Fundamental Aspects of

Operational Risk and Insurance Analytics

FUNDAMENTAL ASPECTS OF Operational Risk and Insurance Analytics

A Handbook of Operational Risk

MARCELO G. CRUZ

GARETH W. PETERS

PAVEL V. SHEVCHENKO

WILEY

Published by John Wiley & Sons, Inc., Hoboken, New Jersey.
Published simultaneously in Canada.

Library of Congress Cataloging-in-Publication Data:

Cruz, Marcelo G.
Fundamental aspects of operational risk and insurance analytics : a handbook of operational risk / Marcelo G. Cruz, GLeonard N. Stern School of Business, New York University, New York, NY, USA, Gareth W. Peters, Department of Statistical Science, University College of London, London, United Kingdom, Pavel V. Shevchenko, Division of Computational Informatics, The Commonwealth Scientific and Industrial Research Organization, Sydney, Australia.
pages cm
Includes bibliographical references and index.
ISBN 978-1-118-11839-9 (hardback)
1. Operational risk. 2. Risk management. I. Peters, Gareth W., 1978– II. Shevchenko, Pavel V. III. Title.
HD61.C778 2014
658.15′5–dc23

2014012662

Printed in the United States of America.
10 9 8 7 6 5 4 3 2 1

To Virginia and Nicholas
Marcel G. Cruz

To my dear wife Chen Mei-Peters, your love, patience support, and encouragement have made this book a reality. To my mother Laraine Peters for teaching me the joy of scientific discovery. To Youxiang Wu, the charity work is complete, thank you for all your support
Gareth W. Peters

To my father Vladimir and mother Galina
Pavel V. Shevchenko

To know, is to know that you know nothing.
That is the meaning of true knowledge.
Socrates

Contents

Preface

Operational risk (OpRisk) has been through significant changes in the past few years with increased regulatory pressure for more comprehensive frameworks. Nowadays, every mid-sized and larger financial institution across the planet has an OpRisk department. However, if we compare the pace of progress of OpRisk to market and credit risks, we would realize that OpRisk is not advancing as fast as its sister risks moved in the past. Market risk management and measurement had its major breakthrough in the early 1990s as J.P. Morgan released publicly its Value-at-Risk (VaR) framework. Only a couple of years after this release, most of the 100 global largest banks had developed a market risk framework and were using, at least to a certain level, VaR methods to measure and manage market risk. A few years later, the Basel Committee allowed banks to use their VaR models for regulatory capital purposes. From the release of JP Morgan's methodology to becoming accepted by Basel and local regulators, it took only about 4 years. This is basically because the methods were widely discussed and the regulators could also see in practice how they would work. As we see it, one of the biggest challenges in OpRisk is to take this area to the same level that market and credit risk management are at. Those two risks are managed proactively and risk managers usually have a say if deals or businesses are approved based on the risk level. OpRisk is largely kept out of these internal decisions at this stage and this is a very worrying issue as quite a few financial institutions have OpRisk as its dominant exposure. We believe that considerable effort in the industry would have to be put into data collection and modeling improvements, and making a contribution to close this gap is the main objective of our book.

Unlike market and credit risks, the methodologies and practices used in OpRisk still vary substantially between banks. Regulators are trying to close the methodological gap by holding meetings with the industry and incentivizing convergence among the different approaches through more individualized guidance. Although some success might be credited to these efforts, there are still considerable challenges and this is where the *Fundamental Aspects of Operational Risk and Insurance Analytics: A Handbook of Operational Risk* can add value to the industry.

In addition, by using this text as a graduate text from which to teach the key components of OpRisk in universities, one will begin to achieve a concensus and understanding of the discipline for junior quantitative risk managers and actuaries. These challenges involve the practical business environment, regulator requirements, as well as the serious and detailed quantitative challenges in the modeling aspects.

This book is a comprehensive treatment of all aspects of OpRisk management and insurance with a focus on the analytical and modeling side but also covering the basic qualitative aspects. The initial chapters cover the building blocks of OpRisk management and measurement. There is broad coverage on the four data elements that need to be used in the

OpRisk framework as well as how a risk taxonomy process should be developed. Considerable focus is given to internal loss data and key risk indicators, as these would be fundamental in developing a risk-sensitive framework similar to market and credit risks. An example is also shown of how OpRisk can be inserted into a firm's strategic decisions. In addition, we cover basic concepts of probability theory and the basic framework for modeling and measuring OpRisk and how loss aggregation should work. We conclude this part of the text with a model to perform stress-testing in OpRisk under the US Comprehensive Capital Analysis and Review (CCAR) program.

We continue by covering more special topics in OpRisk measurement. For example, diverse methods to estimate frequency and severity models are discussed. Another very popular issue in this industry is how to select severity models and this is also comprehensively discussed. One of the biggest challenges in OpRisk is that data used in measurement can be very different, so combining them into a single measure is not trivial. In this part of the book, we show a number of methods to do so.

After the core risk measurement work is done, there are still some issues to address that can potentially mitigate the capital and also indicate how to manage risks. In the third part, we discuss correlation and dependency modeling as well as insurance and risk transfer tools and methods. This is particularly relevant when considering risk mitigation procedures for loss processes that may generate catastrophic losses due to, for instance, nature risk.

This book provides a consistent and comprehensive coverage of all aspects of risk management, more specifically OpRisk–organizational structure, methodologies, policies, and infrastructure–for both financial and nonfinancial institutions. The risk measurement and modeling techniques discussed in the book are based on the latest research. They are presented, however, with considerations based on practical experience of the authors with the daily application of risk measurement tools.

We have incorporated the latest evolution of the regulatory framework. The book offers a unique presentation of the latest OpRisk management techniques and provides one-stop shopping for knowledge in risk management ranging from current regulatory issues, data collection and management to technological infrastructure, hedging techniques, and organizational structure.

It is important to mention that we are publishing at the same time a companion book *Advances in Heavy Tailed Risk Modeling: A Handbook of Operational Risk* (Peters and Shevchenko, 2015), which, although can be seen as an independent tome, covers many important ideas in OpRisk and insurance modeling. This book would be ideally treated as a mathematically detailed companion to this current text, which would go hand in hand with a more advanced graduate course on OpRisk. In this text, we cover in detail significant components of heavy-tailed loss modeling, which is of key importance to many areas of OpRisk.

We would like to thank our families for their patience in our absence while we were writing this book.

Acknowledgments

Dr. Gareth W. Peters also acknowledges the support of the Institute of Statistical Mathematics, Tokyo, Japan and Prof. Tomoko Matsui for extended collaborative research visits and discussions during the development of this book.

Marcelo G. Cruz, Gareth W. Peters, and Pavel V. Shevchenko

New York, London, Sydney

Acronyms

ABC	Approximate Bayesian Computation
ALP	Accumulated Loss Policy
AMA	Advanced Measurement Approach
APT	Arbitrage Pricing Theory
a.s.	almost surely
AUM	Assets under Management
BDSF	Business Disruption and System Failures
BCBS	Basel Committee on Banking Supervision
BCRLB	Bayesian Cramer–Rao Lower Bound
BEICF	Business Environment and Internal Control Factors
BHC	Banking Holding Company
BIS	Bank for International Settlements
CAT	catastrophe bond
CCAR	Comprehensive Capital Analysis and Review
CD	codifference
CLP	Combined Loss Policy
CPI	Consumer Price Index
CRLB	Cramer Rao Lower Bound
CV	covariation
CVaR	Conditional Value-at-Risk
DFT	Discrete Fourier Transform
ES	Expected shortfall
EVI	Extreme Value Index
EVT	Extreme Value Theory
FED	Federal Reserve Bank
FFT	Fast Fourier Transform
GAM	Generalized Additive Models
GAMLSS	Generalized Additive Models for Location Scale and Shape
GAMM	Generalized Additive Mixed Models
GDP	Gross Domestic Product
GLM	Generalized Linear Models
GLMM	Generalized Linear Mixed Models
HILP	Haircut Individual Loss Policy
HMCR	higher moment coherent risk measure
i.i.d.	independent and identically distributed

ILPC	Individual Loss Policy Capped
ILPU	Individual Loss Policy Uncapped
LDA	Loss Distribution Approach
MC	Monte Carlo
MCMC	Markov chain Monte Carlo
MLE	maximum likelihood estimator
MPT	Modern Portfolio Theory
o.d.e.	ordinary differential equation
OpRisk	operational risk
p.g.f.	probability generating function
PMCMC	particle Markov chain Monte Carlo
PPNR	pre-provision net revenue
r.v.	random variable
SCAP	Supervisory Capital Assessment Program
SMC	Sequential Monte Carlo
SRM	spectral risk measure
TCE	tail conditional expectation
TTCE	tempered tail conditional expectation
VaR	Value-at-Risk
Vco	variational coefficient

List of Distributions

Distribution Name	**Distribution Symbol**
Asymmetric Laplace	$AsymmetricLaplace(\cdot)$
Beta	$Beta(\cdot)$
Binomial	$Binomial(\cdot)$
Chi-Squared	$ChiSquared(\cdot)$
Exponential	$Exp(\cdot)$
g-and-h distributions	$T_{g,h}(\cdot)$
g-and-k distributions	$T_{g,k}(\cdot)$
g distributions	$T_g(\cdot)$
Gamma	$Gamma(\cdot)$
Generalized Inverse Gaussian	$GIG(\cdot)$
Generalized Pareto Distribution	$GPD(\cdot)$
Inverse Gaussian	$InverseGaussian(\cdot)$
Inverse Gamma	$InverseGamma(\cdot)$
LogNormal	$LogNormal(\cdot)$
Normal (Gaussian)	$Normal(\cdot)$
Standard Normal	$\Phi(\cdot)$
Negative Binomial	$NegBinomial(\cdot)$
Normal Inverse Gaussian	$NIG(\cdot)$
Pareto	$Pareto(\cdot)$
Poisson	$Poisson(\cdot)$
Tukey Transform h	$T_h(\cdot)$
Tukey Transform k	$T_k(\cdot)$
Tukey Transform j	$T_j(\cdot)$
Tukey Transform hjk	$T_{hjk}(\cdot)$
h distributions	$T_h(\cdot)$
Double h–h distributions	$T_{h,h}(\cdot)$
Generalized Beta	$GB2(\cdot)$
Log-t	$Log\text{-}t(\cdot)$
Generalized Gamma	$GG(\cdot)$
Singh–Maddala or Burr Type III	$BurrIII(\cdot)$
Dagum or Burr Type XII	$BurrXII(\cdot)$
Log-Cauchy	$LogCauchy(\cdot)$

Lomax	$Lomax(\cdot)$
Generalized Hyperbolic	$GH(\cdot)$
Laplace	$Laplace(\cdot)$
Halphen Type A	$Halphe(\cdot)$

CHAPTER ONE

OpRisk in Perspective

1.1 Brief History

Operational risk (OpRisk) is the youngest of the three major risk branches, the others being market and credit risks. The term OpRisk started to be used after the Barings event in 1995, when a rogue trader caused the collapse of a venerable institution by placing bets in the Asian markets and keeping these contracts out of sight of management. At the time, these losses could be classified neither as market nor as credit risks and the term OpRisk started to be used in the industry to define situations where such losses could arise. It took quite some time until this definition was abandoned and a proper definition was established for OpRisk. In these early days, OpRisk had a negative definition as "every risk that is not market and credit", which was not very helpful to assess and manage this risk. Looking back at the history of risk management research, we observe that early academics found the same issue of classifying risk in general, as Crockford (1982) noticed: "*Research into risk management immediately encounters some basic problems of definition. There is still no general agreement on where the boundaries of the subject lie, and a satisfactory definition of risk management is notoriously difficult to formulate*".

Before delving into the brief history of OpRisk it might be useful to first understand how risk management is evolving and where OpRisk fits in this evolution. Risk in general is a relatively new area that began to be studied only after World War II. The concept of risk management came from the insurance industry and this was clear in the early days' definitions. According to Crockford (1982) the term "risk management", in its earliest incarnations, "*encompassed primarily those activities performed to prevent accidental loss*". In one of the first textbooks on risk, Mehr and Hedges (1963) used a definition that reflected this close identification with insurance: "*[T]he management of those risks for which the organization, principles and techniques appropriate to insurance management is useful*". Almost 20 years later, Bannister and Bawcutt (1981) defined risk management as "*the identification, measurement and economic control of risks that threaten the assets and earnings of a business or other enterprise*", which is much closer to the definition used in the financial industry in the twenty-first century.

The association of risk management and insurance came from the regular use of insurance by individuals and corporations to protect themselves against these "accidental losses". It is interesting to see that even early authors on the subject made a case for the separation between risk management and risk-takers (the businesses). Crockford (1982) wrote that "*operational*

Fundamental Aspects of Operational Risk and Insurance Analytics: A Handbook of Operational Risk, First Edition. Marcelo G. Cruz, Gareth W. Peters, and Pavel V. Shevchenko.

convenience continues to dictate that pure and speculative risks should be handled by different functions within a company, even though theory may argue for them being managed as one".

New tools for managing risks started to emerge in the 1950s, in addition to insurance, when many types of insurance coverage became very costly and incomplete; or certainly this "incompletion" started to be better noticed as risk management was beginning to evolve. Several business risks were either impossible or too expensive to insure. Contingent planning activities, an embryo of what is today called Business Continuity Planning (BCP), were developed, and various risk prevention or self-prevention activities and self-insurance instruments against some losses were put in place. Coverage for work-related illnesses and accidents also started to be offered during the 1960s. The 1960s were when a more formal, organized scholarly interest started to blossom in academia on issues related to risk. The first academic journal to show "risk" in their title was the *Journal of Risk and Insurance* in 1964. This journal was actually titled *Journal of Insurance* until then. Other specialized journals followed including *Risk Management*—published by the Risk and Insurance Management Society (RIMS), a professional association of risk managers founded in 1950 and the *Geneva Papers on Risk and Insurance*, published by the Geneva Association since 1976.

Risk management had its major breakthrough as the use of financial derivatives by investors became more spread out. Before the 1970s, derivatives were basically used for commodities and agricultural products; however, in the 1970s but more strongly in the 1980s, the use of derivatives to manage and hedge risks began. In the 1980s, companies began to consider financial risk management of "risk portfolios". Financial risk management has become complementary to pure risk management for many companies. Most financial institutions, particularly investment banks, intensified their market and credit risk management activities during the 1980s. Given this enhanced activity and a number of major losses, it was no surprise that more intense scrutiny drew international regulatory attention. Governance of risk management became essential and the first risk management positions were created within organizations.

A sort of "risk management revolution" was sparked in the 1980s by a number of macroeconomic events that were present during this decade as, for example, fixed currency parities disappeared, the price of commodities became much more volatile, and the price fluctuations of many financial assets like interest rates, stock markets, exchange rates, etc. became much more volatile. This volatility, and the many headline losses that succeeded, revolutionized the concept of financial risk management as most financial institutions had such assets in their balance sheets and managing these risks became a priority for senior management and board of directors. At the same time, the definition of risk management became broader. Risk management decisions became financial decisions that had to be evaluated based on their effect on a firm or portfolio value, rather than on how well they cover certain risks. This change in definition applies particularly to large public corporations, due to the risk these bring to the overall financial system.

These exposures to financial derivatives brought new challenges with regard to risk assessment. Quantifying the risk exposures, given the complexity of these assets, was (and still remains) quite complex and there were no generally accepted models to do so. The first and most popular model to quantify market risks was the famous "Black & Scholes" developed by Black and Scholes (1973) in which an explicit formula for pricing a derivative was proposed—in this case, an equity derivative. The model was so revolutionary that the major finance journals refused to publish it at first. It was finally published in the *Journal of Political Economy* in 1973. An extension of this article was later published by Merton in the *Bell Journal of Economics and Management Science* (Merton, 1973). The impact of the article in the financial industry was significant and the risk coverage of derivatives grew quickly, expanding to many distinct assets like interest rate swaps, currencies, etc.

As risk management started to grow as a discipline, regulation also began to get more complex to catch up with new tools and techniques. It is not a stretch to say that financial institutions have always been regulated one way or another given the risk they bring to the financial system. Regulation was mostly on a country-by-country basis and very uneven, allowing arbitrages. As financial institutions became more globalized, the need for more symmetric regulation that could level the way institutions would be supervised and regulated increased worldwide. The G10, the group of 10 most industrialized countries, started meetings in the city of Basel in Switzerland under the auspices of the Bank for International Settlements (BIS). The so-called Basel Committee on Banking Supervision or Basel Committee was established by the central bank governors of the group of 10 countries at the end of 1974, and continues to meet regularly four times a year. It has four main working groups, which also meet regularly.

The Basel Committee does not possess any formal supranational supervisory authority, and its conclusions cannot, and were never intended to, have legal force. Rather, it formulates broad supervisory standards and guidelines and recommends statements of best practice in the expectation that individual authorities will take steps to implement them through detailed arrangements, statutory or otherwise, which are best suited to their own national systems. In this way, the Committee encourages convergence toward common approaches and common standards without attempting detailed standardization of member countries' supervisory techniques.

The Committee reports to the central bank governors and heads of supervision of its member countries. It seeks their endorsement for its major initiatives. These decisions cover a very wide range of financial issues. One important objective of the Committee's work has been to close gaps in international supervisory coverage in pursuit of two basic principles: that no foreign banking establishment should escape supervision; and that supervision should be adequate. To achieve this, the Committee has issued a long series of documents since 1975 that guide regulators worldwide on best practices that can be found on the website:
www.bis.org/bcbs/publications.htm.

The first major outcome of these meetings was the Basel Accord, now called Basel I, signed in 1988 (see BCBS, 1988). This first accord was limited to credit risk only and required each bank to set aside a capital reserve of 8%, the so-called Cooke ratio, of the value of the securities representing the credit risk in their portfolio. The accord also extended the definition of capital to create reserves encompassing more than bank equity, which were namely:

- **Tier 1 (core capital)**, consisting of common stock, holding in subsidiaries, and some reserves disclosed to the regulatory body;
- **Tier 2 (supplementary capital)**, made up of hybrid capital instruments, subordinated debts with terms to maturity greater than 5 years, other securities, other reserves.

The Basel I Accord left behind one important risk component, which was market risk. In the meantime, JP Morgan released publicly its market risk methodology called Risk Metrics (JP Morgan, 1996), and the popularization of market risk measurement became widespread in the early 1990s. Reacting to that, in 1996 the Basel Committee issued the market risk amendment (BCBS, 1996), which included market risk in the regulatory framework. The acceptance of more sophisticated models like Value at Risk (VaR) as regulatory capital was a significant milestone in risk management. However, this initial rule had a number of limitations as it did not allow diversification, that is, the total VaR of the firm would be the sum of the VaR for all assets without allowing for correlation between these risks.

As the global financial markets became increasingly interconnected and sophisticated as well as financial products, like credit derivatives, it soon became clear to the Basel Committee that a new regulatory framework was needed. In June 1999, the Committee issued a proposal for a revised Capital Adequacy Framework. The proposed capital framework consisted of the following three pillars:

- **Pillar 1.** Minimum capital requirements, which seek to refine the standardized rules set forth in the 1988 Accord;
- **Pillar 2.** Supervisory review of an institution's internal assessment process and capital adequacy;
- **Pillar 3.** Market discipline focused on effective use of disclosure to strengthen market discipline as a complement to supervisory efforts.

Following extensive interaction with banks, industry groups, and supervisory authorities that are not members of the Committee, the revised framework (referred to as Basel II) BCBS (2004) was issued on June 26, 2004; the comprehensive version was published as BCBS (2006). This text serves as a basis for national rule-making and for banks to complete their preparations for the new framework's implementation.

With Basel II, there also came for the first time the inclusion of OpRisk into the regulatory framework. The OpRisk situation was different from the one faced by market and credit risks. For those risks, regulators were looking at the best practice in the industry and issuing regulation mirroring these. The progress in OpRisk during the late 1990s and early 2000s was very slow. Some very large global banks like Lehman Brothers did not have an OpRisk department until 2004, so the regulators were issuing rules without the benefit of seeing how these rules would work in practice. This was a challenge for the industry.

In order to address these challenges, the Basel Committee allowed a few options for banks to assess capital. The framework outlined and presented three methods for calculating OpRisk capital charges in a continuum of increasing sophistication and risk sensitivity: (i) the Basic Indicator Approach (BIA); (ii) the Standardized Approach (SA); and (iii) Advanced Measurement Approaches (AMA). Internationally active banks and banks with significant OpRisk exposures (e.g., specialized processing banks) are expected to use an approach that is more sophisticated than the BIA and that is appropriate for the risk profile of the institution.

Many models have been suggested for modeling OpRisk under Basel II; for an overview, see Chernobai *et al.* (2007, chapter 4), Allen *et al.* (2005), and Shevchenko (2011, Section 1.5). Fundamentally there are two different approaches used to model OpRisk:

- The top–down approach; and
- The bottom–up approach.

The top–down approach quantifies OpRisk without attempting to identify the events or causes of losses while the bottom–up approach quantifies OpRisk on a microlevel as it is based on identified internal events. The top–down approach includes the Risk Indicator models that rely on a number of operational risk exposure indicators to track OpRisks and the Scenario Analysis and Stress Testing Models that are estimated based on the what-if scenarios. The bottom–up approaches include actuarial-type models (referred to as the Loss Distribution

Approach) that model frequency and severity of OpRisk losses. In this book we provide a detailed quantitative discussion on a range of models some of which are appropriate for top-down modelling whilst others are directly applicable to bottom-up frameworks.

1.2 Risk-Based Capital Ratios for Banks

Until the late 1970s, banks in most countries were in general highly regulated and protected entities. This protection was largely a result of the bitter memories of the Great Depression in the US as well as the role that high (or hyper) inflation played in the political developments in Europe in the 1930s, and banks arguably play a significant part in the spreading of inflation. Due to these memories, the activities banks were allowed to undertake were tightly restricted by national regulators and, in return, banks were mostly protected from competitive forces. This cozy relationship was intended to ensure stability of the banking system, and it succeeded in its goals throughout the reconstruction and growth phases, which followed World War II. This agreement held well until the collapse of Bretton Woods[1] (Eichengreen, 2008) in the 1970s. The resulting strain in the banking system was enormous. Banks suddenly were faced with an increasingly volatile environment, but at the same time had very inelastic pricing control over their assets and liabilities, which were subject not just to government regulation but also to protective cartel-like arrangements. The only solution seen by national authorities at this time was to ease regulations on banks. As the banking sector was not used to competitive pressures, the result of the deregulation was that banks started to take too much risk in search of large payoffs. Suddenly banks were overlending to Latin American countries (and other emerging markets); overpaying for expansion (e.g., buying competitors looking for geographic expansion), etc. With the crisis in Latin America in the 1980s, these countries could not repay their debts and banks were once again in trouble. Given that the problems were mostly cross-boundary as the less regulated banks became more international, the only way to address this situation was at the international level and the Basel Committee was consequently established under the auspices of the BIS.

In 1988, the Basel Committee decided to introduce an internationally accepted capital measurement system commonly referred to as Basel I, (BCBS, 1988). This framework was replaced by a significantly more complex capital adequacy framework commonly known as Basel II (BCBS, 2004) and, more recently, the Basel Committee issued the Basel III Accord (BCBS, 2011, 2013), which will add more capital requirements to banks. Table 1.1 shows a summary of key takeaways of the Basel Accords.

Basel I primarily focuses on credit risk and developed a system of risk-weighting of assets. Assets of banks were classified and grouped in five categories according to credit risk, carrying risk weights of 0% for the safest, most liquid assets (e.g., cash, bullion, home country debt like Treasuries) to 100% (e.g., most corporate debt). Banks with an international presence were required to at least hold capital equal to 8% of their risk-weighted assets (RWA). The concept of RWA was kept in all Accords with changes on the weights and in the composition of assets by category. An example of how risk-weighting works can be seen in Table 1.2. In this example, the sum of the assets of this bank is $1015; however, applying the risk-weighting rule established in Basel I, the RWA is actually $675.

[1]The Bretton Woods agreement was established in the summer of 1944 and put in place a system of exchange and interest rate stability which ensured that banks could easily manage their exposures.

TABLE 1.1 Basel framework general summary

Accord	Year	Key points
Basel I	1988	Introduces minimal capital requirement for the banking book. Introduces tier concept for capital requirement. Incorporates trading book into the framework later on through the Market Risk Amendment (MRA).
Basel II	2004	Allows usage of internal models and inputs in risk measurement. Introduces operational risk.
Basel II/III	2010	Increases capital requirement for trading book, with significant increase for correlation trading and securitizations.
Basel III	2010	Motivated by the great financial crisis of 2008, increases capital requirements, introduces leverage constraints and minimum liquidity and funding requirements.

TABLE 1.2 Example of risk-weighted assets calculation under Basel I

Risk-weight (%)	Asset	Amount ($)	RWA ($)
0	Cash	10	0
	Treasury bills	50	0
	Long-term treasury securities	100	0
20	Municipal bonds	20	4
	Items in collection	20	4
50	Residential mortgages	300	150
100	AA+ rated loan	20	20
	Commercial loans, AAA- rated	55	55
	Commercial loans, BB- rated	200	200
	Sovereign loans B- rated	200	200
	Fixed assets	50	50
Not rated	Reserve for loan losses	(10)	(10)
Total		1015	675

Since Basel I, a bank's capital also started to be classified into Tier 1 and Tier 2. Tier 1 capital is considered the primary capital or "core capital"; Tier 2 capital is the supplementary capital. The total capital that a bank holds is defined as the sum of Tier 1 and Tier 2 capitals. Table 1.3 provides a more detailed view of the components of each tier of capital. The key component of Tier 1 capital is the common shareholders equity. This item is so important that a number of banks also report the so-called Tier 1 Common Equity in which only common shareholder equity is considered as Tier 1. As shown in Table 1.3, the Basel Committee made capital requirement much stricter in the latest Basel Accords by changing the definition of some of the current items but also by sending a couple of items to Tier 2 (e.g., trust preferred securities and remaining noncontrolling interest), making it more difficult for banks to comply with these new capital rules.

Another important contribution from Basel I is the concept of capital ratios that remains until today. Basically, a bank needs to assert its capital requirements based on the formula:

$$\text{Capital ratio} = \frac{\text{Eligible capital}}{\text{RWA}}. \tag{1.1}$$

TABLE 1.3 **Tiered capital definition under Basel II and Basel III**

Tier	Capital requirement under Basel II	Basel III capital requirement
Tier 1	(+) Common shareholders equity (+) Partial noncontrolling interest (NCI) (−) Certain deferred tax assets (DTA) (−) Goodwill and intangibles (−) Debt valuation adjustments (DVA) (−) Other deductions = **Tier 1 common** (+) Perpetual preferred stock (+) Trust preferred securities (+) Remaining NCI = **Tier 1 capital**	(+) Common shareholders equity (+) Partial noncontrolling interest (NCI)* (−) Certain deferred tax assets (DTA)* (−) Goodwill and intangibles (−) Debt valuation adjustments (DVA) (−) Other deductions* = **Tier 1 common** (+) Perpetual preferred stock = **Tier 1 capital**
Tier 2	(+) Subordinated debt (+) Allowance for loan loss reserves	(+) Trust preferred securities* (+) Remaining NCI* (+) Subordinated debt (+) Allowance for loan loss reserves

Basel III changes are indicated by ∗.

TABLE 1.4 **Example of capital ratios in some large European banks in 2012**

	UBS	Credit Suisse	Deutsche Bank
Tier 1 capital	40,982	43,547	50,483
Total capital	48,498	49,936	57,015
RWA total	192,505	224,296	333,849
RWA market risk	21,173	29,366	53,058
RWA credit risk	105,807	143,679	229,196
RWA OpRisk	53,277	45,125	51,595
Other risks	6,248	6,126	—
Tier 1 capital ratio	21.3%	19.4%	15.1%
Total capital ratio	25.2%	22.3%	17.1%

Source: Banks annual reports. Figures are in Swiss Francs (CHF) millions for UBS and Credit Suisse and in Euros millions for Deutsche Bank.

Therefore, to find its Tier 1 capital ratio a bank would have to calculate its RWA based on the current Basel rules and also retrieve the elements that compose Tier 1 capital from its balance sheet. Dividing the Tier 1 capital by the RWA would provide the Tier 1 capital ratio. In order to make this process very clear, we show examples on how to calculate each of the steps. Table 1.2 shows an example of RWA calculation using only credit risk-weightings; Table 1.3 provides an overview of capital requirement definitions on the balance sheet; and Table 1.4 shows a real-life example of capital ratios in a few Large European banks that are Basel II approved and, therefore, have to show their capital breakdown.

Basel II discussions started in the late 1990s and ended with the publication of the second Accord, or “Basel II” in 2004 (BCBS, 2004). Basel II was implemented in an era where banks posted record profits and the global macroeconomic scenario did not show many clouds on

TABLE 1.5 New capital charges on Basel III

Capital conservation buffer	Countercyclical capital buffer
2.5% added to the minimum ratios To be built up in good times and available in period of stress Inclusion in target capital ratios by end of transition period (2018) Restriction on distributions (dividends, share buybacks, and bonuses) if the full buffer requirement is not met	Up to 2.5% added to the minimum ratios Declared by any country that is experiencing overheated credit markets—preannouncement of decision by up to 12 months Can be relaxed when the market "cools down" again—takes effect immediately with announcement Restriction on distributions (dividends, share buybacks, and bonuses) if the buffer requirement is not met

TABLE 1.6 Minimum capital requirements

Type of capital	Before Basel III(%)	Basel III(%)	Capital conservation buffer (%)	Countercyclical capital buffer (%)	Total Basel III(%)
Common Equity Tier 1	2	4.5	2.5	0–2.5	9.5
Tier 1	4	6	2.5	0–2.5	11
Total risk-based capital	8	8	2.5	0–2.5	13

Source: BCBS (2013).

the horizon. In this Accord, banks were allowed to use their own internal models to calculate regulatory capital for market, credit, and also operational risk, which was introduced in this Accord. The overall idea of Basel II was that banks would be able to reduce their capital requirements by adopting internal models and following the strict qualification criteria.

In order to calculate the RWA in market and operational risks, where the risk-weighting asset in the example of Table 1.2 would obviously not apply, banks would have to convert the outcomes of their internal models, calculated at the 99.9% quantile, and divide this number by 8% (or multiply by 12.5). Reverse engineering these numbers from Table 1.4, i.e. calculating operational risk capital as RWA OpRisk divided by 12.5, we can see that the operational risk capital at UBS in 2012 was CHF 4264 million, Credit Suisse was CHF 3610 million, and Deutsche Bank was €4127 million.

Unlike Basel I and Basel II, Basel III was motivated by the great banking crisis in 2008 and this motivation made this 3-rd version of the Accord primarily focussed on addressing concerns about a run on the bank risks (i.e., liquidity issues on customers withdrawing resources from a bank due to lack of confidence in its financial health), consequently requiring differing levels of reserves for different forms of bank deposits and other borrowings. Therefore, contrary to what might be expected by its name, Basel III rules do not, for the most part, supersede the guidelines established in Basel I and Basel II but work alongside them. The main changes in the Basel III framework are shown in Table 1.5 and are mostly related to the creation of new capital buffers to ensure banks are enough capitalized in the next crisis.

In addition to the minimum capital ratios already established in the previous Accords (see Table 1.6), Basel III requires that all banking organizations maintain a "capital conservation buffer" consisting of Tier 1 Common Equity capital in an amount equal to 2.5% of risk-weighted assets in order to avoid restrictions on their ability to make capital distributions and to make certain discretionary bonus payments to executive officers. Thus, the capital conservation buffer effectively increases the minimum Tier 1 common equity capital, Tier 1 capital, and total capital requirements for US banking organizations to 7.0%, 8.5%, and 10.5%, respectively. Banking organizations with capital levels that fall within the buffer will be forced to limit dividends, share repurchases or redemptions (unless replaced within the same calendar quarter by capital instruments of equal or higher quality), and make discretionary bonus payments. The limits consist of a sliding scale, so that as the buffer decreases, so does the maximum payout as a percentage of the banking organization's net income over the past four quarters. For large global banks, the capital buffer may be increased during periods of "excessive credit growth" by an incremental "countercyclical capital buffer" of up to 2.5% of risk-weighted assets. In a change from the proposed rules (i.e., presigning the Accord), large global banks would (after completing the "parallel run" process for migrating to the advanced approaches regime) be required to use the lesser of their standardized and advanced approaches risk-based capital ratios as the basis for calculating their capital conservation buffer (and any applicable countercyclical capital buffer). This change will likely increase the capital buffer for at least some large global banks compared to the proposed rules.

Basel III also imposes a Tier 1 minimum leverage ratio of 4.0% for all banking organizations and an additional supplementary Tier 1 leverage ratio of 3.0% for large global banks (BCBS, 2013). The 3.0% supplementary leverage ratio (which, consistent with Basel III schedule, will take effect in January 2018 but be reported beginning in January 2015) incorporates in the denominator certain off-balance sheet exposures that are not included in the standard leverage ratio. Despite significant criticism from the industry, Basel III continues to include in the supplementary leverage ratio derivative exposures based on potential future exposure (without collateral recognition) and 10% of unconditionally cancellable commitments.

1.3 The Basic Indicator and Standardized Approaches for OpRisk

Under the Basel II framework, the simplest method that banks could use to calculate OpRisk capital is the BIA. Banks using the BIA must hold capital for OpRisk equal to the average over the previous 3 years of a fixed percentage (denoted α) of positive annual gross income. Figures for any year in which annual gross income is negative or zero should be excluded from both the numerator and denominator when calculating the average. The capital charge K_{BIA} may be expressed as follows:

$$K_{BIA} = \alpha \frac{1}{n} \sum_{j=1}^{3} \max\{GI(j), 0\}, \quad n = \sum_{j=1}^{3} \mathbb{I}_{\{GI(j)>0\}}, \tag{1.2}$$

where $GI(j), j = 1, 2, 3$ are the annual gross incomes over the last 3 years; $\mathbb{I}_{\{GI(j)>0\}}$ is an indicator function that equals 1 if condition in $\{.\}$ is true and 0 otherwise; n is the number of previous years in which income is positive (expected to be three); and $\alpha = 0.15$ (as of 2013) as established by the Committee (BCBS, 2006, pp. 144–145).

Another simple approach to calculate OpRisk capital under the Basel II framework is the SA where, bank activities are divided into eight business lines: Corporate finance, Trading and sales, Retail banking, Commercial banking, Payment and settlement, Agency services, Asset management, and Retail brokerage. Within each business line, gross income is a broad indicator that serves as a proxy for the scale of business operations and thus the likely scale of OpRisk exposure within each of these business lines. The capital charge for each business line

TABLE 1.7 Coefficients β_i for each business line as determined by Basel II in BCBS (2006, p. 147)

Business line	β_i
Corporate finance	0.18
Trading and sales	0.18
Retail banking	0.12
Commercial banking	0.15
Payment and settlements	0.18
Agency services	0.15
Asset management	0.12
Retail brokerage	0.12

is calculated by multiplying gross income by a factor (denoted β) assigned to that business line. The value of β serves as a proxy for the industry-wide relationship between the OpRisk loss experience for a given business line and the aggregate level of gross income for that business line. It should be noted that in the SA gross income is measured for each business line, not the whole institution, that is, in Corporate finance, the indicator is the gross income generated in the Corporate finance business line.

The total capital charge is calculated as the 3-year average of the simple summation of the regulatory capital charges across each of the business lines in each year. In any given year, negative capital charges (resulting from negative gross income) in any business line may offset positive capital charges in other business lines without limit. However, where the aggregate capital charge across all business lines within a given year is negative, the input to the numerator for that year will be zero. The total capital charge K_{TSA} may be expressed as

$$K_{TSA} = \frac{1}{3}\sum_{j=1}^{3} \max\left(\sum_{i=1}^{8} \beta_i GI_i(j), 0\right), \tag{1.3}$$

where $GI_i(j)$ is the annual gross income of business line i in year j and β_i a fixed coefficient, set by the Committee, relating the level of required capital to the level of gross income for each of the eight business lines. These details can be found in (BCBS, 2006, pp. 146–147); the values of β_i (as of 2013) are presented in Table 1.7.

1.4 The Advanced Measurement Approach

Under the Basel II AMA for OpRisk, banks are allowed to use their own internal models to estimate capital. A bank intending to use the AMA should demonstrate the accuracy of the internal models within the matrix of Basel II risk cells (eight business lines by seven event types) relevant to the bank. The eight business lines are listed in Table 1.7 and the seven event types are as follows:

- Internal fraud;
- External fraud;
- Employment practices and workplace safety;

- Clients, products, and business practices;
- Damage to physical assets;
- Business disruption and system failures;
- Execution, delivery, and process management.

As imagined, given the early stages of bank frameworks, the range of practice was quite broad. In Europe, the methodological focus of most banks was on using scenario analysis while in the US the focus was on internal and external loss data. Understanding the evolutionary nature of OpRisk management as a developing risk management discipline, the Basel Committee provided significant flexibility to banks in the development of an OpRisk measurement and management system. This flexibility was, and continues to be, a critical feature of the AMA. However, substantial efforts are required by national authorities to ensure sufficient consistency in the application of these features. The Basel II framework envisaged that, over time, the OpRisk discipline will mature and converge toward a narrower band of effective risk management and risk measurement practices. Understanding the current range of observed operational risk management and measurement practices, both within and across geographic regions, contributes significantly to the efforts to establish consistent supervisory expectations. Through the analysis of existing practices, and the publication of papers reporting those practices, the Basel Committee expects the maturation of OpRisk practices and supports supervisors in developing more consistent regulatory expectations.

The initial Basel II proposal (BCBS, 2001, Annex 4) suggested three approaches for AMA:

- Internal Measurement Approach (IMA);
- Score Card Approach (SCA);
- Loss Distribution Approach (LDA).

The latest Basel II document (see BCBS, 2006) does not give any guidance for the AMA approach and allows flexibility.

1.4.1 INTERNAL MEASUREMENT APPROACH

Under the IMA, OpRisk events are divided into business lines $i = 1, 2, \ldots$ and event types $j = 1, 2, \ldots$; an exposure indicator EI_{ij} is set for each business line/event type combination (risk cell) to capture the scale of the bank's activities in the risk cell; probability P_{ij} that the event will occur over the next year and average loss AL_{ij} are estimated using internal loss data. Then, the capital charge K_{IMA} is calculated as

$$K_{IMA} = \sum_{i,j} \gamma_{ij} EI_{ij} P_{ij} AL_{ij}, \tag{1.4}$$

where γ_{ij} is the conversion factor translating expected loss, $EI_{ij}P_{ij}AL_{ij}$, for business line/event type risk cell into a capital charge; see BCBS (2001, Annex 4).

1.4.2 SCORE CARD APPROACH

Under a scorecard based approach the bank determines an initial level of OpRisk capital at the firm or business line level, and then attempts to modify the calculated amounts over time on the basis of a qualitative ranking or scoring of each risks evolution.

As stated in the Basel working paper on the regulatory treatment of OpRisk (BCBS 2001, p. 35)

"*These scorecards are intended to bring a forward-looking component to the capital calculations, that is, to reflect improvements in the risk control environment that will reduce both the frequency and severity of future operational risk losses. The scorecards may be based on actual measures of risk, but more usually identify a number of indicators as proxies for particular risk types within business units/lines. The scorecard will normally be completed by line personnel at regular intervals, often annually, and subject to review by a central risk function*".

The SCA approach calculates the capital charge K_{SCA} as

$$K_{SCA} = \sum_{i,j} \omega_{ij} EI_{ij} RS_{ij}, \tag{1.5}$$

where ω_{ij} is the amount of capital per unit of the indicator of exposure, EI_{ij} is the exposure index from a set for each business line/event type combination (risk cell) and RS_{ij} is the risk factor. Under the SCA, a bank assigns a value to each OpRisk event and compares the different OpRisks according to the values. This method relies on experts' assessment in the selection of indicators and their weights (see, e.g., Anders and Sandstedt, 2003). There are a number of references discussing in more detail the nature of scorecard based approaches, see for instance Blunden (2003) and Alexander (2003) and the references therein for more details.

As noted in Alexander (2003), scorecards can be higly subjective and the following important issues are still in the process of being better understood:

- The industry has still been unable to develop industry wide standards for the key risk indicators (KRIs) that should be used for each risk type and underpin the development of scorecard methods;
- They may be inherent biases and moral hazards that must be better understood, modelled and managed before scorecard based methods can be considered reliable. To understand this point, typically, given a set of risk indicators, frequency and severity scores are assigned by a business manager or risk expert in the business that 'owns' the particular operational risk. Hence, one requires a considered design of the management process in order to avoid subjective biases or moral hazard from occurring in the scoring process;
- In addition to the subjectivity of the scores there is also a second problem that under an AMA approach one should figure out a method to map scorecard data to a loss distribution model. This involves the mapping of the scores subjectively to monetary loss amounts.

For these reasons we don't elaborate further on scorecard based approaches. In fact we suggest users of scorecard approaches to consider formulating them under a regression based framework such as those developed in Item Response Theory (IRT), see discussions in Linden and Hambleton (1997).

1.4.3 LOSS DISTRIBUTION APPROACH

The LDA approach is based on modelling annual frequency N and severity $X_1, X_2, \ldots$ of OpRisk events for a risk cell. Then the annual loss for the j-th risk cell is calculated as aggregation of severities over a 1-year time horizon

$$Z^{(j)} = X_1^{(j)} + X_2^{(j)} + \cdots + X_{N^{(j)}}^{(j)} \tag{1.6}$$

and the total loss over all risk cells in a given year is obtained by the following sum over the d risk cells

$$Z = \sum_{j=1}^{d} Z^{(j)}.$$

Then, the regulatory capital is defined as the 0.999 VaR, which is the quantile of the distribution for the next year's annual loss Z:

$$\mathrm{VaR}_q[Z] = \inf\{z \in \mathbb{R} : \mathbb{P}\mathrm{r}[Z > z] \leq 1 - q\} \tag{1.7}$$

at the level $q = 0.999$. For economic capital, banks often use quantile levels in the range $q \in [0.9995, 0.9997]$ depending on a bank's credit rating. The risk cells can be selected at the actual loss generating process level. However, currently, many banks use the LDA for business line/event type risk cells.

Remark 1.1 ***The LDA is considered to be the most comprehensive approach and is a focus of this book. Hereafter, we consider the LDA model only.***

1.4.4 REQUIREMENTS FOR AMA

The qualifying criteria for using the AMA are quite stringent and, in practice, it takes many years of implementation and regulatory exams to validate the approach. The Basel II Accord states that a bank must meet a number of qualitative standards before it is permitted to use an AMA for OpRisk capital (BCBS, 2006, pp. 150–151). In brief, these are as follows:

- The bank must have an independent OpRisk management function responsible for codifying firm-level policies and procedures concerning OpRisk management and controls; the design and implementation of the firm's OpRisk measurement methodology; the design and implementation of a risk-reporting system for OpRisk; and developing strategies to identify, measure, monitor, and control/mitigate OpRisk;
- The bank's internal OpRisk measurement system must be closely integrated into the day-to-day risk management processes. Its output must be an integral part of the process of monitoring and controlling the OpRisk profile. The bank must have techniques for allocating OpRisk capital to major business lines and for creating incentives to improve the management of OpRisk throughout the firm;
- There must be regular reporting of OpRisk exposures and loss experience to business unit management, senior management, and to the board of directors. The bank must have procedures for taking appropriate action according to the information within the management reports;
- The bank's OpRisk management system must be well documented;
- Internal and/or external auditors must perform regular reviews of the OpRisk management processes and measurement systems;
- The validation of the OpRisk measurement system by external auditors and/or supervisory authorities must include:

- ◦ Verifying that the internal validation processes are operating in a satisfactory manner;
- ◦ Making sure that data flows and processes associated with the risk measurement system are transparent and accessible.

In addition to these qualitative factors, Basel II also has quite stringent criteria on AMA acceptance based on a series of quantitative standards (BCBS, 2006, pp. 151–152) as follows:

- Any internal OpRisk measurement system must be consistent with the OpRisk defined by the Committee and the loss event types defined in BCBS (2006);
- The risk measure used for capital charge should correspond to the 99.9% confidence level for a 1-year holding period, that is, $\text{VaR}_{0.999}$ defined in (1.7). Supervisors will require the bank to calculate its regulatory capital requirement $\text{VaR}_{0.999}$ as the sum of expected loss (EL) and unexpected loss (UL), unless the bank can demonstrate that it is adequately capturing EL in its internal business practices. To calculate the minimum regulatory capital as UL, the bank must be able to demonstrate to the satisfaction of its national supervisor that it has measured and accounted for its EL exposure. For illustration, see Figure 1.1. Hereafter, for simplicity, we consider the regulatory capital to be the sum of EL and UL, which is the 99.9% VaR;
- A bank's risk measurement system must be sufficiently "granular" to capture the major drivers of OpRisk affecting the shape of the tail of the loss estimates;
- OpRisk capital charge measures for different risk cells must be added for purposes of calculating the regulatory minimum capital requirement over all risk cells in a bank. However, the bank may be permitted to use internally determined correlations between risk cells, provided it can demonstrate to the satisfaction of the national supervisor that its systems for determining correlations are sound, implemented with integrity, and take into account the uncertainty surrounding any such correlation estimates (particularly in

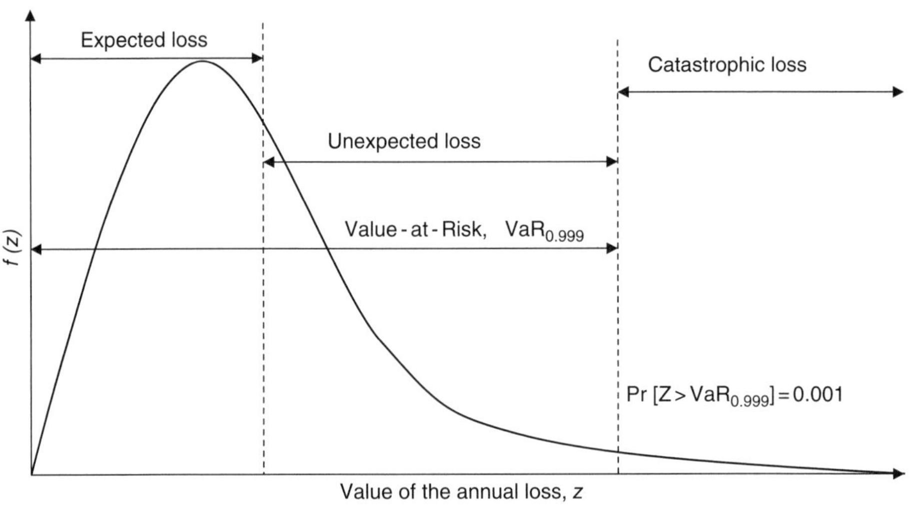

FIGURE 1.1 **Illustration of the expected and unexpected losses in the capital requirements at the 99.9% confidence level for a 1-year holding period; $f(z)$ is the probability density function of the annual loss**

periods of stress). The bank must validate its correlation assumptions using appropriate quantitative and qualitative techniques;

- OpRisk measurement system must be based on the use of internal data, relevant external data, scenario analysis, and factors reflecting the business environment and internal control systems (BEICF). A bank needs to have a credible, transparent, well-documented and verifiable approach for weighting these fundamental elements in its overall OpRisk measurement system. If the estimates of the 99.9% VaR based primarily on internal and external loss event data are unreliable for business lines with a heavy-tailed loss distribution and a small number of observed losses, then scenario analysis and BEICF may play a more dominant role in the risk measurement system. Conversely, OpRisk loss event data may play a more dominant role in the risk measurement system for business lines where estimates of the 99.9% VaR based primarily on such data are deemed reliable.

Given that these rules are quite stringent and were made without benchmarks, unlike market and credit risks, the range of practice can vary significantly from bank to bank. Even banks based in the same block in Midtown Manhattan, just to be very graphic, can have completely different methodologies and frameworks to measure OpRisk. This is very different from market and credit risks where the measurement frameworks are similar across the banks.

The Basel Committee performs surveys on the range of practices for AMA and then issues reports to divulge the results. These reports describe industry practices for some key areas of the governance, data, and modeling components of an AMA framework identifying emerging effective practices as well as practices that are inconsistent with supervisory expectations. The findings from the latest range of practices report (BCBS, 2009a) based on the 2008 Loss Data Collection Exercise (BCBS, 2009b) include the following:

- The absence of definitions in the Basel II text for "gross loss" or "recoveries" and varying loss data collection practices among AMA banks results in differences in the loss amounts recorded for similar events. This practice may lead to potentially large differences in banks' respective capital calculations;
- There was a broad range of practices in the use of loss amount as the AMA input. Most of the 42 participating AMA banks (43%) used "gross loss after all recoveries" (except insurance). "Gross loss before any recoveries" was used by 29%. Other loss amounts used by participating banks include "net loss" (14%) and "other definition" (12%);
- Data collection thresholds vary widely across institutions and types of activity. A bank should be aware of the impact that its choice of thresholds has on OpRisk capital computations;
- There is a broad range of practices for when the loss amounts from legal events are used as a direct input into the model quantifying operational capital, which raises questions of transparency and industry consistency in how these OpRisk exposures are quantified for capital purposes;
- There is considerable diversity across banks in the choice of granularity of their models that may be driven as much by modeler's preferences as by actual differences in OpRisk profiles;
- While it is common for banks to use the Poisson distribution for estimating frequency, there are significant differences in the way banks model severity, including the choice of severity distribution; and

- The combination and weighting of the four elements (internal data, external data, scenario analysis, and BEICF) are significant issues for many banks, given the many possible combination techniques. This is an area where the range of practices is particularly broad both within and across jurisdictions.

1.5 General Remarks and Book Structure

Regulators are trying to close the methodology gap by holding meetings with the industry and they are attempting to incentivize convergence among the different approaches through more individualized guidance. Although some success might be credited to these efforts, there are still considerable challenges and this is where our book *Fundamental Aspects of Operational Risk and Insurance Analytics: A Handbook of Operational Risk* can add value to the industry.

We consider that one of the biggest challenges in OpRisk is to take this risk management branch to the same level that market and credit risk management play. Those two risks are managed proactively and risk managers usually have a say if deals or businesses are approved based on the risk level. OpRisk is mostly kept out of these discussions at this stage and this is an issue as quite a few financial institutions have OpRisk as their dominant exposure. We believe that considerable effort in the industry would have to be put into data collection and modeling improvements, and that is the focus of our work in this book.

Our book can be divided into two parts. The first part (Chapters 1–5) covers the basics, the building blocks, of OpRisk management and measurement. In Chapter 2, there is a broad coverage on the four data elements that need to be used in the OpRisk framework as well as how a risk taxonomy process should be developed. Considerable focus is given to internal loss data and key risk indicators as these would be fundamental in developing a risk-sensitive framework similar to market and credit risks. Subsequently, Chapter 3 shows how OpRisk can be inserted into a firm's strategic decisions and Chapter 4 shows a model to stress-test OpRisk under the US Comprehensive Capital Analysis and Review (CCAR) program. The basic concepts of probability theory and the basic framework for modeling and measuring OpRisk and how loss aggregation should work are considered in Chapter 5.

In the second part of the book (Chapters 6–18), we cover more special topics in OpRisk measurement. For example, diverse methods to estimate frequency and severity models are discussed. Another very popular issue in this industry is how to select severity models and this is also comprehensively discussed in this part. One of the biggest challenges in OpRisk is that data used in measurement can be very different, so combining them into a single measure is not a trivial task. In this part of the book, we show a number of methods to do so. After the core risk measurement work is done, there are still some issues to address that can potentially mitigate the capital found and also on how to manage risks. We also discuss correlation and dependency modeling as well as insurance and risk transfer tools and methods.

We hope this book can be the basis for a number of discussions in the industry. This book can helps novices in the field to learn the building blocks of OpRisk and also suggest new techniques and ideas for those who have been practicing or researching for a while.

Most OpRisk practitioners would say that their focus is always on the tail events as these are the ones that can cause real damage and even force a financial institution into bankruptcy. Realizing this and understanding these tail events and how to model these is a crucial part of OpRisk. Comprehensive treatment of the modeling of heavy-tailed events requires a book-length text and it is a subject covered in the more advanced companion book *Advances in Heavy Tailed Risk Modeling: A Handbook of Operational Risk*, Peters and Shevchenko (2015).

CHAPTER TWO

OpRisk Data and Governance

2.1 Introduction

One of the first and most important phases in any analytical process, and this is certainly no different when developing OpRisk models, is to cast the data into a form amenable to analysis. This is the very first challenge that an analyst or quant faces when determined to model, measure, and even manage OpRisk. At this stage, there is a need to establish how the information available can be modeled to act as an input in the analytical process that would allow proper risk assessment to be used in risk management and mitigation. In risk management, and particularly in OpRisk, this activity is today quite regulated and the entire data process, from collection to maintenance and use, has strict rules, which in a way reduces the variance in the use of the data across the industry.

The OpRisk framework starts by having solid risk taxonomy so risks are properly classified. Firms also need to perform a comprehensive risk mapping across their processes to make sure that no risk is left out of the measurement process. This is a key process to be accomplished and where a number of firms should be paying more attention.

In this chapter, we lay the ground for the basic building blocks of OpRisk management. First we describe how risk taxonomy works, classifying loss events into the major risk categories. Then we describe the four major data elements that should be used to measure and manage OpRisk: internal loss data, external loss data, scenario analysis, and business and control environment factors. When these risk mapping, taxonomy, and data building blocks are reasonably structured, it becomes important to configure the organization of the OpRisk department and a firm's risk governance. Even a very efficient and well-developed OpRisk framework would fail if the proper organization and policies are not in place.

2.2 OpRisk Taxonomy

The term "taxonomy" has become quite popular in the risk management industry. In most conferences and industrial workshops, and most certainly among consultants, the term "risk taxonomy" has become a regular mantra. So, what is risk taxonomy? Taxonomy is actually a term borrowed from biology. One of the missions of the biologist is to discover new species on remote places of the planet and it would make their work easier if they could classify a new

Fundamental Aspects of Operational Risk and Insurance Analytics: A Handbook of Operational Risk, First Edition. Marcelo G. Cruz, Gareth W. Peters, and Pavel V. Shevchenko.

species into a new group based on some characteristics. So taxonomy means the conception, naming, and classifying organisms into groups. It is a common practice in biology to group individuals into species, arranging species into larger groups, and giving those groups names, thus producing a classification. For example, the fact that dolphins live in the sea and look like a fish does not make them a fish as many of their characteristics made biologists classify them as "mammals". Taxonomy basically encompasses description, identification, nomenclature, and classification. Therefore, taxonomy has become an interesting and a popular turn in risk management industry as new risks are being encountered at regular intervals.

Before getting onboard the risk taxonomy bandwagon, a firm must perform a comprehensive risk mapping exercise. This means going through, in excruciating details, every major process of the firm. For example, let us imagine the equity trading process. Analyzing this process would mean going through the risks since the customer places an order until the transaction gets fully settled with exchanges of payment and securities delivered. Those will be the basic risks that unlikely would change, unless there is a change in the process. From this process, a risk manager should also be able to point out where losses are coming from and develop mechanisms to collect them. The outcome of this exercise would be the building block of any risk classification study.

It is interesting to note that even today firms are struggling with basic risk classification, which is the base of the risk management pyramid, the very first building block of a robust risk management framework. Mistakes made in the past years in classifying a risk will have repercussions in the risk management and on the communication of risks, at a minimum, to outside parties like regulators, and might compromise any good work done elsewhere in the framework. There are roughly three ways that firms drive this risk taxonomy exercise: cause-driven, impact-driven, and event-driven. In many firms, risk taxonomy is a mixture of these three making it even more difficult to get it right. Let us discuss these three methods. In the cause-driven method, the risk classification is based on the reasons that cause operational losses. This usually follows the old OpRisk definition (which most firms use in their annual reports) in which OpRisk is defined as a function of "people, systems, and external events". Some risk types in this classification would be, for example, "lack of skills in trade control" or "inappropriate access control to systems". Although there are some advantages in this type of classification, as a "root cause" is pretty much embedded into the risk classification, challenges arise when multiple causes exist or the cause is not immediately clear. If this cause-driven risk classification is applied to a process in which operational losses have high frequency, it would be very difficult for risk managers to classify correctly every single loss, and the attrition with the business and within the department is likely to be high. Another way to perform this classification exercise is through an impact-driven method. In this method, the classification is made according to the financial impact of operational losses. Most firms that follow this type of classification do not invest heavily in OpRisk management; they just use this type to retrieve data from their systems. This is quite common in smaller firms. In this type of classification, it is quite difficult to manage OpRisk as, although the exposures are known, it is difficult to understand what is driving these losses.

The event-driven risk classification is probably the most common one used by large firms. It classifies risk according to OpRisk events. This is the classification used by the Basel Committee. It is interesting to know that during the Basel II discussions, when this type of risk taxonomy was presented, most of the industries were reluctant to accept it. A number of firms, even today, follow their own classification initially and map to the Basel event-type category later. What is interesting in this classification is that the definition is rather broad which should make it easier to accept changes in the process. For example, under "Execution, Delivery, and

Process Management" (EDPM), which is the level-1 event type, there is a category named "Transaction Capture, Execution, and Maintenance" that can be an umbrella for a number of event types. For example, if the equity trading process changes from a old-fashioned phone-based to an online high-frequency trading, using this classification would be easy to define the taxonomy of these risks.

Given how new risks emerge in OpRisk, and also the breadth of its scope, the concept and the ideas behind risk taxonomy in OpRisk sound quite appealing. However, as this is a building block of the OpRisk framework, firms need to be very careful. In the following sections, all seven Basel II event types required for advanced method approach (AMA) are defined and discussed in detail; detailed breakdown into event types at level 1, level 2, and activity groups is provided in BCBS (2006, pp. 305–307).

2.2.1 EXECUTION, DELIVERY, AND PROCESS MANAGEMENT

EDPM loss event type is one of the most prominant in the OpRisk profile of firms or business units with heavy transaction processing and execution businesses. It encompasses losses from failed transaction processing, as well as problems with counterparties and vendors. Table 2.1 describes the Basel event-type breakdown for this risk.

Losses of this event type are quite frequent as these can be due to human errors, miscommunications, and so on, which are very common in an environment where banks have to process millions of transactions per day. A typical example of execution losses might help to illustrate how frequent these losses can be.

TABLE 2.1 **Execution, Delivery & Process Management (EDPM) event-type defined as *losses from failed transaction processing or process management, from relations with trade counterparties and vendors.* Basel II event type classification as provided in BCBS (2006, pp. 305–307)**

Category (level 1)	Categories (level 2)	Activity examples
Execution, Delivery & Process Management	Transaction Capture, Execution and Maintenance	Miscommunication; data entry, maintenance or loading error; missed deadline or responsibility; model/system misoperation; accounting error/entity attribution error; other task misperformance; delivery failure; collateral management failure; reference data maintenance
	Monitoring and Reporting	Failed mandatory reporting obligation; inaccurate external report (loss incurred)
	Customer Intake and Documentation	Client permissions/disclaimers missing; legal documents missing/incomplete
	Customer/Client Account Management	Unapproved access given to accounts; incorrect client records (loss incurred); negligent loss or damage of client assets
	Trade Counterparties	Nonclient counterparty misperformance; misc. nonclient counterparty disputes
	Vendors and Suppliers	Outsourcing; vendor disputes

Consider the following deal: A foreign exchange (FX) trader bought USD 100,000,000 for €90,000,000 (i.e., USD 1 = €0.90) and then sold USD 100,000,000 for €90,050,000 (i.e., USD 1 = €0.9005) with a trading initial profit of €50,000. Both transactions were made almost at the same time, and the trader was obviously very satisfied with a profit of €50,000. In his/her excitement at the successful deal, however, there were some snags in the back-office with some confusion on where to remit the payments of one leg of the deal, and the transaction was finally settled 3 days later than it should have been.

In FX transactions trading tickets are usually larger to compensate for the low margins. Similar situations as described earlier may lead to errors. The counterparties obviously would have demanded a compensation as the settlement has been delayed for 3 days, and the bank would also have paid a penalty, in the form of interest claims of €55,000. Therefore, any error has the potential to be higher than a transactions eventual economic profit.

The overall scenario is alarming. There was a loss of €5000 on the aggregate due to operational errors (€50,000 transaction profit less €55,000 interest claims due for late payment). This is the reality a trading environment faces on the day-to-day. The actions of traders are recognized at the closing of the deal, and errors coming to light at a later time (e.g., mis-pricing, late settlement) are not linked back to the underlying cause. The error goes to an "error account" or the like and, in terms of OpRisk management, those who are responsible for the errors are never identified; even worse is that the real profitability of individual transactions is rarely understood. The cost side (and the OpRisks involved) is in general ignored.

Knowing where these errors occur is very important for OpRisk management. We will see examples like that throughout the book.

Execution, Delivery and Process Management: Misunderstanding a Trading Order: Large US Private Bank, August 2012

Despite the fact that there are currently many options to place orders, where technological devices such as e-mail, Internet, live chats are available, many purchase orders, particularly in private banking, are still being placed by old-fashioned telephone methods. A very common mistake is the misunderstanding of the order, especially frequent when the counterparty is a foreign-language speaker and the communication chain usually goes from client to banker to trader assistant to trader, and in any one of these links there is potential for communication breakdowns to happen.

In a busy afternoon at the end of summer 2012, a client asked his private banker to purchase "USD 100,000 of a particular share". The private banker passed this order to the trader, and at the end of the day the trader passed a bill to the private banker for several million US dollars. The private banker was absolutely stunned to see that they had bought a significant portion of this particular company. As a consequence of this transaction, the share price of this company rose significantly which also generated questions from authorities that suspected some type of *pump-and-dump* scheme. Considering it all, the bank decided to keep the shares and sell it little by little. The operational loss in this case was reflected in the value lost in returning the stocks to the market after the shares returned to their average price.

2.2.2 CLIENTS, PRODUCTS, AND BUSINESS PRACTICES

Loss events under Clients, Products and Business Practices (CPBP) risk type are usually the largest, particularly in the US. These events encompass losses, for example, from disputes with clients and counterparties, regulatory fines from improper business practices, or wrongful advisory activities. Table 2.2 presents the Basel event-type breakdown and definition for this risk type. This is a specific and an important risk type for firms with operations in the US where litigation is very common. As seen in recent regulatory fines imposed on French banks and other foreign banks operating in US jurisdiction, this loss type can also be significant to off-shore entities.

Real OpRisk Events: SBC Warburg (Investment Bank), October 1996

The Securities and Futures Authority in the UK (the former City of London regulator since superseded by the Financial Services Authority) released partial details in March 1997 of an investigation that had commenced in October 1996 into rogue trading in a program trade in SBC Warburg. (A program trade is a transaction where one agent, generally a fund, chooses another agent, generally a bank or a broker, to sell part of its shares in the market in a determined day and hour determined by market prices.) The program trading error that made SBC Warburg the subject of the investigation is thought to have cost it no more than £5 million. Nevertheless, this program trade was one of the largest ever to be awarded to SBC Warburg, and the SFA investigation has clearly embarrassed it. The investigation relates to a mistake made during the execution of a £300 million program trade for an investment trust which caused the price of a

TABLE 2.2 **CPBP event-type defined as *losses arising from an unintentional or negligent failure to meet a professional obligation to specific clients (including fiduciary and suitability requirements) or from the nature or design of a product.* Basel II event type classification as provided in BCBS (2006, pp. 305–307)**

Category (level 1)	Category (level 2)	Activity examples
Clients, Products, and Business Practices	Suitability, Disclosure, and Fiduciary	Fiduciary breaches/guideline violations; suitability/disclosure issues (e.g., KYC); retail customer disclosure violations; breach of privacy; aggressive sales; account churning; misuse of confidential information; lender liability
	Improper Business or Market Practices	Antitrust; improper trade/market practices; market manipulation; insider trading (on firm's account); unlicensed activity; money laundering
	Product Flaws	Product defects (e.g., unauthorised); model errors
	Selection, Sponsorship, and Exposure	Failure to investigate client per guidelines; exceeding client exposure limits
	Advisory Activities	Disputes over performance of advisory activities

number of French stocks to fall sharply. The investigation is being extended whether this bank made a similar error when selling Spanish shares as part of the same program deal.

The SFA investigation focused on a 30-min period on October 30, 1996. At some time around mid-day, SBC Warburg traders learnt that the bank had been awarded three contracts by Kleinwort Benson European Privatization Investment (Kepit) to execute a series of share sales (the so-called program trade) on its behalf. Contracts for programme trades are often awarded just before the deal takes place, and the Kepit deal was no different. It involved SBC Warburg taking the £300 million-worth of shares onto its books just minutes later, at 12:30 pm, and paying Kepit the mid-market prices for each share at that time. In the remaining minutes before the 12:30 pm deadline, SBC Warburg traders sought to sell some of the same shares they were about to get from Kepit in order to reduce the risk (this process is known as short sell, and it is accepted as a normal practice in a program trade, as long as the price does not fall too much).

Elsewhere at SBC Warburg, a trader was running an arbitrage position on Kepit, seeking to make money by exploiting differences between Kepit's own share price and the price of the shares the bank owned. SFA investigators were told that in the minutes before the 12:30 pm deadline, the SBC Warburg trader running the arbitrage position was seen on the trading floor making gestures with his hands for traders to get the price of the shares down. Nevertheless, a mistake by one of the SBC Warburg's Paris-based traders attracted the attention of SFA. Instead of selling as much as he could before 12:30 pm, SFA investigators have been told that the trader misunderstood his instructions and instead attempted to sell at the strike time. The trader also failed to put a so-called down limit on his proposed share sales, effectively turning it into an unlimited sell order.

In the tapes passed to the SFA (all conversations on the trading desk are recorded), the London-based trader is heard talking with a colleague about how the price of the French shares had fallen much further than they had planned. The trader complained that a colleague had just told him, in hindsight after the share prices had collapsed, that they should only have pushed the prices down by 1%. SBC admitted in March 1997 that its short selling had contributed to adverse price movements and dismissed several employees involved in the trade.

2.2.3 BUSINESS DISRUPTION AND SYSTEM FAILURES

Business Disruption and System Failures (BDSF) event type is one the most difficult to spot in a large organization. A system crash, for example, would almost certainly bear some financial loss for a firm, but these losses most likely would be classified as EDPM. An example might help to clarify this point. Suppose that the funding system of a large bank crashes at 9:00 am. Despite all efforts from IT, the system comes back online only by 4:00 pm when money markets are already closed. When the system returns, the bank learns that it needs to fund an extra USD 20 billion on that day. As the markets are already closed, they need to make requests to their counterparties to allow them special conditions; however, the rates in which they capture these funds are higher than the daily average. This extra cost, although due to a system failure and, therefore, should be classified as BDSF, would hardly be captured at all. Table 2.3 presents the formal Basel definition and breakdown of this risk type.

TABLE 2.3 **BDSF event risk type defined as *losses arising from disruption of business or system failures*. Basel II event type classification as provided in BCBS (2006, pp. 305–307)**

Category (level 1)	Category (level 2)	Activity examples
Business Disruption and System Failures	Systems	Hardware; software; telecommunications; utility outage/disruptions

TABLE 2.4 **External fraud event risk type defined as *losses due to acts of a type intended to defraud, misappropriate property, or circumvent the law, by a third party*. Basel II event type classification as provided in BCBS (2006, pp. 305–307)**

Category (level 1)	Category (level 2)	Activity examples
External fraud	Theft and fraud	Theft/robbery; forgery; check kiting
	Systems security	Hacking damage; theft of information (w/monetary loss)

The difficulty to capture this event type is reflected in external databases where, aside damage to physical assets, this risk type has least number of events.

2.2.4 EXTERNAL FRAUDS

External frauds are frauds committed or attempted by third parties or outsiders against the firm. Examples would be system hacking and checque and credit card frauds. External fraud is very common in retail businesses where financial firms deal with millions of clients. Frauds attempted or committed by clients are a daily event in sectors such as retail banking, retail brokerage, and credit card services; see Table 2.4 for Basel II definition and breakdown.

2.2.5 INTERNAL FRAUD

Internal frauds are frauds committed or attempted by a firm's own employees. It is one of the less frequent types of OpRisk loss. Given the sophisticated, controls that most institutions have this would be unlikely. However, events such as traders mismarking positions, particularly in assets that are hard to establish an accepted mark-to-market price are not uncommon. Recently there were a number of large internal frauds in which billions of dollars were lost as traders of a particular bank failed to mention their position. These are usually low-frequency/high-severity events. Table 2.5 presents the formal Basel definition and breakdown of this risk type.

Real OpRisk Events: Model Inputs Fraud, NatWest, March 1997

One of the most famous case in derivatives mispricing was the one that happened at NatWest in 1997. On February 28, 1997, a few days after the bank released its annual results, it announced a loss of approximately USD 150 million caused by a junior trader who has already left the bank. The trader was said to be dealing in long-dated OTC interest rate options, used by companies that borrow at a floating rate and purchase a cap on the interest payments. The major problem in valuing these options is that they are relatively illiquid. The trader calculated the price of the options by providing his own estimates of volatility, which he apparently overestimated, creating fictitious profits that built up in the books over time.

The volatility estimates resulted in the options being underpriced. The trader attracted more clients, booking the requested premium, thereby increasing the apparent profitability of his desk (and, by extension, his remuneration). The loss was realized when the options were exercised.

TABLE 2.5 **Internal fraud event risk type defined as *losses due to acts of a type intended to defraud, misappropriate property or circumvent regulations, the law or company policy, excluding diversity/discrimination events, which involves at least one internal party*. Basel II event type classification as provided in BCBS (2006, pp. 305–307)**

Category (level 1)	Category (level 2)	Activity examples
Internal fraud	Unauthorised/Activity	Transactions not reported (intentional); transaction type unauthorised (w/monetary loss); mismarking of position (intentional)
	Theft and fraud	Fraud/credit fraud/worthless deposits; theft/extortion/embezzlement/robbery; misappropriation of assets, malicious destruction of assets; forgery; check kiting; smuggling; account take-over/impersonation/etc.; tax noncompliance/evasion (wilful); bribes/kickbacks; insider trading (not on firm's account)

2.2.6 EMPLOYMENT PRACTICES AND WORKPLACE SAFETY

Employment Practices and Workplace Safety (EPWS) type of risk is more prominent in the Americas than Europe or Asia as either the labor laws are old-fashioned and/or there is more a culture of litigation against the employers (Table 2.6). For example, some large banks in Brazil would count employment litigation on the tens of thousand and it is one of the main OpRisks for banks. In some lines of business like investment banking employment issues are also quite important. As these line of business mostly provide advisory to large corporations and the key personnel is highly compensated, litigation against some of these key employees and losing them can cost millions of dollars.

TABLE 2.6 **EPWS event risk type defined as *losses arising from acts inconsistent with employment, health or safety laws or agreements, from payment of personal injury claims, or from diversity/discrimination events.* Basel II event type classification as provided in BCBS (2006, pp. 305–307)**

Category (level 1)	Category (level 2)	Activity examples
Employment Practices and Workplace Safety	Employee relations	Compensation, benefit, termination issues; organised labor activity
	Safe environment	General liability (e.g., slip and fall.); employee health and safety rules events; workers compensation
	Diversity and discrimination	All discrimination types

2.2.7 DAMAGE TO PHYSICAL ASSETS

Damage to Physical Assets (DPA) is another OpRisk event type. The most common method to assess the exposure to this risk is through scenario analysis using insurance information. Very few firms actively collect losses on this risk type as these are usually either too small or incredibly large. The formal Basel definition and breakdown of this risk type is presented in Table 2.7.

TABLE 2.7 **DPA event risk type defined as *losses arising from loss or damage to physical assets from natural disaster or other events.* Basel II event type classification as provided in BCBS (2006, pp. 305–307)**

Category (level 1)	Category (level 2)	Activity examples
Damage to physical assets	Disasters and other events	Natural disaster losses; human losses from external sources (e.g., terrorism, vandalism)

2.3 The Elements of the OpRisk Framework

The four elements that should be used in any OpRisk framework are as follows:

- Internal loss data;
- Business environment and internal control factors;
- External loss data;
- Scenario analysis.

We provide a description of each of these elements in the following text.

2.3.1 INTERNAL LOSS DATA

Operational loss means a gross monetary loss (excluding insurance or tax effects) resulting from an operational loss event. An operational loss includes all expenses associated with an operational loss event except for opportunity costs, forgone revenue, and costs related to risk management and control enhancements implemented to prevent future operational losses.

Having a robust historical internal loss database is the basis of any OpRisk framework. These losses need to be classified into the Basel categories (and internal if different than the Basel) and mapped to a firm's business units. Given their importance for the OpRisk framework, the collection and maintenance of these data are heavily regulated. Basel II regulation says that firms need to collect at least 5 years of data, (BCBS, 2006), but most decided not to discard any loss even when these are older than this limit. Since losses are difficult to acquire and take years to build up a reliable and informative loss database, consequently most firms even pay to supplement internal losses (see the external loss database). Hence, it is clear that it would not make sense to discard losses that took place in the own firm unless the business in which this loss took place was sold. There are a number of issues that can come from internal data modeling that are worth comments and are listed below.

Considerable challenges exist in collating a large volume of data, in different formats and from different geographical locations, into a central repository, and ensuring that these data feeds are secure and can be backed up and replicated in case of an accident.

2.3.2 SETTING A COLLECTION THRESHOLD AND POSSIBLE IMPACTS

Most firms set a threshold for loss collection as allowed by Basel. However, this decision can have significant impact in establishing the risk profile of a business unit. This is usually the case in businesses that have heavy transaction execution like asset management or equities. See the example in Table 2.8. If the OpRisk department had chosen USD 100,000 as the threshold, usually under the argument that only tail events drive OpRisk capital, that firm would think that its total loss in that year was USD 49 million. If the threshold choice was USD 20,000, the total losses would be USD 53 million. However, most losses are due to compensating retail clients whose orders are usually ranging from USD 1000 to USD 50,000. The sum of the losses under USD 50,000 is about USD 20 million, which is almost equivalent to the losses

TABLE 2.8 **The impact of threshold choice: losses in a certain year for the asset management division of a bank**

Loss brackets (USD)	Number of losses	Total (USD)	Accumulated total (USD)
> 5,000,000	3	23,750,325	23,750,325
1,000,000–5,000,000	7	13,775,000	37,525,325
500,000–1,000,000	10	8,250,781	45,776,106
100,000 –500,000	12	3,562,177	49,338,283
50,000 –100,000	22	1,723,490	51,061,773
20,000 –50,000	71	2,159,021	53,220,794
< 20,000	1520	17,500,235	70,721,029

above USD 5 million. For this particular firm, setting the loss collection threshold in USD 100,000 would show total losses for the year as USD 49 million. However, if this firm had not set a loss collection threshold they would observe that their actual losses were USD 71 million, a very different risk profile.

A number of OpRisk managers pick their threshold thinking only in terms of OpRisk capital. Disregarding these small losses in many cases can bias the risk profile of a business unit and, of course, this will also have an impact on OpRisk capital.

2.3.3 COMPLETENESS OF DATABASE (UNDER-REPORTING EVENTS)

In gathering data from disparate sources, we need to avoid an OpRisk in collecting the OpRisk data collection. Such risks and subsequent losses may arise, for example, the employee responsible for reporting losses does not send the loss information to the central database, whether accidental or not. The Basel II document BCBS (2006) refers to this scenario with the possible consequence being that an institution that could not prove that loss data is flowing with a high degree of reliability to the central database(s) is likely to be disallowed to employ more advanced techniques for assessing the levels of risk.

The development of filters that capture operational issues and calculate an eventual operational loss is one of the most expensive parts of the entire data collation process, but the outcome can be decisive in making an OpRisk project successful and increasing confidence in the completeness of the loss database.

This OpRisk filter will vary from bank to bank depending on their systems, but in all cases it works like a conduit between systems, collecting every cancellation or alteration made to a transaction or any differences between the attributes of a transaction in one system compared to its attributes in another system. The transaction flow starts at the front-office system that registers the transaction passing it to the accounting and clearing systems. Any discrepancy, alteration, or cancellation must be extracted by the OpRisk filter. Also, abnormal inputs (e.g., a lower volatility in a derivative) can be flagged and investigated. The filter will calculate the OpRisk loss event and several other impacts in the organization.

2.3.4 RECOVERIES AND NEAR MISSES

The Basel II rules (BCBS, 2006) in general do not allow for the use of recoveries to be considered for capital calculation purposes. The issue again is that if firms are trying to estimate losses that can happen once every thousand years, it would not make sense to start applying mitigating factors to reduce the losses and eventually reducing also capital. For this reason, gross losses should be considered for OpRisk calculation purposes.

The only exception is on rapidly recovered loss events but even this exception is not accepted everywhere. Rapidly recovered loss events are OpRisk events that lead to losses recognized in financial statements that are recovered over a short period. For instance, a large internal loss is rapidly recovered when a bank transfers money to a wrong party but recovers all or part of the loss soon thereafter. A bank may consider this to be a gross loss and a recovery. However, when the recovery is made rapidly, the bank may consider that only the loss net of the rapid recovery constitutes an actual loss. When the rapid recovery is full, the event is considered to be a "near miss".

2.3.5 TIME PERIOD FOR RESOLUTION OF OPERATIONAL LOSSES

Some OpRisk events, usually some of the largest, will have a large time gap between the inception of the event and the final closure, due to the complexity of these cases. As an example, most litigation cases that came up from the financial crisis in 2007/2008 were only settled by 2012/2013. These legal cases have their own life cycle and start with a discovery phase in which lawyers and investigators would argue if the other party has a proper case to actually take the action to court or not. At this stage, it is difficult to even come up with an estimate for eventual losses. Even when a case is accepted by the judge it might be several years until lawyers and risk managers are able to estimate properly the losses. Firms can set up reserves for these losses (and these reserves should be included in the loss database), but they usually do that only for a few weeks before the case is settled to avoid disclosure issues (i.e., the counterparty eventually knows the amount reserved and uses this information in their favor). This creates an issue for setting up OpRisk capital because firms would know that they are going to under go a large loss and yet are unable to include it in the database; the inclusion of this settlement would cause some volatility in the capital. The same would happen if a firm set a reserve of, for example, USD 1 billion for a case, and then a few months later, if a judge decides to remove the loss in favor of the firm. For this reason, firms need to have a clear procedure on how to handle those large, long-duration losses.

2.3.6 ADDING COSTS TO LOSSES

As said earlier, an operational loss includes all expenses associated with an operational loss event except for opportunity costs, forgone revenue, and costs related to risk management and control enhancements implemented to prevent future operational losses. Most firms, for example, do not have enough lawyers on payroll (or expertise) to deal with all the cases, particularly some of the largest or those that demand some specific expertise and whose legal fees are quite expensive. There are cases in which the firm wins in the end, maybe due to some external law firms, but the cost can reach tens of millions of dollars. In such cases, though the firms wins a court victory, there will be an operational loss.

2.3.7 PROVISIONING TREATMENT OF EXPECTED OPERATIONAL LOSSES

Unlike credit risk, the calculated expected credit losses might be covered by general and/or specific provisions in the balance sheet. For OpRisk, due to its multidimensional nature, the treatment of expected losses is more complex and restrictive. Recently, with the issuing of IAS37 by the International Accounting Standards Board, Wittsiepe (2008), the rules have become clearer as to what might be subject to provisions (or not). IAS37 establishes three specific applications of these general requirements, namely:

- a provision should not be recognized for future operating losses;
- a provision should be recognized for an onerous contract — a contract in which the unavoidable costs of meeting its obligations exceeds the expected economic benefits;
- a provision for restructuring costs should be recognized only when an enterprise has a detailed formal plan for restructuring and has raised a valid expectation in those affected.

These provisions should not include costs, such as retraining or relocating continuing staff, marketing or investing in new systems and distribution networks; the restructuring does not necessarily entail that.

IAS37 requires that provisions should be recognized in the balance sheet when, and only when, an enterprise has a present obligation (legal or constructive) as a result of a past event. The event must be likely to call upon the resources of the institution to settle the obligation, and, more importantly, it must be possible to form a reliable estimate of the amount of the obligation. Provisions should be measured in the balance sheet at the best estimate of the expenditure required to settle the present obligation at the balance sheet date. Any future changes, like changes in the law or technological changes, may be taken into account where there is sufficient objective evidence that they will occur. IAS37 also indicates that the amount of the provision should not be reduced by gains from the expected disposal of assets (even if the expected disposal is closely linked to the event giving rise to the provision) nor by expected reimbursements (arising from, for example, insurance contracts or indemnity clauses). When and if it is virtually certain that reimbursement will be received should the enterprise settle the obligation, this reimbursement should be recognized as a separate asset.

2.4 Business Environment and Internal Control Environment Factors (BEICFs)

One can see OpRisk as a function of the control environment. If the control environment is fair and under control, large operational losses are not likely to take place and OpRisk is considered to be under control. Therefore, understanding the firm's business processes, mapping the risks on these processes, and assessing the control of these processes are the fundamental roles of an OpRisk manager. A simple example is the equities trading process and is shown in Figure 2.1.

Firms need to be able to assess risk on the many steps of the settlement process and report them regularly. There are a couple of tools that are commonly used by financial firms to perform this task: Risk Control Self-Assessment and Business and Control Environment programs.

2.4.1 RISK CONTROL SELF-ASSESSMENT (RCSA)

These are also knowns as Control Self-Assessment (CSA) in some firms. According to this procedure, firms regularly ask experts about their views on the status of each business process and subprocess. These reviews are usually done every 12 or 18 months and color rated Red/Amber/Green (RAG) according to the perceived status. Some firms go beyond and try to quantify these risks using subjective approaches or through a scorecard. For many firms,

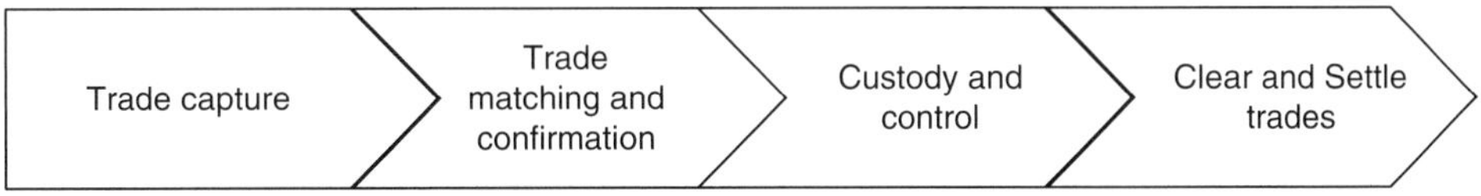

FIGURE 2.1 Equity Settlement Process

RCSA is the anchor of the OpRisk framework and most OpRisk activities are linked to this procedure.

In a broad sense, the RCSA program requires the documentation and assessment of risks embedded in a firm's processes. Levels of risks are derived (usually from a frequency and severity basis), and controls associated with these risks are identified. As risks are usually reported by business units, these processes are aggregated to a certain business unit and rated/assessed.

In the RCSA program, managers first identify and assess inherent risks by making no inferences about controls embedded in the process: controls are assumed to be absent. Under this assumption, managers must carefully identify how risk manifests within the activities in the processes. The following are the usual questions asked by risk managers in this phase:

- **Risk scenarios**. Where are the potential failure points in each of these processes?
- **Exposure**. How big a loss could happen to my operation if a failure happens?
- **Correlation to other risks**. Could a failure altogether change my organization's performance, either financially, its reputation, or affect any other area?

The answers point toward the specific inherent risks embedded within a business unit's process, which must be assessed to determine the likelihood the events could occur (frequency) and severity. The results of this analysis provide a birds' eye view of the inherent risk of a firm's business processes. Management can then use this assessment to prioritize and focus on the most critical risks that must be proactively managed.

Once these inherent risks are understood, controls will be added in the RCSA framework. The effectiveness of these controls are then assessed to understand how efficient these are to mitigate risks. At this stage, the residual risk is also calculated, which is the risk that is left after inherent risks are controlled. Put another way, residual risk is the probability of loss that remains to systems that store, process, or transmit information after security measures or controls have been implemented.

For a firm that has the RCSA program as the core of the OpRisk framework, all other OpRisk initiatives under the firms OpRisk program are usually structured to feed the RCSA. Risk metrics such as key risk indicators (KRIs), internal loss events, and external events would contribute to the risk identification process ensuring the organization has considered all readily available data and benchmark risk assessments.

Once the universe of controls and mitigation measures has been identified, the business unit can partner with various control functions to conduct the control testing phase of the RCSA. Control testing is critical to a mutual understanding of expectations and actions across business units and between the front and back offices.

One significant challenge that arises due to combining RCSA data is interpreting what the data actually means. For example, outputs from a RCSA program might lead a risk manager to conclude that no immediate action is required if the risk exposures are controlled within the tolerances acceptable to the firm. On the other hand, if the RCSA data indicates that the control environment is weakening and threatening the success of a particular business goal, a risk manager might decide to recommend a corrective action. However, weighting those risks across the entire risk universe and naming the most important or "key" might not be an easy and objective task.

There are a number of vendors that provide systems that help to collate these results. The issue with these programs in general is that they make it harder to integrate with the other data inputs that are numeric. Even if these RAG assessments can be converted to a number or rating, there is always a bias embedded that the person who does the assessment would have a motivation to improve their ratings so as to reduce their capital.

2.4.2 KEY RISK INDICATORS

These indicators/factors are mostly quantitative and are used as a proxy for the quality of the control environment of a business. For example, in order to report the quality of the processing systems of an investment bank, we might design factors such as "system downtime" (measuring the number of minutes that a system stayed offline), and "system slow time" (counting the minutes that a system was overload and running slow). These KRIs can be extremely important in OpRisk measurement as they can allow OpRisk models to behave very similarly to those in market and credit risks.

Going back to the equity settlement example, instead of using RAG self-assessment, a better way to assess the quality of these processes is to establish a few KRIs that provides an accurate picture of the control environment as seen in Figure 2.2. As an example, on the trade confirmation stage of the settlement process, if the number of unsigned confirmations older than 30 days increases to over a certain percent of the total population, and the number of

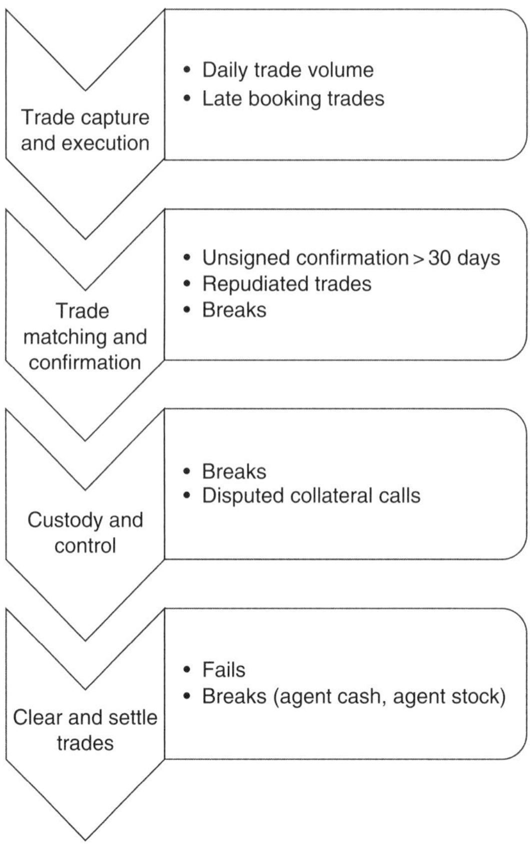

FIGURE 2.2 **Equity Settlement Process**

repudiated trades increases, one might say that this process is facing challenges that need to be addressed.

The process of KRI collection deserves special attention. It is important that these data are absolutely reliable, in order to display relationships between KRIs and losses. Automating the collection straight from the firm's operational systems might help to create a more realistic reflection of the true profile of the infrastructure of a certain business. There are many stages in establishing these links and of course there is a cost associated with the implementation of the KRI program, but probably no other type of data will be more powerful than KRIs for managing and measuring operational risk. It is much easier to explain OpRisk as a function of the control environment in which a firm exists than to say that OpRisk capital is moving up or down because of past losses or changes in scenarios.

The first stage of the KRI collection process is trying to establish assumptions on the OpRisk profile of a certain business. For example, we might assume that execution errors in the equities division can be explained by the trade volume on the day, the number of securities that failed to be received or delivered, the head count available on the trading desk and the back office, and system downtime (measured by minutes offline). The decision to be made is: at what organizational level should this relationship be measured? Equities division as a whole? Should we break down equities division into cash equities, listed derivatives and OTC derivatives, or along any other lines? Should we consider breaking it down along regional lines? All these questions are fundamental for the success of the analysis. The quantitative incorporation of KRI data into OpRisk modeling is discussed in Chapter 16.

If loss data and KRIs are collected at cost center level (the lowest possible level), it becomes possible to perform this disaggregation. In general, the lower the level you model the causal relationship, the better the chances that you will find higher level fits to the model. Put this another way, it is easier to find strong causal relationships, if you model, for example, the US cash equities department than modeling at the global equities division level, as the lower level would better capture local nuances, idiosyncrasies, and trends.

The modeler might also consider using external factors such as equity indexes and interest rates. It is common to find strong relationships between a stock market index and operational losses, for example, higher volatility on stock markets is usually associated with high trading volumes, which in turn is highly associated with execution losses in OpRisk. Table 2.9 presents

TABLE 2.9 **Examples of BEICFs used in few environments**

Business environment	Factor	Description
Systems	System downtime	Number of minutes a system is offline
	System slow time	Number of minutes a system is slow
	Software stability	Number of code lines changed in a program or software in a certain period of time
Information Security	Malware attacks	Number of malware attacks
	Hacking attempts	Number of hacking attempts
People/Organization	Employees	Number of employees
	Employees experience	Average experience of employees
Execution/Processing	Transactions	Number of transactions processed
	Failed transactions	Number of transactions that failed to settle
	Data quality	Ratio of transactions with errors
	Breaks	Number of transactions breaks

few examples of Business Environment and Internal Control Factors (BEICFs) used in few environments.

2.5 External Databases

According to the Basel Accord, OpRisk modelers need to calculate regulatory capital at the 99.9% confidence level, which is equivalent to finding enough capital to protect against losses in the worse year in a 1,000 year period. One way to try to overcome these challenges is through using other firms' loss experiences. This is common in insurance. For example, suppose that a US insurer wants to expand to a new state, say New Jersey. This insurer does not have experience in New Jersey; New Jersey has different characteristics, for example it may have much more cars per square foot than other states and hence the accident ratio is known to be higher. How can this insurer price correctly its premium in New Jersey? The most used alternative is to start with a local database of car accidents. This database is available, with considerable details, for insurance companies to acquire. Obviously, this database would never replace the insurer's own loss experience in their portfolio, but while this loss experience is not available, the best way to start the business is using this external database. As the insurer starts building up their own loss experience, it can start weighting the importance of the external database in their premium through credibility theory methods (which will be discussed later in Chapter 15).

Similarly, banks and other financial firms might struggle to come up with reasonable measures for some types of risk because they were never exposed to large losses, but, despite that, they understand that they are still under the risk that such a loss-would happen eventually. These loss-gathering databases can be very useful in these cases.

There are basically three ways to get hold of these databases as seen in Table 2.10. The best choice for a firm would depend significantly on how their framework is structured and how the modeler expects to use these losses.

TABLE 2.10 **Methods to acquire external data and details**

Type	Details	Pros	Cons
Internally developed	Firm gathers these losses from news feeds and magazines	Cheapest way	It might not be comprehensive enough and miss losses in many industries and jurisdictions
Consortia	The most popular is ORX which has some of the largest banks in the industry	Loss reporting threshold is €20,000	No details on the losses. It can only be used for measurement
Vendors	There are a number of vendors like IBM OpVantage and SAS	More detailed analysis on the loss. It can be used for management or scenarios	Loss threshold is usually high (USD 1 million). Loss details might not be accurate as these were taken from newspapers

2.6 Scenario Analysis

Another important tool in OpRisk management and measurement is scenario analysis. For a significant number of firms, the scenario analysis program is the pillar of their framework. These scenarios estimates are usually gathered through expert opinions, where these experts (or a group of experts) communicate their estimates on how losses can happen on an extreme situation. These experts are commonly guided by information gathered from external data or KRIs and internal loss trends, see for instance discussions on scenario analysis for OpRisk in Rippel and Teplý (2008), Alderweireld *et al.* (2006) and Hoffman (2002).

Though there are different approaches to run a scenario workshop, only three approaches are widiley used: structured workshops, surveys, or individualized discussions. A recent survey in 2012 with the largest US financial firms (the results are not in public domain and reference cannot be provided) shows that information from experts is obtained mainly through structured workshops (Figure 2.3). A comprehensive guide to performing and establishing appropriate statistical structures for surveys in such workshops is provided in detail in O'Hagan *et al.* (2006).

Scenarios can be a useful tool in case of emerging risks where a loss experience would not be available. Financial institutions understanding this challenge are creating many new scenarios for these emerging risks every year. Figure 2.4 presents some other results of this

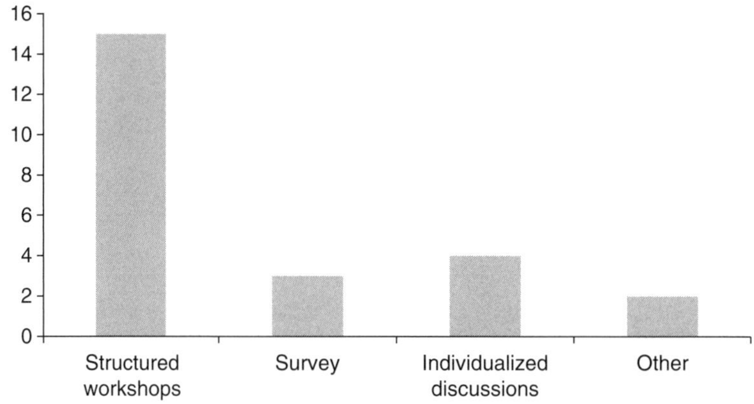

FIGURE 2.3 Survey on how US banks run scenarios

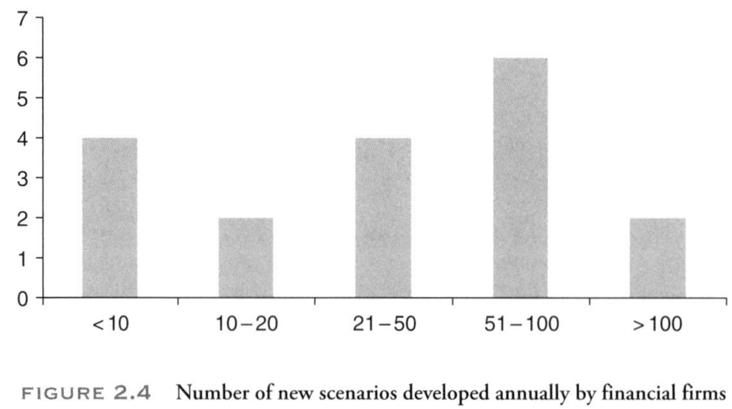

FIGURE 2.4 Number of new scenarios developed annually by financial firms

TABLE 2.11 **Using scenario analysis outcome for measurement**

Loss bracket (in USD thousand)	Loss frequency	Relative frequency
USD 5,000	7	6.9%
1,000–5,000	10	9.8
500–1,000	15	14.7
100–500	30	29.4
50–100	40	39.2
Total	102	

survey about the number of new scenarios developed annually by financial firms showing that most firms develop between 51 and 100 scenarios every year.

In order to make the outcomes of the scenario analysis workshops useful to the OpRisk measurement and quantification efforts, the opinions need to be converted into numbers. There are a few ways to do so, but the most frequent is through gathering estimates on the loss frequencies on predefined severity brackets. These numbers are then converted to empirical distributions, see example in Table 2.11, that are aggregated with internal losses later.

After converting expert opinion into an empirical distribution, the question is how to incorporate this into the OpRisk framework. There are a number of articles on the subject, for example, see recent publications of Dutta and Babbel (2013), Ergashev (2012), and Shevchenko (2011). It will be discussed in detail in Chapters 14 and 15.

Common Issues and Bias in Scenarios. Because scenarios are usually based on expert opinion, they present a number of biases, see for example, a demonstration of such features in the experiments designed by Lin and Bier (2008). This is one of the key limitations of this process as these bias are very difficult to mitigate or avoid. Some of the biases are as follos:

- **Presentation Bias.** This arises when the order in which the information is provided can skew or alter the assessment from the experts; see discussion in Hogarth and Einhorn (1992);
- **Availability bias.** It is related to the over/underestimation of loss events due to respondents' exposure or familiarity to a particular experience or risk. For example, if the expert has a 30 years career in FX trading and had never experimented or seen an individual loss of USD 1 billion or larger, he/she might be unable to accept the risk that such a loss would take place;
- **Anchoring bias.** Anchoring occurs when participants restrict their estimates to being within a range of a given value, which may come from their own experiences, a value they have seen elsewhere (e.g., internally, in the media) or a value provided in the workshop; see discussion in Wright and Anderson (1989);
- **"Huddle" bias or anxiety bias.** It involves the tendency of groups to avoid conflicts and differences of opinion, either because individuals do not want to disrupt the smooth functioning of the group through dissent, or because they are unwilling to disagree openly with the more senior, expert, or powerful people in the room; see discussions in O'Hagan (2005);
- **Gaming.** Conflicts of participants' interests with the goals or consequences of the workshops can cause motivational biases or gaming. Participants may be unwilling to disclose

information or engage meaningfully in the workshop or may seek to influence the outcomes;

- **Over/under confidence bias.** This bias involves over/underestimation of risk due to the available experience and/or literature on the risk being limited;
- **Inexpert opinion.** In many firms, scenario workshops do not attract the expert (or the expert is not identified) and a more junior or someone with much less experience ends up participating in the workshop and providing inaccurate estimates;
- **Context bias.** This bias arises when framing in a certain manner alters the response of experts, that is, color their opinion; see discussion in Fischhoff *et al.* (1978).

A fundamental problem that scenario analysis programs face is the disparity of understanding and opinions on losses sizes and frequencies. To circumvent some of these problems, application of the Delphi technique may be of help. The Delphi technique, as Linstone and Turoff (1975) defined, "*...may be characterized as a method for structuring a group communication process so that the process is effective in allowing a group of individuals, as a whole, to deal with a complex problem*".

The Delphi concept is a spin off from defense research. "Project Delphi" is the name given to an American Air Force project, started in the early 1950s, that made use of expert opinion (see Dalkey and Helmer, 1963). The objective of the original study was to "*obtain the most reliable consensus of opinions within a group of experts*" by a series of intensive questionnaires interspersed with controlled opinion feedback.

Delphi has been tested and broadly used in several applications such as gathering current and historical data not accurately known or available and examining the significance of events. Usually, one or more of the following properties of the problem to be solved leads to the need for employing Delphi.

- The problem does not lend itself to precise analytical techniques but can benefit from subjective judgments on a collective basis;
- The individuals needed to contribute to the examination of a broad or complex problem have no history of adequate communication and may represent diverse backgrounds in respect of experience or expertise;
- Time and cost make frequent group meetings infeasible; and
- More individuals are needed than can effectively interact in a face-to-face exchange.

Therefore, for Delphi to work, it necessary that a group of experts in each business get together in order to estimate OpRisk occurrences at a given confidence level. Consider an example: bank in order to assess transaction execution risk in the fixed income desk decided to get three different perspectives: from the front desk (traders), from the finance, and from the operations. Each one of these areas has a different perspective on what risks would be and how many losses would happen. As the estimates from each of the three areas were very different, a separate scenario workshop was performed in each department and the participants were elicited to estimate extreme losses. At the end, a final number was agreed by the three areas and all recognized that tremendous education took place as traders, for example, did not have the perspective of losses due to settlement failures. Delphi technique (Dalkey and Helmer, 1963) has a number of stages:

1. In the first step, the subject under discussion should be explored with as many individuals contributing additional information;
2. Given the information from step 1, a feedback and a description of the issues are provided to the group;
3. (Optional) Bring out the possible differences found in step 2 and evaluate them; and
4. A final evaluation occurs when all the previously gathered information has been initially analyzed and the evaluations have been fed back to the respondents for consideration.

Finally, we would like to mention that ideas from works on expert elicitation processes were implemented in a freely available toolkit known as the Sheffield Elicitation Framework (SHELF)[1], which is covered under copyright when it comes to commercial usage; see details on the associated website. In agreement with the standard industrial practice of structured workshops, the SHELF framework is developed to be performed with a group elicitation in mind and comprises a framework for eliciting beliefs of one or more experts as a group; SHELF will be discussed further in Chapter 14.

2.7 OpRisk Profile in Different Financial Sectors

After deciding the form of the operational loss data model and the types of losses that need to be reported, it is useful to split the financial institution into different business lines, given that the OpRisk profile is generally very diverse across different businesses within a financial institution. While an asset management unit is more inclined to have legal/liability problems (although still having a few transaction processing problems, in general, asset managers hold their positions longer than treasury), the investment bank arm is more inclined to operational errors in processing transaction. A large investment bank might process over a million transactions a day.

A typical list of business units includes *Corporate Finance*, *Trading and Sales*, *Retail Banking*, *Commercial Banking*, *Payment and Settlement*, *Agency Services*, *Asset Management*, and *Retail Brokerage*. These are business units at level 1 as suggested in Basel II. Detailed breakdown into level 2 business units and activity groups can be found in BCBS (2006, pp. 302). Also it can be appropriate to add extra business unit *Insurance*. Most of these business units are discussed in the following sections.

2.7.1 TRADING AND SALES

It should not come as a surprise that trading and sales OpRisk profile is dominated by "EDPM" or just "Execution". This can be clearly seen Table 2.12, where both frequency and severity execution losses dominate. The business model in trading is quite simple; traders perform trades on behalf of either their own firms or clients, and these trades get settled by exchanging the securities against some form of payments. However, as the products are diverse and complex and settlements deadlines and procedures vary significantly it is not surprising that executing

[1] SHELF is available at http://www.tonyohagan.co.uk/shelf/

TABLE 2.12 **Trading and Sales OpRisk Profile**

Event type	Frequency (%)	Severity (%)
Internal Fraud	1.0	11.0
External Fraud	1.0	0.3
Employment Practices and Workplace safety	3.1	2.3
Clients, Products, and Business Practices	12.7	29.0
Damage to Physical Assets	0.4	0.2
Business Disruption and System Failures	5.0	1.8
Execution, Delivery & Process Management	76.7	55.3

Source: Results from the 2008 Loss Data Collection Exercise for Operational Risk, see BCBS (2009b).

TABLE 2.13 **Corporate Finance OpRisk Profile**

Event type	Frequency (%)	Severity (%)
Internal Fraud	1.6	0.24
External Fraud	5.4	0.12
Employment Practices and Workplace safety	10.1	0.59
Clients, Products, and Business Practices	47.1	93.67
Damage to Physical Assets	1.1	0.004
Business Disruption and System Failures	2.2	0.02
Execution, Delivery & Process Management	32.5	5.36

Source: Results from the 2008 Loss Data Collection Exercise for Operational Risk, see BCBS (2009b).

these transactions is the major OpRisk of this business and, for many trading shops, the major overall risk that they are exposed to.

2.7.2 CORPORATE FINANCE

This business is where financial firms many times behave similar to consulting firms by providing advise to corporations in possible mergers and acquisitions, doing an IPO or even assessing strategic alternatives. The differences to consulting firms are due to the fact that corporate finance in banks constantly offers financing options, so deals are made. Therefore, it is expected that most of the losses fall under the umbrella of "litigation" or disputes with clients for arguably poor advice when, for example, IPOs go wrong; see Table 2.13.

2.7.3 RETAIL BANKING

The OpRisk profile of retail banks is not too dissimilar to that of retail brokerage; see Table 2.14. On the frequency side, most losses are due to external frauds that are daily events for these firms. Execution comes in a far second. However, when looking at severity, the largest risk exposure is due to litigation once again.

TABLE 2.14 **Retail Banking OpRisk Profile**

Event type	Frequency (%)	Severity (%)
Internal Fraud	5.4	6.3
External Fraud	40.3	19.4
Employment Practices and Workplace safety	17.6	9.8
Clients, Products, and Business Practices	13.1	40.4
Damage to Physical Assets	1.4	1.1
Business Disruption and System Failures	1.6	1.5
Execution, Delivery & Process Management	20.6	21.4

Source: Results from the 2008 Loss Data Collection Exercise for Operational Risk, see BCBS (2009b).

2.7.4 INSURANCE

For those not familiar with this industry, this sector can be actually divided into three types given the significant differences: life insurance, health insurance, and property/casualty or "P&C" insurance (or general insurance as known in Europe). To put very simply, life insurers basically charge a premium from individuals in exchange to providing a sum of money when they die. Life insurers also offer retirement and income-protection products. Health insurers provide medical and hospital coverage. P&C insurers offer coverage against damage to properties caused by fire, natural disasters, theft, etc. They also offer protection against liabilities (e.g., directors being sued and professional errors). The actuarial calculation used in the P&C insurance is very similar to the one used in OpRisk capital calculation. Most of operational risk capital techniques, are derived from P&C actuarial techniques, and there are many articles in the *Journal of OpRisk* that were written by P&C actuaries; also Chapters 17 and 18 discuss modeling insurance in detail.

Regarding the sector's overall current financial situation, similar to most of the financial sectors, the effects of the financial crisis still lingers. Life insurers started to feel the consequential effects from the long low-interest rate environment, which affects their profitability and company valuations and also, as consumers struggle, declining sales and revenue. If interest rates continue to stay low, and it appears likely that they will for at least another two years, then life insurers' financial pain will be broader and deeper. On the P&C side, the continuing prospects for weak investment returns and low interest rates over an extended period compel carriers to improve underwriting margins, requiring difficult decisions concerning pricing and operating approaches. Organic growth continues to be a challenge, given the economic situation and the competitive landscape. Individual insurers confront greater competition, driven by an abundance of capital, uncertainty around the timing, and the scope of regulatory changes and the continuing volatility caused by weather-related losses, highlighted recently by Hurricane Sandy in 2012 (in the US, Hurricane Sandy affected 24 states with particularly severe damage in New Jersey and New York). Health insurers in the US, given the advent of the Patient Protection and Affordable Care Act (signed into law by US President Barack Obama on March 23, 2010, and commonly referred to as "Obamacare"), are in much better shape than their counterparts with a better perspective ahead of them.

Regarding risk regulation in this sector, there are significant differences between Europe and the US. In Europe, a process similar to Basel II was developed by insurance regulators, called Solvency 2. Two key themes have dominated regulatory discussions in the past year:

supervisory focus on risk and capital management and concerted efforts to move toward a consistent approach to cross-territory supervision of insurance groups. These initiatives underscore the importance of embedding strong risk management principles throughout an enterprise and moving beyond just "tick the box" compliance, similar to what Basel II has been influencing in the banking industry.

In the US, the regulatory environment also has been changing as State insurance departments and rating agencies, in addition to National Association of Insurance Commissioners (NAIC), are also influencing the direction of solvency regulation. While these varied initiatives place differing degrees of emphasis on capital requirements, reporting standards and risk measures, a common theme, is their intensified focus on clearly articulating an insurer's risk profile. To prepare and address the regulatory pressures to enhance risk management, insurers must significantly enhance their data management, reporting and analytical resources, and their organizations' ability to integrate risk data across disciplines. The US insurance industry is also anticipating potential impacts of Dodd-Frank legislation, including in the systemically important financial institution (SIFI) designation and the Federal Insurance Office's (FIO) pending report to Congress on the state of US insurance regulation, which in practice creates a national insurance regulator.

Regarding OpRisk more specifically, insurers are still in the early stages of the development of their OpRisk frameworks. This comes somehow as a surprise as insurers suffered several large operational losses that were very public and reported in the media. Some of the examples over the last decade[2] are the USD 250 million loss that a large US insurer suffered a few years ago for discrimination (i.e., allegedly pricing their policies differently according to race); a large European reinsurer lost USD 3.5 billion for not having final contracts in place on the 9/11 terror attacks inflicting damages to clients; a large US auto insurer lost USD 1 billion for using low-quality auto parts in vehicle repairs; a large US life insurer lost USD 2 billion for abusive sales practices and illegal sales of securities and the list goes on and on.

Insurers face a number of OpRisks; some of these are mis-selling their products to clients. A number of insurers worldwide got severe penalties for these sales practices. As with any retail sector, insurers are exposed to bad faith claims (i.e., frauds by customers)—Hollywood has a number of movies on these interesting stories. More recently, the issue of unclaimed property has become a concern for insurers as public officials are now focusing much more on the issue than they did in the past. Given these pressures, insurers have been more diligent to catch up with banks in developing more robust OpRisk frameworks. However, they have a long road ahead of them.

2.7.5 ASSET MANAGEMENT

The financial crisis brought to the global asset management industry challenges it has not seen in decades as the industry was accustomed to high margins and substantial profits (particularly in the years 2000–2007 due to the availability of excess liquidity). As the financial markets climbed regularly over the last 30 years, occasional dips notwithstanding, asset managers became used to the steady increases in their assets under management (AUM) and easy profits. However, in the wake of the biggest downturn since the Great Depression, a slow recovery has

[2]To preserve confidentiality, the company names are not mentioned.

left many firms struggling. Even in 2012, most of the growth of the asset management came from market appreciation and not due to increase in flow of resources from clients.

This new environment changed the asset management industry. During the precrisis "golden years" of abundant liquidity, most asset managers were not overly worried about the costs incurred in running their operations and did not pay close attention to the risks involved, since the continuous growth in personal wealth steadily increased their AUM, covering for these expenses. Errors and high operating costs were buried under the increased revenues from a larger asset base and the profits that came from high returns in the world markets. Postcrisis, the situation has changed dramatically. Large asset managers have seen their AUM go down by 30 or 40%, not only because of the drop in asset prices but also because clients are withdrawing funds, either out of necessity to cover debts, because they fear that the stock markets will take a long time to recover, or sometimes even out of concern for the financial well-being of some asset managers. The crisis also showed historic regulatory failures, like the Bernie Madoff case, in which he created a Ponzi scheme, that was discovered during the 2008 financial crisis, and lost USD 6 billion from investors (this case is one of the largest OpRisk events in history). Many investors close to retirement lost their pensions not only because of the market conditions but also because of a lack of caution and risk management from pension fund managers.

This long-lasting dire economic environment forces asset managers to develop a much more careful discipline around costs, risk management, and productivity. Each of these factors has received widespread attention in the specialized media.

The industry has reacted quickly to this new reality. For example, a large independent US asset manager has already put in place several measures to reduce costs, by sharing services in its distribution and administration departments to reduce costs across geographical areas. This same firm has also launched an initiative to reduce its NCE by 20% in 2009, with the development of an inter-company committee to determine the expenses that have to be eliminated.

A European-based global firm decided to reduce the number of products it offered and the development efforts for a few products where it can build competitive advantage on a global scale. This firm also decided to immediately implement a plan, which had been on the shelf for many years, to streamline its operational platforms on a global basis. Currently, each geographical location (and sometimes within the same country) has its own platform with different vendors and frameworks to process securities.

Asset managers are susceptible to all forms of risks, namely market, credit, and OpRisks. However, due to the characteristics of their business (and perhaps helped by a historic disregard for strong controls), OpRisk is typically the largest risk exposure an asset manager has. Market and credit risk associated losses would usually have an indirect impact on the asset manager's revenue, as any loss to the client funds entails lower commissions. However, these losses are usually held by the fund's; clients not the asset manager as financial institution. These market and credit risks losses would impact the quotas and NAVs, so the client would take a direct hit; the asset manager would just have less fee revenue in these cases, an indirect impact. OpRisk can be manifested in many different ways for an asset manager as, for example, in errors in processing transactions or a system failure that can cause severe damage and impact the balance sheet of the asset manager. Asset managers are also regularly sued for poor performance by clients. Consistently failing to comply with local regulations, or with very basic business ethics, can generate very large operational losses and subsequent reputational damage. A number of examples are available in the media for large losses in each of these cases (Table 2.15).

Coming to realize the need to focus in OpRisk, asset managers have been setting up OpRisk departments at a fast speed in the last few years. The higher focus from regulators

TABLE 2.15 **Asset Management OpRisk Profile**

Event type	Frequency (%)	Severity (%)
Internal Fraud	1.5	11.1
External Fraud	2.7	0.9
Employment Practices and Workplace safety	4.3	2.5
Clients, Products, and Business Practices	13.7	30.8
Damage to Physical Assets	0.3	0.2
Business Disruption and System Failures	3.3	1.5
Execution, Delivery & Process Management	74.2	52.8

Source: Results from the 2008 Loss Data Collection Exercise for Operational Risk, see BCBS (2009b).

on hedge funds also made these more sophisticated asset managers to set up better OpRisk procedures around their operations. This new focus on control and risks would actually facilitate a more stabled growth, with less bumps, when the economic environment eventually improves.

2.7.6 RETAIL BROKERAGE

For OpRisk practitioners, this sector is possibly the one of the most interesting. Although we obviously need to consider that risk profiles would vary significantly between institutions given their different business strategies, broker-dealers risk profile is usually dominated by OpRisk, which accounts for at least 60–70% of the total risk capital in these firms. This OpRisk type becomes clear when we review the sector.

Broker-dealers of these days can be roughly classified into online and brick-and-mortar brokers. Although what separation then cannot be precisely defined, the customer focus of these brokers is different. While online brokers tend to compete on the retail, offering the convenience of trading from home or work and charging a reasonable fee for trades and usually offering free online research tools and a few other services, brick-and-mortar brokers are mostly a division of larger financial institution and tend to focus on a wealthier customer base that would pay for high fees they charge, advice from financial advisors, etc.

Over the past decade, the industry had a dramatic transformation with the proliferation of sophisticated, high-speed trading technology that has changed the way broker-dealers trade for their own accounts and as agent for their customers. In addition, customers of these broker-dealers—particularly leading-edge institutions—have themselves begun using technological tools to place orders and to trade on markets with little or no substantive intermediation of their broker-dealers. This, in turn, has given rise to the increased use and reliance on "direct market access" or "sponsored access" arrangements. Under these arrangements, the broker-dealer allows its customers—whether an institution such as a hedge fund, mutual fund, bank or insurance company, an individual, or another broker-dealer—to use the broker-dealer's market participant identifier ("MPID") or other mechanism for the purposes of electronically accessing the exchange. With "direct market access", as commonly understood, the customer's orders first flow through the broker-dealer's systems and then enters the markets, while with "sponsored access", the customer's orders flow directly into the markets without passing through the broker-dealer's systems. In all cases, irrespectively, whether the broker-dealer is trading for its own account, is trading for customers through more traditionally intermediated brokerage arrangements, or is allowing customers direct market access or sponsored access, the broker-dealer with market access is legally responsible for all trading activities that occur under its MPID. In some cases, the broker-dealer providing sponsored access may not utilize any pretrade risk management controls (i.e., "unfiltered" or "naked" access), and thus could be unaware of the trading activity occurring under its market identifier and has no mechanism to control it.

Nowadays, order placement rates can exceed 1000 orders per second with the use of high-speed, automated algorithms. If, for example, an algorithm such as this malfunctions and places repetitive orders with an average size of 300 shares and an average price of USD 20, a two-minute delay in the detection of the problem could result in the entry of, for example, 120,000 orders that values USD 720 million. In sponsored access arrangements, as well as other access arrangements, appropriate pretrade risk controls could prevent this outcome from occurring by blocking unintended orders from being routed to an exchange. Incidents involving algorithmic or other trading errors in connection with market access occur with some regularity. For example, it was reported that, on September 30, 2008, trading in Google became extremely volatile toward the end of the day, dropping 93% in value at one point, due to an influx of erroneous orders onto an exchange from a single market participant. As a result, Nasdaq had to cancel numerous trades, and adjust the closing price for Google and the closing value for the Nasdaq 100 Index. In addition, it was reported that, in September 2009, Southwest Securities announced a USD 6.3 million quarterly loss resulting from deficient market access controls with respect to one of its correspondent brokers that vastly exceeded its credit limits. Despite receiving intra-day alerts from the exchange, Southwest Securities' controls proved insufficient to allow it to respond in a timely manner, and trading by the correspondent continued for the rest of the day, resulting in a significant loss. Another example that highlights the need for appropriate controls in connection with market access occurred in December 2005, when Mizuho Securities, one of Japan's largest brokerage firms, sustained a significant loss due to an erroneous manual order entry that resulted in a trade that, under the applicable exchange rules, could not be canceled. Specifically, it was reported that a trader at Mizuho Securities intended to enter a customer sale order for one share of a security at a price of 610,000 Yen, but the numbers were mistakenly transposed and an order to sell 610,000 shares of the security at a price of 1 Yen was entered instead. A system-driven, pretrade control reasonably designed to reject orders that are not reasonably related to the quoted price of the security would have prevented this order from reaching the market.

As these examples show, broker-dealers are intensively exposed to OpRisk that usually occupies the headlines of most of the newspapers and media. Brokers usually do not hold large proprietary positions and lending, particularly after the 2008 crash, has been limited; therefore, most exposure comes from potentially explosive system issues, execution errors, litigation with retail customers, fraud committed by clients, etc. (Table 2.16)

2.8 Risk Organization and Governance

Developing a solid risk organization is a key part of the framework. Understanding the reporting lines and establishing the position of this organization on the firm would have probably as much importance as having a good measurement system. Also having proper organizational involvement in OpRisk issues where key stakeholders are regularly informed and oversee risk is fundamental for success. Developing a framework in a silo that no one sees or cares is not a desirable situation. The OpRisk manager needs to be integrated to the rest of the organization.

TABLE 2.16 **Asset Management OpRisk Profile**

Event type	Frequency (%)	Severity (%)
Internal Fraud	5.8	18.1
External Fraud	2.3	1.4
Employment Practices and Workplace safety	4.4	6.3
Clients, Products, and Business Practices	66.9	59.5
Damage to Physical Assets	0.1	0.1
Business Disruption and System Failures	0.5	0.2
Execution, Delivery & Process Management	20.0	14.4

Source: Results from the 2008 Loss Data Collection Exercise for Operational Risk, see BCBS (2009b).

In this section, we provide an overview of how risk is organized in financial firms, how policies are structured, and the importance of a solid committee and governance structure. Sound internal governance forms the foundation of an effective OpRisk management framework. Although internal governance issues related to the management of operational risk are not unlike, those encountered in the management of credit or market risk OpRisk management challenges may differ from those in other risk areas.

2.8.1 ORGANIZATION OF RISK DEPARTMENTS

One cannot downplay the role of an organization in any large business. Although many times the focus is on the measurement models with its complex formulas, most of the times the success of implementing an OpRisk framework lies in having the right organization. The organizational design would usually hint at the strength and degree of development of an OpRisk framework at a firm. In the following text, we show a few organizational designs and the beliefs that firms need to have to make them work. Usually firms start with Design 1 and go to Design 4 presented in Figure 2.5.

- **Design 1—Central Risk Function as Coordinator**. In this organizational design, risk management role is more of a facilitator. Usually in this structure, risk management gathers information and reports to the CEO or the Board. Sometimes risk management would add some layer of analysis, but in most cases, the Central Risk group would be a small group. One of the issues with this structure is that the regulators dislike the idea that risk managers report to revenue generating businesses;

  In order for this structure to be successful, one should believe that the Business Units will be responsive to the Central Risk demands even without being part of their reporting line and the control and incentives that such reporting includes (e.g., control over compensation, etc.);
- **Design 2—Matrix reporting—the "dotted lines"**. In this organizational design, a sort of evolution to the previous design, risk managers have a dotted line to the Central Risk function; however, they are appointed by the Business Units and compensation decisions are still taken by these. In order for this to be successful, the Business Units should have a strong risk culture and collaborate very closely with the Central Risk function. This dotted line structure works well when there is a culture of Business Unit independence and distrust of the Central Risk function for some reason or event that happened in the past;

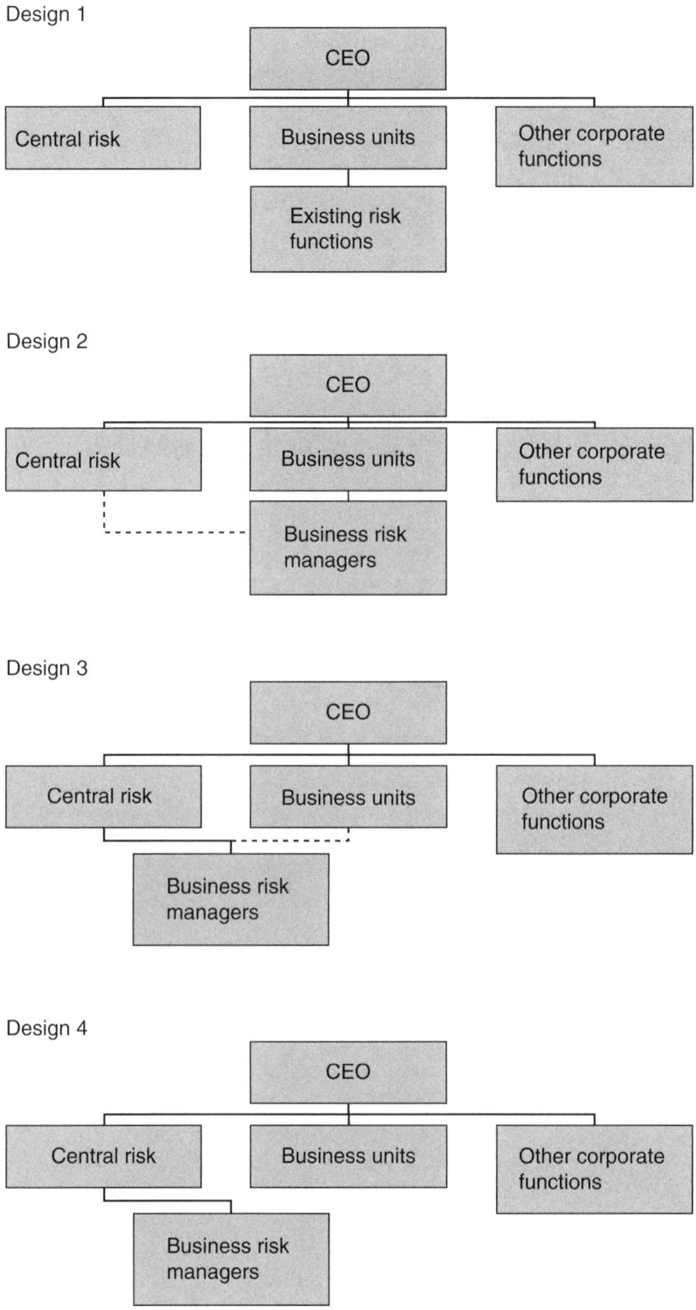

FIGURE 2.5 Organization of risk departments: designs 1–4

- **Design 3—Solid reporting lines to Central Risk Management**. This organization structure is reasonably popular within large firms. Risk Managers still physically work in the Business Units but report to the Central Risk function usually based in the headquarters. The Central Risk function will be better positioned to prioritize risk management efforts across different initiatives. This solid line reporting will also assist in the creation of a more homogenous risk culture and consistent approach across the enterprise;
- **Design 4—Strong Central Risk Management**. Large firms have adopted this structure lately, either by internal agreement or through regulatory pressure. In this structure, the Corporate Chief Risk Officer is the key decision maker in risk management and fully responsible for risk across the firm. Central Risk Management is responsible to monitor and manage all the firm's risks and report to senior management and Board. Such structure makes much easier for the regulator to streamline supervision as they can focus to one particular group instead of being scattered in many business units and geographical areas.

2.8.2 STRUCTURING A FIRM WIDE POLICY: EXAMPLE OF AN OPRISK POLICY

Example of a policy is presented in Table 2.17. A policy defines a firm's operational risk management framework, which includes governance structure, roles and responsibilities, and

TABLE 2.17 **Example of an OpRisk policy**

Content	Description
Executive summary	Defines the rationale and scope of the policy
Policy statements	Provide a quick definition of the standards that will be used across the policy
Risk taxonomy	Categorize OpRisk in different risk types. It can follow the Basel categories, but if it does not, it usually provides a mapping of internal categories to the Basel-defined
Loss collection	Defines what losses or incidents should be reported. Discusses concepts of "near misses" and describes recoveries
Risk assessment	Usually describes other programs used to supplement internal loss data collection like scenario analysis or risk factor analysis
Risk measurement	Describes the basic framework for measuring OpRisk, which types of data are used, and how capital is calculated (overall view of the building blocks not a detailed manual)
Validation	Describes how the risk assessment and measurement are validated, how frequent validation takes place, and which departments are responsible for the validation
Policy assurance and testing	Determines which department(s) in the firm, will be responsible for assurance that the policy is being followed and the reports that assure this firm-wide compliance
Governance	Describes where this policy is situated, which committee approves it, and how the OpRisk governance works
References	Determine on which regulations, external standards, and/or other firm policies this was based upon

standards for OpRisk management and measurement. It also describes the OpRisk management programs, which are the functional activities requiring guidelines for consistent firm wide execution (e.g. loss capture program, risk control self-assessment, and scenario analysis).

2.8.3 GOVERNANCE

Common industry practice for sound OpRisk governance often relies on three lines of defense:

- Business line management;
- An independent corporate OpRisk management function; and
- An independent review (usually internal audit).

Depending on the bank's nature, size and complexity, and the risk profile of a bank's activities, the degree of formality of how these three lines of defense are implemented will vary. In all cases, however, a bank's OpRisk governance function should be fully integrated into the bank's overall risk management governance structure and the regulators closely monitor this.

If OpRisk governance utilizes the three lines of defense model (i.e., the business is the first line of defense, risk management is the second line, and internal audit being the third), the structure and activities of the three lines often varies, depending on the bank's portfolio of products, activities, processes, and systems; the bank's size; and its risk management approach. Strong risk culture and good communications among the three lines of defense, are important characteristics of good OpRisk governance.

The regulators also reinforce the role of the board of directors. In the US and UK it is common that the regulators meet separately with financial firms board of directors regularly to discuss their expectations regarding risk management. The board of directors should take the lead in establishing a strong risk management culture. The board of directors and senior management should establish a corporate culture that is guided by strong risk management and that supports and provides appropriate standards and incentives for professional and responsible behavior. In this regard, it is the responsibility of the board of directors to ensure that a strong OpRisk management culture exists throughout the whole organization and this will be closely monitored by regulators.

CHAPTER THREE

Using OpRisk Data for Business Analysis

The financial crisis that started in 2008 made the financial industry face challenges it had not seen in decades. The industry was accustomed to high margins and substantial profits (particularly in the years 2000–2007, due to the availability of excess liquidity). However, in the wake of the biggest downturn since the Great Depression, a slow recovery left many firms struggling. Even in 2012/2013, the recovery seemed stalled, as the crisis still lingers to a certain extent, and the high regulatory pressure on financial firms not to take risks is putting a cap on their profits; as a result, most firms across the globe are going through severe cost-cutting programs.

This new economic environment is forcing financial firms to develop a much more careful discipline around costs, risk management, and productivity. Each of these factors has received widespread attention in the media. Productivity is a concept usually associated with manufacturing, but it can also play an important role in asset management.

In this chapter, we argue that, within the options available to them for returning to their former profitability levels, financial firms will have to take a very careful look at their cost structure and risk management frameworks. We analyze the cost structure of financial firms and describe strategic/tactical options to reduce costs on an item-by-item basis. In the last section, we describe how a well-tailored and well-implemented risk management program can impact a financial firm bottom line and avoid extreme cost-cutting measures.

To illustrate the impact of the crisis on the financial bottom line in the entire financial industry, we take the example of the asset management industry, which is interesting, as this industry gauges quite well the temperature of the economy; for example, if customers are getting wealthy, they would be investing more and this sector would be performing well. We analyze the impact of the crisis on the 10 largest global asset managers' profitability, measured in basis points. The average profit (operating margin) for an asset manager fell from 38 points at the end of 2007 to 34 at the end of 2008 and in 2011 this figure was at about 28 points. Most players in the industry are also suffering from a substantial decrease in Assets Under Management (AUM) either because of a decrease in asset value or because of client withdrawals. For this reason, their financial bottom line is being severely impacted, and the most tactical way to try to return to a higher level of profitability is via cost-cutting and by developing a robust risk management framework. We examine these two options in detail in the next two main sections.

Fundamental Aspects of Operational Risk and Insurance Analytics: A Handbook of Operational Risk,
First Edition. Marcelo G. Cruz, Gareth W. Peters, and Pavel V. Shevchenko.

3.1 Cost Reduction Programs in Financial Firms

Considering the long-term numb economic environment, a means by which a financial firm may gain some measure of control is to consider cost cutting measures, as shown in Table 3.1, to provide a means to return to their previous levels of profitability. Even if we assume that the economic conditions in the near future will be no worse than those in 2008 and that their revenues will remain at the same level, financial firms may have to cut their current costs by up to 50% to return to their 2006 profitability levels, such is the extent of the current financial crisis. This cost-cutting exercise would need to be accomplished in a much tougher regulatory environment, with regulators keeping close tabs on financial firms to ensure that non-revenue-generating back office functions like risk, legal, and compliance (usually some of the first to be cut in tough times) remain in place. On the positive side, such cost optimization exercises were long overdue. Most financial firms preferred not to face these issues while they were focusing on an expansion of their funds; however, these new lean times are now forcing them to make such adjustments. The industry has indeed been quick to react to this new reality. However, as usual, the "lowest-hanging fruit" is a reduction of headcount. These cuts show companies adapting to the new environment with lower margin products and less demand. While the initial focus was this reduction in headcount, financial firms can optimize their operational

TABLE 3.1 **Economic crisis impact on the fundamentals of the financial industry**

Factor	Description	Impact/reaction
Change in client behavior	Client risk-averse behavior, preferring simpler, transparent products	Development of new products with lower margins
Regulatory pressure	Increasing regulation demands that financial firms enhance transparency through risk disclosure and maintain capital requirements through balance sheet management	Higher compliance costs
Change in industry structure	Sharper differentiation based on chosen business models, increase in the number of independent firms, as well as larger players, due to consolidation	Immediate strategic decisions need to be taken and, based on that, a new focus for tactical decisions
Higher costs in developing robust risk management	Risk management will enter a new paradigm, shifting from client risk reporting to protecting the institution itself, requiring asset managers to develop new tools and techniques	Higher focus on risk management
Pressure on the financial bottom line (revenue, profits, and costs)	Fundamental shift in cost structure (toward more variable costs and "industrial" processes) necessary to address profitability challenges; pressures on revenue and profits due to threat from substitute products; and margin pressure from shifting product mix and lower volumes	Cutting costs

TABLE 3.2 **Examples of cost-cutting (in USD million) in noncompensation costs in three major global asset managers**

	BlackRock	Legg Mason	Franklin Templeton
2007 AUM	USD 1357	USD 999	USD 644
Revenues	USD 4845	USD 4707	USD 4228
Non-compensation expenses (NCE)	USD 1784	USD 1874	USD 923
NCE/revenues	37%	40%	22%
2008 AUM	USD 1154	USD 711	USD 400
Revenues	USD 5112	USD 3935	USD 3711
Non-compensation expenses (NCE)	USD 1613	USD 1706	USD 849
NCE/revenues	32%	43%	23%
Delta variation NCE/revenues	−5%	3%	1%
Decrease in AUM	−USD 203	−USD 288	−USD 244
Variation in revenues	6%	−16%	−12%
Variation in NCE	−USD 171	−USD 168	−USD 74

Source: Company websites.

platforms significantly in order to decrease costs. These platforms were mostly developed and implemented when the industry was growing at double-digit rates per annum and companies were scrambling to keep up with growth and geographical expansion. Cost containment was not the highest priority in those good times; it was consistently of lesser importance than the speed of development.

The cost structure in the financial industry can basically be broken down into two main components, namely, compensation and noncompensation expenses (NCE). It is more difficult for firms to balance the cuts in direct compensation, since, as in any financial service organization, rewarding portfolio managers and investment and quantitative research analysts is key to having good performance and attracting new business. If cuts are too deep in these areas, then firms run the risk of losing their key personnel to competitors who may offer higher renumeration; therefore, they hesitate to make heavy cuts on the revenue-generating side.

Many firms made steep adjustments on the NCE side of their costs in view of the impact that the recession is having on their businesses. A sample of three very large asset managers in the US (see Table 3.2) shows how they have been reducing their NCE. BlackRock, for example, in spite of an unusual increase in revenue from 2007 to 2008, reduced its NCE by USD 171 million, and its NCE/revenue ratio fell from 37% to 32% in 2008. Legg Mason reduced its NCE in absolute value by USD 168 million (very close to BlackRock); however, as its revenue declined 16% in 1 year, its NCE/revenue ratio actually increased to 43% in 2008. Franklin Templeton also had the same problem; although it made a cut in its NCE, its revenue decrease more than offset the NCE cut. This illustrates how deep the cost cuts have to be in order to return to higher levels of profitability. Indeed one may also question the sustainability of such high rates of returns for the financial industry as it continues to mature.

In order to design optimal cost reduction programs, we need to break the NCE down into more detailed categories. The analysis of the cost breakdown of the largest 50 global, US, and European asset managers as a percentage of their total costs in 2008 is shown in Figure 3.1. Occupational expenses (real estate, rents, etc.) represent slightly more than a quarter of the total NCE.

A branch network usually entails a significant real estate cost. A network of branches is useful if the asset manager is focused on retail clients. A branch may be useful in attracting

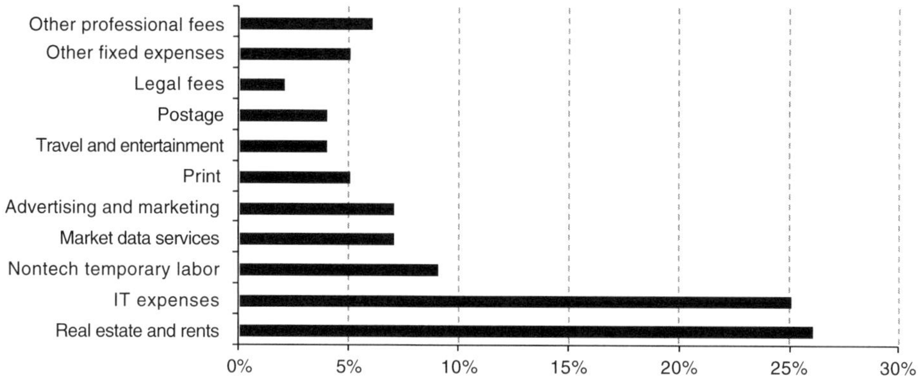

FIGURE 3.1 **Breakdown of noncompensation expenses. Source: Fifty largest asset manager's financial statements from 2008**

new clients by facilitating a first face-to-face contact. (Later on, an investor may communicate with the asset manager through one of the other channels of communication.) If the asset manager is focused on the institutional side, then an extensive network of branches is not necessary. (An asset manager may be content with having a small number of offices only in big cities.) If the asset manager has a large number of smaller individual investors, then a larger and more extensive network of branches may be advantageous. When cutting these costs, a firm needs to bear in mind the strategic consequences when it comes to attracting and retaining clients. The second-largest expense would be in technology and telecommunications. Given their importance, most of the cost savings would have to come from these categories, but cost cutting in these areas is never easy.

Personnel-related cuts are also important, but firms need to be careful to cut only in areas that are directly related to the volume of business. Changes that were not made in previous years because of accelerated growth, like delayering levels of hierarchy inside the firm, should now be a priority, as this can cut headcount by up to 30%. In Table 3.3, we summarize a few cost reduction activities by type of cost, considering the time they would take to be achieved and the average savings they would produce. IT and real estate cost cuts, for example, are extremely relevant, but would take longer to achieve results. Reducing these costs usually demands investment, as breaking or renegotiating leasing contracts, for example, commonly commands fees and charges. The same applies to IT optimization, which needs to be implemented very carefully to avoid serious operational problems in the future.

The industry has been quick to react to this new reality. For example, a large independent US asset manager has already put in place several measures to reduce costs, by sharing services in its distribution and administration departments to reduce costs across geographies. This same firm has also launched an initiative to reduce its NCE by 20% in 2009, with the development of an intercompany committee to determine which expenses will have to be eliminated.

A European-based global firm decided to reduce the amount of product offering and the development efforts for a few products where it can build competitive advantage on a global scale. This firm also decided to immediately implement a plan, which had been on the shelf for many years, to streamline its operational platforms on a global basis. Currently, each geographical location (and sometimes within the same country) has its own platform with

TABLE 3.3 Cost-cutting activities and average savings

Cost	Possible cost-cutting activities (%)	Average savings (%)*	Timeline
IT	Outsourcing programs Re-evaluate IT and telecommunication needs due to the new activity levels Reassess redundancy Server consolidation and right-size laptop and PC ratios	10–20	1 year
Personnel – organization	Cut layers of hierarchy Push activities to lower cost personnel ("empowering")	5–10	3–6 months
Personnel – headcount	Cut headcount across the board, adapting to the new level of activity	10–30	Immediate
Products range	Optimize the product range to better use investment teams, portfolio managers, and research	15–20	3–6 months
Real estate	Close facilities and/or renegotiate leases Increase use of outsourced resources that do not demand real estate use Use shared services Consolidate functions	5–20	6–12 months
Marketing and advertising	Consolidate marketing functions across the firm Shift spending to the most efficient vehicles Cut advertising spending		

* *Average savings, considering only their base cost*

Source: Author's work with asset managers.

TABLE 3.4 Most common types of cost-cutting programs

Type	Definition	Situation
Cost blitz	Companies start cutting costs immediately in a desperate fashion;	Sudden market changes that caught companies unprepared; Quick loss of profitability;
Category specific	Focus on only one category to cut costs–for example, cutting IT costs seen as the solution;	There is an obvious need for cost reduction in this expense category that market conditions aggravate;
Deep dive/transformation programs	A more analytical and holistic way to optimize costs and spending;	As the economic environment keeps deteriorating, companies see the need for a more structural change in their cost management.

different vendors and frameworks to process securities. Another US-based global firm followed the same path, creating and developing global centers of excellence in an attempt to provide their clients with the best possible service.

There are a few ways to perform such cost-cutting programs. Firms tend to go through all of them in recessionary times. These types are shown in Table 3.4.

When suddenly hit by a very serious crisis, as in September 2008 with the demise of Lehman Brothers, a company may often go immediately on a "cost blitz", which may result in a major round of layoffs. As the current situation does not seem to improve, many firms now also have to manage costs in specific categories, such as closing locations that are not profitable and are only viable if experiencing accelerated growth. Some firms still focus all their efforts on the IT category. As the current economic downturn seems to be lasting longer than initially expected, quite a few firms are now cutting their costs dramatically. These transformation programs tend to be longer, but usually present long-lasting results.

On a positive note, these changes are coming at a good time, as the previous fast growth meant that these firms did not use these resources in an optimal way. The crisis is therefore a good opportunity to check all these costs, and, when growth returns, this may stimulate large productivity increases.

3.2 Using OpRisk Data to Perform Business Analysis

As mentioned earlier in this chapter, financial firms are being pushed by regulators to dramatically strengthen their risk management frameworks. This will certainly require not just investments, but also greater management time and attention. However, we show in this section that better risk management, particularly OpRisk management, can also bring opportunities to reduce costs.

Financial firms are susceptible to all forms of risks, namely, market, credit, and OpRisks. Market risks are due to the daily fluctuation of asset prices, and credit risks are due to the possibility that some counterparties with whom the funds do business might default and make a financial asset worthless. Financial firms are particularly subject to OpRisk. In quite a few sectors in the financial industry such as retail brokerage, retail banking, and asset management, OpRisks are predominant. Errors in processing transactions or a system failure can cause severe damage and impact the balance sheet of the financial firm. Consistently failing to comply with local regulations, or with very basic business ethics, can generate very large operational losses and subsequent reputational damage. Clients can also sue for poor performance. OpRisk can be modeled in a few different ways. It particularly affects factors like people (human resources) and IT systems. In what follows, we elaborate on these two risk factors and how good risk management can translate into a positive impact on the bottom line.

3.2.1 THE RISK OF LOSING KEY TALENTS: OPRISK IN HUMAN RESOURCES

As a service sector firm, any type of asset manager needs to hire top talent in order to provide the best return and service for its clients. Human resource talent is needed for the following:

- General management (portfolio managers, etc.);
- Administrative personnel (operations settlements, accountants, etc.);
- Research (equity, bond and currency analysts, risk analysts, etc.);
- Technologists (e.g., IT specialists); and
- Sales force.

As in many financial firms, asset managers have to ensure that they are able to attract and retain, above all, portfolio managers with an established track record and a potential to bring in clients and provide high returns to their funds. Such people are the face of the firm to the outside world and are a basis for attracting clients. Compensation of such personnel is one of the highest costs of any financial firm. Losing top talent is very costly and also increases the susceptibility to OpRisk. There is a learning curve for apprentices and, during this period, the probabilities of error are higher. Asset managers are, therefore, highly exposed to key personnel risk. Particularly in the US, but also in other countries, funds are often named after their portfolio managers. Typically, these portfolio managers develop such a track record and reputation that clients want to invest with them. These funds linked to a name can hold many billions of dollars in investments, and the asset manager may become very dependent on this particular person. The risk of losing such a portfolio manager may represent a loss of revenue of many millions per year in administration and performance fees.

In the front office, sales people need to follow procedures and local regulations to sell pension and other types of funds. Several pension mis-selling cases have occurred in different countries. Probably the most infamous case of pension mis-selling was the situation that arose in Britain between 1988 and 1994, after British regulators decided to allow individuals to buy pensions from private-sector providers. The regulators determined at that time that pension investors should have the choice of who would provide their pension (not necessarily their employer) and that they should be allowed to invest, in effect, in a retail pension fund. Many who decided, or who were persuaded, to buy a retail fund should not have done so. High-pressure tactics by commission-based salespeople led to tens of thousands of people purchasing products that proved to be entirely unsuitable. High fees and charges and poor investment returns combined to shrink the retirement savings of these investors. Many found themselves locked in and unable to switch to more appropriate products without incurring very high exit fees. The result was a nightmare for investors, pension providers, and the government. After a long legal process, the funds were told to reimburse the investor for mis-selling these pensions. Until 2008, an estimated GBP 11.5 billion (nearly USD 20 billion) had been paid in compensation for mis-selling by certain asset managers who operated in this market.

The British experience serves to illustrate what can go wrong when, even with the best intentions, a choice is given to people who are unprepared for it. It also shows how greedy salespeople can exploit unsuspecting consumers, and how something that starts out as a good idea can turn into a major financial liability to asset managers if not properly conducted.

OpRisk can also manifest itself in back office personnel. For example, risk managers, auditors, and accountants play an important role, since they have to guard the firm against the likes of rogue traders, accounting frauds, and Ponzi schemes (like the aforementioned Madoff case). It is important that the reporting lines of the traders and risk managers are kept separate.

3.2.2 OPRISK IN SYSTEMS DEVELOPMENT AND TRANSACTION PROCESSING

Scale plays an important role in asset management. The larger the portfolio, the lower is the cost per transaction. However, the optimal size of a managed fund is often a balance of various trade-offs. For example, while an overall larger scale for an asset manager is preferred because of economies of scale, a small fund would be more agile to move a fund's allocation in reaction to market movements, and would probably be better able to outperform the competition. This is the case with hedge funds. Another aspect that has an impact on the optimal size of a fund is the error rate (OpRisk), which is a function of the transaction frequency. It is to be expected that the probability of error increases with an increasing frequency in the rate of transactions. A larger fund, in order to meet its benchmarks, will have to take bigger bets. So, for each type of fund there is an optimal size and an optimal focus. Historically, several funds that reached a size deemed to be larger than optimal decided to close entry for new clients, such as Fidelity's Magellan.

Financial institutions in general, and asset managers in particular, have traditionally never been as careful with costs as other industries have been. In several industries, like car manufacturing, error rates are extremely low and very well controlled by sophisticated quality control departments, which are usually the most sophisticated areas within an organization except for research (or product) development. On the other hand, in the financial services industry, the most sophisticated departments are located either in the front office or on the revenue side. Financial derivatives are priced taking only market opportunity costs (and rarely transaction costs) into consideration; even if transaction costs are taken into account, the analysis is not very deep. In the portfolio aggregation of these products, the final effects of processing are never considered. In this section, we try to briefly depict how a more sophisticated cost analysis can be developed for financial products based on a traditional microeconomic analysis.

Economic theory postulates that, for a firm to maximize its results, it is necessary that it produces such a quantity that allows equilibrium between the variation of the total cost and the variation of the total revenue. The total (or gross) revenue, R_{gross}, is simply the result of multiplying the price, p, of a certain product by the quantity, K, negotiated, that is, $R_{gross} = p \times K$. In general, the price is a function of quantity, that is, $p = p(K)$, and the marginal revenue, R_{mg}, corresponds to the variation of the total revenue with respect to the quantity sold K. Assuming that the variation of the quantity and the gross revenue can be admitted as infinitesimal (this works in theory, but is unlikely to be the case in business practice), the marginal revenue can be determined by the first derivative of the gross revenue in relation to the quantity sold:

$$R_{mg} = \frac{\partial R_{gross}}{\partial K}. \tag{3.1}$$

In asset management, the increased number of transactions K (the production) will bring an unexpected variable cost, which is an increase in operational error (human and system factors would not perform the same when subject to a higher volume of transactions). The relationship between the number of operational errors and the transaction volume can be estimated through multifactor models. Denote the total cost of the production as C_{gross}, which is a function of K. Then the marginal cost is defined as

$$C_{mg} = \frac{\partial C_{gross}}{\partial K}. \tag{3.2}$$

The entire analysis of revenues, production, and costs based on the (micro)economic theory is complex, and there is vast literature on the subject (see, e.g., Krugman and Wells 2012 and references therein). We will not delve into more detail in this section, but strongly recommend understanding these relationships when developing any growth strategy. It is worth noting that perhaps the most important conclusion from these considerations is that the firm's profit will be maximized when the marginal cost and the marginal revenue are the same, that is, $C_{mg} = R_{mg}$ (when the profit $P(K) = R_{gross} - C_{gross}$ is a concave function of K, this corresponds to the standard condition of maximum $\partial P(K)/\partial K = 0$). In what follows, we present a very simple stylized example to illustrate this concept.

EXAMPLE 3.1 Maximizing profit with respect to the number of trades

Suppose a fund trades a single product with a very tight margin at $\gamma = 0.006\%$ per trade (one trade is A = USD 100,000). Therefore, the gross revenue for K trades (e.g., per day) is

$$R_{gross} = A \times K \times \gamma. \tag{3.3}$$

In general, the fund trader would only see the trades from the revenue side and would be happy to see the revenue growth as the number of trades K increases. Using (3.3), it is easy to see that, for example, the revenue grows from USD 1,200,000 to USD 4,200,000 when the number of trades K increases from USD 200,000 to USD 700,000. This is a very general view, but revenue generators will not bother about the costs incurred to achieve that revenue.

Let us now analyze the costs. We divide the costs into two components: processing cost $C_{process}$ and error cost C_{error}. Assume that the processing cost per trade is δ = USD 5, that is, the total processing cost is $C_{process} = K\delta$. The error cost ε is random, and assume that the expected value of the error cost is $\mu = \mathrm{E}[\varepsilon]$ = USD 9.43. So it would cost USD 5 to process a trade and an additional USD 9.43 to reprocess it in the case of the error on average.

Denote the number of failed trades as K_{failed} and expected number of failed trades as $\lambda = \mathrm{E}[K_{failed}]$. Then the total error cost is εK_{failed} and expected error cost is $C_{error} = \mathrm{E}[\varepsilon K_{failed}] = \mu \times \lambda$ (assuming that ε and K_{failed} are independent). Thus, the total expected gross cost is $C_{gross} = C_{process} + C_{error}$.

Assume a simple linear model for the expected error ratio λ/K with respect to the number of trades K:

$$\frac{\lambda}{K} = \alpha + \beta \times K, \tag{3.4}$$

where $\alpha = 0.0095$ and $\beta = 1.1 \times 10^{-7}$. For this model, one can easily see that the error ratio is about 3.15% when the number of trades is $K = 200,000$; and when the number of trades grows to 700,000, the error ratio climbs to 8.65%!

The total expected profit from K trades is

$$\begin{aligned} P(K) &= R_{gross} - C_{process} - C_{error} \\ &= A \times K \times \gamma - K\delta - \mu K(\alpha + \beta K). \end{aligned} \tag{3.5}$$

It is easy to see that the maximum expected profit $P(K)$ is achieved at $K = K^*$; where $\partial P(K)/\partial K = 0$, this gives a closed form of expression for the optimal number of trades

$$K^* = \frac{A\gamma - \delta - \mu\alpha}{2\mu\beta}. \tag{3.6}$$

The profit $P(K)$ as a function of K is shown in Figure 3.2.

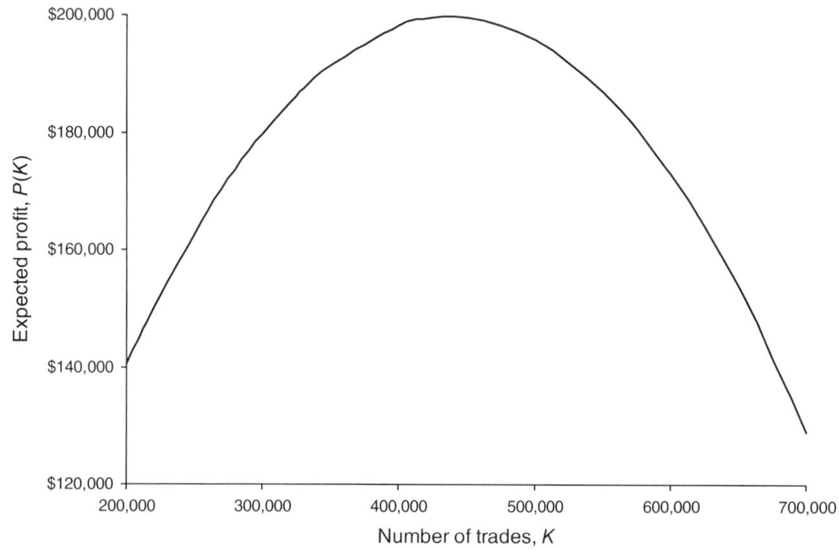

FIGURE 3.2 The profit of the business as a function of the number of trades K; for details, see Example 3.1

It is also easy to see that the maximum condition $\partial P(K)/\partial K = 0$ corresponds to $C_{mg} = R_{mg}$; see formulas (3.1) and (3.2). For the parameter values used in this example, it is easy to calculate using (3.6) that the maximum profit USD 199,762 is achieved at $K^* = 438,838$. If we trade more than K^*, we have declining profits. If the asset manager has any strategy of trading more than that, he/she will also have to take the costs into consideration. This type of modeling also offers us conditions to verify our capacity and see how an improvement in the process (system improvement, training process, hiring employees, etc.) will benefit the organization and increase productivity.

In this example, if the error rate (3.4) and expected error cost ϵ are reduced by 20% (i.e., α, β, and ε are multiplied by 0.8), due to OpRisk reduction (e.g., by training employees and improving systems), the maximum profit USD 334,634 is reached at $K^* = 709,975$ trades. Therefore, the fact that we reduced the OpRisk in a business by 20% increased profit by about 70% and increased our optimal capacity by about 60%, achieving a dramatic productivity gain by managing the OpRisk better.

There are several other factors that affect the costs and risks of transaction processing. Transaction processing can be outsourced (however, usually not offshore, but preferably to some firm relatively close by, so that any form of OpRisk does not increase too much). Another important factor is manual versus automated transaction processing (e.g., society for world interbank financial telecommunication (SWIFT)). Automated transaction processing clearly has a higher productivity than manual transaction processing. However, automated transactions can only be done with regard to standard, plain vanilla transactions, not with regard to more complicated esoteric transactions. Even though one may think that automated processing is more reliable and less susceptible than manual processing to OpRisk, it is not clear that this is actually the case (e.g., automated transactions are still subject to typographical errors, which have often cost managed funds millions). ■

3.3 Conclusions

The financial and economic crisis has changed the financial industry landscape completely, and this now presents many challenges for financial firms all over the world. Senior management and boards at these companies are using the best possible tactics to return to higher levels of profitability, in some cases even in order to survive. The easiest way to control this is through cost reduction programs. Finding the optimal cost structure in the current environment without losing clients for poor quality of service is key. These cost optimization programs in financial firms were overdue. As they were concerned only with expansion in the last few years, there are usually a number of legacy systems that need to be closed (duplicate processing, unnecessary office locations, etc.), which would make these firms leaner and more productive. Costs, productivity, and OpRisk are strongly intertwined. For a firm to optimize its investments and operations, all possible factors and trade-offs have to be taken into account. Such an optimization process is an analytical task that needs to be carefully executed. However, asset managers who survive this crisis will be much stronger when markets recover.

CHAPTER FOUR

Stress-Testing OpRisk Capital and the Comprehensive Capital Analysis and Review (CCAR)

4.1 The Need for Stressing OpRisk Capital Even Beyond 99.9%

Since the Lehman Brothers collapse that culminated in a financial crisis in 2008, banks across the globe have been constantly demanded by regulators, investors, lawmakers, and the public in general to prove their financial health and resilience of their balance sheet under stressed financial conditions. In order to standardize and formalize this process, more formal tests were established by the leading world regulators, which periodically require banks to stress-test their capital base given certain scenarios. On both sides of the Atlantic, this process is similar to that shown in Figure 4.1. It basically requires a firm to develop a set of scenarios or use scenarios developed by the regulators. Regulators would then get the individual results from firms and develop their own systemic stress test to verify if the financial industry can withstand negative scenarios and where regulators need to enforce banks to avoid another situation like the one in 2008.

These scenarios are expressed in stressed macroeconomic factors and financial indicators, and regulators provide these figures on a quarterly basis for a period of 2 or 3 years ahead. For example, in a certain quarter, regulators might establish that the S&P 500 would go down 30% and US unemployment would reach 12% (there are many other factors). Based on this information, banks would then assess the impact of this economic scenario reflected in market and credit losses in their portfolios and how their capital base would behave in this situation. These tests are motivated from the government bailout days in which banks did not have enough capital to cope with extremely negative scenarios and had to be helped by tax payers' money. The novelty is that banks are also required to analyze the impact of this scenario in OpRisk. The relationship between these macroeconomic factors and indicators to market and credit risks is straightforward, but what about OpRisk?

As OpRisk capital is already reported to regulators at 99.9% and considering that the fitted distributions are usually heavy-tailed, it is a regular discussion in the OpRisk community

Fundamental Aspects of Operational Risk and Insurance Analytics: A Handbook of Operational Risk, First Edition. Marcelo G. Cruz, Gareth W. Peters, and Pavel V. Shevchenko.

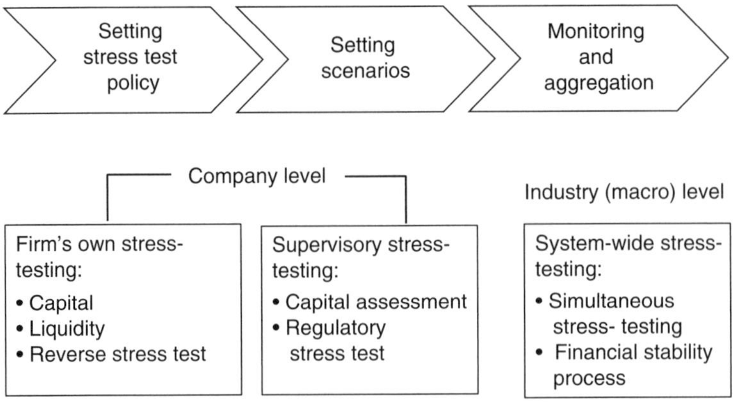

FIGURE 4.1 **Generic stress test framework**

whether or not this large confidence interval is already sufficient to protect against large loss events. Many think that OpRisk capital is large enough and it would be a pointless exercise to estimate it even further down the tail. However, in practice, a number of issues with OpRisk modeling can show the clear need for even deeper estimation in the tail stress of OpRisk events. As an effect of the great financial crisis of 2008, most large financial firms were sued by clients for many reasons, for example, because mortgages were irregularly granted or funds in large asset management were unduly keeping mortgage-backed securities and, therefore, suffered large financial losses in the post crisis. The settlement of these lawsuits is still taking place in 2014 and beyond, and they amount to multibillion dollar amounts. In addition to these large settlements, banks also still continue to face the usual threat of internal frauds. It is understood that banks allocate significant capital against internal frauds but these losses keep exceeding their largest estimates. The same can be applied to business disruptions due to system crashes or pretty much every OpRisk event type. Given that, many firms already have a process to stress-test capital to even higher levels and a number of regulators, particularly the Federal Reserve Bank (FED) in the US, developed a more rigorous process to stress-test OpRisk it seems that the industry is responding with an increased level of caution already indirectly.

4.2 Comprehensive Capital Review and Analysis (CCAR)

Since the great financial crisis in 2008, the regulators in the financial industry have been duly concerned in finding ways to assess the financial health of financial companies on a more regular basis and, more particularly, under stress conditions. In 2009, in the aftermath of the crisis, the FED launched the so-called Supervisory Capital Assessment Program (SCAP) (see FED 2009) as an attempt to try and get a better assessment on how institutions would fare given a number of adverse macroeconomic factors during a period of 2 years ahead. This exercise was very extensive as it involved impacts in the pre-provision net revenue and expenses given a set of scenarios for a number of key macroeconomic factors. However, the main focus was on the impact on bank capital. As the FED noticed at the time, "*capital reassures an institution's depositors, creditors and counterparties – and the institution itself – that an event such as an unexpected surge in losses or an unanticipated deterioration in earnings will not impair its ability to engage in lending to creditworthy borrowers and protect the savings of its depositors*".

Before embarking on a discussion on the recent stress testing of capital being performed in the US, it is useful to observe (Schuermann, 2013, figure 1) relating to the arms race for capital adequacy and its evolution under the Basel Accords. To understand the context of such stress tests, it should be noted that during the 2007/2008 crisis period, banks the failed or came close to failure requiring some form of bailout or assistance in the UK and US were all considered, prior to the crisis, well capitalized by existing regulatory standards. In this initial SCAP, only the 19 largest bank-holding companies (BHCs) were required to participate. Not just commercial or investment banks were participants among these 19 but also a few large insurance and credit card companies. The idea was that the largest financial institutions, deemed capable of impacting the financial system significantly in case they are in financial trouble, had to perform this what-if test to give regulators some assurance. SCAP was a very stringent test and all hypotheses and calculations performed by these institutions were thoroughly audited by the FED and some of them actually failed the test, meaning that under the stress conditions under SCAP, these institutions would not have the expected capital buffer to protect them. The exercise focused not only on the amount of capital but also on the composition of capital held by each of the 19 BHCs. The SCAP's emphasis on what is termed "Tier 1 common capital"[1] reflects the fact that common equity is the first element of the capital structure to absorb losses, offering protection to more senior parts of the capital structure and lowering the risk of insolvency; for more details on bank capital definition, see BCBS (2011). All else equal, more Tier 1 common capital gives a BHC greater permanent loss absorption capacity and a greater ability to conserve resources under stress by changing the amount and timing of dividends and other distributions. This means that institutions would have to be preapproved by the regulators to do any activity that might impact capital, for example, pay dividends, enter a shares buyback program, issue shares, etc.

SCAP was initially designed to be a one-off test; however, it returned in the following year, now named as Comprehensive Capital Analysis and Review (CCAR). It has been run on a yearly basis since then, see for instance the discussion of the results in 2012 available in Federal Reserve (2012) and the more recent summary of findings from such stress tests discussed in Bernanke (2013). These stress tests had to be delivered to the FED around January 7 of the following year so it became a new tradition for risk managers in the US to work extremely long hours during the holidays. The slow period of holiday celebrations became a casualty of the CCAR process as this time of the year became one of the most intense for US-based risk managers. In 2011, the FED created a new program called "Capital Plan Review" that in practice extended CCAR to another 11 institutions and this number is expected to grow in the next few years. As a result of these stress testing exercises several academics have begun to question the outcomes of the tests and to assess them, see for instance the study of Petrella and Resti (2013) and Acharya *et al.* (2014) and the references therein. In addition to the stress tests performed by the FED in the US, there were also a number of stress tests performed in Europe, for instance in 2010 there were reported 91 banks undergoing stress testing in Europe which covered 20 countries. The result of the tests in Europe were alarming with 7 major banks in the set considered failing to meet the capital adequacy standards required under the prescribed stress tests, requiring additional bailouts to stay solvent in excess of €3 billion. In addition, subsequent to this stress testing, there were a number of European countries going into distress

[1]Tier 1 common capital is composed of common shareholders equity + partial noncontrolling interest − certain deferred tax assets − goodwill and intangibles − debt valuation adjustments − other deductions. Tier 1 capital is all of this plus perpetual preferred stocks, trust preferred securities, and remaining noncontrolling interest.

with large scale bank bailouts occurring, such as in Ireland, even when banks that participated in the stress test had passed the capital adequacy standards. This suggests that such exercises need to be further expanded and capital adequacy further explored.

In the recent study of Petrella and Resti (2013) it was noted that since supervisory stress tests can be used to assess the impact of an adverse macroeconomic scenario on the profitability and capitalization of the largest banks in a given economy, then such results should be used to reduce the perceived public opinion that there is an opaqueness in the way tax-payer money may be being used to help support and bailout struggling financial companies. Consequently, as noted in Petrella and Resti (2013) the EU regulators took unprecedented step in releasing the results of the stress test exercises performed to the public in order to help investors distinguish between robust and under capitalized institutions. This involved releasing the results of the 2011 EU region stress test which include around 3,400 data points for each of the 90 participating banks. It was noted in Petrella and Resti (2013) that the important features of the data released included:

- Data on risk-weighted assets and own funds, which also included a breakdown of items recognised as core Tier 1 capital, compulsory deductions, governmental support and other mitigating measures fully committed by 30 April 2011;
- P&L figures which included: net interest income, trading income, impairments, other income/losses and net profit after tax;
- Details on provisions, loss rates and coverage ratios for performing and non-performing exposures. In addition this was separated by retail, corporate, bank and sovereign portfolios;
- Credit exposures by geographic area, counterparty and default status;
- Sovereign exposures by geographic area, accounting treatment (e.g. trading book, fair value option, available for sale, etc.), duration band. This included derivative exposures at fair value.

Based on this stress testing data, academic works such as Petrella and Resti (2013) started to study meaningful questions relating to the impact of such tests on perceived confidence and stability of the financial sector in different regions. For instance they studied questions like:

- Did the stress tests produce relevant information for market participants ("irrelevance hypothesis")?
- If the test's results triggered a market reaction, was this reaction caused by the release of more granular historical data ("zoom hypothesis"); or
- By the resiliency indicators generated by the stress test exercise ("stress hypothesis")?

The outcomes of testing these three hypotheses were that there was evidence obtained to reject the irrelevance hypothesis since the market was shown to significantly react once the disclosure of the results was performed by the EU regulators. In addition, it was shown that the abnormal returns of tested banks could be strongly related to some stress test outputs released by the EU regulators. Finally, with regard to the "zoom" and the "stress hypothesis" it was concluded that these were supported by the analysis post the EU regulators release of information.

CCAR is a comprehensive test not just for OpRisks but also for market, liquidity, and credit risks. As part of the CCAR, the FED assesses institutions' capital adequacy, internal capital adequacy assessment processes, and their plans to make capital distributions, such as dividend payments or stock repurchases. The CCAR includes a supervisory stress test to support the FED's analysis of the adequacy of the firms' capital. Boards of directors of the institutions are required each year to review and approve capital plans before submitting them to the FED. The CCAR process is an intense exercise that involves many top-level executives in BHCs. The general view seen from the industry regarding stress testing is that it possess some important advantages when used as a quantitative tool to assess and determine aggregate capitalization. It delivers a specific annual set of transparent scenarios that are readily understood by a range of members of the financial institution and the executive board and covers not just financial losses but also revenue and costs. Also, it provides regulators with a tool by which they may better understand the country and industry wide risk known as systemic risk. Moreover, it allows for a direct study of practical accounting measures of financial performance such as a lack of capital fungibility.

Before going into specific details of the CCAR process it is worthwhile to first discuss the main constituents of a stress test framework which include:

- defining a risk apetite for the given financial institution;
- given a particular risk appetite, there is a stage of process and governance to be performed. This includes a clear mapping of the role and responsibilities of senior management involved in the exercise;
- the scenario definitions are to be developed and considered/discussed with each business/divisional stakeholder. This can include macro-economic assumptions to be considered, which should be done with historical relationships kept in mind. Then in addition to potential macro-economic scenarios based on historical events, there should be an additional level of expert opinion incorporated to develop additional what-if and plausible scenarios that could be faced in future not yet present in historical realized events;
- there is a stage of credit forecasting for loan losses, provisions and ending reserves. In addition it should consider permanent impairments of investment securities;
- there is a stage of pre-provision net revenue forecasting to consider balance sheet dynamics, net interest income forecasts and other aspects of income and expenses;
- all unaccounted for risks are then considered such as mark-to-market trading losses from given scenarios, operational risk losses and liquidity impacts;
- finally, these items are combined into the final stage of capital assessment which involves a forecasting of the capital position post the stress events.

Having discussed these high level stages, we now discuss in more detail the CCAR process. However, we note at this stage that a set of stress tests is provided each year in the FED CCAR guidelines along with generic templates for reporting of results. In the CCAR process, the FED assesses a BHC's pro forma post-stress capital ratios resulting from the combination of stress performance measures (e.g., revenues, losses, and reserves from the supervisory severely adverse scenario) and the BHC's planned capital actions (e.g., planned dividends, issuance, and repurchases as provided in the BHC baseline scenario) against each minimum regulatory capital ratio and a 5 % Tier 1 common ratio as shown in Table 4.1.

TABLE 4.1 **CCAR regulatory minimum ratios**

Regulatory ratio	Minimum level (%)
Tier 1 common ratio	6
Tier 1 leverage ratio	3 or 4
Tier 1 risk-based capital ratio	4
Total risk-based capital ratio	8

TABLE 4.2 **Types of scenarios in the CCAR process**

Type of scenario	Description
Bank-holding company (BHC) baseline	A baseline scenario for OpRisk defined and built by the own firm
Bank-holding company (BHC) stress	A stress scenario for OpRisk defined and built by the own firm
Supervisory baseline	A baseline scenario provided by the Federal Reserve under the capital plan rule
Supervisory severely adverse	A severely adverse scenario provided by the Federal Reserve under the capital plan rule

The types of scenarios in the CCAR process are presented in Table 4.2 and some elements of process to project preprovision net revenue and capital are presented in Figure 4.2. The results of a BHC's analysis for each scenario should encompass all potential losses and other impacts to net income that the BHC might experience under the scenarios described earlier. In all cases, BHCs should substantiate that their results are consistent with the specified macroeconomic and financial environment, and that the components of their results are internally consistent within each scenario.

The BHC baseline scenario should reflect the BHC's view of the expected path of economy over the planning horizon. A BHC may use the same baseline scenario as the Federal Reserve baseline scenario if the BHC believes the Federal Reserve baseline scenario appropriately represents their view of the most likely outlook for the risk factors salient to the BHC.

The BHC stress scenario should be based on a coherent, logical narrative of a severely adverse economic and financial market environment and potential BHC-specific events. The scenario narrative should detail key events and circumstances that occur in the scenario. BHCs must provide the quarterly trajectories of key macroeconomic and financial variables for its BHC baseline and BHC stress scenario.

A BHC's stress scenario should describe a severely adverse hypothetical combination of circumstances designed with the BHC's particular vulnerabilities in mind. Specifically, and as noted earlier, the BHC stress scenario should be designed to stress factors that affect all of its material exposures and activities, capturing potential exposures from both on- and off-balance sheet positions. In addition, a forward-looking analysis is also required in the BHC stress scenario.

A BHC is required to perform an assessment of the expected uses and sources of capital over the planning horizon assuming both expected and stressful conditions. This assessment must contain the following elements:

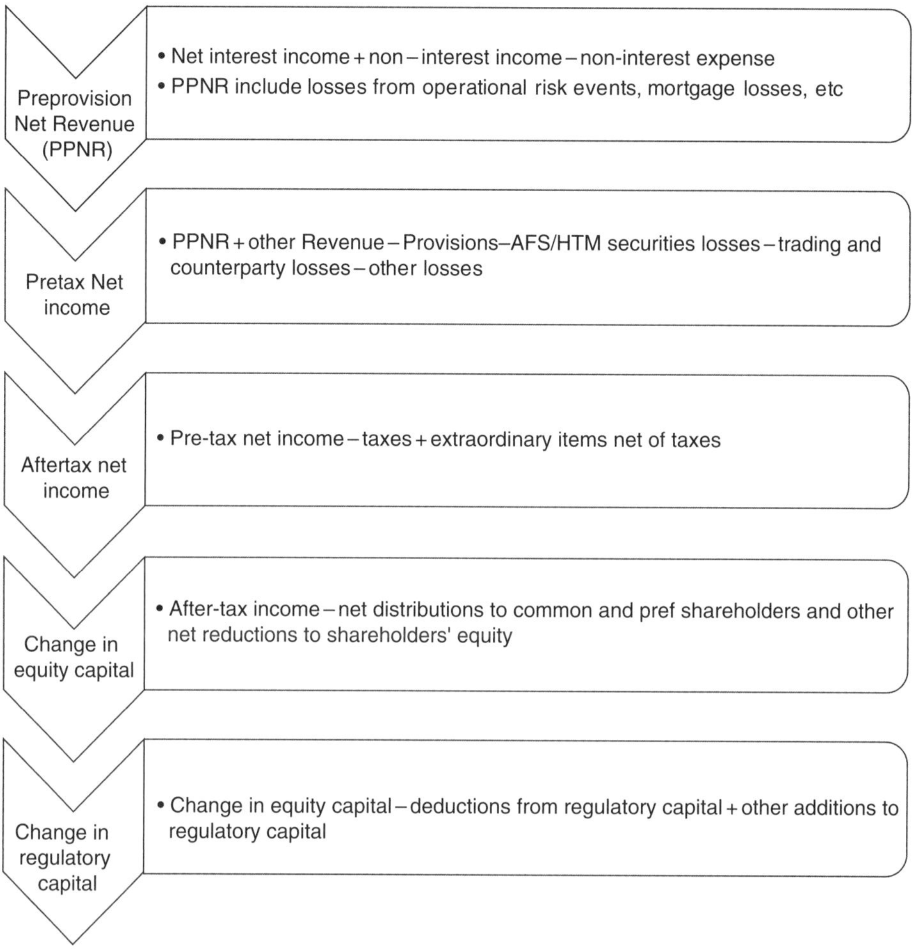

FIGURE 4.2 **Process to project preprovision net revenue and capital**

- Estimates of projected revenues, losses, reserves, and pro forma capital levels, including any regulatory capital ratios (e.g., leverage, Tier 1 risk-based, and total risk-based capital ratios) and any additional capital measures deemed relevant by the BHC, over the planning horizon under expected conditions and under a range of stressed scenarios, including any scenarios provided by the FED and at least one stress scenario developed by the BHC appropriate to its business model and portfolios;
- A calculation of the Tier 1 common ratio over the planning horizon under expected conditions and under a range of stressed scenarios and discussion of how the company will maintain all minimum regulatory capital ratios and a Tier 1 common ratio above 5% under expected conditions and the stressed scenarios required;
- A discussion of the results of the stress tests required by law or regulation, and an explanation of how the capital plan takes these results into account;
- A description of all planned capital actions over the planning horizon.

BHCs should demonstrate that their results are consistent with the macroeconomic and financial environments specified in the scenarios being used, and that the various components of their results are internally consistent. For instance, it might be inconsistent to project a shrinking balance sheet while also projecting large increases in net income in a stress or baseline environment, as this would certainly raise a red flag. BHCs should submit background information on the methodologies supporting their estimates. This material should include a discussion of key approaches and assumptions used to measure BHC-wide exposures and to arrive at stress loss estimates, along with relevant background on positions or business lines that could have a material influence on outcomes.

At the end of this process, the FED can object to a capital plan based on qualitative or quantitative concerns, or both. The FED can make new capital plans from an institution at any time to make improvements in the capital planning process, or if there is a change in condition of an individual institution or in the economy that could potentially lead to a change in a firm's capital position. The outcome of the CCAR is public and by March the industry will know how BHCs fared in the stress test.

It will be interesting to see the impact that the CCAR exercise has in the US financial industry. This exercise has become so important that a number of firms now instead of performing this only annually as required are doing it either semiannually or a few even every quarter. As almost every firm uses the results of CCAR in Pillar 2 instead of Pillar 1, it turns out that the outcome from CCAR tends to be larger and more important than the Basel numbers for US banks, reducing somehow the importance of Basel.

The Macroeconomic Factors and Financial Indicators Used in the Three Scenarios

The main characteristics of the scenarios for the CCAR 2013 are listed in Table 4.3. All scenarios start in the fourth quarter of the current year (e.g., 2014:Q4) and extend through the fourth quarter of 2016 (2016:Q4). The three scenarios are defined over 26 variables. In its description of US economic conditions, each scenario includes the following:

- **Six measures of economic activity and prices.** Real and nominal gross domestic product (GDP), the unemployment rate of the civilian noninstitutional population aged 16 and over, real and nominal disposable personal income, and the Consumer Price Index (CPI);
- **Four aggregate measures of asset prices or financial conditions.** Indexes of house prices, commercial property prices, and equity prices, and US stock market volatility;
- **Four measures of interest rates**. The rate on the 3-month Treasury bill; the yield on the 10-year Treasury bond; the yield on a 10-year Better Business Bureau (BBB) corporate security; and the interest rate associated with a conforming, conventional, fixed-rate, 30-year mortgage.

For the international variables, each scenario includes three variables in four countries/country blocks:

- The three variables for each country/country block are the annualized percent change in real GDP, the annualized percent change in the CPI or local equivalent, and the US dollar/foreign currency exchange rate;
- The four countries/country blocks included are the European area, the UK, developing Asia, and Japan. The European area is defined as the 17 European Union member states that have adopted the € as their common currency, and developing Asia is defined as

TABLE 4.3 **Main characteristics of each scenario for the CCAR 2013**

Scenario	Characteristics	Indicators sample
Supervisory baseline	This scenario follows a contour very similar to the average projections from surveys of economic forecasters. For example, the outlook for real activity and inflation in the baseline is in line with the October and November 2012 consensus projections from *Blue Chip Economic Indicators*. The baseline scenario for the US shows a moderate expansion in economic activity.	Real GDP: +2.75% per year Unemployment rate: edges down in 2013 and falls slowly to 6.75% by the end of 2015 CPI. + 2.25% per year Equity prices: +5.5% per year and equity-market volatility remains low. Nominal house prices: +3% per year
Supervisory adverse	This scenario is characterized by a weakening in economic activity across all of the economies included in the scenario combined with a sudden rise in domestic inflation that brings about a rapid increase in short- and long-term interest rates. This scenario features a moderate recession in the US that begins in the fourth quarter of 2012 and lasts until early 2014.	Real GDP: −2.0% per year Unemployment rate: rises to 9.75% in 2015 CPI: +4% Equity prices: fall 25% by the middle of 2013 and, correspondingly, the equity market volatility index jumps to over 40 (measured by the VIX index) at the start of the scenario Nominal house prices: –6% per year
Severely adverse	The severely adverse scenario is characterized by a substantial weakening in economic activity across all of the economies included in the scenario. In addition, the scenario features a significant further weakening in the US housing market.	Real GDP: −5.0% per year Unemployment rate: rises to 12% in 2015 CPI: +1% Equity prices: fall more than 50% over the course of the recession and, correspondingly, the equity market volatility index jumps above 70 at the start of the scenario Nominal house prices: fall more than 20% over the period

the nominal GDP-weighted aggregate of China, India, Hong Kong Special Administrative Region (SAR), and Taiwan.

Having discussed in some length the CCAR and stress testing exercises and scenarios with associated assumptions performed in the US, we also briefly note some features of the EU stress tests and associated assumptions. In 2011 the EU stress test performed by the European Banking Authority was undertaken on 90 banks which covered in excess of 65% of the total assets in the EU banking system. The stress testing simulation scenarios began witha baseline based on real financial data at the financial close of 2010 and covered forecasted scenarios for two years, 2011 and 2012. Overall, two core scenarios were considered which included: baseline and adverse categories.

The baseline scenarios involved a consideration of a strengthening macroeconomic recovery, where it was assumed that there would be a growth in GDP of 1.7% and 2% in the EU. Alternatively, under the adverse stress scenarios it was assumed that the GDP would instead reduce by 0.4% in 2011 and stay flat in 2012. In addition, it was furthermore assumed that equity prices would drop by 15%; and short-term risk-free rates would increase by 1.40% and long-term ones by 1.25%. Finally, it was assumed that credit spreads for sovereign debts in Europe would also rise, with different increases in each country.

Under these assumptions, the 90 participating banks were requested to utilise their internal capital estimation models, of which OpRisk is a core contributor, to generate values for balance-sheet items and P&L results. There was also imposed a stringent methodology that must be followed according to specifications developed by the European Banking Authority. Then each countries national supervisory body studied each of the firm specific assumptions made and these were cross checked with each countries national supervisors and the European Banking Authority for a uniformity analysis, resulting in additional calibration as was deemed suitable on a case by case basis. For further details see the account provided in Petrella and Resti (2013) as well as a list of outcomes of such stress tests in the EU over the last few years.

4.3 OpRisk and Stress Tests

As OpRisk capital represents a significant chunk of the total capital in most firms, it is obvious that it should be a key part of the stress test exercise. However, if for market and credit risks there is a more apparent relationship between the macroeconomic factors and the key drivers for these risks, this relationship is not clear for OpRisk. Therefore, most banks are heavily using more subjective tools like scenario analysis as the key input in the stress exercise. That does not mean that banks changed their scenario analysis program to deal with CCAR; in reality, new adverse scenarios were added that were very specific to the CCAR questions. Almost every bank developed a special "stress test scenario analysis" program to respond to the regulatory stress test exercise; however, quite a few are moving to integrate the two scenario programs somehow.

In OpRisk, to find a consistent statistical relationship between these factors and indicators and to incorporate them in a sound way into the framework is a significant challenge for a few reasons. The most obvious one is the usual culprit in OpRisk, which is data issues. Although significant progress is under way across the industry to improve the quality of operational loss data, this is still a major challenge. Some of the major issues are as follows:

- **Completeness**. The completeness of internal and external data, while an objective for the industry, is still elusive. Even when using external data to assess correlations, it can be questioned whether banks' loss database in the OpRisk data Exchange Association (ORX) consortia are actually fully comprehensive and, if so, whether they are reporting all losses they suffered;

- **Varying collecting thresholds**. Several banks started with very high collection thresholds and have been reducing these thresholds this so makes it difficult to find a long time series of standard events;
- **Natural scarcity**. Operational losses are sparser and for some risk types losses would not happen at daily or even weekly frequencies, while economic indicators are available daily. The solution in this case is to aggregate losses monthly, quarterly, etc. However, as the aggregation increases, quite a few spurious correlations would appear that would bear no logical support. For example, West Texas Intermediate (WTI) crude oil prices would show a 32% correlation with losses of Business Disruption and System Failures (BDSF) type (aggregated quarterly, using ORX data);
- **Dates**. Operational losses would have many dates associated, for example, "occurrence date" (when losses occur), "impact date" (when losses are realized), and "account date" (when losses are booked to the general ledger). Changing the type of date used would affect correlations.

Another issue is that, for several very important OpRisk types, the lag that exists between a macroeconomic event and the losses can be of many years, way beyond the exercise proposed by the regulators. This is a clear example of litigation losses (mostly under the risk type "Clients, Products, and Business Practices"). For example, only in 2011, banks started to set reserves for litigation originating from the mortgage crisis in the US that took place in 2007/2008. The cycle for a litigation process can take anywhere from 3 to 6 years or even longer. Considering the regulatory stress tests only span for a couple of years ahead, it is very difficult to find a meaningful correlation between a certain macroeconomic scenario and litigation losses within this time frame.

Given these constraints, modelers need to take quite a few cautionary measures. The first one is to break down OpRisks into their Basel risk types. OpRisks are actually an amalgamation of different risk types, and the impact of the macroeconomic factors can vary significantly among them. For execution losses (the "Execution, Delivery, and Process Management" type), a steep decline or volatility in the financial markets usually increase the trading volume, which can increase the execution losses. Using ORX data, this relationship is not so apparent in daily data, but starts to show up on a quarterly aggregation. Considering that execution risk would represent about 20–40% of the total OpRisk, a volatile macroeconomic scenario can potentially have some significance. However, there is no absolutely robust and conclusive correlation using these data. An example of a strong correlation is when this correlation is maintained at any aggregation level. For example, if we analyze the relationship between DJIA and S&P 500, we will find a strong relationship on a daily basis and if we extend this to weekly, monthly, and quarterly bases, the association will hold. If data from ORX were used against any of these macroeconomic factors, this would never happen.

For some risk types, like Employment Practices and Workplace Safety (EPWS), a stress scenario can actually lower the risk. Analyzing unemployment data against employment-related losses in the US, it can be seen that higher unemployment levels reduce the risk, as most employees are more worried about securing their jobs and avoid litigation with employers.

Bearing in mind these difficulties, many banks prefer to use subjective modeling of these correlations and relationships in the preparation of these scenarios. The danger of this method is that we can establish relationships, which although seem logical, might fail to actually be proven with hard data and, therefore, have any connection with reality.

4.4 OpRisk in CCAR in Practice

It has long been the practice of economists and finance professionals to seek out relationships between certain quantifiable factors that are thought to explain the behavior of some variable under study. Regression analysis, the chosen approach to solving this type of problem, is a statistical process for estimating the relationships among variables. It includes many techniques for modeling and analyzing several variables, when the focus is on the relationship between a dependent variable and one or more independent variables.

Regression models most frequently involve the following variables:

- The unknown parameters, denoted as β, which can be either scalar or vector;
- The independent (explanatory) variables, $\boldsymbol{X}$;
- The dependent (or response) variable, Y.

In what follows, we show an example of a study to find the relationship between OpRisk and the macroeconomic factors and financial indicators given by the FED through the use of trade volume (i.e., the count of how many trades are processed in a certain day) as the independent variable and five macroeconomic variables, as supplied by the FED through the CCAR process, as the independent variables. We then analyze the resulting model for stresses in those variables, as prepared by CCAR, in order to test the behavior of trade count under adverse conditions. The study has three steps:

- Find a relationship between a financial firm's internal factor and the FED/Office of the Comptroller of the Currency (OCC) macroeconomic factors;
- Find a relationship between operational losses and a financial institution's internal factor;
- Project stress capital estimates.

We are actually using a "bridge" between the macroeconomic factors and losses because a direct relationship might not be that strong or obvious.

The data used in this example belong to a medium-sized US bank and broker. Due to confidentiality issues we cannot provide much detail about the data, but we can say that losses happen many times every day and most losses would be on the risk event types "Execution, Delivery, and Process Management", and "Business Disruption and System Failures".

For a brokerage, trade volume is the main driver of revenue. As such, it is important to assess how this indicator would behave under economic stress from a revenue perspective. However, large trade volumes also tend to put pressure on the processing platform and more operational losses might happen. As finding a direct relationship between macroeconomic variables and operational losses is complicated by the issues discussed in the previous section, it might be easier to find these relationships with trade volumes, and then we can assess the relationship between trade volume and losses.

In order to relate responses to linear combinations of predictor variables, we use the Generalized Linear Models (GLMs). A detailed discussion on GLM modeling is provided in

Chapter 16. The GLM approach has become popular given its ability to model more than simply continuous dependent variables: for example, rates, proportions, binary, ordinal, and counts are among the many different types of variables that can be incorporated into GLMs. The canonical treatment of GLMs was defined by McCullagh and Nelder (1989). Overall, the model works by considering a mean of the response variable Y to be a function of the independent variables $\boldsymbol{X}$ as

$$\mu = \mathrm{E}[Y] = g^{-1}(\boldsymbol{X\beta}), \quad \text{i.e., } g(\mu) = \boldsymbol{X\beta}, \tag{4.1}$$

where $\eta = \boldsymbol{X\beta}$ is the so-called *linear predictor* and $g(\cdot)$ is a *link* function. The distribution of Y has to belong to the exponential family (that includes Normal, Poisson, Gamma, and many others). For example, in the case of Normal distribution, the link function is identity, $g(\mu) = \mu$; in the case of Poisson distribution, $g(\mu) = \ln \mu$. There are efficient schemes to estimate GLMs; for details and extensions, see Chapter 16.

After testing several variants of the GLM, we decided on the Normal distribution. Table 4.4 shows the results of the best fit (where β_0 corresponds to the explanatory variable $X_0 = 1$). As can be seen in Table 4.4, the Normal model possesses the best combination of a lower Akaike Information Criterion (AIC) and significant p-values as well as a stable R^2; in spite of the R^2 not being the highest, the Normal probability plot of the residuals (Figure 4.3)

TABLE 4.4 **GLM candidates**

			Normal distribution			
$\boldsymbol{\beta}$	**Estimate**	**StdErr**	**tStat**	***p*-value**		
β_0	−0.08992	0.034089	−2.63768	0.016224	**AIC**	−197
β_1	0.005348	0.001772	3.017435	0.007083	**AICc**	−192
β_2	0.005082	0.001439	3.530408	0.002236	**BIC**	−189
β_3	0.000336	9.70E−05	3.467959	0.002576		
β_4	0.000313	8.75E−05	3.573972	0.002025	R^2	0.8064
β_5	0.001329	0.000359	3.706702	0.001497	R^2_{adj}	0.7554
			Gamma distribution			
$\boldsymbol{\beta}$	**Estimate**	**StdErr**	**tStat**	***p*-value**		
β_0	62.47991	10.20403	6.12306	6.9E−06	**AIC**	−196
β_1	−1.67135	0.53389	−3.1305	0.00550	**AICc**	−192
β_2	−1.80615	0.46723	−3.86561	0.00104	**BIC**	−189
β_3	−0.09896	0.02841	−3.48295	0.00249		
β_4	−0.08771	0.02444	−3.58894	0.00195	R^2	0.8366
β_5	−0.36049	0.09528	−3.78339	0.00125	R^2_{adj}	0.7936
			Inverse Gaussian distribution			
$\boldsymbol{\beta}$	**Estimate**	**StdErr**	**tStat**	***p*-value**		
β_0	1876.013	363.2361	5.16472	5.5E−05	**AIC**	−194
β_1	−58.966	19.0674	−3.09253	0.005994	**AICc**	−190
β_2	−67.461	17.3470	−3.88895	0.000987	**BIC**	−187
β_3	−3.392	0.99980	−3.39266	0.003055		
β_4	−2.896	0.83323	−3.47673	0.002525	R^2	0.8389
β_5	−11.698	3.16406	−3.69722	0.001529	R^2_{adj}	0.7965

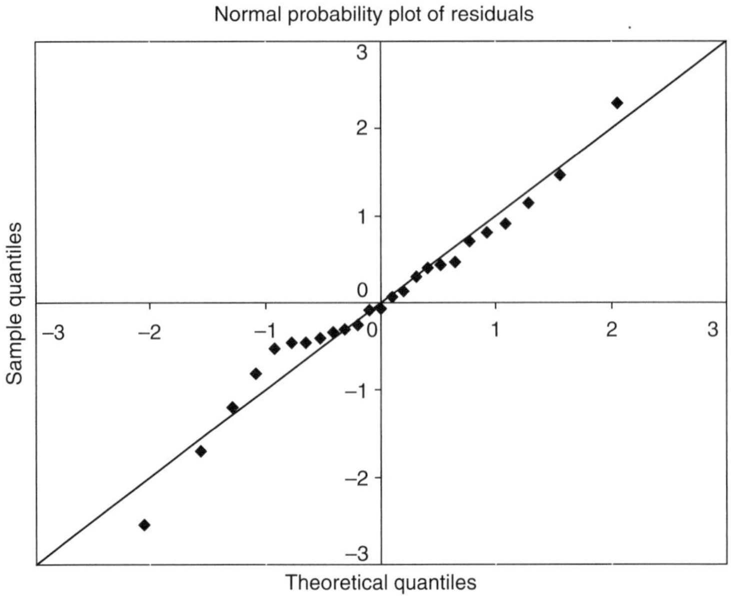

FIGURE 4.3 **Model fit results**

of the model also shows a reasonable fit. Model selection criteria, probability plots, and p-values will be formally defined later in Chapter 8.

In this model, we regressed the quarterly average trade volume from the third quarter 2006 to the third quarter 2012 (the response variable) against the following macroeconomic variables (the independent variables)—as supplied by regulators in the CCAR process:

- X_1, **US unemployment rate**. Quarterly average of monthly data, Bureau of Labor Statistics;
- X_2, **US 10-year Treasury yield.** Quarterly average of the yield on 10-year US Treasury bonds, constructed for the Federal Reserve Board (FRB)/US model by Federal Reserve staff based on the Svensson smoothed term structure model;
- X_3, **US Commercial Real Estate Price Index.** From flow of funds accounts of the US, FRB; the series corresponds to the data for price indexes: Commercial Real Estate Price Index divided by 1000;
- X_4, **US market Volatility Index (VIX).** Chicago Board Options Exchange, converted to quarterly by using the maximum value in any quarter;
- X_5, **developing Asia real GDP growth.** Staff calculations based on Bank of Korea via Haver; Chinese National Bureau of Statistics via CEIC; Indian Central Statistical Organization via CEIC; Census and Statistics Department of Hong Kong via CEIC; and Taiwan Directorate-General of Budget, Accounting, and Statistics via CEIC.

One tool that can be used to assess the goodness of fit is the relative variable impact. This is used to assess how much each variable in the model contributes to the formation of the resulting response variable. The results are presented in Figure 4.4. In this case, we

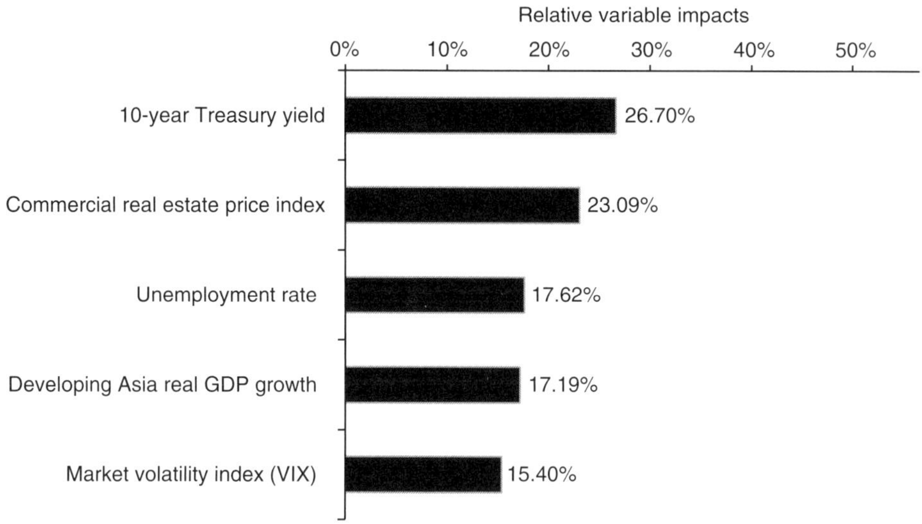

FIGURE 4.4 **Relative impact of the explanatory variables**

see that the 10-year Treasury yield and the Commercial Real-Estate Price Index combined contribute 49.8% (almost half) to the variation in trade volume while the remaining three variables account for 50.2% of the same variation. This calculation was done using the Garson relative contribution method (see Garson, 1991).

As described in previous sections, CCAR defines three scenarios: supervisory baseline, supervisory adverse, and supervisory severely adverse, where a predefined number of macro-economic variables are stressed in adverse direction. With the fitted model we projected the behavior of this bank/brokerage trade volume for each of these scenarios (Figure 4.5).

The economic rationale underlying the behavior of the trade volume for the three scenarios is very straightforward. In the baseline case, trade volume grows at a similar rate to expected economic growth. In the adverse case, trade volume grows at first (because market volatility will cause overall activity to increase) but dips down as the effects of economic weakness are felt. In the severely adverse case, trade volume experiences initial growth driven by increased volatility and deteriorating market conditions but declines to near baseline level as the economy returns to normal.

The last step of this analysis is to find a relationship between trade volume and operational losses so we can estimate losses based on the estimated trade volume. In order to keep the confidentiality of the data, we promised our data, provider to only state that a strong relationship was found and that the $R^2 = 66\%$.

The main objective of this section was to provide a practical view on how models are used by OpRisk analysts in determining the impact of operational losses in the pre provision net revenue and, ultimately, in the capital ratios. Table 4.5 depicts how these models are used in the exercise. The example in this section is quite simplistic and, in more realistic terms, a significantly higher number of models would be used for the determination of the final operational losses impact. For example, a separate model might be developed for litigation risk using models with multiyear temporal lags.

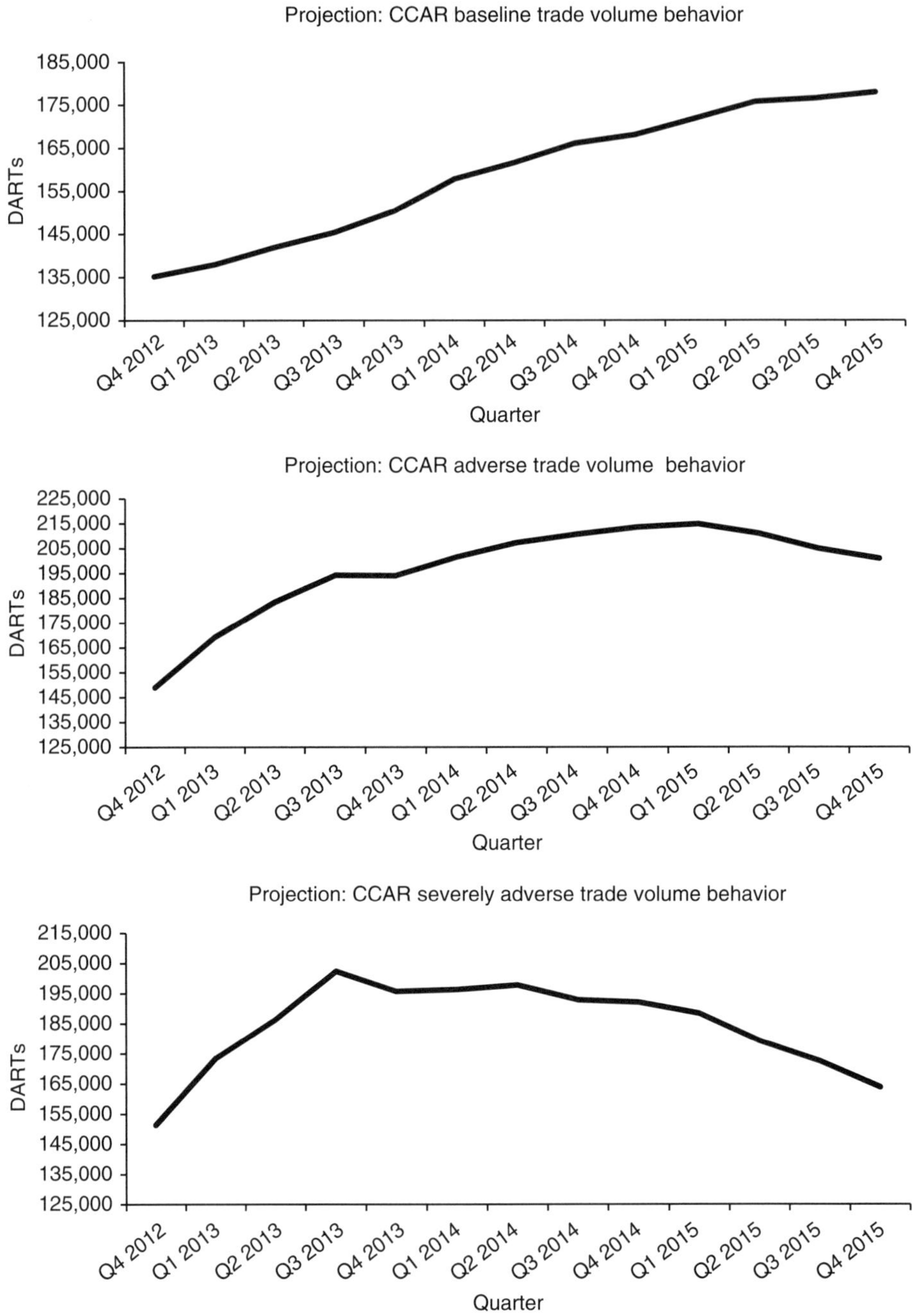

FIGURE 4.5 **DARTs (daily average revenue trades) under different CCAR scenarios**

TABLE 4.5 **Assessing preprovision net revenue (PPNR) impact**

PPNR	Models (example)	Description of model behavior in the example
Net interest income	NA	NA
Noninterest income	Trade volume against a number of CCAR variables (e.g., VIX, unemployment, etc.)	Revenue would grow for the broker as market volatility grows as clients would sell and buy—this is a relationship found by most firms. As the environment settles, the worsened economic environment would lower the trade volume, impacting revenue
Noninterest expenses	Establish how operational losses are related to trade volume growth	As the operational platform is constant in the short term (i.e., no major improvements happen), an increase in trade volume would increase operational losses, particular in transaction execution and systems losses
Total impact	Assess the net impact in PPNR of these variables	

4.5 Reverse Stress Test

Another popular type of stress test in the industry (and mandatory in the UK) is the so-called "reverse stress test". Reverse stress tests require a firm to assess scenarios and circumstances that would render its business model unviable, thereby identifying potential business vulnerabilities. Reverse stress-testing starts from an outcome of business failure and identifies circumstances in which this might occur, for illustrative example see Table 4.6. This is different from general stress and scenario testing, which tests for outcomes arising from changes in circumstances.

In 2009, the financial authority in the UK (then the FSA now the Prudential Regulation Authority) issued the Policy Statement 09/20, which goes into details about the reverse stress test.

Reverse stress-testing is primarily designed to be a risk management tool rather than a highly analytical exercise. It should encourage financial institutions to explore more fully the vulnerabilities and fault lines in their business model and inherent controls, including "tail risks". Based on the analysis of its reverse stress tests, senior management should determine whether it should put in place any mitigating actions at the current time or whether it should put in place triggers for future action should the scenario develop. It is separate but complementary to other stress tests, starting from the outcome of business failure.

Undertaking reverse stress-testing and taking action on its results should also inform contingency planning and enable financial institutions to make decisions that are consistent with both business and capital planning, very similar to the CCAR process in the US but with a more qualitative focus.

The objective is that financial institutions, based on these reverse stress tests, develop mitigation and recovery strategies. This recovery strategy is about the management of a firm taking actions that are aimed at preventing it from failing in circumstances in which it is facing severe stress as identified in the process. In order to avoid failure, the management may need to eventually undertake extreme measures. A recovery plan is one of the outcomes of the reverse stress test and details what options the management may pursue, what would need to

TABLE 4.6 **Reverse engineering stress losses with 1 year horizon**

Loss size	Risk type	Event	Mitigation/ control
USD 4000	CPBP	Litigation settlement on mortgage financing	
USD 1000	Internal Fraud	Trades are not reported in management system	Number of unconfirmed trades over 90 days Daily match of accounting and desk exposures
USD 800	External Fraud	Denial of service attack overload systems and force site to be off the air for a day	Consultancy that monitor and protects against these attacks Independent backup system based in different geography
USD 500	EPWS	Class action lawsuit	
USD 300	EDPM	Known deficiencies in internal transactions operations system can significant delay settlements causing steep losses	Intra-day monitoring of settlement failures Budget allocated to system upgrade

All amounts are in USD million.

happen for each action to be implemented, and the risks to implementing each action. In this way, a recovery plan can build on existing stress and scenario testing requirements, and on management actions that would be taken in response to these events.

In the resolution plan, firms will provide the information that would be necessary for the authorities and central banks to undertake the resolution of the firm and identify the actions that would need to be taken for the authorities to resolve a failing firm in an orderly manner. In the US, a similar process was called "living will". This is a separate process from the reverse stress test which requires a firm to identify and assess the scenarios most likely to cause its current business model to fail and, using these results, to put in place appropriate mitigating action. However, the reverse stress test can be seen as the starting point for resolution plans, as the point at which the risks identified in the reverse stress test crystallize may be the point at which resolution plans are required.

4.6 Stressing OpRisk Multivariate Models—Understanding the Relationship Among Internal Control Factors and Their Impact on Operational Losses

One type of modeling that has become more popular recently, particularly in US banks influenced by the CCAR process, is the multivariate model that relates operational losses to key control and business environment variables and also to external macroeconomic variables. These models are a very powerful tool for risk management as they allow to spotting the factors that are determinant to control losses and bring OpRisk to a similar level to market and credit

TABLE 4.7 **Operational loss data and control environment factors**

Date	Losses	No. losses	Downtime	No. employees	Data quality (%)	No. transactions
July 2	USD 234,412	10,004	3	22	94	250,096
July 3	USD 91,234	7,284	1	24	96	208,111
July 5	USD 2,734,009	17,792	10	19	88	345,611
July 6	USD 545	5,745	0	24	98	185,321
July 9	USD 115,912	9,745	1	24	97	249,876
July 10	USD 1,234	8,075	0	24	98	252,345
July 11	USD 91,233	9,287	1	24	98	250,987
July 12	USD 55,908	8,879	1	24	98	236,765
July 13	USD 12,002	9,079	0	24	98	238,911
July 16	USD 23,456	9,078	0	24	98	237,654
July 17	USD 1,787,634	13,514	8	21	89	293,778
July 18	USD 7,233,704	24,510	16	17	81	415,422
July 19	USD 2,891	8,054	0	24	97	250,912
July 22	USD 122	6,061	0	24	98	191,210
July 23	USD 0	5,360	0	24	99	172,901
July 24	USD 0	5,283	0	24	99	170,415
July 25	USD 200,786	8,387	1	24	95	221,876
July 26	USD 1,456	6,604	0	24	97	200,121
July 27	USD 918	5,934	0	24	98	191,435
July 30	USD 1,234,095	11,438	5	22	95	278,987
July 31	USD 17,654	7,287	0	24	96	238,908
Aug 1	USD 9,871	7,549	0	24	97	235,908
Aug 2	USD 1,095,033	10,988	3	22	97	268,001
Aug 3	USD 1,200	6,492	0	23	99	199,761

Downtime is system downtime in minutes.

risks. GLM and its extensions help to accomplish this task; this will be discussed in detail in Chapter 16.

Here, for illustration, we consider a simple example using data from a retail bank (see Table 4.7) and assume a simple multifactor model for operational daily losses in a particular business or area inside the bank as follows:

$$Y_t = \beta_0 + \beta_1 X_{1,t} + \ldots + \beta_n X_{n,t} + \varepsilon_t, \tag{4.2}$$

where Y_t represents the operational loss for a day t, $(X_{1,t}, \ldots, X_{n,t})$ represent the control environment factors, $(\beta_0, \ldots, \beta_n)$ are model parameters, and ε_t, $t = 1, 2, \ldots$ are independent random variables from zero mean Normal distribution.

Picking the right variables makes this model work very well. Four variables showed significance here: X_1— systems downtime, X_2— number of employees in a department, X_3— number of transactions, and X_4— data quality (% of data moving from front to back office correctly with no need of amendments). These data were provided by a custodian bank. It may be mentioned that the individual losses are mostly very small or even zero. This bank had developed a system to collect every single error even if no losses took place. This was really important in the multivariate analysis.

Running a multivariate regression with these data, we estimate

$$\beta_0 = 3,379,940;\ \beta_1 = 319,294;\ \beta_2 = -35,455;\ \beta_3 = 0.906;\ \beta_4 = -2,945,102 \tag{4.3}$$

with $R^2 = 91\%$. The averages of the explanatory variables from July 2 to August 3 are 2.08 minutes for system downtime; 23.08 for number of employees; 241,055 for number of daily transactions; and 95.83% for data quality. The high R^2 found in this model is surprisingly common within heavy transaction processing environments where the quality of processing is very dependent on system availability, volume, and personnel. Note that the specific model and specific choice of explanatory variables are for illustration only. More advanced analysis should involve a comparison with other distribution types for daily losses (at least within a GLM framework) and consider other possible explanatory variables and their transforms.

We can use this model to stress internal factors and see their impact on losses and OpRisk. For example, we can assume an increase of 30% in the volume of transactions, 20% decrease for the number of employees, etc. As an example, suppose that, in order to increase the profitability of the products traded in the area, the bank decides to increase the daily volume by 30%. Executive management then asks OpRisk management to assess the operational impact of this decision; however, this increase in transactions would not be followed by an increase in headcount, as the bank wants to keep tabs on costs. The average daily number of transactions during the period July 2 to August 3 was 241,055 and average daily loss was USD 622,721. This 30% increase in the number of transactions will mean that the average number of transactions will move to 313,371; then using (4.2) and (4.3) one can calculate that the average daily loss will increase to USD 688,241. As no employees can be hired, we should find ways to improve the system or the average data quality that was on average 95.83%. Using model (4.2) with parameter estimates (4.3), it is easy to find that if an internal quality program is developed and the quality of the input is increased to an average of 98.06%, then it will offset the impact of the 30% growth in the number of transactions and there will be almost no change in the average daily loss. Understanding the level of risk a bank faces given the increase in the number of transactions is quite important and a number of financial institutions are performing this analysis.

The model described is useful however, it is deterministic in explanatory variables. We could let the explanatory variables $X_{1,t}, \ldots, X_{4,t}$ be stochastic and perform a more informative stress test. For example, assuming that $X_{i,t}$, $t = 1, 2, \ldots$ are independent and identically distributed, we can calibrate the distributions for the explanatory variables as follows: system downtime, $X_{1,t} \sim Poisson(\lambda = 2.08)$; number of employees working per day, $X_{2,t} \sim Normal(\mu = 23.08, \sigma = 1.81)$; number of transactions, $X_{3,t} \sim LogNormal(\mu = 12.37, \sigma = 0.208)$; and data quality, $X_{4,t} \sim Beta(\alpha = 20.65, \beta = 0.898)$; for definition of distributions, see Appendix A. Then we could find the quantiles of the explanatory variables that can be used in stress-testing. In addition, given the multifactor model (4.2) and knowing the distributions for each independent (explanatory) variable X_i, we could calculate the unconditional distribution of daily losses and find its high quantiles.

CHAPTER FIVE

Basic Probability Concepts in Loss Distribution Approach

In risk management in general, modelers attempt to assess the uncertain risk exposures or threats using past experiences and other information available. Probability theory seems to be the natural fit for these types of analyses. This chapter provides a description of basic concepts of the probability theory used in this book and introduces relevant notation. There is a range of important concepts that are required to be considered when developing OpRisk models in practical settings. This chapter establishes what will be considered in future chapters as basic presumed knowledge. It covers the following basic concepts:

- Loss Distributional Approach (LDA) modelling;
- Definitions of a probability distribution function and density functions in univariate and multivariate settings, as well as discrete, continuous and mixed type random variables;
- Statements of the Law of Large Numbers and distributional convergence of scaled and translated sums are briefly discussed;
- Then moments and quantile functions for random variables are discussed;
- Following this, the notion of frequency distributional models are discussed for the number of losses in a given year;
- Then naturally, the notion of severity loss models is briefly discussed for the size of each loss event in a given year;
- Next, the compound process is discussed with additional discussion on convolutions and transform methods;
- Finally, a very brief overview of Extreme Value Theory is presented, for a more comprehensive coverage see the companion book (Peters and Shevchenko, 2015, chapter 2).

5.1 Loss Distribution Approach

OpRisks are modeled by random variables representing unknown size of the loss, time of the loss occurrence, number of losses, etc. The value of a random variable is a result of a

Fundamental Aspects of Operational Risk and Insurance Analytics: A Handbook of Operational Risk, First Edition. Marcelo G. Cruz, Gareth W. Peters, and Pavel V. Shevchenko.

measurement (e.g., size of the loss). Specifically, under the Loss Distribution Approach, the OpRisk loss over a 1-one year time horizon is modeled as

$$Z = X_1 + \cdots + X_N,$$

where the number of events per year (frequency) is a random variable N, and the sizes of the loss (severity) when the events occur are $X_1, X_2, \ldots$. It is common to assume that frequency and severity are independent, and severities $X_1, X_2, \ldots$ are independent and identically distributed. These assumptions will be made throughout unless stated otherwise in more advanced chapters of this text. Hereafter, we use the following notation:

- Random variables are denoted by upper case symbols (capital letters) and their realizations are denoted by lower case symbols, for example, random variable X and its realization x;
- Vectors are considered as column vectors and are written in bold, for example, n-dimensional random vector $\boldsymbol{X} = (X_1, X_2, \ldots, X_n)^T$, where superscript "$T$" denotes transposition;
- The realizations of random variables are real numbers, so that $\boldsymbol{x} = (x_1, x_2, \ldots, x_n)^T$ means a point in the n-dimensional Euclidean space of real numbers $\mathbb{R}^n$;
- Operators on random variables are written with square brackets, for example, the variance of a random variable X is denoted as $\mathbb{V}\mathrm{ar}[X]$.

Random variables representing frequency and severity are characterized by distribution functions formally defined as follows.

Definition 5.1 (Univariate distribution function) *The distribution function of a random variable X, denoted as $F_X(x)$, is defined as the probability that X is less than or equal to a number x*

$$F_X(x) = \mathbb{P}\mathrm{r}[X \le x].$$

The support of a random variable X with a distribution function $F_X(\cdot)$ is a set of all points, where $F_X(\cdot)$ is strictly increasing. Often used notation for the survival function or tail distribution function of a random variable X is defined as

$$\overline{F}_X(x) = 1 - F_X(x) = \mathbb{P}\mathrm{r}[X > x].$$

Frequently used notation, $X \sim F_X(x)$, means a random variable X has a distribution function $F_X(x)$. Often, for simplicity of notation, we may drop the subscript and write $X \sim F(\cdot)$. ■

The distribution function has to satisfy the following conditions:

- $F(x)$ is nondecreasing;
- $F(x) \to 1$, as $x \to \infty$;
- $F(x) \to 0$ as $x \to -\infty$;
- $F(x)$ is right continuous, that is, the limiting value of $F(x)$ as x approaches x_0 from the right equals $F(x_0)$.

Most of the standard distributions used throughout the book are formally defined in Appendix A.

Random variables can be classified into different categories (*continuous, discrete*, or *mixed*) according to a set of all possible outcomes (*support*). In particular, severities $X_1, X_2, \ldots$ (the loss sizes) are typically modeled as continuous random variables and frequency N (number of events per time interval) are typically modeled by a discrete random variable; the aggregate loss $Z = X_1 + \cdots + X_N$ is typically a mixed random variable (e.g., if $\Pr[N = 0] > 0$). Formal definitions are as follows.

Definition 5.2 (Continuous random variable) *A continuous random variable X has its support on an interval, a union of intervals, or real line (half-line). The distribution function of a continuous random variable can be written as*

$$F_X(x) = \int_{-\infty}^{x} f_X(y)dy,$$

where $f_X(x)$ is called the continuous probability density function. ■

Definition 5.3 (Discrete random variable) *A discrete random variable X has a finite or countable number of values $x_1, x_2, \ldots$. The distribution function of a discrete random variable has jump discontinuities at $x_1, x_2, \ldots$ and is constant between. The probability function (also called the* probability mass function*) of a discrete random variable is defined as*

$$p_X(x_i) = \Pr[X = x_i], \quad \text{for} \quad i = 1, 2, \ldots$$
$$p_X(x) = 0, \quad \text{for} \quad x \neq x_1, x_2, \ldots.$$ ■

Definition 5.4 (Mixed random variable) *A mixed random variable X is a continuous random variable with positive probability of occurrence on a countable set of exception points. Its distribution function F_X has jumps at these exception points and can be written as*

$$F_X(x) = wF_X^{(d)}(x) + (1 - w)F_X^{(c)}(x),$$

where $0 \leq w \leq 1$, $F_X^{(c)}$ is a continuous distribution function, and $F_X^{(d)}(x)$ is a discrete distribution function. ■

A mixed random variable is common in OpRisk for the loss aggregated over some period of time. This is because typically there is a probability of nonoccurrence loss during a period of time (giving finite probability mass at zero) while the loss amount is a continuous random variable. In general, any distribution function can be represented as a mixture of discrete distribution function, continuous distribution function, and singular continuous distribution function (a continuous distribution function with points of increase on a set of zero Lebesgue measure). The last type of random variable will not be considered in this book. The case of mixed random variables with discrete and continuous components covers all situations encountered in OpRisk practice.

To unify notation for discrete and continuous densities, it may be convenient to write the density functions using the *Dirac δ-function* (also called the impulse δ-function), which is zero everywhere except from the origin, where it is infinite, and its integral over any arbitrary interval containing the origin is equal to 1:

$$\delta(x) = 0, \text{ if } x \neq 0; \ \delta(0) = \infty,$$

$$\int_{-\varepsilon}^{\varepsilon} \delta(x)dx = 1, \text{ for any } \epsilon > 0.$$

This implies that for any function $g(x)$,

$$\int_{a}^{b} g(x)\delta(x - x_0)dx = g(x_0), \quad \text{if } a < x_0 < b, \tag{5.1}$$

and the integral is zero if (a, b) interval does not contain x_0. This definition of δ function is a heuristic definition but it is enough for the purposes of this book; the theory of the Dirac δ-function can be found in many textbooks (see, e.g., Pugachev 1965, section 9).

Then, the density of discrete random variable can be written as

$$f_X(x) = \sum_{i\geq 1} p_X(x_i)\delta(x - x_i), \tag{5.2}$$

and the density of a mixed random variable is

$$f_X(x) = w\sum_{i\geq 1} p_X(x_i)\delta(x - x_i) + (1 - w)f_X^{(c)}(x), \tag{5.3}$$

where $f_X^{(c)}(x)$ is the continuous density function and $p_X(x_i)$ is a probability mass function of a discrete distribution.

Similarly, vectors of random numbers are characterized by multivariate distributions, formally defined as follows.

Definition 5.5 (Multivariate distribution function) *The multivariate distribution function of a random vector* $\boldsymbol{X} = (X_1, X_2, \ldots, X_n)^T$ *is defined as*

$$F_{\boldsymbol{X}}(x_1, x_2, \ldots, x_n) = \Pr[X_1 \leq x_1, X_2 \leq x_2, \ldots, X_n \leq x_n]$$

and the corresponding survival function is

$$\overline{F}_{\boldsymbol{X}}(x_1, x_2, \ldots, x_n) = \Pr[\boldsymbol{X} > \boldsymbol{x}].$$ ■

Often we are interested in convergence of some sequences of random numbers. For example, the well-known probability theorem *Strong Law of Large Numbers* is stated as follows.

Theorem 5.1 (Strong Law of Large Numbers) *Given a sequence of independent and identically distributed random variables* $X_1, X_2, \ldots,$ *which are integrable* $\mathbb{E}[|X_1|] < \infty$,

$$\overline{X}_n = \frac{1}{n}\sum_{i=1}^{n} X_i \ \textit{converges to}\ \mathbb{E}[X_1], \quad \textit{for}\ n \to \infty.$$

The convergence stated is the so-called *almost surely* convergence. This means that the probability of $\lim_{n\to\infty} \overline{X}_n = \mathbb{E}[X_1]$ is 1, that is, there might be a sequence of random numbers that will not satisfy this limit but the probability of that sequence is zero. Often it is written as

$$\overline{X}_n \to \mathbb{E}[X_1], \quad \text{for } n \to \infty \ \textit{almost surely}$$

or

$$\overline{X}_n \to \mathbb{E}[X_1], \quad \text{for } n \to \infty \ \textit{a.s.}$$

EXAMPLE 5.1 **Distribution functions**

Consider the following three functions:

(a)

$$F(x) = 0.05x, \quad 0 \le x \le 20;$$

(b)

$$F(x) = \begin{cases} 0.05x, & 0 \le x \le 5, \\ 0.25, & 5 < x < 10, \\ 0.25 + 0.075(x - 10), & 10 \le x \le 20; \end{cases}$$

(c)

$$F(x) = \begin{cases} 0.025x, & 0 \le x < 10; \\ 0.05x, & 10 \le x \le 20. \end{cases}$$

All these functions are distribution functions and are presented in Figures 5.1a–c, and respectively. Case (a) corresponds to the so-called uniform distribution where all possible values of X have the same chance to occur. Case (b)

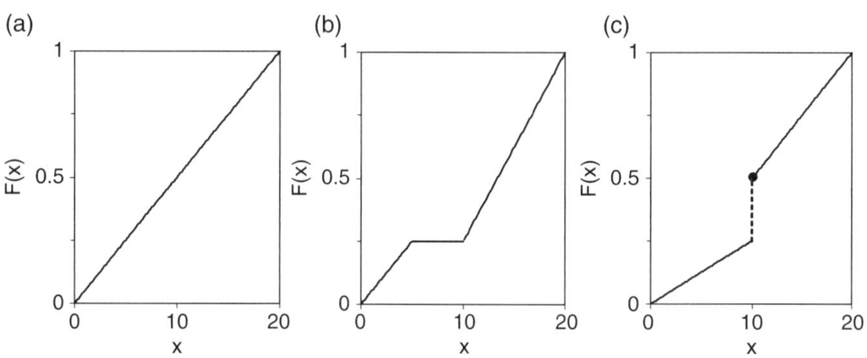

FIGURE 5.1 **Simple examples of distributions: (a) uniform distribution; (b) distribution with a flat piece; (c) distribution with a jump. See Example 5.1 for details**

corresponds to the distribution with a flat piece. Note that there is no chance for X to occur within a flat piece. Finally, Case (c) is a distribution with a jump; note here that the function is right continuous at the point of the jump. ■

EXAMPLE 5.2 Empirical distribution

Often, modelers use parametric distribution functions to model severity and frequency. However, it is also often convenient to use empirical distributions constructed from observed data. For example, a modeler may use empirical distribution to model severities below some large threshold and continuous distribution for severities above the threshold.

Given independent identically distributed realizations $x_1, \ldots, x_n$, empirical distribution is defined as

$$F(x) = \frac{1}{n}\sum_{i=1}^{n} \mathbb{I}_{\{x_i \le x\}}. \tag{5.4}$$

Consider a sample $(0.5; 2; 1; 1.2; 0; 1.5; 1.8; 0.7; 1; 1.9)$. Then the ordered sample is $(0; 0.5; 0.7; 1; 1; 1.2; 1.5; 1.8; 1.9; 2)$. Using (5.4), it is easy to calculate the empirical distribution of the sample, which is presented in Figure 5.2. Note that point 1 is repeated in the sample and thus the jump at this point is 2/10, while for all other points the jump is 1/10.

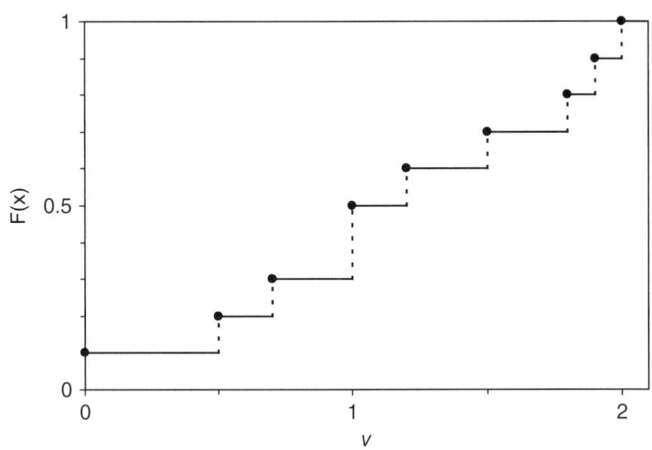

FIGURE 5.2 Empirical distribution. See Example 5.2 for details

■

5.2 Quantiles and Moments

Quantiles and moments of distribution are important characteristics/measures of random variables. Throughout the book we use the following standard definition of a generalized inverse function (also called *quantile function*) for a distribution function.

Definition 5.6 (Quantile function) *Given a distribution function $F_X(x)$, the inverse function F_X^{-1} of F_X is*

$$F_X^{-1}(\alpha) = \inf\{x \in \mathbb{R} : F_X(x) \geq \alpha\} = \sup\{x \in \mathbb{R} : F_X(x) < \alpha\},$$

where $0 < \alpha < 1$. It is also often denoted as $F_X^{\leftarrow}(\alpha)$. ■

Given a probability level α, $F_X^{-1}(\alpha)$ is the α-th quantile of X (often, it is denoted as q_α). The inverse function is defined as the left continuous generalized inverse of the distribution function. This is to handle cases when α corresponds to a flat piece in the distribution (in this case, the quantile corresponds to the left end of the flat piece). In the case when α does not sit on a flat piece, the quantile is the ordinary inverse of $F(x)$. Figure 5.3 illustrates quantiles for the standard and tricky cases such as distribution with flat pieces or jumps. Alternatively, the inverse function can be defined as the right continuous generalized inverse

$$F^{-1+}(\alpha) = \inf\{x : F(x) > \alpha\} = \sup\{x : F(x) \leq \alpha\}. \tag{5.5}$$

That is, the quantile would be to the right end of the flat piece if α corresponds to this flat piece; see Figure 5.3b for an example. We could also define the quantile as a convex combination of left and right continuous generalized inverse distributions. In this book (and in most of the literature), we consider the definition of quantile as $F^{-1}(\alpha)$.

The expected value (*mean*) of a random variable X is denoted as $\mathbb{E}[X]$. A formal construction of the operator $\mathbb{E}[\cdot]$ is somewhat involved but for the purposes of this book we will use the following short definition.

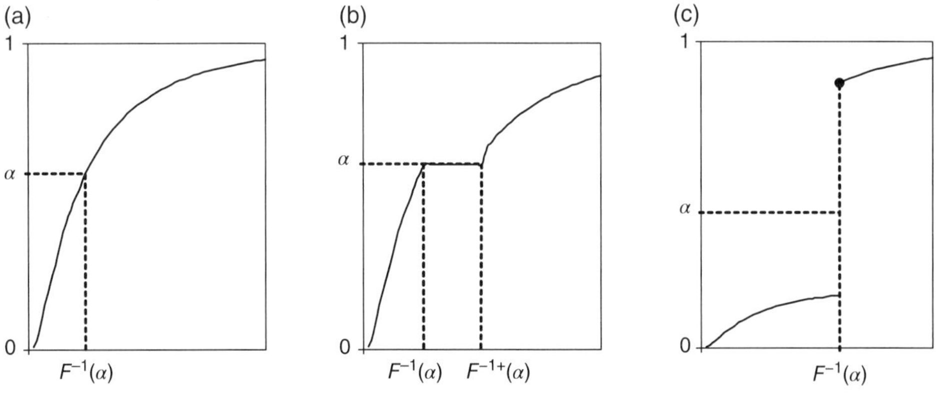

FIGURE 5.3 Calculation of quantiles: (a) continuous distribution; (b) distribution with a flat piece; (c) the case of probability atom in distribution function

Definition 5.7 (Expected value)

- *If X is a continuous random variable with the density function $f_X(x)$, then*

$$\mathbb{E}[X] = \int_{-\infty}^{\infty} x f_X(x) dx. \tag{5.6}$$

- *If X is a discrete random variable with support $x_1, x_2, \ldots$ and probability mass function $p_X(x)$, then*

$$\mathbb{E}[X] = \sum_{j \geq 1} x_j p_X(x_j).$$

- *In the case of a mixed random variable X (see Definition 5.4), the expected value is*

$$\mathbb{E}[X] = w \sum_{j \geq 1} x_j p_X(x_j) + (1 - w) \int_{-\infty}^{\infty} x f_X^{(c)}(x) dx.$$

■

The expected value integral or sum may not converge to a finite value for some distributions. In this case, it is said that the mean does not exist.

The definition of the expected value (5.6) can also be used in the case of the discrete and mixed random variables if their density functions are defined as (5.2) and (5.3), respectively. This gives a unified notation for the expected value of the continuous, discrete, and mixed random variables. Another way to introduce a unified notation is to use Riemann–Stieltjes integral

$$\mathbb{E}[X] = \int_{-\infty}^{\infty} x dF_X(x). \tag{5.7}$$

See Carter and Van Brunt (2000) for a good introduction to this topic.

The expected value is the first moment about the origin (also called the first raw moment). There are two standard types of moments: the raw moments and central moments, defined as follows.

Definition 5.8 (Moments)

- *The k-th moment about the origin (raw moment) of a random variable X is the expected value of X^k, that is, $\mathbb{E}[X^k]$;*
- *The k-th central moment of a random variable X is the expected value of $(X - \mathbb{E}[X])^k$, that is, $\mathbb{E}[(X - \mathbb{E}[X])^k]$.*

■

Typically, k is a nonnegative integer $k = 0, 1, 2, \ldots$. The expected value may not exist for some values of k; then it is said that the k-th moment does not exist. The first four moments are most frequently used and the relevant characteristics are defined as follows:

- **Variance**. The variance of a random variable X is the second central moment

$$\mathbb{V}\text{ar}[X] = \mathbb{E}[(X - \mathbb{E}[X])^2] = \mathbb{E}[X^2] - (\mathbb{E}[X])^2. \tag{5.8}$$

- **Standard deviation**. The standard deviation,

$$\text{stdev}[X] = \sqrt{\mathbb{V}\text{ar}[X]}, \tag{5.9}$$

  is a measure of spread of the random variable around the mean. It is measured in the same units as the mean (i.e., the same units as the values of the random variable).
- **Variational coefficient.** The *variational coefficient* (also called the *coefficient of variation*) is a dimensionless quantity,

$$\text{Vco}[X] = \frac{\text{stdev}[X]}{\mathbb{E}[X]}, \tag{5.10}$$

  which measures the spread relative to the mean;
- **Skewness**. The skewness is a dimensionless quantity that measures an asymmetry of a random variable X and is defined as

$$\gamma_1 = \frac{\mathbb{E}[(X - \mathbb{E}[X])^3]}{(\text{stdev}[X])^3}. \tag{5.11}$$

  For symmetric distributions, the skewness is zero;
- **Kurtosis**. The kurtosis is a dimensionless quantity that measures the flatness of the distribution (tail heaviness) relative to the Normal distribution. It is defined as

$$\gamma_2 = \frac{\mathbb{E}[(X - \mathbb{E}[X])^4]}{(\text{stdev}[X])^4} - 3. \tag{5.12}$$

  For the Normal distribution, kurtosis is zero.

Again, for some distributions these characteristics may not exist. Moreover, central moments can be expressed through the raw moments and vice versa. Detailed discussions, definitions, and relationships for these quantities can be found in most undergraduate statistical texts. To conclude this section, we define the covariance and the linear correlation coefficient that measure the dependence between random variables.

Definition 5.9 (Covariance and linear correlation) *The covariance of random variables X and Y is defined as*

$$\mathbb{C}\text{ov}[X, Y] = \mathbb{E}[(X - \mathbb{E}[X])(Y - \mathbb{E}[Y])] = \mathbb{E}[XY] - \mathbb{E}[X]\mathbb{E}[Y].$$

The linear correlation between X and Y is

$$\rho[X, Y] = \mathbb{C}\text{ov}[X, Y]/\sqrt{\mathbb{V}\text{ar}[X]\mathbb{V}\text{ar}[Y]}.$$

■

These quantities are popular measures of the dependence between X and Y. However, as will be discussed later in the book, the linear correlation can be a bad indicator of dependence for non-Gaussian random variables. Moreover, for some distributions these measures may not exist.

5.3 Frequency Distributions

The most commonly used frequency distributions for the annual number of events N are Poisson, Binomial, and Negative Binomial distributions. An interesting property (that is often used as a criterion to select a frequency distribution) is that a Binomial's variance is less than its mean, it is therefore called under-dispersed; the variance of the Negative Binomial is larger than its mean, it is therefore called over-dispersed; and the Poisson mean equals its variance. The BCBS paper on the range of practices for OpRisk AMA BCBS (2009a) reports that among 42 AMA banks participating in the survey,[1] 93% use the Poisson distribution, 19% use the Negative Binomial, and 7% use other distributions to model frequency. We formally define these distributions as follows.

- A **Poisson distribution function** is denoted as $Poisson(\lambda)$. The random variable N has a Poisson distribution $N \sim Poisson(\lambda)$ if its probability mass function is

$$p(k) = \mathbb{P}\mathrm{r}[N = k] = \frac{\lambda^k}{k!}e^{-\lambda}, \quad \lambda > 0 \tag{5.13}$$

  for all $k \in \{0, 1, 2, \ldots\}$. Expectation, variance, and variational coefficient of a random variable $N \sim Poisson(\lambda)$ are

$$\mathbb{E}[N] = \lambda, \quad \mathbb{V}\mathrm{ar}[N] = \lambda, \quad \mathrm{Vco}[N] = \frac{1}{\sqrt{\lambda}}. \tag{5.14}$$

- The **Binomial distribution function** is denoted as $Binomial(n, p)$. The random variable N has a Binomial distribution $N \sim Binomial(n, p)$ if its probability mass function is

$$p(k) = \mathbb{P}\mathrm{r}[N = k] = \binom{n}{k} p^k(1-p)^{n-k}, \quad p \in (0, 1), \; n \in 1, 2, \ldots \tag{5.15}$$

  for all $k \in \{0, 1, 2, \ldots, n\}$. Expectation, variance, and variational coefficient of a random variable $N \sim Binomial(n, p)$ are

$$\mathbb{E}[N] = np, \; \mathbb{V}\mathrm{ar}[N] = np(1-p), \quad \mathrm{Vco}[N] = \sqrt{\frac{1-p}{np}}. \tag{5.16}$$

  In a common interpretation, N is the number of successes in n independent trials, where p is the probability of a success in each trial;

- A **Negative Binomial distribution** function is denoted as $NegBinomial(r, p)$. The random variable N has a Negative Binomial distribution $N \sim NegBinomial(r, p)$ if its probability mass function is

$$p(k) = \mathbb{P}\mathrm{r}[N = k] = \binom{r+k-1}{k} p^r(1-p)^k, \quad p \in (0, 1), \; r \in (0, \infty) \tag{5.17}$$

[1]Note that banks participating in the survey were able to select more than one answer per question.

for all $k \in \{0, 1, 2, \ldots\}$. Here, the generalized binomial coefficient is

$$\binom{r+k-1}{k} = \frac{\Gamma(k+r)}{k!\Gamma(r)}, \tag{5.18}$$

where $\Gamma(r)$ is the Gamma function. Expectation, variance, and variational coefficient of a random variable $N \sim NegBinomial(r, p)$ are

$$\mathbb{E}[N] = \frac{r(1-p)}{p}, \quad \mathbb{V}\mathrm{ar}[N] = \frac{r(1-p)}{p^2}, \quad \mathbb{V}\mathrm{co}[N] = \frac{1}{\sqrt{r(1-p)}}. \tag{5.19}$$

If r is a positive integer, then in common interpretation, N is the number of failures in a sequence of independent trials until r successes, where p is the probability of a success in each trial.

There are many other discrete distribution types that can be found in many books; for example, for OpRisk context, see Panjer (2006). It can be useful to consider **zero-truncated distributions**: $p^{tr}(k) = p(k)/(1-p(0))$, $k = 1, 2, \ldots$, where $p(k)$ is a discrete distribution defined with $k = 0, 1, \ldots$ such as Poisson or Binomial. These truncated distributions can be used when zero value is impossible. Mixing and splicing methods can also be used to create other distributions from simple distributions (e.g., for special handling of zero values); these methods will be discussed in Section 5.4.3.

5.4 Severity Distributions

There are many standard parametric distributions that can be used for modelling severity. Some of these are listed in Appendix A, and some nonstandard distributions are discussed in Chapter 9. Many statistical books list two-, three-, and four- parameter continuous distributions; for a good review of possible distributions in the context of OpRisk, the reader is referred to Panjer (2006). These standard distributions can be used to create more flexible distributions via mixture and splicing methods discussed in Section 5.4.3.

The BCBS paper on the range of practices for OpRisk AMA BCBS (2009a) reports that among 42 AMA banks participating in the survey[2]

- About 31% banks apply a single severity distribution to model body and tail, with the LogNormal (33%) and Weibull (17%) most widely used;
- About 30% of banks use two distributions for body and tail: LogNormal (19%) and empirical (26%) for modeling the body and LogNormal (14%) and generalized Pareto (31%) for estimating the tail;
- Other distributions used for modeling severity include Gamma, g-and-h, generalized Beta, mixture of LogNormals.

Most of these distributions are formally defined in this section and in Chapter 9.

[2]Note that banks participating in the survey were able to select more than one answer per question.

5.4.1 SIMPLE PARAMETRIC DISTRIBUTIONS

Here, for illustration, we formally define few simple parametric distributions such as LogNormal and Exponential; for many nonstandard distributions the reader is referred to Chapter 9.

5.4.1.1 One-Parameter Distributions.

- **Exponential distribution** function is denoted as $Exp(\theta)$. The random variable X has an Exponential distribution, denoted as $X \sim Exp(\mu, \sigma)$, if its probability density function is

$$f(x) = \frac{1}{\theta} e^{-x/\theta}, \quad \mu > 0 \tag{5.20}$$

  for $x > 0$. The corresponding distribution function is simply $F(x) = 1 - e^{-x/\theta}$. All moments can be calculated as $\mathbb{E}[X^k] = \theta^k \Gamma(k+1)$ for $k > -1$, that is, $\mathbb{E}[X^k] = \theta^k k!$ for integer k. Here, $\Gamma(k)$ is a standard Gamma function formally defined by (A.2) in Appendix A. A very special feature of Exponential distribution is that the expected size of the loss above a threshold does not depend on the threshold;

- **One-parameter Pareto distribution** function is denoted as $Pareto(\xi, x_0)$. The random variable X has a Pareto distribution, denoted as $X \sim Pareto(\xi, x_0)$, if its distribution function is

$$F(x) = 1 - \left(\frac{x}{x_0}\right)^{-\xi}, \quad x \geq x_0, \tag{5.21}$$

  where $x_0 > 0$ and $\xi > 0$. The support starts at x_0, which is typically known and not considered as a parameter. Therefore the distribution is referred to as a single parameter Pareto. The corresponding probability density function is

$$f(x) = \frac{\xi}{x_0} \left(\frac{x}{x_0}\right)^{-\xi-1}. \tag{5.22}$$

  This distribution is heavy-tailed and has only a finite number of moments that can be calculated as $\mathbb{E}[X^k] = \xi x_0^k / (\xi - k)$ for $k < \xi$.

5.4.1.2 Two-Parameter Distributions.

- **LogNormal distribution** function is denoted as $LogNormal(\mu, \sigma^2)$. The random variable X has a LogNormal distribution, denoted as $X \sim LogNormal(\mu, \sigma^2)$, if its probability density function is,

$$f(x) = \frac{1}{x\sqrt{2\pi\sigma^2}} \exp\left(-\frac{(\ln(x) - \mu)^2}{2\sigma^2}\right), \quad \sigma^2 > 0, \ \mu \in \mathbb{R} \tag{5.23}$$

  for $x > 0$. Expectation, variance, and variational coefficient are,

$$\mathbb{E}[X] = e^{\mu + \frac{1}{2}\sigma^2}, \ \mathbb{V}\text{ar}[X] = e^{2\mu + \sigma^2}(e^{\sigma^2} - 1), \ \mathbb{V}\text{co}[X] = \sqrt{e^{\sigma^2} - 1}. \tag{5.24}$$

LogNormal is formally a heavy-tailed distribution (it belongs to the so-called class of subexponential distributions), but it is a very light heavy-tailed distribution with all moments existing;

- **Gamma distribution** function is denoted as $Gamma(\alpha, \beta)$. The random variable X has a gamma distribution, denoted as $X \sim Gamma(\alpha, \beta)$, if its probability density function is

$$f(x) = \frac{x^{\alpha-1}}{\Gamma(\alpha)\beta^{\alpha}} \exp(-x/\beta), \quad \alpha > 0,\ \beta > 0 \tag{5.25}$$

for $x > 0$. The Gamma distribution $Gamma(\alpha, \beta)$ is a light tail distribution. However, if $\ln X \sim Gamma(\alpha, \beta)$, then X is from Log-Gamma distribution, which is a heavy-tailed distribution with a Pareto-type power tail behavior. Expectation, variance, and variational coefficient of a random variable $X \sim Gamma(\alpha, \beta)$ are

$$\mathbb{E}[X] = \alpha\beta, \quad \mathbb{V}\mathrm{ar}[X] = \alpha\beta^2, \quad \mathbb{V}\mathrm{co}[X] = 1/\sqrt{\alpha}. \tag{5.26}$$

- **Pareto distribution** (two-parameter) function is denoted as $Pareto_2(\alpha, \beta)$. The random variable X has a Pareto distribution, denoted as $X \sim Pareto_2(\alpha, \beta)$, if its distribution function is

$$F(x) = 1 - \left(1 + \frac{x}{\beta}\right)^{-\alpha}, \quad x \geq 0, \tag{5.27}$$

where $\alpha > 0$ and $\beta > 0$. The corresponding probability density function is

$$f(x) = \frac{\alpha\beta^{\alpha}}{(x+\beta)^{\alpha+1}}. \tag{5.28}$$

The moments of a random variable $X \sim Pareto_2(\alpha, \beta)$ are

$$\mathbb{E}[X^k] = \frac{\beta^k k!}{\prod_{i=1}^{k}(\alpha - i)}; \quad \alpha > k.$$

Pareto distribution $Pareto_2(\alpha, \beta)$ has a very heavy tail such that the k-th moment and higher do not exist when the tail parameter $\alpha \leq k$;

- **Weibull distribution** function is denoted as $Weibull(\alpha, \beta)$. The random variable X has a Weibull distribution, denoted as $X \sim Weibull(\alpha, \beta)$, if its probability density function is,

$$f(x) = \frac{\alpha}{\beta^{\alpha}} x^{\alpha-1} \exp(-(x/\beta)^{\alpha}), \quad \alpha > 0,\ \beta > 0 \tag{5.29}$$

for $x > 0$. The corresponding distribution function is,

$$F(x) = 1 - \exp\left(-(x/\beta)^{\alpha}\right), \quad \alpha > 0,\ \beta > 0. \tag{5.30}$$

Expectation and variance of a random variable $X \sim Weibull(\alpha, \beta)$ are,

$$\mathbb{E}[X] = \beta\Gamma(1 + 1/\alpha), \quad \mathbb{V}\mathrm{ar}[X] = \beta^2\left(\Gamma(1 + 2/\alpha) - (\Gamma(1 + 1/\alpha))^2\right).$$

5.4.1.3 Three-Parameter Distributions.

- **Generalized Inverse Gaussian (GIG) distribution** function is denoted as $GIG(\omega, \phi, \nu)$. The random variable X has a GIG distribution, denoted as $X \sim GIG(\omega, \phi, \nu)$, if its probability density function is,

$$f(x) = \frac{(\omega/\phi)^{(\nu+1)/2}}{2K_{\nu+1}(2\sqrt{\omega\phi})} x^{\nu} e^{-x\omega - x^{-1}\phi}, \quad x > 0, \tag{5.31}$$

where $\phi > 0, \omega \geq 0$ if $\nu < -1$; $\phi > 0, \omega > 0$ if $\nu = -1$; $\phi \geq 0, \omega > 0$ if $\nu > -1$; and

$$K_{\nu+1}(z) = \frac{1}{2} \int_0^{\infty} u^{\nu} e^{-z(u+1/u)/2} du.$$

$K_{\nu}(z)$ is called a modified Bessel function of the third kind (see, e.g., Abramowitz and Stegun 1965, p. 375). The moments of a random variable $X \sim GIG(\omega, \phi, \nu)$ are not available in a closed form through elementary functions but can be expressed in terms of Bessel functions:

$$\mathbb{E}[X^{\alpha}] = \left(\frac{\phi}{\omega}\right)^{\alpha/2} \frac{K_{\nu+1+\alpha}(2\sqrt{\omega\phi})}{K_{\nu+1}(2\sqrt{\omega\phi})}, \quad \alpha \geq 1, \ \phi > 0, \ \omega > 0.$$

- **Burr distribution**. The random variable X has a Burr distribution, denoted as $X \sim Burr(\alpha, \beta, \gamma)$, if its distribution function is,

$$F(x) = 1 - u^{\alpha}, \quad u = \frac{1}{1 + (x/\beta)^{\gamma}}, \quad x \geq 0. \tag{5.32}$$

The density and moments are expressed in closed form as,

$$f(x) = \frac{\alpha\gamma(x/\beta)^{\gamma}}{x(1 + (x/\beta)^{\gamma})^{\alpha+1}},$$

$$\mathbb{E}[X^k] = \frac{\beta^k \Gamma(1 + k/\gamma)\Gamma(\alpha - k/\gamma)}{\Gamma(\alpha)}, \quad -\gamma < k < \alpha\gamma.$$

This distribution is also known as the Burr Type XII or Singh–Maddala distribution. It is often used to model household income.

5.4.2 TRUNCATED DISTRIBUTIONS

It is often convenient to model data using a truncated version of the standard distributions. For example, a standard distribution $F(x)$ (such as LogNormal, Gamma, etc.) is defined as $x > 0$ with a corresponding density function $f(x)$. However, one may be interested in modeling losses above some threshold $L > 0$ only. Then, one can consider a distribution truncated below L formally defined as

$$F^{tr}(x) = \frac{F(x) - F(L)}{1 - F(L)} \mathbb{I}_{x \geq L} \tag{5.33}$$

with a corresponding truncated density function

$$f^{tr}(x) = \frac{f(x)}{1 - F(L)} \mathbb{I}_{x \geq L}. \tag{5.34}$$

Note that this truncated density is a proper density function, that is, $\int_0^\infty f^{tr}(x)dx = 1$. Similarly, one can model losses below L using distribution truncated above L:

$$F^{tr}(x) = \frac{F(x)}{F(L)} \mathbb{I}_{x \leq L}, \quad f^{tr}(x) = \frac{f(x)}{F(L)} \mathbb{I}_{x \leq L}. \tag{5.35}$$

If there is a need to model losses in a specific range $[L, U]$, one can use distribution $F(x)$ truncated below L and above U:

$$F^{tr}(x) = \frac{F(x) - F(L)}{F(U) - F(L)} \mathbb{I}_{L \leq x \leq U}, \quad f^{tr}(x) = \frac{f(x)}{F(U) - F(L)} \mathbb{I}_{L \leq x \leq U}. \tag{5.36}$$

For example, in OpRisk settings, the lower threshold L may correspond to the data collection loss threshold and the upper threshold U may correspond to the high level separating body and tail losses.

5.4.3 MIXTURE AND SPLICED DISTRIBUTIONS

It is common practice for actuarial scientist and risk managers to consider the class of flexible distributional models known as mixture and spliced distributions. A mixture distribution is just a weighted average of other distributions formally defined as follows.

Definition 5.10 (Mixture distribution) *A random variable X has a mixture distribution if its distribution function is given by*

$$F(x) = w_1 F_1(x) + \cdots + w_k F_k(x),$$

where the weights $w_i > 0$, $w_1 + \cdots + w_k = 1$ and $F_i(x)$ are proper distributions. The density of the mixture is just

$$f(x) = w_1 f_1(x) + \cdots + w_k f_k(x).$$

■

The total number of parameters in the mixture distribution is the number of parameters across all distributions $F_i(x)$ plus $k - 1$ weights. The mixture approach allows to create many possible distributions from simple known distributions. For example, for a risk cell, one may consider a mixture of two different LogNormal distributions: $LogNormal(\mu_1, \sigma_1^2)$ and $LogNormal(\mu_2, \sigma_2^2)$, both defined by (5.23) for $x > 0$, to model a situation when the losses are generated by two different mechanisms. In this case, the total number of parameters is five.

Another closely related approach is splicing together pieces of different distributions (the so-called spliced distribution).

Definition 5.11 (Spliced distribution) *A random variable X has a spliced distribution if its density function is given by*

$$f_X(x) = \begin{cases} w_1 f_1(x), & x_0 \le x < x_1, \\ w_2 f_2(x), & x_1 \le x < x_2, \\ \vdots & \\ w_{k-1} f_{k-1}(x), & x_{k-2} \le x < x_{k-1}, \\ w_k f_k(x), & x_{k-1} \le x < x_k, \end{cases} \tag{5.37}$$

■

where $w_i > 0$ and $f_i(x)$ is a proper density function on (x_{i-1}, x_i) for $i = 1, \ldots, k$, that is, $\int_{x_{i-1}}^{x_i} f_i(x)dx = 1$; also $w_1 + \cdots + w_k = 1$. Typically, the motivation for splicing is to model large losses in the tail using one distribution (e.g., Pareto) and small losses using another distribution (e.g., LogNormal), because it is too restrictive to model both large and small losses using one simple distribution (e.g., Pareto or LogNormal). For example, to model the tail and body losses in a risk cell, one may consider a splicing of two truncated LogNormal distributions: one truncated below level L and another truncated above L (where L is body/tail large threshold level), that is,

$$f(x) = w f_B(x) + (1 - w) f_T(x), \quad 0 < w < 1, \tag{5.38}$$

where

$$f_B(x) = \frac{f_{LN}(x; \mu_B, \sigma_B)}{F_{LN}(L; \mu_B, \sigma_B)} \mathbb{I}_{0<x<L}, \quad f_T(x) = \frac{f_{LN}(x; \mu_T, \sigma_T)}{1 - F_{LN}(L; \mu_T, \sigma_T)} \mathbb{I}_{x \ge L}.$$

Here, $f_{LN}(x; \mu, \sigma)$ and $F_{LN}(x; \mu, \sigma)$ are the density and distribution functions of $LogNormal(\mu, \sigma^2)$ defined in (5.23).

Typically, component densities in the mixture distribution are defined on the same interval while splicing can be viewed as a mixture distribution with component densities defined on nonoverlapping intervals.

5.5 Convolutions and Characteristic Functions

Often we need to calculate the distribution of the sum of independent random variables such as the aggregate loss $X_1 + X_2 + \cdots + X_N$. It can be convenient to calculate these distributions through convolution of corresponding distribution functions.

Definition 5.12 (Convolution) *The convolution of two functions $g(x)$ and $h(x)$ is*

$$g(x) * h(x) = \int h(x - y) g(y) dy.$$

■

The density and distribution functions of the sum of two independent continuous random variables $Y_1 \sim F_1(\cdot)$ and $Y_2 \sim F_2(\cdot)$, with the densities $f_1(\cdot)$ and $f_2(\cdot)$, respectively, can be calculated via convolution as

$$f_{Y_1+Y_2}(y) = f_1(y) * f_2(y) = \int f_2(y - y_1) f_1(y_1) dy_1 \tag{5.39}$$

and

$$F_{Y_1+Y_2}(y) = F_1(y) * F_2(y) = \int F_2(y - y_1) f_1(y_1) dy_1 \tag{5.40}$$

respectively.

This can be generalized to the sum of many independent random variables using n-fold convolutions.

Definition 5.13 (n-fold convolution) *Given distribution functions $F_1(\cdot), \ldots, F_n(\cdot)$, the n-fold convolution is*

$$F_n^{(n)*}(x) = F_{n-1}^{(n-1)*}(x) * F_n(x),$$

calculated recursively as

$$F_k^{(k)*}(x) = F_{k-1}^{(k-1)*}(x) * F_k(x), \quad k = 2, \ldots, n$$

with $F_1^{(1)*}(x) = F_1(x)$. In the case of the same function $F(x) = F_1(x) = \cdots = F_n(x)$, we have

$$F^{(n)*}(x) = F^{(n-1)*}(x) * F(x).$$

■

Using n-fold convolutions, it is easy to calculate the distribution of the sum of independent random variables using the following well-known result.

Proposition 5.1 (Distribution of sum of independent random variables via convolution) *Given $X_1, \ldots, X_n$ are independent random variables with $X_i \sim F_i(\cdot)$, the distribution of the sum $X = X_1 + \cdots + X_n$ is the n-fold convolution*

$$\Pr[X_1 + \cdots + X_n \leq x] = F_n^{(n)*}(x).$$

In the case of independent identically distributed (i.i.d.) $X_1, \ldots, X_n$, where $X_i \sim F(x)$,

$$\Pr[X_1 + \cdots + X_n \leq x] = F^{(n)*}(x). \tag{5.41}$$

Thus, the distribution of the annual loss $X_1 + \cdots + X_N$, where $X_1, \ldots, X_N$ are i.i.d. from the severity distribution $F(\cdot)$ and annual frequency N is random with $\Pr[N = k] = p_k$, can be calculated as

$$\begin{aligned} H(z) = \Pr[Z \leq z] &= \sum_{k=0}^{\infty} \Pr[Z \leq z | N = k] \Pr[N = k] \\ &= \sum_{k=0}^{\infty} p_k F^{(k)*}(z). \end{aligned} \tag{5.42}$$

The k-fold convolution $F^{(k)*}(z)$ is calculated recursively as

$$F^{(k)*}(z) = \int_{0}^{z} F^{(k-1)*}(z-x)f(x)dx$$

with

$$F^{(0)*}(z) = \begin{cases} 1, & z \geq 0, \\ 0, & z < 0. \end{cases}$$

Here the integration limits are 0 and z. This is because we consider nonnegative severities. The obtained formula is analytic. However, closed-form solutions are rare. Panjer recursion and FFT, discussed in Sections 11.4 and 11.6, are very efficient numerical methods to calculate these convolutions.

Another powerful tool to calculate the distribution of the sum of independent random variables is the method of characteristic functions. It is explained in many textbooks on probability theory. In particular, it is often used for calculating aggregate loss distributions. some distributions are defined via characteristic functions and are not available in closed form (e.g., alpha stable distributions).

Definition 5.14 (Characteristic function) *The characteristic function of the density $f(x)$ is defined as*

$$\varphi(t) = \int_{-\infty}^{\infty} f(x)e^{\mathrm{i}tx}dx, \tag{5.43}$$

where $\mathrm{i} = \sqrt{-1}$ *is a unit imaginary number.* ■

If the characteristic function is known, then the original density function can be calculated by inverse Fourier transform

$$f(x) = \frac{1}{2\pi}\int_{-\infty}^{\infty} \varphi(t)\exp(-\mathrm{i}tx)dt. \tag{5.44}$$

The corresponding analogy for discrete distributions is called the *probability generating function*

Definition 5.15 (Probability generating function) *The probability generating function of a discrete distribution with probability mass function $p_k = \mathbb{P}\mathrm{r}[N = k]$ is*

$$\psi(s) = \sum_{k=0}^{\infty} s^k p_k. \tag{5.45}$$

■

Using a well-known property that the characteristic function of the sum of independent random variables is just a product of their characteristic functions, the characteristic function of the annual loss $X_1 + \cdots + X_N$, denoted by $\chi(t)$, can be expressed through the probability generating function of the frequency distribution and characteristic function of the severity distribution as

$$\chi(t) = \sum_{k=0}^{\infty} (\varphi(t))^k p_k = \psi(\varphi(t)). \tag{5.46}$$

Given the characteristic function, the density of the annual loss Z can be calculated via the inverse Fourier transform; this will be discussed in detail in Chapter 11.

5.6 Extreme Value Theory

The topic of extreme value modeling is discussed in detail in companion book *Advances in Heavy Tail Risk Modeling: A Handbook of Operational Risk*, Peters and Shevchenko (2015). However, for completeness of this manuscript, we briefly discuss the main concepts of extreme value theory (EVT).

There are two types of EVT models: traditional *block maxima* and *threshold exceedances*. *Block maxima* EVT is focused on modeling the largest loss per time period of interest. This is used in insurance and in many other fields. For example, it is used in the design of dams for flood control where engineers are interested in quantification of the probability of the annual maximum water level. It is certainly important for operational risk too. However, for capital calculations, the primary focus is to quantify the impact of all losses per year. Modeling of all large losses exceeding a threshold is dealt by EVT *threshold exceedances*. The key result of EVT is that the largest losses or losses exceeding a large threshold can be approximated by the limiting distribution—which is the same regardless of the underlying process. This allows for rational extrapolation to losses beyond those historically observed and estimation of their probability. However, as with any extrapolation method, EVT should be applied with caution.

Typically, to apply EVT (or any other extrapolation method) on a dataset, we assume that there is a single physical process responsible for the observed data and any future losses exceeding the observed levels. This is often the case in physical sciences (e.g., hydrology). However, in assessing operational risk, some people may argue that extreme values are anomalous and are not strongly related to the rest of the data. In addition, multiple processes might be responsible for extreme events within a risk cell and these processes might be different from the processes generating less severe losses. A good discussion on these issues can be found by Cope *et al.* (2009) and Nešlehová *et al.* (2006).

If we assume that a single mechanism is responsible for the losses in dataset and extrapolation can be done, then EVT is a very powerful tool. A detailed presentation of EVT is provided by Embrechts *et al.* (1997), McNeil *et al.* (2005, chapter 7). In this chapter, we summarize the main results relevant to operational risk. It is important to note that EVT is an asymptotic theory. Whether the conditions validating the use of the asymptotic theory are satisfied is often a difficult question to answer. The convergence of some parametric models to the EVT regime is also very slow. For example, this is the case for the by LogNormal and g-and-h distributions studied Mignola and Ugoccioni (October 2005) and Degen *et al.* (2007), respectively. In general, EVT should not preclude the use of other parametric distributions.

5.6.1 EVT—BLOCK MAXIMA

Consider a sequence of n independent random variables $X_1, \ldots, X_n$ from a distribution $F(x)$ representing losses. Denote the maximum loss as

$$M_n = \max(X_1, \ldots, X_n).$$

Because each loss cannot exceed the maximum and due to independence between the losses, the distribution of the maximum is

$$\begin{aligned} F_{M_n}(x) &= \mathbb{P}\mathrm{r}[M_n \leq x] = \mathbb{P}\mathrm{r}[X_1 \leq x, \ldots, X_n \leq x] \\ &= \prod_{i=1}^{n} \mathbb{P}\mathrm{r}[X_i \leq x] = (F(x))^n. \end{aligned} \tag{5.47}$$

Given that $F(x) < 1$ or $F(x) = 1$, it is easy to see that if $n \to \infty$, then the distribution of maximum (5.47) converges to the degenerate distribution, which is either 0 or 1 (i.e., the density concentrates on a single point), which is not very useful information. That is why the study of the largest losses in the limit $n \to \infty$ requires appropriate normalization. This is somewhat similar to the central limit theory stating that the appropriately normalized sum

$$\widetilde{S}_n = (S_n - a_n)/b_n,$$

where $S_n = X_1 + \cdots + X_n$ and $X_1, \ldots, X_n$ are independent and identically distributed random variables with finite variance, converges to the standard Normal distribution as $n \to \infty$. Here, the normalized constants are

$$a_n = n\mathbb{E}[X_1], \quad b_n = \sqrt{n\mathbb{V}\mathrm{ar}[X_1]}.$$

Similarly, the limiting result for the distribution of the normalized maximum $\widetilde{M}_n = (M_n - d_n)/c_n$ shows that for some sequences of $c_n > 0$ and d_n,

$$\lim_{n\to\infty} \mathbb{P}\mathrm{r}[(M_n - d_n)/c_n \leq x] = \lim_{n\to\infty} (F(c_n x + d_n))^n = H(x). \tag{5.48}$$

If $H(x)$ is a nondegenerate distribution, then F is said to be in the *maximum domain of attraction* of H, which is denoted as $F \in MDA(H_\xi)$. Then the well-known Fisher–Trippet, Gnedenko Theorem essentially says that $H(x)$ must be the generalized extreme value (GEV) distribution $H_\xi((x - \mu)/\sigma)$, $\sigma > 0$, $\mu \in \mathbb{R}$ with the standard form

$$H_\xi(x) = \begin{cases} \exp\left(-(1 + \xi x)^{-1/\xi}\right), & \xi \neq 0, \\ \exp(-\exp(-x)), & \xi = 0, \end{cases} \tag{5.49}$$

where $1+\xi x > 0$. The standard GEV will often be referred to as $GEV(\xi)$. If convergence takes place, then it is always possible to choose normalizing sequences c_n and d_n so that the limit will be in the standard form $H_\xi(x)$. The shape parameter ξ determines a type of distribution:

- $\xi > 0$ corresponds to a Fréchet distribution;
- $\xi = 0$ corresponds to a Gumbel distribution;
- $\xi < 0$ corresponds to a Weibull distribution.

The Weibull distribution ($\xi < 0$) has a bounded right tail (i.e., $x \leq -1/\xi$), while Gumbel and Fréchet have an unbounded right tail. In addition, the decay of the Fréchet tail is much slower than the Gumbel tail.

5.6.2 EVT—RANDOM NUMBER OF LOSSES

As earlier mentioned, in OpRisk, the number of losses per time period is not fixed and is modeled as a random variable N with $p_n = \mathbb{P}\mathrm{r}[N = n]$. This has some implications for the use of the previously described EVT.

Distribution of maximum. If the frequency N is random, then, instead of (5.47), the distribution of a maximum M_N is calculated as

$$\begin{aligned} F_{M_N}(x) &= \sum_{n=0}^{\infty} \mathbb{P}\mathrm{r}[M_N \leq x | N = n]\mathbb{P}\mathrm{r}[N = n] \\ &= \sum_{n=0}^{\infty} (F(x))^n \mathbb{P}\mathrm{r}[N = n] = \psi_N(F(x)), \end{aligned} \tag{5.50}$$

where

$$\psi_N(t) = \mathrm{E}[t^N] = \sum_k p_k t^k,$$

is the probability generating function of the frequency distribution. Note that there is a finite probability for zero maximum, that is, $\mathbb{P}\mathrm{r}[M_N = 0] = \psi_N(F(0))$. Typically, severity distribution has $F(0) = 0$ and frequency distribution has a finite probability at zero; in this case, $\mathbb{P}\mathrm{r}[M_N = 0] = \mathbb{P}\mathrm{r}[N = 0]$.

For example, if the annual number of losses $N \sim \mathit{Poisson}(\lambda)$, then $\psi(t) = \exp(-\lambda(1-t))$ and thus the distribution of the maximum loss (per annum) is

$$F_{M_N}(x) = \exp(-\lambda(1 - F(x)). \tag{5.51}$$

The distribution of the maximum loss over m years is

$$(F_{M_N}(x))^m = \exp(-m\lambda(1 - F(x)). \tag{5.52}$$

5.6.3 EVT—THRESHOLD EXCEEDANCES

While it is important to understand and measure maximum possible loss over a 1-year time horizon, the primary focus in operational risk capital charge calculations is quantification of overall impact of all losses. For this purpose, the method of EVT threshold exceedances is very useful. Consider a random variable X, whose distribution is $F(x) = \mathbb{P}r[X \leq x]$. Given a threshold u, the exceedance of X over u is distributed from

$$F_u(y) = \mathbb{P}r[X - u \leq y | X > u] = \frac{F(y+u) - F(u)}{1 - F(u)}. \tag{5.53}$$

As the threshold u increases, the limiting distribution of $F_u(\cdot)$ is given by the Pickands–Balkema-de Haan theorem (see McNeil *et al.* 2005, section 7.2). The theorem essentially states that *if and only if* $F(x)$ is the distribution for which the distribution of the maximum (5.48) is $GEV(\xi)$ given by (5.49), then, as u increases, the excess distribution $F_u(\cdot)$ converges to a generalized Pareto distribution (GPD), $GPD(\xi, \beta)$:

$$G_{\xi,\beta}(y) = \begin{cases} 1 - (1 + \xi y/\beta)^{-1/\xi}, & \xi \neq 0, \\ 1 - \exp(-y/\beta), & \xi = 0. \end{cases} \tag{5.54}$$

Here, the shape parameter ξ is the same as the shape parameter of the GEV distribution H_ξ. More strictly, we can find a function $\beta(u)$ such that

$$\lim_{u \to a} \sup_{0 \leq y \leq a-u} |F_u(y) - G_{\xi,\beta(u)}(y)| = 0, \tag{5.55}$$

where $a \leq \infty$ is the right end point of $F(x)$, ξ is the GPD shape parameter, and $\beta > 0$ is the GPD scale parameter. The domain of GPD is

$$y \in \begin{cases} [0, \infty), & \text{if} \quad \xi \geq 0, \\ [0, -\beta/\xi], & \text{if} \quad \xi < 0. \end{cases} \tag{5.56}$$

The properties of GPD depend on the value of the shape parameter ξ:

- The case $\xi = 0$ corresponds to an exponential distribution with the right tail unbounded;
- If $\xi > 0$, the GPD right tail is unbounded and the distribution is heavy-tailed, so that some moments do not exist. In particular, if $\xi \geq 1/m$, then the m-th and higher moments do not exist. For example, for $\xi \geq 1/2$, the variance and higher moments do not exist. The analysis of operational risk data by Moscadelli (2004) reported even cases of $\xi \geq 1$ for some business lines, that is, infinite mean distributions; also see discussions by Nešlehová *et al.* (2006);
- $\xi < 0$ leads to a bounded right tail, that is, $x \in [0, -\beta/\xi]$. It seems that this case is not relevant to modeling operational risk as all reported results indicate a non-negative shape parameter. One could think though of a risk control mechanism restricting the losses by an upper level and then the case of $\xi < 0$ might be relevant.

The density of GPD is

$$h(x;\xi,\beta) = \begin{cases} \dfrac{1}{\beta}(1+\xi x/\beta)^{-1-1/\xi}, & \xi \neq 0, \\ \dfrac{1}{\beta}\exp(-x/\beta), & \xi = 0, \end{cases} \tag{5.57}$$

where $h(x=0) = 1/\beta$. Note some special cases of negative shape parameter: if $\xi = -1/2$, then $h(x) = \frac{1}{\beta}(1 - \frac{1}{2}x/\beta)$ is a linear function; if $\xi = -1$, then $h(x) = 1/\beta$ is constant; if $\xi < -1$, then $h(x)$ is infinity at the boundary of the domain $-\beta/\xi$. The latter case is certainly not relevant to operational risk in practice and can be excluded during fitting procedures.

The GPD has a special stability property with respect to excesses. Specifically, if $X \sim G_{\xi,\beta}(x)$, $x > 0$, then the distribution of the conditional excesses $X - L|X > L$ over the threshold L is also the GPD with the same shape parameter ξ and changed scale parameter from β to $\beta + \xi L$:

$$\mathbb{P}\mathrm{r}[X - L \leq y|X > L] = G_{\xi,\beta+\xi L}(y), \quad y > 0. \tag{5.58}$$

This stability property implies that if $\xi < 1$, then the mean excess function is

$$e(L) = \mathbb{E}[X - L|X > L] = \frac{\beta + \xi L}{1 - \xi}. \tag{5.59}$$

That is, the mean excess function is linear in L. This is often used as a diagnostic to check that the data follow the GPD model. In particular, it is used in a graphical method (plotting the mean excess of the data versus the threshold) to choose a threshold when the plot becomes approximately "linear".

CHAPTER SIX

Risk Measures and Capital Allocation

OpRisk is a significant risk exposure to most firms and therefore requires effective risk management. Thus, these risks should be modeled, measured, and capital should be held so that a bank can withstand extreme losses. A risk measure is a single number quantifying an exposure to the risk. In particular, risk managers and regulators are interested in assessing the probability that extreme losses may occur and this can be represented through the quantile of the loss distribution (over a specified time horizon). This led to the regulatory requirement for risk capital to be measured as a Value-at-Risk (VaR), which is just a quantile of the loss distribution at some high confidence level (i.e., quantile of the loss distribution at the 0.999 confidence level over a 1-year horizon for OpRisk).

Since VaR had come into widespread use in the financial markets for quantifying market risk, adacemics began to undertake theoretical studies of the properties of such a risk measure and they began to notice that using VaR as a risk measure could sometimes have a poor outcome. Specifically, the diversification principle may fail in some circumstances. The wisdom in choosing the 0.999 VaR as a risk measure for capital is highly contested (see, e.g., Daniélsson *et al.* 2001). Using economic reasoning, a list of axiomatic properties for a good (*coherent*) risk measure was suggested in the seminal paper by Artzner *et al.* (1999). In particular, an alternative risk measure known as expected shortfall (ES) is coherent and considered to be better suited for risk management as it provides information not only about the probability of the default but also about its severity; it can be viewed as an average of losses larger than or equal to the VaR. However, the use of VaR for a capital has good justification from a regulator's point of view when considering minimization of the possible shortfall and cost of the capital. Moreover, it is clear that ES is not so good a measure for OpRisk because some OpRisks exhibit such heavy tails that even the mean (expected loss) may not exist; these are the so-called infinite mean distributions reported in the literature for OpRisk. In addition, ES can be too sensitive to the tail index of heavy-tailed distributions while VaR is more stable.

A lot of research has been done in the area of risk measures. The choice of risk measure is currently discussed in the literature; for example, Basel Committee on Banking Supervision (BCBS, 2012) asks for a possible transition from VaR to ES as the underlying risk measure. We know that in general VaR does not have diversification property, while ES does. However, Gneiting (2011) implies that, while VaR in general is statistically backtestable, ES is not.

Fundamental Aspects of Operational Risk and Insurance Analytics: A Handbook of Operational Risk, First Edition. Marcelo G. Cruz, Gareth W. Peters, and Pavel V. Shevchenko.

Furthermore, while VaR in general has certain robustness properties, ES does not, at least as discussed by Cont *et al.* (2010). So, it is likely that VaR, despite all its shortcomings, will remain in force.

In this chapter, we focus on VaR and ES, which are the most relevant risk measures for OpRisk; their definitions, advantages, and drawbacks are considered throughout Section 6.2.

Typical risk models are parameterized, where the true value of the parameter is unknown and should be estimated. It is expected that the uncertainty in parameters should increase the capital. Accounting for parameter uncertainty in the risk measure is a subject of Section 6.2.9.

A closely related problem is capital allocation. An overall bank capital should be allocated to various levels within a bank. In particular, it is allocated to business line or business line/event type (e.g. execution, delivery and process management event type in asset management business line) level and is often required to be allocated below, that is, to the process level or general manager level. The allocation mechanism should provide incentive to better manage OpRisk. Also, it is desirable that the allocation procedure to different levels shares the same risk factors/drivers. It is a challenging and currently unresolved issue as these two objectives seem to be in conflict. Optimal management of OpRisks considers risks and controls of typical events rather than extreme events that are outside of a risk manager's control. While a bank capital is driven by the risk tail events and the corresponding risk tail measure, risk body events and measure are more meaningful to the business incentives. Moreover, the availability of data below the business line/event type level is typically limited. In this book, we consider the allocation mechanisms to the risk cells (i.e., to the level where OpRisks are modeled, e.g. business line/event type level). Similar to defining a coherent risk measure using a set of axioms, a coherent allocation principle can be defined. It has been demonstrated (using different sets of axioms of economic reasoning) that the capital allocations can be calculated as the gradient of the capital with respect to risk exposures (the so-called the Euler's principle). The subject of risk allocations is considered in Section 6.3.

Before presenting an overview of the different classes of risk measures, we present some motivation and background context of the Basel accords and how they have developed in terms of requirements for capital estimation and risk measure quantification.

6.1 Development of Capital Accords Base I, II and III

In jurisdictions in which active regulation is applied to the banking sector, the modeling of OpRisk has progressively taken a prominent place in financial quantitative measurement. This has occurred as a result of Basel II and now Basel III regulatory requirements. There has been a significant amount of research dedicated to understanding the features of Basel II (see, e.g., Daniélsson *et al.* 2001, Decamps *et al.* 2004, and Kashyap and Stein 2004). In addition, the mathematical and statistical properties of the key risk processes that comprise OpRisk, especially those that contribute significantly to the capital charge required to be held against OpRisk losses, have also been carefully studied; see, for example, the book length discussions in Cruz (2002), King (2001), and Shevchenko (2011).

In January 2001, the Basel Committee on Banking Supervision proposed the Basel II Accord (BCBS, 2002, 2004, 2006), which replaced the 1988 Capital Accord. In 2013, the Basel III Accord was due to start to be considered. Since the initiation of the Basel capital accords, the discipline of OpRisk and its quantification have grown in prominence in the financial sector. Paralleling these developments have been similar regulatory requirements for the insurance industry which are referred to as Solvency 2. In both accords, the primary component of such a regulation revolves around the quantitative modeling of capital.

Under the Basel II/Basel III structures, there is at the core the notion of three pillars, which, by their very nature, emphasize the importance of assessing, modeling, and understanding OpRisk loss profiles. These three pillars are minimum capital requirements (refining and enhancing risk-modeling frameworks), supervisory review of an institution's capital adequacy, and internal assessment processes and market discipline, which deals with disclosure of information.

In the third update to the Basel Accords due for implementation in the period 2013–2018, a global regulatory standard that draws together bank capital adequacy, stress-testing, and market liquidity was developed. It is established as an international best practice for modeling OpRisk by the members of the Basel Committee on Banking Supervision (see Gregoriou 2009 and discussions in Blundell-Wignall and Atkinson 2010).

The Basel III Accord naturally extends the work developed in both the Basel I and Basel II accords, with the new accord arising primarily as a response to the identified issues associated with financial regulation that arose during the recent global financial crisis in the late 2000s. In this regard, the Basel III accord builds on Basel II by strengthening the bank capital requirements as well as introducing additional regulatory requirements on bank liquidity and leverage. The quantification of capital requirements is principally concerned with an evaluation of the risk associated with losses arising from a range of different loss processes in different lines of business.

Banking regulation under Basel II and Basel III specifies that banks are required to hold adequate capital against OpRisk losses. OpRisk is a relatively new category of risk that is additional to more well-established risk areas such as market and credit risks. As such, in its own right, OpRisk attracts a capital charge, which is defined by Basel II (BCBS 2006, p. 144) as "*the risk of loss resulting from inadequate or failed internal processes, people and systems or from external events. This definition includes legal risk, but excludes strategic and reputational risk*". OpRisk is significant in many financial institutions, e.g. see Table 1.4 for examples of capital ratios in some large European banks in 2012.

Before detailing the changes to capital requirements due to come into industry practice under Basel III, it is prudent to recall the Basel definition of Tier 1 capital, which is the key measure of a bank's financial strength from the perspective of the regulatory authority. In particular, the capital accord in Basel II and III states that financial institutions must provide capital above the minimum required amount, known as the floor capital. In addition, this capital as specified in the regulation is comprised of three key components: Tier 1, Tier 2 and Tier 3. Both Tier 1 and Tier 2, capital were first defined in the Basel I capital accord and remained substantially the same in the replacement Basel II and Basel III accords.

Definition 6.1 (Tier 1 capital) *The Tier 1 capital under regulation is comprised of the following main components:*

1. *Paid-up share capital/common stock;*
2. *Disclosed reserves (or retained earnings).*

It may also include nonredeemable noncumulative preferred stock. ■

The Basel Committee also noted the existence of banking strategies to develop instruments in order to generate Tier 1 capital. As a consequence, these must be carefully regulated through the imposition of stringent conditions, with a limit to such instruments at a maximum of 15% of total Tier 1 capital.

Definition 6.2 (Tier 2 capital) *The Tier 2 capital under regulation is comprised of the following main components:*

1. *Undisclosed reserves;*
2. *Asset revaluation reserves;*
3. *General provisions/general loan-loss reserves;*
4. *Hybrid (debt/equity) capital instruments;*
5. *Long-term subordinated debt.*

In this regard, one may consider Tier 2 capital as representing the so-called, supplementary capital. ■

We note at this stage that as a consequence of different legal systems in each jurisdiction, the accord has had to be sufficiently flexible to allow for some interpretation of specific capital components within the context of each regulator's jurisdiction. Depending on the particular jurisdiction in question, the specific country's banking regulator has some discretionary control over how exactly differing financial instruments may count in a capital calculations.

Remark 6.1 *The key reason that Basel III requires financial institutions to hold capital is that it is aimed to provide protection against unexpected losses. This is different to mitigation of expected losses, which are covered by provisions, reserves, and current year profits.*

We note that modifications under the Basel III Accord relative to its predecessor refer to limitations on risk-weighted capital (RWC) and the Tier 1 capital ratio, defined as follows.

Definition 6.3 (Risk-weighted assets (RWA)) *These assets comprise the total of all assets held by the bank weighted by credit risk according to a formula determined by either the jurisdiction's regulatory authority or in some cases the central bank. Most regulators and central banks adhere to the definitions specified by the BCBS guidelines in setting formulae for asset risk weights. Liquid assets such as cash and coins typically have zero risk weight, while certain loans have a risk weight at 100% of their face value. As specified by the BCBS, the total RWA is not limited to credit risk. It contains components for market risk (typically based on VAR) and OpRisk. The BCBS rules for calculating the components of total RWA have also been updated as a result of the recent financial crisis.* ■

Definition 6.4 (Tier 1 capital ratio) *The Tier 1 capital ratio is the ratio of a bank's core equity capital to its total RWA.* ■

Next, we highlight the prominent extensions to the Basel II Accord, established in the Basel III Accord. In particular, the Basel III Accord will require financial institutions to hold for RWA, 4.5% of common equity, which is an increase from the previous 2% under Basel II as well as 6% of Tier 1 capital, itself an increase by 2% relative to Basel II. In addition to these changes to common equity and Tier 1 capital, Basel III also introduces a minimum leverage ratio and two additional required liquidity ratio limits. Finally, of the significant changes, there are also additional capital buffers introduced:

1. A mandatory capital conservation buffer of 2.5%;
2. A discretionary countercyclical buffer, allowing national regulators to require up to another 2.5% of capital during periods of high credit growth.

Against the backdrop of these capital regulatory accord changes and extensions, there is always the base fundamental requirement of risk analysts, actuaries, and quants, which involves the quantitative modeling and reporting of such capital estimates. To quantify the OpRisk capital charge under the current regulatory framework for banking supervision, referred to as Basel II/Basel III, many banks adopt the Loss Distribution Approach (LDA). In this context, we are working with frequency and severity and resulting compound processes. In this chapter, the primary concern involves the development of quantification of risk utilizing different classes of risk measures for LDA models. There are typically three main families of risk measure that are considered for the calculation of OpRisk capital: VaR (LDA annual loss distribution quantile function); ES (LDA annual loss distribution tail conditional expectation), and spectral risk measures. These risk measures and their associated quantitative properties are covered in detail in the remainder of this chapter.

6.2 Measures of Risk

In general, a risk is an event that may or may not occur (i.e., random event) and brings some adverse consequences. It is natural to model OpRisk by a random variable that represents the random amount of loss that a company may experience. It can be assumed that random variables modeling OpRisk losses are non-negative (similar to insurance risk). In general, risk can be defined as a random variable representing future worth, but in OpRisk we focus on losses, not profits.

Given this definition of risk, measuring OpRisk means establishing a correspondence between the random variable representing risk and a non-negative real number. This leads us to the following definition of risk measure.

Definition 6.5 (Risk measure) *A risk measure is a mapping of a random variable representing risk to a real number. Henceforth, denote a general risk measure related to the risk X as $\rho[X]$.* ∎

That is, a risk measure (for OpRisk) is a functional that assigns a single non-negative real number to a non-negative random variable representing risk. No single risk measure can describe all aspects of risk. In this book, we consider risk measures used for setting capital requirements, that is, we focus on measuring the upper tail of a loss distribution. Moreover, the risk measure should describe not only the aspects of the overall risk but also the relative importance of risks within a collection. In particular, risk managers are interested in diversification benefits if risks are merged into a collection. This is typically measured using the diversification coefficient.

Definition 6.6 (Diversification coefficient) *For a collection of risks $X_1, \ldots, X_n$, the diversification coefficient is defined as*

$$D = 1 - \frac{\rho[X_1 + \cdots + X_n]}{\rho[X_1] + \cdots + \rho[X_n]}. \tag{6.1}$$

∎

This coefficient is positive if there are diversification benefits and is negative if diversification fails.

The choice of a risk measure for capital quantification is not a trivial task. The current Basel II requirement is to use VaR as a risk measure. However, VaR has some shortcomings, and other risk measures such as ES are widely discussed in the literature. At the same time, the use of VaR has a justification from the regulator's point of view. In this section, we treat the issue of risk measurement with a particular emphasis on VaR and ES risk measures, which are the most relevant to operational risk.

6.2.1 COHERENT AND CONVEX RISK MEASURES

There are many different risk measures introduced in the literature and practice, and the choice of a risk measure might be difficult. One approach to treat the issue of risk measurement is to start with a list of properties that a risk measure should satisfy. Using economic reasoning, a list of axiomatic properties for a good (*coherent*) risk measure was suggested in the seminal paper by Artzner *et al.* (1999).

Definition 6.7 (A coherent risk measure) *A coherent risk measure, $\rho[X]$, is defined to have the following properties for any two random variables X and Y:*

- *Translation invariance: for any constant c, $\rho[X + c] = \rho[X] + c$;*
- *Monotonicity: if $X \leq Y$ for all possible outcomes, then $\rho[X] \leq \rho[Y]$;*
- *Subadditivity: $\rho[X + Y] \leq \rho[X] + \rho[Y]$;*
- *Positive homogeneity: for any positive constant c, $\rho[cX] = c\rho[X]$.*

■

Note that in OpRisk we define loss to have a positive value while the original paper by Artzner *et al.* (1999) works with the future value of a position (which is negative for losses). As a result, there are changes in sign in some of the axioms (we also set interest rates to zero, i.e., no discounting).

The topic of coherent risk measures has been widely discussed in the literature (see McNeil *et al.* 2005). We list some arguments typically used to explain why the axioms are reasonable requirements.

- *Translation invariance*. This axiom means that adding a fixed amount to a collection of risks will change the capital requirement by the same amount. This is necessary to make sense of ρ as a risk capital, that is, adding cash amount $\rho[X]$ to the risk X gives adjusted loss $\widetilde{X} = X - \rho[X]$, whose capital is $\rho[\widetilde{X}] = \rho[X] - \rho[X] = 0$ and thus the new risk $\widetilde{X}$ is acceptable without further capital requirement (note that loss is defined as a positive value);
- *Monotonicity*. This is perhaps the most obvious axiom: risks that lead to smaller losses in every state require less risk capital. However, some traditional risk measures such as those based on standard deviation can fail this condition (see, e.g., Kalkbrener 2005). This has unpleasant consequences for the capital allocation. For example, if potential losses X are bounded by some level, then the contributory capital of X to the risk collection might exceed this level (see Kalkbrener *et al.* 2004);

- *Subadditivity*. This axiom is the most known because VaR fails to satisfy this condition in some situations. The rationale behind this axiom is to allow diversification benefit. It is easy to see that the diversification coefficient, as defined in (6.1), is positive for subadditive risks and is negative if subadditivity (diversification) fails. Artzner *et al.* (1999) support this by the statement that "a merger does not create extra risk", that is, merging two risks X_1 and X_2 with stand-alone capitals $\rho[X_1]$ and $\rho[X_2]$ will create overall risk $X = X_1 + X_2$ with a capital $\rho[X]$ less than or equal to the sum of stand-alone capitals $\rho[X_1] + \rho[X_2]$. Of course, anyone with experience through a merger can question this statement. So, it might be better to argue that breaking up will increase the capital requirement. If the risk manager wants to restrict $\rho[X_1 + X_2]$ by some level A, then he or she can just impose levels A_1 and A_2 ($A_1 + A_2 \leq A$) such that $\rho[X_1] \leq A_1$ and $\rho[X_2] \leq A_2$. Then subadditivity will ensure that $\rho[X] \leq A_1 + A_2 \leq A$. Note that if a non-subadditive risk measure is used for regulatory capital, then a bank might have an incentive to legally break up into subsidiaries to reduce the overall regulatory capital;
- *Positive homogeneity*. This axiom means that increasing a risk by a factor α should increase the capital by the same factor. Note that the subadditivity axiom implies that $\rho[2X] \leq \rho[X] + \rho[X]$. Thus, positive homogeneity adds an extra condition that $\rho[2X] = \rho[X] + \rho[X]$.

Artzner *et al.* (1999) demonstrated that any coherent risk measure (on a finite set of probability measures) can be written as the so-called scenario-based risk measure.

Definition 6.8 (Scenario-based risk measure) *The scenario-based risk measure for a risky loss random variable X is*

$$\rho[X] = \sup\{\mathbb{E}^Q[X] | Q \in P\}, \tag{6.2}$$

where $\mathbb{E}^Q[\cdot]$ means that the expectation is calculated with respect to probability distribution Q, and P is a nonempty set of probability measures (generalized scenarios). ■

It is straightforward to prove the coherence of this risk measure. The properties of monotonicity, positive homogeneity, and translational invariance can be easily followed by the definition. The subadditivity follows from

$$\begin{aligned}\sup\{\mathbb{E}^Q[X+Y] | Q \in P\} &= \sup\{\mathbb{E}^Q[X] + \mathbb{E}^Q[Y] | Q \in P\} \\ &\leq \sup\{\mathbb{E}^Q[X] | Q \in P\} + \sup\{\mathbb{E}^Q[Y] | Q \in P\}.\end{aligned} \tag{6.3}$$

For a more technical proof that any coherent risk measure can be represented as (6.2), see McNeil *et al.* (2005, section 6.1.4). The earlier defined risk measure is expectation with respect to a worst case scenario because supremum is taken over different distributions Q in a set P. For most of this book, we consider risk measures defined with respect to a single distribution.

It is important to note that the previously listed coherence axioms are accepted by many researchers and practitioners. However, there is no set of desirable axioms universally accepted. One can change a set of axioms and introduce other "coherent" risk measures. In particular, the condition of positive homogeneity has been criticized due to potential liquidity and risk concentration issues. One can argue that doubling the position will lead to a portfolio with more than double risk. It has been suggested to have $\rho[\lambda X] \geq \lambda\rho[X]$, $\lambda > 0$, to penalize for possible lack of liquidity. The condition $\rho[\lambda X] > \lambda\rho[X]$ cannot be satisfied for a subadditive risk measure. As a result, another larger class of *convex* risk measures has been introduced by Föllmer and Schied (2002), in whose work subadditivity and positive homogeneity axioms are replaced by weaker property of *convexity*. It is formally defined as follows.

Definition 6.9 (Convex risk measure) *A convex risk measure $\rho[X]$ is defined to have the following properties for any two random variables X and Y:*

- *Monotonicity: if $X \le Y$ for all possible outcomes, then $\rho[X] \le \rho[Y]$;*
- *Translation invariance: for any constant c, $\rho[X + c] = \rho[X] + c$;*
- *Convexity: $\rho[\lambda X + (1-\lambda)Y] \le \lambda\rho[X] + (1-\lambda)\rho[Y], \quad \lambda \in [0,1]$.*

■

That is, convex risk measure axioms are translation invariance, monotonicity, and convexity.

6.2.2 COMONOTONIC ADDITIVE RISK MEASURES

A desirable property for a risk measure is the so-called comonotonic additivity, which means that diversification (6.1) is zero for risks that are perfectly (and positively) dependent. This is formally defined as follows.

Definition 6.10 (Comonotonic additivity) *The risk measure $\rho[\cdot]$ is comonotonic additive if*

$$\rho[X_1 + \cdots + X_n] = \rho[X_1] + \cdots + \rho[X_n], \tag{6.4}$$

where $X_1, \ldots, X_n$ are comonotonic risks (i.e., perfectly positively dependent risks). The risks are called comonotonic if there exist a random variable Z and nondecreasing functions $h_1, \ldots, h_n$ such that $X_i = h_i(Z)$, $i = 1, \ldots, n$. In particular, comonotonic risks can always be represented as

$$X_1 = F_1^{-1}(U), \ldots, X_n = F_n^{-1}(U),$$

where U is a random variable from the uniform (0,1) distribution and $F_i(\cdot)$ is the distribution of X_i. ■

VaR and expected shortfall risk measures (formally defined later) are comonotonic additive. It is important to note that the risk measure might fail subadditivity but still satisfy comonotonic additivity (e.g., this is the case of VaR in some situations). For coherent risks, $\rho[X_1+\cdots+X_n] \le \rho[X_1]+\cdots+\rho[X_n]$, and thus $\rho[X_1]+\cdots+\rho[X_n]$, is the worst possible case for $\rho[X_1+\cdots+X_n]$. A class of coherent risk measures with such a property is the so-called *spectral risk measures* (see Acerbi 2002, Tasche 2002, and Kusuoka 2001). In fact in (Acerbi, 2002, theorem 7) they show that the class of spectral risk measures can be identified as all the coherent measures which are also law-invariant and comonotonic additive.

6.2.3 VALUE-AT-RISK

The concept of VaR as a quantile of a loss distribution has become a benchmark risk measure and is adopted by Basel regulations for setting the capital requirement. It allows addressing the question of what loss we can experience over a time period (i.e., 1 year for OpRisk) with a given probability and is formally defined as follows.

Definition 6.11 (Value-at-Risk) *The VaR of a random variable* $X \sim F(x)$ *at the* α*-th probability level,* $\mathrm{VaR}_\alpha[X]$*, is defined as the* α*-th quantile of the distribution of* X*:*

$$\begin{aligned}\mathrm{VaR}_\alpha[X] &= F^{-1}(\alpha) = \inf\{x : \Pr[X > x] \leq 1 - \alpha\} = \inf\{x : F(x) \geq \alpha\} \\ &= \sup\{x : F(x) < \alpha\}.\end{aligned} \tag{6.5}$$

That is, VaR is the minimum threshold exceeded by X *with probability at most* $1 - \alpha$*.* ■

The above VaR is defined as the left-continuous generalized inverse of the distribution function. This is to handle cases when α corresponds to a flat piece in the distribution (in this case, VaR corresponds to the left end of the flat piece). In the case when α does not sit on a flat piece, VaR is the ordinary inverse of $F(x)$. Figure 6.1 illustrates VaR for the standard and tricky cases such as a distribution with flat pieces or jumps. Alternatively, VaR can be defined as the right-continuous generalized inverse

$$F^{-1+}(\alpha) = \inf\{x : F(x) > \alpha\} = \sup\{x : F(x) \leq \alpha\} \tag{6.6}$$

that is, VaR would be the right end of the flat piece if α corresponds to this flat piece; see Figure 6.1b as an example. We could also define VaR as a convex combination of left- and right-continuous generalized inverse distributions. In this book (and in most of the literature), we take the definition of VaR as $F^{-1}(\alpha)$.

The VaR has the following obvious properties:

- $\mathrm{VaR}_\alpha[X] \leq \max[X]$ for any $\alpha \in (0, 1)$;
- VaR is *monotonic*: $\mathrm{VaR}_\alpha[X] \leq \mathrm{VaR}_\alpha[Y]$ if $X \leq Y$;

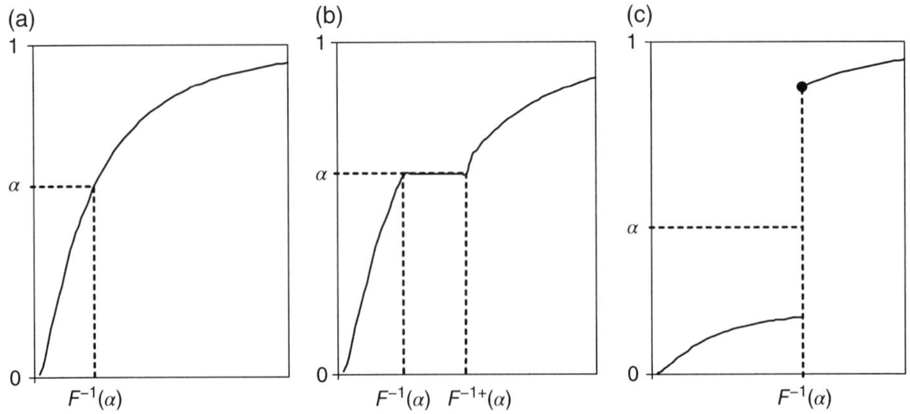

FIGURE 6.1 Calculation of quantiles: (a) continuous distribution; (b) distribution with a flat piece; (c) the case of probability atom in distribution function

- VaR is *translation invariant*, $\mathrm{VaR}_\alpha[X+c] = \mathrm{VaR}_\alpha[X] + c$;
- VaR is *positive homogeneous* $\mathrm{VaR}_\alpha[cX] = c \times \mathrm{VaR}_\alpha[X]$.

The last two properties also follow from the following general relation.

Proposition 6.1 (VaR of transformed random variable) *VaR of a random variable $Y = g(X)$, where $g(\cdot)$ is a nondecreasing function of a random variable X, can be calculated as*

$$\mathrm{VaR}_\alpha[Y] = g(\mathrm{VaR}_\alpha[X]).$$

Proof: Let $F(x)$ be a distribution of X. Then the proof is straightforward from the probability transform $\mathbb{P}\mathrm{r}[X \le F^{-1}(\alpha)] = \mathbb{P}\mathrm{r}[g(X) \le g(F^{-1}(\alpha))]$. ■

It is also easy to see that VaR is *comonotonic additive*, that is, there is no diversification for perfectly dependent risks.

Proposition 6.2 (VaR comonotonic additivity) *If risks $X_1, X_2, \ldots, X_n$ are comonotonic, then*

$$\mathrm{VaR}_\alpha[X_1 + \cdots + X_n] = \mathrm{VaR}_\alpha[X_1] + \cdots + \mathrm{VaR}_\alpha[X_n]. \tag{6.7}$$

Proof: Comonotonic risks can always be represented as $X_i = F_i^{-1}(U)$, where U is a random variable from $Uniform(0,1)$ distribution $F_U(\cdot)$ and $F_i(\cdot)$ is the distribution of X_i. Thus

$$\mathrm{VaR}_\alpha[X_1 + \cdots + X_n] = \mathrm{VaR}_\alpha[F_1^{-1}(U) + \cdots + F_n^{-1}(U)] = \mathrm{VaR}_\alpha[g(U)],$$

where $g(x) = F_1^{-1}(x) + \cdots + F_n^{-1}(x)$. Given that $g(x)$ is a nondecreasing function, we have

$$\mathrm{VaR}_\alpha[g(U)] = g\left(F_U^{-1}(\alpha)\right) = g(\alpha) = F_1^{-1}(\alpha) + \cdots + F_n^{-1}(\alpha),$$

which completes the proof. ■

Remark 6.2 (VaR is not a coherent measure) *It is important to note that the case of perfectly dependent risks is not necessarily an upper bound for* $\mathrm{VaR}_\alpha[X_1 + \cdots + X_n]$ *because subadditivity may fail for VaR. In general, VaR possesses all the properties of a coherent risk measure in Definition 6.7 except subadditivity. For some cases, such as a multivariate Normal distribution, VaR is subadditive. However, in general, the VaR of a sum of risks may be larger than the sum of VaRs of these risks. For examples and discussions, see McNeil et al. (2005); also see Examples 6.1 and 6.2. This has a direct implication for measuring OpRisk. In particular, VaR calculated for individual portfolios (e.g., business lines) may not be summed to produce the upper bound for the VaR of the overall risk.*

A formal Basel II regulatory requirement for OpRisk capital charge refers to a VaR and it can be justified using the following logic. The regulator's objective is to ensure that capital requirement against loss X is large enough so that the shortfall measure $\mathbb{E}[\max[X - \rho[X], 0]]$ is small enough. At the same time, the regulator should avoid requiring too much capital because the capital has a cost for the bank. Thus, the regulatory capital $\rho[X]$ can be determined as the solution of the following minimization problem (balance between low residual risk and low cost capital):

$$\min_{\rho}\{\mathbb{E}[\max(X-\rho,0)]+(1-\alpha)\rho\},\quad 0<\alpha<1. \tag{6.8}$$

The following elegant result justifies the VaR-based regulatory capital.

Proposition 6.3 (VaR as an optimal capital requirement) *If* $\alpha\in(0,1)$ *does not correspond to a flat part of the distribution of* X*, then*

$$\mathrm{VaR}_\alpha[X]=\arg\min_{\rho}\{\mathbb{E}[\max(X-\rho,0)]+(1-\alpha)\rho\}$$

and the solution is unique. In general, including the case when α *corresponds to a flat piece in the distribution of* X*,* $\mathrm{VaR}_\alpha[X]$ *is the lowest* ρ *which is a solution to*

$$\min_{\rho}\{\mathbb{E}[\max(X-\rho,0)]+(1-\alpha)\rho\}.$$

In the case when α *corresponds to a flat piece, the minimum is achieved for any* ρ *that satisfies* $F(\rho)=\alpha$*, that is, the smallest* ρ *in this case corresponds to* VaR_α*; otherwise, the minimum is unique.*

Proof: This result is taken from Dhaene *et al.* (2003) and an elegant proof based on geometrical reasoning is presented by Denuit *et al.* (2005, p. 70). ∎

This result supports the current Basel II regulatory choice of VaR. However, it is important to note that here the VaR is not really used to measure risk but appears as an optimal requirement. The risk controlled here is $\max(X-\rho,0)$, which is measured by $\mathbb{E}[\max(X-\rho,0)]$.

VaR is certainly meaningful when the objective is to avoid the default event while the size of the shortfall is not important. One can argue that for bank management and shareholders, avoiding the default is the primary objective while the size of the shortfall in the event of default is of secondary importance due to limited liability. If a bank has aggregate annual loss X and provision (for this loss) A, then $\mathrm{VaR}_\alpha[X]-A$ is the smallest additional capital required such that the bank may default with a small probability at most $(1-\alpha)$. For $\alpha=0.999$, this means that a bank will have a capital sufficient (on average) to cover losses in 999 out of 1000 years.

As already mentioned, VaR is not a coherent risk measure in general. In particular, under some circumstances, the VaR risk measure may fail a subadditivity property, that is, the diversification

$$D_\alpha=1-\frac{\mathrm{VaR}_\alpha[X_1+\cdots+X_n]}{\mathrm{VaR}_\alpha[X_1]+\cdots+\mathrm{VaR}_\alpha[X_n]} \tag{6.9}$$

may appear to be negative (see Embrechts *et al.* 2009a,b). This may occur even for independent risks when the risks are heavy-tailed. It was shown and discussed by Nešlehová *et al.* (2006) that if independent risks are Pareto type, $X_i\sim F_i(x)=1-x^{-\lambda_i}C_i(x)$, with the tail indexes $0<\lambda_i<1$, then

$$\mathrm{VaR}_\alpha[X_1+\cdots+X_n]>\sum_{i=1}^{n}\mathrm{VaR}_\alpha[X_i], \tag{6.10}$$

at least for sufficiently large α. Here, $C(x)$ is a *slowly varying function*, that is, $C(tx)/C(x) \to 1$ when $x \to \infty$ for all $t > 0$. The case of $0 < \lambda_i \leq 1$ corresponds to infinite mean distribution, that is, $\mathbb{E}[X_i] = \infty$. There are many examples in the literature of subadditivity failure for VaR; for illustration, we calculate two examples presented in Shevchenko (2011, examples 7.2 and 7.3).

EXAMPLE 6.1

Consider two independent risks

$$X \sim Pareto(\beta, 1) \quad \text{and} \quad Y \sim Pareto(\beta, 1),$$

where $Pareto(\beta, a)$ is a distribution function $F(x) = 1 - (x/a)^{-\beta}$. Using the FFT numerical method, we calculate $\mathrm{VaR}_{0.999}[X + Y]$ and diversification $D_{0.999}$ as defined in (6.1). The results for $D_{0.999}$ versus β that demonstrate negative diversification for $\beta < 1$ are presented in Figure 6.2.

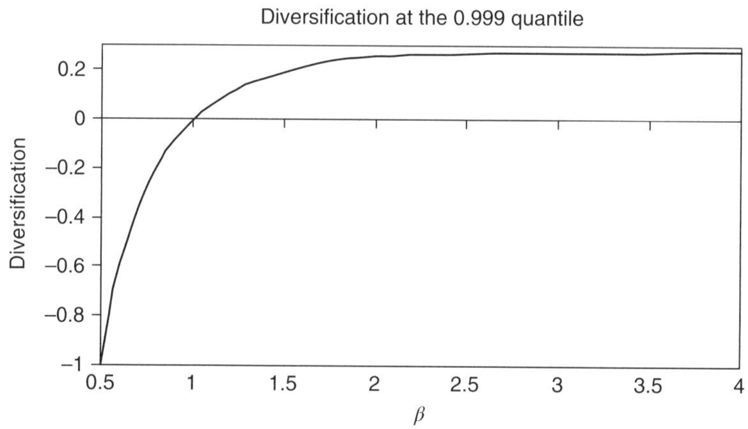

FIGURE 6.2 The diversification coefficient for $X \sim Pareto(\beta, 1)$ and $Y \sim Pareto(\beta, 1)$ versus β; see Example 6.1 for details

■

EXAMPLE 6.2

In the case of VaR, the diversification might be present for some quantile levels and might fail for other levels. In Example 6.1, the diversification is positive for $\beta > 1$. For example, $D_{0.999} \approx 0.27$ for $\beta = 4$; note that in this case mean, variance, and skewness are finite. Results for D_α versus α when $\beta = 4$ are shown in Figure 6.3.

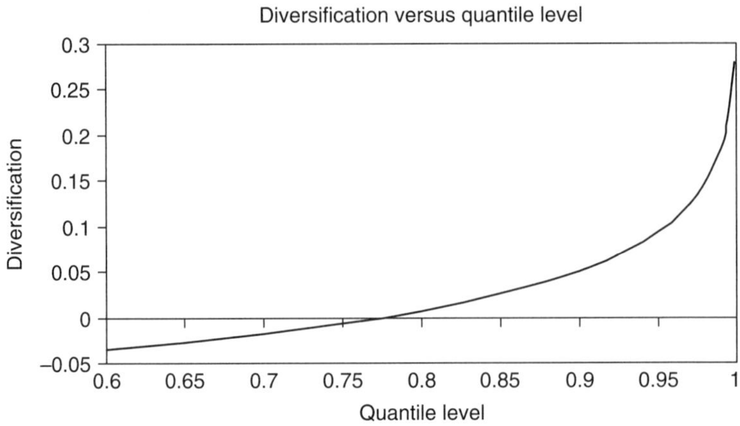

FIGURE 6.3 **The diversification coefficient for $X \sim Pareto(4, 1)$ and $Y \sim Pareto(4, 1)$ versus quantile level; see Example 6.2 for details**

It is easy to see that for high-level quantiles the diversification coefficient is positive but for lower quantiles it becomes negative. ■

6.2.4 EXPECTED SHORTFALL

A VaR at a specified probability level α does not provide any information about the fatness of the distribution upper tail. Often the management and regulators are concerned not only with the probability of default but also with its severity. Therefore, other risk measures are considered such as ES (sometimes referred to as the tail VaR).

Definition 6.12 (Expected shortfall) *The expected shortfall of a random variable $X \sim F(x)$ at the α-th probability level $\mathrm{ES}_\alpha[X]$ is*

$$\mathrm{ES}_\alpha[X] = \frac{1}{1-\alpha}\int_\alpha^1 \mathrm{VaR}_p[X]dp, \tag{6.11}$$

which is the "arithmetic average" of the VaRs of X from α to 1. ■

In general, the following identity is valid

$$\mathrm{ES}_\alpha[X] = \mathrm{VaR}_\alpha[X] + \frac{1}{1-\alpha}\mathbb{E}[\max(X - \mathrm{VaR}_\alpha[X], 0)], \tag{6.12}$$

because

$$\begin{aligned}\mathbb{E}[\max(X - \mathrm{VaR}_\alpha[X], 0)] &= \int_0^1 \max\left(\mathrm{VaR}_q[X] - \mathrm{VaR}_\alpha[X], 0\right) dq \\ &= (1-\alpha)\mathrm{ES}_\alpha[X] - (1-\alpha)\mathrm{VaR}_\alpha[X].\end{aligned} \tag{6.13}$$

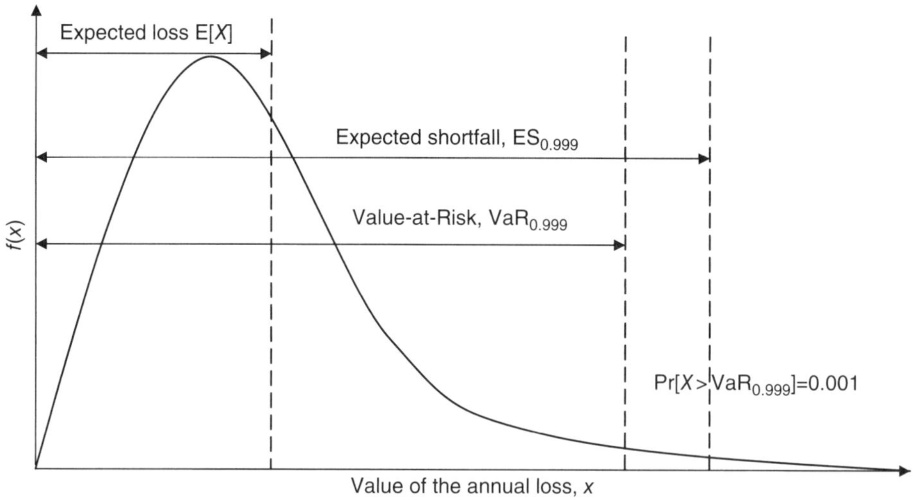

FIGURE 6.4 **Illustration of the 0.999 Value-at-Risk (VaR) and the 0.999 expected shortfall (ES) of the annual loss X with the probability density $f(x)$**

This identity shows that the ES is not less than VaR

$$\mathrm{ES}_\alpha[X] \geq \mathrm{VaR}_\alpha[X]; \tag{6.14}$$

for a simple illustration see Figure 6.4.

In the case of continuous distributions, it can be shown that $\mathrm{ES}_\alpha[X]$ is just expected loss given that the loss exceeds $\mathrm{VaR}_\alpha[X]$.

Proposition 6.4 *For a random variable X with a continuous distribution function $F(x)$, we have*

$$\mathrm{ES}_\alpha[X] = \mathbb{E}[X|X \geq \mathrm{VaR}_\alpha[X]] = \mathbb{E}[X|X > \mathrm{VaR}_\alpha[X]],$$

which is the conditional expected loss given that the loss exceeds $\mathrm{VaR}_\alpha[X]$.

Proof: Using Definition 6.12, the proof is trivial: simply change the integration variable to $x = F_X^{-1}(p)$. ■

For a distribution function $F_X(x)$ discontinuous (i.e., with a jump) at the $\mathrm{VaR}_\alpha[X]$ threshold, we have more general relation expressions and

$$\mathbb{E}[X|X \geq \mathrm{VaR}_\alpha[X]] \leq \mathrm{ES}_\alpha[X] \leq \mathbb{E}[X|X > \mathrm{VaR}_\alpha[X]], \tag{6.15}$$

where the equality on the right side is achieved when $\alpha = \alpha_U = \Pr[X \leq \mathrm{VaR}_\alpha[X]]$; the equality on the left side is achieved when $\alpha = \alpha_L = \Pr[X < \mathrm{VaR}_\alpha[X]]$; and other cases correspond to strict inequalities; for illustration, see Figure 6.5. This is proved in the following proposition.

Proposition 6.5 *For a random variable X the ES can be calculated as*

$$\mathrm{ES}_\alpha[X] = \mathbb{E}\left[X|X > \mathrm{VaR}_\alpha[X]\right] - \frac{\alpha_U - \alpha}{1-\alpha}\mathbb{E}[X - \mathrm{VaR}_\alpha[X]|X > \mathrm{VaR}_\alpha[X]] \tag{6.16}$$

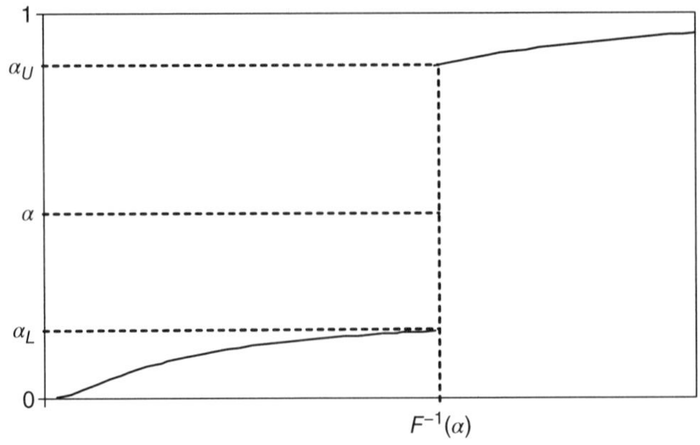

FIGURE 6.5 **An example of distribution $F(x)$ with a jump at $\mathrm{VaR}_\alpha[X]$**

or

$$\mathrm{ES}_\alpha[X] = \mathbb{E}[X|X \geq \mathrm{VaR}_\alpha[X]] + \frac{\alpha - \alpha_L}{1-\alpha}\mathbb{E}[X - \mathrm{VaR}_\alpha[X]|X \geq \mathrm{VaR}_\alpha[X]], \tag{6.17}$$

where $\alpha_L = \Pr[X < \mathrm{VaR}_\alpha[X]]$ *and* $\alpha_U = \Pr[X \leq \mathrm{VaR}_\alpha[X]]$.

Proof: For a distribution continuous at α, we have $\alpha_L = \alpha_U = \alpha$ and the above relations simplify to the correct expression (given by Proposition 6.4)

$$\mathrm{ES}_\alpha[X] = \mathbb{E}[X|X \geq \mathrm{VaR}_\alpha[X]] = \mathbb{E}\left[X|X > \mathrm{VaR}_\alpha[X]\right].$$

If there is a jump in distribution at level α, then $\alpha_L \leq \alpha \leq \alpha_U$ (see Figure 6.5), and relation (6.16) can be proved using simple calculus and splitting the probability atom at α as follows:

$$\begin{aligned}\mathrm{ES}_\alpha[X] &= \frac{1}{1-\alpha}\int_\alpha^1 \mathrm{VaR}_p[X]dp = \frac{1}{1-\alpha}\int_\alpha^{\alpha_U} \mathrm{VaR}_p[X]dp + \frac{1}{1-\alpha}\int_{\alpha_U}^1 \mathrm{VaR}_p[X]dp\\ &= \frac{\alpha_U - \alpha}{1-\alpha}\mathrm{VaR}_\alpha[X] + \frac{1-\alpha_U}{1-\alpha}\mathbb{E}[X|X > \mathrm{VaR}_\alpha[X]].\end{aligned}$$

Also, the relation (6.17) can be proved by the following splitting:

$$\begin{aligned}\mathrm{ES}_\alpha[X] &= \frac{1}{1-\alpha}\int_\alpha^1 \mathrm{VaR}_p[X]dp = \frac{1}{1-\alpha}\int_{\alpha_L}^1 \mathrm{VaR}_p[X]dp - \frac{1}{1-\alpha}\int_{\alpha_L}^\alpha \mathrm{VaR}_p[X]dp\\ &= \frac{1-\alpha_L}{1-\alpha}\mathbb{E}[X|X \geq \mathrm{VaR}_\alpha[X]] - \frac{\alpha - \alpha_L}{1-\alpha}\mathrm{VaR}_\alpha[X].\end{aligned}$$

Given that $\alpha_L \leq \alpha \leq \alpha_U$, it is obvious from relations (6.16) and (6.17) that $\mathbb{E}[X|X \geq \mathrm{VaR}_\alpha[X]] \leq \mathrm{ES}_\alpha[X] \leq \mathbb{E}[X|X > \mathrm{VaR}_\alpha[X]]$ is true (see Acerbi and Tasche 2002, proposition 3.2). ■

This relation might look a bit complicated but is easy to understand from finite sample estimators. Consider a sample of i.i.d. random variables $X_1, \ldots, X_N$ and the corresponding ordered sample $X_{(1,N)} \leq \cdots \leq X_{(N,N)}$. Then, the VaR at level α can be estimated as $X_{(k,N)}$, where $k = \lceil N\alpha \rceil$ is the smallest integer larger or equal to $N\alpha$. Then, in general (i.e., the sample can be from distribution with jumps and there might be repeated values in a sample), the ES (6.11) can be calculated empirically according to results based on the following asymptotic limit of the empirical process of the weighted order statistics:

$$\mathrm{ES}_\alpha[X] = \lim_{N\to\infty} \frac{\sum_{i=k}^{N} X_{(i,N)}}{N-k+1}. \tag{6.18}$$

At the same time, conditional tail expectation $\mathbb{E}[X|X \geq \mathrm{VaR}_\alpha[X]]$ is calculated as a simple average of losses larger than or equal to VaR which can also be empirically estimated and shown to asymptotically satisfy the following relationship in the empirical process limit given by:

$$\mathbb{E}[X|X \geq \mathrm{VaR}_\alpha[X]] = \lim_{N\to\infty} \frac{\sum_{i=1}^{N} X_i \mathbb{I}_{\{X_i \geq \mathrm{VaR}_\alpha[X]\}}}{\sum_{i=1}^{N} \mathbb{I}_{\{X_i \geq \mathrm{VaR}_\alpha[X]\}}}, \tag{6.19}$$

which is clearly different from the ES estimator (6.18) if there are repeated samples at the level α (i.e., there is a jump in distribution at $\mathrm{VaR}_\alpha[X]$ as in Figure 6.5).

The relationship between ES and VaR (6.12) allows to reformulate Proposition 6.3 as follows.

Proposition 6.6 (ES as the minimum of cost function) *ES can be written as*

$$\mathrm{ES}_\alpha[X] = \min_\rho \left\{ \frac{1}{1-\alpha}\mathbb{E}[\max(X-\rho, 0)] + \rho \right\},$$

where the smallest ρ solving the minimization problem is $\mathrm{VaR}_\alpha[X]$.

Proof: This follows directly from Proposition 6.3 and identity (6.12); also see (Rockafellar and Uryasev 2002, theorem 10). ■

Note that this function is valid for continuous and discrete distributions. Moreover the function $\frac{1}{1-\alpha}\mathbb{E}[\max(X-\rho,0)] + \rho$ is convex as a function of ρ.

ES is a *coherent risk measure*. It satisfies coherent risk measure axioms in Definition 6.7, that is, *subadditivity*, *monotonicity*, *translation invariance*, and *positive homogeneity*. It is, *comonotonic additive*.

- Translational invariance, positive homogeneity, and monotonicity follow from corresponding properties of VaR and ES definition:

$$\begin{aligned}\mathrm{ES}_\alpha[X+a] &= \frac{1}{1-\alpha}\int_\alpha^1 \mathrm{VaR}_p[X+a]dp = a + \frac{1}{1-\alpha}\int_\alpha^1 \mathrm{VaR}_p[X]dp \\ &= a + \mathrm{ES}_\alpha[X], \quad \text{for any} \quad a \in \mathbb{R},\end{aligned} \tag{6.20}$$

$$\begin{aligned}\mathrm{ES}_\alpha[\lambda X] &= \frac{1}{1-\alpha}\int_\alpha^1 \mathrm{VaR}_p[\lambda X]dp = \lambda\frac{1}{1-\alpha}\int_\alpha^1 \mathrm{VaR}_p[X]dp\\ &= \lambda \mathrm{ES}_\alpha[X], \quad \text{for any} \quad \lambda > 0 \end{aligned} \tag{6.21}$$

and if $X \leq Y$, then

$$\begin{aligned}\mathrm{ES}_\alpha[X] &= \frac{1}{1-\alpha}\int_\alpha^1 \mathrm{VaR}_p[X]dp\\ &\leq \frac{1}{1-\alpha}\int_\alpha^1 \mathrm{VaR}_p[Y]dp = \mathrm{ES}_\alpha[Y]. \end{aligned} \tag{6.22}$$

- ES is comonotonic additive, that is, for comonotonic (perfectly positively dependent) risks $X_1, \ldots, X_n$, the ES of their sum $X_1 + \cdots + X_n$, is the sum of individual ESs

$$\mathrm{ES}_\alpha[X_1 + \cdots + X_n] = \mathrm{ES}_\alpha[X_1] + \cdots + \mathrm{ES}_\alpha[X_n]. \tag{6.23}$$

This follows from the comonotonic additivity property of VaR

$$\begin{aligned}\mathrm{ES}_\alpha[X_1 + \cdots + X_n] &= \frac{1}{1-\alpha}\int_\alpha^1 \mathrm{VaR}_p[X_1 + \cdots + X_n]dp\\ &= \frac{1}{1-\alpha}\int_\alpha^1 \sum_{i=1}^n \mathrm{VaR}_p[X_i]dp\\ &= \mathrm{ES}_\alpha[X_1] + \cdots + \mathrm{ES}_\alpha[X_n]. \end{aligned} \tag{6.24}$$

ES is not just *subadditive* but also *convex*, and this is proved in the following proposition.

Proposition 6.7 *ES is a subadditive and convex risk measure.*

Proof: Using Proposition 6.6 with $\rho = \lambda \mathrm{VaR}_\alpha[X] + (1-\lambda)\mathrm{VaR}_\alpha[Y]$, $\lambda \in (0,1)$, and convexity of function $\max(x-a, 0)$, we obtain

$$\begin{aligned}\mathrm{ES}_\alpha[\lambda X + (1-\lambda)Y] &= \lambda \mathrm{VaR}_\alpha[X] + (1-\lambda)\mathrm{VaR}_\alpha[Y]\\ &\quad + \frac{1}{1-\alpha}\mathbb{E}[\max(\lambda X + (1-\lambda)Y - \lambda \mathrm{VaR}_\alpha[X] - (1-\lambda)\mathrm{VaR}_\alpha[Y], 0)]\\ &\leq \lambda \mathrm{VaR}_\alpha[X] + (1-\lambda)\mathrm{VaR}_\alpha[Y] + \frac{1}{1-\alpha}\mathbb{E}[\max(\lambda X - \lambda \mathrm{VaR}_\alpha[X], 0)]\\ &\quad + \frac{1-\lambda}{1-\alpha}\mathbb{E}[\max(Y - \mathrm{VaR}_\alpha[Y], 0)]\\ &= \lambda \mathrm{ES}_\alpha[X] + (1-\lambda)\mathrm{ES}_\alpha[Y]. \end{aligned} \tag{6.25}$$

This proves the convexity and in the case $\lambda = 1/2$ (also using positive homogeneity) gives the subadditivity

$$\mathrm{ES}_\alpha[X+Y] \leq \mathrm{ES}_\alpha[X] + \mathrm{ES}_\alpha[Y].$$

■

The fact that ES is subadditive and comonotonic additive implies that the case of perfectly dependent risks is the worst-case scenario for $\mathrm{ES}_\alpha[X_1 + \cdots + X_n]$.

EXAMPLE 6.3 ES for LogNormal distribution.

Assume that loss X is from LogNormal density $f(x; \mu, \sigma)$, i.e., $\ln X$ is from Normal distribution with mean μ and variance σ^2. Then VaR is

$$q_\alpha := \mathrm{VaR}_\alpha[X] = \exp(\mu + \sigma\Phi^{-1}(\alpha))$$

and ES is calculated as follows:

$$\begin{aligned}\mathrm{ES}_\alpha &= \frac{1}{1-\alpha}\int_{q_\alpha}^{\infty} xf(x; \mu, \sigma)dx = \frac{1}{1-\alpha}\int_{(\ln q_\alpha - \mu)/\sigma}^{\infty} e^{\mu+\sigma y}\phi(y)dy \\ &= \frac{1}{1-\alpha}e^{\mu+\frac{1}{2}\sigma^2}\Phi(\sigma - \Phi^{-1}(\alpha)). \end{aligned} \tag{6.26}$$

Here, $\Phi(\cdot)$ and $\Phi^{-1}(\cdot)$ are the standard Normal distribution and its inverse, respectively; $\phi(\cdot)$ is the standard Normal density, and we used the closed form integral

$$\int_a^\infty e^{\gamma x}\phi(x)dx = e^{\frac{1}{2}\gamma^2}\Phi(\gamma - a). \tag{6.27}$$

■

EXAMPLE 6.4 ES for exponential distribution.

Assume that loss X is from exponential distribution $F(x) = 1 - \exp(-\lambda x)$, that is, with the density $f(x) = \lambda\exp(-\lambda x)$. Then VaR is

$$q_\alpha := \mathrm{VaR}_\alpha[X] = F^{-1}(\alpha) = -\frac{1}{\lambda}\ln(1-\alpha)$$

and ES is calculated as

$$\mathrm{ES}_\alpha[X] = \frac{1}{1-\alpha}\int_{q_\alpha}^{\infty} x\lambda e^{-\lambda x}dx = \mathrm{VaR}_\alpha[X] + \frac{1}{\lambda}. \tag{6.28}$$

Note that the difference between ES and VaR does not depend on α; this is because the exponential distribution is memoryless. ■

6.2.5 SPECTRAL RISK MEASURE

Having defined the class of ES risk measures and demonstrated several examples of LDA models under this risk measure, we now observe that ES is a special subclass of a larger class of risk measures known as the Spectral Risk Measures (SRM) as given in Definition 6.14.

It is important at this stage to make the following technical clarification on the use of the notation, ES, Conditional VaR (CVaR), and Tail Conditional Expectation TCE (or CTE). As stated, we treat the ES as defined by the following form:

$$\mathrm{ES}_{\alpha}[X] = \frac{1}{1-\alpha}\int_{\alpha}^{1} \mathrm{VaR}_{p}[X]dp. \tag{6.29}$$

In addition, one often reads about the notions of $\mathrm{CVaR}_{\alpha}[X]$ and $\mathrm{TCE}_{\alpha}[X]$, which has the following definitions:

$$\mathrm{CVaR}_{\alpha}[X] = \mathrm{TCE}_{\alpha}[X] = \mathbb{E}\left[X|X \geq F_X^{-1}(\alpha)\right]. \tag{6.30}$$

In general, one will find that CVaR is not a coherent measure of risk in the most general context. However, in the case of a strictly continuous loss distribution, one will have equivalence between the CVaR and the ES measures, which will be coherent.

Furthermore, we note that in the case that $\alpha = 0$, one can see that at level $\alpha = 0$, $\mathrm{ES}_0[X]$ can be extended to be understood as the worst-case loss scenario given by

$$\mathrm{ES}_0[X] = \text{ess. sup}[X], \tag{6.31}$$

where ess. sup stands for the essential supremum defined in Definition 6.13 below.

Definition 6.13 (Essential Supremum and Essential Infimum) *Consider a measurable function $f : X \mapsto \mathbb{R}$, where X is a measure space with measure μ, the essential supremum is the smallest number α such that the set $\{x : f(x) > \alpha\}$ has measure zero. If no such number exists, then the essential supremum is ∞.* ■

Remark 6.3 *The essential supremum is the generalization to measurable functions of the maximum. The main difference is that the values of a function on a set of measure zero do not affect the essential supremum. The essential supremum of the absolute value of a function $|f|$ is usually denoted $||f||_{\infty}$, and this serves as the norm for L-∞-space.*

Returning to discussion on risk measures, next we turn to the question of how one may utilize the notion of the ES risk measure to extend to a wider class of risk measures. The answer to this question was studied by Acerbi (2002).

To build new risk measures we first consider the result in Proposition 6.8 (see Acerbi 2002, proposition 2.2).

Proposition 6.8 (Linear Combinations of Risk Measures) *Consider the set of risk measures $\{\rho_i\}_{i=1}^{n}$, then any convex combination given by*

$$\rho = \sum_{i=1}^{n} \alpha_i \rho_i \tag{6.32}$$

for weights subject to the restrictions $\{\alpha_i\}_{i=1}^{n}$ *and* $\sum_{i=1}^{n} \alpha_i = 1$ *will produce a risk measure. In addition, if* ρ_α *is a risk measure defined with respect to a parameter* $\alpha \in [a, b]$*, then for any measure* $d\mu_\alpha$ *on* $[a, b]$ *with* $\int_a^b d\mu(\alpha) = 1$, $\rho = \int_a^b d\mu(\alpha)\rho_\alpha$ *is also a risk measure.*

The observation of this result allows one to then define a general family of risk measures based on considering the measure $d\mu(\alpha)$ for $\alpha \in [0, 1]$. As observed by Acerbi (2002), if the measure $d\mu(\alpha)$ is selected to satisfy some basic integrability conditions, then one may now define a class of risk measures based on the ES risk measure as follows:

$$M_\mu[X] = \int_0^1 d\mu(\alpha)(1-\alpha)\mathrm{ES}_\alpha[X] = \int_0^1 d\mu(\alpha) \int_\alpha^1 dpF_X^{-1}(p). \tag{6.33}$$

This will be a risk measure so long as the following condition is satisfied,

$$\int_0^1 (1-\alpha)d\mu(\alpha) = 1. \tag{6.34}$$

Then under the same integrability conditions, one may apply the Fubini–Tonelli theorem to swap orders of integration in the definition of the class of risk measures such that one has

$$M_\mu[X] = \int_0^1 dpF_X^{-1}(p)\phi(p) \equiv M_\phi[X], \tag{6.35}$$

in other words, parameterization in terms of a risk measure $d\mu(\alpha)$ can be transformed into a parameterization in terms of a function ϕ typically termed the "risk spectrum" given by $\phi(p) = \int_p^1 d\mu(\alpha)$ and normalization condition

$$\int_0^1 \phi(p)dp = \int_0^1 d\mu(\alpha)(1-\alpha) = 1. \tag{6.36}$$

This realization led to the definition of the class of Spectral Risk Measures (SRM).

Definition 6.14 (Spectral Risk Measures) *Consider an LDA model for an OpRisk single loss process with annual loss random variable* $Z_N \sim F_{Z_N}(z)$ *with severity distribution* $X_i \sim F_X(x)$ *for all losses* X_i *and frequency distribution* $N \sim F_N(n)$*. The SRM for a weight function* $\phi : [0, 1] \mapsto \mathbb{R}$ *is given by*

$$\mathrm{SRM}_{Z_N}(\phi) = \int_0^1 \phi(s)\mathrm{VaR}_s[Z_N]\,ds \tag{6.37}$$

with $\forall t \in (1, \infty)$ *and function* $\phi(1-1/t) \leq Kt^{-1/\beta+1-\epsilon}$ *for some* $K > 0$ *and* $\epsilon > 0$. ■

Remark 6.4 *Tong and Wu (2012) showed that if an individual has a Constant Absolute Risk Aversion (CARA) utility function with coefficient of absolute risk aversion ξ, then the SRM should be given according to*

$$SRM_{\phi_\kappa}(\kappa) = \int_\kappa^1 \phi_\kappa(s)\text{VaR}_s\, ds,$$

where the weighting function (risk spectrum or risk aversion function) $\phi_\kappa(s)$ is given by

$$\phi_\kappa(s) = (1-\kappa)^{-1}\phi\left(1 - \frac{1-s}{1-\kappa}\right)\mathbb{I}_{[\kappa,1]}(s)$$

with

$$\phi(\kappa) = \frac{\xi e^{-\xi(1-\kappa)}}{1-e^{-\xi}}.$$

Note that if one considers $\phi(t) \equiv 1 \forall t \in [0,1]$, then the SRM resumes to the ES.

Dowd and Blake (2006) explain that the following three properties are required to be satisfied in order for a SRM to be coherent:

1. **Non-negativity**. The risk aversion function $\phi_\kappa(s) \geq 0$ for all $\kappa, s \in [0,1]$;
2. **Normalization**. The risk aversion function $\phi_\kappa(s)$ should be normalized as follows:

$$\int_0^1 \phi_0(s)ds = 1. \tag{6.38}$$

3. **Increasing**. The risk aversion function $\phi_\kappa(s)$ should be increasing such that for any $\kappa \in (0,1)$, one has $\phi_\kappa(s_1) \leq \phi_\kappa(s_2)$ for all $\kappa \leq s_1 \leq s_2 \leq 1$;

These should therefore act as a minimal set of requirements for risk managers to consider when specifying their risk aversion function. The last condition simply implies that larger losses should be no smaller than weights attached to smaller loss amounts. This last point, simple as it may be, is the key to coherency of ES and SRM and also the reason why VaR fails to be a coherent risk measure.

6.2.6 HIGHER-ORDER RISK MEASURES

As observed, the VaR is not a coherent risk measure since the convexity requirement reflects the view that diversification should not increase risk. It has been observed that the VaR will not be a coherent risk measure; consequently, alternatives to VaR have been suggested and investigated such as ES and SRM.

Among the risk measures that satisfy the coherency properties, the notions of CVaR have been already considered. There are also other measures such as Maximum Loss described by

Pflug (2000), as well as other coherent risk measures that are based on one-sided moments such as described by Fischer (2003) and the deviation d-based risk measures discussed by Rockafellar *et al.* (2006).

Other notions that extend the idea of the Average VaR (AVaR or TCE), which in the continuous loss distribution case corresponds to the coherent risk measure ES, have also been considered such as the general higher-order moment representations by Krokhmal (2007). Such higher moment risk measures were also considered in the dual representation under the Kusuoka form by Dentcheva *et al.* (2010). The definition of the class of Higher Moment Coherent Risk (HMCR) measures is given in Definition 6.15 (see Krokhmal 2007). *Note* that throughout the remainder of this section, the loss distribution will be assumed continous such that the AVaR will be equivalent to the ES.

Definition 6.15 (Higher Moment Coherent Risk Measures) *Consider the probability space $(\Omega, \mathcal{F}, \mu)$ with sample space Ω, sigma algebra $\mathcal{F}$, and probability measure μ. Then consider the linear space χ of $\mathcal{F}$-measurable function mappings, that is, loss random variables given by $X : \Omega \mapsto \mathbb{R}$ such as $\chi = \mathcal{L}_p(\Omega, \mathcal{F}, P)$ for some $p \in (1, \infty)$. Then for some $\alpha \in (0, 1)$, consider the function*

$$\phi(X) = \frac{1}{(1-\alpha)} ||(X)^+||_p, \tag{6.39}$$

where the p-norm is defined by $||X||_p = (\mathbb{E}|X|^p)^{1/p}$. Then the HMCRs are defined by

$$HMCR_{p,\alpha}(X) = \min_{\nu \in \mathbb{R}} \left(\nu + \frac{1}{1-\alpha} ||(X-\nu)^+||_p \right), \quad p \geq 1, \ \alpha \in (0,1). \tag{6.40}$$

■

One may make the following remarks about the properties of the HMCR measures.

Remark 6.5 *The HMCR measures satisfy the following properties:*

- *For $p < q$ and loss random variable in the space $X \in \mathcal{L}_q$, one has*

$$HMCR_{p,\alpha}(X) \leq HMCR_{q,\alpha}(X). \tag{6.41}$$

- *The HMCR measures are tail measures or risks, that are coherent measures of risk. They can therefore be seen as generalizations of the convex but not positive homogeneous or translation invariant risk measures considered by Bawa (1975) known as the Lower Partial Moments and given by*

$$LPM_p(X; t) = \mathbb{E}\left[(X-t)^+\right]^p, \quad p \geq 1, \ t \in \mathbb{R}. \tag{6.42}$$

- *The HMCR measures are also related to the Central One-Sided Moment (COSM)-based risk measures considered by Fischer (2003) and defined by the class of coherent measures of semi-$\mathcal{L}_p$ type given by*

$$COSM_{p,\beta}(X) = \mathbb{E}[X] + \beta||(X - \mathbb{E}[X])^+||_p, \quad p \geq 1, \ \beta \geq 0. \tag{6.43}$$

These COSM risk measures are central one-sided moment risk measures whereas the HMCR measures are tail-based risk measures;

- *An advantage of HMCR measures when compared to other coherent risk measures is that their tail cutoff point is adjustable to the chosen level* $\alpha \in (0, 1)$.

Krokhmal (2007) pays particular attention to the second-order HMCR case where $p = 2$, which displays remarkably similar characteristics to the CVaR while at the same time measuring the risk relative to the second-order moments of the loss distributon. Dentcheva *et al.* (2010) consider a dual representation also known as the Kusuoka representation. This involves defining the notion of a Kusuoka measure of risk given in Definition 6.16 (see Kusuoka 2001).

Definition 6.16 (Kusuoka risk measure) *Consider the probability space* $(\Omega, \mathcal{F}, \mu)$ *with sample space* Ω, *sigma algebra* $\mathcal{F}$, *and probability measure* μ. *Then consider the linear space* χ *of* $\mathcal{F}$*-measurable function mappings, that is, loss random variables given by* $X : \Omega \mapsto \mathbb{R}$ *such as* $\chi = \mathcal{L}_p(\Omega, \mathcal{F}, P)$ *for some* $p \in [1, \infty)$. *Then a Kusuoka risk measure denoted by* $\rho(Z)$ *is defined for a convex set of measures* $\mathcal{M}$ *in the set* $\mathcal{P}((0, 1])$ *of probability measures on* $(0, 1]$ *such that for all loss random variables* Z *one has*

$$\rho(Z) = \sup_{m \in \mathcal{M}} \int_0^1 AVaR_\alpha(Z) m(d\alpha), \tag{6.44}$$

where $AVaR_\alpha$ *is the average VaR at level* α. ■

Remark 6.6 *It was shown by Kusuoka (2001) that the Kusuoka representable risk measures are coherent when defined on* $L_\infty(\Omega, \mathcal{F}, P)$, *and then by Dentcheva et al. (2010) showed that this result could be extended to risk measures defined on spaces* $\mathcal{L}_p$. *In addition, it was shown that the HMCR class of risk measures has a Kusuoka representation.*

The AVaR plays a central role in the description of every coherent risk measure via the Kusuoka representation. AVaR is a coherent risk measure, hence it is preferred in stochastic optimization. However, there are other coherent risk measures, generated from AVaR via the Kusuoka representation. To further understand the role played by the AVaR, consider the loss random variable $X \in \mathcal{L}_1(\Omega, \mathcal{F}, P)$ and consider the tail of the severity distribution for this loss random variable given by,

$$F_X(\nu) = \mathbb{P}\mathrm{r}\left[X \leq \nu\right], \tag{6.45}$$

and,

$$F_X^{(k)}(\nu) = \int_\nu^\infty F_X^{(k-1)}(\alpha) d\alpha, \quad k \geq 2. \tag{6.46}$$

Then define the inverse $F_X^{(-1)}(\alpha) = \inf\{\nu : F_X(\nu) \geq \alpha\}$ for $\alpha \in (0,1)$ where the AVaR at a level α is given by

$$\begin{aligned} \text{AVaR}_\alpha(X) &= \frac{1}{1-\alpha} F_X^{(-2)}(\alpha) \\ &= \frac{1}{1-\alpha} \int_\alpha^1 \text{VaR}_s[X] ds, \end{aligned} \tag{6.47}$$

where one can identify $F_X^{(-2)}(\alpha)$ as the absolute Lorenz function. One can then use the result of Kusuoka (2001) to obtain the following alternative extremal representation that generalizes the AVaR. To see this, consider the $\text{AVaR}_\alpha(X)$ given by

$$\begin{aligned} \text{AVaR}_\alpha(X) &= \frac{1}{1-\alpha} \sup_{\nu \in \mathbb{R}} \left\{ \nu\alpha - F_X^{(2)}(\nu) \right\} \\ &= \inf_{\nu \in \mathbb{R}} \left\{ \frac{1}{1-\alpha} \mathbb{E}\left[(\nu - X)^+ \right] - \nu \right\}, \end{aligned} \tag{6.48}$$

which can be generalized to the representation given by the HMCR measure according to

$$\inf_{\nu \in \mathbb{R}} \left\{ \frac{1}{\alpha} ||(\nu - X)^+||_p - \nu \right\}, \quad p > 1. \tag{6.49}$$

Dentcheva *et al.* (2010) demonstrated that the resulting Kusuoka representation can then be obtained by considering the risk measures defined with respect to the convex set of measures $\mathcal{M}$ given by $\mathcal{M} = \mathcal{M}_q$ with $p^{-1} + q^{-1} = 1$ such that

$$\mathcal{M}_q = \left\{ \mu \in \mathcal{P}((0,1]) : \int_0^1 \left\| \int_\alpha^1 \frac{\mu(ds)}{ds} \right\|^q d\alpha \leq c^q \right\}. \tag{6.50}$$

6.2.7 DISTORTION RISK MEASURES

Another popular class of risk measures used in insurance are the so-called *distortion* risk measures introduced by Wang (1996). Distortion risk measures form an important class; they include Value at Risk, Conditional Tail Expectation and Wang's PH transform premium principle. Before definition the distortion risk measure, we first need to introduce the definition of distortion function.

Definition 6.17 (Distortion function) *A distortion function $g(\cdot)$ is a non-decreasing function with $g(0) = 0$ and $g(1) = 1$ such that $g : [0,1] \mapsto [0,1]$.* ■

One can then define the class of distortion risk measures based on the class of distortion funtions as follows in Definition 6.18.

Definition 6.18 (Distortion risk measure) *For a non-negative random variable X from a distribution $F(x)$, the distortion risk measure is defined as*

$$\rho[X] = \int_0^\infty g(\overline{F}(x))dx,$$

where $\overline{F}(x) = 1 - F(x)$ and $g(\cdot)$ is a nondecreasing distortion function such that $g(0) = 0$ and $g(1) = 1$. ■

Remark 6.7 *The distortion risk measure can be interpreted as adjusting the true probability measure to give more weight to higher risk events. Hence, the distortion function $g(\overline{F}(x))$ can be thought of as a risk adjusted decumulative distribution function. Since X is a non-negative random variable, $\rho(X) = \mathbb{E}_g[X]$ where the subscript indicates in this case the change of measure for the expectation.*

The properties and several examples of viable distortion functions $g(\cdot)$ are provided in the discussions in (Wirch, 2001).

This risk measure is *positively homogeneous, translation invariant, monotonic*, and *comonotonic additive.* It is also *subadditive* (i.e., coherent) if distortion function $g(\cdot)$ is concave.

Definition 6.19 *A real-valued function $g(x)$ defined on an interval I is called convex if*

$$g(tx_1 + (1-t)x_2) \leq tg(x_1) + (1-t)g(x_2)$$

for $t \in [0, 1]$ and all $x_1, x_2 \in I$, that is, the graph of the function lies below the line segment joining any two points of the graph. Similarly, the function $g(x)$ is called concave if

$$g(tx_1 + (1-t)x_2) \geq tg(x_1) + (1-t)g(x_2)$$

for $t \in [0, 1]$ and all $x_1, x_2 \in I$, that is, the graph of the function lies above the line segment joining any two points of the graph. ■

The well-known risk measures such as VaR and ES are the distortion risk measures for specific choice of $g(\cdot)$. In particular, for a confidence level $\alpha \in (0, 1)$, VaR corresponds to

$$g(x) = \begin{cases} 1, & \text{if } \ x > 1-\alpha, \\ 0, & \text{otherwise}, \end{cases} \tag{6.51}$$

which is not a concave function; and ES corresponds to a concave distortion function

$$g(x) = \min\left(1, \frac{x}{1-\alpha}\right). \tag{6.52}$$

For more details on distortion risk measures, see Denuit *et al.* (2005, sections 2.6.2 and 2.6.3).

6.2.8 ELICITABLE RISK MEASURES

Recently, the notion of an elicitable risk measure has been introduced by Bellini and Bignozzi (2013) as an adaption to financial risk measures from the more general point estimator setting developed by Gneiting (2011), Osband and Reichelstein (1985), and Lambert *et al.* (2008).

To understand this class of risk measures one must first consider the notion of an elicitable function, which was defined with respect to forecast point estimators under a decision theoretic framework by Lambert *et al.* (2008). Therefore, to understand this class of risk measures we present the general decision theoretic structure introduced before considering this structure for financial risk measures. In particular, we will consider the formal definition of the decision theory framework, the scoring function, the consistency of the scoring function, and then the class of elicitable scoring functions. We start with the notion of a decision theoretic structure, as discussed in detail by, for instance, Berger (1985).

Definition 6.20 (Classical decision theoretic structure) *Consider the following components of the decision theoretic structure:*

- *Space of outcomes of the random process known as the observation domain,* $\mathcal{O}$*;*
- *A class of probability measures* $\mathcal{F}$ *defined on the observation domain;*
- *The action space* $\mathcal{A}$*;*
- *A loss function that maps the cross space of actions and observations for a loss/reward given generically by*

$$L : \mathcal{O} \times \mathcal{A} \mapsto [0, \infty). \tag{6.53}$$

The loss function quantifies the consequence that would be incurred for each possible decision for various possible values of the "state of nature" observed in the observation space of the loss process.

■

Typically, an action in this context is the formation of an estimator, generically denoted by $\hat{\theta}(X)$, which is a function of the random loss process. The loss process itself can be assumed to involve a probability distribution $X \sim F_X(x; \theta)$ with true (unknown) state of nature, characterized, for example, by parameter(s) θ. The loss function then helps one to decide upon the appropriate choice of action that is, to make a decision regarding the estimator. The typical loss functions include the following:

- Squared loss function: $L(\theta, \hat{\theta}) = (\theta - \hat{\theta})^2$;
- Absolute error loss function: $L(\theta, \hat{\theta}) = \|\theta - \hat{\theta}\|$;
- L_p loss function: $L(\theta, \hat{\theta}) = \|\theta - \hat{\theta}\|_p$;
- Binary loss function: $L(\theta, \hat{\theta}) = \mathbb{I}[\theta \neq \hat{\theta}]$.

If one then considers, for a given loss function choice, a decision rule (action) with a small "expected (long-term average) loss" obtained by using the estimator $\hat{\theta}(X)$ for different realizations of the loss process X, then this leads one naturally to the notion of a statistical risk function in statistical decision theory given by

$$R\left(\theta, \hat{\theta}\right) = \mathbb{E}_\theta\left[L\left(\theta, \hat{\theta}\right)\right]. \tag{6.54}$$

Using this general decision theoretic structure, Gneiting (2011) makes a particular choice in which it is assumed that the observation and action spaces coincide, that is, $\mathcal{O} = \mathcal{A} \in \mathbb{R}$;

note that in the remainder of this section we assume these spaces to be the real line. Then one can define a scoring function analogously to the notion of a loss function in a decision theoretic setting.

Then the following assumptions are made about the scoring (loss) function $S(x, y)$:

- [A1] The scoring function $S(x, y)$ is positive $S(x, y) \geq 0$ with equality when $x = y$;
- [A2] The scoring function $S(x, y)$ is a continuous function in x;
- [A3] The partial derivative with respect to the first argument exists and is continuous whenever $x \neq y$.

In addition, it will be desirable to consider scoring (loss) functions that are homogeneous (scale invariant) such that

$$S(cx, cy) = |c|^b S(x, y), \quad \forall x, y \in \mathbb{R}, c \in \mathbb{R}. \tag{6.55}$$

From this general statistical decision theory setup, one may now consider a functional (i.e., a statistical function), which is, for instance, a set valued mapping, denoted T, from a class of probability measures (distributions) $\mathcal{F}$ to a Euclidean space. From this notion, one may define a consistent scoring function as given in Definition 6.21.

Definition 6.21 (Consistent scoring function) *A scoring function $S(x, y)$ is consistent for a functional T relative to a class of measures (distributions) $\mathcal{F}$ if it satisfies the condition*

$$\mathbb{E}_F [S(t, Y)] \leq \mathbb{E}_F [S(x, Y)] \tag{6.56}$$

for all probability distributions $F \in \mathcal{F}$, all $t \in T(F)$, and all $x \in \mathbb{R}$. ■

Note that strict consistency of a scoring function arises when the scoring function is consistent and equality in Equation (6.56) implies that $x \in T(F)$.

One may now observe that the class of the scoring functions, that are consistent for a certain functional T is identical to the class of the loss functions under which the functional is an optimal point forecast.

From the notion of a consistent scoring function, one may now define the concept of an elicitable function, as given in Definition 6.22.

Definition 6.22 (Elicitable function) *A functional T is elicitable with respect to a class of measures (probability distributions) $\mathcal{F}$ if there exists a scoring function (loss function) S which is strictly consistent for the functional T relative to $\mathcal{F}$. That is possibly a set valued functional $T : \mathcal{M}_1(\mathbb{R}) \mapsto \mathbb{R}$, for the set of probability distributions on the real line, $\mathcal{M}_1(\mathbb{R})$, is measurable if it satisfies*

$$T(F) = \arg\min_x \int S(x, y) dF(y), \tag{6.57}$$

where the scoring function satisfies the conditions A1–A3 defined earlier. ■

Simple examples of elicitable functionals include the mean, which minimizes the quadratic score (loss) function; the quantile interval, which is the set of minimizers of a piecewise linear score function; and the expectiles, which minimize the asymmetric piecewise quadratic score.

One may now define an important class of loss distribution functionals that are of direct importance when considering risk measures, as studied by Thomson (1979) and Saerens (2000); see Theorem 6.1.

Theorem 6.1 (Quantile elicitable risk measures) *Consider the class of loss distributions $\mathcal{F}$ on the interval $\mathbb{I} \subseteq \mathbb{R}$ and the value $\alpha \in (0, 1)$. Then the following holds:*

- *The α-quantile function is elicitable relative to the class of loss distributions $\mathcal{F}$;*
- *If the scoring function $S(x, y)$ satisfies conditions A1–A3 on domain $I \times I$, then S is consistent for the α-quantile relative to the class of compactly supported loss distributions on $\mathbb{I}$ if and only if it has the form*

$$S(x, y) = \mathbb{I}(x \geq y)\,(g(x) - g(y))\,, \tag{6.58}$$

for a nondecreasing function g on $\mathbb{I}$.

Remark 6.8 (Elicitabilty and the relationship with backtesting risk measures) *It was observed by Bellini and Bignozzi (2013) for the case of financial risk measures, that it is valuable to consider the elicitability property as it provides a natural methodology to perform backtesting of risk measures. Here, we define the notion of backtesting as the activity of periodically comparing the forecasted risk measure with the realized value of the variable under interest, so as to assess the accuracy of the forecasting methodology.*

A typical example of backtesting involves the VaR measure in which the $VaR_\alpha[X] = q_\alpha(X)$ where $q_\alpha(X)$ denotes the quantile of the loss distribution for a loss random variable $X \sim F$. Then given a model estimated VaR, the question becomes how one may test such an estimated VaR using historical data. A typical approach to backtesting of an estimated VaR measure in OpRisk is to consider counting the number of violations during a fixed time interval (in years for OpRisk) and then comparing this to a theoretical model estimated quantity through a formal binomial hypothesis test. Note that a positive count is recorded for each violation of the historical empirical losses in given sets of periods when compared to the model VaR, that is, in this case, a violation in a year or so period of interest occurs when the realized order statistic for the quantile level α at which one calculates the model-based *VaR* falls below the model estimate.

This concept of backtesting of the risk measure can be generalized to more complex risk measures $\rho(x)$ as long as they satisfy the condition that they are elicitable. Bellini and Bignozzi (2013) observe that an important example of a coherent and backtestable risk measure is the class of expectile risk measures, given in Definition 6.23.

Definition 6.23 (Expectiles) *The κ-level expectile for a random variable Z, denoted by μ_κ, is a parameter that minimizes the expectation given by*

$$\mathbb{E}\left[|\kappa - \mathbb{I}\left[Z < \mu_\kappa\right]|\,(Z - \mu_\kappa)^2\right] \tag{6.59}$$

for $\kappa \in (0, 1)$. ∎

Remark 6.9 *Generally, the κ-level expectile μ_κ is neither the VaR nor the ES and does not have a simple intuitive explanation.*

One can observe that μ_κ occurs at a quantile level of the annual loss Z, denoted by q_κ and typically one has $\kappa < q_\kappa$; hence, one can also see that μ_κ also minimizes the expectation

$$\mathbb{E}\left[\left(q_\kappa - \mathbb{I}\left[Z < \mu_\kappa\right]\right)\left(Z - \mu_\kappa\right)\right]. \tag{6.60}$$

Newey and Powell (1987) then showed that there is a one-to-one relationship between the expectiles and the ES risk measure. In the simple case that $\mathbb{E}[Z] = 0$, one would obtain the relationship

$$\mathrm{ES}_\alpha[Z] = \left(1 + \frac{\kappa}{(1-2\kappa)q_\kappa}\right)\mu_\kappa. \tag{6.61}$$

Note: Typically, this will not be a case of interest in OpRisk settings, and more general expressions may be obtained in the earlier mentioned reference. This relationship provides interesting alternative statistical methods to perform estimation of ES for OpRisk settings based on quantile regressions.

Definition 6.24 (Expectile risk measures) *The expectiles can also be shown to correspond to a class of coherent and elicitable risk measures that correspond to a scoring function given by*

$$S(x,y) = \alpha \mathbb{I}_{y>x}(x-y)^2 + (1-\alpha)\mathbb{I}_{y<x}(x-y)^2. \tag{6.62}$$

■

A second example of a nonstandard elicitable risk measure is the class of measures known as the Λ-VaR as defined by Frittelli *et al.* (2014) and given in Definition 6.25.

Definition 6.25 (Λ-Value-at-Risk) *The elicitable Λ-VaR measure is given by considering the continuous and strictly decreasing mapping $\Lambda : \mathbb{R} \mapsto (0,1)$ satisfying the conditions that $\Lambda(x) \mapsto 1^-$ for $x \to -\infty$ and $\Lambda(x) \mapsto 0^+$ for $x \to \infty$. Then one can define the functional $T(F)$ as the Λ-VaR measure with respect to a class of loss distributions $F \in \mathcal{F}$ according to the definition*

$$T(F) = \inf\left\{m \in \mathbb{R} : F_X(m) \geq \lambda(t)\right\}. \tag{6.63}$$

The corresponding scoring (loss) function for this tail functional (risk measure) is given by

$$S(x,y) = (x-y)^+ - \varphi(x), \tag{6.64}$$

with

$$\varphi(x) = \int_y^x \Lambda(s)ds. \tag{6.65}$$

■

6.2.9 RISK MEASURE ACCOUNTING FOR PARAMETER UNCERTAINTY

Typical risk models are parameterized by a generic parameter vector $\boldsymbol{\theta}$, where the true value of the parameter $\boldsymbol{\theta}$ is unknown and should be estimated. It is expected that the uncertainty in parameters should increase the capital. A convenient way to deal with this problem is to model parameter $\boldsymbol{\theta}$ by a random variable vector $\boldsymbol{\Theta}$ with its own distribution. Consider risk X with conditional density $f(x|\boldsymbol{\Theta})$, where $\boldsymbol{\Theta}$ is a random variable vector from $\pi(\boldsymbol{\theta})$. Then the predictive density of X is

$$f(x) = \int f(x|\boldsymbol{\theta})\pi(\boldsymbol{\theta})d\boldsymbol{\theta} \tag{6.66}$$

and we can calculate a capital using some risk measure $\rho[X]$ based on this distribution.

Denote a risk measure based on the conditional distribution $F(x|\boldsymbol{\theta})$ as $\rho[X|\boldsymbol{\Theta}]$. It is possible to get the following useful result for a risk measure that can be represented as a distortion risk measure (see Definition 6.18)

$$\rho[X] \geq \mathbb{E}_{\boldsymbol{\Theta}}[\rho[X|\boldsymbol{\Theta}]] \tag{6.67}$$

if the distortion function $g(\cdot)$ is concave. This can be proved using Jensen's inequality[1] as follows:

$$\begin{aligned}\rho[X] &= \int_0^\infty g(\overline{F}(y))dy = \int_0^\infty g\left(\int \overline{F}(y|\boldsymbol{\theta})\pi(\boldsymbol{\theta})d\boldsymbol{\theta}\right) dy \\ &\geq \int \pi(\boldsymbol{\theta})d\boldsymbol{\theta} \int_0^\infty g(\overline{F}(y|\boldsymbol{\theta}))dy = \mathbb{E}_{\boldsymbol{\Theta}}[\rho[X|\boldsymbol{\Theta}]].\end{aligned} \tag{6.68}$$

One can also consider risk measure $\rho[X|\boldsymbol{\Theta}]$ as a function of a random variable vector $\boldsymbol{\Theta}$; find the distribution of $\rho[X|\boldsymbol{\Theta}]$; and form a predictive interval $[L, U]$ to contain the true value with a probability γ:

$$\Pr[L \leq \rho[X|\boldsymbol{\Theta}] \leq U] = \gamma, \tag{6.69}$$

or form a one-sided predictive interval $\Pr[\rho[X|\boldsymbol{\Theta}] \leq U] = \gamma$. Then it can be argued that the conservative estimate of the capital accounting for parameter uncertainty should be based on the upper bound of the constructed predictive interval. However, it might be difficult to justify a particular choice of confidence γ. One should answer the question as to whether it is conservative enough to use, for example, $\gamma = 0.95$ for estimation of 0.999 quantile.

Modeling parameter $\boldsymbol{\theta}$ by a random variable vector $\boldsymbol{\Theta}$ corresponds to the Bayesian inference approach. In this case, $\pi(\boldsymbol{\theta})$ would be a posterior distribution for given observed data. The frequentist analogy is to replace parameter $\boldsymbol{\theta}$ by its point estimator $\hat{\boldsymbol{\theta}}$, which is treated as random.

Remark 6.10

- *Note that VaR is a distorted risk measure with a function $g(\cdot)$ given by (6.51), which is neither concave nor convex. ES is a distorted risk measure with $g(\cdot)$ given by (6.52), which is concave. Thus, inequality (6.67) is guaranteed for ES. However, it is not true in general for VaR; see Example 6.5;*

[1] Jensen's inequality for a random variable Z states that $\varphi(\mathbb{E}[Z]) \leq \mathbb{E}[\varphi(Z)]$ if $\phi(\cdot)$ is a convex function and inequality is reversed if $\phi(\cdot)$ is concave.

- *Assuming that the distribution parameter has some distribution itself is also known as mixing. As a result, the unconditional distribution typically has a heavier tail than the conditional one; see Example 6.5.*

EXAMPLE 6.5 Exponential distribution with the Gamma distributed parameter

Assume that for a given parameter λ, loss X is from the exponential distribution $F(x|\lambda) = 1 - \exp(-\lambda x)$, that is, with the density $f(x|\lambda) = \lambda \exp(-\lambda x)$. Also assume that parameter λ is modeled by a random variable Λ from the Gamma distribution $Gamma(\alpha, 1/\beta)$ with the density denoted $\pi(\lambda)$. Then the unconditional density of loss X is

$$\begin{aligned} f(x) &= \int_0^\infty f(x|\lambda)\pi(\lambda)d\lambda \\ &= \frac{\beta^\alpha}{\Gamma(\alpha)} \int_0^\infty \lambda e^{-\lambda x} \lambda^{\alpha-1} e^{-\lambda\beta} \\ &= \frac{\beta^\alpha}{\Gamma(\alpha)} \int_0^\infty \lambda^\alpha e^{-\lambda(x+\beta)} dx \\ &= \frac{\beta^\alpha}{\Gamma(\alpha)} \frac{\Gamma(\alpha+1)}{(x+\beta)^{\alpha+1}} \\ &= \frac{\alpha\beta^\alpha}{(x+\beta)^{\alpha+1}}, \end{aligned} \tag{6.70}$$

which is a density of Pareto distribution $Pareto(\alpha, \beta)$

$$F(x) = 1 - \left(1 + \frac{x}{\beta}\right)^{-\alpha}.$$

Thus,

$$\mathrm{VaR}_q[X] = \beta\left(\exp\left(-\frac{1}{\alpha}\ln(1-q)\right) - 1\right). \tag{6.71}$$

The CVaR, $\mathrm{VaR}_q[X|\Lambda]$, is just the inverse of the exponential distribution

$$\mathrm{VaR}_q[X|\Lambda] = -\frac{1}{\lambda}\ln(1-\alpha)$$

Given that Λ is from $Gamma(\alpha, 1/\beta)$, $Q_q(\Lambda) \equiv \mathrm{VaR}_q[X|\Lambda]$ is from inverse Gamma distribution with the shape parameter α and scale parameter $-\beta\ln(1-q)$. Then it is easy to find the quantiles of $Q_q(\Lambda)$ and other characteristics such as the mean:

$$\mathbb{E}[Q_q(\Lambda)] = -\frac{\beta\ln(1-q)}{\alpha-1}, \quad \alpha > 1. \tag{6.72}$$

Comparing (6.72) and (6.71), it is easy to see that $\mathrm{VaR}_q[X]$ is not always larger than $\mathbb{E}[\mathrm{VaR}_q[X|\Lambda]]$. It depends on the shape α and quantile level q. It is easy to find that

$$\mathrm{VaR}_q[X] \geq \mathbb{E}[\mathrm{VaR}_q[X|\Lambda]] \quad \text{if} \quad q \in [q_c(\alpha), 1), \tag{6.73}$$

where $q_c(\alpha) = 1 - \exp(\alpha y_c)$ and y_c is a solution of $h(y) = \exp(-y) + y\alpha/(\alpha-1) - 1 = 0, y < 0$. Conversely,

$$\mathrm{VaR}_q[X] < \mathbb{E}[\mathrm{VaR}_q[X|\Lambda]] \quad \text{if} \quad q \in (0, q_c(\alpha)). \tag{6.74}$$

The function $q_c(\alpha)$ is presented in Figure 6.6. One can see that for large α, only small quantile levels will break the inequality (6.73). However, for $\alpha \to 1$, only large quantiles $q_c \to 1$ will have loading for parameter uncertainty.

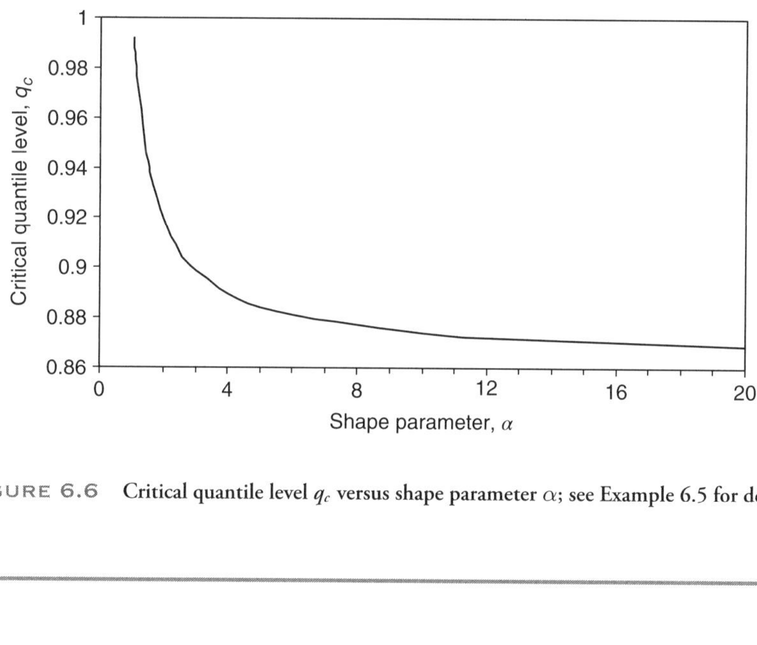

FIGURE 6.6 **Critical quantile level q_c versus shape parameter α; see Example 6.5 for details**

■

6.3 Capital Allocation

Bank capital should be allocated to the various levels within a bank. The allocated capital can be used by risk managers as a mechanism to provide incentives for better risk management. Risk allocation is closely related to the choice of risk measure but in addition should account for diversification in a risk collection. In this section, we treat the issue of risk allocation with a focus on cases of VaR and ES risk measures.

Consider a collection of risks $X_1, \ldots, X_n$. If the risk measure $\rho[\cdot]$ is chosen, then risk capital can be quantified for each risk $\rho_i = \rho[X_i]$. If these risks are combined into one business (collections), then the total capital for the business is $\rho[X_1 + \cdots + X_n]$, which is less than or equal to $\rho_1 + \cdots + \rho_n$ for coherent risk measures. After the total capital is measured by $\rho[\cdot]$, it is important to answer the question as to how much a risk cell i contributes to the total capital.

Calculation of the bank overall capital $\rho[X]$, where

$$X = X_1 + \cdots + X_n,$$

is the annual loss in a bank over the next year and should be followed by an important procedure of allocation of the capital into risk cells in such a way that

$$\rho[X] = \sum_{i}^{n} \Pi_i. \tag{6.75}$$

Here, Π_i denotes the capital allocated to the i-th risk cell. To formalize this statement, define the allocation principle as follows.

Definition 6.26 (Allocation principle) *An allocation principle is a mapping of a collection of risks X_i, $i = 1, 2, \ldots, n$ into unique allocations $\Pi_i = \Pi_i[X_1, \ldots, X_n]$, $i = 1, \ldots, n$ such that*

$$\sum_{i=1}^{n} \Pi_i = \rho[X].$$

■

Capital allocation Π_i can be used for performance measurement providing incentives for a business to improve its risk management practices. Naive choice $\Pi_i = \rho[X_i]$ is certainly not appropriate because it disregards risk diversification. Moreover, the sum of $\rho[X_i]$ adds up to $\rho[X]$ only in the case of perfect positive dependence between risk cells. In this section, we present two popular methods, the *Euler principle* and *marginal contribution*, to allocate the capital.

6.3.1 COHERENT CAPITAL ALLOCATION

Similar to defining a coherent risk measure using a set of axioms, a coherent allocation principle can be defined. A set of axioms (argued to be necessary properties of a reasonable allocation principle) are introduced by Denault (2001).

Definition 6.27 (Coherent allocation axioms set 1) *An allocation principle is coherent if it satisfies the following three properties:*

- ***No undercut**. For any subset M of $\{1, \ldots, n\}$*

$$\sum_{i \in M} \Pi_i \leq \rho\left[\sum_{i \in M} X_i\right].$$

- ***Symmetry**. If for any subset M of $\{1, \ldots, n\}$ that excludes risks i and k,*

$$\rho\left[X_i + \sum_{j \in M} X_j\right] = \rho\left[X_k + \sum_{j \in M} X_j\right],$$

  then $\Pi_i = \Pi_k$. That is, if by joining a subset M, risks i and k make the same contribution to the risk capital, then $\Pi_i = \Pi_k$;

- ***Riskless allocation**. If X_i is riskless, that is, $X_i = \alpha$, then $\Pi_i = \rho[\alpha] = \alpha$.*

■

Remark 6.11

- *While the risk measure of the i-th risk $\rho_i = \rho[X_i]$ does not depend on other risks, the contribution of this risk to the total risk $\rho[X_1 + \cdots + X_n]$ is $\Pi_i = \Pi_i[X_1, \ldots, X_n]$, which depends on all other risks;*
- *Often we are interested in* ***non-negative allocation*** *that satisfies $\Pi_i \geq 0$ for $i = 1, \ldots, n$;*
- *Note that we define loss as a positive random variable, that is, cash amount corresponds to a negative value.*

The proposition is that the three axioms in Definition 6.27 are necessary conditions of the fairness of allocation principle. These conditions can be justified as follows:

- *No-undercut* ensures that no risk can undercut the proposed allocation. An undercut is the situation when a capital allocation to a risk is higher than the amount of capital this risk would face if it were an entity separate from a collection of risks. If a risk joins the collection of risks (or any subset of the collection), then the capital of the collection increases by no more than the capital of this risk. In addition, the property ensures that the coalitions of risks cannot create an undercut;
- *Symmetry* ensures that risk allocation depends only on its contribution to the risk within a collection and nothing else;
- *Riskless allocation* means that riskless position should be allocated a capital exactly the same as its risk measure. It also means that allocated capital decreases by the amount of increase in a cash position.

A different set of axioms is considered by Kalkbrener (2005), who assumes that capital allocation Π_i depends on X_i and X only; we denote this allocation as $\rho[X_i, X] \equiv \Pi_i$.

Definition 6.28 (Coherent allocation axioms set 2)

- ***Linear aggregation.*** *The risk capital of the portfolio (collection) of risks equals the sum of the contributory risk capital of its individual risks, that is,*

$$\rho[X] = \rho[X_1, X] + \cdots + \rho[X_n, X].$$

- ***Diversification.*** *The risk capital $\rho[X, Y]$ of X considered as a subportfolio of Y does not exceed the risk capital $\rho[X]$ of X considered as a stand-alone portfolio;*
- ***Continuity.*** *Small changes to the portfolio of risks only have a limited effect on the risk capital of its subportfolios. More formally, the risk capital $\rho[X, Y + \epsilon X]$ converges to $\rho[X, Y]$ if ϵ converges to 0.*

■

Both sets of axioms lead to the same result that allocation should be done using Euler's principle if the utilized risk measure is coherent, which is the subject of the next section.

6.3.2 EULER ALLOCATION

Typically, the allocated capital is calculated as

$$\Pi_i = \left.\frac{\partial \rho[X + hX_i]}{\partial h}\right|_{h=0}, \quad \text{subject to} \quad \rho[X] = \sum_{i=1}^{n} \Pi_i; \tag{6.76}$$

(see Litterman 1996, Tasche 1999, 2008, and McNeil *et al.* 2005, section 6.3). These are called the Euler allocations and represent capital allocation per unit of exposure X_i. They are consistent with axioms of coherent allocations; see Definition 6.27 and 6.28, for coherent risk measures. However, only positive homogeneity and differentiability of the risk measure are required for formula (6.76), which is based on the following well-known Euler theorem for homogeneous functions.

Definition 6.29 (Homogeneous function) *A function $f(\boldsymbol{u}) = f(u_1, \ldots, u_n)$, $\boldsymbol{u} \in \mathbb{R}^n$, is called homogeneous of degree τ if for all $\lambda > 0$*

$$f(\lambda u_1, \ldots, \lambda u_n) = \lambda^{\tau} f(u_1, \ldots, u_n).$$

■

Homogeneous functions that are continuous and differentiable have several properties relevant to risk modeling summarized later; for a proof, see Tasche (2002 later, 2008).

Theorem 6.2 (Euler's theorem for homogeneous functions) *If function $f(\boldsymbol{u})$ is a continuously differentiable function, then $f(\boldsymbol{u})$ is homogeneous of degree τ if and only if it satisfies*

$$\tau f(\boldsymbol{u}) = \sum_{i=1}^{n} u_i \frac{\partial f(\boldsymbol{u})}{\partial u_i}.$$

Proposition 6.9 *Function $f(\boldsymbol{u})$, which is homogeneous of degree 1, that is, $f(\lambda \boldsymbol{u}) = \lambda f(\boldsymbol{u})$, is convex*

$$f(t\boldsymbol{u} + (1-t)\boldsymbol{v}) \leq tf(\boldsymbol{u}) + (1-t)f(\boldsymbol{v}), \quad t \in [0, 1], \quad \boldsymbol{u}, \boldsymbol{v} \in \mathbb{R}^n$$

if and only if it is subadditive, that is,

$$f(\boldsymbol{u} + \boldsymbol{v}) \leq f(\boldsymbol{u}) + f(\boldsymbol{v}).$$

In addition, a continuously differentiable homogeneous function of degree 1 is subadditive (and convex) if and only if,

$$\sum_{i=1}^{n} u_i \frac{\partial f(\boldsymbol{u} + \boldsymbol{v})}{\partial u_i} \leq f(\boldsymbol{u}).$$

Consider random variables $X_1, \ldots, X_n$ representing risks (losses) with the total loss $X = X_1 + \cdots + X_n$. The capital for this risk collection is determined by risk measure $\rho[X]$. Introducing weight variables $\boldsymbol{u} = (u_1, \ldots, u_n)$, so that,

$$X(\boldsymbol{u}) = u_1 X_1 + \cdots + u_n X_n,$$

that is, $X = X(1, \ldots, 1)$, we can consider the risk measure as a function of $\boldsymbol{u}$,

$$f_\rho(\boldsymbol{u}) = \rho[X(\boldsymbol{u})],$$

and determine the finally required risk measure as,

$$\rho[X] = \rho[X(1, \ldots, 1)].$$

It is obvious that the function f_ρ corresponding to risk measure ρ is homogeneous of degree τ if ρ is homogeneous of degree τ; the risk measure is homogeneous of degree τ if for all $\lambda > 0$, $\rho[\lambda X] = \lambda^\tau \rho[X]$. This correspondence allows translating properties of homogeneous functions (such as Euler theorem) to the homogeneous risk measures. In OpRisk, we are interested in the case of homogeneous functions (homogeneous risk measures) of degree 1, which is the case for risk measures such as VaR and ES. Now it is easy to prove the Euler allocation formula (6.76), formally given by the following theorem.

Theorem 6.3 (Euler allocation principle) *If risk measure $\rho[\cdot]$ is positive homogeneous of degree 1 (i.e., $\rho[\lambda X] = \lambda\rho[X], \lambda > 0$) and differentiable, then*

$$\rho[X] = \sum_{i=1}^{n} \Pi_i^{Euler}, \tag{6.77}$$

where

$$\Pi_i^{Euler} = \left.\frac{\partial\rho[X + hX_i]}{\partial h}\right|_{h=0}. \tag{6.78}$$

Proof: Consider $X(\boldsymbol{u}) = u_1X_1 + \cdots + u_nX_n$, where $\boldsymbol{u} \in \mathbb{R}^J$. Then risk measure $\rho[X(\boldsymbol{u})]$ can be considered as a function of $\boldsymbol{u}$, $f_\rho(\boldsymbol{u}) = \rho[X(\boldsymbol{u})]$, which is a homogenous function of degree 1. Applying Euler's theorem for homogeneous functions, Theorem 6.2 with $\boldsymbol{u} = (1, \ldots, 1)$ gives (6.78). It is also easy to prove this result directly. Consider $f_\rho(\lambda\boldsymbol{u}) = \rho[\lambda X(\boldsymbol{u})], \lambda > 0$. Then using the homogeneity property $\rho[\lambda X] = \lambda\rho[X]$,

$$\frac{df_\rho(\lambda\boldsymbol{u})}{d\lambda} = \rho[X(\boldsymbol{u})].$$

On the other hand, using the standard rule of derivative calculus,

$$\frac{df_\rho(\lambda\boldsymbol{u})}{d\lambda} = \sum_{i=1}^{n} \frac{\partial f_\rho(\lambda\boldsymbol{u})}{\partial(\lambda u_i)} u_i = \sum_{i=1}^{n} \frac{\partial f_\rho(\boldsymbol{u})}{\partial u_i} u_i,$$

where the last equality follows from the homogeneity property. Thus,

$$\rho[X(\mathbf{1})] = \sum_{i=1}^{n} \left.\frac{\partial\rho[X_1 + \cdots + X_i + hX_i]}{\partial h}\right|_{h=0},$$

which completes the proof. ∎

Remark 6.12

- *Note that for Euler allocations, it is only required that the risk measure be homogeneous of degree 1, which is the case for VaR and ES. Formally, subadditivity is not required for the Euler theorem to hold. However, allocations will be coherent in a sense of Definition 6.27 or 6.28 if the risk measure is coherent; this has been demonstrated by Denault (2001), who used game-theoretic considerations, and also by Kalkbrener (2005);*
- *Tasche (1999) showed that Euler allocation is the only allocation principle compatible with the return on risk-adjusted capital (RORAC; i.e., expected return divided by risk capital) measure of performance in portfolio management.*

Another property of homogeneous functions allows us to get a useful result. A continuously differentiable homogeneous function of degree 1 is subadditive (and convex) if and only if,

$$\sum_{i=1}^{n} u_i \frac{\partial f(\boldsymbol{u}+\boldsymbol{v})}{\partial u_i} \leq f(\boldsymbol{u}),$$

(see Tasche 2002b, proposition 2.5). Substituting $u_k = 1, v_k = 0$ if $k = i$ and $u_k = 0, v_k = 1$ if $k \neq i$ (for $k = 1, \ldots, n$), we obtain

$$\Pi_i^{Euler} \leq \rho[X_i]. \tag{6.79}$$

Risk contributions calculated as Euler contributions will never exceed the stand-alone risk capital if the risk measure is positive homogeneous and subadditive. Often violations of subadditivity property are observed through violation of (6.79).

6.3.3 STANDARD DEVIATION

Standard deviation is a risk measure in classical portfolio theory. It is frequently used in finance; however, it is not well suited for heavy-tailed distribution and nonsymmetric distributions in OpRisk. Here, we consider this risk measure to illustrate the concept of risk allocation. Formally, a capital based on standard deviation risk measure can be defined as

$$\rho[X] = \gamma\sqrt{\mathbb{V}\mathrm{ar}[X]} + \mathbb{E}[X], \tag{6.80}$$

where γ is a non-negative real number (that can be chosen to correspond to some confidence level). The Euler capital allocation (6.78) in this case is

$$\Pi_i^{Euler} = \gamma \frac{\mathbb{C}\mathrm{ov}[X_i, X]}{\sqrt{\mathbb{V}\mathrm{ar}[X]}} + \mathbb{E}[X_i]. \tag{6.81}$$

If $\mathbb{V}\mathrm{ar}[X] = 0$, then $\rho_i^{Euler} = \mathbb{E}[X_i]$. It is easy to see that these risk allocations depend not only on the distribution of X_i but also on the dependence between the risk X_i and overall risk

$X = X_1 + \cdots + X_n$. Formula (6.81) can be easily proven by considering $X(\boldsymbol{u}) = u_1X_1 + \cdots + u_nX_n$, calculating $\mathbb{V}\mathrm{ar}[X(\boldsymbol{u})] = \sum_{ij} u_i u_j \mathbb{C}\mathrm{ov}[X_i, X_j]$ and

$$\frac{\partial\left(\gamma\sqrt{\mathbb{V}\mathrm{ar}[X(\boldsymbol{u})]} + \mathbb{E}[X(\boldsymbol{u})]\right)}{\partial u_i} = \gamma\frac{\mathbb{C}\mathrm{ov}[X_i, X(\boldsymbol{u})]}{\sqrt{\mathbb{V}\mathrm{ar}[X(\boldsymbol{u})]}} + \mathbb{E}[X_i],$$

and setting $\boldsymbol{u} = (1, \ldots, 1)$.

It is obvious that the standard deviation risk measure (6.80) is

- *Translation invariant*: $\rho[X + a] = \rho[X] + a$;
- *Positively homogeneous*: $\rho[\lambda X] = \lambda\rho[X]$, $\lambda > 0$;
- *Subadditive*: $\rho[X + Y] \leq \rho[X] + \rho[Y]$.

However, in general, *it is not monotonic*, that is, $X \leq Y \Rightarrow \rho[X] \leq \rho[Y]$ is not valid in general (see, e.g., Kalkbrener 2005). This has unpleasant consequences for the allocation. For example, if potential losses X are bounded by some level, then contributory capital of X to the portfolio Y might exceed this level (see Kalkbrener *et al.* 2004).

6.3.4 EXPECTED SHORTFALL

If there is no jump in the distribution of X at confidence level α, that is, $\mathbb{P}\mathrm{r}[X = \mathrm{VaR}_\alpha[X]] = 0$, then Euler's allocations (6.78) for $\mathrm{ES}_\alpha[\cdot]$ can be easily calculated, that is, the derivatives in (6.76) are

$$\Pi_i^{Euler} = \left.\frac{\partial \mathrm{ES}_\alpha[X + hX_i]}{\partial h}\right|_{h=0} = \mathbb{E}[X_i|X \geq \mathrm{VaR}_\alpha[X]]; \tag{6.82}$$

(see McNeil *et al.* 2005, section 6.3). It is trivial to verify that,

$$\sum_{i=1}^{n} \mathbb{E}[X_i|X \geq \mathrm{VaR}_\alpha[X]] = \mathbb{E}[X|X \geq \mathrm{VaR}_\alpha[X]] = \mathrm{ES}_\alpha[X].$$

In general, that is, if there are jumps in distribution of X at α level,

$$\left.\frac{\partial \mathrm{ES}_\alpha[X + hX_i]}{\partial h}\right|_{h=0} = \frac{1}{1-\alpha}\left(\mathbb{E}\left[X_i \mathbb{I}_{X \geq \mathrm{VaR}_\alpha[X]}\right] + \beta_X \mathbb{E}\left[X_i \mathbb{I}_{X = \mathrm{VaR}_\alpha[X]}\right]\right), \tag{6.83}$$

where

$$\beta_X = \frac{\mathbb{P}\mathrm{r}[X \leq \mathrm{VaR}_\alpha[X]] - \alpha}{\mathbb{P}\mathrm{r}[X = \mathrm{VaR}_\alpha[X]]}, \quad \text{if} \quad \mathbb{P}\mathrm{r}[X = \mathrm{VaR}_\alpha[X]] > 0,$$

(see, e.g., Kalkbrener 2005). This expression is equivalent to (6.82) if $\mathbb{P}\mathrm{r}[X = \mathrm{VaR}_\alpha[X]] = 0$.

Typically, the Euler allocations should be calculated numerically. Assume that the total capital is quantified using Monte Carlo methods and $\mathbb{P}\mathrm{r}[X = \mathrm{VaR}_\alpha[X]] = 0$. That is, a sample of independent and identically distributed annual losses $x_k^{(j)}$, $k = 1, \ldots, K$ is simulated for each risk cell i (here, the dependence between risk cells is allowed). Then, a sample $x^{(1)}, \ldots, x^{(K)}$, where $x^{(k)} = \sum_{i=1}^{n} x_i^{(k)}$, can be calculated and $\mathrm{VaR}_\alpha[X]$ is estimated in the usual way by sorting the samples and taking a sample (after sorting) with the index $\lceil n\alpha \rceil$. Denote this estimate by $\widehat{\mathrm{VaR}}_\alpha[X]$. Then the Euler allocations in the case of ES (6.82) are estimated via

$$\mathbb{E}[X_i|X \geq \mathrm{VaR}_\alpha[X]] \approx \frac{\sum_{k=1}^{K} x_i^{(k)} \mathbb{I}_{\{x^{(k)} \geq \widehat{\mathrm{VaR}}_\alpha[X]\}}}{\sum_{k=1}^{K} \mathbb{I}_{\{x^{(k)} \geq \widehat{\mathrm{VaR}}_\alpha[X]\}}}. \tag{6.84}$$

For simplicity, in (6.84) we assumed that there are no repeated samples $x^{(k)}$ at $\widehat{\mathrm{VaR}}_\alpha[X]$. A more general formula (6.83) can be estimated using Monte Carlo samples but it is not typically required in OpRisk models.

6.3.5 VALUE-AT-RISK

Although the VaR is not subadditive and differentiable in general, the derivatives of $\mathrm{VaR}_\alpha[\cdot]$ to calculate risk allocations (6.76),

$$\lim_{h \to 0} \left. \frac{\mathrm{VaR}_\alpha[X + hX_i] - \mathrm{VaR}_\alpha[X]}{h} \right|_{h=0} \tag{6.85}$$

may exist for some risk collections. Under some technical conditions, it can be calculated as,

$$\left. \frac{\partial \mathrm{VaR}_\alpha[X + hX_i]}{\partial h} \right|_{h=0} = \mathbb{E}[X_i|X = \mathrm{VaR}_\alpha[X]] =: \Pi_i^{Euler}. \tag{6.86}$$

For precise conditions when this is true, see Tasche (1999). Here we just note that it is easy to verify that these contributions add up to the total risk,

$$\sum_{j=1}^{J} \mathbb{E}[X_i|X = \mathrm{VaR}_\alpha[X]] = \mathbb{E}[X|X = \mathrm{VaR}_\alpha[X]] = \mathrm{VaR}_\alpha[X].$$

In the case of VaR, the Euler allocation can be difficult to estimate using the Monte Carlo sample, because $\Pr[X = \mathrm{VaR}_\alpha[X]] = 0$ in the case of continuous distributions. To handle this problem, the condition $X = \mathrm{VaR}_\alpha[X]$ can be replaced by $|X - \mathrm{VaR}_\alpha[X]| < \epsilon$ for some $\epsilon > 0$ large enough to have $\Pr[|X - \mathrm{VaR}_\alpha[X]| < \epsilon] > 0$. However, this condition will be satisfied by only a few Monte Carlo simulations and important sampling techniques are needed to get an accurate estimation (see Glasserman 2005).

It can be somewhat easier to calculate the Euler allocations using the finite difference approximation,

$$\left. \frac{\partial \rho[X + hX_i]}{\partial h} \right|_{h=0} \approx \frac{\rho[X + \Delta X_i] - \rho[X]}{\Delta} \tag{6.87}$$

with some small suitable $\Delta \neq 0$. Note that the choice of Δ depends on the numerical accuracy of the estimator for $\rho[\cdot]$ and curvature of the $\rho[\cdot]$ with respect to h. So, Δ should be neither very small nor too large. This is a typical problem with estimating derivatives via finite difference, and details can be found in many books on numerical recipes (see, e.g., Press *et al.* 2002, section 5.7).

Another approach is to allocate VaR using ES,

$$\mathrm{ES}_{\beta(X)}[X] = \mathrm{VaR}_{\alpha}[X], \tag{6.88}$$

if $\mathbb{E}[X] \leq \mathrm{VaR}_{\alpha}[X]$. This technique was proposed in several papers (Overbeck 2000, Bluhm *et al.* 2002; see also Kalkbrener 2005). That is, we calculate $\mathrm{VaR}_{\alpha}[X]$ and then find a confidence level β such that (6.88) is satisfied; then we allocate capital $\mathrm{ES}_{\beta}[X]$ into risk cells using ES allocations

$$\Pi_i = \mathbb{E}[X_i | X \geq \mathrm{VaR}_{\beta}[X]]$$

or using ES allocations in (6.83) for the general case. Even if marginal distributions and dependence are known, closed-form solutions are rarely available in OpRisk for VaR, ES, and their allocations. However, all these quantities can easily be calculated using Monte Carlo by following logical steps:

- Simulate all risks $X_1, \ldots, X_n$ and find corresponding $X = X_1 + \cdots + X_n$. For K simulations we have $x_i^{(k)}$, $k = 1, \ldots, K$, $i = 1, \ldots, n$ and $x^{(k)} = x_1^{(k)} + \cdots + x_n^{(k)}$, $k = 1, \ldots, K$. Sort sample $x^{(k)}$ in increasing order; for simplicity of notation, assume that $x^{(k)}$ denotes the ordered sample and samples $x_1^{(k)}, \ldots, x_n^{(k)}$ are reordered correspondingly;
- Using the ordered sample, estimate $\widehat{\mathrm{VaR}}_{\alpha}[X] = x^{\lceil \alpha K \rceil}$;
- Find the confidence level $\hat{\beta} = k_{\beta}/K$ such that $\widehat{\mathrm{ES}}_{\beta}[X] \approx \widehat{\mathrm{VaR}}_{\alpha}[X]$. This can be simply achieved by calculating

$$\widehat{\mathrm{ES}}_{\beta}[X] = \frac{\sum_{k=k_{\beta}}^{K} x^{(k)}}{K - k_{\beta} + 1}. \tag{6.89}$$

  for different values of $\beta \leq \alpha$, that is, for different values of $k_{\beta} \leq \lceil \alpha K \rceil$;
- Once β and the corresponding index k_{β} are found, the allocations $\mathbb{E}[X_i | X \geq \mathrm{VaR}_{\beta}[X]]$ are estimated as

$$\frac{\sum_{k=k_{\beta}}^{K} x_i^{(k)}}{K - k_{\beta} + 1}. \tag{6.90}$$

  Note that, here, $x_i^{(k)}$ is not originally simulated or ordered in increasing order sample; index k here corresponds to the index in the ordered sample $x^{(k)}$.

EXAMPLE 6.6 Allocation of VaR using ES for LogNormal distribution

Assume that overall loss $X = X_1 + \cdots + X_n$ is from LogNormal density $f(x; \mu, \sigma)$, that is, $\ln X$ is from Normal distribution with mean μ and variance σ^2. Then

$$\mathrm{VaR}_\alpha[X] = \exp(\mu + \sigma\Phi^{-1}(\alpha))$$

and ES is

$$\mathrm{ES}_\alpha[X] = \frac{1}{1-\alpha} e^{\mu+\frac{1}{2}\sigma^2}\Phi(\sigma - \Phi^{-1}(\alpha)), \tag{6.91}$$

See Example 6.3 for details. To allocate $\mathrm{VaR}_\alpha[X]$ via ES, the following equation should be solved for β:

$$\mathrm{ES}_\beta[X] = \mathrm{VaR}_\alpha[X].$$

For example, assume that $\mu = 10, \sigma = 2, \alpha = 0.999$. Then numerical root finding gives $\beta \approx 0.9961$, where $\mathrm{ES}_\beta[X] \approx \mathrm{VaR}_\alpha[X] \approx 10,643,550$. This capital can then be allocated to risk cells as $\mathbb{E}[X_i|X \geq \mathrm{VaR}_\beta[X]]$, but this requires information on individual risks and their dependence. ■

6.3.6 ALLOCATION BY MARGINAL CONTRIBUTIONS

Another popular way to allocate capital is based on marginal risk contribution

$$\rho_i^{marg} = \rho[X] - \rho[X - X_i], \tag{6.92}$$

which is the difference between total risk (across all risk cell) and total risk without risk cell i. This can be viewed as some crude approximation of Euler allocation derivatives (6.87), but of course the risk measure differentiability is not required to calculate marginal contribution. The sum of marginal contributions may not add up to $\rho[X]$. In particular, if the risk measure is subadditive, continuously differentiable, and homogeneous of degree 1, then it can be shown that

$$\rho_i^{marg} \leq \Pi_i^{Euler}, \quad \sum_{i=1}^{n} \rho_i^{marg} \leq \rho[X] \tag{6.93}$$

(see Tasche 2008). One can define

$$\Pi_i^{marg} = \frac{\rho_i^{marg}}{\sum_{j=1}^{J} \rho_j^{marg}} \rho[X], \tag{6.94}$$

to ensure that allocated capitals add up to $\rho[X]$, that is, $\rho[X] = \Pi_1^{marg} + \cdots + \Pi_n^{marg}$.

6.3.7 NUMERICAL EXAMPLE

To illustrate the allocation procedure using the previously described Euler allocation principle, consider the following simple example.

Assume that there are four risk cells where the annual losses X_i are independent random variables from the LogNormal distribution $LogNormal(0, \sigma_i^2)$ with $\sigma_1 = 1.25$, $\sigma_2 = 1.5$, $\sigma_3 = 1.75$, and $\sigma_4 = 2$, respectively. Results based on 4×10^6 Monte Carlo simulations are given in Tables 6.1 and 6.2 for VaR and ES risk measures correspondingly.

Value-at-Risk results.
Monte Carlo estimate of the capital, measured as VaR of the total loss, is

$$\hat{C} = \widehat{\text{VaR}}_{0.999}\left[\sum_i X_i\right] = 552.$$

In Table 6.1, we present VaRs of individual risk cells $\text{VaR}_{0.999}[X_i]$, $i = 1, \ldots, 4$; and marginal and Euler risk allocations Π_i^{marg} and Π_i^{Euler}, respectively. Marginal contributions ρ_i^{marg} and normalized marginal contributions Π_i^{marg} are calculated using (6.92) and (6.94), respectively. The standard errors of Monte Carlo estimates (due to finite number of simulations) for the capital and individual VaRs are on the order of 1%. Π_i^{Euler} estimated using finite difference

TABLE 6.1 Allocation of VaR capital $C = \text{VaR}_{0.999}[X_1 + \cdots + X_4] \approx 552$ by marginal and Euler contributions Π_i^{marg} and Π_i^{Euler}, respectively; here, $X_i \sim LogNormal(0, \sigma_i^2)$

i	σ_i	$\text{VaR}_{0.999}[X_i]$	ρ_j^{marg}	$\Pi_i^{marg}(\%)$	$\hat{\Pi}_i^{Euler}$	$\widetilde{\Pi}_i^{Euler}(\%)$
1	1.25	48	2	0.5	1	0.2
2	1.5	102	8	2.2	6	1.1
3	1.75	226	60	16	106	18.9
4	2.0	480	303	81.3	448	79.8
Total		856	373	100	561	100

Estimated Π_i are given in absolute terms and as a percentage of the total capital C. See Section 6.3.7 for details.

TABLE 6.2 Allocation of ES capital $C = \text{ES}_{0.999}[X_1 + \cdots + X_4] \approx 1118$ by marginal and Euler contributions Π_i^{marg} and Π_i^{Euler}, respectively; here, $X_i \sim LogNormal(0, \sigma_i^2)$

i	σ_i	$\text{ES}_{0.999}[X_i]$	ρ_i^{marg}	$\Pi_i^{marg}(\%)$	Π_i^{Euler}	$\Pi_i^{Euler}(\%)$
1	1.25	72	2	0.3	2	0.2
2	1.5	171	7	0.9	15	1.3
3	1.75	419	71	9.4	156	14.0
4	2	1036	675	89.4	945	84.5
Total		1698	755	100	1118	100

Estimated Π_i are given in absolute terms and as a percentage of the total capital C. See Section 6.3.7 for details.

approximation (6.87) with $\Delta = 0.02$ is denoted as $\hat{\Pi}^{Euler}$. Due to finite difference approximation, $\sum_i \hat{\Pi}_i^{Euler} = 561$ is slightly different from $\mathrm{VaR}_{0.999}[\sum_i X_i] \approx 552$, so the final estimate for capital allocations using Euler principle is

$$\widetilde{\Pi}_i^{Euler} = \frac{\hat{\Pi}_i^{Euler}}{\sum_j \hat{\Pi}_j^{Euler}} \mathrm{VaR}_{0.999}\left[\sum_j X_j\right],$$

which is presented in Table 6.1 as percentage of the total capital $\hat{C}$. The total diversification

$$1 - \frac{\mathrm{VaR}_{0.999}\left[\sum_i X_i\right]}{\sum_i \mathrm{VaR}_{0.999}[X_i]} \tag{6.95}$$

is approximately 35%, which indicates that VaR is subadditive for given distributions and 0.999 quantile level. It is easy to observe that both marginal and Euler allocations are significantly less than corresponding $\mathrm{VaR}_{0.999}[X_i]$ as expected from (6.79) for subadditive risk measure. Due to finite difference approximation and Monte Carlo errors, one can observe that inequality (6.93) does not hold for the first and second risks whose contributions are very small but is satisfied for the third and fourth risks where the errors are not material.

Finally, it is important to note that the relative importance of risk cells cannot be measured by simple ratios

$$\Pi_i^{naive} = \frac{\mathrm{VaR}_{0.999}[X_i]}{\sum_j \mathrm{VaR}_{0.999}[X_j]}, \quad i = 1, \ldots, 4,$$

which are 6, 12, 26, and 56%, respectively. These are referred to as "naive" allocation and compared with the Euler allocation in Figure 6.7. "Naive" allocations are quite different from Euler allocations; at the same time the Euler allocations are more skewed in a sense that risk cells with large losses get relatively larger allocation, that is, the relative difference between risks increases when risks are considered as a collection (when compared to the "naive" allocations); this feature (typical in practice) can be observed in Figure 6.7.

Expected shortfall results.
Monte Carlo estimate of the capital, measured as ES of the total loss, is

$$\hat{C} = \widehat{\mathrm{ES}}_{0.999}\left[\sum_i X_i\right] = 1118.$$

In Table 6.2, we present ES of individual risk cells $\mathrm{ES}_{0.999}[X_i], i = 1, \ldots, 4$; and marginal and Euler risk allocations Π_i^{marg} and Π_i^{Euler}, respectively. Marginal contributions ρ_i^{marg} and normalized marginal contributions Π_i^{marg} are calculated using (6.92) and (6.94), respectively. The standard errors of Monte Carlo estimates (due to finite number of simulations) for the capital and individual ESs are on the order of 1%. Π_i^{Euler}, $i = 1, \ldots, 4$, were estimated using (6.82) and thus their sum is exactly the same as $\widehat{\mathrm{ES}}_{0.999}[\sum_i X_i] = 1118$.

The total diversification

$$1 - \frac{\mathrm{ES}_{0.999}[\sum_i X_i]}{\sum_i \mathrm{ES}_{0.999}[X_i]} \tag{6.96}$$

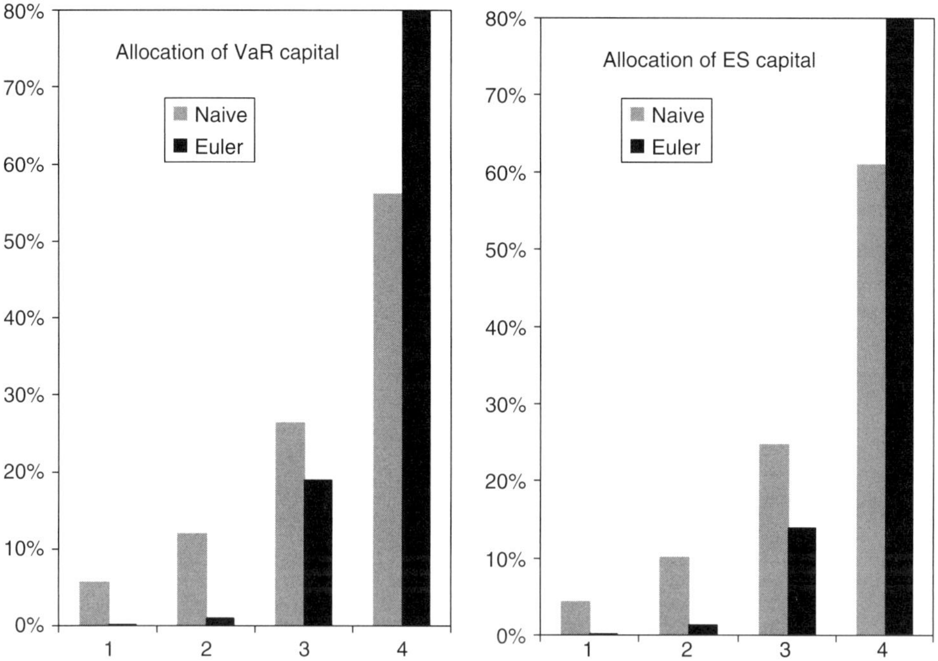

FIGURE 6.7 **Capital allocation weights via "Naive" and Euler allocations to four risk cells (1, 2, 3, and 4) considered in numerical example in Section 6.3.7. Left figure—allocation of VaR capital, right figure—allocation of ES capital**

is approximately 34%, which conforms with subadditivity of ES. It is easy to observe that both marginal and Euler allocations are significantly less than corresponding $\mathrm{ES}_{0.999}[X_i]$ as expected from (6.79) for subadditive risk measure. In addition, marginal allocations are less than corresponding Euler allocations, which is consistent with the inequality (6.93). Finally, it is important to note that the relative importance of risk cells cannot be measured by simple allocation weights

$$\Pi_i^{naive} = \frac{\mathrm{ES}_{0.999}[X_i]}{\sum_i \mathrm{ES}_{0.999}[X_i]}, \quad i = 1, \ldots, 4,$$

which are 4, 10, 25, and 61%, respectively and referred to as "naive" allocation in Figure 6.7. Similar to the results for VaR, "naive" allocations for ES can be quite different from Euler allocations Π_i^{Euler}. At the same time, for Euler allocations, risk cells with large losses get relatively larger allocation (when compared to the "naive" allocations); see Figure 6.7.

CHAPTER SEVEN

Estimation of Frequency and Severity Models

Estimation of the frequency and severity distributions is a challenging task for low-frequency/high-severity losses, due to very limited data for these risks. The main tasks involved in fitting the frequency and severity distributions using data are as follows:

- Finding the best point estimates for the distribution parameters;
- Quantification of the parameter uncertainties;
- Assessing the model quality (model error).

In general, these tasks can be accomplished by undertaking either a frequentist or a Bayesian approach. In this chapter, we present key aspects of each of these approaches. In addition, we note that such modeling paradigms can be performed in both parametric and non-parametric modeling frameworks, but here we focus primarily on a parametric modeling approach, typically adopted in OpRisk. In the context of parameteric modelling we cover components of estimation based on key statistical methods such as Maximum Likelihood Estimation (MLE), Expectation Maximization (EM) algorithm, Bayesian posterior inference methods such as Markov chain Monte Carlo (MCMC), Sequential Monte Carlo Samplers (SMC Samplers) as well as estimation in the presence of truncations. For a comprehensive overview of the non-parametric case, see a book-length review for Bayesian approaches Ghosh and Ramamoorthi (2003); Hjort *et al.* (2010), and for frequentist approaches, Van der Vaart (2000).

7.1 Frequentist Estimation

Fitting distribution parameters using data via the frequentist approach is a classical problem described in many textbooks. For the purposes of this book, we detail important components of several methods that will be of practical use in OpRisk modelling. We note that, under the frequentist approach, one says that the model parameters are fixed while their estimators have

Fundamental Aspects of Operational Risk and Insurance Analytics: A Handbook of Operational Risk, First Edition. Marcelo G. Cruz, Gareth W. Peters, and Pavel V. Shevchenko.

associated uncertainties that typically converge to zero when a sample size increases. Several popular methods to fit parameters (finding point estimators for the parameters) of the assumed distribution include the following:

- Method of moments: finding the parameter estimators to match the observed moments;
- Matching certain quantiles of the empirical distribution;
- Maximum likelihood method: finding parameter values that maximize the joint density of observed data;
- Estimating parameters by minimizing a certain distance between empirical and theoretical distributions, for example, Anderson–Darling or other statistics.

A *point estimator* is a function of a data sample. Notationally, an *estimator* is a function of the sample while an *estimate* is the realized value of an estimator for a realization of the data sample. For example, given a vector of random variables $\boldsymbol{X} = (X_1, X_2, \ldots, X_K)^T$, the estimator is a function of $\boldsymbol{X}$ while the estimate is a function of the realization $\boldsymbol{x}$.

Given a sample $\boldsymbol{X} = (X_1, X_2, \ldots, X_K)^T$ from a density $f(\boldsymbol{x}|\theta)$, we try to find a point estimator $\hat{\Theta}$ for a parameter θ. In most cases, different methods will lead to different point estimators. One of the standard ways to evaluate an estimator is to calculate its *mean squared error.*

Definition 7.1 (Mean squared error) *The mean squared error (MSE) of an estimator $\hat{\Theta}$ for a parameter θ is defined as*

$$\mathrm{MSE}_{\hat{\Theta}}(\theta) = \mathbb{E}\left[(\hat{\Theta} - \theta)^2\right].$$

■

Any increasing function of the discrepancy $|\hat{\Theta} - \theta|$ can be used as a measure of the accuracy of the estimator but MSE is the most popular due to tractability and clear interpretation. In particular, it can be written according to the following decomposition,

$$\mathrm{MSE}_{\hat{\Theta}}(\theta) = \mathbb{V}\mathrm{ar}[\hat{\Theta}] + \left(\mathbb{E}[\hat{\Theta}] - \theta\right)^2, \tag{7.1}$$

where the first term is due to the uncertainty (variability) of the estimator and the second term is due to the bias. The latter is defined as follows.

Definition 7.2 (Bias of a point estimator) *The bias of a point estimator $\hat{\Theta}$ for a parameter θ is*

$$\mathrm{Bias}_{\hat{\Theta}}(\theta) = \mathbb{E}[\hat{\Theta}] - \theta.$$

■

An estimator with zero bias, that is, $\mathbb{E}[\hat{\Theta}] = \theta$, is called *unbiased*. The MSE of an unbiased estimator is reduced to $\mathrm{MSE}_{\hat{\Theta}}(\theta) = \mathbb{V}\mathrm{ar}[\hat{\Theta}]$.

EXAMPLE 7.1

Consider a sample of independent random variables $N_1, N_2, \ldots, N_M$ from a *Poisson*(λ) distribution, with a mean given by $\mathbb{E}[N_m] = \lambda$, and an estimator of this population parameter based on M samples given by $\hat{\Lambda} = \frac{1}{M}\sum_{m=1}^{M} N_m$ (in this case, it is a maximum likelihood estimator (MLE); see Section 7.1.1). Then

$$\mathbb{E}[\hat{\Lambda}] = \frac{1}{M}\mathbb{E}\left[\sum_{m=1}^{M} N_m\right] = \lambda.$$

Thus, the estimator $\hat{\Lambda}$ is an unbiased estimator of λ. ■

It is important for the point estimator of a parameter to be a *consistent* estimator, that is, converge to the "true" value of the parameter in probability as the sample size increases. Formally, a property of consistency is defined for a sequence of estimators as follows.

Definition 7.3 (Consistent estimator) *For a sample $X_1, X_2, \ldots,$ a sequence of estimators*

$$\hat{\Theta}_n = \hat{\Theta}_n(X_1, \ldots, X_n), \quad n = 1, 2, \ldots$$

for the parameter θ is a consistent sequence of estimators if for every $\epsilon > 0$

$$\lim_{n\to\infty} \Pr[|\hat{\Theta}_n - \theta| < \epsilon] = 1.$$

■

We note that consistency is related to bias since a consistent estimator has the property that it is convergent and asymptotically unbiased, therefore it converges to the correct value asymptotically as the sample size increases. However, a consistent estimator may have that the individual estimators in the sequence (i.e. for finite sample sizes) in a consistent sequence may be biased, so long as the bias converges to zero as the sample size increases. A more informative estimation of the parameter (in comparison with the point estimator) is based on a confidence interval specifying the range of possible values.

Definition 7.4 (Confidence interval) *Given a data realization $\boldsymbol{X} = \boldsymbol{x}$, the* $1-\alpha$ confidence interval *for a parameter θ is $[L(\boldsymbol{x}), U(\boldsymbol{x})]$ such that*

$$\Pr[L(\boldsymbol{X}) \le \theta \le U(\boldsymbol{X})] \ge 1 - \alpha.$$

That is, the random interval $[L, U]$, where $L = L(\boldsymbol{X})$ and $U = U(\boldsymbol{X})$, contains the true value of parameter θ with at least probability $1-\alpha$. ■

Typically, it is difficult to construct a confidence interval exactly. However, often it can be found approximately using Gaussian distribution approximation in the case of large data samples (see e.g., Section 7.1.1). Specifically, if a point estimator $\hat{\Theta}$ is distributed according to a *Normal*$(\theta, \sigma^2(\theta))$ distribution, then

$$\mathbb{P}\mathrm{r}\left[-F_N^{-1}(1-\alpha/2) \leq \frac{\hat{\Theta}-\theta}{\sigma(\theta)} \leq F_N^{-1}(1-\alpha/2)\right] = 1-\alpha,$$

where $F_N^{-1}(\cdot)$ is the inverse of the standard Normal distribution *Normal*$(0, 1)$. Note that $\sigma(\theta)$ depends on θ. For a given data realization, typically $\sigma(\theta)$ is replaced by $\sigma(\hat{\theta})$ to approximate a confidence interval by

$$\left[\hat{\theta} - F_N^{-1}(1-\alpha/2)\sigma(\hat{\theta}), \hat{\theta} + F_N^{-1}(1-\alpha/2)\sigma(\hat{\theta})\right]. \tag{7.2}$$

7.1.1 PARAMETERIC MAXIMUM LIKELIHOOD METHOD

The most popular approach to fit the parameters of the assumed distribution is the maximum likelihood method. Given the model parameters $\boldsymbol{\theta} = (\theta_1, \theta_2, \ldots, \theta_K)^T$, assume that the joint density of data $\boldsymbol{X} = (X_1, X_2, \ldots, X_n)^T$ is $f(\boldsymbol{x}|\boldsymbol{\theta})$. Then the *likelihood function* is defined as the joint density $f(\boldsymbol{x}|\boldsymbol{\theta})$ considered as a function of parameter $\boldsymbol{\theta}$.

Definition 7.5 (Likelihood function) *For a sample $\boldsymbol{X} = \boldsymbol{x}$ from the joint density $f(\boldsymbol{x}|\boldsymbol{\theta})$ with the parameter vector $\boldsymbol{\theta}$, the* likelihood function *is a function of $\boldsymbol{\theta}$:*

$$L_{\boldsymbol{x}}(\boldsymbol{\theta}) = f(\boldsymbol{x}|\boldsymbol{\theta}). \tag{7.3}$$

The log likelihood function *is $\ell_{\boldsymbol{x}} = \ln L_{\boldsymbol{x}}(\boldsymbol{\theta})$.* ■

Often it is assumed that $X_1, X_2, \ldots, X_n$ are independent with a common density $f(x|\boldsymbol{\theta})$; then the likelihood function is $L_{\boldsymbol{x}}(\boldsymbol{\theta}) = \prod_{i=1}^{n} f(x_i|\boldsymbol{\theta})$.

The MLE $\hat{\boldsymbol{\Theta}}^{\mathrm{MLE}} = \hat{\boldsymbol{\Theta}}(\boldsymbol{X})$ of the parameters $\boldsymbol{\theta}$ are formally defined as follows.

Definition 7.6 (Maximum likelihood estimator) *For a sample $\boldsymbol{X}$, $\hat{\boldsymbol{\Theta}}(\boldsymbol{X})$ is the* MLE, *if for each realization $\boldsymbol{x}$, $\hat{\boldsymbol{\theta}}(\boldsymbol{x})$ is a value of parameter $\boldsymbol{\theta}$ maximizing the likelihood function $L_{\boldsymbol{x}}(\boldsymbol{\theta})$ or equivalently maximizing the log likelihood function $\ell_{\boldsymbol{x}} = \ln L_{\boldsymbol{x}}(\boldsymbol{\theta})$.* ■

An important property of MLEs is their convergence to the true value in probability as the sample size increases, that is, MLEs are *consistent* estimators, under the weak regularity conditions on the likelihood discussed later.

Theorem 7.1 *For a sample $X_1, X_2, \ldots, X_n$ of independent and identically distributed random variables from $f(x|\boldsymbol{\theta})$ and corresponding MLE $\hat{\boldsymbol{\Theta}}_n$, under the suitable regularity conditions, as the sample size n increases,*

$$\lim_{n\to\infty} \Pr[|\hat{\boldsymbol{\Theta}}_n - \boldsymbol{\theta}| \geq \epsilon] = 0, \quad \textit{for every} \quad \epsilon > 0. \tag{7.4}$$

The required regularity conditions are as follows:

- The parameter should be identifiable: $\boldsymbol{\theta} \neq \tilde{\boldsymbol{\theta}} \Rightarrow f(x|\boldsymbol{\theta}) \neq f(x|\tilde{\boldsymbol{\theta}})$;
- The true parameter should be an interior point of the parameter space;
- The support of $f(x|\boldsymbol{\theta})$ should not depend on $\boldsymbol{\theta}$;
- $f(x|\boldsymbol{\theta})$ should be differentiable in $\boldsymbol{\theta}$.

Asymptotically, for large sample size, under stronger conditions (that further require $f(x|\boldsymbol{\theta})$ to be differentiable three times with respect to $\boldsymbol{\theta}$ and to have continuous and bounded third derivatives), the MLEs are distributed according to a Normal distribution.

Theorem 7.2 *Under the suitable regularity conditions, for a sample $X_1, X_2, \ldots, X_n$ of independent and identically distributed random variables from $f(x|\boldsymbol{\theta})$, $\boldsymbol{\theta} = (\theta_1, \theta_2, \ldots, \theta_K)^T$, and corresponding MLE $\hat{\boldsymbol{\Theta}}_n$:*

$$\sqrt{n}(\hat{\boldsymbol{\Theta}}_n - \boldsymbol{\theta}) \to Normal\left(\mathbf{0}, [\boldsymbol{I}(\boldsymbol{\theta})]^{-1}\right), \tag{7.5}$$

as the sample size n increases. Here, $[\boldsymbol{I}(\boldsymbol{\theta})]^{-1}$ is the inverse matrix of the expected Fisher information matrix for one observation $\boldsymbol{I}(\boldsymbol{\theta})$, whose matrix elements are given by

$$\begin{aligned} \boldsymbol{I}(\boldsymbol{\theta})_{km} &= \mathbb{E}\left[\frac{\partial}{\partial\theta_k}\ln f(X_1|\boldsymbol{\theta})\frac{\partial}{\partial\theta_m}\ln f(X_1|\boldsymbol{\theta})\right] \\ &= -\mathbb{E}\left[\frac{\partial^2}{\partial\theta_k\partial\theta_m}\ln f(X_1|\boldsymbol{\theta})\right]. \end{aligned} \tag{7.6}$$

That is, $\hat{\boldsymbol{\Theta}}^{\text{MLE}}$ converges to $\boldsymbol{\theta}$ as the sample size increases and asymptotically $\hat{\boldsymbol{\Theta}}^{\text{MLE}}$ is Normally distributed with the mean $\boldsymbol{\theta}$ and covariance matrix $n^{-1}\boldsymbol{I}(\boldsymbol{\theta})^{-1}$. For precise details on regularity conditions and proofs, see Lehmann (1983, theorems 6.2.1 and 6.2.3); these can also be found in many other books such as Van der Vaart (2000), Casella and Berger (2002, p. 516), Stuart *et al.* (1999, chapter 18), Ferguson (1996, part 4), and Lehmann and Casella (1998, section 6.3).

In practice, this asymptotic result is often used even for small samples and for the cases that do not formally satisfy the regularity conditions. Note that the mean and covariances depend on the unknown parameters $\boldsymbol{\theta}$ and are usually estimated by replacing $\boldsymbol{\theta}$ with $\hat{\boldsymbol{\theta}}^{\text{MLE}}$ for a given realization of data. Often in practice, the expected Fisher information matrix is approximated by the *observed information matrix*

$$\hat{\boldsymbol{I}}(\hat{\boldsymbol{\theta}})_{km} = -\frac{1}{n}\sum_{i=1}^{n}\left.\frac{\partial^2 \ln f(x_i|\boldsymbol{\theta})}{\partial\theta_k\partial\theta_m}\right|_{\boldsymbol{\theta}=\hat{\boldsymbol{\theta}}} = -\frac{1}{n}\left.\frac{\partial^2 \ln L_{\boldsymbol{x}}(\boldsymbol{\theta})}{\partial\theta_k\partial\theta_m}\right|_{\boldsymbol{\theta}=\hat{\boldsymbol{\theta}}}, \tag{7.7}$$

for a given realization of data. This should converge to the expected information matrix by the law of large numbers. It has been suggested by Efron and Hinkley (1978) that the use of the observed information matrix leads to a better inference in comparison with the expected information matrix.

Though very useful and widely used, these asymptotic approximations are usually not accurate enough for small samples, that is, the distribution of parameter errors can be materially different from Normal and MLEs may have significant bias. Moreover, as for any asymptotic results, *a priori*, one cannot decide on a sample size that is large enough to use the asymptotic approximation.

To assess the quality of the fit, there are several popular goodness-of-fit tests including Kolmogorov–Smirnov, Anderson–Darling, and Chi-square tests. In addition, the likelihood ratio test and Akaike's information criterion are often used to compare models; these are discussed in detail in Chapter 8.

Usually maximization of the likelihood (or minimization of some distances in other methods) must be done numerically. Popular numerical optimization algorithms include simplex method, Newton methods, expectation maximization (EM) algorithm, and simulated annealing. It is worth mentioning that the last is attempting to find a global maximum while other methods find a local maximum. Moreover, EM is usually more stable and robust than the standard deterministic methods such as simplex or Newton methods.

Again, detailed descriptions of the earlier-mentioned methodologies can be found in many textbooks; for application in an OpRisk context, see Panjer (2006).

7.1.2 MAXIMUM LIKELIHOOD METHOD FOR TRUNCATED AND CENSORED DATA

When performing maximum likelihood for OpRisk models, one has to be aware of potential data truncations and censoring as defined generically below.

Definition 7.7 (Censored loss processes) *A general definition of data censoring in OpRisk is that loss data are censored when the number of observations that fall in a given set is known, but the specific values of the observations are unknown; data are said to be censored from below when the set comprises all numbers less than a specific value.* ■

Definition 7.8 (Truncated loss processes) *A general definition of data truncation in OpRisk is that loss data are said to be truncated when observations that fall in a given set are excluded and the number of such observations is also unknown; data are said to be truncated from below when the set comprises all numbers less than a specific value.* ■

This would result in two potential modifications to the MLE given as follows.

Proposition 7.1 (Data Lower Truncated Likelihood) *Given a data-generating model $X \sim F_X(x; \boldsymbol{\theta})$ for i.i.d. data with a lower truncation threshold of x_L, the resulting truncated log likelihood for n observations is given by*

$$l(\boldsymbol{\theta}; \boldsymbol{x}) = -n \ln\left(1 - F_X(x_L; \boldsymbol{\theta})\right) + \sum_{i=1}^{n} \ln f\left(x_i; \boldsymbol{\theta}\right). \tag{7.8}$$

Proposition 7.2 (Data Left Censored Likelihood) *Given a data-generating model $X \sim F_X(x; \boldsymbol{\theta})$ for i.i.d. data with a left censoring threshold of x_c, the resulting left censored log likelihood for n observations (n_u uncensored and $n_c = n - n_u$ censored) is given by*

$$l(\boldsymbol{\theta}; \boldsymbol{x}) = \underbrace{n_c \ln\left(F_X(x_c; \boldsymbol{\theta})\right)}_{censored} + \sum_{i=1}^{n} \underbrace{\ln f\left(x_i; \boldsymbol{\theta}\right)}_{uncensored} \mathbb{I}\left[x_i \in \mathcal{X}_u\right], \tag{7.9}$$

where $\mathcal{X}_u$ denotes the set of uncensored observed losses and $\mathcal{X}_c$ denotes the set of censored observed losses.

Modeling truncated data will be considered in detail in Section 7.9.

7.1.3 EXPECTATION MAXIMIZATION AND PARAMETER ESTIMATION

The EM algorithm is a general iterative method to estimate model parameters maximizing the likelihood of observed data when some of the variables/data are hidden (missing or not observed). Under the Bayesian framework, it can be used to estimate parameters maximizing the posterior of the parameters given observed data, though such approaches will be discussed later.

For simplicity, here we consider maximization of the data likelihood; extension to posterior maximization in a Bayesian framework is trivial. The algorithm is very convenient and efficient when maximization of the likelihood is simplified for the case of complete data (i.e., if hidden variables are known) in comparison with the maximization of the observed data likelihood. Often it is used for cases with truly missing data such as data truncation, but it may also be convenient to artificially introduce hidden variables if the resulting maximization of the complete likelihood is simplified by the addition of such variables.

Each iteration of the algorithm consists of two steps: an expectation step (E-step), and a maximization step (M-step). In the E-step, the hidden variables are estimated given the observed data and the current estimate of the model parameters. This is achieved using the conditional expectation, explaining the choice of terminology. In the M-step, the likelihood function is maximized under the assumption that the missing data are known. The algorithm convergence is guaranteed because the likelihood increases at each iteration.

Dempster *et al.* (1977) presented a proof of general results of the algorithm and introduced the term *EM algorithm*. However, this idea was in use for many years. The reader is also referred to a book by McLachlan and Krishnan (1997) devoted entirely to EM and applications. The algorithm is particularly suitable for situations with missing data (e.g., truncated data or censored data). In OpRisk, it has been used by Bee (2005b) to fit a model accounting for data truncation (typically data below some level are not reported in OpRisk). It is also often used to fit mixture distributions; for a general approach to fitting mixtures, see McLachlan and Krishnan (1997). Truncated mixtures are considered by Sansom and Thompson (1998). McLachlan and Jones (1988) describe the EM algorithm for data grouped into intervals which may also be truncated.

Maximizing of the log-likelihood function can be accomplished by other methods such as gradient-based optimization algorithms or simplex-type algorithms. Often EM is preferred due to its stability and convergence properties. However, it is important to note that in general EM is guaranteed to converge to a local maximum (not global maximum), which is typical across optimization algorithms. In such cases it may be wise to run the EM algorithm multiple times from different starting points randomly selected in the parameter space, then keep the solution maximizing the likelihood.

Consider observed data $\boldsymbol{X}$, unobserved data (or factors) $\boldsymbol{Y}$, and a parameter vector $\boldsymbol{\theta}$ for a chosen model with the density of the observed data $f(\boldsymbol{X}|\boldsymbol{\theta})$ and joint density of observed and unobserved data (complete dataset) $f(\boldsymbol{X}, \boldsymbol{Y}|\boldsymbol{\theta})$. Denote the likelihood of complete data $(\boldsymbol{X}, \boldsymbol{Y})$ as $L_{\boldsymbol{X},\boldsymbol{Y}}(\boldsymbol{\theta}) = f(\boldsymbol{X}, \boldsymbol{Y}|\boldsymbol{\theta})$. Then the marginal likelihood of observed data $L_{\boldsymbol{X}}(\boldsymbol{\theta}) = f(\boldsymbol{X}|\boldsymbol{\theta})$ can be calculated from a complete likelihood as

$$L_X(\theta) = f(\boldsymbol{X}|\boldsymbol{\theta}) = \int f(\boldsymbol{X}, \boldsymbol{y}|\theta) d\boldsymbol{y}. \tag{7.10}$$

The maximum likelihood method estimates parameters $\boldsymbol{\theta}$ by maximizing the marginal likelihood $L_X(\boldsymbol{\theta})$. Often, this is more difficult in comparison with maximizing $L_{X,Y}(\theta)$. Starting with the initial guess for parameters $\boldsymbol{\theta}^0$, the EM algorithm proceeds as follows. For parameter estimates at iteration t, $\boldsymbol{\theta}^t$,

- **E-step:** calculate condition expectation

$$Q(\boldsymbol{\theta}|\boldsymbol{\theta}^t) = \mathbb{E}\left[\ln f(\boldsymbol{X}, \boldsymbol{Y}|\boldsymbol{\theta})|\boldsymbol{X}, \boldsymbol{\theta}^t\right] = \int \ln f(\boldsymbol{X}, \boldsymbol{y}|\boldsymbol{\theta}) f(\boldsymbol{y}|\boldsymbol{X}, \boldsymbol{\theta}^t) d\boldsymbol{y}. \tag{7.11}$$

- **M-step:** maximize $Q(\boldsymbol{\theta}|\boldsymbol{\theta}^t)$ with respect to $\boldsymbol{\theta}$

$$\boldsymbol{\theta}^{t+1} = \arg\max_{\boldsymbol{\theta}} Q(\boldsymbol{\theta}|\boldsymbol{\theta}^t). \tag{7.12}$$

The E- and M-steps are repeated until the change in $Q(\boldsymbol{\theta}^{t+1}|\boldsymbol{\theta}^t)$ (or estimated parameters $\boldsymbol{\theta}^t$) is less than a user prescribed accuracy or tolerance level. In the M-step, we choose $\boldsymbol{\theta}^t$ as the value of $\boldsymbol{\theta}$ that maximizes $Q(\boldsymbol{\theta}|\boldsymbol{\theta}^t)$. If this maximization is difficult, then it can be replaced with finding $\boldsymbol{\theta}^t$ that simply increases $Q(\boldsymbol{\theta}|\boldsymbol{\theta}^t)$, that is, $Q(\boldsymbol{\theta}^{t+1}|\boldsymbol{\theta}^t) \geq Q(\boldsymbol{\theta}^t|\boldsymbol{\theta}^t)$; this is the so-called Generalized Expectation Maximization (GEM) algorithm.

Remark 7.1 (EM for exponential family likelihoods) *If the data are generated from the exponential family distribution, the E-step and M-step are simplified. The E-step reduces to computing the expectation of the complete-data sufficient statistics given the observed data. In the M-step, the conditional expectations of the sufficient statistics computed in the E-step can be directly substituted for the sufficient statistics that occur in the expressions obtained for the complete-data MLEs (i.e., explicit maximization of the expected log likelihood can be avoided).*

As a general algorithm available for complex maximum likelihood computations, the EM algorithm has several appealing properties relative to other iterative algorithms such as Newton–Raphson. First, it is typically easily implemented because it relies on complete-data computations: the E-step of each iteration only involves taking expectations over complete-data conditional distributions. The M-step of each iteration only requires complete-data MLE, for which simple closed-form expressions are already available. Second, it is numerically stable: each iteration is required to increase the log likelihood $\ln L_X(\boldsymbol{\theta})$ in each iteration, and if $\ln L_X(\boldsymbol{\theta})$ is bounded, the sequence $\ln L_X(\boldsymbol{\theta}^t)$ converges to a stationary value. If the sequence converges, it does so to a local maximum or saddle point of the likelihood and to the unique MLE if the likelihood is unimodal. A disadvantage of EM is that its rate of convergence can be extremely slow if a lot of data are missing: Dempster *et al.* (1977) show that convergence is linear with rate proportional to the fraction of information about $\boldsymbol{\theta}$ in $\ln L_{X,Y}(\boldsymbol{\theta})$ that is observed.

In OpRisk settings, one would typically consider EM-type methods for estimation of model parameters under a likelihood-based procedure when there is potential for data censoring or truncation; see definitions by Klugman *et al.* (1998).

EXAMPLE 7.2 OpRisk Application of EM for LDA Model Estimation

The estimation of a Loss Distribution Approach (LDA) model in OpRisk, comprised of estimation of the model parameters for the frequency and severity models in principle, is straightforward statistical estimation. However, in practice for OpRisk modeling, as noted in Bee (2005b), a major difficulty may arise from the fact that loss data are usually left-censored or, more frequently, left-truncated; according to whether data are truncated or censored, specific inferential procedures are needed. Consider the model for the severity in which one considers losses given by $\{X_i\}_{i=1}^n$ with each of the i.i.d. losses, given by $X_i \sim LogNormal(\mu, \sigma^2)$ with a total of n_c and $n_u = n - n_c$ censored and uncensored losses respectively. Denote by $\boldsymbol{x}_{1:n_c} = (x_1, x_2, \ldots, x_{n_c})$ the unobserved censored losses and $\boldsymbol{y}_{1:n_u} = (y_1, y_2, \ldots, y_{n_u})$ as the observed uncensored losses given by $x_i > x_c$ for all $i \in \{1, 2, \ldots, n\}$. One can then define the complete data and observed data likelihoods, for log-transformed data $\widetilde{X}_i = \ln X_i \sim Normal(\mu, \sigma^2)$ and $\widetilde{Y}_i = \ln X_i \sim Normal(\mu, \sigma^2)$, as follows:

1. The complete log-transformed data likelihood is given by

$$
\begin{aligned}
L\left(\boldsymbol{\theta}; \tilde{\boldsymbol{x}}_{1:n_u}, \tilde{\boldsymbol{y}}_{1:n_c}\right) &= \prod_{i=1}^{n_u} f\left(\tilde{x}_i|\boldsymbol{\theta}\right) \prod_{i=1}^{n_c} f\left(\tilde{y}_i|\boldsymbol{\theta}\right) \\
&= \left[\prod_{i=1}^{n_u} \frac{1}{\sqrt{2\pi\sigma^2}} \exp\left(-\frac{1}{2}\left(\frac{\tilde{x}_i - \mu}{\sigma}\right)^2\right)\right] \\
&\times \left[\prod_{j=1}^{n_c} \frac{1}{\sqrt{2\pi\sigma^2}} \exp\left(-\frac{1}{2}\left(\frac{\tilde{y}_j - \mu}{\sigma}\right)^2\right)\right].
\end{aligned} \tag{7.13}
$$

2. The observed log-transformed data likelihood is given by

$$
\begin{aligned}
L_{obs}\left(\boldsymbol{\theta}; \tilde{\boldsymbol{x}}_{1:n_u}\right) &\propto \left(F\left(\tilde{x}_c; \boldsymbol{\theta}\right)\right)^{n_c} \prod_{i=1}^{n_u} f\left(\tilde{x}_i|\boldsymbol{\theta}\right) \\
&\propto \Phi\left(\frac{\tilde{x}_c - \mu}{\sigma}; \boldsymbol{\theta}\right)^{n_c} \prod_{i=1}^{n_u} \left[\frac{1}{\sqrt{2\pi\sigma^2}} \exp\left(-\frac{1}{2}\left(\frac{\tilde{x}_i - \mu}{\sigma}\right)^2\right)\right].
\end{aligned} \tag{7.14}
$$

To perform the inference in this case one can then utilize the EM algorithm with the following two steps, which are updated at iteration τ:

1. **E-Step**. Take the conditional expectation (w.r.t. censored data) of the complete data likelihood function conditional on the observed (uncensored data) and model parameters at iteration τ given by $\hat{\boldsymbol{\theta}}^{(\tau)}$ producing

$$
\begin{aligned}
&\mathbb{E}\left[L\left(\tilde{\boldsymbol{x}}_{1:n_u}, \tilde{\boldsymbol{Y}}_{1:n_c}; \boldsymbol{\theta}\right)\middle|\tilde{\boldsymbol{x}}_{1:n_u}, \boldsymbol{\theta}^{(\tau)}\right] \\
&= \left[\prod_{i=1}^{n_c} f\left(\tilde{x}_i|\boldsymbol{\theta}\right)\right]\left[\prod_{j=1}^{n_c}\mathbb{E}\left[f\left(\widetilde{Y}_i;\boldsymbol{\theta}\right)\middle|\tilde{x}_{1:n_u}, \boldsymbol{\theta}^{(\tau)}\right]\right] \\
&= \left[\prod_{i=1}^{n_c} f\left(\tilde{x}_i|\boldsymbol{\theta}\right)\right]\prod_{j=1}^{n_c}\mathbb{E}\left[f\left(\widetilde{Y}_i;\boldsymbol{\theta}\right)\middle|\tilde{x}_{1:n_u}, \boldsymbol{\theta}^{(\tau)}\right],
\end{aligned}
\tag{7.15}
$$

where the distribution of the missing censored data is given by the right-truncated Gaussian,

$$
f(\tilde{y}_i;\boldsymbol{\theta}) = \frac{1}{\Phi\left(\frac{x_c-\mu}{\sigma}\right)}\frac{1}{\sqrt{2\pi}\sigma}\exp\left[-\frac{1}{2}\left(\frac{\tilde{y}_i-\mu}{\sigma}\right)^2\right]\mathbb{I}\left[\tilde{y}\leq c\right]. \tag{7.16}
$$

As detailed by Bee (2005b), the truncated conditional expectation $\mathbb{E}\left[f(\widetilde{Y}_i;\boldsymbol{\theta})\middle|\tilde{x}_{1:n_u}, \boldsymbol{\theta}^{(\tau)}\right]$ is simplified by the fact that the complete log-transformed likelihood is linear in $\widetilde{Y}_i$ and $\widetilde{Y}_i^2$, which produces a first and second moment given by

$$
\begin{aligned}
\mathbb{E}\left[\widetilde{Y}_i\middle|\tilde{x}_{1:n_u}, \boldsymbol{\theta}^{(\tau)}\right] &= \mu^{(\tau)} - \sigma^{(\tau)}\alpha\left(\frac{x_c-\mu^{(\tau)}}{\sigma^{(\tau)}}\right) \\
\mathbb{E}\left[\widetilde{Y}_i^2\middle|\tilde{x}_{1:n_u}, \boldsymbol{\theta}^{(\tau)}\right] &= \left(\sigma^{(\tau)}\right)^2\left[1-\left(\frac{x_c-\mu^{(\tau)}}{\sigma^{(\tau)}}\right)\alpha\left(\frac{x_c-\mu^{(\tau)}}{\sigma^{(\tau)}}\right)\right. \\
&\quad \left. -\left(\alpha\left(\frac{x_c-\mu^{(\tau)}}{\sigma^{(\tau)}}\right)\right)^2\right] + \left[\mathbb{E}\left[\widetilde{Y}_i\middle|\tilde{x}_{1:n_u}, \boldsymbol{\theta}^{(\tau)}\right]\right]^2
\end{aligned}
$$

with $\alpha\left(\frac{x_c-\mu^{(\tau)}}{\sigma^{(\tau)}}\right) = \phi\left(\frac{x_c-\mu^{(\tau)}}{\sigma^{(\tau)}}\right)\left[\Phi\left(\frac{x_c-\mu^{(\tau)}}{\sigma^{(\tau)}}\right)\right]^{-1}$.

2. **M-Step.** Then one takes the MLEs for the log-transformed likelihoods, which are Gaussian and conditional upon the sufficient statistics estimated for the missing (censored) data from the E-step giving

$$
\begin{aligned}
\mu^{(\tau)} &= \frac{1}{N}\left(\sum_{i=1}^{n_u} x_i + n_c\mathbb{E}\left[\widetilde{Y}_i\middle|\tilde{x}_{1:n_u}, \boldsymbol{\theta}^{(\tau)}\right]\right); \\
\left(\sigma^{(\tau)}\right)^2 &= \frac{1}{N}\left(\sum_{i=1}^{n_u} x_i^2 + n_c\mathbb{E}\left[\widetilde{Y}_i^2\middle|\tilde{x}_{1:n_u}, \boldsymbol{\theta}^{(\tau)}\right]\right).
\end{aligned}
\tag{7.17}
$$

These two steps are then iterated progressively until convergence. ■

The EM algorithm for the truncated Poisson–LogNormal LDA model is also developed by Bee (2005b, section 3.3). We also note that the utilization of such EM steps in LDA model parameter estimation has been explored, with regard to the impact on Value-at-Risk (VaR) and Expected Shortfal (ES) estimations by Chernobai *et al.* (2006). Again these results are developed for both the truncated and censored cases.

7.1.4 BOOTSTRAP FOR ESTIMATION OF PARAMETER ACCURACY

A popular method often used in practice to estimate parameter uncertainties is the so-called *bootstrap*. This method is based on a simple idea: that we can learn about the characteristics of a sample by taking resamples from the original sample with replacement and calculating the parameter estimates for each resampled set to asses the parameter variability. The bootstrap method was originally developed by Efron in the 1970s. For a good introduction to the method we refer the reader to Efron and Tibshirani (1993). Often the bootstrap estimators are reasonable and consistent. Two types of bootstrapping, *nonparametric bootstrap* and *parametric bootstrap*, are commonly used in practice.

Nonparametric bootstrap. Suppose we have a sample of independent and identically distributed random variables $\boldsymbol{X} = (X_1, X_2, \dots, X_K)^T$ and there is an estimator $\hat{\Theta}(\boldsymbol{X})$. Then:

- Draw M independent samples

$$\boldsymbol{X}^{(m)} = (X_1^{(m)}, X_2^{(m)}, \dots, X_K^{(m)})^T, \quad m = 1, \dots, M$$

  with replacement from the original sample $\boldsymbol{X}$. That is $X_k^{(m)}$, $k = 1, \dots, K$, $m = 1, \dots, M$ are independent and identically distributed, and drawn from the empirical distribution of the original sample $\boldsymbol{X}$;
- Calculate estimator $\hat{\Theta}^{(m)} = \hat{\Theta}(\boldsymbol{X}^{(m)})$ for each resample $m = 1, \dots, M$;
- Calculate

$$\widehat{\mathbb{V}\mathrm{ar}}[\hat{\Theta}] = \frac{1}{M-1} \sum_{m=1}^{M} \left(\hat{\Theta}^{(m)} - \mu\right)^2, \quad \text{where} \quad \mu = \frac{1}{M} \sum_{m=1}^{M} \hat{\Theta}^{(m)}. \tag{7.18}$$

Parametric bootstrap. Suppose we have a sample of independent and identically distributed random variables $\boldsymbol{X} = (X_1, X_2, \dots, X_K)^T$ from $f(x|\theta)$ and we can calculate some estimator $\hat{\Theta}(\boldsymbol{X})$ (e.g., MLE) for θ. Then:

- Draw M independent samples

$$\boldsymbol{X}^{(m)} = (X_1^{(m)}, X_2^{(m)}, \dots, X_K^{(m)})^T, \quad m = 1, \dots, M,$$

  where $X_k^{(m)}$, $k = 1, \dots, K$, $m = 1, \dots, M$ are independent and identically distributed from $f(x|\hat{\theta})$;
- Calculate estimator $\hat{\Theta}^{(m)} = \hat{\Theta}(\boldsymbol{X}^{(m)})$ for each resample $m = 1, \dots, M$;
- Calculate $\widehat{\mathbb{V}\mathrm{ar}}[\hat{\Theta}] = \frac{1}{M-1} \sum_{m=1}^{M} \left(\hat{\Theta}^{(m)} - \mu\right)^2$, where $\mu = \frac{1}{M} \sum_{m=1}^{M} \hat{\Theta}^{(m)}$.

The obtained $\widehat{\mathbb{V}\text{ar}}[\hat{\Theta}]$ is used as an estimator for $\mathbb{V}\text{ar}[\hat{\Theta}]$. Typically, for independent and identically distributed samples, this estimator is consistent, that is,

$$\widehat{\mathbb{V}\text{ar}}[\hat{\Theta}] \to \mathbb{V}\text{ar}[\hat{\Theta}], \quad \text{as} \quad M \to \infty \quad \text{and} \quad K \to \infty, \tag{7.19}$$

though in more general situations it may not occur.

Remark 7.2 *More accurate treatment of nonparametric bootstrap estimators involves an estimator given by*

$$\widehat{\mathbb{V}\text{ar}}^*[\hat{\Theta}] = \frac{1}{N-1}\sum_{m=1}^{N}\left(\hat{\Theta}^{(m)} - \mu\right)^2, \quad \mu = \frac{1}{N}\sum_{m=1}^{N}\hat{\Theta}^{(m)},$$

where $N = K^K$ is the total number of nondistinct resamples. N is very large even for small K, for example, for $K = 10$, $N = 10^{10}$. Calculations of the variance estimators (7.18) with $M \ll N$ is considered an approximation for $\widehat{\mathbb{V}\text{ar}}^*$ variances. Then, convergence of bootstrap estimators is considered in two steps: $\widehat{\mathbb{V}\text{ar}}[\hat{\Theta}] \to \mathbb{V}\text{ar}^*[\hat{\Theta}]$ as $M \to \infty$; and $\mathbb{V}\text{ar}^*[\hat{\Theta}] \to \mathbb{V}\text{ar}[\hat{\Theta}]$ as $K \to \infty$.

7.1.5 INDIRECT INFERENCE–BASED LIKELIHOOD ESTIMATION

In cases in which one considers statistical models for OpRisk data that may not produce a tractable likelihood distribution or density form that can be written down analytically or perhaps evaluated pointwise. Then in such cases, there are also numerous estimation procedures available often based on simulation based methods. Under a likelihood-based inference, there is the method known in econometrics as indirect inference (see, e.g., Gourieroux *et al.* 2006, Gallant and Tauchen 1996, and the book-length coverage by Gourieroux and Monfort 1997).

At its most fundamental level, indirect inference is a technique of parameter estimation for simulation models, that is, models for which one can generate data given (unknown) parameters but not evaluate the density for the data-generating model. One would then like to compare the simulated data with the observed data to decide on the model parameter estimations.

To achieve this via indirect inference one introduces a new model, called the "auxiliary model", which is misspecified and typically not even generative, but is easily fit to the data via, say, standard closed-form estimators for the MLE of the parameters of the auxiliary model. This auxiliary model has its own parameter vector $\boldsymbol{\beta}$, with estimator $\hat{\boldsymbol{\beta}}$. These parameters of the auxiliary model describe aspects of the distributions of the observations. The idea of indirect inference is then to simply try to match aspects of the estimated parameters on the observed data $\boldsymbol{x}$ given by $\hat{\boldsymbol{\beta}}(\boldsymbol{x})$ and the simulated data $\boldsymbol{x}^*$ using parameters of the actual model $\boldsymbol{\theta}$ given by $\hat{\boldsymbol{\beta}}(\boldsymbol{x}^*)$. The indirect inference estimator is then obtained by the following algorithm.

Algorithm 7.1 (Indirect Inference–Based Estimation)

1. *Initialize parameter vector of intractable model $\boldsymbol{\theta}_0$ and simulate initial synthetic data from intractable model $\boldsymbol{X}^* \sim F(\boldsymbol{x}; \boldsymbol{\theta}_0)$;*

2. *Develop an artificial simplified auxiliary model for which it is possible to make inference on parameters of auxiliary model using observed data to obtain true data reference auxiliary model parameter estimator* $\hat{\boldsymbol{\beta}}(\boldsymbol{x})$;
3. *Estimate auxiliary model parameters using synthetic simulated data* $\hat{\boldsymbol{\beta}}_0(\boldsymbol{x}^*(\boldsymbol{\theta}_0))$;
4. *Estimate Mahlanobis distance or Euclidean distances between auxiliary parameter vectors by*

$$D\left(\hat{\boldsymbol{\beta}}(\boldsymbol{x}),\hat{\boldsymbol{\beta}}_0(\boldsymbol{x}^*(\boldsymbol{\theta}_0))\right)=\sqrt{\left(\hat{\boldsymbol{\beta}}(\boldsymbol{x})-\hat{\boldsymbol{\beta}}_0(\boldsymbol{x}^*(\boldsymbol{\theta}_0))\right)^T\Sigma^{-1}\left(\hat{\boldsymbol{\beta}}(\boldsymbol{x})-\hat{\boldsymbol{\beta}}_0(\boldsymbol{x}^*(\boldsymbol{\theta}_0))\right)}.$$

5. *Set optimal parameter vector* $\hat{\boldsymbol{\theta}}=\boldsymbol{\theta}_0$ *with distance* $D_{min}=D\left(\hat{\boldsymbol{\beta}}(\boldsymbol{x}),\hat{\boldsymbol{\beta}}_0(\boldsymbol{x}^*(\boldsymbol{\theta}_0))\right)$;
6. *Repeat until convergence or until you reach J total iterations; for iterations* $j=j+1$, *carry out the following steps:*
   a) *Generate proposed parameter vector* $\boldsymbol{\theta}_j$ *from a proposal mechanism, for instance a genetic algorithm mutation stage, to perturb the parameters in the parameter space;*
   b) *Given parameter vector* $\boldsymbol{\theta}_j$, *generate synthetic data from intractable model* $\boldsymbol{X}^*\sim F\left(\boldsymbol{x};\boldsymbol{\theta}_j\right)$;
   c) *Calculate auxiliary model parameters from synthetic data* $\hat{\boldsymbol{\beta}}_j\left(\boldsymbol{x}^*\left(\boldsymbol{\theta}_j\right)\right)$;
   d) *Calculate distance metric*

$$D\left(\hat{\boldsymbol{\beta}}(\boldsymbol{x}),\hat{\boldsymbol{\beta}}_j\left(\boldsymbol{x}^*\left(\boldsymbol{\theta}_j\right)\right)\right)=\sqrt{\left(\hat{\boldsymbol{\beta}}(\boldsymbol{x})-\hat{\boldsymbol{\beta}}_j\left(\boldsymbol{x}^*\left(\boldsymbol{\theta}_j\right)\right)\right)^T\Sigma^{-1}\left(\hat{\boldsymbol{\beta}}(\boldsymbol{x})-\hat{\boldsymbol{\beta}}_j\left(\boldsymbol{x}^*\left(\boldsymbol{\theta}_j\right)\right)\right)}.$$

   e) *If* $D_{min}>D\left(\hat{\boldsymbol{\beta}}(\boldsymbol{x}),\hat{\boldsymbol{\beta}}_j\left(\boldsymbol{x}^*\left(\boldsymbol{\theta}_j\right)\right)\right)$, *then update the optimal parameter estimate* $\hat{\boldsymbol{\theta}}=\boldsymbol{\theta}_j$.

Several theoretical properties are known about the estimators obtained from such a data-generative procedure (see discussions by Smith 2008 and Genton and Ronchetti 2003). Under several assumptions (see Gourieroux and Monfort, 1997), it can be shown that the indirect inference procedure produces a point estimator of the model parameters which is both consistent and asymptotically Normal under standard regularity conditions. In addition, indirect inference can be shown to be asymptotically efficient when the model is correctly specified for the observed data.

As a consequence, it has recently been proposed as a viable method even to tackle problems in which the likelihood is tractable but perhaps the model parameter estimation is non-robust to model misspecification. For example, consider the observed i.i.d. observations $X_1, X_2, \ldots, X_n$ from a parametric model (family), $\mathcal{M}=\{P(\boldsymbol{\theta}):\boldsymbol{\theta}\in\Theta\}$ of probability measure $P(\boldsymbol{\theta})$, which are indexed by the parameter set $\boldsymbol{\Theta}\subseteq\mathbb{R}^d$. If the model is correctly specified for the given data, that is, $P=P(\boldsymbol{\theta}_0)$ for a unique $\boldsymbol{\theta}_0\in\Theta$, then the MLE is typically a desirable estimator for $\boldsymbol{\theta}_0$ since it is asymptotically efficient under well-known regularity conditions (see Van der Vaart 2000).

However, as discussed by Nickl and Pötscher (2010), if the model class $\mathcal{M}$ is misspecified, that is, we select an inappropriate parametric model family to consider modeling the observed data, then we would like the estimated parameter $\boldsymbol{\theta}_0$ to be robust to such misspecifications. Such cases are considered in the review article of Huber (1972) and Daszykowski *et al.* (2007) where they discuss how to obtain an estimator that is robust to model misspecifications of the form discussed earlier. That is, for an estimator of $\boldsymbol{\theta}_0$ that is robust to perturbations of $P(\boldsymbol{\theta}_0)$

in a metric $D(\cdot,\cdot)$, it would be advisable to utilize, instead, minimum distance estimators. For example if $\widetilde{P}_n$ is a suitable D-consistent estimator of P, then it is better to estimate $\boldsymbol{\theta}$ by the minimizer over Θ of

$$Q_n(\boldsymbol{\theta}) := D\left(\widetilde{P}_n, P(\boldsymbol{\theta})\right). \tag{7.20}$$

Beran and Millar (1987) showed that if the distance is selected specifically, such that D is the Hellinger distance, and if $\widetilde{P}_n$ is a kernel density estimator, then the resulting minimum-distance estimator is both robust and simultaneously asymptotically efficient. Therefore, in such cases, they will outperform the MLE in this sense.

Examples where indirect inference has been applied to parameter estimation in interesting models for OpRisk include recent work on estimation of α-stable model parameters (see Garcia *et al.*, 2011).

7.2 Bayesian Inference Approach

There is a broad literature covering Bayesian inference and its applications for the insurance industry. For a good generic introduction to the Bayesian inference method, see Berger (1985) and Robert (2001), and in OpRisk settings, see Peters and Sisson (2006); Shevchenko (2011). This approach is well suited for OpRisk. It is sketched later to introduce notation and concepts it will also be discussed in detail in Chapter 15.

Consider a random vector of data $\boldsymbol{X} = (X_1, X_2, \ldots, X_n)^T$ whose density, for a given vector of parameters $\boldsymbol{\theta} = (\theta_1, \theta_2, \ldots, \theta_K)^T$, is $f_{X|\Theta}(\boldsymbol{x}|\boldsymbol{\theta})$. In the Bayesian approach, both data and parameters are considered to be random. A convenient interpretation is to think that the parameter vector is a random vector with some distribution and the true value (which is deterministic but unknown) of the parameter is a realization of this random vector. Then the joint density of the data and parameters is

$$f_{X,\Theta}(\boldsymbol{x},\boldsymbol{\theta}) = f_{X|\Theta}(\boldsymbol{x}|\boldsymbol{\theta})\pi_\Theta(\boldsymbol{\theta}) = \pi_{\Theta|X}(\boldsymbol{\theta}|\boldsymbol{x})f_X(\boldsymbol{x}), \tag{7.21}$$

where

- $\pi_\Theta(\boldsymbol{\theta})$ is the density of parameters (a so-called *prior density*);
- $\pi_{\Theta|X}(\boldsymbol{\theta}|\boldsymbol{x})$ is the density of parameters given data $\boldsymbol{X} = \boldsymbol{x}$ (a so-called *posterior density*);
- $f_{X,\Theta}(\boldsymbol{x},\boldsymbol{\theta})$ is the joint density of the data and parameters;
- $f_{X|\Theta}(\boldsymbol{x}|\boldsymbol{\theta})$ is the density of the data given parameters $\boldsymbol{\Theta} = \boldsymbol{\theta}$. This is the same as a likelihood function see (7.3) if considered as a function of $\boldsymbol{\theta}$ for a given $\boldsymbol{x}$, that is, $L_X(\boldsymbol{\theta}) = f_{X|\Theta}(\boldsymbol{x}|\boldsymbol{\theta})$;
- $f_X(\boldsymbol{x})$ is the marginal density of $\boldsymbol{X}$. If $\pi_\Theta(\boldsymbol{\theta})$ is continuous, then

  $$f_X(\boldsymbol{x}) = \int f_{X|\Theta}(\boldsymbol{x}|\boldsymbol{\theta})\pi_\Theta(\boldsymbol{\theta})d\boldsymbol{\theta}$$

  and if $\pi_\Theta(\boldsymbol{\theta})$ is a discrete probability mass function, then the integration should be replaced by a corresponding summation.

Remark 7.3 *Typically,* $\pi_\Theta(\boldsymbol{\theta})$ *depends on a set of further parameters, the so-called* hyper-parameters, *omitted here for simplicity of notation. The choice and estimation of the prior will be discussed in detail in Chapter 15.*

Using (7.21), the well-known Bayes's theorem, Bayes (1763) gives the following.

Theorem 7.3 (Bayes's theorem) *The posterior density can be calculated as*

$$\pi_{\Theta|X}(\boldsymbol{\theta}|\boldsymbol{x}) = f_{X|\Theta}(\boldsymbol{x}|\boldsymbol{\theta})\pi_{\Theta}(\boldsymbol{\theta})/f_X(\boldsymbol{x}). \tag{7.22}$$

Here, $f_X(\boldsymbol{x})$ plays the role of a normalization constant and the posterior can be viewed as a combination of prior knowledge (contained in $\pi_{\Theta}(\boldsymbol{\theta})$) with information from the data (contained in $f_{X|\Theta}(\boldsymbol{x}|\boldsymbol{\theta})$).

Given that $f_X(\boldsymbol{x})$ is a normalization constant, the posterior is often written up to proportionality according to

$$\pi_{\Theta|X}(\boldsymbol{\theta}|\boldsymbol{x}) \propto f_{X|\Theta}(\boldsymbol{x}|\boldsymbol{\theta})\pi_{\Theta}(\boldsymbol{\theta}), \tag{7.23}$$

where "$\propto$" means "is proportional to" with a constant of proportionality independent of the parameter $\boldsymbol{\theta}$. Typically, in closed-form calculations, the right-hand side of the equation is calculated as a function of $\boldsymbol{\theta}$ and then the normalization constant is determined by integration over $\boldsymbol{\theta}$.

Using the posterior $\pi_{\Theta|X}(\boldsymbol{\theta}|\boldsymbol{x})$, one can easily construct a probability interval for Θ, which is the analogue for confidence intervals (see Definition 7.4) under the frequentist approach.

Definition 7.9 (Credibility interval) *Given a data realization $\boldsymbol{X} = \boldsymbol{x}$, if $\pi_{\Theta|X}(\theta|\boldsymbol{x})$ is the posterior density of Θ and*

$$\Pr[a \leq \Theta \leq b|\boldsymbol{X} = \boldsymbol{x}] = \int_a^b \pi_{\Theta|X}(\theta|\boldsymbol{x})d\theta \geq 1 - \alpha,$$

then the interval $[a, b]$ contains the true value of parameter θ with at least probability $1 - \alpha$. The interval $[a, b]$ is called a credibility interval *(sometimes referred to as* predictive interval *or* credible interval*) for parameter θ.* ■

Remark 7.4

- *The inequality in Definition 7.9 is to cover the case of discrete posterior distributions;*
- *Typically, one chooses the smallest possible interval $[a, b]$. One can also consider one-sided intervals, for example, $\Pr[\Theta \leq b|\boldsymbol{X} = \boldsymbol{x}]$;*
- *Extension to the multivariate case, that is, parameter vector $\boldsymbol{\theta}$, is trivial;*
- *Though the Bayesian credibility interval looks similar to the frequentist confidence interval (see Definition 7.4), these intervals are conceptually different. To determine a confidence (probability to contain the true value), the bounds of the frequentist confidence interval are considered to be random (functions of random data) while bounds of the Bayesian credibility interval are functions of a data realization. For some special cases, the intervals are the same (for given data realization) but in general they are different, especially in the case of strong prior information.*

If the data $X_1, X_2, \ldots$ are conditionally (given $\boldsymbol{\Theta} = \boldsymbol{\theta}$) independent, then the posterior can be calculated iteratively, that is, the posterior distribution calculated after $k - 1$ observations can be treated as a prior distribution for the k-th observation. Thus, the loss history over many years is not required, making the model easier to understand and manage, and allowing experts to adjust the priors at every step.

For simplicity of notation, the density and distribution subscripts indicating random variables will often be omitted, for example, $\pi_\Theta(\boldsymbol{\theta})$ will be written as $\pi(\boldsymbol{\theta})$.

7.2.1 CONJUGATE PRIOR DISTRIBUTIONS

Sometimes the posterior density can be calculated in closed form, which is very useful in practice when Bayesian inference is applied. This is the case for the so-called conjugate prior distributions, where the prior and posterior distributions are of the same type.

Definition 7.10 (Conjugate prior) *Let F denote a class of density functions $f(\boldsymbol{x}|\boldsymbol{\theta})$, indexed by $\boldsymbol{\theta}$. A class U of prior densities $\pi(\boldsymbol{\theta})$ is said to be a conjugate family for F and $F - U$ is called a conjugate pair, if the posterior density $\pi(\boldsymbol{\theta}|\boldsymbol{x}) = f(\boldsymbol{x}|\boldsymbol{\theta})\pi(\boldsymbol{\theta})/f(\boldsymbol{x})$, where $f(\boldsymbol{x}) = \int f(\boldsymbol{x}|\boldsymbol{\theta})\pi(\boldsymbol{\theta})d\boldsymbol{\theta}$ is in the class U for all $f \in F$ and $\pi \in U$.* ∎

Formally, if the family U contains all distribution functions, then it is conjugate to any family F. However, to make a model useful in practice it is important that U should be as small as possible while containing realistic distributions. In Chapter 15, we present $F - U$ conjugate pairs (Poisson–Gamma, LogNormal–Normal, Pareto–Gamma) that are useful and illustrative examples of modeling frequencies and severities in OpRisk. Several other pairs (Binomial–Beta, Gamma–Gamma, Exponential–Gamma) can be found, for example, in Bühlmann and Gisler (2005). In all these cases, the prior and posterior distributions have the same type and the posterior distribution parameters are easily calculated using the prior distribution parameters and observations (or recursively).

In general, if the posterior cannot be found in closed form or is difficult to evaluate, one can use Gaussian approximation, Markov chain Monte Carlo methods, or Sequential Monte Carlo methods, discussed next.

7.2.2 GAUSSIAN APPROXIMATION FOR POSTERIOR (LAPLACE TYPE)

For a given data realization $\boldsymbol{X} = \boldsymbol{x}$, denote the mode of the posterior $\pi(\boldsymbol{\theta}|\boldsymbol{x})$ by $\hat{\boldsymbol{\theta}}$. If the prior is continuous at $\hat{\boldsymbol{\theta}}$, then a Gaussian approximation for the posterior is obtained by a second-order Taylor series expansion around $\hat{\boldsymbol{\theta}}$:

$$\ln \pi(\boldsymbol{\theta}|\boldsymbol{x}) \approx \ln \pi(\hat{\boldsymbol{\theta}}|\boldsymbol{x}) + \frac{1}{2}\sum_{i,j} \left.\frac{\partial^2 \ln \pi(\boldsymbol{\theta}|\boldsymbol{x})}{\partial\theta_i \partial\theta_j}\right|_{\boldsymbol{\theta}=\hat{\boldsymbol{\theta}}} (\theta_i - \hat{\theta}_i)(\theta_j - \hat{\theta}_j). \tag{7.24}$$

Under this approximation, $\pi(\boldsymbol{\theta}|\boldsymbol{x})$ is a multivariate Normal distribution with the mean $\hat{\boldsymbol{\theta}}$ and covariance matrix

$$\boldsymbol{\Sigma} = \boldsymbol{I}^{-1}, \ (\boldsymbol{I})_{ij} = -\left.\frac{\partial^2 \ln \pi(\boldsymbol{\theta}|\boldsymbol{x})}{\partial\theta_i \partial\theta_j}\right|_{\boldsymbol{\theta}=\hat{\boldsymbol{\theta}}}. \tag{7.25}$$

Remark 7.5 *In the case of improper constant priors, this approximation is comparable to the Gaussian approximation for the MLEs (Equation 7.5). Note also that in the case of constant priors, the mode of the posterior and the MLE are the same. This is also true if the prior is uniform within a bounded region, provided that the MLE is within this region.*

7.2.3 POSTERIOR POINT ESTIMATORS

Once the posterior density $\pi(\theta|\boldsymbol{x})$ is found, for given data $\boldsymbol{X}$, one can define point estimators of Θ. The mode and mean of the posterior are the most popular point estimators. These Bayesian estimators are typically referred to as the Maximum a Posteriori (MAP) estimator and the Minimum Mean Square Estimator (MMSE), formally defined as follows:

$$\text{MAP}: \quad \hat{\Theta}^{\text{MAP}} = \arg\max_{\theta} \left[\pi(\boldsymbol{\theta} \mid \boldsymbol{X})\right], \tag{7.26}$$

$$\text{MMSE}: \quad \hat{\Theta}^{\text{MMSE}} = \mathbb{E}\left[\Theta|\boldsymbol{X}\right]. \tag{7.27}$$

The median of the posterior is also often used as a point estimator for Θ. Note also that if the prior $\pi(\theta)$ is constant and the parameter range includes the MLE, then the MAP of the posterior is the same as the MLE (see Remark 7.5).

More formally, the choice of point estimators is considered using a *loss function* $l(\theta, \hat{\theta})$ that measures the cost (loss) of a decision to use a particular point estimator $\hat{\Theta}$. For example:

- Quadratic loss. $l(\theta, \hat{\theta}) = (\theta - \hat{\theta})^2$;
- Absolute loss: $l(\theta, \hat{\theta}) = |\theta - \hat{\theta}|$;
- All or nothing loss: $l(\theta, \hat{\theta}) = 0$ if $\theta = \hat{\theta}$ and $l(\theta, \hat{\theta}) = 1$ otherwise;
- Asymmetric loss function: e.g. $l(\theta, \hat{\theta}) = \hat{\theta} - \theta$ if $\hat{\theta} > \theta$ and $l(\theta, \hat{\theta}) = -2(\hat{\theta} - \theta)$ otherwise.

Then the value of $\hat{\Theta}$ that minimizes $\mathbb{E}[l(\Theta, \hat{\Theta})|\boldsymbol{X}]$ is called a Bayesian point estimator of Θ. Here, the expectation is calculated with respect to the posterior $\pi(\theta|\boldsymbol{X})$. In particular:

- The posterior mean is a Bayesian point estimator in the case of a quadratic loss function;
- In the case of an absolute loss function, the Bayesian point estimator is the median of the posterior;
- All or nothing loss function gives the mode of the posterior as the point estimator.

Remark 7.6 $\hat{\Theta} = \hat{\Theta}(\boldsymbol{X})$ *is a function of data* $\boldsymbol{X}$ *and thus it is referred to as estimator. For a given data realization* $\boldsymbol{X} = \boldsymbol{x}$*, we get* $\hat{\Theta} = \hat{\theta}$*, which is referred to as a point estimate.*

In addition, one may be interested in reporting a marginal posterior confidence interval or a measure of precision for the posterior point estimators defined for the i-th static parameter with posterior distribution $\boldsymbol{\Theta}_i \sim F(\boldsymbol{\Theta}_i)$. To achieve this one would typically utilize credibility intervals, see Definition 7.9.

Though the point estimators and interval estimators are useful, for quantification of OpRisk annual loss distribution and capital we recommend the use of the whole posterior, as discussed in the following chapters.

7.2.4 RESTRICTED PARAMETERS

In practice, it is not unusual to restrict parameters. In this case, the posterior distribution will be a truncated version of the posterior distribution in the unrestricted case. That is, if $\boldsymbol{\theta}$ is restricted to some range $[\boldsymbol{\theta}_L, \boldsymbol{\theta}_H]$, then the posterior distribution will have the same type as in the unrestricted case but truncated outside this range.

For example, we choose the LogNormal distribution $LogNormal(\mu, \sigma^2)$ to model the data $\boldsymbol{X} = (X_1, \ldots, X_n)^T$ and we choose a prior distribution for μ to be the Normal distribution $Normal(\mu_0, \sigma_0^2)$. This case will be considered in section 13.2.4. However, if we know that μ cannot be negative, we restrict $Normal(\mu_0, \sigma_0^2)$ to non-negative values only.

Another example is the Pareto–Gamma case, where the losses are modeled by $Pareto(\xi, L)$ and the prior distribution for the tail parameter ξ is $Gamma(\alpha, \beta)$ (see section 13.2.5). The prior is formally defined for $\xi > 0$. However, if we do not want to allow infinite mean predicted loss, then the parameter should be restricted to $\xi > 1$.

These cases can be easily handled by using the truncated versions of the prior–posterior distributions. Assume that $\pi(\theta)$ is the prior whose corresponding posterior density is $\pi(\theta|\boldsymbol{x}) = f(\boldsymbol{x}|\theta)\pi(\theta)/f(\boldsymbol{x})$, where θ is unrestricted. If the parameter is restricted to $a \leq \theta \leq b$, then we can consider the prior,

$$\pi^{\mathrm{tr}}(\theta) = \frac{\pi(\theta)}{\Pr[a \leq \theta \leq b]}\mathbb{I}_{\{a \leq \theta \leq b\}}, \quad \Pr[a \leq \theta \leq b] = \int_a^b \pi(\theta)d\theta, \tag{7.28}$$

for some a and b with $\Pr[a \leq \theta \leq b] > 0$. $\Pr[a \leq \theta \leq b]$ plays the role of normalization and thus the posterior density for this prior is simply

$$\pi^{\mathrm{tr}}(\theta|\boldsymbol{x}) = \frac{\pi(\theta|\boldsymbol{x})}{\Pr[a \leq \theta \leq b|\boldsymbol{x}]}\mathbb{I}_{\{a \leq \theta \leq b\}}, \quad \Pr[a \leq \theta \leq b|\boldsymbol{x}] = \int_a^b \pi(\theta|\boldsymbol{x})d\theta. \tag{7.29}$$

Remark 7.7 *It is obvious that if $\pi(\theta)$ is a conjugate prior, then $\pi^{\mathrm{tr}}(\theta)$ is a conjugate prior too.*

7.2.5 NONINFORMATIVE PRIOR

Sometimes there is no prior knowledge about the model parameter $\boldsymbol{\theta}$, or we would like to rely on data only and avoid an impact from any subjective information. In this case, we need a *noninformative prior* (sometimes called *vague prior*) that attempts to represent a near-total absence of prior knowledge. A natural noninformative prior is the uniform density

$$\pi(\boldsymbol{\theta}) \propto \text{const} \quad \text{for all } \boldsymbol{\theta}. \tag{7.30}$$

If parameter $\boldsymbol{\theta}$ is restricted to a finite set, then this $\pi(\boldsymbol{\theta})$ corresponds to a proper uniform distribution. For example, the parameter p in a Binomial distribution $Binomial(n, p)$ is restricted to the interval $[0, 1]$. Then one can choose a noninformative constant prior, which is the uniform distribution $Uniform(0, 1)$.

However, if the parameter $\boldsymbol{\theta}$ is not restricted, then a constant prior is not a proper density (since $\int f(\boldsymbol{\theta})d\boldsymbol{\theta} = \infty$). Such a prior is called an *improper prior*. For example, the parameter

μ (mean) of the Normal distribution $Normal(\mu, \sigma^2)$ is defined on $(-\infty, \infty)$. Then, for any constant $c > 0$, $\pi(\mu) = c$ is not a proper density because $\int \pi(\mu)d\mu = \infty$. It is not a problem to use improper priors as long as the posterior is a proper distribution. Moreover, as noted in previous sections, if the prior $\pi(\boldsymbol{\theta})$ is constant and the parameter range includes the MLE, then the mode of the posterior is the same as the MLE (see Remark 7.5).

A constant prior is often used as a noninformative prior, though it can be criticized for a lack of invariance under transformation. For example, if a constant prior is used for parameter θ and the model is reparameterized in terms of $\widetilde{\theta} = \exp(\theta)$, then the prior density for $\widetilde{\theta}$ is proportional to $1/\widetilde{\theta}$. Thus, we cannot choose a constant prior for both θ and $\widetilde{\theta}$. In this case, one typically argues that some chosen parameterization is the most intuitively reasonable and absence of prior information corresponds to a constant prior in this parameterization. One can propose noninformative priors through consideration of problem transformations. This has been considered in many studies starting with Jeffreys (1961). For discussion on this topic, see Berger (1985, section 3.3). Here, we just mention that for scale densities of the form $\theta^{-1}f(x/\theta)$, the recommended noninformative prior for a scale parameter $\theta > 0$ is given by,

$$\pi(\theta) \propto \frac{1}{\theta}, \tag{7.31}$$

which is an improper prior because $\int_0^\infty \pi(\theta)d\theta = \infty$.

7.3 Mean Square Error of Prediction

To illustrate the difference between the frequentist and Bayesian approaches, consider the so-called (conditional) mean squared error of prediction (MSEP), which is often used for prediction of uncertainty.

Consider a sample $X_1, X_2, \ldots, X_n, \ldots$ and assume that, given data,

$$\boldsymbol{X} = (X_1, X_2, \ldots, X_n)^T,$$

we are interested in prediction of a random variable R, which is some function of $X_{n+1}, X_{n+2}, \ldots$. Assume that $\hat{R}$ is a predictor for R and an estimator for $\mathbb{E}[R|\boldsymbol{X}]$. Then, the conditional MSEP is defined by

$$\mathrm{MSEP}_{R|\boldsymbol{X}}\left(\hat{R}\right) = \mathbb{E}\left[\left(R - \hat{R}\right)^2 |\boldsymbol{X}\right]. \tag{7.32}$$

It allows for a good interpretation if decoupled into *process variance* and *estimation error* as

$$\begin{aligned} \mathrm{MSEP}_{R|\boldsymbol{X}}\left(\hat{R}\right) &= \mathbb{V}\mathrm{ar}[R|\boldsymbol{X}] + \left(\mathbb{E}[R|\boldsymbol{X}] - \hat{R}\right)^2 \\ &= \text{Process variance} + \text{estimation error}. \end{aligned} \tag{7.33}$$

It is clear that the estimator $\hat{R}$ that minimizes conditional MSEP is $\hat{R} = \mathbb{E}[R|\boldsymbol{X}]$. Assume that the model is parameterized by the parameter vector $\boldsymbol{\theta} = (\theta_1, \ldots, \theta_K)^T$. Then under the frequentist and Bayesian approaches we get the following estimators of MSEP.

Frequentist Approach. Unfortunately, in the frequentist approach, $\mathbb{E}[R|\boldsymbol{X}]$ is unknown and the second term in (7.33) is often estimated by $\mathbb{V}\mathrm{ar}[\hat{R}]$ (see Wüthrich and Merz (2008, section 6.4.3)). Under the frequentist approach, $\mathbb{V}\mathrm{ar}[R|\boldsymbol{X}]$ and $\mathbb{E}[R|\boldsymbol{X}]$ are functions of parameter $\boldsymbol{\theta}$ and can be denoted as $\mathbb{V}\mathrm{ar}_{\boldsymbol{\theta}}[R|\boldsymbol{X}]$ and $\mathbb{E}_{\boldsymbol{\theta}}[R|\boldsymbol{X}]$, respectively. Typically, these are estimated as $\widehat{\mathbb{V}\mathrm{ar}}_{\boldsymbol{\theta}}[R|\boldsymbol{X}] = \mathbb{V}\mathrm{ar}_{\hat{\boldsymbol{\Theta}}}[R|\boldsymbol{X}]$ and $\hat{\mathbb{E}}_{\boldsymbol{\theta}}[R|\boldsymbol{X}] = \mathbb{E}_{\hat{\boldsymbol{\Theta}}}[R|\boldsymbol{X}]$, where $\hat{\boldsymbol{\Theta}}$ is a point estimator of $\boldsymbol{\theta}$ obtained by maximum likelihood or other methods. Also, typically, one chooses $\hat{R} = \mathbb{E}_{\hat{\boldsymbol{\Theta}}}[R|\boldsymbol{X}]$, so that now $\hat{R}$ is a function of $\hat{\boldsymbol{\Theta}}$ that we denote as $\hat{R}(\hat{\boldsymbol{\Theta}})$. The parameter uncertainty term $\mathbb{V}\mathrm{ar}_{\boldsymbol{\theta}}[\hat{R}]$ is usually estimated using the first-order Taylor expansion of $\hat{R}(\hat{\boldsymbol{\Theta}})$ around $\boldsymbol{\theta}$

$$\hat{R}(\hat{\boldsymbol{\theta}}) \approx \hat{R}(\boldsymbol{\theta}) + \sum_i \left. \frac{\partial \hat{R}(\hat{\boldsymbol{\theta}})}{\partial \hat{\theta}_i} \right|_{\hat{\boldsymbol{\theta}}=\boldsymbol{\theta}} (\hat{\theta}_i - \theta_i)$$

leading to

$$\mathbb{V}\mathrm{ar}_{\theta}[\hat{R}(\hat{\boldsymbol{\Theta}})] \approx \sum_{i,j} \left. \frac{\partial \hat{R}}{\partial \hat{\theta}_i} \right|_{\hat{\boldsymbol{\theta}}=\boldsymbol{\theta}} \left. \frac{\partial \hat{R}}{\partial \hat{\theta}_j} \right|_{\hat{\boldsymbol{\theta}}=\boldsymbol{\theta}} \mathbb{C}\mathrm{ov}[\hat{\Theta}_i, \hat{\Theta}_j].$$

Estimating $\boldsymbol{\theta}$ by $\hat{\boldsymbol{\Theta}}$ gives the final estimator

$$\widehat{\mathbb{V}\mathrm{ar}}_{\theta}[\hat{R}(\hat{\boldsymbol{\Theta}})] = \mathbb{V}\mathrm{ar}_{\hat{\boldsymbol{\Theta}}}[\hat{R}(\hat{\boldsymbol{\Theta}})].$$

Note that if the point estimators are unbiased, that is, $\mathbb{E}[\hat{\Theta}_i - \theta_i] = 0$, then $\mathbb{E}[\hat{R}(\hat{\boldsymbol{\theta}})] \approx \hat{R}(\boldsymbol{\theta})$. Finally, the estimator for conditional MSEP is

$$\begin{aligned}\widehat{\mathrm{MSEP}}_{R|X}\left[\hat{R}\right] &= \widehat{\mathbb{V}\mathrm{ar}}[R|\boldsymbol{X}] + \widehat{\mathbb{V}\mathrm{ar}}[\hat{R}] \\ &= \text{Process variance} + \text{estimation error.}\end{aligned} \tag{7.34}$$

These estimators are typically consistent and unbiased in the limit of a large sample size.

Bayesian Approach. Under the Bayesian inference approach, where the unknown parameters $\boldsymbol{\theta}$ are modeled as random variables $\boldsymbol{\Theta}$, $\mathbb{V}\mathrm{ar}\left[R|\,\boldsymbol{X}\right]$ can be decomposed as

$$\begin{aligned}\mathbb{V}\mathrm{ar}\left[R|\boldsymbol{X}\right] &= \mathbb{E}\left[\mathbb{V}\mathrm{ar}\left[R|\,\boldsymbol{\Theta},\boldsymbol{X}\right]|\,\boldsymbol{X}\right] + \mathbb{V}\mathrm{ar}\left[\mathbb{E}\left[R|\,\boldsymbol{\Theta},\boldsymbol{X}\right]|\,\boldsymbol{X}\right] \\ &= \text{Average process variance } + \text{ parameter estimation error,}\end{aligned} \tag{7.35}$$

which equals $\mathrm{MSEP}_{R|X}\left[\hat{R}\right]$ if we choose $\hat{R} = \mathbb{E}[R|\boldsymbol{X}]$. Estimation of the terms involved requires knowledge of the posterior distribution for $\boldsymbol{\Theta}$ that can be obtained either analytically or approximated accurately using Markov chain Monte Carlo methods discussed in the next section.

We also note that under the Bayesian setting it is often of interest to consider the posterior predictive distribution for K additional losses (or annual losses), as defined generically by

$$f\left(\boldsymbol{x}_{n+1:n+K}|\boldsymbol{x}_{1:n}\right) = \int f\left(\boldsymbol{x}_{n+1:n+K}|\boldsymbol{\theta}\right)\pi(\boldsymbol{\theta}|\boldsymbol{x}_{1:n})\; d\boldsymbol{\theta} \tag{7.36}$$

if $\boldsymbol{x}_{1:n}$ and $\boldsymbol{x}_{n+1:n+K}$ are independent given $\boldsymbol{\theta}$. If they are not independent then $f(\boldsymbol{x}_{n+1:n+K}|\boldsymbol{\theta})$ should be replaced by $f(\boldsymbol{x}_{n+1:n+K}|\boldsymbol{\theta}, \boldsymbol{x}_{1:n})$.

7.4 Standard Markov Chain Monte Carlo (MCMC) Methods

There are typically three main reasons why OpRisk practitioners may utilize the following sets of Monte Carlo procedures when undertaking estimation in OpRisk settings.

1. The first involves working with the posterior distribution for the parameters of an LDA model structure to obtain point estimators, posterior credible intervals for the model parameters, and estimation of functions with respect to the posterior, which is the conditional distribution of the LDA model parameters given the observed loss data (as discussed in previous sections);
2. The second involves estimation with respect to annual loss models for a risk or group of risk processes, such as integrals with respect to the LDA model(s). This could include quantities such as predictive distributions for losses in future years, distributions for capital allocations, distributions for joint loss processes with dependence or insurance features, or perhaps sensitivity analysis in derived LDA quantities to changes in model parameters;
3. The third area of interest involves the estimation of risk measures and capital under a particular LDA model.

The first case arises when, for example, the posterior distribution is not known in closed form (i.e., up to normalization) or perhaps one wants to integrate functions with respect to the posterior to find point estimates or predictive interval summaries of the conditional distribution. This is particularly important when one is outside of the class of conjugate Bayesian models. Hence, often practitioners would resort to numerical estimation procedures via sampling. However, when the posterior is not easily sampled from by simple exact methods such as inversion and transformation, then estimation of quantities of interest empirically by direct simulation in a basic Monte Carlo strategy is also problematic. In general, Markov chain Monte Carlo methods (hereafter referred to as MCMC methods) and Sequential Monte Carlo (hereafter referred to as SMC) methods can be used in such settings where direct sampling and basic Monte Carlo procedures will not be possible.

A range of standard as well as more advanced MCMC and SMC methods of relevance are presented here, and fundamentally all approaches to be discussed aim to achieve the same common goal of obtaining samples efficiently from the posterior distribution, which are as close to independent as possible. The biggest difference between each approach to be discussed relates to the accuracy that one can perform such inference given a computational budget such as total samples, total simulation time, etc.

Typically, there will be a trade-off for practitioners related to the complexity of the sampling algorithm they wish to consider versus the reduction in variance in estimated quantities of interest. Therefore, we provide a range of methods one may consider suiting those interested in very simple approaches with nonrestrictive simulation budgets through to those looking for state-of-the-art sampling methods that will provide sample estimates accurately for constrained computational budgets or make challenging goals such as rare event estimation possible in reasonable computational budgets.

Given the different estimation goals discussed, there is no unique way to present the following sampling algorithms based on Markov chain and Importance Sampling methodologies. Therefore, throughout the following sections, we will present the sampling problems in general as generating a sequence of L samples $\boldsymbol{\Theta}^{(1)}, \boldsymbol{\Theta}^{(2)}, \ldots, \boldsymbol{\Theta}^{(L)}$, which will be understood in the following algorithmic descriptions to be generic random vectors that will correspond to quantities of interest in the inference, which are not necessarily always static LDA model parameters as will be demonstrated in the examples provided.

7.4.1 MOTIVATION FOR MARKOV CHAIN METHODS

In the following sections, we will generally detail examples for the most common situation encountered in practice which involves obtaining samples from a posterior conditional distribution for the OpRisk model parameters, given observed loss data. However, we note with some examples alternative situations that are also of interest and can be addressed by the Monte Carlo methods presented; these include situations in which we may assume we know the model parameters and our target is to sample the distribution of the annual loss, or joint distribution of multiple risk annual losses to obtain estimates of tail functionals, such as would be required in risk measure estimation for capital purposes.

Note: we make the following note on notation used throughout this book, regarding the dirac-delta function. In general we will utilise a range of representations for this special function including $\delta(x - x_0)$, δ_{x_0} and $\delta_{x_0}(x)$ which will be all variations on presenting the same dirac-delta function. In addition, we will also utilise $\delta_{x_0}(dx)$ to represent the dirac-delta measure.

First, consider the generic question of quantification of the probability of a particular event or set of events that are measurable with respect to outcomes of the model, denoted by $A \subset \mathbb{R}^d$ some measurable subset of the support of the posterior for the OpRisk model. Now consider computation of quantities such as $\mathbb{P}\mathrm{r}[\boldsymbol{\Theta} \in A] = \int_A \pi\left(\boldsymbol{\Theta}|\boldsymbol{x}_{1:T}\right) d\boldsymbol{\Theta}$.

Consider a sequence of samples $(\boldsymbol{\Theta}^{(i)})_{1\leq i\leq L}$ of independent copies of the random variable $\boldsymbol{\Theta}$. In this situation, the traditional Monte Carlo approximation of quantities such as $\mathbb{P}\mathrm{r}[\boldsymbol{\Theta} \in A]$ (which is the most simple special case of the inference problems previously defined) is given by the empirical measures with L samples

$$\hat{\pi}\left(\boldsymbol{\Theta}|\boldsymbol{x}_{1:T}\right) = \frac{1}{L}\sum_{1\leq i\leq L} \delta_{\boldsymbol{\Theta}^{(i)}} \longrightarrow \pi\left(\boldsymbol{\Theta}|\boldsymbol{x}_{1:T}\right), \quad L \to \infty.$$

Under this most basic of Monte Carlo estimators, the convergence of the empirical measure is understood as a weak convergence of empirical measures in the following sense for any bounded and measurable test function ψ on $\mathbb{R}^d$:

$$\begin{aligned}\hat{\pi}(\psi) &:= \int \psi(\boldsymbol{\Theta})\, \hat{\pi}\left(d\boldsymbol{\Theta}|\boldsymbol{x}_{1:T}\right) = \frac{1}{L}\sum_{1\leq i\leq L} \psi\left(\boldsymbol{\Theta}^{(i)}\right) \\ \longrightarrow \pi(\psi) &:= \int \psi(\boldsymbol{\Theta})\, \pi\left(d\boldsymbol{\Theta}|\boldsymbol{x}_{1:T}\right) = \mathbb{E}_{\pi(\boldsymbol{\Theta}|\boldsymbol{x}_{1:T})}(\psi(\boldsymbol{\Theta})), \quad L \to \infty.\end{aligned} \tag{7.37}$$

It is often highly informative for practitioners to consider the marginal behavior of sub-blocks of the parameter vector $\boldsymbol{\Theta} \in \mathbb{R}^d$. From the samples obtained under a basic Monte Carlo strategy, we observe that through the use of indicator functions for cells in $\mathbb{R}^d$, one can study visualizations for the shape of the posterior distribution marginally simply by plotting the histograms of the samples $\boldsymbol{\Theta}^{(i)}$ in every dimension.

In the specific choice of test functions given by the indicator function on a set of events A denoted $\psi = \mathbb{I}_A$, in the notational convention, the empirical measure of the posterior $\hat{\pi}\left(\mathbb{I}_A|\boldsymbol{x}_{1:T}\right)$ and the true posterior measure $\pi\left(\mathbb{I}_A|\boldsymbol{x}_{1:T}\right)$ is denoted by $\hat{\pi}\left(A|\boldsymbol{x}_{1:T}\right)$ and $\pi\left(A|\boldsymbol{x}_{1:T}\right)$. Hence, for indicator function $\psi = \mathbb{I}_A$, one has by the a.s. convergence of the empirical measure of the test function ψ, in Equation 7.37, the resulting empirical estimator is an unbiased estimator given by,

$$\mathbb{E}\left[\hat{\pi}\left(A|\boldsymbol{x}_{1:T}\right)\right] = \pi\left(A|\boldsymbol{x}_{1:T}\right), \tag{7.38}$$

and a variance in this estimator given by,

$$\mathbb{V}\text{ar}\left(\hat{\pi}\left(A|\boldsymbol{x}_{1:T}\right)\right) = \frac{1}{L}\,\pi\left(A|\boldsymbol{x}_{1:T}\right)\left(1 - \pi\left(A|\boldsymbol{x}_{1:T}\right)\right). \tag{7.39}$$

This assumes that these samples are attainable through techniques such as generic methods based on the inverse transform exact sampling, or accept–reject Monte Carlo sampling methods; (see Glasserman 2004, section 2.2).

Corollary 7.1 (The inverse transform) *If $U \sim Uniform(0, 1)$, then the distribution of the random variable $X = F^{-1}(U)$ is $F(x)$.*

That is, to simulate X from the distribution $F(x)$ using the inverse transform, generate $U \sim Uniform(0, 1)$ and calculate $X = F^{-1}(U)$.

Corollary 7.2 *Simulating X from the density $f(x)$ is equivalent to simulating (X, U) from the uniform distribution on (x, u), where $0 \leq u \leq f(x)$.*

This means that to simulate X from the density $f(x)$, generate (X, U) from the uniform distribution under the curve of $f(x)$. The latter is typically done through the accept–reject algorithm (sometimes called rejection sampling).

Corollary 7.3 (Accept–reject method) *Assume that the density $f(x)$ is bounded by M (i.e., $f(x) \leq M$) and defined on the support $a \leq x \leq b$. Then, to simulate X with the density $f(x)$*

- *Draw $X \sim Uniform(a, b)$ and $U \sim Uniform(0, M)$;*
- *Accept the sample of X if $U \leq f(X)$, otherwise repeat the previous steps.*

If another density $g(x)$ such that $Mg(x) \geq f(x)$ can be found for constant M, then to simulate X with the density $f(x)$

- *Draw X from $g(x)$ and $U \sim Uniform(0, Mg(X))$;*
- *Accept the sample of X if $U \leq f(X)$, otherwise repeat the previous steps.*

The inverse method cannot be used if the normalization constant is unknown, and the accept–reject method cannot be used if you cannot easily find the bounds for the density or one is working in high dimension where rejection probabilities may be high and will scale nonlinearly with the dimension of the parameter space of the posterior. These difficulties are often arise for posterior densities; therefore, unfortunately in practice, one cannot typically easily generate i.i.d. samples from the target posterior distribution, due to the intractability of the inverse of the distribution function (i.e., the quantile function is not known in closed form). In such cases, which incidentally correspond to the majority of situations in practice, one must resort to alternative statistical approaches to provide samples. There are numerous such examples, which include Importance Sampling (IS), MCMC, SMC, Sequential Monte Carlo Samplers (SMC Samplers), Particle Markov chain Monte Carlo (PMCMC), and their adaptive versions. Each of these classes of algorithms is significantly different in its attributes and in the problems for which it is appropriate to utilize each one when making inference in OpRisk modeling. It is the intention of the following sections to introduce practitioners to a subset of the many possible choices that are selected as their performance is efficient and widely applicable for the types of problems discussed in the context of OpRisk Bayesian modeling. We begin with MCMC general properties, then present a special case of what is known as an auxiliary variable sampler illustrated by the Slice sampler of Neal (2003). This is widely applicable to many Bayesian inference problems and is now a standard package in statistical software such as R and Matlab.

For a good introduction on estimation (sampling) of the posterior $\pi(\boldsymbol{\theta}|\boldsymbol{x})$ numerically using MCMC methods, see Robert and Casella (2004). MCMC has almost unlimited applicability though its performance depends on the problem particulars. The idea of MCMC methods is based on a simple observation that to obtain an acceptable approximation to some integrals depending on a distribution of interest $\pi(\boldsymbol{\theta}|\boldsymbol{x})$, it is enough to sample a sequence (Markov chain) $\{\boldsymbol{\theta}^{(1)}, \boldsymbol{\theta}^{(2)}, \ldots\}$, whose limiting density is the density of interest $\pi(\boldsymbol{\theta}|\boldsymbol{x})$. This idea appeared as early as the original Monte Carlo method but became very popular and practical in the last few decades only when fast computing platforms became available.

A Markov chain is a sequence of random variables defined as follows.

Definition 7.11 (Markov chain) *A sequence of random variables,*

$$\{\boldsymbol{\Theta}^{(0)}, \boldsymbol{\Theta}^{(1)}, \ldots, \boldsymbol{\Theta}^{(l)}, \ldots\},$$

is a first-order Markov chain *if, for any l, the conditional distribution of $\boldsymbol{\Theta}^{(l+1)}$ given $\boldsymbol{\Theta}^{(i)}$, $i = 0, 1, \ldots, l$ is the same as the conditional distribution of $\boldsymbol{\Theta}^{(l+1)}$ given $\boldsymbol{\Theta}^{(l)}$. A conditional probability density of $\boldsymbol{\Theta}^{(l+1)}$ given $\boldsymbol{\Theta}^{(l)}$ is called* transition kernel *of the chain and is usually denoted as $K(\boldsymbol{\Theta}^{(l)}, \boldsymbol{\Theta}^{(l+1)})$.* ■

Remark 7.8 *The challenge is to construct a Markov chain sampling procedure that from any location in the state space will produce samples (as close to uncorrelated and i.i.d. as possible) for a given distribution of interest–either a posterior distribution for model parameters or, for example, a distribution for the annual loss of a single risk process or multiple, dependent risk processes.*

One way to achieve such goals is to utilize an MCMC approach. MCMC methods produce an *ergodic Markov chain* with a *stationary distribution* (which is also a *limiting distribution*) that is designed to match the distribution one wishes to obtain samples from. These chains are also *recurrent* and *irreducible* (see details in Meyn *et al.* 2009 or Robert and Casella 2004). In simpler terms, one would like to design MCMC samplers that have certain desirable properties related to their mixing rates; crudely put, the ability of the Markov chain to explore the state space of the target distribution from any initial starting point, and how quickly the chain reaches the stationary regime, i.e. obtains samples from the desired distribution of interest. In addition, it would be nice to be able to rely upon the convergence of MCMC sample estimators of functionals on the state space as provided by some form of ergodic theorem. In addition to this, it is often of interest to ascertain knowledge around the accuracy and rates of convergence such as through the existence of the Central Limit Theorem (CLT) and knowledge of the asymptotic variance (see discussions by Jones 2004 and the references therein).

Theorem 7.4 (Strong Law of Large Numbers) *Consider a sequence of non-negative random variables $Y_1, Y_2, \ldots$, which are i.i.d. with a mean $\mathbb{E}[Y_1] = \mu$. Then the following convergence in probability holds for all $\epsilon > 0$:*

$$\lim_{L\to\infty} \Pr\left[\left|\frac{Y_1 + Y_2 + \cdots + Y_L}{L} - \mu\right| > \epsilon\right] = 0. \tag{7.40}$$

With this theorem in mind, consider the definition of the ergodic theorem for an MCMC sampler. Ergodic theorems concern the limiting behavior of averages over time; therefore, an ergodic theorem is basically a result describing the limiting behavior of a sequence such as

$$\frac{1}{L}\sum_{l=0}^{L-1} f\left(X^{(l)}\right) \tag{7.41}$$

as $L \to \infty$. The formulation of the ergodic theorem considered in any given example will depend on the class of functions f (e.g. integrable, $L^2, \ldots$ continuous), and the notion of convergence used (e.g., pointwise convergence, L^2 convergence, ...uniform convergence). In the case of pointwise ergodic theorems, one would typically consider the Birkhoff Khinchin theorem (see, e.g., Kornfeld *et al.* 1982); in the case of a mean ergodic theorem for Hilbert spaces, one may consider Von Neumann's mean ergodic theorem (see, e.g., Birkhoff 1931 and Cohen 1940 and references therein). For a summary, see the book-length review of Krengel and Brunel (1985).

For simplicity consider first a discrete state space and discrete time setting, and define the proportion of time that the Markov chain spends in a state i before L is achieved, denoted by $T_i(L)$ and defined by the sum of indicator events

$$T_i(L) = \sum_{l=0}^{L-1} \mathbb{I}\left[X^{(l)} = i\right]. \tag{7.42}$$

Then, if we normalize this sequence by the length of the chain L to get the proportion of time spent in a state i, we can state one version of the ergodic theorem as follows, based on definitions of irreducibility given by, for example, Meyn *et al.* (2009).

Theorem 7.5 (Ergodic Theorem: Discrete Time and Discrete State Space) *Consider a transition matrix for a Markov chain, denoted by P, which is irreducible, and let π be any distribution from which we wish to obtain samples. Furthermore, we assume that the Markov chain $\left(X^{(L)}\right)_{L\geq 0}$ that corresponds to this Markov(P, π) structure has for all $\epsilon > 0$ the property*

$$\lim_{L\to\infty} \Pr\left[\left|\frac{T_i(L)}{L} - \frac{1}{\mu_i}\right| > \epsilon\right] = 0, \tag{7.43}$$

where $\mu_i = \mathbb{E}\left[T_i\right]$ is the expected return time to state i. Then, if the chain is also positive recurrent, one can state that for any bounded function defined on the state space $X \in \mathcal{X}$ given by $f : \mathcal{X} \to \mathbb{R}$, one has for all $\epsilon > 0$ the following convergence in probability:

$$\lim_{n\to\infty} \Pr\left[\left|\frac{1}{L}\sum_{l=0}^{L-1} f\left(X^{(l)}\right) - \sum_{\mathcal{X}} f\left(x^{(l)}\right)\pi\left(x^{(l)}\right)\right| > \epsilon\right] = 0, \tag{7.44}$$

where π is the unique invariant distribution of the Markov chain.

In OpRisk, we are typically interested in more general state space settings, such as continuous supports. So we generalize the previous notions to consider a general Markov transition kernel $P(x, dy)$ on a general state space $(\mathcal{X}, \mathbb{B}(\mathcal{X}))$ with Borel sigma algebra $\mathcal{B}(\mathcal{X})$ for an associated discrete time Markov chain $\boldsymbol{X} = \left(X^{(l)}\right)_{l\geq 0}$. Furthermore, we denote the l-step Markov transition by $P^{(l)}(x, dy)$ and then for $i \in \mathbb{L}$, $x \in \mathcal{X}$, and a measurable set A, we can define $P^{(l)}(x, A) = \mathbb{P}\text{r}\left(X^{(l+i)} \in A | X^{(i)} = x\right)$. If we then consider the class of Borel functions given by $f : \mathcal{X} \rightarrow \mathbb{R}$, then the following operator notation is well defined: $Pf(x) = \int f(y)P(x, dy)$ and $\triangle f(x) = Pf(x) - f(x)$. Furthermore, if we assume the Markov chain $\boldsymbol{X}$ to be Harris ergodic, that is, aperiodic, ψ-irreducible, and positive Harris recurrent (see Meyn *et al.* 2009) with invariant distribution π on some general state space $\mathcal{X}$, then these assumptions are sufficient to show the strong convergence in total variation norm, given for every initial distribution $\nu(\cdot)$ on $\mathbb{B}(\mathcal{X})$ as $l \rightarrow \infty$, by

$$||P^{(l)}(\nu, \cdot) - \pi(\cdot)||_{TV} \rightarrow 0, \tag{7.45}$$

where we define $P^{(l)}(\lambda, A) = \int_{\mathcal{X}} P^{(l)}(x, A)\nu(dx)$ and $\|\cdot\|_{TV}$ denotes the Total variation norm between probability measures, given in Definition 7.12.

Definition 7.12 (Total Variation Distance Norm) *Consider two probability measures μ and ν. Then the total variation distance between probability measures μ and ν can be defined by*

$$\|\mu - \nu\|_{TV} = \sup\left\{|\mu(A) - \nu(A)| : A \in \Sigma\right\} \tag{7.46}$$

and its values are non-trivial. ■

The meaning of this distance is that it measures the largest possible difference between the probabilities that the two probability distributions can assign to the same event. In the context of Markov chains, this will be the maximum difference between probabilities on all Borel measurable events arising from the l-step Markov chain with initial distribution ν and the target stationary distribution π.

In other words, a weaker version of this result tells us that if we consider a class of Borel functions f that we use to define a sample average, from L samples,

$$\overline{f}_L = L^{-1} \sum_{l=0}^{L-1} f\left(X^{(l)}\right)$$

and the target functional $\mathbb{E}_\pi[f] = \int_{\mathcal{X}} f(x)\pi(dx)$. Then, if we assume that $\mathbb{E}_\pi\left[|f|\right] < \infty$, a generalization of the previous ergodic theorem will guarantee that our sample average will converge $\overline{f}_L \rightarrow \mathbb{E}_\pi[f]$ with probability 1 as $L \rightarrow \infty$.

To extend these results by considering the *rate* at which these sample estimators, constructed by the Markov chain samples, will converge as well as the accuracy of such sample estimators, one needs to consider the existence of a Central Limit Theorem (CLT). The CLT condition or result is important as it would state the following convergence in distribution holds asymptotically in the length of the Markov chain (i.e. number of samples L), according to

$$\sqrt{L}\left(\overline{f}_L - \mathbb{E}_\pi[f]\right) \xrightarrow{d} Normal(0, \sigma_f^2) \tag{7.47}$$

as $L \to \infty$ and where the asymptotic variance is given by

$$\sigma_f^2 := \mathbb{V}\mathrm{ar}_\pi \left[f(X_0)\right] + 2\sum_{i=1}^{\infty} \mathbb{C}\mathrm{ov}_\pi \left[f(X_0) f(X_i)\right] < \infty. \tag{7.48}$$

The existence of a CLT result is far from certain, however it is highly informative and crucial to sensible implementation of MCMC methods, since knowledge of the behaviour of the sample average, for function f, using MCMC samples, informs directly the performance of $\overline{f}_L$ as an estimator of $\mathbb{E}_\pi[f]$ and its accuracy in large samples through σ_f^2. As noted, in developing a CLT result, generally the first step is to verify its existence which is typically obtained through consideration of discussions on minorization conditions for the Markov transition kernel. Briefly, one can consider conditions on the Markov chain that would result in some form of bound on the following total variation norm

$$||P^{(l)}(x,\cdot) - \pi(\cdot)||_{TV} \leq M(x)g(n) \tag{7.49}$$

for some non-negative function $M(x)$ and a non-negative decreasing function $g(l)$ for $l \in \mathbb{Z}^+$. That is some bound on the maximum difference between probabilities assigned to all measurable events from the Markov chain after l iterations and the target stationary distribution π. If this upper bound is in the form of a decreasing function of the length of the chain l then one can utilise this convergence rate knowledge to verify the existance of a CLT result. Typically, one would consider cases such as geometric ergodicity of the Markov chain $\boldsymbol{X}$ in which $g(l) = t^l$ for some $t < 1$, or uniform ergodicity in which M is bounded and $g(l) = t^l$ again for some $t < 1$, or alternatively polynomial ergodicity on the order of $m \geq 0$, where $g(l) = l^{-m}$. These minorization conditions can be translated to conditions on the Markov chain transition kernel. For instance, a minorization condition will hold for a particular set A if there exists a probability measure, say Q, taking support on the Borel sets $\mathbb{B}(X)$ such that for some positive integer l_0 and a positive constant ϵ on has the lower bound

$$P^{(l)}(x,A) \geq \epsilon Q(A), \quad \forall x \in A, \quad A \in \mathcal{B}(X). \tag{7.50}$$

From this minorization condition on the transition kernel of the Markov chain one typically then formulates some form or drift condition to verify different forms of ergodicity mixing, such as polynomial, uniform or geometric. The minorization conditions required of a Markov chain to achieve these convergence results and the implications on the resulting sequence of estimators with regard to the existence of a CLT are summarized in the tutorial article of Jones (2004) and the references therein. In particular from the related drift conditions there are known results relating to the existence of the CLT for a Markov chain, see (Jones, 2004, theorem 1).

Throughout the presentation of this section, we assume that one is utilizing MCMC methods to sample from a target distribution, which, for illustration purposes, is going to be the posterior distribution of the model parameters for an LDA risk framework given observed losses and the total number of losses over time. Of course, as already discussed, this can be more general to also include cases in which we just wish to sample from complex distributions efficiently; to highlight this fact, we will also provide several examples of such applications throughout the following presentation.

For the purposes of this book, we make the following further general remark regarding the types of Markov chain methods that will be discussed and developed in future sections of this chapter.

Remark 7.9

- *We are interested in the case where the chain stationary distribution corresponds to, for example, the posterior density $\pi(\boldsymbol{\theta}|\boldsymbol{x})$ or perhaps, as a second example, the distribution of multiple risk processes $\pi(Z^{(1)}, Z^{(2)}, \ldots, Z^{(d)})$ for which we know the parameters and wish to obtain samples from this multivariate intractable distribution subject to tail dependence features after restriction to a joint tail event;*
- *The* ergodic *property means that the distribution of $\boldsymbol{\Theta}^{(l)}$ converges to a* limiting distribution *$\pi(\boldsymbol{\theta}|\boldsymbol{x})$ for almost any starting value of $\boldsymbol{\Theta}^{(0)}$. Therefore, for large l, $\boldsymbol{\Theta}^{(l)}$ is approximately distributed from $\pi(\boldsymbol{\theta}|\boldsymbol{x})$ regardless of the starting point. Of course, the problem is to decide what is large l. This can formally be accomplished by running diagnostic tests on the stationarity of the chain;*
- *A Markov chain is said to have a* stationary distribution *if there is a distribution $\pi(\boldsymbol{\theta}|\boldsymbol{x})$ such that if $\boldsymbol{\Theta}^{(l)}$ is distributed from $\pi(\boldsymbol{\theta}|\boldsymbol{x})$, then $\boldsymbol{\Theta}^{(l+1)}$ is distributed from $\pi(\boldsymbol{\theta}|\boldsymbol{x})$ too;*
- *A Markov chain is* irreducible *if it is guaranteed to visit any set $\mathcal{A}$ of the support of $\pi(\boldsymbol{\theta}|\boldsymbol{x})$. This property implies that the chain is* recurrent, *that is, that the average number of visits to an arbitrary set $\mathcal{A}$ is infinite and even* Harris recurrent. *The latter means that the chain has the same limiting behavior for* every *starting value rather than* almost every *starting value;*
- *Markov chains considered in MCMC algorithms are almost always* homogeneous, *that is, the distribution of $\boldsymbol{\Theta}^{(l_0+1)}, \boldsymbol{\Theta}^{(l_0+2)}, \ldots, \boldsymbol{\Theta}^{(l_0+k)}$ given $\boldsymbol{\Theta}^{(l_0)}$ is the same as the distribution of $\boldsymbol{\Theta}^{(1)}, \boldsymbol{\Theta}^{(2)}, \ldots, \boldsymbol{\Theta}^{(k)}$ given $\boldsymbol{\Theta}^{(0)}$ for any $l_0 \geq 0$ and $k > 0$. We detail a few special cases of adaptive MCMC algorithms in which we do not make this assumption; for a general overview of relevance to OpRisk, see Del Moral et al. (2013);*
- *Another important stability property is called* reversibility, *which means that the direction of the chain does not matter. That is, the distribution of $\boldsymbol{\Theta}^{(l+1)}$ conditional on $\boldsymbol{\Theta}^{(l+2)} = \boldsymbol{\theta}$ is the same as the distribution of $\boldsymbol{\Theta}^{(l+1)}$ conditional on $\boldsymbol{\Theta}^{(l)} = \boldsymbol{\theta}$. The chain is* reversible *if the transition kernel satisfies the* detailed balance condition

$$K(\boldsymbol{\theta}, \boldsymbol{\theta}')\pi(\boldsymbol{\theta}|\boldsymbol{x}) = K(\boldsymbol{\theta}', \boldsymbol{\theta})\pi(\boldsymbol{\theta}'|\boldsymbol{x}). \tag{7.51}$$

The detailed balance condition is not necessary but sufficient condition for $\pi(\boldsymbol{\theta}|\boldsymbol{x})$ to be stationary density associated with the transitional kernel $K(\cdot\ ,\ \cdot)$, which usually can easily be checked for MCMC algorithms.

Of course, the samples $\boldsymbol{\Theta}^{(1)}, \boldsymbol{\Theta}^{(2)}, \ldots$ are not independent. However, the independence is not required if we have to calculate some functionals of $\pi(\boldsymbol{\theta}|\boldsymbol{x})$, because the Ergodic theorem implies that for large L, the average

$$\frac{1}{L}\sum_{l=1}^{L} g(\boldsymbol{\Theta}^{(l)}) \tag{7.52}$$

converges to $\mathbb{E}[g(\boldsymbol{\Theta})|\boldsymbol{X} = \boldsymbol{x}]$ (if this expectation is finite), where expectation is calculated with respect to $\pi(\boldsymbol{\theta}|\boldsymbol{x})$.

In the following we illustrate a few example functionals that typically arise in practice for OpRisk settings. Consider multiple risk processes denoted by $Z^{(1)}, Z^{(1)}, \ldots, Z^{(d)}$. Assume that each risk process $Z^{(i)}$ satisfies marginally an LDA structure with frequency distribution

$N^{(i)} \sim F_{N^{(i)}}(n)$ and severity distribution for the j-th loss $X_j^{(i)} \sim F_{X_j^{(i)}}(x)$. Then assume that the d risk processes are dependent upon each other and this dependence is modeled in the following fashion where jointly the annual loss d-variate random vector has distribution $\left(Z^{(1)}, \ldots, Z^{(d)}\right) \sim C\left(F_{Z^{(1)}}(z_1), \ldots, F_{Z^{(d)}}(z_d)\right)$. Given this multivariate risk process model, OpRisk practitioners may be interested in obtaining samples (e.g., by MCMC methods) from the joint density given by

$$f_{Z^{(1)},\ldots,Z^{(d)}}(z_1, \ldots, z_d) = c\left(F_{Z^{(1)}}(z_1), \ldots, F_{Z^{(d)}}(z_d)\right) \prod_{i=1}^{d} f_{Z^{(i)}}(z_i), \tag{7.53}$$

These L samples, $\left\{\left(Z^{(1)}, \ldots, Z^{(d)}\right)^{(l)}\right\}_{l=1:L}$ can then be utilized to estimate quantities such as the following:

- **Joint tail functionals.** In this case for some measurable and bounded test function $\varphi:\mathbb{R}^{+d} \mapsto \mathbb{R}$ one has the Monte Carlo estimator

$$\int_{\mathbb{R}+} \cdots \int_{\mathbb{R}+} \varphi(z_1, \ldots, z_d) f_{Z^{(1)},\ldots,Z^{(d)}}(z_1, \ldots, z_d) dz_1 \ldots dz_d \approx \frac{1}{L} \sum_{l=1}^{L} \varphi\left(\left[z^{(1)}, \cdots, z^{(d)}\right]^{(l)}\right). \tag{7.54}$$

  Examples of such functionals include marginal distributions, conditional distributions, tail functionals such as quantiles, and tail expectations;
- **Conditional constrained tail functionals:** In many settings, one is interested in calculating for some measurable and bounded test function $\varphi : \mathbb{R}^{+d} \mapsto \mathbb{R}$ a Monte Carlo estimator of quantities such as

$$\int_{\mathbb{R}+} \cdots \int_{\mathbb{R}+} \varphi(z_1, \ldots, z_d) f_{Z^{(1)},\ldots,Z^{(d)}|R\left(Z^{(1)},\ldots,Z^{(d)}\right)}\left(z_1, \ldots, z_d \middle| R\left(Z^{(1)}, \ldots, Z^{(d)}\right)\right) dz_1 \ldots dz_d$$
$$\approx \frac{1}{L} \sum_{l=1}^{L} \varphi\left(\left[z^{(1)}, \ldots, z^{(d)}\right]^{(l)}\right) \mathbb{I}_{R\left(\left[z^{(1)},\ldots,z^{(d)}\right]^{(l)}\right)}\left[\left(Z^{(1)}, \ldots, Z^{(d)}\right)\right], \tag{7.55}$$

  subject to some constraints on the joint sequence of losses denoted generically by a constraint function $R(Z^{(1)}, \ldots, Z^{(d)})$. Examples of constraints of interest in OpRisk settings include choices such as the following:
  - Example 1: to calculate marginal, joint, and groups of marginal tail functionals for the joint loss distribution restricted to certain tail events of interest,

$$R\left(Z^{(1)}, \ldots, Z^{(d)}\right) = \mathbb{I}_{Z^{(1)}>\rho_1,\ldots,Z^{(d)}>\rho_d}\left[\left(Z^{(1)}, \ldots, Z^{(d)}\right)\right]. \tag{7.56}$$

  - Example 2: constraints on linear combinations of the marginal annual losses would allow one to obtain estimators for functionals of jointly constrained risk processes under constraints such as

$$R\left(Z^{(1)}, \ldots, Z^{(d)}\right) = \sum_{i=1}^{d} Z^{(i)} > \rho_T. \tag{7.57}$$

There are several possibilities one may consider in this context, and in general this is relevant for insurance settings as well as coherent capital allocation methods.

- **Quantile function estimation**. Given samples $\left\{\left(Z^{(1)}, \ldots, Z^{(d)}\right)^{(l)}\right\}_{l=1:L}$ from the dependent joint risk process, one can transform these samples via a measurable and bounded test function $\varphi : \mathbb{R}^{+^d} \mapsto \mathbb{R}$ that corresponds to the new univariate samples $\left\{W^{(j)}\right\}_{l=1:L}$ with $W^{(j)} = \varphi\left(\left(Z^{(1)}, \ldots, Z^{(d)}\right)^{(j)}\right)$ having density denoted $f_W(w)$. Now sorting these samples to obtain the order statistics $\left\{W_{(l,L)}\right\}_{l=1:L}$ one can obtain an estimator of the resulting quantile via

$$\hat{q}_\alpha(W) = W_{(l,L)}, \quad \text{with } \frac{l-1}{L} < \alpha \leq \frac{l}{L}. \tag{7.58}$$

This is of relevance for risk measure estimations and capital estimation of, for example, the institutional capital, such as when one considers

$$\varphi\left(Z^{(1)}, \ldots, Z^{(d)}\right) = \sum_{i=1}^{d} Z^{(i)}.$$

Note that in this case we can state two things about the accuracy of such a quantile estimator (see discussions by Flegal *et al.* 2012) depending on whether the samples obtained are independent or autocorrelated. In the case of independent samples, one has the following convergence in distribution between the obtained sample estimated quantiles $\hat{q}_\alpha(W)$ and the target theoretical quantile $q_\alpha(W)$ at level α for the chosen constraint function φ, asymptoticly according to:

$$\sqrt{L}\left(\hat{q}_\alpha(W) - q_\alpha(W)\right) \rightarrow Normal\left(0, \frac{\alpha(1-\alpha)}{\left[f_W(q_\alpha(W))\right]^2}\right), \quad \text{as } L \rightarrow \infty. \tag{7.59}$$

If the samples are autocorrelated, as would typically be the case if the samples $\left\{\left(Z^{(1)}, \ldots, Z^{(d)}\right)^{(l)}\right\}_{l=1}^{L}$, used to construct the constrained samples $\left\{W^{(l)}\right\}_{l=1}^{L}$, were obtained by a Markov chain procedure, then Flegal *et al.* (2012, theorem 1) state that under polynomial mixing (of order 3) of the MCMC sampler, the following convergence in distribution is satisfied in Theorem 7.6.

Theorem 7.6 *If there exists an $\epsilon > 0$ such that W is polynomially ergodic of order $2.5 + \epsilon$, and if W has density satisfying that its derivative f'_W is positive and bounded in the neighborhood of the quantile $q_\alpha(W)$, then as $L \rightarrow \infty$, one obtains the following convergence in distribution:*

$$\sqrt{L}\left(\hat{q}_\alpha(W) - q_\alpha(W)\right) \rightarrow Normal\left(0, \frac{\sigma^2\left(q_\alpha(W)\right)}{\left[f_W(q_\alpha(W))\right]^2}\right), \quad as \quad L \rightarrow \infty. \tag{7.60}$$

with $\sigma^2(y)$ defined by

$$\begin{aligned}\sigma^2(y) &= \mathbb{V}\mathrm{ar}_{f\left(Z^{(1)},\ldots,Z^{(d)}\right)\left(Z^{(1)},\ldots,Z^{(d)}\right)}\left(\mathbb{I}\left[W^{(0)}<y\right]\right)\\ &\quad + 2\sum_{k=1}^{\infty}\mathbb{C}\mathrm{ov}_{f\left(Z^{(1)},\ldots,Z^{(d)}\right)\left(Z^{(1)},\ldots,Z^{(d)}\right)}\left(\mathbb{I}\left[W^{(0)}<y\right],\mathbb{I}\left[W^{(k)}<y\right]\right).\end{aligned} \tag{7.61}$$

Remark 7.10 *Note that this variance expression is then approximated using the MCMC samples in order to use this information to form confidence intervals for the resulting quantile estimator, which will be of the form $\hat{q}_\alpha(W) \pm t_* \frac{\sigma^2(q_\alpha(W))}{\sqrt{L}}$ for some desired t distributed confidence interval at level $*$ with t-score t_*. See examples in Flegal et al. (2012).*

Having defined the notions required to understand Markov chain methods at a rudimentary level, we now summarize how we utilize these concepts to proceed with MCMC algorithms (see further discussion by Roberts 1995). The MCMC approach constructs an ergodic Markov chain $\left\{\boldsymbol{\Theta}^{(1)},\ldots,\boldsymbol{\Theta}^{(L)}\right\}$, taking values in $\mathbb{R}^d$. This Markov chain is constructed to have the property that it has a limiting, invariant distribution is the target distribution of interest $\pi\left(d\boldsymbol{\Theta}|\boldsymbol{x}_{1:T}\right)$. This invariant distribution is the target distribution, that, in the cases considered in this chapter, will correspond to the posterior for the OpRisk model. For the Markov chain samples to be used as samples from the target distribution, it is necessary that there exist a unique invariant distribution of the Markov chain corresponding to the posterior of interest for the OpRisk model. A detailed review of the properties of more general state space Markov chain theory can be found in, for example, Meyn *et al.* (2009) and Del Moral (2004).

To achieve this construction of a Markov chain with desired stationary distribution, the majority of methods developed in the statistics literature have focused on the case in which the Markov chain created satisfies the condition of reversibility, whereby the following holds:

$$\pi\left(d\boldsymbol{\Theta}^{(i)}|\boldsymbol{x}_{1:T}\right)Q(\boldsymbol{\Theta}^{(i)},d\boldsymbol{\Theta}^{(j)}) = \pi\left(d\boldsymbol{\Theta}^{(j)}|\boldsymbol{x}_{1:T}\right)Q(\boldsymbol{\Theta}^{(j)},d\boldsymbol{\Theta}^{(i)}), \tag{7.62}$$

where $\boldsymbol{\Theta}^{(i)}$ and $\boldsymbol{\Theta}^{(j)}$ represent states of the Markov chain and $Q(\boldsymbol{\Theta}^{(j)},d\boldsymbol{\Theta}^{(i)})$ denotes the Markov transition representing the probability of starting in state $\boldsymbol{\Theta}^{(j)}$ and transition to a neighborhood of the state $\boldsymbol{\Theta}^{(i)}$.

Under this condition, there is a wide range of methods that one may utilize to construct the desired Markov chain, which in a large number of instances involves the careful design of the transition kernel $Q(\boldsymbol{\Theta}^{(i)},d\boldsymbol{\Theta}^{(j)})$. The transition kernel for the class of MCMC methods of interest in this section is typically given by

$$\begin{aligned}Q\left(\boldsymbol{\Theta}^{(l)},d\boldsymbol{\Theta}^{(l+1)}\right) &= q\left(\boldsymbol{\Theta}^{(l)},d\boldsymbol{\Theta}^{(l+1)}\right)\alpha\left(\boldsymbol{\Theta}^{(l)},d\boldsymbol{\Theta}^{(l+1)}\right)\\ &\quad + \left[1-\int q\left(\boldsymbol{\Theta}^{(l)},\boldsymbol{z}\right)\alpha\left(\boldsymbol{\Theta}^{(l)},\boldsymbol{z}\right)d\boldsymbol{z}\right]\mathbb{I}\left[\boldsymbol{\Theta}^{(l+1)}=\boldsymbol{\Theta}^{(l)}\right],\end{aligned} \tag{7.63}$$

where the design of transition density $q\left(\boldsymbol{\Theta}^{(l)},d\boldsymbol{\Theta}^{(l+1)}\right)$ is of direct interest for reducing variance in Monte Carlo estimates. The first component

$$q\left(\boldsymbol{\Theta}^{(l)},d\boldsymbol{\Theta}^{(l+1)}\right)\alpha\left(\boldsymbol{\Theta}^{(l)},d\boldsymbol{\Theta}^{(l+1)}\right) \tag{7.64}$$

corresponds to the probability of starting in a state $\boldsymbol{\Theta}^{(l)}$ at iteration l of the Markov chain and moving to some state $\boldsymbol{\Theta}^{(l+1)}$ with acceptance of such a proposed sampled move typically denoted generically by acceptance probability $\alpha\left(\boldsymbol{\Theta}^{(l)}, d\boldsymbol{\Theta}^{(l+1)}\right)$. The remainder of the kernel, that is, the second term, corresponds to rejecting such a proposed move and remaining in the current state for the next iteration of the Markov chain.

Algorithms of this form are generally considered as special cases of the general framework established by Metropolis *et al.* (1953) and extended by Hastings (1970). It is instructive to first present the basic Metropolis–Hastings (MH) MCMC sampler and the univariate Gibbs sampler prior to explaining how more recent advances in these methods can be utilized effectively for OpRisk model inference.

7.4.2 METROPOLIS–HASTINGS ALGORITHM

The MH algorithm is almost a universal algorithm used to generate a Markov chain with a stationary distribution $\pi(\boldsymbol{\theta}|\boldsymbol{x})$. It has been developed by Metropolis *et al.* (1953) in mechanical physics and generalized by Hastings (1970) in a statistical setting. It can be applied to a variety of problems since it requires the knowledge of the distribution of interest up to a normalizing constant only, i.e. as is typically the case in practice, the normalizing constant does not need to be known for the posterior for one to apply the following methods. Given a density $\pi(\boldsymbol{\theta}|\boldsymbol{x})$, known up to a normalization constant, and a conditional density $q(\boldsymbol{\theta}^*|\boldsymbol{\theta})$, the method generates the chain $\{\boldsymbol{\theta}^{(1)}, \boldsymbol{\theta}^{(2)}, \dots\}$ using the following algorithm.

Algorithm 7.2 (Metropolis–Hastings algorithm)

1. *Initialize* $\boldsymbol{\theta}^{(l=0)}$ *with any value within a support of* $\pi(\boldsymbol{\theta}|\boldsymbol{x})$;
2. *For* $l = 1, \dots, L$
   a) *Set* $\boldsymbol{\theta}^{(l)} = \boldsymbol{\theta}^{(l-1)}$;
   b) *Generate a proposal* $\boldsymbol{\theta}^*$ *from* $q(\boldsymbol{\theta}^*|\boldsymbol{\theta}^{(l)})$;
   c) *Accept a proposal with the acceptance probability*

$$\alpha(\boldsymbol{\theta}^{(l)}, \boldsymbol{\theta}^*) = \min\left\{1, \frac{\pi(\boldsymbol{\theta}^*|\boldsymbol{x})q(\boldsymbol{\theta}^{(l)}|\boldsymbol{\theta}^*)}{\pi(\boldsymbol{\theta}^{(l)}|\boldsymbol{x})q(\boldsymbol{\theta}^*|\boldsymbol{\theta}^{(l)})}\right\}, \tag{7.65}$$

   that is, simulate U *from the uniform distribution function* $Uniform(0, 1)$ *and set* $\boldsymbol{\theta}^{(l)} = \boldsymbol{\theta}^*$ *if* $U < \alpha(\boldsymbol{\theta}^{(l)}, \boldsymbol{\theta}^*)$. *Note that the normalization constant of the posterior does not contribute here.*
3. *Next* l *(i.e., do an increment,* $l = l + 1$*, and return to step 2).*

Remark 7.11

- *The density* $\pi(\boldsymbol{\theta}|\boldsymbol{x})$ *is called the* target *or* objective density*;*
- $q(\boldsymbol{\theta}^*|\boldsymbol{\theta})$ *is called the* proposal density *and will be discussed shortly.*

For further discussions on this algorithm and tutorials on its properties and variants, see examples by Chib and Greenberg (1995), Gilks *et al.* (1996), and Andrieu *et al.* (2003).

7.4.3 GIBBS SAMPLER

The Gibbs sampler is a technique for generating random variables from a distribution indirectly, without having to calculate the density. The method takes its name from the Gibbs random fields in image-processing models starting with the paper of Geman and Geman (1984). Its roots can be traced back to the 1950s; see Robert and Casella (2004) for a brief summary of the early history.

To illustrate the idea of the Gibbs sampler, consider the case of two random variables X and Y that have a joint bivariate density $\pi(x, y)$. Assume that simulation of X from $\pi(x)$ cannot be done directly but we can easily sample X from the conditional density $\pi(x|y)$ and Y from the conditional density $\pi(y|x)$. Then, the Gibbs sampler generates samples as follows.

Algorithm 7.3 (Gibbs sampler, bivariate case)

1. *Initialize* $y^{(l=0)}$ *with an arbitrary value within a support of* Y;
2. *For* $l = 1, \ldots, L$
   (a) *Simulate* $x^{(l)}$ *from* $\pi\left(x|y^{(l-1)}\right)$;
   (b) *Simulate* $y^{(l)}$ *from* $\pi\left(y|x^{(l)}\right)$.
3. *Next* l *(i.e., do an increment,* $l = l + 1$*, and return to step 2).*

Under quite general conditions, $\pi(x, y)$ is a stationary distribution of the chain $\{(x^{(l)}, y^{(l)}), l = 1, 2, \ldots\}$; and the chain is ergodic with a limiting distribution $\pi(x, y)$, that is, the distribution of $X^{(l)}$ converges to $\pi(x)$ and the distribution of $Y^{(l)}$ converges to $\pi(y)$ for large l.

Gibbs sampling can be thought of as a practical implementation of the fact that knowledge of the conditional distributions is sufficient to determine a joint distribution (if it exists!). The generalization of the Gibbs sampling to a multidimensional case is as follows. Consider a random vector $\boldsymbol{X}$ with a joint density $\pi(\boldsymbol{x})$. Denote full conditionals $\pi_i(x_i|\boldsymbol{x}_{-i}) = \pi(x_i|x_1, \ldots, x_{i-1}, x_{i+1}, \ldots, x_N)$. Then, do the following steps.

Algorithm 7.4 (Gibbs sampler, multivariate case)

1. *Initialize* $x_2^{(l=0)}, \ldots, x_N^{(l=0)}$ *with an arbitrary value;*
2. *For* $l = 1, \ldots, L$
   1) *Simulate* $x_1^{(l)}$ *from* $\pi_1\left(x_1|x_2^{(l-1)}, \ldots, x_N^{(l-1)}\right)$;
   2) *Simulate* $x_2^{(l)}$ *from* $\pi_2\left(x_2|x_1^{(l)}, x_3^{(l-1)}, \ldots, x_N^{(l-1)}\right)$;
   ⋮
   N) *Simulate* $x_N^{(l)}$ *from* $\pi_N\left(x_N|x_1^{(l)}, \ldots, x_{N-1}^{(l)}\right)$.
3. *Next* l.

Again, under general conditions, the joint density $\pi(\boldsymbol{x})$ is a stationary distribution of the generated chain $\{\boldsymbol{x}^{(l)}, l = 1, 2, \ldots\}$; and the chain is ergodic, that is, $\pi(\boldsymbol{x})$ is a limiting distribution of the chain.

For detailed analysis of properties and justification of this algorithm, see analyses by Casella and George (1992), Liu *et al.* (1995), Chan (1993) and Smith and Roberts (1993).

In many cases, such as in the univariate Gibb's sampler framework, the full conditional posterior distributions may not be sampled via inversion. To handle this complication, there have been several developments in which adaptive rejection sampling has been utilized to sample from each successive full conditional posterior distribution. Well-known examples of these include Gilks and Wild (1992); Gilks *et al.* (1994, 1995), and Gelfand (2000). Another approach to tackle this challenge in general is to adopt a mixed strategy in which one utilizes combinations of Gibbs steps for some "blocks" of parameters and MH within Gibbs steps for other parameter blocks.

7.4.4 RANDOM WALK METROPOLIS–HASTINGS WITHIN GIBBS

The *Random Walk Metropolis–Hastings* (RW-MH) *within the Gibbs* algorithm is easy to implement and often efficient if the likelihood function can be easily evaluated. It is referred to as *single-component Metropolis–Hastings* by Gilks *et al.* (1996, section 1.4). The algorithm is not well known among OpRisk practitioners and we would like to mention its main features; see Peters and Sisson (2006); Shevchenko and Temnov (2009) for application in the context of OpRisk and Peters *et al.* (2009a) for application in the context of a similar problem in insurance.

The RW-MH within the Gibbs algorithm creates a reversible Markov chain with a stationary distribution corresponding to our target posterior distribution π. Denote by $\boldsymbol{\theta}^{(l)}$ the state of the chain at iteration l. The algorithm proceeds by proposing to move the i-th parameter from the current state $\theta_i^{(l-1)}$ to a new proposed state θ_i^* sampled from the MCMC proposal transition kernel denoted generically here by density f with distribution F. Typically, the parameters are restricted by simple ranges, $\theta_i \in [a_i, b_i]$, and proposals are sampled from the Normal distribution. Then, the logical steps of the algorithm are as follows.

Algorithm 7.5 (RW-MH within Gibbs)

1. *Initialize* $\theta_i^{(l=0)}$, $i = 1, \ldots, I$ *by for example, using MLEs;*
2. *For* $l = 1, \ldots, L$
   a) *Set* $\boldsymbol{\theta}^{(l)} = \boldsymbol{\theta}^{(l-1)}$;
   b) *For* $i = 1, \ldots, I$.
      i. *Sample proposal* θ_i^* *from the transition kernel, for example, from the truncated Normal density*

$$f^{\text{tr}}(\theta_i^* | \theta_i^{(l)}, \sigma_i) = \frac{f(\theta_i^* | \theta_i^{(l)}, \sigma_i)}{F(b_i | \theta_i^{(l)}, \sigma_i) - F(a_i | \theta_i^{(l)}, \sigma_i)}, \tag{7.66}$$

where $f(x|\mu, \sigma)$ *and* $F(x|\mu, \sigma)$ *are the Normal density and its distribution with mean* μ *and standard deviation* σ;

ii. *Accept proposal with the acceptance probability*

$$\alpha(\boldsymbol{\theta}^{(l)}, \boldsymbol{\theta}^*) = \min\left\{1, \frac{\pi(\boldsymbol{\theta}^*|\boldsymbol{x}) f^{\text{tr}}(\theta_i^{(l)}|\theta_i^*, \sigma_i)}{\pi(\boldsymbol{\theta}^{(l)}|\boldsymbol{x}) f^{\text{tr}}(\theta_i^*|\theta_i^{(l)}, \sigma_i)}\right\}, \tag{7.67}$$

where $\boldsymbol{\theta}^* = (\theta_1^{(l)}, \ldots, \theta_{i-1}^{(l)}, \theta_i^*, \theta_{i+1}^{(l-1)}, \ldots)$, *that is, simulate* U *from Uniform*$(0, 1)$ *and set* $\theta_i^{(l)} = \theta_i^*$ *if* $U < \alpha(\boldsymbol{\theta}^{(l)}, \boldsymbol{\theta}^*)$. *Note that the normalization constant of the posterior does not contribute here.*

c) *Next i.*

3. *Next l.*

This procedure builds a set of correlated samples from the target posterior distribution. One of the most useful asymptotic properties is the convergence of ergodic averages constructed using the Markov chain samples to the averages obtained under the posterior distribution. The chain has to be run until it has sufficiently converged to the stationary distribution (the posterior distribution) and then one obtains samples from the posterior distribution. General properties of this algorithm, including convergence results, can be found in Robert and Casella (2004, sections 6–10). The RW-MH algorithm is simple in nature and easy to implement. However, for a bad choice of the proposal distribution, in the case above the tuning of the proposal f variance parameters, the algorithm gives a very slow convergence to the stationary distribution. There have been several recent studies regarding the optimal scaling of the proposal distributions to ensure optimal convergence rates (see Bedard and Rosenthal, 2008). The suggested asymptotic acceptance rate optimizing the efficiency of the process is 0.234. Usually, it is recommended that the σ_i in (7.66) be chosen to ensure that the acceptance probability is roughly close to 0.234. This requires some tuning of the σ_i prior to the final simulations.

7.5 Standard MCMC Guidelines for Implementation

There are several numerical issues when implementing MCMC. For the majority of standard MCMC algorithms, one must consider the following practical advice. In practice, an MCMC run consists of three stages: *tuning*, *burn-in*, and *sampling* stages. It is also important to assess the numerical errors of the obtained estimators due to finite number of MCMC iterations.

7.5.1 TUNING, BURN-IN, AND SAMPLING STAGES

Tuning. The use of MCMC samples can be very inefficient for an arbitrary chosen proposal distribution. Typically, parameters of a chosen proposal distribution are adjusted to achieve a reasonable acceptance rate for each component. There have been several studies regarding the optimal scaling of proposal distributions to ensure optimal convergence rates. Gelman *et al.* (1997), Bedard and Rosenthal (2008), and Roberts and Rosenthal (2001) were the first authors to publish theoretical results for the optimal scaling problem in RW-MH algorithms with Gaussian proposals. For the d-dimensional target distributions with independent and identically distributed components, the asymptotic acceptance rate optimizing the efficiency of the process is 0.234 independent of the target density. Though for most problems the posterior parameters are not independent Gaussian, it provides a practical guide.

There is no need to be very precise in this stage. In practice, the chains with acceptance rate between 0.2 and 0.8 work well. Typically, turning is easy. In an ad hoc procedure, one can initialize the proposal distribution parameters with the values corresponding to the proposal with a very small variability, and start the chain. This will lead to a very high acceptance rate. Then run the chain and gradually change the parameters toward the values that correspond to the proposal with a large uncertainty. This will gradually decrease the acceptance rate. Continue this procedure until the acceptance rate is within the 0.2–0.8 range. For example, for Gaussian proposal choose a very small standard deviation parameter. Then increase the standard deviation in small steps and measure the average acceptance rate over the completed iterations until the rate is within the 0.2–0.8 range. One can apply a reverse procedure, that is, start with parameter values corresponding to a very uncertain proposal resulting in a very low acceptance rate. Then gradually change the parameters toward the values corresponding to the proposal with small variability. Many other alternative ways can be used in this context.

Gaussian proposals are often useful with the covariance matrix given by (7.25), that is, using Gaussian approximation for the posterior, or just MLE observed information matrix (7.7) in the case of constant prior. An alternative approach is to utilize a new class of adaptive MCMC algorithms recently proposed in the literature (see Atchadé and Rosenthal 2005 and Rosenthal 2007).

Burn-in stage. Subject to regularity conditions, the chain converges to the stationary target distribution. The number of iterations required for the chain to converge should be discarded and called *burn-in* iterations. Again, we do not need to identify this quantity precisely. Rough approximations of the order of magnitude work well. Visual inspections of the chain trace plot is the most commonly used method. If the chain is run long enough, then the impact of these *burn-in* iterations on the final estimates is not significant. There are many formal *convergence diagnostics* that can be used to determine the length of *burn-in* (for a review, see Cowles and Carlin 1996).

Sampling stage. Consider the chain $\{\boldsymbol{\theta}^{(1)}, \boldsymbol{\theta}^{(2)}, \ldots, \boldsymbol{\theta}^{(L)}\}$ and the number of *burn-in* iterations is L_b. Then, $\boldsymbol{\theta}^{(L_b+1)}, \boldsymbol{\theta}^{(L_b+2)}, \ldots, \boldsymbol{\theta}^{(L)}$ are considered as dependent samples from the target distribution $\pi(\boldsymbol{\theta}|\boldsymbol{x})$ and used for estimation purposes. For example, $\mathbb{E}[g(\boldsymbol{\Theta})|\boldsymbol{X} = \boldsymbol{x}]$ is estimated as

$$\mathbb{E}[g(\boldsymbol{\Theta})|\boldsymbol{X} = \boldsymbol{x}] = \int g(\boldsymbol{\theta})\pi(\boldsymbol{\theta}|\boldsymbol{x})d\boldsymbol{\theta} \approx \frac{1}{L - L_b} \sum_{l=L_b+1}^{L} g(\boldsymbol{\theta}^{(l)}). \tag{7.68}$$

Typically, when we calculate the posterior characteristics using MCMC samples, we assume that the samples are taken after burn-in and L_b is dropped in corresponding formulas to simplify notation.

In addition to visual inspection of MCMC, checking that after the burn-in period the samples are mixing well over the support of the posterior distribution, it is useful to monitor the serial correlation of the MCMC samples. For a given chain sample $\theta_i^{(1)}, \ldots, \theta_i^{(L)}$, the autocorrelation at lag k is estimated as

$$\widehat{\mathrm{ACF}}[\theta_i, k] = \frac{1}{(L - k)\hat{s}^2} \sum_{l=1}^{L-k} (\theta_i^{(l)} - \hat{\mu})(\theta_i^{(l+k)} - \hat{\mu}), \tag{7.69}$$

where $\hat{\mu}$ and $\hat{s}^2$ are the mean and variance of a sample $\theta_i^{(1)}, \ldots, \theta_i^{(L)}$. In well-mixed MCMC samples, the autocorrelation falls to near zero quickly and stays near zero at larger lags. It is useful to find a lag $k^{\max}$ where the autocorrelations seem to have *"died out"*, that is, fallen to near zero (for some interesting discussion on this issue, see e.g., Kass *et al.* 1998). It is not unusual to choose a $k_i^{\max}$ for each component such that the autocorrelation at lag $k_i^{\max}$ has reduced to less than 0.01.

EXAMPLE 7.3

To illustrate the described stages, consider a dataset of the annual counts $\boldsymbol{n} = (9, 12, 7, 9)$ simulated from $Poisson(10)$. Then, we obtain the chain $\lambda^{(0)}, \lambda^{(1)}, \ldots$ using the RW-MH algorithm with the Gaussian proposal distribution for the $Poisson(\lambda)$ model and constant prior on a very wide range $[0.1, 100]$. Figure 7.1 shows the chains in the case of different starting values $\lambda^{(0)}$ and different standard deviations σ_{RW} of the Gaussian proposal. One can see that after the burn-in stage indicated by the vertical broken line, the chain looks stationary. Figure 7.1a and b were obtained when $\sigma_{RW} = \widehat{\text{stdev}}[\hat{\lambda}^{\text{MLE}}] \approx 1.521$, leading to the acceptance probability of approximately 0.7, while Figure 7.1c and d were obtained when $\sigma_{RW} = 0.4$ and $\sigma_{RW} = 30$, leading to the acceptance probability of about 0.91 and 0.10, respectively. The MLE was calculated in the usual way as $\widehat{\text{stdev}}[\hat{\lambda}^{\text{MLE}}] = (\sum_{i=1}^{m} n_i/m)^{1/2}/\sqrt{m}$, where $m = 4$. The impact of the value of σ_{RW} is easy to see: the chains on Figures 7.1c and d are *mixing* slowly (moving slowly around the support of the posterior) while the chains on Figure 7.1a and b are mixing rapidly. Slow mixing means that much longer chain should be run to get good estimates. ■

7.5.1.1 MCMC Convergence Diagnostics.

In all cases of using MCMC algorithms in practice, one has to decide on how many samples to draw from the Markov chain mechanism, that is, how long a Markov chain to run in the burn-in and sampling phase. In principle, this will dependent on a few different factors such as the precision in the estimated inferential quantities that one wishes to achieve, since we have seen that the accuracy of Monte Carlo estimates increases with the number of samples.

In addition, one must remember that when using MCMC samplers, if one initializes the sampler from an arbitrary point in the parameter space, it will take a certain number of iterations before the Markov chain reaches what can be considered the stationary regime, that is, begins to sample from the true posterior target distribution (the reason for the burn-in phase). Therefore, it is common in practice to do two things: the first is to discard the initial samples that may not have come from the stationary distribution (know as discarding "burn-in"), and the second is to monitor the mixing (exploration of the Markov chain) around the support of the posterior distribution. This helps to ensure that we are not using samples that are too autocorrelated. Both these tasks require some version of monitoring, and generally there have been statistical approaches developed to monitor these aspects as the Markov chain progresses, which are known as convergence diagnostics (see the review by Mengersen *et al.* 1999).

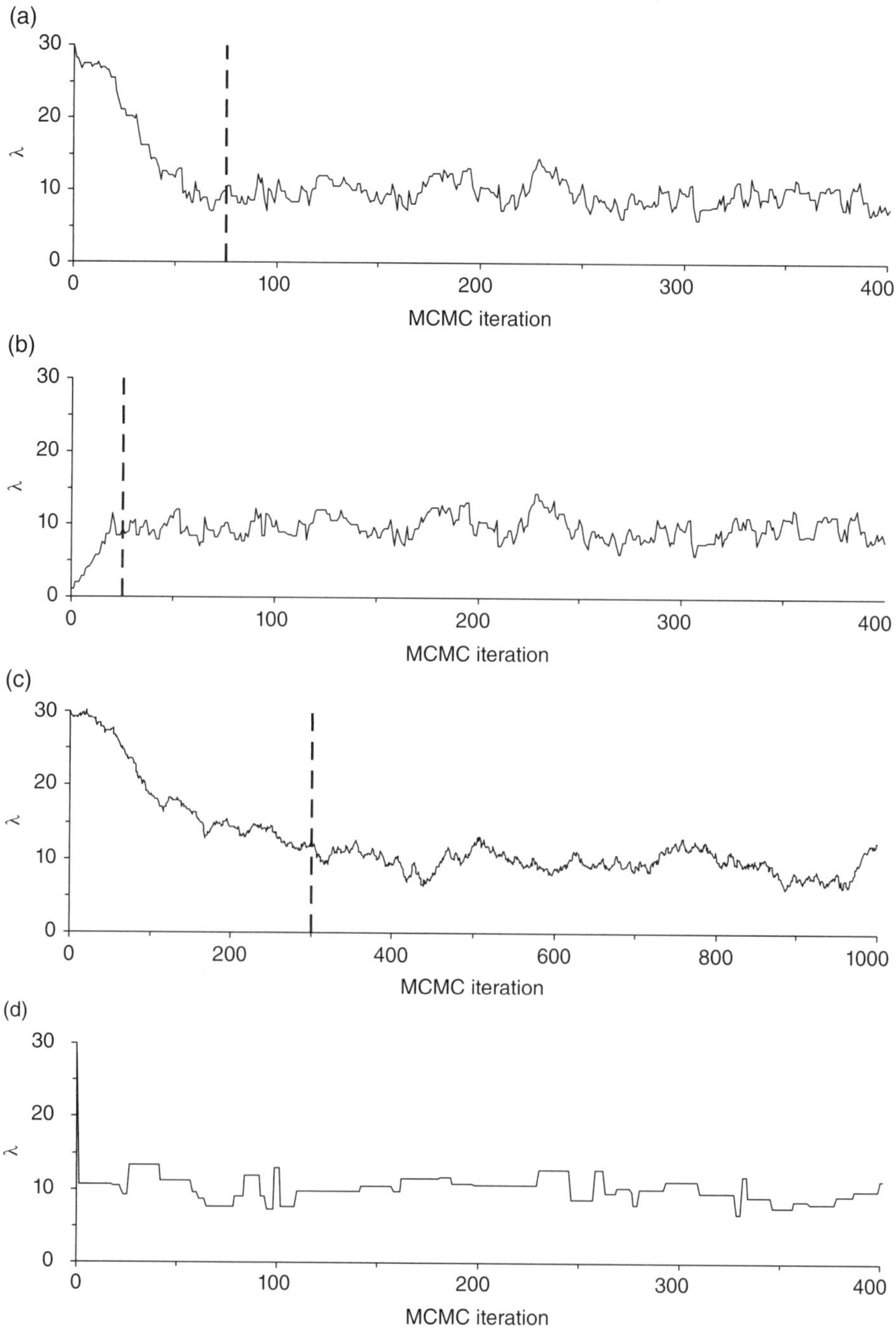

FIGURE 7.1 MCMC chains of λ parameter of *Poisson*(λ) model in the case of different starting points $\lambda^{(0)}$ and different standard deviations of the Gaussian proposal distribution: (a) starting point $\lambda^{(0)} = 30$ and $\sigma_{RW} = 1.521$; (b) $\lambda^{(0)} = 1$ and $\sigma_{RW} = 1.521$; (c) $\lambda^{(0)} = 30$ and $\sigma_{RW} = 0.4$; (d) $\lambda^{(0)} = 30$ and $\sigma_{RW} = 30$. The *burn-in* stage is to the left of the vertical broken line. The dataset consisting of the annual number of events $(9, 12, 7, 9)$ over 4 years was simulated from *Poisson*(10)

Hence, we stress that when using an MCMC algorithm, it is crucial to carefully monitor the convergence diagnostics of the Markov chain. This is more important in general MCMC contexts and approximate Bayesian computation settings due to the possibility of extended rejections where the Markov chain can stick in a given state for long periods.

If the total chain has length L, the initial burn-in stage will correspond to the first L_b samples and we define $\widetilde{L} = L - L_b$. Note that in this particular section, we will denote by $\{\Theta^{(l)}\}_{l=1:\widetilde{L}}$ the Markov chain of the i-th parameter after burn-in; for simplicity of notation, the parameter index i is dropped. The diagnostics we consider are given as follows.

- *Geweke et al. (1991); Cowles and Carlin (1996) time series diagnostic*;
  1. Split the Markov chain samples into two subsequences,
  $$\{\Theta^{(l)}\}_{l=1:L_1} \quad \text{and} \quad \{\Theta^{(l)}\}_{l=L^*:\widetilde{L}},$$
  such that $L^* = \widetilde{L} - L_2 + 1$, and with ratios $L_1/\widetilde{L}$ and $L_2/\widetilde{L}$ fixed such that $(L_1 + L_2)/\widetilde{L} < 1$ for all $\widetilde{L}$;
  2. Evaluate $\hat{\mu}_{L_1}$ and $\hat{\mu}_{L_2}$ corresponding to the sample means of each subsequence;
  3. Evaluate consistent spectral density estimates for each subsequence, at frequency 0, denoted $\widehat{SD}_{L_1}$ and $\widehat{SD}_{L_2}$. The spectral density estimator is the classical non-parametric periodogram or power spectral density estimator; for details of the power spectral density, see Oppenheim *et al.* (1989);
  4. Evaluate convergence diagnostic given by
  $$Z_{\widetilde{L}} = \frac{\hat{\mu}_{L_1} - \hat{\mu}_{L_2}}{L_1^{-1}\widehat{SD}_{L_1} + L_2^{-1}\widehat{SD}_{L_2}}.$$
  According to the CLT, as $\widetilde{L} \to \infty$ one has $Z_{\widetilde{L}} \to Normal(0,1)$ if the sequence $\{\Theta^{(l)}\}_{l=1:\widetilde{L}}$ is stationary.
- *Gelman and Rubin (1992), and Brooks and Gelman (1998), R-statistic diagnostic.* This approach to convergence analysis requires that one run multiple parallel independent Markov chains each starting at randomly selected initial starting points (e.g., consider running five chains). For comparison purposes, we split the total computational budget of $\widetilde{L}$ into $L_1 = L_2 = \cdots = L_5 = \widetilde{L}/5$. The convergence diagnostic for parameter Θ is calculated using the following steps:
  1. Generate five independent Markov chain sequences, producing the chains for parameter Θ denoted $\{\Theta_k^{(l)}\}_{l=1:L_k}$ for $k \in \{1, \ldots, 5\}$;
  2. Calculate the sample means $\hat{\mu}_{L_k}$ for each sequence and the overall mean $\hat{\mu}_{\widetilde{L}}$;
  3. Calculate the variance of the sequence means
  $$\frac{1}{4}\sum_{k=1}^{5}\left(\hat{\mu}_{L_k} - \hat{\mu}_{\widetilde{L}}\right)^2 =: B/L_k.$$
  4. Calculate the within-sequence variances $\hat{s}^2_{L_k}$ for each sequence;
  5. Calculate the average within-sequence variance, $\frac{1}{5}\sum_{k=1}^{5}\hat{s}^2_{L_k} =: W$;

6. Estimate the target posterior variance for parameter Θ by the weighted linear combination $\hat{\sigma}^2_{\widetilde{L}} = \frac{L_k - 1}{L_k} W + \frac{1}{L_k} B$. This estimate is unbiased for samples that are from the stationary distribution. In the case in which not all subchains have reached stationarity, this overestimates the posterior variance for a finite $\widetilde{L}$ but asymptotically, $\widetilde{L} \to \infty$, it converges to the posterior variance;
7. Improve on the Gaussian estimate of the target posterior given by $Normal(\hat{\mu}_{\widetilde{L}}, \hat{\sigma}^2_{\widetilde{L}})$ by accounting for sampling variability in the estimates of the posterior mean and variance. This can be achieved by making a Student-t approximation with location $\hat{\mu}_{\widetilde{L}}$, scale $\sqrt{\hat{V}}$, and degrees of freedom df, each given respectively by: $\hat{V} = \hat{\sigma}^2_{\widetilde{L}} + B/\widetilde{L}$ and $df = 2(\hat{V})^2/\widehat{\mathrm{Var}}(\hat{V})$, where the variance is estimated as

$$\begin{aligned}\widehat{\mathrm{Var}}\left(\hat{V}\right) &= \frac{1}{5}\left(\frac{L_1 - 1}{L_1}\right)^2 \widehat{\mathrm{Var}}\left[\hat{s}^2_{L_k}\right] + \left(\frac{6}{\sqrt{2\widetilde{L}}}\right)^2 B^2 \\ &\quad + \frac{12(L_1 - 1)}{25L_1}\widehat{\mathrm{Cov}}\left(\hat{s}^2_{L_k}, \hat{\mu}_{\widetilde{L}}\right) - \frac{24(L_1 - 1)}{25L_1}\hat{\mu}_{\widetilde{L}}\widehat{\mathrm{Cov}}\left[\hat{s}^2_{L_k}, \hat{\mu}_{\widetilde{L}}\right].\end{aligned} \tag{7.70}$$

Note that the covariance terms are estimated empirically using the within-sequence estimates of the mean and variance obtained for each sequence;
8. Calculate the convergence diagnostic $\hat{R} = \hat{V} df / W(df - 2)$, where as $\widetilde{L} \to \infty$ one can prove that $\hat{R} \to 1$. This convergence diagnostic monitors the scale factor by which the current distribution for Θ may be reduced if simulations are continued for $\widetilde{L} \to \infty$.

7.5.2 NUMERICAL ERROR

Due to the finite number of iterations, MCMC estimates have numerical error that reduces as the chain length increases. Consider the estimator

$$\hat{\Omega} = \hat{\mathbb{E}}[g(\boldsymbol{\Theta})|\boldsymbol{X} = \boldsymbol{x}] = \frac{1}{L}\sum_{l=1}^{L} g(\boldsymbol{\Theta}^{(l)}). \tag{7.71}$$

If the samples $\boldsymbol{\Theta}^{(1)}, \ldots, \boldsymbol{\Theta}^{(L)}$ are independent and identically distributed then the standard error of $\hat{\Omega}$ (due to the finite L) is estimated using

$$\mathrm{stdev}[\hat{\Omega}] = \mathrm{stdev}[g(\boldsymbol{\Theta})|\boldsymbol{X} = \boldsymbol{x}]/\sqrt{L},$$

where $\mathrm{stdev}[g(\boldsymbol{\Theta})|\boldsymbol{X}]$ is estimated by the standard deviation of the sample $g(\boldsymbol{\Theta}^{(l)})$, $l = 1, \ldots, L$. This formula is only approximate for MCMC samples due to serial correlations between the samples. Of course, one can keep every $k_{\max}$ sample from the chain to get approximately independent samples, but it is always a suboptimal approach (see MacEachern and Berliner 1994).

Effective sample size. If there is only one parameter θ, then one of the popular approaches is to calculate *effective sample size*, $T_{\mathit{eff}} = T/\tau$, where τ is autocorrelation time

$$\tau = 1 + 2\sum_{k=1}^{\infty} \mathrm{ACF}[\theta, k]. \tag{7.72}$$

To estimate τ, it is necessary to cut off the sum in (7.72) at a value of $k = k^{\max}$, where the autocorrelations seem to have fallen to near zero. Then the standard error of the $\hat{\Omega}$ (7.71) is estimated using

$$\mathrm{stdev}[\hat{\Omega}] = \frac{\mathrm{stdev}[g(\Theta)]}{\sqrt{L/\tau}}$$

(see Ripley 1987, and Neal 1993).

Batch sampling. Probably the most popular approach to estimating the numerical error of the MCMC posterior averages is the so-called *batch sampling* (see Gilks *et al.* 1996, section 3.4.1). Consider MCMC posterior samples $\boldsymbol{\Theta}^{(1)}, \ldots, \boldsymbol{\Theta}^{(L)}$ of $\boldsymbol{\Theta}$ with the length $L = K \times N$, and an estimator $\widehat{\Omega} = \sum_{l=1}^{L} g(\boldsymbol{\Theta}^{(l)})$ of $\mathbb{E}[g(\boldsymbol{\Theta})]$. If N is sufficiently large, the means

$$\hat{\Omega}_j = \frac{1}{N}\sum_{i=(j-1)N+1}^{j\times N} g(\boldsymbol{\Theta}^{(i)}), \quad j = 1, \ldots, K \tag{7.73}$$

are approximately independent and identically distributed. Then the overall estimator and its variance are

$$\hat{\Omega} = \frac{1}{K}(\hat{\Omega}_1 + \cdots + \hat{\Omega}_K),$$
$$\mathbb{V}\mathrm{ar}[\hat{\Omega}] = \frac{1}{K^2}(\mathbb{V}\mathrm{ar}[\hat{\Omega}_1] + \cdots + \mathbb{V}\mathrm{ar}[\hat{\Omega}_K]) = \frac{\sigma^2}{K},$$

where $\sigma^2 = \mathbb{V}\mathrm{ar}[\hat{\Omega}_1] = \cdots = \mathbb{V}\mathrm{ar}[\hat{\Omega}_K]$. In the limit of large K, by the CLT (we also assume that σ^2 is finite), the distribution of $\hat{\Omega}$ is Normal with the standard deviation $\sigma/\sqrt{K}$. The latter is referred to as the standard error of $\hat{\Omega}$. Finally, σ^2 can be estimated using sample variance

$$\hat{\sigma}^2 = \frac{1}{K-1}\sum_{j=1}^{K}(\hat{\Omega}_j - \hat{\Omega})^2. \tag{7.74}$$

Note that K is the number of quasi-independent bins, and $N = L/K$ is the size of each bin or batch. Typically, in practice $K \geq 20$ and $N \geq 100k^{\max}$, where $k^{\max} = \max(k_1^{\max}, k_2^{\max}, \ldots)$ is the maximum of the cutoff lags over components. In general, we would like to run the chain until the numerical error is not significant. So, one can set N using $k^{\max}$ identified during tuning and burning stages, for example, set $N = 100k^{\max}$, then run the chain in batches until the numerical error of the estimates is less than the desired accuracy.

7.5.3 MCMC EXTENSIONS: REDUCING SAMPLE AUTOCORRELATION

Sometimes, in the developed Bayesian models, there is a strong correlation between the model parameters in the posterior. In extreme cases, this can cause slow rates of convergence in the Markov chain to reach the ergodic regime, translating into longer Markov chain simulations. In such a situation, several approaches can be tried to overcome this problem. The following are suggestions that are widely used in practice.

- **Hybrid Samplers**. The first involves the use of a mixture transition kernel for the Markov chain, where one combines local and global moves. Local moves are from Markov transition kernels that sample the next chain transition based on local information of the current chain's state, whereas global moves sample the next state of the Markov chain independently of the current state location. Global moves can produce a wider exploration potential of the state space, whereas local moves produce a local exploration with a higher chance of acceptance of a proposed move. For example, one can perform local moves via a univariate slice sampler and global moves via an independent MH sampler with adaptive learning of its covariance structure. Such an approach is known as a hybrid sampler (see comparisons in Brewer *et al.* 1996); the slice sampler will be discussed later. Alternatively, for the global move if determination of level sets in multiple dimensions is not problematic for the model under consideration, then some of the multivariate slice sampler approaches designed to account for correlation between parameters can be incorporated (see Neal 2003 for details);
- **Transformations of Parameters (change of variable)**. Another approach to breaking correlation between parameters in the posterior is via the transformation of the parameter space. If the transformation is effective, this will reduce correlation between parameters of the transformed target posterior. Sampling can then proceed in the transformed space, and then samples can be transformed back to the original space. It is not always straightforward to find such transformations;
- **Change of Target Distribution (distortion of target)**. A third alternative is based on *simulated tempering*, introduced by Marinari and Parisi (1992) and discussed extensively by Geyer and Thompson (1995). In particular, a special version of simulated tempering, first introduced by Neal (1996), can be utilized in which one considers a sequence of target distributions $\{\pi_l\}$ constructed such that they correspond to the objective posterior in the following way:

$$\pi_l = (\pi(\boldsymbol{\theta}|\boldsymbol{x}))^{\gamma_l} \tag{7.75}$$

  with sequence $\{\gamma_l\}$. Then one can use the standard MCMC algorithms (e.g., slice sampler) and replace π with π_l. Running a Markov chain such that at each iteration l we target the posterior π_l and then only keeping samples from the Markov chain corresponding to situations in which $\gamma_l = 1$ can result in significant improvement in exploration around the posterior support. This can overcome slow mixing arising from a univariate sampling regime. The intuition for this is that for values of $\gamma_l << 1$ the target posterior is almost uniform over the space, resulting in large moves being possible around the support of the posterior, then as γ_l returns to a value of 1, several iterations later, it will be in potentially new unexplored regions of the posterior support.

For example, one can utilize a sine function

$$\gamma_l = \min\left(\sin\left(\frac{2\pi}{K}l\right) + 1, 1\right)$$

with large K (e.g., $K = 1000$), which has its amplitude truncated to ensure it ranges between 0 and 1. That is, the function is truncated at $\gamma_l = 1$ for extended iteration periods for our simulation index l to ensure the sampler spends significant time sampling from the actual posterior distribution.

In the application of tempering, one must discard many simulated states of the Markov chain, whenever $\gamma_l \neq 1$. There is, however, a computational way to avoid discarding these samples (see Gramacy *et al.*, 2010);

- **Adaptive Transition Kernels and Mixed Samplers.** Finally, we note that there are several alternatives to an MH within a Gibbs sampler such as a basic Gibbs sampler combined with *adaptive rejection sampling* (ARS) (Gilks and Wild 1992). Note that ARS requires distributions to be log-concave. Alternatively, an adaptive version of this known as the adaptive Metropolis rejection sampler could be used (see Gilks *et al.* 1995).

Remark 7.12 *Knowing which of theses strategies is most appropriate for a given application is a combination of three factors: careful consideration of the properties of the target distribution to determine which approach may be possible to implement efficiently; consideration of the total computational budget and desired precision in the estimation target; and some trial of competing methods prior to full simulation. It is still a challenge to definitively state that a particular MCMC approach or sampling procedure will work universally for all problems in an efficient manner and so some trial and error is required.*

It is the intention of the following sections to provide more advanced techniques that we recommend be adopted only if one has already tried the simple MCMC procedures discussed previously and found them to be inefficient for the given sampling challenge. Therefore, these more advanced methods will provide a significant increase in sampler "performance" at an increased cost of complexity of understanding and implementation.

7.6 Advanced MCMC Methods

In this section, we will survey a few examples of more recently developed MCMC methods aimed at improving the performance of the fundamental approaches discussed in the section presenting the standard MCMC algorithms. The first methods we present are a class of auxiliary variable MCMC methods in which we focus on the special case of the univariate Gibbs sampler algorithm, with discussions and references to more recent advanced multivariate versions. Following this we present the framework of adaptive MCMC methods, illustrating the properties briefly in order to explain to practitioners how such an approach can be implemented. The adaptive strategies chosen to be presented will be based on nontrivial modifications to the standard algorithms to obtain the adaptive Metropolis algorithm and the Reimann–Manifold Hamiltonian Monte Carlo algorithms. Then we introduce briefly the family of sequential IS methods known in the statistics literature as SMC Samplers, which are direct competitors to MCMC methods (see Del Moral *et al.* 2006, Peters *et al.* 2009 and

Peters 2005). We discuss these briefly and then present particular detailed examples of such algorithms are also presented in the companion book by Peters and Shevchenko (2015) under the topic of rare-event simulation.

7.6.1 AUXILIARY VARIABLE MCMC METHODS: SLICE SAMPLING

In this section, we explain the general class of auxiliary variable methods that are available to practitioners to improve the efficiency of MCMC algorithms in exploring complex posterior supports. In general, there are different classes of such algorithms ranging from slice samplers by, for example, Neal (2003) and for OpRisk, Peters *et al.* (2009); auxiliary variable techniques to remove intractability in likelihood models such as by Godsill (2000), West (1987), and Peters *et al.* (2011b); data augmentation schemes (Tanner and Wong, 1987) such as those used to tackle complicated dependence structures by Peters *et al.* (2012b); and the general summary of such methods by Higdon (1998).

7.6.2 GENERIC UNIVARIATE AUXILIARY VARIABLE GIBBS SAMPLER: SLICE SAMPLER

In this section, we focus on settings in which the approach of auxiliary variables can be utilized to improve sampling performance. Often, the full conditional distributions in Gibbs samplers do not take standard explicit closed forms and typically the normalizing constants are not known in closed form. Therefore, this will exclude straightforward simulation using the inversion method (see Corollary 7.1) or basic rejection sampling (see Corollaries 7.2 and 7.3). In this case, for sampling, one may adopt a Metropolis–Hastings within a Gibbs algorithm (described in Section 7.4.4). This typically requires tuning of the proposal for a given target distribution, which becomes computationally expensive, especially for high-dimensional problems. To overcome this problem one may use an adaptive Metropolis–Hastings within a Gibbs sampling algorithm (see Atchadé and Rosenthal 2005 and Rosenthal 2009). An alternative approach, which is more efficient in some cases, is known as a univariate *slice sampler* (see Neal 2003). The latter was developed with the intention of providing a "black box" approach for sampling from a target distribution, which may not have a simple form.

The slice sampling methodology we develop will be automatically tailored to the desired target posterior. As such, it does not require pretuning and in many cases will be more efficient than an MH within Gibbs sampler. The reason for this, pointed out by Neal (2003), is that an MH within Gibbs has two potential problems. The first arises when an MH approach attempts moves that are not well adapted to local properties of the density, resulting in slow mixing of the Markov chain. Second, the small moves arising from the slow mixing typically lead to traversal of a region of posterior support in the form of a Random Walk. Therefore, L^2 steps are required to traverse a distance that could be traversed in only L steps if moving consistently in the same direction. A univariate slice sampler can adaptively change the scale of the moves proposed avoiding problems that can arise with the MH sampler when the appropriate scale of proposed moves varies over the support of the distribution.

We will utilize the notations $\boldsymbol{\Theta}_{(-i)} = (\Theta_1, \ldots, \Theta_{i-1}, \Theta_{i+1}, \ldots, \Theta_d)$ with $\boldsymbol{\theta} \in \mathbb{R}^d$. The intuition behind slice sampling arises from the fact that sampling from a univariate distribution, in this case given by say the i-th component full conditional of the posterior for the

Bayesian model, given by $\pi\left(\Theta_i|\boldsymbol{\Theta}_{(-i)}, \boldsymbol{x}_{1:T}\right)$, can always be achieved by sampling uniformly from the region under the distribution $\pi\left(\Theta_i|\boldsymbol{\Theta}_{(-i)}, \boldsymbol{x}_{1:T}\right)$.

The procedure in Algorithm 7.6 is repeated for each of the d elements of the posterior parameter vector $\boldsymbol{\Theta}$ to obtain the l-th sample from the univariate slice sampler. Then typically, such a procedure is repeated L times to obtain sufficient draws for resulting estimation challenges. We note that in applying Algorithm 7.6 we actually discard the auxiliary variable sample u_j^l for each of the dimensions, $j \in \{1, 2, \ldots, d\}$, just keeping the resulting correlated samples $\Theta_j^{(l)}$ which when combined together in a vector $\left(\Theta_1^{(l)}, \Theta_2^{(l)}, \ldots, \Theta_d^{(l)}\right)$ will make a draw $\boldsymbol{\Theta}^{(l)}$ from $\pi\left(\boldsymbol{\Theta}|\boldsymbol{x}_{1:T}\right)$. Neal (2003) demonstrates that a Markov chain $(\boldsymbol{U}, \boldsymbol{\Theta})$ constructed in this way will have stationary distribution defined by a uniform distribution under $\pi\left(\boldsymbol{\Theta}|\boldsymbol{x}_{1:T}\right)$ and therefore discarding the vector of auxiliary variables $\boldsymbol{U}$ at each of the l iterations allows one to obtain the marginal of $\boldsymbol{\Theta}$ which will produce samples from the desired stationary distribution $\pi\left(\boldsymbol{\Theta}|\boldsymbol{x}_{1:T}\right)$. Additionally, Mira and Tierney (2003), and Mira *et al.* (2002) proved that the slice sampler algorithm, assuming a bounded target distribution $\pi\left(\boldsymbol{\Theta}|\boldsymbol{x}_{1:T}\right)$ with bounded support is uniformly ergodic.

Similar to a deterministic scan Gibbs sampler, the simplest way to apply the slice sampler to a multivariate distribution is by considering each of the univariate full conditional distributions either in turn under a deterministic scan Gibbs sampler; or alternatively under a random scan Gibbs sampler in which the dimension of $\boldsymbol{\Theta}$ to be updated at the l-th iteration of the slice sampler is randomly selected. Discussions that relate to the benefits provided by Random Walk behavior suppression, as achieved by the slice sampler, are presented in the context of nonreversible Markov chains in for instance Diaconis *et al.* (2000).

A single iteration of the slice sampler algorithm for a toy example is presented in Figure 7.2. The intuition behind the slice sampling arises from the fact that sampling from a univariate density $\pi(\theta)$ can always be achieved by sampling uniformly from the region under the density $\pi(\theta)$, where, for instance, $\pi(\theta)$ could be a posterior distribution $\pi\left(\theta|\boldsymbol{x}_{1:T}\right)$

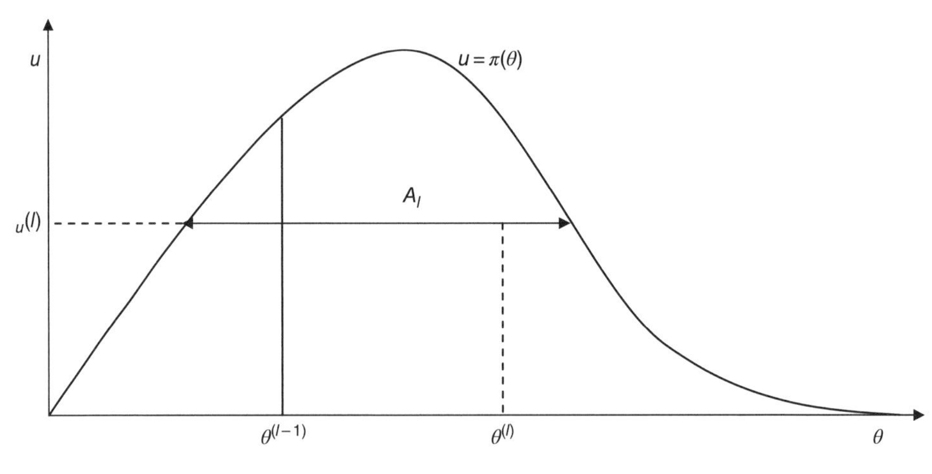

FIGURE 7.2 **Markov chain created for Θ and auxiliary random variable U, $\left(u^{(1)}, \theta^{(1)}\right), \ldots, \left(u^{(l-1)}, \theta^{(l-1)}\right), \left(u^{(l)}, \theta^{(l)}\right), \ldots$ has a stationary distribution with the desired marginal density $\pi(\theta)$**

or any desired target distribution. The example we present is for a single univariate distribution; if one has a multivariate posterior, then the algorithm is applied iteratively via either a random scan or a deterministic scan over each of the univariate full conditional distributions. This is basically then a version of an auxiliary variable Gibbs sampler as discussed above with the sampling stage at each iteration replaced with the two steps of Algorithm 7.6 below for each full conditional target distribution. Again, we note that this algorithm only requires that the target posterior distribution and univariate full conditional posterior distributions are only required to be known up to a normalization constant. That is, the normalization constant is not required to be known to apply this method.

Algorithm 7.6 (Univariate slice sampler)

1. *Initialize* $\theta^{(0)}$ *by any value within the support of* $\pi(\theta)$*;*
2. *For* $l = 1, 2, \ldots, L$
   a) *Sample a value* $u^{(l)} \sim Uniform\left(0, \pi\left(\theta^{(l-1)}\right)\right)$*;*
   b) *Sample a value* $\theta^{(l)}$ *uniformly from the level set* $A_l = \left\{\theta : \pi(\theta) > u^{(l)}\right\}$*, that is,* $\theta^{(l)} \sim Uniform(A_l)$.
3. *Next* l.

As noted above, in general to apply this univariate procedure to a multivariate posterior distribution, with $\Theta \in \mathbb{R}^d$, one would use at iteration l, of a deterministic scan Gibbs sampler, for the i-th element, having updated $i-1$ elements l times the full conditional posterior choice

$$\pi\left(\theta_i|\theta_1^{(l)}, 2\theta_2^{(l)}, \ldots, \theta_{i-1}^{(l)}, \theta_{i+1}^{(l-1)}, \ldots, \theta_d^{(l-1)}, \boldsymbol{x}_{1:T}\right). \tag{7.76}$$

The sampling in Algorithm 7.6 above would then be applied for each $i \in \{1, 2, \ldots, d\}$ to obtain the complete l-th sample of the Markov chain from $\pi\left(\boldsymbol{\theta}|\boldsymbol{x}_{1:T}\right)$.

There are many approaches that could be used in the determination of the level sets A_l for the density $\pi(\cdot)$ (see Neal 2003, section 4). For example, one can use a stepping out and a shrinkage procedure (see Neal 2003, figure 1, p. 713).

The basic idea is that given a sampled vertical level $u^{(l)}$, the level sets A_l can be found by positioning an interval of width w randomly around $\theta^{(l-1)}$. This interval is expanded in step sizes of width w until both ends are outside the slice. Then a new state is obtained by sampling uniformly from the interval until a point in the slice A_l is obtained. Points that fail can be used to shrink the interval. Developing such a procedure can be rather intricate in practice to implement if the full conditional posterior distributions are multi-modal. Thankfully, there are now efficient Slice Sampler packages available in standard softwares such as R, Matlab and Python.

Additionally, it is important to note that we only need to know the target full conditional posterior up to normalization (see Neal 2003, p. 710). To make more precise the intuitive description of the slice sampler presented earlier, we briefly detail the argument made by Neal on this point. Suppose we wish to sample a random vector $\boldsymbol{\Theta}$ whose density $\pi(\boldsymbol{\theta})$ is proportional to some function $f(\boldsymbol{\theta})$. This can be achieved by sampling uniformly from the $(n+1)$-dimensional region that lies under the plot of $f(\boldsymbol{\theta})$. This is formalized by introducing the auxiliary random variable U and defining a joint distribution over $\boldsymbol{\Theta}$ and U (which is uniform over the region $\{(\boldsymbol{\Theta}, U) : 0 < u < f(\boldsymbol{\theta})\}$ below the surface defined by $f(\boldsymbol{\theta})$) given by

$$\pi(\boldsymbol{\theta}, u) = \begin{cases} 1/Z, & \text{if} \quad 0 < u < f(\boldsymbol{\theta}), \\ 0, & \text{otherwise,} \end{cases} \tag{7.77}$$

where $Z = \int f(\boldsymbol{\theta})\, d\boldsymbol{\theta}$. Then the target marginal density for $\boldsymbol{\Theta}$ is given by

$$\pi(\boldsymbol{\theta}) = \int_0^{f(\boldsymbol{\theta})} \frac{1}{Z} du = \frac{f(\boldsymbol{\theta})}{Z} \tag{7.78}$$

as required.

Remark 7.13 *We note that such an algorithm aims to improve the mixing of the Markov chain around the support of the posterior through the use of the auxiliary variables, which means that the joint Markov chain on the parameters and auxiliary variables has a stationary distribution that is uniform in some domain for the volume under the posterior, such that the marginal distribution obtained by discarding the Markov chain samples for the auxiliary variables are actually samples from the true posterior. The challenge with the implementation of this algorithm is to obtain at each slice the level sets (or an approximation of the level sets) of the posterior. This is typically done numerically through a stepping in and stepping out routine, followed by a rejection sample. The interested reader is referred to Neal (2003, section 4).*

The simplest way to apply the slice sampler in a multivariate case is by applying the univariate slice sampler for each fully conditional distribution within the Gibbs sampler; for example, in the OpRisk context, see Peters *et al.* (2009).

Recently, several extensions have been developed for the slice sampler algorithm with a view to generalizing it to multivariate block Gibbs samplers known as reflective slice samplers, hyper rectangle slice samplers, and "crumbs" see the approaches presented by Tibbits *et al.* (2011), Mira *et al.* (2002), Murray *et al.* (2010), Thompson and Neal (2010a), Thompson (2011), Roberts and Rosenthal (2002), and Thompson and Neal (2010).

7.6.3 ADAPTIVE MCMC

As has now been demonstrated in the previous few sections, MCMC sampling has gained wide recognition in all areas of modeling and statistical estimation as an essential tool for performing inference in Bayesian models (see reviews and discussions by Gilks *et al.* 1996 and Brooks 1998). In this section, we discuss two recently developed classes of algorithms known as forms of adaptive MCMC (see a review by Andrieu and Thoms 2008).

As discussed in the section on MH algorithms, standard MCMC algorithms that do not incorporate adaptation often require a degree of "tuning" of the parameters controlling the algorithms' performance. This is typically performed by offline simulations to assess performance of the mixing of the resulting Markov chain followed by numerical investigation of the convergence rates to stationarity of the chain for different algorithmic settings of the proposal distribution. For example, the variant of the MH algorithm, the RW-MH algorithm with the widely used multivariate Gaussian proposal, has mixing performance that is controlled through specification of the Markov chain proposal distributions covariance matrix. Tuning this matrix for optimal performance can be computationally expensive and inefficient (see detailed discussions by Gilks *et al.* 1996, Brooks 1998, and Chib and Greenberg 1995). Optimal performance of an MCMC algorithm is typically either specified by the convergence rate of the Markov chain to stationarity or through the related quantity, the acceptance probability of the rejection step in the MCMC algorithm. In this regard, theoretically optimal results have been derived for several classes of statistical models, which now act as guides for more complicated sampling problems (see discussions by Roberts and Rosenthal 2001).

The potential in OpRisk modeling to have high dimensionality in the posterior parameter space provides a significant challenge for standard MCMC algorithms with respect to the design of an efficient proposal mechanism for the Markov chain. Therefore, it is desirable to automate this proposal construction for the MCMC sampler, avoiding computationally expensive tuning processes. Hence, we develop an adaptive version of the RW-MH algorithm. The incorporation of an adaptive proposal mechanism in an MCMC algorithm has been demonstrated to improve the performance of the sampling algorithm relative to standard MCMC approaches (see reviews of several examples of this improvement by Andrieu and Thoms 2008). The improvement is achieved by learning the structure of the Markov chain proposal distribution online in an automated fashion, avoiding offline tuning of the MCMC proposal mechanism.

There are several classes of adaptive MCMC algorithms and each class has several adaptation strategies (Roberts and Rosenthal 2009, Atchadé and Rosenthal 2005, Andrieu and Thoms 2008). These approaches can be classified as either internal adaptation mechanisms, including controlled MCMC methods, or external adaptation strategies (see discussion by Atchadé and Rosenthal, 2005).

Remark 7.14 *The distinguishing feature of adaptive MCMC algorithms, when compared to standard MCMC, is that the Markov chain is generated via a sequence of transition kernels. Adaptive algorithms get their name from the fact that they utilize a combination of time or state inhomogeneous proposal kernels. Each proposal in the sequence is allowed to depend on the past history of the Markov chain generated, resulting in many possible variants.*

When using inhomogeneous Markov kernels, it is particularly important to ensure that the generated Markov chain is ergodic, with the appropriate stationary distribution. Several recent papers proposing theoretical conditions that must be satisfied to ensure ergodicity of adaptive algorithms include Atchadé and Rosenthal (2005) and Haario *et al.* (2001, 2006). The papers by Roberts and Rosenthal (2007), Łatuszyński *et al.* (2013), and Bai *et al.* (2009) consider properties such as the ergodicity of the adaptive MCMC under conditions such as *Diminishing Adaptation* and *Bounded Convergence.*

Designing an adaptation strategy that satisfies these conditions guarantees asymptotic convergence of the law of the Markov chain samples to the target posterior and ensures that the Weak Law of Large Numbers holds for bounded test functions of the parameter space (interested readers are referred to Roberts and Rosenthal 2009 for details).

Technical Notes Regarding Adaptive MCMC. The practitioner reading this section may wish to skip the following technical notes relating to the validity of the adaptive MCMC algorithm. Primarily the focus of this section involves discussing under what types of conditions one way achieve a sensible notion of sample average from a Markov chain in which the transition matrix is constantly changing, that is, "adapting". However, if a practitioner wishes to adopt or try such methods with confidence that they are theoretically justified, they may go directly past this section to the description of the algorithms.

In stating these conditions more precisely, we consider the distance between two probability distributions generically denoted by ν and μ, that are formed from measures that we assume admit a density with respect to, say, the Lebesgue measure. We can again recall this distance in a slightly different form from where we denoted it by $|| \cdot ||_{TV}$, where we define this distance generically according to

$$d(\mu, \nu) = \sup \left\{ \left| \int \varphi d\mu - \int \varphi d\nu \right| : \varphi \in \mathcal{D} \right\} \tag{7.79}$$

for some test function φ in a class of functions denoted $\mathcal{D}$, for example, the class of all bounded and k-th order differentiable functions, etc. In the case of considering the Total Variation distance $||\mu - \nu||_{TV}$, we consider the space of Borel sets $\mathcal{B}$ and define the distance

$$d(\mu, \nu) = ||\mu - \nu||_{TV} := \sup_{A \in \mathcal{B}} |\nu(A) - \mu(A)| \tag{7.80}$$

with the class of functions given by the indicators on the Borel sets $\mathcal{D} = \{\mathbb{I}_A : A \in \mathcal{B}\}$. This distance in the case of two probability measures ν and μ will clearly be between 0 and 1 and will provide a comparison of convergence between two probability measures, which will imply weak convergence (convergence in distribution).

One can formally consider this distance and utilize it to develop a condition that will succinctly state one of the required conditions for an adaptive Markov chain to satisfy ergodicity. This condition is known as diminishing adaptation and is given as follows.

Diminishing Adaptation.

$$\lim_{n \to \infty} \sup_{\boldsymbol{\Theta} \in \mathbb{R}^d} \| Q_{\Gamma_{n+1}}(\boldsymbol{\Theta}, \boldsymbol{\Theta}') - Q_{\Gamma_n}(\boldsymbol{\Theta}, \boldsymbol{\Theta}') \|_{TV} = 0 \text{ in prob.},$$

where $\boldsymbol{\Theta}$ is the vector of parameters in the Bayesian model and the measures ν and μ are selected to be the Markov transition kernel $Q_{\Gamma_{n+1}}$ at a random time denoted by index Γ_{n+1} when the $n+1$-th update of the kernel (in the learning phase) was applied.

The second condition required for a transition kernel to satisfy ergodicity is known as bounded convergence and is given as follows.

Bounded Convergence.

$$\{M_\epsilon(\Theta^{(l)}, \Gamma_j)\}_{j=0}^{\infty} \text{ is bounded in prob., } \epsilon > 0$$

with convergence time defined as $M_\epsilon(\theta, \gamma) = \inf \left\{ l \geq 1 : \| Q_\gamma^l(\theta, \cdot) - P(\cdot)\|_{TV} \leq \epsilon \right\}$.

Creating an adaptive MCMC sampler with a proposal distribution that satisfies these technical conditions ensures the following:

- Asymptotic convergence:

$$\lim_{l \to \infty} \| \mathcal{L}(\Theta^{(l)}) - \pi(\cdot)\|_{TV} = 0 \text{ in prob.},$$

where $\pi(\cdot)$ is the target posterior distribution–intended stationary distribution of the Markov chain and $\mathcal{L}(\cdot)$ denotes the law of the random variable (distribution);

- Weak Law of Large Numbers (for all bounded functions g)

$$\lim_{L\to\infty} \frac{1}{L}\sum_{l=1}^{L} g(\Theta^{(l)}) = \pi(g) = \int g(\theta)\pi(\theta)\,d\theta.$$

In the following, we discuss one particular illustrative choice of transition kernel that satisfies these conditions and has been used successfully in several applications (see Andrieu and Thoms 2008, Peters *et al.* 2010, Korostil *et al.* 2012, and Roberts and Rosenthal 2009). The algorithm we present is one of many possibilities in this literature and is known as the adaptive Metropolis algorithm. It involves utilizing an MCMC proposal distribution, parameterized by parameter vector or matrix $\mathbf{\Psi}$, and learning the appropriate values for $\mathbf{\Psi}$ recursively utilizing the previous samples of the Markov chain that have been accepted under the MCMC accept–reject mechanism. This is achieved online, adapting according to the support of the posterior distribution, thereby allowing the Markov chain to discover and explore the regions of the posterior distribution that have the most mass. Through this online adaptive learning mechanism, the Markov chain proposal distribution can significantly improve the acceptance rate of the Markov chain, enabling efficient mixing and improving the exploration of the posterior support by the Markov chain.

To provide practitioners with perhaps the simplest version of an adaptive MCMC algorithm proposal that could be considered, we present the internal adaptation strategy based on the adaptive Metropolis algorithm detailed by Roberts and Rosenthal (2009). This is a variant of the approach proposed by Haario *et al.* (2001), which develops an RW-MH that estimates the global covariance structure from the past samples.

Under an adaptive Metropolis algorithm, the proposal distribution is based on a Gaussian mixture kernel detailed by Roberts and Rosenthal (2009). The proposal, $q\left(\mathbf{\Theta}^{(j-1)}, \mathbf{\Theta}^{(j)}\right)$, involves an adaptive Gaussian-mixture Metropolis proposal, one component of which has a covariance structure that is adaptively learnt online as the algorithm explores the posterior distribution. For iteration j of the Markov chain the proposal is

$$q_j\left(\mathbf{\Theta}^{(j-1)}, \cdot\right) = \gamma \mathit{Normal}\left(\mathbf{\Theta}^{(*)}; \mathbf{\Theta}^{(j-1)}, \frac{(2.38)^2}{d}\Sigma_j\right) + (1-\gamma)\,\mathit{Normal}\left(\mathbf{\Theta}^{(*)}; \mathbf{\Theta}^{(j-1)}, \frac{(0.1)^2}{d}I_{d,d}\right). \quad (7.81)$$

Here, $\mathbf{\Psi}_j = \Sigma_j$ is the current empirical estimate of the proposal parameters; in this case, the posterior covariance between the parameters of $\mathbf{\Theta}$, estimated using samples from the Markov chain up to time $j-1$. Small positive constant γ is usually taken as equal to 0.05 (Roberts and Rosenthal 2009). The theoretical motivation for the choices of scale factors 2.38, 0.1 and dimension d are all provided by Roberts and Rosenthal (2009) and are based on optimality conditions presented by Roberts and Rosenthal (2001).

We note that the update of the covariance matrix can be done recursively online via the following recursion (as detailed by Atchadé *et al.* 2011) and presented in the following algorithm. In the following sequence of steps for the j-th iteration of the adaptive Metropolis algorithm, we will update the state of the Markov chain from $\mathbf{\Theta}^{(j-1)}$ to parameter vector $\mathbf{\Theta}^{(j)}$ according to the steps in Algorithm 7.7.

Algorithm 7.7 (Adaptive Metropolis Algorithm)

1. *Initialize the parameter vector* $\boldsymbol{\theta}^{(0)} \in \mathbb{R}^d$ *and the covariance matrix of the proposal* $\Psi^{(0)} = \Sigma^{(0)} = \frac{(0.1)^2}{d} I_{d,d}$;
2. *For* $l = 1, \ldots, L$
   a) *If* $l > 1$ *then update the adaptive Metropolis proposal covariance matrix recursively using previous samples from the Markov chain created via*

$$\mu_{l+1} = \mu_l + \frac{1}{l+1}\left(\boldsymbol{\Theta}^{(l-1)} - \mu_l\right),$$
$$\Sigma_{l+1} = \Sigma_l + \frac{1}{l+1}\left(\left(\boldsymbol{\Theta}^{(l-1)} - \mu_l\right)\left(\boldsymbol{\Theta}^{(l-1)} - \mu_l\right)^T - \Sigma_l\right).$$

   b) *Sample a proposed vector of parameters* $\boldsymbol{\theta}^{(*)} \sim q\left(\boldsymbol{\theta}^{(l-1)}, \cdot\right)$ *from an adaptive MCMC proposal* $\left(q_l\left(\boldsymbol{\Theta}^{(l-1)}, \cdot\right)\right)$ *constructed using previous Markov chain samples* $\left\{\boldsymbol{\Theta}^{(0)}, \ldots, \boldsymbol{\Theta}^{(l-1)}\right\}$ *as detailed by the mixture proposal in Equation (7.81);*
   c) *Accept the proposed new Markov chain state comprised of* $\boldsymbol{\theta}^{(*)}$ *with acceptance probability given by*

$$\alpha\left(\boldsymbol{\theta}^{(l-1)}, \boldsymbol{\theta}^*\right) = \min\left(1, \frac{\pi\left(\boldsymbol{\theta}^* | \boldsymbol{x}_1, \ldots, \boldsymbol{x}_T\right) q\left(\boldsymbol{\theta}^{(l-1)}, \boldsymbol{\theta}^*\right)}{\pi\left(\boldsymbol{\theta}^{(l-1)} | \boldsymbol{x}_1, \ldots, \boldsymbol{x}_T\right) q\left(\boldsymbol{\theta}^*, \boldsymbol{\theta}^{(l-1)}\right)}\right), \tag{7.82}$$

   where we evaluate this acceptance probability utilizing the expressions detailed previously. If there is acceptance, then $\boldsymbol{\theta}^{(l)} = \boldsymbol{\theta}^{(*)}$, *else one sets* $\boldsymbol{\theta}^{(l)} = \boldsymbol{\theta}^{(l-1)}$.
3. *Next* l.

7.6.4 RIEMANN–MANIFOLD HAMILTONIAN MONTE CARLO SAMPLER (AUTOMATED LOCAL ADAPTION)

By this stage, we have clearly established the fact that the design of the proposal distribution for the Markov chain that is created for an MCMC method with target posterior $\pi\left(\boldsymbol{\Theta} | \boldsymbol{x}_{1:T}\right)$ can directly effect the ability to make accurate inference. In addition, we have discovered that the transition kernel for the class of MCMC methods of interest is typically given by

$$\begin{aligned} Q\left(\boldsymbol{\Theta}^{(l)}, d\boldsymbol{\Theta}^{(l+1)}\right) &= q\left(\boldsymbol{\Theta}^{(l)}, d\boldsymbol{\Theta}^{(l+1)}\right) \alpha\left(\boldsymbol{\Theta}^{(l)}, d\boldsymbol{\Theta}^{(l+1)}\right) \\ &\quad + \left[1 - \int q\left(\boldsymbol{\Theta}^{(l)}, \boldsymbol{z}\right) \alpha\left(\boldsymbol{\Theta}^{(l)}, \boldsymbol{z}\right) d\boldsymbol{z}\right] \mathbb{I}\left[\boldsymbol{\Theta}^{(l+1)} = \boldsymbol{\Theta}^{(l)}\right], \end{aligned} \tag{7.83}$$

where the (adaptive) design of $q\left(\boldsymbol{\Theta}^{(l)}, d\boldsymbol{\Theta}^{(l+1)}\right)$ is of direct interest for reducing variance in Monte Carlo estimates. In the following sections, we will primarily focus on how to adaptively modify $q\left(\boldsymbol{\Theta}^{(l)}, d\boldsymbol{\Theta}^{(l+1)}\right)$ to improve the mixing of the resulting MCMC samplers.

Next we present another more advanced class of algorithms known as Hamiltonian Monte Carlo (HMC) methods. The extension to this class of algorithms we discuss is a recently introduced class of MCMC samplers by Girolami and Calderhead (2011), which was developed to help automate the design of the proposal distributions within the Markov kernel; in this case, this will be achieved through the use of what is known as Riemann–Manifold Hamiltonian Monte Carlo (RM-HMC; see a detailed tutorial overview by Neal 2010).

We first define the basic principles on which RM-HMC is discussed in detail by Duane *et al.* (1987), Girolami and Calderhead (2011), and Neal (2010). The context of the RM-HMC algorithm derives from the design of a Markov chain proposal obtained from a discretized Langevin diffusion with two components: a stochastic discretized diffusion component and a second component based on a discretized deterministic component constructed from gradient information of the target density. This first class of algorithm was known as the Metropolis-Adjusted Langevin Algorithm (MALA) method of Stramer and Tweedie (1999) and adaptive versions by Marshall and Roberts (2012). Alternative approaches of a similar nature were also developed and are generally known as Hybrid Monte Carlo (hybrid MC) proposals as they also involve a combination of deterministic and stochastic components obtained from discretization of a physical stochastic process. Such hybrid MC algorithms typically produce an ergodic Markov chain in which large traversals of the posterior support are accepted with high probability.

7.6.4.1 Sampling the Posterior Density via Establishing Related Hamiltonian Mechanics. This section provides basic details to aid the understanding of how one develops such an HMC proposal mechanism to efficiently explore the support of the target posterior distribution. It is instructive to consider the following nonstandard formulation of a posterior distribution, which we repose as the equations of motion under Hamiltonian mechanics. This involves specification of a system of partial differential equations that can be solved to provide the building blocks of the HMC algorithms. Consider the random vector of posterior parameters $\boldsymbol{\Theta} \in \mathbb{R}^d$ with $\boldsymbol{\Theta} \sim \pi\left(\boldsymbol{\Theta}|\boldsymbol{x}_{1:T}\right)$ and an independent auxiliary random vector denoted by $\boldsymbol{Z} \in \mathbb{R}^d$ with $\boldsymbol{Z} \sim \mathit{Normal}(0, \Sigma)$. Now consider construction of the negative log joint density given by the equivalent interpretation as a Hamiltonian $H(\boldsymbol{\Theta}, \boldsymbol{z})$ of an energy-conserving physical dynamic system described by

$$H(\boldsymbol{\Theta}, \boldsymbol{z}) = -\ln \pi\left(\boldsymbol{\Theta}|\boldsymbol{x}_{1:T}\right) + \frac{1}{2}\ln(2\pi)^d|\Sigma| + \frac{1}{2}\boldsymbol{z}^T\Sigma^{-1}\boldsymbol{z}, \tag{7.84}$$

with $-\ln \pi\left(\boldsymbol{\Theta}|\boldsymbol{x}_{1:T}\right)$ the accumulated potential energy at location $\boldsymbol{\Theta}$, the term $\frac{1}{2}\boldsymbol{z}^T\Sigma^{-1}\boldsymbol{z}$ representing the kinetic energy, and $\boldsymbol{z}$ the momentum and mass matrix Σ (see discussions by Girolami and Calderhead, 2011) and Neal (2010). To understand why this interpretation may be of relevance to the design of an MCMC sampler one needs to consider the score function of the joint distribution of random vectors $\boldsymbol{\Theta}$ and $\boldsymbol{Z}$ given by

$$\frac{\partial H}{\partial \boldsymbol{z}} = \Sigma^{-1}\boldsymbol{z}, \quad -\frac{\partial H}{\partial \boldsymbol{\Theta}} = \nabla_{\boldsymbol{\Theta}} \ln \pi\left(\boldsymbol{\Theta}|\boldsymbol{x}_{1:T}\right). \tag{7.85}$$

This deterministic system of partial differential equations can be used to re-interpret, with respect to the joint distribution of the two random vectors, a dynamical system with artificial "time" unit τ given by

$$\frac{d\Theta}{d\tau} = \frac{\partial H}{\partial \boldsymbol{z}} \quad \text{and} \quad \frac{d\boldsymbol{z}}{d\tau} = -\frac{\partial H}{\partial \Theta}.$$

By linking the joint distribution of random vectors $\boldsymbol{\Theta}$ and $\boldsymbol{Z}$ to a physical system evolution, one may now construct from these artificial dynamics a dynamic proposal mechanism for a Markov chain sampler (see Algorithm 7.8).

Remark 7.15 *It should be noted that the numerical integrator should provide a dynamic solution, which is interpreted as a transformation mapping from the parameter vector* $(\boldsymbol{\Theta}, \boldsymbol{z})$ *to the newly "proposed" parameter vector* $(\boldsymbol{\Theta}', \boldsymbol{z}')$*. If this mapping is time-reversible (in the artificial time* τ*) and volume-preserving, then it can be utilized to design an MH Markov chain reject–accept sampler, that is, it may used as a proposal mechanism in an MCMC sampler.*

Fortunately, by constructing such a Hamiltonian dynamical system for utilization in the proposal as an efficient and perhaps adaptive MCMC sampler, one automatically satisfies the following important properties (see proofs by Neal 2010):

1. **Reversible proposals**. Hamiltonian mechanics preserve the reversibility of a Markov chain constructed with a proposal that utilizes such a dynamic system to explore the support of the posterior. In other words, one may define a mapping from the state of the system at time τ_1 given by $(\boldsymbol{\theta}(\tau_1), \boldsymbol{z}(\tau_1))$ to a new state at time $\tau_2 > \tau_1$, denoted $(\boldsymbol{\theta}(\tau_2), \boldsymbol{z}(\tau_2))$, which is one-to-one and therefore invertible;
2. **Invariance of the Hamiltonian system**. Designing a proposal from a Hamiltonian system of equations will create dynamics that are invariant within the Hamiltonian system. That is, the dynamics preserve the structure of the Hamiltonian system;
3. **Volume preservation (Liouville theorem)**. It is well known that a Hamiltonian system is volume-preserving. The consequence of this is that using this dynamic system to construct a proposal in MCMC will result in an acceptance probability that does not require a Jacobian mass transform. This is a significant advantage of such a transformational proposal, making evaluation of the proposal in the MCMC acceptance probability numerically easier and more numerically well behaved.

7.6.4.2 Sampling the Posterior via Discretization of the Hamiltonian Mechanics.

Utilising this Hamiltonian dynamic system for the MCMC proposal therefore requires a numerical solver for the two partial differential equations (PDEs) in order to generate the proposal at each iteration of the HMC. Therefore, the challenge lies in finding numerical integrators that are both time-reversible and volume-preserving. Fortunately, one such explicit class of integrators is; the symplectic class, a particular example from this class is the leapfrog integrator (see Duane *et al.* 1987). This was utilized to define an HMC solution, where one iteration of the HMC algorithm therefore involves drawing randomly a realized vector $\boldsymbol{z}$ and then iterating the leapfrog integrator defined by deterministic recursions for step size ϵ in Algorithm 7.8.

Hence, the generation of an MCMC proposal under the HMC system would proceed as follows.

Algorithm 7.8 (Hybrid Monte Carlo Algorithm Proposal)

1. *Sample a realization of auxiliary random vector* $\boldsymbol{Z} \sim Normal(0, \Sigma)$;
2. *Perform numerical integration to solve* $\frac{d\boldsymbol{\Theta}}{d\tau} = \frac{\partial H}{\partial \boldsymbol{z}}$ *and* $\frac{d\boldsymbol{z}}{d\tau} = -\frac{\partial H}{\partial \boldsymbol{\Theta}}$, *using the sampled value of the auxiliary variable, thus providing an evolution equation in the joint distribution space for random vectors* $\boldsymbol{\Theta}$ *and* $\boldsymbol{Z}$ *which are characterized as follows:*
   a) $\boldsymbol{z}(\tau + \epsilon/2) = \boldsymbol{z}(\tau) + \epsilon \nabla_{\boldsymbol{\Theta}} \ln \pi \left(\boldsymbol{\Theta}|\boldsymbol{x}_{1:T}\right)\big|_{\boldsymbol{\Theta}=\boldsymbol{\Theta}(\tau)} /2$;
   b) $\boldsymbol{\Theta}(\tau + \epsilon) = \boldsymbol{\Theta}(\tau) + \epsilon\Sigma^{-1}\boldsymbol{z}(\tau + \epsilon/2)$;
   c) $\boldsymbol{z}(\tau + \epsilon) = \boldsymbol{z}(\tau + \epsilon/2) + \frac{\epsilon}{2}\nabla_{\boldsymbol{\Theta}} \ln \pi \left(\boldsymbol{\Theta}|\boldsymbol{x}_{1:T}\right)\big|_{\boldsymbol{\Theta}=\boldsymbol{\Theta}(\tau+\epsilon)}$.

Iteration of this algorithm generates a sequence of random proposals of initial value for $\boldsymbol{Z}$ followed by a deterministic trajectory solution for n steps of size ϵ via a leapfrog integration iteration for the proposal. Taking the last point $\boldsymbol{\Theta}_T = \boldsymbol{\Theta}_*$ as the proposal, one then accepts this proposed point under the MCMC accept–reject mechanism with the following probability, which involves the Hamiltonian energy functions:

$$\alpha(\boldsymbol{z}, \boldsymbol{\Theta}; \boldsymbol{z}_*, \boldsymbol{\Theta}_*) = \min\left(1, -H(\boldsymbol{z}_*, \boldsymbol{\Theta}_*) + H(\boldsymbol{z}, \boldsymbol{\Theta})\right). \tag{7.86}$$

It was recently realized that one could further adapt this Hamiltonian proposal through observing that the behavior of the simulated trajectory was directly affected by "tuning" the matrix Σ; therefore, one could try to find a way to learn efficient choices for Σ to improve the acceptance probability of a move by adapting Σ to local structure of the target distribution posterior. Therefore, the MCMC proposal constructed in this fashion is then tuned via the selection of the mass matrix Σ, the number of iteration steps n, and the step size ϵ. In general, one may summarize this HMC algorithm according to the Langevin discretized diffusion recursion

$$\boldsymbol{\Theta}(\tau + \epsilon) = \underbrace{\boldsymbol{\Theta}(\tau) + \frac{\epsilon^2}{2}\Sigma^{-1}\frac{\epsilon}{2}\nabla_{\boldsymbol{\Theta}} \ln \pi \left(\boldsymbol{\Theta}|\boldsymbol{x}_{1:T}\right)\Big|_{\boldsymbol{\Theta}=\boldsymbol{\Theta}(\tau)}}_{\text{Preconditioned deterministic innovation}} + \underbrace{\epsilon\Sigma^{-2}\boldsymbol{z}(\tau)}_{\text{Stochastic innovation}}. \tag{7.87}$$

The adaptive MCMC development of this algorithm is discussed extensively by Girolami and Calderhead (2011) and involves primarily the adaption of the mass matrix Σ. The other algorithmic parameters to consider involve the number of steps n and the step size ϵ – these may typically be effectively estimated from acceptance probabilities of the MCMC chain. The following two key points were noted by Girolami and Calderhead (2011) to consider in order to improve the performance of the HMC algorithm.

Remark 7.16 *Stochastic transitions that account for local geometric structure of the target distribution when making proposals to different regions of the distributional support can improve the Markov chain exploration and mixing. One way to achieve this is to replace the HMC global covariance matrix proposal* Σ *(mixing matrix) with a position specific version.*

Remark 7.17 *Under the HMC algorithm described, the deterministic component of the Langevin proposal involves the gradient of the target distribution, which is preconditioned by the inverse global mass matrix. It was noted that adapting this mass matrix* Σ *to local structure of the target distribution would improve mixing performance. This can be achieved by exploiting a Riemannian structure of the target distribution parameter space using a localized metric tensor.*

To address these remarks the approach of Girolami and Calderhead (2011) was to develop an RM-HMC algorithm. Here we briefly discuss this adaptive HMC algorithm and present the details so that it may be utilized to make inference in model estimation.

In the RM-HMC setting, one considers locally adapting the generic Hamiltonian given by $-\ln \pi(\boldsymbol{\Theta}, \boldsymbol{z}) = -\ln \pi(\boldsymbol{\Theta}) - \ln \pi(\boldsymbol{z})$ in the HMC setting. This is achieved by interpreting the family of parameterized probability densities for d-dimensional random vector $\boldsymbol{\Theta}$ given by $\ln p(\boldsymbol{\Theta})$ as defining a Riemannian manifold that has an associated metric tensor, which may, for example, be selected to be the Fisher information matrix for the target distribution model given by $\mathcal{I}(\boldsymbol{\Theta}) = \mathbb{E}\left[\nabla_{\boldsymbol{\Theta}} \ln \pi(\boldsymbol{\Theta}) \nabla_{\boldsymbol{\Theta}} \ln \pi(\boldsymbol{\Theta})^T\right]$.

Under this modified specification using the alternate metric tensor, say, the Fisher information matrix $\mathcal{I}(\boldsymbol{\Theta})$, one obtains a Hamiltonian equation given by

$$H(\boldsymbol{\Theta}, \boldsymbol{z}) = -\ln \pi\left(\boldsymbol{\Theta} | \boldsymbol{x}_{1:T}\right) + \frac{1}{2} \ln (2\pi)^d |\mathcal{I}(\boldsymbol{\Theta})| + \frac{1}{2} \boldsymbol{z}^T \mathcal{I}(\boldsymbol{\Theta})^{-1} \boldsymbol{z}. \tag{7.88}$$

Under this formulation, one can sample the auxiliary variable vectors in the RM-HMC scheme for $\boldsymbol{z}$ given by a conditionally Gaussian distribution $\boldsymbol{Z} \sim \mathit{Normal}(\boldsymbol{z}; \boldsymbol{0}, \mathcal{I}(\boldsymbol{\Theta}))$. It is now clear that such a modification through the Riemannian structure of the target distribution allows one to utilize a locally adapted proposal; however, the consequence of this structure is that the Hamiltonian is no longer separable. The consequence of this loss of separability is that the symplectic integration procedure previously proposed for the standard HMC algorithm will be required to be modified, as detailed by Girolami and Calderhead (2011) and shown later.

Consider designing an RM-HMC algorithm to move from state $(\boldsymbol{z}_0, \boldsymbol{\Theta}_0)$, that is, some previous RM-HMC state, to the proposed state $(\boldsymbol{z}_*, \boldsymbol{\Theta}_*)$. If one defines the metric tensor for the local adaption of the covariance Σ in which $\Sigma\left(\boldsymbol{\Theta}_{j-1}\right) = \mathcal{I}(\boldsymbol{\Theta})$ and the integration step size is given by ϵ and total number of iterations by T, then the full symplectic integrator for the RM-HMC algorithm is now adjusted to the following five steps:

$$\begin{aligned}
\boldsymbol{z}_1 &= \boldsymbol{z}_0 - \frac{\epsilon}{2} \nabla_{\boldsymbol{\Theta}} \left(-\ln \pi\left(\boldsymbol{\Theta} | \boldsymbol{x}_{1:T}\right) + \frac{1}{2} \ln (2\pi)^d |\mathcal{I}(\boldsymbol{\Theta})| \right) \Big|_{\boldsymbol{\Theta}_0} \\
\boldsymbol{z}_2 &= g\left(\boldsymbol{\Theta}_0, \boldsymbol{z}_1, \frac{\epsilon}{2}\right) \\
\boldsymbol{\Theta}_* &= \boldsymbol{\Theta}_0 + \epsilon \mathcal{I}(\boldsymbol{\Theta})^{-1} \\
\boldsymbol{z}_3 &= g\left(\boldsymbol{\Theta}_*, \boldsymbol{z}_2, \frac{\epsilon}{2}\right) \\
\boldsymbol{z}_* &= \boldsymbol{z}_3 - \frac{\epsilon}{2} \nabla_{\boldsymbol{\Theta}} \left(-\ln \pi\left(\boldsymbol{\Theta} | \boldsymbol{x}_{1:T}\right) + \frac{1}{2} \ln (2\pi)^d |\mathcal{I}(\boldsymbol{\Theta})| \right) \Big|_{\boldsymbol{\Theta}_*}
\end{aligned}$$

with the vector valued function g defined as presented in the technical appendix by Girolami and Calderhead (2011) for both the scalar and multivariate parameter $\boldsymbol{\Theta}$ cases.

This is then iteratively applied in Algorithm 7.8 with the new symplectic integration scheme (five steps), and the final proposed state after T iterations of the proposal is accepted with the MH-rejection scheme with acceptance probability again presented with the appropriate exponential of the difference between the nonseparable Hamiltonian (see Equation (7.88)) at the old proposed state minus the Hamiltonian at the newly proposed Markov chain state obtained from the symplectic integrator proposal.

7.7 Sequential Monte Carlo (SMC) Samplers and Importance Sampling

SMC methods have emerged out of the fields of engineering, probability and statistics in recent years. Variants of the methods sometimes appear under the names of particle filtering or interacting particle systems (e.g., Ristic *et al.* 2004, Doucet *et al.* 2001, Del Moral 2004), and their theoretical properties have been extensively studied by Crisan and Doucet (2002), Del Moral (2004), Chopin (2004), and Künsch (2005). In the OpRisk context, such algorithms have been developed for insurance and OpRisk applications (see Peters *et al.* 2009 and Del Moral *et al.* 2013).

The standard SMC algorithm involves finding a numerical solution to a set of filtering recursions, such as filtering problems arising from nonlinear/non-Gaussian state space models. Under this framework, the SMC algorithm samples from a (often naturally occurring) sequence of distributions π_t, indexed by $t = 1, \ldots, T$. Each distribution is defined on the support $E^t = E \times E \times \cdots \times E$ for some generic space denoted E. This context is not typically of interest to OpRisk settings; however, this class of algorithms was adapted to tackle the same class of problems typically addressed by MCMC methods where one has instead a sequence of distributions $\{\pi_l\}_{l \geq 1}$ each defined on fixed support E; NOTE: not a product space $E_l = E \times E \cdots \times E$ but a fixed space E. Del Moral *et al.* (2006), Peters (2005), and Peters *et al.* (2009) generalize the SMC algorithm to the case where the target distributions π_t are all defined on the same support E. This generalization, termed the SMC *sampler*, adapts the SMC algorithm to the more popular setting in which the state space E remains static, that is, the settings we have discussed earlier with regard to the MCMC algorithms.

In short, the SMC sampler generates weighted samples (termed *particles*) from a sequence of distributions π_t, for $t = 1, \ldots, T$, where π_T may be of particular interest. We refer to π_T as the target distribution such as a posterior distribution for model parameters in an LDA model.

Procedurally, particles obtained from an arbitrary initial distribution π_1, with a set of corresponding initial weights, are sequentially propagated through each distribution π_t in the sequence via three processes, involving mutation (or move), correction (or importance weighting), and selection (or resampling). The final weighted particles at distribution π_T are considered weighted samples from the target distribution π. The mechanism is similar to sequential IS (resampling), see details by Liu 2008 and Doucet *et al.* 2001, with one of the crucial differences being the framework under which the particles are allowed to move, resulting in differences in the calculation of the importance weights of the particles.

One of the major difficulties with SMC-type algorithms is particle depletion, in which the weights of the majority of the particles gradually decrease to zero, while a few particle weights dominate the population. This severely increases the variability of Monte Carlo estimates of expectations under π. In this chapter, we develop an algorithm that incorporates the partial rejection control (PRC) strategy of Liu (1998) into the SMC sampler framework. A particular motivation for this stems from the recent developments in "likelihood-free" (or approximate Bayesian) computation to be discussed in Section 7.8, where an extremely high proportion of mutated particles are expected to have very small, or exactly zero, posterior weights (see discussions in this context by Peters *et al.* 2009).

In this chapter, we survey some basic developments in the SMC samplers and SMC samplers PRC algorithm, in which the PRC mechanism is built directly into the mutation kernel of the SMC sampler. This choice of algorithm allows a particle mutation to be rejected if the resulting importance weight is below a certain threshold. This turns out to be very valuable for a range of estimation and sampling problems in OpRisk. We also discuss implementation issues arising from the inclusion of the PRC stage, including estimation for the resultant kernel normalizing constant. The theoretical properties and justifications for this class of algorithms is provided by Peters *et al.* (2009).

7.7.1 MOTIVATING OPRISK APPLICATIONS FOR SMC SAMPLERS

The context of the SMC sampler algorithm involves sampling from a sequence of distributions $\{\pi_t(d\boldsymbol{\theta})\}_{t=1}^{T}$. This has many applications in practice for OpRisk modeling and includes settings such as the following:

1. **Tempering (on the data)**. In this case, the sequence of distributions is constructed as $\pi_t(d\boldsymbol{\theta}) = \pi(d\boldsymbol{\theta}|\boldsymbol{x}_{1:t})$; see discussions and motivation for this type of application by Chopin (2002). Alternatively, versions of tempering one could also consider would involve sequences of distributions given by $\pi_t(d\boldsymbol{\theta}) \propto [\pi(d\boldsymbol{\theta})]^{\gamma_t} [\pi(d\boldsymbol{\theta})]^{1-\gamma_t}$ for some schedule of increasing powers, $\gamma_t \in [0,1]$, with $0 \leq \gamma_0 \leq \gamma_1 \leq \cdots \leq \gamma_L = 1$;
2. **Progressively constrained distributions and rare events**. A second common application for such a sequence of distributions would be to move from a simple and tractable posterior distribution $\pi_0(d\boldsymbol{\theta})$ to a distribution of interest (such as constrained or truncated distribution) $\pi_T(d\boldsymbol{\theta})$. This could be achieved through a progressive sequence of distributions; see applications in rare-event simulations by Johansen (2009). In this context, one may consider $\pi_t(d\boldsymbol{\theta}) = \pi_t(d\boldsymbol{\theta} \in A_t)$ with a successively contracting sequence of sets $A_0 \supseteq A_1 \supseteq \cdots \supseteq A_T$. Examples may include $A_0 = \mathbb{R}^+$ and $A_n = [a_n, \infty)$ with $a_1 \leq a_2 \leq \cdots \leq a_T < \infty$;
3. **Stochastic optimization and parameter estimation**. A third application of such sequences of distributions would involve the notion of simulated annealing in which the sequence of distributions is given by $\pi_t(d\boldsymbol{\theta}) \propto [\pi(d\boldsymbol{\theta})]^{\gamma_t}$ for some sequence of increasing powers $\{\gamma_t\}_{t=0}^{T}$.

In the following three examples, we provide details of a few of the particular applications that are of relevance and important to OpRisk settings where such sequences of distributions can be used effectively in an OpRisk setting to simulate via SMC samplers algorithm key quantities.

EXAMPLE 7.4 Sampling Rare Events in Compound Processes Tails

Consider a single risk LDA model defined by the compound process $Z = \sum_{n=1}^{N} X_n$ with a random number of losses given by frequency distribution $N \sim F_N$ and a severity distribution $X_n \sim F_X$ for each i.i.d. loss. In many settings in OpRisk, one is interested in quantifying tail functionals, which would require being able to obtain draws from the tail of the compound process given by $Z|Z > A \sim F_{Z|Z>A}$ given in terms of the frequency and severity models for some $A > 0$ by

$$\pi\left(z|\sum_{n=1}^{N} X_n > A\right) = f_Z\left(z\left|\sum_{n=1}^{N} X_n > A\right.\right)$$
$$= \frac{\sum_{n=1}^{\infty} \mathbb{P}\mathrm{r}(N=n) f_X^{(n)*}(x)}{1-\left(\mathbb{P}\mathrm{r}(N=0)+\sum_{n=1}^{\infty} \mathbb{P}\mathrm{r}(N=n) F_X^{(n)*}(A)\right)} \mathbb{I}\left[\sum_{n=1}^{N} X_n > A\right]. \tag{7.89}$$

In general, it will be difficult to sample directly via standard Monte Carlo simulation techniques from such a distribution when A is very large, such as $A = \mathrm{VaR}_{1-\alpha}[Z]$ for some small α. Therefore, it is sensible and computationally more efficient to consider constructing a sequence of such distributions, defined by a decreasing sequence of levels $\{\alpha_t\}$ that are progressively moving the focus of the samples obtained toward the tails of the target distribution. One example of such a sequence involves the following choice:

$$\pi_t\left(z\left|\sum_{n=1}^{N} X_n > \mathrm{VaR}_{1-\alpha_t}[Z]\right.\right)$$
$$= \frac{\sum_{n=1}^{\infty} \mathbb{P}\mathrm{r}(N = n) f_X^{(n)*}(x)}{1-\left(\mathbb{P}\mathrm{r}(N = 0)+\sum_{n=1}^{\infty} \mathbb{P}\mathrm{r}(N = n) F_X^{(n)*}(\mathrm{VaR}_{1-\alpha_t}[Z])\right)} \tag{7.90}$$
$$\times \mathbb{I}\left[\sum_{n=1}^{N} X_n > \mathrm{VaR}_{1-\alpha_t}[Z]\right].$$

Specific examples of this type of sequence construction are detailed extensively in Peters and Shevchenko (2015).

Poisson–Inverse Gaussian LDA Model Tail Estimation. In Figure 7.3, we show the sequence of such target distributions for a simple LDA model with Poisson–Inverse Gaussian frequency and severity models. In such a framework, we consider the losses in the risk process generated from shape-scale Inverse-Gaussian severity models (with strictly positive support) and closed under convolution given by

$$X_i \sim F_X(x;\mu,\gamma) = \Phi\left(\sqrt{\frac{\gamma}{x}}\left(\frac{x}{\mu}-1\right)\right) + \exp\left(\frac{2\gamma}{\mu}\right)\Phi\left(-\sqrt{\frac{\gamma}{x}}\left(\frac{x}{\mu}+1\right)\right) \tag{7.91}$$

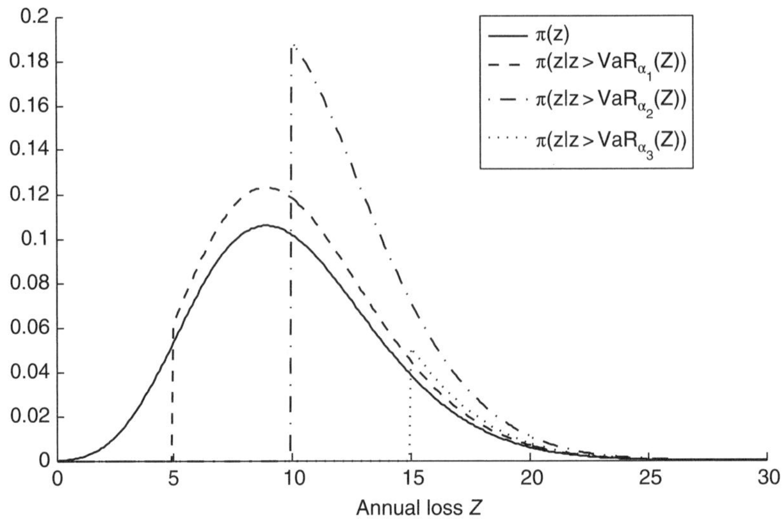

FIGURE 7.3 **Plot of the sequence of truncated annual loss distributions for a Poisson($\lambda = 2$) and Inverse Gaussian($\mu = 1, \gamma = 2$) with a truncation of $N \in \{0, 1, 2 \ldots, 20\}$. The sequence of truncations correspond to $\alpha_t \in \{0, \alpha_1, \alpha_2, \alpha_3\}$, which produce quantile values of $q_{\alpha_t}(Z) \in \{0, 5, 10, 15\}$**

with density

$$f_X(x; \mu, \gamma) = \left[\frac{\gamma}{2\pi x^3}\right]^{\frac{1}{2}} \exp\left(\frac{-\gamma(x-\mu)^2}{2\mu^2 x}\right). \tag{7.92}$$

Then each target distribution is attainable in closed form using the fact that

$$S_n = \sum_{i=1}^{n} X_i \sim F_{S_n}^{(n)*}(x) = F_X(x; n\mu, n^2\gamma), \tag{7.93}$$

which allows one to specify uniquely the sequence of distributions according to

$$\begin{aligned}
&\pi_t\left(z \,\middle|\, \sum_{n=1}^{N} X_n > \mathrm{VaR}_{1-\alpha_t}[Z]\right) \\
&= \frac{1}{g\left(\mathrm{VaR}_{1-\alpha_t}[Z]\right)} \left[\sum_{n=1}^{\infty} \mathbb{P}\mathrm{r}(N = n) \left\{\left[\frac{n^2\gamma}{2\pi z^3}\right]^{\frac{1}{2}} \exp\left(\frac{-n^2\gamma(z-n\mu)^2}{2n^2\mu^2 z}\right)\right\}\right] \\
&\quad \times \mathbb{I}\left[\sum_{n=1}^{N} X_n > \mathrm{VaR}_{1-\alpha_t}[Z]\right]
\end{aligned} \tag{7.94}$$

with the normalizing constant for a given truncation threshold given by

$$\begin{aligned} &g\left(\mathrm{VaR}_{1-\alpha_t}[Z]\right) \\ &= 1 - \mathbb{P}\mathrm{r}[N=0] - \sum_{n=1}^{\infty} \mathbb{P}\mathrm{r}[N=n]\Phi\left(\sqrt{\frac{n^2\gamma}{\mathrm{VaR}_{1-\alpha_t}[Z]}}\left(\frac{\mathrm{VaR}_{1-\alpha_t}[Z]}{n\mu}-1\right)\right) \\ &\quad - \sum_{n=1}^{\infty} \mathbb{P}\mathrm{r}[N=n]\exp\left(\frac{2n^2\gamma}{n\mu}\right)\Phi\left(-\sqrt{\frac{n^2\gamma}{\mathrm{VaR}_{1-\alpha_t}[Z]}}\left(\frac{\mathrm{VaR}_{1-\alpha_t}[Z]}{n\mu}+1\right)\right). \end{aligned} \quad (7.95)$$

In practice, one would also place an upper bound on the total number of losses N_T that may occur in a given year for this risk process and then these Poisson-weighted mixtures would have finite numbers of terms. ■

EXAMPLE 7.5 Multivariate Risk Process with Copula Dependence

Consider d risk processes each characterized by a compound process $Z^{(i)} = \sum_{n=1}^{N^{(i)}} X_n^{(i)}$ for $i \in \{1, 2, \ldots, d\}$ with frequency distributions $\left\{F_N^{(i)}\right\}_{i=1}^d$ and severity distributions $\left\{F_X^{(i)}\right\}_{i=1}^d$. Furthermore, assume that one wishes to model dependence between the risk process for the random vector of annual losses $\boldsymbol{Z} = \left(Z^{(1)}, Z^{(2)}, \ldots, Z^{(d)}\right)$ given by the multivariate copula model generically denoted by the distribution function given three elements, the copula dependence function, the marginal single risk process annual loss distributions, and the mapping from the marginal annual loss positive random variables to the unit cube, as denoted by $C(U_1, U_2, \ldots, U_d)$, $\{F_{Z^{(i)}}(z_i)\}_{i=1}^d$, and $G_i(z)$, respectively. Note that we set $U_i = G_i\left(Z^{(i)}\right)$ for each risk process $i \in \{1, 2, \ldots, d\}$ for mappings G_i, which are monotonic and strictly increasing functions. That is, one can consider any G_i for $i \in \{1, 2, \ldots, d\}$, which is a nonunique transform selected to map the i-th marginal risk process annual loss random variable from $\mathbb{R}^+$ to $[0, 1]$, that is $G_i : \mathbb{R}^+ \mapsto [0, 1]$, s.t. G_i is a monotonically increasing function. Note that a natural choice for $G_i(\cdot)$ is the annual loss distribution $F_{Z^{(i)}}(\cdot)$. The resulting joint distribution function is then given by

$$\mathbb{P}\mathrm{r}\left(Z^{(1)} \leq z_1, \ldots, Z^{(d)} \leq z_d\right) = C\left(G_1(z_1), \ldots, G_d(z_d)\right). \quad (7.96)$$

Now if one differentiates this distribution to get the density, it produces in general the result

$$\pi(\boldsymbol{z}) = c\left(G_1(z_1), \ldots, G_d(z_d); \boldsymbol{\Psi}\right) \prod_{j=1}^{d} f_{Z^{(j)}}(z_j) \left|\frac{dG_j^{-1}(u_j)}{du_j}\right|, \quad (7.97)$$

where $z_j = G_j^{-1}(u_j)$. It is common practice to work with transformations that avoid the need for the Jacobian terms mentioned earlier, which would correspond to mapping to the unit cube each marginal risk process annual loss random variable by the marginal annual loss distribution, that is, $G_i(z) = F_{Z^{(i)}}(z)$. In this case, one would obtain for the density of the joint risk process the expression

$$\begin{aligned}\pi(\boldsymbol{z}) &= f_Z(\boldsymbol{z})\\ &= c\left(F_{Z^{(1)}}(z_1), \ldots, F_{Z^{(d)}}(z_d); \boldsymbol{\Psi}\right)\prod_{j=1}^{d} f_{Z^{(j)}}(z)\\ &= c\left(F_{Z^{(1)}}(z_1), \ldots, F_{Z^{(d)}}(z_d); \boldsymbol{\Psi}\right)\prod_{j=1}^{d}\left[\sum_{n=1}^{\infty}\Pr(N^{(j)} = n)\, f^{(j)}_{X^{(n)*}}(x)\right].\end{aligned} \tag{7.98}$$

Then, in general sampling, such a multivariate distribution is very challenging when using standard Monte Carlo approaches. Here we demonstrate how to construct an SMC sampler solution where one must decide upon a suitable sequence of intermediate distributions $\{\pi_t\}_{t=1}^{T}$, which are easier to sample, and such that the sequence will progressively target the original distribution of interest $\pi_T(\boldsymbol{z}) = \pi(\boldsymbol{z})$. Next we provide a few examples one may consider for achieving this goal.

1. **Annealing via power schedule: from independence to dependence**
   One of many possible examples of such a sequence could involve the following annealing scheme, which would start from the case of completely independent risks, for which it is trivial to generate Monte Carlo draws from the distribution and progress through to the copula-coupled processes via the sequence

   $$\pi_t(\boldsymbol{z}) = \left\{c(G_1(z_1), \ldots, G_d(z_d); \boldsymbol{\Psi})\right\}^{\gamma_t}\prod_{j=1}^{d} f_{Z^{(j)}}(z_j) \tag{7.99}$$

   with $\{\gamma_t\}$ some schedule starting with $\gamma_0 = 0$ through to a maximum $\gamma_T = 1$;
2. **Annealing via copula parameter schedule: from independence to dependence**
   An alternative approach to constructing such a sequence of distributions would be to consider the sequence defined by annealing on the copula parameter, as opposed to the previous example where one anneals on the power of the copula density, in which one defines the sequence by

   $$\pi_t(\boldsymbol{z}) = \left\{c(G_1(z_1), \ldots, G_d(z_d); \boldsymbol{\Psi}_t)\right\}\prod_{j=1}^{d} f_{Z^{(j)}}(z_j), \tag{7.100}$$

   where now one may define a sequence of tempered copula models index by parameter vector sequence $\{\boldsymbol{\Psi}_t\}$. For example, in the Archimedean copula families, popular in practice, this could correspond to a sequence on a univariate parameter ranging from independence through to the final value $\boldsymbol{\Psi} = \boldsymbol{\Psi}_T$. In the case of, for example, the Student-t copula model, this could be based on

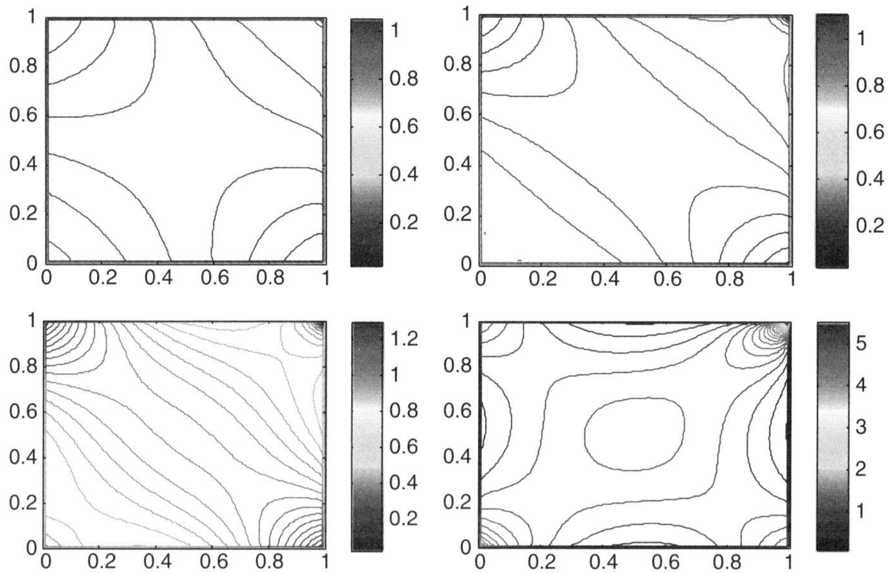

FIGURE 7.4 **Top left subplot: target distribution π_0 at a low temperature, where the distribution is fairly flat and simple to sample. Top right subplot: target distribution π_{t_1} at an intermediate temperature, where the distribution is still fairly flat and simple to sample. Bottom left subplot: target distribution π_{t_2} at an intermediate temperature, where the distribution is increasingly concentrated. Bottom right subplot: target distribution π_T final distribution, which is the target distribution. (For color detail please see color plate section.)**

a sequence of Student-t copulas, which would have a progressively decreasing degree of freedom parameters ν_t such that $\nu_t > \nu_{t-1}$ and $\nu_T = \nu$ are the actual model d.f. Details of different copula models can be found in Chapter 10–12.

In Figure 7.4, ignoring the marginals we show the sequence of such intermediate distributions that are constructed to allow the SMC sampler to progress successively through the case of independence to dependence for a mixture of Clayton and Gumbel copulas in a bivariate example, with two risk processes. The model considered is given by

$$\pi_T(\boldsymbol{z}) = 0.4\underbrace{c\left(G_1(z_1), G_2(z_2); 2\right)}_{Frank} + 0.4\underbrace{c\left(G_1(z_1), G_2(z_2); -3\right)}_{Frank} + 0.2\underbrace{c\left(G_1(z_1), G_2(z_2); 1.6\right)}_{Gumbel}.$$

■

In the next example, we consider a challenging multivariate set of annual losses under constraints, making for a challenging rare event problem.

EXAMPLE 7.6 Constrained Dependent Multivariate Risk Processes

Consider d risk processes, each characterized by a compound process $Z^{(i)} = \sum_{n=1}^{N^{(i)}} X_n^{(i)}$ for $i \in \{1, 2, \ldots, d\}$ with frequency distributions $\{F_N^{(i)}\}_{i=1}^d$ and severity distributions $\{F_X^{(i)}\}_{i=1}^d$. Furthermore, assume that one wishes to model dependence between the risk process for the random vector of annual losses $\boldsymbol{Z} = (Z^{(1)}, Z^{(2)}, \ldots, Z^{(d)})$ given by the multivariate copula model generically denoted by the density, this time subject to constraints. For example, one may be interested in considering the constraints on each of the d risk processes written as a joint restriction on the aggregate of the d annual losses according to

$$\sum_{i=1}^{d} Z^{(i)} \geq \text{VaR}_{1-\alpha}\left[\sum_{i=1}^{d} Z^{(i)}\right] \tag{7.101}$$

for some tail aggregate process events as characterized by α. Such constraints arise naturally in OpRisk when considering capital allocations under a Euler principle (see discussions and details in Chapter 6). In this case, one is interested in a joint distribution given by

$$\begin{aligned}\pi(\boldsymbol{z}) &= f_Z\left(\boldsymbol{z} \middle| \sum_{i=1}^{d} Z^{(i)} \geq \text{VaR}_{1-\alpha}\left[\sum_{i=1}^{d} Z^{(i)}\right]\right) \\ &= c\left(G_1(z_1), \ldots, G_d(z_d) \middle| \sum_{i=1}^{d} Z^{(i)} \geq \text{VaR}_{1-\alpha}\left[\sum_{i=1}^{d} Z^{(i)}\right]; \boldsymbol{\Psi}\right) \\ &\quad \times \prod_{j=1}^{d} f_{Z^{(j)}}(z_j) \left|\frac{dG_j^{-1}(u_j)}{du_j}\right| \mathbb{I}\left[\sum_{i=1}^{d} Z^{(i)} \geq \text{VaR}_{1-\alpha}\left[\sum_{i=1}^{d} Z^{(i)}\right]\right],\end{aligned} \tag{7.102}$$

where $z_j = G_j^{-1}(u_j)$. In general, sampling such a distribution is very challenging and cannot be efficiently done via standard Monte Carlo methods. Again we propose such a problem is ideal for SMC sampler solutions. As discussed already in previous examples, to utilize this solution technique one must select a sequence of intermediate distributions $\{\pi_t\}_{t=1}^T$ which are easier to sample such that the sequence will progressively target the original distribution of interest $\pi_T(\boldsymbol{z}) = \pi(\boldsymbol{z})$.

In this case, it is natural to consider the schedule given by a sequence of tail quantiles $0 \leq \alpha_1 \leq \cdots \leq \alpha_{t-1} \leq \alpha_t \leq \cdots \leq \alpha_T$. If one considers how this constraint region looks in the bivariate risk process setting, using the models of Example 7.4, which would produce risk processes with Poisson–Inverse Gaussian LDA models, coupled with a bivariate Frank copula model, then the sequence of constraints given by $(\alpha_1, \alpha_2, \alpha_3, \alpha_4) = (0.25, 0.5, 0.75, 0.95)$ would produce the following constrained distributions, where we show both the joint distribution and the constrained copula dependence distributions. Note that depending on the mappings $G_1(Z^{(1)})$ and $G_2(Z^{(2)})$ from the space of the annual losses $(Z^{(1)}, Z^{(2)})$ in space $\mathbb{R}^+ \times \mathbb{R}^+$ to $[0, 1] \times [0, 1]$ in the copula, the constraint region will take different shapes. For simplicity, we consider marginal annual loss processes with finite support for the maximum losses achieved in an interval $[0, Z_{\max}]$ and we take as G_1 and G_2 uniform distribution functions, each on the support $[0, Z_{\max}]$. This means the resulting density is given by defining the uniform distribution transform for each $i \in \{1, 2\}$ by $U_i = G_i\left(Z^{(i)}\right) = \frac{Z^{(i)}}{Z_{\max}}$ and inverse transform function given by $Z^{(i)} = G_i^{-1}(U_i) = U_i Z_{\max}$, which results in density

$$\pi(\boldsymbol{z}) = c\left(u_1, u_2 \left| \sum_{i=1}^{2} Z^{(i)} \geq \mathrm{VaR}_{1-\alpha}\left[\sum_{i=1}^{2} Z^{(i)}\right] ; \boldsymbol{\Psi}\right.\right)$$
$$\times Z_{\max}^2 \prod_{j=1}^{2} f_{Z^{(j)}}(z_j)\, \mathbb{I}\left[\sum_{i=1}^{d} Z^{(i)} \geq \mathrm{VaR}_{1-\alpha}\left[\sum_{i=1}^{d} Z^{(i)}\right]\right] \qquad (7.103)$$

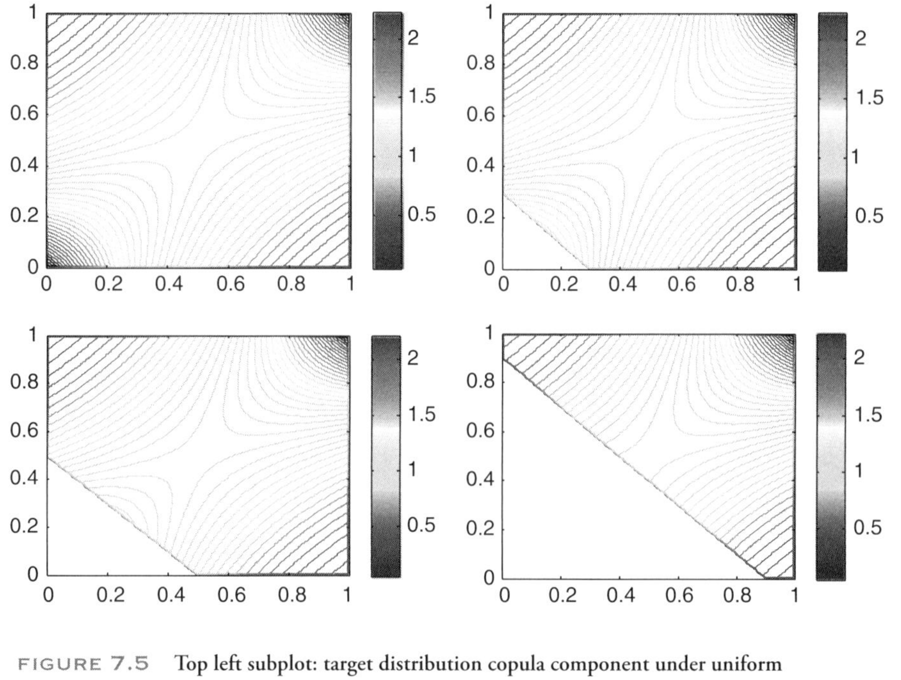

FIGURE 7.5 Top left subplot: target distribution copula component under uniform distribution function transformation for π_0 at little truncation, where the distribution is fairly flat and simple to sample. Top right subplot: target distribution copula component under uniform distribution function transformation for π_{t_1} at an intermediate truncation. Bottom left subplot: target distribution copula component under uniform distribution function transformation for π_{t_2} at an intermediate truncation. Bottom right subplot: target distribution copula component under uniform distribution function transformation for π_T final distribution, which is the target distribution. (For color detail please see color plate section.)

The choice of mapping as performed by a Uniform distribution function results in the constraint regions remaining linear as presented in Figures 7.5 and 7.6; however, generally, one would chose alternative mappings with noncompact supports.

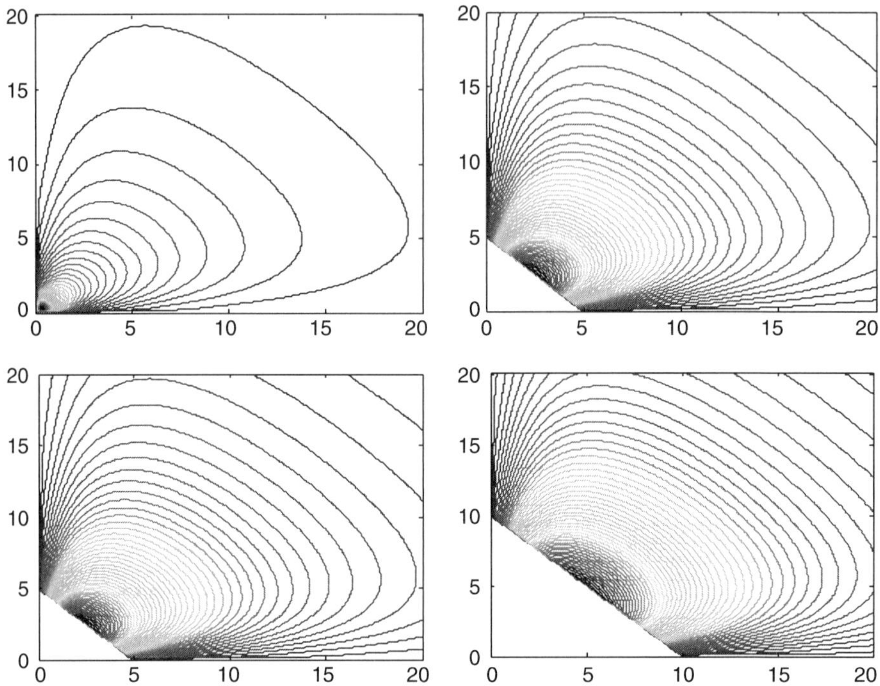

FIGURE 7.6 **Top left subplot: target distribution π_0 at little truncation, where the distribution is fairly flat and simple to sample. Top right subplot: target distribution π_{t_1} at an intermediate truncation. Bottom left subplot: target distribution π_{t_2} at an intermediate truncation. Bottom right subplot: target distribution π_T final distribution, which is the target distribution. (For color detail please see color plate section.)**

7.7.2 SMC SAMPLER METHODOLOGY AND COMPONENTS

To address such sampling challenges for a sequence of distributions $\{\pi_t(d\boldsymbol{\theta})\}_{t=1}^{T}$, the aim is to develop a large collection of N-weighted random samples at each time t denoted by $\left\{W_t^{(i)}, \boldsymbol{\Theta}_t^{(i)}\right\}_{i=1}^{N}$ such that $W_t^{(i)} > 0$ and $\sum_{i=1}^{N} W_t^{(i)} = 1$. These importance weights and samples, denoted by $\left\{W_t^{(i)}, \boldsymbol{\Theta}_t^{(i)}\right\}_{i=1}^{N}$, are known as particles (hence the name often given to such algorithms as particle filters or interacting particle systems). For such approaches to be sensible we would require that the empirical distributions constructed through these samples should converge asymptotically $(N \to \infty)$ to the target distribution π_t for each time t. This means that for any π_t integrable function, denoted, for example, by $\phi(\boldsymbol{\theta}) : E \rightarrow \mathbb{R}'$ one would have the following convergence:

$$\sum_{i=1}^{N} W_t^{(i)} \phi\left(\boldsymbol{\theta}_t^{(i)}\right) \overset{a.s.}{\rightarrow} \mathbb{E}_{\pi_t}\left[\phi(\boldsymbol{\Theta})\right]. \tag{7.104}$$

The sequential nature of such algorithms arises from the fact that they iteratively construct the sets of weighted particles recursively through a sequential IS framework (see many examples of such algorithms in Doucet *et al.* 2000, Oh and Berger 1993, Givens and Raftery 1996, Gilks and Berzuini 2002, Neal 2001 and the tutorial of Doucet and Johansen 2009).

In the SMC Samplers algorithm, a particular variant of SMC algorithms, a modification of the SMC algorithm, is developed. Consider a generic sequence of distributions given by $\pi_t(\boldsymbol{\theta}), t = 1, \ldots, T$, with $\boldsymbol{\theta} \in E$, where the final distribution π_T is the distribution of interest. By introducing a sequence of backward kernels L_k, a new distribution

$$\widetilde{\pi}_t(\boldsymbol{\theta}_1, \ldots, \boldsymbol{\theta}_t) = \pi_t(\boldsymbol{\theta}_t) \prod_{k=1}^{t-1} L_k\left(\boldsymbol{\theta}_{k+1}, \boldsymbol{\theta}_k\right) \tag{7.105}$$

may be defined for the *path* of a particle $(\boldsymbol{\theta}_1, \ldots, \boldsymbol{\theta}_t) \in E^t$ through the sequence $\pi_1, \ldots, \pi_t$. The only restriction on the backward kernels is that the correct marginal distributions $\int \widetilde{\pi}_t(\boldsymbol{\theta}_1, \ldots, \boldsymbol{\theta}_t) d\boldsymbol{\theta}_1, \ldots, d\boldsymbol{\theta}_{t-1} = \pi_t(\boldsymbol{\theta}_t)$ are available. Within this framework, one may then work with the constructed sequence of distributions, $\widetilde{\pi}_t$, under the standard SMC algorithm.

In summary, the SMC Sampler algorithm involves three stages:

1. *Mutation*, whereby the particles are moved from $\boldsymbol{\theta}_{t-1}$ to $\boldsymbol{\theta}_t$ via a mutation kernel $M_t(\boldsymbol{\theta}_{t-1}, \boldsymbol{\theta}_t)$;
2. *Correction*, where the particles are reweighted with respect to π_t via the incremental importance weight (Eq. 7.106);
3. *Selection*, where according to some measure of particle diversity, commonly the effective sample size, the weighted particles may be resampled in order to reduce the variability of the importance weights.

In more detail, suppose that at time $t-1$, the distribution $\widetilde{\pi}_{t-1}$ can be approximated empirically by $\widetilde{\pi}_{t-1}^N$ using N-weighted particles. These particles are first propagated to the next distribution $\widetilde{\pi}_t$ using a mutation kernel $M_t(\boldsymbol{\theta}_{t-1}, \boldsymbol{\theta}_t)$, and then assigned new weights $W_t = W_{t-1} w_t(\boldsymbol{\theta}_1, \ldots \boldsymbol{\theta}_t)$, where W_{t-1} is the weight of a particle at time $t-1$ and w_t is the incremental importance weight given by

$$w_t\left(\boldsymbol{\theta}_1, \ldots, \boldsymbol{\theta}_t\right) = \frac{\widetilde{\pi}_t\left(\boldsymbol{\theta}_1, \ldots, \boldsymbol{\theta}_t\right)}{\widetilde{\pi}_{t-1}\left(\boldsymbol{\theta}_1, \ldots, \boldsymbol{\theta}_{t-1}\right) M_t\left(\boldsymbol{\theta}_{t-1}, \boldsymbol{\theta}_t\right)} = \frac{\pi_t\left(\boldsymbol{\theta}_t\right) L_{t-1}\left(\boldsymbol{\theta}_t, \boldsymbol{\theta}_{t-1}\right)}{\pi_{t-1}\left(\boldsymbol{\theta}_{t-1}\right) M_t\left(\boldsymbol{\theta}_{t-1}, \boldsymbol{\theta}_t\right)}. \tag{7.106}$$

The resulting particles are now weighted samples from $\widetilde{\pi}_t$. Consequently, from Eq. (7.106), under the SMC Sampler framework, one may work directly with the marginal distributions $\pi_t(\boldsymbol{\theta}_t)$ such that $w_t(\boldsymbol{\theta}_1, \ldots, \boldsymbol{\theta}_t) = w_t(\boldsymbol{\theta}_{t-1}, \boldsymbol{\theta}_t)$. While the choice of the backward kernels L_{t-1} is essentially arbitrary, their specification can strongly affect the performance of the algorithm, as will be discussed in the following subsections.

The basic version of the SMC Sampler algorithm therefore proceeds explicitly as given in Algorithm 7.9.

Remark 7.18 *In all cases in which we utilize the incremental importance sampling weight correction, the arguments in the expressions only need to be known up to normalization. That is, it is perfectly acceptable to only be able to evaluate the sequence of target distributions* $\{\pi_t\}$ *up to normalization constant. This is tue as long as the same normalization constant is present for all particles, since the renormalization step will correct for this lack of knowledge in the importance weighting. In practice, this is critical to the application of such methods.*

Algorithm 7.9 (Sequential Monte Carlo Sampler)

1. *Initialize the particle system;*
   a) *Set* $n = 1$;
   b) *For* $i = 1, \ldots, N$, *draw initial particles* $\boldsymbol{\Theta}_1^{(i)} \sim p(\boldsymbol{\theta})$;
   c) *Evaluate incremental importance weights* $\left\{w_1\left(\boldsymbol{\Theta}_1^{(i)}\right)\right\}$ *using Equation (7.106) and normalize the weights to obtain* $\left\{W_1^{(i)}\right\}$.

   Iterate the following steps through each distribution in sequence $\{\pi_t\}_{t=2}^{T}$.
2. *Resampling*
   a) *If the effective sampling size* $(ESS) = \frac{1}{\sum_{i=1}^{N}\left(w_t^{(i)}\right)^2} < N_{eff}$ *is less than a threshold* N_{eff}, *then resample the particles via the empirical distribution of the weighted sample either by multinomial or stratefied methods; see discussions on unbiased resampling schemes by Künsch (2005) and Del Moral (2004).*
3. *Mutation and correction*
   a) *Set* $t = t + 1$, *if* $t = T + 1$, *then stop;*
   b) *For* $i = 1, \ldots, N$ *draw samples from mutation kernel* $\boldsymbol{\Theta}_t^{(i)} \sim M_t\left(\Theta_{t-1}^{(i)}\right)$;
   c) *Evaluate incremental importance weights* $\left\{w_1\left(\boldsymbol{\Theta}_1^{(i)}\right)\right\}$ *using Equation (7.106) and normalize the weights to obtain* $\left\{W_1^{(i)}\right\}$ *via*

$$W_t^{(i)} = W_{t-1}^{(i)} \frac{w_t^{(i)}\left(\boldsymbol{\Theta}_{t-1}, \boldsymbol{\Theta}_t\right)}{\sum_{j=1}^{N} W_{t-1}^{(i)} w_t^{(i)}\left(\boldsymbol{\Theta}_{t-1}, \boldsymbol{\Theta}_t\right)}. \tag{7.107}$$

7.7.2.1 Choice of Mutation Kernel and Backward Kernel.

There are many choices for mutation kernel and backward kernel that could be considered when designing an SMC Sampler algorithm. In this section, we survey a few possible choices and note an important difference between the SMC Sampler and MCMC methods in the following remark.

Remark 7.19 *In the MCMC methods presented previously, the proposal kernel was typically selected to ensure the resulting Markov chain satisfied reversibility and detailed balance conditions, or in the case of the adaptive proposals, some notion of eventual non-adaption (diminishing adaptation and bounded convergence). Unlike the MCMC methods, in the case of the SMC Sampler algorithms, the mutation kernel is significantly more flexible with regard to choice and with regard to adaption strategies. It is clear that the optimal choice of mutation kernel would be the next* *distribution in the sequence* $M_t(\boldsymbol{\theta}_{t-1}, \boldsymbol{\theta}_t) = \pi_t(\boldsymbol{\theta}_t)$*, as this would minimize the variance of the incremental weights, though clearly the context of the application of SMC Samplers is one in which the target distribution cannot be sampled directly via inversion or rejection sampling methods, so this choice is not practical.*

Some examples of possible choices of the mutation kernel are given as follows:

1. **Independent kernels**. In this setting, one would select a mutation kernel given for all $t \in \{1, 2, \ldots, T\}$ by $M_t(\boldsymbol{\theta}_{t-1}, \boldsymbol{\theta}_t) = M_t(\boldsymbol{\theta}_t)$;
2. **Local Random Walks**. In this setting, the kernel would be selected for all $t \in \{1, 2, \ldots, T\}$ to be of the form $M_t(\boldsymbol{\theta}_{t-1}, \boldsymbol{\theta}_t)$, where the mutation from $\boldsymbol{\theta}_{t-1}$ to $\boldsymbol{\theta}_t$ follows a local Random Walk based around, say, a Gaussian smoothing kernel as given by Givens and Raftery (1996);
3. **Markov chain Monte Carlo Kernels**. In this setting, the kernel would be selected for all $t \in \{1, 2, \ldots, T\}$ to be an MCMC kernel of invariant distribution π_t. As noted by Del Moral *et al.* (2006) and Peters (2005), this option is suitable if the Markov chain kernel is mixing rapidly or if the sequence of distributions is such that π_{t-1} is close to π_t, which is often the case by design. Then the use of an MCMC kernel would result in running for each stage, N inhomogeneous Markov chains. Then one must correct for the fact that one is not targeting the correct distribution under these Markov chains, which is achieved using IS: $\hat{\pi}_{t-1}^N = \sum_{i=1}^N W_{t-1}^{(i)} \delta_{\boldsymbol{\theta}_{t-1}^{(i)}}(\boldsymbol{\theta})$ and running L iterations of the Markov chain for each particle, where each of the N chains will target $\sum_{i=1}^N W_{t-1}^{(i)} \prod_{l=1}^L M_l\left(\boldsymbol{\theta}_{l-1}^{(i)}, \boldsymbol{\theta}_l\right)$, which is not in general π_t, then with an IS correction, such an approach is accurate and unbiased (i.e., targets the distribution of interest at time t given by π_t;
4. **Gibbs Sampler kernels**. If the sequence of target distributions $\{\pi_t\}_{t\geq 0}$ is such that its support is multivariate, then it may also be possible to sample from the full conditional distributions in the sequence of distributions. This approach allows one to undertake a Gibbs step, which would involve a kernel for update of the k-th element given in the form

$$M_t(\boldsymbol{\theta}_{t-1}, d\boldsymbol{\theta}_t) = \delta_{\boldsymbol{\theta}_{t-1,-k}}(d\boldsymbol{\theta}_{t,-k})\, \pi_t(\boldsymbol{\theta}_{t,k} | \boldsymbol{\theta}_{t,-k}) \tag{7.108}$$

   with $\boldsymbol{\theta}_{t,-k} = (\boldsymbol{\theta}_{t,1}, \boldsymbol{\theta}_{t,2}, \ldots, \boldsymbol{\theta}_{t,k-1}, \boldsymbol{\theta}_{t,k+1}, \ldots, \boldsymbol{\theta}_{t,J})$, where there are J parameters in the OpRisk model target posterior. If the full conditionals are not available, one could approximate them accurately at each stage and then correct for the approximation error through IS;
5. **Mixture kernels**. It is always possible to consider a mixture kernel choice given by

$$M_t(\boldsymbol{\theta}_{t-1}, \boldsymbol{\theta}_t) = \sum_{m=1}^M \alpha_{t,m}(\boldsymbol{\theta}_{t-1})\, M_{t,m}(\boldsymbol{\theta}_{t-1}, \boldsymbol{\theta}_t), \tag{7.109}$$

   with $\alpha_{t,m}(\boldsymbol{\theta}_{t-1}) > 0$ and $\sum_{m=1}^M \alpha_{t,m}(\boldsymbol{\theta}_{t-1}) = 1$. One special case of this type of kernel would be an independent kernel constructed by a kernel density estimate of $M_{t,m}(\boldsymbol{\theta}_{t-1}, \boldsymbol{\theta}_t) = M_t(\boldsymbol{\theta}_{t-1}, \boldsymbol{\theta}_t)$ for all m and $\alpha_{t,m}(\boldsymbol{\theta}_{t-1}) = W_{t-1}^{(i)}$ with $M = N$;

6. **Partial Rejection Control kernels.** In this case, one aims to construct a mutation kernel in the SMC Sampler that guarantees all sampled particles have importance weights with a "fitness" exceeding a user-specified threshold at each time t, denoted by c_t such that $w_t^{(i)} \geq c_t,\ \forall i \in \{1, 2, \ldots, N\}$. To achieve this, one modifies any of the earlier mutation kernels to take the form given by

$$M_t^*\left(\boldsymbol{\theta}_{t-1}^i, \boldsymbol{\theta}_t\right) = r(c_t, \boldsymbol{\theta}_{t-1}^{(i)})^{-1}\left[\min\left\{1, W_{t-1}^{(i)}\frac{w_t\left(\boldsymbol{\theta}_{t-1}^{(i)}, \boldsymbol{\theta}_t\right)}{c_t}\right\} M_t\left(\boldsymbol{\theta}_{t-1}^{(i)}, \boldsymbol{\theta}_t\right)\right]. \tag{7.110}$$

The quantity $r(c_t, \boldsymbol{\theta}_{t-1}^{(i)})$ denotes the normalizing constant for particle $\boldsymbol{\theta}_{t-1}^{(i)}$, given by

$$r(c_t, \boldsymbol{\theta}_{t-1}^{(i)}) = \int \min\left\{1, W_{t-1}^{(i)}\frac{w_t\left(\boldsymbol{\theta}_{t-1}^{(i)}, \boldsymbol{\theta}_t\right)}{c_t}\right\} M_t\left(\boldsymbol{\theta}_{t-1}^{(i)}, \boldsymbol{\theta}_t\right) d\boldsymbol{\theta}_t. \tag{7.111}$$

Note that $0 < r(c_t, \boldsymbol{\theta}_{t-1}) \leq 1$ if (w.l.o.g.) the mutation kernel M_t is normalized, so that $\int M_t(\boldsymbol{\theta}_{t-1}, \boldsymbol{\theta}_t) d\boldsymbol{\theta}_t = 1$, and if the PRC threshold $0 \leq c_t < \infty$ is finite. The sequence of PRC thresholds is then user-specified to ensure a certain particle "fitness" at each stage of the SMC Sampler. We will detail more explicitly this example in a future section.

Proposition 7.3 (Optimal Backward Kernel) *Given any of the possible mutation kernels $M_t(\boldsymbol{\theta}_{t-1}, \boldsymbol{\theta}_t)$, one can define the optimal backward kernel in the SMC Sampler as the one that minimizes the variance of the incremental (unnormalized) IS weights, given by Peters (2005) and Del Moral et al. (2006) by*

$$L_{t-1}^{opt}\left(\boldsymbol{\theta}_t, \boldsymbol{\theta}_{t-1}\right) = \frac{\nu_{t-1}\left(\boldsymbol{\theta}_{t-1}\right) M_t\left(\boldsymbol{\theta}_{t-1}, \boldsymbol{\theta}_t\right)}{\nu_t\left(\boldsymbol{\theta}_t\right)}, \tag{7.112}$$

where one defines the sequence of integrated distributions on the path space by

$$\nu_t\left(\boldsymbol{\theta}_t\right) = \int \cdots \int \pi_1\left(\boldsymbol{\theta}_1\right) \prod_{l=1}^{t} M_l\left(\boldsymbol{\theta}_{l-1}, \boldsymbol{\theta}_l\right) d\boldsymbol{\theta}_1 d\boldsymbol{\theta}_2 \cdots d\boldsymbol{\theta}_t. \tag{7.113}$$

This optimal choice is difficult to utilize in practice as it involves knowledge of the ability to draw from each of the distributions in the sequence.

This choice of optimal backward kernel is easily understood by interpreting it as the choice of kernel in which one would perform IS on the space E rather than the product space E^t. The resulting incremental IS weight for the optimal choice of backward kernel is simply

$$w_t\left(\boldsymbol{\theta}_{1:t}\right) = \frac{\pi_t\left(\boldsymbol{\theta}_t\right)}{\nu_t\left(\boldsymbol{\theta}_t\right)}. \tag{7.114}$$

In addition, we also note some examples of possible choices of the backward kernel given along with the corresponding incremental IS weights.

1. **Mixture Backward kernel**. Given a mixture mutation kernel in Equation (7.109), the equivalent backward kernel is given by

$$L_{t-1,m}(\boldsymbol{\theta}_t, \boldsymbol{\theta}_{t-1}) = \sum_{m=1}^{M} \beta_{t-1,m}(\boldsymbol{\theta}_t) L_{t-1,m}(\boldsymbol{\theta}_t, \boldsymbol{\theta}_{t-1}) \tag{7.115}$$

with $\beta_{t,m}(\boldsymbol{\theta}_t) > 0$ and $\sum_{m=1}^{M} \beta_{t,m}(\boldsymbol{\theta}_t) = 1$. In this case, the incremental IS weight can be written in the following form, with respect to an index auxiliary random variable for the mixture I_t that was sampled:

$$w_t(\boldsymbol{\theta}_{t-1}, \boldsymbol{\theta}_t, i_t) = \frac{\pi_t(\boldsymbol{\theta}_t) \beta_{t-1,i_t}(\boldsymbol{\theta}_t) L_{t-1,i_t}(\boldsymbol{\theta}_t, \boldsymbol{\theta}_{t-1})}{\pi_{t-1}(\boldsymbol{\theta}_{t-1}) \alpha_{t,i_t}(\boldsymbol{\theta}_{t-1}) M_{t,i_t}(\boldsymbol{\theta}_{t-1}, \boldsymbol{\theta}_t)}. \tag{7.116}$$

2. **Approximate Optimal Backward kernel**. One of the best possible approximations to the optimal backward kernel is to consider replacing $\nu_t(\boldsymbol{\theta}_t)$ with $\pi_t(\boldsymbol{\theta}_t)$, to get

$$L_{t-1}^{opt}(\boldsymbol{\theta}_t, \boldsymbol{\theta}_{t-1}) = \frac{\pi_{t-1}(\boldsymbol{\theta}_{t-1}) M_t(\boldsymbol{\theta}_{t-1}, \boldsymbol{\theta}_t)}{\int \pi_{t-1}(d\boldsymbol{\theta}_{t-1}) M_t(\boldsymbol{\theta}_{t-1}, \boldsymbol{\theta}_t)}, \tag{7.117}$$

which would give an incremental importance weight of

$$w_t(\boldsymbol{\theta}_{t-1}, \boldsymbol{\theta}_t) = \frac{\pi_{t-1}(\boldsymbol{\theta}_{t-1})}{\int \pi_{t-1}(\boldsymbol{\theta}_{t-1}) M_{t,i_t}(\boldsymbol{\theta}_{t-1}, \boldsymbol{\theta}_t) d\boldsymbol{\theta}_{t-1}}. \tag{7.118}$$

Note that if resampling has occurred at time $t-1$, then this kernel is already equivalent to the optimal choice and therefore its particle approximation is already the optimal option. In general, if using the optimal backward kernel, one would still need to typically be able to approximate the univariate integrals, usually done via approximation using the particles at time $t-1$ as follows:

$$\int \pi_{t-1}(\boldsymbol{\theta}_{t-1}) M_t(\boldsymbol{\theta}_{t-1}, \boldsymbol{\theta}_t) d\boldsymbol{\theta}_{t-1} \approx \sum_{i=1}^{N} W_{t-1}^{(i)} M_t\left(\boldsymbol{\theta}_{t-1}^{(i)}, \boldsymbol{\theta}_t\right). \tag{7.119}$$

Note that this results in an $O(N^2)$ algorithm, which is not ideal computationally;

3. **MCMC Backward kernel**. A generic approximation of the "approximate optimal backward kernel" in Equation (7.117) is often selected as an MCMC kernel in which one uses for the mutation kernel M_t an invariant MCMC kernel for target distribution π_t and the backward kernel given by

$$L_{t-1}(\boldsymbol{\theta}_{t-1}, \boldsymbol{\theta}_t) = \frac{\pi_t(\boldsymbol{\theta}_{t-1}) M_t(\boldsymbol{\theta}_{t-1}, \boldsymbol{\theta}_t)}{\pi_t(\boldsymbol{\theta}_t)}. \tag{7.120}$$

This choice is a good approximation whenever the sequence of distributions π_{t-1} and π_t is close for all t, since this choice simply correspond to the time-reversed Markov kernel of the mutation kernel M_t. In addition, we note that you cannot adopt this choice for examples such as the successive sequence of constrained distributions as in the rare-event setting. When this backward kernel is utilized, one obtains an incremental importance weight given by a very simple form

$$w_t(\boldsymbol{\theta}_{t-1}, \boldsymbol{\theta}_t) = \frac{\pi_{t-1}(\boldsymbol{\theta}_t)}{\pi_{t-1}(\boldsymbol{\theta}_{t-1})}. \tag{7.121}$$

7.7.3 INCORPORATING PARTIAL REJECTION CONTROL INTO SMC SAMPLERS

It is well known that the performance of SMC methods is strongly dependent on the mutation kernel. If M_t is poorly chosen, such that it does not place particles in regions of the support of π_t with high density, then many IS weights will be close to zero. This leads to sample degeneracy, as a few well-located particles with large weights dominate the particle population, resulting in large variance for estimates made using these samples.

Liu (2008) and Liu *et al.* (1998) introduced a method, known as PRC strategy, to overcome particle degeneracy in a sequential IS setting. Under this mechanism, when the weight of a particle at distribution π_t falls below a finite threshold $c_t \geq 0$, the particle is probabilistically discarded. It is replaced with a particle drawn from the previous distribution π_{t-1}, which is then mutated to π_t. This new particle's weight is then compared to the threshold, with this process repeating until a particle is accepted. This concept was extended into an understanding of the resulting mutation kernel and developed under an SMC Sampler framework by Peters *et al.* (2009). This approach is termed *partial* rejection, as the replacement particle is drawn from π_{t-1}, not π_1 (see Liu 2008).

As demonstrated by Peters *et al.* (2009), under the SMC sampler framework, one may modify this approach and incorporate the partial rejection mechanism directly within the mutation kernel. Hence, at time $t-1$, the particle $\boldsymbol{\theta}_{t-1}$ is moved via the mutation kernel $M_t(\boldsymbol{\theta}_{t-1}, \boldsymbol{\theta}_t)$ and weighted according to (16.55). This particle is accepted with probability p, determined by the particle's weight and the weight threshold c_t. If rejected, a new particle is obtained via the mutation kernel M_t, until a particle is accepted.

For the sequence of distributions π_t, $t = 1, \ldots, T$, the mutation and backward kernels M_t and L_{t-1}, a sequence of weight thresholds c_t, and PRC normalizing constants $r(c_t, \boldsymbol{\theta}_{t-1})$ (defined later), the SMC sampler PRC algorithm is given in Algorithm 7.10.

Algorithm 7.10 (SMC Sampler PRC Algorithm)

1. *Initialization:*
   Set $t = 1$.
   For $i = 1, \ldots, N$, *sample* $\boldsymbol{\theta}_1^{(i)} \sim \pi_1(\boldsymbol{\theta})$, *and set weights* $W_1^{(i)} = \frac{1}{N}$;
2. *Resample:*
   Normalize the weights $\sum_i W_t^{(i)} = 1$. *If* $[\sum_i (W_t^{(i)})^2]^{-1} < H$ *resample* N *particles with respect to* $W_t^{(i)}$ *and set* $W_t^{(i)} = \frac{1}{N}, i = 1, \ldots, N$.
3. *Mutation and correction:*
   Set $t = t + 1$ *and* $i = 1$:
   (a) *Sample* $\boldsymbol{\theta}_t^{(i)} \sim M_t(\boldsymbol{\theta}_{t-1}^{(i)}, \boldsymbol{\theta}_t)$ *and set weight for* $\boldsymbol{\theta}_t^{(i)}$ *to*
   $$W_t^{(i)} = W_{t-1}^{(i)} \frac{\pi_t(\boldsymbol{\theta}_t^{(i)}) L_{t-1}(\boldsymbol{\theta}_t^{(i)}, \boldsymbol{\theta}_{t-1}^{(i)})}{\pi_{t-1}(\boldsymbol{\theta}_{t-1}^{(i)}) M_t(\boldsymbol{\theta}_{t-1}^{(i)}, \boldsymbol{\theta}_t^{(i)})}.$$
   (b) *With probability* $1 - p^{(i)} = 1 - \min\{1, W_t^{(i)}/c_t\}$, *reject* $\boldsymbol{\theta}_t^{(i)}$ *and go to (a).*
   (c) *Otherwise, accept* $\boldsymbol{\theta}_t^{(i)}$ *and set* $W_t^{(i)} = W_t^{(i)} r(c_t, \boldsymbol{\theta}_{t-1}^{(i)})/p^{(i)}$.
   (d) *Increment* $i = i + 1$. *If* $i \leq N$, *go to (a).*
   (e) *If* $t < T$, *go to Resample.*

Remark 7.20 *In the SMC Sampler PRC algorithm, we present the general framework in which we consider adaptive resampling. The derivation of the resulting normalizing constant for the PRC mechanism can be addressed under both adaptive and non-adaptive resampling schemes, which can be found in Peters et al. (2009, section 2.3). However, as they discuss, it will be shown to be computationally convenient when estimating the normalizing constant under PRC to consider the special case of H=N, thereby resampling at each iteration t.*

Algorithm 7.10, without the mutation and correction steps (b) and (c), is equivalent to the standard SMC Sampler algorithm. In the resample stage, the degeneracy of the particle approximation is quantified through the usual estimate of the effective sample size, $1 \leq [\sum_i (W_t^{(i)})^2]^{-1} \leq N$ (see Liu and Chen 1998). The addition of a rejection step at each time t effectively modifies the mutation kernel M_t. We denote the resulting kernel by M_t^*, to the choice presented in Equation (7.110). Thus, the SMC sampler PRC algorithm can be considered as an SMC sampler algorithm with the mutation kernel $M_t^* (\boldsymbol{\theta}_{t-1}, \boldsymbol{\theta}_t)$, and the correction weight

$$W_t = W_{t-1} \frac{\pi_t (\boldsymbol{\theta}_t) L_{t-1} (\boldsymbol{\theta}_t, \boldsymbol{\theta}_{t-1})}{\pi_{t-1} (\boldsymbol{\theta}_{t-1}) M_t^* (\boldsymbol{\theta}_{t-1}, \boldsymbol{\theta}_t)}. \tag{7.122}$$

Remark 7.21

- ***Estimation of the normalizing constant.*** *As the normalizing constant $r(c_t, \boldsymbol{\theta}_{t-1})$ in the weight calculation (7.122) in general depends on $\boldsymbol{\theta}_{t-1}$, it must be evaluated as it will not disappear in the renormalization across all weights for each particle. Where no analytic solution can be found, approximating (7.111) may be achieved by, for example, quadrature methods if the sample space E is relatively low-dimensional or Monte Carlo methods if E is high-dimensional;*
- ***Exact kernel selection normalization.*** *This is an alternative approach that restricts the mutation and backward kernel choices admitting an exact solution for the normalizing constant. Furthermore, this approach provides a computationally efficient approach to dealing with the PRC normalizing constant. This involves selecting kernels M_t and L_{t-1} such that $r(c_t, \boldsymbol{\theta}_{t-1}) = r(c_t)$ will be constant for all particles $\boldsymbol{\theta}_{t-1}$. In this case, the value of $r(c_t)$ may be absorbed into the proportionality constant of the weights, and safely ignored. Equation (7.111) suggests that this can be achieved if $M_t(\boldsymbol{\theta}_{t-1}, \boldsymbol{\theta}_t)$, W_{t-1}, and $w(\boldsymbol{\theta}_{t-1}, \boldsymbol{\theta}_t)$ are independent of $\boldsymbol{\theta}_{t-1}$.*

Specifying mutation kernels M_t such that $M_t(\boldsymbol{\theta}_{t-1}, \boldsymbol{\theta}_t) = M_t(\boldsymbol{\theta}_t)$ amounts to choosing a *global* kernel that is the same for all particles $\boldsymbol{\theta}_{t-1}$. The particle-dependent weight W_{t-1} can be set to $1/N$ for all particles following a resampling (or preselection) step; hence, setting $H = N$ will induce resampling at each iteration. Finally, consider for a moment the backward kernel of the form

$$L_{t-1}^{opt}(\boldsymbol{\theta}_t, \boldsymbol{\theta}_{t-1}) = \frac{\pi_{t-1}(\boldsymbol{\theta}_{t-1}) M_t(\boldsymbol{\theta}_{t-1}, \boldsymbol{\theta}_t)}{\int \pi_{t-1}(\boldsymbol{\theta}_{t-1}) M_t(\boldsymbol{\theta}_{t-1}, \boldsymbol{\theta}_t) d\boldsymbol{\theta}_{t-1}}. \tag{7.123}$$

Under the backward kernel (7.123), the incremental weight can be approximated by

$$\begin{aligned} w_t(\boldsymbol{\theta}_{t-1}, \boldsymbol{\theta}_t) &= \pi_t(\boldsymbol{\theta}_t) \Big/ \int \pi_{t-1}(\boldsymbol{\theta}_{t-1}) M_t(\boldsymbol{\theta}_{t-1}, \boldsymbol{\theta}_t) d\boldsymbol{\theta}_{t-1} \\ &\approx \pi_t(\boldsymbol{\theta}_t) \Big/ \sum_{i=1}^{N} W_{t-1}^{(i)} M_t(\boldsymbol{\theta}_{t-1}^{(i)}, \boldsymbol{\theta}_t). \end{aligned} \tag{7.124}$$

Under a global mutation kernel $M_t(\boldsymbol{\theta}_t)$, and following a resampling step, the incremental weight under this backward kernel reduces to $w_t(\boldsymbol{\theta}_{t-1}, \boldsymbol{\theta}_t) = \pi_t(\boldsymbol{\theta}_t)/M_t(\boldsymbol{\theta}_t)$, which is independent of $\boldsymbol{\theta}_{t-1}$.

Remark 7.22 *One such example of a global mutation kernel one may consider involves* $M_t\left(\left\{\boldsymbol{\theta}_{t-1}^{(i)}\right\}_{i=1:N}, \boldsymbol{\theta}_t\right) = \sum_{i=1}^{N} W_{t-1}^{(i)} M_t(\boldsymbol{\theta}_{t-1}^{(i)}, \boldsymbol{\theta}_t)$. *Thus, the weight calculation in (7.122) becomes*

$$\begin{aligned} W_t &\propto \pi_t(\boldsymbol{\theta}_t) \Big/ \left[\min\left\{1, \frac{w(\boldsymbol{\theta}_{t-1}, \boldsymbol{\theta}_t)}{Nc_t}\right\} M_t(\boldsymbol{\theta}_t)\right] \\ &= \begin{cases} \pi_t(\boldsymbol{\theta}_t)/M_t(\boldsymbol{\theta}_t), & \textit{if } \min\left\{1, \frac{w(\boldsymbol{\theta}_{t-1}, \boldsymbol{\theta}_t)}{Nc_t}\right\} = 1, \\ Nc_t, & \textit{otherwise.} \end{cases} \end{aligned}$$

It is instructive to consider the implications of this finding. Firstly, the resulting acceptance probability for each particle will range over the interval $(0, 1)$. To see this consider two illustrative scenarios, the first involving the trivial case of simply setting the user-controlled threshold to $c_t = 1/N$, thereby ensuring that as N increases, the acceptance probability does not necessarily decrease. This may not always be desirable since it reduces the threshold condition that particles must satisfy for large particle systems. The second nontrivial setting is to consider the incremental weight expression obtained in Equation (7.124). Under these choices for mutation and backward kernel, and assuming resampling in the setting $H = N$, we obtain an expression for the PRC probability of acceptance given by

$$\begin{aligned} \min\left\{1, \frac{w(\boldsymbol{\theta}_{t-1}, \boldsymbol{\theta}_t)}{Nc_t}\right\} &= \min\left\{1, \frac{\pi_t(\boldsymbol{\theta}_t)}{Nc_t \sum_{i=1}^{N} (1/N) M_t(\boldsymbol{\theta}_{t-1}^{(i)}, \boldsymbol{\theta}_t)}\right\} \\ &= \min\left\{1, \frac{\pi_t(\boldsymbol{\theta}_t)}{c_t \sum_{i=1}^{N} M_t(\boldsymbol{\theta}_{t-1}^{(i)}, \boldsymbol{\theta}_t)}\right\}. \end{aligned} \tag{7.125}$$

Note that under this setting, the SMC sampler PRC algorithm can be considered as a sequence of IS strategies with partial rejection control.

Finally, we observe that there are several variants of the SMC Sampler algorithm available in the context of interacting SMC Samplers: Annealed Importance Sampling and Population Monte Carlo, Island models, and transdimensional SMC Samplers (see examples in Jasra *et al.* 2007, 2008, Neal 2001, and Cappé *et al.* 2004.)

7.7.4 FINITE SAMPLE (NONASYMPTOTIC) ACCURACY FOR PARTICLE INTEGRATION

In this section, we detail some properties of the class of SMC algorithms discussed earlier, in particular, what is known about the accuracy of such methods. In addition, we also present examples for estimators of quantiles of annual loss distributions from such approaches, of direct interest to capital estimation. We begin by presenting some recent examples of concentration inequalities for particle methods that are finite sample result (see discussion and references by Del Moral *et al.* 2013).

The exponential concentration inequalities presented here are satisfied under some regularity conditions on the particle weights and the mutation kernel M_n when defined on some general state space E_n; see specific probabilistic details of these conditions by Del Moral (2004).

Using the concentration analysis of mean field particle models, the following exponential estimate can be obtained (see discussion by Del Moral 2004) and references therein. Note in the following when the N particle approximation to a distribution or density, such as π, is used we will denote it by π^N.

Theorem 7.7 (Finite Sample Exponential Concentration Inequality) *For any $x \geq 0$, $t \geq 0$, and any population size $N \geq 1$, the probability of the event is*

$$\mathbb{P}\mathrm{r}\left(\left|\pi_t^N(\varphi) - \pi_t(\varphi)\right| \leq \frac{c_1}{N}\left(1 + x + \sqrt{x}\right) + \frac{c_2}{\sqrt{N}}\sqrt{x}\right) \geq 1 - e^{-x}, \tag{7.126}$$

where one defines the N particle sample estimator as follows:

$$\pi_t^N(\varphi) = \sum_{i=1}^{N} W_t^{(i)} \varphi\left(\boldsymbol{\theta}_t^{(i)}\right)$$

and

$$\pi_t(\varphi) = \int \varphi\left(\boldsymbol{\theta}_t\right) \pi_t\left(\boldsymbol{\theta}_t\right)\, d\boldsymbol{\theta}_t. \tag{7.127}$$

In the case of a stable SMC algorithm, that is, one that is insensitive to initial conditions, such as those we discussed earlier, the constants c and (c_1, c_2) do not depend on the time parameter. One can also bound the difference between the particle estimate of the target distribution and the true distribution as follows. Consider that for any $\boldsymbol{\theta} = (\theta_i)_{1\leq i\leq d}$ and any $(-\infty, x] = \prod_{i=1}^{d}(-\infty, \theta_i]$ cells in $E_t = \mathbb{R}^d$, we let

$$F_t(x) = \pi_t\left(\mathbb{I}_{(-\infty,x]}\right) \quad \text{and} \quad F_t^N(x) = \pi_t^N\left(\mathbb{I}_{(-\infty,x]}\right).$$

Using these definitions of the empirical particle constructed distribution function and the target distribution function at sequence number t in the sequence of distribution $\{\pi_1, \pi_2, \ldots, \pi_T\}$, we can state the following corollary for the distribution functions for sequence of densities π_t given previously.

Corollary 7.4 *For any $y \geq 0$, $t \geq 0$, and any population size $N \geq 1$, the probability of the following event*

$$\sqrt{N}\, \left\|F_t^N - F_t\right\| \leq c\sqrt{d\,(y+1)}$$

is greater than $1 - e^{-y}$.

This concentration inequality ensures that the particle repartition function F_t^N converges to F_t, almost surely for the uniform norm. We complete this section with an example of a nonasymptotic estimate for a risk measure estimation via SMC Sampler output.

EXAMPLE 7.7 SMC Samplers Estimators for Risk Measures

Consider the single risk measure, where $d = 1$. Then let $F_t^{\leftarrow}$ be the generalized inverse on $[0, 1]$ of the function F_t, which is the annual loss distribution for the LDA model under consideration; that is, we have

$$F_t^{\leftarrow}(\alpha) := \inf\{\theta \in \mathbb{R} \,:\, F_t(x) \geq \alpha\}. \tag{7.128}$$

Now let $F_t^{\leftarrow}(\alpha) = q_t(\alpha)$ be the quantile, of order α, and we denote by $\zeta_t^{(i)}$ the order particle statistic associated with the particle system θ_t^i at time t; that is, we have

$$\zeta_t^{(1)} := \theta_t^{\sigma(1)} \leq \zeta_t^{(2)} := \theta_t^{\sigma(2)} \leq \cdots \leq \zeta_t^{(N)} := \theta_t^{\sigma(N)}$$

for some random permutation σ. We also denote by $q_t^N(\alpha) := \zeta_t^{1+\lfloor N\alpha \rfloor}$ the α particle quantile. By construction, we have

$$\begin{aligned}
\left|F_t\left(q_t^N(\alpha)\right) - F_t(q_t(\alpha))\right| &\leq \left|F_t\left(q_t^N(\alpha)\right) - F_t^N(q_t^N(\alpha))\right| + \left|F_t^N(q_t^N(\alpha)) - \alpha\right| \\
&\leq \left\|F_t^N - F_t\right\| + \left(\frac{1 + \lfloor N\alpha \rfloor}{N} - \alpha\right) \\
&\leq \left\|F_t^N - F_t\right\| + 1/N.
\end{aligned} \tag{7.129}$$

This clearly implies that $q_t^N(\alpha)$ converges almost surely to $q_t(\alpha)$, as N tends to ∞. In addition, for any $y \geq 0$, $n \geq 0$, and any population size $N \geq 1$, the probability of the following event

$$\sqrt{N}\,\left|F_n\left(q_n^N(\alpha)\right) - \alpha\right| \leq c\sqrt{d\,(y+1)} + \frac{1}{\sqrt{N}}$$

is greater than $1 - e^{-y}$. ■

7.8 Approximate Bayesian Computation (ABC) Methods

Here we present a class of estimation methods that generalize the classes of applicable models for the posterior to those which have intractable likelihoods. That is, we generalize now to classes of Monte Carlo algorithms that can tackle settings in which the posterior distribution may be constructed from a likelihood model for which the density of the observations (likelihood of the parameters) cannot be evaluated pointwise in closed form. This arises surprisingly often in settings related to heavy-tailed models; see discussions in the context of OpRisk Peters and Sisson (2006) or in financial modeling by Peters *et al.* (2010, 2011b).

The standard MCMC methods described previously assume that the likelihood of the data for given model parameters can be easily evaluated. If this is not the case, but synthetic data are easily simulated from the model for given parameters, then the so-called *approximate Bayesian computation* (ABC) methods can be utilized to estimate the model. For example, this is the case when the severity is modeled by the α-stable or g-and-h distributions that can easily be simulated but the density is not available in closed form (see the discussion in the OpRisk context by Peters and Sisson 2006, and Peters *et al.* 2010, 2008).

ABC methods are relatively recent developments in computational statistics (see Beaumont *et al.* 2002 and Tavaré *et al.* 2003). For applications in the context of OpRisk and insurance, see Peters and Sisson (2006) and Peters *et al.* (2010).

Consider the data $\boldsymbol{X}$ and denote the model parameters by $\boldsymbol{\theta}$. Then the posterior from which we wish to draw samples is $\pi(\boldsymbol{\theta}|\boldsymbol{x}) \propto f(\boldsymbol{x}|\boldsymbol{\theta})\pi(\boldsymbol{\theta})$. The purpose of ABC is to sample from the posterior $\pi(\boldsymbol{\theta}|\boldsymbol{x})$ without evaluating the computationally intractable $f(\boldsymbol{x}|\boldsymbol{\theta})$. The logical steps of the simplest ABC algorithm are as follows.

Algorithm 7.11 (Rejection Sampling ABC)

1. *Choose a small tolerance level* ϵ;
2. *For* $l = 1, 2, \ldots$
   a) *Draw* $\boldsymbol{\theta}^*$ *from the prior* $\pi(\cdot)$;
   b) *Simulate a synthetic dataset* $\boldsymbol{x}^*$ *from the model given parameters* $\boldsymbol{\theta}^*$, *that is, simulate from* $f(\cdot|\boldsymbol{\theta}^*)$;
   c) *Rejection condition: calculate a distance metric* $\rho(\boldsymbol{x}, \boldsymbol{x}^*)$ *that measures a difference between* $\boldsymbol{x}$ *and* $\boldsymbol{x}^*$. *Accept the sample, that is, set* $\boldsymbol{\theta}^{(l)} = \boldsymbol{\theta}^*$ *if* $\rho(\boldsymbol{x}, \boldsymbol{x}^*) \leq \epsilon$; *otherwise return to step (a).*
3. *Next* l.

It is easy to show that, if the support of the distributions on $\boldsymbol{x}$ is discrete and the rejection condition $\rho(\boldsymbol{x}, \boldsymbol{x}^*) \leq \epsilon$ is a simple condition of accepting the proposal only if $\boldsymbol{x}^* = \boldsymbol{x}$, then the obtained $\boldsymbol{\theta}^{(1)}, \boldsymbol{\theta}^{(2)}, \ldots$ are exact samples from $\pi(\boldsymbol{\theta}|\boldsymbol{x})$. For more general cases, the obtained samples $\boldsymbol{\theta}^{(l)}$ are from

$$\pi_{ABC}(\boldsymbol{\theta}|\boldsymbol{x}, \epsilon) \propto \int \pi(\boldsymbol{\theta})\pi(\boldsymbol{x}^*|\boldsymbol{\theta})g_\epsilon(\boldsymbol{x}|\boldsymbol{x}^*)d\boldsymbol{x}^*, \tag{7.130}$$

where a weighting function $g_\epsilon(\boldsymbol{x}|\boldsymbol{x}^*)$ is used. The previous rejection algorithm considers a weight function such as the hard decision choice

$$g_\epsilon(\boldsymbol{x}|\boldsymbol{x}^*) \propto \begin{cases} 1, & \text{if } \ \rho(\boldsymbol{x}, \boldsymbol{x}^*) \leq \epsilon, \\ 0, & \text{otherwise.} \end{cases} \tag{7.131}$$

As $\epsilon \to 0$, for appropriate choices of distance $\rho(\cdot,\cdot)$,

$$\pi_{ABC}(\boldsymbol{\theta}|\boldsymbol{x},\epsilon) \to \pi(\boldsymbol{\theta}|\boldsymbol{x}).$$

Of course, for a finite ϵ we obtain an approximation for $\pi(\boldsymbol{\theta}|\boldsymbol{x})$.

To improve the computational efficiency, $\rho(\boldsymbol{x},\boldsymbol{x}^*)$ is often replaced by $\rho(S(\boldsymbol{x}),S(\boldsymbol{x}^*))$, where $S(\boldsymbol{x})$ is a summary statistic of the data sample. Other weighting functions can be used. In general, the procedure is simple: given a realization of the model parameters, a synthetic dataset $\boldsymbol{x}^*$ is simulated and compared to the original dataset $\boldsymbol{x}$. Then the summary statistic $S(\boldsymbol{x}^*)$ is calculated for the simulated dataset $\boldsymbol{x}^*$ and compared to the summary statistic of the observed data $S(\boldsymbol{x})$; and a distance $\rho(S(\boldsymbol{x}),S(\boldsymbol{x}^*))$ is calculated. Finally, a greater weight is given to the parameter values producing $S(\boldsymbol{x}^*)$ close to $S(\boldsymbol{x})$ according to the weighting function $g_\epsilon(\boldsymbol{x}|\boldsymbol{x}^*)$. The obtained sample is from $\pi_{ABC}(\boldsymbol{\theta}|\boldsymbol{x},\epsilon)$, which converges to the target posterior $\pi(\boldsymbol{\theta}|\boldsymbol{x})$ as $\epsilon \to 0$, assuming that $S(\boldsymbol{x})$ is a *sufficient statistic*[1] and the weighting function converges to a point mass on $S(\boldsymbol{x})$. The tolerance ϵ is typically set as small as possible for a given computational budget. One can calculate the results for subsequently reduced values of ϵ until further reduction does not make material difference for the model outputs. The described ABC can be viewed as a general augmented model

$$\pi(\boldsymbol{\theta},\boldsymbol{x},\boldsymbol{x}^*) = \pi(\boldsymbol{x}|\boldsymbol{x}^*,\boldsymbol{\theta})\pi(\boldsymbol{x}^*|\boldsymbol{\theta})\pi(\boldsymbol{\theta}),$$

where $\pi(\boldsymbol{x}|\boldsymbol{x}^*,\boldsymbol{\theta})$ is replaced by $g(\boldsymbol{x}|\boldsymbol{x}^*)$.

To improve the performance of the ABC algorithm, it can be combined with MCMC, producing the stationary distribution $\pi_{ABC}(\boldsymbol{\theta}|\boldsymbol{x},\epsilon)$. For example, the MCMC–ABC can be implemented as follows.

Algorithm 7.12 (MCMC–ABC)

1. *Initialize* $\boldsymbol{\theta}^{(l=0)}$;
2. *For* $l = 1,\ldots,L$
   a) *Draw proposal* $\boldsymbol{\theta}^*$ *from the proposal density* $q(\cdot|\boldsymbol{\theta}^{(l-1)})$;
   b) *Simulate a synthetic dataset* $\boldsymbol{x}^*$ *from the model given parameters* $\boldsymbol{\theta}^*$;
   c) *Accept the proposal with the acceptance probability*

$$p(\boldsymbol{\theta}^{(l-1)},\boldsymbol{\theta}^*) = \min\left\{1, \frac{\pi(\boldsymbol{\theta}^*)q(\boldsymbol{\theta}^{(l-1)}|\boldsymbol{\theta}^*)}{\pi(\boldsymbol{\theta}^{(l-1)})q(\boldsymbol{\theta}^*|\boldsymbol{\theta}^{(l-1)})}\mathbb{I}_{\{\rho(S(\boldsymbol{x}),S(\boldsymbol{x}^*))\le\epsilon\}}\right\},$$

   that is, simulate U *from the Uniform(0,1) and set* $\boldsymbol{\theta}^{(l)} = \boldsymbol{\theta}^*$ *if* $U \le p(\boldsymbol{\theta}^{(l-1)},\boldsymbol{\theta}^*)$ *otherwise set* $\boldsymbol{\theta}^{(l)} = \boldsymbol{\theta}^{(l-1)}$. *Here,* $\mathbb{I}_{\{.\}}$ *is a standard indicator function.*
3. *Next* l.

Various summary statistics of the dataset $x_1,\ldots,x_N$ are used in practice. For example, the statistic $S(\boldsymbol{x})$ can be defined as the following vectors:

[1]A sufficient statistic is a function of the dataset $\boldsymbol{x}$ that summarizes all the available sample information about $\boldsymbol{\theta}$; for a formal definition, see Berger (1985, section 1.7).

- $\boldsymbol{S} = (\widetilde{\mu}, \widetilde{\sigma})$, where $\widetilde{\mu}$ and $\widetilde{\sigma}$ are empirical mean and standard deviation of the dataset $\boldsymbol{x}$, respectively;
- $\boldsymbol{S} = (x_1, \ldots, x_N)$, that is, all data points in the dataset;
- $\boldsymbol{S}\left(q_1(\boldsymbol{x}), \ldots, q_p(\boldsymbol{x})\right)$ a vector of empirical quantities summarizing the empirical distribution function at a fixed set of probabilities.

Popular choices for the distance metrics, $\rho(\boldsymbol{S}, \boldsymbol{S}^*)$, include the following:

- Euclidean distance: $\rho(\boldsymbol{S}, \boldsymbol{S}^*) = \sum_{l=1}^{L}(S_l - S_l^*)^2$;
- L_1-distance $\rho(\boldsymbol{S}, \boldsymbol{S}^*) = \sum_{l=1}^{L}|S_l - S_l^*|$.

We also note that there are efficient SMC Sampler versions of these ABC algorithms developed by Peters *et al.* (2009) and Del Moral *et al.* (2012).

7.9 OpRisk Estimation and Modeling for Truncated Data

Accurate modeling of the severity and frequency distributions is the key to estimating a capital charge. One of the challenges in modeling OpRisk is the lack of complete data—often a bank's internal data are not reported below a certain level (typically on the order of € 10,000). These data are said to be left-truncated. Generally speaking, missing data increase uncertainty in modeling. Sometimes, a threshold level is introduced to avoid difficulties with collection of too many small losses. Industry data in external databases from vendors and consortia of banks are available above some thresholds: Algo OpData provides publicly reported operational risk losses above USD 1 million and ORX provides OpRisk losses above € 20,000 reported by ORX members. The OpRisk data from Loss Data Collection Exercises (LDCE) over many institutions are truncated too. For example, Moscadelli (2004) analyzed 2002 LDCE and Dutta and Perry (2006) analyzed 2004 LDCE, where the data were mainly above € 10,000 and USD 10,000, respectively. Fitting data reported above a constant threshold is a well-known and studied problem. However, in practice, the losses are scaled for business and other factors before the fitting and thus the threshold varies across the scaled data sample. Moreover, the actual threshold might be unknown for some external databases and should be treated as stochastic (see, e.g., Baud *et al.* 2003 by De Fontnouvelle *et al.* 2006).

The reporting level may also change when a bank changes its reporting policy. In this section, we consider the cases of constant, time-varying, unknown, and stochastic thresholds. We also discuss the approaches to fit these models and the impact of ignoring data truncation on the estimation of the 0.999 quantile of the annual loss distribution.

Often, modeling of missing data is done assuming a parametric distribution for losses below and above the threshold. Then fitting is accomplished using losses reported above the threshold via the maximum likelihood method (see, e.g., Frachot *et al.* 2004b) or the EM algorithm (see, e.g., Bee 2005b). In practice, often the missing data are ignored completely. This may lead to a significant underestimation or overestimation of the capital. The impact of data truncation in OpRisk was discussed in the literature (see Baud *et al.* 2003, Chernobai *et al.* 2006, Mignola and Ugoccioni 2006, Luo *et al.* 2007, and Ergashev *et al.* 2012). Typically, the case of a constant threshold is discussed in research studies, though in practice, a threshold level varies across observations (see Shevchenko and Temnov 2009). One of the reasons for a varying threshold appearing in OpRisk loss data is that the losses are scaled for inflation and other factors before fitting to reflect changes in risk over time. The reporting level may also change from time to time within a bank when the reporting policy is changed. The problem with multiple thresholds also appears when the different companies report losses into the same database using different threshold levels (see Baud *et al.* 2003).

Of course, for risks with heavy-tailed severities, the impact of the data threshold should not be important in a limit of high quantiles. However, it should be quantified first before making such a conclusion and to justify a chosen reporting level. For light-tailed risks too the impact can be significant.

In this section, we consider the case of a single risk cell and use the following notation and assumptions.

Model Assumptions 7.1 *Consider a single risk cell where*

- *The annual loss in a risk cell in year m is*

$$Z_m = \sum_{i=1}^{N_m} X_i(m). \tag{7.132}$$

- N_m *is the number of events (frequency) and* $X_i(m)$, $i = 1, \ldots, N_m$ *are the severities of the events in year* m;
- *If convenient, we may index severities* $X_i(m)$ *and their event times* $T_i(m)$, $i = 1, \ldots, N_m$, $m = 1, 2, \ldots$ *(ordered in time) as* X_j *and* T_j, $j = 1, 2, \ldots$, *respectively, where* $T_1 < T_2 < \cdots$;
- *The severities of the events* X_j, $j = 1, 2, \ldots$, *occurring at times* T_j, $j = 1, 2, \ldots$, *respectively are modeled as independent and identically distributed random variables from a continuous distribution* $F(x|\boldsymbol{\beta})$, $0 < x < \infty$, *whose density is denoted as* $f(x|\boldsymbol{\beta})$. *Here,* $\boldsymbol{\beta}$ *are the severity distribution parameters;*
- N_m, $m = 1, 2, \ldots$ *are independent and identically distributed random variables from a discrete frequency distribution with probability mass function* $p(n|\lambda) = \Pr[N_m = n]$, *where* λ *is a frequency parameter (or a vector of parameters);*
- *The severities* $X_i(m)$ *and frequencies* N_m *of the events are independent;*
- $\boldsymbol{\gamma} = (\lambda, \boldsymbol{\beta})$ *is a vector of frequency and severity distribution parameters.*

7.9.1 CONSTANT THRESHOLD - POISSON PROCESS

If we assume that loss events follow a homogeneous Poisson process with the intensity parameter λ, then $N_1, N_2, \ldots$ are independent and identically distributed random variables from the Poisson distribution, *Poisson*(λ), with

$$\Pr[N_m = n] = p(n|\lambda) = \frac{\lambda^n}{n!}\exp(-\lambda), \quad \lambda > 0,\ n = 0, 1, \ldots \tag{7.133}$$

and the event interarrival times $\delta T_j = T_j - T_{j-1}, j = 1, 2, \cdots$ (where $T_0 < T_1 < T_2 < \cdots$ are the event times and $T_0 = t_0$ is the start of the observation period) are independent exponentially distributed random variables with the density and distribution functions

$$g(\tau|\lambda) = \lambda\exp(-\lambda\tau) \quad \text{and} \quad G(\tau|\lambda) = 1 - \exp(-\lambda\tau), \tag{7.134}$$

respectively.

If the losses, originating from severity $f(x|\boldsymbol{\beta})$ and frequency $p(n|\lambda)$ densities, are recorded above a known reporting level (truncation level) L, then the density of the losses above L is left-truncated density,

$$f_L(x|\boldsymbol{\beta}) = \frac{f(x|\boldsymbol{\beta})}{1 - F(L|\boldsymbol{\beta})}; \quad L \leq x < \infty. \tag{7.135}$$

The events of the losses above L follow the Poisson process with the intensity,

$$\theta(\boldsymbol{\gamma}, L) = \lambda(1 - F(L|\boldsymbol{\beta})), \tag{7.136}$$

the so-called *thinned* Poisson process, and the annual number of events above the threshold is distributed from $Poisson(\theta)$.

The series of the annual counts or event times can be used for estimating frequency distribution. These cases are considered separately here.

Proposition 7.4 (Likelihood for Annual Counts and Truncated Severities) *For independent losses from Poisson process with intensity λ and severity density $f(x|\boldsymbol{\beta})$, consider a corresponding random vector $\boldsymbol{Y}$ of the events recorded above the threshold L over a period of T years consisting of the annual frequencies $\widetilde{N}_m$, $m = 1, \ldots, T$ and severities $\widetilde{X}_j$, $j = 1, \ldots, J$, $J = \widetilde{N}_1 + \ldots + \widetilde{N}_T$. Then, for given model parameters $\boldsymbol{\gamma}$, the joint density of $\boldsymbol{Y}$ at $\widetilde{N}_m = \tilde{n}_m$ and $\widetilde{X}_j = \tilde{x}_j$ can be written as*

$$h(\boldsymbol{y}|\boldsymbol{\gamma}) = \prod_{j=1}^{J} f_L(\tilde{x}_j|\boldsymbol{\beta}) \prod_{m=1}^{T} p(\tilde{n}_m|\theta(\boldsymbol{\gamma}, L)). \tag{7.137}$$

That is, the log likelihood function for this model is $\ell_{\boldsymbol{y}}(\boldsymbol{\gamma}) = \ln h(\boldsymbol{y}|\boldsymbol{\gamma})$.

Proof: The proof is obvious because severities and frequencies are independent. ■

From (7.137), the MLEs for model parameters $\hat{\boldsymbol{\gamma}}$ can be found as a solution of

$$\frac{\partial \ell_{\boldsymbol{Y}}(\boldsymbol{\gamma})}{\partial \lambda} = (1 - F(L|\boldsymbol{\beta})) \sum_{m=1}^{T} \frac{\partial}{\partial \theta} \ln p(\widetilde{N}_m|\theta(\boldsymbol{\gamma}, L)) = 0, \tag{7.138}$$

$$\begin{aligned}\frac{\partial \ell_{\boldsymbol{Y}}(\boldsymbol{\gamma})}{\partial \boldsymbol{\beta}} &= \sum_{j=1}^{J} \frac{\partial}{\partial \boldsymbol{\beta}} \ln f_L(\widetilde{X}_j|\boldsymbol{\beta}) \\ &\quad - \lambda \frac{\partial F(L|\boldsymbol{\beta})]}{\partial \boldsymbol{\beta}} \sum_{m=1}^{T} \frac{\partial}{\partial \theta} \ln p(\widetilde{N}_m|\theta(\boldsymbol{\gamma}, L)) = 0.\end{aligned} \tag{7.139}$$

It is easy to see that the MLEs $\hat{\boldsymbol{\beta}}$ for the severity parameters can be found marginally (independently from frequency) by maximizing

$$\sum_{j=1}^{J} \ln f_L(\widetilde{X}_j|\boldsymbol{\beta}) \tag{7.140}$$

and then Equation (7.138) gives the MLE for the intensity

$$\hat{\lambda} = \frac{1}{1 - F(L|\hat{\boldsymbol{\beta}})} \times \frac{1}{T} \sum_{m=1}^{T} \widetilde{N}_m. \tag{7.141}$$

Proposition 7.5 (Likelihood for Event Times and Truncated Severities) *For independent losses from Poisson process with intensity λ and severity density $f(x|\boldsymbol{\beta})$, consider the data $\boldsymbol{Y}$ of the events above a constant threshold over the time period $[t_0, t_E]$ consisting of the event interarrival times $\delta\widetilde{T}_j = \widetilde{T}_j - \widetilde{T}_{j-1}, j = 1, \ldots, J$ (where $\widetilde{T}_j, j = 1, 2, \ldots$ are the event times and $\widetilde{T}_0 = t_0$) and the severities $\widetilde{X}_j$, $j = 1, \ldots, J$. Then the joint density (for given γ) of $\boldsymbol{Y}$ at $\delta\widetilde{T}_j = \widetilde{\tau}_j$ and $\widetilde{X}_j = \widetilde{x}_j$ is*

$$\begin{aligned} h(\boldsymbol{y}|\gamma) &= (1 - G(t_E - \tilde{t}_J|\theta(\gamma, L))) \prod_{j=1}^{J} f_L(\tilde{x}_j|\boldsymbol{\beta}) g(\widetilde{\tau}_j|\theta(\gamma, L)) \\ &= \lambda^J \exp(-\theta(\gamma, L)(t_E - t_0)) \prod_{j=1}^{J} f(\tilde{x}_j|\boldsymbol{\beta}). \end{aligned} \tag{7.142}$$

Here, $1 - G(t_E - \tilde{t}_J|\theta(\gamma, L))$ is the probability that no event will occur within $(\tilde{t}_J, t_E]$. The log likelihood function for this model is $\ell_{\boldsymbol{y}}(\gamma) = \ln h(\boldsymbol{y}|\gamma)$.

Proof: The proof is obvious using the distribution of interarrival time (7.134) and the fact that severities and frequencies are independent, severities are independent. ■

From (7.142), the MLEs $\hat{\gamma}$ can be found as a solution of

$$\begin{aligned} \frac{\partial \ell_{\boldsymbol{Y}}(\gamma)}{\partial \lambda} &= \frac{J}{\lambda} - (1 - F(L|\boldsymbol{\beta}))(t_E - t_0) = 0, \\ \frac{\partial \ell_{\boldsymbol{Y}}(\gamma)}{\partial \boldsymbol{\beta}} &= \lambda(t_E - t_0)\frac{\partial F(L|\boldsymbol{\beta})}{\partial \boldsymbol{\beta}} + \sum_{j=1}^{J} \frac{\partial}{\partial \boldsymbol{\beta}} \ln f(\widetilde{X}_j|\boldsymbol{\beta}) = 0. \end{aligned} \tag{7.143}$$

This gives the MLE for the intensity parameter

$$\hat{\lambda} = \frac{J}{\left[1 - F(L|\hat{\boldsymbol{\beta}})\right](t_E - t_0)}, \tag{7.144}$$

which is equivalent to (7.141) if the start and end of the observation period correspond to the beginning and end of the first and last years, respectively. Substituting $\hat{\lambda}$ into the second equation in (7.143), it is easy to see that the severity MLEs $\hat{\boldsymbol{\beta}}$ can be obtained by maximizing $\sum_{j=1}^{J} \ln f_L(\widetilde{X}_j|\boldsymbol{\beta})$.

Remark 7.23

- *If the start and end of the observation period correspond to the beginning and end of the first and last years, respectively, then the inferences based on the likelihoods (7.142) and (7.137) are equivalent. This is because the likelihoods, in this case, are different by a factor that does not depend on the model parameters;*

- *The MLE errors are typically estimated using asymptotic Gaussian approximation via the inverse of the Fisher information matrix (see Section 7.1.1). The latter is often estimated by the observed information matrix. That is,*

$$\mathbb{C}\mathrm{ov}[\hat{\gamma}_i, \hat{\gamma}_j] \approx (\hat{\boldsymbol{I}}^{-1})_{ij}, \quad (\hat{\boldsymbol{I}})_{ij} = -\left.\frac{\partial^2 \ell_{\boldsymbol{y}}(\boldsymbol{\gamma})}{\partial \gamma_i \partial \gamma_j}\right|_{\boldsymbol{\gamma}=\hat{\boldsymbol{\gamma}}}. \tag{7.145}$$

Whether the sample size is large enough to use this asymptotic approximation is a difficult question that should be addressed in a practical solution. Also, the regularity conditions required for this approximation are mild but difficult to prove.

Detailed illustrative examples of fitting truncated data in the case of constant threshold using maximum likelihood and Bayesian MCMC methods are given by Shevchenko (2011, examples 5.1 and 5.2, pp. 184–188).

7.9.2 NEGATIVE BINOMIAL AND BINOMIAL FREQUENCIES

Negative Binomial and Binomial are other distributions often used to model frequencies. The mean of a Binomial is less than the variance; the mean of the Negative Binomial is larger than its variance; and Poisson mean equals its variance. This property is often used as a criterion to choose a frequency distribution, and is known as under and over dispersion of the counting distribution (frequency distribution).

Another convenient property of these distributions is that their type is preserved in the case of loss truncation as given by the following proposition.

Proposition 7.6 (Frequency of Truncated Losses) *Consider independent losses $X_1, X_2, \ldots, X_N$ with a common distribution $F(x)$ over some time period. Assume that the losses are independent of the loss frequency N. Denote the frequency of the losses above the reporting level L as N_L. Then*

(a) *If $N \sim Poisson(\lambda)$, $N_L \sim Poisson(\lambda(1 - F(L))$;*

(b) *If $N \sim NegBinomial(r, p)$, where the parameter $p = 1/(1 + q)$, then $N_L \sim NegBinomial(r, \tilde{p})$ with $\tilde{p} = 1/(1 + \tilde{q})$, where $\tilde{q} = q(1 - F(L))$;*

(c) *If $N \sim Binomial(n, p)$, then $N_L \sim Binomial(n, \tilde{p})$, where $\tilde{p} = p(1 - F(L))$.*

Proof: The proof is trivial using a more general result given by Equation (7.150) derived later. ∎

In general, the relation between the distributions of N and N_L can be calculated as follows. Assume that the probability function for the number of events N is known to be $p_n = \mathbb{P}\mathrm{r}[N = n]$ and its probability-generating function is

$$\psi_N(t) = \mathbb{E}[t^N] = \sum_k p_k t^k. \tag{7.146}$$

Consider a compound sum $S = M_1 + \cdots + M_N$, where N is a discrete random variable with probability-generating function $\psi_N(t)$, and M_i are independent discrete random variables with probability-generating function $\psi_M(t)$. Utilizing the fact that the probability-generating function of the sum of independent random variables is the product of the individual probability-generating functions, the probability-generating function of S can be found as

$$\begin{aligned}\psi_S(t) &= \sum_k \mathbb{P}\mathrm{r}[S = k]t^k \\ &= \sum_k \sum_n \mathbb{P}\mathrm{r}[M_1 + \cdots + M_n = k|N = n]\mathbb{P}\mathrm{r}[N = n]t^k \\ &= \sum_n \mathbb{P}\mathrm{r}[N = n](\psi_M(t))^n \\ &= \psi_N(\psi_M(t)). \end{aligned} \tag{7.147}$$

The number of events above the threshold can be written as

$$N_L = I_1 + \cdots + I_N,$$

where I_j are independent and identically distributed indicator random variables

$$I_j = \begin{cases} 1, \ \mathbb{P}\mathrm{r}[I_j = 1] = 1 - F(d), & \text{if} \quad X_j > u, \\ 0, \ \mathbb{P}\mathrm{r}[I_j = 0] = F(d), & \text{if} \quad X_j \le u, \end{cases} \tag{7.148}$$

with probability-generating function

$$\psi_I(t) = F(L) + t(1 - F(L)) = 1 + (1 - F(L))(t - 1).$$

The probability-generating function of the number of events above the threshold L can then be calculated as

$$\psi_{N_L}(t) = \psi_N(\psi_I(t)). \tag{7.149}$$

Moreover, if the distribution of N is parameterized by some θ and its probability-generating function has a special form $\psi_N(t;\theta) = g(\theta(t-1))$, that is, t and θ appear in $\psi_N(t;\theta)$ as $\theta(t-1)$ only, then

$$\psi_{N_L}(t;\theta) = g(\theta(1 - F(L))(t - 1)) = \psi_N(t; \theta(1 - F(L))). \tag{7.150}$$

That is, both N and N_L have the same distribution type with different parameter θ. Specifically, if N is distributed from $P(\cdot|\theta)$, then N_L is distributed from $P(\cdot|\widetilde{\theta})$, where $\widetilde{\theta} = \theta(1 - F(L))$. It can be checked directly that this relationship holds for Poisson, Binomial, and Negative Binomial. This property of Poisson distribution has already been used in Section 7.9.1. For more details and examples, see Panjer (2006, sections 5.7 and 7.8.2).

7.9.3 IGNORING DATA TRUNCATION

Often, the data below a reported level are simply ignored in the analysis, arguing that the high quantiles are mainly determined by the low-frequency/heavy-tailed severity risks. However, even if the impact is small, often it should be estimated to justify the reporting level. There are several ways to ignore truncation discussed here.

Assume that the true model is based on the annual number of events N and severities X_j coming from distributions $P(\cdot|\boldsymbol{\lambda})$ and $F(\cdot|\boldsymbol{\beta})$, respectively. Here, $P(\cdot|\boldsymbol{\lambda})$ can be different from Poisson and $\boldsymbol{\lambda}$ denotes all frequency parameters. The density of the distribution $F(\cdot|\boldsymbol{\beta})$ is $f(\cdot|\boldsymbol{\beta})$. If it is further assumed that severities are independent and identically distributed, and independent of frequency. Then the frequency $\widetilde{N}$ and losses $\widetilde{X}_j$ above the threshold L are from $\widetilde{P}(\cdot|\boldsymbol{\theta})$ and

$$F_L(x|\boldsymbol{\beta}) = \frac{F(x|\boldsymbol{\beta}) - F(L|\boldsymbol{\beta})}{1 - F(L|\boldsymbol{\beta})}, \quad x \geq L,$$

respectively. Note that $\boldsymbol{\theta}$ is a function of $\boldsymbol{\lambda}$, $\boldsymbol{\beta}$, and L (see Section 7.9.2). The corresponding truncated severity density is

$$f_L(x|\boldsymbol{\beta}) = \frac{f(x|\boldsymbol{\beta})}{1 - F(L|\boldsymbol{\beta})}, \quad x \geq L.$$

Denote the data above the threshold as $\widetilde{\boldsymbol{Y}}$. Then fitting of the correct model proceeds as follows.

"True model". Using the frequency $\widetilde{P}(\cdot|\boldsymbol{\theta})$ and severity $F_L(x|\boldsymbol{\beta})$ distributions of the truncated data $\widetilde{\boldsymbol{Y}}$, fit the model parameters $\boldsymbol{\lambda}$ and $\boldsymbol{\beta}$, using the likelihood of the observed data $\widetilde{\boldsymbol{Y}}$ via the MLE or Bayesian inference methods as described in Section 7.9.1. Then calculate the annual loss as

$$Z^{(0)} = \sum_{i=1}^{N} X_i, \quad N \sim P(\cdot|\boldsymbol{\lambda}), \quad X_i \overset{i.i.d.}{\sim} F(\cdot|\boldsymbol{\beta}). \tag{7.151}$$

Of course, it is assumed that data below the threshold are generated from the same process as for data above. To simplify the fitting procedure or to avoid making the assumptions about data below the level, several approaches are popular in practice. In particular *"naive model"*, *"shifted model"*, and *"truncated model"* are defined.

"Naive model". Using truncated data $\widetilde{\boldsymbol{Y}}$, fit frequency distribution $\widetilde{P}(\cdot|\boldsymbol{\theta})$ and severity $F(\cdot|\boldsymbol{\beta}_U)$ assuming that there is no truncation. Then calculate the annual loss as

$$Z^{(U)} = \sum_{i=1}^{N} X_i, \quad N \sim \widetilde{P}(\cdot|\boldsymbol{\theta}), \quad X_i \overset{i.i.d.}{\sim} F(\cdot|\boldsymbol{\beta}_U). \tag{7.152}$$

"Shifted model". Using truncated data $\widetilde{\boldsymbol{Y}}$, fit frequency $\widetilde{P}(\cdot|\boldsymbol{\theta})$ and severity $F_L^{(S)}(x) = F(x - L|\boldsymbol{\beta})$. Then calculate the annual loss as

$$Z^{(S)} = \sum_{i=1}^{N} X_i, \quad N \sim \widetilde{P}(\cdot|\boldsymbol{\theta}), \quad X_i \overset{i.i.d.}{\sim} F_L^{(S)}(\cdot|\boldsymbol{\beta}_S). \tag{7.153}$$

"Truncated model". Using truncated data $\widetilde{\boldsymbol{Y}}$, fit frequency $\widetilde{P}(\cdot|\boldsymbol{\theta})$ and severity $F_L(x|\boldsymbol{\beta})$. Then calculate the annual loss as

$$Z^{(T)} = \sum_{i=1}^{N} X_i, \quad N \sim \widetilde{P}(\cdot|\boldsymbol{\theta}), \quad X_i \overset{i.i.d.}{\sim} F_L(\cdot|\boldsymbol{\beta}). \tag{7.154}$$

Denote the 0.999 quantiles of the annual losses under the "true", "naive", "shifted" and "truncated" models as $Q^{(0)}$, $Q^{(U)}$, $Q^{(S)}$, and $Q^{(T)}$, respectively. The bias introduced into the 0.999 quantile of the annual loss distribution from use of the wrong model can be quantified by the relative difference

$$\delta^{(\star)} \equiv \frac{Q^{(\star)} - Q^{(0)}}{Q^{(0)}}, \quad (\star) = \text{“}U,\text{” “}T\text{”, “}S\text{”}. \tag{7.155}$$

Calculation of the annual loss quantile using the incorrect model (wrong frequency and severity distributions) will induce a bias. One may think that the bias is not significant and use one of the mentioned methods.

Each of the "naive model", "shifted model", and "truncated model" is biased for finite truncation, that is, their quantile estimates will never converge to the true value as the data sample size increases.

The difference (bias) between $Q^{(0)}$ and $Q^{(S)}$, and between $Q^{(0)}$ and $Q^{(U)}$ was studied by Luo *et al.* (2007) and Ergashev *et al.* (2012). The difference between $Q^{(T)}$ and $Q^{(0)}$ was studied by Mignola and Ugoccioni (2006). The "naive model" was analyzed by Chernobai *et al.* (2006) and Frachot *et al.* (2004b).

Example for Poisson—LogNormal case. To demonstrate the impact of ignoring data truncation consider N and X_i modeled by the *Poisson*(λ) and *LogNormal*(μ, σ^2) with the probability mass $p(\cdot|\lambda)$ and the density $f(x|\mu, \sigma)$, $0 < x < \infty$, respectively. The density of a left-truncated LogNormal distribution is

$$f_L(x|\mu, \sigma) = \frac{f(x|\mu, \sigma)}{1 - F(L|\mu, \sigma)}, \quad L \le x < \infty \tag{7.156}$$

Assuming that losses originating from $f(x|\mu, \sigma)$ and $p(k|\lambda)$ are recorded above known reporting level L, the data above L are counts from *Poisson*(θ), $\theta = \lambda(1 - F(L|\mu, \sigma))$, and losses from $f_L(x|\mu, \sigma)$. Then the following models are calculated.

- ***"True model"*** is obtained by using λ, μ, and σ in (7.151);
- ***"Shifted model".*** Suppose that the shifted LogNormal density

$$f_L^{(S)}(x|\mu_s, \sigma_s) = \frac{1}{(x - L)\sqrt{2\pi\sigma_s^2}} \exp\left(-\frac{(\ln(x - L) - \mu_s)^2}{2\sigma_s^2}\right), \tag{7.157}$$

where $L \le x < \infty$, is fitted to the truncated data using the method of maximum likelihood. In the limit of large sample size, the parameters of this distribution μ_S and σ_S can be determined using the first two moments, that is, expressed in terms of the true parameters μ and σ as follows:

$$\mu_S = \int_L^{\infty} \ln(x - L) f_L^{(T)}(x|\mu, \sigma)dx, \tag{7.158}$$

$$\sigma_S^2 = \int_L^{\infty} [\ln(x - L)]^2 f_L^{(T)}(x|\mu, \sigma)dx - \mu_S^2. \tag{7.159}$$

These integrals can be efficiently calculated using Gaussian quadrature or just using standard adaptive integration routines available from most of software packages e.g. In this model, the frequency is modeled by $Poisson(\theta)$, that is, the losses below L are ignored. Finally, θ, μ_S, and σ_S are used in (7.153);

- ***"Naive model".*** This model is based on the untruncated LogNormal with density $f(x|\mu_U, \sigma_U)$ defined by (2) and fitted to data above the threshold L using the method of maximum likelihood. Similar to the "shifted model", in the limit of large sample size, parameters μ_U and σ_U can be determined via the true parameters μ and σ as follows (see Chernobai *et al.* 2006):

$$\mu_U = \int_L^{\infty} \ln(x) f_L^{(T)}(x|\mu, \sigma)dx, \tag{7.160}$$

$$\sigma_U^2 = \int_L^{\infty} (\ln x)^2 f_L^{(T)}(x|\mu, \sigma)dx - \mu_U^2. \tag{7.161}$$

Unlike the "shifted model", these integrals can be evaluated in closed form. The frequency under the "naive model" is modeled by $Poisson(\theta)$, that is, the losses below the threshold, are ignored when the intensity of loss events is estimated. Finally, θ, μ_U, and σ_U are used in (7.152);

- ***"Truncated model"*** is obtained by using θ, μ, and σ in (7.154).

Figure 7.7 shows the relative bias in the 0.999 annual loss quantile (7.155) versus a fraction of truncated points $\Psi = F(L|\mu, \sigma) \times 100\%$, for the cases of light- and heavy-tailed severities. In this example, the parameter values are chosen the same as some cases considered in Luo *et al.* (2007). In particular, we show the results for $(\theta = 10, \sigma = 1)$ and $(\theta = 10, \sigma = 2)$. The latter corresponds to the heavier-tailed severity. Here, the calculated bias is due to the model error only, that is, the bias corresponds to the limiting case of a very large data sample. Also note that the actual value of the scale parameter μ is not relevant because only relative quantities are calculated. "*Naive model*" and "*shifted model*" are easy to fit but induced bias can be very large. Typically, "*naive model*" leads to a significant underestimation of the capital, even for a heavy-tailed severity; "*shifted model*" is better than "*naive model*" but worse than "*truncated model*"; the bias from "*truncated model*" is less for heavier-tailed severities.

The biases introduced by the "naive" and "shifted" models, studied in this example, are the biases in the limit of large sample size. The parameters fitted using real data are estimates that have statistical fitting errors due to finite sample size. The true parameters are not known. The impact of parameter uncertainty on quantile estimates can be taken into account using a

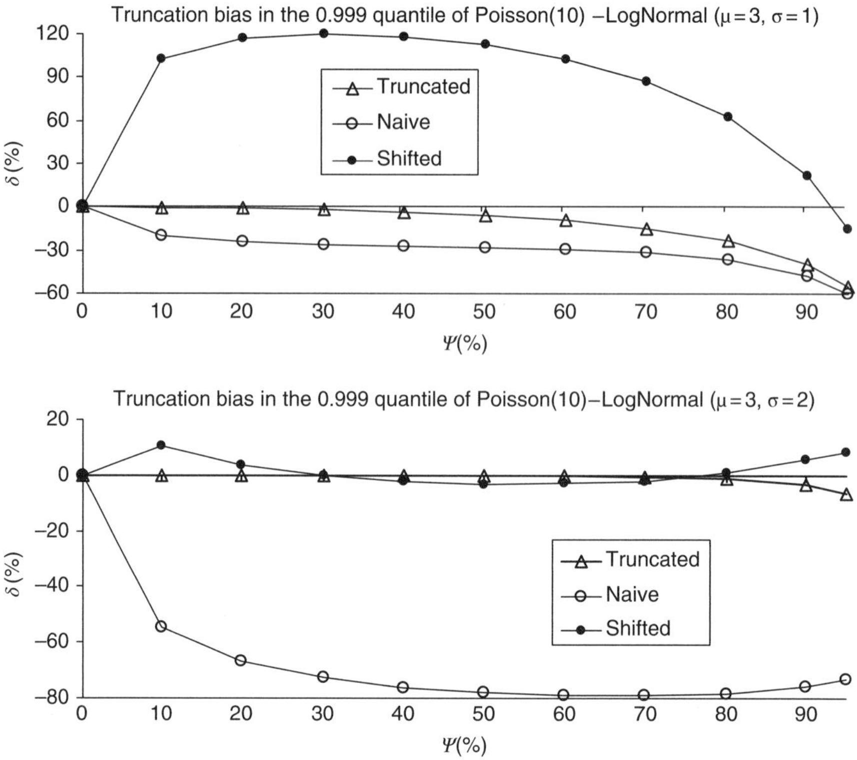

FIGURE 7.7 **Relative bias in the 0.999 quantile of the annual loss versus percentage of truncated points for several models ignoring truncation in the case of light-tailed severities from *LogNormal*$(\mu = 3, \sigma = 1)$ (top figure) and heavy-tailed severities from *LogNormal*$(\mu = 3, \sigma = 2)$ (bottom figure). The annual counts above the truncation level are from *Poisson*(10)**

Bayesian framework. The problem with the use of the simplified models that ignore data truncation, such as "naive" and "shifted" models, is not just the introduced bias but underestimation of extra capital required to cover parameter uncertainty. Typically, these simplified models lead to smaller fitting errors. It is not difficult to find a realistic example where the "shifted model" overestimating the quantile leads to underestimation when the parameter uncertainty is taken into account; for an example, see Luo *et al.* (2007, section 6, table 1). "Naive model" typically underestimates the capital even if the parameter uncertainty is taken into account This is because the "shifted" and "naive" models lead to smaller fitting errors in comparison to the "unbiased model". Of course, as the number of observations increases, the impact of parameter uncertainty diminishes. However, for modest fitting errors 5–10% (often, in modeling OpRisk data, the errors are larger) the impact of parameter uncertainty is significant.

7.9.4 THRESHOLD VARYING IN TIME

Often, in practice, a modeler should handle a reporting threshold varying in time. This might be due to scaling losses by some factors (inflation, business factors, etc.) that scales the reporting threshold too or changes in reporting policy in time. As a result, the losses in the fitted sample will have different threshold levels. Consider the following model assumptions.

Model Assumptions 7.2

- *In the absence of a threshold, the events follow a homogeneous Poisson process with the intensity* λ *and the severities* X_j *are independent with a common distribution* $F(\cdot|\boldsymbol{\beta})$*; denote* $\boldsymbol{\gamma} = (\lambda, \boldsymbol{\beta})$*;*
- *The losses are reported above the known time-dependent level* $L(t)$*. Denote the severities and arrival times of the reported losses as* $\widetilde{X}_j$ *and* $\widetilde{T}_j$*,* $j = 1, \ldots, J$*, respectively, and* t_0 *is the start of the observation period.*

Under these assumptions, the events above $L(t)$ follow a nonhomogeneous Poisson process with the intensity given by,

$$\theta(\boldsymbol{\gamma}, L(t)) = \lambda(1 - F(L(t)|\boldsymbol{\beta})). \tag{7.162}$$

Furthermore, denote by $\Lambda(t, h)$ the following integral

$$\Lambda(t, h) = \int_{t}^{t+h} \theta(\boldsymbol{\gamma}, L(x))dx. \tag{7.163}$$

Then, given that $(j-1)$-th event occurred at $\tilde{t}_{j-1}$, the interarrival time for the j-th event $\delta\widetilde{T}_j = \widetilde{T}_j - \widetilde{T}_{j-1}$ is distributed from

$$G_j(\tau|\boldsymbol{\gamma}) = 1 - \exp(-\Lambda(\tilde{t}_{j-1}, \tau)) \tag{7.164}$$

with the density

$$g_j(\tau|\boldsymbol{\gamma}) = \theta(\boldsymbol{\gamma}, L(\tilde{t}_{j-1} + \tau)) \exp(-\Lambda(\tilde{t}_{j-1}, \tau)). \tag{7.165}$$

The implied number of events in year m is $Poisson(\Lambda(s_m, 1))$-distributed, where s_m is the time of the beginning of year m, and the number of events over the nonoverlapping periods are independent.

Proposition 7.7 (Likelihood for Event Times and Truncated Severities) *Under Model Assumptions 7.2, for given parameters* $\boldsymbol{\gamma}$*, the joint density of the data* $\mathbf{Y}$ *of the events above* $L(t)$ *over the time period* $[t_0, t_E]$*, consisting of the interarrival times* $\delta\widetilde{T}_j = \widetilde{T}_j - \widetilde{T}_{j-1}$ *and severities* $\widetilde{X}_j$*,* $j = 1, \ldots, J$ *above* $L(t)$*, can be written as*

$$\begin{aligned} h(\boldsymbol{y}|\boldsymbol{\gamma}) &= (1 - G_J(t_E - \tilde{t}_J|\boldsymbol{\gamma})) \prod_{j=1}^{J} f_{L(\tilde{t}_j)}(\tilde{x}_j|\boldsymbol{\beta}) g_j(\tilde{\tau}_j|\boldsymbol{\gamma}) \\ &= \lambda^J \exp(-\Lambda(t_0, t_E - t_0)) \prod_{j=1}^{J} f(\tilde{x}_j|\boldsymbol{\beta}). \end{aligned} \tag{7.166}$$

Here, explicitly,

$$\Lambda(t_0, t_E - t_0) = \lambda \int_{t_0}^{t_E} [1 - F(L(x)|\boldsymbol{\beta})]dx.$$

Then, the likelihood function for the model is $\ell_{\boldsymbol{y}}(\boldsymbol{\gamma}) = \ln h(\boldsymbol{y}|\boldsymbol{\gamma})$.

Proof: This follows from independence between frequencies and severities, independence between severities, distribution of interarrival times (7.164), and its density (7.165). ∎

The MLEs for model parameters $\boldsymbol{\gamma}$ can be found by solving the maximum likelihood equations

$$\frac{\partial \ell_{\boldsymbol{y}}(\boldsymbol{\gamma})}{\partial \lambda} = \frac{J}{\lambda} - \int_{t_0}^{T} [1 - F(L(x)|\boldsymbol{\beta})]dx = 0, \tag{7.167}$$

$$\frac{\partial \ell_{\boldsymbol{y}}(\boldsymbol{\gamma})}{\partial \boldsymbol{\beta}} = -\frac{\partial}{\partial \boldsymbol{\beta}}\Lambda(t_0, T - t_0) + \sum_{j=1}^{J} \frac{\partial}{\partial \boldsymbol{\beta}} \ln f(\tilde{x}_j|\boldsymbol{\beta}) = 0. \tag{7.168}$$

The first equation gives

$$\hat{\lambda} = \frac{J}{\int_{t_0}^{T} [1 - F(L(x)|\hat{\boldsymbol{\beta}})]dx}, \tag{7.169}$$

which can be substituted into (7.166) and maximization will be required with respect to $\boldsymbol{\beta}$ only. The likelihood contains integral over the severity distribution. If integration is not possible in closed form, then it can be calculated numerically (which can be done efficiently using standard routines available in many numerical packages). For convenience, one can assume that a threshold is constant between the reported events $L(t) = L(t_j)$, $\tilde{t}_{j-1} < t \leq \tilde{t}_j$ and $L(t) = L(t_E)$ for $\tilde{t}_j < t \leq t_E$, so that

$$\int_{t_0}^{t_E} [1 - F(L(x)|\boldsymbol{\beta})]dx = [1 - F(L(t_E)|\boldsymbol{\beta})](t_E - \tilde{t}_J) + \sum_{j=1}^{J} [1 - F(L(\tilde{t}_j)|\boldsymbol{\beta})]\tau_j. \tag{7.170}$$

Of course, this assumption is reasonable if the intensity of the events is not small. Typically, scaling is done on the annual basis and one can assume a piece-wise constant threshold per annum and the integral is replaced by a simple summation.

Proposition 7.8 (Likelihood for Annual Counts and Truncated Severities) *Under Model Assumptions 7.2, the joint density of data* $\boldsymbol{Y}$ *of the events above the reporting threshold* $L(t)$ *consisting of the annual counts* $\widetilde{N}_m$, $m = 1, \ldots, T$ *and severities* $\widetilde{X}_j$, $j = 1, \ldots, J$ $(J = \widetilde{N}_1 + \ldots + \widetilde{N}_T)$ *can be written as*

$$h(\mathbf{y}|\gamma) = \prod_{j=1}^{J} f_{L(t_j)}(\tilde{x}_j|\boldsymbol{\beta}) \prod_{m=1}^{T} p(\tilde{n}_m|\Lambda(s_m, 1)), \tag{7.171}$$

where $p(\cdot|\Lambda(s_m, 1))$ *is probability mass function of Poisson*$(\Lambda(s_m, 1))$*. Then, the likelihood function for the model is* $\ell_{\mathbf{y}}(\gamma) = \ln h(\mathbf{y}|\gamma)$.

Proof: This follows from independence between frequencies and severities, independence between severities, and intensity of nonhomogenous Poisson process (7.162). ■

Usually, in practice, scaling is done on an annual basis. Thus, we can consider the case of a piece-wise constant threshold per annum such that for year m:

$$L(t) = L_m, \quad \theta_m = \theta(\gamma, L(t)) = \lambda(1 - F(L_m|\boldsymbol{\beta})), \quad s_m \leq t < s_m + 1,$$

where s_m is the time of the beginning of year m. The joint density in this case is

$$h(\mathbf{y}|\gamma) = \prod_{j=1}^{J} f_{L(\tilde{t}_j)}(\tilde{x}_j|\boldsymbol{\beta}) \prod_{m=1}^{T} p(\tilde{n}_m|\theta_m)) \tag{7.172}$$

and equations to find MLEs using the likelihood function $\ell_{\mathbf{y}}(\gamma) = \ln h(\mathbf{y}|\gamma)$ are

$$\frac{\partial \ell_{\mathbf{y}}(\gamma)}{\partial \lambda} = \sum_{m=1}^{T} [1 - F(L_m|\boldsymbol{\beta})] \frac{\partial}{\partial \theta_m} \ln p(\tilde{n}_m|\theta_m) = 0, \tag{7.173}$$

$$\frac{\partial \ell_{\mathbf{y}}(\gamma)}{\partial \boldsymbol{\beta}} = \sum_{j=1}^{J} \frac{\partial}{\partial \boldsymbol{\beta}} \ln f_{L(\tilde{t}_j)}(\tilde{x}_j|\boldsymbol{\beta}) - \lambda \sum_{m=1}^{T} \frac{\partial F(L_m|\boldsymbol{\beta})]}{\partial \boldsymbol{\beta}} \frac{\partial}{\partial \theta_m} \ln p(\tilde{n}_m|\theta_m) = 0. \tag{7.174}$$

The first equation gives

$$\hat{\lambda} = \frac{\sum_{m=1}^{T} \tilde{n}_m}{\sum_{m=1}^{T} (1 - F(L_m|\hat{\boldsymbol{\beta}}))}. \tag{7.175}$$

The MLEs of the severity parameters should be estimated jointly with the intensity. Given that the intensity MLE can be expressed in terms of the severity parameter MLEs via the given equation, one can substitute (7.175) into the likelihood function (7.172) and find severity parameter MLEs by maximizing the obtained likelihood profile.

Often the MLEs for severity parameters calculated marginally (i.e., by simply maximizing $\sum \ln f_{L(t_j)}(\tilde{x}_j|\boldsymbol{\beta})$) do not differ materially from the results of the joint estimation if the variability of the threshold is not extremely fast. In addition, marginal estimation does not allow for quantification of the covariances between frequency and severity parameters required to account for parameter uncertainty. For an illustrative example of fitting truncated data with time-varying threshold using maximum likelihood and Bayesian MCMC methods, the reader is referred to Shevchenko (2011, example 5.4, pp. 199–200).

7.9.5 UNKNOWN AND STOCHASTIC TRUNCATION LEVEL

One of the most significant problems in fitting OpRisk models using publicly available data is handling the issue that not all losses are reported above the threshold (typical threshold for external databases of public data is USD 1 million). Moreover, the truncation level for different losses is unknown. It is expected that there will be a positive relationship between the loss amount and the probability of its reporting. In this case, the dataset is a biased sample containing a disproportionate number of very large losses. One can say that an operational loss is publicly reported only if it exceeds some unobserved truncation point. This can be modeled as an unknown deterministic truncation level or stochastic truncation level.

Unknown deterministic truncation level. Baud *et al.* (2002) consider the case of unknown deterministic truncation level L. In this case, it is considered as an additional parameter to be estimated along with the parameters characterizing the loss distribution. The log likelihood function is identical to the one given in the previous sections for known truncation level, except that it is now an explicit function of both severity distribution parameters and L. Now the maximum likelihood approach corresponds to maximization of the likelihood with respect to distribution parameters and L. Furthermore, it can be immediately observed that the MLE for L is just the smallest observed loss in the fitted dataset. In practice, Baud *et al.* (2002) suggest the following procedure to account for possible contamination with untruncated or badly recorded data.

- Estimate severity parameters $\hat{\boldsymbol{\theta}}$ for each L ranging from 0 to ∞;
- Plot $\hat{\boldsymbol{\theta}}$ as a function of L;
- Truncation level estimator $\hat{L}$ is eventually calculated as the threshold beyond which $\hat{\boldsymbol{\theta}}$ remains approximately flat as a function of L.

As a result, the loss parameters are eventually estimated with fewer data than available, that is, losses above the highest threshold. In order to avoid this loss of information, one can consider modeling truncation level as a stochastic variable.

Stochastic truncation level. Stochastic truncation problem was reviewed in many papers (see, e.g., Amemiya 1984, Maddala 1983). Application of this technique to operational loss data is considered by De Fontnouvelle *et al.* (2006) and Baud *et al.* (2002). Note that here we refer to a threshold level as a known reporting level in the database while the truncation level is unknown.

Let X and Y be random variables with joint density $h(x,y)$ and marginal densities $f(x)$ and $g(y)$, respectively. Denote corresponding distribution functions as $F(x)$ and $G(y)$. Here, X is randomly truncated if it is observed only when it exceeds the unobserved truncation level Y. If X and Y are statistically independent, then the joint density $h(x,y)$ is the product of the marginal densities $f(x)$ and $g(y)$. The joint density of X and Y, conditional on X being observed, can be written as

$$\begin{aligned} h(x,y|x>y) &= \frac{f(x)g(y)}{\mathbb{P}\mathrm{r}[X>Y]} \\ &= \frac{f(x)g(y)}{\int_{-\infty}^{\infty}\int_{-\infty}^{x} f(x)g(y)dydx} \\ &= \frac{f(x)g(y)}{\int_{-\infty}^{\infty} f(x)G(x)dx}. \end{aligned} \tag{7.176}$$

The marginal density of X, conditional on $X > Y$, is obtained by integrating out the unobserved variable y

$$f(x|x > y) = \frac{f(x)G(x)}{\int_{-\infty}^{\infty} f(x)G(x)dx}. \quad (7.177)$$

Thus, the likelihood of the observed data $x_1, \ldots, x_n$ can be written as

$$L_{X|X>Y}(\boldsymbol{\theta}) = \prod_{i=1}^{n} \frac{f(x_i)G(x_i)}{\int_{-\infty}^{\infty} f(x)G(x)dx}. \quad (7.178)$$

Here, $\boldsymbol{\theta}$ are parameters of severity and truncation level distributions that can be estimated using maximum likelihood or Bayesian MCMC methods. Of course, in practice, we fit data above some known threshold (e.g., USD 1 million); then one can consider the above formulas for the loss exceedances above the threshold or log of the loss minus log of the threshold. This approach was used by De Fontnouvelle *et al.* (2006) to fit SAS OpRisk Global Data and Fitch Risk/OpVantage OpVar Loss Database.

There are many factors that affect whether a loss is publicly reported or not. It may depend on the type of loss, the business line, legal proceedings related to the loss, executives and reporters deciding whether to report the loss or not, etc. Thus, one can argue that the truncation level should be normally distributed. However, De Fontnouvelle *et al.* (2006) found that assumption of Normal distribution often leads to nonconvergence of the numerical optimization of the maximum likelihood and recommend using a logistic distribution (for truncation level of log losses)

$$G(x) = \frac{1}{1 + \exp(-(x - \tau)/\beta)}. \quad (7.179)$$

Here, the location parameter τ corresponds to the amount with a 50% chance of being reported and a scale parameter β regulates the increase (decrease) of the probability of reporting as the loss amount increases (decreases).

CHAPTER EIGHT

Model Selection and Goodness-of-Fit Testing for Frequency and Severity Models

In this chapter, we present details on statistical approaches to performing model selection. We separate the sections first into diagnostic tools that may be adopted to make quantitative assessments for model selection purposes. This includes analysis of the presence of heavy-tailed features of the data. Then the focus of the next few sections is on individual risk process model selections under a Loss Distribution Approach (LDA) structure, for the severity model and the frequency models. This can be achieved under a number of different frameworks such as information criteria, frequentist hypothesis testing, and Bayesian model selection approaches. A particular focus in these sections involves the suitable modifications to classical hypothesis tests that should be considered when performing model selection on heavy-tailed models, for instance, for the severity distribution. This is important to consider as it can have a substantial impact on the choice of the model and therefore on the capital. The last sections of this chapter involve the model selection of dependence features between multiple risks, such as model selection for the copula distribution, which may be used to link multiple risk processes as discussed in detail in Chapters 10–12.

8.1 Qualitative Model Diagnostic Tools

In general, it is often practical to utilize a range of model diagnostic tools to make qualitative judgments on the suitability of a particular choice of severity or frequency model. In this regard, there are a range of possible tools one can consider, each providing different interpretations as to the suitability of a particular aspect of the fitted model. In this section, we discuss the popular Quantile–Quantile plot (Q–Q plot) and analog Probability–Probability plots (P–P plot) as well as diagnostics for heavy-tailed behavior such as Mean Excess (ME) plots and Hill plots.

A Q–Q plot is a graphical method of comparing two probability distributions by plotting their quantiles against each other. One first selects a set of intervals for the quantiles

Fundamental Aspects of Operational Risk and Insurance Analytics: A Handbook of Operational Risk, First Edition. Marcelo G. Cruz, Gareth W. Peters, and Pavel V. Shevchenko.

to be plotted and then a point (x, y) on the plot corresponds to one of the quantiles of the second distribution (y-coordinate) plotted against the same quantile of the first distribution (x-coordinate). Typically, one considers the empirical quantiles from a sample to be plotted on the y-axis versus the hypothesized model quantiles on the x-axis.

The main step in constructing a Q–Q plot is calculating or estimating the quantiles to be plotted. In the case in which the distribution function(s) for one or both of the axes is based on a theoretical distribution that is continuous, all quantiles are uniquely defined and can be obtained by inverting the distribution function. If there is an atom in the support of the distribution function, that is, a discontinuity in the distribution for one or both of the axes, and a theoretical probability distribution is considered, then one should take care to observe that the definition used for the quantile at such points may utilise an interpolated quantile.

Typically, the Q–Q plot is based on data, such as losses, for which there can be multiple choices for quantile estimators. The approach adopted with regard to forming Q–Q plots when quantiles must be estimated or interpolated is called selection of "plotting positions", which literally means selecting which quantile levels to plot.

Remark 8.1 (Properties of Q–Q plots) *The following is worth noting when considering the interpretation of a Q–Q plot. The points plotted are always nondecreasing when viewed from left to right. In the case in which the two distributions compared are identical, the resulting Q–Q plot would follow the 45° line $y = x$. However, if there is a mismatch between the Q–Q plot curve and the line $y = x$, then this provides qualitative evidence of features of the data that are not in agreement with the proposed parametric model.*

Examples of such qualitative analysis are provided as follows:

1. If the overall trend in the Q–Q plot tends to be flatter (gradient less than 1) than the line $y = x$, then this implies that the hypothesized distribution plotted on the x-axis will be more dispersed than the distribution plotted on the vertical axis, which relates to the population distribution from which the sample was obtained;
2. If the overall trend of the Q–Q plot tends to be steeper (gradient greater than 1) than the line $y = x$, then this implies that the population distribution plotted on the vertical axis is more dispersed than the hypothesized distribution plotted on the horizontal axis;
3. If the Q–Q plot is curved, arched, or S-shaped, then this can indicate that relative to the hypothesized parametric distribution the population distribution from which the sample is obtained has different skew characteristics. It can also indicate that one of the distributions has heavier tails than the other.

An example is provided in Figure 8.1 where we consider a sample of $n = 100$ losses from a distribution F (unknown population distribution) and we compare this to the hypothesized severity model we are considering to use as a model. In these examples, we will consider a LogNormal model for the hypothesized model and simulate data from three different cases. The first is the ideal case where we simulate the data also from a LogNormal with the same parameters, the second is the case where the data come from a Weibull, and the third is the case where the data come from an exponential distribution.

Similar in concept, one can also plot what are known as P–P plots, in which a comparison between the empirical cumulative distribution function of a data set of size n samples, denoted by $\hat{F}_n$, is compared with a specified theoretical cumulative distribution function F. We note

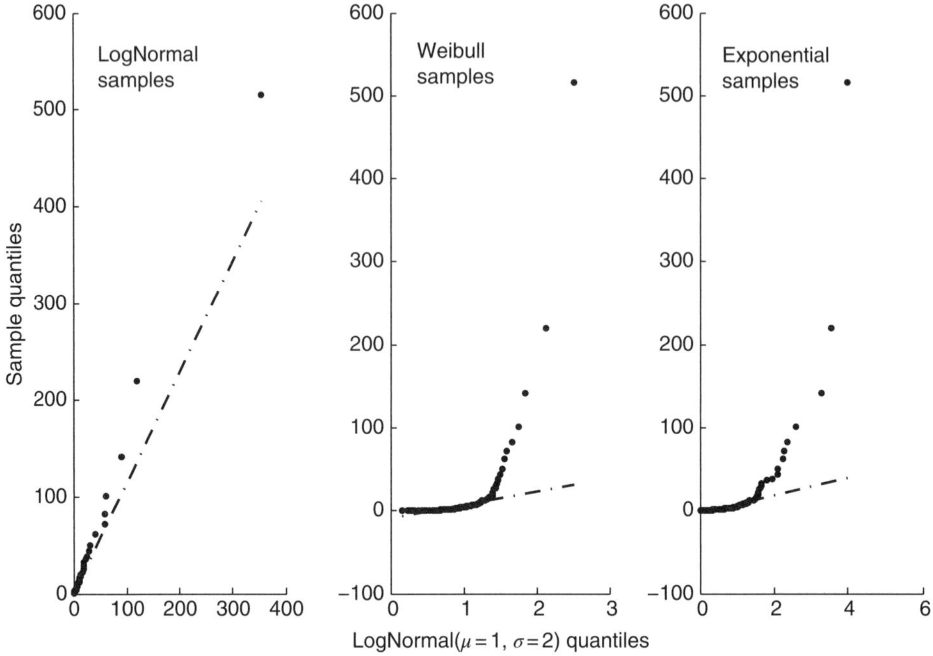

FIGURE 8.1 **In each plot, the x-axis corresponds to quantiles from a *LogNormal*$(\mu = 1, \sigma = 2)$ model. Left subplot: generated data are from a *LogNormal*$(\mu = 1, \sigma = 2)$ model. Middle subplot: generated data are from a *Weibull*$(\alpha = 1, \beta = 2)$ model. Right subplot: generated data are from an *Exp*$(\mu = 1)$ model**

some basic differences in the way that P–P plots and Q–Q plots are constructed. A Q–Q plot doesn't require information on the location or scale parameters of F to be known. The reason for this is that a linear relationship in the plotted Q–Q points indicates that the specified family describes the data distribution. However, the location and scale parameters do not affect the fact that this linear relationship will be present as they only affect the slope and intercept. However, when constructing a P–P plot one must be careful as it requires the location and scale parameters of F to be specified in order to evaluate the distribution at the ordered values. This is important since, on a P–P plot, changes in location or scale do not necessarily preserve linearity. Hence, one is advised to utilize the Q–Q plot when the intention is to assess the suitability of a family of parametric models from which the data may have been drawn.

The advantage of a P–P plot, when it is appropriate to utilize one, is that they are discriminating in regions of high probability density. To understand this point, we note that in these regions the empirical and theoretical cumulative distributions are changing more rapidly than where there is low probability; see further discussions on these plots by Wilk and Gnanadesikan (1968).

8.2 Tail Diagnostics

In many cases in which one is fitting a heavy-tailed severity model to data, one may be interested in diagnostic tools to assess the suitability of such a model feature. There are several qualitative plots that are available for such analysis such as the Mean Excess (ME) plot and the Hill plot, see Kratz and Resnick (1996).

When considering heavy-tailed features under a parametric model, it is often natural to consider the distribution of exceedances above a threshold such as under a Peaks Over Threshold (POT) estimation framework. The choice of threshold for which to consider the Generalized Pareto Distribution (GPD) is a challenging quantity to assess. Selecting the threshold and assessing the appropriateness of a heavy-tailed GPD model for data is often aided by an ME plot analysis. In the special case of a GPD model $GPD(\xi, \beta)$, the ME function takes a closed-form expression in terms of the extreme value index (EVI) parameter ξ, which for a level u threshold is given by

$$M(u) = \mathbb{E}\left[X - u | X > u\right] = \frac{\beta}{1-\xi} + \frac{\xi}{1-\xi}u, \tag{8.1}$$

which clearly indicates that the sample estimated ME should form a linear relationship if the GPD heavy-tailed model is suitable to describe the tail behavior of the data. In fact, any heavy-tailed subexponential model (though it may not have a parametric form for the ME function) will produce an upward-sloping ME plot, where as light-tailed or exponentially tailed models will produce a downward or decreasing ME plot; see discussion and references by Ghosh and Resnick (2010) for details.

As mentioned, for plotting the ME plot we consider the sample ME function defined by Equation (8.2), which represents the sum of excesses over a threshold u divided by the number of data points that exceed the threshold u. It approximates the ME function describing the expected exceedance amount for a particular threshold u given an exceedance occurred. If the empirical ME function estimate has a positive slope for large thresholds u, then this indicates that the observed data are consistent with a GPD with a positive tail index parameter (Beirlant *et al.* 2004, chapter 1). The sample ME is then given by

$$e_n(u) = \frac{\sum_{i=1}^{n}(X_i - u)\mathbb{I}_{\{X_i > u\}}}{\sum_{i=1}^{n}\mathbb{I}_{\{X_i > u\}}} \tag{8.2}$$

which estimates the conditional expectation $e(u) = \mathbb{E}\left[(X-u)|X > u\right]$.

We note that the ME plot is only one of a large set of widely used tools for extreme value model selection. Other diagnostic tools include the Hill plot, the Pickands plot, and the moment estimator plot, see discussion by Ghosh and Resnick (2010).

In addition, when performing qualitative assessments of the tail thickness of the data-generating distribution, it is common to consider the Hill plot. The Hill plot represents the estimated inverse tail index as a function of the upper-order statistics k. In other words, it considers the order statistics of the data set of length n given by $\left\{X_{(i,n)}\right\}_{i=1}^{n}$ and takes the m upper-order statistics to obtain the Hill estimator of the extreme value tail index given by

$$H_{m,n} = \left(\frac{1}{m}\sum_{i=1}^{m}\ln\frac{X_{(i,n)}}{X_{(m+1,n)}}\right)^{-1}, \quad 1 \le m \le n. \tag{8.3}$$

The Hill plot is then the plot of points $\{(k, H_{k,n}) : 1 \le k \le n\}$. This plot provides feedback on the suitability of the selected threshold utilized in the estimation of the Extreme Value Theory (EVT) models, in particular the POT's method, see Beirlant *et al.* (2004) for details. For further detailed discussions on the suitability of such a plot to certain model assumptions, see discussion by Ghosh and Resnick (2010, section 4).

8.3 Information Criterion for Model Selection

In this section, we provide background on alternative popular model selection and penalization approaches developed in the statistical literature, which include the frequentist Akaike Information Criterion (AIC) and small sample results of AICc. We also present the Bayesian equivalent quantities given by the Bayesian Information Criterion (BIC) and the Deviance Information Criterion (DIC). Hence, we discuss methods of model comparison based on ideas of information theory, model complexity and accuracy or bias, and variance trade-off. However, these are not measures of absolute suitability of a model choice as would be obtained from goodness-of-fit (GOF) testing under a formal hypothesis test. In other words, if all the candidate models for the severity fit poorly, these criteria will not give any warning of this possibility. We therefore also comment on the importance of estimation of both the tail properties of a heavy-tailed severity model in an LDA framework, as well as the assessment of the suitability of the model more generally—these are related but not equivalent concepts.

8.3.1 AKAIKE INFORMATION CRITERION FOR LDA MODEL SELECTION

The AIC is a measure of the relative GOF of OpRisk LDA model components such as the severity or frequency distribution under consideration. One can interpret this criterion as providing an entropy-based trade-off between bias and variance in an OpRisk model construction. To understand the connection to the concept of entropy we observe that the AIC is based on the Kullback–Leibler (KL) divergence given in Definition 8.1.

Definition 8.1 (Kullback–Leibler divergence) *The Kullback–Leibler divergence, in this case for two densities f and g, is given by the following two components:*

$$KL(f \,||\, g) = \underbrace{-\int f(\boldsymbol{x}) \ln g(\boldsymbol{x}) d\boldsymbol{x}}_{\text{Cross Entropy } \triangle(f||g)} + \underbrace{\int f(\boldsymbol{x}) \ln f(\boldsymbol{x}) d\boldsymbol{x}}_{\text{Entropy}} \tag{8.4}$$

■

If one now considers $f(\boldsymbol{x};\boldsymbol{\theta})$ and $g(\boldsymbol{x};\boldsymbol{\psi})$ to be the likelihoods of two competing models and notes that the integrals are taken with respect to the observed losses, then the cross entropy term represents the expected negative log-likelihood of data coming from f under g. Now, since we don't know the true model generating process of the losses, we assume we have a set of possible candidates and we rank them by their AIC score given in Definition 8.2.

Definition 8.2 (Akaike Information Criterion for severity models in OpRisk) *Consider a given model for severity loss random variable $X \sim F_X(x;\boldsymbol{\theta})$ parameterized by $(k \times 1)$ dimensional vector $\boldsymbol{\theta}$. Given observed losses one obtains the value of the maximum likelihood estimation (MLE) of the severity model parameters $\hat{\boldsymbol{\Theta}}^{\text{MLE}}(\boldsymbol{x})$, which is a function of the observed data. Then the AIC is given by*

$$AIC(\hat{\boldsymbol{\Theta}}^{\text{MLE}}(\boldsymbol{x})) = \underbrace{-2l(\hat{\boldsymbol{\theta}}^{\text{MLE}};\boldsymbol{x})}_{\text{Maximum of Likelihood}} + \underbrace{2k}_{\text{Model Complexity Penalty}} \tag{8.5}$$

■

Given several possible candidate severity models for an OpRisk LDA model, the preferred model is the one with the minimum AIC value.

Remark 8.2 (Cautionary comments on application of AIC) *The following general issues are required to be considered for the application of AIC:*

1. *AIC is asymptotic; it requires conventional large-sample properties;*
2. *The maximum number of parameters in the severity model should not exceed 2kn, where n is the number of observations. This is because larger values will weaken the bias correction;*
3. *There are cases when AIC decreases monotonically, that is, there is no solution. In most of these cases, the culprit is poor selection of model class;*
4. *If an AIC score difference between two severity models has a magnitude of between* 1 *to* 2 *or more, then the difference is significant;*
5. *In some cases, AIC has been shown to be inconsistent.*

In general, in OpRisk settings where the sample sizes are low, it may be better to consider the small sample AIC with a different bias correction given in Definition 8.3 (see discussions in Burnham and Anderson 2002). In this text of model selection, they advise to utilize AICc, rather than AIC, when the number of losses n is small or the number of severity model parameters k is large. Note further that the AICc will converge to AIC as the sample size grows and for small sample sizes we see that the bias is reduced by keeping AIC with a greater penalty for extra parameters.

Definition 8.3 (Small-sample Akaike Information Criterion AICc) *The correction to AIC for small sample sizes, known as the AICc is given by*

$$AICc\left(\hat{\Theta}^{\text{MLE}}(\boldsymbol{x})\right) = AIC\left(\hat{\Theta}^{\text{MLE}}(\boldsymbol{x})\right) + \frac{2k(k+1)}{n-k-1}, \tag{8.6}$$

where n denotes the sample size and k the number of parameters in the severity model. ■

8.3.1.1 Understanding How the AIC Criterion is Obtained. To understand how the AIC criterion is obtained, we consider a hypothetical true data-generating (losses) severity model denoted by $h(\boldsymbol{x}|\boldsymbol{\theta}^*)$ with true parameters $\boldsymbol{\theta}^*$. Furthermore, consider a class of models $\mathcal{M}_k = \{f(\boldsymbol{x}|\boldsymbol{\theta}_k)\,|\boldsymbol{\theta}_k \in \boldsymbol{\Omega}(k)\}$, where each member of this class is a data-generating density that is parameterized by a k-dimensional parameter vector (risk profile) $\boldsymbol{\theta}_k$. For each model in this class of models, denote the log-likelihood at the MLE by $l\left(\hat{\boldsymbol{\theta}}_k^{\text{MLE}}\right)$. Then one may define the expected log likelihood, with respect to the true model parameters $\boldsymbol{\theta}^*$, for a given model in class $\mathcal{M}_k$ by

$$\mathbb{E}_{\boldsymbol{\theta}^*}\left[\ln f(\boldsymbol{X}|\boldsymbol{\theta}_k)\right] = \int h(\boldsymbol{x}|\boldsymbol{\theta}^*)\ln f(\boldsymbol{x}|\boldsymbol{\theta}_k)\,d\boldsymbol{x}, \tag{8.7}$$

where this expectation is interpreted as that taken with respect to the hypothetical true data-generating distribution $h(\boldsymbol{x}|\boldsymbol{\theta}^*)$. In addition, one may define this expectation at a particular point corresponding to the expected maximized log-likelihood given by

$$\mathbb{E}_{\boldsymbol{\theta}^*}\left[\ln f\left(\boldsymbol{X}|\hat{\boldsymbol{\theta}}_k^{\mathrm{MLE}}\right)\right] = \int h(\boldsymbol{x}|\boldsymbol{\theta}^*)\ln f\left(\boldsymbol{x}|\hat{\boldsymbol{\theta}}_k^{\mathrm{MLE}}\right)d\boldsymbol{x}, \tag{8.8}$$

where the point estimator of the MLE $\hat{\boldsymbol{\theta}}_k^{\mathrm{MLE}}$ is clearly dependent on one fixed realization of the observations. The use of $\mathbb{E}_{\boldsymbol{\theta}^*}\left[\ln f\left(\boldsymbol{X}|\hat{\boldsymbol{\theta}}_k^{\mathrm{MLE}}\right)\right]$ to estimate $\mathbb{E}_{\boldsymbol{\theta}^*}\left[\ln f\left(\boldsymbol{X}|\boldsymbol{\theta}_k\right)\right]$ is critical to the derivation of the AIC. That is, the AIC is derived by making an estimate of the expected log-likelihood using the maximized log-likelihood function. However, since the MLE will depend on one realization of the data $\boldsymbol{X}$, it will produce a biased estimator of the mean expected log-likelihood w.r.t. to the loss data and this bias is asymptotically given by k, the number of parameters in the model, see Akaike (1981).

To proceed with the understanding of how the AIC is obtained, we consider improving the estimator of the expected maximum log-likelihood. This will be achieved by considering the mean expected maximum log-likelihood, where the single realization of observation vector $\boldsymbol{X}$ is observed and then averaging over the MLE estimator from i.i.d. observation vectors $\boldsymbol{Y}$, each assumed to come from the same hypothetical true distribution as the observed losses $\boldsymbol{X}$, thus producing the following expectations to be considered:

$$\mathbb{E}_{\boldsymbol{Y}|\boldsymbol{\theta}^*}\mathbb{E}_{\boldsymbol{X}|\boldsymbol{\theta}^*}\left[\ln f\left(\boldsymbol{X}|\boldsymbol{\theta}_k^{\mathrm{MLE}}(\boldsymbol{Y})\right)\right]. \tag{8.9}$$

In the work of Akaike, it was postulated that as the mean expected maximum log-likelihood increases, the model provides an improving fit. It can also be shown that the estimator of this mean expected log-likelihood is estimated by the maximum likelihood function, with a bias, and that the bias can be obtained easily as the number of free parameters in the model.

Another way of seeing this is to note that one can show that the AIC score relates to the cross entropy between the unknown "true" data-generating model h with true parameters $\boldsymbol{\theta}^*$ and a model under consideration f, where we denote the cross entropy by $\triangle(h\,\|\,f)$, which is given in the following way for an $(n \times 1)$ vector of observed losses $\boldsymbol{x}$ according to the following expectation,

$$\mathbb{E}_{\boldsymbol{\theta}^*}\left[\hat{\boldsymbol{\Theta}}^{\mathrm{MLE}}(\boldsymbol{x})\right] = \mathbb{E}_{\boldsymbol{\theta}^*}\left[\triangle\left(\boldsymbol{\theta}^*||\hat{\boldsymbol{\Theta}}^{\mathrm{MLE}}(\boldsymbol{x})\right)\right] + o_n(1). \tag{8.10}$$

This shows that the AIC score, given for model $\mathcal{M}_k$ by

$$AIC\left(\hat{\boldsymbol{\Theta}}^{\mathrm{MLE}}(\boldsymbol{x})\right) = -\ln f\left(\boldsymbol{x}|\hat{\boldsymbol{\theta}}_k^{\mathrm{MLE}}\right) + k, \tag{8.11}$$

is an asymptotically unbiased estimator of the cross-entropy risk and can only be accurately applied in the large sample size setting, something that is not often available in OpRisk modeling. One way to show this property of the AIC score estimator is by taking the case in which the true model $\mathcal{M}_{k^*}$ is nested in the class of models considered and there are two sources of error in the model selection:

1. Discrepancy from approximation. This is the main source of error when underfitting where the number of parameters in the fitted model $\mathcal{M}_k$ is such that $k < k^*$;
2. Discrepancy from estimation. This is the main source of error when overfitting where the number of parameters in the fitted model $\mathcal{M}_k$ is such that $k \geq k^*$.

To complete this derivation, one shows that under particular regularity conditions discussed in Akaike (1981) the mean of the AIC score with respect to the true data-generating model is given by the mean cross entropy $\triangle\left(\boldsymbol{\theta}^* || \hat{\boldsymbol{\Theta}}^{\mathrm{MLE}}(\boldsymbol{x})\right)$ up to the first order. In other words, one must show that

$$
\begin{aligned}
&\mathbb{E}_{\boldsymbol{\theta}^*}\left[\ln f\left(\boldsymbol{X}|\hat{\boldsymbol{\theta}}_k^{\mathrm{MLE}}\right) + k\right] \\
&= \mathbb{E}_{\boldsymbol{\theta}^*}\left[\triangle\left(\boldsymbol{\theta}^* || \hat{\boldsymbol{\Theta}}^{\mathrm{MLE}}(\boldsymbol{x})\right)\right] + o_n(1) \\
&= \triangle\left(\boldsymbol{\theta}^* || \boldsymbol{\Theta}_0\right] + \frac{1}{2}\left[\hat{\boldsymbol{\Theta}}^{\mathrm{MLE}}(\boldsymbol{x}) - \boldsymbol{\Theta}_0\right]^T J\left(\boldsymbol{\Theta}_0\right)\left[\hat{\boldsymbol{\Theta}}^{\mathrm{MLE}}(\boldsymbol{x}) - \boldsymbol{\Theta}_0\right] + o_n(1) \\
&= \mathbb{E}_{\boldsymbol{\theta}^*}\left[\ln f\left(\boldsymbol{X}|\hat{\boldsymbol{\theta}}_k^{\mathrm{MLE}}\right)\right] + \frac{1}{2}\left[\hat{\boldsymbol{\Theta}}^{\mathrm{MLE}}(\boldsymbol{x}) - \boldsymbol{\Theta}_0\right]^T H\left(\hat{\boldsymbol{\Theta}}^{\mathrm{MLE}}(\boldsymbol{x})\right)\left[\hat{\boldsymbol{\Theta}}^{\mathrm{MLE}}(\boldsymbol{x}) - \boldsymbol{\Theta}_0\right] \\
&\quad + \frac{1}{2}\left[\hat{\boldsymbol{\Theta}}^{\mathrm{MLE}}(\boldsymbol{x}) - \boldsymbol{\Theta}_0\right]^T J\left(\boldsymbol{\Theta}_0\right)\left[\hat{\boldsymbol{\Theta}}^{\mathrm{MLE}}(\boldsymbol{x}) - \boldsymbol{\Theta}_0\right] + o_n(1),
\end{aligned}
\tag{8.12}
$$

where the matrices J and H are given at location $\boldsymbol{\theta}_0$ by,

$$
\begin{aligned}
J\left(\boldsymbol{\Theta}_0\right) &= \left[\left.\frac{\partial^2 \triangle\left(\boldsymbol{\theta}^* || \boldsymbol{\Theta}\right]}{\partial \boldsymbol{\theta} \partial \boldsymbol{\theta}'}\right|_{\boldsymbol{\theta}=\boldsymbol{\theta}_0}\right] \\
H\left(\hat{\boldsymbol{\Theta}}^{\mathrm{MLE}}(\boldsymbol{x})\right) &= \left[\left.\frac{\partial^2 \ln f\left(\boldsymbol{X}|\boldsymbol{\theta}\right)}{\partial \boldsymbol{\theta} \partial \boldsymbol{\theta}'}\right|_{\boldsymbol{\theta}=\hat{\boldsymbol{\theta}}_k^{\mathrm{MLE}}}\right]
\end{aligned}
\tag{8.13}
$$

Then one notes that under the same regularity conditions utilised to obtain this expansion, the following holds

$$
\begin{aligned}
\frac{1}{2}\mathbb{E}_{\boldsymbol{\theta}^*}\left[\left[\hat{\boldsymbol{\Theta}}^{\mathrm{MLE}}(\boldsymbol{x}) - \boldsymbol{\Theta}_0\right]^T J\left(\boldsymbol{\Theta}_0\right)\left[\hat{\boldsymbol{\Theta}}^{\mathrm{MLE}}(\boldsymbol{x}) - \boldsymbol{\Theta}_0\right]\right] &= \frac{k}{2} + o_n(1) \\
\frac{1}{2}\mathbb{E}_{\boldsymbol{\theta}^*}\left[\left[\hat{\boldsymbol{\Theta}}^{\mathrm{MLE}}(\boldsymbol{x}) - \boldsymbol{\Theta}_0\right]^T H\left(\hat{\boldsymbol{\Theta}}^{\mathrm{MLE}}(\boldsymbol{x})\right)\left[\hat{\boldsymbol{\Theta}}^{\mathrm{MLE}}(\boldsymbol{x}) - \boldsymbol{\Theta}_0\right]\right] &= \frac{k}{2} + o_n(1).
\end{aligned}
\tag{8.14}
$$

After substitution of these results one obtains the required result for the AIC criterion expression.

8.3.2 DEVIANCE INFORMATION CRITERION

There are also a number of other information crtiria, some of which are particular relevant for Baysian modelling. One such example, commonly used in practice, is the Deviance Information Criterion (DIC). For a dataset $\boldsymbol{X} = \boldsymbol{x}$ generated by the model with the posterior density $\pi(\boldsymbol{\theta}|\boldsymbol{x})$, define the deviance by

$$
D(\boldsymbol{\theta}) = -2\ln \pi(\boldsymbol{x}|\boldsymbol{\theta}) + C,
\tag{8.15}
$$

where the constant C is common to all candidate models. Then the DIC is calculated as

$$
\begin{aligned}
DIC &= 2\mathbb{E}[D(\boldsymbol{\Theta})|\boldsymbol{X} = \boldsymbol{x}] - D(\mathbb{E}[\boldsymbol{\Theta}|\boldsymbol{X} = \boldsymbol{x}]) \\
&= \mathbb{E}[D(\boldsymbol{\Theta})|\boldsymbol{X} = \boldsymbol{x}] + \left(\mathbb{E}[D(\boldsymbol{\Theta})|\boldsymbol{X} = \boldsymbol{x}] - D(\mathbb{E}[\boldsymbol{\Theta}|\boldsymbol{X} = \boldsymbol{x}])\right),
\end{aligned}
\tag{8.16}
$$

where

- $\mathbb{E}[\cdot|\boldsymbol{X} = \boldsymbol{x}]$ is the expectation with respect to the posterior density of $\boldsymbol{\Theta}$;
- The expectation $\mathbb{E}[D(\boldsymbol{\Theta})|\boldsymbol{X} = \boldsymbol{x}]$ is a measure of how well the model fits the data; the smaller this is, the better the fit;
- The difference $\mathbb{E}[D(\boldsymbol{\Theta})|\boldsymbol{X} = \boldsymbol{x}] - D(\mathbb{E}[\boldsymbol{\Theta}|\boldsymbol{X} = \boldsymbol{x}])$ can be regarded as the effective number of parameters. The larger this difference, the easier it is for the model to fit the data.

The DIC criterion favors the model with a better fit but at the same time penalizes the model with more parameters. Under this setting the model with the smallest DIC value is the preferred model.

DIC is a Bayesian alternative to BIC (*Schwarz's criterion* Schwarz 1978) and AIC (Akaike, 1983). For more details on these criteria, see, for example, Robert (2001, chapter 7).

8.4 Goodness-of-Fit Testing for Model Choice (*How to Account for Heavy Tails!*)

It is also natural under a frequentist modeling perspective to consider performing a Goodness of Fit (GOF) hypothesis test. This is a formal hypothesis testing procedure for assessing the statistical significance of whether the observed loss process was likely to have been generated from the statistical model considered. Measures of GOF typically summarize the discrepancy between observed loss values and the loss values expected under the model in question. In this section, we will consider several possible tests, such as the Kolmogorov–Smirnov, test, the Chi-squared test, and heavy-tailed tests of particular relevance to OpRisk when considering the appropriateness of particular tail properties of the severity model.

Stated more formally, one can say that a GOF test for a set of n i.i.d. random variables $X_1, \ldots, X_n$ with an unspecified distribution function $G_X(x)$ aims to inform a decision between whether the samples follow a null distribution $F_X(x; \boldsymbol{\theta})$, where $\boldsymbol{\theta}$ contains possibly unknown model parameters, or an alternative. This can be stated according to the two following hypotheses:

$$\mathcal{H}_0 : G_X(x) = F_X(x; \boldsymbol{\theta}).$$
$$\mathcal{H}_1 : G_X(x) \neq F_X(x; \boldsymbol{\theta}).$$

To be more precise, we will first recall some basic inference definitions. Generally, when performing such inferential procedures, one should distinguish between simple hypotheses and compound hypotheses as detailed below:

- **Simple hypothesis**. A hypothesis that completely specifies the probability distribution;
  - Example 1. The parameter of this Binomial distribution is $p = 0.6$;
  - Example 2. The distribution is a Normal one of average $\mu = 4$ and standard deviation $\sigma = 1$.
- **Compound hypothesis**. A hypothesis that does not completely specify the distribution. **Note**. In this case the alternative hypothesis cannot be directional; it must measure deviations in all directions;

- Example 1. The parameter p of this Binomial distribution is greater than 0.1;
- Example 2. These two distributions have the same mean and common standard deviation of $\sigma = 1$.

In addition, we need to consider how we handle Type I and Type II errors that specify the possible mistakes we can make in our decision.

Definition 8.4 (Type I and Type II errors) *A Type I error occurs when the null hypothesis $\mathcal{H}_0$ is rejected though it should have been accepted, and a Type II error occurs when the alternative hypothesis $\mathcal{H}_1$ is rejected though it should have been accepted. Note that this is equivalent in logic to the case in which the null hypothesis $\mathcal{H}_0$ is accepted though it should have been rejected.* ∎

We also note that a hypothesis test will partition the space of observations into two regions denoted by $\mathcal{R}$ and $\mathcal{A}$. These can then be considered to help define the attributes of a given test according to the characteristics that define a given testing procedure, known as the significance of the test and the power of the test. These characteristics are directly related to the decision errors of Type I and II, which are specified formally according to the following definition.

Definition 8.5 (Power and significance of hypothesis test) *The significance of a hypothesis test refers to the Type I errors and is defined by*

$$S = 1 - \alpha = 1 - \mathbb{P}\mathrm{r}\left(x \in \mathcal{R}|\mathcal{H}_0\right) = 1 - \int_{\mathcal{R}} \mathbb{P}\mathrm{r}\left(x|\mathcal{H}_0\right) dx. \tag{8.17}$$

The power of a hypothesis test refers to the Type II errors and is defined by

$$P = 1 - \beta = 1 - \mathbb{P}\mathrm{r}\left(x \in \mathcal{A}|\mathcal{H}_1\right) = 1 - \int_{\mathcal{A}} \mathbb{P}\mathrm{r}\left(x|\mathcal{H}_1\right) dx. \tag{8.18}$$

∎

To select a test and study its properties one will typically encounter a trade-off between α and β. It is therefore standard practice to set *a priori* the significance to a fixed value ($\alpha = 0.01; 0.05; \ldots$) and then to find the most powerful test, where β is as small as possible. In general, the results will correspond to the class of testing procedures referred to as the Neyman Pearson tests, which apply to the setting of a simple null hypothesis against a simple alternative hypothesis.

8.4.1 CONVERGENCE RESULTS OF THE EMPIRICAL PROCESS FOR GOF TESTING

In characterizing the decision rule for the hypothesis test, one can either specify the decision boundary (critical values) for a given acceptable level of precision or significance level α, or one can specify the probability of events exceeding these critical values for any given value α (a p-value). Both specifications are equivalent and require knowledge of the distribution of the statistic used to make the decision in the test under the assumption that the null hypothesis is correct.

In the context of GOF testing, typically the distribution for the statistic is based on a functional or transformation that maps an empirical process and its limiting (large sample) asymptotic process to a random vector or random variable; this random vector/variable is known as the test statistic. It is typically obtained conditional on the assumption that the nominal claim is correct. We will see that while evaluation of the statistic in practice requires the knowledge of the model, the actual distribution of this statistic turns out to be model-free (distribution-free) and can be evaluated, stated, or tabulated once for any desired model.

The distribution of this test statistic in a distribution-free GOF test is based entirely on the limiting process of the empirical process under study. For example, if the statistic were the maximal vertical distance between the empirical distribution and the null distribution over the support of the null distribution (KS test), then we are talking about a p-value for such a test using this statistic which is obtained by first understanding the limiting behavior of the empirical distribution process as the sample size increases. This understanding of the limiting behavior of the empirical process can then be used to study functionals of the empirical and limiting process and in particular the tail events of the distributions of such functions (which are then random variables/vectors) in order to obtain well-defined p-values. With these p-values one can then probabilistically characterize events under the null hypothesis, using the given statistic that would lead one to decide against the nominal claim, based on the evidence from the observation of the process.

Therefore, it will also be beneficial to observe the following results that are based on comparisons between properties of the empirical distribution function and a hypothesized distribution function which regularly arise in the context of GOF testing. We first detail two fundamental results of relevance to these tests: the Glivenko–Cantelli theorem (Cantelli 1933) and the Dvoretzky–Kiefer–Wolfowitz (DKW) inequality (see Dvoretzky *et al.* 1956 and Birnbaum and McCarty 1958).

Definition 8.6 (Empirical distribution function) *Given a continuous distribution, let $X_1, X_2, \ldots, X_n$ be a sequence of i.i.d. random variables from this distribution with observed realizations $x_1, x_2, x_3, \ldots, x_n$. Then the empirical distribution function denoted by $\hat{F}_n$ is defined according to*

$$\hat{F}_n(x) = \frac{1}{n} \sum_{i=1}^{n} \mathbb{I}\left[x_i \leq x\right]. \tag{8.19}$$

■

In Figure 8.2, we present an example of the empirical distribution function for a small sample size generated from an example of a Gamma distribution. As expected for each location in which there is a Dirac mass, there will be a jump in probability of $1/n$. Throughout this chapter we will study different functions of this basic quantity in the context of hypothesis testing.

One can also bound the probability of all measurable events for a given distribution function F, from which n i.i.d. samples are assumed to be drawn, the accuracy of the sample estimated empirical distribution function probabilities and the true distribution via the results in Theorems 8.1 and 8.2

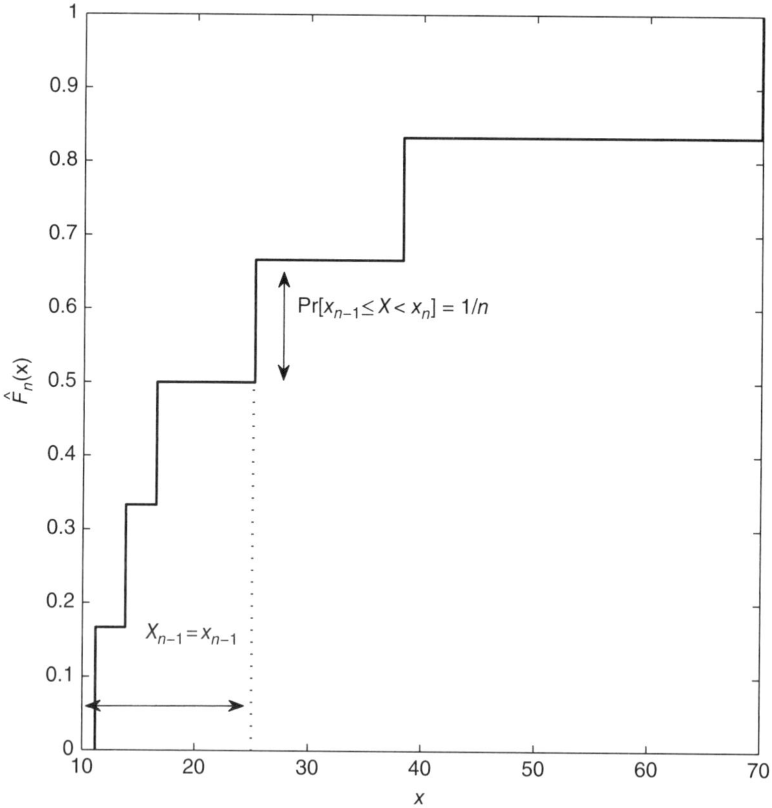

FIGURE 8.2 **Example of an empirical distribution function given by $\hat{F}_n$ with $n = 6$**

Theorem 8.1 (Glivenko–Cantelli theorem) *Given a sample size n of i.i.d. real-valued sample realizations of random variables $X_1, X_2, \ldots, X_n$ with distribution F, the following uniform convergence holds. For every fixed x, $\hat{F}_n(x)$ is a sequence that converge to $F(x)$ almost surely by the strong law of large numbers, that is, $\hat{F}_n$ converges to F pointwise. In addition, this convergence is uniform,*

$$||\hat{F}_n - F||_\infty = \sup_{x \in \mathbb{R}} |\hat{F}_n(x) - F(x)| \to 0, \quad \textit{almost surely}. \tag{8.20}$$

One can strengthen this result of uniform convergence with information on the rate via the following inequality of Dvoretzky-Keifer-Wolfowitz (DKW).

Theorem 8.2 (Dvoretzky–Keifer–Wolfowitz inequality) *Given a sample size n of i.i.d. real-valued sample realizations of random variables $X_1, X_2, \ldots, X_n$ with distribution F, the following inequality holds*

$$\mathbb{P}\mathrm{r}\left(\sup_{x \in \mathbb{R}} \left(\hat{F}_n(x) - F(x)\right) > \epsilon\right) \leq \exp(-2n\epsilon^2) \tag{8.21}$$

for every $\epsilon \geq \sqrt{\frac{1}{2n} \ln 2}$.

Remark 8.3 *Therefore, the DKW inequality predicts how close an empirically determined distribution function will be to the distribution function from which the empirical samples are drawn. We will also see that the DKW inequality can inform the tail of the KS test statistic under the null hypothesis.*

8.4.1.1 Convergence of the empirical distribution function for simple hypotheses. We first start by assuming a parametric family of models in which the parameters of the population distribution are fixed and known; they do not require estimation. This is in agreement with a hypothesis test in which the hypotheses are stated in a simple form. Then it is directly relevant to understand further the convergence of the empirical distribution function to the true population distribution function pointwise, as this will be utilized in the GOF tests. We therefore consider the following additional results that are discussed by Chicheportiche and Bouchaud (2012), Mason and Schuenemeyer (1983), and the right censored analysis by Fleming *et al.* (1980). Start by defining the Bernoulli random variables $Y_i(x) = \mathbb{I}[X_i \leq x]$ and denoting $u = F(x)$ and $\nu = F(y)$; we note the following properties of the mean and covariance of these random variables

$$\mathbb{E}[Y_i(x)] = F(x),$$
$$\mathbb{E}\left[Y_i(x)Y_j(x')\right] = \begin{cases} F(\min(x,x')), & i = j, \\ F(x)F(x'), & i \neq j, \end{cases} \tag{8.22}$$

as discussed by Chicheportiche and Bouchaud (2012).

Now we consider constructing sample estimators for the centered sample mean, $\overline{\boldsymbol{Y}}$, of the random vector of Bernoulli random variables $\boldsymbol{Y} = [Y_1(x), \ldots, Y_n(x)]$, given by Equation (8.23). This sample estimator is a measure of the difference between the true distribution and the empirical distribution at a point x, which as we will see shortly, is used to construct a GOF test statistic and is defined by one of the following representations:

$$\overline{\boldsymbol{Y}}(x) = \frac{1}{n}\sum_{k=1}^{n} Y_k(x) - F(x) \tag{8.23}$$

for any x in the support of $F(x)$, or equivalently for any $u \in [0,1]$ by

$$\overline{\boldsymbol{Y}}(u) = \frac{1}{n}\sum_{k=1}^{n} Y_k\left(F^{-1}(u)\right) - u. \tag{8.24}$$

In addition, one can show the covariance between the sample means of two quantile levels as given in Equation (8.25):

$$\mathbb{C}\text{ov}\left(\overline{Y}(u), \overline{Y}(v)\right) = \frac{1}{n}\left(\min\{u,v\} - uv\right)\left[1 + D_N(u,v)\right], \tag{8.25}$$

for $u = F(x)$, $v = F(x')$ and where one defines

$$D_N(u,v) = \frac{1}{n}\sum_{j,k\neq j}^{n} \frac{\mathbb{C}\text{ov}\left(Y_j(x), Y_k(x')\right) - uv}{\min\{u,v\} - uv}, \tag{8.26}$$

which quantifies the departure from independence in which case one would have $\mathbb{C}\text{ov}\left(Y_j(x), Y_k(x')\right) = uv$ and $D_N(u,v) = 0$.

Now, given the manner in which the random variable $\overline{Y}(u)$ is constructed for a given quantile level u, one can state the asymptotic behavior of an appropriately scaled version of this random variable, as detailed in Theorem 8.3

Theorem 8.3 (Convergence of the empirical distribution function process) *According to the Lindeberg–Lévy Central limit theorem, given a sample size n of i.i.d. real-valued random variables $X_1, X_2, \ldots, X_n$ with distribution F, the following convergence in distribution holds as $n \to \infty$:*

$$\sqrt{n}\left(\overline{Y}(u) - y(u)\right) \xrightarrow{d} Normal\left(0, \sigma^2(u, \nu)\right) \tag{8.27}$$

with covariance function

$$\sigma(u, \nu) = \min(u, \nu) - u\nu. \tag{8.28}$$

One can think about this covariance function (kernel) as characterizing the Brownian motion $y(u)$, which satisfies $y(0) = y(1) = 0$ and therefore forms what is known as a Brownian bridge.

An example of such Brownian bridge sample paths are constructed to illustrate the concept in Figure 8.3. As expected, each realization of the bridge trajectory involves a smooth continous function with variability as a function of the distance and tied down points at $u = 0$ and $u = 1$.

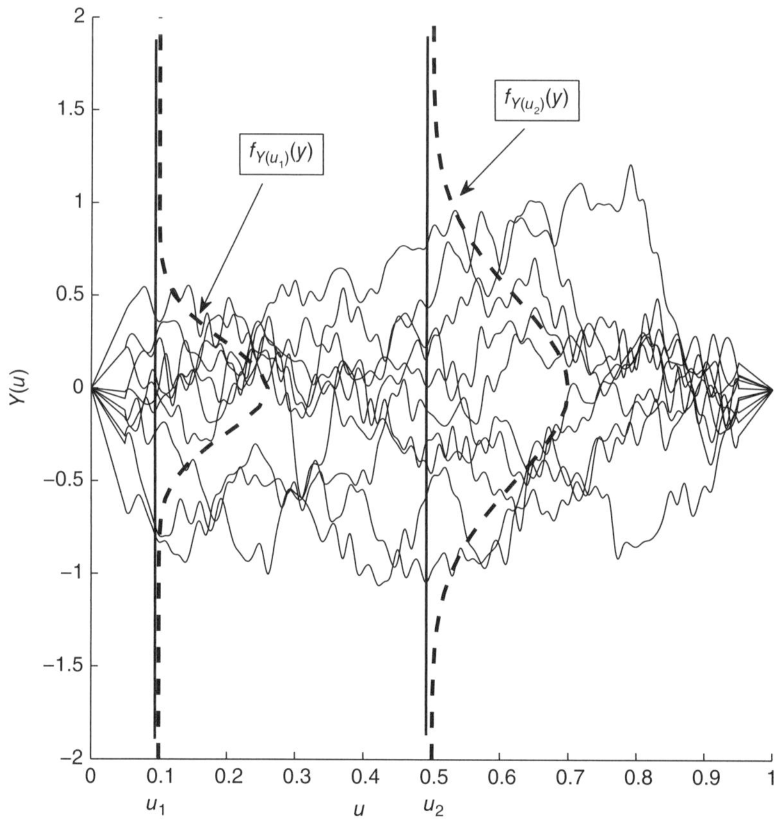

FIGURE 8.3 **Example of realizations of a Brownian bridge formed from the random emprical process convergence of $\sqrt{n}\left(\overline{Y}(u) - y(u)\right)$ as the number of samples $n \to \infty$**

Remark 8.4 *The key insight of this result realized by Kolmogorov when forming the GOF test that takes his name was that this limiting process and the resulting law of any functional of the limiting process y is not explicitly a function of the data-generating distribution F. This is precisely what makes it possible to design "universal" GOF tests.*

8.4.1.2 Convergence of the empirical distribution function for compound hypotheses.

In the case of the compound hypothesis setting, in which the population distribution contains a set of unknown parameters, one must consider carefully the convergence of the empirical distribution function to the population distribution under the additional component of estimation of the population parameters and the effect this may have. This exact problem was studied by Durbin (1973). Consider the setting involving i.i.d. observations $X_1, \ldots, X_n$ from a continuous distribution function $F(x; \boldsymbol{\theta})$ in which $\boldsymbol{\theta} = [\boldsymbol{\theta}_1, \boldsymbol{\theta}_2]$ is p-dimensional. Assume that under the null one has a statement about $0 < q \leq p$ of the parameters $\boldsymbol{\theta}_1$ given by $\mathcal{H}_0 : \boldsymbol{\theta}_1 = \boldsymbol{\theta}_{\mathcal{H}_0}$ and the remaining $p - q$ parameters, denoted by sub-vector of parameters $\boldsymbol{\theta}_2$, are unknown and must be estimated from the sample data according to an estimator with n samples denoted by $\hat{\boldsymbol{\theta}}_{2,n}$. For convenience, we refer to the null hypothesis values for vector of parameters $\boldsymbol{\theta}_1$ by the notation $\boldsymbol{\theta}_{1,0}$ and analgous notation holds for $\boldsymbol{\theta}_2$.

In this case, one can show that the sample process is not going to display the same weak convergence to a tied down Brownian motion discussed earlier for the simple hypothesis setting. Instead, we can make certain assumptions about the properties of the estimator $\hat{\boldsymbol{\theta}}_n = \left[\boldsymbol{\theta}_{1,0}, \hat{\boldsymbol{\theta}}_{2,n}\right]$ to obtain a weak convergence of the empirical distribution function for a sample of size n with estimated parameters $\hat{F}_n\left(x; \boldsymbol{\theta}_1, \hat{\boldsymbol{\theta}}_{2,n}\right)$ to the population distribution $F\left(x; \boldsymbol{\theta}_1, \boldsymbol{\theta}_2\right)$.

Durbin (1973) considered the analogous estimated sample process to that defined earlier in terms of the Bernoulli random variables, where now the difference in $\overline{Y}(u)$ will be studied in terms of the estimated sample process $Z(u) = \sqrt{n}\left(\overline{Y}(u) - y(u; \hat{\boldsymbol{\theta}}_{2,n})\right)$. In terms of the alternative hypothesis, the author assumes that it can be defined according to a sequence of alternative hypothesis with the special form

$$\mathcal{H}_n : \boldsymbol{\theta}_1 = \boldsymbol{\theta}_{1,0} + \sqrt{n}\boldsymbol{\gamma} \tag{8.29}$$

for a given vector $\boldsymbol{\gamma}$ and sample size n. Then, with this structure, one can show the weak convergence of the empirical distribution function to the population distribution under specific conditions on the decomposition and regularity of the estimator $\hat{\boldsymbol{\theta}}_{2,n}$ (see details in Durbin 1973, p. 281, assumptions 1 and 2). More precisely, it can be shown that $\overline{Y}(u)$ converges to a Gaussian process with a modified mean and covariance, as stated in Theorem 8.4. It is assumed that the estimator for the unknown parameters after appropriate scaling and translation by the true unknown parameters will satisfy the following structural form of decomposition given by,

$$\sqrt{n}\left(\hat{\boldsymbol{\theta}}_{2,n} - \boldsymbol{\theta}_{2,0}\right) = \frac{1}{\sqrt{n}}\sum_{i=1}^{n} l(x_i, \boldsymbol{\theta}_n) + A\boldsymbol{\gamma} + \epsilon_{1,n} \tag{8.30}$$

where assumptions on the random function $l(x_i, \boldsymbol{\theta}_n)$ and other terms are specified in detail by the conditions presented by Durbin (1973).

Theorem 8.4 (Convergence of the empirical distribution function for a compound hypothesis) *Assume that $\hat{\boldsymbol{\theta}}_{2,n}$ satisfies conditions (A1) and (A2) (Durbin 1973, p. 281, assumptions 1 and 2); then under the sequence of alternative hypothesis $\{\mathcal{H}_n\}$, $\overline{Z}(u)$ converges weakly to a Gaussian process, where the convergence is understood to be in the space of right continuous functions with left limits on $[0, 1]$. The resulting mean and covariance functions of the Gaussian process are given by*

$$\begin{aligned}\mathbb{E}\left[Z(u)\right] &= \gamma^T\left(\boldsymbol{g}_1(u) - A^T\boldsymbol{g}_2(u)\right)\\ \mathbb{C}\mathrm{ov}\left[Z(u), Z(\nu)\right] &= \min(u,\nu) - u\nu - \underbrace{\boldsymbol{h}(u)^T\boldsymbol{g}_2(\nu) - \boldsymbol{h}(\nu)^T\boldsymbol{g}_2(u) + \boldsymbol{g}_2(u)^T L\boldsymbol{g}_2(\nu)}_{\textit{modification to covariance function}}\end{aligned} \tag{8.31}$$

where one defines the vector-valued functions

$$\boldsymbol{g}_1(u) = \left.\frac{\partial F(x,\boldsymbol{\theta})}{\partial \boldsymbol{\theta}_1}\right|_{x=x(u,\theta)} \quad \textit{and} \quad \boldsymbol{g}_2(u) = \left.\frac{\partial F(x,\boldsymbol{\theta})}{\partial \boldsymbol{\theta}_2}\right|_{x=x(u,\theta)} \tag{8.32}$$

and

$$\boldsymbol{h}(u) = h\left(u, \boldsymbol{\theta}_{\mathcal{H}_0}\right) \quad \textit{with} \quad h\left(u,\boldsymbol{\theta}\right) = \int_{-\infty}^{x(u,0)} l(x,\boldsymbol{\theta})dF(x;\boldsymbol{\theta}) \tag{8.33}$$

and the finite non-negative definite matrix sequence $L\left(\boldsymbol{\theta}_n\right) = \mathbb{E}\left[l(x,\boldsymbol{\theta}_n)l(x,\boldsymbol{\theta}_n)^T \middle| \boldsymbol{\theta} = \boldsymbol{\theta}_n\right]$ converges to the resulting matrix L, i.e. $L\left(\boldsymbol{\theta}_n\right) \to L\left(\boldsymbol{\theta}_0\right) = L$ as $n \to \infty$.

Often in the case of heavy-tailed models we may also be interested in hypothesis testing on nominal claims relating to the characteristic function and therefore we will be concerned with convergence of the empirical characteristic function (ECF).

8.4.1.3 Convergence of the ECF for simple and compound hypotheses.

It will also be beneficial to observe the following results which are based on comparisons between properties of the ECF (Definition 8.7) as detailed by Parzen (1962) and a hypothesized distribution characteristic function, which regularly arises in the context of GOF testing in the work of for example Heathcote (1972), Press (1972), and Koutrouvelis and Kellermeier (1981).

Definition 8.7 (Empirical characteristic function) *Given a continuous distribution, let X_1, $X_2, \ldots, X_n$ be a sequence of i.i.d. random variables from this distribution with observed realizations $x_1, x_2, \ldots, x_n$. Then the ECF $\hat{\phi}$ is defined according to*

$$\hat{\phi}_{X,n}(t) = \frac{1}{n}\sum_{j=1}^{n} \exp\left(\mathrm{i}tX_j\right). \tag{8.34}$$

■

To illustrate the ECF, consider Example 8.1.

EXAMPLE 8.1 Understanding the Empirical Characteristic Function

Consider a continuous distribution F that is LogNormal and let $X_1, X_2, \ldots, X_n$ be a sequence of i.i.d. random variables from this distribution with observed realizations $x_1, x_2, x_3, \ldots, x_n$. Then the ECF $\hat{\phi}$ can be thought of as taking the realization of each random sample and considering it as orbiting the unit circle in the complex plane, that is, the transformation of the i-th sample via the mapping $\exp(itX_i)$ will produce a set of points on the unit disk in the complex plane (as depicted in red circles, see Figure 8.4). These points are then averaged to get an estimated reconstruction of the characteristic function as depicted in the dashed black line for the estimate and the solid black line for the true characteristic function.

Basically, we can consider the ECF as the expected orbit or mean of the random variable orbits. For large sample sizes, the ECF converges to the distribution characteristic function as formalized in the figure.

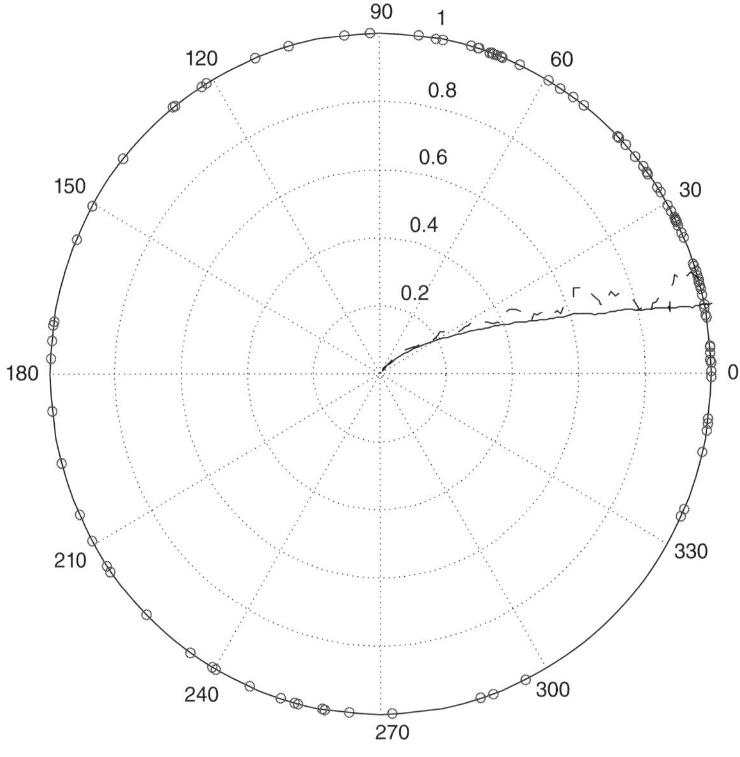

FIGURE 8.4 Red circles depict the project's observation realizations $x_1, x_2, x_3, \ldots, x_n$ on the unit disk in the complex plane for a severity model $LogNormal(\mu = 1, \sigma = 2)$. In the dashed black line we see the ECF estimated for the model from the data and the solid black line demonstrates the true characteristic function. (see insert for color representation of the figure.)

Theorem 8.5 (Convergence in probability of empirical characteristic function) *Given a sample size n of i.i.d. real-valued sample realizations of random variables $X_1, X_2, \ldots, X_n$ with distribution F, the following convergence in probability applies according to the Strong Law of Large Numbers for any fixed $T < \infty$:*

$$\mathbb{P}\mathrm{r}\left(\lim_{n\to\infty} \sup_{|t|\leq T} \left|\hat{\phi}_{X,n}(t) - \phi_X(t)\right| = 0 \right) = 1 \;\; a.s. \tag{8.35}$$

Other results related to the convergence of the ECF to the population characteristic function are discussed in detail by Feuerverger and Mureika (1977). As was done with the empirical distribution function, one can also consider the definition of a stochastic process representation of the ECF, given in Definition 8.8.

Definition 8.8 (Empirical characteristic function process) *Given a continuous distribution, let $X_1, X_2, \ldots, X_n$ be a sequence of i.i.d. random variables from this distribution with observed realizations $x_1, x_2, x_3, \ldots, x_n$. Then the stochastic process given by*

$$R_n(t) = \sqrt{n}\left(\hat{\phi}_{X,n}(t) - \phi_X(t)\right) \tag{8.36}$$

is a random complex process in t with the following mean and covariance function characteristics:

$$\begin{aligned} \mathbb{E}\left[R_n(t)\right] &= 0, \\ \mathbb{E}\left[R_n(t_1)R_n(t_2)\right] &= \phi_X(t_1 + t_2) - \phi_X(t_1)\phi_X(t_2). \end{aligned} \tag{8.37}$$

■

One can then consider the convergence of the appropriately scaled and translated ECF process and one can state the following weak convergence result in Theorem 8.6.

Theorem 8.6 (Weak convergence of the ECF process) *Consider a continuous distribution and let $X_1, X_2, \ldots, X_n$ be a sequence of i.i.d. random variables from this distribution; then the stochastic process given by*

$$R_n(t) = \sqrt{n}\left(\hat{\phi}_{X,n}(t) - \phi_X(t)\right) \tag{8.38}$$

is a random complex process in t, which converges weakly to a process $R(t)$ in every finite interval, where $R(t)$ is a zero mean complex valued Gaussian process satisfying the symmetry condition that $R(t) = R(-t)$ with covariance structure of the real and imaginary components as follows:

$$\begin{aligned} \mathbb{C}\mathrm{ov}\left[\mathcal{R}\mathrm{e}\left[R_n(t_1)\right]\mathcal{R}\mathrm{e}\left[R_n(t_2)\right]\right] &= \frac{1}{2}\left[\mathcal{R}\mathrm{e}\left[\phi_X(t_1+t_2)\right] + \mathcal{R}\mathrm{e}\left[\phi_X(t_1-t_2)\right]\right] \\ &\quad - \mathcal{R}\mathrm{e}\left[\phi_X(t_1)\right]\mathcal{R}\mathrm{e}\left[\phi_X(t_2)\right], \\ \mathbb{C}\mathrm{ov}\left[\mathcal{R}\mathrm{e}\left[R_n(t_1)\right]\mathcal{I}\mathrm{m}\left[R_n(t_2)\right]\right] &= \frac{1}{2}\left[\mathcal{I}\mathrm{m}\left[\phi_X(t_1+t_2)\right] + \mathcal{I}\mathrm{m}\left[\phi_X(t_1-t_2)\right]\right] \\ &\quad - \mathcal{R}\mathrm{e}\left[\phi_X(t_1)\right]\mathcal{I}\mathrm{m}\left[\phi_X(t_2)\right], \\ \mathbb{C}\mathrm{ov}\left[\mathcal{I}\mathrm{m}\left[R_n(t_1)\right]\mathcal{I}\mathrm{m}\left[R_n(t_2)\right]\right] &= \frac{1}{2}\left[-\mathcal{R}\mathrm{e}\left[\phi_X(t_1+t_2)\right] + \mathcal{R}\mathrm{e}\left[\phi_X(t_1-t_2)\right]\right] \\ &\quad - \mathcal{I}\mathrm{m}\left[\phi_X(t_1)\right]\mathcal{I}\mathrm{m}\left[\phi_X(t_2)\right]. \end{aligned} \tag{8.39}$$

Remark 8.5 (ECF central limit theorem) *It will also be useful to note the following result that will specify the conditions required for the complex valued ECF to converge weakly to a Gaussian process (see discussion by Feigin and Heathcote 1976). The real component of the ECF process*

$$\begin{aligned}\mathcal{R}\mathrm{e}\left\{R_n(t)\right\} &= \frac{1}{n}\left(\mathcal{R}\mathrm{e}\left\{\hat{\phi}_{X,n}(t)\right\} - \mathcal{R}\mathrm{e}\left\{\phi_X(t)\right\}\right)\\ &= \frac{1}{n}\left(\sum_{i=1}^{n}\cos\left(tX_i\right) - \mathbb{E}\left[\cos(tX)\right]\right)\\ &= \frac{1}{n}\left(U_n(t) - u(t)\right)\end{aligned} \tag{8.40}$$

will converge in distribution to a zero mean Gaussian random variable for any t that satisfies

$$1 + u(2t) - 2u^2(t) > 0. \tag{8.41}$$

The imaginary component of the ECF process

$$\begin{aligned}\mathcal{I}\mathrm{m}\left\{R_n(t)\right\} &= \frac{1}{n}\left(\mathcal{I}\mathrm{m}\left\{\hat{\phi}_{X,n}(t)\right\} - \mathcal{I}\mathrm{m}\left\{\phi_X(t)\right\}\right)\\ &= \frac{1}{n}\left(\sum_{i=1}^{n}\sin\left(tX_i\right) - \mathbb{E}\left[\sin(tX)\right]\right)\\ &= \frac{1}{n}\left(V_n(t) - v(t)\right)\end{aligned} \tag{8.42}$$

will converge in distribution to a zero mean Gaussian random variable for any t that satisfies

$$1 + v(2t) - 2v^2(t) > 0. \tag{8.43}$$

In general, it is possible to work with multiple t values to obtain convergence to a multivariate Gaussian for the real and imaginary components. It is also common in practice to work with one value of t that may be selected in order to maximize the power of the resulting test.

Again, as was the case in the empirical distribution function process for a compound hypothesis, the behavior of the ECF process when parameters of the model are estimated is studied by Koutrouvelis and Kellermeier (1981).

We are now in a position to state some general results for generic GOF tests and their properties, which are based on the empirical process convergence results considered in this section just completed.

8.4.2 OVERVIEW OF GENERIC GOF TESTS—OMNIBUS DISTRIBUTIONAL TESTS

In this section, we briefly mention the notion of generic GOF tests, which include approaches based on P–P plots, χ^2 tests, empirical distribution function, and ECF tests. In all cases of formal inference, we consider the following generic steps appropriate to formally set out the test and its outcomes.

Generic Structure of a Hypothesis Test Procedure

1. Set up suitable notation for the random variables and distributions being tested;
2. Make a statement of the null and alternative hypotheses in terms of population distribution/parameters;
3. State the test statistic and its observed value as well as the distribution of the test statistic;
4. State a formal mathematical expression for the p-value;
5. State the range of values within which the p-value falls (and a statement of how these are obtained);
6. State the conclusion of the test in plain language (relevant to the experimental context).

With this generic framework, we can now state some well-known examples of hypothesis tests and their properties.

- **P–P plots**. As discussed earlier, in addition to plotting P–P plots as a qualitative diagnostic tool, one can also perform a hypothesis test on the relevance of a regression relationship formed by regressing the percentiles of the data against the percentiles under the null hypothesis;
- **Pearson's χ^2 GOF test**. In this type of universal test, the observations are binned into a partition of the observed random variables' support. Then the χ^2 test statistics compare the observed counts from the realized data sample with those one would expect to see under the null hypothesis distributions' support on the given partition. These comparisons are then summed over all partitions to obtain the observed value of the test statistic (denoted d), which under the null hypothesis will have asymptotically a χ^2 distribution. This allows for the calculation of a p-value $(p = \mathbb{P}\text{r}\,(D \geq d|\mathcal{H}_0))$ in order to make a decision at a given level of significance.
  If the data are divided into k bins, then the test statistic under the null is defined as

$$\chi^2 = \sum_{i=1}^{k} \frac{(O_i - E_i)^2}{E_i}, \tag{8.44}$$

where O_i is the observed frequency for bin i obtained from the counts on the empirical loss data and E_i is the expected frequency for bin i using the null hypothesis claim on the distribution for data generation. The expected frequency is calculated by

$$E_i = n\,(F(X_u(i)) - F(X_l(i)))\,, \tag{8.45}$$

where $X_u(i)$ and $X_l(i)$ are the upper and lower limits, respectively, of the i-th partition (bin). This test statistic follows, approximately, a chi-square distribution with $(k - p)$ degrees of freedom, where k is the number of nonempty cells and p is the number of estimated parameters (including location and scale parameters and shape parameters) for the distribution +1. For example, for a two-parameter LogNormal distribution, $p = 3$. Then under the null, this statistic will produce a p-value given by

$$p = \mathbb{P}\text{r}\left(\chi^2 \geq c|\mathcal{H}_0\right) = 1 - F(c; n - p - 1), \tag{8.46}$$

where $F(c; n-p-1)$ is a χ^2 distribution with $n-p-1$ degrees of freedom and a distribution given by

$$F(c; n-p-1) = \frac{\gamma\left(\frac{n-p-1}{2}, \frac{c}{2}\right)}{\Gamma\left(\frac{n-p-1}{2}\right)} \tag{8.47}$$

with the lower incomplete Gamma function $\gamma(x, y) = \int_0^y t^{x-1} \exp(-t) dt$.

Note 1. This test is sensitive to the choice of bins and there is no optimal choice for the bin width as it will be distribution-specific.

Note 2. The asymptotic chi-square distributional approximation under the null is valid when the expected frequency is sufficiently large. Hence, it should not be applied for small samples, and if some of the counts are less than 5, you may need to combine some bins in the tails.

Note 3. If one is considering a compound hypothesis where parameters of the model under the null must be estimated, there is a well-established correction for the χ^2 p-values due to the fact that the resulting test statistic is no longer asymptotically χ^2 (see Snedecor 1989, Chernoff and Lehmann 1954 and LeCam *et al.* 1983). More precisely, when estimating the parameters for the test, it is possible to utilize an MLE (or equivalent) estimator based on either the cell frequencies or the original observations. If the observations are utilized in the estimation of the parameters, then the resulting test statistic constructed would be a function of the parameter estimates $\chi^2\left(\hat{\boldsymbol{\theta}}_n\right)$, which under the null is no longer asymptotically χ^2 distributed. In particular, when the test statistic is evaluated by using the MLE procedure and it does not coincide with a minimum chi-squared estimation, then the resulting asymptotic distribution of the test statistic can be shown to lie somewhere between a chi-squared distribution with $n-p-1$ and $n-1$ degrees of freedom (see Chernoff and Lehmann 1954, p. 580, theorem 1).

Note 4. Pearson's χ^2 GOF test is the best known of several chi-squared tests (Yates, likelihood ratio, portmanteau test in time series, etc.)

Note 5. The chi-square GOF test can be applied to discrete distributions such as the Binomial and the Poisson. The KS and Anderson–Darling (AD) tests are restricted to continuous distributions.

Note 6. The disadvantage is that you must evaluate the distribution function and there will also be a loss of information from the grouping of observations;

- **Empirical distribution function GOF tests.** In this type of universal test, one measures the distance between the empirical distribution and the null distribution. In a general sense, one can consider measuring a limit distance between distributions under a norm $\|\cdot\|$ over the space of continuous bridges. Examples include norm-2 on the limiting bridge process,

$$|y|_2 = \int_0^1 y(u)^2 du, \tag{8.48}$$

or the norm-sup on the limiting bridge process,

$$|y|_{\infty} = \sup_{u\in[0,1]} |y(u)|. \tag{8.49}$$

These general ideas translate in practice into the evaluation of the quadratic measures (Cramer-von-Mises (CVM) family) and (AD) tests given for a null hypothesis of distribution F given according to the test statistic given by

$$Q = \int_{-\infty}^{\infty} w(x)\left(\hat{F}(x) - F(x;\boldsymbol{\theta})\right)^2 dF(x;\boldsymbol{\theta}), \tag{8.50}$$

for some weight function such as the quadratic Cramer-von-Mises statistic when $w(x) = 1$ or the AD statistic when $w(x) = F(x;\boldsymbol{\theta})\,(1 - F(x;\boldsymbol{\theta}))^{-1}$. There is also the vertical maximum distance measure given by the supremum norm for example, KS test and their weighted versions:

$$D = \sup_x \left|\hat{F}(x) - F(x;\boldsymbol{\theta})\right|. \tag{8.51}$$

- **Empirical characteristic function GOF tests**. In this type of universal test, one measures the distance between the empirical characteristic function, given for an i.i.d. sample $X_1, \ldots, X_n$ by

$$\hat{\phi}_{X,n}(t) = \frac{1}{n}\sum_{j=1}^{n} \exp\left(\mathrm{i}tX_j\right), \tag{8.52}$$

and the null characteristic function, given by

$$\phi_X(t) = \int_{-\infty}^{\infty} \exp\left(\mathrm{i}tX\right) dF_X(x). \tag{8.53}$$

Typically, one still utilizes distance-based measures such as

$$\sup_t \left|\hat{\phi}_{X,n}(t) - \phi_X(t)\right|. \tag{8.54}$$

In addition to GOF tests based on the characteristic function for the distributional form, there are also interesting tests for general attributes of the distribution under the null that utilize the characteristic function and ECF. For example, one can utilize the fact that a characteristic function is real if and only if the corresponding distribution function is symmetric about the origin; Feuerverger and Mureika (1977) suggest that such a result could consider testing for symmetry through the consideration of a statistic such as

$$\int_{-\infty}^{\infty} \left(\mathcal{I}\mathrm{m}\left[\hat{\phi}_{X,n}(t)\right]\right)^2 dF(t). \tag{8.55}$$

Other tests based on the characteristic function have been discussed by Feuerverger and Mureika (1977), who note that for testing the symmetry of a distribution function about the origin, it suffices to test if the characteristic function is real. In addition, other tests that have been proposed based on the ECF have included the work of Heathcote (1972) and Feigin and Heathcote (1976), who studied the case of simply hypothesis testing for the null specification that $\mathcal{H}_0 : \phi_X(t) = \phi_{X_0}(t)$, under a test statistic that could be constructed from the real or imaginary component at a single value of t, which was specifically selected. This test was generalized by Koutrouvelis (1980) to multiple points $t_1, \ldots, t_m$ under a test statistic constructed on these points, which comprised of measuring the Mahlanobis distance between the vector of the ECF evaluated at the points after it had been suitably translated and scaled by the null hypothesis mean and covariance functions evaluated at these points for the null characteristic function. The points must be selected to ensure that the inverse covariance function at these points is not singular.

8.4.3 KOLMOGOROV–SMIRNOV GOODNESS-OF-FIT TEST AND WEIGHTED VARIANTS: TESTING IN THE PRESENCE OF HEAVY TAILS

There are many situations where OpRisk practitioners need to assess the adequacy of assumptions or hypotheses regarding the distribution from which their observed losses may have been sampled. This would typically be in the form of addressing a question such as the following "*What is an appropriate model for the observed losses in a given business unit and risk event type?*"

To address such questions one may adopt a hypothesis-testing procedure based on a KS test, which is a formal inference procedure for verifying that a sample comes from a population with some known distribution (one-sample test) or alternatively for considering whether two populations have the same distribution (two-sample test). We will proceed below under the setting in which the simple hypothesis is assumed.

Remark 8.6 (Relevance of heavy-tailed GOF testing to OpRisk severity models) *When assessing the tail behavior of the loss process severity model in an LDA structure, the assessment of the heavy-tailed behavior under a particular model is often the focus, see detailed discussion in our companion book Peters and Shevchenko (2015). However, the point of the following section is to make clear that the analysis of the tail index (heaviness or fatness) of the right severity tail under a parametric model assumption is not the complete analysis. In particular, the presence of heavy tails does not tell you that the correct model has been considered; instead, one should also consider a formal inferential procedure to assess this question regarding the appropriate model structure.*

We present in detail the standard GOF testing for general LDA severity models, then we explain why in OpRisk settings one should consider nonstandard modifications to the basic statistical GOF tests, in particular, how such modifications allow one to correctly account for the right tail behavior appropriately when informing a decision of a nominal claim via a p-value. The tail-weighted variants are particularly important for testing model hypotheses regarding the right tail of a heavy-tailed severity model.

The one-sample KS test is defined for observed loss realizations $x_1, x_2, \ldots, x_n$ of a set of continuous i.i.d. random loss variables $X_1, X_2, \ldots, X_n$, for which it is hypothesized that their sampling distribution function is F. The test is performed according to the following stages in Algorithm 8.1 that are based on the results in Theorem 8.7.

Theorem 8.7 (Kolmogorov–Smirnov's approximation of null distribution) *Consider a null hypothesis for the data-generating distribution F_0, which one assumes is a continuous distribution. Let $X_1, X_2, \ldots, X_n$ be a sequence of i.i.d. random variables with distribution F_0. Then the following holds:*

1. *The test statistic is evaluated as,*

$$D_n = \sup_{x \in \mathbb{R}} \left| \hat{F}_n(x) - F_0(x) \right| \tag{8.56}$$

and it depends only on the sample size n, and second the maximum will always occur at one of the sample points in the unweighted test;

2. *If $n \to \infty$, the distribution $\sqrt{n}D_n$ is asymptotically Kolmogorov's distribution given by*

$$Q(x) = 1 - 2\sum_{k=1}^{\infty}(-1)^{k-1}\exp(-2k^2x^2), \tag{8.57}$$

where this defines the following probability limit

$$\lim_{n \to \infty} \mathbb{P}\mathrm{r}\left(\sqrt{n}D_n \leq x\right) = Q(x). \tag{8.58}$$

Remark 8.7 *The Kolmogorov distribution can be shown to be formally the distribution of the random variable*

$$K = \sum_{t \in [0,1]} |B(t)|, \tag{8.59}$$

where $B(t)$ is the Brownian bridge (see Kolmogorov 1933, Smirnov 1948, Anderson and Darling 1952; and Massey 1951).

The KS GOF hypothesis test then proceeds as further detailed.

Algorithm 8.1 (Kolmogorov–Smirnov One-Sample Test)

1. ***Step 1****. Set up suitable notation for the random variables and distributions being tested and make a statement of the null and alternative hypotheses in terms of population distribution/parameters. Determine hypothesis for GOF testing where null claims loss data are from a hypothesized distribution $F_0(x)$*

$$H_0 : F(x) = F_0(x),\ \forall x \tag{8.60}$$

versus an alternative claim that the observed losses are not realizations from F_0

$$H_A : F(x) \neq F_0(x),\ \forall x. \tag{8.61}$$

2. ***Step 2****. State the test statistic and its observed value and when possible state the distribution of the test statistic or its approximation. Under the null hypothesis calculate the KS test statistic given by*

$$\begin{aligned} D_n &= \sup_{x \in \mathbb{R}} \left| \hat{F}_n(x) - F_0(x) \right| \\ &= \max_{1 \leq i \leq n} \left(F_0(X_i) - \frac{i-1}{n}, \frac{i}{n} - F_0(X_i) \right), \end{aligned} \tag{8.62}$$

where $\hat{F}$ is the empirical cumulative distribution defined according to

$$\hat{F}_n(x) = \frac{1}{n}\sum_{i=1}^{n} \mathbb{I}\left[x_i \le x\right], \tag{8.63}$$

and the supremum occurs at one of the observed values x_i to the left of x_i. This procedure will produce an observed realization of the test statistic based on the observed data samples $\{x_i\}_{i=1}^{n}$ under the null hypothesis, denoted by d_n.

3. ***Step 3**. State a formal mathematical expression for the p-value. Determine the p-value for the test under the null hypothesis given by considering*

$$p\text{-}value = \mathbb{P}\mathrm{r}\left[|D_n| \ge d_n | H_0\right].$$

To obtain the p-value one first needs to obtain an approximation of the distribution of the test statistic under the null. This can be done in two cases, depending on the size of the sample:
***Small-sample p-value evaluation**. If the sample size n is small, one can perform evaluation of the p-value for making a decision on the test via the following simple simulation procedure, where $\{X_i\}_{i=1}^{n}$ are the samples from the experiment and $j = 1, \ldots, J$ is the index of the simulated test statistic realizations $\left\{d_n^{(j)}\right\}$ obtained by the following procedures:*

- *Simulate a set of samples $\left\{U_i^{(j)}\right\}_{i=1}^{n}$ with $U_i \sim Uniform(0,1)$ that is, distribution $F(u) = u$;*
- *Construct the empirical distribution function $\hat{F}^{(j)}$ using the samples $\left\{U_i^{(j)}\right\}_{i=1}^{n}$;*
- *Evaluate for each set of samples $\left\{U_i^{(j)}\right\}_{i=1}^{n}$ the maximum distance between the distribution $F(u) = u$ and the empirical distribution function for the generated sample $\hat{F}^{(j)}$ in the verticle direction, to get $d_n^{(j)}$. Repeat many times $j \in \{1, 2, \ldots, J\}$ to get an estimate of the distribution for the test statistic under the null D_n, that is, the null distribution of the test statistic D_n is then approximated by the samples $\left\{d_n^{(j)}\right\}$ known by simulation.*

Given the empirical estimator for the distribution of the test statistic under the null $\hat{F}_{D_n}(x)$, use this to evaluate the p-value.
***Large-sample p-value evaluation**. If the sample size n is large, one can perform evaluation of the p-value for making a decision on the test via the asymptotic result for the KS distribution function. If $n \to \infty$, the distribution $\sqrt{n}D_n$ is asymptotically Kolmogorov's distribution given by*

$$Q(x) = 1 - 2\sum_{k=1}^{\infty}(-1)^{k-1}\exp(-2k^2x^2), \tag{8.64}$$

where this defines the following probability limit

$$\lim_{n\to\infty} \Pr\left(\sqrt{n}D_n \le x\right) = Q(x). \tag{8.65}$$

4. ***Step 4***. *State the range of values within which the p-value falls (and a statement of how these are obtained. If the p-value is significantly lower than a given level of testing significance, typically 5%, then one has sufficient evidence from the observed loss data to reject the claim of the null hypothesis in favor of the alternative;*
5. ***Step 5***. *State the conclusion of the test in plain language (relevant to the experimental context).*

Remark 8.8 *When applying the standard KS GOF test specified earlier, it is well known that these tests will overweight the quantiles around the median and downweight the quantiles in the tails. In the case of a heavy-tailed model, this is not ideal, as in such cases, it is precisely the null assumption on the tail decay of the statistical model that is of most interest for testing and practical features of the use of the model.*

EXAMPLE 8.2 Kolmogorov–Smirnov Test and Heavy-Tailed Severity Model

Consider a GPD for the severity model with tail index (shape) parameter $k = 1$, scale parameter $\sigma = 1$, and threshold (location) parameter $\theta = 0$. This is a heavy-tailed loss model in the sense that the mean is not finite when $K \geq 1$ under this GPD model specification. We simulate loss data realizations $\{X_i\}_{i=1}^{n}$ for sample sizes of $n = 10, 20, 50, 100$. Then we evaluate the test statistic for a standard KS test in the following three cases:

- CASE 1. Nominal claimed distribution F_0 is the GPD with exact parameters $GPD(k = 1, \sigma = 1, \theta = 0)$—that is, no model or parameter misspecification;
- CASE 2. Nominal claimed distribution F_0 is the exponential distribution with parameter $GPD(k = 0, \sigma = 1, \theta = 0)$—that is, (in this special cases) there is a parameter misspecification producing a light-tailed nominal claim when really the data are from a heavy-tailed population distribution;
- CASE 3. Nominal claimed distribution F_0 is the LogNormal distribution with parameters $LogNormal(\mu = 0, \sigma = 1)$—that is, model and parameter misspecification.

Using these nominal claim models, the test statistics distribution under each sample size is simulated via Monte Carlo and plotted in Figure 8.5. In Tables 8.1, 8.2 and 8.3, we also show the quantile (critical values) for the distribution of the KS test statistic for each case as a function of sample size. The results of this analysis demonstrate that as the sample size increases, one should expect the maximum absolute distance between the empirical distribution function and the null hypothesis (verticle direction) to reduce if in fact the nominal claim is correct. In Case 1, the true distribution that generated the sample loss data was used for the nominal claim and therefore, as expected, the distribution of the test statistic, as the sample size increases, produces critical values at each level of significance closer to zero, making it less and less likely that the nominal claim will be rejected by the test. In Case 2, the nominal claim involves the correct distribution; however, the parameters are estimated incorrectly and, importantly, this results in the nominal claim for the GPD severity distribution resembling an exponential distribution that is much more lightly tailed than the true data-generating distribution. In this case, there is still a strong chance that one would reject the nominal claim even for a large sample size, as expected especially when the tails are so poorly matched by the nominal claim. The same occurs for the case when the nominal claim is the wrong distributional family, but the tails are still subexponential as occurs in Case 3 with the LogNormal example. It is clear from the reported critical values in the table that the misspecification of the tails can have a big effect on the performance of this test, as shown with the LogNormal example, where it is hard to distinguish this model from the true model that was used in Case 3.

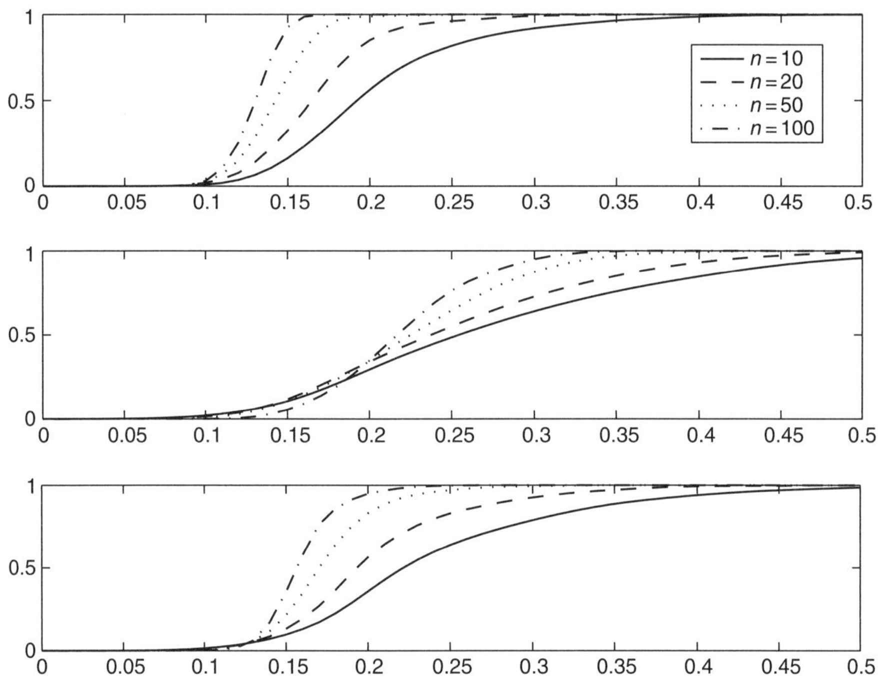

FIGURE 8.5 **In each subplot, the distribution of the KS test statistic is displayed under the assumption that the null hypothesis is correct for data sizes $n = 10, 20, 50, 100$. The top subplot shows the distribution of the KS test statistic under CASE 1. The middle subplot shows the distribution of the KS test statistic under CASE 2. The bottom subplot shows the distribution of the KS test statistic under CASE 3**

■

To overcome these problems in the supremum norm context, one can develop the weighted KS GOF test as given in Proposition 8.1. One, then, still measures the largest verticle distance between the empirical distribution function and the distribution; however, weights are now attributed to each deviation as a function of the quantile level.

TABLE 8.1 Assessing the heavy tailed feature of loss data under a GOF test based on KS. True population loss model is $GPD(k = 1, \sigma = 1, \theta = 0)$. Nominal claim is GPD with correct population parameters, test is performed at a number of sample sizes n and a range of significance levels $1 - \alpha$

Case 1: $H_0 : GPD(k = 1, \sigma = 1, \theta = 0)$				
$1 - \alpha$	$n = 10$	$n = 20$	$n = 50$	$n = 100$
80%	0.2444	0.1913	0.1590	0.1422
90%	0.2854	0.2094	0.1687	0.1470
95%	0.3327	0.2345	0.1750	0.1517
97.5%	0.3745	0.2710	0.1810	0.1542
99%	0.4017	0.2947	0.2037	0.1606
99.5%	0.4392	0.3198	0.2294	0.1668

TABLE 8.2 Assessing the heavy tailed feature of loss data under a GOF test based on KS. True population loss model is $GPD(k = 1, \sigma = 1, \theta = 0)$. Nominal claim is Exponential with incorrect population parameters, test is performed at a number of sample sizes n and a range of significance levels $1 - \alpha$. Nominal claim is light tailed, true population distribution is heavy tailed

Case 2: $H_0 : GPD(k = 0, \sigma = 1, \theta = 0) = Exp(\sigma = 1)$				
$1 - \alpha$	$n = 10$	$n = 20$	$n = 50$	$n = 100$
80%	0.3683	0.3268	0.2783	0.2551
90%	0.4339	0.3756	0.3077	0.2821
95%	0.4840	0.4153	0.3297	0.2980
97.5%	0.5310	0.4559	0.3536	0.3125
99%	0.5806	0.4844	0.3760	0.3306
99.5%	0.5981	0.5080	0.4069	0.3350

TABLE 8.3 Assessing the heavy tailed feature of loss data under a GOF test based on KS. True population loss model is $GPD(k = 1, \sigma = 1, \theta = 0)$. Nominal claim is LogNormal and the test is performed at a number of sample sizes n and a range of significance levels $1 - \alpha$. Nominal claim is heavy-tailed but misspecified relative to the true population distribution which is also heavy-tailed

Case 3: $H_0 : LogNormal(\mu = 0, \sigma = 1)$				
$1 - \alpha$	$n = 10$	$n = 20$	$n = 50$	$n = 100$
80%	0.3024	0.2376	0.1953	0.1720
90%	0.3575	0.2817	0.2127	0.1872
95%	0.4127	0.3195	0.2314	0.1989
97.5%	0.4623	0.3588	0.2558	0.2132
99%	0.5119	0.3880	0.2766	0.2276
99.5%	0.5399	0.4175	0.3114	0.2393

Proposition 8.1 (Weighted supremum norm tests) *Consider a null hypothesis for the data-generating distribution F_0, which one assumes is a continuous distribution. Let $X_1, X_2, \ldots, X_n$ be a sequence of i.i.d. random variables with distribution F_0 (null distribution). The resulting test statistic then takes the form*

$$\widetilde{D}_n = \sup_{x \in \mathbb{R}} \left| w(F_0(x)) \left(\hat{F}_n(x) - F_0(x) \right) \right| \tag{8.66}$$

for some weight function $w(u)$. Now, the weighted fluctuation analysis for this statistic can be studied under the null to find the tail distribution to obtain the p-values. As $n \to \infty$, the distribution $\sqrt{n}\widetilde{D}n$ should be considered; this is nontrivial and involves studying the behavior of the Brownian bridge for quantile levels $u \in [0, 1]$ given by

$$\widetilde{y}(u) = \sqrt{w(u)}y(u) \tag{8.67}$$

such that $y(0) = y(1) = 0$.

The distribution of the limiting fluctuation process has been studied for different weight functions in different regimes for the number of samples obtained. For example, one could focus attention on the left or right or both tails using indicator functions on tail regions with weights such as $w(u; a) = \mathbb{I}[u \geq a]$ for the right tail and $w(u; b) = \mathbb{I}[u \leq b]$ for the left tail. It should be noted that in OpRisk settings it will typically be the case that one would only be interested in the right tail behavior and the suitability of a fitted model in this region.

Another popular choice for the weight function, studied by Noé and Vandewiele (1968), Niederhausen (1981), Borokov and Sycheva (1968), and Chicheportiche and Bouchaud (2012), involves $w(u) = 1/\mathbb{V}\text{ar}(y(u))$. This choice is made in order to allocate equal weight to all quantile levels. Noé and Vandewiele (1968) studied the supremum norm over the interval $[0, 1]$ and derived the distribution for the test statistic under one-and two-sided simple hypotheses, which were then studied and numerically tabulated Niederhausen (1981, example 4) via a basis expansion using Sheffer polynomials. They noted that this case was studied by Borokov and Sycheva (1968), who obtained an exact distribution for finite samples as well as the asymptotic distribution for the case of a test statistic defined in Equation (8.68):

$$\widetilde{D}_n = \sup_{a \leq F_0(x) \leq b} \frac{\left(\hat{F}_n(x) - F_0(x) \right)}{\sqrt{F_0(x)\left(1 - F_0(x)\right)}}. \tag{8.68}$$

In the case of the GOF tests we typically consider, we do not wish to differentiate our analysis between positive and negative vertical deviations between the empirical distribution function and the null distribution, and for this reason we are interested in the distribution for maximum absolute deviations. In this case, the tractability of the results obtained by Borokov and Sycheva (1968) disappears. To address this Chicheportiche and Bouchaud (2012) obtained an interesting result for a large-sample analysis for this variance-weighted KS test as detailed in Proposition 8.2. Then the general result for any finite sample size is obtained recursively according to Niederhausen (1981).

Proposition 8.2 (Kolmogorov–Smirnov's variance–weighted test for the tails) *Consider a null hypothesis for the data-generating distribution F_0, which one assumes is a continuous distribution. Let $X_1, X_2, \ldots, X_n$ be a sequence of i.i.d. random variables with distribution F_0. Then one may characterize the distribution for the law of the weighted supremum of a Brownian bridge given by,*

$$D(a,b) = \sup_{u\in[a,b]} |\tilde{y}(u)| = \sup_{u\in[a,b]} \left|\sqrt{w(u;a,b)}y(u)\right| \tag{8.69}$$

such that $y(0) = y(1) = 0$ with probability 1 and the weight applied to interval $a \in]0,1[$ and $b \in [a,1[$ given by

$$w(u;a,b) = \begin{cases} \frac{1}{u(1-u)}, & a \le u \le b, \\ 0, & \text{otherwise} \end{cases} \tag{8.70}$$

according to a large-sample asymptotic result (see Chicheportiche and Bouchaud 2012, equation 13), which considers

$$\mathbb{P}\mathrm{r}\left[D(a,b) \le d | a = \ln n, b = 1 - \ln n\right] = \widetilde{A}(d)n^{\theta_0(d)} \tag{8.71}$$

with the expressions for $\widetilde{A}(d)$ and $\theta_0(d)$ provided explicitly in general by Chicheportiche and Bouchaud (2012), and the right and left tail asymptotic behavior that one cares about for the two-sided test given by

$$\begin{aligned} \theta_0(d) \overset{d\to\infty}{\to} \to 0, \quad \theta_0(d) \overset{d\to 0}{\to} \to \frac{\pi^2}{4d^2} - \frac{1}{2}, \\ \widetilde{A}(d) \overset{d\to\infty}{\to} 1, \quad \widetilde{A}(d) \overset{d\to 0}{\to} \frac{16}{\pi^2\sqrt{2\pi}} d. \end{aligned} \tag{8.72}$$

Remark 8.9 *Note that the critical value for this test, that is, the decision boundary for a level of significance $\alpha = 5\%$, will produce a value $d^*(n)$, which is a function of the sample size n. Thankfully, the critical values of the test have been tablulted by Chicheportiche and Bouchaud (2012) as follows:*

Sample size n	10^3	10^4	10^5	10^6
Critical value $d^*(n)$	3.439	3.529	3.597	3.651

It should be noted that in general in OpRisk settings one would not be in the setting of large n sample size as earlier; therefore, it would be advisable to evaluate the law of $\mathbb{P}\mathrm{r}[D(a,b) \le d | a = 1, b = 0]$ exactly for small n. Niederhausen (1981, section 2) demonstrates how to calculate the null distribution of the weighted KS test for any sample size. This is achieved by utilizing the well-known generalized representation of this problem as a class of bivariate Renyi statistics, which, under suitable choices of parameters, can be molded into the modified variance-weighted KS test discussed earlier.

Then the evaluation of the tail probability for the p-values of this test is specified as a special member of a more general distribution given by the family of Renyi distributions. Given this Reyni family of distributions, one can write the probability of a particular tail event according to a system of differential equations (see discussion by Steinbrecher and Shaw 2008). However, in this particular context, the resulting boundary conditions are challenging to work with and typically one adopts a piecewise solution, which is presented with regard to a Sheffer polynomial recursive solution approach to obtain the p-value as specified by Niederhausen (1981, section 2, example 4 and table 1).

The tail-weighted KS GOF hypothesis test then proceeds as further detailed.

Algorithm 8.2 (Tail-Weighted Kolmogorov–Smirnov One-Sample Test)

1. ***Step 1***. *Set up suitable notation for the random variables and distributions being tested and make a statement of the null and alternative hypotheses in terms of population distribution/parameters. Determine hypothesis for GOF testing where null claims loss data are from a hypothesised distribution $F_0(x)$*

$$H_0 : F(x) = F_0(x), \ \forall x \tag{8.73}$$

versus an alternative claim that the observed losses are not realizations from F_0

$$H_A : F(x) \neq F_0(x), \ \forall x. \tag{8.74}$$

2. ***Step 2***. *State the test statistic and its observed value and when possible state the distribution of the test statistic or its approximation. Under the null hypothesis calculate the tail-weighted KS test statistic given by*

$$D_n = \sup_{a \leq F_0(x) \leq b} \frac{\left(\hat{F}_n(x) - F_0(x)\right)}{\sqrt{F_0(x)\left(1 - F_0(x)\right)}} \tag{8.75}$$

with the weight applied to interval $a \in]0, 1[$ and $b \in [a, 1[$ and where $\hat{F}_n$ is the empirical cumulative distribution defined according to

$$\hat{F}_n(x) = \frac{1}{n} \sum_{i=1}^{n} \mathbb{I}\left[x_i \leq x\right], \tag{8.76}$$

and the supremum occurs at one of the observed values x_i to the left of x_i. This procedure will produce an observed realization of the test statistic based on the observed data samples $\{x_i\}_{i=1}^{n}$ under the null hypothesis, denoted by d_n;

3. ***Step 3***. *State a formal mathematical expression for the p-value. Determine the p-value for the test under the null hypothesis given by considering*

$$p\text{-}value = \mathbb{P}\mathrm{r}\left[|D_n| \geq d_n | H_0\right].$$

To obtain the p-value one first needs to obtain an approximation of the distribution of the test statistic under the null. In this case, one can either use the large-sample results for the critical values, as a function of sample size for testing at $\alpha = 5\%$ *significance given in the remark before Equation (8.9) or perform a simulation estimation.*
Small-sample p-value evaluation. *If the sample size n is small, one can perform an evaluation of the p-value for making a decision on the test via the following simple simulation procedure, where* $\{X_i\}_{i=1}^n$ *are the samples from the experiment and* $j = 1, \ldots, J$ *is the index of the simulated test statistic realizations* $\left\{d_n^{(j)}\right\}$ *obtained by the following procedures:*

   - *Simulate a set of samples* $\left\{U_i^{(j)}\right\}_{i=1}^n$ *with* $U_i \sim Uniform(0,1)$ *that is, distribution* $F(u) = u$;
   - *Transform the samples* $\left\{U_i^{(j)}\right\}_{i=1}^n$ *to samples from the Null distribution* $X_i^{(j)} = F_0^{-1}\left(U_i^{(j)}\right)$;
   - *Construct the empirical distribution function* $\hat{F}^{(j)}$ *using the samples* $\left\{X_i^{(j)}\right\}_{i=1}^n$;
   - *Evaluate the realized test statistic*

$$d_n^{(j)} = \sup_{a \leq F_0(x) \leq b} \frac{\left(\hat{F}_n^{(j)}(x) - F_0(x)\right)}{\sqrt{F_0(x)\left(1 - F_0(x)\right)}} \tag{8.77}$$

   for each set of samples $\left\{U_i^{(j)}\right\}_{i=1}^n$ *to get* $d_n^{(j)}$. *Repeat many times* $j \in \{1, 2, \ldots, J\}$ *to get an estimate of the distribution for the test statistic under the null* D_n, *that is, the null distribution of the test statistic* D_n *is then approximated by the samples* $\left\{d_n^{(j)}\right\}$ *known by simulation.*

Given the empirical estimator for the distribution of the test statistic under the null, $\hat{F}_{D_n}(x)$, *use this to evaluate the p-value.*

4. ***Step 4***. *State the range of values within which the p-value falls (and a statement of how these are obtained. If the p-value is significantly lower than a given level of testing significance, typically 5%, then one has sufficient evidence from the observed loss data to reject the claim of the null hypothesis in favor of the alternative.*
5. ***Step 5***. *State the conclusion of the test in plain language (relevant to the experimental context).*

In the following example, we show the resulting estimated empirical distributions for the distribution of a KS and tail-weighted KS test statistic as a function of sample size.

EXAMPLE 8.3 Standard versus Tail-Weighted Kolmogorov–Smirnov Tests

Consider a GPD for the severity model with tail index (shape) parameter $k = 1$, scale parameter $\sigma = 1$, and threshold (location) parameter $\theta = 0$. This is a heavy-tailed loss model in the sense that the mean is not finite when $K \geq 1$ under this GPD model specification. We simulate loss data realizations $\{X_i\}_{i=1}^{n}$ for sample sizes of $n = 10, 20, 50, 100$. Then we evaluate the test statistic for a standard KS test and the tail-weighted KS test in the case that the nominal claimed distribution F_0 is the GPD with exact parameters $GPD(k = 1, \sigma = 1, \theta = 0)$—that is, no model or parameter misspecification. Using these nominal claim models, the test statistics distribution under each sample size is simulated via Monte Carlo and ploted in Figure 8.6.

It is clear from the results in the bottom subplot that when one accounts more for the tails of the distribution in deciding whether to reject the nominal claim or not (as with the tail-weighted KS test), one requires a larger number of samples to be able to reject the nominal claim as would be the case when the tails are not taken into account. This will be particularly the case for the heavy-tailed loss models. In Table 8.4, we also show the quantile (critical values) for the distribution of the KS test statistic as a function of sample size.

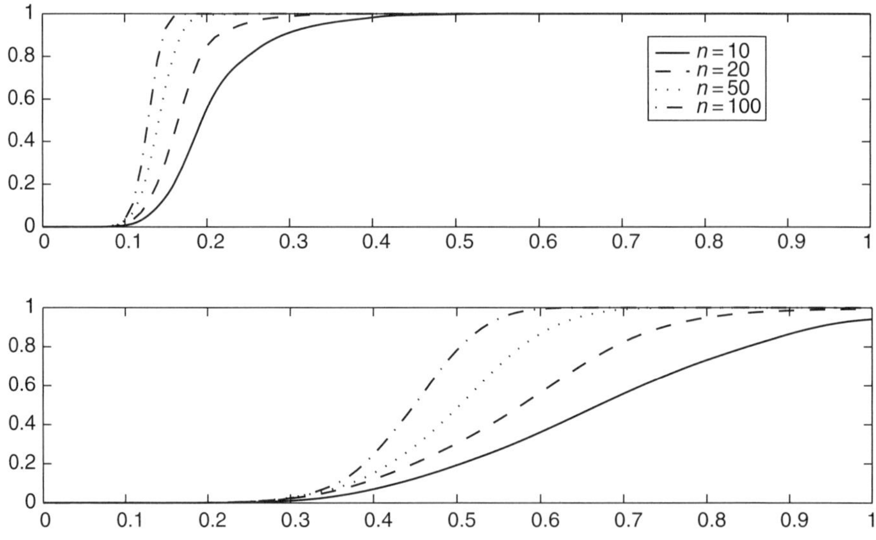

FIGURE 8.6 In each subplot, the distribution of the KS test statistic is displayed under the assumption that the null hypothesis is correct for data sizes $n = 10, 20, 50, 100$. The top subplot shows the distribution of the standard KS test statistic. The bottom subplot shows the distribution of the tail-weighted KS test statistic

TABLE 8.4 **Assessing the heavy tailed feature of loss data under a GOF test based on KS. True population loss model is $GPD(k=1, \sigma=1, \theta=0)$. Nominal claim is GPD with correct population parameters, test is performed at a number of sample sizes n and a range of significance levels $1-\alpha$. The top table presents the results from a standard KS test, the bottom table presents the results from a weighted KS test, with equal contribution from all quantiles in the distribution**

$H_0 : GPD(k = 1, \sigma = 1, \theta = 0)$

$1-\alpha$	$n = 10$	$n = 20$	$n = 50$	$n = 100$
80%	0.2444	0.1913	0.1590	0.1422
90%	0.2854	0.2094	0.1687	0.1470
95%	0.3327	0.2345	0.1750	0.1517
97.5%	0.3745	0.2710	0.1810	0.1542
99%	0.4017	0.2947	0.2037	0.1606
99.5%	0.4392	0.3198	0.2294	0.1668

$H_0 : GPD(k = 1, \sigma = 1, \theta = 0)$

$1-\alpha$	$n = 10$	$n = 20$	$n = 50$	$n = 100$
80%	0.8465	0.6821	0.5723	0.5029
90%	0.9097	0.7391	0.6101	0.5302
95%	1.0410	0.7922	0.6390	0.5488
97.5%	1.3434	0.8446	0.6662	0.5657
99%	1.7852	0.9097	0.6872	0.5843
99.5%	1.9195	1.0100	0.7066	0.6021

8.4.4 CRAMER-VON-MISES GOODNESS-OF-FIT TESTS AND WEIGHTED VARIANTS: TESTING IN THE PRESENCE OF HEAVY TAILS

The one-sample Cramer-von-Mises (CvM) tests are defined for observed loss realizations $x_1, x_2, \ldots, x_n$ of a set of continuous i.i.d. random loss variables $X_1, X_2, \ldots, X_n$, for which it is hypothesized that their sampling distribution function is F. To test this hypothesis one can consider the weighted 2-norm distributional test statistic given by,

$$Q = \int_{-\infty}^{\infty} w(x) \left(\hat{F}_n(x) - F(x;\boldsymbol{\theta})\right)^2 dF(x;\boldsymbol{\theta}) \tag{8.78}$$

for some weight function such as the quadratic CvM statistic when $w(x) = 1$ or the Anderson–Darling (AD) statistic when $w(x) = F(x;\boldsymbol{\theta})\,(1 - F(x;\boldsymbol{\theta}))^{-1}$. In this section, we present both the AD test and its tail-weighted variant, as well as the CvM test and its right tail-weighted variant.

8.4.4.1 Weighted Anderson–Darling Goodness-of-Fit Tests for Heavy Tails.

In the case of the AD form of the CvM test statistic Q, the test is performed according to the following stages in Algorithm 8.3. As it turns out two of the more important results one can derive for the asymptotic expansions of the right tail of the AD test statistic, under the nominal claim, are based on a result derived by Zolotarev (1961); we provide this result in Theorem 8.8.

Theorem 8.8 (Distribution of positively weighted quadratic Gaussian infinite sums) *Consider $Z_1, Z_2, \ldots$ as i.i.d. random variables with $Z_i \sim Normal(0,1)$. Then the limiting distribution of the positively weighted quadratic sequence given by*

$$Y_n = \sum_{j=1}^{n} \lambda_j Z_j^2 \sim F_n, \tag{8.79}$$

with $\lambda_i \geq 0$ for all $i \in \{1, 2, \ldots, n\}$ and decreasing $\lambda_1 > \lambda_2 > \cdots > \lambda_n$ for all n, satisfies $F_n \to F$ as $n \to \infty$ with the distribution right tail given explicitly by the product

$$1 - F(x) = \left[\prod_{i=2}^{\infty} \left(1 - \frac{\lambda_i}{\lambda_1}\right)^{-\frac{1}{2}} / \Gamma(1/2) \right] \left(\frac{x}{2\lambda_1}\right)^{-\frac{1}{2}} \exp\left(-\frac{x}{2\lambda_1}\right) [1 + \epsilon(x)]. \tag{8.80}$$

See details in Zolotarev (1961).

In Figure 8.7, the distribution for the random sum

$$Y_n = \sum_{j=1}^{n} \lambda_j Z_j^2 \sim F_n \tag{8.81}$$

is displayed as a function of sample size n with two different weight functions for $\{\lambda_i\}_{i=1}^{n}$ given by the following:

- Case 1: weight function $\lambda_j = \frac{1}{j(j+1)}$, top subplot;
- Case 2: weight function $\lambda_j = \frac{1}{j}$, bottom subplot.

It is clear that the weight function also has a strong influence on the asymptotic convergence rate as a function of the number of summand terms. This is clearly the reason why in the literature it is typically preferred to utilize a weight function such as Case 1, as the convergence even for small sample size to the asymptotic distribution is more rapid. In practice, it should be pointed out that there will be a direct relationship between the λ_i weights and the weight function $w(x)$ chosen in the GOF citerion. This small illustration shows that in general one can expect a large influence on the rate of convergence, and therefore the suitability of asymptotic results for the distribution of the test statistic under the null claim, for different choices of weight function. In other words, one should carefully consider the sample size before applying the asymptotic results for the tail of the distribution of the test statistic under the nominal claim. In cases where it is suspected that such an asymptotic may not be utilized accurately, a closed-form expression for the finite sample test statistic distribution under the null will be noted and in other cases, it is also possible to adopt other approximations such as the saddle point. If all these cases fail, then one can resort to numerical procedures to reconstruct the distribution of the test statistic numerically.

One can then utilize this result to obtain the asymptotic behavior of the distribution of the AD test.

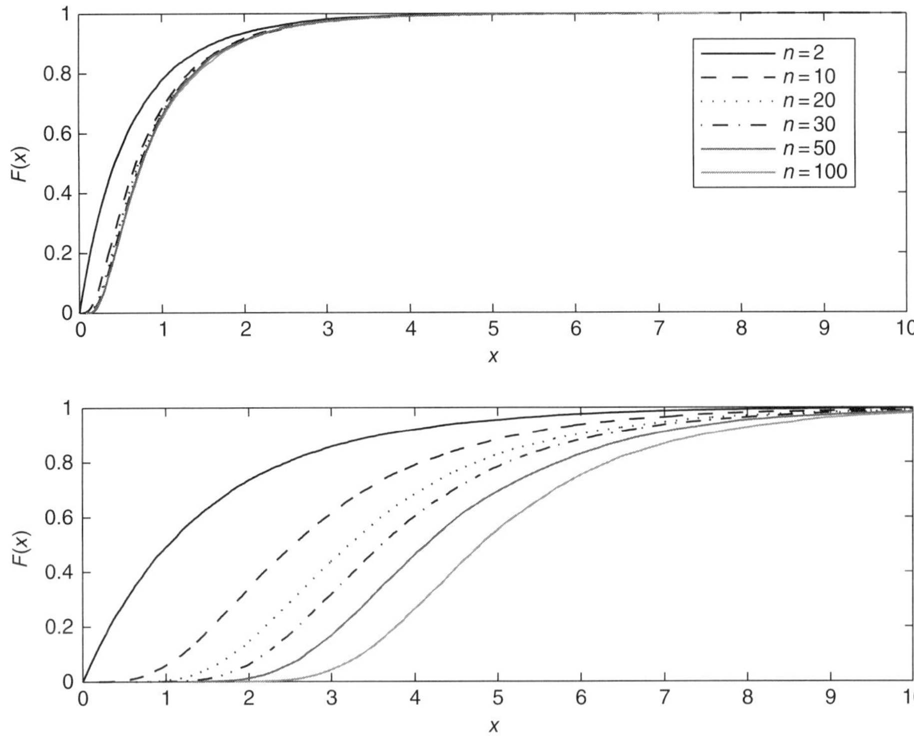

FIGURE 8.7 Distribution of the test statistic as a function of the number of random variables $\{\lambda_i\}_{i=1}^{n}$ for $n \in \{2, 10, 20, 30, 50, 100\}$. Top subplot is for weight function Case 1 and the bottom subplot is for weight function Case 2. (see insert for color representation of the figure.)

Proposition 8.3 (Characterizing the Anderson–Darling null distribution) *Consider a null hypothesis for the data-generating distribution F_0, which one assumes is a continous distribution. Let $X_1, X_2, \ldots, X_n$ be a sequence of i.i.d. random variables with distribution F_0. Then the following holds:*

1. ***Test statistic.** The test statistic is given by*

$$Q_n = n \int_{-\infty}^{\infty} \frac{\left(\hat{F}_n(x) - F(x)\right)^2}{F(x)\left(1 - F(x)\right)} dF(x)$$
$$= -n - \sum_{i=1}^{n} \frac{2i-1}{n} \left(\ln\left(F\left(X_{(i,n)}\right)\right) + \ln\left(1 - F\left(X_{(n+1-i,n)}\right)\right)\right), \tag{8.82}$$

which is constructed using the order statistics $X_{(1,n)} \leq X_{(2,n)} \leq \cdots \leq X_{(n,n)}$ and can be shown to be strictly contained in the interval $Q \in [0, 8]$ assymptotically for large samples n (see Lewis 1961);

2. ***Characteristic function****. The characteristic function of the AD test statistic Q for large samples $n \to \infty$ is given by*

$$\phi(t) = \left[\frac{-2\pi \mathrm{i} t}{\cos\left(\frac{\pi}{2\sqrt{1+8\mathrm{i}t}}\right)} \right]^{\frac{1}{2}}, \tag{8.83}$$

which one can then observe is equivalent to the characteristic function of an infinite weighted sum of independent chi-squared random variables with positive weights

$$\lambda_i = \frac{1}{j(j+1)}. \tag{8.84}$$

This means one can use the result of Zolotarev (1986) given in Theorem 8.8 to obtain the tail distribution of the AD test statistic;

3. ***Large-sample asymptotic tail expansion****. An asymptotic expansion of the large-sample right tail of the AD test statistic under the null is attainable in closed form. If the sampe size $n \to \infty$, then the upper right tail of the distribution $\sqrt{n}Q_n$ is asymptotically approximated as $x \to \infty$ by*

$$\mathbb{P}\mathrm{r}\left(Q > x\right) = \left[\prod_{i=2}^{\infty} \left(1 - \frac{\lambda_i}{\lambda_1}\right)^{-\frac{1}{2}} / \Gamma(1/2) \right] \left(\frac{x}{2\lambda_1}\right)^{-\frac{1}{2}} \exp\left(-\frac{x}{2\lambda_1}\right) \left[1 + \epsilon(x)\right], \tag{8.85}$$

where $\lambda_i = \frac{1}{i(i+1)}$ and $\epsilon(x) \to 0$ (see Sinclair and Spurr 1988);

4. ***Cumulant-generating function****. The cummulant-generating function of the distribution for Q can be obtained in closed form according to*

$$\kappa(t) = -\frac{1}{2} \sum_{j=1}^{\infty} \ln\left(1 - 2\lambda_j t\right). \tag{8.86}$$

5. ***Saddle point approximation****. One can then obtain a saddle point approximation of the distribution for the AD test statistic via the derivatives of the cummulants when they exist and are analytic, which are given by*

$$\begin{aligned} \frac{d}{dt}\kappa(t) &= \sum_{j=1}^{\infty} \frac{\lambda_j}{(1 - 2\lambda_j t)}, \\ \frac{d^2}{dt^2}\kappa(t) &= \sum_{j=1}^{\infty} \left[\frac{\lambda_j}{(1 - 2\lambda_j t)} \right]^2, \end{aligned} \tag{8.87}$$

giving a saddle point approximation for the right tail at location x according to

$$\mathbb{P}\mathrm{r}\left(Q > x\right) = 1 - \Phi\left(\hat{w}\right) + \phi\left(\hat{w}\right)\left[\hat{u}^{-1} - \hat{w}^{-1}\right] \tag{8.88}$$

with $\hat{t}$ the unique and existing solution to the inverse problem $\frac{d}{dt}k(t) = x$ and

$$\begin{aligned} \hat{w} &= \sqrt{2\left[\hat{t}x - \kappa\left(\hat{t}\right)\right]} sgn(\hat{t}) \\ \hat{u} &= \hat{t}\left\{\frac{d^2}{dt^2}\kappa(\hat{t})\right\}^{\frac{1}{2}}. \end{aligned} \tag{8.89}$$

See details in Giles (2001), Daniels (1954), and Lugannani and Rice (1980).

These details are then sufficient to perform the AD GOF hypothesis test, which is summarized in the following algorithm.

Algorithm 8.3 (Anderson–Darling One-Sample Test)

- ***Step 1.*** *Set up suitable notation for the random variables and distributions being tested and make a statement of the null and alternative hypotheses in terms of population distribution/parameters. Determine hypothesis for GOF testing where null claims loss data are from a hypothesized distribution $F_0(x)$*

$$H_0 : F(x) = F_0(x), \ \forall x \tag{8.90}$$

versus an alternative claim that the observed losses are not realizations from F_0

$$H_A : F(x) \neq F_0(x), \ \forall x. \tag{8.91}$$

- ***Step 2.*** *State the test statistic and its observed value and when possible state the distribution of the test statistic or its approximation. Under the null hypothesis calculate the AD test statistic given by*

$$\begin{aligned} Q_n &= n \int_{-\infty}^{\infty} \frac{\left(\hat{F}_n(x) - F(x)\right)^2}{F(x)\left(1 - F(x)\right)} dF(x) \\ &= -n - \sum_{i=1}^{n} \frac{2i-1}{n}\left(\ln\left(F\left(X_{(i,n)}\right)\right) + \ln\left(1 - F\left(X_{(n+1-i,n)}\right)\right)\right), \end{aligned} \tag{8.92}$$

which is constructed using the order statistics $X_{(1,n)} \leq X_{(2,n)} \leq \cdots \leq X_{(n,n)}$ and $\hat{F}_n$ is the empirical cumulative distribution defined according to

$$\hat{F}_n(x) = \frac{1}{n}\sum_{i=1}^{n} \mathbb{I}\left[x_i \leq x\right]. \tag{8.93}$$

This procedure will produce an observed realization of the test statistic based on the observed data samples $\{x_i\}_{i=1}^{n}$ under the null hypothesis, denoted by d_n;

- ***Step 3.*** *State a formal mathematical expression for the p-value. Determine the p-value for the test under the null hypothesis given by considering*

$$p\text{-}value = \mathbb{P}\mathrm{r}\left[|Q_n| \geq q_n | H_0\right].$$

To obtain the p-value one first needs to obtain an approximation of the distribution of the test statistic under the null. This can be done in two cases, depending on the size of the sample:
Small-sample p-value evaluation*. If the sample size n is small, one can perform evaluation of the p-value for making a decision on the test via the following simple simulation procedure, where* $\{X_i\}_{i=1}^n$ *are the samples from the experiment and* $j = 1, \ldots, J$ *is the index of the simulated test statistic realizations* $\left\{q_n^{(j)}\right\}$ *obtained by the following procedures:*

- *Simulate a set of samples* $\left\{U_i^{(j)}\right\}_{i=1}^n$ *with* $U_i \sim \text{Uniform}(0,1)$*, that is, distribution* $F(u) = u$*;*
- *Transform the samples* $\left\{U_i^{(j)}\right\}_{i=1}^n$ *to samples from the null distribution* $X_i^{(j)} = F_0^{-1}\left(U_i^{(j)}\right)$*;*
- *Evaluate for each set of samples* $\left\{X_i^{(j)}\right\}_{i=1}^n$ *the test statistic*

$$q_n^{(j)} = -n - \sum_{i=1}^{n} \frac{2i-1}{n}\left(\ln\left(F\left(X_{(i,n)}^{(j)}\right)\right) + \ln\left(1 - F\left(X_{(n+1-i,n)}^{(j)}\right)\right)\right). \quad (8.94)$$

Repeat many times $j \in \{1, 2, \ldots, J\}$ *to get an estimate of the distribution for the test statistic under the null* Q_n*, that is, the null distribution of the test statistic* Q_n *is then approximated by the samples* $\left\{q_n^{(j)}\right\}$ *known by simulation.*

Given the empirical estimator for the distribution of the test statistic under the null, $\hat{F}_{Q_n}(x)$*, use this to evaluate the p-value.*
Large-sample p-value evaluation*. If the sample size n is large, one can perform evaluation of the p-value for making a decision on the test via the asymptotic expansion of the large-sample right tail of the AD test statistic under the null; whereas, if the sample size* $n \to \infty$*, then the upper right tail of the distribution* $\sqrt{n}Q_n$ *is asymptotically approximated as* $x \to \infty$ *by*

$$\mathbb{P}\mathrm{r}(Q > x) = \left[\prod_{i=2}^{\infty} \frac{1}{\Gamma(1/2)}\left(1 - \frac{\lambda_i}{\lambda_1}\right)^{-\frac{1}{2}}\right]\left(\frac{x}{2\lambda_1}\right)^{-\frac{1}{2}} \exp\left(-\frac{x}{2\lambda_1}\right)[1 + \epsilon(x)], \quad (8.95)$$

where $\lambda_i = \frac{1}{i(i+1)}$ *and* $\epsilon(x) \to 0$*;*

- ***Step 4****. State the range of values within which the p-value falls (and a statement of how these are obtained. If the p-value is significantly lower than a given level of testing significance, typically 5%, then one has sufficient evidence from the observed loss data to reject the claim of the null hypothesis in favor of the alternative;*
- ***Step 5****. State the conclusion of the test in plain language (relevant to the experimental context).*

In the following example, a small case study illustrating the properties of the standard AD test is illustrated for a few simple distribution models in OpRisk.

EXAMPLE 8.4 Anderson–Darling GOF Test Example

A sample of $J = 1000$ random numbers for a Normal, double exponential, Cauchy, and LogNormal distribution was considered. In each case, the AD test was utilized to see if the data had come from a model with exponential tail decay, in this simple example a Gaussian distribution. In this case, the nominal and alternative claims were given by the following:

- H_0: the data are Gaussian distributed;
- H_A: the data are not Gaussian distributed.

In the case of the Gaussian sample, the estimated test statistic is given by $q_{500} = 0.2576$; in the case of the double exponential sample, the estimated test statistic is given by $q_{500} = 5.8492$; in the case of the Cauchy sample, the estimated test statistic is given by $q_{500} = 288.7863$; in the case of the Cauchy sample, the estimated test statistic is given by $q_{500} = 83.3935$. In this case, when looking at the GOF test, at a significance level of $\alpha = 0.05$, the resulting critical value for the test under the null is given by 0.75 so that the nominal claim is rejected if the test statistic exceeds this critical value. ■

There have also been versions of the AD test developed to tackle situations in which the tails of the distribution of the observed sample are also given more importance. For example, Sinclair *et al.* (1990) developed a modification to the AD test statistic to obtain two test statistics for testing suitablity of the null claim about the upper or the lower tails. We will focus here on the upper tail test, which is of most relevance for OpRisk settings (see Proposition 8.4). In particular, such a test will be directly of interest for those wishing to assess the validity of a nominal claim relating to the presence of a particular tail behavior believed to be present in the data.

Proposition 8.4 (Upper tail modified Anderson–Darling test for heavy tails) *Consider a null hypothesis for the data-generating distribution F_0, which one assumes is a continous distribution. Let $X_1, X_2, \ldots, X_n$ be a sequence of i.i.d. random variables with distribution F_0. Then the following holds:*

1. ***Upper tail modified test statistic.*** *The modified AD test statistic given below will give heavier weight to the upper tail compared to the classical AD test statistic. The evaluation of the statistic for n samples involves*

$$\begin{aligned}\tilde{Q}_n &= n \int_{-\infty}^{\infty} \frac{\left(\hat{F}(x) - F(x)\right)^2}{(1 - F(x))} dF(x) \\ &= \frac{n}{2} - 2\sum_{i=1}^{n} F\left(X_{(i,n)}\right) - \sum_{i=1}^{n}\left[2 - \frac{2i-1}{n}\right] \ln\left(1 - F\left(X_{(i,n)}\right)\right),\end{aligned} \tag{8.96}$$

which is constructed using the order statistics $X_{(1,n)} \leq X_{(2,n)} \leq \cdots \leq X_{(n,n)}$ (see Sinclair et al. 1990, equation 2.5);

2. ***Modified characteristic function***. *The characteristic function of this modified AD test statistic $\tilde{Q}$ for large samples $n \rightarrow \infty$ is given by*

$$\phi(t) = \left[\frac{\sqrt{2it}}{J_1\left(2\sqrt{2it}\right)}\right]^{\frac{1}{2}}, \tag{8.97}$$

which, as noted by Sinclair et al. (1990), can be written in terms of an infinite number of eigen values λ_i, where each is obtained as a real solution to $J_1\left(2\sqrt{\lambda}\right) = 0$ for $\lambda \neq 0$, that is, the roots of the Bessel function. These eigenvalues are denoted by Abramowitz and Stegun (1965, section 9.5, p. 370) according to $j_{1,1} < j_{1,2} < \cdots < j_{1,i} < \ldots$ giving eigenvalues

$$\lambda_i = \frac{j_{1,i}^2}{4}. \tag{8.98}$$

Using these eigenvalues one can re-express the characteristic function of the modified test statistic according to the equivalent form of an infinite weighted sum of independent χ-squared random variables with one degree of freedom according to

$$\phi(t) = \prod_{j=1}^{\infty}\left[1 - \frac{2it}{\lambda_j}\right]^{-\frac{1}{2}}, \tag{8.99}$$

where the weights are given by the inverse of the eigen values λ_i^{-1};

3. ***Large-sample asymptotic tail expansion***. *The distribution of the large sample $n \rightarrow \infty$ as $\tilde{Q}_n \rightarrow \tilde{Q}$ is given by Sinclair et al. (1990) and MacNeill (1974) according to*

$$\mathbb{P}\mathrm{r}\left(Q > x\right) = \left[\prod_{i=1}^{\infty}\frac{1}{\Gamma(\frac{1}{2})}\left(1 - \frac{\lambda_i}{\lambda_1}\right)^{-\frac{1}{2}}\right]\left(\frac{x}{2\lambda_1}\right)^{-\frac{1}{2}}\exp\left(-\frac{x}{2\lambda_1}\right)\left[1 + \epsilon(x)\right], \tag{8.100}$$

with $\epsilon(x) \rightarrow 0$ as $x \rightarrow \infty$.

Remark 8.10 *It is interesting to note that the asymptotic result for the distribution tail for the upper tail modified AD test takes the same basic structural form as the same expression obtained for the standard AD GOF test distribution with no modification to the right tail. Of course, the key component that differentiates the two cases and changes the resulting p-values accordingly is the different expressions for the resulting eigenvalues λ_i used in the representation of the characteristic function.*

These details are then sufficient to perform the AD GOF hypothesis test, which is summarized in the following algorithm.

Algorithm 8.4 (Tail-weighted Anderson–Darling One-Sample Test)

- ***Step 1***. *Set up suitable notation for the random variables and distributions being tested and make a statement of the null and alternative hypotheses in terms of population* *distribution/parameters. Determine hypothesis for GOF testing where null claims loss data are from a hypothesized distribution* $F_0(x)$

$$H_0 : F(x) = F_0(x), \ \forall x \tag{8.101}$$

versus an alternative claim that the observed losses are not realizations from F_0

$$H_A : F(x) \neq F_0(x), \ \forall x. \tag{8.102}$$

- ***Step 2***. *State the test statistic and its observed value and when possible state the distribution of the test statistic or its approximation. Under the null hypothesis calculate the tail-weighted AD test statistic given for n samples involving*

$$\tilde{Q}_n = \frac{n}{2} - 2\sum_{i=1}^{n} F\left(X_{(i,n)}\right) - \sum_{i=1}^{n}\left[2 - \frac{2i-1}{n}\right] \ln\left(1 - F\left(X_{(i,n)}\right)\right) \tag{8.103}$$

which is constructed using the order statistics $X_{(1,n)} \leq X_{(2,n)} \leq \cdots \leq X_{(n,n)}$. *This procedure will produce an observed realization of the test statistic based on the observed data samples* $\{x_i\}_{i=1}^{n}$ *under the null hypothesis, denoted by* q_n;
- ***Step 3***. *State a formal mathematical expression for the p-value. Determine the p-value for the test under the null hypothesis given by considering*

$$p\text{-}value = \mathbb{P}\text{r}\left[|Q_n| \geq q_n | H_0\right].$$

To obtain the p-value one first needs to obtain an approximation of the distribution of the test statistic under the null. This can be done in two cases, depending on the size of the sample: ***Small-sample p-value evaluation***. *If the sample size n is small, one can perform evaluation of the p-value for making a decision on the test via the following simple simulation procedure, where* $\{X_i\}_{i=1}^{n}$ *are the samples from the experiment and* $j = 1, \ldots, J$ *is the index of the simulated test statistic realizations* $\left\{q_n^{(j)}\right\}$ *obtained by the following procedures:*

- *Simulate a set of samples* $\left\{U_i^{(j)}\right\}_{i=1}^{n}$ *with* $U_i \sim Uniform(0,1)$, *that is, distribution* $F(u) = u$;
- *Transform the samples* $\left\{U_i^{(j)}\right\}_{i=1}^{n}$ *to samples from the null distribution* $X_i^{(j)} = F_0^{-1}\left(U_i^{(j)}\right)$;
- *Evaluate for each set of samples* $\left\{X_i^{(j)}\right\}_{i=1}^{n}$ *the test statistic*

$$q_n^{(j)} = \frac{n}{2} - 2\sum_{i=1}^{n} F\left(X_{(i,n)}^{(j)}\right) - \sum_{i=1}^{n}\left[2 - \frac{2i-1}{n}\right] \ln\left(1 - F\left(X_{(i,n)}^{(j)}\right)\right). \tag{8.104}$$

Repeat many times $j \in \{1, 2, \ldots, J\}$ *to get an estimate of the distribution for the test statistic under the null* Q_n *that is, the null distribution of the test statistic* Q_n *is then approximated by the samples* $\left\{q_n^{(j)}\right\}$ *known by simulation.*

Given the empirical estimator for the distribution of the test statistic under the null, $\hat{F}_{Q_n}(x)$, use this to evaluate the p-value.

Large-sample p-value evaluation*. If the sample size n is large, one can perform evaluation of the p-value for making a decision on the test via the asymptotic expansion of the large-sample right tail of the AD test statistic under the null; whereas, if the sample size $n \to \infty$, then the upper right tail of the distribution $\sqrt{n}Q_n$ is asymptotically approximated as $x \to \infty$ by*

$$\mathbb{P}\mathrm{r}(Q > x) = \left[\prod_{i=2}^{\infty} \frac{1}{\Gamma(1/2)}\left(1 - \frac{\lambda_i}{\lambda_1}\right)^{-\frac{1}{2}}\right]\left(\frac{x}{2\lambda_1}\right)^{-\frac{1}{2}} \exp\left(-\frac{x}{2\lambda_1}\right)[1 + \epsilon(x)], \tag{8.105}$$

with $\epsilon(x) \to 0$ as $x \to \infty$;

- ***Step 4****. State the range of values within which the p-value falls (and a statement of how these are obtained. If the p-value is significantly lower than a given level of testing significance, typically 5%, then one has sufficient evidence from the observed loss data to reject the claim of the null hypothesis in favor of the alternative;*
- ***Step 5****. State the conclusion of the test in plain language (relevant to the experimental context).*

8.4.4.2 Weighted Cramer-von-Mises Goodness-of-Fit Tests for Heavy Tails.

An alternative test one may consider is the CvM test (see details in Csorgo and Faraway 1996 and Brown 1982). The test statistic considered in the CvM test is given in Proposition 8.5 and is based on the CvM family of test statistics given by,

$$Q = n \int_{-\infty}^{\infty} w(x)\left(\hat{F}(x) - \hat{F}(x;\boldsymbol{\theta})\right)^2 dF(x;\boldsymbol{\theta}), \tag{8.106}$$

where $w(x) = 1$.

Proposition 8.5 (Characterizing the Cramer-von-Mises null distribution) *Consider a null hypothesis for the data-generating distribution F_0, which one assumes is a continous distribution. Let $X_1, X_2, \ldots, X_n$ be a sequence of i.i.d. random variables with distribution F_0. Then the following holds*

1. ***Test statistic****. The test statistic is given by*

$$\begin{aligned} Q_n &= n \int_{-\infty}^{\infty} \left[\hat{F}(x) - F(x)\right]^2 dF(x) \\ &= n\omega^2 = \frac{1}{12n} + \sum_{i=1}^{n}\left[\frac{2i-1}{2n} - F\left(X_{(i,n)}\right)\right]^2, \end{aligned} \tag{8.107}$$

which is constructed using the order statistics $X_{(1,n)} \leq X_{(2,n)} \leq \cdots \leq X_{(n,n)}$ and can be shown to be strictly contained in the interval $Q \in \left[\frac{1}{12n}, \frac{n}{3}\right]$ (see Csorgo and Faraway 1996);

2. ***Characteristic function***. *The characteristic function of the AD test statistic Q for large samples $n \to \infty$ is given by*

$$\phi(t) = \left[\frac{(-2\pi i t)^{\frac{1}{2}}}{\sinh(-2it)^{\frac{1}{2}}}\right]^{\frac{1}{2}} \tag{8.108}$$

(see discussions by Mises 1947 and Smirnov 1936);

3. ***Large-sample distribution***. *If the sampe size $n \to \infty$, then the distribution $\sqrt{n}Q_n$ is asymptotically given by inversion of the characteristic function, as first performed by Smirnov (1936) and detailed by Csorgo and Faraway (1996, equation 1.3) to produce, for $x \geq 0$,*

$$\mathbb{P}\mathrm{r}(Q \leq x) = \frac{1}{\pi^{\frac{3}{2}} x^{\frac{1}{2}}} \sum_{k=0}^{\infty} \frac{\Gamma\left(k+\frac{1}{2}\right)}{k!}(4k+1)^{\frac{1}{2}} \exp\left(-\frac{(4k+1)^2}{16x}\right) K_{\frac{1}{4}}\left(\frac{4k+1}{2x^{\frac{1}{2}}}\right) \tag{8.109}$$

for $K_\nu(x)$, the modified Bessel function of the third kind. Evaluation of finite sample approximations of this distribution has been studied by Götze (1979) and summarized by Csorgo and Faraway (1996);

4. *The result by Prokhorov (1968, theorem 1) is utilized by Csorgo and Faraway (1996) to illustrate that one may obtain a bound for the right tail of the finite sample distribution for $Q_n \sim F_{Q_n}(x)$ given for any sample size $n \geq 1$ by*

$$\sup\{1 - F_{Q_n}(x)\} \leq C\exp(-Kx) \tag{8.110}$$

with $x \geq \frac{8}{\pi^2}$, $C = 1 + \frac{1}{\pi\sqrt{2}}\exp\left(\frac{5}{12}\right)$, and $K = \frac{3}{32}\exp(-2)$. Knott (1974) and later Csorgo and Faraway (1996) showed that one can evaluate the distribution for the small-sample n distribution of the CvM statistic in a bounded range according to

$$F_{Q_n}(x) = \frac{n!\pi^{\frac{n}{2}}}{\Gamma\left(\frac{n}{2}+1\right)}\left(x - \frac{1}{12n}\right)^{\frac{n}{2}} \tag{8.111}$$

for $x \in \left[\frac{1}{12n}, \frac{n+3}{12n^2}\right]$.

These details are then sufficient to perform the CvM GOF hypothesis test, which is summarized in the following algorithm.

Algorithm 8.5 (Tail-Weighted Cramers-Von-Mise-One-Sample Test)

- ***Step 1***. *Set up suitable notation for the random variables and distributions being tested and make a statement of the null and alternative hypotheses in terms of population distribution/parameters. Determine hypothesis for GOF testing where null claims loss data are from a hypothesized distribution $F_0(x)$*

$$H_0 : F(x) = F_0(x),\ \forall x \tag{8.112}$$

versus an alternative claim that the observed losses are not realizations from F_0

$$H_A : F(x) \neq F_0(x), \ \forall x. \tag{8.113}$$

- ***Step 2***. *State the test statistic and its observed value and when possible state the distribution of the test statistic or its approximation. Under the null hypothesis calculate the tail-weighted CvM test statistic given for n samples involving*

$$Q_n = n\omega^2 = \frac{1}{12n} + \sum_{i=1}^{n} \left[\frac{2i-1}{2n} - F\left(X_{(i,n)}\right) \right]^2, \tag{8.114}$$

which is constructed using the order statistics $X_{(1,n)} \leq X_{(2,n)} \leq \cdots \leq X_{(n,n)}$. *This procedure will produce an observed realization of the test statistic based on the observed data samples* $\{x_i\}_{i=1}^{n}$ *under the null hypothesis, denoted by* q_n*;*

- ***Step 3***. *State a formal mathematical expression for the p-value. Determine the p-value for the test under the null hypothesis given by considering*

$$p\text{-}value = \mathbb{P}\mathrm{r}\left[|Q_n| \geq q_n | H_0\right].$$

To obtain the p-value one first needs to obtain an approximation of the distribution of the test statistic under the null. This can be done in two cases, depending on the size of the sample: ***Small-sample p-value evaluation***. *If the sample size n is small, one can perform evaluation of the p-value for making a decision on the test via either first using the following simple simulation procedure, where* $\{X_i\}_{i=1}^{n}$ *are the samples from the experiment and* $j = 1, \ldots, J$ *is the index of the simulated test statistic realizations* $\left\{q_n^{(j)}\right\}$ *obtained by the following procedures:*

  - *Simulate a set of samples* $\left\{U_i^{(j)}\right\}_{i=1}^{n}$ *with* $U_i \sim$ *Uniform*$(0,1)$ *that is, distribution* $F(u) = u$*;*
  - *Transform the samples* $\left\{U_i^{(j)}\right\}_{i=1}^{n}$ *to samples from the null distribution* $X_i^{(j)} = F_0^{-1}\left(U_i^{(j)}\right)$*;*
  - *Evaluate for each set of samples* $\left\{X_i^{(j)}\right\}_{i=1}^{n}$ *the test statistic*

$$q_n^{(j)} = \frac{n}{2} - 2\sum_{i=1}^{n} F\left(X_{(i,n)}^{(j)}\right) - \sum_{i=1}^{n}\left[2 - \frac{2i-1}{n}\right] \ln\left(1 - F\left(X_{(i,n)}^{(j)}\right)\right). \tag{8.115}$$

  Repeat many times $j \in \{1, 2, \ldots, J\}$ *to get an estimate of the distribution for the test statistic under the null* Q_n *that is, the null distribution of the test statistic* Q_n *is then approximated by the samples* $\left\{q_n^{(j)}\right\}$ *known by simulation.*

Alternatively, if the smaple size is appropriate for the quantile of the test statistic distribution to fall in the interval for $\left[\frac{1}{12n}, \frac{n+3}{12n^2}\right]$ *at the desired level of significance, then the p-value can be obtained using the representation*

$$F_{Q_n}(x) = \frac{n!\pi^{\frac{n}{2}}}{\Gamma\left(\frac{n}{2}+1\right)}\left(x - \frac{1}{12n}\right)^{\frac{n}{2}}. \tag{8.116}$$

Given the empirical estimator for the distribution of the test statistic under the null, $\hat{F}_{Q_n}(x)$, use this to evaluate the p-value.

Large-sample p-value evaluation. *If the sample size n is large, one can perform evaluation of the p-value for making a decision on the test via the asymptotic expansion of the large-sample right tail of the tail-weighted CvM test statistic under the null; whereas if the sample size $n \to \infty$, then the upper right tail of the distribution $\sqrt{n}Q_n$ is asymptotically approximated as $x \to \infty$ by*

$$\mathbb{P}\mathrm{r}\left(Q \leq x\right) = \frac{1}{\pi^{\frac{3}{2}} x^{\frac{1}{2}}} \sum_{k=0}^{\infty} \frac{\Gamma\left(k+\frac{1}{2}\right)}{k!} (4k+1)^{\frac{1}{2}} \exp\left(-\frac{(4k+1)^2}{16x}\right) K_{\frac{1}{4}}\left(\frac{4k+1}{2x^{\frac{1}{2}}}\right) \tag{8.117}$$

for $K_\nu(x)$, the modified Bessel function of the third kind;

- ***Step 4.*** *State the range of values within which the p-value falls (and a statement of how these are obtained. If the p-value is significantly lower than a given level of testing significance, typically 5%, then one has sufficient evidence from the observed loss data to reject the claim of the null hypothesis in favor of the alternative;*
- ***Step 5.*** *State the conclusion of the test in plain language (relevant to the experimental context).*

We finish this section on model selection for components of an LDA model structure in a single risk by discussing briefly how one may undertake Bayesian model selection.

8.5 Bayesian Model Selection

Consider a model M with parameter vector $\boldsymbol{\theta}$. The model likelihood with data $\boldsymbol{x}$ can be found by integrating out the parameter $\boldsymbol{\theta}$

$$\pi(\boldsymbol{x}|M) = \int \pi(\boldsymbol{x}|\boldsymbol{\theta}, M)\pi(\boldsymbol{\theta}|M)d\boldsymbol{\theta}, \tag{8.118}$$

where $\pi(\boldsymbol{\theta}|M)$ is the prior density of $\boldsymbol{\theta}$ in the model indexed by M. Given a set of K competing models $(M_1, \ldots, M_K)$ with parameters $\boldsymbol{\theta}_{[1]}, \ldots, \boldsymbol{\theta}_{[K]}$ respectively, the Bayesian alternative to traditional hypothesis testing is to evaluate and compare the posterior probability ratio between the models. Assuming we have some prior knowledge about the model probability $\pi(M_i)$, we can compute the posterior probabilities for all models using the model likelihoods

$$\pi(M_i|\boldsymbol{x}) = \frac{\pi(\boldsymbol{x}|M_i)\ \pi(M_i)}{\sum_{k=1}^{K} \pi(\boldsymbol{x}|M_k)\ \pi(M_k)}. \tag{8.119}$$

Consider two competing models M_1 and M_2, parameterized by $\boldsymbol{\theta}_{[1]}$ and $\boldsymbol{\theta}_{[2]}$, respectively. The choice between the two models can be based on the posterior model probability ratio, given by

$$\frac{\pi(M_1|\boldsymbol{x})}{\pi(M_2|\boldsymbol{x})} = \frac{\pi(\boldsymbol{x}|M_1)\ \pi(M_1)}{\pi(\boldsymbol{y}|M_2)\ \pi(M_2)} = \frac{\pi(M_1)}{\pi(M_2)} B_{12}, \tag{8.120}$$

where $B_{12} = \pi(\boldsymbol{x}|M_1)/\pi(\boldsymbol{x}|M_2)$ is the Bayes factor, the ratio of the posterior odds of model M_1 to that of model M_2. As shown by Lavin and Scherrish (1999), an accurate interpretation of the Bayes factor is that the ratio B_{12} captures the change of the odds in favor of model M_1 as we move from the prior to the posterior. Jeffreys (1961) recommended a scale of evidence for interpreting the Bayes factor, which was later modified by Wasserman (1997). A Bayes factor $B_{12} > 10$ is considered strong evidence in favor of M_1. Kass and Raftery (1995) give a detailed review of the Bayes factor.

Typically, the integral (8.118) required by the Bayes factor is not analytically tractable, and sampling-based methods must be used to obtain estimates of the model likelihoods. There are quite a few methods in the literature for direct computation of the Bayes factor or indirect construction of the Bayesian model selection criterion, both based on Markov chain Monto Carlo (MCMC) outputs. The popular methods are direct estimation of the model likelihood, thus the Bayes factor; indirect calculation of an asymptotic approximation as the model selection criterion; and direct computation of the posterior model probabilities, as discussed later. Popular model selection criteria, based on simplifying approximations, include DIC and Bayesian information criterion BIC; see, e.g., Robert (2001, chapter 7).

In general, given a set of possible models $(M_1, \ldots, M_K)$, the model uncertainty can be incorporated in the Bayesian framework by considering the joint posterior for the model and the model parameters $\pi(M_k, \boldsymbol{\theta}_{[k]}|\boldsymbol{x})$, where $\boldsymbol{\theta}_{[k]}$ is a vector of parameters for model k. Subsequently, calculated posterior model probabilities $\pi(M_k|\boldsymbol{x})$ can be used to select an optimal model as the model with the largest probability or average over possible models according to the full joint posterior.

Accurate estimation of the required posterior distributions usually involves the development of a Reversible Jump MCMC framework. This type of Markov chain sampler is complicated to develop and analyze. It goes beyond the scope of this book but interested readers can find details in Green (1995). In the case of a small number of models, Congdon (2006) suggests running a standard MCMC (e.g., Random Walk Metropolis Hastings (RW-MH)) for each model separately and using the obtained MCMC samples to estimate $\pi(M_k|\boldsymbol{x})$. Peters *et al.* (2009a) adopted this method for modeling claims-reserving problem in the insurance literature with an appropriate modification. They used the following modified version for the special case of nested models and utilized the Markov chain results for each model, in the case of equiprobable nested models, and calculated the posterior model probabilities $\pi(M_i|\boldsymbol{x})$ as

$$\pi(M_i|\boldsymbol{x}) = \frac{1}{L}\sum_{l=1}^{L} \frac{f\left(\boldsymbol{x}|M_i, \boldsymbol{\theta}_{[i]}^{(l)}\right)}{\sum_{j=1}^{K} f\left(\boldsymbol{x}|M_j, \boldsymbol{\theta}_{[j]}^{(l)}\right)}, \tag{8.121}$$

where $\boldsymbol{\theta}_{[i]}^{(l)}$ is the MCMC posterior sample at Markov chain step l for model M_i, $f(\boldsymbol{x}|M_i, \boldsymbol{\theta}_{[i]}^{(l)})$ is the joint density of the data $\boldsymbol{x}$ given the parameter vector $\boldsymbol{\theta}_{[i]}^{(l)}$ for model M_i, and L is the total number of MCMC steps after the burn-in period.

8.5.1 RECIPROCAL IMPORTANCE SAMPLING ESTIMATOR

Given MCMC samples $\boldsymbol{\theta}^{(l)}$, $l = 1, \ldots, L$ from the posterior distribution obtained through MCMC, Gelfand and Dey (1994) proposed the *reciprocal importance sampling estimator* (RISE) to approximate the model likelihood

$$\hat{p}_{RI}(\boldsymbol{x}) = \left[\frac{1}{L} \sum_{l=1}^{L} \frac{h(\boldsymbol{\theta}^{(l)})}{\pi(\boldsymbol{x}|\boldsymbol{\theta}^{(l)})\, \pi(\boldsymbol{\theta}^{(l)})} \right]^{-1}, \tag{8.122}$$

where h plays the role of an importance sampling density roughly matching the posterior. Gelfand and Dey (1994) suggested the multivariate Normal or t distribution density with mean and covariance fitted to the posterior sample.

The RISE estimator can be regarded as a generalization of the *harmonic mean estimator* suggested by Newton and Raftery (1994). The latter is obtained from the RISE estimator by setting $h = 1$. However, we strongly advise against such choices since it is known that the harmonic mean estimator will produce an estimator with infinite variance as discussed by Wolpert and Schmidler (2012). Other estimators include the *bridge sampling* proposed by Meng and Wong (1996), and *Chib's candidate estimator* by Chib (1995). In addition, there are also alternative approaches recently proposed that are efficient to implement in convex likelihood models such as the nested sampling framework of Skilling (2006).

In a recent comparison study by Miazhynskaia and Dorffner (2006), these estimators were employed as competing methods for Bayesian model selection on GARCH-type models, along with the reversible jump MCMC. It was demonstrated that the RISE estimator (either with Normal or t importance sampling density), the bridge sampling method, and Chib's algorithm gave statistically equal performance in model selection. Also, the performance more or less matched the much more involved reversible jump MCMC; however, it should be clearly noted that the relative computational costs and efficiency in general between each of these approaches will differ depending on the complexity of the model selection task. For this reason, we also present the details of the simplest form of the Chib estimator for model evidence (see Carlin and Chib 1995).

8.5.2 CHIB ESTIMATOR FOR MODEL EVIDENCE

The version of the Chib estimator that we propose for practitioners to utilize for the estimation of the Bayes factors is still based on sample output from the posterior model, typically from Markov chain samples. An important statistical property of this estimator from Chib is that it satisfies a standard Gaussian Central Limit Theorem and has finite variance. Therefore, we can estimate not only the model evidence but also, for a given set of simulations, we can report a measure of uncertainty in our model selections through an assessment of the accuracy of our evidence estimation.

In the following, we provide a brief description of the simplest form of the Chib estimator, the single block estimator. Under this approach, one proceeds to evaluate the evidence for the i-th model, denoted by $\ln \hat{p}\,(\boldsymbol{x}|M_i)$, according to the log decomposition as a function of the posterior with generic vector of parameters $\boldsymbol{\theta}$ and data $\boldsymbol{x}$ as follows:

$$\ln \hat{p}\,(\boldsymbol{x}|M_i) = \ln p\,(\boldsymbol{x}|\boldsymbol{\theta}^*, M_i) + \ln p\,(\boldsymbol{\theta}^*|M_i) - \ln \hat{p}\,(\boldsymbol{\theta}^*|\boldsymbol{x}, M_i)\,, \tag{8.123}$$

where $\boldsymbol{\theta}^*$ represents a point estimator for the parameters obtained from the MCMC output, such as the posterior mean (minimum mean square error (MMSE)) or the posterior mode (maximum a posteriori (MAP)) estimators. Here the estimator of $\hat{p}\,(\boldsymbol{\theta}^*|\boldsymbol{x}, M_i)$ obtained via Chib's approach is given for J samples from the proposal $\left\{\boldsymbol{\theta}^{(j)} : \boldsymbol{\theta}^{(j)} \sim q\,(\boldsymbol{\theta}, \boldsymbol{\theta}^*|\boldsymbol{x})\right\}_{j=1:J}$ and M samples from the MCMC output (i.e., correlated draws from the posterior) by

$$\hat{p}\left(\boldsymbol{\theta}^*|\boldsymbol{x}, M_i\right) = \frac{\frac{1}{M}\sum_{m=1}^{M} \alpha\left(\boldsymbol{\theta}^{(m)}, \boldsymbol{\theta}^*|\boldsymbol{x}\right) q\left(\boldsymbol{\theta}^{(m)}, \boldsymbol{\theta}^*|\boldsymbol{x}\right)}{\frac{1}{J}\sum_{j=1}^{J} \alpha\left(\boldsymbol{\theta}^*, \boldsymbol{\theta}^{(j)}|\boldsymbol{x}\right)}. \tag{8.124}$$

Here, the function $\alpha\left(\boldsymbol{\theta}, \boldsymbol{\theta}'|\boldsymbol{x}\right)$ represents the standard Metropolis–Hastings acceptance probability given by

$$\alpha\left(\boldsymbol{\theta}^*, \boldsymbol{\theta}^{(j)}|\boldsymbol{x}\right) = \min\left\{1, \frac{p\left(\boldsymbol{x}|\boldsymbol{\theta}'\right) p\left(\boldsymbol{\theta}'\right) q\left(\boldsymbol{\theta}', \boldsymbol{\theta}\right)}{p\left(\boldsymbol{x}|\boldsymbol{\theta}\right) p\left(\boldsymbol{\theta}\right) q\left(\boldsymbol{\theta}, \boldsymbol{\theta}'\right)}\right\}. \tag{8.125}$$

The proposal often considered is a multivariate student-t distribution for $q\left(\boldsymbol{\theta}, \boldsymbol{\theta}'\right)$ given for location parameter vector $\boldsymbol{\theta} \in \mathbb{R}^p$ and covariance matrix $\Sigma \in SP^+(p)$, where SP^+ is the space of symmetric and positive definite matrices in $\mathbb{R}^p$, for parameter $\boldsymbol{\theta}' \in \mathbb{R}^p$ by a probability density of

$$q\left(\boldsymbol{\theta}, \boldsymbol{\theta}'\right) = \frac{\Gamma\left(\frac{(n+p)}{2}\right)}{\Gamma\left(\frac{n}{2}\right) n^{p/2}\pi^{p/2}|\Sigma|^{1/2}\left[1 + \frac{1}{n}\left(\boldsymbol{\theta}' - \boldsymbol{\theta}\right)^T \Sigma^{-1}\left(\boldsymbol{\theta}' - \boldsymbol{\theta}\right)\right]^{(n+p)/2}}, \tag{8.126}$$

where one could estimate the covariance of the proposal, Σ, from the empirical sample covariance obtained from the samples out of the MCMC output for the M_i-th model according to

$$\hat{\Sigma} = \frac{1}{M-1}\sum_{m=1}^{M}\left(\boldsymbol{\theta}^{(m)} - \bar{\boldsymbol{\theta}}\right)\left(\boldsymbol{\theta}^{(m)} - \bar{\boldsymbol{\theta}}\right)^T, \tag{8.127}$$

where $\bar{\boldsymbol{\theta}}$ is the empirical mean of the parameters for the model.

8.6 SMC Sampler Estimators of Model Evidence

Del Moral *et al.* (2012) note that one can also utilize specially developed evidence estimators in the sequential Monto Carlo (SMC) samplers output, based on bridge sampling estimators of Gelman and Meng (1998) as follows. The particle estimate of target distribution evidence for π_t and π_{t-1} for each time step is given by using $\left\{W_t^{(i)}, \boldsymbol{\Theta}_t^{(i)}\right\}_{i=1}^{N}$ to approximate the normalizing constant ratios (model evidence that is denoted here as Z_n)

$$\frac{Z_t}{Z_{t-1}} = \frac{\int \pi_t\left(\boldsymbol{\theta}_t\right) d\boldsymbol{\theta}_t}{\int \pi_{t-1}\left(\boldsymbol{\theta}_{t-1}\right) d\boldsymbol{\theta}_{t-1}} \tag{8.128}$$

using the particle estimator given by

$$\widehat{\frac{Z_t}{Z_{t-1}}} = \sum_{i=1}^{N} W_{t-1}^{(i)} w_t\left(\boldsymbol{\Theta}_{t-1}^{(i)}, \boldsymbol{\Theta}_t^{(i)}\right). \tag{8.129}$$

Then one notes that to estimate the ratio Z_t/Z_1 one forms a product estimate based on these local ratio SMC sampler approximations; see discussions in Chapter 7 for SMC sampler algorithms.

8.7 Multiple Risk Dependence Structure Model Selection: Copula Choice

In this section, we discuss how to perform model selection for the dependence structure linking multiple risk processes, with a particular emphasis on copula model representations. For a model choice of copula using frequentist GOF testing, see Klugman and Parsa (1999) and Panjer (2006, section 14.5). One can also use the AIC to choose a copula. However, formally, it does not hold for copulas fitted using data marginally transformed into $[0,1]^d$; a proper correction, referred to as copula information criterion, has been derived by Grønneberg and Hjort (2008). Under the Bayesian approach, model choice can be made using Bayesian criteria presented in Section 8.5; for a case study of t-copula choice, see Luo and Shevchenko (2012).

To proceed, in this section we first discuss how to perform model selection purely on the dependence component of the model that is used to combine or relate multiple risk processes LDA models. The detailed presentation of such model structures is presented in Chapters 10–12. The generic parametric copula model considered in this section will be denoted by distribution C. In this section, the presentation of such model selection approaches will assume that one has already made inference on the appropriate models for the marginal (each individual loss processes) LDA structures. On this point, it is important to observe the following fact: if one truly wants to test the hypothesis given by

$$H_0: \quad C \in \mathcal{C}_0 \tag{8.130}$$

that is, that the dependence structure between the risk processes is well represented by a particular parameteric multivariate distribution in the copula family $\mathcal{C}_0$, then the option of modeling the marginal LDA models by parameteric families first is no longer strictly viable. To understand why this is the case, one must realize that such a procedure would actually correspond to a different much more restricted null hypothesis, corresponding to $H_0 \cap H_0'$ where H_0' relates to the assumption of the structure of the marginal models, producing a hypothesis for the full parameteric model of the multiple risk processes and not just their dependence features.

Hence, this indicates that when testing purely for the dependence structure between multiple risk process LDA models, one should consider the marginal distributions, such as the annual losses between d risk processes $F_{Z_1}(z_1), F_{Z_2}(z_2), \ldots, F_{Z_d}(z_d)$ or the equivalent quantitites for the d severity of frequency models, as infinite dimensional nuisance parameters (i.e., nuisance functions). Having recognized this fact, one needs to also utilize the property of all copula distributions, that they are multivariate distributions for d risk processes on support $[0,1]^d$, which are invariant to strictly increasing transformations of their components. The implications of this for the perspective of hypothesis testing and model selection is that one may instead make inference on the evidence against H_0 based on the maximally invariant statistics with respect to these types of transformations, that is, one may work directly on the ranks or order statistics. This leads one to the notion of pseudo observations for copula GOF hypothesis testing as detailed in Definition 8.9.

Definition 8.9 (Pseudo observations for copula GOF testing) *Consider the d-dimensional multivariate loss data* $\{\boldsymbol{Z}_i\}_{i=1}^n$ *for n loses from the d risk processes, with* $\boldsymbol{Z}_i = \left(Z_i^{(1)}, \ldots, Z_i^{(d)}\right)$*. Convert the individual loss data into pseudo observations based on scaled ranks by considering new data* $\{\boldsymbol{U}_i\}_{i=1}^n$ *with the j-th component of the i-th random vector observation given by*

$$U_i^{(j)} = \frac{R_i^{(j)}}{n+1} = \frac{n}{n+1}\hat{F}_{Z^{(j)}}\left(Z_i^{(j)}\right), \tag{8.131}$$

where $\boldsymbol{U}_i \in [0,1]^d$ *for all* $i \in \{1,2,\ldots,n\}$ *and* $R_i^{(j)}$ *is the rank of the j-th component of vector* $\boldsymbol{Z}_i$ *among the samples, that is, the rank of* $Z_i^{(j)}$ *among* $\left\{Z_1^{(j)}, \ldots, Z_n^{(j)}\right\}$*. This transformation of each margin through the normalized ranks is known as the empirical marginal transformation. Therefore, these pseudo observations can then be interpreted as draws from the underlying copula C acting as the dependence structure between the multiple risk processes.* ■

Remark 8.11 *The pseudo observations discussed earlier are not mutually independent of each other and therefore the components are only approximately uniform on* $[0,1]$ *in each margin. They will only be exactly uniform if the exact model is considered. If a model selection or hypothesis-testing procedure is developed which ignores these features, it will suffer from a lack of power and may fail to hold its nominal level.*

EXAMPLE 8.5 Pseudo Data for GOF Dependence Structure Analysis

The aim of this example is to illustrate that one can recover samples that closely resemble the true samples from the copula density for a multivariate distribution, via the pseudo data in practical applications in OpRisk. Consider a multivariate loss model for $\boldsymbol{Z}_i = \left(Z_i^{(1)}, Z_i^{(2)}, \ldots, Z_i^{(d)}\right)$ with the following density:

$$f_Z(z_1, \ldots, z_d) = c_\rho\left(F_{Z_1}(z_1), \ldots, F_{Z_d}(z_d)\right) \prod_{i=1}^{d} f_{Z_i}(z_i), \tag{8.132}$$

where we consider each marginal loss process $F_{Z_i}(z_i)$ to be given by annual loss density from the compound sum LDA model

$$Z_i = \sum_{j=1}^{N} X_j \tag{8.133}$$

with $N \sim$ *Poisson*(λ_i) and i.i.d. losses for risk process i given by $X_j \sim$ *InverseGaussian*(μ_i, γ_i). Then the marginal for the i-th loss process is a Poisson-weighted mixture of inverse Gamma distributions given by

$$f_{Z_i}(z) = \sum_{n=1}^{\infty} \exp(-\lambda_i) \frac{\lambda_i^n}{n!} \left\{ \left[\frac{n^2\gamma_i}{2\pi z^3}\right]^{\frac{1}{2}} \exp\left(-\frac{n^2\gamma_i (z - n\mu_i)^2}{2n^2\mu_i^2 z}\right) \right\}. \tag{8.134}$$

Consider the case with $d = 2$, $\lambda_1 = \lambda_2 = 3$, $\mu_1 = 2, \mu_2 = 1$, and $\gamma_1 = 1, \gamma_2 = 2$. Furthermore, assume the annual losses are jointly dependent with a Frank copula dependence structure for the density $c_\rho\left(F_{Z_1}(z_1), \ldots, F_{Z_d}(z_d)\right)$ with dependence parameter $\rho = -1$ and density given by

$$c(u_1, u_2) = \frac{\rho\left[1 - \exp(-\rho)\right]\exp\left(-\rho\left(u_1 + u_2\right)\right)}{\left(\left[1 - \exp(-\rho)\right] - \left(1 - \exp\left(-\rho u_1\right)\right)\left(1 - \exp\left(-\rho u_2\right)\right)\right)^2}. \tag{8.135}$$

Under this model it is assumed that a total of 100 samples have been obtained and the true copula density contour plots are drawn, followed by the true joint density of (Z_1, Z_2). The pseudo data are then plotted over the top of the copula contours; see Figure 8.8. Clearly, the pseudo data display behavior consistent with the Frank copula model from which they are approximately drawn. It should also be noted that the accuracy of this pseudo data transformation by the rank will diminish (i.e., how representative the pseudo data are of the true copula) and be affected by the sample size available. ■

Having briefly explained the need to be cautious when performing specialized GOF tests specifically for the copula dependence structures and the definition of pseudo data that may be used to undertake such testing procedures, we next provide a brief summary of the various approaches that have been adopted in the literature for undertaking such testing (see detailed discussions in Berg 2009). Before proceeding with the overview of these different testing procedures, we need to introduce a few additional basic concepts. The first is the transformation of a random vector known as Rosenblatt's probability integral transformation, as detailed in Definition 8.10 (see Bickel and Rosenblatt 1973).

Definition 8.10 (Rosenblatt's probability integral transformation) *Rosenblatt's probability integral transformation (PIT) of a copula distribution C is a mapping $\mathcal{T} : (0,1)^d \mapsto (0,1)^d$ such that every vector $\boldsymbol{u} = (u_1, u_2, \ldots, u_d) \in (0,1)^d$ is assigned to a new vector under the mappint $\mathcal{T}(u_1, u_2, \ldots, u_d) = (e_1, e_2, \ldots, e_d)$ such that the following holds:*

$$\begin{aligned} e_1 &= u_1 \\ e_i &= \frac{\partial^{i-1} C\left(u_1, \ldots, u_i, 1, \ldots, 1\right)}{\partial u_1 \ldots \partial u_{i-1}} \Big/ \frac{\partial^{i-1} C\left(u_1, \ldots, u_{i-1}, 1, \ldots, 1\right)}{\partial u_1 \ldots \partial u_{i-1}}, \quad \forall i \in \{2, \ldots, d\}. \end{aligned} \tag{8.136}$$

Under this transformation, the original random vector $\boldsymbol{U}$ is distributed from a copula C that is, $\boldsymbol{U} \sim C$ if and only if the resulting transformed random vector $\boldsymbol{E}$ is distributed from a uniform distribution $\boldsymbol{E} \sim [0,1]^d$, that is, the resulting copula of $\boldsymbol{E}$ is the independence copula with uniform marginals. ■

Remark 8.12 *Under the application of Rosenblatt's transformation of a random vector with dependence features given by copula distribution C, one may transform a hypothesis test from the nominal claim that*

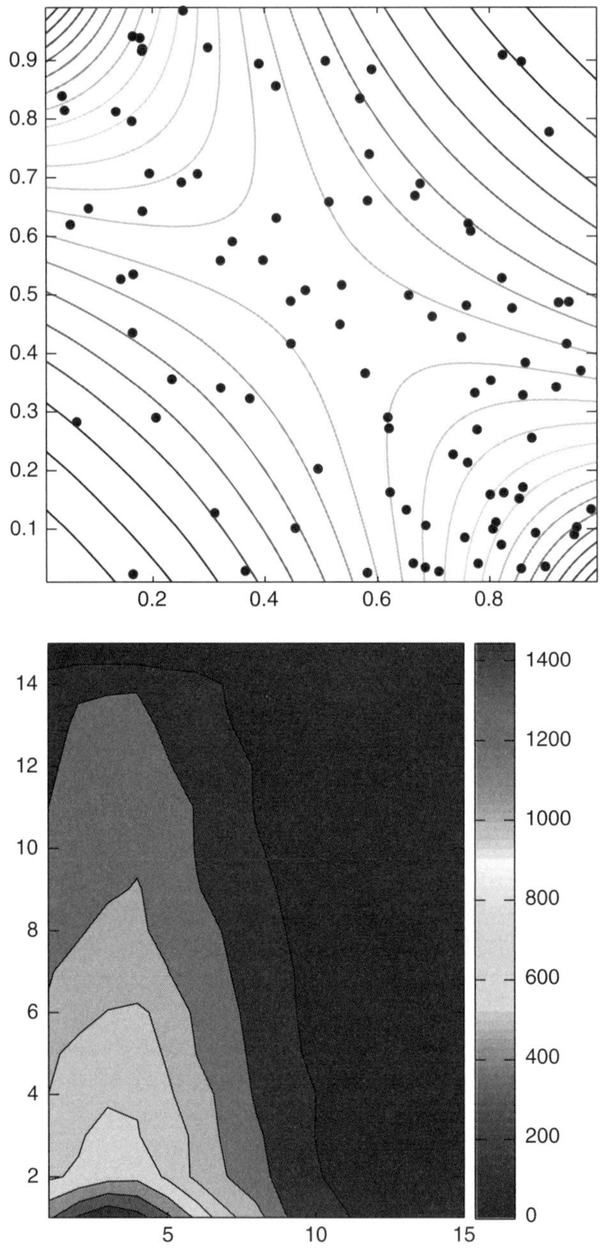

FIGURE 8.8 Top subplot: this plot shows the true copula contours used in this model, that is, a Frank copula, and the points correspond to the pseudo data obtained by transformation through the empirical marginals (i.e., using the marginal scaled ranks). Bottom subplot: this plot shows the contours of the joint loss process density for $\left(Z^{(1)}, Z^{(2)}\right)$. (see insert for color representation of the figure.)

$$H_0 : \;\; \boldsymbol{U} \sim C \in \mathcal{C}_0 \tag{8.137}$$

to an equivalent omnibus claim that

$$H_0^* : \;\; \mathcal{T}(\boldsymbol{U}) \sim U[0,1]^d \tag{8.138}$$

for some parameter vector $\boldsymbol{\rho}$ that parameterizes the copula C.

In the following examples, the required Rosenblatt PIT is provided for several examples. In each case, one makes use of the multivariate Faa Di Bruno composite derivative expressions to obtain simple closed-form expressions for the PIT transforms for Archimedean families. This first involves recognizing that the Archimedean copula family has a particular distribution form given by a composite function comprised of $\psi(\cdot)$ and linear combinations of its inverse $\psi^{-1}(\cdot)$:

$$C(u_1, \ldots, u_n) = \psi\left(\sum_{i=1}^{n} \psi^{-1}(u_i)\right) \tag{8.139}$$

See detailed discussions in Chapters 10–12 on dependence modeling. It should then be noted that to find the evaluation of the distribution given by $C(u_1, u_2, \ldots, u_{n-k}, 1, 1, \ldots, 1)$, one obtains

$$C(u_1, u_2, \ldots, u_{n-k}, 1, 1, \ldots, 1) = \psi\left(\sum_{i=1}^{n-k} \psi^{-1}(u_i)\right), \tag{8.140}$$

since $\psi^{-1}(1) = 0$ in order for ψ to be a well-defined generator for the Archimedean family; see details in Chapters 10–12. Hence, this means that taking the derivatives for the terms e_i under Rosenblatt's PIT for the Archimedean family of copula models will result in terms, for $\forall i \in \{2, \ldots, d\}$, given by the ratio of the integrated density in dimension i with respect to argument u_i and the density in dimension $i-1$ according to

$$\begin{aligned} e_i &= \frac{\partial^{i-1} C(u_1, \ldots, u_i, 1, \ldots, 1)}{\partial u_1 \ldots \partial u_{i-1}} \Bigg/ \frac{\partial^{i-1} C(u_1, \ldots, u_{i-1}, 1, \ldots, 1)}{\partial u_1 \ldots \partial u_{i-1}} \\ &= \frac{\int_0^{u_i} c(u_1, \ldots, s)\, ds}{c(u_1, \ldots, u_{i-1})} \\ &= \frac{\left[\prod_{j=1}^{i-1} (\psi^{-1})'(u_j)\right] \int_0^{u_i} \psi^i\{\psi^{-1}(s) + \sum_{i=1}^{i-1} \psi^{-1}(u_i)\}(\psi^{-1})'(s) ds}{\psi^{(d)}\{\sum_i = 1^{i-1} \psi^{-1}(u_i)\} \prod_{j=1}^{i-1} (\psi^{-1})'(u_j)} \\ &\times \frac{\int_0^{u_i} \psi^i\{\psi^{-1}(s) + \sum_{i=1}^{i-1} \psi^{-1}(u_i)\}(\psi^{-1})'(s) ds}{\psi^{(i-1)}\{\sum_{i=1}^{i-1} \psi^{-1}(u_i)\}}. \end{aligned} \tag{8.141}$$

One can then utilize closed-form expressions for the derivatives of these generators for any dimension; see details in Chapter 10. We present examples for the bivariate setting in a few popular models in the Archimedean family.

EXAMPLE 8.6 Rosenblatt's Probability Integral Transform Clayton Copula

Consider the Clayton copula given by distribution and density

$$C^C(u_1, \ldots, u_n) = \left(1 - n + \sum_{i=1}^{n} u_i^{-\rho^C}\right)^{-1/\rho^C}, \quad (8.142)$$

$$c^C(u_1, \ldots, u_n) = \left(1 - n + \sum_{i=1}^{n} (u_i)^{-\rho^C}\right)^{-n-\frac{1}{\rho^C}} \prod_{i=1}^{n}\left((u_i)^{-\rho^C-1}((i-1)\rho^C + 1)\right), \quad (8.143)$$

where $\rho^C \in [0, \infty)$ is the dependence parameter. Hence, given a random vector $\boldsymbol{U} = (U_1, U_2)$ distributed from this copula model, one can obtain the Rosenblatt PIT as for the *bivariate case:*

$$\begin{aligned} e_1 &= u_1, \\ e_2 &= \frac{\frac{\partial}{\partial u_1}\left[\left(-1 + \sum_{i=1}^{2} u_i^{-\rho^C}\right)^{-1/\rho^C}\right]}{\frac{\partial}{\partial u_1}[u_1]} \\ &= u_1^{-\rho^C-1}\left(-1 + \sum_{i=1}^{2} u_i^{-\rho^C}\right)^{-1/\rho^C-1}. \end{aligned}$$

EXAMPLE 8.7 Rosenblatt's Probability Integral Transform Gumbel Copula

Consider the Gumbel copula given by distribution

$$C^G(u_1, \ldots, u_d) = \exp\left(-\left[\sum_{i=1}^{d}(-\ln(u_i))^{\rho^G}\right]^{\frac{1}{\rho^G}}\right), \quad (8.144)$$

where $\rho^G \in [1, \infty)$ is the dependence parameter. In the bivariate case, the explicit expression for the Gumbel copula density is given by

$$\begin{aligned} c(u_1, u_2) &= \frac{\partial^2}{\partial u_1 \partial u_2} C(u_1, u_2) \\ &= C(u_1, u_2)\, u_1^{-1} u_2^{-1} \left[\sum_{i=1}^{2}(-\ln u_i)^{\rho}\right]^{2\left(\frac{1}{\rho}-1\right)} (\ln u_1 \ln u_2)^{\rho-1} \\ &\times \left[1 + (\rho - 1)\left[\sum_{i=1}^{2}(-\ln u_i)^{\rho}\right]^{-\frac{1}{\rho}}\right]. \end{aligned}$$

Hence, given a random vector $\boldsymbol{U} = (U_1, U_2)$ distributed from this copula model, one can obtain the Rosenblatt PIT as for the *bivariate case:*

$$
\begin{aligned}
e_1 &= u_1 \\
e_2 &= \frac{\frac{\partial}{\partial u_1}\left[\exp\left(-\left[\sum_{i=1}^{2}(-\ln(u_i))^{\rho^G}\right]^{\frac{1}{\rho^G}}\right)\right]}{\frac{\partial}{\partial u_1}[u_1]} \\
&= \frac{\partial}{\partial u_1}\left[\exp\left(-\left[\sum_{i=1}^{2}(-\ln(u_i))^{\rho^G}\right]^{\frac{1}{\rho^G}}\right)\right] \\
&= \frac{1}{u_i}(-\ln(u_i))^{\rho^G-1}\left[\sum_{i=1}^{2}(-\ln(u_i))^{\rho^G}\right]^{\frac{1}{\rho^G}-1} \\
&\quad \times \exp\left(-\left[\sum_{i=1}^{2}(-\ln(u_i))^{\rho^G}\right]^{\frac{1}{\rho^G}}\right).
\end{aligned}
\tag{8.145}
$$

■

8.7.1 APPROACHES TO GOODNESS-OF-FIT TESTING FOR DEPENDENCE STRUCTURES

The following are popular approaches that have been proposed to perform copula dependence GOF testing and model selection.

- **Rosenblatt's transformation test**. The Rosenblatt transformation has been proposed for copula GOF testing by several authors including Genest *et al.* (2006), Dobrić and Schmid (2007), Berg and Bakken (2005), and Berg (2009). In this case, the pseudo observations $\{\boldsymbol{U}_i\}_{i=1}^n$ given in Definition 8.9 are transformed through Rosenblatt's transformation $\mathcal{T}(\cdot)$ to obtain new observations $\{\boldsymbol{E}_i\}_{i=1}^n$ with each $\boldsymbol{E}_i = \mathcal{T}(\boldsymbol{U}_i)$ and the null hypothesis being tested is then transformed to become

$$H_0^*: \ \mathcal{T}(\boldsymbol{U}) \sim U[0,1]^d. \tag{8.146}$$

  The resulting test statistic under this new nominal claim can then be considered under two different sets of assumptions. The simplest would be to assume that the pseudo observations are mutually independent and uniformly distributed on $(0,1)^d$ if $\boldsymbol{U}_i \sim C \in \mathcal{C}_0$. Of course, as noted by several authors (see discussion by Genest *et al.* 2009b), the approximate uniformity of the transformed observations $\{\boldsymbol{E}_i\}_{i=1}^n$ on the space $(0,1)^d$ allows one to utilize the following transform on each marginal component followed by the convolution of the approximately i.i.d. components to make up a GOF test as follows: first, construct new aggregate univariate random variables given by

$$\chi_i = \sum_{j=1}^{d} \left\{ \Phi^{-1}\left(E_i^{(j)}\right) \right\}^2, \quad \forall i \in \{1, 2, \ldots, n\}. \tag{8.147}$$

Next, consider the distribution of the set $\{\chi_i\}_{i=1}^n$ and in particular the empirical distribution function

$$\hat{F}_n(x) = \frac{1}{n} \sum_{i=1}^{n} \mathbb{I}_{\chi_i \leq x}, \quad x \geq 0. \tag{8.148}$$

If the assumption of uniformity of the sample $\{\boldsymbol{E}_i\}_{i=1}^n$ were valid, then one can construct the empirical process convergence to a Brownian bridge, as discussed previously by considering the process as $n \to \infty$ given by

$$\sqrt{n}\left(\hat{F}_n(x) - F(x)\right), \tag{8.149}$$

which would then allow one (if the copula parameter was assumed to be known—not a compound test) to perform, for instance, an AD test to test the nominal claim

$$H_0: \quad \boldsymbol{U} \sim C \in \mathcal{C}_0 \tag{8.150}$$

using the test statisic given by

$$A_n = -n - \frac{1}{n} \sum_{i=1}^{n} (2i - 1) \ln\left[F\left(\chi_{(i,n)}\right)\right] + \ln\left[1 - F\left(\chi_{(n+1-i,n)}\right)\right] \tag{8.151}$$

with $\chi_{(i,n)}$ the i-th order statistic where $\chi_{(1,n)} \leq \cdots \leq \chi_{(n,n)}$. As noted by Breymann *et al.* (2003) and Dobrić and Schmid (2007), the assumption underpinning this asymptotic convergence may not apply in many settings and, consequently, the test statistic and p-value must be adjusted or calculated numerically via a bootstrap procedure as discussed by Genest *et al.* (2009b, appendix C);

- **Rosenblatt's weighted transformation test**. Malevergne *et al.* (2003) and Berg (2009) considered the pseudo data samples $\{\boldsymbol{u}_i\}_{i=1}^n$ and applied the Rosenblatt transformation to obtain samples $\{\boldsymbol{e}_i\}_{i=1}^n$, which under the nominal claim copula $C_{\hat{\rho}_n}$, will produce samples that are from the independence copula. They note that when the pseudo data are obtained from the rank data, this assumption on uniformity is not strictly achieved. The Rosenblatt transformed data are then transformed further to produce a univariate sample given by the weighted transformation

$$\widetilde{\chi}_i(\Gamma) = \sum_{j=1}^{n} \Gamma\left(e_i^{(j)}; \boldsymbol{\psi}\right), \tag{8.152}$$

where $\Gamma(\cdot; \boldsymbol{\psi})$ is a weighting function parameterized by parameter vector $\boldsymbol{\psi}$. This function can be used to focus testing on different regions of the unit cube such as particular quadrants of interest or the tails of the dependence copula tails in different regions of the unit hyper cube. Examples of such weighting functions include the following:

$$\Gamma_1\left(e_i^{(j)};\psi\right) = \left|e_i^{(j)} - 0.5\right|; \\ \Gamma_2\left(e_i^{(j)};\psi\right) = \Phi^{-1}\left(e_i^{(j)}\right)^2. \tag{8.153}$$

In the case of weighting function $\Gamma_1(\cdot)$, the resulting dimension-reduced data $\widetilde{\chi}_i(\Gamma_1)$ will be a χ_d^2 distribution for all data samples $i \in \{1, 2, \ldots, n\}$. If the dimension-reduced data are obtained with the second weighting function $\widetilde{\chi}_i(\Gamma_2)$, then one does not have a simple closed-form distribution for these random variables and hence a double bootstrap procedure is required.

In general, for any choice of weighting function $\Gamma(\cdot;\psi)$, the resulting random variables $\{\widetilde{\chi}_i(\Gamma)\}_{i=1}^n$, are each distributed under the null according to the distribution $F[\widetilde{\chi}_i(\Gamma)]$, producing the process

$$S_1(w) = \mathbb{P}\mathrm{r}\left(F[\widetilde{\chi}_1(\Gamma)] \leq w\right), \quad w \in [0, 1]. \tag{8.154}$$

Under the null, Berg (2009) showed that one can empirically estimate the process $S_1(w)$ using the sample estimate

$$\hat{S}_1(w) = \frac{1}{n+1}\sum_{i=1}^{n} \mathbb{I}\left[F[\widetilde{\chi}_{1i}(\Gamma)] \leq w\right]. \tag{8.155}$$

Then using this emprical process estimator, one can obtain an estimate of the resulting CvM test statistic

$$T_1 = n\int_0^1 \left(\hat{S}_1(w) - S_1(w)\right)^2 dS_1(w), \tag{8.156}$$

which is empirically evaluated using the statistic

$$\hat{T}_1 = \frac{n}{3} + \frac{n}{n+1}\sum_{j=1}^{n} \hat{S}_1\left(\frac{j}{n+1}\right)^2 - \frac{n}{(n+1)^2}\sum_{i=1}^{n}(2j+1)\hat{S}_1\left(\frac{j}{n+1}\right). \tag{8.157}$$

- **Empirical Copula Distribution Functions**. As discussed by Genest *et al.* (2009b), two copula GOF tests based on emprical copula distribution functions are summarized from those developed by Fermanian *et al.* (2004) and Tsukahara (2005). The first test developed involves the approximation of the copula distribution using the pseudo data and the d-variate empirical distribution function given by

$$\hat{C}_n(\boldsymbol{u}) = \frac{1}{n}\sum_{i=1}^{n} \mathbb{I}_{\left(U_i^{(1)} \leq u_1, U_i^{(2)} \leq u_2, \ldots, U_i^{(d)} \leq u_d\right)} \tag{8.158}$$

for $\boldsymbol{u} = (u_1, u_2, \ldots, u_d) \in [0, 1]^d$, which is known as the emprical copula, though technically it is not strictly a copula distribution (see discussions by Deheuvels 1979). Under conditions discussed by Fermanian *et al.* (2004) and Tsukahara (2005), the empirical copula distribution will converge as $n \to \infty$ such that $\hat{C}_n \to C$ as a consistent estimator whether the nominal claim holds or not. One may therefore test the hypothesis

$$H_0: \quad \boldsymbol{U} \sim C \in \mathcal{C}_0 \tag{8.159}$$

using a distance-based measure comparing the empirical copula $\hat{C}_n$ and an estimate $C_{\hat{\rho}_n}$ obtained under the nominal claim. To develop such a hypothesis test, Genest and Rémillard (2008) considered rank-based versions of the CvM and KS tests with statistics given by considering the process $\mathbb{C}_n = \sqrt{n}\left(\hat{C}_n - C_{\hat{\rho}_n}\right)$, which produces a CvM statistic

$$S_n = \int_{[0,1]^d} \mathbb{C}_n(\boldsymbol{u})^2 d\hat{C}_n(\boldsymbol{u}) \tag{8.160}$$

or a KS statistic

$$T_n = \sup_{\boldsymbol{u} \in [0,1]^d} \left|\mathbb{C}_n(\boldsymbol{u})\right|. \tag{8.161}$$

As discussed by Genest and Rémillard (2008), one can be sure that under particular regularity conditions on the parameteric copula family $\mathcal{C}_0$ and the sequence of parameter estimators $\{\hat{\rho}_n\}$ as $n \to \infty$, the tests based on S_n and T_n are consistent in that they will reject the nominal claim if the true copula for the data is not in the nominal class. The evaluation of the empirical test statistic for the CvM test was shown by Berg (2009) to be estimated by

$$\hat{S}_n = \sum_{i=1}^{n} \left[\hat{C}_n(\boldsymbol{u}_i) - C_{\hat{\rho}_n}(\boldsymbol{u}_i)\right]^2. \tag{8.162}$$

It is generally not possible to find the asymptotic distribution of the test statistics to find the tabulation of the p-values for the decision rule on these tests since the distribution will depend heavily on the nominal claims class of copula distributions $\mathcal{C}_0$. The p-values for these test statistics can be obtained via a bootstrap procedure (see discussions in Genest *et al.* 2009b, appendix A);

- **Empirical Copula Distribution and Rosenblatt's Transformation**. This idea basically follows equivalently the idea proposed in the empirical copula process test, except that there are two transformations applied to the data: the first is that the data are transformed under a rank-based transform to obtain the pseudo data, and the second is to apply Rosenblatt's transformation. The empirical copula process considered then should be compared to the resulting independence copula as the nominal claim, rather than a particular copula model family;
- **Kendall's Tau Transformation Tests**. Under Kendall's tau transformation testing approach to inference on the copula linking d risk processes, the approach considered by Genest *et al.* (2006) involves the transformation of the data $\boldsymbol{Z}$ via a transformation

$$\boldsymbol{Z} \mapsto V = C(U_1, U_2, \ldots, U_d) \tag{8.163}$$

with $U_i = F_{Z^{(i)}}\left(Z^{(i)}\right)$. This transform relates to Kendall's tau since the expectation of V is the affine transformation of the multivariate version of Kendall's coefficient of concordance as discussed in Chapter 10 and by Barbe *et al.* (1996). As discussed by Berg (2009), the resulting empirical process for the case of Kendall's tau tests is to consider the pseudo data $\{\boldsymbol{U}_i\}_{i=1}^n$ to construct the process

$$K(w) = \mathbb{P}\mathrm{r}\left(C\left(U_1, U_2, \ldots, U_d\right) \leq w\right), \quad w \in [0, 1]. \tag{8.164}$$

Then the nominal claim is that $K(w) = K_{\hat{\rho}_n}(w)$, which is copula-specific (see details in Chapter 10). The resulting empirical estimate is given by

$$\hat{K}_n(w) = \frac{1}{n+1}\sum_{i=1}^{n}\mathbb{I}\left[\hat{C}_n\left(\boldsymbol{u}_i\right) \leq w\right] \tag{8.165}$$

and the resulting CvM test statistic is given by

$$S_n = n\int_0^1 \left(\hat{K}_n(w) - K_{\hat{\rho}_n}(w)\right)^2 d\hat{K}_n(w), \tag{8.166}$$

which can be evaluated empirically by the following expression:

$$\hat{S}_n = \sum_{i=1}^{n}\left(\hat{K}_n\left(\frac{j}{n+1}\right) - K_{\hat{\rho}_n}\left(\frac{j}{n+1}\right)\right)^2. \tag{8.167}$$

There are also a number of other tests available based on Spearman's rho, Shih's test, and other alternatives; see detailed discussions and references in Berg (2009) and Genest *et al.* (2009b).

We finish this section by discussing how one would calculate the p-values in a double bootstrap procedure, which will be required for most tests that do not admit a distributional form under the nominal claim for the test statistic that is noncopula family–specific. There are several approaches one may adopt; we provide briefly the details of the standard example detailed by Genest *et al.* (2009b, appendix A).

8.7.2 DOUBLE PARAMETERIC BOOTSTRAP FOR COPULA GOF

The following procedure allows one to calculate the p-value of tests based on, for instance, the CvM test statistic via a double bootstrap procedure. This is particularly useful in cases where one may not be able to calculate the copula distribution in closed form, but generation of data from this model is trivial and efficient.

Algorithm 8.6 (Double Parameteric Bootstrap for Copula GOF)

- ***Step 1****. Compute the emprical copula according to the expression to obtain the approximation of the copula distribution using the pseudo data and the d-variate empirical distribution function given by*

$$\hat{C}_n(\boldsymbol{u}) = \frac{1}{n}\sum_{i=1}^{n}\mathbb{I}_{\left(U_i^{(1)}\le u_1, U_i^{(2)}\le u_2,\ldots,U_i^{(d)}\le u_d\right)} \tag{8.168}$$

for $\boldsymbol{u} = (u_1, u_2, \ldots, u_d) \in [0,1]^d$, *which is known as the empirical copula. Then make an estimation of the copula parameter* $\hat{\boldsymbol{\rho}}_n$ *using the pseudo data;*

- ***Step 2***. *If the copula family under the nominal claim can be evaluated in closed form, then evaluate the test statistic* S_n *given by*

$$S_n = \int_{[0,1]^d} \mathbb{C}_n(\boldsymbol{u})^2 d\hat{C}_n(\boldsymbol{u}) \tag{8.169}$$

via the empirical approximation

$$\hat{S}_n = \sum_{i=1}^{n}\left[\hat{C}_n(\boldsymbol{u}_i) - C_{\hat{\boldsymbol{\rho}}_n}(\boldsymbol{u}_i)\right]^2. \tag{8.170}$$

- ***Step 3***. *If the copula family under the nominal claim cannot be evaluated pointwise in closed form, then perform the following steps:*
  1. *Generate random sample* $\{\boldsymbol{U}_i^*\}_{i=1}^m$ *as i.i.d. draws from the distribution* $C_{\hat{\boldsymbol{\rho}}_n}$;
  2. *Evaluate the empirical copula to approximate* $C_{\hat{\boldsymbol{\rho}}_n}$ *using the estimator*

$$\hat{C}^*_{\hat{\boldsymbol{\rho}}_n}(\boldsymbol{u}) = \frac{1}{m}\sum_{i=1}^{m}\mathbb{I}\left[\boldsymbol{U}_i^* \le \boldsymbol{u}\right]. \tag{8.171}$$

  3. *Approximate the test statistic using the two empirically estimated copula distributions via the original pseudo data* $\{\boldsymbol{U}_i\}_{i=1}^n$,

$$\hat{S}_n = \sum_{i=1}^{n}\left[\hat{C}_n(\boldsymbol{U}_i) - \hat{C}^*_{\hat{\boldsymbol{\rho}}_n}(\boldsymbol{U}_i)\right]^2. \tag{8.172}$$

- ***Step 4***. *Then perform a large number of repetitions (J) of the following steps for* $j \in \{1, \ldots, J\}$:
  1. *Generate random sample* $\left\{\boldsymbol{V}^*_{i,j}\right\}_{i=1}^n$ *as i.i.d. draws from the distribution* $C_{\hat{\boldsymbol{\rho}}_n}$ *and evaluate their ranks given by* $\left\{\boldsymbol{R}^*_{i,j}\right\}_{i=1}^n$;
  2. *Compute the pseudo data using the ranks to obtain* $\left\{\widetilde{\boldsymbol{U}}^*_{i,j}\right\}_{i=1}^n$ *where each sampel is obtained by*

$$\widetilde{\boldsymbol{U}}^*_{i,j} = \frac{1}{n+1}\boldsymbol{R}^*_{i,j}. \tag{8.173}$$

  3. *Evaluate the empirical copula given by*

$$\hat{C}^*_{n,j}(\boldsymbol{u}) = \frac{1}{n}\sum_{i=1}^{n}\mathbb{I}\left[\widetilde{\boldsymbol{U}}^*_{i,j} \le \boldsymbol{u}\right] \tag{8.174}$$

and evaluate the resulting copula parameter estimate $\hat{\boldsymbol{\rho}}^*_{n,j}$ *using the pseudo data samples* $\left\{ \widetilde{\boldsymbol{U}}^*_{i,j} \right\}_{i=1}^{n}$;

- ***Step 5**. If the nominal claim copula distribution admits a parametric form that can be evaluated pointwise, then evaluate the test statistic for each* $j \in \{1, 2, \ldots, J\}$ *according to*

$$\hat{S}^*_{n,j} = \sum_{i=1}^{n} \left[\hat{C}^*_{n,j} \left(\widetilde{\boldsymbol{U}}^*_{i,j} \right) - C_{\hat{\boldsymbol{\rho}}^*_{n,j}} \left(\widetilde{\boldsymbol{U}}^*_{i,j} \right) \right]^2 . \tag{8.175}$$

- ***Step 6**. If the nominal claim copula is not available in closed parameteric form to be evaluated pointwise, then proceed as follows for* $j \in \{1, 2, \ldots, J\}$:
    - *Generate random sample* $\left\{ \boldsymbol{V}^{**}_{i,j} \right\}_{i=1}^{n}$ *as i.i.d. draws from the distribution* $C_{\hat{\boldsymbol{\rho}}_n}$;
    - *Evaluate the empirical copula given by*

$$\hat{C}^{**}_{n,j}(\boldsymbol{u}) = \frac{1}{n} \sum_{i=1}^{n} \mathbb{I} \left[\widetilde{\boldsymbol{V}}^{**}_{i,j} \leq \boldsymbol{u} \right] . \tag{8.176}$$

    - *Evaluate the approximation of the test statistic according to*

$$\hat{S}^*_{n,j} = \sum_{i=1}^{n} \left[\hat{C}^*_{n,j} \left(\widetilde{\boldsymbol{U}}^*_{i,j} \right) - \hat{C}^{**}_{n,j} (\widetilde{\boldsymbol{U}}^*_{i,j}) \right]^2 . \tag{8.177}$$

- ***Step** 7. Evaluate the p-value for the CvM test using the empirical estimator given by*

$$p = \frac{1}{J} \sum_{j=1}^{J} \mathbb{I} \left[\hat{S}^*_{n,j} > \hat{S}_n \right] . \tag{8.178}$$

CHAPTER NINE

Flexible Parametric Severity Models: Basics

9.1 Motivation for Flexible Parametric Severity Loss Models

In Chapter 5, we provided a description of standard loss distribution models. In the case of severity models, this has included LogNormal, Gamma, Weibull, Pareto, and Generalized Pareto models. In the case of frequency-based models, the families of Poisson, Binomial, and Negative Binomial have been considered. In this chapter, we provide a more flexible set of models that should be considered by OpRisk practitioners, especially in the modeling of heavy-tailed loss processes.

In the following subsections, we will first introduce important members of the general family of heavy-tailed loss models for the severity distribution; some of these will also be members of the subexponential family of models or models with different properties of tail variation as well as flexible skew and kurtosis characteristics. It is typical when modeling such severity distributions to consider families of models that have members which take positive support and are typically unimodal and left skewed. The models presented in the following sections introduce several families of parametric statistical models that are of direct interest in the areas of OpRisk and insurance modeling. The focus will be on severity models under a Loss Distribution Approach (LDA) structure and the properties of the considered parametric families that make them amenable to heavy-tailed modeling in OpRisk.

The intention of this section will be to define the key properties of each of these subclasses of models and explain how they are of relevance to modeling OpRisk loss processes. We will then consider properties of subexponential family members when incorporated into compound process models in an LDA framework, illustrating in the process how such models can be successfully incorporated into OpRisk models.

In particular, we will provide detailed discussions on several important families of severity models for capturing features of heavy-tailed loss processes. The models covered will include important basic model choices that have been proposed in OpRisk modeling scenarios in practice as well as in academic literature:

Fundamental Aspects of Operational Risk and Insurance Analytics: A Handbook of Operational Risk, First Edition. Marcelo G. Cruz, Gareth W. Peters, and Pavel V. Shevchenko.

- Generalized hyperbolic (GIG);
- Normal inverse-Gaussian (NIG);
- Generalized inverse Gaussian (GIG);
- Inverse Gaussian (IG) and the related family of Halphen severity models; and
- Elongation transform quantile function g-and-h severity distribution models.

This chapter will aim to make such models accessible and better understood by practitioners so they may consider application of such models in practice. For a detailed discussion on the properties of heavy-tailed models and additional results relating to their characterization, estimation, and modeling properties, we refer the interested reader to the advanced coverage in the companion book Peters and Shevchenko (2015).

9.2 Context of Flexible Heavy-Tailed Loss Models in OpRisk and Insurance LDA Models

In this section, we will seek to first provide the motivation for such families of models based on empirical studies on OpRisk banking losses, we will characterize each family of models as well as present relevant and useful modeling properties of these models. Then we will discuss parameter estimation under each model as well as properties that may be of relevance for each model when it comes to incorporation of these models into compound process LDA frameworks. In several cases, we will provide examples that illustrate how one may incorporate such models into LDA OpRisk modeling settings. The resulting features of the loss process will be examined analytically and numerically in the process.

Before entering into the detail of such models, we find it important to understand the motivation and justifications for considering such models in an OpRisk modeling framework. A bank adopting an Advanced Measurement Approach (AMA) must develop a comprehensive internal risk quantification system. This approach is the most flexible from a quantitative perspective, as banks may use a variety of methods and models, which they believe are most suitable for their operating environment and culture, provided they can convince the local regulator (BCBS, 2006, pp. 150–152). The key quantitative criterion is that a bank's models must sufficiently account for potentially high-impact rare events. As discussed previously, the idea of the LDA involves modeling the severity and frequency distributions over a predetermined time horizon, typically annual, as specified in, for instance, the Australian regulators documents, the prudential standard APS115 (see APRA 2008).

The fitting of frequency and severity distributions, as opposed to simply fitting a single parametric annual loss distribution, involves making the mathematical choice of working with compound distributions. This would seem to complicate the matter, since it is well known that, for most situations, analytical expressions for the distribution of a compound random variable are not attainable. However, as demonstrated in this section, there are particular model choices that can overcome this complication as they have severity distributions which will produce loss processes closed under convolution. These models include members of the Generalized Hypergometric family of severity models and also the α-Stable family of severity models; see detailed discussions on such models in Peters and Shevchenko (2015).

Typically, the reason for modeling severity and frequency distributions separately and then constructing a compound process is because some factors affect the frequency, while others may affect only the severity, see discussions in Panjer (2006). Some of the key points relating to why this is important in most practical settings are provided here in brief:

1. The expected number of operational losses will change as the company grows. Typically, growth needs to be accounted for in forecasting the number of OpRisk losses in future years, based on previous years;
2. Economic inflationary effects can be directly factored into the size of losses through scaling of the severity distribution;
3. Insurance and the impacts of altering policy limits and excesses are more easily understood by directly altering severity distributions;
4. Changing recording thresholds for loss events and the impact this will have on the number of losses required to be recorded is more transparent.

This can easily be understood when modeling is performed for frequency and severity separately. Alternative modeling approaches that also consider utilization of some of the heavy-tailed distributions discussed in this chapter have been proposed under a semi-linear credibility theory and Extreme Value Theory (EVT)-based framework, see discussions Lu *et al.* (2012). However, the most popular choices for frequency distributions in practical settings are Poisson, Binomial, and Negative Binomial. The typical choices of severity distribution include exponential, Weibull, LogNormal, Generalized Pareto, and the g-and-h family of distributions (Dutta and Perry 2006, Peters and Sisson 2006) and recently the α-Stable family (Peters *et al.* 2010).

Remark 9.1 *From the perspective of capital calculation, the most important processes to model accurately are those that have relatively infrequent losses. However, when these losses do occur, they are distributed as a very heavy-tailed severity distribution such as members of the subexponential family. Therefore, the intention of the following sections is to present families of models suitable for such severity distribution modeling, as well as their properties and estimators for the parameters that specify these models.*

The importance of the focus on particular heavy-tailed processes is highlighted in numerous reviews on OpRisk modeling. For instance, it was reported by Gagan (2008) that the total loss associated with OpRisk has reached as high as USD 96 billion in the US during the financial crisis in 2008. There have also been numerous OpRisk loss events that have been highlighted in the media to support such enormous aggregate figures. Some of the lesser reported cases have recently come to light with the paper of Lu *et al.* (2012), who paint similar pictures in Chinese banking sectors as have been observed in US and European markets. For example, they state that typical examples of large OpRisk loss events in recent years in the Chinese banking sector include the Guangdong branch of the Industrial and Commercial Bank of China (ICBC), which in 2003 lost 740 million yuan; the Jinzhou branch of the Bank of Communications in 2004, which lost 22.1 million yuan; the Heilongjiang branch of the Bank of China (BOC) in 2005, which lost 100 million yuan; the Guangdong branch of BOC in 2006, which lost 400 million yuan; and the Qilu Bank in 2010, which lost 100 million yuan.

These single loss events are significant and indicate the importance of models for loss processes which will capture such extreme loss events adequately when undertaking capital estimation.

9.3 Empirical Analysis Justifying Heavy-Tailed Loss Models in OpRisk

In this section, we summarize some of the key findings in an instrumental paper on empirical features of OpRisk data in US banking institutions from the Federal Reserve Bank of Boston (see Dutta and Perry 2006). In addition, we discuss and compare these findings with those of the Chinese banking system recently reported by Lu *et al.* (2012).

The first study of US banking institutions considered the 2004 Loss Data Collection Exercise (LDCE) survey data and narrowed down the number of suitable candidate data sets from all institutions surveyed to just seven institutions for which it was deemed that sufficient numbers of reported losses were acquired. The somewhat heuristic selection criterion that the authors utilized was that a total of at least 1000 reported total losses was required, and in addition each institution was required to have consistent and coherent risk profiles relative to each other, which would cover a range of business types and risk types as well as asset sizes for the institutions.

The second study on the Chinese banking sector utilized less reliable data sources as they adopted the approach of Feng *et al.* (2012), who collected loss data of Chinese commercial banks through the national media, covering 1990–2010. In the process of collecting data for banks, which include the four major state-owned commercial banks (SOCBs), nine joint-stock commercial banks (JSCBs), 35 city commercial banks (CCBs), 74 urban and rural credit cooperatives (URCCs), and 13 China Postal Savings (CPS) subsidiaries. The authors also note that the highest single OpRisk loss amount is up to 7.4 billion yuan, whereas the lowest amount is 50,000 yuan. In addition, losses measured in foreign currency were converted back via the real exchange rate when the loss occurred to convert it to the equivalent amount in yuan. Details of the incidence bank, incidence bank location, type of OpRisk loss, amount of loss, incident time and time span, and the sources of OpRisk events were noted.

Starting with the first study, the work of Dutta and Perry (2006), we note that in this paper the authors explored a number of key statistical questions relating to the modelling of OpRisk data in practical banking settings. As noted by Dutta and Perry (2006), a key concern for banks and financial institutions, when designing an LDA model, is the choice of model to use for modeling the severity (dollar value) of operational losses. In addition, a key concern for regulatory authorities is the question of whether institutions using different severity-modeling techniques can arrive at very different (and inconsistent) estimates of their exposure. They find, not surprisingly, that using different models for the same institution can result in materially different capital estimates. However, on the more promising side for LDA modeling in OpRisk, they find that there are some models that yield consistent and plausible results for different institutions even when their data differ in some core characteristics related to collection processes. This suggests that OpRisk data display some regularity across institutions which can be modeled. In this analysis, the authors note that they were careful to consider both the modeling of aggregate data at the enterprise level, which would group losses from different business lines and risk types, as well as modeling the attributes of the individual business line and risk types under the recommended business lines of Basel II/Basel III.

Data were collected from seven institutions, with each institution selected as it had at least 1000 loss events in total, and these data were part of the 2004 LDCE. Using these data, the authors performed a detailed statistical study of attributes of the data and flexible distributional models that could be considered for OpRisk models. Based on these seven data sources, over a range of different business units and risk types, they found that to fit all of the various data sets one would need to use a model that is flexible enough in its structure. Dutta and Perry (2006) considered modeling via several different means: parametric distributions, Extreme Value Theory (EVT) models, and nonparametric empirical models. In this chapter, we focus on the parametric models.

Dutta and Perry (2006) focused on models considered by financial institutions in Quantitative Impact Study 4 (QIS-4) submissions; these included one-two-, and four-parameter models. The one- and two-parameter distributions for the severity models included exponential, gamma, Generalized Pareto, loglogistic, truncated LogNormal, and Weibull. The four-parameter distributions included models such as the Generalized Beta Distribution of Second Kind (GB2) and the g-and-h distribution. These models were also considered Peters and Sisson (2006) for modeling severity models in OpRisk under a Bayesian framework. In this chapter, we consider these models as well as generalizations of these families of severity models.

Dutta and Perry (2006) discusss the importance of fitting distributions that are flexible but appropriate for the accurate modeling of OpRisk data, focus on the following five simple attributes in deciding upon a suitable statistical model for the severity distribution:

1. *Good fit.* Statistically, how well does the model fit the data?
2. *Realistic.* If a model fits well in a statistical sense, does it generate a loss distribution with a realistic capital estimate?
3. *Well-specified.* Are the characteristics of the fitted data similar to the loss data and logically consistent?
4. *Flexible.* How well is the model able to reasonably accommodate a wide variety of empirical loss data?
5. *Simple.* Is the model easy to apply in practice, and is it easy to generate random numbers for the purposes of loss simulation?

Their criterion was to regard any technique that is rejected as a poor statistical fit for the majority of institutions to be inferior for modeling OpRisk. The reason for this consideration was related to their desire to investigate the ability to find aspects of uniformity or universality in the OpRisk loss process that they studied. They concluded from the analysis undertaken that such an approach would suggest that OpRisk can be modeled and there is regularity in the loss data across institutions. While this approach combined elements of expert judgment and statistical hypothesis testing, it was partially heuristic and not the most formal statistical approach to address such problems. However, it does represent a plausible attempt given the limited data sources and resources, as well as the competing constraints mentioned in the measurement criterion they considered.

We note that an alternative purely statistical approach to such model selection processes was proposed for OpRisk modeling in the work of Peters and Sisson (2006), whose approach to model selection was to consider a Bayesian model selection based on Bayesian methodology of the Bayes Factor and information criterion for penalized model selection such as the Bayesian Information Criterion.

In both approaches, it is generally acknowledged that accurate selection of an appropriate severity model is paramount to appropriate modeling of the loss processes and therefore to the accurate estimation of capital.

Returning to the findings from the seven sources of OpRisk data studied by Dutta and Perry (2006), they found that the Exponential, Gamma, and Weibull distributions are rejected as good fits to the loss data for virtually all institutions at the enterprise, business line, and event-type levels. This was decided based on formal one-sample statistical goodness-of-fit tests for these models.

When considering the g-and-h distribution, they did not perform the standard hypothesis test for the goodness-of-fit, opting instead for a comparison of Quantile–Quantile (Q–Q) plots and diagnostics based on the five criteria posed earlier. In all situations, they found that the g-and-h distribution fit as well as other distributions on the Q–Q plot. The next most preferred distributions were the GB2, loglogistic, truncated LogNormal, and Generalized Pareto models, indicating the importance of considering flexible severity loss models. In addition, they noted that the EVT models fitted under a Peaks Over Threshold (POT) framework were also generally suitable fits for the tails, consistent with the discussions and findings for OpRisk data in the Chinese banking sector reported by Lu *et al.* (2012).

Having motivated the need for flexible families of loss distribution in OpRisk, in the following sections we present, several different families of severity distributional models along with a description of their features. This will allow practitioners and researchers to gain a deeper understanding of the flexible classes of models that are available from statistics to utilize in their LDA modeling exercises and more importantly the features and properties of such models that make them appropriate for OpRisk settings.

In general, when considering models for severity distributions in OpRisk, it is useful to recognize that distributional families typically fall into two broad classes of models: those with general forms for the density or distribution functions; and those that are defined by a family of transformations of a base distribution, and hence by their quantile function.

9.4 Quantile Function Heavy-Tailed Severity Models

In this section, we discuss a popular distributional family for severity models in OpRisk which can only be specified via the transformation of another standard random variable such as a Gaussian. Examples of OpRisk severity models defined through their quantile functions include the Johnson family with base distribution given by Gaussian or logisitic and the Tukey family with base distribution Gaussian or logistic. The concept of constructing skew and heavy-tailed distributions through the use of a transformation of a Gaussian random variable was originally proposed in the work of Tukey (1977a) and is therefore aptly named the family of Tukey distributions. This family of distributions was then extended by Hoaglin (1985) and Jorge and Boris (1984). Within this family of distributions, two particular subfamilies have received the most attention in the literature; these correspond to the g-and-h and the g-and-k distributions. The first of these families has been studied in a few contexts in OpRisk (see Dutta and Perry 2006, Peters and Sisson 2006, Degen *et al.* 2007, and Jiménez and Arunachalam 2011, and the references therein for applications).

Before presenting details of the g-and-h and the g-and-k distributions, we first discuss the general family of Tukey distributions. Basically, Tukey suggested several nonlinear transformations of a standard Gaussian random variable, denoted here by $W \sim Normal(0, 1)$ so as not to be confused with the annual loss that we denote throughout by Z. The g-and-h transformations involve a skewness transformation g and a kurtosis transformation h. If one replaces the kurtosis transformation of the type h with the type k, one obtains the g-and-k family of distributions discussed by Rayner and MacGillivray (2002). If the h transformation is replaced by the j transformation, one obtains the g-and-j transformations of Fischer and Klein (2004).

We begin with the generic specification of the Tukey transformation given in Definition 9.1. These types of transformations were labelled elongation transformations, where the notion of elongation was noted to be closely related to tail properties such as heavy-tailedness (see discussions by Hoaglin 1985). In considering such a class of elongation transformations to obtain a distribution, one is comparing the tail strength of the new distribution with that of the base distribution (such as a Gaussian or logistic). In this regard, one can think of tail strength or heavy-tailedness as an absolute concept, whereas the notion of elongation strength is a relative concept. In the following, we will first consider relative elongation compared to a base distribution for a generic random variable W. It should be clear that such a measure of relative tail behavior is independent of location and scale. Other properties, that such an elongation transform $T(\cdot)$, should satisfy are that it should preserve symmetry $T(w) = T(-w)$, and the base distribution should not be significantly transformed in the center, such that $T(w) = w + O(w^2)$ for w around the mode. Then, to increase the tails of the resulting distribution relative to the base, it is important to assume that T is strictly monotonically increasing transform that is convex, that is, one has the transform satisfying for positive $w > 0$ that $T'(w) > 0$ and $T''(w) > 0$. One such transformation family satisfying these properties includes the Tukey transformations.

Definition 9.1 (Tukey transformations) *Consider a Gaussian random variable $W \sim Normal(0, 1)$ and a transformation $T(w)$ given by*

$$X = W\,T(W)^{\theta}, \tag{9.1}$$

for a parameter $\theta \in \mathbb{R}$. ■

Typically, in the OpRisk setting, it will be desirable when working with such severity models to enforce a constraint that the tails of the resulting distribution after transformation are heavier than the Gaussian distribution. In this case, one should consider a transformation $T(w)$, which is positive, symmetric, and strictly monotonically increasing for positive values of $W \geq 0$. In addition, it will be desirable to obtain this property of heavy tails relative to the Gaussian to also consider setting the parameter $\theta \geq 0$. As discussed, a series of kurtosis transformations were proposed in the literature. The Tukey h, k, and j transforms are provided in Definition 9.2.

Definition 9.2 (Tukey's kurtosis transformations, h, k and j types)
The h-type of transformation, denoted by $T_h(w)$, is given by

$$T_h(w) = \exp\left(w^2\right). \tag{9.2}$$

The k-type of transformation, denoted by $T_k(w)$, is given by

$$T_k(w) = 1 + w^2. \tag{9.3}$$

The j-type of transformation, denoted by $T_j(w)$, is given by

$$T_j(w) = \frac{1}{2}\left[\exp(w) + \exp(-w)\right]. \tag{9.4}$$

■

EXAMPLE 9.1 Shape of Tukey Base Elongation Transforms for Kurtosis: h, j, k

In Figure 9.1, we plot the simple base transforms for the h, j and k Tukey elongation transforms.

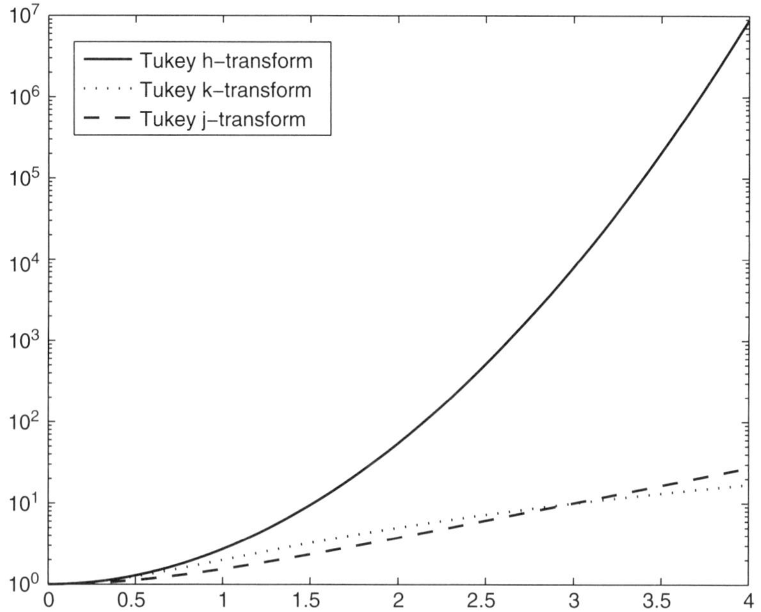

FIGURE 9.1 **Base transforms for the Tukey elongation kurtosis transforms**

These plots demonstrate that the base j and k transforms have a similar effect on the tails of the base distribution, which is distinct in the kurtosis introduced by the h transform. ■

To nest all these transformations within one class of transformations, the work of Fischer (2010) proposed a power series representation denoted by the subscript a given in Equation (9.5). This suggestion, though it nested the other families of distributions, is not practical for use as it involves the requirement of estimating a very large (infinite) number of parameters a_i to obtain the data-generating mechanism:

$$T_a(w) = \sum_{i=0}^{\infty} a_i w^{2i}. \tag{9.5}$$

As a consequence, this nesting structure was replaced with the general transformation given by Fischer (2010) which took the form given in Equation (9.6):

$$T_{hjk}(w; \alpha, \beta, \gamma) = \left(1 + \frac{(w^2+\gamma)^{\alpha} - \gamma^{\alpha}}{\beta}\right)^{\beta}, \quad \alpha > 0, \beta \geq 1, \gamma > 0. \tag{9.6}$$

Then it is clear that the original h, k, and j transformations are recovered with $T_h(w) = T_{hjk}(w; 1, \infty, \gamma)$, $T_k(w) = T_{hjk}(w; 1, 1, \gamma)$, and $T_j(w) \approx T_{hjk}(w; 0.5, \infty, 0.5)$.

EXAMPLE 9.2 Tukey Elongation Transform Density Shapes

Consider the Tukey elongation transform of a base reference random variable $W \sim Normal(0,1)$ given by the generic transform super class of Fischer (2010) according to

$$T_{hjk}(w;\alpha,\beta,\gamma) = \left(1 + \frac{(w^2+\gamma)^{\alpha} - \gamma^{\alpha}}{\beta}\right)^{\beta}, \quad \alpha > 0, \beta \geq 1, \gamma > 0. \quad (9.7)$$

The plots in Figures 9.2–9.4 show the distributions of this general transform relative to the base Gaussian distribution without truncation, scaling, or translation parameters — purely the elongation transform effects. In the first set of plots (Figure 9.2), we consider the effect of the parameter α in the generalized transform.

In the second set of plots (Figure 9.3), we consider the effect of the parameter β in the generalized transform.

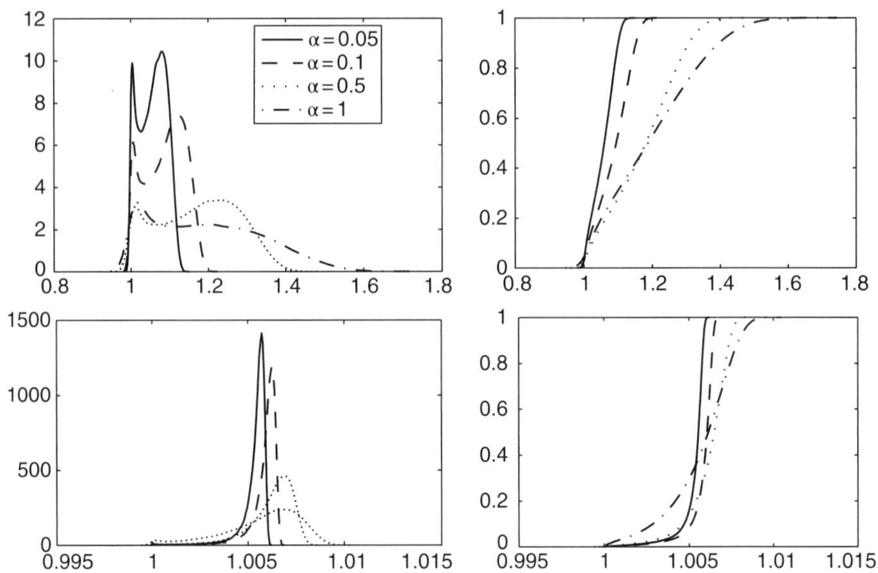

FIGURE 9.2 Top left subplot: The density of the T_{hjk} transform with base standard Gaussian distribution and parameters $\alpha \in \{0.05, 0.1, 0.5, 1\}$, $\beta = 0.1$, and $\gamma = 0.1$. Top right subplot: The distribution function of the T_{hjk} transform with base standard Gaussian distribution and parameters $\alpha \in \{0.05, 0.1, 0.5, 1\}$, $\beta = 0.1$, and $\gamma = 0.1$. Bottom left subplot: The density of the T_{hjk} transform with base standard Gaussian distribution and parameters $\alpha \in \{0.05, 0.1, 0.5, 1\}$, $\beta = 0.001$, and $\gamma = 0.001$. Bottom right subplot: The distribution of the T_{hjk} transform with base standard Gaussian distribution and parameters $\alpha \in \{0.05, 0.1, 0.5, 1\}$, $\beta = 0.001$, and $\gamma = 0.001$

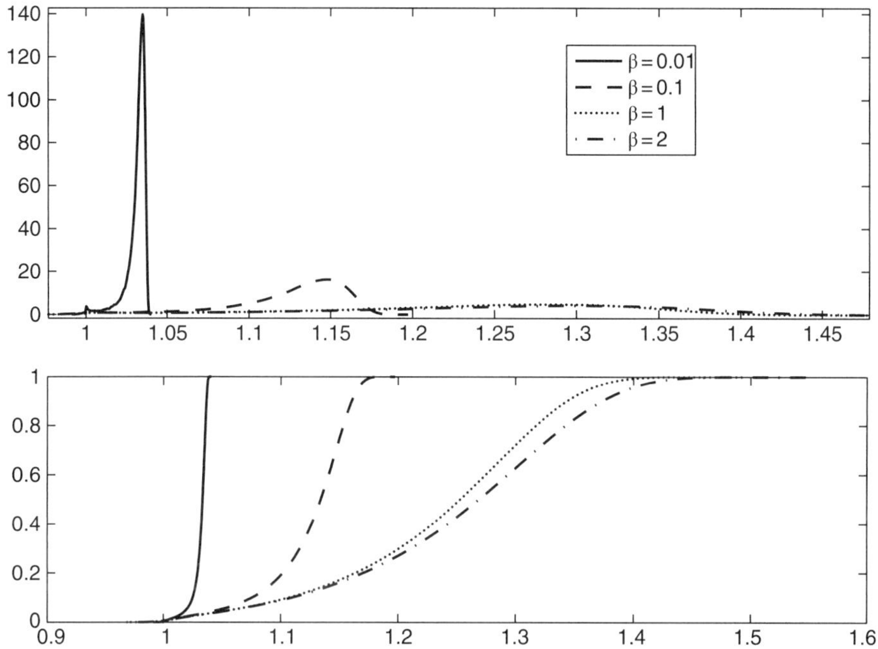

FIGURE 9.3 **Top subplot: The density of the T_{hjk} transform with base standard Gaussian distribution and parameters $\alpha = 0.05$, $\beta \in \{0.01, 0.1, 1, 2\}$, and $\gamma = 0.001$. Bottom subplot: The distribution of the T_{hjk} transform with base standard Gaussian distribution and parameters $\alpha = 0.05$, $\beta \in \{0.01, 0.1, 1, 2\}$, and $\gamma = 0.001$**

In the third set of plots (Figure 9.4), we consider the effect of the parameter γ in the generalized transform. ■

These examples simply demonstrate the flexible distributional shapes that can be obtained with the basic elongation transform given by the generalized transform of Fischer (2010) for different sets of parameter values. In terms of practical severity models, we will now continue to parameterize these transforms to provide sufficient parameters that may make these models suitable for OpRisk loss modeling, leading to, for example, the g-and-h distribution family.

Under the T_{hjk} superclass of transformations, one can state the following basic properties. Assuming that $W \sim Normal(0, 1)$ will produce the severity random variable $X = K(W) = WT_{hjk}(W)^{\theta}$, the severity density $f_X(\cdot)$ and quantile functions $Q_X(\cdot)$, for loss random variable X, are given by

$$\begin{aligned} f_X(x) &= \frac{1}{Q_X'\left(Q_X^{-1}(x)\right)} \\ &= \frac{\phi\left(K^{-1}(x)\right)}{K'\left(K^{-1}(x)\right)}, \quad \inf\{x : x \in S\} < x < \sup\{x : x \in S\} \qquad (9.8) \\ Q_X(\alpha) &= K\left(Q_W(\alpha)\right), \quad \alpha \in [0, 1], \end{aligned}$$

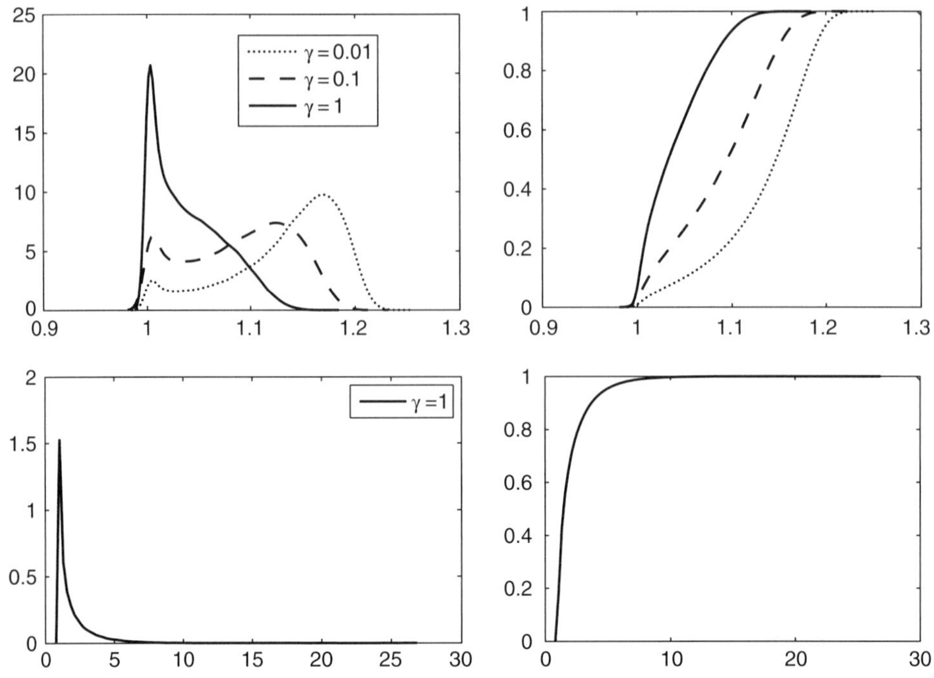

FIGURE 9.4 Top left subplot: The density of the T_{hjk} transform with base standard Gaussian distribution and parameters $\alpha = 0.1$, $\beta = 0.1$, and $\gamma \in \{0.01, 0.1, 1\}$. Top right subplot: The distribution function of the T_{hjk} transform with base standard Gaussian distribution and parameters $\alpha = 0.1$, $\beta = 0.1$, and $\gamma \in \{0.01, 0.1, 1\}$. Bottom left subplot: The density of the T_{hjk} transform with base standard Gaussian distribution and parameters $\alpha = 1$, $\beta = 1$, and $\gamma = 1$. Bottom right subplot: The distribution of the T_{hjk} transform with base standard Gaussian distribution and parameters $\alpha = 1$, $\beta = 1$, and $\gamma = 1$

with S the appropriate support of the random variable X and

$$K'(w) = T_{hjk}(w)^{\theta-1}\left(T_{hjk}(w) + \theta w T'_{hjk}(w)\right).$$

Other generalizations to the Tukey family include those of Rayner and MacGillivray (2002), who propose the generalized forms for the quantile functions of the g-and-h and g-and-k families of distributions given in Equations (9.9) and (9.10). The generalized g-and-h and g-and-k families have $b > 0$ and c is a constant to ensure proper distributions are obtained.

$$Q_X(\alpha; a, b, g, h) = a + bQ_W(\alpha)\left(1 + c\frac{1-\exp(-gQ_W(\alpha))}{1+\exp(-gQ_W(\alpha))}\right)\exp\left(\frac{1}{2}hQ_W(\alpha)^2\right) \quad (9.9)$$

$$Q_X(\alpha; a, b, g, k) = a + bQ_W(\alpha)\left(1 + c\frac{1-\exp(-gQ_W(\alpha))}{1+\exp(-gQ_W(\alpha))}\right)\left(1 + Q_W(\alpha)^2\right)^k. \quad (9.10)$$

Next, we explain the properties of specific subfamilies of distributions, showing how these results are derived for the basic g-and-h family.

9.4.1 G-AND-H SEVERITY MODEL FAMILY IN OPRISK

The family of g-and-h distributions was first studied by Tukey (1977b) and then considered in a number of works such as those by Hoaglin (1985), Azzalini (1985), and Fischer *et al.* (2007). The multivariate versions of this model have been discussed by Field and Genton (2006).

The advantage of the g-and-h family for modeling severity in an OpRisk LDA framework is the fact that it provides a very flexible range of skew, kurtosis, and heavy-tailed features while also being specified as a rather simple transformation of standard Gaussian random variates, making simulation of the annual loss under such a model efficient and simple. It is important to note that the support of the g-and-h density incudes the entire real line, as such, one must be cautious in OpRisk settings to manage the treatment of the parameter settings to restrict the probability of negative values as much as possible. This can be achieved either by truncation or by restriction of the parameter values. In some subfamily members, the g-and-h family automatically takes a positive support such as the Double h–h subfamily. In addition, it has been shown that the g-and-h distribution can approximate most members of the Personian family of distributions up to a desired level of accuracy.

9.4.1.1 g-and-h, g, h, and h–h Family Transformations.

The g-and-h family can be considered as composed of three transformations that can produce subfamilies of non-Gaussian distributions for severity based on the g-distributions, the h-distributions, and the g-and-h distributional families. The basic specifications in which g and h components are treated as constants are given in Definitions 9.3, 9.4, and 9.5 in terms of transformations of Gaussian random variables.

Definition 9.3 (g-and-h Distributional family) *Let $W \sim Normal(0, 1)$ be a standard Gaussian random variable. Then the loss random variable X has severity distribution given by the g-and-h distribution with parameters $a, b, g, h \in \mathbb{R}$, denoted $X \sim GH(a, b, g, h)$, if X is given by (for $g \neq 0$)*

$$X = T_{g,h}(W; a, b, g, h) := a + b\frac{\exp(gW) - 1}{g}\exp\left(\frac{hW^2}{2}\right). \tag{9.11}$$

■

The parameters a and b are linear transformations whereas the parameters g and h can be significantly extended to polynomials as discussed later, and play an important role in the skewness and kurtosis properties of the g-and-h family.

Remark 9.2 *In general, one may consider the constants g and h to be more flexibly selected as polynomials, which would include higher orders of W^2. These polynomials could take the form, for example, of any integers p and q:*

$$\begin{aligned} g(w) &:= \alpha_0 + \alpha_1 w + \cdots + \alpha_p w^p, \\ h(w) &:= \beta_0 + \beta_1 w + \cdots + \beta_q w^q. \end{aligned} \tag{9.12}$$

The addition of these polynomial terms can provide additional degrees of freedom to improve the ability to fit data. These have been shown to be significant when modeling certain types of OpRisk data, as demonstrated by Dutta and Perry (2006) and Peters and Sisson (2006).

EXAMPLE 9.3 g-and-h Elongation Transform Density Shapes (Base Gaussian Distribution)

Consider the g-and-h elongation transform of a base reference random variable $W \sim Normal(0, 1)$ given by

$$X = T_{g,h}(W; a, b, g, h) := a + b\frac{\exp(gW) - 1}{g} \exp\left(\frac{hW^2}{2}\right). \tag{9.13}$$

The plots in Figure 9.5 show the distributions of this general transform relative to the base Gaussian distribution without truncation, scaling, or translation parameters — purely the elongation transform effects. In the first set of plots (Figure 9.5), we consider the effect of the parameters g and h in the generalized transform.

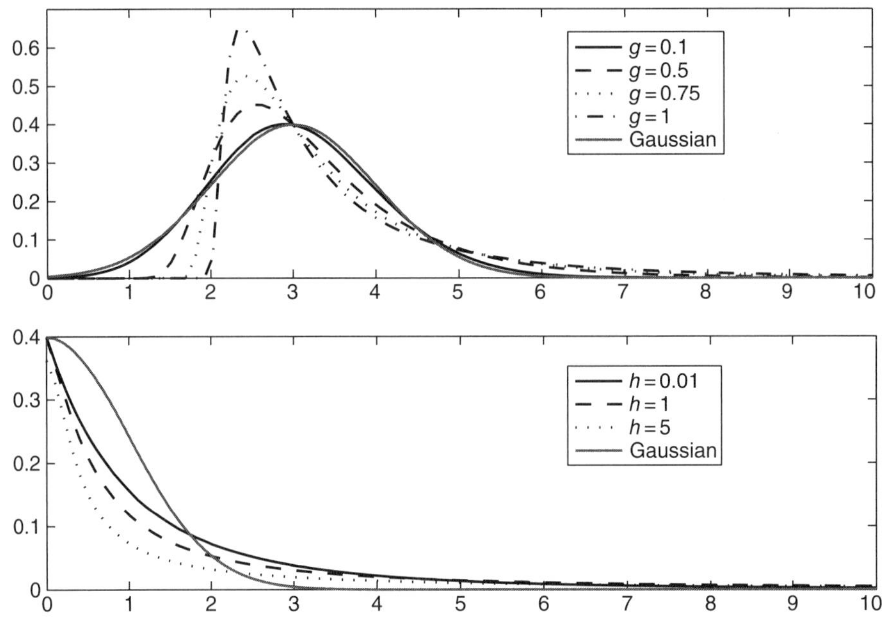

FIGURE 9.5 Top subplot: This plot shows the effect of the skewness parameter g on the elongation transformed severity distribution versus the base Gaussian distribution with $g \in \{0.1, 0.5, 0.75, 1\}$. In this case, the other parameters were set to $a = 3$, $b = 1$, and $h = 0.001$. Bottom subplot: This plot shows the effect of the kurtosis parameter h on the elongation transformed severity distribution versus the base Gaussian distribution with $h \in \{0.01, 1, 5\}$. In this case, the other parameters were set to $a = 0$, $b = 1$, and $g = 1$

In the second example (Figure 9.6), we demonstrate the effect of changing the base distribution to a LogNormal density model instead of the Gaussian distribution.

EXAMPLE 9.4 g-and-h Elongation Transform Density Shapes: Base Gaussian versus Logistic Distribution

Consider the g-and-h elongation transform of one of two base reference random variables $W \sim Normal(0, 1)$ or $W \sim LogNormal(0, 1)$ given by

$$X = T_{g,h}(W; a, b, g, h) := a + b\frac{\exp(gW) - 1}{g} \exp\left(\frac{hW^2}{2}\right). \tag{9.14}$$

The plot in Figure 9.6 shows the distributions of this general transform relative to the base Gaussian distribution without truncation, scaling, or translation parameters — purely the elongation transform effects. In the first set of plots, we consider the effect of the parameter g and h in the generalized transform.

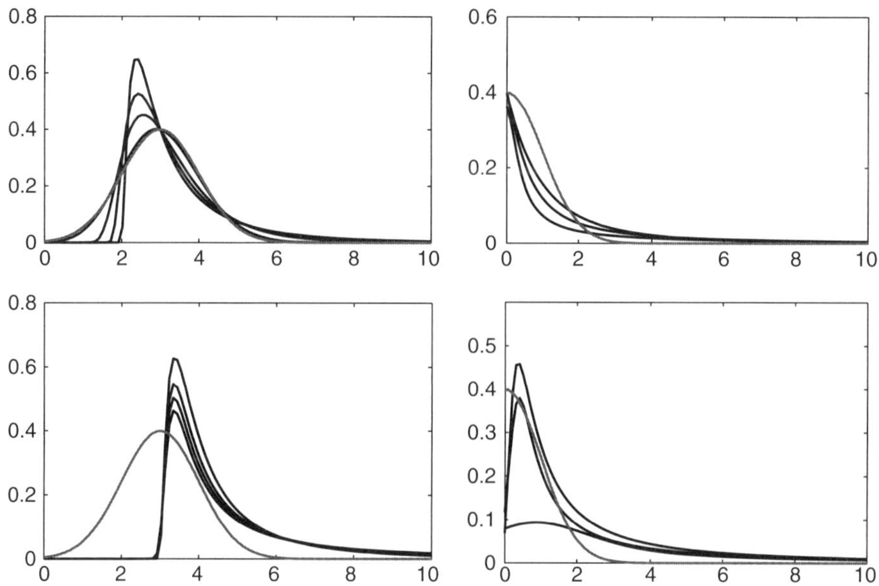

FIGURE 9.6 Top left subplot: This plot shows the effect of the skewness parameter g on the elongation transformed severity distribution versus the base Gaussian distribution with $g \in \{0.1, 0.5, 0.75, 1\}$. In this case, the other parameters were set to $a = 3$, $b = 1$, and $h = 0.001$. Top right subplot: This plot shows the effect of the kurtosis parameter h on the elongation transformed severity distribution versus the base Gaussian distribution with $h \in \{0.01, 1, 5\}$. In this case, the other parameters were set to $a = 0$, $b = 1$, and $g = 1$. Bottom left subplot: This plot shows the effect of the skewness parameter g on the elongation transformed severity distribution versus the base LogNormal(0,1) distribution with $g \in \{0.1, 0.5, 0.75, 1\}$. In this case, the other parameters were set to $a = 3$, $b = 1$, and $h = 0.001$. Bottom right subplot: This plot shows the effect of the kurtosis parameter h on the elongation transformed severity distribution versus the base LogNormal(0,1) distribution with $h \in \{0.01, 1, 5\}$. In this case, the other parameters were set to $a = 0$, $b = 1$, and $g = 1$

Remark 9.3 *Dutta and Perry (2006) recommended that a reasonable range of parameter values for the parameters g and h was to restrict them to be* $g, h > 0$ *and in particular the enterprise level modeling they performed involved the ranges* $g \in [1.79, 2.30]$ *and* $h \in [0.10, 0.35]$.

Within this family of g-and-h distributions, one can also define the subfamilies of distributions given by the g and the h families. Again, we present these models in their simplest form, with constant g or h, though in practice one may include polynomials in W for such models.

Definition 9.4 (g Distributional Family) *Let* $W \sim Normal(0, 1)$ *be a standard Gaussian random variable. Then the loss random variable* X *has severity distribution given by the g distribution with parameters* $a, b, g \in \mathbb{R}$, *denoted* $X \sim G(a, b, g)$, *if* X *is given by (for* $g \neq 0$*)*

$$X = T_g(W; a, b, g) := a + b\frac{\exp(gW) - 1}{g}. \tag{9.15}$$

■

Remark 9.4 *Note that the g-distribution subfamily corresponds (in the case that g is a constant) to a scaled LogNormal distribution.*

Definition 9.5 (h Distributional Family) *Let* $W \sim Normal(0, 1)$ *be a standard Gaussian random variable. Then the loss random variable* X *has severity distribution given by the h distribution with parameters* $a, b, h \in \mathbb{R}$, *denoted* $X \sim H(a, b, h)$, *if* X *is given by*

$$X = T_h(W; a, b, h) := a + bW\exp\left(\frac{hW^2}{2}\right). \tag{9.16}$$

■

In addition, one may obtain an asymmetric class of *h–h* distributions studied by Morgenthaler and Tukey (2000, section 2.2), Headrick and Pant (2012a and 2012b). The asymmetric *h–h* distribution transformation is given in Definition 9.6.

Definition 9.6 (Double *h–h* Distributional Family) *Let* $W \sim Normal(0, 1)$ *be a standard Gaussian random variable. Then the loss random variable* X *has severity distribution given by the unit h–h distribution with parameters* $h_l, h_r \in \mathbb{R}$, *denoted* $X \sim HH(h_l, h_r)$, *if* X *is given by*

$$X = T_{h,h}(W; h_l, h_r) := \begin{cases} W\exp\left(\frac{1}{2}h_l W^2\right), & W \leq 0, \\ W\exp\left(\frac{1}{2}h_r W^2\right), & W \geq 0, \end{cases} \tag{9.17}$$

for $h_r \geq 0$ *and* $h_l \geq 0$. ■

In addition to these families of Tukey transformations discussed, there have been modified g-and-h families developed based on L-moments. The L-moment Tukey transformation families developed by Headrick and Pant (2012b) are based on transformation of a random variable which involves a logistic distribution that is, the distribution of random variable W is changed from the standard Gaussian to the following form $W \sim F(w; \mu, s)$, which has density, distribution, and quantile functions

$$\begin{aligned} f(w) &= \frac{\exp\left(-(w-\mu)/s\right)}{s\left(1+\exp\left(-(w-\mu)/s\right)\right)^2}, \\ F(w) &= \frac{1}{1+\exp\left(-(w-\mu)/s\right)}, \\ Q_W(\alpha) &= \mu + s\ln\left(\frac{\alpha}{1-\alpha}\right), \quad \alpha \in [0,1], \end{aligned} \tag{9.18}$$

for all $w \in \mathbb{R}$, $\mu \in \mathbb{R}$, and $s \in \mathbb{R}^+$. The motivation for modifying the distribution transformed under the Tukey structure was related to the fact that inference on the parameters was to be performed with L-moments and L-correlation. The four classes of modified Tukey quantile function transformations are then given in Definition 9.7

Definition 9.7 (L-Moment Tukey Transforms) *Let $W \sim Logistic(\mu = 0, s = 1)$ be a standard logistic distributed random variable. Then the loss random variable X has severity distribution given by the L-moment Tukey family as follows:*

1. *The $\gamma-\kappa$ Tukey family transformation is given by*

$$X = T_{\gamma,\kappa}(W) = \gamma^{-1}\left(\exp(\gamma W) - 1\right)\exp(\kappa|W|). \tag{9.19}$$

   This is the analog of the g-and-h Tukey transform for the logistic distribution case for $\gamma \neq 0$ and $\kappa \geq 0$;

2. *The $\kappa_L-\kappa_R$ Tukey family transformation is given by*

$$X = T_{\kappa_L,\kappa_R}(W) = \begin{cases} W\exp\left(\kappa_L|W|\right), & W \leq 0 \\ W\exp\left(\kappa_R|W|\right), & W \geq 0. \end{cases} \tag{9.20}$$

   This is the analog of the double h–h Tukey transform for the logistic distribution case for $\kappa_L \geq 0$, $\kappa_R \geq 0$, and $\kappa_L \neq \kappa_R$.

■

Algorithm 9.1 (Simulating Losses from a g-and-h Severity Model)

1. *Draw a standard Gaussian random variate: $Z_i \sim Normal(0,1)$;*
2. *Given p, q, and coefficients $\{\alpha_i\}_{i=0}^p$ and $\{\beta_i\}_{i=0}^q$, evaluate the polynomials*

$$\begin{aligned} g\left(W_i\right) &= \alpha_0 + \alpha_1 W_i + \cdots + \alpha_p W_i^p \\ h\left(W_i\right) &= \beta_0 + \beta_1 W_i + \cdots + \beta_q W_i^q. \end{aligned} \tag{9.21}$$

3. *Then, given parameters a, b and polynomials $g\left(W_i\right)$ and $h\left(W_i\right)$, evaluate transformation*

$$X_i = a + b\frac{\exp\left(g\left(W_i\right)W_i\right) - 1}{g\left(W_i\right)}\exp\left(\frac{h\left(W_i\right)W_i^2}{2}\right).$$

9.4.1.2 g-and-h, g, h, and h–h Family Distribution, Density, and Statistical Properties. Next we discuss properties of the g-and-h family and in particular different ways that people have sought to evaluate and present the distribution and density functions for the g-and-h family. In general, it will be informative for this section to remind the reader of the following basic property.

Proposition 9.1 *If X is a continuous random variable distributed according to distribution $X \sim F(x)$, which is monotonically increasing on support $Supp\{F(x)\} = \{x : 0 < F(x) < 1\}$, then, in this general case, one can show that the quantile function $Q_X(\alpha) = F^{-1}(\alpha)$ for $\alpha \in [0, 1]$ determines the relationship between the random variable X and any other continuous random variable with monotonically increasing distribution, say $W \sim G(w)$. The relationship is then specified through the transformation*

$$X \stackrel{d}{=} F^{-1}(G(W)). \tag{9.22}$$

When the random variable of W is standard Gaussian as utilized in the g-and-h family, one can show that for any continously differentiable transformation $X = T(W)$, X will have a density given in Equation (9.23) with respect to the standard Gaussian density $\phi(\cdot)$. In this case, one can also observe that when the transform $T(\cdot)$ increases rapidly, the resulting density is heavy-tailed. For instance, a linear growth in the function $T(\cdot)$ results in tail behavior for the distribution of random variable X being equivalent to a Gaussian:

$$f_X(x) = \frac{\phi\left(T^{-1}(x)\right)}{T'\left(T^{-1}(x)\right)}. \tag{9.23}$$

As observed previously, the Tukey family has a transformation $T(\cdot)$ given in Definition 9.1:

$$X = T(W) = W\left(\frac{h}{2}W^2\right). \tag{9.24}$$

Note: The original Tukey h-type transformation had $\theta = 1$ and an addition scaling of $\frac{1}{2}$ as indicated earlier. This transformation has the property that its derivative

$$\frac{d}{dw}T(w) = \left(1 + hw^2\right)\exp\left(\frac{1}{2}hw^2\right) \geq 1 \tag{9.25}$$

for all $h \geq 0$.

In addition, in the following discussions, it will be useful to recall the following properties of the g-and-h family of distributions (see discussions in Dutta and Babbel 2002):

1. The g-and-h transformation can be shown to be strictly monotonically increasing in its argument, that is, for all $w_1 \leq w_2$ one has $T_{gh}(w_1) \leq T_{gh}(w_2)$;
2. If $a = 0$, then the g-and-h transformation satisfies the condition $T_{-g,h}(W) = -T_{g,h}(-W)$.

Degen *et al.* (2007) observed that one can specify the distribution function of the g-and-h distribution as given in Definition 9.8 via a composite function. Note that the scale parameters

a and b can be dropped without loss of generality. The particular representation developed by these authors was convenient to allow one to obtain a closed-form representation of the Value-at-Risk for such a random variable.

Definition 9.8 (g-and-h Distribution Function (constant g and h with $h > 0$)) *Consider the g-and-h distributed random variable $X \sim GH(a = 0, b = 1, g, h)$ with constant parameters g and $h > 0$. The distribution function can be specified according to the following composite function:*

$$F_X(x; g, h) = \Phi\left(k^{-1}(x)\right), \tag{9.26}$$

where $\Phi(\cdot)$ is the standard Gaussian distribution and the function $k(x)$ is specified by

$$k(x) = \frac{\exp(gx) - 1}{g} \exp\left(\frac{hx^2}{2}\right). \tag{9.27}$$

■

In this parametrization, the parameter g will control the skew of the distribution both in terms of the sign and the magnitude, while the parameter h will control heaviness of the tails and is related directly to the kurtosis. This will be discussed further when the regular variation properties of this model are explored. Under the restricted parametrization for the distribution given in Definition 9.8, one can obtain the quantile function given in Definition 9.9.

Definition 9.9 (g-and-h Quantile Function (constant g and h with $h > 0$)) *Consider the g-and-h distributed random variable $X \sim GH(a = 0, b = 1, g, h)$ with constant parameters g and $h > 0$. The quantile function for a level $\alpha \in [0, 1]$ can be specified according to the following representation:*

$$Q_X(\alpha) := q_X(\alpha; g, h) = F_X^{-1}(\alpha; g, h) = k\left(\Phi^{-1}(\alpha)\right), \tag{9.28}$$

where $\Phi(\cdot)$ is the standard Gaussian distribution and the function $k(x)$ is specified by

$$k(x) = \frac{\exp(gx) - 1}{g} \exp\left(\frac{hx^2}{2}\right). \tag{9.29}$$

■

Headrick *et al.* (2008) approach the problem of specification of the distribution and density for generic parameterizations of the g-and-h family from an analytic geometry perspective; this can be seen to be an analogous representation of the approach discussed earlier by Degen *et al.* (2007). In fact, the representations they obtained were equivalent with equivalent parameter restrictions. We briefly mention these results here as they allow for an alternative perspective on how one obtains the results in Definitions 9.8 and 9.9. To proceed with the representation developed by Headrick *et al.* (2008), one needs to define $\phi(w) = f_W(w)$ and $\Phi(w) = F_W(w)$, respectively, to be the curves that characterize the standard Gaussian density and distribution. Then consider that w is actually comprised of two components (a vector) with an additional auxiliary variable such that $w = (x, y)$ will produce the following mappings of the curves $f_W(w)$ and $F_W(w)$ according to the following relationships:

$$
\begin{aligned}
f_W(w) &: w \to \mathbb{R}^2 := f_W\left(w, f_W(w)\right), \\
F_W(w) &: w \to \mathbb{R}^2 := F_W\left(w, F_W(w)\right).
\end{aligned}
\tag{9.30}
$$

Given these mappings, one may then define analytically the form of the quantile function for the g-and-h distribution according to the expression given in Equation (9.1), which we reiterate below to present the quantile function notation $Q_X(w)$ of the g-and-h loss random variable given by

$$
q(w; g, h) = \frac{\exp(gw) - 1}{g} \exp\left(\frac{hw^2}{2}\right). \tag{9.31}
$$

Here, $q(w; g, h)$ is a strictly increasing monotonic function in w with the parameter restrictions $g \neq 0$ and $h > 0$. Using these definitions Headrick *et al.* (2008) then provide a specification for the density and distribution functions for the g-and-h family as detailed in Definition 9.10.

Definition 9.10 (g-and-h Distribution and Density Functions) *Consider the g-and-h distributed random variable $X \sim GH(a = 0, b = 1, g, h)$ with constant parameters g and $h > 0$. The density and distribution functions associated to the quantile function $q(w; g, h)$ can be specified according to the following composite functions based on the auxiliary variable $w = (x, y)$:*

$$
\begin{aligned}
f \circ q &: q(w; g, h) \to \mathbb{R}^2 := f_{q(W;g,h)}(w) = f_{q(W;g,h)}\left(q(w; g, h), \frac{f_W(w)}{q'(w; g, h)}\right), \\
F \circ q &: q(w; g, h) \to \mathbb{R}^2 := F_{q(W;g,h)}(w) = F_{q(W;g,h)}\left(q(w; g, h), F_W(w)\right),
\end{aligned}
\tag{9.32}
$$

where $q'(z; g, h)$ denotes the derivative of the quantile function given by

$$
\begin{aligned}
q'(z; g, h) &:= \frac{d}{dz}\left[\frac{\exp(gz) - 1}{g} \exp\left(\frac{hz^2}{2}\right)\right] \\
&= \exp\left(gz + \frac{hz^2}{2}\right) + \frac{h}{g} z \exp\left(\frac{hz^2}{2}\right)(\exp(gz) - 1).
\end{aligned}
\tag{9.33}
$$

■

One advantage of the specification of the distribution and density functions with regard to a particular quantile function is that the statistical properties of these distributions can now be easily studied. For instance, the mode and moments of the distribution can be characterized. The result in Proposition 9.2 provides the mode for the g-and-h distribution in Definition 9.10.

Proposition 9.2 (Mode and Median of the g-and-h Density) *Consider the g-and-h distributed random variable $X \sim GH(a = 0, b = 1, g, h)$ with constant parameters g and $h > 0$. The mode of the density is located at the value $\widetilde{w} = Mode[W]$, which produces a maximum value of the density at $f_{q(W;g,h)}(\widetilde{w})$ and can be found as the solution to the following equation when $w = \widetilde{w}$, which is selected to satisfy*

$$
\frac{d}{dw}\left[\frac{f_W(w)}{q'(w; g, h)}\right] = 0. \tag{9.34}
$$

The median of the g-and-h distributed random variable is then given by $w_{0.5} = Median[W]$ *and will correspond to the median being the limit of* $\lim_{w\to 0} q(w; g, h) = 0$. *Therefore, one sees that in the general g-and-h distribution the median of the data set will be the parameter a.*

These can be found numerically and for the mode the solution obtained will be guaranteed to be a globally optimal solution (see discussion by Headrick *et al.* 2008). It may also be noted that in the case of the h-type and double h-type Tukey distributions, the median and mode are at the origin (for $a = 0$).

In addition, one may now express the moments of the g-and-h distributed random variable according to the results in Proposition 9.3. It was noted by Dutta and Babbel (2002) that since the g-distribution is a horizontally shifted LogNormal distribution, then the moments of the g-distribution take the same form as those of a LogNormal model with appropriate adjustment for the translation. The h-distributional family is symmetric (except the double h–h family); consequently, all odd-order moments for the h-subfamily are zero.

Proposition 9.3 (Moments of the g-and-h Density) *Consider the g-and-h distributed random variable* $X \sim GH(a = 0, b = 1, g, h)$ *with constant parameters g and* $h > 0$. *The r-th integer moment of the distribution in Equation (9.32) is given with respect to the standard Normal distribution and the r-th power of the quantile function* $q(w)$ *by*

$$\mathbb{E}[X^r] = \mathbb{E}[q(W; g, h)^r] = \int_{-\infty}^{\infty} q(w; g, h)^r f_W(w) dw, \tag{9.35}$$

which will exist if $h \in \left[0, \frac{1}{r}\right)$. *One can also observe more generally that under the g-and-h transform the following identity holds with regard to powers of the standard Gaussian,* $W \sim Normal(0, 1)$, *such that*

$$\begin{aligned} X^n &= T_{g,h}(W; a, b, g, h)^n \\ &= \left(a + bT_{g,h}(W; a = 0, b = 1, g, h)\right)^n \\ &= \sum_{i=0}^{n} \frac{n!}{(n-i)!i!} a^{n-i} b^i T_{g,h}(W; a = 0, b = 1, g, h)^i, \end{aligned} \tag{9.36}$$

which will produce moments given by

$$\begin{aligned} \mathbb{E}[X^n] &= \mathbb{E}\left[\left(a + bT_{g,h}(W; a = 0, b = 1, g, h)\right)^n\right] \\ &= \sum_{i=0}^{n} \frac{n!}{(n-i)!i!} a^{n-i} b^i \mathbb{E}\left[T_{g,h}(W; a = 0, b = 1, g, h)^i\right]. \end{aligned} \tag{9.37}$$

Furthermore, it was shown by Dutta and Babbel (2002) that when it exists one can obtain the general expression

$$\mathbb{E}\left[T_{g,h}(W; a = 0, b = 1, g, h)^i\right] = \frac{\sum_{r=0}^{i} (-1)^r \frac{i!}{(i-r)!r!} \exp\left(\frac{(i-r)^2 g^2}{2(1-ih)}\right)}{\sqrt{(1-ih)} g^i}, \tag{9.38}$$

which would produce the following four moments in closed form:

$$\mathbb{E}[X] = \mathbb{E}[q(W;g,h)] = \left[\exp\left(\frac{g^2}{2-2h}\right) - 1\right]\left[g\sqrt{1-h}\right]^{-1}$$
$$\mathbb{E}\left[X^2\right] = \mathbb{E}\left[q(W;g,h)^2\right] = \left[1 - 2\exp\left(\frac{g^2}{2-4h}\right) + \exp\left(\frac{2g^2}{1-2h}\right)\right]\left[g^2\sqrt{1-2h}\right]^{-1}$$
$$\mathbb{E}\left[X^3\right] = \mathbb{E}\left[q(W;g,h)^3\right] = \left[3\exp\left(\frac{g^2}{2-6h}\right) + \exp\left(\frac{9g^2}{2-6h}\right)\right.$$
$$\left. - 3\exp\left(\frac{2g^2}{1-3h}\right) - 1\right]\left[g^3\sqrt{1-3h}\right]^{-1}$$
$$\mathbb{E}\left[X^4\right] = \mathbb{E}\left[q(W;g,h)^4\right] = s(g,h)\exp\left(\frac{8g^2}{1-4h}\right)\left[g^4\sqrt{1-4h}\right]^{-1}.$$

with the function $s(g,h)$ being given by

$$s(g,h) = \left(1 + 6\exp\left(\frac{6g^2}{4h-1}\right) + \exp\left(\frac{8g^2}{4h-1}\right) - 4\exp\left(\frac{7g^2}{8h-2}\right) - 4\exp\left(\frac{15g^2}{8h-2}\right)\right).$$

Remark 9.5 *These results allow one to perform model estimation via moment matching of model moments to empirical moments of the loss data.*

As a consequence, one can easily then find the skew, kurtosis, and coefficient of variations for the g-and-h distribution as well as the subfamilies for the g-distributions and h-distributions. Note that one can also develop variations of the g-and-h distribution density and distribution functions will which avoid the restrictions specified in Definitions 9.8 and 9.10. In addition, there are numerous authors who have studied the generalized properties of quantile-based functionals of asymmetry and kurtosis (see examples in Definition 9.11; also see Balanda and MacGillivray 1988, 1990, Rayner and MacGillivray 2002, and Balanda and MacGillivray 1988).

Definition 9.11 (Generalized Skewness and Kurtosis Functionals in OpRisk) *In considering the generalizations of the skewness and kurtosis for transformation-based quantile function severity models, one can utilize the generalized specifications given for the skewness functional, for a given distribution $F_X(x)$ with respect to its quantile function $Q_X(x)$ by*

$$\gamma_F = \frac{Q_X(\alpha) + Q_X(1-\alpha) - 2Q_X\left(\frac{1}{2}\right)}{Q_X(\alpha) - Q_X(1-\alpha)}, \quad \alpha \in (0,1). \tag{9.39}$$

In addition, there is the spread functional given by

$$S_F = Q_X(\alpha) - Q_X(1-\alpha), \quad \alpha \in (0,1). \tag{9.40}$$

■

Such measures were discussed by Balanda and MacGillivray (1990) and it can be shown that $|\gamma_F(\alpha)| \leq 1$. In the case of the g-and-h family of severity models, one would obtain the forms given in Definition 9.12.

Definition 9.12 (Generalized Skewness and Kurtosis for g-and-h Family) *Consider the g-and-h distributed random varaible $X \sim GH(a = 0, b = 1, g, h)$ with constant parameters g and $h > 0$ and a quantile function for a level $\alpha \in [0, 1]$ given by*

$$Q_X(\alpha; g, h) = F_X^{-1}(\alpha; g, h) = k\left(\Phi^{-1}(\alpha)\right), \tag{9.41}$$

where $\Phi(\cdot)$ is the standard Gaussian distribution and the function $k(x)$ is specified by

$$k(x) = \frac{\exp(gx) - 1}{g} \exp\left(\frac{hx^2}{2}\right). \tag{9.42}$$

Then the generalized skewness and kurtosis are given by

$$\begin{aligned}
S_F &= Q_X(\alpha) - Q_X(1-\alpha) \\
&= \frac{\exp\left(g\Phi^{-1}(\alpha)\right) - 1}{g} \exp\left(\frac{1}{2} h \Phi^{-1}(\alpha)^2\right) \\
&\quad - \frac{\exp\left(g\Phi^{-1}(1-\alpha)\right) - 1}{g} \exp\left(\frac{1}{2} h \Phi^{-1}(1-\alpha)^2\right), \\
\gamma_F &= \frac{Q_X(\alpha) + Q_X(1-\alpha) - 2Q_X\left(\frac{1}{2}\right)}{Q_X(\alpha) - Q_X(1-\alpha)} \\
&= \frac{\frac{\exp\left(g\Phi^{-1}(\alpha)\right)-1}{g} \exp\left(\frac{1}{2} h\Phi^{-1}(\alpha)^2\right)}{S_F} + \frac{\frac{\exp\left(g\Phi^{-1}(1-\alpha)\right)-1}{g} \exp\left(\frac{1}{2} h\Phi^{-1}(1-\alpha)^2\right)}{S_F} \\
&\quad - 2\frac{\frac{\exp\left(g\Phi^{-1}(0.5)\right)-1}{g} \exp\left(\frac{1}{2} h\Phi^{-1}(0.5)^2\right)}{S_F}.
\end{aligned}$$

■

9.4.2 TAIL PROPERTIES OF THE G-AND-H, G, H, AND H–H SEVERITY IN OPRISK

In terms of the tail behavior of the g-and-h family of distributions, the properties of such severity models have been studied by numerous authors such as Morgenthaler and Tukey (2000) and Degen *et al.* (2007). In particular, the tail property (index of regular variation) for the g-and-h family of distributions was first studied for the h-distribution by Morgenthaler and Tukey (2000) and later for the g-and-h distribution by Degen *et al.* (2007) (see Proposition 9.4). In addition, the second-order regular variation properties of the g-and-h family of distributions was studied by Degen *et al.* (2007). In order to study the properties of regular variation of the g-and-h family of loss distribution models it is first important to recall some basic definitions. First, we note that a postive measurable function $f(\cdot)$ is regularly varying if it satisfies the conditions in Definition 9.13, see discussion in Karatzas and Shreve (1991).

Definition 9.13 (Regularly Varying Function) *A positive measurable function $f(\cdot)$ is regularly varying (at infinity) with an index $\alpha \in \mathbb{R}$ if it satisfies:*

- *It is defined on some neighbourhood $[x_0, \infty)$ of infinity; and*
- *It satisfies the following limiting relationship*

$$\lim_{x\to\infty} \frac{f(\lambda x)}{f(x)} = \lambda^{\alpha}, \quad \forall \lambda > 0. \tag{9.43}$$

■

We note that when $\alpha = 0$, then the function $f(\cdot)$ is said to be slowly varying (at infinity). From this definition one can show that a random variable has a regularly varying distribution if it satisfies the condition in Definition 9.14, see further discussion in detail in Peters and Shevchenko (2015).

Definition 9.14 (Regularly Varying Random Variable) *A loss random variable X with distribution $F_X(x)$ taking positive support is said to be regularly varying with index $\alpha \geq 0$ if the right tail distribution $\overline{F}_X(x) = 1 - F_X(x)$ is regularly varying with index $-\alpha$.* ■

The following important features can be noted about regularly varying distributions as shown in Theorem 9.1, see detailed discussion in Bingham *et al.* (1989).

Theorem 9.1 (Properties of Regularly Varying Distributions) *Given a loss distribution $F_X(x)$ satisfying $F_X(x) < 1$ for all $x \geq 0$, the following conditions on $F_X(x)$ can be used to verify that it is regularly varying such that $F_X(x) \in RV_{\alpha}$:*

- *If $F_X(x)$ is absolutely continuous with density $f_X(x)$ such that for some $\alpha > 0$ one has the limit*

$$\lim_{x\to\infty} \frac{x f_X(x)}{\overline{F}_X(x)} = \alpha. \tag{9.44}$$

Then $f_X(x)$ is regularly varying with index $-(1+\alpha)$ and consequently $\overline{F}_X(x)$ is regularly varying with index $-\alpha$;

- *If the density $f_X(x)$ for loss distribution $F_X(x)$ is assumed to be regularly varying with index $-(1+\alpha)$ for some $\alpha > 0$. Then the following limit,*

$$\lim_{x\to\infty} \frac{x f_X(x)}{\overline{F}_X(x)} = \alpha, \tag{9.45}$$

will also be satisfied if $\overline{F}_X(x)$ is regularly varying with index $-\alpha$ for some $\alpha > 0$ and the density $f_X(x)$ will be ultimately monotone.

Many additional properties are described for such heavy tailed distribution and density functions. Here we will utilise the above stated conditions to assess the regular variation properties of the right tail of the g-and-h family of loss models. In particular we will see if a single distributional parameter characterizes the heavy tailed feature as captured by the notion of regular variation index, or if the relationship is more complex.

Proposition 9.4 (Index of Regular Variation of g-and-h Distribution) *Consider the random variable $W \sim Normal(0, 1)$ and a loss random variable X, which has severity distribution given by the g-and-h distribution with parameters $a, b, g, h \in \mathbb{R}$, denoted $X \sim GH(a, b, g, h)$, with $h > 0$ and density (distribution) $f(x)$ (and $F(x)$). Then the index of regular variation is obtained by considering the following limit*

$$\lim_{x\to\infty} \frac{xf(x)}{\overline{F}(x)} = \lim_{x\to\infty} \frac{\phi(u)\,(\exp(gu) - 1)}{(1 - \Phi(u))\,(g\exp(gu) + hu(\exp(gu) - 1))} = \frac{1}{h} \tag{9.46}$$

for $u = k^{-1}(x)$ where the function $k(x)$ is given by

$$k(x) = \frac{\exp(gx) - 1}{g} \exp\left(\frac{hx^2}{2}\right). \tag{9.47}$$

Hence, one can state that $\overline{F} \in RV_{-\frac{1}{h}}$.

The asymptotic tail behavior of the h-family of Tukey distributions was studied by Morgenthaler and Tukey (2000 proposition 1) and is given in Proposition 9.5.

Proposition 9.5 (h-Type Tail Behaviour) *Consider the h-type transformation, where $W \sim Normal(0, 1)$ is a standard Gaussian random variable and the loss random variable X has severity distribution given by the h-distribution with parameters $a, b, h \in \mathbb{R}$, denoted $X \sim H(a, b, h)$ according to*

$$X = T_h(W; a, b, h) := a + bW \exp\left(\frac{hW^2}{2}\right). \tag{9.48}$$

Then the asymptotic tail index of the h-type distribution is then given by $1/h$. This is equivalent to the g-and-h family for $g \neq 0$.

This shows that the h-type family has a Pareto heavy-tailed property, hence the restriction that moments will only exist on the order of less than $1/h$. The g-family of distributions can be shown to be subexponential in the tail behavior but not regularly varying. It was shown Degen *et al.* (2007, theorem 2.2) that one can obtain an explicit form for the function of slow variation in the g-and-h family as detailed in Theorem 9.2.

Theorem 9.2 (Slow Variation Representation of g-and-h Severity Models) *Consider the random variable $W \sim Normal(0, 1)$ and a loss random variable X, which has severity distribution given by the g-and-h distribution with parameters $a, b, g, h \in \mathbb{R}$, denoted $X \sim GH(a, b, g, h)$, with $g > 0$ and $h > 0$ and density (distribution) $f(x)$ (and $F(x)$). Then $\overline{F}(x) = x^{-1/h}L(x)$ for some slowly varying function $L(x)$ given as $x \to \infty$ by*

$$L(x) = \frac{h}{\sqrt{2\pi}g^{1/h}} \frac{\left[\exp\left(\frac{g}{h}\sqrt{g^2 + 2h\ln(gx)} - \frac{g^2}{h}\right) - 1\right]^{1/h}}{\sqrt{g^2 + 2h\ln(gx)} - g} \left(1 + O\left(\frac{1}{\ln x}\right)\right). \tag{9.49}$$

From this explicit Karamata representation developed by Degen *et al.* (2007), it was also shown that one can obtain the second-order regular variation properites of the g-and-h family.

The implications of these findings are that the g-and-h distribution, under the parameter restrictions $g > 0$ and $h > 0$, belongs to the domain of attraction of an Extreme Value Distribution, such that $X \sim GH(a, b, g, h)$ with distribution F satisfying $F \in MDA(H_\gamma)$ where $\gamma = h > 0$. As a consequence, by the Pickands–Balkema–de Haan Theorem, discussed in detail in companion book Peters and Shevchenko (2015), one can state that there exists an Extreme Value Index (EVI) constant γ and a positive measurable function $\beta(\cdot)$ such that the following result between the excess distribution of the g-and-h (denoted by $F_u(x) = \mathbb{P}\mathrm{r}(X - u \leq x | X > u)$ and the generalized Pareto distribution (GPD) is satisfied in the tails

$$\lim_{u \uparrow \infty} \sup_{x \in (0,\infty)} \left| F_u(x) - G_{\gamma,\beta(u)}(x) \right| = 0. \tag{9.50}$$

For discussion on the rate of convergence in the tails, see Raoult and Worms (2003) and the application of this theorem to the g-and-h case by Degen *et al.* (2007 lemma 3.1) where it is shown that the order of covergence is given by $O\left(A \exp\left(V^{-1}(u)\right)\right)$ for functions

$$\begin{aligned} V(x) &:= \overline{F}^{-1}\left(\exp(-x)\right), \\ A(x) &:= \frac{V''(\ln x)}{V'(\ln x)} - \gamma. \end{aligned} \tag{9.51}$$

Hence, the conclusion from this analysis regarding the tail convergence of the excess distribution of the g-and-h family toward the GPD $G_{\gamma,\beta(u)}(x)$ is given explicitly by

$$\frac{\ln L(x)}{\ln x} \sim \sqrt{2}\frac{g}{h^{\frac{3}{2}}} \frac{1}{\sqrt{\ln(x)}} = O\left(\frac{1}{\sqrt{\ln\left(k^{-1}(x)\right)}} \right), \quad x \to \infty. \tag{9.52}$$

Remark 9.6 *The implications of this slow rate of convergence are that when data for severities are obatained from a loss process, if a goodness-of-fit test suggests that one may not reject the null hypothesis that these data came from a g-and-h distribution (under a composite test as described in Chapter 8), then one should avoid performing estimation of the extreme quantiles, such as those used to measure the capital via the Value-at-Risk, via methods based on Peaks Over Threshold (POT) or Extreme Value Theory (EVT) based penultimate approximations.*

9.4.3 PARAMETER ESTIMATION FOR THE G-AND-H SEVERITY IN OPRISK

There have been many proposed methods for performing parameter estimation in the g-and-h family of distributions. In this section, we survey a few of the possibilities. Dutta and Babbel (2002) suggest a method of parameter estimation for constant g and h parameters based on percentile matching. This involves recognizing the relationships between data percentiles and the g-and-h parameters given in Proposition 9.6.

Proposition 9.6 (g-and-h Distribution Percentile Matching Estimation) *Consider a g-and-h distributed random variable $X \sim GH(a, b, g, h)$ and a sample of n loss data points with order* *statistics* $\left\{X_{(i,n)}\right\}_{i=1}^{n}$ *that will be used to fit the g-and-h distribution. Then the location parameter is given by the following percentile:*

$$\hat{a} = Median\left\{X_1, X_2, \ldots, X_n\right\} = X_{\left(\left\lfloor \frac{n}{2} \right\rfloor, n\right)}. \tag{9.53}$$

Then the value of the g parameter is given by the following estimator (based on the p-th percentile given by $p = \frac{i}{n}$ *for one or more* $i \in \{1, 2, \ldots, n\}$*) given by*

$$g_p = -\frac{1}{W_p} \ln\left(\frac{X_{1-p} - X_{0.5}}{X_{0.5} - X_p}\right) \tag{9.54}$$

with $W_p = \inf\{w \ : \ \Phi(w) > p\}$ *the p-th percentile of a standard Normal and* $X_p = \inf\{x \ : F(x) > p\}$ *denotes the p-th percentile of the sample loss data. It would then be suggested to take a robust estimate of a set of* g_p*, for a range of percentiles taken from* $p \in \{p_L, \ldots, p_U\}$ *and then to form the median*

$$\hat{g} = Median\left\{\hat{g}_L, \ldots, \hat{g}_U\right\}. \tag{9.55}$$

Then, given $\hat{g}$*, one can estimate the h parameter and the b parameters using the relationship*

$$h_p = \frac{2}{W_p^2} \ln\left(\frac{1}{b}\frac{g\left(X_p - X_{1-p}\right)}{\exp\left(gW_p\right) - \exp\left(-gW_p\right)}\right). \tag{9.56}$$

From this, the estimates of $\hat{g}$ *and* $\hat{a}$ *would then allow one to select a range of percentiles taken from* $p \in \{p_L, \ldots, p_U\}$ *and then regress* $Y_p = \ln\left(\frac{g\left(X_p - X_{1-p}\right)}{\exp\left(gW_p\right) - \exp\left(-gW_p\right)}\right)$ *against the quadrative of the corresponding percentiles of standard Gaussian with scaling given by* $\frac{W_p^2}{2}$*, the resulting intercept of the regression would be an estimate of* $\ln(b)$ *and the resulting gradient would be an estimate of h. Given a selection of k percentile levels in* $[p_L, p_U]$*, the resulting estimators would each be given by*

$$\hat{h} = \frac{\sum_{p=1}^{k}\left(\frac{W_p^2}{2} - \frac{1}{2k}\sum_{j=1}^{k} W_j^2\right)\left(Y_p - \frac{1}{k}\sum_{j=1}^{k} Y_k\right)}{\sum_{p=1}^{k}\left(\frac{W_p^2}{2} - \frac{1}{k}\sum_{j=1}^{k}\frac{W_j^2}{2}\right)^2}$$
$$\hat{b} = \frac{1}{k}\sum_{j=1}^{k} Y_k - \hat{h}\frac{1}{2k}\sum_{j=1}^{k} W_j^2. \tag{9.57}$$

In addition, one could perform moment-based matching to estimate the parameters. In this regard, there have been two approaches proposed: one which utilizes the expressions derived earlier for the first four moments that define a system of four nonlinear equations that are solved numerically via root search (see Mahbubul *et al.* 2008 for the discussion on the solutions for *g* and *h*) and the other (Jiménez and Arunachalam, 2011, section 2.2.2) for the simple moment-based estimators for location and scale parameters *a* and *b*.

Remark 9.7 *Note, this simple matching moment-based approach is not particularly recommended as in cases in OpRisk, where these models are of major interest, one is typically not only interested* *in light-tailed models for the severity but instead the heavy-tailed high kurtosis severity models are considered. In such cases, it may be that the population moments may not even be finite, though the sample moments will always be finite. The consequence of this is that no matter how many data are utilised to estimate the parameters, they will always be biased, or perhaps not well defined under the system of moment conditions.*

The more robust alternative to simple moment matching, especially when the tails of the empirical distribution of the data suggest heavy-tailed features, such as in situations where parameter estimates of h will be large positive values, then it will be numerically more robust to consider an L-moments based approach such as those proposed by Headrick and Pant (2012b). In the advanced text Peters and Shevchenko (2015), we detail extensively the properties of L-moment estimators for EVT models; we therefore defer the reader to this section for details on the estimators we present in Proposition 9.7. Before presenting these results, we briefly recall the definition of the sample L-moments (see Greenwood *et al.* 1979).

Definition 9.15 (Sample L-Moments) *Consider a sample of n observed losses denoted by random variables $\{X_i\}_{i=1}^n$ with associated order statistics in increasing order $\{X_{(i,n)}\}_{i=1}^n$. Then the first four sample L-moments from the data are given by*

$$\begin{aligned} l_1 &= m_0, \\ l_2 &= 2m_1 - m_0, \\ l_3 &= 6m_2 - 6m_1 + m_0, \\ l_4 &= 20m_3 - 30m_2 + 12m_1 - m_0, \end{aligned} \tag{9.58}$$

with the sample probability moments m_i's given by

$$\begin{aligned} m_0 &= \frac{1}{n}\sum_{j=1}^{n} X_{(j,n)}, \\ m_i &= \frac{1}{n}\sum_{j=i+1}^{n} \frac{(j-1)(j-2)\dots(j-i)}{(n-1)(n-2)\dots(n-i)} X_{(j,n)}. \end{aligned} \tag{9.59}$$

■

Next we will define the population L-moments in terms of the parameters of the L-moment $\gamma - \kappa$ Tukey family as well as the asymmetric L-moment $\kappa_L - \kappa_R$ Tukey family, which can be matched to the sample-estimated L-moments and then utilized as a system of nonlinear equations to be solve numerically via root search for the resulting L-moment parameter estimates.

Proposition 9.7 (L-Moment Estimators for the L-Moment $\gamma-\kappa$ Tukey Family) *Consider a γ-and-κ distributed random variable $X \sim F(\gamma, \kappa)$ and a sample of n loss data points with order statistics $\{X_{(i,n)}\}_{i=1}^n$ that will be used to fit the γ-and-κ distribution. Then under the restrictions that $\gamma + \kappa < 1$, $\kappa < 1$, and $1 + \gamma > \kappa$, which allow the first two L-moments to be* *finite, one obtains the following two equations for the population's first two L-moments λ_1 and λ_2 given by:*

$$\begin{aligned}\lambda_1 &= \frac{(-\gamma-\kappa)h_1+(\gamma-\kappa)h_2+(-\gamma+\kappa)h_3+2\kappa h_4+(\gamma+\kappa)h_5-2\kappa h_6}{2\gamma}\\ \lambda_2 &= \frac{2\gamma-(\gamma+\kappa)^2 h_1+(\gamma-\kappa)^2(h_1-h_3)+(\gamma+\kappa)^2 h_5}{2\gamma},\end{aligned} \tag{9.60}$$

where $h_1, h_2, \ldots, h_6$ are defined with respect to the Harmonic number functions with the following arguments according to

$$\begin{aligned} h_1 &= H\left[\frac{1}{2}(-1-\gamma-\kappa)\right], \quad h_2 = H\left[\frac{1}{2}(-1+\gamma-\kappa)\right]\\ h_3 &= H\left[\frac{1}{2}(\gamma-\kappa)\right], \quad h_4 = H\left[\frac{1}{2}(-1-\kappa)\right]\\ h_5 &= H\left[-\frac{1}{2}(\gamma+\kappa)\right], \quad h_6 = H\left[-\frac{1}{2}\kappa\right],\end{aligned} \tag{9.61}$$

with the harmonic number functions defined for any $x > 0$ by

$$H[x] := x\sum_{k=1}^{\infty}\frac{1}{k(x+k)}. \tag{9.62}$$

One can then estimate sample L-moments that can be matched to the population moments to solve numerically for the parameters.

Remark 9.8 *As noted by Headrick and Pant (2012b, p. 9), expressions are also developed for the population L-skewness τ_3 and L-kurtosis τ_4 should one wish to utilize these for L-moment matching parameter estimation.*

Analogously, the solutions for the first two population L-moments for the class of $\kappa_L - \kappa_R$ Tukey transformations were detailed by Headrick and Pant (2012b) and can be used to perform parameter estimation, as detailed in Proposition 9.8.

Proposition 9.8 (L-Moment Estimators for the L-Moment $\kappa_L - \kappa_R$ Tukey Family) *Consider the asymmetric κ_L-and-κ_R distributed random variable $X \sim F(\kappa_L, \kappa_R)$ and a sample of n loss data points with order statistics $\{X_{(i,n)}\}_{i=1}^{n}$ that will be used to fit theκ_L-and-κ_R distribution. Then under the restrictions that $\kappa_L < 1$ and $\kappa_R < 1$, which allow the first two L-moments to be finite, one obtains the following two equations for the population's first two L-moments λ_1 and λ_2 given by*

$$\begin{aligned}\lambda_1 &= \frac{1}{4}\left[2p_5-2p_6-2p_7+2p_8-\kappa_L p_9+\kappa_L p_{10}+\kappa_R p_{11}-\kappa_R p_{12}\right]\\ \lambda_2 &= \frac{1}{4}\left[4+\kappa_L\left(-4p_5+4p_6+\kappa_L\left(p_9-p_{10}\right)\right)+4+\kappa_R\left(-4p_7+4p_8+\kappa_R\left(p_{11}-p_{12}\right)\right)\right],\end{aligned} \tag{9.63}$$

where $p_5, p_6, \ldots, p_{12}$ are defined with respect to the polygamma functions with the following arguments according to

$$
\begin{aligned}
p_5 &= P\left[0, \frac{1}{2} - \frac{\kappa_L}{2}\right], \quad p_6 = P\left[0, 1 - \frac{\kappa_L}{2}\right], \quad p_7 = P\left[0, \frac{1}{2} - \frac{\kappa_R}{2}\right] \\
p_8 &= P\left[0, 1 - \frac{\kappa_R}{2}\right], \quad p_9 = P\left[1, \frac{1}{2} - \frac{\kappa_L}{2}\right], \quad p_{10} = P\left[1, 1 - \frac{\kappa_L}{2}\right] \\
p_{11} &= P\left[1, \frac{1}{2} - \frac{\kappa_R}{2}\right], \quad p_{12} = P\left[1, 1 - \frac{\kappa_R}{2}\right],
\end{aligned}
\tag{9.64}
$$

with the polygamma functions defined by

$$
P[m, x] := (-1)^{m+1} m! \sum_{k=0}^{\infty} \frac{1}{(x+k)^{m+1}}. \tag{9.65}
$$

One can then estimate sample L-moments that can be matched to the population L-moments to solve numerically for the parameters.

There are also approaches based on numerical maximum likelihood applied to the estimation of parameters in the g-and-h family of models (see discussions by Rayner and MacGillivray 2002).

Often in practice, the amount of observed data may not be large, however there may be reasonable expert opinion available. As such, it is often beneficial to adopt a Bayesian estimation framework, as detailed in the next section.

9.4.4 BAYESIAN MODELS FOR THE G-AND-H SEVERITY IN OPRISK

In this section, we consider two different approaches to constructing Bayesian models for the g-and-h family of severity models. This is particularly important if one wishes to develop an LDA modeling structure that would calibrate such models using a combination of expert opinions as well as collected loss data, as required by Basel II/Basel III standards. The two frameworks we develop can be considered approximations in that the posterior distribution obtained will be approximate up to any desired level of precision, as specified by the modeler. In particular, we first consider an approach based on the work of Peters and Sisson (2006), which utilizes an Approximate Bayesian Computation (ABC) formulation, which was the first application of such statistical techniques in finance and risk; then we consider a specially designed conjugate Bayesian formulation based on Askey orthogonal polynomials (see detailed discussion in Chapter 17).

9.4.4.1 Approximate Bayesian Computation and the g-and-h Severity Model.

The basic concept of ABC methods is covered in Chapter 7. Here we briefly review the concept of ABC methods that is growing in popularity in statistics (see Peters and Sisson 2006, Peters *et al.* 2012a, Beaumont *et al.* 2009, Csillery *et al.* 2010, Del Moral *et al.* 2012, and Sisson *et al.* 2010 and the references therein). In particular, we describe the basic Markov chain Monte Carlo (MCMC) sampling methodology first developed by Peters and Sisson (2006) for the g-and-h distribution. In this work, it was recognized that one could exploit the efficiency of simulation from the g-and-h distribution, which only required simulation of a single standard Gaussian random variate for each observation in order to apply ABC methods.

Consider the parameters of the severity model, upon which one wishes to build a Bayesian model, given in the case of the g-and-h model by vector

$$\boldsymbol{\Theta} = \left\{a, b, \alpha_0, \alpha_1, \ldots, \alpha_p, \beta_0, \beta_1, \ldots, \beta_q\right\} \in \Omega,$$

where it is assumed we condition upon a choice of p and q for the dimension of the $g(z)$ and $h(z)$ polynomials, which can be selected by a desired model selection criterion such as Bayesian Information Criterion (BIC), Bayes Factors, or Deviance Information Criterion (DIC). Then, we will denote the prior distribution generically for these severity model parameters by the joint density $\pi\left(a, b, \alpha_0, \ldots, \alpha_p, \beta_0, \beta_1, \ldots, \beta_q\right)$. These priors can be elicited in a number of methods *a priori*; see discussions on the different approaches for instance in O'Hagan (1998).

Given the prior distribution, essentially, the ABC methods first reduce the observed loss data in a year, denoted by the n losses given by $\boldsymbol{x} = x_{1:n}$, to a low-dimensional vector of summary statistics denoted by $t_{\boldsymbol{x}} = T(\boldsymbol{x}) \in \mathcal{T}$, where $\dim(\boldsymbol{\Theta}) \leq \dim(t_{\boldsymbol{x}}) << n$. Then, the true posterior $\pi(\boldsymbol{\theta}|\boldsymbol{x})$ is replaced with a new posterior given by $\pi(\boldsymbol{\theta}|t_{\boldsymbol{x}})$, which would theoretically match exactly the true posterior in two cases if $t_{\boldsymbol{x}} = \boldsymbol{x}$ or if $t_{\boldsymbol{x}}$ is sufficient for $\boldsymbol{\theta}$, otherwise it is an approximation $\pi(\boldsymbol{\theta}|t_{\boldsymbol{x}}) \approx \pi(\boldsymbol{\theta}|\boldsymbol{x})$. The new target posterior, still assumed to be computationally intractable (with regard to evaluation of the density pointwise), is embedded within an augmented model from which a Monte Carlo sampling scheme is viable, such as MCMC or Sequential Monte Carlo (SMC) (see the Estimation section Chapter 7).

The secret to all ABC methods is the replacement of the evaluation of the intractable likelihood model with the simulation of auxiliary data given a set of model parameters $\boldsymbol{\Theta}$. Hence, the auxilairy data will be denoted by vector $\boldsymbol{X}^* = X^*_{1:n}$ and the i-th sample is obtained by $X^*_i = x^*_i$ through simulation in the case of the g-and-h model by $X_i \sim GH\left(a, b, \alpha_0, \alpha_1, \ldots, \alpha_p, \beta_0, \beta_1, \ldots, \beta_q\right)$ using the algorithm specified in the previous subsection. The auxiliary data are then also summarized by the summary statistic $t_{\boldsymbol{x}^*} \in \mathcal{T}$ for the given simulated realization.

Specifically, under the ABC method, one then expresses the joint posterior of the model parameters $\boldsymbol{\Theta}$ and auxiliary data $\boldsymbol{X}^*$ conditional upon the observed data $\boldsymbol{X}$ according to the kernel-based representation given in Equation (9.66):

$$\pi\left(\boldsymbol{\theta}, \boldsymbol{X}^* | \boldsymbol{x}\right) \propto K_h\left(t_{\boldsymbol{x}^*} - t_{\boldsymbol{x}}\right) f\left(\boldsymbol{X}|\boldsymbol{\theta}\right) \pi(\boldsymbol{\theta}), \tag{9.66}$$

where $\boldsymbol{X}^* \sim f\left(\boldsymbol{X}|\boldsymbol{\theta}\right)$ and $f(\cdot|\boldsymbol{\theta})$ is in this case the g-and-h model. Then under this ABC posterior framework, the marginal posterior distribution as given by

$$\pi_M\left(\boldsymbol{\theta}|\boldsymbol{x}\right) = c_M \int K_h\left(t_{\boldsymbol{x}^*} - t_{\boldsymbol{x}}\right) f\left(\boldsymbol{X}|\boldsymbol{\theta}\right) \pi(\boldsymbol{\theta}) d\boldsymbol{X}^*, \tag{9.67}$$

where $c_M^{-1} = \int_\Omega \int K_h\left(t_{\boldsymbol{x}^*} - t_{\boldsymbol{x}}\right) f\left(\boldsymbol{X}|\boldsymbol{\theta}\right) \pi(\boldsymbol{\theta}) d\boldsymbol{X}^* d\boldsymbol{\theta}$, normalizes the posterior such that it is a well-defined density (see discussions by Reeves and Pettitt 2005 and Peters and Hübner 2009).

The function $K_h\left(t_{\boldsymbol{x}^*} - t_{\boldsymbol{x}}\right)$ is a standard kernel function with scale parameter $h \geq 0$, which weights the intractable posterior with high density in regions in which $t_{\boldsymbol{x}^*} \approx t_{\boldsymbol{x}}$ where the auxiliary and observed data sets are similar. Therefore, $\pi_M\left(\boldsymbol{\theta}|\boldsymbol{x}\right) \approx \pi\left(\boldsymbol{\theta}|\boldsymbol{x}\right)$ forms an approximation of the intractable posterior through standard smoothing arguments (see Marin *et al.* 2012). As $h \to 0$, so that $K_h\left(t_{\boldsymbol{x}^*} - t_{\boldsymbol{x}}\right)$ becomes a point mass at the origin (i.e., $t_{\boldsymbol{x}^*} = t_{\boldsymbol{x}}$)

and is zero elsewhere, if $t_{\boldsymbol{x}}$ is sufficient for $\boldsymbol{\theta}$, then the intractable posterior marginal is recovered exactly where $\pi_M(\boldsymbol{\theta}|\boldsymbol{x}) = \pi(\boldsymbol{\theta}|\boldsymbol{x})$. Note typically setting h too small is computationally impractical; see disucssions in the context of g-and-h models in this regard by Peters and Sisson (2006). There has been a reasonable amount of discussion on the different possible choices one may adopt in practice (see Peters *et al.* 2009) and discussions therein.

Typically in practice, it is common to consider a generlization of this scheme in which the joint posterior distribution in Equation (9.66) is augmented with more than one auxiliary summary vector, by considering for $S \geq 1$ the auxiliary posterior

$$\pi\left(\boldsymbol{\theta}, \boldsymbol{X}^{*,1:S}\middle|\boldsymbol{x}\right) \propto \widetilde{K}_h\left(t_{\boldsymbol{x}^*}^{1:S}, t_{\boldsymbol{x}}\right) f\left(\boldsymbol{X}^{1:S}|\boldsymbol{\theta}\right)\pi(\boldsymbol{\theta}), \tag{9.68}$$

where $t_{\boldsymbol{x}^*}^{1:S} = \left(t_{\boldsymbol{x}^*}^1, t_{\boldsymbol{x}^*}^2, \ldots, t_{\boldsymbol{x}^*}^S\right)$ and for all $i \in \{1, 2, \ldots, S\}$ one has i.i.d. data sets generated from the g-and-h intractable likelihood $\boldsymbol{X}^{*,i} \sim f(\boldsymbol{X}|\boldsymbol{\theta})$. By construction, the auxiliary data are conditionally independent given $\boldsymbol{\theta}$, which gives

$$f\left(\boldsymbol{X}^{*,1:S}|\boldsymbol{\theta}\right) = \prod_{s=1}^{S} f\left(\boldsymbol{X}^{*,s}|\boldsymbol{\theta}\right). \tag{9.69}$$

In addition, as discussed by Del Moral *et al.* (2012), one may select the kernel $\widetilde{K}_h\left(t_{\boldsymbol{x}^*}^{1:S}, t_{\boldsymbol{x}}\right)$ according to

$$\widetilde{K}_h\left(t_{\boldsymbol{x}^*}^{1:S}, t_{\boldsymbol{x}}\right) = \frac{1}{S}\sum_{s=1}^{S} K_h\left(t_{\boldsymbol{x}^*}^s - t_{\boldsymbol{x}}\right), \tag{9.70}$$

which will result in a joint posterior given by

$$\pi_M\left(\boldsymbol{\theta}, \boldsymbol{X}^{*,1:S}\middle|\boldsymbol{x}\right) = c_M\left[\frac{1}{S}\sum_{s=1}^{S} K_h\left(t_{\boldsymbol{x}^*}^s - t_{\boldsymbol{x}}\right)\right]\left[\prod_{s=1}^{S} f\left(\boldsymbol{x}^s|\boldsymbol{\theta}\right)\right]\pi(\boldsymbol{\theta}), \tag{9.71}$$

with the normalizing constant $c_M > 0$. In this case, by construction one again obtains the appropriate marginal target distribution

$$\int \pi_M\left(\boldsymbol{\theta}, \boldsymbol{X}^{*,1:S}\middle|\boldsymbol{x}\right)\, d\boldsymbol{X}^{*,1:S} = \pi(\boldsymbol{\theta}|\boldsymbol{x}). \tag{9.72}$$

Working with such posterior ABC distributions, one needs to typically obtain samples via a Monte Carlo sampling strategy. In this regard, there are two approaches one may adopt to sample from the target posterior:

1. The first involves treating the summary quantities (statistics) for the auxiliary data, $t_{\boldsymbol{x}^*}^{1:S}$, as parameters in an augmented statespace model. This approach therefore involves sampling directly on the augmented model, say $\pi_M\left(\boldsymbol{\theta}, \boldsymbol{X}^{*,1:S}\middle|\boldsymbol{x}\right)$, by obtaining joint samples on the product space $\left(\boldsymbol{\theta}, \boldsymbol{X}^{*,1:S}\right) \in \Theta \times \mathcal{X}^S$. Then one *a posteriori* marginalizes over the samples $\boldsymbol{X}^{*,1:S}$ by simply discarding these realizations from the sampler output;
2. The second approach involves sampling the lower-dimensional ABC posterior given by $\pi_M(\boldsymbol{\theta}|\boldsymbol{x})$. Within the Monte Carlo sampler this would involve approximation

of the Monte Carlo integral, by draws at each iteration of sample for $\boldsymbol{\theta}$, using $\boldsymbol{X}^{*,1}, \boldsymbol{X}^{*,2}, \ldots, \boldsymbol{X}^{*,S} \sim f(\boldsymbol{x}|\boldsymbol{\theta})$ to obtain

$$\begin{aligned} \pi_M(\boldsymbol{\theta}|\boldsymbol{x}) &\propto \pi(\boldsymbol{\theta}) \int_{\mathcal{X}} \widetilde{K}_h(t_{\boldsymbol{x}^*}, t_{\boldsymbol{x}}) f(\boldsymbol{X}|\boldsymbol{\theta})\, d\boldsymbol{X} \\ &\approx \frac{1}{S}\pi(\boldsymbol{\theta}) \sum_{s=1}^{S} \widetilde{K}_h(t^s_{\boldsymbol{x}^*}, t_{\boldsymbol{x}}) := \hat{\pi}_M(\boldsymbol{\theta}|\boldsymbol{x}) \end{aligned} \tag{9.73}$$

in lieu of posterior evaluation of $\pi_M(\boldsymbol{\theta}|\boldsymbol{x})$.

To proceed with the design of a Monte Carlo sampling strategy to obtain samples from the posterior $\pi_M(\boldsymbol{\theta}|\boldsymbol{x}) \approx \pi(\boldsymbol{\theta}|\boldsymbol{x})$, which is based upon a simple MCMC approach, there are several more advanced strategies available; see, for instance, the SMC samplers–Partial Rejection Control method proposed by Peters *et al.* (2012a). An in complete list of such approaches are provided here:

1. Marginal versus augmented auxiliary ABC posterior: Sisson *et al.* (2010), Sisson and Fan (2011);
2. Rejection, MCMC: Beaumont *et al.* (2002), Marjoram *et al.* (2003);
3. SMC samplers PRC: Peters *et al.* (2012a);
4. SMC samplers: Sisson *et al.* (2007).

The ABC–MCMC algorithm for the g-and-h family then proceeds as follows given the order of the g and h polynomials p, q.

Algorithm 9.2 (ABC–MCMC for the Bayesian Posterior g-and-h Severity Model)

1. *Initialize the g-and-h model parameters (Markov chain sate):*

$$\boldsymbol{\theta}^{(0)} = \left[\alpha^{(0)}_{1:p}, \beta^{(0)}_{1:p}, a^{(0)}, b^{(0)}\right]$$

and draw synthetic data realizations and evaluate summary statistics

$$t^1_{\boldsymbol{x}^*}, t^2_{\boldsymbol{x}^*}, \ldots, t^S_{\boldsymbol{x}^*} \sim f(t|\boldsymbol{\theta}^{(0)}).$$

This involves for the j-th summary vector $t^j_{\boldsymbol{x}^*}$ drawing n samples from the g-and-h model according to the following step:

a) *Draw a standard Gaussian random variate: $Z_i \sim Normal(0,1)$;*

b) *Given p, q. and coefficients $\left\{\alpha_i^{(0)}\right\}_{i=0}^{p}$ and $\left\{\beta_i^{(0)}\right\}_{i=0}^{q}$, evaluate the polynomials*

$$\begin{aligned} g(W_i) &= \alpha_0^{(0)} + \alpha_1^{(0)} W_i + \ldots + \alpha_p^{(0)} W_i^p \\ h(W_i) &= \beta_0^{(0)} + \beta_1^{(0)} W_i + \ldots + \beta_q^{(0)} W_i^q. \end{aligned} \tag{9.74}$$

c) *Then given parameters $a^{(0)}$, $b^{(0)}$, and polynomials $g(W_i)$ and $h(W_i)$, evaluate transformation*

$$X_i^{(j)} = a^{(0)} + b^{(0)} \frac{\exp\left(g(W_i) W_i\right) - 1}{g(W_i)} \exp\left(\frac{h(W_i) W_i^2}{2}\right).$$

d) *Then given synthetic data samples $\left\{X_i^{(j)}\right\}_{i=1}^{n}$ evaluate the summary statistic $t_{\boldsymbol{x}^*}^{j}$.*

2. *For iterations $k \geq 1$ perform one update on the ABC–MCMC algorithm as follows:*

a) *Generate a proposal vector of parameters for the new state of the Markov chain $\boldsymbol{\theta} \sim q\left(\boldsymbol{\theta}^{(k)}, \boldsymbol{\theta}\right)$ using an MCMC proposal for the g-and-h parameters such as a local random walk or a mixture of local and global proposals for $q\left(\boldsymbol{\theta}^{(k)}, \boldsymbol{\theta}\right)$;*

b) *Draw synthetic data realizations independently from the model with the proposed parameters $\boldsymbol{\theta}$ such that*

$$t_{\boldsymbol{x}^*}^{1}, t_{\boldsymbol{x}^*}^{2}, \ldots, t_{\boldsymbol{x}^*}^{S} \sim f(t|\boldsymbol{\theta}).$$

This involves for the j-th summary vector $t_{\boldsymbol{x}^*}^{j}$ drawing n samples from the g-and-h model according to the following steps:

i. *Draw a standard Gaussian random variate: $Z_i \sim Normal(0, 1)$;*

ii. *Given p, q, and coefficients $\{\alpha_i\}_{i=0}^{p}$ and $\{\beta_i\}_{i=0}^{q}$, evaluate the polynomials*

$$\begin{aligned} g(W_i) &= \alpha_0 + \alpha_1 W_i + \ldots + \alpha_p W_i^p \\ h(W_i) &= \beta_0 + \beta_1 W_i + \ldots + \beta_q W_i^q. \end{aligned} \tag{9.75}$$

iii. *Then given parameters a, b. and polynomials $g(W_i)$ and $h(W_i)$ evaluate transformation*

$$X_i^{(j)} = a + b \frac{\exp\left(g(W_i) W_i\right) - 1}{g(W_i)} \exp\left(\frac{h(W_i) W_i^2}{2}\right).$$

iv. *Then given synthetic data samples $\left\{X_i^{(j)}\right\}_{i=1}^{n}$ evaluate the summary statistic $t_{\boldsymbol{x}^*}^{j}$.*

c) *With probability*

$$min\left\{1, \frac{\frac{1}{S}\sum_s K_h\left(t_{\boldsymbol{x}^*}^{s} - t_{\boldsymbol{x}}\right) \pi(\boldsymbol{\theta}) q\left(\boldsymbol{\theta}, \boldsymbol{\theta}^{(k)}\right)}{\frac{1}{S}\sum_s K_h\left(t_{\boldsymbol{x}^*}^{s} - t_{\boldsymbol{x}}\right) \pi\left(\boldsymbol{\theta}^{(k)}\right) q\left(\boldsymbol{\theta}^{(k)}, \boldsymbol{\theta}\right)}\right\}$$

accept the proposed state and set $\boldsymbol{\theta}^{(k+1)} = \boldsymbol{\theta}$ and keep track of the new sampled values $\{t_{\boldsymbol{x}^*}^{s}\}_{s=1}^{S}$. Otherwise set $\boldsymbol{\theta}^{(k+1)} = \boldsymbol{\theta}^{(k)}$.

d) *Increment $k = k + 1$.*

Remark 9.9 *The typical summary statistic one may consider involves a summary of the empirical distribution function*

$$\hat{F}_n(x) = \frac{1}{n}\sum_{i=1}^{n} \mathbb{I}_{x \leq x_i} \tag{9.76}$$

such as a set of sample quantiles $\{\hat{q}_i\}_{i=1}^{p}$ *with* $\hat{q}_i = x_{(i,n)}$. *In addition, one typically has a wide choice of kernel choice such as soft and hard decision kernels (see discussions by Peters et al. 2010).*

9.5 Generalized Beta Family of Heavy-Tailed Severity Models

McDonald and Xu (1995) present a general representation of the families of the Generalized Beta form, which nests both the Generalized Beta of the first and second kinds (GB1 and GB2). This is a five-parameter family of models, which is also related to the Exponential Generalized Beta family. The standard Beta distributions of the first and second kinds are some of the most widely utilized distributions in statistical applications as they include nested subfamilies of models such as the power distirbutions, uniform distribution, gamma, Lomax, F, Chi-square, and exponential distributions (see discussions by Johnson *et al.* 1970). Generalizations such as the Generalized-F distribution, the Feller-Pareto, Generalized Beta Prime, and Transformed Beta distributions have been proposed by various authors. They are all members of the GB2 family of models to be presented next. First, we present the global family of the five-parameter Generalized Beta distributions of McDonald and Xu (1995) given in Definition 9.16

Definition 9.16 (Generalized Beta Distribution Severity Models) *A loss random variable X has a Generalized Beta distribution* $X \sim GB(x; a, b, c, p, q)$ *if the density is given by*

$$f_X(x; a, b, c, p, q) = \frac{|a| x^{ap-1}\left(1 - (1-c)\left(\frac{x}{b}\right)^a\right)^{q-1}}{b^{ap} B(p,q)\left(1 + c\left(\frac{x}{b}\right)^a\right)^{p+q}}, \quad 0 < x^a < \frac{b^a}{1-c}, \tag{9.77}$$

with $c \in [0, 1]$, $a \neq 0$, *and* $b, p, q > 0$ *and where* $B(p, q)$ *is the Beta function.* ∎

Remark 9.10 *One can obtain the GB1 family by setting* $c = 0$ *and the GB2 family by setting* $c = 1$.

Next, we present the GB2 subfamily as these have been shown to be particulary relevant to OpRisk modeling scenarios.

9.5.1 GENERALIZED BETA FAMILY TYPE II SEVERITY MODELS IN OPRISK

In this section, we introduce a family of severity models known as the GB2 family that has been utilized in OpRisk settings successfully (see discussions by Dutta and Perry 2006 and Peters and Sisson 2006). The GB2 family, like the previously discussed quantile transformation models, will also allow for a wide range of flexible skew and kurtosis distributions. In the case of the GB2 family, it admits a parametric specification for its distribution and density functions. There is a detailed account of this family, with some emphasis on financial and actuarial modeling in the works of Bookstabber and McDonald (1987), McDonald (1996), McDonald and Xu (1995), Cummins *et al.* (1990), and the book-length review by Gupta and Nadarajah (2004). It is also worth noting that a restricted form of the GB2 family was also studied for $a > 0$ where it was termed the generalized F distribution (see Kalbfleisch and Prentice 2011).

The density and distribution for the four-parameter GB2 family is given in Definition 9.17 and automatically has the required support for a loss distribution, with positive support. The GB2 family is parameterized by four parameters:

1. a is the location parameter that also determines the rate at which the tails approach the x-axis; hence, large values of a imply a strong peakedness for the GB2 model density function;
2. b is the scale parameter and it affects the height of the density;
3. q determines the kurtosis of the distribution and the product aq directly affects the kurtosis;
4. p when combined with q affects the skewness of the distribution.

One can obtain expressions for the GB2 family distribution and density functions as solutions to the differential equation given by

$$\frac{d\ln f_X(x)}{dx} = \frac{ap - 1 - (aq+1)\left(\frac{x}{b}\right)^a}{x\left(1 + \left(\frac{x}{b}\right)^a\right)}, \tag{9.78}$$

where the solution will produce the closed-form distributional form given in Definition 9.17. The differential equation representation is interesting to consider since it demonstrates any possible relationships between the GB2 and other distributional families also specified in such an integro-differential form. As a result of this differential equation representation, it can be seen that the GB2 family is neither contained in nor contains other well-known distributional families such as the Pearsonian family (Pearson 1894, 1895). It therefore warrants consideration as a unique positively supported class of a heavy-tailed flexible skew–kurtosis model for OpRisk.

Definition 9.17 (Generalized Beta Family of the Second Kind (GB2) Severity Models) *A severity random variable $X \sim GB2(x; a, b, p, q)$ if its distribution is given by*

$$X \sim F_X(x; a, b, p, q) = \frac{z(x)^p}{pB(p,q)}\, {}_2F_1\left[p, 1-q, 1+p; z(x)\right], \quad x > 0 \tag{9.79}$$

where ${}_2F_1\left[a, b, c; x\right]$ is a hypergeometric function given with respect to Pochhammer notation $(x)_n$ by

$${}_2F_1\left[a, b, c; x\right] = \sum_{n=0}^{\infty} \frac{(a)_n (b)_n x^2}{(b)_n n!} \tag{9.80}$$

and

$$z(x) := \frac{\left(\frac{x}{b}\right)^a}{\left(1 + \left(\frac{x}{b}\right)^a\right)}. \tag{9.81}$$

The density of the GB2 model is also closed form and given by

$$f_X(x; a, b, p, q) = \frac{|a| x^{ap-1}}{b^{ap} B(p, q) \left[1 + \left(\frac{x}{b}\right)^a\right]^{p+q}}, \quad x > 0 \tag{9.82}$$

where $B(p, q)$ is the Beta function. ■

In Example 9.5, there are plots of the GB2 distribution displayed for a range of different parameter settings.

EXAMPLE 9.5 GB2 Severity Model Density Shapes

In the Figure 9.7 plots, the GB2 severity density is plotted for a range of parameter values to illustrate the skewness and kurtosis properties this model offers, the location and scale parameters are set to $a = 0$ and $b = 1$, and for the shape, skewness and kurtosis parameters p and q are considered.

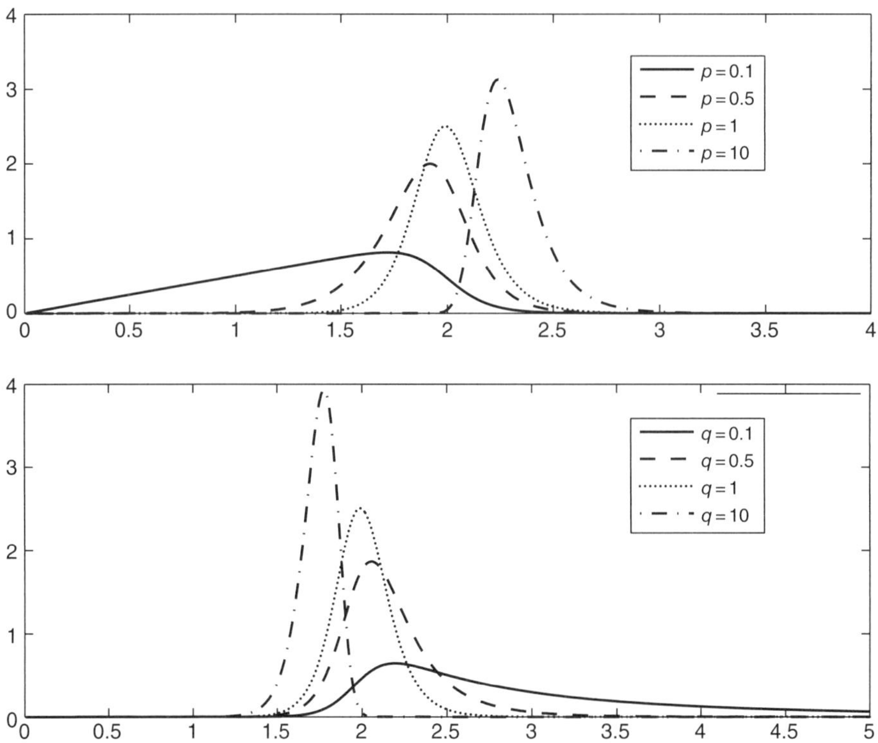

FIGURE 9.7 Top subplot: the top subplot shows the effect of the parameter $p \in \{0.1, 0.5, 1, 10\}$, showing the skewness that results from decreasing the value of p. Bottom subplot: the bottom subplot shows the effect of the parameter $q \in \{0.1, 0.5, 1, 10\}$, showing the kurtosis that results from decreasing the parameter q

■

The tail properties of the GB2 model's density function are given in Proposition 9.9, where the regular variation feature will therefore also limit the existence of moments for certain parameter ranges.

Proposition 9.9 (Regular Variation of the GB2 Density Function) *If a loss random variable has GB2 distribution $X \sim GB2(x; a, b, p, q)$, then the right tail of the GB2 density function is regularly varying at infinity with index $-aq - 1$.*

The mode of the GB2 model is given by the closed-form location specified in Proposition 9.10.

Proposition 9.10 (Mode of GB2 Severity Model) *If a loss random variable has GB2 distribution $X \sim GB2(x; a, b, p, q)$, then the mode is given by the expression*

$$Mode[X] = \begin{cases} b\left(\dfrac{ap-1}{aq+1}\right)^{\frac{1}{a}}, & ap > 1, \\ 0, & otherwise. \end{cases} \tag{9.83}$$

The moments of the GB2 model are given in Proposition 9.11 (see Bookstabber and McDonald 1987). It is clear that as $a \to \infty$ the variance will decrease to zero and the mean of the distribution will tend toward b, which will therefore asymptotically become the location of a dirac mass, where the distribution will collapse in the limit.

Proposition 9.11 (Moments of the GB2 Family of Severity Models) *If a loss random variable has GB2 distribution $X \sim GB2(x; a, b, p, q)$, then the integer moments $\mathbb{E}[X^r]$ exist if $-ap < r < aq$ for all $r \in \mathbb{J}^+$. The first moment (mean) is given by*

$$\mathbb{E}[X] = b\frac{B\left(p+\frac{1}{a}, q-\frac{1}{a}\right)}{B(p,q)}. \tag{9.84}$$

The moment-generating function of the GB2 family is given by

$$M_X(t) = \sum_{k=0}^{\infty} \frac{B\left(p+\frac{k}{a}, q-\frac{k}{a}\right)}{B(p,q)} \frac{t^k b^k}{k!}, \tag{9.85}$$

so in general the r-th integer moment when it exists is given by

$$\mathbb{E}[X^r] = b^r\frac{B\left(p+\frac{r}{a}, q-\frac{r}{a}\right)}{B(p,q)}. \tag{9.86}$$

9.5.2 SUB FAMILIES OF THE GENERALIZED BETA FAMILY TYPE II SEVERITY MODELS

The GB2 family has a wide range of skew–kurtosis subfamilies, which are also well-known distributions; for example, a first layer of nested distributions (in the sense that one of the four parameters in the GB2 family is constrained) includes the following families:

1. Log-t distribution family when $a \to 0$;
2. Generalized Gamma distribution family when $q \to \infty$;
3. Beta distribution of the second kind if $a = 1$;
4. Singh–Maddala or Burr type XII distributions if $p = 1$;
5. Dagum or Burr type III distributions if $q = 1$.

When one constrains two parameters or more, one can also obtain the following nested distributional families:

1. Log-Cauchy severity models are obtained when $a \to 0$ and $q = \frac{1}{2}$;
2. LogNormal severity models are obtained when $a \to 0$ and $q \to \infty$ simultaneously;
3. Weibull severity models are obtained when $p = 1$ and $q \to \infty$;
4. Gamma severity models are obtained when $a = 1$ and $q \to \infty$;
5. Lomax severity models are obtained when $p = 1$ and $a = 1$;
6. Exponential distribution with $a = 1, p = 1$, and $q \to \infty$;
7. Generalized Log-Logistic with $a = 1$, $b = 1, p = 1$, and $q = 1$.

It may also be noted that the GB2 model has some nested family members that also overlap with the Pearson family, such as when one sets $a = 1$ to recover the subfamily of Beta distributions of the second kind.

It is also useful to know how to simulate from any member of the GB2 family, as this is critical for many applications in OpRisk when using the GB2 as a severity distribution model; this can be achieved as follows.

Algorithm 9.3 (Simulating Losses from a GB2 Severity Model.)

1. *Draw a standard Gamma random variate:* $Y_1 \sim Gamma(p, 1)$;
2. *Draw a standard Gamma random variate:* $Y_2 \sim Gamma(q, 1)$;
3. *Construct the GB2 distributed random variable* $X \sim GB2(x; a, b, p, q)$ *according to the transformation*

$$X = b\left(\frac{Y_1}{Y_2}\right)^{\frac{1}{a}}. \tag{9.87}$$

9.5.3 MIXTURE REPRESENTATIONS OF THE GENERALIZED BETA FAMILY TYPE II SEVERITY MODELS

It is also worth noting that the GB2 family can be represented as characterizing a large family of mixed-type distributions. A mixed-type distribution is formally defined in Definition 9.18.

Definition 9.18 (Mixed-Type Distributions) *A mixed distribution is one that is generated from two distinct distributions, the first known as the structural distribution and the second known* *as a mixing density. Consider a severity random variable* $X \sim F_X(x;\theta_1,\phi)$ *parameterized by* θ_1 *and* ϕ *and with a density that satisfies the relationship*

$$\begin{aligned} f_X\left(x;\theta_1,\phi\right) &= f\left(x;\theta_1,\theta_2\right) \oslash_{\theta_2} g\left(\theta_2;\phi\right) \\ &:= \int \underbrace{f\left(x;\theta_1,\theta_2\right)}_{\textit{Structural distribution}} \;\; \underbrace{g\left(\theta_2;\phi\right)}_{\textit{Mixing density}} \; d\,\theta_2, \end{aligned} \tag{9.88}$$

where we denote the mixing integral operator by $\oslash_{\theta_2}$ *with respect to argument* θ_2. ∎

In the case of the GB2 family of models, one can show the following mixed-type distributional properties given in Proposition 9.12. This will result in many more flexible families of distributions, some of which are nested in the GB2 family.

Proposition 9.12 (Mixed-Type GB2 Severity Models) *If* $X \sim GB2(a,b,p,q)$, *then the density function satisfies the following mixing property:*

$$f_X\left(x;a,b,p,q\right) = f_X\left(x;a,\theta,p,q\right) \oslash_{\theta} f_X\left(\theta;a,b,\theta_2,\theta_3\right). \tag{9.89}$$

The resulting density takes the form

$$f_X(x;a,b,p,q) = \begin{cases} \dfrac{|a|\left(\dfrac{x}{b}\right)^{a\theta_2-1} B\left(q+\theta_2,p+\theta_3\right)}{bB(p,q)B\left(\theta_2,\theta_3\right)}, & 0 < x \leq b. \\[2ex] \dfrac{|a|\left(\dfrac{b}{x}\right)^{a\theta_3+1} B\left(q+\theta_2,p+\theta_3\right)}{bB(p,q)B\left(\theta_2,\theta_3\right)}, & x > b. \end{cases} \tag{9.90}$$

Remark 9.11 *One can show that the GB2 family is itself a mixture class, since the mixture between the Generalized Gamma distribution as a base distribution when mixed with an Inverse Generalized Gamma distribution as a mixing distribution will produce a GB2 distribution. In addition, one can also show that the GB2 family can be used to characterize generalized LogNormal–Gamma mixtures.*

The following is a list of popular mixture representations of members in the GB2 family of severity distributions:

1. GB2 comes from a mixture between Generalized Gamma as structural distribution and Inverse Generalized Gamma as mixing distribution:
$$GB2(x;a,b,p,q) = GG(x;a,\theta,p) \oslash_{\theta} IGG(\theta;a,b,q).$$
2. Beta distribution of the second kind B2 comes from a mixture between Gamma as structural distribution and Inverse Gamma as mixing distribution:
$$B2(x;b,p,q) = Gamma(x;\theta,p) \oslash_{\theta} IGamma(\theta;b,q).$$
3. GB2 ($p = 1$ Singh–Maddala distribution) comes from a mixture between Weibull as structural distribution and Inverse Generalized Gamma as mixing distribution:
$$GB2(x;a,b,p=1,q) = SM(x;a,b,q) = Weibull(x;a,\theta) \oslash_{\theta} IGG(\theta;a,b,q).$$

4. GB2 ($q = 1$ Dagum distribution) comes from a mixture between Generalized Gamma as structural distribution and Inverse Weibull as mixing distribution:

$$GB2(x; a, b, p, q = 1) = Dagum(x; a, b, p) = GG(x; a, \theta, p) \oslash_{\theta} IWeib(\theta; a, b).$$

5. GB2 ($p = 1, a = 1$, Lomax distribution) comes from a mixture between Exponential as structural distribution and Inverse Gamma as mixing distribution:

$$GB2(x; a, b, p = 1, q) = Lomax(x; b, q) = Exp(x; \theta) \oslash_{\theta} IGamma(\theta; b, p).$$

Other useful properties one can observe about the GB2 model family that have been used to great effect in certain applications of the GB2 family are the following:

1. Closure under multiplication of two independent GB2-distributed random variables (see Proposition 9.13);
2. Closure under inversion (see Proposition 9.14 and Venter 1983).

Proposition 9.13 (GB2 Closure Under Multiplication) *Given two i.i.d. random variables $X_i \sim GB2(a, b, p, q)$ for $i \in \{1, 2\}$, one has the product random variable $Y = \prod_{i=1}^{2} X_i$ with density given by*

$$f_Y(y) = \begin{cases} \dfrac{|a| \left(\frac{y}{b^2}\right)^{ap} B(p+q, p+q)}{yB(p, q)^2}, & 0 < y < b^2, \\ \dfrac{|a| \left(\frac{y}{b^2}\right)^{-aq} B(p+q, p+q)}{yB(p, q)^2}, & y \geq b^2. \end{cases} \tag{9.91}$$

Then to get closure under multiplication of the GB2 family, one needs to impose some additional parameter restrictions such as would occur for the LogNormal model. In general, one also has the property that if $X \sim GB2(a, b, p, q)$, then $X^r \sim GB2\left(\frac{a}{r}, b^r, p, q\right)$.

Proposition 9.14 (GB2 Closure Under Inversion) *Given a loss random variable $X \sim GB2$ (a, b, p, q), one observes that the inverse loss random variable $Y = 1/X$ has a distribution given by $Y \sim GB2\left(a, \frac{1}{b}, q, p\right)$.*

9.5.4 ESTIMATION IN THE GENERALIZED BETA FAMILY TYPE II SEVERITY MODELS

The estimation of the GB2 model parameters proceeds typically via maxiumum likelihood estimation (MLE) or method of moments (see Chapter 7). Venter (1983) provides the system of equations for the likelihood estimation with n-samples from a severity model with GB2 distribution (see Proposition 9.15).

Proposition 9.15 (Maximum Likelihood Estimation GB2 Severity Model Parameters) *Given a severity model with i.i.d. losses distributed as $X_i \sim GB2(a, b, p, q)$, one has the following system of nonlinear equations for the parameters when performing MLE:*

$$
\begin{aligned}
\frac{n}{a} + p\sum_{i=1}^{n} \ln\left(\frac{x_i}{b}\right) &= (p+q)\sum_{i=1}^{n} \ln\left(\frac{x_i}{b}\right)\left[\left(\frac{b}{x_i}\right)^a + 1\right]^{-1}, \\
np &= (p+q)\sum_{i=1}^{n}\left[\left(\frac{b}{x_i}\right)^a + 1\right]^{-1}, \\
n\psi(p+q) + a\sum_{i=1}^{n} \ln\left(\frac{x_i}{b}\right) &= n\psi(p) + \sum_{i=1}^{n} \ln\left[\left(\frac{x_i}{b}\right)^a + 1\right], \\
n\psi(p+q) &= n\psi(q) + \sum_{i=1}^{n} \ln\left[\left(\frac{x_i}{b}\right)^a + 1\right],
\end{aligned}
\tag{9.92}
$$

where $\psi(\cdot)$ is the digamma function. This system is solved by first solving the first two linear equations in p and q in terms of a and b followed by a Newton method for the second two equations for solving for a and b. Note that the Fisher Information matrix for the confidence intervals of such MLE estimates is also known in closed form (see Brazauskas 2002).

9.6 Generalized Hyperbolic Families of Heavy-Tailed Severity Models

The family of distributions known as the Generalized Hyperbolic (GH) class was studied extensively by Barndorff-Nielsen (1977, 1978a) and the book-length review by Barndorff-Nielsen and Blaesild (1981). Since their introduction these models have found many applications where initially they were very influential only in areas of physics and biology. Then, more recently, they have become influential models in areas of financial mathematics; in this context, notable examples include the works of Eberlein and Keller (1995), Cont (2001), Eberlein (2001), and the thesis of Prause (1999). The widespread interest in the GH family of models has primarily arisen due to their flexibility for skew–kurtosis characteristics as well as the tractability and closed-form expressions for the density, characteristic function, cummulants, and Levy measure.

In this section, we will first discuss some basic properties of the GH family before presenting more details on two relevant subfamilies for the context of OpRisk given by the GIG family and the NIG family. These two families are particularly interesting for OpRisk settings as they display the property of closure under convolution, making specification of the annual loss process in an LDA model comprising these models for the severity model particularly tractable as they admit closed-form representations for the annual loss distribution and density.

A loss random variable X has a severity distributional model that is from the GH family if its distribution satisfies the following definition for the distribution in Definition 9.19 (see Barndorff-Nielsen and Stelzer 2005). The parameters of the GH family have the following influence on the properties of the resulting distribution:

1. α is the shape parameter;
2. β is the skewness parameter;
3. μ is the location parameter;
4. δ is the scale parameter;

5. The special parameter ν characterizes which subclass the model represents and in particular the tail properties of the resulting subfamily.

The density is then given by the following expression in Definition 9.19 in terms of parameters α, β, γ, and δ. It is also common practice to consider an alternative parameterization in the scale invariant form where one uses instead the reparametrization involving

$$\begin{aligned} \rho &= \frac{\beta}{\alpha}, \quad \chi = \rho\eta, \\ \eta &= \left(1 + \delta\sqrt{\alpha^2 - \beta^2}\right)^{\frac{1}{2}}, \end{aligned} \tag{9.93}$$

where χ is the skewness-type measure (assymmetry) and η is the kurtosis-type measure (steepness), which satisfy $0 < |\chi| < \eta < 1$. We will see later that for particular cases of the model parameter ν such as in the NIG family these parameters are known as steepness and assymmetry and can be used to characterize all distributional members by what is known as the "shape triangles", which are the analog of the classical skewness and kurtosis plots for the GH members.

Definition 9.19 (Generalized Hyperbolic Severity Models) *A loss random variable X has a GH distribution $X \sim GH(x; \nu, \alpha, \beta, \mu, \delta)$ if it has a density given by*

$$f_X(x) = \frac{\overline{\gamma}^{\nu}\overline{\alpha}^{\frac{1}{2}-\nu}}{\sqrt{2\pi}\delta K_{\nu}(\overline{\gamma})}\left(1 + \frac{(x-\mu)^2}{\delta^2}\right)^{\frac{\nu}{2}-\frac{1}{4}} K_{\nu-\frac{1}{2}}\left(\overline{\alpha}\sqrt{1 + \frac{(x-\mu)^2}{\delta^2}}\right) e^{\beta(x-\mu)}, \quad x \in \mathbb{R}$$

with parameters $\nu \in \mathbb{R}$, $0 \leq |\beta| \leq \alpha$, $\mu \in \mathbb{R}$, and $\delta \in \mathbb{R}^+$. In addition, one defines

$$\gamma = \sqrt{\alpha^2 - \beta^2}, \quad \overline{\alpha} = \delta\alpha, \quad \overline{\beta} = \delta\beta, \quad \overline{\gamma} = \delta\gamma$$

with $K_{\nu}(\cdot)$ the modified Bessel function of the third kind. ■

The cummulant-generating function of the GH family of severity models is also known in closed form as specified in Proposition 9.16, which allows one to utilize a result from Barndorff-Nielsen (1978b, corollary 7.1) to obtain expressions for the mean and variance in closed from, as detailed in Proposition 9.17.

Proposition 9.16 (Cummulant-Generating Function GH Severity Models) *A loss random variable X with a GH distribution $X \sim GH(x; \nu, \alpha, \beta, \mu, \delta)$ has a cummulant-generating function given by*

$$\ln\left(\mathbb{E}\left[\exp(tX)\right]\right) = \frac{\nu}{2}\ln\left(\frac{\gamma}{\alpha^2 - (\beta+t)^2}\right) + \ln\left(\frac{K_{\nu}\left(\delta\sqrt{\alpha^2 - (\beta+t)^2}\right)}{K_{\nu}\left(\delta\sqrt{\alpha^2 - \beta^2}\right)}\right) + t\mu. \tag{9.94}$$

Consequently, using this result one may show that the mean and variance of a loss random variable with GH severity are given as follows, as well as the integer centralized moments.

Proposition 9.17 (Mean and Variance of GH Severity Models) *A loss random variable X with a GH distribution $X \sim GH(x; \nu, \alpha, \beta, \mu, \delta)$ has a mean and variance given by the following expressions*

$$\mathbb{E}[X] = \mu + \beta \frac{\delta K_{\nu+1}(\overline{\gamma})}{\gamma K_{\nu}(\overline{\gamma})}$$
$$\mathbb{V}\mathrm{ar}[X] = \delta^2 \left(\frac{K_{\nu+1}(\overline{\gamma})}{\overline{\gamma} K_{\nu}(\overline{\gamma})} + \frac{\beta^2}{\gamma^2} \left(\frac{K_{\nu+2}(\overline{\gamma})}{K_{\nu}(\overline{\gamma})} - \left(\frac{K_{\nu+1}(\overline{\gamma})}{K_{\nu}(\overline{\gamma})} \right)^2 \right) \right). \tag{9.95}$$

In addition, the integer centralized and absolute centralized moments are given by the series expansion (see Barndorff-Nielsen and Stelzer 2005, theorem 2):

$$\mathbb{E}[(X-\mu)^r] = \frac{2^{\lceil \frac{r}{2} \rceil} \overline{\gamma}^{\nu} \delta^{2\lceil \frac{r}{2} \rceil} \beta^{r \bmod 2}}{\sqrt{\pi} K_{\nu}(\overline{\gamma}) \overline{\alpha}^{\nu + \lceil \frac{r}{2} \rceil}} \sum_{k=0}^{\infty} \frac{2^k \overline{\beta}^{2k} \Gamma\left(k + \lceil \frac{r}{2} \rceil + \frac{1}{2}\right)}{\overline{\alpha}^k (2k + (r \bmod 2))!} K_{\nu+k+\lceil \frac{r}{2} \rceil}(\overline{\alpha}),$$
$$\mathbb{E}[|X-\mu|^r] = \frac{2^{\frac{r}{2}} \overline{\gamma}^{\nu} \delta^r}{\sqrt{\pi} K_{\nu}(\overline{\gamma}) \overline{\alpha}^{\nu + \frac{r}{2}}} \sum_{k=0}^{\infty} \frac{2^k \overline{\beta}^{2k} \Gamma\left(k + \frac{r}{2} + \frac{1}{2}\right)}{\overline{\alpha}^k (2k)!} K_{\nu+k+\frac{r}{2}}(\overline{\alpha}). \tag{9.96}$$

The GH family of severity models also has the translation and scale invariance closure properties given in Proposition 9.18.

Proposition 9.18 (Scale and Translation Properties of GH Severity Models) *Given a loss random variable X with a GH distribution $X \sim GH(x; \nu, \alpha, \beta, \mu, \delta)$, the scaled and translated random variable is also distributed according to a GH distribution as given by*

$$aX + b \sim GH\left(x; \nu, \frac{\alpha}{|a|}, \frac{\beta}{a}, \delta|a|, a\mu + b\right). \tag{9.97}$$

9.6.1 TAIL PROPERTIES AND INFINITE DIVISIBILITY OF THE GENERALIZED HYPERBOLIC SEVERITY MODELS

To understand the asymptotic tail behavior of the GH family of severity models, it is important to first consider the tail behavior of the modified Bessel function of the third kind $K_{\nu}(\cdot)$ given in Proposition 9.19 (see Gil *et al.* 2002, p. 401, and for details on evaluation, see Lozier and Olver 1994).

Proposition 9.19 (Tail Behavior of Modified Bessel Function of the Third Kind) *Consider the modified Bessel function of the third kind given by $K_{\nu}(x)$, which can be represented asymptotically according to the series expansion*

$$K_{\nu}(x) \sim \left(\frac{\pi}{2x}\right)^{\frac{1}{2}} \exp(-x) \sum_{k=0}^{\infty} \frac{(\nu, k)}{(2x)^k} \tag{9.98}$$

with (ν, k) representing the Hankel symbol given by

$$(\nu, k) = \frac{1}{\pi k!} (-1)^k \cos(\nu\pi) \Gamma\left(\frac{1}{2} + \nu + k\right) \Gamma\left(\frac{1}{2} - \nu + k\right), \quad x \to \infty, \tag{9.99}$$

and the special case of $\nu = 1$, which simplifies the asymptotic behavior as $|x| \to \infty$ to be given by

$$K_1(x) \sim \frac{\pi}{2x} \exp(-x). \tag{9.100}$$

The GH family of severity models exhibits a range of tail behaviors that are characterized as semiheavy in nature. In general, the following asymptotic expression can be shown for the tails of the GH family given in Proposition 9.20, (see Barndorff-Nielsen and Stelzer 2005).

Proposition 9.20 (Tail Behavior of the Generalized Hyperbolic Severity Models) *A loss random variable X with a GH distribution $X \sim GH(x; \nu, \alpha, \beta, \mu, \delta)$ has a tail behavior characterized by the expression*

$$f_X(x; \nu, \alpha, \beta, \mu = 0, \delta) \sim C|x|^{\nu-1} \exp\left[(\beta - \alpha)x\right], \quad x \to \infty, \tag{9.101}$$

for some constant C.

In the following example, we consider the special case when $\nu = -\frac{1}{2}$, which produces the result shown in Proposition 9.21.

Proposition 9.21 ***(Tail Behavior of the Normal Inverse Gamma Models: (GH with $\nu = -\frac{1}{2}$))*** *A loss random variable X with a GH distribution $X \sim GH(x; \nu = -\frac{1}{2}, \alpha, \beta, \mu, \delta)$ has a tail behavior characterized by the expression*

$$f_X\left(x; \nu = -\frac{1}{2}, \alpha, \beta, \mu = 0, \delta\right) \sim |x|^{-\frac{3}{2}} \exp\left(\beta x - \alpha|x|\right), \quad |x| \to \infty. \tag{9.102}$$

In the case that $\alpha - |\beta| << 1$, this asymptotic tail behavior has the same form as a Cauchy distribution tail decay

$$f_X\left(x; \nu = -\frac{1}{2}, \alpha, \beta, \mu = 0, \delta\right) \sim |x|^{-2}. \tag{9.103}$$

In addition to having semiheavy tails, the GH family of severity models is also important for modeling in OpRisk as it displays properties of infinite divisibility as characterized in Proposition 9.22, though it is not closed under convolution in general.

Proposition 9.22 (Infinite Divisibility of GH Severity Models) *A loss random variable X with a GH distribution $X \sim GH(x; \nu, \alpha, \beta, \mu, \delta)$ is infinitely divisible, meaning that it can be represented such that for every positive integer n, there exist n i.i.d. random variables with sum*

$$S_n = \sum_{i=1}^{n} Y_i, \quad \textit{such that } X \stackrel{d}{=} S_n. \tag{9.104}$$

Again in the case in which $\nu = -\frac{1}{2}$, one obtains the subfamily of NIG which is not only infinitely divisible but, under appropriate parameter restrictions, is also closed under convolutions as shown in Proposition 9.23.

Proposition 9.23 (Closure Under Convolution of GH Severity Model with $\nu = -\frac{1}{2}$) *Given two i.i.d. loss random variables $X_1 \sim GH\left(x; \nu = -\frac{1}{2}, \alpha, \beta, \mu_1, \delta_1\right)$ and $X_2 \sim GH\left(x; \nu = -\frac{1}{2}, \alpha, \beta, \mu_2, \delta_2\right)$, the sum of the two random variables*

$$X = X_1 + X_2 \sim GH\left(x; \nu = -\frac{1}{2}, \alpha, \beta, \mu_1 + \mu_2, \delta_1 + \delta_2\right).$$

9.6.2 SUBFAMILIES OF THE GENERALIZED HYPERBOLIC SEVERITY MODELS

As noted by Barndorff-Nielsen and Stelzer (2005), the GH family of models contains several well-known subclasses of parametric severity models given by different values of ν such as in the following cases:

1. $\nu = 1$ one obtains the subfamily of hyperbolic distributions (see Example 9.6);
2. $\nu = -\frac{1}{2}$ one obtains the subfamily of NIG distributions;
3. Other distributional subfamilies include Gaussian, Exponential, Laplace, Variance-Gamma and Student-t.

EXAMPLE 9.6 Examples of Flexible Hyperbolic Distributions (GH with $\nu = 1$)

In the Figure 9.8 plots and we explore the density shapes for the case of the subfamily of GH distributions given by the hyperbolic distributions where $\nu = 1$.

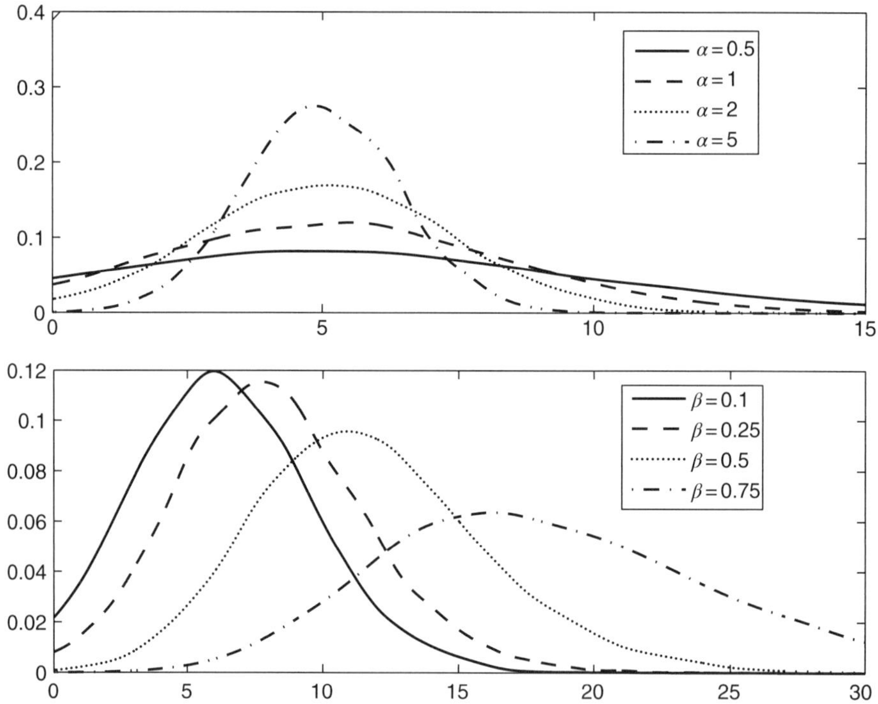

FIGURE 9.8 Top subplot: parameters used in this subplot study the effect of the shape parameter ranges $\alpha = [0.5, 1, 2, 5]$, skewness parameter $\beta = 0$, location parameters $\mu = 5$, and scale parameter $\delta = 10$. Bottom subplot: parameters used in this subplot study the effect of the skewness parameter ranges $\beta = [0.1, 0.25, 0.5, 0.75]$, shape parameter $\alpha = 1$, location parameters $\mu = 5$, and scale parameter $\delta = 10$

One of the representations of the GH family of models that is particularly useful, especially from the perspective of simulating draws from the GH family, is the Normal variance–mean mixture representation presented in Proposition 9.24 (see Barndorff-Nielsen and Stelzer 2005).

Proposition 9.24 (Variance–Mean Mixture Representation of GH Family) *Consider independent random variables $X \sim GH(\nu, \alpha, \beta, \mu, \delta)$, the GIG distributed random variable $V \sim GIG(\nu, \delta, \gamma)$ such that $\gamma = \sqrt{\alpha^2 - \beta^2}$ and the standard Normal distributed random variable $\epsilon \sim Normal(0, 1)$, which produces the following distributional equality*

$$X \overset{d}{=} \mu + \beta V + \sqrt{V}\epsilon. \tag{9.105}$$

The representation of the Variance–Mean mixture of the GH family results in the following general algorithm for simulation from any of the GH severity models. In some subfamilies such as the NIG case, there are even simpler samplers available, as discussed later.

Algorithm 9.4 (Simulating Losses from a GH Severity Model)

1. *Draw a GIG random variate $V \sim GIG(\nu, \delta, \gamma)$, where $\gamma = \sqrt{\alpha^2 - \beta^2}$. This is achieved as follows via a rejection envelope method (see Atkinson 1982):*
   a) *Choose the envelope distribution function $g(v)$ to sample via inversion from, for example $V = g^{-1}(U)$ for $U_1 \sim Uniform(0, 1)$. In the GIG distribution case, the envelope function and its domain $[0, \infty)$ is partitioned as follows:*

$$g(v) = \begin{cases} k_1 d_1(v), & x \in [0, t] \\ k_2 d_2(v), & x \in (t, \infty) \end{cases} \tag{9.106}$$

Where we select t as the mode of the GIG distribution given by

$$t = m(\nu, \delta, \gamma) = \begin{cases} \dfrac{\nu - 1 + \sqrt{(1-\nu)^2 + \gamma\delta}}{\gamma}, & \gamma > 0 \\ \dfrac{\gamma}{2(1-\nu)}, & \gamma = 0, \end{cases} \tag{9.107}$$

and the envelope functions are given by simple functions to sample from

$$d_1(v) = \exp(sv), \quad d_2(v) = \exp(-pv), \tag{9.108}$$

where s and p are selected numerically to maximize the objective function expression given by

$$\begin{aligned} &\left(\frac{\exp(st) - 1}{s}\right) m(\nu, \delta, \gamma + 2s)^{\nu - 1} \exp\left(-\frac{1}{2}[\delta m(\nu, \delta, \gamma + 2s)^{-1} + \gamma m(\nu, \delta, \gamma + 2s)]\right) \\ &+ \left(\frac{\exp(-pt)}{p}\right) m(\nu, \delta, \gamma - 2p)^{\nu - 1} \\ &\exp\left(-\frac{1}{2}\left[\delta m(\nu, \delta, \gamma - 2p)^{-1} + \gamma m(\nu, \delta, \gamma - 2p)\right]\right). \end{aligned}$$

b) *Now denote the target distribution (GIG density) by* $f(v) = ce(v)$ *and define the function* $h_i(v) = \frac{e(v)}{d_i(v)}$ *with maximum in each partition of the domain given by* $S_i = \sup h_i(v)$. *One can then accept the generated value according to the following condition, after sampling a second independent uniform variate* $U_2 \sim Uniform(0, 1)$

$$U_2 \leq \frac{h_i(V)}{S_i}. \tag{9.109}$$

Repeat until one can accept a draw V.

2. *Draw a standard Normal random variate:* $\epsilon \sim Normal(0, 1)$*;*
3. *Construct the GH distributed random variable* $X \sim GH(x; \nu, \alpha, \beta, \mu, \delta)$ *according to the transformation*

$$X = \mu + \beta V + \sqrt{V}\epsilon. \tag{9.110}$$

Next we present a special subfamily of the GH class of severity models known as the NIG severity distribution.

9.6.3 NORMAL INVERSE GAUSSIAN FAMILY OF HEAVY-TAILED SEVERITY MODELS

The NIG distribution was recently introduced in the financial literature to capture non-Gaussian residuals observed in the financial time series by Barndorff-Nielsen (1997) and Barndorff-Nielsen and Shephard (2001). The NIG model takes its name from the fact that it represents a normal variance–mean mixture that occurs as the marginal distribution for a random variable X when considering a pair of random variables (X, Z), where Z is distributed as an IG $Z \sim InverseGaussian(\delta, \sqrt{\alpha^2 - \beta^2})$, and X conditional on Z is $(X|Z = z) \sim Normal(\mu + \beta z, z)$. The resulting density function for the NIG model is given in Definition 9.20.

Definition 9.20 (Normal Inverse Gaussian (NIG) — Scale Invariant) *A random variable* $X \sim NIG(\alpha, \beta, \mu, \delta)$ *is characterized by the density function*

$$f_X(x; \alpha, \beta, \mu, \delta) = \frac{\alpha\delta}{\pi}\frac{\exp[p(x)]}{q(x)} K_1[\alpha q(x)], \tag{9.111}$$

where $K_1[\cdot]$ *is a modified Bessel function of the second kind with index 1 (see Olver 1960), with*

$$p(y) = \delta\sqrt{\alpha^2 - \beta^2} + \beta(y - \mu)$$

and

$$q(y) = \left((y - \mu)^2 + \delta^2\right)^{1/2}. \qquad \blacksquare$$

As with the GH family, for the NIG subfamily under this parametrization, the parameters have the constraints $\mu \in \mathcal{R}, \delta > 0, 0 \leq |\beta| \leq \alpha$. The parameter α is inversely related to the heaviness of the tails, where a small α corresponds to heavier tails. The skewness is directly controlled by the parameter β, where negative (positive) values of β result in a left (right) skew and $\beta = 0$ is the symmetric model. The translation (or location) of the distribution is given by the parameter μ and the scale of the distribution is given by the parameter δ.

An alternative parametrization proposed by Eriksson *et al.* (2009), which is scale-invariant and may be considered in further studies, is obtained by setting $\overline{\alpha} = \delta\alpha$ and $\overline{\beta} = \delta\beta$, which is defined in Definition 9.21.

Definition 9.21 (Normal Inverse Gaussian (NIG) — scale invariant) *A random variable $X \sim NIG\left(\overline{\alpha}, \overline{\beta}, \mu, \delta\right)$ is characterized by the density function*

$$f_{\mathrm{NIG}}(x; \overline{\alpha}, \overline{\beta}, \mu, \delta) = \frac{\overline{\alpha} K_1\left[\frac{\overline{\alpha}}{\delta}\sqrt{\delta^2 + (x-\mu)^2}\right]}{\pi\sqrt{\delta^2 + (x-\mu)^2}} \exp\left(\sqrt{\overline{\alpha}^2 - \overline{\beta}^2} + \frac{\overline{\beta}}{\delta}(x-\mu)\right) \quad (9.112)$$

■

When considering the NIG severity model subfamily of the GH distributions, it is more convenient to simulate the severity losses via the following algorithm.

Algorithm 9.5 (Simulating Losses from a Normal Inverse Gaussian Severity Model)

1. *Draw an IG random variate: $Z \sim InverseGaussian(\delta, \sqrt{\alpha^2 - \beta^2})$. This is achieved as follows via a transformation and rejection stage:*
   a) *Draw a standard Normal random variate: $V \sim Normal(0, 1)$;*
   b) *Evaluate $Y = V^2$;*
   c) *Evaluate $D = \delta + \frac{\delta^2 Y}{2\sqrt{\alpha^2-\beta^2}} - \frac{\delta}{2\sqrt{\alpha^2-\beta^2}}$;*
   d) *Sample a uniform random variate $U \sim Uniform(0, 1)$ and perform rejection stage where one accepts $Z = D$ if $U < \frac{\delta}{\delta+D}$, otherwise set $Z = \frac{\sqrt{\alpha^2-\beta^2}}{D}$.*
2. *Draw a conditional Normal random variate $X \sim Normal(\mu + \beta Z, Z)$.*

In addition, the NIG has the following features that relate it to other distributions:

1. If one restricts $\beta = 0$ and μ is arbitrary, the NIG model asymptotically approaches the popular Gaussian model $X \sim Normal\left(\mu, \frac{\delta}{\alpha}\right)$ as $\alpha \to \infty$ or $\delta \to \infty$;
2. If one restricts $\alpha = \beta = 0$ with μ and δ as arbitrary, the NIG model approaches the Cauchy distribution;
3. The NIG model can also approximate the skewness and kurtosis of the LogNormal, Student's *t*, and Gamma distributions, among others (see Hosack *et al.* 2012 and Hanssen and Oigard 2001).

To better understand the flexible features of the NIG model, it is convenient to consider the steepness–asymmetry specification. The shape of the NIG distribution can be conveniently summarized with a graphical representation called the NIG shape triangle (Barndorff-Nielsen and Shephard 2001). This plot uses indices of steepness and asymmetry, which are analogous to kurtosis and skewness, given by

$$\text{Steepness} = \left(1 + \delta\sqrt{\alpha^2 - \beta^2}\right)^{-1/2},$$
$$\text{Asymmetry} = \frac{\beta}{\alpha} \times \text{Steepness},$$

with $0 < \text{Steepness} < 1$ and $-1 < \text{Asymmetry} < 1$. Distributions with $\text{Asymmetry} = 0$ are symmetric, and the Gaussian and Cauchy distributions occur as limiting cases for $(\text{Asymmetry}, \text{Steepness})$ near (0,0) and (0,1), respectively. Figure 9.9 provides a graphical representation of NIG probability example density functions.

The expressions for the mean, variance, skewness, and kurtosis for the NIG model are given conveniently in terms of the model parameters in the following closed, form expressions in Proposition 9.25.

Proposition 9.25 (Moments of NIG Severity Models) *A loss random variable $X \sim NIG(\alpha, \beta, \mu, \delta)$ is characterized sufficiently by the first four moments*

$$\begin{aligned}
\mathbb{E}(X) &= \mu + \frac{\delta\left(\frac{\beta}{\alpha}\right)}{\left(1 - \left(\frac{\beta}{\alpha}\right)^2\right)^{1/2}}, \\
\mathbb{V}\text{ar}(X) &= \frac{\delta}{\alpha\left(1 - \left(\frac{\beta}{\alpha}\right)^2\right)^{3/2}}, \\
\mathbb{S}\text{kew}(X) &= \frac{3\left(\frac{\beta}{\alpha}\right)}{(\delta\alpha)^{1/2}\left(1 - \left(\frac{\beta}{\alpha}\right)^2\right)^{1/4}}, \\
\mathbb{K}\text{urt}(X) &= 3\frac{4\left(\frac{\beta}{\alpha}\right)^2 + 1}{\delta\alpha\left(1 - \left(\frac{\beta}{\alpha}\right)^2\right)^{1/2}}.
\end{aligned} \tag{9.113}$$

In general, the moment-generating function is given by the expression

$$\mathbb{E}\left[\exp(tX)\right] = \exp\left(\delta\sqrt{\alpha^2 - \beta^2} - \delta\sqrt{\alpha^2 - (\beta + t)^2} + \mu t\right). \tag{9.114}$$

One may then solve these equations numerically to perform parameter estimation via matching of the population moments and the sample estimated moments.

9.6.3.1 Parameter Estimation for Normal Inverse Gaussian Severity Models. In this section, we consider the approach to parameter estimation based on Method of Moments (MOM) and a variant of this approach, which ensures the strict ϵ-positivity of the support of the resulting NIG severity distribution. Other approaches, which

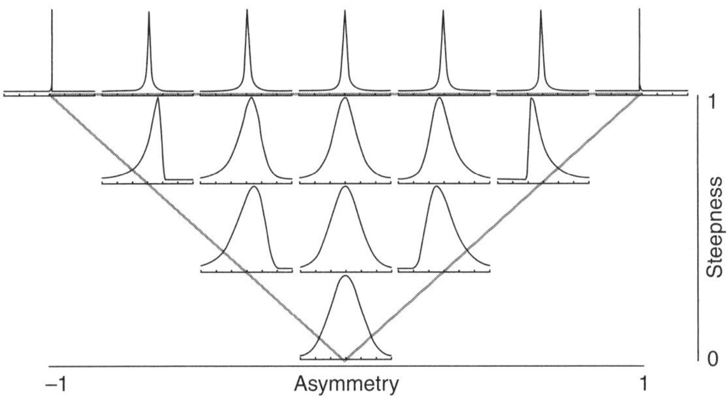

FIGURE 9.9 NIG triangle characterizing the flexibility of the skewness and kurtosis properties of the NIG family of models

are numerically slower to implement but provide accurate results, include maximum likelihood (see Chapter 7).

Under the MOM approach, if the parameters are unconstrained, then one may achieve any of the possible skew and kurtosis characteristics of the NIG family to be obtained, with the computational constraint that no closed-form solution for the parameter estimates can be obtained from algebraic manipulation of the system of equations produced by matching distribution expressions for moments with empirical sample moments. Hence, for the unconstrained case, one must resort to numerical root-finding solutions in four parameters and care should be taken with the numerical procedures adopted. Alternatively, one may restrict to a subfamily of the NIG distributions, through constraining of the existence of the first four cummulants, as detailed by Eriksson *et al.* (2004). These expressions for the parameters of the NIG distribution in terms of its mean, variance, skewness, and excess kurtosis under these constraints are then achieved as shown in Proposition 9.26.

Proposition 9.26 ***(Closed-Form Parameter Estimation for NIG Severity Models via MOM)*** *Consider that i.i.d. distributed loss random variables $X_i \sim NIG(\alpha, \beta, \mu, \delta)$ with sample mean, sample variance, sample skewness, and sample excess kurtosis, denoted by $\hat{\mathcal{M}}$, $\hat{\mathcal{V}}$, $\hat{\mathcal{S}}$, and $\hat{\mathcal{K}}$, respectively, can be utilized to estimate the model parameters with a constraint imposed. Assume that the following constraint applies to the kurtosis $3\hat{\mathcal{K}} > 5$ and the skewness $\hat{\mathcal{S}}^2 > 0$, then the method of moment estimators for the parameters are given in closed form under these constraints by*

$$\begin{aligned}
\hat{\alpha} &= 3\hat{\rho}^{1/2}(\hat{\rho}-1)^{-1}\hat{\mathcal{V}}^{-1/2}|\hat{\mathcal{S}}|^{-1}, \\
\hat{\beta} &= 3(\hat{\rho}-1)^{-1}\hat{\mathcal{V}}^{-1/2}\hat{\mathcal{S}}^{-1}, \\
\hat{\mu} &= \hat{\mathcal{M}} - 3\hat{\rho}^{-1}\hat{\mathcal{V}}^{1/2}\hat{\mathcal{S}}^{-1}, \\
\hat{\delta} &= 3\hat{\rho}^{-1}(\hat{\rho}-1)^{1/2}\hat{\mathcal{V}}^{1/2}|\hat{\mathcal{S}}|^{-1},
\end{aligned} \tag{9.115}$$

where $\hat{\rho} = 3\hat{\mathcal{K}}\hat{\mathcal{S}}^{-2} - 4 > 1$.

In the case where one wishes to ensure ϵ-strict positivity of support when estimating the model parameters, one may wish to select $\beta = \alpha - |\epsilon|$ for $\epsilon \in n.e.(0)$ where ϵ is a small value in the neighborhood of the origin. In this case, one has the estimating equations for the MOM yield of the closed-from expressions given in Proposition 9.27.

Proposition 9.27 (Parameter Estimation of NIG Model with ϵ-Positive Support) *Consider that i.i.d. distributed loss random variables $X_i \sim NIG(\alpha, \beta, \mu, \delta)$ with sample mean, sample variance, sample skewness, and sample excess kurtosis, denoted by $\hat{\mathcal{M}}$, $\hat{\mathcal{V}}$, $\hat{\mathcal{S}}$, and $\hat{\mathcal{K}}$, respectively, can be utilized to estimate the model parameters with a constraint imposed. In this case, the constraint is selected to ensure the estimated model has ϵ-positive support where $\beta = \alpha - |\epsilon|$ for $\epsilon \in n.e.(0)$ where ϵ is a small value in the neighborhood of the origin. The parameters are estimated via the following closed-form expressions:*

$$
\begin{aligned}
\hat{\alpha} &= \frac{\epsilon}{\left(1 - \sqrt{\frac{\mathcal{S}^2}{3\mathcal{K} - 4\mathcal{S}^2}}\right)}, \\
\hat{\beta} &= \hat{\alpha} - \epsilon, \\
\hat{\delta} &= \frac{9\left(\hat{\alpha} - \epsilon\right)^2}{\hat{\alpha}^2 \mathcal{S}^2 \sqrt{2\hat{\alpha}\epsilon - \epsilon^2}}, \\
\hat{\mu} &= \mathcal{E} - \frac{\hat{\delta}\left(\frac{\hat{\alpha}-\epsilon}{\hat{\alpha}}\right)}{\sqrt{1 - \left(\frac{\hat{\alpha}-\epsilon}{\hat{\alpha}}\right)^2}}.
\end{aligned}
\tag{9.116}
$$

Next, we briefly discuss the related class of GIG distributions, where it was first formed as part of a family of distributions known as the Halphen family. The GIG submembers of this family (Halphen Type A distribution) were shown to be instrumental in constructing the GH family and the NIG family.

9.7 Halphen Family of Flexible Severity Models: GIG and Hyperbolic

The history of the Halphen family of distributions is very interesting as pointed out by the series of two expository papers by Perreault *et al.* (1999a,b) and an article by Seshadri (2004). The Halphen severity distribution is an interesting family of distributions that was first proposed by a French statistician, Etienne Halphen. His publication of this work was complicated by the fact that his country was in a war period and he had an early death. Consequently, the official record of his work first appeared in his papers (Halphen 1941, 1953) on harmonic distributions, which were renamed in the 1970s as the hyperbola distribution family (subfamily of the GH) in the pioneering works of Rukhin (1974) and Barndorff-Nielsen and Halgreen (1977), who seem to have rediscovered this family independently of Halphen's original works. In fact, as pointed out by Perreault *et al.* (1999a), it was George Morlat, Halphen's colleague, who eventually published some of Halphen's work (Morlat 1951). This work would have remained hidden as it was written in lesser-known journals to the statistical community and was published in French, except for the exposition developed on this family by Perreault *et al.* (1999a), which itself was not written for a wide statistical audience as it was published in a hydrological engineering journal. This made sense as the primary motivation for Halphen for this work was originally related to the modeling of river flows. Since this early work there have only been a few applications of this family of models (under the name of Halphen family; see Guillot 1964 and Cam 1949); however, there have been many more under the name of GIG and GH distributions.

In this section, we will provide a brief introduction to the family of positive supported Halphen Type A, Type B, and Type IB distributions for the modeling of OpRisk severity models in an LDA structure. These families will be shown to contain several important subfamilies such as the GIG, GH, Inverese Gamma, Gamma, and Normal distributions. The family of Halphen distributions has recently become popular in the hydrological literature for modeling flows, but remains, however, largely under utilized in the statistics literature even though it provides comparative performance to other models of extreme events such as the Generalized Extreme Value (GEV) models discussed in Peters and Shevchenko (2015); also see discussions by El Adlouni *et al.* (2009). We start with a general characterization of this family of distributions.

One can characterize the Halphen family according to the differential equation in Proposition 9.28, where four generic parameters q, a_0, a_1, and a_2 are specified. When considering each subfamily of the Halphen distributions, the reparametrization in terms of parameters is as follows:

- $m > 0$ is a scale parameter;
- ν is a shape parameter with a range of admissible parameter values that will depend on the subfamily (Type A, $\nu \in \mathbb{R}$; Type B, $\nu > 0$; and Type IB, $\nu > 0$);
- α is a shape parameter with a range of admissible parameter values that will depend on the subfamily (Type A, $\alpha > 0$; Type B, $\alpha \in \mathbb{R}$ and Type IB, $\alpha \in \mathbb{R}$).

The specification of the family via this o.d.e. representation is particularly useful as it allows one to show the mode and antimode behaviors of each of the subfamilies of distributions (see discussions in Perreault *et al.* 1999a).

Proposition 9.28 (Halphen Family of Severity Models) *A loss random variable $X \sim Halphen\,(q, a_0, a_1, a_2)$ is characterized generically by the ordinary differential equation*

$$\frac{1}{f(x)}\frac{df(x)}{dx} = \frac{a_0 + a_1 x + a_2 x^2}{x^q}, \tag{9.117}$$

which gives rise to the following three subfamilies of distributions known as Type A, Type B, and Type IB:

- ***Type A subfamily.*** *Set the parameters to $q = 2$, $a_0 = \alpha m$, $a_1 = \nu - 1$, and $a_2 = -\alpha/m$;*
- ***Type B subfamily.*** *Set the parameters to $q = 1$, $a_0 = 2\nu - 1$, $a_1 = \alpha/m$, and $a_2 = -2/m^2$;*
- ***Type IB subfamily.*** *Set the parameters to $q = 3$, $a_0 = 2m^2$, $a_1 = -\alpha m$, and $a_2 = -(2\nu + 1)$.*

With this specification one can show the following model flexibility of each of the subfamilies of the Halphen distribution with regard to modes and antimodes simply by finding the roots of the equation $a_0 + a_1 x + a_2 x^2$ (see Proposition 9.29 and details by Perreault *et al.* 1999a).

Proposition 9.29 (Halphen Severity Model Mode and AntiModes) *One can classify each subfamily of the Halphen distributions via the existence of no mode, one mode or a mode and antimode relationship. It can be shown that under this representation, the mode and antimode properties of the distribution are obtained by equating $df/dx = 0$, which shows that such modes and antimodes are simply solutions to the quadratic equation $a_0 + a_1 x + a_2 x^2 = 0$ given by*

$$\text{Mode}(X) = -\frac{a_1}{2a_2} \pm \sqrt{\left(\frac{a_1}{2a_2}\right)^2 - \frac{a_0}{a_2}}. \tag{9.118}$$

As a result, the types of Halphen distribution can be characterized according to the existence of no real solutions (no mode), one real solution (one mode), and two real solutions (mode and antimode) subfamilies, labeled Type I, Type II, and Type III, respectively.

- ***Type I.*** *No modes with conditions $\frac{a_0}{a_2} \geq 0$ and $\frac{a_1}{2a_2} \geq 0$;*
- ***Type II.*** *A mode and an antimode with conditions $\frac{a_0}{a_2} > 0$, $\left(\frac{a_1}{2a_2}\right)^2 > \frac{a_0}{a_2}$ and $\frac{a_1}{2a_2} < 0$;*
- ***Type III.*** *One mode with conditions $\frac{a_0}{a_2} < 0$; or $\frac{a_0}{a_2} = 0$ and $\frac{a_1}{2a_2} < 0$.*

Before presenting details of each of the subfamilies of models in the Halphen class, we first detail some general properties of the tail behavior of the Halphen Type A, Type B, and Type IB distributions. As discussed by Perreault *et al.* (1999a), the Halphen family has the tail properties specified in Proposition 9.30.

Proposition 9.30 (Halphen Family Distributional Tail Properties) *Consider a loss random variable $X \sim$ Halphen(q, a_0, a_1, a_2), then the following tail behaviors are possible in each subfamily.*

1. ***Type A subfamily.*** *Set the parameters to $q = 2$, $a_0 = \alpha m$, $a_1 = \nu - 1$, and $a_2 = -\alpha/m$ and the tail properties of the Halphen Type A, Gumbel, and Gamma distributions are characterized by the return period with quantiles x proportional to the log of the return period such that $x \propto \ln T$;*
2. ***Type B subfamily.*** *Set the parameters to $q = 1$, $a_0 = 2\nu - 1$, $a_1 = \alpha/m$, and $a_2 = -2/m^2$ and the tail properties of the Halphen Type B and Gaussian distributions are characterized by the return period with quantiles x proportional to the squareroot of the log of the return period such that $x \propto \sqrt{\ln T}$;*
3. ***Type IB subfamily.*** *Set the parameters to $q = 3$, $a_0 = 2m^2$, $a_1 = -\alpha m$, and $a_2 = -(2\nu + 1)$ and the tail properties of the Halphen Type IB distributions are characterized by the return period with quantiles x proportional to the power of the return period such that $x \propto T^{\frac{1}{2\nu}}$.*

In these properties $T(x)$ denotes the return periods given by $T(x) = \overline{F}^{-1}(x)$.

At this stage, it will be relevant to introduce a special function introduced by Halphen (1955) that plays an equivalent role in normalization for the Type B and Type IB distributions as the modified Bessel function of the second kind (third kind) in the Type A (Generalized Inverse Gaussian) subfamily (see Proposition 9.31).

Proposition 9.31 (Exponential Factorial Function) *The exponential factorial function* $ef_\nu(\alpha)$ *is given by the integral equation*

$$ef_\nu(\alpha) = 2\int_0^\infty x^{2\nu-1}\exp\left[-x^2+\alpha x\right]dx, \quad \nu > 0 \tag{9.119}$$

or by the series expansion

$$ef_\nu(\alpha) = \sum_{r=0}^{\infty}\frac{\alpha^r}{r!}\Gamma\left(\nu+\frac{r}{2}\right). \tag{9.120}$$

The exponential factorial function is plotted in Example 9.7 for a few parameter settings.

EXAMPLE 9.7 Features of the Exponential Factorial Function

In this example, we plot the value of the log of the exponential factorial function as a function of ν and α for a range of parameter settings (Figure 9.10).

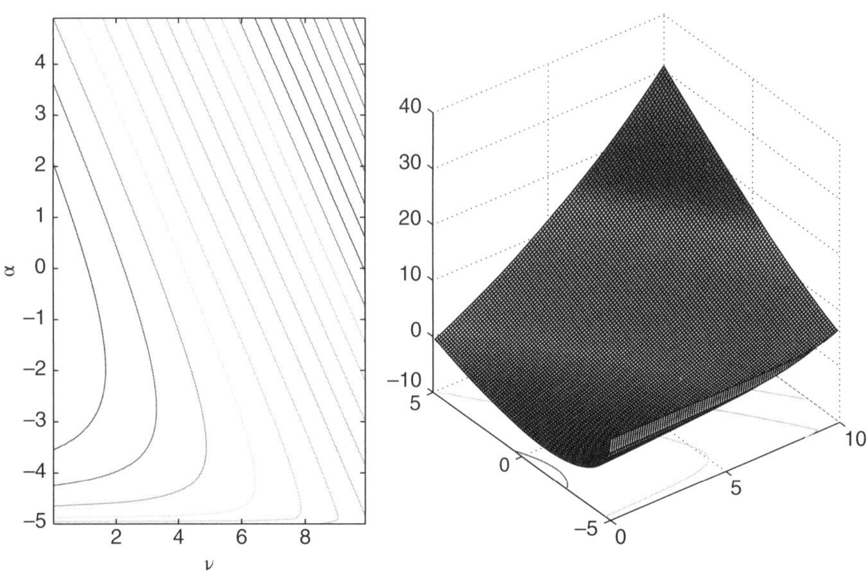

FIGURE 9.10 Log of the exponential factorial function for a range of parameters ν and α

To conclude this general discussion on the Halphen family of distributions, we discuss some basic sampling procedures for drawing loss random variates from a Halphen family distribution as discussed in detail by El Adlouni and Bobée (2007), where an acceptance–rejection method is proposed for each of the Type A, Type B, and Type IB subfamilies.

Algorithm 9.6 (Simulating Losses from a Halphen Type A Severity Model)

1. *Construct the instrumental distribution for aceptance–rejection sampling given by a Gamma density*

$$g(x;\lambda,\delta) = \frac{\delta^{\lambda}}{\Gamma(\lambda)} x^{\lambda-1}\exp(-\delta x) \tag{9.121}$$

with scale $\delta > 0$ and shape $\lambda > 0$, where if $X \sim$ HalphenA (m, ν, α), one selects the shape and scale of the instrumental distribution by

$$\begin{aligned} \lambda &= \frac{\mathbb{E}[X]^2}{2\mathbb{V}\text{ar}(X)} = \frac{K_{\nu}^2(2\alpha)}{K_{\nu}(2\alpha)K_{\nu+2}(2\alpha) - K_{\nu+1}^2(2\alpha)} \\ \delta &= \frac{\lambda}{\mathbb{E}[X]} = \frac{\lambda K_{\nu}(2\alpha)}{mK_{\nu+1}(2\alpha)}. \end{aligned} \tag{9.122}$$

2. *Numerically find the maximum point M such that the following inequality with the Halphen Type A density $f(x)$ and the instrumental Gamma density $g(x)$ satisfy $f(x) \leq Mg(x)$ where one should select M as the minimum value such that $Mg(x)$ is an envelope for $f(x)$ over its entire support;*
3. *Draw a Gamma random variate: $X \sim$ Gamma(λ, δ);*
4. *Draw a Uniform random variate: $U \sim$ Uniform$(0, 1)$;*
5. *Accept Draw X if $U \leq \frac{f(X;m,\nu,\alpha)}{Mg(X;\lambda,\delta)}$, otherwise repeat.*

In the same manner, one can also design an acceptance–rejection algorithm for sampling loss random variates from the Halphen Type B and Type IB distributions as follows. The only real difference compared to the Type A samplers is in the specification of the shape and scale parameters that will ensure the dominance of the instrumental distribution to act as an envelope function of the support of the Type B and Type IB distributions.

Algorithm 9.7 (Simulating Losses from a Halphen Type B Severity Model)

1. *Construct the Instrumental distribution for acceptance–rejection sampling given by a Gamma density*

$$g(x;\lambda,\delta) = \frac{\delta^{\lambda}}{\Gamma(\lambda)} x^{\lambda-1}\exp(-\delta x) \tag{9.123}$$

with scale $\delta > 0$ and shape $\lambda > 0$, where if $X \sim$ HalphenA (m, ν, α), one selects the shape and scale of the instrumental distribution by one of the two following specifications:

If $\nu \leq 1$,

$$\lambda = \frac{1}{5} ef^2_{\nu-\frac{1}{2}}(\alpha) \left[ef_\nu(\alpha) ef_{\nu-1}(\alpha) - ef^2_{\nu-\frac{1}{2}}(\alpha) \right]^{-1}$$
$$\delta = \sqrt{\frac{\lambda ef^2_\nu(\alpha)}{5m^2}} \left[ef_\nu(\alpha) ef_{\nu-1}(\alpha) - ef^2_{\nu-\frac{1}{2}}(\alpha) \right]^{-\frac{1}{2}}. \quad (9.124)$$

If $\nu \leq 1$,

$$\lambda = \frac{1}{2} ef^2_{\nu-\frac{1}{2}}(\alpha) \left[ef_\nu(\alpha) ef_{\nu-1}(\alpha) - ef^2_{\nu-\frac{1}{2}}(\alpha) \right]^{-1}$$
$$\delta = \sqrt{\frac{\lambda ef^2_\nu(\alpha)}{2m^2}} \left[ef_\nu(\alpha) ef_{\nu-1}(\alpha) - ef^2_{\nu-\frac{1}{2}}(\alpha) \right]^{-\frac{1}{2}}. \quad (9.125)$$

2. *Numerically find the maximum point M such that the following inequality with the Halphen Type A density $f(x)$ and the instrumental Gamma density $g(x)$ satisfy $f(x) \leq Mg(x)$ where one should select M as the minimum value such that $Mg(x)$ is an envelope for $f(x)$ over its entire support;*
3. *Draw a Gamma random variate: $X \sim Gamma(\lambda, \delta)$;*
4. *Draw a Uniform random variate: $U \sim Uniform(0, 1)$;*
5. *Accept draw X if $U \leq \frac{f(X;m,\nu,\alpha)}{Mg(X;\lambda,\delta)}$, otherwise repeat.*

To obtain samples from the Halphen Type IB distribution one simply uses the fact that if $X \sim HalphenB(x; m, \alpha, \nu)$, then $1/X \sim HalphenIB\left(x; \frac{1}{m}, \alpha, \nu\right)$. Next we present some properties of the there subfamilies classified by Type A, Type B, and Type IB. Before presenting this, we note that several authors have developed parameter estimation procedures for the Halphen family including Maximum Likelihood, MOM, and mixed methods (see discussions by Chebana *et al.* 2008 and Perreault *et al.* 1999b).

9.7.1 HALPHEN TYPE A: GENERALIZED INVERSE GAUSSIAN FAMILY OF FLEXIBLE SEVERITY MODELS

Here we discuss the important subfamily of the Halphen system given by the Halphen Type A distributions. As discussed by Morlat (1951) and Perreault *et al.* (1999a), the original specification of a (reparameterized) form of the GIG distribution better known in statistics from the work of Good (1953) was actually developed almost a decade earlier by Halphen (1941). The reparameterized form of the Halphen Type A family (relative to the typically used GIG parameterization in modern statistics) is shown in Proposition 9.32.

Proposition 9.32 (Halphen Type A (GIG) Severity Models) *A loss random variable X has a Halphen Type A distribution $X \sim HalphenA(x; m, \nu, \alpha)$ if it has a density given by*

$$f(x; m, \nu, \alpha) = \frac{1}{2m^\nu K_\nu(2\alpha)} x^{\nu-1} \exp\left[-\alpha\left(\frac{x}{m} + \frac{m}{x}\right)\right], \quad x > 0 \quad (9.126)$$

for $m > 0$, $\alpha > 0$, and $\nu \in \mathbb{R}$.

In the following example, the shape of the Halphen Type A severity distribution is plotted for a range of shape and scale parameters to demonstrate how each parameter changes the characteristics of the severity model.

EXAMPLE 9.8 Examples of Flexible Halphen Severity Distributions (Type A)

In the plots shown in Figure 9.11 plots, we explore the density shapes for the case of the subfamily of Halphen Type A distributions (GIG subfamily) given by a range of scale parameters m and shape parameters ν and α.

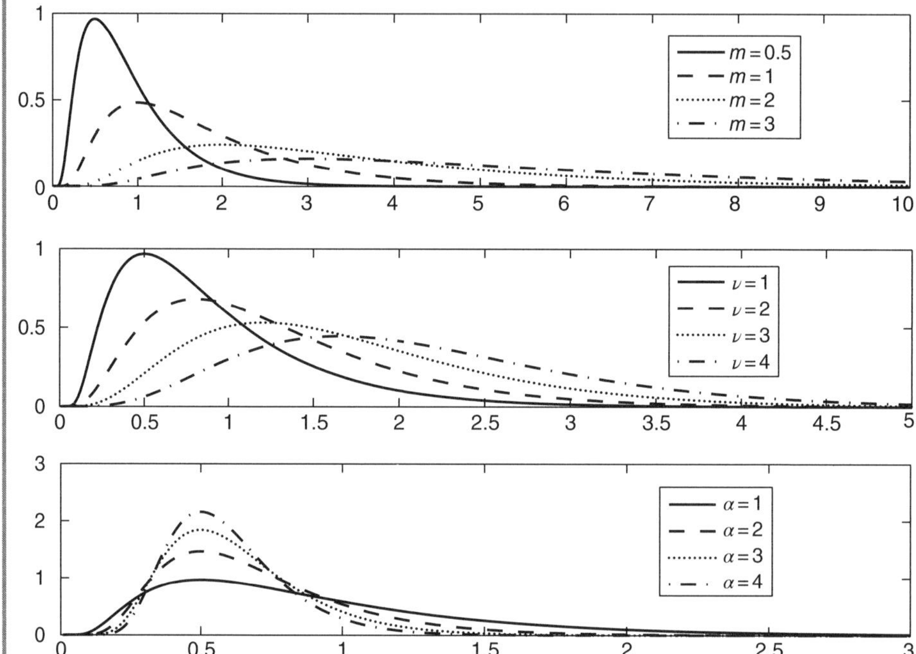

FIGURE 9.11 **Halphen Type A distributions (GIG subfamily). Top subplot: parameters used in this subplot study the effect of the scale parameter ranges $m = [0.5, 1, 2, 3]$ with shape parameters set to $\nu = 1$ and $\alpha = 1$. Middle subplot: parameters used in this subplot study the effect of the shape parameter ranges $\nu = [1, 2, 3, 4]$ with shape parameter set to $\alpha = 1$ and scale parameter $m = 0.5$. Bottom subplot: parameters used in this subplot study the effect of the shape parameter ranges $\alpha = [1, 2, 3, 4]$ with shape parameter set to $\nu = 1$ and scale parameter $m = 0.5$**

It is trivial to see that if one considers the reparametrization in terms of parameters ν, δ, and γ, which are typically used in specifying the GIG family of severity distributions, then the setting

1. ν parameter unchanged;
2. $\alpha = \frac{\sqrt{\delta\gamma}}{2}$;
3. $m = \frac{\gamma}{\sqrt{\delta\gamma}}$;

will reproduce the typical parametrization of the GIG family shown in works such as that of Good (1953). In the remainder of this section, we will present some properties of the GIG family when parameterized under the Halphen Type A structure or the more familiar structure of the GIG family, making explicit the distinction by persisting with the two sets of parameters, and making explicit which is used in each discussion.

As previously demonstrated, the GIG subfamily of the Halphen family also plays an important role in simulation of losses from a GH severity distribution model. However, in this section, we will also study the properties of the GIG family of models that are useful for OpRisk modeling in their own right. The GIG family is a subfamily of the GH models with strictly positive support with many usefull properties for OpRisk LDA structures. The family of GIG models is parametrized by $\nu \in \mathbb{R}$, $\gamma \in \mathbb{R}^+$, and $\delta \in \mathbb{R}^+$ such that $\gamma + \delta > 0$ and has density given in Definition 9.22 (see details by Jørgensen 1982). There are a few special subfamilies nested in the GIG model that include the ($\delta = 0$) case, which produces the Gamma distribution family, the ($\gamma = 0$), which yields the Inverse Gamma family, and the ($\nu = -\frac{1}{2}$) giving the Inverse Gaussian family.

Definition 9.22 (Generalized Inverse Gaussian Severity Model Density) *A loss random variable X has a GIG distribution $X \sim GIG(x; \nu, \delta, \gamma)$ if it has a density given by*

$$f_X(x) = \frac{\left(\frac{\gamma}{\delta}\right)^{\frac{\nu}{2}}}{2K_\nu\left(\sqrt{\delta\gamma}\right)} x^{\nu-1} \exp\left(-\frac{1}{2}\left(\delta x^{-1} + \gamma x\right)\right), \quad x > 0 \tag{9.127}$$

with domain of variation of the parameters divided into three cases depending on the value of ν as follows:

1. $\delta > 0$ *and* $\gamma \geq 0$, *if* $\nu < 0$;
2. $\delta > 0$ *and* $\gamma > 0$, *if* $\nu = 0$;
3. $\delta \geq 0$ *and* $\gamma > 0$, *if* $\nu > 0$.

■

In the form of the Halphen Type A distribution, it can be shown, based on the conditions in Proposition 9.29, that the Halphen Type A and therefore also the reparameterized GIG family of models are unimodal (see Proposition 9.33).

Proposition 9.33 (Mode of Halphen Type A (GIG) Severity Models) *A loss random variable X with a Halphen Type A distribution $X \sim HalphenA(x; m, \nu, \alpha)$ has a single mode at location*

$$Mode[X] = m\left[\frac{\nu - 1}{2\alpha} + \sqrt{\left(\frac{\nu - 1}{2\alpha}\right)^2 + 1}\right]. \tag{9.128}$$

In the parametrization of the GIG distribution $X \sim GIG(x; \nu, \delta, \gamma)$, the mode is located at

$$Mode[X] = \frac{\gamma}{\sqrt{\delta\gamma}}\left[\frac{\nu - 1}{\sqrt{\delta\gamma}} + \sqrt{\left(\frac{\nu - 1}{\sqrt{\delta\gamma}}\right)^2 + 1}\right]. \tag{9.129}$$

In addition, as was originally the intention of Halphen in developing this family, the Halphen Type A family of distributions may be written in exponential family form with sufficient statistics for each of the parameters given in Proposition 9.34. Note that this result is just a reparametrized version of the well-known result for the exponential family representation of the GIG family.

Proposition 9.34 (Exponential Family Representation of Halphen Type A) *A loss random variable X with a Halphen Type A distribution $X \sim HalphenA(x; m, \nu, \alpha)$ belongs to a three-parameter exponential family presented in the following form:*

$$f(x; m, \nu, \alpha) = \exp\left((\nu - 1)\ln x - \frac{\alpha}{m}x - \alpha m\frac{1}{x} - \ln\left[2m^{\nu}K_{\nu}(2\alpha)\right]\right), \tag{9.130}$$

which results in the following sufficient statistics from a sample of n i.i.d. losses $\{X_i\}_{i=1}^{n}$ from the Halphen Type A distribution

$$\begin{aligned} T_1(X_{1:n}) &= \sum_{i=1}^{n} \ln X_i = n \ln G, \\ T_2(X_{1:n}) &= \sum_{i=1}^{n} X_i = nA, \\ T_3(X_{1:n}) &= \sum_{i=1}^{n} X_i^{-1} = nH^{-1}, \end{aligned} \tag{9.131}$$

for the arithmetic mean A, the geometric mean G, and the harmonic mean H.

It is clear that one could then perform estimation of the parameters if it were possible to obtain the population means (arithmetic, geometric, and harmonic) in terms of the parameters. Fortunately, this is possible as illustrated by Perreault *et al.* (1999a) who show that one may obtain the results given in Proposition 9.35.

Proposition 9.35 ***(Arithmetic, Geometric, and Harmonic Population Moments Halphen Type A)*** *A loss random variable X with a Halphen Type A distribution $X \sim HalphenA(x; m, \nu, \alpha)$ has the following closed-form expressions for the Arithmetic (A), the Geometric (G), and Harmonic (H) means:*

$$\begin{aligned} A &= \mathbb{E}[X] = m\frac{K_{\nu+1}(2\alpha)}{K_{\nu}(2\alpha)} \\ H &= \mathbb{E}\left[\frac{1}{X}\right] = \frac{1}{m}\frac{K_{\nu-1}(2\alpha)}{K_{\nu}(2\alpha)}. \end{aligned} \tag{9.132}$$

Note that the Geometric mean is given by the moment of order quasi-zero (see Kendall et al. 1994), hence it is given, for i.i.d. random variables $\{X_i\}_{i=1}^{n}$, by the limit

$$G := \lim_{r \to 0}\left[\frac{1}{n}\sum_{i=1}^{n} X_i^r\right]^{\frac{1}{r}}. \tag{9.133}$$

The log of G is given by

$$\ln G = \mathbb{E}[\ln X] = \ln m + K_{\nu}^{-1}(2\alpha)\frac{\partial K_{\nu}(2\alpha)}{\partial \nu}. \tag{9.134}$$

In terms of the GIG parameterization of the Halphen Type A family, the moments of the GIG family can be characterized by the moment-generating function given in Proposition 9.36.

Proposition 9.36 (Moment-Generating Function Generalized Inverse Gaussian Family) *A loss random variable X with a GIG distribution $X \sim GIG(x; \nu, \delta, \gamma)$ has a moment-generating function given by*

$$\mathbb{E}\left[\exp(tX)\right] = \left(\frac{\gamma}{\gamma - 2t}\right)^{\frac{\nu}{2}} \frac{K_\nu\left(\sqrt{\delta(\gamma - 2t)}\right)}{K_\nu\left(\sqrt{\delta\gamma}\right)}. \tag{9.135}$$

If one then differentiates the Moment Generating Function (MGF) and evaluates it at the origin, the following moment results are obtained in Proposition 9.37.

Proposition 9.37 (Moments of Generalized Inverse Gaussian Severity Models) *A loss random variable X with a GIG distribution $X \sim GIG(x; \nu, \delta, \gamma)$ has integer moments $r \in \mathbb{J}^+$ given in closed form by*

$$\mathbb{E}\left[X^r\right] = \left(\frac{\delta}{\gamma}\right)^{\frac{r}{2}} \frac{K_{\nu+r}\left(\sqrt{\delta\gamma}\right)}{K_\nu\left(\sqrt{\delta\gamma}\right)}, \tag{9.136}$$

which gives a mean and variance according to the expressions

$$\begin{aligned}\mathbb{E}[X] &= \frac{\sqrt{\delta}K_{\nu+1}\left(\sqrt{\gamma\delta}\right)}{\sqrt{\gamma}K_\nu\left(\sqrt{\gamma\delta}\right)}\\ \mathbb{V}\mathrm{ar}[X] &= \frac{\delta}{\gamma}\left[\frac{K_{\nu+2}\left(\sqrt{\gamma\delta}\right)}{K_\nu\left(\sqrt{\gamma\delta}\right)} - \left(\frac{K_{\nu+1}\left(\sqrt{\gamma\delta}\right)}{K_\nu\left(\sqrt{\gamma\delta}\right)}\right)^2\right].\end{aligned} \tag{9.137}$$

The log moment is given by evaluating the following expression at $r = 0$:

$$\mathbb{E}\left[\ln X\right] = \frac{d\mathbb{E}\left[X^r\right]}{dr}. \tag{9.138}$$

With regard to the tail properties of the GIG family of distributions, these were considered by Embrechts (1983). In this work, the case $\nu < 0$ was considered and it was shown that all GIG distributions under this restriction satisfy an asymptotic convolution property, which allowed Embrechts (1983) to show particular tail properties of the distribution such as those presented in Proposition 9.38 (see Embrechts 1983, theorem 1).

Proposition 9.38 (Subexponential Tail Behavior of Generalized Inverse Gaussian) *Consider a loss random variable X with a GIG distribution $X \sim GIG(x; \nu, \delta, \gamma)$ with parameter restrictions $\nu < 0$, $\delta > 0$, and $\gamma \geq 0$; then the distribution function $F_X(x)$ is a member of the subexponential family of distributions with $F_X(x) \in \mathcal{S}\left(\frac{\delta}{2}\right)$ such that the following limiting tail behavior is satisfied for $n \geq 2$*

$$\lim_{x\to\infty} \frac{\overline{F_X^{(n)*}}(x)}{\overline{F}_X(x)} = n\left[f\left(-\frac{\delta}{2}\right)\right]^{n-1}, \tag{9.139}$$

where $F^{(n)*}$ is the n-fold convolution and the function $f\left(-\frac{\delta}{2}\right)$ represents the Laplace–Stieltjes transform of distribution F that satisfies

$$\lim_{x\to\infty} \frac{\overline{F_X^{(2)*}}(x)}{\overline{F}_X(x)} = 2f\left(-\frac{\delta}{2}\right) < \infty. \tag{9.140}$$

9.7.1.1 Subfamilies of the Halphen Type A (Generalized Inverse Gaussian) Severity Models. The GIG family of severity models also contains as a special subfamily the IG (Wald) severity models with parameter settings $\nu = -\frac{1}{2}$, $\gamma = \frac{\lambda}{\mu}$, and $\delta = \lambda$ giving the IG family with parameters λ and μ. The IG severity models are also members of the exponential family (see discussions by Johnson *et al.* 1970). The resulting distribution and density are provided in Definition 9.23.

Definition 9.23 (Inverse Gaussian Severity Model Distribution and Density) *Consider a loss random variable with IG severity model $X \sim InverseGaussian(\lambda, \mu)$, then the density function is given by*

$$f_X(x;\lambda,\mu) = \left[\frac{\lambda}{2\pi x^3}\right]^{\frac{1}{2}} \exp\left(\frac{-\lambda(x-\mu)^2}{2\mu^2 x}\right), \quad x > 0, \tag{9.141}$$

for $\lambda > 0$ and $\mu > 0$ and the distribution function by

$$F_X(x;\lambda,\mu) = \Phi\left(\sqrt{\frac{\lambda}{x}}\left(\frac{x}{\mu}-1\right)\right) + \exp\left(\frac{2\lambda}{\mu}\right)\Phi\left(-\sqrt{\frac{\lambda}{x}}\left(\frac{x}{\mu}+1\right)\right), \tag{9.142}$$

where $\Phi(\cdot)$ is the standard Normal distribution function. ■

One of the key features for risk and insurance modeling of the IG subfamily is the fact that it is a unimodal severity distribution with positive support which also has the property of closure under convolution as shown in Proposition 9.39 (see discussions by Folks and Chhikara 1978).

Proposition 9.39 (Closure Under Convolution of the Inverse Gaussian Severity Model) *Given n i.i.d. loss random variables $X_i \sim InverseGaussian\,(x;\mu,\lambda)$, the sum of these loss random variables is also Inverse Gaussian distributed as follows:*

$$S_n = \sum_{i=1}^{n} X_i \sim InverseGaussian\,(\mu, n\lambda)\,. \tag{9.143}$$

One other useful property of the IG subfamily of severity models is the closure under scaling property given by Propositions 9.40 and 9.41 (see Folks and Chhikara 1978).

Proposition 9.40 (Closure Under Scaling of IG Severity Models) *Given a loss random variable $X \sim InverseGaussian\,(x;\mu,\lambda)$, the scaled loss random variable is also IG-distributed as follows:*

$$kX \sim InverseGaussian\,(k\mu, k\lambda) \tag{9.144}$$

for some constant $k > 0$.

Proposition 9.41 (Moments and Negative Moments of IG Severity Models) *Given a loss random variable $X \sim InverseGaussian(x; \mu, \lambda)$, the integer moments and negative moments are related as follows:*

$$\mathbb{E}\left[X^{-r}\right] = \frac{1}{\mu^{2r+1}}\mathbb{E}\left[X^{r+1}\right] \tag{9.145}$$

for any $r \in \mathbb{J}^+$ and where the moment-generating function is given by

$$\mathbb{E}\left[\exp(tX)\right] = \exp\left(\frac{\lambda}{\mu}\right)\left[1 - \sqrt{1 - \frac{2\mu^2 t}{\lambda}}\right] \tag{9.146}$$

this gives the following mean and variance:

$$\mathbb{E}\left[X\right] = \mu\mathbb{E}\left[X\right] = \frac{\mu^3}{\lambda}. \tag{9.147}$$

9.7.2 HALPHEN TYPE B AND IB FAMILIES OF FLEXIBLE SEVERITY MODELS

In this section, we propose the use of the Halphen Type B and Type IB families of distribution as a flexible choice of severity models for OpRisk loss processes. As detailed previously, the Type B Halphen family of severity models has a very special feature — that it can display a mode and antimode relationship (members of the Type B family may also be classified into Type I, Type II, or Type III with respect to the mode). The density of the Halphen Type B family is characterized as shown in Definition 9.24.

Definition 9.24 (Halphen Type B Severity Models) *A loss random variable X has a Halphen Type B distribution $X \sim HalphenB(x; m, \nu, \alpha)$ if it has a density given by*

$$f(x; m, \alpha, \nu) = \frac{1}{m^{2\nu} ef_{\nu}(\alpha)} x^{2\nu-1} \exp\left(-\left(\frac{x}{m}\right)^2 + \alpha\frac{x}{m}\right) \tag{9.148}$$

with m representing the shape, and the first and second scale parameters given by α and ν and $x > 0$, $m > 0$, $\alpha \in \mathbb{R}$, $\nu > 0$, and $ef_{\nu}(\alpha)$ is the exponential factorial function given by either the integral, series, or special function representations:

$$\begin{aligned} ef_{\nu}(\alpha) &= 2\int_0^{\infty} x^{2\nu-1}\exp\left(-x^2 + \alpha x\right)dx, \quad \nu > 0 \\ &= \Gamma(\nu) + \frac{\alpha}{1!}\Gamma\left(\nu + \frac{1}{2}\right) + \frac{\alpha^2}{2!}\Gamma\left(\nu + 1\right) + \cdots + \frac{\alpha^r}{r!}\Gamma\left(\nu + \frac{r}{2}\right) + \cdots \\ &= \Gamma(\nu)M\left(\nu, \frac{1}{2}, \frac{\alpha^2}{4}\right) + \alpha\Gamma\left(\nu + \frac{1}{2}\right)M\left(\nu + \frac{1}{2}, \frac{3}{2}, \frac{\alpha^2}{4}\right), \end{aligned} \tag{9.149}$$

where the confluent hypergeometric function $M(a, b, z)$ is given by Abramowitz and Stegun (1965) as follows:

$$M(a, b, z) = \frac{\Gamma(b)}{\Gamma(b-a)\Gamma(a)}\int_0^1 e^{zt} t^{a-1}(1-t)^{b-a-1}\,dt. \tag{9.150}$$

■

The density of the Halphen Type IB family is characterized by the fact that if $X \sim HalphenB(x; m, \alpha, \nu)$ then $1/X \sim HalphenIB\left(x; \frac{1}{m}, \alpha, \nu\right)$. In Example 9.9, we demonstrate the flexible features of these families of distributions.

In the following example, the shape of the Halphen Type B severity distribution is plotted for a range of shape and scale parameters to demonstrate how each parameter changes the characteristics of the severity model.

EXAMPLE 9.9 Examples of Flexible Halphen Severity Distributions (Type B)

In the plots shown in Figure 9.12, we explore the density shapes for the case of the subfamily of Halphen Type B distributions given by a range of scale parameters m and shape parameters ν and α.

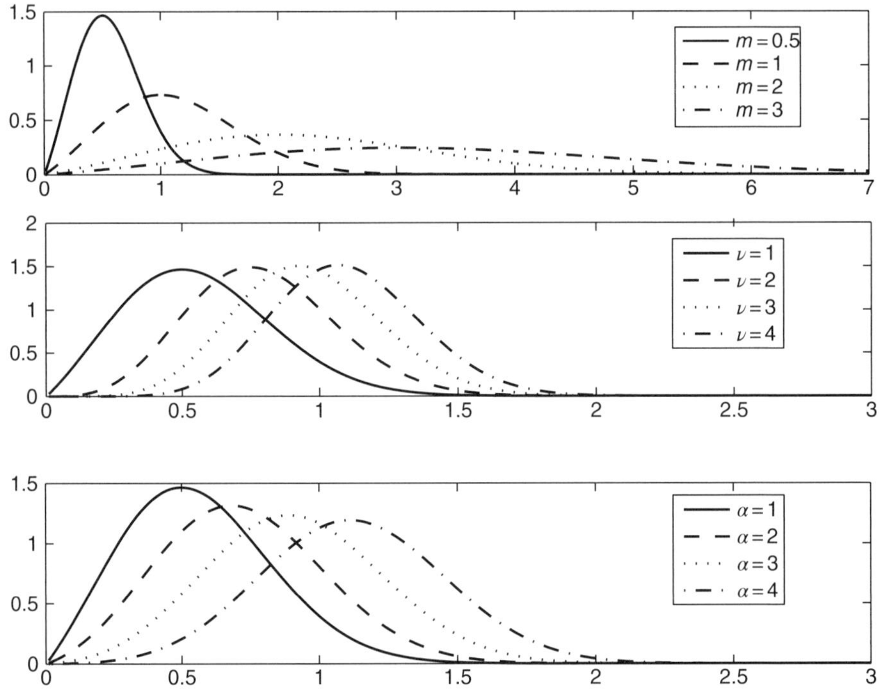

FIGURE 9.12 Halphen Type B distributions. Top subplot: parameters used in this subplot study the effect of the scale parameter ranges $m = [0.5, 1, 2, 3]$ with shape parameters set to $\nu = 1$ and $\alpha = 1$. Middle subplot: parameters used in this subplot study the effect of the shape parameter ranges $\nu = [1, 2, 3, 4]$ with shape parameter set to $\alpha = 1$ and scale parameter $m = 0.5$. Bottom subplot: parameters used in this subplot study the effect of the shape parameter ranges $\alpha = [1, 2, 3, 4]$ with shape parameter set to $\nu = 1$ and scale parameter $m = 0.5$

We now focus on features of the Halphen Type B family. The Type B family has the following mode and antimode features:

- No mode if $\alpha \leq 0$ and $\nu \leq 0.5$;
- Two modes (i.e., mode and antimode) if $\alpha < 0, \nu < 0.5$, and $\left(-\frac{\alpha}{4}\right)^2 < 0.5 - \nu$ and they are located at

$$\begin{aligned} \text{Mode}(X) &= m\left[\frac{\alpha}{4} + \sqrt{\left(\frac{\alpha}{4}\right)^2 + \nu - 0.5}\right], \\ \text{Antimode}(X) &= m\left[\frac{\alpha}{4} - \sqrt{\left(\frac{\alpha}{4}\right)^2 + \nu - 0.5}\right]. \end{aligned} \tag{9.151}$$

- One mode if either $\nu > 0.5$ or if $\nu = 0.5$ and $\alpha > 0$, then the mode is located at

$$\text{Mode}(X) = m\left[\frac{\alpha}{4} + \sqrt{\left(\frac{\alpha}{4}\right)^2 + \nu - 0.5}\right]. \tag{9.152}$$

In addition, as was the case for the Type A Halphen family, the Type B and Type IB Halphen families also have the feature that the distributions may be written in exponential family form with sufficient statistics for each of the parameters given in Proposition 9.42 (see Perreault *et al.* 1999a).

Proposition 9.42 ***(Exponential Family Representation of Halphen Type B Severity Models)*** *A loss random variable X with a Halphen Type B distribution $X \sim HalphenB(x; m, \nu, \alpha)$ belongs to a three-parameter exponential family presented in the following form:*

$$f(x; m, \nu, \alpha) = \exp\left((2\nu - 1)\ln x - \frac{1}{m^2}x^2 - \frac{\alpha}{m}x + \ln\left[\frac{2}{m^{2\nu}}ef_\nu(\alpha)\right]\right), \tag{9.153}$$

which results in the following sufficient statistics from a sample of n i.i.d. losses $\{X_i\}_{i=1}^n$ from the Halphen Type A distribution

$$\begin{aligned} T_1(X_{1:n}) &= \sum_{i=1}^n \ln X_i = n \ln G, \\ T_2(X_{1:n}) &= \sum_{i=1}^n X_i^2 = nQ, \\ T_3(X_{1:n}) &= \sum_{i=1}^n X_i = nA, \end{aligned} \tag{9.154}$$

for the arithmetic mean A, the geometric mean G, and the quadratic mean Q.

It is clear that one could then perform estimation of the parameters if it were possible to obtain the population means (arithmetic, geometric, and quadratic) in terms of the parameters. Fortunately, this is possible as illustrated by Perreault *et al.* (1999a) who show that one may obtain the following results in Proposition 9.43.

Proposition 9.43 ***(Arithmetic, Geometric, and Harmonic Population Moments Halphen Type B)*** *A loss random variable X with a Halphen Type B distribution $X \sim HalphenB(x; m, \nu, \alpha)$ has the following closed-form expressions for the Arithmetic (A), the Geometric (G), and Quadratic (Q) means:*

$$\begin{aligned} A &= \mathbb{E}[X] = m\frac{ef_{\nu+\frac{1}{2}}(\alpha)}{ef_{\nu}(\alpha)}, \\ Q &= \mathbb{E}\left[X^2\right] = m^2\frac{ef_{\nu+1}(\alpha)}{ef_{\nu}(\alpha)}. \end{aligned} \tag{9.155}$$

Note that the Geometric mean is given by the moment of order quasi-zero (see Kendall et al. 1994), hence it is given, for i.i.d. random variables $\{X_i\}_{i=1}^n$, by the limit

$$G := \lim_{r\to 0}\left[\frac{1}{n}\sum_{i=1}^{n} X_i^r\right]^{\frac{1}{r}}. \tag{9.156}$$

The log of G is given by

$$\ln G = \mathbb{E}[\ln X] = \ln m + 2ef_{\nu}^{-1}(\alpha)\frac{\partial ef_{\nu}(\alpha)}{\partial \nu}. \tag{9.157}$$

The estimation of these models has also been performed by other methods such as the Generalized MOM approach specified in Proposition 9.44 (see Perreault *et al.* 1999b).

Proposition 9.44 ***(Generalized Method of Moments Parameter Estimation Halphen Type B)*** *Given a loss random variable X with a Halphen Type B distribution $X \sim HalphenB(x; m, \nu, \alpha)$ and a sample of loss data $\{X_i\}_{i=1}^n$, the parameters can be estimated via the Generalized MOM by first estimating ν, substituting the estimate $\hat{\nu}$ to find m, and finally estimating α through substitution of the estimates $\hat{\nu}$ and $\hat{m}$ according to the following equations:*

$$\begin{aligned} \hat{\nu} &= \frac{1}{2}\frac{\mathbb{E}[X]\mathbb{E}\left[X^{-1}\right]\left(\mathbb{E}\left[X^3\right]\mathbb{E}[X] - \mathbb{E}\left[X^2\right]^2\right) - \mathbb{V}\mathrm{ar}[X]\mathbb{E}[X]^2}{\left(1 - \mathbb{E}[X]\mathbb{E}\left[X^{-1}\right]\right)\left(\mathbb{E}\left[X^2\right]^2 - \mathbb{E}\left[X^3\right]\mathbb{E}[X]\right) - \mathbb{V}\mathrm{ar}[X]^2}, \\ \hat{m} &= \sqrt{\frac{2\mathbb{V}\mathrm{ar}[X]}{2\nu\left(1 - \mathbb{E}[X]\mathbb{E}\left[X^{-1}\right]\right)} + \mathbb{E}[X]\mathbb{E}\left[X^{-1}\right]}, \\ \hat{\alpha} &= \frac{m\left(2\nu\left(\mathbb{E}[X] - \mathbb{E}\left[X^2\right]\mathbb{E}\left[X^{-1}\right]\right) + \mathbb{E}\left[X^2\right]\mathbb{E}\left[X^{-1}\right]\right)}{\mathbb{V}\mathrm{ar}[X]}. \end{aligned} \tag{9.158}$$

CHAPTER TEN

Dependence Concepts

10.1 Introduction to Concepts in Dependence for OpRisk and Insurance

This chapter is the first component of three chapters (Chapters 10–12) covering dependence modelling in OpRisk frameworks. These three chapters jointly cover a detailed account of the fundamental concepts that OpRisk practitioners should consider when developing dependence models for LDA OpRisk loss processes. In particular we first present an overview of dependence modelling approaches for LDA models which includes discussion on:

- Which components of the LDA model can dependence be added, such as between severities, between frequencies or between annual losses. This can be done explicitly via a parametric model such as a copula specification or via common shock frameworks, both of which are discussed in detail in the following chapters;
- A case study that provides an understanding of the basic impacts that dependence has in multiple risk LDA models. For instance, adding dependence between frequencies can induce dependence between annual losses etc. We provide some theoretical bounds on impacts of dependence for simple Poisson-LogNormal LDA models.

Having performed this case study, next we develop a mathematical description of the various notions of dependence that have been developed in the statistics literature, these include:

- Parametric model based Copula dependence;
- Multivariate Upper Negative (positive) Dependence, Lower Negative (positive) Dependence and Negative (positive) Dependence;
- Multivariate Negative and Positive Quadrant Dependence
  Key concepts for determining if a parameter of a multivariate distribution (copula) is directly a dependence parameter;
- Multivariate Association, Comonotonicity and Stochastic Ordering
  Associated to key concepts such as increasing positive dependence used in analysis of mixing of Markov chains, time series etc;

Fundamental Aspects of Operational Risk and Insurance Analytics: A Handbook of Operational Risk,
First Edition. Marcelo G. Cruz, Gareth W. Peters, and Pavel V. Shevchenko.

- Regression Dependence: Positive and Negative;
- Extreme Dependence, Tail Dependence and Intermediate Tail Dependence
  Crucial to the study of joint extreme dependence.

Having developed a clear understanding of the different notions of dependence, we then introduce the concept of concordance, and present detailed discussion on different measures of dependence that aim to capture the dependence concepts mentioned above. This includes, linear and non-linear measures of dependence as well as measures suitable explicitly for heavy tailed loss processes.

Then in Chapter 11 we consider the statistical modelling of dependence in OpRisk, it is dedicated to considerations involving a careful detailed discussion on many families of parametric copula that are of direct relevance to OpRisk practitioners - explaining the specification and features of the models, the estimation of the parameters in such models via Inference Functions for the Margins (IFM), and the sampling from such models in an LDA framework. The copula models include:

- Gaussian copula;
- student-T copula; skew student-T copula; grouped student-T copula and generalised student-T copula;
- Archimedean copulas: Frank, Clayton, Gumbel, Joe; Mixture Archimedean copula; Heirarchical Archimedean copulas; Nested Archimedean copulas; Outer and Inner power transformed Archimedean copula;
- Levy copula; Max-stable models and Self-Chaining copula;
- Common factor models and factor copulas.

Finally, in Chapter 12 different LDA models for OpRisk settings are developed completely with a range of different copula models. In addition, it is demonstrated how to perform combining of different sources of information under an LDA framework which includes dependence structures of different forms.

10.2 Dependence Modeling Within and Between LDA Model Structures

The aim of this chapter is to address the issue of dependence modelling within and between many OpRisks and to consider the effect of such dependence modelling on the aggregated total loss distribution. The LDA model discussed throughout this book so far has focused on the case of a single risk. This chapter considers modeling of dependence between the risks. Recall, that in this text, the LDA for a bank's total loss in year t is calculated as

$$Z_t = \sum_{j=1}^{J} Z_t^{(j)}, \tag{10.1}$$

where $Z_t^{(j)}$ is the annual loss in the j-th risk cell (business line/event type) modeled as a compound random variable,

$$Z_t^{(j)} = \sum_{s=1}^{N_t^{(j)}} X_s^{(j)}(t). \tag{10.2}$$

Here,

- $t = 1, 2, \ldots, T, T+1$ is discrete time (in annual units) with $T+1$ corresponding to the next year. For simplicity of notation in this chapter, this subscript is often dropped;
- The superscript j is used to identify the risk cell. Formally for OpRisk, $J = 56$ (eight business lines times seven event types), but this may differ depending on the financial institution and type of problem;
- The annual number of events $N_t^{(j)}$ is a random variable distributed according to a frequency distribution $P_j(\cdot|\lambda_t^{(j)})$, typically Poisson, which also depends on parameter(s) $\lambda_t^{(j)}$ that can be time dependent;
- The severities, in year t, are represented by random variables $X_s^{(j)}(t)$, $s \geq 1$, distributed according to a severity distribution $F_j(\cdot|\psi_t^{(j)})$ with parameter(s) $\psi_t^{(j)}$;
- The index j on the distributions $P_j(\cdot)$ and $F_j(\cdot)$ reflects the fact that distribution type can be different for different risks. For simplicity of notation, often we shall omit this j if the parameter index is presented, that is, using $P(\cdot|\lambda_t^{(j)})$ and $F(\cdot|\psi_t^{(j)})$;
- The variables $\lambda_t^{(j)}$ and $\psi_t^{(j)}$ generically represent distribution (model) parameters of the j-th risk that we refer to hereafter as the risk profiles;
- Typically, it is assumed that given $\lambda_t^{(j)}$ and $\psi_t^{(j)}$, the frequency and severities of the j-th risk are independent, and the severities within the j-th risk are also independent.

Modeling dependence between different risk cells and factors is an important challenge in OpRisk management. If initially one thinks of such dependence structures in terms of correlation, then in this case the difficulties associated with such dependence modeling approaches are well known and, hence, regulators typically take a conservative approach when considering correlation in risk models. For example, the Basel II OpRisk regulatory requirement for the Advanced Measurement Approach, BCBS (BCBS, 2006, p. 152), states as follows[1]:

> Risk measures for different OpRisk estimates must be added for purposes of calculating the regulatory minimum capital requirement. However, the bank may be permitted to use internally determined correlations in OpRisk losses across individual OpRisk estimates, provided it can demonstrate to the satisfaction of the national supervisor that its systems for determining correlations are sound, implemented with integrity, and take into account the uncertainty surrounding any such correlation estimates (particularly in periods of stress). The bank must validate its correlation assumptions using appropriate quantitative and qualitative techniques.

[1]The original text is available free of charge on the BIS website www.BIS.org/bcbs/publ.htm.

The current risk measure specified by regulatory authorities is value-at-risk (VaR) at the 0.999 level for a 1-year holding period. In this case, simple summation over VaRs corresponds to an assumption of perfect dependence between risks. This can be very conservative as it ignores any diversification effects. If the latter are allowed in the model, it is expected that the capital may reduce, providing a strong incentive to model dependence in the banking industry. At the same time, limited data do not allow for reliable estimates of correlations and there are attempts to estimate these using expert opinions. In such a setting, a transparent dependence model is very important from the perspective of model interpretation, understanding of model sensitivity, and with the aim of minimizing possible model risk. In this chapter, we will discuss how to formulate such dependence models and how to study and understand their features in the context of OpRisk.

Remark 10.1 *One should note at this stage that VaR is not a coherent risk measure; see definition 6.7 in section 6.2.1. This means that, in principle, dependence modeling could also increase VaR; see Embrechts et al. (2009a,b). This issue will be discussed in section 12.8.*

The pitfalls with the use of linear correlation as a measure of dependence and its limitations are now widely known, and consequently the use of more general dependence modeling concepts based around parametric models known as copula functions has become more prominent. Copula models are especially being increasingly used to model dependence structures in financial risk management. This was not the case until the publication of the highly influential paper by Embrechts *et al.* (2002), which was first available as a RiskLab (ETH Zurich) report in early 1999. These will be discussed throughout this chapter. A textbook reference for modeling dependence between financial risks is McNeil *et al.* (2005), which also contains an extensive bibliography on this subject.

10.2.1 WHERE CAN ONE INTRODUCE DEPENDENCE BETWEEN LDA MODEL STRUCTURES?

Before we proceed to discuss copula modeling structures, we will first observe that there is first an important modeling question to be addressed that is not just what types of dependence features should I consider in my model, but between which components of the risk models should I consider modeling the dependence structure and what effect might this have on my overall institutional risk model.

Conceptually, under model (10.2), the dependence between the annual losses $Z_t^{(j)}$ and $Z_t^{(i)}, i \neq j$, can be introduced in several ways:

- Modeling dependence between frequencies $N_t^{(j)}$ and $N_t^{(i)}$ directly through copula methods; see Frachot *et al.* (2004a), Bee (2005a), and Aue and Klakbrener (2006). Here, we note that the use of copula methods, in the case of discrete random variables, needs to be done with care;
- Common shocks; see Lindskog and McNeil (2003) and Powojowski *et al.* (2002). The approach of common shocks is proposed as a method to model events affecting many cells at the same time. Formally, this leads to a dependence between frequencies of the risks if superimposed with cell internal events. Dependence between severities occurring at the same time is considered in Lindskog and McNeil (2003);

- Modeling dependence between the k-th severities or between k-th event times of different risks; see Chavez-Demoulin *et al.* (2006) (e.g., first, second, etc. losses/event times of the j-th risk are correlated to the first, second, etc., losses/event times of the i-th risk, respectively). This can be difficult to interpret especially when one considers high-frequency versus low-frequency risks;
- Modeling dependence between annual losses directly via copula methods; see Giacometti *et al.* (2008) and Embrechts and Puccetti (2008). However, this may create irreconcilable problems with modeling insurance for OpRisk that directly involves event times. Additionally, it will be problematic to quantify these correlations using historical data, and the LDA model (10.2) will lose its structure. One can, however, consider dependence between losses aggregated over shorter periods such as monthly or quarterly; we will discuss such contexts in this chapter with regard to self-chaining copula structures;
- Using the multivariate compound Poisson model based on Lévy copulas as suggested in Böcker and Klüppelberg (2008, 2009);
- Using structural models with common (systematic) factors that can lead to the dependence between severities and frequencies of different risks and within risk; see section 12.5;
- Modeling dependence between severities and frequencies from different risks and within risk using dependence between risk profiles, as considered in Peters *et al.* (2009);
- In the general case, when no information about the dependence structure is available, Embrechts and Puccetti (2006) work out bounds for aggregated operational risk capital; see also Embrechts *et al.* (2009a).

In the remainder of the chapter, we detail and describe the main concepts, approaches, and issues behind some of these approaches. The choice of appropriate dependence structures is crucial and determines the amount of diversification—it is still an open challenging problem.

Remark 10.2 (Dependence on Macroeconomic Factors) *It is important to note that there is empirical evidence, as reported in Allen and Bali (2004), that some OpRisks are dependent on macroeconomic variables such as GDP, unemployment, equity indices, interest rates, foreign exchange rates, regulatory environment variables, and others. For example, some OpRisks typically increase during economic downturns, high unemployment, and low interest rates. This will be discussed further in section 12.5.*

10.2.2 UNDERSTANDING BASIC IMPACTS OF DEPENDENCE MODELING BETWEEN LDA COMPONENTS IN MULTIPLE RISKS

When modeling dependence between different aspects of OpRisk LDA structures and multiple risks as discussed earlier, a natural question that arises involves understanding the amount of "induced" correlation or dependence present in the annual losses. The answer to this question in OpRisk models has been given in numerous papers such as Frachot *et al.* (2004b) and Peters *et al.* (2009). In some cases, it is clear what the resulting dependence may be, such as when dependence is introduced across the annual losses; however, in most cases described earlier, this induced dependence and behavior or influence on capital calculation on an institutional level is not transparent and must be studied carefully.

To motivate an understanding from a practical perspective of the impact that dependence can have on the annual loss between two LDA risk processes, we consider a class of models in which one can obtain closed form results and bounds. The example considered involves two risk processes each with an LDA structure with a frequency, which is Poisson, and the severity distribution, which is LogNormal, as given by the following model definition.

Definition 10.1 (Multiple Risk Model: Poisson–LogNormal LDAs) *Consider two risk processes with the following LDA structures.*

1. *Consider two loss processes for annual losses given by* $\left\{Z_t^{(1)}\right\}_{t=1}^T$ *and* $\left\{Z_t^{(2)}\right\}_{t=1}^T$ *for T years, which are given by the compound process in year t according to*

$$Z_t^{(\cdot)} = \sum_{i=1}^{N_t^{(\cdot)}} X_i^{(\cdot)}(t). \tag{10.3}$$

   a) *Assume for each given loss process* $\left\{Z_t^{(\cdot)}\right\}_{t=1:T}$ *there is independence between the number of events and the severities;*
   b) *Assume for each given loss process* $\left\{Z_t^{(\cdot)}\right\}_{t=1:T}$ *the severities are independent and identically distributed loss random variables.*

2. *Assume that the severity distribution is such that* $X_i^{(j)}(t) \sim F_X^{(j)}(x)$ *with LogNormal severity model given by*

$$F_X^{(j)}(x; \mu^{(j)}, \sigma^{(j)}) = \frac{1}{2} + \frac{1}{2} erf\left[\frac{\ln x - \mu^{(j)}}{\sqrt{2}\sigma^{(j)}}\right] \tag{10.4}$$

   with log scale $\sigma^{(j)} > 0$ *and shape* $\mu^{(j)} \in \mathbb{R}$*;*

3. *Assume that the frequency distribution is marginally given by* $N_t^{(j)} \sim F_N^{(j)}(n)$ *with Poisson frequency model given by*

$$F_N^{(j)}(n; \lambda^{(j)}) = \frac{\left(\lambda^{(j)}\right)^n}{n!} \exp\left(-\lambda^{(j)}\right). \tag{10.5}$$

■

If one were to consider a relaxation of the model assumptions given earlier with regard to the independence of the severity loss random variables, then the first result we note is studied in Asmussen and Rojas-Nandayapa (2008) where they consider the single risk setting and assume that there is a dependence structure between the individual loss amounts such that if $(Y_1, Y_2, \ldots, Y_n)$ have a joint n-dimensional Gaussian distribution with a general covariance structure (that does not include perfect negative or positive dependence between any pair of losses), where the individual loss amounts are then given by marginal LogNormal distribution $X_i \sim Exp(Y_i)$ such that $X_i \sim LogNormal(\mu_i, \sigma_i^2)$, then the partial sum of n such losses given by

$$S_n = X_1 + \cdots + X_n, \tag{10.6}$$

can be shown to have an asymptotic behavior as $n \to \infty$ in which the tail of the aggregate loss distribution $\mathbb{P}r[S_n > x]$ can be shown to have the same asymptotic tail behavior as the independent case, where asymptotically in $x \to \infty$ one can show that the partial sum will have the property that it is asymptotically equivalent (ie. equivalent in the tails of the distribution as the loss $x \to \infty$) to the distribution of the tail of the maximum loss, given by the order statistic $X_{(n,n)}$ (i.e. n-th largest loss out of n losses) according to the relationship,

$$\mathbb{P}\mathrm{r}\left[S_n > x\right] \sim m_n \overline{F}_{X_{(n,n)}}\left(x; \mu_m, \sigma_m\right), \tag{10.7}$$

where $\mu_m = \max\{\mu_1, \ldots, \mu_n\}$ and $\sigma_m = \max\{\sigma_1, \ldots, \sigma_n\}$ and $m_n = \sharp\{k : \mu_k = \mu_m, \sigma_k = \sigma_m\}$ and $\sharp$ is used to denote the number of elements in a set or its cardinality.

Remark 10.3 *Of course, the result could also be applied to multiple loss processes in the case that each individual loss processes annual loss was modeled by a LogNormal model.*

If we now go back to consider the severity loss random variables as being independent and instead consider dependence on the number of losses over time, then the second result we present was studied in Frachot *et al.* (2004b) and involves incorporation of dependence between the frequency distributions of the two risk cells. This can be achieved in two ways, either on the actual annual counts of losses or on the intensity in the loss process frequency distributions as studied in Peters *et al.* (2009). In this case, we consider the setting in which dependence is introduced between the counts of losses between the two risk processes, which can be shown to induce a closed form result for the dependence between the annual loss that is directly a function of the linear correlation considered between the frequency counts as shown in Proposition 10.1.

Proposition 10.1 (Multiple Risk-Induced Correlation in Annual Loss) *Consider the two risk processes specified by LDA models with Poisson frequency and LogNormal severity as detailed in model Definition 10.1. If the correlation between the annual counts of each risk process is considered, as measured by corr* $\left(N_t^{(1)}, N_t^{(2)}\right)$, *then the induced correlation between the annual loss random variables for the Poisson–LogNormal LDA models is given by*

$$corr\left(Z_t^{(1)}, Z_t^{(2)}\right) = corr\left(N_t^{(1)}, N_t^{(2)}\right) \exp\left(-\frac{1}{2}\left(\sigma^{(1)}\right)^2 - \frac{1}{2}\left(\sigma^{(2)}\right)^2\right). \tag{10.8}$$

Furthermore, one may construct such a correlation between the frequencies in numerous ways such as considering three independent Poisson random variables Y, Y_1, and Y_2 with intensity parameters λ, $\lambda^{(1)} - \lambda$, and $\lambda^{(2)} - \lambda$, which are used to construct the annual loss frequency random variables $N^{(i)} = Y + Y_i$ for $i \in \{1, 2\}$ and the resulting correlation between the frequency counts given by

$$corr\left(N_t^{(1)}, N_t^{(2)}\right) = \frac{\lambda}{\sqrt{\lambda^{(1)}\lambda^{(2)}}} \leq R = \sqrt{\frac{\min\left(\lambda^{(1)}, \lambda^{(2)}\right)}{\max\left(\lambda^{(1)}, \lambda^{(2)}\right)}}. \tag{10.9}$$

Remark 10.4 *In fact in Brunel (2013), it is demonstrated that if the frequency distribution is assumed to be of mixed type, where the intensity of each Poisson frequency model is stochastic such that*

$$\lambda^{(i)} = \mu + \sigma G_i, \tag{10.10}$$

for uncorrelated standard Gaussian random variables G_i, then one can obtain the bound on the correlation on the frequency counts given by the expression

$$corr\left(N_t^{(1)}, N_t^{(2)}\right) \leq R = \exp\left(-\frac{\sigma}{2}|G_1 - G_2|\right) \tag{10.11}$$

and $R \sim F_R$ with F_R a truncated LogNormal distribution. This case was studied further in a dynamic Cox process setting with an autoregressive component in Peters et al. (2009).

Other than these simple model structures, to understand the impact that dependence will have, one must be more precise on the form of dependence and generally this will result in a need to study such features numerically, as is discussed toward the end of this chapter.

To generalize the discussion beyond simple considerations of dependence through correlation measures, the next few sections will introduce more general concepts of dependence modeling followed by models to represent these dependence concepts parametrically, either within a single risk process or between multiple risk processes.

10.3 General Notions of Dependence

In this section, we first discuss the concepts of dependence. Then, we proceed to introduce parametric models as well as measures of the strength of these notions of dependence. In the statistics literature, there have been many notions of dependence that have been discussed, including (but not exhaustive)

- multivariate upper negative dependence, lower negative dependence and negative dependence;
- multivariate association;
- multivariate negative and positive quadrant dependence;
- commonotonicity and stochastic ordering;
- negative regression dependence;
- parametric copula dependence.

Before detailing each of these different notions of dependence, we first present a brief introduction to the notion of a copula which is used throughout this chapter. Since, it is ubiquitous in modelling dependence, and appears numerous times throughout the following sections, it will be beneficial to readers to see a brief introduction here to copulas, before a more comprehensive discussion of the properties and theoretical derivation of the copula is presented in Section 10.4 where Sklar's theorem is formally presented.

One can consider copulas to be synonymous with a model based characterization of dependence. Indeed, Fisher (1997) observed that *"Copulas [are] of interest to statisticians for two main reasons:*

1. *as a way of studying scale-free measures of dependence;*
2. *as a starting point for constructing families of multivariate distributions, sometimes with a view to simulation".*

- Copula theory can be traced back to Hoeffding's work on standardised distributions on the square $\left[-\frac{1}{2}, \frac{1}{2}\right] \times \left[-\frac{1}{2}, \frac{1}{2}\right]$.
- Following this work, the term copula was first coined as a mathematical concept in Abel Sklar's theorem, covered in detail in Section 10.4, which showed that one-dimensional distributions can be joined by a copula function to form multivariate distributions.

In general one can consider that a d-dimensional copula is a multivariate distribution C with uniform $[0, 1]$ margins such that $C : [0, 1]^d \to [0, 1]$ and C satisfies:

- $C(u_1, \ldots, u_d) = 0$ whenever $u_i = 0$ for at least one $i \in \{1, \ldots, d\}$;
- $C(u_1, \ldots, u_d) = u_i$ if $u_j = 1$ for all $j = 1, \ldots, d$ and $j \neq i$;
- C is quasi-monotone on its support $[0, 1]^d$ i.e. for every hyperrectangle $B = \prod_{i=1}^d [x_i, y_i] \subseteq [0, 1]^d$ the C-volume of B is non-negative.

To understand this last condition on volumes, note that it requires that for every $\boldsymbol{a}$ and $\boldsymbol{b}$ in $[0, 1]^d$, such that for each $a_i < b_i$ for all $i \in \{1, 2, \ldots, n\}$ the condition on the volume for copula C is satisfied: $V_C([\boldsymbol{a}, \boldsymbol{b}]) \geq 0$. NOTE: The volume of an d-box is given by

$$\begin{aligned} V_C([\boldsymbol{a}, \boldsymbol{b}]) &= \sum \operatorname{sgn}(\boldsymbol{v}) C(\boldsymbol{v}) \\ &= \triangle_{a_1}^{b_1} \triangle_{a_2}^{b_2} \cdots \triangle_{a_d}^{b_d} C(\boldsymbol{v}) \end{aligned} \tag{10.12}$$

where the sum is taken over all vertices $\boldsymbol{v}$ of the d-box $[\boldsymbol{a}, \boldsymbol{b}]$ and $\operatorname{sgn}(\boldsymbol{v}) = 1$ if $v_k = a_k$ for an even number of k's of $\operatorname{sgn}(\boldsymbol{v}) = -1$ if $v_k = a_k$ for an odd number of k's. In addition one defines the notation

$$\triangle_{a_k}^{b_k} C(\boldsymbol{u}) = C(u_1, u_2, \ldots, u_{k-1}, b_k, u_{k+1}, \ldots, u_d) - C(u_1, u_2, \ldots, u_{k-1}, a_k, u_{k+1}, \ldots, u_d). \tag{10.13}$$

Hence, one can now consider the definition of a copula informally as follows. Consider random vector $\boldsymbol{X} \in \mathbb{R}^d$ with continuous distribution F. Then to every $\boldsymbol{X}$ one can associate a d-copula $C : [0, 1]^d \mapsto [0, 1]$, defined by

$$F(x_1, \ldots, x_d) = C(F_1(x_1), \ldots, F_d(x_d)), \tag{10.14}$$

where F_i is the marginal distribution of X_i. It will also be useful to consider the notion of a survival copula, which is informally defined as follows

$$\begin{aligned} \mathbb{P}\mathrm{r}[X_1 > x_1, X_2 > x_2] &= \overline{F}(x_1, x_2) \\ &= 1 - F_{X_1}(x_1) - F_{X_2}(x_2) + F(X_1, X_2) \\ &= \overline{F}_{X_1}(x_1) + \overline{F}_{X_2}(x_2) - 1 + C(F_{X_1}(x_1), F_{X_2}(x_2)) \\ &= \overline{F}_{X_1}(x_1) + \overline{F}_{X_2}(x_2) - 1 + C\left(1 - \overline{F}_{X_1}(x_1), 1 - \overline{F}_{X_2}(x_2)\right) \end{aligned} \tag{10.15}$$

Hence, one can define for instance in $d = 2$ the mapping $\widetilde{C} : [0, 1]^2 \mapsto [0, 1]$ by

$$\widetilde{C}(1 - u, 1 - u) = 1 - 2u - C(u, u) \tag{10.16}$$

to be the survival copula of C i.e. $\overline{F}(x_1, x_2) = \widetilde{C}(\overline{F}_{X_1}(x_1), \overline{F}_{X_2}(x_2))$.

There are many different parametric families of copula models that are discussed in future sections of this chapter, below we present some important base copula models that act as bounds for all other families of copula model and are therefore instructive to present at this stage.

Definition 10.2 (Frechet-Hoffding Copula Bounds) *The Frechet-Hoffding Upper Bound copula is given by*

$$M^d(u_1, \ldots, u_d) = \min\{u_1, \ldots, u_d\}.$$

The Frechet-Hoffding Lower Bound copula is given by

$$W^d(u_1, \ldots, u_d) = \max\left\{1 - d + \sum_{i=1}^{d} u_i, 0\right\}.$$

One has the following bounds on all copulas

$$W^d(u_1, \ldots, u_d) \leq C(F_1(x_1), \ldots, F_d(x_d)) \leq M^d(u_1, \ldots, u_d).$$

■

We can also note the following properties of such copula model bounds.

- Probability Mass M^d is distributed uniformly along the line segment $u_1 = \ldots = u_d$ running from $(0, \ldots, 0)$ to $(1, \ldots, 1)$ in $[0, 1]^d$;
- For all d-copula distributions $C \leq M^d$ and M^d can be thought of as a state of 'maximal concordance'.

Note: the notion of concordance will be formally defined shortly in this subsection. In addition it is worthwhile to observe the following remark regarding the Frechet-Hoffding Lower Bound copula.

Remark 10.5 *Consider the Frechet-Hoffding Lower Bound copula W^d in d-dimensions. Then for $d \geq 3$ the function W^d is not strictly a copula, this can be seen by calculating $W^d([1/2, 1] \times [1/2, 1] \times \cdots \times [1/2, 1])$ which may not produce $V_C([\boldsymbol{a}, \boldsymbol{b}]) \geq 0$. Recall the definition of a Volume of a d-box:*

$$V_C([\boldsymbol{a}, \boldsymbol{b}]) = \sum sgn(\boldsymbol{v})C(\boldsymbol{v}) = \triangle_{a_1}^{b_1} \triangle_{a_2}^{b_2} \cdots \triangle_{a_d}^{b_d} C(\boldsymbol{v}),$$

where the sum is taken over all vertices $\boldsymbol{v}$ of the n-box $[\boldsymbol{a}, \boldsymbol{b}]$ and $sgn(\boldsymbol{v}) = 1$ if $v_k = a_k$ for an even number of k's of $sgn(\boldsymbol{v}) = -1$ if $v_k = a_k$ for an odd number of k's and we used

$$\triangle_{a_k}^{b_k} C(\boldsymbol{u}) = C(u_1, u_2, \ldots, u_{k-1}, b_k, u_{k+1}, \ldots, u_d) - C(u_1, u_2, \ldots, u_{k-1}, a_k, u_{k+1}, \ldots, u_d).$$

Applying this to the copula W^d for the d-box $[1/2, 1]^d$ produces

$$\begin{aligned} W^d\left([1/2,1]^d\right) &= \max\{1+1+\ldots+1-d+1,0\} \\ &- d\max\{1/2+1+\ldots+1-d+1,0\} \\ &+ C_2^n \max\{1/2+1/2+1+\ldots+1-d+1,0\} \\ &\ldots \\ &+ \max\{1/2+\ldots+1/2-d+1,0\} \\ &= 1-d/2+0+\ldots+0. \end{aligned}$$

Hence, for $d \geq 3$ the function W^d is not strictly a copula.

One may then ask the question, so how is W^d the best possible lower bound on copulas? The answer to this question is that it should be understood in the following sense, W^d is best possible lower bound, where Nelsen (1999) showed that for any $d \geq 3$ and any $\boldsymbol{u} \in [0, 1]^d$, there is a d-copula C, which depends on $\boldsymbol{u}$, such that

$$C(\boldsymbol{u}) = W^d(\boldsymbol{u}). \tag{10.17}$$

One last special copula also valuable is the Independence Copula.

Definition 10.3 (Independence Copula) *The independence copula is given by*

$$\Pi^d(u_1, \ldots, u_d) = u_1 u_2 \cdots u_d.$$

■

Having informally defined the basic idea of a copula distribution, we also present a few key properties, that often come in useful in practice, when working with copulas in OpRisk settings. These properties discussed pertain to the invariance properties of such dependence models to certain types of transformation of the underlying random vector.

Proposition 10.2 (Copula Invariance to Strictly Increasing Transformations) *If $X_1, \ldots, X_d$ are continuous random variables with copula $C_{X_1,\ldots,X_d}$. Then if $T_1(X_1), \ldots, T_d(X_d)$ are strictly increasing on $Ran(X_1), \ldots, Ran(X_d)$, then $C_{T_1(X_1),\ldots,T_d(X_d)} = C_{X_1,\ldots,X_d}$. Copula $C_{X_1,\ldots,X_d}$ is invariant under strictly increasing transforms.*

Proof: The proof of this invariance is sketched as follows.

- Consider marginal distributions $F_1, \ldots, F_d$ for continuous r.v.'s $X_1, \ldots, X_d$ and joint copula $C_{X_1,\ldots,X_d}$;
- Let $G_1, \ldots, G_d$ be the distributions of $T_1(X_1), \ldots, T_d(X_d)$ respectively with joint copula $C_{T_1(X_1),\ldots,T_d(X_d)}$;
- $T_i(\cdot)$ is strictly increasing for each i, hence

$$G_i(x) = \Pr(T_i(X_i) \leq x) = \Pr\left(X_i \leq T_i^{-1}(x)\right) = F_i\left(T_i^{-1}(x)\right) \tag{10.18}$$

for any $x \in \mathrm{Ran}(X_i)$.

Hence, one may now show that

$$\begin{aligned}
&C_{T_1(X_1),\ldots,T_d(X_d)}\left(G_1(x_1),\ldots,G_d(x_d)\right)\\
&=\mathbb{P}\mathrm{r}\left(T_1(X_1)\le x_1,\ldots,T_d(X_d)\le x_d\right)\\
&=\mathbb{P}\mathrm{r}\left(X_1\le T_1^{-1}(x_1),\ldots,X_d\le T_d^{-1}(x_d)\right)\\
&=C_{X_1,\ldots,X_d}\left(F_1(T_1^{-1}(x_1)),\ldots,F_d(T_d^{-1}(x_d))\right)\\
&=C_{X_1,\ldots,X_d}\left(G_1(x_1),\ldots,G_d(x_d)\right)
\end{aligned} \tag{10.19}$$

Since $X_1,\ldots,X_d$ are continous, $\mathrm{Ran}G_1 = \ldots = \mathrm{Ran}G_d = [0,1]$. Hence it follows that $C_{T_1(X_1),\ldots,T_d(X_d)} = C_{X_1,\ldots,X_d}$ on $[0,1]^d$. ■

One can also state some useful analogous properties related to copula invariance to strictly monotone transformations, as detailed briefly in Proposition 10.3.

Proposition 10.3 (Copula Invariance to Strictly Monotone Transformations) *If X_1 and X_2 are continuous random variables with copula C_{X_1,X_2}. Then if $T_1(X_1)$ and $T_2(X_2)$ are strictly monotone on $Ran(X_1)$ and $Ran(X_2)$, then:*

- *If $T_1(\cdot)$ is strictly increasing and $T_2(\cdot)$ strictly decreasing, then*

$$C_{T_1(X_1),T_2(X_2)}(u,v) = u - C_{X_1,X_2}(u,1-v).$$

- *If $T_1(\cdot)$ is strictly decreasing and $T_2(\cdot)$ strictly increasing, then*

$$C_{T_1(X_1),T_2(X_2)}(u,v) = v - C_{X_1,X_2}(1-u,v).$$

- *If $T_1(\cdot)$ and $T_2(\cdot)$ are strictly decreasing, then*

$$C_{T_1(X_1),T_2(X_2)}(u,v) = u + v - 1 + C_{X_1,X_2}(1-u,1-v).$$

Having detailed a very brief preliminary discussion on some key aspects of the concept of a copula, we now proceed to discuss different notions of dependence. The dependence concepts discussed include:

- Stochastic Ordering and Properties Implied by a Stochastic Order;
- Multivariate Negative and Positive Dependence: Upper Negative and Lower Negative Dependence;
- Multivariate Negative and Positive Association;
- Quadrant Dependence: Pairwise Negative and Posative Quandrant Dependence;
- Lower and Upper Orthant Dependence;
- Tail Increasing and Tail Decreasing, Tail Increase/Decrease in Sequence;
- Stochastic Increase and Stochastic Decrease;
- Regression Dependence: Bivariate and Multivariate;
- Comonotonicity;
- Multivariate Total Positivity of Order 2, see Nelsen (1999).

We start with the concept of negative dependence first proposed in Block *et al.* (1982) and then studied further in Ghosh (1981), with the specification taken as given in Definition 10.4

Definition 10.4 (Multivariate Negative Dependence in LDA Single Risk Models) *Consider a sequence of loss random variables in an OpRisk loss model $\{X_i\}_{i\geq 1}$. The sequence can be called lower or upper negatively dependent as follows:*

- ***Lower Negative Dependence (LND).*** *A sequence of loss random variables have LND if for each $n \geq 1$ and all $X_1, X_2, \ldots, X_n$ one has*

$$\mathbb{P}\mathrm{r}\left[X_1 \leq x_1, X_2 \leq x_2, \ldots, X_n \leq x_n\right] \leq \prod_{i=1}^{n} \mathbb{P}\mathrm{r}\left[X_i < x_i\right]. \tag{10.20}$$

- ***Upper Negative Dependence (UND).*** *A sequence of loss random variables have UND if for each $n \geq 1$ and all $X_1, X_2, \ldots, X_n$ one has*

$$\mathbb{P}\mathrm{r}\left[X_1 > x_1, X_2 > x_2, \ldots, X_n > x_n\right] \leq \prod_{i=1}^{n} \mathbb{P}\mathrm{r}\left[X_i > x_i\right]. \tag{10.21}$$

- ***Negative Dependence (ND).*** *A sequence of loss random variables have negative dependence if for each $n \geq 1$ and all $X_1, X_2, \ldots, X_n$ it is both LND and UND.*

■

The concept of UND is clearly directly relevant to OpRisk modeling as it involves explicitly the concept of a lower bound on the joint probability of a large loss occurring in all the n risk processes given by the product of the probability that such an event happens in each loss process marginally. The difference between such probabilities is then the influence of the dependence structure introduced—this will be explored in more detail when we build models for such events.

In addition, if one considers a simple example in which each of the n loss processes has the same risk model marginally, that is, $X_i \sim F$ for all $i \in \{1, 2, \ldots, n\}$ and furthermore the thresholds $x_1 = x_2 = \cdots = x_n = x$. Then, in the independence case clearly the lower bound decays with $x \to \infty$ as $\overline{F}_X(x)^n$, which can rapidly go to zero. However, in the UND case, this probability of joint exceedance $\overline{F}_{X_1,\ldots,X_n}(x, \ldots, x)$ will still decay to zero as $x \to \infty$ but clearly the rate that this occurs will depend on the type of upper negative dependence that is present, that is, on the model used to capture the notion of UND. In the following sections, different parametric copulas will be considered that will demonstrate UND structure.

Remark 10.6 (LND and UND Versus Negative Association) *The notion of lower and upper negative dependence is a weaker notion of dependence than the more familiar idea of negative association.*

To make precise this remark, we recall in Definition 10.5 the concept of negative association of random variables; see, for instance, Joag-Dev and Proschan (1983).

Definition 10.5 (Multivariate Negative Association in LDA Single Risk Models) *Consider a sequence of loss random variables in an OpRisk loss model* $\{X_1, \ldots, X_n\}$. *The sequence is called negatively associated (NA) if for every pair of disjoint subsets* A_1, A_2 *of* $\{1, \ldots, n\}$ *then one has*

$$\mathbb{C}\text{ov}\left[f_1\left(X_i; i \in A_1\right), f_2\left(X_j; j \in A_2\right)\right] \leq 0 \tag{10.22}$$

whenever f_1 *and* f_2 *are increasing functions.* ■

That is, one can consider negative association as the situation in which the values of one variable tend to increase as the other variable increases.

Remark 10.7 *Several practically important multivariate distributions possess the property of being negatively associated such as multinomial, multivariate hypergeometric, Dirichlet and Dirichlet compound multinomial distributions.*

The following properties of negatively associated random sequences of loss random variables given in Proposition 10.4 are also useful to recall; see Joag-Dev and Proschan (1983).

Proposition 10.4 (Properties of Negatively Associated Loss Random Variables) *Consider a sequence of loss random variables in an OpRisk loss model* $\{X_i\}_{i\geq 1}$ *that satisfy that they are negatively associated, then the following properties apply*

- *A subset of two or more negatively associated random variable losses is negatively associated;*
- *A set of independent random variable losses is negatively associated;*
- *Increasing functions defined on disjoint subsets of a set of negatively associated random variable losses are negatively associated;*
- *Unions of independent sets of negatively associated random variable losses are negatively associated.*

Another notion of dependence that is of significance in the following sets of results relates to the notion of pairwise quadrant dependence given by Definition 10.6; see Lehmann (1966) and the associated pairwise positive quadrant dependence (PPQD). This is simply a less restrictive notion of negative or positive dependence as discussed previously.

Definition 10.6 (Pairwise Negative Quadrant Dependence) *A pair of loss random variables* X_i *and* X_j *are said to be pairwise negative quadrant dependent (PNQD) if for all* $x, y \in \mathbb{R}$ *one has*

$$\mathbb{P}\text{r}\left[X_i \leq x, X_j \leq y\right] \leq \mathbb{P}\text{r}\left[X_i \leq x\right] \mathbb{P}\text{r}\left[X_j \leq y\right]. \tag{10.23}$$

■

Definition 10.7 (Pairwise Positive Quadrant Dependence) *A pair of loss random variables* X_i *and* X_j *are said to pairwise positive quadrant dependence (PPQD) if for all* $x, y \in \mathbb{R}$ *one has*

$$\mathbb{P}\text{r}\left[X_i \leq x, X_j \leq y\right] \geq \mathbb{P}\text{r}\left[X_i \leq x\right] \mathbb{P}\text{r}\left[X_j \leq y\right]. \tag{10.24}$$

■

One can also observe that if X_i and X_j are PQD then one has the following property for the copula between these two loss random variables,

$$C\left(F_{X_i}(x), F_{X_j}(y)\right) \geq F_{X_i}(x)F_{X_j}(y) \tag{10.25}$$

for all $\left(F_{X_i}(x), F_{X_j}(y)\right)$ in the unit square.

Remark 10.8 *Intuitively, X and Y are PQD if the probability that they are simultaneously small (or simultaneously large) is at least as great as it would be were they independent.*

One can also observe the following features of loss random variables which display a PQD relationship:

- Like independence, quadrant dependence (positive or negative) is a property of the copula of continuous random variables, and consequently is invariant under strictly increasing transformations of the random variables;
- If X and Y are PQD, then the graph of the copula of X and Y given by C lies on or above the graph of the independence copula Π ie. $C(u, v) \geq uv$ for all $(u, v) \in [0, 1]^2$;
- Many examples of copula model families exist that satisfy quadrant dependence. Examples include: many totally ordered one-parameter families of copulas have subfamilies of PQD copulas and NQD copulas.
    - Example: the Mardia family, the Farlie-Gumbel-Morgenstein FGM family, the Ali-Mikhail-Haq AMH family, or the Frank Archimedean family satisfy that they are PQD for copula parameter $\rho \geq 0$ and NQD for $\rho \leq 0$ with $\rho = 0$ giving $C = \Pi$.

We also observe that the notion of PQD(X, Y) can be rewritten conditionally. To see this consider the following representations:

$$\begin{aligned} &\mathbb{P}\mathrm{r}\left[X \leq x, Y \leq y\right] \geq \mathbb{P}\mathrm{r}\left[X \leq x\right]\mathbb{P}\mathrm{r}\left[Y \leq y\right], \text{ or as} \\ &\mathbb{P}\mathrm{r}\left[X \leq x | Y \leq y\right] \geq \mathbb{P}\mathrm{r}\left[X \leq x\right], \text{ or as} \\ &\mathbb{P}\mathrm{r}\left[X \leq x | Y \leq y\right] \geq \mathbb{P}\mathrm{r}\left[X \leq x | Y \leq \infty\right] \end{aligned} \tag{10.26}$$

One can now also observe that a stronger condition than Quadrant dependence is to require that for each $x \in \mathbb{R}$, the conditional distribution function $\mathbb{P}\mathrm{r}\left[X \leq x | Y \leq y\right]$ is a non-increasing function of y.

Remark 10.9 *This stronger condition leads to the notion of Tail Decreasing and Tail Increasing, Esary et al. (1972).*

In addition to these notions of pairwise quadrant dependence, one may also consider the concepts such as positive lower orthant dependence, left tail decreasing in sequence, and multivariate left tail decreasing as discussed in (Hua and Joe, 2011, definition 4) and given below in Definitions 10.8, 10.10 and 10.11. Note that, analogous definitions for positive upper orthant dependence, right tail increasing in sequence, and multivariate right tail increasing can be defined.

Definition 10.8 (Positive Lower Orthant Dependence) *A random vector is said to have the dependence structure in its distribution known as positive lower orthant dependence (PLOD) if its distribution satisfies*

$$\mathbb{P}\mathrm{r}\left[X_1 \leq x_1, \ldots, X_d \leq x_d\right] \geq \prod_{i=1}^{d} \mathbb{P}\mathrm{r}\left[X_i \leq x_i\right]. \tag{10.27}$$

■

Clearly, the situation of PLOD dependence is opposite in the inequality sign compared to LND, defined earlier.

In addition, one can also say the following about orthant dependent loss random variables. Consider two d-copulas C_1 and C_2 then the following relationship between orthant dependencies and concordance holds:

- C_1 is more **Positive Lower Orthant Dependent** than C_2 if for all $\boldsymbol{u} \in [0,1]^d$ one has $C_1(\boldsymbol{u}) \geq C_2(\boldsymbol{u})$;
- C_1 is more **Positive Upper Orthant Dependent** than C_2 if for all $\boldsymbol{u} \in [0,1]^d$ one has $\overline{C}_1(\boldsymbol{u}) \geq \overline{C}_2(\boldsymbol{u})$;
- C_1 is more **Positive Orthant Dependent** than C_2, or C_1 is more concordant than C_2 if for all $\boldsymbol{u} \in [0,1]^d$, both $C_1(\boldsymbol{u}) \geq C_2(\boldsymbol{u})$ and $\overline{C}_1(\boldsymbol{u}) \geq \overline{C}_2(\boldsymbol{u})$ holds.

Having defined the notion of orthant dependence, it is now natural to start to consider the concept of Tail Increase and Tail Decrease given in the following definition.

Definition 10.9 (Tail Increasing and Tail Decreasing Loss Random Variables) *In the case of two random variables X and Y one can define the following:*

- *Y is **left tail decreasing** in X ie. LTD($Y|X$) if $\mathbb{P}\mathrm{r}[Y \leq y | X \leq x]$ is a non-increasing function of x for all y;*
- *X is **left tail decreasing** in Y ie. LTD($X|Y$) if $\mathbb{P}\mathrm{r}[X \leq x | Y \leq y]$ is a non-increasing function of y for all x;*
- *Y is **right tail increasing** in X ie. RTI($Y|X$) if $\mathbb{P}\mathrm{r}[Y > y | X > x]$ is a non-decreasing function of x for all y;*
- *X is **right tail increasing** in Y ie. RTI($X|Y$) if $\mathbb{P}\mathrm{r}[X > x | Y > y]$ is a non-decreasing function of y for all x.*

■

To see the relationship between the notions of dependence already presented and the concept of tail monotonicity, we note that these four tail monotonicity conditions each implies positive quadrant dependence. Analogously, negative dependence properties, known as left tail increasing and right tail decreasing, are defined by exchanging the words "nonincreasing" and "nondecreasing", see discussions in detail in Kimeldorf *et al.* (1989).

From these notions of Tail Increase and Tail Decrease one can also define the analogous concept for sequences as follows.

Definition 10.10 (Left Tail Decreasing in Sequence) *A random vector is said to have the dependence structure in its distribution known as left tail decreasing in sequence (LTDS) if its distribution satisfies*

$$\mathbb{P}\mathrm{r}\left[X_i \leq x_i | X_1 \leq x_1, \ldots, X_{i-1} \leq x_{i-1}\right] < \mathbb{P}\mathrm{r}\left[X_{i-1} \leq x_{i-1} | X_1 \leq x_1, \ldots, X_{i-2} \leq x_{i-2}\right] \tag{10.28}$$

for all $i \in \{1, 2, \ldots, d\}$. ■

Definition 10.11 (Multivariate Left Tail Decreasing) *A random vector is said to have the dependence structure in its distribution known as multivariate left tail decreasing if its distribution satisfies that the random vector* $(X_{i_1}, \ldots, X_{i_d})$ *is LTDS for all possible permutations* $(i_1, \ldots, i_d)$ *of* $(1, \ldots, d)$. ■

In addition, the notion of Tail Increase and Tail Decrease can be captured directly in terms of properties of the parametrization of the precedence structure through a copula model as follows.

Proposition 10.5 (Copula Conditions for Tail Increase or Decrease) *Consider loss r.v.'s X and Y with copula C then:*

- *LTD($Y|X$) holds iff for any* $v \in [0, 1]$ *one has that* $C(u, v)/u$ *is nonincreasing in u, or equivalently one has that*

$$\frac{\partial C(u, v)}{\partial u} \leq \frac{C(u, v)}{u}, \quad \textit{almost all } u; \tag{10.29}$$

- *LTD($X|Y$) holds iff for any* $u \in [0, 1]$ *one has that* $C(u, v)/v$ *is nonincreasing in v, or equivalently one has that*

$$\frac{\partial C(u, v)}{\partial v} \leq \frac{C(u, v)}{v}, \quad \textit{almost all } v; \tag{10.30}$$

- *RTI($Y|X$) holds iff for any* $v \in [0, 1]$ *one has that* $[1 - u - v + C(u, v)]/(1 - u)$ *is nonincreasing in u, or equivalently one has that*

$$\frac{\partial C(u, v)}{\partial u} \geq \frac{[v - C(u, v)]}{1 - u}, \quad \textit{almost all } u; \tag{10.31}$$

- *RTI($X|Y$) holds iff for any* $u \in [0, 1]$ *one has that* $[1 - u - v + C(u, v)]/(1 - v)$ *is nonincreasing in v, or equivalently one has that*

$$\frac{\partial C(u, v)}{\partial v} \leq \frac{[u - C(u, v)]}{1 - v}, \quad \textit{almost all } v; \tag{10.32}$$

Yet another notion one may consider for modeling dependence structures that is discussed later is known as regression dependence. The influence that regression dependence has on compound process tail asymptotics was considered in Ko and Tang (2008, assumption 2.1). In this study, they considered the bivariate setting and the general assumption on dependence given in Definition 10.15, which is known as regression dependence. Such a dependence assumption is based upon setting up a conditional probability in a ratio, such that if one of the variables were independent, then the ratio should collapse to one since it would appear both in the numerator and the denominator. This general dependence representation allows one to capture limited positive as well as negative dependence features, in particular limited positive and negative quadrant dependence. To present this concept of regression dependence for the bivariate and multivariate settings, it is first instructive to recall the notion of comonotonicity and stochastic order. The notion of stochastic order involves the quantification of 'one random variable being "bigger" than another'.

Definition 10.12 (Stochastic Ordering Notations) *The most basic definition of stochastic ordering (partial ordering) that allows one to compare two random variables X_1 and X_2 involves a statement on the tails of their distribution (also denoted as equivalently their distribution) given by $X_1 \preceq X_2$ (also denoted as $X_1 \leq_{st} X_2$) if and only if*

$$\overline{F}_{X_1}(x) \leq \overline{F}_{X_2}(x), \quad \forall x \in \mathbb{R}. \tag{10.33}$$

Of course, it is equally valid to state that $X_1 \leq_{st} X_2$ if $F_{X_1} \leq F_{X_2}$; this statement when applicable can also be extended to the density function. To be precise, the following are all equivalent definitions:

- $X_1 \leq_{st} X_2 \Leftrightarrow F_{X_1}(x) \geq F_{X_2}(x), \quad \forall x \in \mathbb{R}$;
- $X_1 \leq_{st} X_2 \Leftrightarrow \Pr[X_1 \geq x] \leq \Pr[X_2 \geq x], \quad \forall x \in \mathbb{R}$;
- $X_1 \leq_{st} X_2 \Leftrightarrow \mathbb{E}_{X_1}[f(x)] \geq \mathbb{E}_{X_2}[f(x)]$, *for all increasing (nondecreasing) functions f.*

This concept of stochastic ordering is considered partial ordering since it can be shown to be reflexive, transitive, and antisymmetric. The following properties of stochastic ordering apply:

- *If $X_1 \leq_{st} X_2$ and a function $g(\cdot)$ is nondecreasing, then $g(X_1) \leq_{st} g(X_2)$;*
- *Consider random vectors $(X_1, \ldots, X_n)$ and $(Y_1, \ldots, Y_n)$ such that for all $i \in \{1, 2, \ldots, n\}$ one has $X_i \leq_{st} Y_i$ then for any function $g : \mathbb{R}^n \mapsto \mathbb{R}$, which is nondecreasing one has $g(X_1, \ldots, X_n) \leq_{st} g(Y_1, \ldots, Y_n)$. Note: this is useful in OpRisk since it applies for the choice of interest in insurance and risk given by loss aggregation $g(X_1, \ldots, X_n) = \sum_{i=1}^{n} X_i$;*
- *Reflexive: $F_{X_i} \leq_{st} F_{X_i}$;*
- *Transitive: $F_{X_i} \leq_{st} F_{X_j}$ and $F_{X_j} \leq_{st} F_{X_k}$ then $F_{X_i} \leq_{st} F_{X_k}$;*
- *Antisymmetric: $F_{X_i} \leq_{st} F_{X_j}$ and $F_{X_j} \leq_{st} F_{X_i}$ would imply that $F_{X_i} = F_{X_j}$ which is another statement of stochastic equivalence that is, that $X_i \sim F_{X_i}$ and $X_j \sim F_{X_j}$ then $X_i =_{st} X_j$ when $F_{X_i} \sim F_{X_j}$.*

■

We observe the following remark about a stochastic order.

Remark 10.10 *We note that a stochastic order can be considered an antisymmetric preorder since it is a binary relation that is reflexive and transitive. It is not a complete ordering since there exist random variables (distributions) which cannot be ordered through this ordering.*

Definition 10.13 (Comonotonicity) *The notion of comonotonicity involves the perfect positive dependence between the components of a random vector. This means that they can be represented as increasing functions of a single random variable. In general, one can define a random vector* $(X_1, \ldots, X_n)$ *as comonotonic if its multivariate distribution satisfies*

$$\mathbb{P}\mathrm{r}\left[X_1 \leq x_1, \ldots, X_n \leq x_n\right] = \min_{i \in \{1, \ldots, n\}} \mathbb{P}\mathrm{r}\left[X_i \leq x_i\right]. \tag{10.34}$$

■

Remark 10.11 (Comonotonicity and Stochastically Decreasing) *The concept of stochastic ordering with regard to stochastically decreasing variables involves a dependence relation imposed that excludes any extremely positive dependence structures such as those arising from comonotonic (nondecreasing support) loss random variables.*

Having briefly discussed stochastic ordering and monotonicity, one can now proceed to also define the notion of regression dependence or Stochastic Increase/Decrease. It is based upon setting up a conditional probability in a ratio, such that if one of the variables were independent, then the ratio should collapse to unity. In this way, regression dependence captures limited positive and negative dependence features, in particular quadrant dependence and is defined as follows, see details in Shaked (1977).

Definition 10.14 (Stochastic Increase and Decrease Dependence) *Consider loss random variables X and Y, then:*

- *Postive Dependence: Y is **Stochastically Increasing** in X, $SI(Y|X)$ if $\mathbb{P}\mathrm{r}[Y > y|X = x]$ is non-decreasing function of x for all y;*
- *Postive Dependence: X is **Stochastically Increasing** in Y, $SI(X|Y)$ if $\mathbb{P}\mathrm{r}[X > x|Y = y]$ is non-decreasing function of y for all x;*
- *Negative Dependence: Y is **Stochastically Decreasing** in X, $SD(Y|X)$ if $\mathbb{P}\mathrm{r}[Y > y|X = x]$ is non-increasing function of x for all y;*
- *Negative Dependence: X is **Stochastically Decreasing** in Y, $SD(X|Y)$ if $\mathbb{P}\mathrm{r}[X > x|Y = y]$ is non-decreasing function of y for all x.*

■

One may now proceed to define the notion of regression dependence, in particular negative regression dependence between two loss random variables, by the expression presented in Definition 10.15. In general, this can be seen as another way to express Stochastic Increase or Decrease.

Definition 10.15 (Negative Regression Dependence: Bivariate) *Consider two loss random variables X_1 and X_2 in which X_2 is stochastically decreasing in X_1 denoted as $SD(X_2|X_1)$. Then, define the dependence relationship for variables with index $(i, j) \in \{(1, 2), (2, 1)\}$ by*

$$\lim_{x \to \infty} \frac{\mathbb{P}\mathrm{r}\left[X_j > x - y | X_i = y\right]}{\mathbb{P}\mathrm{r}\left[X_j > x - y\right]} = O(1), \tag{10.35}$$

which will hold uniformly for all $y \in [x_0, x]$ *for some large* $x_0 > 0$ *such that*

$$\limsup_{x \to \infty} \int_{x_0}^{x - x_0} \frac{\mathbb{P}\mathrm{r}\left[X_j > x - y | X_i = y\right]}{\mathbb{P}\mathrm{r}\left[X_j > x - y\right]} dy < \infty. \tag{10.36}$$

In this setting, the following relationship will hold, for all $x_0 \geq 0$, *such that* $F_{X_1}(x_0) > 0$, *uniformly for all* $y > x_0$ *according to*

$$\mathbb{P}\mathrm{r}\left[X_2 > x - y | X_1 = y\right] \leq \frac{\overline{F}_{X_2}(x - y)}{F_{X_1}(x_0)}. \tag{10.37}$$

■

Definition 10.16 (Negative Regression Dependence: Multivariate) *Consider* n *loss random variables with marginal distributions* $X_i \sim F_{X_i}$ *and a joint dependence that is the analog of the bivariate negative regression dependence for* $n \geq 2$ *that is captured by the following relationship on the conditional distributions of the partial sums:*

$$\frac{\mathbb{P}\mathrm{r}\left[\sum_{s=1}^{j-1} X_s > x - y | X_j = y\right]}{\mathbb{P}\mathrm{r}\left[\sum_{s=1}^{j-1} X_s > x - y\right]} = O(1), \tag{10.38}$$

which will hold uniformly for all $y \in [x_0, x]$ *that should exist for some large* $x_0 > 0$ *such that this order of asymptotic convergence is satisfied for all* $j \in \{2, \ldots, n\}$. ■

Analogously to negative regression dependence, one may also define positive regression dependence as follows, and detailed in Kimeldorf *et al.* (1989).

Definition 10.17 (Positive Regression Dependence) *The loss random variable* X_1 *is positive regression dependent on loss random variable* X_2 *if it holds that* X_1 *is Stochastically Increasing in loss random variable* X_2, *such that* $\mathbb{P}\mathrm{r}(X_1 > x | X_2 = y)$ *is a non-decreasing function of* x *for all* y. ■

One can also link the notion of Statistic Increase/Decrease to copula properties as detailed in Proposition 10.6 below.

Proposition 10.6 (Copula Specification of Stochastic Increase and Decrease Dependence) *Consider continuous loss random variables* X *and* Y *with copula* C, *then:*

- Y *is Stochastically Increasing in* X, *SI*$(Y|X)$ *iff for any* $v \in [0, 1]$ *one has that* $\frac{\partial C(u,v)}{\partial u}$ *is non-increasing in* u, *i.e.* $C(u, v)$ *is a concave function of* u;
- X *is Stochastically Increasing in* Y, *SI*$(X|Y)$ *iff for any* $u \in [0, 1]$ *one has that* $\frac{\partial C(u,v)}{\partial v}$ *is non-increasing in* v, *i.e.* $C(u, v)$ *is a concave function of* v;

We can also observe the following properties:

- If $SI(Y|X)$, then one as $LTD(Y|X)$ and $RTI(Y|X)$;
- If $SI(X|Y)$, then one as $LTD(X|Y)$ and $RTI(X|Y)$;

Note that Geluk and Tang (2009) considered a related approach to specification of the dependence in the severity loss random variables under two assumptions, (A1) and (A2), given later. They consider these dependence relationships to be their analogs of the proposed Negative Regression Dependence given in Definition 10.16. They show that such a relationship can be satisfied by a family of copula models such as the Farlie–Gumbel–Morgenstern (FGM) distribution that is characterized in the following section after recalling the notion of a copula model. The two dependence specifications they developed involved for n-losses the conditions:

(A1) $\lim_{x_i \wedge x_j \to \infty} \mathbb{P}r\left[|X_i| > x_i | X_j > x_j\right] = 0,\ \forall i,j \in \{1,2,\ldots,n\},\ i \neq j;$

(A2) $\exists x_0, c > 0\ s.t.\ \mathbb{P}r\left[|X_i| > x_i | X_j = x_j\right] \leq c\overline{F}_{X_i}(x_i), \forall i,j \in \{1,\ldots,n\},\ i \neq j,\ x_j > x_0.$

The next notion of dependence between loss random variables to be discussed is one which implies the majority of the other dependence relationships discussed above and involves the concept of Bivariate Total Positivity of Order 2, given in the following definition.

Definition 10.18 (Bivariate Total Positivity Order 2) *Tow loss random variables (X_1, X_2) have total positive dependence of order 2 if their joint distirbution $F(x,y)$ satisfies that :*

$$det \begin{bmatrix} F(x,y) & F(x,y') \\ F(x',y) & F(x'y') \end{bmatrix} \geq 0$$

whenever $x \leq x'$ and $y \leq y'$. ∎

One can then generalize this notion of Total Positivity of Order 2 to the multivariate d-dimensional setting as follows.

Definition 10.19 (Multivariate Total Positivity Order 2) *Random vector $(X_1, \ldots, X_d)$ with density f has total positivity dependence of order 2 (MTP2) if:*

$$f(\boldsymbol{x} \vee \boldsymbol{y}) f(\boldsymbol{x} \wedge \boldsymbol{y}) \geq f(\boldsymbol{x}) f(\boldsymbol{y}) \tag{10.39}$$

for all $\boldsymbol{x}, \boldsymbol{y} \in \mathbb{R}^d$, see Nelson (1992). ∎

One can show the following properties of MTP2 loss random vectors:

- If a random vectors density is MTP2 then so are all of its marginal densities of order 2 and higher;
- If the above inequality expression has its inequality sign reversed, then the density f is said to be multivariate reverse rule of order 2 (MRR2) which is a weak negative dependence concept. Unlike MTP2, the property of MRR2 is not closed under marginalization!

One can also develop the following properties of MTP2 random vectors based on power transforms of the loss random variables distributions. First, we recall some basic properties of loss distributions under power transformation.

Proposition 10.7 (Powers of Univaraite Distributions) *Consider a univariate distribution F and tail $\overline{F}$, if $\gamma > 0$ then F^{γ} and $\overline{F}^{\gamma}$ are distributions (tail functions).*

In the general multivariate setting, one also has the following results.

Proposition 10.8 (Powers of Multivaraite Distributions, Max-ID and Min-ID) *Consider a random vector $\boldsymbol{X} \in \mathbb{R}^d$ with multivariate distribution F and tail $\overline{F}$.*

- *If $\gamma > d - 1$ then F^{γ} ($\overline{F}^{\gamma}$) are distributions (tail functions);*
- *If F^{γ} is a distribution for $\gamma > 0$ then F is max-infinitely divisible (max-id);*
- *If $\overline{F}^{\gamma}$ is a tail function for $\gamma > 0$ then F is min-infinitely divisible (min-id).*

Now, we may make the following links between MTP2 and power transforms of the distribution of the loss random vector.

- If random vector $\boldsymbol{X} \in \mathbb{R}^d$ has a distribution F which is max-id then for all $m \in \mathbb{N}$ one has $F^{1/m}$ is a distribution;
  - If $\left(X_{i1}^m, \ldots, X_{id}^m\right)$, $i = 1, \ldots, d$ are i.i.d. with distribution $F^{1/m}$, then

$$\boldsymbol{X} \stackrel{d}{=} \left(\max_i X_{i1}^m, \ldots, \max_i X_{id}^m\right) \tag{10.40}$$

  where max is over all indices $1, \ldots, d$.
- In bivariate case: F is max-id iff F is TP2;
- In bivariate case: F is min-id iff $\overline{F}$ is TP2.

The sixth concept of dependence between a random vector of risk processes that one may wish to consider involves the idea of upper and lower orthant tail dependence, as discussed, for example, in Li (2009). This notion of dependence is used to describe relative deviations of upper or lower orthant tail probabilities of say a loss random vector, from orthant probabilities comprising a subset of the random vectors components. As such it represents a notion of dependence in extreme values that the joint loss process (sub-)vectors may take. This notion of dependence can be thought of in several ways: by defining what will be known as the tail dependence coefficients that are limiting extreme upper and lower tail dependencies or by defining intermediate behaviors known as tail functions or intermediate tail dependencies. The tail dependence coefficients by their limiting definition become asymptotically independent of the marginal loss process model behavior and only dependent on the joint extreme dependence features of the loss random vector, whereas the intermediate tail dependence components retain the influence of the marginal loss process as well as the intermediate tail dependence. Formal definitions of these notions will be provided in the following sections.

The seventh approach involves the specification of a particular parametric form of dependence via, for instance, a copula model specification that was briefly introduced above and will be discussed in detail in the following sections. We simply note here the comments of

Fisher (1997), where it was observed that "*Copulas [are] of interest to statisticians for two main reasons:*

1. *as a way of studying scale-free measures of dependence;*
2. *as a starting point for constructing families of bivariate distributions, sometimes with a view to simulation*".

In addition to parametric models for dependence between random variables, there are notions of dependence between Lévy measures for compound processes tail measures that are known generically as Lévy copulae, also discussed in detail later.

10.4 Dependence Measures

Having discussed briefly different concepts of dependence, before embarking on discussions regarding parametric copula models that aim to capture these notions of dependence parametrically, we first discuss the different approaches that have been adopted to measure and quantify these notions of dependence. Measuring the dependence between random variables has long been of interest to statisticians and practitioners alike. A history of the development of dependency measures can be found in Mari and Kotz (2001). It is important to note that, in general, the dependence structure between two random variables can *only* be captured in full by their joint probability distribution, and thus any scalar quantity extracted from this structure must be viewed as a representation of some feature of the dependence. Scarsini (1984) gives the following intuitive definition of dependence:

> Dependence is a matter of association between X and Y along any measurable function, i.e. the more X and Y tend to cluster around the graph of a function, either $y = f(x)$ or $x = g(y)$, the more they are dependent.

The choice of dependence measure is influenced by the type of dependence one seeks to study, such as lower left quadrant, upper right quadrant, etc. However, in nontrivial multivariate distributions, it is not possible to capture all of the possible combinations of dependence patterns within a single dependence measure. In what follows, we discuss a few standard measures of dependence that are utilized widely in practice.

Before presenting basic measures of dependence that are widely used, it will be instructive to first provide a general overview of the concept of concordance measures. This will be presented first under a general axiomatic framework, then re-expressed under a copula based framework. Then several examples of concordance measurs satisfying different aspects of the axioms proposed will be discussed. Before proceeding, we require some basic definitions of relevance to be presented. The first of these is the notion of permutation and symmetry.

- **Symmetries:** a symmetry of $[0,1]^d$ is a one-to-one, onto map $\phi : [0,1]^d \mapsto [0,1]^d$ of form $\phi(x_1,\ldots,x_d) = (u_1,\ldots,u_d)$ where for each i one has $u_i = x_{k_i}$ or $1 - x_{k_i}$ and where $(k_1,\ldots,k_d)$ is a permutation of $(1,\ldots,n)$;
- **Permutation:** the map ϕ is a permutation if for each i one has $u_i = x_{k_i}$;
- **Reflection:** the map ϕ is a reflection if for each i one has $u_i = x_i$ or $u_i = 1 - x_i$.

- **Elementary reflections:** an elementary reflection of the i-th component, denoted σ_i is given by

$$\sigma_i(x_1, \ldots, x_d) = (x_1, \ldots, x_{i-1}, 1 - x_i, x_{i+1}, \ldots, x_d)$$

- **Symmetry Length:** the length of a symmetry is denoted by $|\phi|$ and corresponds to the number elementary reflections required to obtain it.

Having made these basic specifications of symmetry, permutation and reflection, we can now discuss the measurement or quantification of dependence between loss random variables. Measuring the dependence between random variables has long been of interest to statisticians and practitioners. Indeed, Scarsini (1984) provides the following intuitive definition of dependence which aligns with the notions of dependence previously discussed:

> "Dependence is a matter of association between X and Y along any measurable function, i.e. the more X and Y tend to cluster around the graph of a function, either $y = f(x)$ or $x = g(y)$, the more they are dependent".

The choice of dependence measure is influenced by the type of dependence one seeks to study, such as lower left quadrant, upper right quadrant etc. This leads one to consideration of the general notion of concordance between loss random variables.

Definition 10.20 (Concordance Between Loss Random Variables) *Informally, a pair of random variables are concordant if 'large' values of one tend to be associated with 'large' values of the other and 'small' values of one with 'small' values of the other. Analogous definitions of discordance are available in reverse directions.* ■

There are numerous ways of mathematically trying to quantify this statement, so consequently, many measures of concordance are available. Scarsini (1984) proposed a set of axioms for general concordance measures, which will be denoted here by κ, which are given in Proposition 10.9

Proposition 10.9 (Multivariate Concordance Measures) *A general concordance measures κ is a function attaching to all d-tuples of continuous r.v.'s $(X_1, X_2, \ldots, X_d)$ defined on a common probability space, when $d \geq 2$, a real number $\kappa\,(X_1, X_2, \ldots, X_d)$ satisfying:*

- ***Normalization:*** *$\kappa\,(X_1, X_2, \ldots, X_d) = 1$ if each X_i is a.s. an increasing function of every other X_j and $\kappa\,(X_1, X_2, \ldots, X_d) = 0$ if $X_1, \ldots, X_d$ are independent;*
- ***Monotonicity:*** *If $X_1, \ldots, X_d$ is less concordent than $Y_1, \ldots, Y_d$ then $\kappa\,(X_1, X_2, \ldots, X_d) < \kappa\,(Y_1, Y_2, \ldots, Y_d)$;*
- ***Continuity:*** *If F_k is the joint distribution of $(X_{k1}, \ldots, X_{kd})$ and F the distribution of $(X_1, \ldots, X_d)$ and one has convergence in the sequence $F_k \to F$ as $k \to \infty$, then $\kappa\,(X_{k1}, \ldots, X_{kd}) \to \kappa\,(X_1, \ldots, X_d)$;*
- ***Permutation Invariance:*** *If $(i_1, \ldots, i_d)$ is a permutation of $(1, \ldots, n)$ then $\kappa\,(X_{i_1}, \ldots, X_{i_d}) = \kappa\,(X_1, \ldots, X_n)$;*
- ***Duality:*** *$\kappa\,(-X_1, \ldots, -X_n) = \kappa\,(X_1, \ldots, X_n)$;*
- ***Reflection Symmetry:*** *$\sum_{\epsilon_1, \ldots, \epsilon_d = \pm 1} \kappa\,(\epsilon_1 X_1, \ldots, \epsilon_d X_d) = 0$ where the sum is over 2^d vectors of the form $(\epsilon_1 X_1, \ldots, \epsilon_d X_d)$ with $\epsilon_i \in \{-1, 1\}$;*

- ***Transition:*** *There exists a sequence* $\{r_d\}$ *for* $d \geq 2$ *such that every* d*-tuple of continous r.v.'s* $(X_1, \dots, X_d)$ *satisfies*

$$r_{d-1}\kappa\left(X_2, \dots, X_d\right) = \kappa\left(X_1, \dots, X_d\right) + \kappa\left(-X_1, X_2, \dots, X_d\right)$$

These general axioms for concordance measures were also recently re-specified in terms of copula models by Taylor (2007) in the following axioms for general concordance measures κ via copula C as detailed in Proposition 10.10

Proposition 10.10 (Multivariate Concordance Measures via Copula) *Consider a sequence of maps* $\kappa_d : Cop(d) \mapsto \mathbb{R}$ *and a sequence of numbers* $\{r_d\}$*, such that if* A, B, C *and* C_m *are* d*-copulas and* $n \geq 2$ *then:*

- *Normalization:*$\kappa\left(M^d\right) = 1$ *and* $\kappa\left(\Pi^d\right) = 0$;
- *Monotonicity: If* $A <_{st} B$ *and* $\overline{A} \leq_{st} \overline{B}$ *then* $\kappa_d\left(A\right) \leq \kappa_d\left(B\right)$;
- *Continuity: If* $C_m \to C$, *then* $\kappa_d\left(C_m\right) \to \kappa_d\left(C\right)$ *as* $m \to \infty$;
- *Permutation Invariance: If* $(i_1, \dots, i_d)$ *is a permutation of* $(1, \dots, d)$ *then* $\kappa\left(C(u_{i_1}, \dots, u_{i_d})\right) = \kappa\left(c(u_1, \dots, u_d)\right)$;
- *Duality:* $\kappa_d\left(c(1 - u_1, \dots, 1 - u_d\right) = \kappa_d\left(c(u_1, \dots, u_d)\right)$;
- *Reflection Symmetry:* $\sum_{\Psi \in \mathcal{R}_d} \kappa_d\left(C^{\Psi}\right) = 0$, *where* Ψ *is a reflection,* $\Psi \in \mathcal{R}_d$ *is an element of the subgroup of reflections in the group of symmetries under composition* $\mathcal{S}([0, 1]^d)$;
- *Transition:*

$$r_n\kappa_d\left(C\right) = \kappa_{n+1}\left(E\right) + \kappa_{n+1}\left(E(1 - u_1, u_2, \dots, u_d)\right)$$

whenever E *is an* $(d+1)$*-copula s.t.* $C(u_1, \dots, u_d) = E(1, u_1, \dots, u_d)$.

One can also state the following theorem regarding the properties of concordance measures that satisfy these axioms, see details in Taylor (2007).

Theorem 10.1 (Properties of Concordance Measures Satisfying Proposition 10.10) *Consider the* d*-copula that is permutation symmetric ie.* $C^{\varsigma} = C$ *for all permuations* ς *of* $[0, 1]^d$. *Then for all measures of concordance* κ *and for all symmetries* Ψ *and* ς *of* $[0, 1]^d$ *one has*

$$\kappa_d(C^{\Psi}) = \kappa_d(C^{\varsigma}) \tag{10.41}$$

whenever $|\Psi| = |\varsigma|$ *or* $|\Psi| + |\varsigma| = d$

Recall: symmetry length $|\cdot|$ corresponds to the number elementary reflections required to obtain it.

Corollary 10.1 *For all* $d \geq 2$ *and for all symmetries* Ψ *and* ς *of* $[0, 1]^d$ *such that* $|\Psi| = |\varsigma|$ *or* $|\Psi| + |\varsigma| = d$ *one has*

$$\kappa_d(M^{\Psi}) = \kappa_d(M^{\varsigma}). \tag{10.42}$$

where M *is the* d*-Frechet-Hoffding Upper Bound copula under permutation.*

We now present several examples of concordance measure that are widely used in practice and are of relevance in many settings in OpRisk modelling.

10.4.1 LINEAR CORRELATION

Arguably the most widely known and utilized concordance measure of dependence, Pearson's product moment correlation coefficient, was developed by Karl Pearson, see Pearson (1896), building on Sir Francis Galton's approach using the median and semi-interquartile range, see Galton (1889). Pearson's correlation coefficient is a measure of how well the two random variables can be described by a linear function and is defined as detailed in Definition 10.21. Pearson's correlation coefficient, otherwise sometimes known as Pearson's product moment correlation coefficient, detailed below, is an extension of the median and semi-interquartile range discussed by Galton in 1889. It acts as a measure of how well the two random variables can be described by a linear function.

Definition 10.21 (Pearson's Correlation Coefficient) *Consider two random variables X and Y with finite second moments $\mathbb{E}[X^2] < \infty$ and $\mathbb{E}[Y^2] < \infty$, then the definition of Pearson's correlation coefficient is given by the ratio of the covariance to the variation of each random variable according to*

$$\rho := \frac{\operatorname{Cov}[X, Y]}{\sqrt{\operatorname{Var}[X]\operatorname{Var}[Y]}}. \tag{10.43}$$

■

Hence, we see that the correlation coefficient is what is known as a linear correlation that is a measure of *linear dependence* between random variables. The notion of linear correlation arises from the fact that such a measure of dependence is invariant under strictly increasing linear transformations

$$\rho[\alpha_i + \beta_i X_i, \alpha_j + \beta_j x_j] = \rho[X_i, X_j], \ \beta_i, \beta_j > 0. \tag{10.44}$$

Hence, perfect linear dependence gives $\rho = +1$ or $\rho = -1$.

Remark 10.12 *A weakness of linear correlation is its noninvariance under nonlinear monotonic transformations of the random variables.*

The problems with using the linear correlation coefficient as a measure of dependence between OpRisks can be summarised as follows:

- It is defined if variances of X_i and X_j are finite. As has already been discussed, some OpRisks are modeled by heavy-tailed distributions with infinite variance and even cases of infinite mean are possible in some loss processes to do with disasters, such as losses resulting from say natural disasters;
- It is not invariant under strictly increasing nonlinear transformations $T(\cdot)$ and $\widetilde{T}(\cdot)$. In general, $\rho[T(X_i), \widetilde{T}X_j)] \neq \rho[X_i, X_j]$;
- Independence between random variables implies that linear correlation is zero. However, in general, zero linear correlation does not imply independence. For example, if

$X \sim Normal(0, 1)$ and $Y = X^2$, then $\rho[X, Y] = 0$ while it is obvious that there is as strong dependence between X and Y. Zero linear correlation and independence are equivalent only in the case of a multivariate Normal distribution as a joint distribution for random variables;

- The linear correlation is bounded to the region $[\rho_{\min}, \rho_{\max}]$, where $-1 \leq \rho_{\min} \leq \rho_{\max} \leq 1$. For example, if $X \sim LogNormal(0, 1)$ and $Y \sim LogNormal(0, \sigma^2)$, then the minimum and maximum bounds for correlation are plotted in Figure 10.1a as functions of σ; for more details, see McNeil *et al.* (2005, example 5.26). Figure 10.1b presents the correlation bounds for the case of $X \sim Pareto(2.1, 1)$ and $Y \sim Pareto(\beta, 1)$, where $Pareto(\beta, a) = 1 - (x/a)^{-\beta}$; for more details, see Nešlehová *et al.* (2006, example 3.1).

As noted earlier, such a pairwise measure of dependence is no longer valid when the second moment of the loss random variables is not finite, as will be the case in heavy-tailed loss

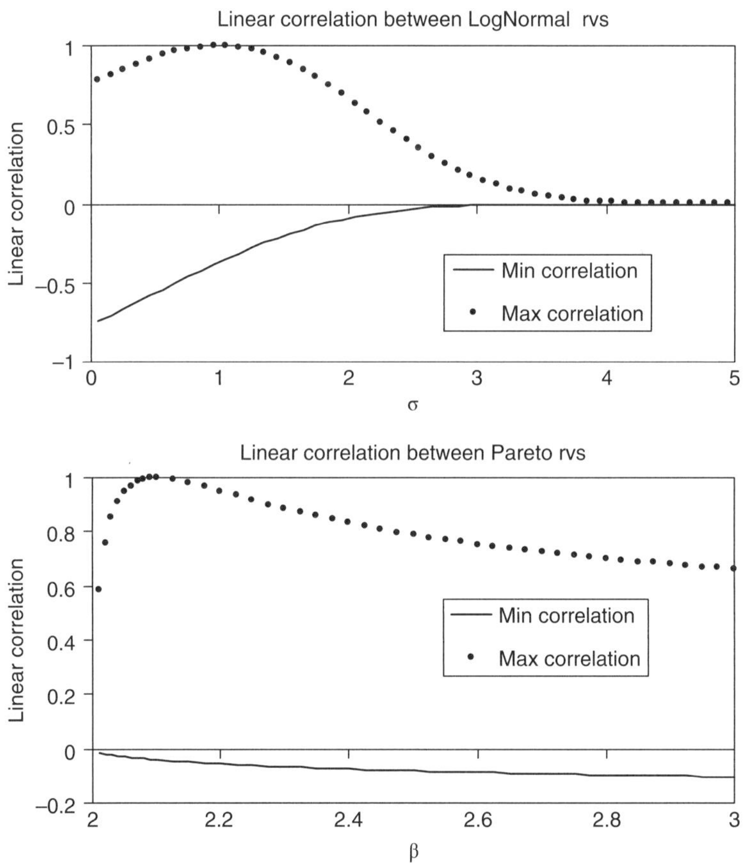

FIGURE 10.1 Upper figure: the minimum and maximum possible linear correlation between the random variables $X \sim LogNormal(0, 1)$ and $Y \sim LogNormal(0, \sigma^2)$. Bottom figure: the minimum and maximum possible linear correlation between the random variables $X \sim Pareto(2.1, 1)$ and $Y \sim Pareto(\beta, 1)$

process models. Instead, it is common to move to notions of co-difference and co-variation (see Definition 10.22, see Kokoszka and Taqqu (1994) and Nowicka-Zagrajek and Wyłomańska (2008)). In particular, we will illustrate the definition of these measures of the important class of heavy-tailed models known as alpha-stable models; see extensive discussions in the advanced companion to this book.

Definition 10.22 (Co-Difference and Co-Variation) *Consider two loss random variables X_1 and X_2 that are jointly from a heavy-tailed model that is symmetric α-Stable $(S\alpha S)$ with a tail index $\alpha \in (1,2)$ such that the second moments $\mathbb{E}[X_1]$ and $\mathbb{E}[X_1]$ are not finite. Then, the co-variation and co-difference are defined by*

1. ***Co-Difference***. *The co-difference between two loss random variables that are jointly $S\alpha S$ distributed is given by*

$$CD(X_1, X_2) = \ln \mathbb{E}[\exp(iX_1 - iX_2)] - \ln \mathbb{E}[\exp(iX_1)] - \ln \mathbb{E}[\exp(-iX_2)]. \tag{10.45}$$

2. ***Co-Variation***. *The co-variation between two loss random variables that are jointly $S\alpha S$ distributed is given by*

$$CV(X_1, X_2) = \int_{\mathbb{S}^2} s_1 s_2^{\alpha-1} \Gamma(d\mathbf{s}), \tag{10.46}$$

where $\mathbb{S}^2$ is the unit 2-sphere defined by

$$\mathbb{S}^2 = \left\{x \in \mathbb{R}^3 : ||x|| = r\right\}, \tag{10.47}$$

which is a two-dimensional manifold in three-dimensional Euclidean space, that is, the 2-sphere is the two-dimensional surface of a (three-dimensional) ball in three-dimensional space

■

Remark 10.13 *It is worth observing that in contrast to the co-difference, the covariation is not symmetric in its arguments. In addition, in the case in which the tail exponent $\alpha = 2$, then one has recovered a joint distribution for the losses in which the second moment is finite and then one can show the following relationship between co-difference, co-variation, and covariance:*

$$\mathbb{C}\text{ov}(X_1, X_2) = 2CV(X_1, X_2) = CD(X_1, X_2). \tag{10.48}$$

We note the following properties of Covariation and Codifference measures of concordance.

Proposition 10.11 (Properties of Covariation and Codifference) *The Codifference and Covariation satisfy the following properties:*

- *In contrast to the co-difference, the covariation is not symmetric in its arguments;*
- *If $\alpha > 1$ then the covariation induces a norm on the linear sub-space of jointly $S\alpha S$ random variables*

$$||\boldsymbol{X}||_\alpha = [CV(X_1, X_2)]^{1/\alpha}$$

- *The codifference can be written*

$$CD(X_1, X_2) = ||X_1||_\alpha^\alpha + ||X_2||_\alpha^\alpha - ||X_1 - X_2||_\alpha^\alpha$$

10.4.2 RANK CORRELATION MEASURES

Following from the notions of pairwise dependence based explicitly on the loss random variables, one can also define notions of correlation and dependence based on the rank of loss random variables. Rank correlation measures the relationship between the *rankings* of variables, that is, after assigning the labels "first", "second", "third", etc., to different observations of a particular variable. The coefficient lies in the interval $[1, 1]$, where $+1$ indicates the agreement between the two rankings is perfect, that is, the same; -1 indicates the disagreement between the two rankings is perfect, that is, one ranking is the reverse of the other; 0 indicates the rankings are completely independent. Due to this scale invariance, rank correlations thus provide an approach for fitting copulae to data.

One of the most popular choices for measuring the rank correlation is known as Spearman's rho. Charles Spearman introduced the nonparametric measure of dependence, Spearman's rank correlation coefficient, in Spearman (1904). This measure assesses how well the dependence between two random variables can be described by a monotonic function. As such it is equivalent to the Pearson's correlation coefficient between the ranked variables as detailed in Definition 10.23.

Definition 10.23 (Spearman's Rank Correlation Coefficient) *Consider two sets of order statistics for two loss processes $\left\{X_{(i,n)}\right\}_{i=1}^{n}$ and $\left\{Y_{(i,n)}\right\}_{i=1}^{n}$. Then, the Spearman's rank correlation is given by*

$$\rho := \frac{\sum_{i=1}^{n}(x_i - \bar{x})(y_i - \bar{y})}{\sqrt{\sum_i (x_i - \bar{x})^2 \sum_i (y_i - \bar{y})^2}}, \tag{10.49}$$

where x_i, y_i are the ranks and $\bar{x} = \frac{1}{n}\sum_{i=1}^{n} x_i$ and $\bar{y} = \frac{1}{n}\sum_{i=1}^{n} y_i$. ■

Spearman's rank correlation is often referred to as *Spearman's rho* and can be seen from its definition to be a simple scalar measure of dependence that depends on the copula of two random variables but not on their marginal distributions. An equivalent way to consider understanding Spearman's rank correlation for two random variables X_1 and X_2 with marginal distributions $F_1(x_1)$ and $F_2(x_2)$ is to consider its representation given by

$$\rho_S[X_1, X_2] = \rho[F_1(X_1), F_2(X_2)]. \tag{10.50}$$

Here we see that Spearman's rank correlation is simply the linear correlation of the probability transformed random variables. For multivariate case $(X_1, \ldots, X_d)$, Spearman's rho matrix is defined by the matrix coefficients $\rho_S[X_i, X_j] = \rho[F_i(X_i), F_j(X_j)]$. The main properties can be summarized as follows:

- The range for possible values of $\rho_S[X_1, X_2]$ is $[-1, 1]$;
- For independent random variables, $\rho_S[X_1, X_2] = 0$. However, zero Spearman's rank correlation does not necessarily imply independence;

- $\rho_S[X_1, X_2] = 1$ if X_1 and X_2 are comonotonic (perfect positive dependence); and $\rho_S[X_1, X_2] = -1$ if X_1 and X_2 are countermonotonic (perfect negative dependence). Note that this is not the case for the linear correlation coefficient $\rho[X_1, X_2]$;
- In the case of bivariate Gaussian copula with correlation parameter ρ, the following relation is true:

$$\rho_S[X_1, X_2] = \frac{6}{\pi} \arcsin\left(\frac{1}{2}\rho\right) \approx \rho; \tag{10.51}$$

see McNeil *et al.* (2005, theorem 5.36). This relationship between Spearman's rank correlation and the Gaussian copula correlation parameter is often used to calibrate the Gaussian copula. The error in approximating the right-hand side of the aforementioned equation by ρ itself is very small:

$$\left| \frac{6}{\pi} \arcsin\left(\frac{1}{2}\rho\right) - \rho \right| \leq (\pi - 3)|\rho|/\pi \leq 0.0181.$$

One can also express Spearman's Rank correlation via a copula specification, see Definition 10.24.

Definition 10.24 (Spearman's Rank Correlation via Copula) *The bivariate Spearman's Rank Correlation can be expressed explicitly via the bivaraite copula C according to*

$$\rho = 12 \int_{[0,1]} \int_{[0,1]} u_1 u_2 dC(u_1, u_2) - 3. \tag{10.52}$$

■

In addition, a general multivariate extension of Spearman's Rank Correlation is developed for d-dimensional loss random vectors and given below, see details in Nelsen (2002).

Definition 10.25 (Multivariate Generalized Spearman's Rho via Copula) *Consider the d-copula given by C and the permuted copula C^σ, then the generalized Spearman's Rho concordance measure of dependence is given according to*

$$\rho_d(C) = \alpha_d \left(\int_{[0,1]^d} (C + C^\sigma)\, d\Pi^d - \frac{1}{2^{d-1}} \right) \tag{10.53}$$

where one has $\alpha_d = \frac{(d+1)2^{d-1}}{2^d-(d+1)}$ and Π^d is the d-Independence Copula. ■

Another widely utilized rank correlation measure was developed by Maurice Kendall and is known as Kendall's τ rank correlation coefficient as detailed in Definition 10.26; see Kendall (1938). It should also be noted that Gustav Fechner proposed a similar measure in the context of time series in 1897; see Kruskal (1958).

Definition 10.26 (Kendall's Tau) *Let (X_1, Y_1) and (X_2, Y_2) be two independent pairs of random variables from a joint distribution function F, then Kendall's rank correlation is given by*

$$\tau := \Pr[(X_1 - X_2)(Y_1 - Y_2) > 0] - \Pr[(X_1 - X_2)(Y_1 - Y_2) < 0]. \tag{10.54}$$

■

Again, one may also define Kendall's tau rank correlation for random variables X_1 and X_2 according to the following equivalent form:

$$\begin{aligned}\rho_\tau[X_1, X_2] &= \mathbb{P}\mathrm{r}[(X_1 - \widetilde{X}_1)(X_2 - \widetilde{X}_2) > 0] - \mathbb{P}\mathrm{r}[(X_1 - \widetilde{X}_1)(X_2 - \widetilde{X}_2) < 0] \\ &= \mathbb{E}[\mathrm{sign}((X_1 - \widetilde{X}_1)(X_2 - \widetilde{X}_2))], \end{aligned} \tag{10.55}$$

where $(\widetilde{X}_1, \widetilde{X}_2)$ and (X_1, X_2) are independent random vectors from the same distribution. It can also be written as

$$\rho_\tau[X_i, X_j] = \mathbb{C}\mathrm{ov}[\mathrm{sign}(X_i - \widetilde{X}_i)\mathrm{sign}(X_j - \widetilde{X}_j)]. \tag{10.56}$$

Similar to Spearman's rank correlation, Kendall's tau rank correlation is a simple scalar measure of dependence that depends on the copula of two random variables but not on their marginal distributions.

- The range for possible values of $\rho_\tau[X_1, X_2]$ is $[-1, 1]$;
- For independent random variables $\rho_\tau[X_1, X_2] = 0$, although zero Kendall's tau does not necessarily imply independence;
- $\rho_\tau[X_1, X_2] = 1$ if X_1 and X_2 are comonotonic (perfect positive dependence); and $\rho_\tau[X_1, X_2] = -1$ if X_1 and X_2 are countermonotonic (perfect negative dependence);
- In the case of the bivariate Gaussian copula with correlation parameter ρ, the following relation is true:

$$\rho_\tau[X_1, X_2] = \frac{2}{\pi} \arcsin(\rho) \approx \rho; \tag{10.57}$$

  see McNeil *et al.* (2005, theorem 5.36). This relationship is also true for a general class of normal variance mixture distributions such as t-copula (it is often used to calibrate t-copula). Strictly speaking, it is true for the bivariate case only. That is, for the multivariate case $(X_1, \ldots, X_d)$, if Kendall's tau rank correlation is found for all pairs $\rho_\tau[X_i, X_j]$, then the correlation matrix coefficients ρ_{ij} calculated using (10.57) may not form a positive definite matrix. If this is the case, then the eigenvalue method can be used to adjust the correlation coefficients so that the matrix is well defined; see McNeil *et al.* (2005, example 5.54 and algorithm 5.5). The remaining degrees-of-freedom parameter ν in the t-copula is estimated, for example, by the maximum likelihood method.

Given data $(\mathcal{Y} = y_1, \ldots, y_N)$, which can be ranked, the estimation of Kendall's tau can be performed using the estimator

$$\hat{\tau} = 1 - \frac{2S(\pi, \sigma)}{N(N-1)/2}, \tag{10.58}$$

where π and σ denote two distinct orderings of the data $\mathcal{Y}$ and $S(\pi, \sigma)$ denotes the minimum number of adjacent transpositions needed to bring π to σ. This shows that Kendall's τ is based on the number of transpositions, that is, interchanges of consecutive elements, necessary to rearrange π into σ.

Remark 10.14 (Spearman's Rho versus Kendall's Tau) *Spearman's ρ and Kendall's τ share a lot of common properties; however, Spearman's ρ is a measure of average quadrant dependence,* *while Kendall's τ is a measure of average likelihood ratio dependence, see discussions in Fredricks and Nelsen (2007). Hence, this means that Kendall's τ will penalizes rank displacements by the distance of the displacement, where as Spearman's ρ will penalize by the square of the distance. It is also worth noting the observation of Newson (2002) that confidence intervals for Spearman's ρ are typically less reliable and less interpretable than confidence intervals for Kendall's τ-parameters.*

One can also relate Kendall's tau measure of concordance to a copula specification. Consider the concordance function κ quantifying the difference in probabilities of concordance and discordance for bi-variate loss random vectors (X_1, Y_1) and (X_2, Y_2) which are specified as follows:

- Assume X_1 and X_2 have common continuous marginal F_X;
- Assume Y_1 and Y_2 have common continuous marginal F_Y;
- Assume (X_1, Y_1) and (X_2, Y_2) have different copula C_1 and C_2 respectively.

Then in Nelsen (2002), it was proposed to consider an alternative copula specified concordance function κ for the equivalent Kendall's tau measuring the probability of concordance and discordance given by

$$\begin{aligned}\kappa &= \mathbb{P}\mathrm{r}\left[(X_1 - X_2)(Y_1 - Y_2) > 0\right] - \mathbb{P}\mathrm{r}\left[(X_1 - X_2)(Y_1 - Y_2) < 0\right] \\ &= 4\int_0^1\int_0^1 C_2(u,v)\,dC_1(u,v) - 1.\end{aligned} \tag{10.59}$$

One can show in under this concordance-discordance measure the results:

- $\kappa(C_1, C_2) \in [-1, 1]$;
- $\kappa(C, \Pi^d) \in [-1/3, 1/3]$;
- $\kappa(C, M^d) \in [0, 1]$;
- $\kappa(C, W^d) \in [-1, 0]$.

Recall: M^d - Frechet-Hoffding Upper-Bound; W^d - Frechet-Hoffding Lower-Bound; and Π^d - independence copula.

We finish this section by briefly also mentioning a third correlation measure that is related to medial correlation known as Blomqvist's β given in Definition 10.27; see Blomqvist (1950). For generalizations of Blomqvist's beta to higher dimensions, see discussions in Nelsen (2002), Joe (1990), and Dolati and Úbeda-Flores (2006).

Definition 10.27 (Blomqvist's Beta) *Consider two random variables X_1 and X_2, then Blomqvist's beta is given by*

$$\begin{aligned}\rho_\beta\left[X_1, X_2\right] := \mathbb{P}\mathrm{r}\left[(X_1 - med(X_1))\,(X_2 - med(X_2)) > 0\right] \\ - \mathbb{P}\mathrm{r}\left[(X_1 - med(X_1))\,(X_2 - med(X_2)) < 0\right],\end{aligned} \tag{10.60}$$

where $med(X_i)$ is the median of random variable X_i. ■

One can also make the following comments regarding the properties of Blomqvist's Beta measure of concordance:

- The empirical version $\hat{\rho}_\beta$ of Blomqvist's beta is a suitably scaled version of the proportion of points whose components are either both smaller, or both larger, than their respective sample medians;
- The computation of $\hat{\rho}_\beta$ involves only $O(n)$ operations, as opposed to $O(n^2)$ for the empirical versions of Kendall's tau and Spearman's rho.

In addition, Blomqvist's Beta can also be specified with regard to a copula as follows.

Definition 10.28 (Blomqvist's Beta via Copula) *The bivariate Blomqvist's Beta can be expressed explicitly via the bivaraite copula C according to*

$$\beta = 4C\left(\frac{1}{2}, \frac{1}{2}\right) - 1. \tag{10.61}$$

■

Remark 10.15 *Recently in Genest et al. (2013) they proposed the inversion of this copula based representation of Blomqvist's Beta to perform explicit parameter estimation for several copula models.*

As with the other popular measures of concordance specified above, there is also a generalization of Blomqvist's Beta to multivariate settings, see discussions in Nelsen (2002).

Definition 10.29 (Generalized Blomqvist's Beta via Copula) *Consider an d-copula C, then the generalized Blomqvist's Beta is given by*

$$\beta_d(C) = \alpha_d \left(C(\frac{1}{2}, \dots, \frac{1}{2}) - \frac{1}{2^d} \right), \tag{10.62}$$

where $\alpha_d = \frac{2^d}{2^{d-1}-1}$ ■

To complete the basic specification of concordance measures widely used in practice, we finish with a brief discussion on the notion of intermediate directional dependence in the 3-dimensional context, see details in Nelsen (2002). We will denote such dependence measures as rho-directional dependence.

Definition 10.30 (3-Copula ρ-Directional Dependence) *Consider a loss random vector* $\boldsymbol{X} = (X_1, X_2, X_3)$ *with* $\boldsymbol{X} \in \mathbb{R}^3$ *and associated 3-dimensional copula* $C_{\boldsymbol{X}}$. *Then for any direction* $(\alpha_1, \alpha_2, \alpha_3)$ *characterised by the vector components* $\alpha_i \in \{-1, 1\}$ *for* $i \in \{1, 2, 3\}$, *one has the* ρ*-directional dependence given by*

$$\begin{aligned}\rho^{(\alpha_1,\alpha_2,\alpha_3)}_{X_1,X_2,X_3} &= \frac{\alpha_1\alpha_2\rho_{X_1,X_x} + \alpha_2\alpha_3\rho_{X_2,X_3} + \alpha_3\alpha_1\rho_{X_3,X_1}}{3} \\ &+ \alpha_1\alpha_2\alpha_3\frac{\rho^{+}_{X_1,X_2,X_3} - \rho^{-}_{X_1,X_2,X_3}}{2}\end{aligned} \tag{10.63}$$

with pairwise Spearman's rho and

$$\rho^{+}_{X_1,X_2,X_3}(C_X) = 8\int_{[0,1]^3} \overline{C}_X(u,v,w)dudvdw - 1,$$
$$\rho^{-}_{X_1,X_2,X_3}(C_X) = 8\int_{[0,1]^3} C_X(u,v,w)dudvdw - 1. \tag{10.64}$$

■

Remark 10.16 *The eight vectors which characterize directions* $(\alpha_1, \alpha_2, \alpha_3)$ *where* $\alpha_i \in \{-1, 1\}$ *for* $i \in \{1, 2, 3\}$ *in* $[0, 1]^3$ *allow one to utilise the* ρ*-directional dependence to measure directional dependence in different quadrants.*

To better understand the notion of rho-directional dependence we note that for example, if $\rho_X^{(-1,-1,1)}$ or $\rho_X^{(1,1,-1)}$ are positive, then there will be positive dependence in the direction of $(-1,-1,1)$ or $(1,1,-1)$, hence one would expect large (small) values of X_1 and X_2 to occur with small (large) values of X_3, ie. $\rho_{X_1,X_2} > 0$ with $\rho_{X_1,X_3} < 0$ and $\rho_{X_2,X_3} < 0$.

10.5 Tail Dependence Parameters, Functions, and Tail Order Functions

We begin this section with a discussion on tail dependence coefficients, introducing this concept and how to interpret the properties of models with this feature. Then we move to relaxing this definition to nonasymptotic cases by considering the case of tail dependence functions and the related tail order functions. The concept of tail dependence parameters, tail dependence functions, or tail order functions each play a crucial role in both copula modeling as well as extreme value theory.

10.5.1 TAIL DEPENDENCE COEFFICIENTS

Tail dependence provides one approach to quantification of the dependence in extremes of a multivariate distribution. Traditionally this notion of dependence was considered from a pairwise construction due to tractability of expressions for the pairwise construction when applied to copula models. However, there is no reason to restrict this notion to just pairwise analysis and later we consider first the pairwise definition and then the generalized definition for d-variate random vectors.

The importance of thinking about tail dependence was succinctly summarized in the questions posed in Charpentier (2003) as detailed later:

1. If one considers data that are is taken from a multivariate distribution anywhere in its support, then through the measures of dependence just discussed and dependence concepts previously detailed in this chapter, it is possible to obtain all the overall dependence structure between say two loss random variables $X^{(1)}$ and $X^{(2)}$.

   However, it is interesting to question whether dependence properties still hold if focusing only on extremes of the distribution in any particular quadrant?

For instance, if the correlation between $X^{(1)}$ and $X^{(2)}$ is positive, is it reasonable to assume that the correlation between extreme values of $X^{(1)}$ and extreme values of $X^{(2)}$ will still be positive or even present at all?

2. Another way to think of this is to consider the case in which $X^{(1)}$ and $X^{(2)}$ may exhibit a positive dependence. Then in what types of models and under what conditions can one assume that the same dependence property will hold for $X^{(1)}$ and $X^{(2)}$ given $X^{(1)}$ is higher than a given threshold and $X^{(2)}$ is also higher than a given threshold?

To understand the quantification of dependence in such cases, that is, cases in which each marginal loss random variable is considered in an extreme region of support of the distribution (which may happen in a number of different ways) one can start to define the notion of tail dependence. The notion of bivariate tail dependence coefficient is defined as the conditional probability that a random variable exceeds a certain threshold given that the other random variable in the joint distribution has exceeded this threshold as detailed formally in Definition 10.31. Note that the notion of copula distributions will be discussed in detail in the following sections; we simply provide an introduction to the notation used for such copula distributions for random vector $\boldsymbol{X} \in \mathbb{R}^d$, which take support on $[0,1]^d$ where the multivariate distribution of random vector $\boldsymbol{X}$ is given by copula and marginal distributions according to

$$\mathbb{P}\mathrm{r}\left[X_1 < x_1, \ldots, X_d < x_d\right] = C\left(F_{X_1}\left(x_1\right), \ldots, F_{X_d}\left(x_d\right)\right). \tag{10.65}$$

Clearly, the copula is obtained after transforming each of the marginal components of random vector $\boldsymbol{X} \in \mathbb{R}^d$ to the unit cube via marginal transformations $U_i = F_{X_i}(X_i)$.

Definition 10.31 (Bivariate Tail Dependence Coefficient) *Consider two random variables X_1 and X_2 with distributions $F_i, i = 1,2$ and a distribution describing their dependence on the unit cube known as a copula C. We define the coefficient of upper tail dependence by*

$$\lambda_u := \lim_{u \uparrow 1} \mathbb{P}r\left[X_2 > F_2^{-1}(u) \,|\, X_1 > F_1^{-1}(u)\right] = \lim_{u \uparrow 1} \frac{1 - 2u + C(u,u)}{1-u} \tag{10.66}$$

and similarly we define the coefficient of lower tail dependence by

$$\lambda_l := \lim_{u \downarrow 0} \mathbb{P}r\left[X_2 \leq F_2^{-1}(u) \,|\, X_1 \leq F_1^{-1}(u)\right] = \lim_{u \downarrow 0} \frac{C(u,u)}{u}. \tag{10.67}$$

■

Note that $\widetilde{C}(1-u, 1-u) = 1 - 2u - C(u,u)$ is known as the survival copula of C. The aforementioned relationships show that the upper tail dependence coefficients of copula C is also equal to the lower tail dependence coefficient of the survival copula of C. Analogously, the lower tail dependence coefficient of copula C is equivalent to the upper tail dependence coefficient of the survival copula.

Remark 10.17 *Similar to rank correlations, the tail dependence coefficient is a simple scalar measure of dependence that depends on the copula of two random variables but not on their marginal distributions.*

Both λ_u and λ_l belong to the range $[0, 1]$, provided that the aforementioned limits exist. Essentially, these coefficients are measures of the dependence in the tails of bivariate distribution. For OpRisk purposes, the upper tail dependence (a chance that X_1 is very large if X_2 is very large) is of primary importance.

As shown in Definition 10.31, when the marginal loss distributions $F_1(\cdot)$and $F_2(\cdot)$ are continuous, then the tail dependence coefficients can be expressed in terms of the unique copula $C(u_1, u_2)$ between loss random variables X_1 and X_2. Therefore, it will not come as a surprise, when different copula parametric models are discussed later, that the tail dependence will be directly a function of the copula parameter. This will allow one to obtain a relationship between the parameters in the parametric density that will represent the joint dependence of the loss random variables and the extreme strength of this dependence.

Given the definition of pairwise tail dependence, one can state the following properties of such a quantification of dependence; see Proposition 10.12.

Proposition 10.12 (Properties of Tail Dependence Coefficient) *Consider two loss random variables with marginal loss distributions $X_i \sim F_{X_i}$ and a joint dependence modeled by the copula C, then defining the constant*

$$c = \lim_{x \to \infty} \frac{\overline{F}_{X_2}(x)}{\overline{F}_{X_1}(x)} \tag{10.68}$$

one can show the following features of upper tail dependence:

1. *The following bounds apply to the upper tail dependence*

$$c\lambda_u \leq \hat{\lambda} \leq \min(c, \lambda_u) \tag{10.69}$$

with

$$\hat{\lambda} = \lim_{x \to \infty} \frac{1 - F_{X_1}(x) - F_{X_2}(x) + C\left(F_{X_1}(x), F_{X_2}(x)\right)}{1 - F_{X_1}(x)}. \tag{10.70}$$

2. *The following relationship between the maximum of a sum of two random variables and the tail dependence holds*

$$\mathbb{P}\mathrm{r}\left[\max\left\{X_1, X_2\right\} > x\right] \sim \left(1 + c - \hat{\lambda}\right)\overline{F}_{X_1}(x) \tag{10.71}$$

and the tail result given by

$$\lim_{x \to \infty} \mathbb{P}\mathrm{r}\left[X_1 > x \middle| \max\left\{X_1, X_2\right\} > x\right] = \frac{1}{1 + c - \hat{\lambda}}. \tag{10.72}$$

3. *The following worst, case bounds can be obtained:*

$$\overline{F}_{X_1}(x) << \mathbb{P}\mathrm{r}\left[X_1 + X_2 > x\right] << (1 + c)\overline{F}_{X_1}\left(\frac{x}{2}\right). \tag{10.73}$$

4. *Considering the identically distributed losses* $X_i \sim F_X(x)$ *with a copula distribution* $C(u_1, u_2) = C(F_X(x), F_X(y))$*, one can obtain the following upper and lower bounds:*

$$\begin{aligned} \lambda_u \leq \liminf_{x\to\infty} \frac{\mathbb{P}\mathrm{r}\left[c_1X_1 + c_2X_2 > x\right]}{\mathbb{P}\mathrm{r}\left[X_1 > \frac{x}{c_1+c_2}\right]}, \\ \limsup_{x\to\infty} \frac{\mathbb{P}\mathrm{r}\left[c_1X_1 + c_2X_2 > x\right]}{\mathbb{P}\mathrm{r}\left[X_1 > \frac{x}{c_1+c_2}\right]} \leq 2 - \lambda_u, \end{aligned} \tag{10.74}$$

for constants c_1 *and* c_2 *satisfying* $y = c_1x/(c_1 + c_2)$.

5. *The tail dependence coefficients are invariant to strictly increasing transformations of the margins.*

One can also relate the notion of upper tail dependence to negative regression dependence as follows.

Remark 10.18 (Upper Tail Dependence and Negative Regression Dependence) *One can show that if loss random variables satisfy the definition of negative regression dependence, then they will always have upper tail dependence of zero as measured by the tail dependence measure*

$$\lambda_u = \lim_{u\uparrow 1} \mathbb{P}\mathrm{r}\left[U_2 > u \middle| U_1 > u\right] \tag{10.75}$$

for marginally uniform $U_1 = F_{X_1}(x)$ *and* $U_2 = F_{X_2}(x)$.

One can also show the following properties of the concordance measures known as tail dependence or extremal dependence:

- Tail dependence provides one approach to quantification of the dependence in extremes of a multivariate distribution;
- The notion of bivariate tail dependence coefficient is defined as the conditional probability that a random variable exceeds a certain threshold given that the other random variable in the joint distribution has exceeded this threshold;
- The tail dependence coefficients are invariant to strictly increasing transformations of the margins;
- If a random vector satisfies the definition of negative regression dependence then it will always have upper tail dependence of zero.

Remark 10.19 *Similar to rank correlations, the tail dependence coefficient is a simple scalar measure of dependence that depends on the copula not the marginals.*

There have also been developments that have made extensions of the notion of tail dependence to arbitrary d-variate cases for $d > 2$ as recently studied in a number of papers; see, for instance, De Luca and Rivieccio (2012). Under the specification provided in Definition 10.32, one may quantify the tail dependence present between subvector partitions of the multivariate random vector with regard to joint tail dependence behaviors; see discussions in Li (2009).

Definition 10.32 (Multivariate Tail Dependence) *Let $X = (X_1, \dots, X_d)^T$ be a d-dimensional random vector with marginal distribution functions $F_1, \dots, F_d$ and copula C.*

1. *One may define the coefficient of multivariate upper tail dependence (upper orthant dependence) by*

$$\lambda_u^{1,\dots,h|h+1,\dots,d} = \lim_{\nu\to 1-} \mathbb{P}\mathrm{r}\left(X_1 > F^{-1}(\nu), \dots, X_h > F^{-1}(\nu) | X_{h+1} > F^{-1}(\nu), \dots, X_d > F^{-1}(\nu)\right) = \lim_{\nu\to 1-} \frac{\overline{C}_n(1-\nu,\dots,1-\nu)}{\overline{C}_{n-h}(1-\nu,\dots,1-\nu)},$$

where $\overline{C}$ is the survival copula of C;

2. *One may define the coefficient of multivariate lower tail dependence (lower orthant dependence) by*

$$\lambda_l^{1,\dots,h|h+1,\dots,d} = \lim_{\nu\to 0+} \mathbb{P}\mathrm{r}\left(X_1 < F^{-1}(\nu), \dots, X_h < F^{-1}(\nu) | X_{h+1} < F^{-1}(\nu), \dots, X_d < F^{-1}(\nu)\right) = \lim_{\nu\to 0+} \frac{C_n(\nu,\dots,\nu)}{C_{n-h}(\nu,\dots,\nu)}.$$

Here, h is the number of variables conditioned on (from the d variables considered).

■

It is important to observe that the tail dependence coefficients of a copula C represent the conditional tail probabilities that components of U_is will go to extreme values at the same rate and hence they can only describe components of the extreme dependence that is independent of the marginal distributions. To address this property, we will discuss the properties of tail dependence functions and tail order.

10.5.1.1 Estimation of Tail Dependence Coefficients.

In Frahm *et al.* (2005), they discuss the estimation of the tail dependence coefficient in a number of different scenarios ranging from complete lack of knowledge of the copula density family in which case the estimation is nonparametric, through to the fully parametric cases.

Assume that the marginal distributions $\{F_{X_i}(\cdot;\theta_i)\}_{i\in\{1,\dots,d\}}$ are known and furthermore that the joint distribution $F_{\mathbf{X}}(\cdot;\phi)$ admits a tail dependence that is nonzero in either the upper or lower coefficient and generically can be written as $\lambda_u = f_u(\phi)$ or $\lambda_l = f_l(\phi)$, respectively, for some known function $f_u(\cdot)$ or $f_l(\cdot)$. Then MLE estimation of the $\hat{\phi}$ will produce an MLE estimator for the tail dependence coefficient $\hat{\lambda}_u = f_u(\hat{\phi})$ or $\hat{\lambda}_l = f_l(\hat{\phi})$. If the standard regularity conditions on the MLE are satisfied, see Chapter 7, then the MLE estimator of the tail dependence coefficients will inherit the asymptotic (in the sample size) consistency and normality properties of the MLE estimator for the model parameter vector ϕ. Of course, there will be bias present in the estimated tail dependence coefficients if there is missspecification of either the marginal or the joint distribution copula.

If one relaxes these assumptions to only assume that the copula distribution family $C(\cdot;\phi)$ is from a particular known parametric model class, however, the marginals distributions are unknown. Then the estimation of the tail dependence coefficient can proceed by first transforming the data to the unit cube via the empirical distribution functions; see discussions on pseudo data in Chapter 8. Then the copula distribution can be estimated on the unit cube via the pseudo data via MLE to get the copula parameter estimate $\hat{\phi}$, where the consistency and asymptotic normality of such a copula parameter estimator has been studied in Genest *et al.* (1995) and Shih and Louis (1995). Then if there is a known functional mapping between the copula parameter vector and the tail dependence coefficient that is sufficiently smooth, then asymptotic consistency and normality will be inherited by the estimator of the tail dependence coefficient.

Finally, one may also consider the completely nonparametric setting in which no presumed knowledge of the copula distribution parametric family or the marginal distribution parametric families is assumed, that is, one has the following empirical copula and empirical marginal distribution functions given n random vectors of loss data $\{\boldsymbol{X}_i\}_{i\in\{1,2,\ldots,n\}}$,

$$\hat{F}_n(\boldsymbol{X}) = \hat{C}_n\left(\hat{F}_{1,n}(x_1), \ldots, \hat{F}_{d,n}(x_d)\right), \tag{10.76}$$

where the empirical copula is given by

$$\hat{C}_n(u_1, \ldots, u_d) = \frac{1}{n}\sum_{i=1}^{n}\mathbb{I}\left(\frac{R_{1i}}{n} \leq u, \ldots, \frac{R_{di}}{n} \leq u\right), \tag{10.77}$$

where R_{ji} is the rank of the variable in its marginal dimension that makes up the pseudo data.

In the case of a bivariate distribution, the nonparametric estimation of the tail dependence coefficients can be performed under one of following estimators that have been proposed in the literature:

- **Nonparametric Estimation of Upper Tail Dependence (Estimator 1):**

$$\hat{\lambda}_u^{(1)} = 2 - \frac{\ln \hat{C}_n\left(\frac{n-k}{n}, \frac{n-k}{n}\right)}{\ln\left(\frac{n-k}{n}\right)}, \quad k \in \{0, 1, 2, \ldots, n\}. \tag{10.78}$$

  This estimator is the nonparametric estimator counterpart of an alternative tail dependence coefficient given in Definition 10.33;

- **Nonparametric Estimation of Upper Tail Dependence (Estimator 2):**

$$\hat{\lambda}_u^{(2)} = 2 - \frac{1 - \hat{C}_n\left(\frac{n-k}{n}, \frac{n-k}{n}\right)}{1 - \left(\frac{n-k}{n}\right)}, \quad k \in \{0, 1, 2, \ldots, n\}. \tag{10.79}$$

  This estimator is the nonparametric estimator for the upper tail dependence studied in Joe *et al.* (1992);

- **Nonparametric Estimation of Upper Tail Dependence (Estimator 3):**

$$\hat{\lambda}_u^{(3)} = 2 - 2\exp\left[\frac{1}{n}\sum_{i=1}^{n}\ln\left(\frac{\sqrt{\ln\frac{1}{U_{1,i}}\ln\frac{1}{U_{2,i}}}}{\ln\left(\frac{1}{\max\{U_{1,i},U_{2,i}\}^2}\right)}\right)\right]. \tag{10.80}$$

This estimator is the nonparametric estimator for the upper tail dependence based on extreme value theory results and EVT copula as studied in Capéraà *et al.* (1997).

One can define analogously the lower tail dependence estimators.

We conclude this section with a basic description of asymptotic independence to round of the discussion of extreme dependence and independence. Basically, this case corresponds to the situation in which the extremes of the distributions are asymptotically independent and one would consequently find the tail dependence coefficient to be zero.

When performing modeling, one must be careful not to impose tail dependence in situations where it may not be suitable to incorporate (this will be tackled by model selection) since it will tend to overestimate the chance of extremal joint events. This statement, when looking at the definitions of upper and lower tail dependence coefficients, is equivalent to saying do not arbitrarily impose copula models for dependence with specified copula parameters, as this will artificially inflate the chance of tail dependence when the limits exist.

Coles *et al.* (1999) consider this concept in the context of multivariate EVT (also see detailed discussion in Peters and Shevchenko, 2015), where they discuss the fact that applying extreme value models based on nonzero tail dependence to cases in which actually no extreme dependence is justified will result in an overestimation of probabilities of extreme joint events. Examining this class of distributions at finite levels, that is, nonasymptotic levels, allows for a more useful measure of extremal dependence that is specified in Definition 10.33, see Coles *et al.* (1999).

Definition 10.33 ($\bar{\chi}$ - Measure of Extremal Dependence) *A modified measure of extreme dependence is given by the following quantity:*

$$\bar{\chi} := \frac{2\ln \mathbb{P}\mathrm{r}(U > u)}{\ln \mathbb{P}\mathrm{r}(U > u, V > v)} - 1 = \frac{2\ln(1-u)}{\ln \bar{C}(u,u)} - 1, \tag{10.81}$$

where $-1 < \bar{\chi}(u) \leq 1$ *for all* $0 \leq u \leq 1$. ■

This measure of dependence is particular useful for multivariate EVT-based models for which it was first developed. The reason for its utility is that $\bar{\chi}$ increases with dependence strength and equals unity for asymptotically dependent variables. In addition, in the case of a multivariate Gaussian model, the dependence measure $\bar{\chi}$ is equal to the correlation, providing a benchmark for interpretation in general models of dependence. Coles *et al.* (1999) thus argue that using this new measure in addition to the tail dependence measure gives a more complete summary of extremal dependence. Since this modified proposal for studying tail dependence, there have been a number of new works on tail functions and intermediate tail dependence that we briefly highlight in the following sections.

10.5.1.2 Tail Dependence and Heavy Tailedness in Elliptical Families.

In this section, we draw attention to recent studies, see Schmidt (2002), on the relationship between tail dependence and heavy tailedness of marginal distributions, specifically in the family of elliptically contoured distributions. Elliptically contoured multivariate distributions are characterized by Definition 10.35.

Definition 10.34 (Spherically Contoured Distributions) *A random vector* $\boldsymbol{X} \in \mathbb{R}^d$ *in* d*-dimensions is said to have spherical contours in its quantile function if it satisfies the equality in distribution*

$$X \stackrel{d}{=} AX \tag{10.82}$$

for every orthogonal matrix $A \in \mathbb{R}^{d\times d}$. Analogously, any random vector with spherical quantile level sets will have a representation given by

$$X \stackrel{d}{=} R_d U \tag{10.83}$$

with random variable $R_d \geq 0$ such that $R_d \sim F_{R_d}$ and independent of the random vector U, which is uniformly distributed on the sphere in $\mathbb{R}^d$. ∎

From the definition of the class of spherical distributions, one may construct a definition of elliptically contoured quantile function distributions.

Definition 10.35 (Elliptically Contoured Distributions) *Consider a random vector $X \in \mathbb{R}^d$ in d-dimensions with parameters in the distribution function given by $\mu \in \mathbb{R}^d$ and $\Sigma \in \mathbb{R}^{d\times d}$. Then the random vector is said to have an elliptical distribution function if it satisfies the equality in distribution*

$$X \stackrel{d}{=} \mu + A^T Y \stackrel{d}{=} \mu + R_d A^T U \tag{10.84}$$

for a k-dimensional ($k \leq d$) spherically distributed random vector Y, matrix $A \in \mathbb{R}^{k\times d}$ satisfying $A^T A = \Sigma$ and $rank(\Sigma) = k$, random variable $R_d \geq 0$ independent of the random vector U, which is uniformly distributed on the sphere in $\mathbb{R}^d$. ∎

Under the constraint of each marginal distribution having to jointly satisfy the expression in Equation 10.84, one can show some interesting connections between the heavy-tailedness of the marginal distributions and the existence or strength of the tail-dependence coefficient. These relationships are captured by the result in Theorem 10.2; see Schmidt (2002, theorem 5.2).

Theorem 10.2 (Elliptical Family Tail Dependence and Marginal Heavy-Tailedness) *Consider an elliptically distributed random vector $X \in \mathbb{R}^d$ in d-dimensions ($d \geq 2$) with parameters in the distribution function given by $\mu \in \mathbb{R}^d$ and $\Sigma \in \mathbb{R}^{d\times d}$. Then the following results hold if the random vector X has a tail-dependent bivariate margin:*

- *The tail distribution function of the random variable R_d given by $\overline{F}_{R_d}$ must satisfy the O-regular variation condition for any $t \geq 1$ given by*

$$0 < \liminf_{r\to\infty} \frac{\overline{F}_{R_d}(rt)}{\overline{F}_{R_d}(r)} \leq \liminf_{r\to\infty} \frac{\overline{F}_{R_d}(rt)}{\overline{F}_{R_d}(r)} < \infty. \tag{10.85}$$

- *The tail distribution function of each marginal of the k-dimensional random vector Y each denoted by $\overline{F}_{Y_i}$ for $i \in \{1, 2, \ldots, k\}$ must also be O-regularly varying and therefore satisfy the condition for any $t \geq 1$ given by*

$$0 < \liminf_{y\to\infty} \frac{\overline{F}_{Y_i}(yt)}{\overline{F}_{Y_i}(y)} \leq \liminf_{y\to\infty} \frac{\overline{F}_{Y_i}(yt)}{\overline{F}_{Y_i}(y)} < \infty. \tag{10.86}$$

for all tail distribution marginals of random vector Y.

Conversely, if $\overline{F}_{R_d}$ *is regularly varying and therefore satisfies the condition that*

$$\lim_{r\to\infty} \frac{\overline{F}_{R_d}(rt)}{\overline{F}_{R_d}(r)} = t^{\alpha} \tag{10.87}$$

for any $t \geq 0$ *and a value of* $\alpha \in \mathbb{R}$, *then all bivariate marginal distributions of random vector* $\boldsymbol{X}$ *will be tail dependent.*

This result is significant for OpRisk since it demonstrates the direct relationship between marginal heavy-tailed distributions and the existence of tail dependence in the family of spherical and elliptical distributions. The results of this Theorem 10.2 were applied for several possible families of loss distributions, which will have a multivariate elliptical distributions with a summary given by

- If $\boldsymbol{X} \in \mathbb{R}^d$ has a multivariate spherically contoured Pearson type VII distribution (such as the multivariate Cauchy distribution), then all bivariate margins of the random vector $\boldsymbol{X}$ will have tail dependence coefficients that are nonzero;
- If $\boldsymbol{X} \in \mathbb{R}^d$ has a multivariate elliptically contoured logistic distribution, then all bivariate marginal distributions are tail independent;
- If $\boldsymbol{X} \in \mathbb{R}^d$ has a multivariate elliptically contoured generalized hyperbolic distribution, then all bivariate marginal distributions are tail independent.

One can estimate the tail dependence for the elliptical family of distributions as discussed in Frahm *et al.* (2005) via the Pickand's dependence function given in Definition 10.36 which plays an important role in extreme value copulas as will be defined later.

Definition 10.36 (Pickand's Dependence Function) *The Pickand's dependence function is a function* $A : [0,1] \mapsto [0.5,1]$, *which is convex such that it satisfies that* $\max(t, 1-t) \leq A(t) \leq 1$ *for every* $t \in [0,1]$. ■

Given an elliptical bivariate distribution $(X_1, X_2) \stackrel{d}{=} \boldsymbol{\mu} + R_2\Lambda[U_1, U_2]$ with R_2 independent of $\boldsymbol{U}$. Then if the tail distribution of the Euclidean norm $D = ||(X_1, X_2)||_2$ given by $\overline{F}_D$ is regularly varying with $\overline{F}_D \in \mathrm{RV}_{\alpha}$ and $\alpha > 0$, then on can show that the tail dependence coefficient can be estimated via the expression linking the tail regularity to the tail dependence coefficient via the Student-t distribution and the correlation ρ in Equation 10.88; see discussion in Frahm *et al.* (2005),

$$A\left(\frac{1}{2}\right) = t_{\alpha+1}\left(\sqrt{\alpha+1}\sqrt{\frac{1-\rho}{1+\rho}}\right). \tag{10.88}$$

Having seen the relationship between the heavy-tailed properties of the marginal distributions and the existence of tailed dependence in the special case of elliptical distributions, we next consider more general notions of intermediate tail dependence. In particular, we will note some recent results regarding the role played by the heavy tailedness of the marginal distributions in more general settings.

10.5.2 TAIL DEPENDENCE FUNCTIONS AND ORDERS

The study of tail dependence functions and tail orders can aid in the selection of an appropriate parametric model for OpRisk loss distributions. An early definition of tail dependence functions was provided in Klüppelberg *et al.* (2008), where they define the tail dependence function given in Definition 10.37.

Definition 10.37 (Tail Dependence Function of a Multivariate Distribution) *Consider a loss random variable* $\boldsymbol{X} \in \mathbb{R}^d$ *for* $d \geq 2$*, then the tail dependence function is given by*

$$\lambda\left(x_1, x_2, \ldots, x_d\right) = \lim_{t \to 0} \frac{1}{t} \mathbb{P}\mathrm{r}\left[\overline{F}_{X_1}(X_1) \leq tx_1, \ldots, \overline{F}_{X_d}(X_d) \leq tx_d\right]. \tag{10.89}$$

■

Joe *et al.* (2010) studied properties of tail dependence functions and conditional tail dependence functions, which they defined via the joint distribution on the unit d-dimensional hyper-cube known as a copula, for a multivariate loss distribution. Note that detailed discussions on copulas are provided in the following sections. The definition adapted in Joe *et al.* (2010) for the upper and lower tail dependence functions differs to that provided in Definition 10.37 via the fact that each marginal can go to the limit at different rates according to the functions:

- Lower tail dependence function. The tail dependence function for the copula distribution $C\left(u_1, \ldots, u_d\right)$ is given by

$$\lambda_l\left(\boldsymbol{t}; C\right) = \lim_{u \downarrow 0} \frac{C\left(ut_1, \ldots, ut_d\right)}{u}, \quad \forall \boldsymbol{t} = \left(t_1, \ldots, t_d\right) \in \mathbb{R}_+^d. \tag{10.90}$$

- Upper tail dependence function. The tail dependence function for the copula distribution $C\left(u_1, \ldots, u_d\right)$ is given by

$$\lambda_u\left(\boldsymbol{t}; C\right) = \lim_{u \downarrow 0} \frac{\overline{C}\left(ut_1, \ldots, ut_d\right)}{u}, \quad \forall \boldsymbol{t} = \left(t_1, \ldots, t_d\right) \in \mathbb{R}_+^d, \tag{10.91}$$

  with survival copula distribution $\overline{C}\left(u_1, \ldots, u_d\right) = C\left(1 - u_1, \ldots, 1 - u_d\right)$.

The existence of such limits in the definition of the tail dependence functions can be linked to existence of multivariate regular variation on the copula distribution tails. Recall the definition of regular variation given in Definition 10.38.

Definition 10.38 (Regular Variation) *A measurable function* $f : \mathbb{R}_+ \to \mathbb{R}_+$ *is regularly varying at* ∞ *with index* α *that is,* $f \in RV_\alpha$ *if for any* $a > 0$ *one has*

$$\lim_{x \to \infty} \frac{f(xa)}{f(x)} = a^\alpha. \tag{10.92}$$

Any function that is regularly varying of order $\alpha = 0$ *is said to be slowly varying and will be denoted by* $l(x) \in RV_0$*. Then any regularly varying function can be written according to the decomposition*

$$f(x) = x^\alpha l(x). \tag{10.93}$$

■

Joe *et al.* (2010) show the following properties of the tail dependence function for the lower tail dependence, with analogous results for the upper tail dependence achieved through the link between these expressions and the survival copula.

- A tail dependence function $\lambda_l(\boldsymbol{t}; C)$ satisfies the properties that $\lambda_l(a\boldsymbol{t}; C) = a\lambda_l(\boldsymbol{t}; C)$ for any $a \geq 0$;
- A tail dependence function satisfies that $\lambda_l(\boldsymbol{t}; C) = 0$ for all $\boldsymbol{t} \geq \boldsymbol{0}$ if and only if $\lambda_l(\boldsymbol{k}; C) = 0$ for some positive $\boldsymbol{k}$;
- Setting $\boldsymbol{t}$ in $\lambda_l(\boldsymbol{t}; C)$ to one returns the lower tail dependence coefficient, $\lambda_l(\boldsymbol{1}; C) = \lambda_l$.

Having defined the notion of tail dependence functions, one can also consider the idea of tail orders that will also be linked to the regular variation of the tails of the copula distribution. In Hua and Joe (2011), they defined the concept of tail order functions as given in Definition 10.39.

Definition 10.39 (Tail Order Parameters) *If a d-dimensional copula distribution C can be written according to a slowly varying function decomposition given by the asymptotic representation as $u \downarrow 0$ according to*

$$C(u, u, \ldots, u) \sim u^{\kappa_l(C)} l(u), \quad u \to 0^+, \tag{10.94}$$

then the power $\kappa_l(C)$ for copula C is denoted the lower tail order. The upper tail order is analogously defined according to

$$\overline{C}(1-u, \ldots, 1-u) \sim u^{\kappa_u(C)} l(u), \quad u \to 0^+ \tag{10.95}$$

with $\kappa_u(C)$ the upper tail order for copula C. ■

Remark 10.20 *Assuming that the slowly varying function $l(u)$ is nonzero that is, $\lim_{s\to 0^+} l(s) = h \in (0, \infty)$, then with $\kappa_l(C) = 1$ ($\kappa_u(C) = 1$) one obtains the standard definitions of lower (and upper) tail dependence coefficients.*

In Hua and Joe (2011), they also define the notion of a tail order function for the upper and lower tail orders according to Definition 10.40.

Definition 10.40 (Tail Order Functions) *Consider a d-dimensional copula C such that*

$$C(u, u, \ldots, u) \sim u^{\kappa_l(C)} l(u), \quad u \to 0^+, \tag{10.96}$$

for some slowly varying function $l(u) \in RV_0$. Then the lower tail order function is given by

$$\lambda_l^o(\boldsymbol{t}; C, \kappa_l(C)) = \lim_{u\to 0^+} \frac{C(ut_1, ut_2, \ldots, ut_d)}{u^{\kappa_l(C)} l(u)}, \tag{10.97}$$

Analogously, the upper tail order function is then given by

$$\lambda_u^o(\boldsymbol{t}; \overline{C}, \kappa_u(C)) = \lim_{u\to 0^+} \frac{\overline{C}(ut_1, ut_2, \ldots, ut_d)}{u^{\kappa_u(C)} l(u)}, \tag{10.98}$$

where $\overline{C}$ is the survival copula. ■

Note that these definitions of tail order functions are directly related to the case of tail dependence functions specified in Definitions 10.90 and 10.91. It is clear that the tail dependence function and tail order functions will be asymptotically directly related to each other since the regular variation decomposition in numerator and denominator that arises will cancel each other eventually. Put another way, one can see that given the slowly varying function is nonzero (i.e., the copula is regularly varying in the tails) such that $\lim_{s \to 0+} l(s) = h \in (0, \infty)$, then one can obtain the direct link between the tail dependence functions and the tail order functions as $\lambda_l(\boldsymbol{t}; C) = h\lambda_l^o(\boldsymbol{t}; C, 1)$.

Remark 10.21 (Copula Reflection Asymmetry and Tail Order Coefficients) *The definition of upper and lower tail dependence functions and tail order functions aids in the understanding of the attributes of particular dependence features obtained from a copula distribution C. For instance, one can utilize the tail orders to assess the degree of reflection symmetry or asymmetry in the tails of the copula. Consider the copula* $C(u_1, \ldots, u_d)$ *and the reflected copula of* $(1 - U_1, \ldots, 1 - U_d)$ *given by* $C_R(u_1, \ldots, u_d)$. *One can consider reflection symmetry as the case in which* $C_R \equiv C$. *If there are inequalities, such that* $C(u, u, \ldots, u) \geq C_R(1 - u, \ldots, 1 - u)$ *for all* $u \in (0, u_0)$ *with* $u_0 \in (0, 0.5]$, *then the copula has greater mass in the lower tail and the reflection asymmetry indicates skewness toward the lower tail and vice versa if the inequality is the other direction. In practice, it is difficult to compare these copula densities, hence the tail orders* $\kappa_l(C)$ *and* $\kappa_u(C)$ *provide an alternative way to assess reflection asymmetry.*

- ***Upper Tail Skewness.*** *If* $\kappa_l(C) > \kappa_u(C)$, *then the copula C has upper tail skewness and the smaller the magnitude of* $\kappa_l(C)$ *the slower the convergence to zero in the tails;*
- ***Lower Tail Skewness.*** *If* $\kappa_l(C) < \kappa_u(C)$, *then the copula C has lower tail skewness and the smaller the magnitude of* $\kappa_u(C)$ *the slower the convergence to zero in the tails.*

Hua and Joe (2011, proposition 2) also relate the lower tail order $\kappa_l(C)$ and upper tail order $\kappa_u(C)$ coefficients for a copula C to the existence of dependence features such as PLOD, LTDS, and MLTD discussed in the previous section, as detailed in Proposition 10.13.

Proposition 10.13 (Tail Order Coefficients and PLOD, LTDS, and MLTD) *Consider a copula distribution function* $C(u_1, u_2, \ldots, u_d)$ *that has lower tail order* $\kappa_l(C) \geq 1$, *then one can show the following properties related to positive lower orthant dependence (PLOD), lower tail decreasing in sequence (LTDS), and multivariate left tail decreasing (MLTD) as a function of the tail order coefficients magnitude:*

- *A d-dimensional copula C will be PLOD eventually if* $\kappa_l(C) \leq d$;
- *Under regularity conditions on the copula C, the marginal copula distributions will preserve the order of the tail orders such that marginals will have smaller tail orders. Consider sets* $S_1 \subset S_2 \subseteq [0,1]^d$ *such that the size of* S_1 *satisfies* $|S_1| = k \geq 2$ *and* $|S_2| = p \in [k, d]$, *then one has that*

$$\kappa_l(C(s_{11}, \ldots, s_{1k}, u_{k+1}, \ldots, u_d)) - \kappa_l(C(s_{21}, \ldots, s_{2p}, u_{p+1}, \ldots, u_d)) \geq 0. \tag{10.99}$$

If $\kappa_l(C) = 1$, *then for any* $S \subset [0,1]^d$ *with* $|S| = k \geq 2$ *one has* $\kappa_l(C(s_1, \ldots, s_k, \ldots, u_d) = 1$;

One can also show that if the copula C satisfies that it is MLTD, then

$$\kappa_l\left(C\left(s_{11},\ldots,s_{1k},u_{k+1},\ldots,u_d\right)\right)-\kappa_l\left(C\left(s_{21},\ldots,s_{2p},u_{p+1},\ldots,u_d\right)\right)\leq|S_2|-|S_1|.$$

- *Analogous results also hold for the case of upper tail order coefficients $\kappa_u(C)$ and the corresponding notions of positive upper orthant dependence and multivariate right tail increasing.*

Having discussed briefly the quantification of dependence through different criteria, we now move to models that can be constructed parametrically to capture such notions of dependence and will allow OpRisk practitioners to incorporate these features into their multirisk LDA structures.

10.5.3 A LINK BETWEEN ORTHANT EXTREME DEPENDENCE AND SPECTRAL MEASURES: TAIL DEPENDENCE

In this section, we finish the chapter with a brief description of some interesting links between the notion of tail dependence and the characterization of multivariate dependence in heavy tailed loss distributions as specified by the spectral measure. In particular we will consider basic links between orthant extreme dependence and the location of mass and quantity of mass on the unit sphere when the spectral measure characterizing a loss random vector is presented in polar co-ordinates.

As an illustration, we will consider the bivariate example for the upper tail dependence:

$$\begin{aligned}\lambda_u&=\lim_{u\uparrow 1}\mathbb{P}r\left(X_1>F_{X_1}^{-1}(u)|X_2>F_{X_2}^{-1}(u)\right)\\&=\lim_{u\uparrow 1}\frac{1-2u+C(u,u)}{1-u}.\end{aligned}\tag{10.100}$$

Now, observe that for a set A in the d-unit sphere $A\subset\mathbb{S}_d$ one can define the cone generated by A to be

$$\text{Cone}(A)=\left\{\boldsymbol{x}\in\mathbb{R}^d:||\boldsymbol{x}||>0,\frac{\boldsymbol{x}}{||\boldsymbol{x}||}\in A\right\}=\{r\boldsymbol{a}:r>0,\boldsymbol{a}\in A\}.\tag{10.101}$$

Next, of relevance to the examples discussed in this section, we observe that if one selects the set A to be the upper right quadrant mapped out by the angle $[0,\pi/2]$ that makes the cone Cone(A) correspond to an arc on the top right quadrant.

We now consider a particular class of heavy tailed loss process considered in detail in Peters and Shevchenko (2015) corresponding to the class of infinitely divisible multivariate α-stable loss random vectors. This involves considering the class of random vectors $\boldsymbol{X}\in\mathbb{R}^d$ which have an infinitely divisible law. Such loss random vectors allow us to make the following representation of their joint characteristic function, according the Levy-Khintchine formula as given in the followin definition.

Definition 10.41 (Levy-Khintchine Formula of Characteristic Function for Infinitely Divisible Loss Random Variables) *A probabilty law μ of a real-valued random vector is inifinitely divisible with characteristic exponent Ψ, given by*

$$\int_{\mathbb{R}^d} \exp\left(\mathrm{i} < \boldsymbol{\theta}, \boldsymbol{x} >\right) \mu(d\boldsymbol{x}) = \exp\left(-\Psi(\boldsymbol{\theta})\right), \ for\ \boldsymbol{\theta} \in \mathbb{R}^d \tag{10.102}$$

iff there exists a triple $(a, \Sigma, W(d\boldsymbol{x}))$, *where* $\boldsymbol{a} \in \mathbb{R}^d$, $\Sigma \in SPD(\mathbb{R}^d)$ *and* $W(d\boldsymbol{x})$ *is a measure concentrated on* $\mathbb{R}^d \setminus \{0\}$ *satisfying* $\int_{\mathbb{R}^d} (1 \wedge ||\boldsymbol{x}||^2)\, W(d\boldsymbol{x}) < \infty$, *s.t.*

$$\Psi(\boldsymbol{\theta}) = \mathrm{i} < \boldsymbol{a}, \boldsymbol{\theta} > + \frac{1}{2}\boldsymbol{\theta}\Sigma\boldsymbol{\theta}^T + \int_{\mathbb{R}^d} \left(1 - e^{\mathrm{i}<\boldsymbol{\theta},\boldsymbol{x}>} + \mathrm{i} < \boldsymbol{\theta}, \boldsymbol{x} > \mathbb{I}_{||\boldsymbol{x}||<1}\right) W(d\boldsymbol{x}) \tag{10.103}$$

■

We note the following additional remarks regarding this representation of the characteristic function of the multivariate loss random vector:

- Measure $W(d\boldsymbol{x})$ is known as the Levy measure and it is unique;
- Spectral measure can be shown to be directly linked to aspects of dependence of the random vector.

Next, we observe that one can map between the spectral measure $W(d\boldsymbol{x})$ defined on $\mathbb{R}^d$ and the spectral measure in polar co-ordinates on unit hyper-sphere $\Gamma(d\boldsymbol{s})$ on $\mathbb{S}_d$ as shown in the pure-jump process setting of Tempered Stable models, see detailed discussions in for instance Rosiński (2007). Then in polar co-ordinates, it was proven in Araujo and Evarist (1980) that there is a link between spectral measure and extreme regional (quadrant etc.) types of dependence as shown in the following Proposition 10.14

Proposition 10.14 (Spectral Measure to Quadrant Extreme Dependence) *Consider a set* $A \subset \mathbb{S}_d$, *and define the cone generated by* A *to be*

$$Cone(A) = \left\{\boldsymbol{x} \in \mathbb{R}^d : ||\boldsymbol{x}|| > 0, \frac{\boldsymbol{x}}{||\boldsymbol{x}||} \in A\right\} = \{r\boldsymbol{a} : r > 0, \boldsymbol{a} \in A\}, \tag{10.104}$$

then

$$\lim_{r\to\infty} \frac{\mathbb{P}r\left(\boldsymbol{X} \in Cone(A), ||\boldsymbol{X}|| > r\right)}{\mathbb{P}r\left(||\boldsymbol{X}|| > r\right)} = \frac{\Gamma(A)}{\Gamma(\mathbb{S}_d)}. \tag{10.105}$$

To further interpret this relationship we note that the mass that the spectral measure, in polar co-ordinates, $\Gamma(\cdot)$ assigns to A determines the tail behavior of $\boldsymbol{X}$ in the direction of A. From the perspective of risk management and insurance, a special case of this relationship has been studied, see for instance in Embrechts *et al.* (2009b), where they considered this type of result from Araujo and Evarist (1980) in elliptical families under context of multivariate regular variation. We recall the basic definition of multivariate regular variation of a function below.

Definition 10.42 (Multivariate Regular Variation) *A random vector* $\boldsymbol{X} = (X_1, \ldots, X_d)$ *is multivariate regularly varying with index* $-\beta < 0$ *if there exists*

- *a probability measure* μ;
- *a measurable function* $b : (0, \infty) \mapsto (0, \infty)$ *with* $\lim_{t \to \infty} b(t) = \infty$; *and*
- *a scalar* $q = q(b)$

such that for all $r > 0$

$$\lim_{t \to \infty} t\mathbb{P}r\left(||\boldsymbol{X}|| > rb(t), \frac{\boldsymbol{X}}{||\boldsymbol{X}||} \in B \right) = qr^{-\beta}\mu(B) \tag{10.106}$$

for any Borel set $B \subset \left\{(x_1, \ldots, x_d) \in \mathbb{R}^d | \ ||\boldsymbol{x}|| = 1\right\}$. *Then* $\boldsymbol{X}$ *is said to be* $MRV_d(-\beta)$. ■

Remark 10.22 *It can then be shown Barbe et al. (2006) and Resnick et al. (2004) that for* $\boldsymbol{X} \in MRV_d(-\beta)$ *for* $\beta > 0$ *one has*

$$q\left(\beta, ||\cdot||\right) = \lim_{x \to \infty} \frac{\mathbb{P}r\left(||\boldsymbol{X}|| > x\right)}{\mathbb{P}r(X_1 > x)} > 0. \tag{10.107}$$

This will have implications for extremal quadrant/orthant dependence.

Embrechts *et al.* (2009b) linked this result to the quantiles as detailed in the next Lemma 10.1

Lemma 10.1 (Multivariate Regular Variation Expressed Via Quantiles) *Consider a loss random vector* $\boldsymbol{X}$ *such that* $\boldsymbol{X} = (X_1, \ldots, X_d) \in MRV_d(-\beta)$ *with* $\beta > 0$ *and identically distributed marginals. Then for a measurable function* $\varphi : \mathbb{R}^d \mapsto \mathbb{R}$,

$$\lim_{x \to \infty} \frac{\mathbb{P}r\left(\varphi(\boldsymbol{X}) > x\right)}{\mathbb{P}r\left(X_1 > x\right)} = q_\varphi \in (0, \infty) \tag{10.108}$$

which implies that for quantile functions Q *at level* α *one has*

$$\lim_{\alpha \uparrow 1} \frac{Q_\alpha\left(\varphi(\boldsymbol{X})\right)}{Q_\alpha(X_1)} = q_\varphi. \tag{10.109}$$

In addition, Resnick *et al.* (2004) made connections between Multivariate Regular Variation and spectral measure of a random vector as follows:

- Consider the random d-vector $\boldsymbol{X} \in \mathbb{R}_+^d$ which has a distribution which satisfies $\boldsymbol{X} \in MVR(-\beta)$ with $\beta > 0$.
- Define the positive part of unit d-sphere with respect to an arbitrary norm $||\cdot|| : \mathbb{R}^d \mapsto \mathbb{R}_+$ according to

$$\boldsymbol{S}_{+,||\cdot||}^{d-1} = \left\{\boldsymbol{x} \in \mathbb{R}_+^d | \ ||\boldsymbol{x}|| = 1\right\}. \tag{10.110}$$

- Define the Radon measure (i.e. finite for all compact sub-sets) by $\mu_\beta(B)$ for all $B \subset [0, \infty]^d \setminus \{0\}$ relatively compact with $\mu_\beta(\partial B) = 0$.

Then one can show the following relationship between such a measure and the limiting behaviour of a MRV random vector:

$$\lim_{t\to\infty} t\mathbb{P}\mathrm{r}\left(\frac{\boldsymbol{X}}{b(t)} \in B\right) = \mu_\beta(B). \tag{10.111}$$

To further relate

$$\lim_{t\to\infty} t\mathbb{P}\mathrm{r}\left(\frac{\boldsymbol{X}}{b(t)} \in B\right) = \mu_\beta(B), \tag{10.112}$$

to the spectral measure in the case of r.v. which satisfies $\boldsymbol{X} \in MVR(-\beta)$, first choose the sets B according to:

$$B = \left\{\boldsymbol{x} \in [0,\infty]^d \middle|\ ||\boldsymbol{x}|| > r, \frac{\boldsymbol{x}}{||\boldsymbol{x}||} \in G\right\}$$

for $r > 0$ and a Borel set $G \in \boldsymbol{S}^{d-1}_{+,||\cdot||}$. Then by the definition of MVR one has the constant q (depending on β and norm $||\cdot||$) given by:

$$q(\beta, ||\cdot||)r^{-\beta}\mu(G) = \nu_\beta\left\{\boldsymbol{x} \in [0,\infty]^d \middle|\ ||\boldsymbol{x}|| > r, \frac{\boldsymbol{x}}{||\boldsymbol{x}||} \in G\right\}. \tag{10.113}$$

Then setting $\beta = 1$ and $r = 1$ one can express the spectral measure as

$$\Gamma_{||\cdot||}(G) = \mu_1\left\{\boldsymbol{x} \in [0,\infty]^d \middle|\ ||\boldsymbol{x}|| > 1, \frac{\boldsymbol{x}}{||\boldsymbol{x}||} \in G\right\} \tag{10.114}$$

which gives according to Barbe *et al.* (2006) the constant function

$$q(\beta, ||\cdot||) = \mu_1\left\{\boldsymbol{x} \in [0,\infty]^d \middle|\ ||\boldsymbol{x}^{1/\beta}|| > 1\right\}. \tag{10.115}$$

With these relationships one has the following theorem from Barbe *et al.* (2006).

Theorem 10.3 (Multivariate Regular Variation and Spectral Measure Representation) *Let the $\mathbb{R}^d_+$ valued random vector $\boldsymbol{X}$ with i.i.d. marginals satisfy $\boldsymbol{X} \in MVR(-\beta)$ with $\beta > 0$, then*

$$q(\beta, ||\cdot||) = \lim_{x\to\infty} \frac{\mathbb{P}r(||\boldsymbol{X}|| > x)}{\mathbb{P}r(X_1 > x)} = \int_{\boldsymbol{S}^{d-1}_{+,||\cdot||}} ||\boldsymbol{x}^{\frac{1}{\beta}}||^\beta \Gamma_{||\cdot||}(d\boldsymbol{x}). \tag{10.116}$$

Remark 10.23 *Note that the existence of such limits in the definition of the tail dependence functions can be linked to existence of multivariate regular variation on the copula distribution tails.*

CHAPTER ELEVEN

Dependence Models

In this chapter we build upon the notions of dependence modelling in OpRisk described in Chapter 10 by presenting a variety of parametric models that practitioners may consider for construction of LDA dependence frameworks. We discuss specifically many families of parametric copula that are of direct relevance to OpRisk practitioners - explaining the specification and features of the models, the estimation of the parameters in such models via Inference Functions for the Margins (IFM), and the sampling from such models in an LDA framework. The copula models include:

- Gaussian copula;
- Student-T copula; skew Student-T copula; grouped Student-T copula and generalised Student-T copula;
- Archimedean copulas: Frank, Clayton, Gumbel, Joe; Mixture Archimedean copula; Heirarchical Archimedean copulas; Nested Archimedean copulas; Outer and Inner power transformed Archimedean copula;
- Levy copula; Max-stable models and Self-Chaining copula;
- Common factor models and factor copulas.

We then conclude this chapter with several examples of LDA models with dependence incorporated, including between frequency and severity models as well as common factor formulations. These act as small illustrative case studies for practitioners to see a complete development of such models that can be utilized and extended for practical application.

11.1 Introduction to Parametric Dependence Modeling Through a Copula

The extensive interest in copula modeling can largely be attributed to the flexibility they offer to modeling a wide range of practical applications, particularly in financial mathematics, risk and insurance; see discussions in Genest *et al.* (2009a). The origins of copula theory can be traced back to Hoeffding's work on standardized distributions on the square

Fundamental Aspects of Operational Risk and Insurance Analytics: A Handbook of Operational Risk, First Edition. Marcelo G. Cruz, Gareth W. Peters, and Pavel V. Shevchenko.

$\left[-\frac{1}{2}, \frac{1}{2}\right] \times \left[-\frac{1}{2}, \frac{1}{2}\right]$, which was undertaken in Hoeffding (1994a, b). Following from this work, the term copula was first coined as a mathematical concept in Abel Sklar's theorem, see Sklar (1959), which proposed that one-dimensional distribution functions may be joined together by a copula function to form multivariate distribution functions. Several recently written discussions on the properties and origins of copula modeling have been developed; see, for instance, the monograph of Nelsen (1999), and the work of Schweizer (1991) and Sklar (1996).

It is reasonable to argue that the explosion of interest in copula modeling, beginning in the 1980s, was in most part due to advances in quantitative risk management methodology in the financial and insurance world. The creation of more complex derivative products and new guidelines on regulation (see discussion in McNeil *et al.* 2005, chapter 1) contributed heavily to the need for risk management developments.

To understand the formal definition of a copula distribution, see Definition 11.1 and then we present the representation result in Theorem 11.1, see Sklar (1959). Sklar's theorem highlights one of the key attractions for practitioners for the use of copula models that involves the separation of a multivariate distribution into its marginal distributions and the dependence structure between the margins.

Definition 11.1 (Copula Distribution) *A d-dimensional copula is a multivariate cumulative distribution function C with uniform* $[0, 1]$ *margins such that* $C : [0, 1]^d \to [0, 1]$ *and the distribution C satisfies the following:*

- $C(u_1, \ldots, u_d) = 0$ *whenever* $u_i = 0$ *for at least one* $i \in \{1, \ldots, d\}$;
- $C(u_1, \ldots, u_d) = u_i$ *if* $u_j = 1$ *for all* $j = 1, \ldots, d$ *and* $j \neq i$;
- *C is quasi-monotone on its support* $[0, 1]^d$.

■

The definition of a n-dimensional copula distribution, denoted generically by $C(u_1, u_2, \ldots, u_n)$, was given as any distribution taking support on the unit d-hypercube that satisfies the following two conditions; see Roger (2006, definition 2.10.6):

1. For every vector $\boldsymbol{u} = (u_1, u_2, \ldots, u_n) \in [0, 1]^n$, one can show that $C(\boldsymbol{u}) = 0$ if at least one coordinate of $\boldsymbol{u}$ is 0;
2. In addition for every $\boldsymbol{a}$ and $\boldsymbol{b}$ in $[0, 1]^n$, such that $\boldsymbol{a} \leq \boldsymbol{b}$ such that for each $a_i < b_i$ for all $i \in \{1, 2, \ldots, n\}$ the following condition on the volume for copula C is satisfied, $V_C([\boldsymbol{a}, \boldsymbol{b}]) \geq 0$. In this notation, the volume of an n-box is given by

$$\begin{aligned} V_C([\boldsymbol{a}, \boldsymbol{b}]) &= \sum \operatorname{sgn}(\boldsymbol{v}) C(\boldsymbol{v}) \\ &= \triangle_{a_1}^{b_1} \triangle_{a_2}^{b_2} \cdots \triangle_{a_n}^{b_n} C(\boldsymbol{v}), \end{aligned} \tag{11.1}$$

where the sum is taken over all vertices $\boldsymbol{v}$ of the n-box $[\boldsymbol{a}, \boldsymbol{b}]$ and $\operatorname{sgn}(\boldsymbol{v}) = 1$ if $v_k = a_k$ for an even number of ks of $\operatorname{sgn}(\boldsymbol{v}) = -1$ if $v_k = a_k$ for an odd number of ks. In addition, one defines the notation

$$\begin{aligned} \triangle_{a_k}^{b_k} C(\boldsymbol{u}) = C(u_1, u_2, \cdots, u_{k-1}, b_k, u_{k+1}, \ldots, u_n) \\ - C(u_1, u_2, \ldots, u_{k-1}, a_k, u_{k+1}, \ldots, u_n). \end{aligned} \tag{11.2}$$

Sklar's Theorem (11.3) provides the foundation to the study of copulae by proving that any multivariate distribution with continuous margins has a unique copula representation.

Theorem 11.1 (Sklar's Theorem) *Consider a d-dimensional distribution H with marginals $F_1, \ldots, F_d$. There exists a copula C, such that*

$$H(x_1, \ldots, x_d) = C(F_1(x_1), \ldots, F_d(x_d)) \tag{11.3}$$

for all $x_i \in (-\infty, \infty)$, $i \in 1, \ldots, d$. Furthermore, if F_i is continuous for all $i = 1, \ldots, d$, then C is unique; otherwise, C is uniquely determined only on $RanF_1 \times \cdots \times RanF_d$, where $RanF_i$ denotes the range of the distribution F_i.

It is also interesting to consider the alternative statement of Sklar's theorem introduced in McNeil and Nešlehová (2009) with regard to survival functions of a multivariate distribution. To present this representation, we first provide Definition 11.2 for survival functions, see McNeil and Nešlehová (2009, lemma 1).

Definition 11.2 (Multivariate Survival Functions) *A survival function $\overline{H}$ of a probability distribution H is a mapping $\overline{H} : \mathbb{R}^d \mapsto [0, 1]$ if and only if it satisfies*

- $\overline{H}(-\infty, \ldots, -\infty) = 1$ *and* $\overline{H}(\boldsymbol{x}) = 0$ *if* $x_i = \infty$ *for at least one index* $i \in \{1, 2, \ldots, d\}$*;*
- $\overline{H}$ *is a right continuous function such that for all* $\boldsymbol{x} \in \mathbb{R}^d$ *one has*

$$\forall \epsilon > 0,\ \exists \delta > 0,\ \forall \boldsymbol{y} \geq \boldsymbol{x}\ \ ||\boldsymbol{y} - \boldsymbol{x}||_1 < \delta \Rightarrow |\overline{H}(\boldsymbol{y}) - \overline{H}(\boldsymbol{x})| < \epsilon. \tag{11.4}$$

- $\overline{H}(-\boldsymbol{x})$ *is quasi-monotone on* $\mathbb{R}^d$.

■

One can then re-express Sklar's theorem in terms of survival functions of a multivariate distribution according to the following result in Theorem 11.2; see discussion in McNeil and Nešlehová (2009, theorem 2.1).

Theorem 11.2 (Sklar's Theorem Expressed via Survival Function) *Considering a d-dimensional survival function $\overline{H}$ with marginal distribution survival functions $\overline{F}_{X_i}$ for $i \in \{1, 2, \ldots, d\}$, then there exists a copula $\overline{C}$, called the survival copula of $\overline{H}$ such that*

$$\overline{H}(\boldsymbol{x}) = \overline{C}\left(\overline{F}_{X_1}(x_1), \ldots, \overline{F}_{X_d}(x_d)\right), \quad \forall \boldsymbol{x} \in \mathbb{R}^d, \tag{11.5}$$

or conversely one has

$$\overline{C}(\boldsymbol{u}) = \overline{H}\left(\overline{F}_{X_1}^{-1}(u_1), \ldots, \overline{F}_{X_d}^{-1}(u_d)\right), \quad \forall \boldsymbol{u} \in \mathcal{D}, \tag{11.6}$$

with $\mathcal{D} = \left\{\boldsymbol{u} \in [0, 1]^d : \boldsymbol{u} \in ran\overline{F}_{X_1} \times \ldots \times ran\overline{F}_{X_d}\right\}$. The survival copula $\overline{C}$ is uniquely determined on the support $\mathcal{D}$. Conversely, given a copula $\overline{C}$ and marginal survival functions $\overline{F}_{X_i}$ for $i \in \{1, 2, \ldots, d\}$, then the multivariate survival function $\overline{H}$ is uniquely given by Equation 11.5.

Considering Sklar's theorem, it is readily apparent that copula models provide a mechanism to model the marginal behavior of each loss process (severity, frequency, annual loss, or even second-order characteristics such as intensity function) and then separately to focus on developing hypotheses regarding the possible dependence structures between these values.

The modeling of dependence through this flexible copula framework has exploded in recent years, fuelled in part by the flexibility such a bottom up modeling framework provides as well as significant progress in simulation and estimation of such models in complex settings; see excellent book-length reviews in Nelsen (1997), Joe (1997), Cherubini *et al.* (2004), Denuit *et al.* (2005), and the recent addition on vine copulae Dorota (2010) and the tutorial introductions in Meucci (2011), Genest and Favre (2007), Schmidt (2006), Bouyé *et al.* (2000), Embrechts *et al.* (2003), Frees and Valdez (1998). It is therefore not the intention of this chapter to review all these topics, instead we focus on key components of this literature of relevance directly to OpRisk practitioners.

The motivation for copula modeling is pervasive in risk management, and many papers have been written espousing the attributes of such a modeling approach. In addition there have been several healthy skeptical articles highlighting areas where copula modeling needs a stronger foundation or exploration to explain some existing deficiencies in such modeling perspectives. We will discuss these later; first, we note the work of Embrechts *et al.* (2002) in which the authors argue for copula approaches over linear correlation for the modeling of dependency for risk management. In particular, the authors point out the pitfalls of using linear correlation in the non-Gaussian world of finance and insurance. Hence, *beyond elliptical multivariate models* we have the following fallacies:

- **Fallacy 1**. Marginal distributions and correlation determine the joint distribution;
- **Fallacy 2**. Given marginal distributions F_1 and F_2 for X and Y, all linear correlations between -1 and 1 can be attained through suitable specification of the joint distribution;
- **Fallacy 3**. The worst case VaR (quantile) for a linear portfolio $X + Y$ occurs when $\rho(X, Y)$ is maximal, that is, X and Y are comonotonic.

As noted earlier, growth of copula literature continues for a range of different disciplines, such as hydrology—Genest and Favre (2007), climate research—Schoelzel *et al.* (2008), ecology—Hossack *et al.* (2014) and neuroscience—Onken *et al.* (2009) to name but a few. There has also been some healthy scepticism of the copula framework. The most notable is Mikosch (2006a), who cited a concern that copulae were being viewed as the solution to all problems in stochastic dependence modeling, whereas in his view "copulas do not contribute to a better understanding of multivariate extremes". There were numerous responses from leaders in the copula field to Mikosch's attack, such as Genest and Rémillard (2006); Embrechts (2006); Joe (2006); de Vries and Zhou (2006); Lindner (2006); Peng (2006) and Segers (2006)—leading to a rejoinder by Mikosch, see Mikosch (2006b). Embrechts (2009) sums up the responses best in his personal review of copulae shortly after:

> Copulas form a most useful concept for a lot of applied modeling, they do not yield, however, a panacea for the construction of useful and well-understood multivariate dfs, and much less for multivariate stochastic processes. But none of the copula experts makes these claims.

Without entering into this debate, we simply highlight some of the pros and cons of copula modeling that one should be aware of when embarking on the application of such models in practical OpRisk settings

Modeling with a Copula Pros:

- Separating out the modeling of the marginals and the dependence structure allows for more flexibility in the complete multivariate model;
- The dependence structure as summarized by a copula is invariant under increasing and continuous transformations of the marginals;
- The tail characteristics within the dependence structure can be explicitly modeled using well-known and interpretable parametric models, for example, Archimedean copulae;
- High-dimensional copulae can be reduced to the composition of lower-dimensional building block copulae, for example, pair-copula constructions, to create extremely flexible models of complex dependence structures.

Modeling with a Copula Cons:

- Which copula to choose? Sometimes it is not easy to say which parametric copula fits a dataset best since some copulae may provide a better fit near the center and others near the tails. However, by focusing on models with suitable characteristics for the application at hand and using goodness-of-fit tests and model selection criteria, for example, AIC, BIC, or CIC, one can overcome this issue;
- As with any statistical model, ignorance on the behalf of practitioners can lead to dangerous oversimplification and reliance on inappropriate models.

Thus, when applying these models in practice it is of the utmost importance to carefully consider the assumptions one is making. The key focus in this research is on combining suitable marginal models, that is, with the capacity to mode skewness and tail-heaviness flexibly, with a model of the dependence structure that captures the upper and lower multivariate tail characteristics asymmetrically.

Theorem 11.3 (Negative Regression Dependence via a Copula: Bivariate Case) *Consider two loss random variables X_1 and X_2 with marginal loss distributions $X_i \sim F_{X_i}$ and a copula dependence given by the copula distribution C. The joint distribution of the loss random variables $F_{X_1,X_2}(x_1, x_2)$ will satisfy that X_1 and X_2 have negative regression dependence if one rewrites the definition in one of two ways in terms of the copula distribution:*

1. *If the first derivative of the copula distribution exists, then one can express the conditions on the copula to satisfy negative regression dependence by considering for $u_1 = F_{X_1}(y)$ and $u_2 = F_{X_2}(x - y)$ the relationship*

$$\frac{1 - \frac{\partial^2}{\partial u_1 \partial u_2} C(u_1, u_2)}{\overline{F}_{X_2}(x - y)} = O(1). \tag{11.7}$$

This holds uniformly for all $y \in [x_0, x]$ for some large $x_0 > 0$;

2. *The alternative way of expressing the conditions on a copula distribution to satisfy negative regression dependence involves the copula density when it exists given by the mixed second-order partial derivatives:*

$$c(u_1, u_2) = \frac{\partial^2}{\partial u_1 \partial u_2} C(u_1, u_2), \tag{11.8}$$

which when the copula density exists and it is uniformly bounded by a constant $M > 0$ *for all* $(u_1, u_2) \in [c, 1] \times [c, 1]$ *then the expression of the copula density conditions required for negative regression dependence according to the relationship can be given by*

$$\frac{\int_{x-y}^{\infty} c\left(F_{X_1}(y), F_{X_2}(z)\right) dF_{X_2}(z)}{\overline{F}_{X_2}(x-y)} \leq M. \tag{11.9}$$

We complete this section on basic introduction to the notion of a copula model by discussing the stochastic ordering of copula models on the unit hyper-cube as detailed in Definition 11.3; see Nelsen (1999, p. 34).

Definition 11.3 (Copula Stochastic Orderings) *A copula distribution* C_1 *is said to be smaller that another copula distribution* C_2 *written in stochastic ordering* $C_1 \prec C_2$ *(or* C_2 *is larger than* C_1 *with* $C_2 \succ C1$*) if the following holds*

$$\forall (u_1, u_2, \ldots, u_n, \ldots, u_N) \in [0, 1]^N, \quad C_1(u_1, \ldots, u_n, \ldots, u_N) \leq C_2(u_1, \ldots, u_n, \ldots, u_N). \tag{11.10}$$

■

It is precisely this stochastic ordering that gives rise to the general notion of concordance, which has already been partially introduced in the section on measures of dependence where definitions for dependence measures such as rank correlations were introduced. More formally, the notion of concordance as discussed in Scarsini (1984) and Joe (1990). The notion of concordance the simple bivariate case involves considering two random variables X_1 and X_2 that are concordant if large values of X_1 occur when large values of X_2 occur and vice versa. In Scarsini (1984) and then extended in Joe (1990), the general definition of concordance is provided in multivariate settings, as detailed in Definition 11.3.

Remark 11.1 *A multivariate concordance ordering is defined by the ordering of the multivariate distribution functions and the corresponding survival functions.*

Using these multivariate notions of concordance, Joe (1990) provides several interesting extensions to multivariate measures of dependence such as multivariate extensions to Spearman's rho, Kendall's tau, and Blomqvist's beta.

In defining the stochastic ordering of copulas, it is often useful to note two important members of the copula distributions on the unit cube; these are known as the Frechet bounds copulae and they often have their distributions denoted by C^- and C^+ for the lower and upper bound copula, respectively. The Frechet bound copulae distributions are detailed in Definition 11.4; see discussion in Bouyé *et al.* (2000).

Definition 11.4 (Copula Frechet Bound Distributions) *The lower copula Frechet bound distribution is given by*

$$C^{-}\left(u_1,\ldots,u_n,\ldots,u_N\right) = \max\left(\sum_{n=1}^{N} u_n - N + 1, 0\right) \tag{11.11}$$

and the upper copula Frechet bound distribution is given by

$$C^{+}\left(u_1,\ldots,u_n,\ldots,u_N\right) = \min\left(u_1,u_2,\ldots,u_n,\ldots,u_N\right). \tag{11.12}$$

Then one can show that all copula models satisfy the following stochastic ordering with respect to the Frechet bound copulae,

$$C^{-} \prec C \prec C^{+}. \tag{11.13}$$

Note that clearly C^{-} is no longer strictly a copula for $N > 2$. ■

Another useful generic property of copula models is the invariance of such models to strictly increasing transformations as detailed in Proposition 11.1.

Proposition 11.1 (Copula Distribution Invariance) *The copula distribution of a random vector $(X_1, X_2, \ldots, X_d)$ is invariant under strictly increasing transformations such that*

$$C_{X_1,\ldots,X_d} = C_{h_1(X_1),\ldots,h_d(X_d)}, \quad \text{if} \quad \partial_x h_i\left(X_i\right) > 0. \tag{11.14}$$

We finish this section by mentioning the notion of concordance that will lead to the generic copula representation of the dependence measures such as rank correlations being expressed with respect to a copula distribution. The notion of a measure of concordance is presented in Definition 11.5; see Nelsen (1999).

Definition 11.5 (Measures of Concordance and Copulae) *A measure of concordance, denoted $\kappa(X_1, X_2)$, between to random variables X_1 and X_2 with a copula dependence distribution C, satisfies the following conditions:*

1. *The concordance measure between two random variables X_1 and X_2 given by $\kappa(X_1, X_2)$ must be defined for every pair of continuous random variables X_1 and X_2;*
2. *The concordance measure*

   $$-1 = \kappa(X, -X) \leq \kappa_C(X_1, X_2) \leq \kappa(X, X) = 1, \tag{11.15}$$

   where C is the copula between random variables X_1 and X_2;
3. *The concordance measure between two random variables X_1 and X_2 given by $\kappa(X_1, X_2)$ must satisfy the following symmetry condition:*

   $$\kappa(X_1, X_2) = \kappa(X_2, X_1). \tag{11.16}$$

4. *If the two random variables X_1 and X_2 are independent and follow an independence (product copula) specification, then the measure of their concordance given by $\kappa(X_1, X_2)$ must satisfy that*

$$\kappa(X_1, X_2) = 0, \quad \text{if } X_1 \perp X_2. \tag{11.17}$$

5. *The concordance measure between two random variables X_1 and X_2 given by $\kappa(X_1, X_2)$ must satisfy the following sign conditions:*

$$\kappa(-X_1, X_2) = \kappa(X_1, -X_2) = -\kappa(X_1, X_2). \tag{11.18}$$

6. *The ordering of concordance between two random variables X_1 and X_2 under two different copulae distributions C_1 and C_2 such that $\kappa_{C_1}(X_1, X_2) \leq \kappa_{C_1}(X_1, X_2)$ implies the stochastic order of the two copulae distributions such that $C_1 \prec C_2$.*

■

There are several excellent discussions and multivariate extensions of the notion of concordance that may be found in Nelsen (2002), Joe (1990), and Dolati and Úbeda-Flores (2006).

Using the notion of measures of concordance, one can observe that the rank measures given by Kendall's tau, Spearman's rho, and Blomqvist's beta discussed previously satisfy these conditions and can be given with respect to a copula distribution for two random variables X_1 and X_2 by

$$\begin{aligned}
\tau &= 4 \int_{[0,1]} \int_{[0,1]} C(u_1, u_2)\, dC(u_1, u_2) - 1, \\
\rho &= 12 \int_{[0,1]} \int_{[0,1]} u_1 u_2 dC(u_1, u_2) - 3, \\
\beta &= 4C\left(\frac{1}{2}, \frac{1}{2}\right) - 1.
\end{aligned} \tag{11.19}$$

See Schweizer and Wolff (1981).

As mentioned one may generalize these measures of association to higher dimensions; next we consider what was termed a directional measure of association ρ-coefficient in Nelsen and Úbeda-Flores (2012). Considering a three-dimensional random vector $\boldsymbol{X} = (X_1, X_2, X_3)$, then one may define the 3-copula ρ-directional dependence by the following result in Definition 11.6.

Definition 11.6 (3-Copula ρ-Directional Dependence) *Consider a random vector $\boldsymbol{X} = (X_1, X_2, X_3)$ with $\boldsymbol{X} \in \mathbb{R}^3$ and associated three-dimensional copula $C_{\boldsymbol{X}}$. Then for any direction $(\alpha_1, \alpha_2, \alpha_3)$ characterized by the vectors $\alpha_i \in \{-1, 1\}$ for $i \in \{1, 2, 3\}$, one has the ρ-directional dependence given by*

$$\rho_{X_1,X_2,X_3}^{(\alpha_1,\alpha_2,\alpha_3)} = \frac{\alpha_1\alpha_2\rho_{X_1,X_x} + \alpha_2\alpha_3\rho_{X_2,X_3} + \alpha_3\alpha_1\rho_{X_3,X_1}}{3} + \alpha_1\alpha_2\alpha_3 \frac{\rho^{+}_{X_1,X_2,X_3} - \rho^{-}_{X_1,X_2,X_3}}{2} \tag{11.20}$$

with pairwise Spearman's rho and

$$
\begin{aligned}
\rho^{+}_{X_1,X_2,X_3}(C_X) &= 8 \int_{[0,1]^3} \overline{C}_X(u,v,w)dudvdw - 1, \\
\rho^{-}_{X_1,X_2,X_3}(C_X) &= 8 \int_{[0,1]^3} C_X(u,v,w)dudvdw - 1.
\end{aligned}
\tag{11.21}
$$

■

Remark 11.2 *The eight vectors that characterize directions* $(\alpha_1, \alpha_2, \alpha_3)$ *where* $\alpha_i \in \{-1, 1\}$ *for* $i \in \{1, 2, 3\}$ *in* $[0, 1]^3$ *allow one to utilize the* ρ*-directional dependence to measure directional dependence in different quadrants. For instance, if* $\rho_{\mathbf{X}}^{(-1,-1,1)}$ *or* $\rho_{\mathbf{X}}^{(1,1,-1)}$ *are positive, then there will be positive dependence in the direction of* $(-1, -1, 1)$ *or* $(1, 1, -1)$*, hence one would expect large (small) values of* X_1 *and* X_2 *to occur with small (large) values of* X_3*, that is,* $\rho_{X_1,X_2} > 0$ *with* $\rho_{X_1,X_3} < 0$ *and* $\rho_{X_2,X_3} < 0$.

11.2 Copula Model Families for OpRisk

There is a vast collection of different parametric copulae in the literature, each with associated dependence features. The monograph Nelsen (1999) provides a detailed mathematical background of many important copulae, including a well-explained introduction to the basis copula function building blocks for many families of copulae. In addition, there are many useful papers reviewing the different families of copulae available to the practitioner, such as (Bouyé *et al.*, 2000; Schmidt, 2006; Trivedi and Zimmer, 2007; Durante and Sempi, 2010). Hence, in this chapter, we will only focus on a small fraction of copula models that have been found to be useful for OpRisk practitioners in practice and have well studied and convenient properties.

In the case of multivariate distributions that may contain some marginals with discrete support, Genest and Neslehova (2007) discuss the issues associated with modeling via copulas under such situations. The main consequence will be evident directly from Sklar's Theorem (11.3), from which it is readily apparent that the copula representation of a multivariate distribution will no longer be guaranteed to be unique when the marginal distributions are not continuous.

As was noted in the introduction to copula functions, they have become popular and flexible tools in modeling multivariate dependence among risks. In general, a copula is a d-dimensional multivariate distribution on $[0, 1]^d$ with uniform marginal distributions. Given a copula function $C(u_1, \ldots, u_d)$, the joint distribution of random variables $Y_1, \ldots, Y_d$ with marginal distributions $F_1(y_1), \ldots, F_d(y_d)$ can be constructed as

$$
F(y_1, \ldots, y_d) = C(F_1(y_1), \ldots, F_d(y_d)). \tag{11.22}
$$

The well-known theorem due to Sklar, published in 1959, says that one can always find a unique copula $C(\cdot)$ for a joint distribution with given continuous marginals. Note that in the case of discrete distributions this copula may not be unique. Given (11.22), the joint density can be written as

$$f(y_1, \ldots, y_d) = c(F_1(y_1), \ldots, F_d(y_d)) \prod_{i=1}^{d} f_i(y_i), \tag{11.23}$$

where $c(\cdot)$ is a copula density and $f_1(y_1), \ldots, f_d(y_d)$ are marginal densities. The copula density $c(F_1(y_1), \ldots, F_d(y_d))$ is given by

$$c(u_1, u_2, \ldots, u_n, \ldots, u_d) = \frac{\partial C(u_1, u_2, \ldots, u_n, \ldots, u_d)}{\partial u_1 \partial u_2 \cdots \partial u_n \cdots \partial u_d}. \tag{11.24}$$

As a short note, it will often be useful when evaluating the copula density to recall the univariate and multivariate chain rule differentiations for composite functions known as Faà di Bruno's Formula; see Faa di Bruno (1857) and discussions in, for example, Constantine and Savits (1996) and Roman (1980). Before stating Faà di Bruno's Formula for differentiation of multivariate composite functions via a generalized chain rule, it will be convenient notationally to present such results with respect to Bell polynomials.

Definition 11.7 (Bell Polynomial) *The Bell polynomial with arguments n and k is given by*

$$B_{n,k}(x_1, x_2, \ldots, x_{n-k+1}) = \sum \frac{n!}{j_1! j_2! \cdots j_{n-k+1}!} \left(\frac{x_1}{1!}\right)^{j_1} \left(\frac{x_2}{2!}\right)^{j_2} \cdots \left(\frac{x_{n-k+1}}{(n-k+1)!}\right)^{j_{n-k+1}}, \tag{11.25}$$

where the sum is taken over all sequences j_1, j_2, j_{n-k+1} *of non-negative integers such that* $j_1 + j_2 + \cdots = k$ *and* $j_1 + 2j_2 + 3j_3 + \cdots = n$. ■

These polynomials are then utilized to simplify the expressions for the differentiation of composite functions in Faà di Bruno's formula as detailed next; see Riordan (1946) and Mihoubi (2008) for details.

Definition 11.8 (Univariate Faà di Bruno's Formula Composite Functions) *If f and g are functions with a sufficient number of derivatives, then*

$$\frac{d^n}{dx^n} f(g(x)) = \sum_{k=0}^{n} f^{(k)}(g(x)) B_{n,k}\left(g'(x), g''(x), \ldots, g^{n-k+1}(x)\right), \tag{11.26}$$

where $B_{n,k}$ are the Bell polynomials, defined earlier. ■

The multivariate Faà di Bruno's Formula is difficult and involved; in the following, we show an example of the bivariate result that is of relevance to this chapter in the form given in Definition 11.9 and then the general result; see detailed discussions in Leipnik and Pearce (2007).

Definition 11.9 (Bivariate Faà di Bruno's Formula Composite Functions) *Consider the first-order multivariate chain rule for* $G(\boldsymbol{z}) = F(\boldsymbol{u}(\boldsymbol{z}))$ *with scalar function* F, $\boldsymbol{u}(\boldsymbol{z}) = (u_1(\boldsymbol{z}), \ldots, u_M(\boldsymbol{z}))$ *and* $\boldsymbol{z} = (z_1, \ldots, z_N)$. *Then one has*

$$\frac{\partial G}{\partial z_k} = \sum_{j=1}^{M} \frac{\partial F(\boldsymbol{u})}{\partial u_j} \frac{\partial u_j(\boldsymbol{z})}{\partial z_k} = (D_{\boldsymbol{u}})(D_{\boldsymbol{zu}})_{\cdot k} \tag{11.27}$$

where $(D_{\boldsymbol{zu}})_{\cdot k}$ *is the k-th column of the* $(M \times N)$ *first derivative matrix* $D_{\boldsymbol{zu}}$. *Now we can generalize this higher-order derivatives by supposing* $F(u_1, u_2)$ *has continuous derivatives up to order* $(p+1, p+1)$ *and* $u_i(z_1, z_2)$ *for* $i \in \{1, 2\}$ *have continuous derivatives up to order* $(p_1 + 1, p_2 + 1)$ *on appropriate domains. Then define the sets*

$$\begin{aligned} A(\boldsymbol{p}) &= (\{0, 1, \ldots, p_1\} \times \{0, 1, \ldots, p_2\}) \notin (\{0\} \times \{0\}), \\ C(\boldsymbol{p}) &= \{(m, m') \in A(p, p) : m + m' \leq p\}, \end{aligned} \tag{11.28}$$

and define the function

$$B_{\phi}(u_i(\boldsymbol{z})) = \prod_{\boldsymbol{n} \in A(p)} \left\{ \frac{1}{\phi(\boldsymbol{n})!} \left(\frac{D_z^{\boldsymbol{n}} u_i(z)}{n_1! n_2!} \right)^{\phi(\boldsymbol{n})} \right\}, \quad i \in \{1, 2\}, \tag{11.29}$$

where $\phi : A(\boldsymbol{p}) \mapsto \{0, 1, 2, \ldots, m\}$. *Then, one has the bivariate composite function chain rule if* $G(z_1, z_2) = F(u_1(z_1, z_2), u_2(z_1, z_2))$ *given by*

$$\frac{\partial^p G(\boldsymbol{z})}{\partial z_1^{p_1} \partial z_2^{p_2}} = p_1! p_2! \sum_{\boldsymbol{m} \in C(\boldsymbol{p})} \boldsymbol{D}_{\boldsymbol{u}}^{\boldsymbol{m}} \boldsymbol{F} \sum_{(\phi, \phi') \in V(\boldsymbol{m})} B_{\phi}(u_1(\boldsymbol{z}) B_{\phi'}(u_2(\boldsymbol{z})) \tag{11.30}$$

with

$$V(m, m') = \{(\phi, \phi') : \phi \in T(m), \phi' \in T(m'), \tau_i(\phi, \phi') = p_i \textit{ for } i \in \{1, 2\}\}, \tag{11.31}$$

with $T(m)$ *defined with respect to the family of maps* $U(m)$ *with regard to* ϕ *such that*

$$T(m) = \left\{ \phi \in U(m) : \sum_{\boldsymbol{n} \in A(\boldsymbol{p})} \phi(\boldsymbol{n}) = m \right\} \tag{11.32}$$

and finally $\tau_i(\phi, \phi')$ *given by*

$$\tau_i(\phi, \phi') = \sum_{\boldsymbol{n} \in A(\boldsymbol{p})} n_i [\phi(\boldsymbol{n}) + \phi'(\boldsymbol{n})], \quad i \in \{1, 2\}. \tag{11.33}$$

■

As in Leipnik and Pearce (2007), one can then extend this notation to the general multivariate setting to obtain the chain rule for general multivariate composite functions using notation extensions:

- Define $\boldsymbol{m} = (m_1, \ldots, m_M)$, $\boldsymbol{\phi} = (\phi_1, \ldots, \phi_M)$, $\boldsymbol{p} = (p_1, \ldots, p_N)$ with $\sum_{i=1}^{N} = p$;
- Define the sets

$$A(\boldsymbol{p}) = (\{0, \ldots, p_1\} \times \cdots \times \{0, \ldots, p_N\}) \notin (\{0\} \times \ldots \times \{0\}),$$
$$C(\boldsymbol{p}) = \left\{ \boldsymbol{m} \in A(p, \ldots, p) : \sum_{l=1}^{M} m_l \leq p \right\}. \tag{11.34}$$

- Define ϕ_i mapping as

$$\phi_i : A(\boldsymbol{p}) \mapsto \{0, 1, \ldots, m_i\}. \tag{11.35}$$

- Define $V(\boldsymbol{m})$ as the set of mappings given by

$$V(\boldsymbol{m}) = \left\{ \phi \,\middle|\, \left\{ \begin{array}{ll} \sum_{\boldsymbol{n} \in A(\boldsymbol{p})} \phi_i(\boldsymbol{n}) = m_i, & i \in \{1, \ldots, M\} \\ \sum_{\boldsymbol{n} \in A(\boldsymbol{p})} n_l \sum_{j=1}^{M} \phi_j(\boldsymbol{n}) = p_l, & l \in \{1, \ldots, N\} \end{array} \right\} \right\}. \tag{11.36}$$

The multivariate Faà di Bruno's formula is then given according to Definition 11.10; see Leipnik and Pearce (2007, theorem 4.2).

Definition 11.10 (Multivariate Faà di Bruno's Formula Composite Functions) *Consider a function $F(\boldsymbol{u})$, which is in the class $C^{(p+1)}$ and $u_i(\boldsymbol{z})$ for $i \in (1, 2, \ldots, M)$ each have continuous derivatives to order $(p_1 + 1, \ldots, p_N + 1)$ on appropriate domains. Then one may define*

$$B_{\phi_i}(u_i(\boldsymbol{z})) = \prod_{\boldsymbol{n} \in A(\boldsymbol{p})} \left\{ \frac{1}{\phi_i(\boldsymbol{n})!} \left(\frac{\boldsymbol{D}_{\boldsymbol{z}}^{\boldsymbol{n}} u_i(\boldsymbol{z})}{\prod_{l=1}^{M} n_l!} \right)^{\phi_i(\boldsymbol{n})} \right\}, \quad i \in \{1, 2, \ldots, M\} \tag{11.37}$$

and if $G(\boldsymbol{z}) = F(\boldsymbol{u}(\boldsymbol{z}))$ and $\boldsymbol{p} \neq \boldsymbol{0}$ then one has the mixed derivatives given by

$$\frac{\partial^p G(\boldsymbol{z})}{\partial z_1^{p_1} \ldots \partial z_N^{p_N}} = \left(\prod_{j=1}^{N} p_j! \right) \sum_{\boldsymbol{c} \in C(\boldsymbol{p})} \boldsymbol{D}_{\boldsymbol{u}}^{\boldsymbol{m}} F \sum_{\phi \in V(\boldsymbol{m})} \prod_{l=1}^{M} B_{\phi_l}(u_l(\boldsymbol{z})). \tag{11.38}$$

■

Remark 11.3 *Using these composite function chain rules, one can obtain density representations for most copula families when they exist.*

Returning back to general copula distribution families, we note that there are many different copulas discussed in the literature and these can be found in many textbooks; for example, see McNeil *et al.* (2005, section 5). Next, for illustration of the concept and notation, we give definitions for the Gaussian, Clayton, Gumbel, and t copulas (Clayton and Gumbel copulas belong to a so-called family of the Archimedean copulas). An important difference between these three copulas is that they each display different tail dependence properties. The Gaussian copula has no upper and lower tail dependence, the Clayton copula will produce greater lower tail dependence as ρ increases, whereas the Gumbel copula will produce greater upper tail dependence as ρ increases.

For a general description of copulas and their properties in the context of financial risk modeling, see McNeil *et al.* (2005, chapter 5) and Panjer (2006, chapter 8); multivariate extreme value copulas are described in McNeil *et al.* (2005, sections 7.5 and 7.6). Before proceeding to definitions for different parametric copula models, we make the following observation regarding simulation from copula models. This will be required for practitioners to work with copula models in OpRisk settings, where Monte Carlo simulation is required for simulation of bank capital. In general any copula distribution can be simulated via the following procedure in Algorithm 11.1. In general we note that this is not the optimal approach to simulate from many families of copula models, we will provide more specialized methods when we present each particular parametric copula model.

- Consider general d-copula C, let the k-dim marginals of C be given by

$$C_k(u_1, \dots, u_k) = C(u_1, \dots, u_k, 1, \dots, 1), \ k = 2, \dots, d-1, \tag{11.39}$$

  with $C_1(u_1) = u_1$ and $C_d(u_1, \dots, u_d) = C(u_1, \dots, u_d)$;
- Let $U_1, \dots, U_d$ have joint distribution C. Then the conditional distribution of U_k given $U_1, \dots, U_{k-1}$ is given by

$$\begin{aligned} C_k\left(u_k | u_1, \dots, u_{k-1}\right) &= \mathbb{P}\mathrm{r}\left(U_k \leq u_k | U_1 = u_1, \dots, U_{k-1} = u_{k-1}\right) \\ &= \frac{\partial^{k-1} C_k(u_1, \dots, u_k)}{\partial u_1 \dots \partial u_{k-1}} \Big/ \frac{\partial^{k-1} C_{k-1}(u_1, \dots, u_{k-1})}{\partial u_1 \dots \partial u_{k-1}}. \end{aligned}$$

Algorithm 11.1 (General Copula Simulation Method)

Step 1 *Simulate a random variate* u_1 *from Uniform*$(0, 1)$*;*

Step 2 *Simulate a random variate* u_2 *from* $C_2(\cdot | u_1)$*;*

$\vdots$

Step d *Simulate a random variate* u_d *from* $C_d(\cdot | u_1, \dots, u_{d-1})$.

We begin the discussion by introducing families of elliptical copulae. In general, elliptical copulae arise naturally from their respective elliptical distributions following Sklar's theorem. Although elliptical copulae have no closed form, they have the property that the dependence structure is fully described by the correlation. This family of distributions was discussed previously in the section on tail dependence, where the relationship between the regular variation of the tails of the elliptical distribution marginals was related to the extremal dependence. In this section, we discuss the works of Fang *et al.* (2002), where they provide an alternative representation of the family of elliptical distributions as detailed in Definition 11.11 that they refer to as the meta-elliptical family.

Definition 11.11 (Elliptical Distribution) *The density function of an elliptical distributions (if it exists) is given by*

$$f(x) = |\Sigma|^{-\frac{1}{2}} g\left[(x-\mu)^T \Sigma^{-1} (x-\mu)\right] \quad x \in \mathbb{R}^n, \tag{11.40}$$

where Σ *(dispersion) is a symmetric positive semidefinite matrix,* $\mu \in \mathbb{R}^n$ *(location) and* g *(density generator) is a* $[0,\infty) \rightarrow [0,\infty)$ *function. Note the scale function (density generator)* $g(\cdot)$ *is uniquely determined by the distribution of the random variable* R *in the elliptical distribution representation*

$$\boldsymbol{X} \stackrel{d}{=} \boldsymbol{\mu} + RA\boldsymbol{U}. \tag{11.41}$$

where $R \geq 0$ *is a positive valued random variable,* A *is a* $d \times d$ *constant matrix such that the covariance of random vector* $\boldsymbol{X}$ *is given by* $AA^T = \Sigma$*, and* $\boldsymbol{U}$ *is uniformly distributed on the unit sphere in* $\mathbb{R}^d$*.* ■

Note that in general if one considers setting $\boldsymbol{\mu} = \boldsymbol{0}$ and the correlation matrix corresponding to Σ given by R with (i,j)-th element $\rho_{ij} \in (-1,1)$ for $i \neq j$ and $\rho_{ii} = 1$. Then one can write the marginal density and distributions of the elliptical family of distributions according to the following integral representations:

$$\begin{aligned} q_g(x_i) &= \frac{\pi^{\frac{(d-1)}{2}}}{\Gamma\left(\frac{d-1}{2}\right)} \int_{x^2}^{\infty} \left(y - x^2\right)^{\frac{(d-1)}{2}-1} g(y)dy, \\ Q_g(x_i) &= \frac{1}{2} + \frac{\pi^{\frac{(d-1)}{2}}}{\Gamma\left(\frac{d-1}{2}\right)} \int_0^x \int_{x^2}^{\infty} \left(y - x^2\right)^{\frac{(d-1)}{2}-1} g(y)dy. \end{aligned} \tag{11.42}$$

In Fang *et al.* (2002), they then utilize this elliptical family construction to define the meta-elliptical densities as given in Definition 11.12.

Definition 11.12 (Meta-Elliptical Distributions) *Consider a random vector* $\boldsymbol{X} = (X_1, \ldots, X_d)$ *such that each marginal random variable* X_i *has continuous density* f_{X_i} *and distribution* F_{X_i}*. Furthermore, assume an elliptically distributed random vector* $\boldsymbol{Z}$ *with characteristics* $\boldsymbol{\mu_Z} = \boldsymbol{0}$*, correlation matrix* R*, and density generator function* $g(\cdot)$ *that satisfies the relationship*

$$Z_i = Q_g^{-1}\left(F_{X_i}(x_i)\right), \quad \forall i \in \{1, 2, \ldots, d\}. \tag{11.43}$$

Then the resulting distribution is given by

$$f_{\boldsymbol{X}}(x_1, \ldots, x_d) = \phi\left(Q_g^{-1}\left(F_{X_1}(x_1)\right), \ldots, Q_g^{-1}\left(F_{X_d}(x_d)\right)\right) \prod_{i=1}^{d} f_{X_i}(x_i), \tag{11.44}$$

where the function $\phi(\cdot)$ *is what Fang* et al. *(2002) defined as a density weighting function, otherwise known as a copula. In this case a prelude to a Gaussian copula is given by*

$$\phi(z_1, \ldots, z_d) = |R|^{-\frac{1}{2}} \frac{g\left(\boldsymbol{z}^T \Sigma^{-1} \boldsymbol{z}\right)}{\prod_{i=1}^{d} q_g(z_i)}. \tag{11.45}$$

■

Some examples of density generator functions $g(\cdot)$ in the bivariate case include the following:

- Symmetric Kotz-type distributions:

$$g(x_1, x_2) = \frac{sr^{N/s}\left(x_1^2 + s_2^2 - 2\rho x_1 x_2\right)^{N-1}}{\pi\Gamma(N/s)\left(1-\rho^2\right)^{N-\frac{1}{2}}} \exp\left(-r\frac{x_1^2 + x_2^2 - 2\rho x_1 x_2}{1-\rho^2}\right) \tag{11.46}$$

with parameters $r > 0$, $s > 0$, and $N > 0$;

- Symmetric bivariate Pearson type VII distributions:

$$g(x_1, x_2) = \frac{N-1}{\pi m\sqrt{1-\rho^2}}\left(1 + \frac{1}{m(1-\rho^2)}\left(x_1^2 + x_2^2 - 2\rho x_1 x_2\right)\right)^{-N} \tag{11.47}$$

with parameters $N > 1$ and $m > 0$.

In the family of elliptical copulae, two important subfamilies involve the Gaussian copula and the Students-t copula as detailed next.

11.2.1 GAUSSIAN COPULA

The d-dimensional Gaussian copula is obtained by transformation of the multivariate Normal distribution

$$C(u_1, \ldots, u_d) = F_N^{\mathbf{\Sigma}}\left(F_N^{-1}(u_1), \ldots, F_N^{-1}(u_d)\right) \tag{11.48}$$

and its density is

$$c(u_1, \ldots, u_d) = \frac{f_N^{\mathbf{\Sigma}}\left(F_N^{-1}(u_1), \ldots, F_N^{-1}(u_d)\right)}{\prod_{i=1}^{d} f_N\left(F_N^{-1}(u_i)\right)}. \tag{11.49}$$

Here, $F_N(\cdot)$ and $f_N(\cdot)$ are the standard Normal distribution and its density, respectively; $f_N^{\mathbf{\Sigma}}(\cdot)$ and $F_N^{\mathbf{\Sigma}}(\cdot)$ are the standard multivariate Normal density and distribution, respectively, with zero means, unit variances, and correlation matrix $\mathbf{\Sigma}$.

Simulation of the random variates from a Gaussian copula is very simple and can be done as follows.

Algorithm 11.2 (Simulation from Gaussian Copula)

1. *Simulate d-variate $(x_1, \ldots, x_d)^T$ from the standard multivariate normal distribution Normal $(\mathbf{0}, \mathbf{\Sigma})$ with zero means, unit variances, and correlation matrix $\mathbf{\Sigma}$;*
2. *Calculate $u_1 = F_N(x_1), \ldots, u_d = F_N(x_d)$. Obtained $(u_1, \ldots, u_d)^T$ is a d-variate from a Gaussian copula.*

11.2.1.1 Fitting a Gaussian Copula. In order to fit a Gaussian copula to a set of n samples of d-variate data $\{(X_{1,i}, \ldots, X_{d,i})\}_{i\in\{1,\ldots,n\}}$, one can perform such parameter estimation of the Gaussian copula via a method-of-moments procedure based on rank correlation estimates. One could calculate the standard marginal pairwise standard linear correlation coefficients for the pseudo observations, where $\rho_S\left(X_i, X_j\right) = \rho\left(F_{X_i}(X_i), F_{X_j}(X_j)\right)$ using the empirical estimates of the marginal distributions $\hat{F}_{i,n}$ to obtain pseudo data

$$\left\{\left(\hat{F}_{i,n}\left(X_{i,k}\right), \hat{F}_{j,n}\left(X_{i,k}\right)\right)\right\}_{k\in\{1,\dots,n\}}.$$

Alternative estimators that can be used are the rank correlations such as the pairwise empirical estimators for the rank correlations:

$$\begin{aligned}
\hat{\rho}_S &= \frac{12}{n\left(n^2-1\right)}\sum_{k=1}^{n}\left(\text{rank}\left(X_{k,i} - \frac{1}{2}(n+1)\right)\right)\left(\text{rank}\left(X_{k,j} - \frac{1}{2}(n+1)\right)\right)\\
\hat{\rho}_K &= \frac{2!(n-2)!}{n!}\sum_{1\leq k\leq l\leq n}\text{sgn}\left(\left(X_{(k,i)} - X_{(l,i)}\right)\left(X_{(k,j)} - X_{(l,j)}\right)\right),
\end{aligned} \tag{11.50}$$

where, for instance, one would then obtain the pairwise linear correlations by the transformation for Spearman's rho

$$\rho_S\left(X_i, X_j\right) = \frac{6}{\pi}\text{arcsin}\frac{1}{2}\rho_{ij}, \tag{11.51}$$

or for Kendall's tau by

$$\rho_K\left(X_i, X_j\right) = \frac{2}{\pi}\text{arcsin}\rho_{ij}. \tag{11.52}$$

Then one must ensure that the resulting pairwise correlations obtained from transformation of the rank correlations actually produce a valid correlation matrix. This is typically achieved by a type of matrix regularization as detailed in Algorithm 11.3

Algorithm 11.3 (Method of Moments Fitting Gaussian Copula)

- *Estimate the rank correlation, either $\rho_S\left(X_i, X_j\right)$ or $\rho_K\left(X_i, X_j\right)$, for each marginal pair of variables. Then transform to the linear correlation measure;*
- *Construct the estimated sample pseudo correlation matrix $\hat{R}^*$ with (i,j)-th element given by, for instance, using Kendall's tau, $\hat{r}^*_{ij} = \sin\left(\frac{\pi}{2}\hat{\rho}_K\left(X_i, X_j\right)\right)$;*
- *The pseudo correlation matrix $\hat{R}^*$ must be made positive definite with unit diagonal entries and off-diagonal entries in the range $[-1, 1]$. To achieve this, complete the following steps:*
    - *Evaluate the spectral decomposition $\hat{R}^* = E\Lambda E^T$ where Λ is a diagonal matrix of eigenvalues and E is an orthogonal matrix whose columns are eigenvectors of $\hat{R}^*$;*
    - *Replace any negative eigenvalues in Λ by a small value $\delta > 0$ to obtain the regularized matrix $E\tilde{\Lambda}E^T$, which then is turned into a regularized correlation matrix $\hat{R}$ given by application of the correlation matrix transform $\mathcal{T}\left[E\tilde{\Lambda}E^T\right]$ with $\mathcal{T}$ given by*

$$\mathcal{T}[\Sigma] = \left(\triangle(\Sigma)\right)^{-1}\Sigma\left(\triangle(\Sigma)\right)^{-1} \tag{11.53}$$

with $\triangle(\Sigma) := diag\left(\sqrt{\sigma_{11}}, \dots, \sqrt{\sigma_{dd}}\right)$;

11.2.1.2 Multivariate Dispersion Models: MDM and Gaussian Copulas.

In treating the multivariate versions of these distribution models, one can adopt one of two main approaches. Before presenting these approaches, we recall briefly that a loss random variable X_i has a dispersion model severity generically denoted by $X_i \sim DM(\mu, \sigma^2)$ with respect to position μ and dispersion σ^2 if it has a density given by

$$f_X\left(x;\mu,\sigma^2\right) = a\left(x;\sigma^2\right)\exp\left(-\frac{1}{2\sigma^2}d(x;\mu)\right) \tag{11.54}$$

for unit deviance function $d(x;\mu) \geq 0$. We note that an exponential dispersion family model is achieved when the deviance function takes the linear form

$$d(x;\mu) = xd_1(\mu) + d_2(x) + d_3(\mu) \tag{11.55}$$

for suitable functions d_1, d_2, and d_3. In addition, we recall that one can show that continuous distributions such as Gaussian, Exponential, Gamma, and Inverse-Gaussian are members of this family of models. In addition, the discrete support members include Poisson, Negative Binomial, and Binomial distributions.

In developing a multivariate version of this severity model, such as would be suitable for modeling joint dependence between multiple losses in different risk cells, or even a single risk cell in which the losses in a given year were considered dependent, there are two approaches proposed in the literature. The first is due to Jørgensen and Lauritzen (2000), where they consider a multivariate extension to the dispersion model, though it has the undesirable property that the marginal severity models are no longer closed in the class of dispersion models.

The second approach rectifies this and is due to the work of Xue-Kun Song (2000). In this case, a Gaussian copula is utilized to combine with the marginal dispersion models for each individual loss to obtain a multivariate dispersion model given by

$$f_{X_1,X_2,\ldots,X_d}(\boldsymbol{x};\boldsymbol{\mu},\boldsymbol{\sigma}^2,\boldsymbol{\rho}) = \frac{1}{|\boldsymbol{\rho}|^{\frac{1}{2}}}\exp\left(-\frac{1}{2}\boldsymbol{\zeta}^T\left(\boldsymbol{\rho}^{-1} - \mathbb{I}\right)\boldsymbol{\zeta}\right)\prod_{i=1}^{d} f_{X_i}\left(x_i;\mu_i,\sigma_i^2\right) \tag{11.56}$$

with $\boldsymbol{\mu} = \left(\mu_1,\ldots,\mu_d\right)^T$, $\boldsymbol{\sigma}^2 = \left(\sigma_1^2,\ldots,\sigma_d^2\right)^T$, $\boldsymbol{\zeta} = \left(\zeta_1,\ldots,\zeta_d\right)^T$ and each element of $\boldsymbol{\zeta}$ given by

$$\zeta_i = \boldsymbol{\Phi}^{-1}\left(\boldsymbol{F}_i\left(x_i;\mu_i,\sigma_i^2\right)\right). \tag{11.57}$$

As noted in Bouyé *et al.* (2000), in this instance, the model preserves the feature that the univariate margins remain as dispersion models.

11.2.2 *T*-COPULA

In practice, one of the most popular copula in modeling multivariate financial data is perhaps the t-copula, implied by the multivariate t-distribution; see Embrechts *et al.* (2002), Fang *et al.* (2002), and Demarta and McNeil (2005). This is due to its simplicity in terms of simulation and calibration, combined with its ability to model tail dependence, which is often observed in financial returns data.

Student's t-copula retains much of the simplicity of the Gaussian copula, such as in simulation and calibration, but also allows for the modeling of tail dependence between variables. The behavior of the model at the four corners of the unit cube is quite different from that of the Gaussian copula, while toward the center they are more similar in nature. Although T-copula and Gaussian copula distributions may share the same correlation matrix, in such cases, the extreme events are much more likely under the t-copula. This copula has often been referred to as the "desert island copula" by Dr. Paul Embrechts due to its excellent fit to multivariate financial return data. However, its simplest specification does not allow for asymmetry in the tails, that is, differing upper and lower tail dependence in a portfolio of loss processes. This can be rectified with grouped and generalized grouped Students-t copula models.

The t-copulas are most easily described and understood by a stochastic representation, as discussed next. We introduce notation and definitions as follows:

- $\boldsymbol{Z} = (Z_1, \ldots, Z_n)'$ is a random vector from the standard n-variate Normal distribution $F_N^{\boldsymbol{\Sigma}}(\boldsymbol{z})$ with zero mean vector, unit variances, and correlation matrix $\boldsymbol{\Sigma}$;
- $\boldsymbol{U} = (U_1, U_2, \ldots, U_n)'$ is defined on $[0, 1]^n$ domain;
- V is a random variable from the $Uniform(0, 1)$ distribution independent of $\boldsymbol{Z}$;
- $W = G_\nu^{-1}(V)$, where $G_\nu(\cdot)$ is the distribution function of $\sqrt{\nu/S}$ with S distributed from the Chi-square distribution with ν degrees of freedom, that is, random variables W and $\boldsymbol{Z}$ are independent;
- $t_\nu(\cdot)$ is the standard univariate t-distribution and $t_\nu^{-1}(\cdot)$ is its inverse.

Then we have the following representations:

Standard t-copula. The random vector

$$\boldsymbol{X} = W \times \boldsymbol{Z}, \tag{11.58}$$

is distributed from a multivariate t-distribution and random vector

$$\boldsymbol{U} = (t_\nu(X_1), \ldots, t_\nu(X_n))^T \tag{11.59}$$

is distributed from the standard t-copula.

Skew t-copula The standard t-copula is sometimes criticized for its restriction relating to tail symmetry. To resolve this issue several different parameterizations of skew t-copula have been developed, see discussions in Demarta and McNeil (2005) and Allen and Satchell (2013) and the detailed references therein. To specify the skew t-copula it will be beneficial to first recall the definition of the generalized multivariate Hyperbolic distribution and its sub-family the multivariate skew-t distribution. The reason for this is that the skew t-copula is the implicitly defined copula that produces the multivariate dependence in this family of distributions, when the marginals are also in this family.

Definition 11.13 (Generalized Multivariate Hyperbolic Distribution) *A d-dimensional random vector $\boldsymbol{X}$ is distributed according to a multivariate hyperbolic distribution if it has density given by*

$$f_X(\boldsymbol{x}) = c\frac{K_{\lambda-d/2}\left(\sqrt{(\chi+(\boldsymbol{x}-\boldsymbol{\mu})^T\Sigma^{-1}(\boldsymbol{x}-\boldsymbol{\mu}))(\psi+\boldsymbol{\gamma}^T\Sigma^{-1}\boldsymbol{\gamma})}\right)\exp\left((\boldsymbol{x}-\boldsymbol{\mu})^T\Sigma^{-1}\boldsymbol{\gamma}\right)}{\left[\sqrt{(\chi+(\boldsymbol{x}-\boldsymbol{\mu})^T\Sigma^{-1}(\boldsymbol{x}-\boldsymbol{\mu}))(\psi+\boldsymbol{\gamma}^T\Sigma^{-1}\boldsymbol{\gamma})}\right]^{d/2-\lambda}}, \tag{11.60}$$

where $K_\nu(\cdot)$ is the modified Bessel function of the second kind and c is a normalizing constant given by

$$c = \frac{\left(\sqrt{\chi\psi}\right)^{-\lambda}\left(\psi+\boldsymbol{\gamma}^T\Sigma^{-1}\boldsymbol{\gamma}\right)^{d/2-\lambda}\psi^\lambda}{(2\pi)^{d/2}|\Sigma|^{1/2}K_\lambda\left(\sqrt{\chi\psi}\right)}, \tag{11.61}$$

where $\boldsymbol{\mu}$ is the d-dimensional location vector, Σ is a $d \times d$ positive definite symmetric covariance matrix, $\boldsymbol{\gamma}$ is a d-dimensional skewness vector and χ and ψ are constants. ■

From this family of distributions one can obtain the following sub-families:

- If $\lambda = (d+1)/2$ one obtains the hyperbolic family of distributions;
- If $\lambda = -1/2$ then one obtains the Normal Inverse Gaussian distribution family; and
- If one selects $\lambda = -\nu/2$ for some degree of freedom parameter, then one obtains the GH skewed-t distribution. Then, if one in addition selects $\boldsymbol{\gamma} = \mathbf{0}$, one recovers the family of multivariate skew-t distributions.

One may also define analogously the univariate skew t-distribution according to the distribution for random variable X, which is distributed according to a multivariate hyperbolic distribution if it has density given by

$$f_X(x) = c\frac{K_{(\nu+1)/2}\left(\sqrt{\left(\nu+\frac{(x-\mu)^2}{\sigma^2}\right)\frac{\gamma^2}{\sigma^2}}\right)\exp\left((x-\mu)\frac{\gamma}{\sigma^2}\right)}{\left[\sqrt{\left(\nu+\frac{(x-\mu)^2}{\sigma^2}\right)\frac{\gamma^2}{\sigma^2}}\right]^{-(\nu+1)/2}\left[1+\frac{(x-\mu)^2}{\nu\sigma^2}\right]^{(\nu+1)/2}}. \tag{11.62}$$

Then if one considers the multivariate skew-t density with marginal distributions given by the marginal skew-t density, the ratio of the multivariate skew-t distribution over the product of the marginal skew-t densities, is according to Sklar's theorem, the implicitly defined skew t-copula density.

In addition, it will be useful for simulation and estimation purposes to also observe that under this specification of the multivariate skew-t distribution, one can obtain the following location-scale mixture representation for a skew-t distributed d-dimensional random vector according to

$$\boldsymbol{X} \stackrel{d}{=} \boldsymbol{\mu} + \boldsymbol{\gamma}W + \sqrt{W}A\boldsymbol{Z}, \tag{11.63}$$

where W is a random variable with $W \sim InvGamma(\nu/2, \nu/2)$ independently distributed from random vector $\boldsymbol{Z}$ which has distribution $\boldsymbol{Z} \sim Normal(\mathbf{0}, \Sigma)$.

One can simulate from the skew-t copula then by the following algorithmic procedure:

Algorithm 11.4 (Simulation from Skew-t-copula)

1. *Given covariane matrix* Σ*, sample a normal d-dimensional random vector* $\boldsymbol{Z} \sim Normal(\mathbf{0}, \Sigma)$*;*
2. *Given degrees of freedom parameter* ν*, sample a realization of the Inverse-Gamma random variable* $W \sim InvGamma(\nu/2, \nu/2)$*;*
3. *Given parmeter vectors* $\boldsymbol{\mu}$ *and* $\boldsymbol{\gamma}$ *and matrix A, create a new random vector*

$$\boldsymbol{X} = \boldsymbol{\mu} + \boldsymbol{\gamma} W + \sqrt{W} A \boldsymbol{Z}; \tag{11.64}$$

4. *Evaluate for each marginal element of the d-dimensional random vector* $\boldsymbol{X}$*, for* $i \in \{1, 2, \ldots, d\}$*, the transformed random variable* $U_i = F(X_i)$*. This is achieved by solving the integral below, numerically for each marginal,typically under a univariate Guass-Quadrature rule,*

$$U_i = F(X_i) = \int_{-\infty}^{X_i} c \frac{K_{(\nu+1)/2}\left(\sqrt{\left(\nu + \frac{(s-\mu)^2}{\sigma^2}\right)\frac{\gamma^2}{\sigma^2}}\right)\exp\left((s-\mu)\frac{\gamma}{\sigma^2}\right)}{\left[\sqrt{\left(\nu + \frac{(s-\mu)^2}{\sigma^2}\right)\frac{\gamma^2}{\sigma^2}}\right]^{-(\nu+1)/2}\left[1 + \frac{(s-\mu)^2}{\nu\sigma^2}\right]^{(\nu+1)/2}} ds. \tag{11.65}$$

Grouped *t*-copula. The standard t-copula is sometimes criticized due to the restriction of having only one parameter for the degrees of freedom ν, which may limit its ability to model tail dependence in multivariate cases. To overcome this problem, Daul *et al.* (2003) proposed the use of the grouped t-copula, where risks are grouped into classes and each class has its own t-copula with a specific degrees-of-freedom parameter. Specifically, partition $\{1, 2, \ldots, n\}$ into m nonoverlapping subgroups of sizes $n_1, \ldots, n_m$. Then the copula of the distribution of the random vector

$$\boldsymbol{X} = (W_1 Z_1, \ldots, W_1 Z_{n_1}, W_2 Z_{n_1+1}, \ldots, W_2 Z_{n_1+n_2}, \ldots, W_m Z_n)^T, \tag{11.66}$$

where $W_k = G_{\nu_k}^{-1}(V)$, $k = 1, \ldots, m$, is the grouped t-copula. That is,

$$\boldsymbol{U} = (t_{\nu_1}(X_1), \ldots, t_{\nu_1}(X_{n_1}), t_{\nu_2}(X_{n_1+1}), \ldots, t_{\nu_2}(X_{n_1+n_2}), \ldots, t_{\nu_m}(X_n))^T$$

is a random vector from the grouped t-copula. Here, the copula for each group is a standard t-copula with its own degrees-of-freedom parameter.

Generalized t-Copula with Multiple Degrees-of-Freedom Parameters. It is not always obvious how the risk factors should be divided into sub-groups. An adequate choice of grouping configurations requires substantial additional effort if there is no natural grouping, for example, by sector or class of asset. The described grouped t-copula can be generalized so that each group will have only one member; see Luo and Shevchenko (2010). The generalized t-copula has the advantages of a grouped t-copula with flexible modeling of multivariate dependencies. At the same time, it overcomes the difficulties with *a priori* choice of groups. Specifically, the copula of the random vector

$$\boldsymbol{X} = (W_1 Z_1,\ W_2 Z_2, \ldots,\ W_n Z_n)^T \tag{11.67}$$

is said to have a t-copula with multiple degrees-of-freedom parameters, which we denote as $\widetilde{t}_{\boldsymbol{\nu}}$-copula, that is,

$$\boldsymbol{U} = (t_{\nu_1}(X_1), t_{\nu_2}(X_2),\ \ldots, t_{\nu_n}(X_n))^T \tag{11.68}$$

is a random vector distributed according to this copula. Note that all W_i are perfectly dependent.

Given the stochastic representation, simulation of the $\widetilde{t}_{\boldsymbol{\nu}}$-copula is straightforward. In the case of a standard t-copula $\nu_1 = \cdots = \nu_n = \nu$; and in the case of grouped t-copula, the corresponding subsets have the same degrees-of-freedom parameter. Note that the standard t-copula, and grouped t-copula are special cases of $\widetilde{t}_{\boldsymbol{\nu}}$-copula.

From the stochastic representation (11.67), it is easy to show that the $\widetilde{t}_{\boldsymbol{\nu}}$-copula distribution has the following explicit integral expression:

$$C_{\boldsymbol{\nu}}^{\boldsymbol{\Sigma}}(\boldsymbol{u}) = \int_0^1 F_N^{\boldsymbol{\Sigma}}(z_1(u_1,s), \ldots, z_n(u_n,s))ds \tag{11.69}$$

and its density is

$$\begin{aligned} c_{\boldsymbol{\nu}}^{\boldsymbol{\Sigma}}(\boldsymbol{u}) &= \frac{\partial^n C_{\boldsymbol{\nu}}^{\boldsymbol{\Sigma}}(\boldsymbol{u})}{\partial u_1 \ldots \partial u_n} \\ &= \frac{1}{\prod_{k=1}^{n} f_{\nu_k}(x_k)} \int_0^1 f_N^{\boldsymbol{\Sigma}}\left(z_1(u_1,s), \ldots, z_n(u_n,s)\right) \prod_{k=1}^{n} (w_k(s))^{-1} ds. \end{aligned} \tag{11.70}$$

Here,

- $z_k(u_k,s) = t_{\nu_k}^{-1}(u_k)/w_k(s),\ k = 1,2,\ \ldots, n$;
- $w_k(s) = G_{\nu_k}^{-1}(s)$;
- $f_N^{\boldsymbol{\Sigma}}(z_1, \ldots, z_n) = \exp(-\frac{1}{2}\boldsymbol{z}^T \boldsymbol{\Sigma}^{-1} \boldsymbol{z}) / \left((2\pi)^{n/2} (\det \boldsymbol{\Sigma})^{1/2}\right)$ is the standard multivariate Normal density;
- $x_k = t_{\nu_k}^{-1}(u_k),\ k = 1,2,\ \ldots, n$;
- $f_\nu(x) = (1 + x^2/\nu)^{-(\nu+1)/2} \frac{\Gamma((\nu+1)/2)}{\Gamma(\nu/2)\sqrt{\nu\pi}}$ is the univariate t density.

The multivariate density (11.70) involves a one-dimensional integration that should be done numerically. This makes the calculation of the copula density more demanding computationally in comparison with the standard t-copula. However, it is still practical, because fast and accurate algorithms are available for the one-dimensional numerical integration. If all degrees-of-freedom parameters are equal (i.e., $\nu_1 = \cdots = \nu_n = \nu$), then it is easy to show that the copula defined by (11.69) becomes the standard t-copula; see Luo and Shevchenko (2010) for a proof.

For the Gaussian and Students-t-copula models, one can find detailed discussions on tail dependence in McNeil *et al.* (2005, section 5.2.3). Here, we just mention that the tail dependence coefficient can be very useful for comparing different copulas. In particular:

- For the bivariate Gaussian copula, defined by (11.48): $\lambda_l = \lambda_u = 0$, if the correlation coefficient of the copula $\rho < 1$;
- For the bivariate t-copula, defined by stochastic representation (11.58) and (11.59):

$$\lambda_l = \lambda_u = 2t_{\nu+1}\left(-\sqrt{\frac{(\nu+1)(1-\rho)}{1+\rho}}\right), \tag{11.71}$$

  which is positive if $\rho > -1$. Here, ρ is a correlation coefficient parameter of the t-copula and ν is a copula degrees-of-freedom parameter.

Before presenting a detailed account of one of the most widely used families of copula models, the Archimedean family, we first briefly introduce the Farlie-Gumbel-Morgenstern (FGM) copula family. This is also relatively widely used for OpRisk settings and was discussed in Geluk and Tang (2009), where it is was shown that the notion of negative regression dependence, discussed earler, in the multivariate setting was satisfied by the family of FGM copula models presented in Definition 11.14; see Johnson *et al.* (2002).

Definition 11.14 (Farlie–Gumbel–Morgenstern Copula Models) *Consider a loss process with n loss random variables with marginal distributions $X_i \sim F_{X_i}(x)$ and a joint distribution given by a copula C. The joint dependence will be in the family of FGM copulas with parameters $\left\{a_{i,j}\right\}_{i,j=1}^{n}$, which satisfy constraints provided in Johnson et al. (2002) and have a distribution given by*

$$F_{X_1,X_2,\ldots,X_n}(x_1,x_2,\ldots,x_n) = \prod_{i=1}^{n} \overline{F}_{X_i}(x_i)\left(1 + \sum_{1\le i<j\le n} a_{i,j}\overline{F}_{X_i}(x_i)\overline{F}_{X_j}(x_j)\right). \tag{11.72}$$

■

11.2.3 ARCHIMEDEAN COPULAS

Generally, Archimedean copulae are not derived from a well-known parametric multivariate distribution; nevertheless, they can be stated explicitly in a simple form. Many Archimedean copulae have been proposed in the literature, see Nelsen (1999), with many further copulae available as extensions and combinations of these base copulae. Archimedean copulae are attractive to researchers and practitioners due to their directly interpretable tail dependence features and parsimonious representations.

Remark 11.4 *It should be noted that throughout the literature there are multiple parameterizations and representations of Archimedean copula distribution families. In particular, one can find representations of the general family of Archimedean copulas where the distribution is written with one of two forms with respect to a composite of two functions (which are inverse of each other)*

$$C(u_1, \ldots, u_n) = \psi\left(\sum_{i=1}^{n} \psi^{-1}(u_i)\right)$$

or

$$C(u_1, \ldots, u_n) = \psi^{-1}\left(\sum_{i=1}^{n} \psi(u_i)\right).$$

One should always be careful to check the particular realization being utilized.

We begin this section with a basic definition of the bivariate Archimedean copula and then this is generalized to the d-variate copula case. This is followed by a detailed account of the required properties of the generator function of this family of parametric dependence models.

The family of Archimedean copula models has the following useful properties (as detailed in a simple bivariate setting) presented in Lemma 11.1.

Lemma 11.1 *Let C be an Archimedean copula with generator ψ. Then according to Nelsen (1999, lemma 4.1.2 and theorem 4.1.5), the following properties hold:*

1. *C is an Archimedean copula if it can be represented by*

$$C(u, v) = \psi^{[-1]}(\psi(u) + \psi(v)),$$

   where ψ is the generator of this copula and is a continuous, strictly decreasing function from $[0, 1]$ to $[0, \infty]$ such that $\psi(1) = 0$ and $\psi^{[-1]}$ is the pseudo inverse of ψ;
2. *C is symmetric, $C(u, v) = C(v, u)\ \forall (u, v) \in [0, 1] \times [0, 1]$;*
3. *C is associative, $C(C(u, v), w) = C(u, C(v, w))\ \forall (u, v, w) \in [0, 1]^3$;*
4. *If $c > 0$ is any constant, then $c\psi$ is a generator of C.*

According to Denuit *et al.* (2005, definition 4.7.6), the extension of the Archimedean copula family to d-dimensions is achieved by considering the strictly monotone generator function ψ such that $\psi : (0, 1] \to \mathbb{R}^+$ with $\psi(1) = 0$, then the resulting Archimedean copula can be expressed as detailed next.

Definition 11.15 (*d*-Dimensional Archimedean Copula) *A d-dimensional copula C is called Archimedean if for some generator ψ it can be represented as*

$$C(\boldsymbol{u}) = \psi\{\psi^{-1}(u_1) + \cdots + \psi^{-1}(u_d)\} = \psi\{t(\boldsymbol{u})\} \quad \forall \boldsymbol{u} \in [0, 1]^d, \tag{11.73}$$

where $\psi^{-1} : [0, 1] \to [0, \infty]$ is the inverse generator with $\psi^{-1}(0) = inf\{t : \psi(t) = 0\}$. ■

Note the shorthand notation $t(\mathbf{u}) = \psi^{-1}(u_1) + \cdots + \psi^{-1}(u_d)$ that will be used occasionally in the remainder of this section.

As we will see later, it is necessary to have formulae for computing the copula densities as this will be useful in many settings such as performing parameter estimation or calculating tail dependence or performing Rosenblatt's probability integral transforms for goodness-of-fit testing. For instance, if one seeks to fit these models using a maximum likelihood approach or a Bayesian approach, both of which require the model likelihood. Equation (11.74) provides such a formula in a generic form for each member of the family of Archimedean copulae. In this regard, one can show the following result regarding existence of a copula density function for an Archimedean family given in Proposition 11.2; see McNeil and Nešlehová (2009).

Proposition 11.2 (Existence of an Archimedean Copula Density) *A d-dimensional Archimedean copula with generator given by ψ will admit a density function if and only if the $(d-1)$-th derivative $\psi^{(d-1)}$ exists and is absolutely continuous on $(0, \infty)$. The density is then given by*

$$c(\boldsymbol{u}) = \frac{\psi^{(d)}\left(\psi^{-1}\left(u_1\right) + \ldots + \psi^{-1}\left(u_d\right)\right)}{\psi^{(1)}\left(\psi^{-1}\left(u_1\right)\right)\psi^{(1)}\left(\psi^{-1}\left(u_2\right)\right)\cdots\psi^{(1)}\left(\psi^{-1}\left(u_d\right)\right)} \tag{11.74}$$

for almost all $\boldsymbol{u} \in (0,1)^d$.

This section first introduces families of exchangeable Archimedean copulae models. To understand this notion, we first recall the definition of exchangeable random variables and sequences in Definition 11.16.

Definition 11.16 (Exchangeable Random Vectors and Sequences) *An exchangeable sequence of random variables is a finite or infinite sequence $X_1, X_2, X_3, \cdots$ of random variables such that for any finite permutation $\pi(\cdot)$ of subsets of the indices $1, 2, 3, \ldots$, then the resulting joint probability distribution of the permuted sequence $X_{\pi(1)}, X_{\pi(2)}, X_{\pi(3)}, \ldots$ is the same as the joint probability distribution of the original sequence.* ■

The first introduction to Archimedean copulas presented later in terms of generator functions will be clearly seen to be representative of exchangeable multivariate random vectors. This is clearly reflected in the representation of a d-dimensional Archimedean copula model that has been shown to be given by

$$C\left(u_1, \ldots, u_d\right) = \psi\left(\psi^{-1}\left(u_1\right) + \cdots + \psi^{-1}\left(u_d\right)\right), \tag{11.75}$$

where ψ is a decreasing function known as the generator for the given copula; see Frees and Valdez (1998). It is clear that such an expression is exchangeable as one could easily swap the order of any subset of variables without a change in the resulting distribution mass.

11.2.3.1 Properties of Archimedean Copulae Generators.

It was shown in the Ph.D. thesis of Ling (1964) that the generator ψ will produce a bivariate copula distribution if and only if it is a convex function. Then in Kimberling (1974) it was shown that in order for the generator ψ to generate any Archimedean copula distribution in any dimension d then it must be a completely monotone function; see Theorem 11.4.

Theorem 11.4 (Completely Monotone Generators and Existence of Archimedean Copulae) *If a generator ψ that is a mapping $\psi : [0, \infty] \mapsto [0, 1]$ is continuous and strictly decreasing such that $\psi(0) = 1$ and $\psi(\infty) = 0$, that is, $\psi \in C^{\infty}(0, \infty)$ and one has that $(-1)^k \psi^{(k)}(x) \geq 1$ for $k = 1, \ldots$ then this class of generators can create Archimedean copulae models in any dimension. This class of completely monotone generators for Archimedean copula in any dimension are denoted by ψ_{∞}.*

Note that it is useful to note the following relevant properties of completely monotone functions in Lemma 11.2 see, for instance, discussion in Hofert (2008).

Lemma 11.2 (Properties of Completely Monotone Functions) *A completely monotone function satisfies the following properties:*

- *Closure under multiplication and positive affine transformations (i.e., linear additive combinations with positive coefficients);*
- *If a function f is a Laplace–Stieltjes transform, then the function f^{α} is completely monotone for any power $\alpha \in (0, \infty)$ if and only if the derivative $(-\ln f)'$ is completely monotone;*
- *If a function f is completely monotone and a second function g is non-negative with its first derivative g' completely monotone, then the composite function $f \circ g$ is a completely monotone function;*
- *If a function f is non-negative and its derivative f' is completely monotone, then the reciprocal of the function f given by $1/f$ is a completely monotone function;*
- *If a function f is continuous on $[0, \infty]$, satisfying $\frac{d^k}{dx^k} f(x) \geq 0$ for any integer $k \in \mathbb{J}$ and $x \in (0, \infty)$ and a function g is completely monotone, then the composite function $f \circ g$ is a completely monotone function.*

The requirement for complete monotonicity is only required to create copula of any dimension, so this was then further relaxed for d-variate Archimedean copula in further studies to include only the positivity of derivatives for $k = 1, 2, \ldots, d$ for a d-variate Archimedean copula; see discussion in McNeil and Nešlehová (2009), where it was shown that one only requires the necessary and sufficient conditions on the generator function to be a d-monotone function as given in Definition 11.17 in order to create Archimedean copulae models up to dimension d.

Definition 11.17 (D-Monotone Functions) *A real function $g(\cdot)$ is d-monotone in a range (a, b) for $a, b \in \mathbb{R}$ and $d \geq 2$ if it is differentiable on this range up to order $d - 2$ and the derivatives satisfy the condition that*

$$(-1)^k g^{(k)}(x) \geq 0, \quad k = 0, 1, \ldots, d - 2 \tag{11.76}$$

for any $x \in (a, b)$ and $(-1)^{d-2} g^{(d-2)}$ is nonincreasing and convex in (a, b). ■

One can then conclude that a function ψ is said to generate an Archimedean copula if it satisfies the following properties.

Definition 11.18 (Archimedean Generator) *An Archimedean generator is a continuous, decreasing function* $\psi : [0, \infty] \rightarrow [0, 1]$ *that satisfies the following conditions:*

1. $\psi : [0, \infty) \mapsto [0, 1]$ *with* $\psi(0) = 1$ *and* $\lim_{t \to \infty} \psi(t) = 0$;
2. ψ *is a continuous function;*
3. ψ^{-1} *is given by* $\psi^{-1}(t) = \inf \{u : \psi(u) \leq t\}$;
4. ψ *is strictly decreasing on* $[0, \inf \{t : \psi(t) = 0\}] = [0, \psi^{-1}(0)]$.

■

McNeil and Nešlehová (2009) discuss the class of generators, denoted by ψ_∞, which represent all the generators for Archimedean copulae models that produce valid copula distributions in any dimension, that is, those that are completely monotone functions. In this context, they note two representations of such generators: the first based on Bernstein–Widder theorem and the Laplace transform; and the second based on the Williamson d-transform. We discuss these two representations in the following subsections.

11.2.4 ARCHIMEDEAN COPULA GENERATORS AND THE LAPLACE TRANSFORM OF A NON-NEGATIVE RANDOM VARIABLE

In understanding the first representation for the completely monotone generator, it will be instructive to first recall the theorem of Bernstein–Widder; see, for instance, a proof in Pollard (1944) or Feller (1966). This theorem links a completely monotone function to a Laplace transform representation.

Theorem 11.5 (Bernstein–Widder Theorem) *Consider a real function* $f(x)$ *such that it satisfies*

$$f(0) = f(0+), \quad (-1)^k f^{(k)}(x) \geq 0, \quad x \in (0, \infty), \forall k = 0, 1, \ldots. \tag{11.77}$$

Then the function $f(x)$ *admits the following representation as a Laplace transform*

$$f(x) = \int_0^\infty \exp(-xt) d\alpha(t) \tag{11.78}$$

for $x \geq 0$ *and* $\alpha(t)$ *an increasing and bounded function.*

For an Archimedean generator ψ, one can then use this result to link the existence of distributions in all dimensions to the range of complete monotonicity of the generator, see Proposition 11.3.

Proposition 11.3 (Complete Monotonicity and Generator Support) *A generator* ψ *for an Archimedean copula belongs to the class of generators* ψ_∞ *if and only if it is completely monotone on* $[0, \infty)$.

Remark 11.5 *One can see from the combination of Theorem 11.5 and Proposition 11.3 that a generator* ψ *of an Archimedean copula is completely monotone only when it is formed from the*

Laplace transform of a non-negative random variable Z. It can then be shown that the resulting Archimedean copula for such a generator $\psi \subset \psi_\infty$ in d-dimensions is given by the survival copula coming from the survival function, which is expressed via the generator of the l_1-norm according to

$$\begin{aligned}\overline{H}(x_1, x_2, \ldots, x_d) &= \psi\left(||\max(\boldsymbol{x}, \boldsymbol{0})||_1\right) \\ &= \mathbb{E}\left[\exp\left(-||\max(\boldsymbol{x}, \boldsymbol{0})||_1 Z\right)\right] \\ &= \mathbb{E}\left[\exp\left(-Z\sum_{i=1}^{d}\max(x_i, 0)\right)\right], \end{aligned} \tag{11.79}$$

which correspond to a survival function of a random vector $\boldsymbol{X} = \frac{1}{Z}\boldsymbol{E}$ with $\boldsymbol{E}$ a vector of i.i.d. exponential random variables that are independent of Z; see discussion in McNeil and Nešlehová (2009).

One important result of this representation is the ability to simulate exactly Archimedean copula random variates, as discussed in Marshall and Olkin (1988) and shown in Algorithm 11.5.

Algorithm 11.5 (Simulation from Archimedean Copula via Laplace Transform)

1. *Sample a random variable $V \sim F$ where the distribution F is given by the inverse Laplace transform of the generator ψ such that $F = \mathcal{L}^{-1}[\psi]$;*
2. *Sample d i.i.d. draws from a uniform distribution $U_i \sim Uniform(0, 1)$ for $i \in \{1, 2, \ldots, d\}$;*
3. *Construct via transformation the d-variate random vector $\boldsymbol{U} = (U_1, \ldots, U_d)$, which is drawn from the Archimedean copula characterized by generator ψ given by*

$$X_i = \psi\left(-\frac{1}{V}\ln(U_i)\right), \quad i \in \{1, 2, \ldots, d\}. \tag{11.80}$$

The following results in Table 11.1 from Hofert (2008, table 1) demonstrate examples of popular Archimedean copula models for which closed form distributions of such inverse Laplace transforms of the generator are known.

As noted in Hofert (2008), it is then a trivial consequence to obtain other Archimedean copula model simulation schemes based on, for instance, those presented in Table 11.1 via exponential tilting results presented in Theorem 11.6.

Theorem 11.6 (Exponential Tilting of Generator Inverse Laplace Transforms) *Consider an Archimedean copula generator ψ in the family of completely monotone Archimedean generators $\psi \in \psi_\infty$ with a known distribution for the inverse Laplace transform given by $F = \mathcal{L}^{-1}[\psi]$. Then, define a new generator $\widetilde{\psi}(x)$ in terms of $\psi(x)$ according to*

$$\widetilde{\psi}(x) = \frac{\psi(x + h; \rho)}{\psi(h; \rho)}, \quad \forall x \in [0, \infty]. \tag{11.81}$$

Then the following holds:

- *$\widetilde{\psi}(x)$ is completely monotone on $x \in [0, \infty]$ and $\widetilde{\psi}(0) = 1$;*

TABLE 11.1 Generators and inverse Laplace transforms for several copulas from Archimedean family

Archimedean Family	ρ Range	Generator $\psi(x;\rho)$	Distribution of $\mathcal{L}^{-1}[\psi]$
Ali–Mikhail–Haq	$[0,1)$	$\psi(x)=\frac{1-\rho}{\exp(x)-\rho}$	$y_k=(1-\rho)\rho^{k-1}, k\in\mathbb{J}$
Clayton	$(0,\infty)$	$\psi(x)=(1+x)^{-1/\rho}$	$\Gamma(1/\rho,1)$
Frank	$(0,\infty)$	$\psi(x)=-\frac{1}{\rho}\ln\left(e^{-x}(e^{-\rho}-1)+1\right)$	$y_k=\frac{(1-e^{-\rho})^k}{k\rho}, k\in\mathbb{J}$
Gumbel	$[1,\infty)$	$\psi(x)=\exp\left(-x^{1/\rho}\right)$	$S_{\frac{1}{\rho}}\left(1,\cos\left(\frac{\pi}{2\rho}\right)^{\rho},0;S1\right)$
Joe	$[1,\infty)$	$\psi(x)=1-\left(1-e^{-x}\right)^{\frac{1}{\rho}}$	$y_k=(-1)^{k+1}\frac{(1/\rho)!}{k!(1/\rho-k)!}, k\in\mathbb{J}.$

Note: $S_\alpha(\beta,\gamma,\delta;S1)$ is the univariate α-stable distribution with S1 parametrization of Nolan.

- *The distribution of the inverse Laplace transform for the new generator* $\widetilde{F}=\mathcal{L}^{-1}\left[\widetilde{\psi}(x)\right]$ *is given in terms of the distribution F by*

$$\widetilde{F}(x)=\frac{1}{\psi(h)}\left(F(0)+\int_0^x \exp(-hu)dF(u)\right),\quad x\in[0,\infty). \tag{11.82}$$

- *If the distribution F admits a density f, the* $\widetilde{F}$ *admits the exponential tilted density given by*

$$\widetilde{f}(x)=\frac{1}{\psi(h)}\exp(-hx)f(x),\quad x\in[0,\infty). \tag{11.83}$$

11.2.5 ARCHIMEDEAN COPULA GENERATORS, L_1-NORM SYMMETRIC DISTRIBUTIONS AND THE WILLIAMSON TRANSFORM

The second representation developed in McNeil and Nešlehová (2009) utilizes the fact that the random vector discussed in the previous subsection given by $\boldsymbol{X}=\frac{1}{Z}\boldsymbol{E}$ can be re-represented by utilizing the fact that if one transforms the vector of i.i.d. exponential random variables according to

$$\boldsymbol{S}_d=\frac{\boldsymbol{E}}{||\boldsymbol{E}||_1}, \tag{11.84}$$

then $\boldsymbol{S}_d$ will be distributed according to a Uniform distribution on the d-dimensional simplex given by the space $\mathcal{S}_d$

$$\mathcal{S}_d=\left\{\boldsymbol{x}\in\mathbb{R}_+^d : ||\boldsymbol{x}||_1=1\right\}. \tag{11.85}$$

In addition, since $\boldsymbol{S}_d$ and Z are independent, then one can write the random vector $\boldsymbol{X}=R\boldsymbol{S}_d$ with random variable R given by $R=\frac{1}{Z}||\mathbb{E}||_1$. The implications of this result for the transformed distribution indicates that the random vector $\boldsymbol{X}$ admits a representation in terms of a mixture of Uniform distributions on simplices.

The significance of this result is that although only completely monotone Archimedean generators will admit representations as survival copulas of random vectors following a particular frailty model, it is clear from the aforementioned result that even only d-monotone Archimedean generators will produce representations as survival copulae of random vectors with l_1-norm symmetric distributions. As observed in McNeil and Nešlehová (2009), in the case of completely monotone generators of Archimedean copulae one could form a link between the Laplace transform of a particular frailty model and the generator via the Bernstein–Widder theorem. In the case of the d-monotone (not completely monotone) generator functions, one can form an analogous link between d-variate Archimedean copulas and the l_1-norm symmetric distributions via a special class of Mellin–Stieltjes integral transforms known as Williamson transforms; see Definition 11.19 and Williamson (1956) and McNeil and Nešlehová (2009, proposition 3.1).

Definition 11.19 (Williamson d-Transforms) *The Williamson transform of a positive random variable X with distribution F is a real function on $[0, \infty)$ given for any integer $d \geq 2$ by*

$$f(x) = \mathcal{W}_d\left[F_X(x)\right] = \int\limits_{(x,\infty)} \left(1 - \frac{x}{t}\right)^{d-1} dF(t) = \begin{cases} \mathbb{E}\left[\left(1 - \frac{x}{X}\right)_+^{d-1}\right], & \text{if } x > 0 \\ 1 - F(0), & \text{if } x = 0. \end{cases} \tag{11.86}$$

The Williamson d-transform $\mathcal{W}_d$ will consist of real functions f on $[0, \infty)$ that are d-monotone on $[0, \infty)$ and satisfy boundary conditions that $\lim_{x\to\infty} f(x) = 0$ and $f(0) = p$ for $p \in [0, 1]$. Furthermore, any non-negative random variable's distribution function can be uniquely defined by its Williamson d-transform $f = \mathcal{W}_d\left[F_X(x)\right]$ such that $F_X(x) = \mathcal{W}_d^{-1}\left[f(x)\right]$ with the inverse given by

$$F_X(x) = \mathcal{W}_d^{-1}\left[f(x)\right] = 1 - \sum_{k=0}^{d-2} \frac{(-1)^k x^k f^{(k)}(x)}{k!} - \frac{(-1)^{d-1} x^{d-1} f_+^{(d-1)}(x)}{(d-1)!}. \tag{11.87}$$

■

Remark 11.6 *It was therefore observed in McNeil and Nešlehová (2009) that the d-monotone Archimedean copula generators ψ will consist of Williamson d-Transforms of distribution functions F from non-negative loss random variables that satisfy $F(0) = 0$.*

In addition, in Williamson (1956), the result in Proposition 11.4 completes the link between l_1-norm symmetric distributions and Archimedean copulas; see McNeil and Nešlehová (2009).

Proposition 11.4 (l_1-Norm Symmetric Distributions and Williamson d-Transforms) *Consider the d-dimensional random vector $\boldsymbol{X}$ with representation as a l_1-norm symmetric distribution $\boldsymbol{X} \stackrel{d}{=} R\boldsymbol{S}_d$ with radial distribution F_R. Then one has the following relationship between the multivariate survival function of $\boldsymbol{X}$ and the Williamson d-transform:*

- *$\overline{H}(\boldsymbol{x})$ is given by*

$$\overline{H}(\boldsymbol{x}) = \mathcal{W}_d\left[F_R\left(||\max\left(\boldsymbol{x}, \boldsymbol{0}\right)||_1\right)\right] + F_R(0)\mathbb{I}\left[\boldsymbol{x} < \boldsymbol{0}\right], \quad \boldsymbol{x} \in \mathbb{R}^d. \tag{11.88}$$

If in addition $F_R(0) = 0$, then $\boldsymbol{X}$ has an Archimedean survival copula with generator given by $\psi = \mathcal{W}_d\left[F_R(r)\right]$;

- *The density $\boldsymbol{X}$ exists if and only if R has a density, which is given with regard to the density of R denoted $f_R(r)$ by*

$$h\left(||\boldsymbol{x}||_1\right) = \Gamma(d)||\boldsymbol{x}||^{1-d} f_R\left(||\boldsymbol{x}||_1\right). \tag{11.89}$$

- *If $\Pr\left[\boldsymbol{X} = \boldsymbol{0}\right] = 0$, then one has that $R \stackrel{d}{=} ||\boldsymbol{X}||_1$ and $\boldsymbol{S}_d \stackrel{d}{=} \boldsymbol{X}/||\boldsymbol{X}||_1$.*

An important result of this Simplectic representation is the ability to simulate exactly Archimedean copula random variates, as discussed in McNeil and Nešlehová (2009) and shown in Algorithm 11.6.

Algorithm 11.6 (Simulation from Archimedean Copula via Williamson d-Transform)

1. *Sample a random variable $R \sim F_R$ where the distribution F_R is given by the inverse Williamson d-transform of the generator ψ such that $F_R = \mathcal{W}_d^{-1}\left[\psi\right]$, which is given by*

$$F_R(x) = \mathcal{W}_d^{-1}\left[f(x)\right] = 1 - \sum_{k=0}^{d-2} \frac{(-1)^k x^k \psi^{(k)}(x)}{k!} - \frac{(-1)^{d-1} x^{d-1} \psi_+^{(d-1)}(x)}{(d-1)!}. \tag{11.90}$$

2. *Sample independently of R the random vector $\boldsymbol{S}_d$ given by transformation of d i.i.d. exponential random variates with $E_i \sim Exp(1)$ such that*

$$\boldsymbol{S}_d \stackrel{d}{=} \left(\frac{E_1}{\sum_{i=1}^d E_i}, \ldots, \frac{E_d}{\sum_{i=1}^d E_i}\right). \tag{11.91}$$

3. *Construct via transformation the d-variate random vector $\boldsymbol{U} = (U_1, \ldots, U_d)$, which is drawn from the Archimedean copula characterized by generator ψ given by*

$$U_i = \psi\left(R \frac{E_i}{\sum_{i=1}^d E_i}\right), \quad i \in \{1, 2, \ldots, d\}. \tag{11.92}$$

Having presented two general representations of Archimedean copula distribution generators and how to simulate from such representations, we now utilize these specifications of the Archimedean generator to formally introduce the definition of an Archimedean copula distributions first under one-parameter subfamilies, then two-parameter cases with inner and outer-power transforms, followed by generalized forms. Before looking more closely at a few subfamilies of Archimedean families, we note the following result about the level sets (quantile function) of an Archimedean copula with regard to the generator; see McNeil and Nešlehová (2009) and Genest and Mackay (1986).

Definition 11.20 (Level Sets of Archimedean Copulae) *The level sets of a d-variate Archimedean copula C are given by the set $L(s) = \left\{\boldsymbol{u} \in [0,1]^d : C(\boldsymbol{u}) = s\right\}$ for $s \in [0,1]$, which are characterized by*

$$L(s) = \begin{cases} \left\{ \boldsymbol{u} \in [0,1]^d : \sum_{i=1}^d \psi^{-1}(u_i) = \psi^{-1}(s) \right\}, & \text{if } s \in (0,1], \\ \left\{ \boldsymbol{u} \in [0,1]^d : \sum_{i=1}^d \psi^{-1}(u_i) = \psi^{-1}(0) \right\}, & \text{if } s = 1. \end{cases} \tag{11.93}$$

■

11.2.5.1 One-Parameter Archimedean Members. Next, we provide some explicit distribution and density representations for some widely utilized subfamilies of Archimedean copulae families; see Lemma 11.3 for the one parameter versions of the Archimedean copulae.

Lemma 11.3 *From the results in Nelsen (1999, section 4.3, table 4.1), the distribution and density functions of the Clayton, Gumbel, and Frank copulae subfamilies are given by:*

1. ***Clayton Copula.*** *The distribution and density are given respectively as*

$$C^C(u_1, \ldots, u_n) = \left(1 - n + \sum_{i=1}^n u_i^{-\rho^C}\right)^{-1/\rho^C}, \tag{11.94}$$

$$c^C(u_1, \ldots, u_n) = \left(1 - n + \sum_{i=1}^n (u_i)^{-\rho^C}\right)^{-n-\frac{1}{\rho^C}} \prod_{i=1}^n \left((u_i)^{-\rho^C-1}\left((i-1)\rho^C + 1\right)\right), \tag{11.95}$$

where $\rho^C \in [0, \infty)$ is the dependence parameter. The generator and inverse generator for the Clayton copula are given by

$$\psi_C(t) = \left(t^{-\rho} - 1\right); \quad \psi_C^{-1}(s) = (1+s)^{-\frac{1}{\rho}}. \tag{11.96}$$

The Clayton copula does not have upper tail dependence. Its lower tail dependence is $\lambda_L = 2^{-1/\rho^C}$;

2. ***Gumbel Copula.*** *The distribution function is given by*

$$C^G(u_1, \ldots, u_d) = \exp\left(-\left[\sum_{i=1}^d \left(-\ln(u_i)\right)^{\rho^G}\right]^{\frac{1}{\rho^G}}\right), \tag{11.97}$$

where $\rho^G \in [1, \infty)$ is the dependence parameter. The generator and inverse generator for the Gumbel copula are given by

$$\psi_G(t) = (-\ln t)^{\rho}; \quad \psi_G^{-1}(s) = \exp\left(-s^{\frac{1}{\rho}}\right). \tag{11.98}$$

The Gumbel copula does not have lower tail dependence. The upper tail dependence of the Gumbel copula is $\lambda_U = 2 - 2^{1/\rho^G}$. In the bivariate case, the explicit expression for the Gumbel copula density is given by

$$
\begin{aligned}
c(u_1, u_2) &= \frac{\partial^2}{\partial u_1 \partial u_2} C(u_1, u_2) \\
&= C(u_1, u_2)\, u_1^{-1} u_2^{-1} \left[\sum_{i=1}^{2} (-\ln u_i)^{\rho} \right]^{2\left(\frac{1}{\rho}-1\right)} (\ln u_1 \ln u_2)^{\rho-1} \\
&\times \left[1 + (\rho - 1) \left[\sum_{i=1}^{2} (-\ln u_i)^{\rho} \right]^{-\frac{1}{\rho}} \right].
\end{aligned}
$$

3. ***Frank Copula***. *The distribution function is given by*

$$
C^F(u_1, \ldots, u_n) = \frac{1}{\rho} \ln \left(1 + \frac{\prod_{i=1}^{n} (e^{\rho^F u_i} - 1)}{(e^{\rho^F} - 1)^{n-1}} \right), \tag{11.99}
$$

where $\rho^F \in \mathbb{R}/\{0\}$ is the dependence parameter. The Frank copula does not have upper or lower tail dependence. In the bivariate case, one can represent the copula density for the Frank distribution as follows:

$$
\begin{aligned}
c(u_1, u_2) &= \frac{\partial^2}{\partial u_1 \partial u_2} C(u_1, u_2) \\
&= \frac{\rho \left[1 - \exp(-\rho)\right] \exp\left(-\rho \left(u_1 + u_2\right)\right)}{\left(\left[1 - \exp(-\rho)\right] - \left(1 - \exp\left(-\rho u_1\right)\right)\left(1 - \exp\left(-\rho u_2\right)\right)\right)^2}.
\end{aligned}
$$

In general, it will be of practical use to be able to evaluate the copula density pointwise and it has already been demonstrated that this will in general require up to d-th order derivatives for a d-variate Archimedean copula of mixed types. Hence, one may combine the following derivative results for the different Archimedean copula models discussed and their generators with the formula for composite differentiation in Definition 11.8 (Table 10.2).

TABLE 11.2 **Archimedean copula generator functions, inverse generator functions, and generator function d-th derivatives**

Family	ψ	ψ^{-1}	$(-1)^d \psi^{(d)}$
Clayton	$(1+t)^{-1/\rho}$	$(s^{-\rho} - 1)$	$\frac{\Gamma(d+1/\rho)}{\Gamma(1/\rho)} (1+t)^{-(d+1/\rho)}$
Frank	$-\frac{1}{\rho} \ln \left[1 - e^{-t}(1 - e^{-\rho})\right]$	$-\ln \frac{e^{-s\rho}-1}{e^{-\rho}-1}$	$\frac{1}{\rho} Li_{-(d-1)}\{(1 - e^{-\rho})e^{-t}\}$
Gumbel	$\exp\left(-t^{1/\rho}\right)$	$(-\ln s)^{\rho}$	$\frac{\psi_{\rho}(t)}{t^d} P^G_{d,1/\rho}\left(t^{1/\rho}\right)$

The densities for the one-parameter copulae in this table can be calculated using Equation (11.74). For details of the results contained in this table, see Hofert *et al.* (2012).

We note the following definitions are utilized:

$$
\begin{aligned}
a_{dk}^{G}\left(\frac{1}{\rho}\right) &= \frac{d!}{k!}\sum_{i=1}^{k}\binom{k}{i}\binom{i/\rho}{d}(-1)^{d-i}, \quad k \in \{1,\ldots,d\}, \\
Li_s(z) &= \sum_{k=1}^{\infty}\frac{z^k}{k^s}, \\
P_{d,\frac{1}{\rho}}^{G}\left(t^{\frac{1}{\rho}}\right) &= \sum_{k=1}^{d} a_{dk}^{G}\left(\frac{1}{\rho}\right)\left(t^{\frac{1}{\rho}}\right)^k.
\end{aligned} \tag{11.100}
$$

Hence, using these closed from results combined with the knowledge of composite function differentiation via Fa di Bruno, one can then work out the required multivariate densities for implementation of likelihood and Bayesian estimation methods using the definition given by

$$
\begin{aligned}
c\left(u_1,u_2,\ldots,u_n,\ldots,u_d\right) &= \frac{\partial C\left(u_1,u_2,\ldots,u_n,\ldots,u_d\right)}{\partial u_1 \partial u_2 \cdots \partial u_n \cdots \partial u_d} \\
&= \frac{\psi^{(d)}\left(\psi^{-1}\left(u_1\right)+\cdots+\psi^{-1}\left(u_d\right)\right)}{\psi^{(1)}\left(\psi^{-1}\left(u_1\right)\right)\psi^{(1)}\left(\psi^{-1}\left(u_2\right)\right)\cdots\psi^{(1)}\left(\psi^{-1}\left(u_d\right)\right)}.
\end{aligned} \tag{11.101}
$$

Remark 11.7 *It is also worth noting that in the case of the bivariate Clayton copula model one can show that this subfamily is comprehensive in its coverage in the sense that its dependence properties can range from the Frechet–Hoeffding lower bound of perfect negative dependence through to the Frechet–Hoeffding upper bound corresponding to perfect positive dependence.*

It is also worth noting that occasionally an alternative parametrization of the multivariate Clayton copula is presented according to the distribution

$$
C_\theta\left\{u_1,u_2,\ldots,u_d\right) = \max\left\{u_1^{1-\theta}+u_2^{1-\theta}+\cdots+u_d^{1-\theta}-d+1,0\right\}^{\frac{1}{1-\theta}} \tag{11.102}
$$

with $\theta \in [0,\infty)\setminus\{1\}$. In addition, in the case that $\theta \in [0,1)$, one must have the condition on the dimension given by $d \leq \lfloor (1-\theta)^{-1} \rfloor + 1$ to ensure that the resulting function is a valid distribution; see discussions in Nelsen (1999).

11.2.5.2 Two-Parameter Archimedean Members via Outer or Inner Power Transforms. In addition to the commonly utilized one-parameter versions of the Archimedean copula families described earlier it is also worthwhile noting that practitioners can gain additional flexibility in these simple copula families through addition of a second parameter. This second parameter can be incorporated through what is known as an outer-power or an inner-power transform; see Feller (1966).

Definition 11.21 (Outer-Power Copula Transform) *The copula family generated by $\tilde{\psi}(x) = \psi\left(x^{\frac{1}{\beta}}\right)$ is called an outer-power family, where $\beta \in [1, \infty)$ and $\psi \in \Psi_{\infty}$ (the class of completely monotone Archimedean generators).* ■

The proof of this follows from Feller (2008), that is, the composition of a completely monotone function with a non-negative function that has a completely monotone derivative is again completely monotone. Such copula model transforms were also studied in Nelsen (1997), where they are referred to as a beta family associated with the inverse generator ψ^{-1}.

As has been noted earlier, in performing the estimation of these transformed copula models via likelihood-based inference, it will be of great benefit to be capable of performing evaluation pointwise of the copula densities. In the case of the outer-power transformed models, this will require the utilization of a specific multivariate chain rule result widely known as the Faà di Bruno's formula; see Faa di Bruno (1857) and discussions in, for example, Constantine and Savits (1996) and Roman (1980). To understand how such a result is required, consider the following remark.

Remark 11.8 *The generator derivatives for the outer-power transforms can be calculated using the base generator derivatives and the following multidimensional extension to the chain rule for the outer-power versions. The densities for the outer-power copulae in Table 11.2 can thus be calculated using Equation (11.74).*

Hence, one may combine the following derivative results (presented in Table 11.3) for the different Archimedean copula models discussed and their generators with the formula for composite differentiation in Definition 11.8. Using these closed from results combined with the knowledge of composite function differentiation via Fa di Bruno one can then work out the required multivariate densities for implementation of likelihood and Bayesian estimation methods.

Next we briefly introduce the alternative mechanism for adding additional flexibility through the inner-power transform.

Definition 11.22 (Inner-power copula) *The copula family generated by $\tilde{\psi}(x) = \psi^{\frac{1}{\alpha}}(x)$ is called an inner power family, where $\alpha \in (0, \infty)$ and $\psi \in \Psi_{\infty}$ (the class of completely monotone Archimedean generators).* ■

Inner power transforms produce a family of generators associated with the base generator, for example, the Clayton generator is the inner power transform of the base generator $\psi(x) = (1 + x)^{-1}$. The lower tail dependence of the transformed copula is $\lambda_L^{1/\alpha}$, while the upper tail dependence remains unchanged. Inner-power copula model transforms were also studied in Nelsen (1997), where they are referred to as an alpha family associated with the inverse generator ψ^{-1}.

We conclude this brief introduction to Archimedean copula families by mentioning one additional family the Ali–Mikhail–Haq copula. In particular, in the context of modeling dependence compound process with unaffected compound process first-and second-order tail asymptotics, a particular member of the Archimedean family will be of interest, the Ali-Mikhail-Haq (AMH) copula model. The AMH copula is detailed in Definition 11.23; see Kumar (2010). Note that we have seen this copula mentioned previously when introducing the generators and Laplace transform representations of the generator of Archimedean copulae models.

TABLE 11.3 Archimedean copula generator functions, inverse generator functions, and generator function d-th derivatives

Family	ψ	ψ^{-1}	$(-1)^d \psi^{(d)}$
OP-Clayton	$\left(1+t^{1/\beta}\right)^{-1/\rho}$	$(s^{-\rho}-1)^{\beta}$	$\frac{1}{t^d}\sum_{k=1}^{d} a_{dk}^G\left(\frac{1}{\beta}\right)\frac{\Gamma\left(k+\frac{1}{\rho}\right)}{\Gamma\left(\frac{1}{\rho}\right)}\left(1+t^{1/\beta}\right)^{-(k+1/\rho)}\left(t^{1/\beta}\right)^k$
OP-Frank	$-\frac{1}{\rho}\ln\left[1-e^{-t^{1/\beta}}(1-e^{-\rho})\right]$	$\left[-\ln\frac{e^{-s\rho}-1}{e^{-\rho}-1}\right]^{\beta}$	$\frac{1}{t^d}\sum_{k=1}^{d} a_{dk}^G\left(\frac{1}{\beta}\right)\frac{1}{\rho}Li_{-(k-1)}\left\{(1-e^{-\rho})e^{-t^{1/\beta}}\right\}\left(t^{1/\beta}\right)^k$
OP-Gumbel	$\exp\left(-t^{1/(\beta\rho)}\right)$	$(-\ln s)^{\rho\beta}$	$\frac{1}{t^d}\sum_{k=1}^{d} a_{dk}^G\left(\frac{1}{\beta}\right)\frac{\psi_\rho\left(t^{1/\beta}\right)}{t^{k/\beta}}P_{k,1/\rho}^G\left(t^{1/(\rho\beta)}\right)\left(t^{1/\beta}\right)^k$

Definition 11.23 (Ali-Mikhail-Haq Copula Models) *The AMH copula distribution in the bivariate case is given by*

$$C(u_1, u_2) = \frac{u_1 u_2}{1 - \rho(1 - u_1)(1 - u_2)} \tag{11.103}$$

for copula parameter $\rho \in [-1, 1]$. The AMH copula parameter ρ has the following relationship to Kendall's τ and Spearman's ρ rank correlations

$$\begin{aligned}\tau &= \frac{3\rho - 2}{3\rho} - \frac{2(1-\rho)^2 \ln(1-\rho)}{3\rho^2},\\ \rho &= \frac{12(1+\rho)di\ln(1-\rho) - 24(1-\rho)\ln(1-\rho)}{\rho^2} - \frac{3(\rho+12)}{\rho}\end{aligned} \tag{11.104}$$

with $di\ln(x) = \int_1^x \frac{\ln t}{1-t} dt$. ■

11.2.5.3 Truncation Invariant Archimedean Members. The focus of this section is to discuss in which settings and under what conditions can a copula distribution be truncation invariant, that is the truncation of the marginal random variables will not affect the dependence structure between the random variables. It should be noted that in this section we consider cases in which the variables are all truncated, perhaps by different amounts. The truncation invariance to special values of the random vectors has also been studied such as diagonal-invariance and curve-invariance; see Charpentier and Segers (2007). Here we focus on the class of truncation, invariant copula for complete truncation, which is interesting as such models will not alter the measure of dependence such as Kendall's tau or Spearman's rho.

We start this section based on the bivariate discussion on survival copula functions in Oakes (2005). Consider two loss random variables X_1 and X_2 with joint distribution $F_{X_1,X_2}(x_1, x_2)$ and joint survival function $\overline{F}_{X_1,X_2}$. Define the marginal survival functions for each variable respectively by $\overline{F}_{X_1}(x_1) = \overline{F}_{X_1,X_2}(x_1, 0)$ and $\overline{F}_{X_2}(x_2) = \overline{F}_{X_1,X_2}(0, x_2)$. Under the conditions of Sklar's theorem, one can show that there will be a unique copula denoted by C_{X_1,X_2} the links the marginal survival functions to the joint as follows:

$$\overline{F}_{X_1,X_2}(x_1, x_2) = C\left(\overline{F}_{X_1}(x_1), \overline{F}_{X_2}(x_2)\right). \tag{11.105}$$

Now consider the conditional joint survival function of loss random variables X_1, X_2 as would be of interest in joint tail conditional expectations and risk measures in OpRisk, given by $\overline{F}_{X_1,X_2}(x_1, x_2 | X_1 > x, X_2 > y)$, which is given by

$$\overline{F}_{X_1,X_2}(x_1, x_2 | X_1 > x, X_2 > y) = \frac{\overline{F}_{X_1,X_2}(x_1, x_2)}{\overline{F}_{X_1,X_2}(x, y)}, \quad x_1 > x,\ x_2 > y. \tag{11.106}$$

The marginal survival functions are then also easily obtained as

$$\begin{aligned}\overline{F}_{X_1,X_2}(x_1, y | X_1 > x, X_2 > y) &= \frac{\overline{F}_{X_1,X_2}(x_1, y)}{\overline{F}_{X_1,X_2}(x, y)}, \quad x_1 > x,\ x_2 = y,\\ \overline{F}_{X_1,X_2}(x, x_2 | X_1 > x, X_2 > y) &= \frac{\overline{F}_{X_1,X_2}(x, x_2)}{\overline{F}_{X_1,X_2}(x, y)}, \quad x_1 = x,\ x_2 > y.\end{aligned} \tag{11.107}$$

Analogously one may also define a unique copula, denoted by $\widetilde{C}$, for the truncated joint survival function given by

$$\overline{F}_{X_1,X_2}(x_1, x_2 | X_1 > x, X_2 > y) = \widetilde{C}\left(\frac{\overline{F}_{X_1,X_2}(x_1, y)}{\overline{F}_{X_1,X_2}(x, y)}, \frac{\overline{F}_{X_1,X_2}(x, x_2)}{\overline{F}_{X_1,X_2}(x, y)}\right). \quad (11.108)$$

Now for the copula dependence structure to be invariant to truncation, this would imply that one would have $C = \widetilde{C}$ for all measurable events.

Remark 11.9 *Having this property is beneficial for simulation and also estimation. For instance, it would tell one that truncation of the individual loss random variables will not affect the measure or quantified levels of association that depend on the copula model, such as Kendall's tau and Spearman's rho.*

The characterization of such copula models was studied in Oakes (2005) where it was shown that in the bivariate case the Clayton family of Archimedean copula will satisfy this conditional invariance property.

In the context of multivariate survival functions, this property was also studied in Javid (2009), where they extended such results to multivariate settings for d-dim with $d > 2$. In this article, they demonstrate that products of algebraically independent Archimedean multivariate Clayton copulas and standard uniform distributions are the only truncation invariant copulas. To generalize the aforementioned result, consider the multivariate extensions of the above quantities that would produce a multivariate survival distribution and multivariate conditional survival distributions given by

$$\begin{aligned}
&\overline{F}_{X_1,\ldots,X_d}(x_1, \ldots, x_d) \\
&= C\left(\overline{F}_{X_1}(x_1), \ldots, \overline{F}_{X_d}(x_d)\right) \overline{F}_{X_1,\ldots,X_d}(x_1, \ldots, x_d | \{X_1 > y_1, \ldots, X_d > y_d\}) \\
&= \widetilde{C}\left(\frac{\overline{F}_{X_1,\ldots,X_d}(x_1, y_2, \ldots, y_d)}{\overline{F}_{X_1,\ldots,X_d}(y_1, y_2, \ldots, y_d)}, \ldots, \frac{\overline{F}_{X_1,\ldots,X_d}(y_1, y_2, \ldots, x_d)}{\overline{F}_{X_1,\ldots,X_d}(y_1, y_2, \ldots, y_d)}\right).
\end{aligned} \quad (11.109)$$

In the d-variate case, the notion of truncation invariance in the copula structure would imply, as discussed earlier in the trivariate case, that $\widetilde{C} = C$ for all measurable sets. As discussed in Javid (2009), the necessary condition for truncation invariance is given by considering for each $i \in \{1, 2, \ldots, d\}$ the variable $a_i = \overline{F}_{X_i}(x_i)$ and $b_i = \overline{F}_{X_i}(y_i)$, where $x_i > y_i$, $a_i \leq b_i$ and one has $C = \widetilde{C}$ which implies the condition

$$\frac{C(a_1, a_2, \ldots, a_d)}{C(b_1, b_2, \ldots, b_d)} = C\left(\frac{C(a_1, b_2, \ldots, b_d)}{C(b_1, b_2, \ldots, b_d)}, \frac{C(b_1, a_2, \ldots, b_d)}{C(b_1, b_2, \ldots, b_d)}, \ldots, \frac{C(b_1, b_2, \ldots, a_d)}{C(b_1, b_2, \ldots, b_d)}\right). \quad (11.110)$$

For detailed examples in the general d-variate case of Archimedean copulae that will satisfy this condition, see Javid (2009).

We provide a trivariate example that was studied in Sungur (1999) where they also discuss the notion of truncation invariant dependence structure within the context of the Archimedean copula family. They considered a slightly different structure, where one considers the conditional probability for a trivariate random vector $\boldsymbol{X} = (X_1, X_2, X_3)$, which is transformed through the marginals of each component to the unit cube $[0, 1]^3$ that will produce a random vector $\boldsymbol{U} = (U_1, U_2, U_3)$ that is now uniquely characterized by the copula dependence structure, generically denoted by C.

Now consider the conditional distribution of two of these components given the third is constrained to some set, event, or region of the hypercube. For instance, this may yield the example given by the following copula identity

$$\mathbb{P}\mathrm{r}\left[U_1 \leq u_1, U_2 \leq u_2 | U_3 \leq u_3\right] = \frac{C\left(u_1, u_2, u_3\right)}{u_3}, \tag{11.111}$$

corresponding to the joint distribution of (U_1, U_2) conditional on the restriction of the third component where U_3 is truncated to remove to the interval $[u_3, 1]$. If one denotes the truncated vector by notation $\widetilde{\boldsymbol{U}}$, then one may consider under what conditions does the resulting copula for the marginal random vector having removed U_3 to produce $\boldsymbol{U}_{-3} = (U_1, U_2)$ correspond to the copula for the truncated distribution $\widetilde{\boldsymbol{U}}_{-3} = (U_1, U_2) \,|\, \{U_3 \leq u_3\}$. This question was addressed in Sungur (1999) and also in Oakes (2005) and relates directly to considering which members of the Archimedean family of copulae will be truncation invariant. The first result that was shown relates to the member of the Archimedean copula family in the three-dimensional case known as the Cook and Johnson Copula as detailed in Definition 11.24; see details originally developed in the bivariate setting in Cook and Johnson (1981).

Definition 11.24 (Cook and Johnson Archimedean Copula (Three-Dimensional)) *A random vector $\boldsymbol{U} \in [0, 1]^3$ has a distribution in the Cook–Johnson subfamily of Archimedean copula distribution functions if the distribution is given by*

$$C\left(u_1, u_2, u_3\right) = \left(u_1^{-\rho} + u_2^{-\rho} + u_3^{\rho} - 2\right)^{-\frac{1}{\rho}}. \tag{11.112}$$

with $\rho > 0$ and Archimedean generator given by

$$\psi_{CJ}\left(t\right) = \rho^{-1}\left(t^{-\rho} - 1\right). \tag{11.113}$$

■

One can then show that this one-parameter Archimedean family of copula model satisfies the conditions required to be truncation invariant. In general, one can state the conditions for truncation invariance, for instance, in three dimensions according to the following result in Theorem 11.7; see details in Sungur (1999).

Theorem 11.7 (Trivariate Truncation Invariant Copula Representation) *Consider random variables $\{X_i\}_{i=1}^3$ that form a random vector $\boldsymbol{X}$ with a copula distribution C_{X_1,X_2,X_3}. The dependence structure of a random pair of components (X_j, X_k) over the right-sided marginal truncation region $T_{X_i} = \{x_i : x_i > a_i\}$ given by $C_{\widetilde{X}_j,\widetilde{X}_k}\left(u_j, u_k\right)$ is independent of a_i iff C_{X_1,X_2,X_3} takes the following form:*

$$C_{X_1,X_2,X_3}\left(u_1, u_2, u_3\right) = C_{X_j,X_k}\left(\frac{C_{X_j,X_i}\left(u_j, u_i\right)}{u_i}, \frac{C_{X_k,X_i}\left(u_k, u_i\right)}{u_i}\right) u_i, \quad i \neq j \neq k \in \{1, 2, 3\}. \tag{11.114}$$

In addition, one can state the following general result for any d-variate Archimedean copula to be truncation invariant as detailed in Theorem 11.8; see discussions in Sungur (1999).

Theorem 11.8 (d-Variate Truncation Invariant Archimedean Copulae Generators) *Consider a d-dimensional random vector $\boldsymbol{X}$ with an Archimedean copula given by*

$$C_X(u_1, u_2, \ldots, u_d) = \psi^{[-1]}(\psi(u_1) + \cdots + \psi(u_d)). \tag{11.115}$$

The copula will be truncation dependence invariant iff its distribution takes either of the following forms

$$c_X(u_1, u_2, \ldots, u_d) = \prod_{i=1}^{n} u_i, \tag{11.116}$$

or the form

$$C_X(u_1, u_2, \ldots, u_d) = \left(\sum_{i=1}^{n} u_i^{-\rho} - n + 1\right)^{-\frac{1}{\rho}}. \tag{11.117}$$

The only two forms of Archimedean generator that will satisfy these forms are given by

$$\psi(x) = \gamma \ln x \;\; \text{or} \; \psi(x) = \delta(x^{\gamma} - 1) \tag{11.118}$$

for some parameters γ or δ that will ensure the generator is a valid generator for an Archimedean copula.

11.2.6 HIERARCHICAL AND NESTED ARCHIMEDEAN COPULAE

The idea of nested Archimedean copula models is to try to relax the restrictive symmetry enforced by working with an exchangeable Archimedean copula. Allowing for asymmetry, while still working with exchangeable Archimedean copula families can be achieved through composition of different copulas known as "nesting" or in other cases as a "hierarchical" structure. It is important to consider when composite Archimedean copulas will produce a valid density; this was considered in McNeil (2008), see the result in Theorem 11.9.

Theorem 11.9 (Composite Archimedean Generators) *Consider completely monotone Archimedean generators $\psi_i \in \psi_\infty$ for $i \in \{0, 1, \ldots, d\}$ such that the composite function formed by $\psi_k^{-1} \circ \psi_{k+1}$ have completely monotone derivatives for any $k \in \{0, 1, \ldots, d-3\}$, then $C(u_1, \ldots, u_d; \psi_0, \psi_1, \ldots, \psi_{d-2})$ is a copula.*

From this result, one can construct many types of copulae model using the standard Archimedean copula generators. The class of hierarchical archimedean copula (HAC) models were considered in Savu and Trede (2006) and involve the joining of two or more standard bivariate or higher-dimensional Archimedean copulas by another Archimedean copula in such a manner that the resulting dependence structure is well defined and interpretable. Effectively, the approach involves developing multilevel hierarchical Archimedean copulae families. An example of such as structure was developed in Joe (1997) to produce a fully nested d-dimensional copula that no longer has d-exchangeability but can be reduced to have partial exchangeability that allows for greater flexibility in the dependence structures that can be captured. The fully nested d-variate Archimedean copula case involves considering $d-1$ generators and constructing component wise the composite copula given by

$$C(u_1, u_2, \ldots, u_d) = \psi_{d-1}^{-1}\left(\psi_{d-1} \circ \psi_{d-2}^{-1}\left[\ldots\left(\psi_2 \circ \psi_1^{-1}\left[\psi_1(u_1) + \psi_2(u_2)\right] + \psi_2(u_3)\right)\right.\right.$$
$$\left.\left. + \cdots + \psi_{d-2}(u_{d-1})\right] + \psi_{d-1}(u_d)\right). \tag{11.119}$$

In this structure, one achieves $d(d-1)/2$ distinct bivariate margins and $d-1$ copulas with the corresponding parameters, the required conditions on the generators and composite functions of the generators, and inverse generators obtained in Joe (1997).

The second approach that has been proposed involves a mixture of exchangeable and fully nested copulas, which is known as a partially nested model. As discussed in Savu and Trede (2006), such a family of copula models is defined for any dimension $d \geq 4$ where the four-dimensional case involves

$$C(u_1, u_2, \ldots, u_4) = \psi^{-1}\left(\psi \circ \psi_{12}^{-1}\left[\psi_{12}(u_1) + \psi_{12}(u_2)\right] + \psi \circ \psi_{34}^{-1}\left[\psi_{34}(u_3) + \psi_{34}(u_4)\right]\right) \tag{11.120}$$

with generators ψ, ψ_{12}, and ψ_{34}; see discussions in Savu and Trede (2006). In this case, the random variables U_1 and U_2 are exchangeable, as are U_3 and U_4. The remaining pairs are not exchangeable in this construction.

The most general approach to HAC copulas developed in Savu and Trede (2006) involves the developing of a multivariate tree–based copula structure for a d-variate copula with a framework of L levels indexed by $l \in \{0, 1, 2, \ldots, L\}$. Each of the levels involves n_l distinct components:

- **Level** $l = 0$. One has at the lowest base level components $u_1, \ldots, u_d$;
- **Level** $l = 1$. One has n_1 standard multivariate Archimedean copulae that group the variables $u_1, \ldots, u_d$ into copulae $C_{1,j}$ for $j \in \{1, 2, \ldots, n_1\}$ given by

$$C_{1,j}\left(\boldsymbol{u}_{1,j}\right) = \psi_{1,j}^{-1}\left(\sum_{\boldsymbol{u}_{1,j}} \psi_{1,j}\left(\boldsymbol{u}_{1,j}\right)\right) \tag{11.121}$$

  with $\psi_{1,j}$ the generator for copula $C_{1,j}$ and $\boldsymbol{u}_{1,j}$ the associated subset of variables from $u_1, \ldots, u_d$ that are grouped. Note that the copulae $C_{1,1}, \ldots, C_{1,n_1}$ may be different Archimedean families such as Frank, Gumbel, or Clayton, etc.;
- **Level** $l \in \{2, \ldots, L-1\}$. One has n_l generalized Archimedean copulas comprising the aggregation of the copulas from $l-1$. The multivariate Archimedean copulae that group the $l-1$ copulae grouped by $C_{l,j}$ for $j \in \{1, 2, \ldots, n_l\}$ are given by

$$C_{l,j}\left(\boldsymbol{C}_{l-1,j}\right) = \psi_{l,j}^{-1}\left(\sum_{\boldsymbol{C}_{l,j}} \psi_{l,j}\left(\boldsymbol{C}_{l,j}\right)\right) \tag{11.122}$$

with $\psi_{l,j}$ the generator for copula $C_{l,j}$ and $\boldsymbol{C}_{1,j}$ the associated subset of all copulas from level $l-1$ combined in copula $C_{l,j}$;

- **Level** $l = L$. There is just one copula linking the remaining copula groups given by $C_{L,1}$.

The technical conditions that will guarantee this structure produces a valid hierarchical copula structure can be obtained in Savu and Trede (2006, section 3).

In practice, one would typically work with partial nesting in an Archimedean copula model such as given by the model

$$\begin{aligned} C\left(u_1, \ldots, u_d\right) &= C\left(C\left(u_{1,1}, \ldots, u_{1,d_1}; \psi_1\right), \ldots, C\left(u_{s,1}, \ldots, u_{s,d_s}; \psi_s\right); \psi_0\right) \\ &= \psi_0\left(\sum_{i=1}^{s} \psi_0^{-1}\left(\psi_i\left(\sum_{j=1}^{d_i} \psi_i^{-1}\left(u_{i,j}\right)\right)\right)\right) \end{aligned} \tag{11.123}$$

with $d = \sum_{i=1}^{s} d_i$.

One can simulate from a nested Archimedean copula model via an algorithm originally developed in McNeil (2008); see Algorithm 11.7. This can of course be trivially modified to sample from partially nested Archimedean copula models.

Algorithm 11.7 (Sampling from Nested Archimedean Copula Models)

1. *Sample* $V_0 \sim F_0 = \mathcal{L}^{-1}\left[\psi_0\right]$;
2. *Sample* $(X_2, \ldots, X_d)$ *from* $C\left(u_2, \ldots, u_d; \psi_{0,1}\left(\cdot; V_0\right), \ldots, \psi_{0,d-2}\left(\cdot; V_0\right)\right)$;
3. *Sample* $X_1 \sim$ *Uniform*$(0, 1)$;
4. *Return* $(U_1, \ldots, U_d)$ *where* $U_i = \psi_0\left(-\frac{1}{V_0}\ln\left(X_i\right)\right)$ *for* $i \in \{1, 2, \ldots, d\}$.

11.2.7 MIXTURES OF ARCHIMEDEAN COPULAE

In this section, we observe the fact that in practice it is often highly beneficial to consider constructing mixture models for the copula dependence as detailed in Lemma 11.4. Such models allow for asymmetric features in the tail dependence as well as flexible models with additional degrees of freedom when modeling higher-dimensional multivariate random vectors. That is, the advantage of this approach is that one may consider asymmetric dependence relationships in the upper tails and the lower tails in the multivariate model. In addition, one can perform a type of model selection purely by incorporating into the estimation the mixture weights associated with each dependence hypothesis. That is, the data can be utilized to decide the strength of each dependence feature as interpreted directly through the estimated mixture weight attributed to the feature encoded in the particular mixture component from the Archimedean family.

Lemma 11.4 *Consider copula distributional members* $C_i\left(u_1, u_2, \ldots, u_n\right) \in \mathcal{A}^n$, *where* $\mathcal{A}^n$ *defines the space of all possible n-variate distributional members of the Archimedean family of copula models, specified in Lemma 11.3. Any finite mixture distribution constructed from* *such copula components that admit tractable density functions* $c_i(u_1, u_2, \ldots, u_n)$, *denoted by* $\tilde{c}(u_1, u_2, \ldots, u_n) = \sum_{i=1}^{m} w_i c_i(u_1, u_2, \ldots, u_n)$, *such that* $\sum_{i=1}^{m} w_i = 1$, *is also the density of a copula distribution.*

The proof of Lemma 11.4 is provided next.

Proof: The proof of Lemma 11.4 requires one to demonstrate that the resulting distribution function

$$\begin{aligned}\widetilde{C}(u_1, u_2, \ldots, u_n) &= \int\limits_{[0,u_1]\times[0,u_2]\times\cdots\times[0,u_n]} \tilde{c}(x_1, x_2, \ldots, x_n)\, dx_{1:n} \\ &= \sum_{i=1^m} w_i \int\limits_{[0,u_1]\times[0,u_2]\times\cdots\times[0,u_n]} c_i(x_1, x_2, \ldots, x_n)\, dx_{1:n} \\ &= \sum_{i=1^m} w_i C_i(u_1, u_2, \ldots, u_n)\end{aligned}$$

satisfies the two conditions of a n-variate copula distribution given in (Definition 2.10.6) of Nelsen (1999). The first of these conditions requires that for every $\boldsymbol{u} = (u_1, u_2, \ldots, u_n) \in [0,1]^n$, one can show that $\widetilde{C}(\boldsymbol{u}) = 0$ if at least one coordinate of $\boldsymbol{u}$ is 0. Clearly since we have shown that $\widetilde{C}(\boldsymbol{u}) = \sum_{i=1^m} w_i C_i(\boldsymbol{u})$ and given each member $C_i(u_1, u_2, \ldots, u_n) \in \mathcal{A}^n$ is defined to be in the family of Archimedean copulas each of which therefore satisfies this condition for all such points $\boldsymbol{u}$, then it is trivial to see that the probability weighted sum of such points also satisfies this first condition. Secondly, one must show that for every $\boldsymbol{a}$ and $\boldsymbol{b}$ in $[0,1]^n$, such that $\boldsymbol{a} \le \boldsymbol{b}$ (i.e., $a_i < b_i\ \forall i \in \{1, 2, \ldots, n\}$) the following condition on the volume for copula $\widetilde{C}$ is satisfied, $V_{\widetilde{C}}([\boldsymbol{a}, \boldsymbol{b}]) \ge 0$. As in Nelsen (1999), we adopt the notation for the n-box, $[\boldsymbol{a}, \boldsymbol{b}]$, representing $[a_1, b_1] \times [a_2, b_2] \times \cdots \times [a_n, b_n]$ and we define the n-box volume for copula distribution $\widetilde{C}$ by (Definition 2.10.1, p. 43) of Nelsen (1999) giving

$$\begin{aligned}V_{\widetilde{C}}([\boldsymbol{a}, \boldsymbol{b}]) &= \sum \operatorname{sgn}(\boldsymbol{c}) \widetilde{C}(\boldsymbol{c}) \\ &= \triangle_{a_1}^{b_1} \triangle_{a_2}^{b_2} \cdots \triangle_{a_n}^{b_n} \widetilde{C}(\boldsymbol{c}),\end{aligned}$$

where the domain $\mathrm{Dom}\widetilde{C}$ of the mixture copula $\widetilde{C}$ satisfies $[\boldsymbol{a}, \boldsymbol{b}] \subseteq \mathrm{Dom}\widetilde{C}$. In addition, we note that this sum is understood to be taken over all vertices $\boldsymbol{c}$ of n-box $[\boldsymbol{a}, \boldsymbol{b}]$ and $\operatorname{sgn}(\boldsymbol{c}) = 1$ if $c_k = a_k$ for an even number of ks or $\operatorname{sgn}(\boldsymbol{c}) = -1$ if $c_k = a_k$ for an odd number of ks. Equivalently, we consider

$$\triangle_{a_k}^{b_k} \widetilde{C}(\boldsymbol{t}) = \widetilde{C}(t_1, t_2, \ldots, t_{k-1}, b_k, t_{k+1}, \ldots, t_n) - \widetilde{C}(t_1, t_2, \ldots, t_{k-1}, a_k, t_{k+1}, \ldots, t_n). \tag{11.124}$$

In the case of the mixture copula, we can expand the volume of the n-box $[\boldsymbol{a}, \boldsymbol{b}]$ as follows:

$$V_{\widetilde{C}}([\boldsymbol{a}, \boldsymbol{b}]) = \sum \operatorname{sgn}(\boldsymbol{c}) \widetilde{C}(\boldsymbol{c}) = \sum_{i=1}^{m} \sum w_i \operatorname{sgn}(\boldsymbol{c}) C_i(\boldsymbol{c}) = \sum_{i=1}^{m} \sum w_i V_{C_i}([\boldsymbol{a}, \boldsymbol{b}]).$$

Hence, we see that since each component $C_i(u_1, u_2, \ldots, u_n)$ is a member of the set of Archimedean copula distributions $\mathcal{A}^n$, therefore for each component we have that $V_{C_i}([\boldsymbol{a}, \boldsymbol{b}]) \ge 0$ for all $i \in \{1, 2, \ldots, m\}$. ■

Remark 11.10 *We note that the tail dependence of a mixture copula can be obtained as the linear weighted combination of the tail dependence of each component in the mixture weighted by the appropriate mixture weight, as discussed in, for example, Nelsen (1999) and Peters et al. (2012b).*

11.2.8 MULTIVARIATE ARCHIMEDEAN COPULA TAIL DEPENDENCE

We conclude this section on discussion on the Archimedean copula family by making explicit expressions for the general multivariate tail dependence measure introduced earlier. In the case of the Archimedean copula families discussed, one may obtain several useful closed form expressions for quantification of this tail dependence measure with respect to the copula parameter.

The explicit generalized multivariate expressions for Archimedean copulae, Equations (11.125) and (11.126), were derived in De Luca and Rivieccio (2012) and are presented in Definition 11.25 for the upper tail dependence and in Definition 11.26 for the corresponding lower tail dependence.

Definition 11.25 (Generalized Archimedean Upper Tail Dependence)
Let $X = (X_1, \ldots, X_d)^T$ be a d-dimensional random vector with marginal distribution functions $F_1, \ldots, F_d$. The coefficient of upper tail dependence is defined as

$$\begin{aligned}\lambda_u^{1,\ldots,h|h+1,\ldots,d} &= \lim_{\nu\to 1-} P\left(X_1 > F^{-1}(\nu), \ldots, X_h > F^{-1}(\nu) | X_{h+1} > F^{-1}(\nu), \ldots, X_d > F^{-1}(\nu)\right)\\ &= \lim_{t\to 0+} \frac{\sum_{i=1}^{d}\left(\binom{d}{d-i} i(-1)^i \left[(\psi^{-1})'(it)\right]\right)}{\sum_{i=1}^{d-h}\left(\binom{d-h}{d-h-i} i(-1)^i \left[(\psi^{-1})'(it)\right]\right)},\end{aligned} \tag{11.125}$$

where $(\psi^{-1})'$ is the derivative of the inverse generator. Here, h is the number of variables conditioned on (from the d considered). ■

Definition 11.26 (Generalized Archimedean Lower Tail Dependence)
Let $X = (X_1, \ldots, X_d)^T$ be a d-dimensional random vector with marginal distribution functions $F_1, \ldots, F_d$. The coefficient of lower tail dependence is defined as

$$\begin{aligned}\lambda_l^{1,\ldots,h|h+1,\ldots,d} &= \lim_{\nu\to 0+} P\left(X_1 < F^{-1}(\nu), \ldots, X_h < F^{-1}(\nu) | X_{h+1} < F^{-1}(\nu), \ldots, X_d < F^{-1}(\nu)\right)\\ &= \lim_{t\to\infty} \frac{d}{d-h}\frac{(\psi^{-1})'(dt)}{(\psi^{-1})'((d-h)t)},\end{aligned} \tag{11.126}$$

where $(\psi^{-1})'$ is the derivative of the inverse generator. Here, h is the number of variables conditioned on (from the d considered). ■

TABLE 11.4 **Kendall's tau and tail dependence coefficients**

Family	τ	λ_L	λ_U
Clayton	$\frac{\rho}{\rho+2}$	$2^{-\frac{1}{\rho}}$	0
Frank	$1+\frac{4D_1(\rho)-1}{\rho}$	0	0
Gumbel	$\frac{(\rho-1)}{\rho}$	0	$2-2^{\frac{1}{\rho}}$

The exact nonlinear transformations between the copula parameter ρ and Kendall's rank correlation τ for the Clayton, Frank, and Gumbel copulae can be seen in Table 11.4.

Remark 11.11 *In defining the mapping for the Frank copula between the copula parameter and the upper tail dependence, one utilizes the Debye function of order one given by*

$$D_1 = \frac{1}{\rho}\int_0^{\rho} \frac{t}{\exp(t)-1}\,dt. \tag{11.127}$$

Next we also briefly mention the known closed from results for the tail dependence of the AMH copula family. The lower and upper tail dependence of the AMH copula family is given in Kumar (2010) according to the result in Proposition 11.5.

Proposition 11.5 (Upper and Lower Tail Dependence for AMH Copula) *Considering the bivariate copula distribution in the AMH family with parameter $\rho \in [-1,1]$, then the upper and lower tail dependence are given by*

$$\begin{aligned}\lambda_l &= \lim_{u\downarrow 0}\frac{C(u,u)}{u} = \begin{cases}0.5, & \text{if } \rho=1,\\ 0, & \text{if } \rho<1,\end{cases}\\ \lambda_u &= \lim_{u\uparrow 1}\frac{1-2u+C(u,u)}{1-u} = 0.\end{aligned} \tag{11.128}$$

This result shows that under an AMH copula two loss random variables will be asymptotically dependent only if the copula parameter is on the boundary $\rho = 1$, otherwise they are asymptotically independent. Clearly, this shows that in practice one must be careful to undertake estimation of such parameters with care as we will see later that one value may admit an asymptotic compound process expansion, whereas others will not.

11.3 Copula Parameter Estimation in Two Stages: Inference for the Margins

The inference function for margins (IFM) technique introduced in Joe (2005) provides a computationally faster method for estimating parameters than full-maximum likelihood, that is, simultaneously maximizing all model parameters and produces in many cases a more stable likelihood estimation procedure. An alternative approach to copula model parameter estimation that is popular in the literature is known as the maximum partial likelihood estimator

(MPLE) detailed in Genest *et al.* (1995). We begin this section with a brief description of the MPLE and then introduce the IFM methodology.

There are two considerations to be made when applying these methodology. The first is what assumptions are suitable for the marginal distributions, and the second is what is the suitable class of copula dependence models. In terms of the marginal distributions, one has really two choices:

- **Parameteric Marginal Models.** To utilize particular parameteric families of marginal distributions, where each possible parameter model is characterized by model index $\mathcal{M}_j$ for J total model classes under consideration with $j = \{1, 2, \ldots, J\}$ and for each marginal random variable (loss process) one has parameter vector for the j-th model given by $\boldsymbol{\theta}_i(\mathcal{M}_j)$, which produces the set of marginal models under consideration for the case of d-loss processes given by

$$\left\{F_{X_i}\left(x_i; \boldsymbol{\theta}_i\left(\mathcal{M}_j\right), \mathcal{M}_j\right)\right\}_{i=1}^{d}.$$

- **Nonparametric Marginal Models (Empirical Distribution Function).** One may alternatively choose to utilize a nonparametric marginal model for some or all of the loss processes in which case one would assume that for the i-th marginal loss process the following empirical distribution function approximation is considered:

$$\hat{F}_{X_i}\left(x_i\right) = \frac{1}{n_i}\sum_{j=1}^{n_i}\mathbb{I}\left[X_{i,j} \leq x_i\right]. \tag{11.129}$$

This is only sensible if sufficient loss data n_i is available to make an accurate representation.

11.3.1 MPLE: COPULA PARAMETER ESTIMATION

In the MPLE estimation procedure, the copula parameters are estimated based on marginal distributions obtained from the emprical distribution functions of each individual loss process where for the i-th marginal loss process the following empirical distribution function approximation is considered for a given data realization:

$$\hat{F}_{X_i}\left(x_i\right) = \frac{1}{n_i}\sum_{j=1}^{n_i}\mathbb{I}\left[X_{i,j} \leq x_i\right]. \tag{11.130}$$

The resulting likelihood estimation for the generic copula parameter ρ is achieved by maximizing the log likelihood given by

$$l(\rho) = \sum_{i=1}^{n}\ln\left[c_\rho\left(\hat{F}_{X_1}\left(x_{1,i}\right), \ldots, \hat{F}_{X_d}\left(x_{d,i}\right)\right)\right]. \tag{11.131}$$

Note that in practice it is wise to rescale each of the marginals by $n/(n+1)$ to avoid numerical problems if the i-th value u_i in the copula approaches the boundaries 0 or 1 causing the copula density to grow unboundedly large. Note that this can be rewritten in terms of ranks or pseudo data samples. The properties of the estimator for the parameter ρ that maximizes this pseudo

Color Insert

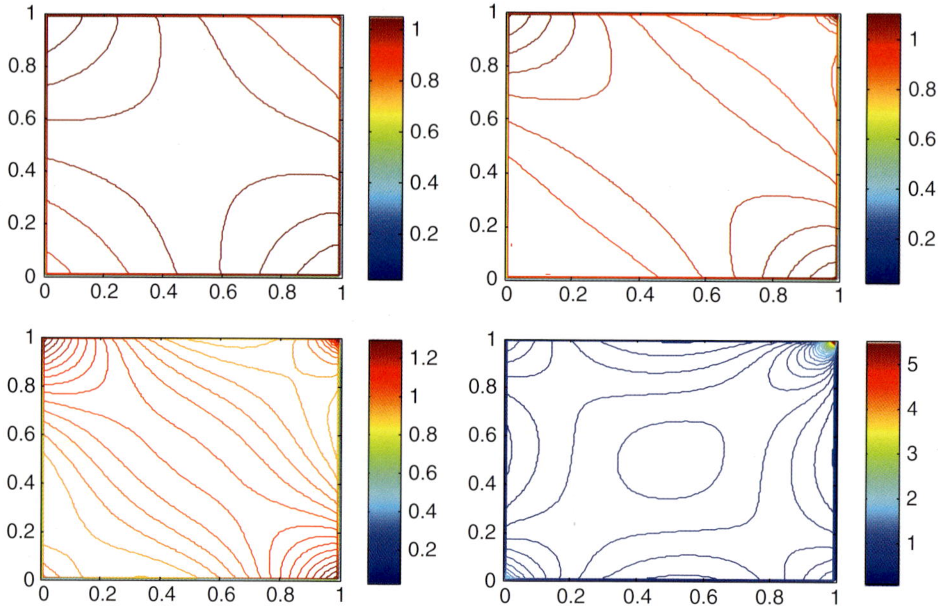

FIGURE 7.4 Top left subplot: target distribution π_0 at a low temperature, where the distribution is fairly flat and simple to sample. Top right subplot: target distribution π_{t_1} at an intermediate temperature, where the distribution is still fairly flat and simple to sample. Bottom left subplot: target distribution π_{t_2} at an intermediate temperature, where the distribution is increasingly concentrated. Bottom right subplot: target distribution π_T final distribution, which is the target distribution.

Fundamental Aspects of Operational Risk and Insurance Analytics: A Handbook of Operational Risk, First Edition. Marcelo G. Cruz, Gareth W. Peters, and Pavel V. Shevchenko.

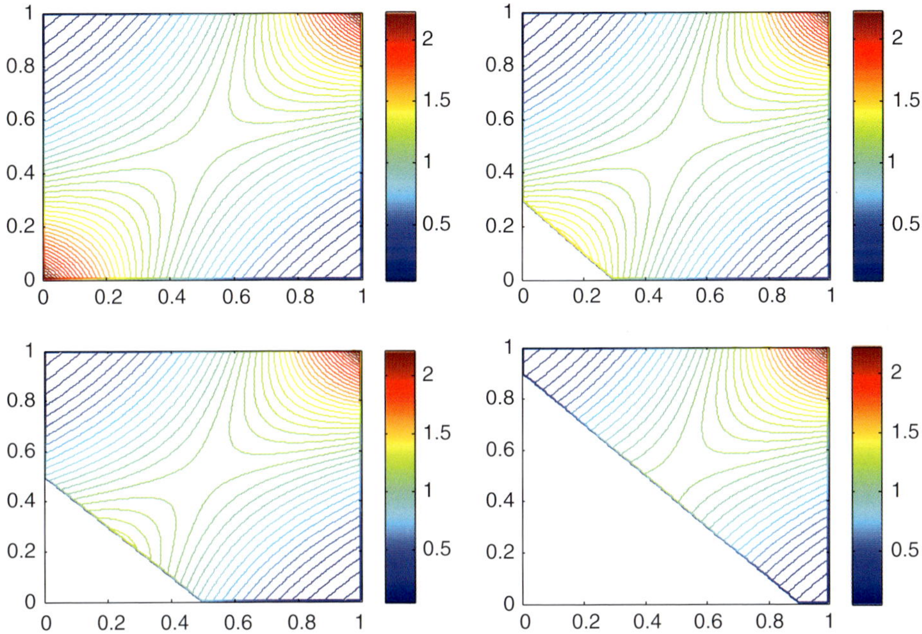

FIGURE 7.5 Top left subplot: target distribution copula component under uniform distribution function transformation for π_0 at little truncation, where the distribution is fairly flat and simple to sample. Top right subplot: target distribution copula component under uniform distribution function transformation for π_{t_1} at an intermediate truncation. Bottom left subplot: target distribution copula component under uniform distribution function transformation for π_{t_2} at an intermediate truncation. Bottom right subplot: target distribution copula component under uniform distribution function transformation for π_T final distribution, which is the target distribution.

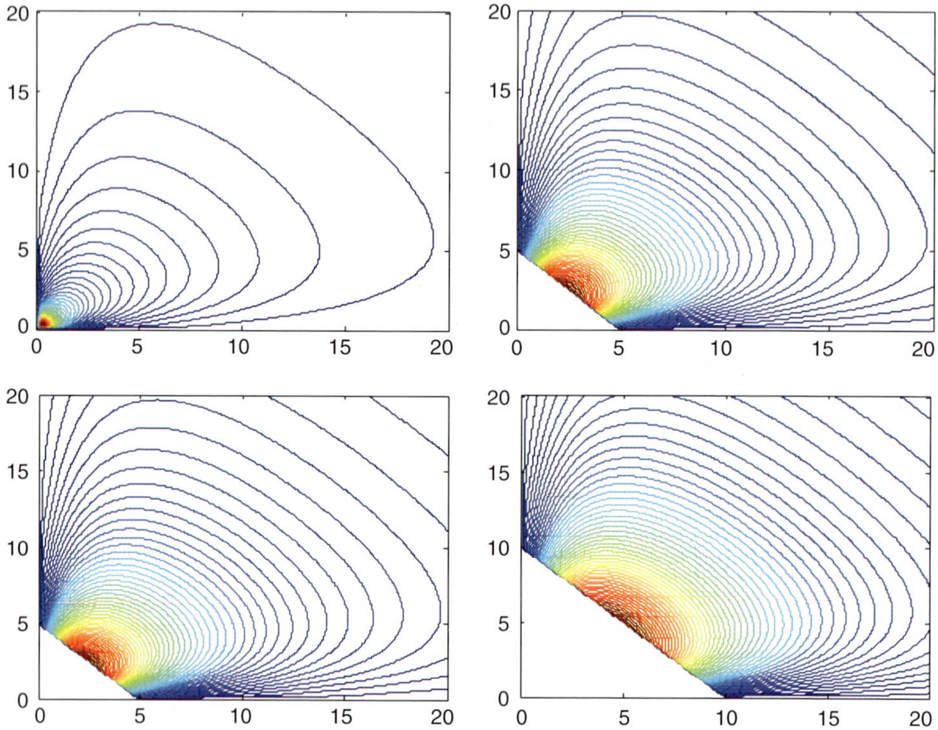

FIGURE 7.6 Top left subplot: target distribution π_0 at little truncation, where the distribution is fairly flat and simple to sample. Top right subplot: target distribution π_{t_1} at an intermediate truncation. Bottom left subplot: target distribution π_{t_2} at an intermediate truncation. Bottom right subplot: target distribution π_T final distribution, which is the target distribution.

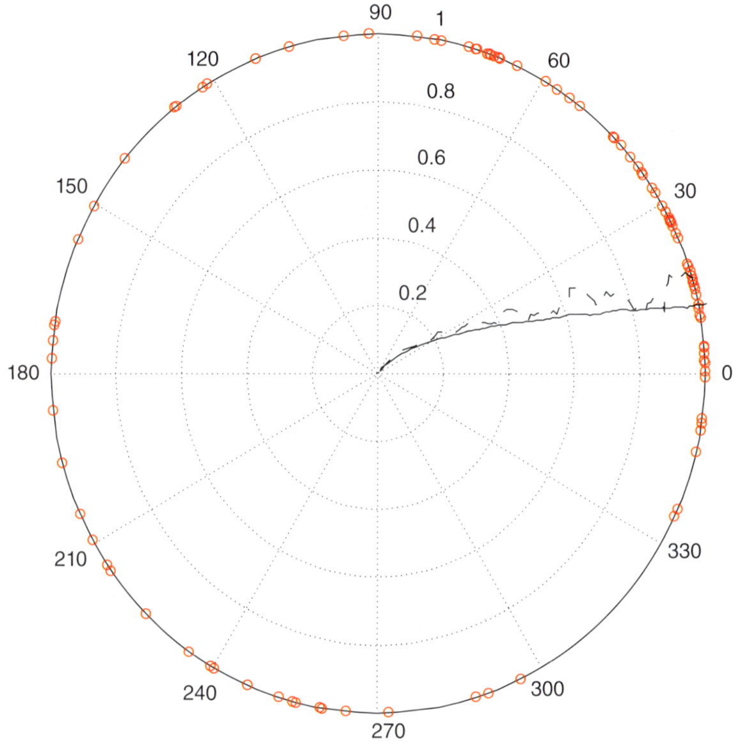

FIGURE 8.4 Red circles depict the project's observation realizations $x_1, x_2, x_3, \ldots, x_n$ on the unit disk in the complex plane for a severity model *LogNormal*$(\mu = 1, \sigma = 2)$. In the dashed black line we see the ECF estimated for the model from the data and the solid black line demonstrates the true characteristic function.

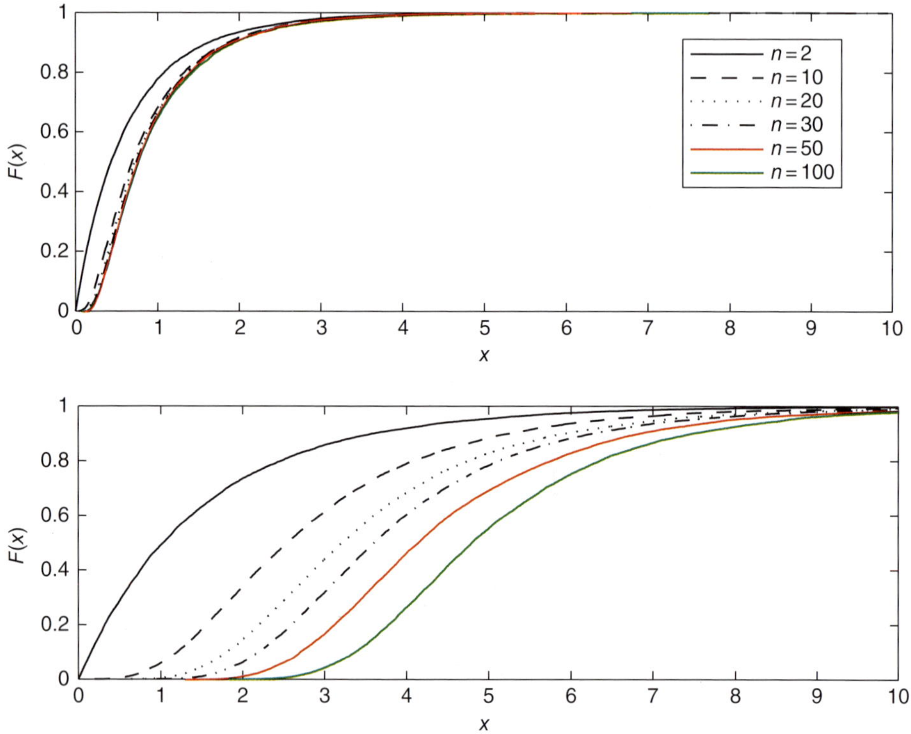

FIGURE 8.7 Distribution of the test statistic as a function of the number of random variables $\{\lambda_i\}_{i=1}^{n}$ for $n \in \{2, 10, 20, 30, 50, 100\}$. Top subplot is for weight function Case 1 and the bottom subplot is for weight function Case 2.

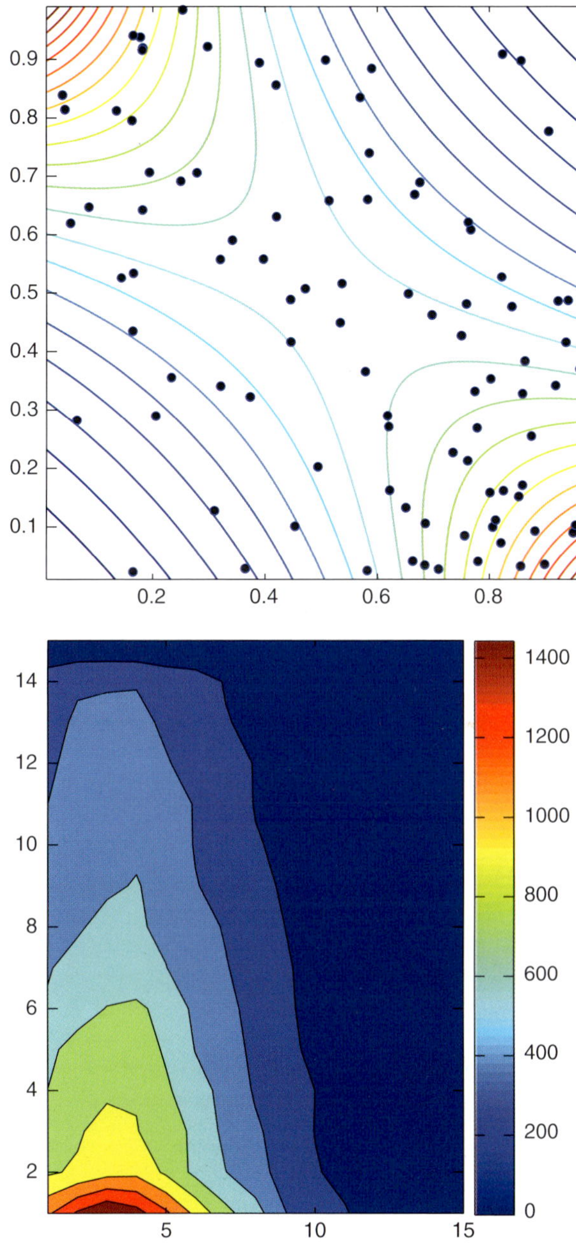

FIGURE 8.8 Top subplot: this plot shows the true copula contours used in this model, that is, a Frank copula, and the points correspond to the pseudo data obtained by transformation through the empirical marginals (i.e., using the marginal scaled ranks). Bottom subplot: this plot shows the contours of the joint loss process density for $\left(Z^{(1)}, Z^{(2)}\right)$.

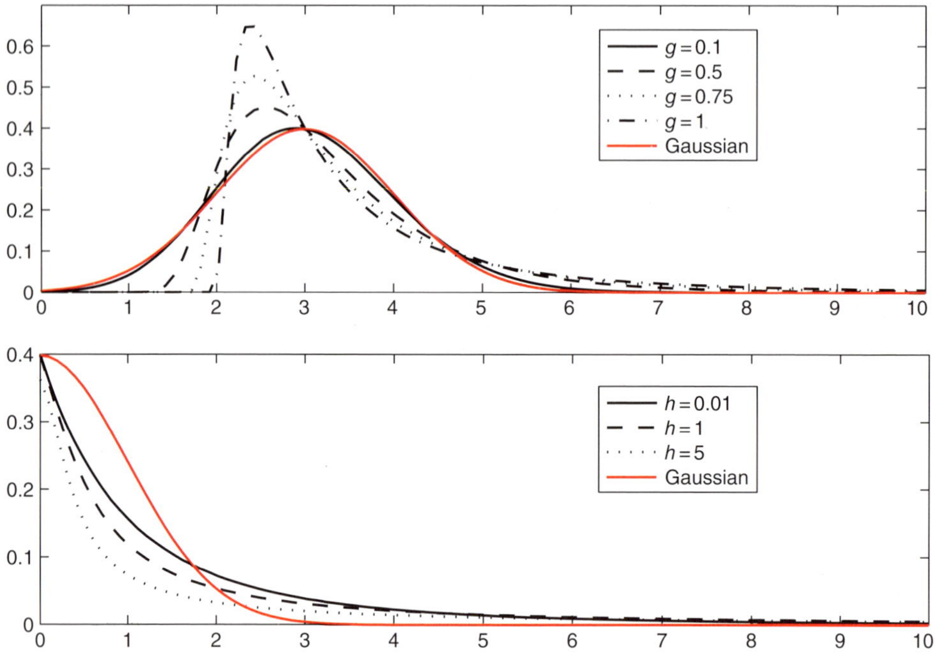

FIGURE 9.5 Top subplot: This plot shows the effect of the skewness parameter g on the elongation transformed severity distribution versus the base Gaussian distribution with $g \in \{0.1, 0.5, 0.75, 1\}$. In this case, the other parameters were set to $a = 3$, $b = 1$, and $h = 0.001$. Bottom subplot: This plot shows the effect of the kurtosis parameter h on the elongation transformed severity distribution versus the base Gaussian distribution with $h \in \{0.01, 1, 5\}$. In this case, the other parameters were set to $a = 0$, $b = 1$, and $g = 1$.

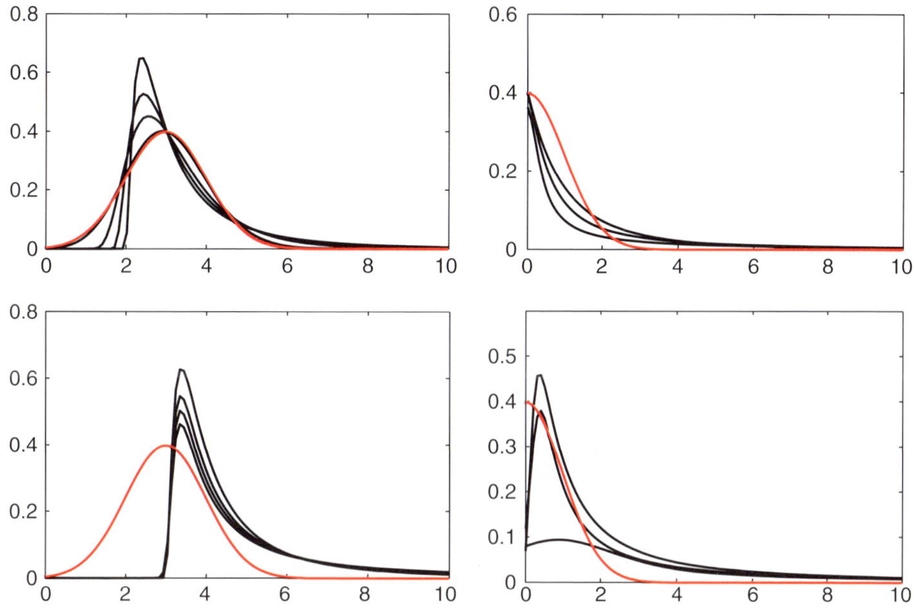

FIGURE 9.6 Top left subplot: This plot shows the effect of the skewness parameter g on the elongation transformed severity distribution versus the base Gaussian distribution with $g \in \{0.1, 0.5, 0.75, 1\}$. In this case, the other parameters were set to $a = 3$, $b = 1$, and $h = 0.001$. Top right subplot: This plot shows the effect of the kurtosis parameter h on the elongation transformed severity distribution versus the base Gaussian distribution with $h \in \{0.01, 1, 5\}$. In this case, the other parameters were set to $a = 0$, $b = 1$, and $g = 1$. Bottom left subplot: This plot shows the effect of the skewness parameter g on the elongation transformed severity distribution versus the base *LogNormal*$(0, 1)$ distribution with $g \in \{0.1, 0.5, 0.75, 1\}$. In this case, the other parameters were set to $a = 3$, $b = 1$, and $h = 0.001$. Bottom right subplot: This plot shows the effect of the kurtosis parameter h on the elongation transformed severity distribution versus the base *LogNormal*$(0, 1)$ distribution with $h \in \{0.01, 1, 5\}$. In this case, the other parameters were set to $a = 0$, $b = 1$, and $g = 1$.

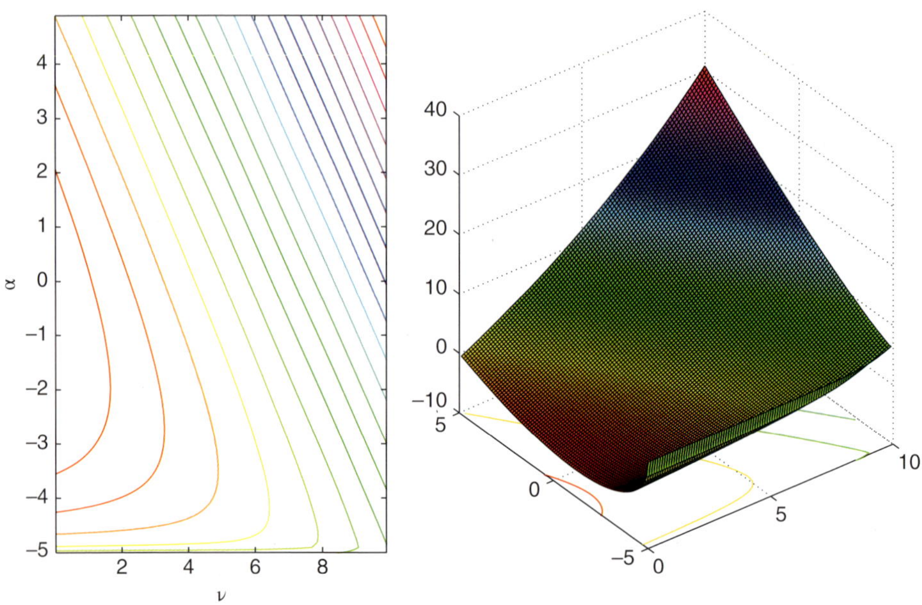

FIGURE 9.10 Log of the exponential factorial function for a range of parameters ν and α.

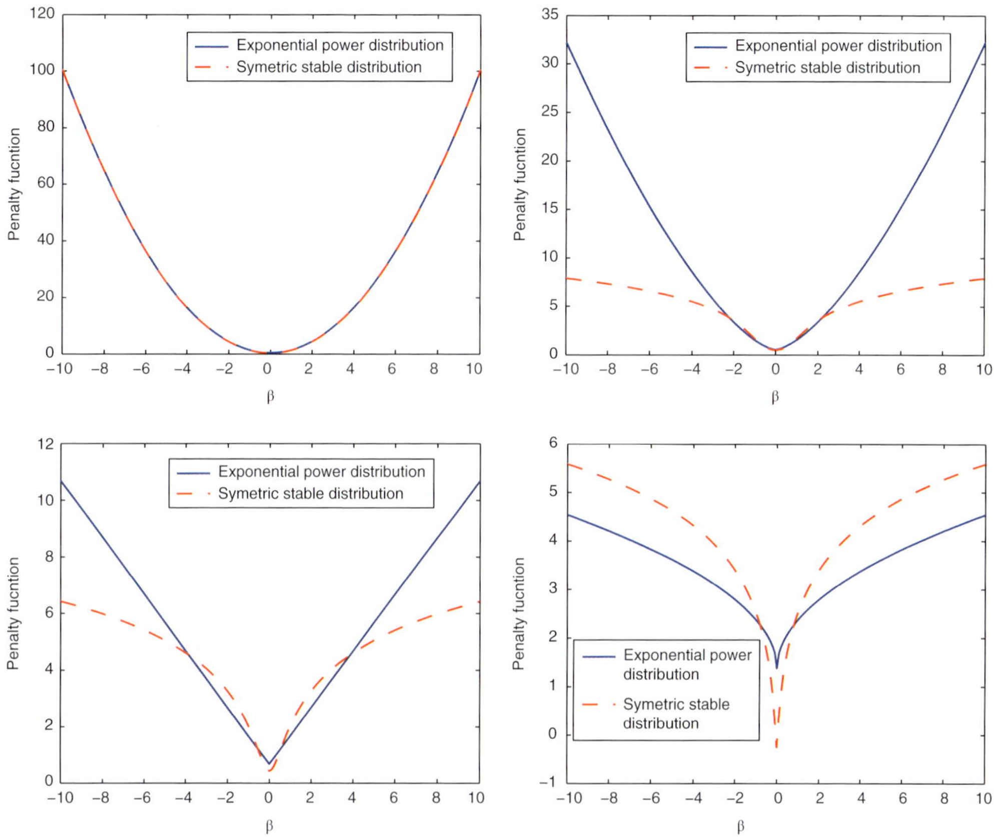

FIGURE 16.1 Comparison of the penalty term induced by the log prior of the regression coefficient to be either the exponential power distribution or the α-stable distribution ($\gamma_{EP} = 2\gamma_{\alpha} = 1$). Top left $q = 2$, Top right $q = 1.5$, Bottom Left $q = 1$, Bottom Right $q = 0.5$.

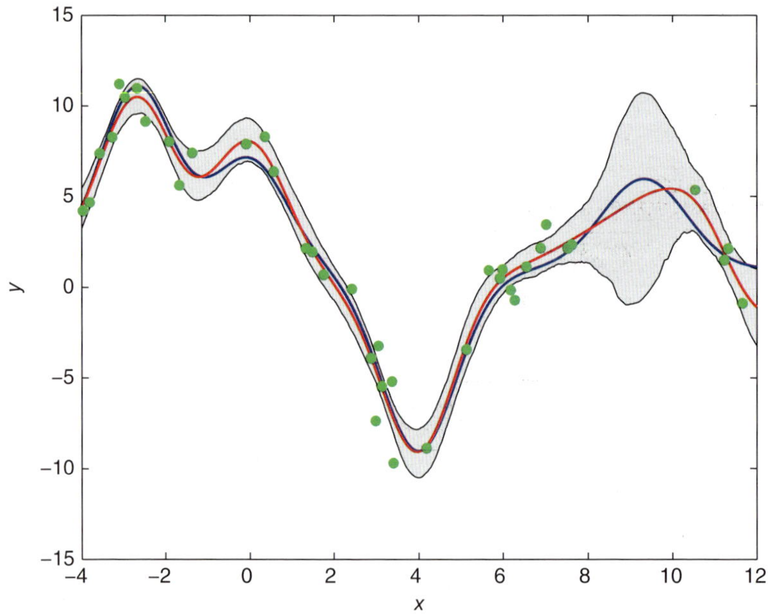

FIGURE 16.2 Normal regression with *EP* prior ($q = 1$): true function in blue - observed responses in green-filled circles, posterior mean from SMC under model $\mathcal{M}_3$ in red and confidence region in gray (5–95% percentiles).

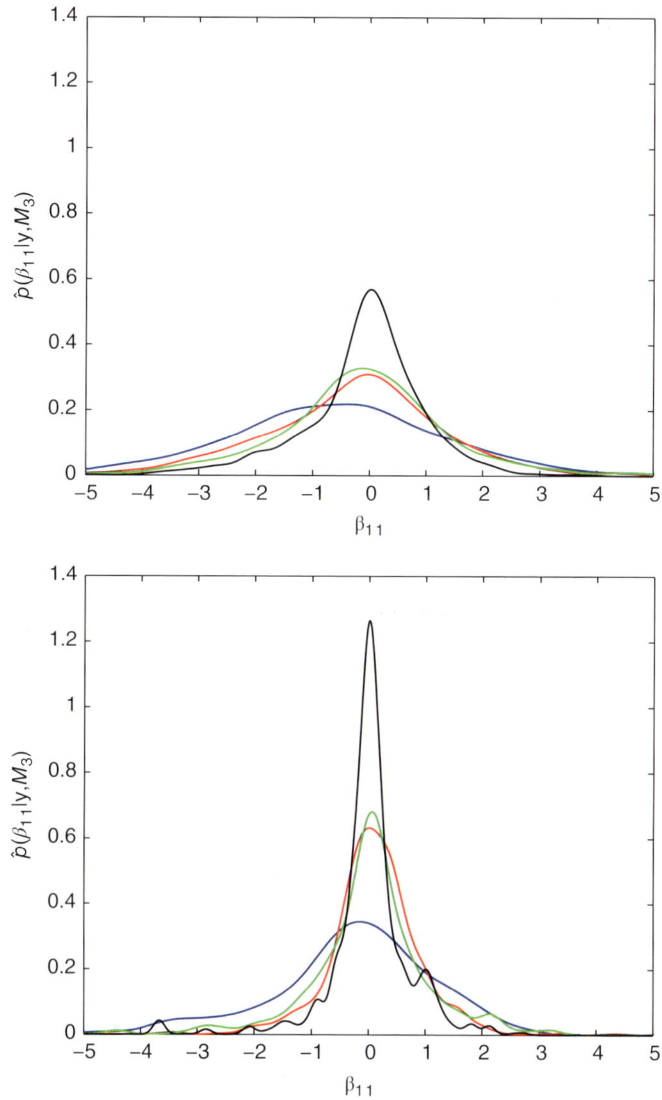

FIGURE 16.3 Comparison of the shrinkage results obtained with the two different priors as q decreases (blue: $q = 1.5$, red: $q = 1$, green: $q = 0.8$, black: $q = 0.5$). Top plot is EP prior and bottom plot is a symmetric α-Stable prior.

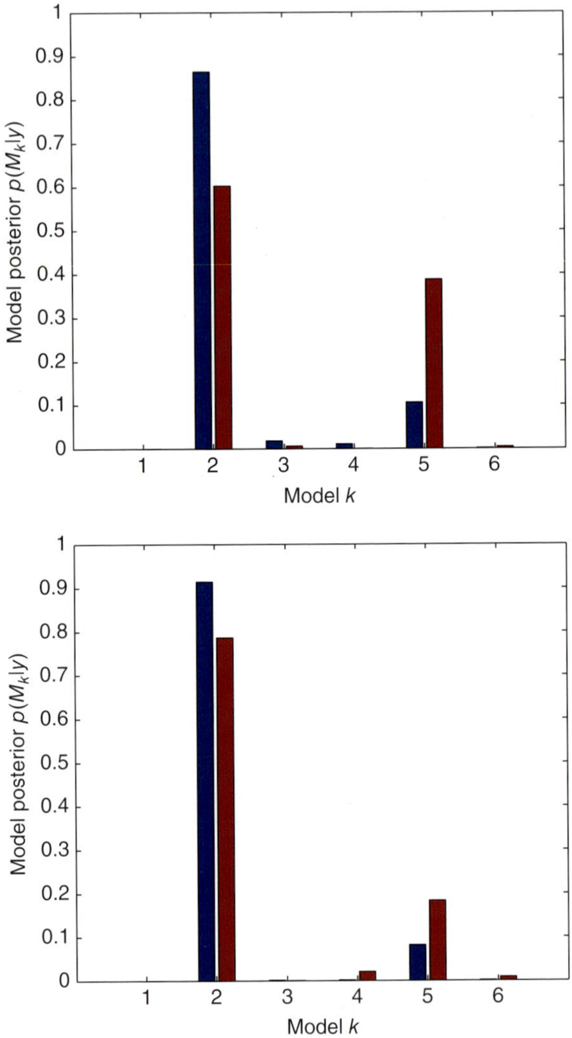

FIGURE 16.4 Comparison of the approximation of the model posterior (blue: α, red: EP).

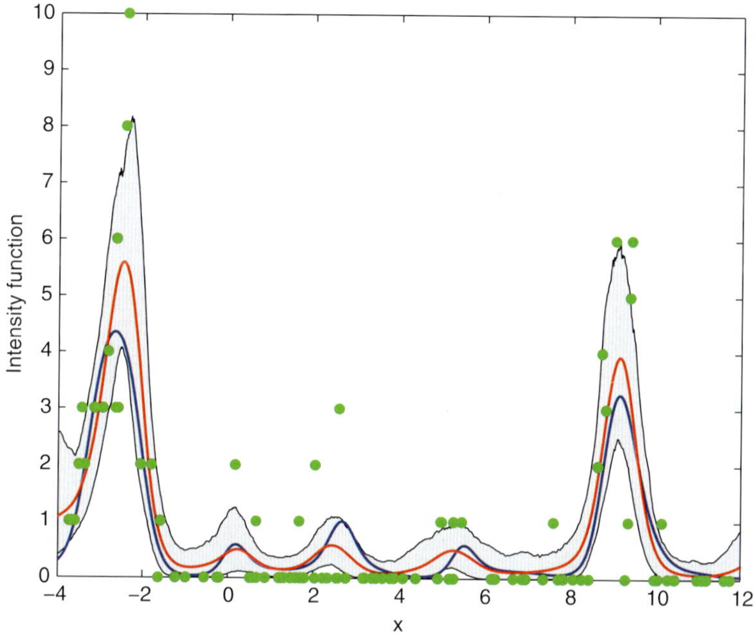

FIGURE 16.5 Poisson regression with α prior ($q = 1$): true function in blue, observed count responses in green-filled circles, posterior mean from SMC under model $\mathcal{M}_3$ in red, and confidence region in gray (5–95% percentiles).

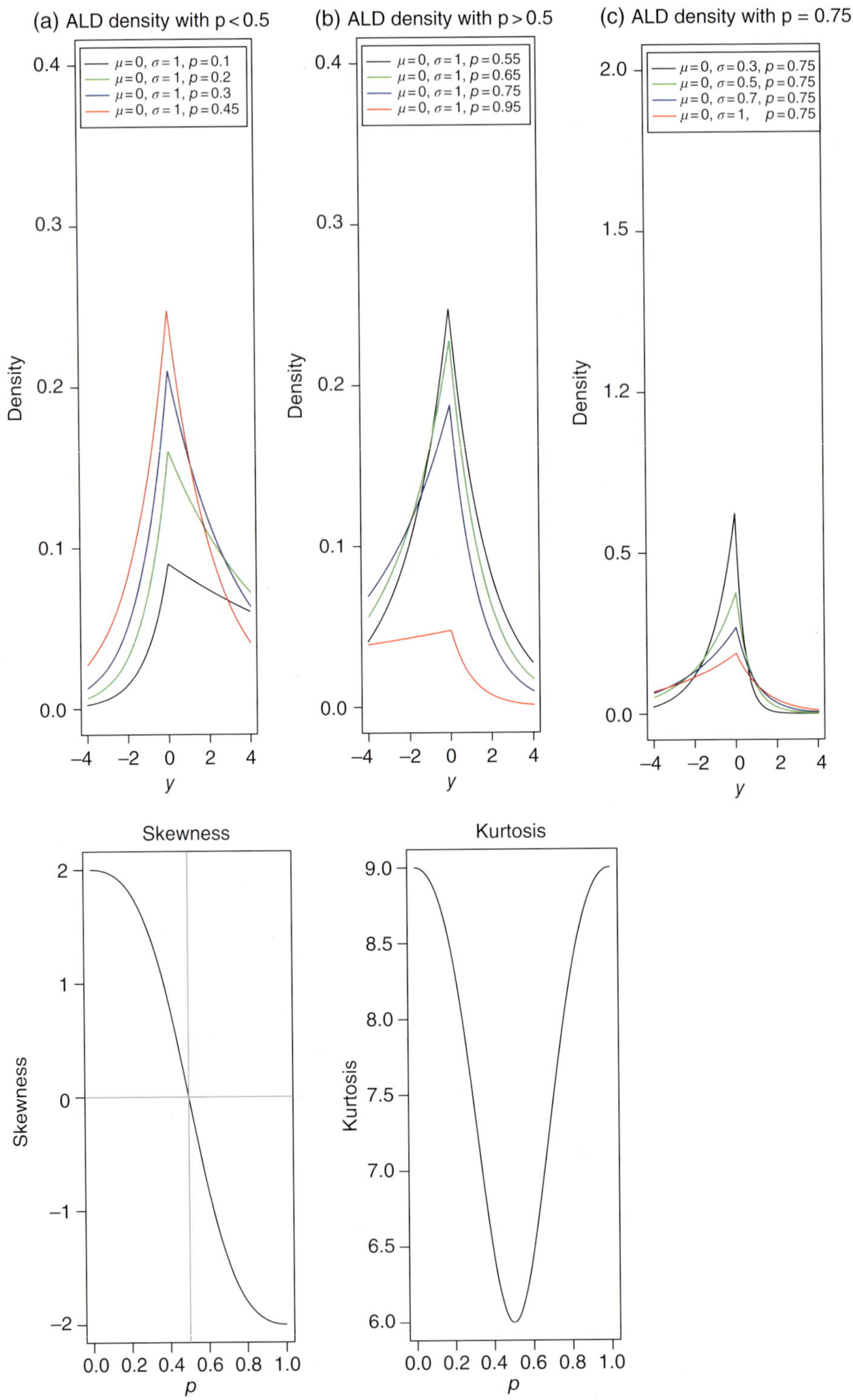

FIGURE 16.7 Top: Assymetric Laplace densities for a range of parameter values. Bottom: ALD skewness and kurtosis.

data likelihood have been shown in Genest *et al.* (1995) under mild regularity conditions on the copula family considered to be a unique solution with the MPLE estimator also satisfying asymptotic normality.

11.3.2 INFERENCE FUNCTIONS FOR MARGINS (IFM): COPULA PARAMETER ESTIMATION

Here, we consider the likelihood-based estimation in two stages via inference on the margins, which is studied with regard to the asymptotic relative efficiency of the two-stage estimation procedure compared with maximum likelihood estimation in Joe (2005) and in Hafner and Manner (2010). It can be shown that the IFM estimator is consistent under weak regularity conditions. However, it is not fully efficient for the copula parameters. Nevertheless, it is widely used for its ease of implementation and efficiency in large data settings. For details on what can go wrong in the estimation of the copula parameter when the marginals are poorly selected in the two-stage IFM procedure, refer to Kim *et al.* (2007).

11.3.2.1 Stage 1: Fitting the Marginal Distributions via MLE. In the first stage, one has to fit the marginal distributions and select the most appropriate marginal model for each individual loss process. For instance, in the case of a LogNormal model, this is achieved trivially since we may utilize the well-known analytic expressions for the MLE estimates:

$$\begin{aligned}\hat{\mu} &= \frac{1}{N}\sum_j \ln\left(x_j\right),\\ \hat{\sigma} &= \sqrt{\frac{1}{N}\sum_j \left(\ln x_j\right)^2 - \hat{\mu}^2}.\end{aligned} \tag{11.132}$$

If one is considering a log generalized Gamma distribution (l.g.g.d.), the estimation for the three model parameters can be significantly more challenging due to the fact that a wide range of model parameters, especially for k, can produce similar resulting density shapes; see discussions in Lawless (1980). To overcome this complication and to make the estimation efficient, it is proposed to utilize a combination of profile likelihood methods over a grid of values for k and perform profile likelihood based MLE estimation for each value of k, then for the other two parameters b and u. The differentiation of the profile likelihood for a given value of k produces the system of two equations given by

$$\begin{aligned}&\exp(\tilde{\mu}) = \left[\frac{1}{n}\sum_{i=1}^{n}\exp\left(\frac{y_i}{\tilde{\sigma}\sqrt{k}}\right)\right]^{\tilde{\sigma}\sqrt{k}},\\ &\frac{\sum_{i=1}^{n} y_i \exp\left(\frac{y_i}{\tilde{\sigma}\sqrt{k}}\right)}{\sum_{i=1}^{n}\exp\left(\frac{y_i}{\tilde{\sigma}\sqrt{k}}\right)} - \bar{y} - \frac{\tilde{\sigma}}{\sqrt{k}} = 0,\end{aligned} \tag{11.133}$$

with n the number of observations, $y_i = \ln x_i$ and the parameter transformations $\tilde{\sigma} = \frac{b}{\sqrt{k}}$ and $\tilde{\mu} = u + b\ln k$. The second equation is solved directly via a simple root search for the estimation of $\tilde{\sigma}$ and then substitution into the first equation provides the estimation of $\tilde{\mu}$. Note that, for each value of k we select in the grid, we get the pair of parameter estimates $\tilde{\mu}$ and $\tilde{\sigma}$, which can then be plugged back into the profile likelihood to make it purely a function of k, with the estimator for k then selected as the one with the maximum likelihood score.

11.3.2.2 Stage 2: Fitting the Mixture Copula via MLE.

In the second stage of the esitmation, one aims to estimate the copula model parameters, given the fixed marginal distributions $\left\{F_{X_i}\left(x_i;\hat{\boldsymbol{\theta}}\left(\mathcal{M}_i\right)\right)\right\}_{i=1}^{d}$, where $\mathcal{M}_i$ represents the model selected for the i-th loss process. The resulting log-likelihood that will be optimized in general for the copula parameter, generically denoted by ρ, is given by

$$l(\rho)=\sum_{i=1}^{n}\ln\left[c_\rho\left(F_{X_1}\left(x_{1,i};\hat{\boldsymbol{\theta}}\left(\mathcal{M}_1\right)\right),\ldots,F_{X_d}\left(x_{d,i};\hat{\boldsymbol{\theta}}\left(\mathcal{M}_d\right)\right)\right)\right]. \tag{11.134}$$

To illustrate this, we consider the following mixture copula model given by components of the Archimedean families of Clayton, Frank, and Gumbel (C-F-G) models with the copulae parameters $(\rho_{Clayton}, \rho_{Frank}, \rho_{Gumbel})$ and the copulae mixture parameter weights

$$(\lambda_{Clayton},\ \lambda_{Frank},\ \lambda_{Gumbel}).$$

These parameters will be estimated in the second stage of the IFM procedure by using maximum likelihood on the data after conditioning on the selected marginal distribution models and their corresponding estimated parameters obtained in stage 1. These models are utilized to transform the data using the distribution function with the MLE parameters ($\hat{\mu}$ and $\hat{\sigma}$) if the LogNormal model is used or ($\hat{k}$, $\hat{u}$, and $\hat{b}$) if the l.g.g.d. is considered.

Therefore, in this second stage of MLE estimation, one aims to estimate either the one-parameter mixture of C-F-G components with parameters

$$\boldsymbol{\theta}=(\rho_{Clayton},\ \rho_{Frank},\ \rho_{Gumbel},\ \lambda_{Clayton},\ \lambda_{Frank},\ \lambda_{Gumbel})$$

or, for instance, the two-parameter mixture of outer-power transformed mixture components OC-OF-OG components with parameters

$$\boldsymbol{\theta}=(\rho_{Clayton},\rho_{Frank},\rho_{Gumbel},\lambda_{Clayton},\lambda_{Frank},\lambda_{Gumbel},\beta_{Clayton},\beta_{Frank},\beta_{Gumbel}).$$

This can be achieved in each case by the conditional maximum likelihood procedure. To achieve this, we need to maximize the log likelihood expressions for the mixture copula models, which in this framework are given generically by the following function for which we need to find the mode,

$$l(\boldsymbol{\theta})=\sum_{i=1}^{n}\ln c^{C-F-G}(F_1(X_{i,1};\hat{\mu}_1,\hat{\sigma}_1),\ldots,F_d(X_{i,d};\hat{\mu}_d,\hat{\sigma}_d))+\sum_{i=1}^{n}\sum_{j=1}^{d}\ln f_j(X_{i,j};\hat{\mu}_j,\hat{\sigma}_j) \tag{11.135}$$

with respect to the parameter vector $\boldsymbol{\theta}$.

For example in the case of the Clayton–Frank–Gumbel mixture copula, we need to maximize on the log-scale the following expression.

$$
\begin{aligned}
l(\boldsymbol{\theta}) = \sum_{i=1}^{n} \ln \big[& \lambda_C \left(c^C_{\rho_C} \left(F_1 \left(X_{i,1}; \hat{\mu}_1, \hat{\sigma}_1 \right) \ldots, F_d \left(X_{i,d}; \hat{\mu}_d, \hat{\sigma}_d \right) \right) \right) \\
& + \lambda_F \left(c^F_{\rho_F} \left(F_1 \left(X_{i,1}; \hat{\mu}_1, \hat{\sigma}_1 \right) \ldots, F_d \left(X_{i,d}; \hat{\mu}_d, \hat{\sigma}_d \right) \right) \right) \\
& + \lambda_G \left(c^G_{\rho_G} \left(F_1 \left(X_{i,1}; \hat{\mu}_1, \hat{\sigma}_1 \right) \ldots, F_d \left(X_{i,d}; \hat{\mu}_d, \hat{\sigma}_d \right) \right) \right) \big].
\end{aligned}
\tag{11.136}
$$

This optimization is achieved via a gradient descent iterative algorithm that can be quite robust given the likelihood surfaces considered in these models, even when real data are utilized, see discussion in Ames *et al.* (2013). In some cases, for the mixture copula models, it would also be suitable to utilize an EM-based estimation algorithm as discussed in Chapter 7.

CHAPTER TWELVE

Examples of LDA Dependence Models

In this third chapter on dependence modelling in OpRisk, we utilise the theory and models developed in the previous two chapters to construct a range of OpRisk LDA models that are directly applicable to practitioners. These include:

- Multiple risk LDA compound Poisson processes and Levy copulas;
- Multiple risk LDA models with dependence between frequencies via copula;
- Multipel risk LDA models with dependence between event times via copula;
- Multiple risk LDA models with dependence between severities via copula;
- Multiple risk LDA models with common shock process dependence features and self chaining copula models;
- Multiple risk LDA models with dependence between annual (aggregate) losses via copula; and
- Multiple risk LDA models with dependence in the risk profiles of the LDA model frequency and severity parameters.

We then conclude the chapter with a complete model of multiple risk LDA models with multiple data sources combined and dependence structures incorporated. We demonstrate the properties of such a model and show how to make inference with this model under a Bayesian formulation with MCMC samplers via a Slice sampler. A numerical example is developed and the predictive posterior distribution specified.

12.1 Multiple Risk LDA Compound Poisson Processes and Lévy Copula

Characterizing multivariate Lévy processes has been an active topic in recent years for financial mathematics, risk, and insurance. In general, there are three well-known methods to construct a multidimensional Lévy process:

Fundamental Aspects of Operational Risk and Insurance Analytics: A Handbook of Operational Risk, First Edition. Marcelo G. Cruz, Gareth W. Peters, and Pavel V. Shevchenko.

1. Subordination of multidimensional Brownian motion;
2. Linear transformation of independent Lévy processes;
3. Multi-dimensional Lévy measure constructions.

As noted in the thesis of Chen (2008), the first approach, though widely studied, tends to have a feature that is not really desirable for OpRisk modeling settings. The problem associated with construction of a multivariate Lévy process via subordination of a multidimensional Brownian motion is that under such a construction the heavy tail behaviors of the joint process are restricted to be highly similar in all marginals. For instance, the widely considered marginal Lévy process given by a variance Gamma family, when constructed into a multivariate Lévy process in this fashion, will produce a joint process in which kurtosis are almost identical in all marginal processes. This restrictive feature makes the flexibility of the model insufficient to capture the diversity in attributes of a range of OpRisk LDA risk processes in different risk types and business structures. This effect is not surprising since under such a construction, one imposes the property that all marginal processes will share the same subordinator that is the source of all heavy-tail behavior.

The second approach mentioned based on linear transformation of independent Lévy processes for each risk process LDA model will produces Lévy processes with dependence. This approach is effectively what will be discussed in a section on common factor–based models that induce dependence between the individual risk Lévy processes for each risk process. The key to this approach is to construct the marginal processes according to a idiosyncratic process plus some common process. The dependence comes from the common process while the idiosyncratic process makes it possible to match some pre-specified marginals. One drawback of such an approach, apart from the fact that it is not always trivial to figure out the joint dependence induced in the multivariate Lévy process, is that the separation of marginal models and dependence structure that copulas offer is lost. That is to say, under such an approach, one cannot separate the dependence part from the marginals. If the dependence is changed by changing the common process, then consequently the entire marginal process is also changed.

In this section, we focus on the third of these approaches that aligns with the approach adopted in Böcker and Klüppelberg (2008, 2009) to model dependence in frequency and severity between different risks at the same time using a new concept of Lévy copulas; see in Cont and Tankov (2004, sections 5.4–5.7). It is assumed that each risk follows to a univariate compound Poisson process (that belongs to a class of Lévy processes). Then, the idea is to introduce the dependence between risks in such a way that any conjunction of different risks constitutes a univariate compound Poisson process. It is achieved using the multivariate compound Poisson processes based on Lévy copulas. Note that if dependence between frequencies or annual losses is introduced via copula as in (12.16) or (12.34), then the conjunction of risks does not follow to a univariate compound Poisson.

More specifically, in this section, we consider the d risk processes each under an LDA structure given generically by the notation for the i-th risk process in year t according to

$$Z_t^{(i)} = \sum_{n=1}^{N_t^{(i)}} X_n^{(i)}(t). \tag{12.1}$$

Then we are interested in the aggregate of the risk processes for the bank or business unit level total annual loss given by

$$Z_t^{(T)} = \sum_{i=1}^{d} Z_t^{(i)}. \tag{12.2}$$

We note that when the individual loss process are compound Poisson processes each with Poisson intensity function $\lambda^{(i)} > 0$ then the resulting total aggregate loss process is again a compound Poisson process if the loss processes are considered independent. These results are somewhat elementary and have been discussed elsewhere in the book, so in this section the study of the influence of dependence on elements of this model structure is considered. In particular, the question of what dependence structure will be preserving of the compound Poisson model structure is considered. In particular, the LDA models for each risk process will be restricted to being Lévy processes such that they form for d risk processes a multivariate d-dimensional Lévy process. In general at a time t, the d-variate vector of annual losses $\boldsymbol{Z}_t = \left(Z_t^{(1)}, \ldots, Z_t^{(d)}\right)$ for a multivariate Lévy process is uniquely defined by the law of $\boldsymbol{Z}_t$ at a fixed time t. Hence, one could define a copula between each marginal component, but it would be highly beneficial to do so in such manner so as to preserve the Lévy process structure of the random vector and its marginals. This turns out to be not so simple; one cannot just apply any copula to the random vector if the preservation of the Lévy process structure is the goal. So the question naturally arises how does one formulate such a class of dependence structures?

In Kallsen and Tankov (2006), they note that if one considers a bivariate setting and considers two infinitely divisible measures ν and μ, then even in this simple general setting it is highly nontrivial to characterize the class of copulas that will preserve the property of infinite divisibility, some notable exceptions being the Gaussian and α-stable laws.

Instead, the idea of Kallsen and Tankov (2006) is to develop a notion of copula dependence (not strictly a copula in the sense presented earlier) that will instead be based directly on the Lévy triplet representation of the multivariate Lévy processes; see detailed discussions in Peters and Shevchenko (2015). In particular, they note that considering the drift, volatility, and Lévy measure in the triplet (γ, a, ν) of the process one can utilize this characterization of the process in a time-independent manner (rather than the finite dimensional distributions specification that may be time dependent) to capture the dependence structure of the d-dimensional Lévy process. In fact, this is not a new concept and has been utilized previously in the literature on stable processes; see discussions in Samorodnitsky and Taqqu (1997). In this context, they observe that the location parameter in the triplet γ is not required in the considerations of the dependence structure of the multivariate Lévy process. Furthermore, they observe that the dependence structure of the Brownian motion component of the Lévy process is characterized completly by the covariance matrix arising from the volatility component of the triplet a. Hence, the remaining aspect of the multi-variate Lévy process, the Lévy spectral measure ν, must hold the required properties that are sought, namely, the valid properties of the dependence structure between the marginal Lévy processes to ensure the resulting multivariate process is still a Lévy process with Lévy process marginals.

If one observes that the continous component and jump components of a Lévy process are stochastically independent, then the remaining component of the triplet, the Lévy measure ν, must charcterize the dependence in the jump component of the Lévy process.

In particular, we consider the model structures proposed in Cont and Tankov (2004), Kallsen and Tankov (2006), Barndorff-Nielsen and Lindner (2004), and specifically in the context of OpRisk the multivariate models proposed in Böcker and Klüppelberg (2010). In each of these studies, the authors consider the family of Lévy processes.

For brevity, we recall that a Lévy measure ν on the space $\mathbb{R}^n$ is a measure with no atom at the origin and statisfies the condition

$$\int_{\mathbb{R}^n} \left(|\boldsymbol{x}|^2 \wedge 1 \right) \nu(d\boldsymbol{x}) < \infty \tag{12.3}$$

with Euclidean norm $|\boldsymbol{x}|^2 = x_1^2 + \cdots + x_n^2$. Furthermore, it will be convenient to also denote the marginal Lévy measures $\nu_1, \ldots, \nu_n$ that are each one-dimensional Lévy measures.

A positive Lévy measure that will be the focus of this section on Lévy copulae is a measure with strictly positive support given by the product space $\mathbb{R}^n = [0, \infty)^m$. To formally introduce a Lévy measure on this space one must first introduce the image measure of ν, which will be denoted by χ and corresponds to the following mapping for Borel sets B in $[0, \infty]^n$ given by

$$\chi(B) := (Q\nu)(B) = \nu\left(Q^{-1}(B)\right), \tag{12.4}$$

where the function Q is a bijection mapping given by

$$Q := Q_n : [0, \infty]^n \mapsto [0, \infty]^n, \quad (x_1, \ldots, x_n) \mapsto \left(x_1^{-1}, \ldots, x_n^{-1}\right); \tag{12.5}$$

see detailed discussion in Barndorff-Nielsen and Lindner (2004).

Note that since ν is a Lévy measure, one has that the measure χ is also finte on any closed rectangles in $[0, \infty]^n$ that do not contain the point $(\infty, \ldots, \infty)$. From these two measures, one can define a volume function with respect to the Lévy measure ν according to

$$F_\nu : [0, \infty]^n \mapsto [0, \infty] \tag{12.6}$$

and given explicitly with respect to the image measure of ν given by χ according to the function

$$F(x_1, \ldots, x_n) := \begin{cases} \chi\left([0, x_1] \times \cdots \times [0, x_n]\right), & (x_1, \ldots, x_n) \neq (\infty, \ldots, \infty) \\ \infty, & (x_1, \ldots, x_n) = (\infty, \ldots, \infty). \end{cases} \tag{12.7}$$

It should also be noted that for each marginal Lévy measure ν_i one can associate the image measure $\chi_i := Q_i\nu_i$ that allows one to define the volume function F_i for ν_i, which is given for all $x_i \in [0, \infty]$ by

$$F_i(x_i) = F(\infty, \infty, \ldots, \infty, x_i, \infty, \ldots, \infty). \tag{12.8}$$

Given these components, one may now formally define the Lévy copula as given in Definition 12.1; see details in Kallsen and Tankov (2006) and Barndorff-Nielsen and Lindner (2004).

Definition 12.1 (Lévy Copula) *Consider a positive n-dimensional Lévy measure ν with marginal one-dimensional Lévy measures $\{\nu_i\}_{i=1}^n$ and a volume function F with marginal volume functions $\{F_i\}_{i=1}^n$. Then there exists a positive Lévy copula $\widetilde{C}$ such that*

$$F(x_1, \ldots, x_n) = \widetilde{C}\left(F_1(x_1), F_2(x_2), \ldots, F_n(x_n)\right), \quad \forall x_1, \ldots, x_n \in [0, \infty]. \tag{12.9}$$

The Lévy copula $\widetilde{C}$ is then uniquely determined on $RanF_1 \times \cdots \times RanF_n$. Conversely, if one has copula $\widetilde{C}$ that is a positive Lévy copula and $\{F_i\}_{i=1}^{n}$ are volume functions of the one-dimensional positive Lévy measures $\{\nu_i\}_{i=1}^{n}$, then Equation (12.9) defines a positvie measure ν with volume function F and marginal Lévy measures $\{\nu_i\}_{i=1}^{n}$. ■

Remark 12.1 *It is clear that this definition of a Lévy copula is the Lévy measure analog of Sklar's theorem for standard copula distributions for continous marginals. In the aforementioned definition one can consider the Lévy copula as a transformation of special Lévy measures.*

We note that this is not the only way to consider caracterizing the Lévy copula. In Böcker and Klüppelberg (2010), they adopt a different approach to presenting the definition of a Lévy copula by first considering the restriction of the aggregate compound process $Z_t^{(T)}$ as defined previously to be a Lévy process and therefore represented by a Lévy–Khinchine characteristic function representation according to

$$\mathbb{E}\left[\exp\left(\mathrm{i}\left\langle \boldsymbol{\theta}, \boldsymbol{Z}_t \right\rangle\right)\right] = \exp\left\{ t \int_{\mathbb{R}_+^d} \left(\exp(\mathrm{i}\left\langle \boldsymbol{\theta}, \boldsymbol{x} \right\rangle - 1\right) \nu(dx) \right\}, \quad \boldsymbol{\theta} \in \mathbb{R}^d, \tag{12.10}$$

with ν a measure on $\mathbb{R}_+^d = [0,\infty)^d$, which is the positive Lévy measure of $\boldsymbol{Z}_t$ with

$$\left\langle \boldsymbol{\theta}, \boldsymbol{Z}_t \right\rangle = \sum_{i=1}^{d} \theta_i Z_t^{(i)}, \tag{12.11}$$

where $\boldsymbol{Z}_t = \left(Z_t^{(1)}, \ldots, Z_t^{(d)}\right)$.

Remark 12.2 *The Lévy measure ν is clearly independent of time and can be utilized to capture the dependence between components of the Poisson compound processes (Lévy processes) characterized by the vector of annual losses $\boldsymbol{Z}_t = \left(Z_t^{(1)}, \ldots, Z_t^{(d)}\right)$. The Lévy copula for this Lévy measure effectively models the dependence between the jumps that occur between the different Lévy processes.*

In Böcker and Klüppelberg (2010), the Lévy copula is represented through specification of the Tail measure as detailed in Definition 12.2.

Definition 12.2 (Tail Measure) *Consider the multivariate d-dimensional Lévy process $\boldsymbol{Z}$ that is restricted to have a positive spectrum in $\mathbb{R}^d$ with Lévy measure ν. The tail integral of the Lévy measure is defined to be the mapping $\overline{\nu} : [0,\infty]^d \mapsto [0,\infty]^d$, which satisfies for $\mathbf{z} = (z_1, \ldots, z_d)$ the following properties:*

1. *The tail measure is given by $\overline{\nu} = \nu\left([z_1,\infty) \times \cdots \times [z_d,\infty)\right)$ with $\mathbf{z} \in [0,\infty)^d$ with $\overline{\nu}(\mathbf{0})$ given by the finite limit for Poisson processes according to*

$$\overline{\nu}(\mathbf{0}) = \lim_{z_1 \downarrow 0, \cdots, z_d \downarrow 0} \nu\left([z_1,\infty) \times \ldots \times [z_d,\infty)\right). \tag{12.12}$$

2. *The tail measure $\overline{\nu}$ is zero if one of its arguments is ∞;*

3. *The i-th marginal tail measure is defined as follows:*

$$\overline{\nu}(0, \ldots, z_i, \ldots, 0) = \overline{\nu}_i(z_i) \tag{12.13}$$

for $\mathbf{z} \in \mathbb{R}_+^d$ *with* $\overline{\nu}_i(z_i) = \nu_i([z_i, \infty))$ *the tail integral of component* z_i.

■

Using the tail integral one may now define the Lévy measure in terms of the tail integral as detailed in Definition 12.3, which is equivalent to that provided in Definition 12.1, see Böcker and Klüppelberg (2010).

Definition 12.3 (Lévy Copula via Tail Measure) *A d-dimensional Lévy copula of a spectrally positive Lévy process is a measure defining function* $\widetilde{C} : [0,\infty]^d \mapsto [0,\infty]$ *such that for all* $z_1, \ldots, z_d \in [0,\infty]$ *it satisfies the following tail measure condition:*

$$\overline{\nu}(z_1, \ldots, z_d) = \widetilde{C}(\overline{\nu}_1(z_1), \ldots, \overline{\nu}_d(z_d)). \tag{12.14}$$

If the marginal tail integrals $\{\overline{\nu}_i\}_{i=1}^d$ *are continous, then the Lévy copula* $\overline{C}$ *is unique, otherwise it is unique on the range* $Ran\overline{\nu}_1 \times \cdots \times Ran\overline{\nu}_d$. *Conversely, if* $\widetilde{C}$ *is a Lévy copula and* $\{\overline{\nu}_i\}_{i=1}^d$ *are marginal tail integrals of spectrally positive Lévy processes, then Equation (12.14) defins the tail integral of a d-dimensional spectrally positive Lévy Process.* ■

Here, we would like to mention that in the case of a compound Poisson process, Lévy measure is the expected number of losses per unit of time with a loss amount in a pre-specified interval,

$$\overline{\nu}_j(x) = \lambda_j \mathbb{P}\mathrm{r}[X_j > x].$$

Then the multivariate Lévy measure can be constructed from the marginal measures and a Lévy copula $\widetilde{C}$ as

$$\overline{\nu}(x_1, \ldots, x_d) = \widetilde{C}(\overline{\nu}_1(x_1), \ldots, \overline{\nu}_d(x_d)). \tag{12.15}$$

This is somewhat similar to (11.22) in a sense that the dependence structure between different risks can be separated from the marginal processes. However, it is quite a different concept. In particular, a Lévy copula for processes with positive jumps is $[0,\infty)^d \to [0,\infty)$ mapping while a standard copula (11.22) is $[0,1]^d \to [0,1]$ mapping. Also, a Lévy copula controls dependence between frequencies and dependence between severities (from different risks) at the same time.

The interpretation of this model is that dependence between different risks is due to the loss of events occurring at the same time. An important implication of this approach is that a bank's total loss can be modeled as a compound Poisson process with some intensity and independent severities. If this common severity distribution is subexponential, then a closed-form approximation (13.75) can be used to estimate the VaR of the total annual loss.

We conclude this section on discussion of Lévy copula by describing recent results related to pair copula constructions of multivariate Lévy copula models. As discussed previously for distributional copula models, a challenge faced by building such models involves finding flexible but still applicable models for higher dimensions. To overcome this problem, the concept of pair copula constructions has been developed in distributional copula models such as was described for the Student t copula constructions. In the context of nondistributional copula constructions, there has been a few works looking at pairwise constructions and tree-based couplings of pairwise Lévy copula models; see, for instance, Grothe and Nicklas (2013) for details of such copula constructions in the context of Lévy processes.

12.2 Multiple Risk LDA: Dependence Between Frequencies via Copula

The most popular approach in practice is to consider a dependence between the annual counts of different risks via a copula. Assuming a J-dimensional copula $C(\cdot)$ and the marginal distributions $P_j(\cdot)$ for the annual counts $N_t^{(1)}, \ldots, N_t^{(J)}$ leads to a model

$$N_t^{(1)} = P_1^{-1}(U_t^{(1)}), \ldots, N_t^{(J)} = P_J^{-1}(U_t^{(J)}), \tag{12.16}$$

where $U_t^{(1)}, \ldots, U_t^{(J)}$ are *Uniform*$(0,1)$ random variables from a copula $C(\cdot)$ and $P_j^{-1}(\cdot)$ is the inverse marginal distribution of the counts in the j-th risk. Here, t is discrete time (typically in annual units but shorter steps might be needed to calibrate the model). Usually, the counts are assumed to be independent between different t steps.

The approach allows us to model both positive and negative dependence between counts. As reported in the literature, the implied dependence between annual losses even for a perfect dependence between counts is relatively small and as a result the impact on capital is small too. Some theoretical reasons for the observation that frequency dependence has only little impact on the OpRisk capital charge are given in Böcker and Klüppelberg (2008).

As an example, in Figure 12.1, we plot Spearman's rank correlation between the annual losses of two risks, $Z^{(1)}$ and $Z^{(2)}$, induced by the Gaussian copula dependence between frequencies. Marginally, the frequencies $N^{(1)}$ and $N^{(2)}$ are from the *Poisson*$(\lambda = 5)$ and *Poisson*$(\lambda = 10)$ distributions, respectively, and the severities are from *LogNormal*$(\mu = 1, \sigma = 2)$ distributions for both risks.

12.3 Multiple Risk LDA: Dependence Between the *k*-th Event Times/Losses

Theoretically, one can introduce dependence between the k-th severities or between the k-th event interarrival times or between the k-th event times of different risks. For example, first, second, etc. losses of the j-th risk are correlated to the first, second, etc. losses of the i-th risk, respectively while the severities within each risk are independent. The actual dependence can be done via a copula similar to (12.16); for an accurate description, we refer to Chavez-Demoulin *et al.* (2006). Here, we would like to note that a physical interpretation of such models can be difficult. Also, an example of dependence between annual losses induced by dependence between the k-th interarrival times is presented in Figure 12.1.

We begin this section with discussion on common shock processes that induces indirect dependence and then we present a more explicit parameteric dependence model via the notion

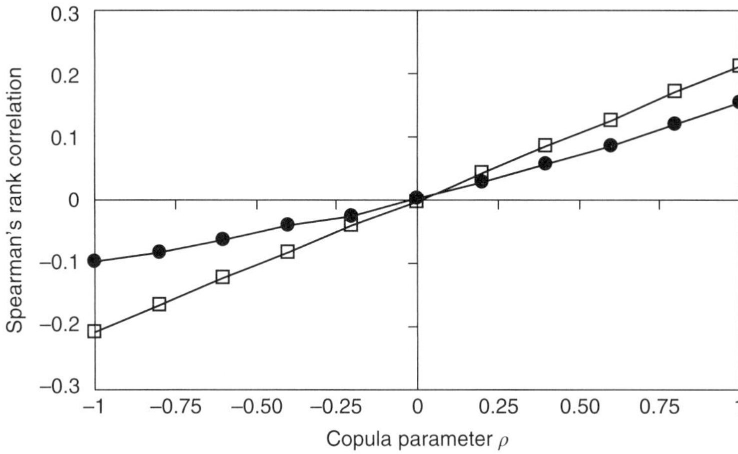

FIGURE 12.1 Spearman's rank correlation between the annual losses $\rho_S[Z^{(1)}, Z^{(2)}]$ vs. the Gaussian copula parameter ρ: (□)—copula between counts $N^{(1)}$ and $N^{(2)}$; (•)—copula between inter-arrival times of two Poisson processes. Marginally, the frequencies are from *Poisson*(5) and *Poisson*(10), respectively, and the severities are from *LogNormal*$(\mu = 1, \sigma = 2)$ for both risks

of dependence through self-chaining copula models for interarrival times of losses in multiple loss processes.

12.3.1 COMMON SHOCK PROCESSES

Modeling OpRisk events affecting many risk cells can be done using common shock process models; see Johnson *et al.* (1997, section 37). In particular, consider J risks with the event counts

$$N_t^{(j)} = N_t^{(C)} + \widetilde{N}_t^{(j)},$$

where $\widetilde{N}_t^{(j)}, j = 1, \ldots, J$ and $N_t^{(C)}$ are generated by independent Poisson processes with the intensities $\widetilde{\lambda}_j$ and λ_C, respectively. Then, $N_t^{(j)}, j = 1, \ldots, J$ are Poisson distributed marginally with the intensities

$$\lambda_j = \widetilde{\lambda}_j + \lambda_C$$

and are dependent via the common events $N_t^{(C)}$. The linear correlation and covariance between risk counts are

$$\rho[N_t^{(i)}, N_t^{(j)}] = \lambda_C / \sqrt{\lambda_i \lambda_j}$$

and

$$\mathbb{C}\text{ov}[N_t^{(i)}, N_t^{(j)}] = \lambda_C,$$

respectively.

Only a positive dependence between counts can be modeled using this approach. Note that the covariance for any pair of risks is the same though the correlations are different. More flexible dependence can be achieved by allowing a common shock process to contribute to the k-th risk process with some probability p_k; then

$$\mathbb{C}\text{ov}[N_t^{(i)}, N_t^{(j)}] = \lambda_C p_i p_j.$$

This method can be generalized to many common shock processes; see Lindskog and McNeil (2003) and Powojowski *et al.* (2002). It is also reasonable to consider the dependence between the severities in different risk cells that occurred due to the same common shock event.

12.3.2 MAX-STABLE AND SELF-CHAINING COPULA MODELS

In this section, we consider the setting in which one has Poisson processes for multiple risk processes in LDA structures; however, there is an interest in not just modeling the distribution of the number of counts, but instead to model the distribution of the arrival times explicitly. Such settings arise typically in OpRisk models when considering applications or particular insurance policies; see discussion in Chapter 17.

In particular, it may be of interest to study the relationship between dependence and the possibility to sample final multivariate survival in a long time-interval as a sequence of iterations of local multivariate survivals along a partition of the total time interval. This challenge was addressed in Brigo and Chourdakis (2012), where it was shown that indeed it is possible to achieve this goal under a form of multivariate lack of memory. This modified definition of memorylessness for multivariate settings can then be linked to properties of the survival times copula. The authors denote such a structure as the "self-chaining-copula".

To be more precise about the model formulation, consider modeling the loss times for two loss processes in a year, where the i-th loss event for risk j occurs at random time $\tau_i^{(j)} \sim F^{(j)}$. Now consider splitting the analysis into two possibilities, either a single period (i.e., a year interval or time) versus multiple period (i.e., monthly in the year) analysis. In general one may consider N periods of duration T given by $[0, T), [T, 2T) \ldots, [(N-1)T, NT]$ such that the final period is of duration NT. Furthermore, consider memoryless marginal event times for each j where the $\tau_i^{(j)}$ are i.i.d. for all i and exponentially distributed satisfying the standard marginal definitions of memorylessness given by

$$\mathbb{P}\text{r}\left[\tau_i^{(j)} > T | \tau_i^{(j)} > S\right] = \mathbb{P}\text{r}\left[\tau_i^{(j)} > T - S\right]. \tag{12.17}$$

In Brigo and Chourdakis (2012), the challenge is stated in the following way:

> Is it possible to iterate a simulation of the joint survival of arrival times always in the same way in all sub-intervals $[0, T), [T, 2T), \ldots, [(N-1)T, NT]$ and also in the same way as one simulates joint survival of arrival times in a single sampling run $[0, NT]$?

To proceed with the specification of a family of self-chaining copula models, it is important to present the modified definition of memorylessness termed the common periods lack of memory (CPLM). In the bivariate case, this is given in Definition 12.4; see discussions in Brigo and Chourdakis (2012).

Definition 12.4 (Common Periods Lack of Memory) *Consider two loss processes, where for a given year one has interarrival times for the i-th loss in the j-th processes at times of loss events given by random variable $\tau_i^{(j)}$ with distribution $\tau_i^{(j)} \sim F^{(j)}$. Assume that individual loss process event times are memoryless and exponentially distributed such that*

$$\mathbb{P}\mathrm{r}\left[\tau_j^{(i)} > T|\tau_j^{(i)} > S\right] = \mathbb{P}\mathrm{r}\left[\tau_j^{(j)} > T - S\right], \tag{12.18}$$

for any j and any $0 \leq S \leq T$. The bivariate lack of memory property in common periods considered involves how to draw i.i.d. random vectors $\boldsymbol{\tau}_i = \left(\tau_i^{(1)}, \tau_i^{(2)}\right)$ such that

$$\mathbb{P}\mathrm{r}\left[\tau_i^{(1)} \geq kT, \tau_i^{(2)} \geq kT|\tau_i^{(1)} \geq hT, \tau_i^{(2)} \geq hT\right] = \mathbb{P}\mathrm{r}\left[\tau_i^{(1)} \geq (k-h)T, \tau_i^{(2)} \geq (k-h)T\right] \tag{12.19}$$

with $k > h$ integers. ■

As noted in Brigo and Chourdakis (2012), this is not a standard definition of memorlessness in bivariate cases, the standard definition would involve

$$\mathbb{P}\mathrm{r}\left[\tau_i^{(1)} \geq lT, \tau_i^{(2)} \geq mT|\tau_i^{(1)} \geq nT, \tau_i^{(2)} \geq pT\right] = \mathbb{P}\mathrm{r}\left[\tau_i^{(1)} \geq (l-n)T, \tau_i^{(2)} \geq (m-p)T\right] \tag{12.20}$$

with $\max(n,p) \leq \max(l,m)$. For the purposes of consideration in this chapter, it is not suitable to consider such an assumption since it would imply independence within the random vector $\boldsymbol{\tau}_i$.

12.3.2.1 Multivariate Exponential Distributions.

In the statistics literature, there are two common definitions of the multivariate exponential distributions that admit exponential marginal distributions while the joint distribution satisfies the CPLM definition earlier.

In the bivariate case, one of the widely used choices is known as the Marshall–Olkin distribution given in Marshall and Olkin (1967). One possible drawback of this model is that there is a nonzero probability of the event $\tau_i^{(1)} = \tau_i^{(2)}$, which is generally undesirable. Others who have studied generalized multivariate exponential distributions that remove this property include the three choices of model proposed in Gumbel (1960) and the characterizations of this family of models and its generalized forms in Nair and Nair (1988), and Paulson (1973), and Lu and Bhattacharyya (1991).

The third choice of models was selected for its flexibility in Brigo and Chourdakis (2012) and is characterized in Definition 12.5 by the joint survival function.

Definition 12.5 (Gumbel's Joint Survival Function of Multivariate Exponentials) *Consider marginal exponential distributions with positive intensities $\lambda^{(j)}$ for $j \in \{1,2\}$ and a joint distribution with dependence parameter denoted by $\theta \in [1,\infty)$ with a joint survival function given by*

$$\overline{F}\left(\tau_i^{(1)} > t_1, \tau_i^{(2)} > t_2\right) = \exp\left(-\left(\left(\lambda^{(1)}t_1\right)^{\theta} + \left(\lambda^{(2)}t_2\right)^{\theta}\right)^{\frac{1}{\theta}}\right). \tag{12.21}$$

■

This choice of multivariate exponential distribution satisfies that both marginal distributions are exponential and the joint distribution satisfies the condition of CPLM. In addition, one can show that the rank correlation measure of Kendall is given for this model by

$$\tau^K = 1 - \frac{1}{\theta}, \tag{12.22}$$

with $\theta = 1$ producing independence and $\theta \to \infty$ producing a comonotonic case.

If one now considers the general d-variate cases, one can reexpress the multivariate joint distribution of d-loss process characterizing the random vector $\boldsymbol{\tau}_i = \left(\tau_i^{(1)}, \ldots, \tau_1^{(d)}\right)$ with marginal exponential distributions $\left\{F^{(i)}\right\}_{i=1}^{d}$ by a joint distribution written in terms of a copula, which would be required to still produce the required exponential marginal distributions as well as the joint property of CPLM.

As has been discussed throughout this chapter, the representation can be related to a copula model in the general d-variate case in which if one considers d-loss process and aims to simulate $\boldsymbol{\tau}_i = \left(\tau_i^{(1)}, \ldots, \tau_1^{(d)}\right)$ with marginal exponential distributions $\left\{F^{(i)}\right\}_{i=1}^{d}$ and a joint distribution characterized by the following survival times copula:

$$\begin{aligned}
\mathbb{P}\mathrm{r}\left[\tau_i^{(1)} \geq t_1, \ldots, \tau_i^{(d)} \geq t_d\right] &= \mathbb{P}\mathrm{r}\left[\overline{F}^{(1)}\left(\tau_i^{(1)}\right) \leq \overline{F}^{(1)}(t_1), \ldots, \overline{F}^{(d)}\left(\tau_i^{(d)}\right) \leq \overline{F}^{(d)}(t_d)\right] \\
&= \mathbb{P}\mathrm{r}\left[U_i^{(1)} \leq \overline{F}^{(1)}(t_1), \ldots, U_i^{(d)} \leq \overline{F}^{(d)}(t_d)\right] \\
&= C(u_1, \ldots, u_d).
\end{aligned} \tag{12.23}$$

In Brigo and Chourdakis (2012), it was discussed that if the copula C is to satisfy the CPLM condition so that one can perform simulation of N intervals of duration T that is consistent with the simualtion over a single interval $[0, NT]$, then this would require the condition that the joint copula in Equation (12.23) should satisfy the condition

$$\left(C\left(u_1^{\frac{1}{N}}, \ldots, u_d^{\frac{1}{N}}\right)\right)^N = C(u_1, \ldots, u_d), \quad u_i \in [0, 1], N \in \mathbb{J}, \tag{12.24}$$

such a copula that satisfies this condition was termed a "self-chaining copula".

It is intersting to observe that such a condition is also used to define the class of max-stable copula models that are obtained by considering a random vector $\boldsymbol{X} = (X_1, \ldots, X_d)$ with copula C. Then consider k i.i.d. copies of this random vector $\{X_{i,1}, \ldots, X_{i,d}\}_{i=1}^{k}$ and the marginal maxima, given for the j-th marginal component by the order statistic

$$M_{(k,k),j} = \max\left\{X_{1,j}, X_{2,j}, \ldots, X_{k,j}\right\}. \tag{12.25}$$

Then the resulting copula for the vector of maxima is given by

$$C(\boldsymbol{u}) = C_{M_{(k,k),1}, M_{(k,k),2}, \ldots, M_{(k,k),d}}(u_1, \ldots, u_d) = C\left(u_1^k, \ldots, u_d^k\right)^{\frac{1}{k}}. \tag{12.26}$$

Then it is said that the orginal copula C is max-stable if and only if it satisfies this condition, that is, $C(\boldsymbol{u}) = C\left(u_1^k, \ldots, u_d^k\right)^{\frac{1}{k}}$. This is equivalent to the notion of self-chaining considered in Brigo and Chourdakis (2012).

At this stage, it will also be useful to define the directly related concept of extreme value copula models; see discussions in, for instance, Pickands (1981) and the review in Gudendorf and Segers (2010). Extreme value copulas arise as possible limits of component-wise maxima of independent, identically distributed samples, and are defined in Definition 12.6.

Definition 12.6 (Bivariate Extreme Value Copula) *A copula C is an extreme value copula if and only if there exists a real-valued function A on the interval $[0, 1]$ such that the copula is defined by*

$$C(u, v) = \exp\left(\ln(uv)A\left(\frac{\ln(v)}{\ln(uv)}\right)\right), \quad u \in [0, 1], v \in [0, 1], \tag{12.27}$$

where function A is the Pickand's dependence function formalized in Definition 10.36 above. ∎

Remark 12.3 *The Pickand's dependence function A can be interpreted in terms of exponential distributions by considering the pair (U, V) with joint distribution given by the extreme value copula C. Consider random variables under transformation*

$$\begin{aligned} D &= -\ln U, \\ E &= -\ln V \end{aligned} \tag{12.28}$$

and define for $t \in [0, 1]$ the function

$$\zeta(t) = \min\left\{\frac{D}{1-t}, \frac{E}{t}\right\} \tag{12.29}$$

with $\zeta(0) = D$ and $\zeta(1) = E$. One can then show that for $x \geq 0$ one has that $\zeta(t)$ follows an exponential distribution with distribution

$$\mathbb{P}\text{r}\left[\zeta(t) \leq x\right] = 1 - \exp\left(-A(t)x\right). \tag{12.30}$$

It was then shown in Brigo and Chourdakis (2012) that two classes of copula that satisfy this self-chaining property corespond to the Gumbel–Hougaard copula and the Marshal–Olkin copula. In general, they show the following conditions that self-chaining copula models must satisfy; see Definition 12.7.

Definition 12.7 (Characterization of Bivariate Self-Chaining Copula Models) *Self-chaining copulas that satisfy the condition that*

$$\left(C\left(u_1^{\frac{1}{N}}, u_d^{\frac{1}{N}}\right)\right)^N = C\left(u_1, u_d\right), \quad u_i \in [0, 1], N \in \mathbb{J}, \tag{12.31}$$

in the bivariate case where $d = 2$ are characterized as the solution to the partial differential equation given by

$$\frac{\partial}{\partial u}C(u,v)u\ln(u) + \frac{\partial}{\partial v}C(u,v)v\ln(v) = C(u,v)\ln C(u,v), \tag{12.32}$$

which also satisfy the conditions for a copula distribution given by Definition 11.1:

- $C(u_1, u_2) = 0$ *whenever* $u_i = 0$ *for at least one* $i \in \{1, 2\}$;
- $C(u_1, u_2) = u_i$ *if* $u_i = 1$ *for all* $j = 1, \ldots, d$ *and* $j \neq i$;
- C *is quasi-monotone on its support* $[0, 1]^2$.

■

In general, one can also show the result in Theorem 12.1 for self-chaining Archimedean copula models; see Brigo and Chourdakis (2012, theorem 7.3).

Theorem 12.1 (Self-Chaining Archimedean Copula Models) *The self-chaining Archimedean copula models correspond to those for which the generator ψ produces a frailty distribution satisfying:*

- *Infinite divisibility;*
- *Strictly semistable, that is, its measure satisfies*

$$\mu(x)^a = \mu(bz) \tag{12.33}$$

with $a > 0$, $a \neq 1$ and $b > 0$.

12.4 Multiple Risk LDA: Dependence Between Aggregated Losses via Copula

Dependence between aggregated losses can be introduced in a manner similar to (12.16). In this approach, one can model the aggregated losses as

$$Z_t^{(1)} = F_1^{-1}(U_t^{(1)}), \ldots, Z_t^{(J)} = F_J^{-1}(U_t^{(J)}), \tag{12.34}$$

where $U_t^{(1)}, \ldots, U_t^{(J)}$ are *Uniform*$(0, 1)$ random variables from a copula $C(\cdot)$ and $F_j^{-1}(\cdot)$ is the inverse marginal distribution of the aggregated loss of the j-th risk.

Note that the marginal distribution $F_j(\cdot)$ should be calculated using the frequency and severity distributions. Typically, the data are available over several years only and a short time step t (e.g., quarterly) is needed to calibrate the model.

This approach is probably the most flexible in terms of the range of achievable dependencies between risks, for example, perfect positive dependence between the annual losses is achievable. However, it may create difficulties with incorporation of insurance into the overall model. This is because an insurance policy may apply to several risks with the cover limit applied to the aggregated loss recovery; see Chapter 17. One can overcome this problem with an approximate solution described in the following example.

EXAMPLE 12.1 Simulation of Multivariate Compound Processes with Annual Copula Dependence

In the aforementioned descriptions we noted that the modeling of dependence between annual losses directly via copula methods theoretically can create irreconcilable problems with modeling insurance for OpRisk that directly involves event times. That is, for any "adjustments" that must be made on an event basis, the modeling of dependence through a copula distribution on the annual loss can be challenging. In this example, we will demonstrate an approximate Monte Carlo procedure that will allow one to overcome this challenge. Consider a bivariate annual loss process $\left\{Z_t^{(i)}\right\}_{t\geq 0, i\in\{1,2\}}$ such that the joint distribution $F_{\mathbf{Z}_t}\left(z_t^{(1)}, z_t^{(2)}\right)$ is characterized uniquely by its copula dependence distribution $C(\boldsymbol{u})$ and its marginals $F_{Z_t^{(i)}}$ according to the relationship

$$F_{\mathbf{Z}_t}\left(z_t^{(1)}, z_t^{(2)}\right) = C\left(F_{Z_t^{(1)}}\left(z_t^{(1)}\right), F_{Z_t^{(2)}}\left(z_t^{(2)}\right)\right), \tag{12.35}$$

where the copula distribution will be explored in extensive details later. Each marginal distribution is assumed comprising an LDA model that involves a compound process distribution $F_{Z_t^{(i)}}$ for the compound process

$$Z_t^{(i)} = \sum_{n=0}^{N_t^{(i)}} X_n^{(i)}(t) \tag{12.36}$$

with frequency distribution model $N_t^{(i)} \sim F_{N^{(i)}}$ and severity distribution model $X_n^{(i)}(t) \sim F_{X^{(i)}}$. Then, in order to work with such a model where dependence is between the annual losses, while still making adjustments to the marginal distributions, for example, under an insurance mitigation policy one can adopt the following approximation procedure for each annual year $t \in \{1, 2, \ldots, T\}$:

1. For each marginal loss process $i \in \{1, 2\}$, simulate J Monte Carlo draws from the compound process

$$Z_t^{(i)} = \sum_{n=0}^{N_t^{(i)}} X_n^{(i)}(t) \tag{12.37}$$

   via simulation for each $j \in \{1, 2, \ldots, J\}$ the following steps:

   - Draw a random variable for the number of losses from the frequency models $N_t^{(i,j)} \sim F_{N^{(i)}}$ and the corresponding event times in the year $\left\{\tau_s^{(i,j)}(t)\right\}_{s\in\left\{1,2,\ldots,N_t^{(i,j)}\right\}}$ that match the desired frequency model. Store these times for each j Monte Carlo draw;

- Draw a vector of loss random variables from the severity model

$$\boldsymbol{X}^{(i)}(t) = \left[X_1^{(i,j)}(t), \ldots, X_{N_t^{(i,j)}}^{(i,j)}(t)\right]$$

as i.i.d. draws from the severity distribution model $X_n^{(i)}(t) \sim F_{X^{(i)}}$. Store these loss random variates for each j Monte Carlo draw;
- Evaluate the j-th realization of the compound process random sum

$$Z_t^{(i,j)} = \sum_{n=0}^{N_t^{(i,j)}} X_n^{(i,j)}(t). \tag{12.38}$$

2. Using the J Monte Carlo draws for each marginal loss process $\left\{Z_t^{(i,j)}\right\}_{j\in\{1,2,\ldots,J\}}, i \in \{1,2\}$ construct emipirical distribution functions for each marginal annual loss process

$$\hat{F}_{Z_t^{(i)},J}(z) = \frac{1}{J}\sum_{j=1}^{J} \mathbb{I}\left[Z_t^{(i,j)} \leq z\right] \tag{12.39}$$

and the empirical quantile function by ordering the annual loss random variables $0 \leq Z_t^{i,(1,J)} < Z_t^{i,(2,J)} \leq \cdots \leq Z_t^{i,(J,J)}$ to obtain

$$\hat{Q}_{Z^{(i)},J}(\alpha) = \inf\left\{j:\ Z_t^{i,(j,J)} \leq \alpha\right\}. \tag{12.40}$$

3. Simulate a random vector on $[0,1]^2$ with uniform marginals and joint distribution given by the copula model $\boldsymbol{U} \sim C$;
4. Draw a copula-dependent annual loss random variable from each marginal by using the empirical quantile function

$$Z_t^{(i)} = \hat{Q}_{Z^{(i)},J}(U_i) \tag{12.41}$$

and record the index of the order statistic that corresponds to the U_i, denoted by $k(i)$;
5. To apply insurance or any event-based adjustment to the annual loss process, take the draw for index $k(i)$ given by $\left\{\tau_s^{(i,k(i))}(t)\right\}_{s\in\left\{1,2,\ldots,N_t^{(i,k(i))}\right\}}$ and

$$\boldsymbol{X}^{(i)}(t) = \left[X_1^{(i,k(i))}(t), \ldots, X_{N_t^{(i,k(i))}}^{(i,k(i))}(t)\right]$$

and make the appropriate insurance required adjustments to the severities.

■

12.5 Multiple Risk LDA: Structural Model with Common Factors

Common (systematic) factors are useful for identifying dependent risks and for reducing the number of required correlation coefficients that must be estimated; for example, see McNeil *et al.* (2005, section 3.4). Structural models with common factors to model dependence are widely used in credit risk; see industry examples in McNeil *et al.* (2005, section 8.3.3). For OpRisk, these models are qualitatively discussed in Marshall (2001, sections 5.3 and 7.4), and there are unpublished examples of practical implementation. As an example, assume a Gaussian copula for the annual counts of different risks and consider one common (systematic) factor Ω_t affecting the counts as follows:

$$\begin{aligned} Y_t^{(j)} &= \rho_j \Omega_t + \sqrt{1-\rho_j^2}\, W_t^{(j)}, \quad j = 1, \ldots, J; \\ N_t^{(1)} &= P_1^{-1}\left(F_N(Y_t^{(j)})\right), \ldots, N_t^{(J)} = P_J^{-1}\left(F_N(Y_t^{(J)})\right). \end{aligned} \tag{12.42}$$

Here, $W_t^{(1)}, \ldots, W_t^{(J)}$ and Ω_t are independent random variables from the standard Normal distribution. All random variables are independent between different time steps t. Given Ω_t, the counts are independent; unconditionally, the risk profiles are dependent if the corresponding ρ_j are nonzero. In this example, one should identify J correlation parameters ρ_j only instead of $J(J-1)/2$ parameters of the full correlation matrix.

Extension of this approach to many factors $\Omega_{t,k}$, $k = 1, \ldots, K$ is easy:

$$Y_t^{(j)} = \sum_{k=1}^{K} \rho_{jk} \Omega_{t,k} + \sqrt{1 - \sum_{k=1}^{K} \rho_{jk}\rho_{jm} \mathbb{C}\text{ov}[\Omega_{t,k}\Omega_{t,m}]}\, W_t^{(j)}, \tag{12.43}$$

where $(\Omega_{t,1}, \ldots, \Omega_{t,K})^T$ is from the standard multivariate Normal distribution with zero means, unit variances, and some correlation matrix.

This approach can also be extended to introduce a dependence between both severities and frequencies. For example, in the case of one factor, one can structure the model as follows:

$$\begin{aligned} Y_t^{(j)} &= \rho_j \Omega_t + \sqrt{1-\rho_j^2}\, W_t^{(j)}, \quad j = 1, \ldots, J; \\ N_t^{(j)} &= P_j^{-1}\left(F_N(Y_t^{(j)})\right), \quad j = 1, \ldots, J; \\ R_s^{(j)}(t) &= \widetilde{\rho}_j \Omega_t + \sqrt{1-\widetilde{\rho}_j^2}\, V_s^{(j)}(t), \quad s = 1, \ldots, N_t^{(j)}, \quad j = 1, \ldots, J; \\ X_s^{(j)}(t) &= F_j^{-1}\left(F_N(R_s^{(j)}(t))\right), \quad s = 1, \ldots, N_t^{(j)}, \quad j = 1, \ldots, J. \end{aligned}$$

Here $W_t^{(j)}$, $V_s^{(j)}(t)$, $s = 1, \ldots, N_t^{(j)}$, $j = 1, \ldots, J$ and Ω_t are independent random variables from the standard Normal distribution. Again, the logic is that there is a factor affecting severities and frequencies within a year such that conditional on this factor, severities, and frequencies are independent. The factor is changing stochastically from year to year so that unconditionally there is dependence between frequencies and severities. Also note that in such setup there is a dependence between severities within a risk category.

Often, common factors are unobservable and practitioners use generic intuitive definitions such as changes in political, legal, and regulatory environments, economy, technology, system security, system automation, etc. Several external and internal factors are typically considered so that some of the factors affect frequencies only (e.g., system automation), some factors affect severities only (e.g., changes in legal environment), and some factors affect both the frequencies and the severities (e.g., system security).

It is possible to derive a full joint distribution for all data (frequencies and severities) given model parameters; however, in general, it will not have a closed form because the latent variables (factors) should be integrated out. Thus, standard methods cannot be used to maximize corresponding likelihood function and one should use more technically involved methods, for example, a slice sampler used in Peters *et al.* (2009).

The common factor models are supported by empirical evidence, reported in Allen and Bali (2004), that some OpRisks are dependent on macroeconomic variables such as GDP, unemployment, equity indices, interest rates, foreign exchange rates, and regulatory environment variables.

12.6 Multiple Risk LDA: Stochastic and Dependent Risk Profiles

Consider the LDA for risk cells $j = 1, \ldots, J$:

$$Z_j(t) = \sum_{s=1}^{N_j(t)} X_j^{(s)}(t), \quad t = 1, 2, \ldots, \tag{12.44}$$

where $N_j(t) \sim P(\cdot|\lambda_t^{(j)})$ and $X_j^{(s)}(t) \sim F(\cdot|\psi_t^{(j)})$. It is realistic to consider that the risk profiles $\boldsymbol{\lambda}_t = (\lambda_t^{(1)}, \ldots, \lambda_t^{(J)})$ and $\boldsymbol{\psi}_t = (\psi_t^{(1)}, \ldots, \psi_t^{(J)})$ are not constant but changing in time stochastically due to changing risk factors (e.g., changes in business environment, politics, regulations). That is, we may model risk profiles $\boldsymbol{\lambda}_t = \left(\lambda_t^{(1)}, \ldots, \lambda_t^{(J)}\right)$ and $\boldsymbol{\psi}_t = \left(\psi_t^{(1)}, \ldots, \psi_t^{(J)}\right)$ by random variables $\boldsymbol{\Lambda}_t = \left(\Lambda_t^{(1)}, \ldots, \Lambda_t^{(J)}\right)$ and $\boldsymbol{\Psi}_t = \left(\Psi_t^{(1)}, \ldots, \Psi_t^{(J)}\right)$, respectively.

Now consider a sequence $(\boldsymbol{\Lambda}_1, \boldsymbol{\Psi}_1), \ldots, (\boldsymbol{\Lambda}_{T+1}, \boldsymbol{\Psi}_{T+1})$. It is naive to assume that risk profiles of all risks are independent. Intuitively these are dependent, for example, due to changes in politics, regulations, law, economy, and technology (sometimes called drivers or external risk factors) that jointly impact on many risk cells. One can model this by assuming some copula $C(\cdot)$ and marginal distributions for the risk profiles $\boldsymbol{\Lambda}_t$ and $\boldsymbol{\Psi}_t$ (as developed in Peters *et al.*, 2009) that gives the following joint distribution of the risk profiles

$$F(\boldsymbol{\lambda}_t, \boldsymbol{\psi}_t) = C\left(G_1(\lambda_t^{(1)}), \ldots, G_J(\lambda_t^{(J)}), H_1(\psi_t^{(1)}), \ldots, H_J(\psi_t^{(J)})\right),$$

where $G_j(\cdot)$ and $H_j(\cdot)$ are the marginal distributions of $\lambda_t^{(j)}$ and $\psi_t^{(j)}$, respectively.

Dependence between the risk profiles will induce a dependence between the annual losses. This general model can be used to model the dependencies between the annual counts; between the severities of different risks; between the severities within a risk; and between the frequencies and severities. The likelihood of data (counts and severities) can be derived but involves a multidimensional integral with respect to latent variables (risk profiles). Advanced MCMC methods (such as the slice sampler method described in Section 7.6.1 and used in Peters *et al.*, 2009) can be used to fit the model.

Stochastic modeling of risk profiles may appeal to intuition. For example, consider the annual number of events for the j-th risk modeled as random variables from the Poisson distribution *Poisson* $\left(\Lambda_t^{(j)} = \lambda_t^{(j)}\right)$. Conditional on $\Lambda_t^{(j)}$, the expected number of events per year is $\Lambda_t^{(j)}$. The latter is not only different for different banks and different risks but also changes from year to year for a risk in the same bank. In general, the evolution of $\Lambda_t^{(j)}$, can be modeled as having deterministic (trend, seasonality) and stochastic components. In actuarial mathematics, this is called a mixed Poisson model.

Remark 12.4 *The use of common (systematic) factors is useful to identify dependent risks and to reduce the number of required correlation coefficients that must be estimated. For example, assuming a Gaussian copula between risk profiles, consider one common factor Ω_t affecting all risk profiles as follows:*

$$
\begin{aligned}
Y_t^{(i)} &= \rho_i \Omega_t + \sqrt{1-\rho_i^2}\, W_t^{(i)}, \quad i = 1, \ldots, 2J; \\
\Lambda_t^{(j)} &= G^{-1}(F_N(Y_t^{(j)})), \quad \Psi_t^{(j)} = H^{-1}(F_N(Y_t^{(j+J)})),\ j = 1, \ldots, J,
\end{aligned}
$$

where $W_t^{(1)}, \ldots, W_t^{(2J)}$ and Ω_t are independent random variables from the standard Normal distribution and all random variables are independent between different time steps t. Given Ω_t, all risk profiles are independent but unconditionally the risk profiles are dependent if the corresponding ρ_i are nonzero. One can consider many factors: some factors affect frequency risk profiles, some factors affect severity risk profiles, and some factors affect both frequency and severity risk profiles.

As an example, consider the following possible model setup for stochastic and dependent risk profiles, proposed in Peters *et al.* (2009).

Model Assumptions 12.1 *Consider J risks each with a general model (10.2) for the annual loss in year t, $Z_t^{(j)}$, and each modeled by severity $X_s^{(j)}(t)$ and frequency $N_t^{(j)}$. The frequency and severity risk profiles are modeled by random vectors*

$$\boldsymbol{\Lambda}_t = (\Lambda_t^{(1)}, \ldots, \Lambda_t^{(J)}) \quad \textit{and} \quad \boldsymbol{\Psi}_t = (\Psi_t^{(1)}, \ldots, \Psi_t^{(J)})$$

respectively and parameterized by risk characteristics

$$\boldsymbol{\theta}_\Lambda = (\theta_\Lambda^{(1)}, \ldots, \theta_\Lambda^{(J)}) \quad \textit{and} \quad \boldsymbol{\theta}_\Psi = (\theta_\Psi^{(1)}, \ldots, \theta_\Psi^{(J)})$$

correspondingly. Additionally, the dependence between risk profiles is parameterized by $\boldsymbol{\theta}_\rho$. Assume that, given $\boldsymbol{\theta} = (\boldsymbol{\theta}_\Lambda, \boldsymbol{\theta}_\Psi, \boldsymbol{\theta}_\rho)$:

1. *The random vectors*

$$\left(\mathbf{\Psi}_1, \mathbf{\Lambda}_1, N_1^{(j)}, X_s^{(j)}(1); \quad j = 1, \ldots, J, s \geq 1\right)$$

$$\vdots$$

$$\left(\mathbf{\Psi}_{T+1}, \mathbf{\Lambda}_{T+1}, N_{T+1}^{(j)}, X_s^{(j)}(T+1); \quad j = 1, \ldots, J, \quad s \geq 1\right)$$

are independent. That is, between different years the risk profiles for frequencies and severities as well as the number of losses and actual losses are independent;

2. *The vectors* $(\Psi_1, \Lambda_1)^T, \ldots, (\Psi_{T+1}, \Lambda_{T+1})^T$ *are independent and identically distributed from a joint distribution with marginal distributions* $\Lambda_t^{(j)} \sim G\left(\cdot|\theta_\Lambda^{(j)}\right)$, $\Psi_t^{(j)} \sim H\left(\cdot|\theta_\Psi^{(j)}\right)$ *and 2J-dimensional copula* $C(\cdot|\boldsymbol{\theta}_\rho)$;
3. *Given* $\mathbf{\Lambda}_t = \boldsymbol{\lambda}_t$ *and* $\mathbf{\Psi}_t = \boldsymbol{\psi}_t$, *the compound random variables* $Z_t^{(1)}, \ldots, Z_t^{(J)}$ *are independent with* $N_t^{(j)}$ *and* $X_1^{(j)}(t), X_2^{(j)}(t), \ldots$ *independent; frequencies* $N_t^{(j)} \sim P\left(\cdot|\lambda_t^{(j)}\right)$; *and independent severities* $X_s^{(j)}(t) \sim F\left(\cdot|\psi_t^{(j)}\right), s \geq 1$.

Calibration of the aforementioned model requires estimation of $\boldsymbol{\theta}$. It can be treated within a Bayesian framework as a random variable $\boldsymbol{\Theta}$ to incorporate expert opinions and external data into the estimation procedure (in Section 12.7, we describe the estimation procedure for frequencies). Also note that for simplicity of notation, we assumed one severity risk profile $\Psi_t^{(j)}$ and one frequency risk profile $\Lambda_t^{(j)}$ per risk—extension is trivial if more risk profiles are required.

In general, a copula can be introduced between all risk profiles. For illustration, consider the bivariate case ($J = 2$). That is, we assume that the model assumptions 12.1 are fulfilled for the aggregated losses

$$Z_t^{(1)} = \sum_{s=1}^{N_t^{(1)}} X_s^{(1)}(t) \quad \text{and} \quad Z_t^{(2)} = \sum_{s=1}^{N_t^{(2)}} X_s^{(2)}(t). \tag{12.45}$$

As marginals, for $j = 1, 2$, we choose

- $N_t^{(j)} \sim Poisson(\lambda_t^{(j)})$ and $X_s^{(j)}(t) \sim LogNormal(\mu_j(t), \sigma_j^2(t))$;
- $\lambda_t^{(1)} \sim Gamma(2.5, 2)$, $\lambda_t^{(2)} \sim Gamma(5, 2)$, $\mu_j(t) \sim Normal(1, 1)$, $\sigma_j(t) = 2$;
- The dependence between $\lambda_t^{(1)}$, $\lambda_t^{(2)}$, $\mu_1(t)$, and $\mu_2(t)$ is a Gaussian copula.

The parameters in the marginal distributions correspond to $\boldsymbol{\theta}_\Lambda$ and $\boldsymbol{\theta}_\Psi$ in model assumptions 12.1. Here, we assume the parameters are known *a priori*. In Section 12.7, we will demonstrate the Bayesian inference model and associated methodology to perform an estimation of the model parameters.

Given marginal and copula parameters $(\boldsymbol{\theta}_\Lambda, \boldsymbol{\theta}_\Psi, \boldsymbol{\theta}_\rho)$, the simulation of the annual losses for year $t = T + 1$, when risk profiles are dependent via a copula, can be accomplished using the following procedure.

Algorithm 12.1

1. *Simulate 2J-variate* $u_1, \ldots, u_J, v_1, \ldots, v_J$ *from a 2J dimensional copula* $C(\cdot|\boldsymbol{\theta}_\rho)$;
2. *Calculate* $\lambda_t^{(j)} = G^{-1}\left(u_j|\theta_\Lambda^{(j)}\right)$ *and* $\psi_t^{(j)} = H^{-1}\left(v_j|\theta_\Psi^{(j)}\right), j = 1, \ldots, J$;
3. *Sample* $n_t^{(j)}$ *from* $P\left(\cdot|\lambda_t^{(j)}\right), j = 1, \ldots, J$;
4. *Sample independent* $x_s^{(j)}(t), s = 1, \ldots, n_t^{(j)}, j = 1, \ldots, J$ *from* $F\left(\cdot|\psi_t^{(j)}\right)$;
5. *Calculate annual losses* $z_t^{(j)} = \sum_{s=1}^{n_t^{(j)}} x_s^{(j)}(t), j = 1, \ldots, J$;
6. *Repeat steps 1–5 K times to get K random samples of the annual losses* $z_t^{(j)}$.

Using the aforementioned simulation procedure, we can examine the strength of dependence between the annual losses if there is a dependence between the risk profiles. Figure 12.2 shows the induced dependence between the annual losses $Z_t^{(1)}$ and $Z_t^{(2)}$ versus the copula dependence parameter for three cases:

- Only $\lambda_t^{(1)}$ and $\lambda_t^{(2)}$ are dependent;
- Only $\mu_1(t)$ and $\mu_2(t)$ are dependent;
- The dependence between $\lambda_t^{(1)}$ and $\lambda_t^{(2)}$ is the same as between $\mu_1(t)$ and $\mu_2(t)$.

In all cases the dependence is the Gaussian copula (11.49) denoted as $C(u_1, u_2|\rho)$ and parameterized by one parameter ρ, which controls the degree of dependence. In the case of the

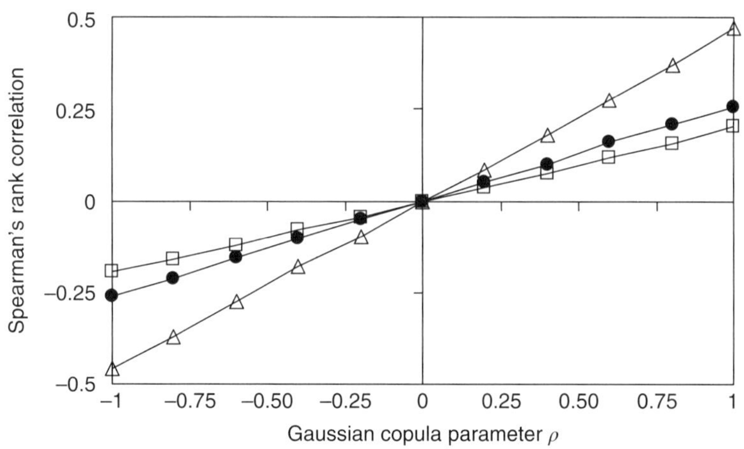

FIGURE 12.2 Spearman's rank correlation $\rho_S[Z^{(1)}, Z^{(2)}]$ between annual losses versus the Gaussian copula parameter ρ: (□)—copula for the frequency profiles $\Lambda_t^{(1)}$ and $\Lambda_t^{(2)}$; (•)—copula for the severity profiles $\Psi_t^{(1)}$ and $\Psi_t^{(2)}$ that correspond to μ_1 and μ_2 in the severity distribution, respectively; (△)—copula for λ_1 and λ_2 and the same copula for $\Psi_t^{(1)}$ and $\Psi_t^{(2)}$

Gaussian copula, ρ is a nondiagonal element of correlation matrix $\boldsymbol{\Sigma}$ in (11.49). The parameter ρ corresponds to θ_ρ in Model Assumptions 12.1.

In each of these examples, we vary the parameter of the copula model ρ from weak to strong dependence. The annual losses are not Gaussian distributed and to measure the dependence between the annual losses we use a nonlinear rank correlation measure, Spearman's rank correlation, $\rho_S[Z_t^{(1)}, Z_t^{(2)}]$. The Spearman's rank correlation between the annual losses was estimated using $10,000$ simulated years for each value of ρ. These numerical experiments show that the range of possible dependence between the annual losses of different risks induced by the dependence between risk profiles is very wide and should be flexible enough to model dependence in practice. Note that the degree of induced correlation can be further extended by working with more flexible copula models at the expense of estimation of a larger number of model parameters.

12.7 Multiple Risk LDA: Dependence and Combining Different Data Sources

Basel II OpRisk models have to combine information from internal data, external data, and expert opinions as discussed in Chapter 15. We should also note that experts in financial institutions often attempt to specify not only frequency and severity distributions but also correlations between risks.

Combining of expert opinions with internal and external data is a difficult problem and complicated ad hoc procedures are used in practice. Some prominent risk professionals in industry have argued that statistically consistent combining of these different data sources is one of the most pertinent and challenging aspects of OpRisk modeling.

A Bayesian model to combine three data sources (internal data, external data, and expert opinion) for the case of a single risk cell is presented in Lambrigger *et al.* (2007). Then Peters *et al.* (2009) extended this to a multivariate case. The main idea was to utilize Bayesian inference to estimate the parameters of the model through the combination of expert opinions and observed loss data (internal and external).

To illustrate the approach, consider modeling frequencies only. The estimation procedure is presented for frequencies only. However, it is not difficult to extend the actual procedure to include severities. Here we extend a single risk cell frequency model to the general multiple risk cell setting. This will involve formulation of the multivariate posterior distribution.

Model Assumptions 12.2 (Multiple Risk Cell Frequency Model) *Consider J risk cells. Assume that every risk cell j has a fixed, deterministic volume $V^{(j)}$.*

1. *The risk characteristic $\boldsymbol{\Theta}_\Lambda = (\Theta_\Lambda^{(1)}, \ldots, \Theta_\Lambda^{(J)})$ has a J-dimensional prior density $\pi(\boldsymbol{\theta}_\Lambda)$. The copula parameters $\boldsymbol{\theta}_\rho$ are modeled by a random vector $\boldsymbol{\Theta}_\rho$ with the prior density $\pi(\boldsymbol{\theta}_\rho)$; $\boldsymbol{\Theta}_\Lambda$ and $\boldsymbol{\Theta}_\rho$ are independent;*
2. *Given $\boldsymbol{\Theta}_\Lambda = \boldsymbol{\theta}_\Lambda$ and $\boldsymbol{\Theta}_\rho = \boldsymbol{\theta}_\rho$: the vectors $(\boldsymbol{\Lambda}_1, \boldsymbol{N}_1), \ldots, (\boldsymbol{\Lambda}_{T+1}, \boldsymbol{N}_{T+1})$ are independent and identically distributed; and the intensities $\boldsymbol{\Lambda}_t = (\Lambda_t^{(1)}, \ldots, \Lambda_t^{(J)})$ have a J-dimensional conditional density with marginal distributions*

$$\Lambda_t^{(j)} \sim G\left(\cdot|\theta_\Lambda^{(j)}\right) = Gamma\left(\alpha^{(j)}, \alpha^{(j)}/\theta_\Lambda^{(j)}\right)$$

and the copula $c(\cdot|\boldsymbol{\theta}_\rho)$. Thus, the joint density of $\boldsymbol{\Lambda}_t$ is given by

$$\pi(\boldsymbol{\lambda}_t|\boldsymbol{\theta}_\Lambda,\boldsymbol{\theta}_\rho) = c\left(G(\lambda_t^{(1)}|\theta_\Lambda^{(1)}),\ldots,G(\lambda_t^{(J)}|\theta_\Lambda^{(J)})|\boldsymbol{\theta}_\rho\right)\prod_{j=1}^{J}\pi(\lambda_t^{(j)}|\theta_\Lambda^{(j)}), \quad (12.46)$$

where $\pi\left(\cdot|\theta_\Lambda^{(j)}\right)$ denotes the marginal density;

3. *Given $\boldsymbol{\Theta}_\Lambda = \boldsymbol{\theta}_\Lambda$ and $\boldsymbol{\Lambda}_t = \boldsymbol{\lambda}_t$, the frequencies are independent with*

$$N_t^{(j)} \sim Poisson(V^{(j)}\lambda_t^{(j)}), \quad j = 1,\ldots,J.$$

4. *There are expert opinions $\boldsymbol{\Delta}_k = (\Delta_k^{(1)},\ldots,\Delta_k^{(J)})$, $k = 1,\ldots,K$. Given $\boldsymbol{\Theta}_\Lambda = \boldsymbol{\theta}_\Lambda$: $\boldsymbol{\Delta}_k$ and $(\boldsymbol{\Lambda}_t, \boldsymbol{N}_t)'$ are independent for all k and t; and $\Delta_k^{(j)}$ are all independent with*

$$\Delta_k^{(j)} \sim Gamma(\xi^{(j)}, \xi^{(j)}/\theta_\Lambda^{(j)}).$$

Prior Structure $\pi(\boldsymbol{\theta}_\Lambda)$ and $\pi(\boldsymbol{\theta}_\rho)$. In the following examples, *a priori,* the risk characteristics $\Theta_\Lambda^{(j)}$ are independent gamma distributed: $\Theta_\Lambda^{(j)} \sim Gamma(a^{(j)}, b^{(j)})$ with hyper-parameters $a^{(j)} > 0$ and $b^{(j)} > 0$. This means that *a priori* the risk characteristics for the different risk classes are independent. That is, if the company has a bad risk profile in risk class *j*, then the risk profile in risk class *i* is not necessarily bad. Dependence is then modeled through the dependence between the intensities $\Lambda_t^{(1)}, \ldots, \Lambda_t^{(J)}$. If this is not appropriate then, of course, this can easily be changed by assuming dependence within $\boldsymbol{\Theta}_\Lambda$. In the following simulation experiments, we consider cases when the copula is parameterized by a scalar θ_ρ. Additionally, we are interested in obtaining inferences on θ_ρ implied by the data only so we use noninformative constant prior on the range $[-1, 1]$ in the case of Gaussian copula.

Posterior Density. The marginal posterior density of random vector $(\boldsymbol{\Theta}_\Lambda, \Theta_\rho)$ given data of counts $\boldsymbol{N}_{1:T} = \boldsymbol{n}_{1:T}$ and expert opinions $\boldsymbol{\Delta}_{1:K} = \boldsymbol{\delta}_{1:K}$ is

$$\begin{aligned}
\pi(\boldsymbol{\theta}_\Lambda,\theta_\rho|\boldsymbol{n}_{1:T},\boldsymbol{\delta}_{1:K}) &= \prod_{t=1}^{T}\int \pi(\boldsymbol{\theta}_\Lambda,\theta_\rho,\boldsymbol{\lambda}_t|\boldsymbol{n}_{1:T},\boldsymbol{\delta}_{1:K})\,d\boldsymbol{\lambda}_t \\
&\propto \prod_{t=1}^{T}\left(\int\prod_{j=1}^{J}\exp\left\{-V^{(j)}\lambda_t^{(j)}\right\}\frac{(V^{(j)}\lambda_t^{(j)})^{n_t^{(j)}}}{n_t^{(j)}!}\,\pi(\boldsymbol{\lambda}_t|\boldsymbol{\theta}_\Lambda,\theta_\rho)\,d\boldsymbol{\lambda}_t\right) \\
&\times\prod_{k=1}^{K}\prod_{j=1}^{J}\left(\frac{(\xi^{(j)}/\theta_\Lambda^{(j)})^{\xi^{(j)}}}{\Gamma(\xi^{(j)})}(\delta_k^{(j)})^{\xi^{(j)}-1}\exp\left\{-\delta_k^{(j)}\xi^{(j)}/\theta_\Lambda^{(j)}\right\}\right) \\
&\times\prod_{j=1}^{J}\frac{(b^{(j)})^{a^{(j)}}}{\Gamma(a^{(j)})}(\theta_\Lambda^{(j)})^{a^{(j)}-1}\exp\left\{-b^{(j)}\theta_\Lambda^{(j)}\right\}\pi(\theta_\rho).
\end{aligned} \quad (12.47)$$

Here, for convenience, we use notation $x_{1:M} = \{x_1, x_2, \ldots, x_M\}$. For example,

$$\boldsymbol{N}_{1:T} = \left\{\left(N_1^{(1)},\ldots,N_1^{(J)}\right),\left(N_2^{(1)},\ldots,N_2^{(J)}\right),\ldots,\left(N_T^{(1)},\ldots,N_T^{(J)}\right)\right\}$$

are the annual number of losses for all risk profiles and years; and

$$\boldsymbol{\Delta}_{1:K} = \left\{\left(\Delta_1^{(1)},\ldots,\Delta_1^{(J)}\right),\left(\Delta_2^{(1)},\ldots,\Delta_2^{(J)}\right),\ldots,\left(\Delta_K^{(1)},\ldots,\Delta_K^{(J)}\right)\right\}$$

are the expert opinions on mean frequency intensities for all experts and risk profiles.

12.7.1 BAYESIAN INFERENCE USING MCMC

Posterior (12.47) involves integration and sampling from this distribution is difficult. The common trick is to sample from the desired target posterior distribution $\pi(\boldsymbol{\theta}_\Lambda,\theta_\rho,\boldsymbol{\lambda}_{1:T}|\boldsymbol{n}_{1:T},\boldsymbol{\delta}_{1:K})$. Then marginally taken samples of $\boldsymbol{\Theta}_\Lambda$ and Θ_ρ are samples from $\pi\left(\boldsymbol{\theta}_\Lambda,\theta_\rho|\boldsymbol{n}_{1:T},\boldsymbol{\delta}_{1:K}\right)$, which can be used to make inferences for required quantities.

Sampling from $\pi\left(\boldsymbol{\theta}_\Lambda,\theta_\rho,\boldsymbol{\lambda}_{1:T}|\boldsymbol{n}_{1:T},\boldsymbol{\delta}_{1:K}\right)$ via closed-form inversion or rejection sampling is still not an option. To accomplish this task, one can develop a specialized MCMC method. One possible way is to use Gibbs sampling methodology. This requires the knowledge of full conditional distributions that can be derived for this particular model, (see Peters *et al.*, 2009, appendix B), as

$$\begin{aligned}
\pi(\theta_\Lambda^{(j)}|\boldsymbol{\theta}_\Lambda^{(-j)},\boldsymbol{\lambda}_{1:T},\boldsymbol{n}_{1:T},\boldsymbol{\delta}_{1:K},\theta_\rho) &\propto \pi(\boldsymbol{\lambda}_{1:T}|\boldsymbol{\theta}_\Lambda^{(-j)},\theta_\Lambda^{(j)},\theta_\rho)\pi(\boldsymbol{\delta}_{1:K}|\boldsymbol{\theta}_\Lambda^{(-j)},\theta_\Lambda^{(j)}) \\
&\quad\times\pi(\boldsymbol{\theta}_\Lambda^{(-j)}|\theta_\Lambda^{(j)})\pi(\theta_\Lambda^{(j)}), \qquad (12.48)\\
\pi(\lambda_t^{(j)}|\boldsymbol{\theta}_\Lambda,\boldsymbol{\lambda}_{1:T}^{(-t,-j)},\boldsymbol{n}_{1:T},\boldsymbol{\delta}_{1:K},\theta_\rho) &\propto \pi(\boldsymbol{n}_{1:T}|\boldsymbol{\lambda}_{1:T}^{(-t,-j)},\lambda_t^{(j)}) \\
&\quad\times\pi(\boldsymbol{\lambda}_t^{(-j)},\lambda_t^{(j)}|\boldsymbol{\theta}_\Lambda,\theta_\rho), \qquad (12.49)\\
\pi\left(\theta_\rho|\boldsymbol{\theta}_\Lambda,\boldsymbol{\lambda}_{1:T},\boldsymbol{n}_{1:T},\boldsymbol{\delta}_{1:K}\right) &\propto \pi(\boldsymbol{\lambda}_{1:T}|\boldsymbol{\theta}_\Lambda,\theta_\rho)\pi\left(\theta_\rho\right). \qquad (12.50)
\end{aligned}$$

Here, $\boldsymbol{\lambda}_{1:T}^{(-i,-j)}$, $\boldsymbol{\theta}_\Lambda^{(-j)}$ and $\boldsymbol{\lambda}_t^{(-j)}$ are the exclusion operators:

- $\boldsymbol{\lambda}_{1:T}^{(-2,-1)} = \left\{\left(\lambda_1^{(1)},\ldots,\lambda_1^{(J)}\right),\left(\lambda_2^{(2)},\ldots,\lambda_2^{(J)}\right),\ldots,\left(\lambda_T^{(1)},\ldots,\lambda_T^{(J)}\right)\right\}$ are frequency intensities for all risk profiles and years, excluding risk profile 1 from year 2;
- $\boldsymbol{\theta}_\Lambda^{(-j)} = \left\{\theta_\Lambda^{(1)},\ldots,\theta_\Lambda^{(j-1)},\theta_\Lambda^{(j+1)},\ldots,\theta_\Lambda^{(J)}\right\}$; and similar for $\boldsymbol{\lambda}_t^{(-j)}$.

These full conditionals do not take standard explicit closed forms and typically the normalizing constants are not known in closed form. Therefore, this will exclude straightforward inversion or basic rejection sampling being used to sample from these distributions. One may adopt a Metropolis–Hastings within Gibbs sampler to obtain samples; see Section 7.4.4. To utilize such algorithm, it is important to select a suitable proposal distribution. Quite often in high-dimensional problems such as ours, this requires tuning of the proposal for a given target distribution. Hence, one incurs a significant additional computational expense in tuning the proposed distribution parameters offline so that mixing of the resulting Markov chain is sufficient. An alternative not discussed here would include an adaptive Metropolis–Hastings within Gibbs sampling algorithm; see Atchadé and Rosenthal (2005) and Rosenthal (2009). Here, we take a different approach that utilizes the full conditional distributions, known as a univariate slice sampler described in Section 7.6.1. Note that we only need to know the target full conditional posterior up to normalization. This is important in this example since solving the normalizing constant in this model is not possible analytically.

Algorithm 12.2 (Slice sampling)

1. *For* $l = 0$*, initialize the parameter vector* $(\boldsymbol{\theta}_{\Lambda,0}, \boldsymbol{\lambda}_{1:T,0}, \theta_{\rho,0})$ *randomly or deterministically;*
2. *Repeat while* $l \leq L$
   a) *Set* $(\boldsymbol{\theta}_{\Lambda,l}, \boldsymbol{\lambda}_{1:T,l}, \theta_{\rho,l}) = (\boldsymbol{\theta}_{\Lambda,l-1}, \boldsymbol{\lambda}_{1:T,l-1}, \theta_{\rho,l-1})$*;*
   b) *Sample* j *uniformly from set* $\{1, 2, \ldots, J\}$*;*
      Sample new parameter value $\widetilde{\theta}_{\Lambda}^{(j)}$ *from the full conditional posterior distribution* $\pi\left(\theta_{\Lambda}^{(j)}|\boldsymbol{\theta}_{\Lambda,l}^{(-j)}, \boldsymbol{\lambda}_{1:T,l}, \boldsymbol{n}_{1:T}, \boldsymbol{\delta}_{1:K}, \theta_{\rho,l}\right)$.
      Set $\theta_{\Lambda,l}^{(j)} = \widetilde{\theta}_{\Lambda}^{(j)}$.
   c) *Sample* j *uniformly from set* $\{1, 2, \ldots, J\}$ *and* t *uniformly from set* $\{1, \ldots, T\}$*;*
      Sample new parameter value $\widetilde{\lambda}_t^{(j)}$ *from the full conditional posterior distribution* $\pi\left(\lambda_t^{(j)}|\boldsymbol{\theta}_{\Lambda,l}, \boldsymbol{\lambda}_{1:T,l}^{(-t,-j)}, \boldsymbol{n}_{1:T}, \boldsymbol{\delta}_{1:K}, \theta_{\rho,l}\right)$.
      Set $\lambda_{t,l}^{(j)} = \widetilde{\lambda}_t^{(j)}$.
   d) *Sample new parameter value* $\widetilde{\theta}_{\rho}$ *from the full conditional posterior distribution* $\pi\left(\theta_{\rho}|\boldsymbol{\theta}_{\Lambda,l}, \boldsymbol{\lambda}_{1:T,l}, \boldsymbol{n}_{1:T}, \boldsymbol{\delta}_{1:K}\right)$.
      Set $\theta_{\rho,l} = \widetilde{\theta}_{\rho}$.
3. $l = l + 1$ *and return to 2.*

The sampling from the full conditional posteriors in stage 2 uses a univariate slice sampler. *For example, to sample the next iteration of the Markov chain from* $\pi\left(\theta_{\Lambda}^{(j)}|\boldsymbol{\theta}_{\Lambda,l}^{(-j)}, \boldsymbol{\lambda}_{1:T,l}, \boldsymbol{n}_{1:T}, \boldsymbol{\delta}_{1:K}, \theta_{\rho}\right)$:

- *Sample* u *from a uniform distribution*

$$\textit{Uniform}\left(0, \pi\left(\theta_{\Lambda,l}^{(j)}|\boldsymbol{\theta}_{\Lambda,l}^{(-j)}, \boldsymbol{\lambda}_{1:T,l}, \boldsymbol{n}_{1:T}, \boldsymbol{\delta}_{1:K}, \theta_{\rho}\right)\right). \tag{12.51}$$

- *Sample* $\widetilde{\theta}_{\Lambda}^{(j)}$ *uniformly from the intervals (level set)*

$$A = \left\{\theta_{\Lambda}^{(j)} : \pi\left(\theta_{\Lambda}^{(j)}|\boldsymbol{\theta}_{\Lambda,l}^{(-j)}, \boldsymbol{\lambda}_{1:T,l}, \boldsymbol{n}_{1:T}, \boldsymbol{\delta}_{1:K}, \theta_{\rho}\right) > u\right\}. \tag{12.52}$$

The level sets A *are determined, for example, by a stepping out and a shrinkage procedure, the details of which can be found in Neal (2003, figure 1, p. 713); see also Section 7.6.1.*

12.7.2 NUMERICAL EXAMPLE

Consider the model with Model Assumptions 12.2 in the case of two risks with dependent intensities and set risk cell volumes $V^{(1)} = V^{(2)} = 1$. Here, we estimate $\Theta_{\Lambda}^{(1)}, \Theta_{\Lambda}^{(2)}$, and Θ_{ρ} jointly. We set the true values of $\Theta_{\Lambda}^{(1)}$ and $\Theta_{\Lambda}^{(2)}$ to be $\theta_{true}^{(1)} = 5$ and $\theta_{true}^{(2)} = 10$, respectively. Also, we assume a Gaussian copula with $\rho = 0.9$, that is, the true value of Θ_ρ is 0.9. For the expert opinions on the true parameters, assume opinion that underestimates risk profile 1, $\Delta_1^{(1)} = 2$ and opinion that overestimates the risk profile 2, $\Delta_1^{(2)} = 13$. The model parameters were set as follows:

- $\xi^{(1)} = \xi^{(2)} = 2, \alpha^{(1)} = 2, \alpha^{(2)} = 2$ – parameters of the conditional distributions for the intensities and expert opinions;
- $a^{(1)} = a^{(2)} = 2, b^{(1)} = 2.5, b^{(2)} = 5$ – parameters of the prior distribution for $\Theta_\Lambda^{(1)}$ and $\Theta_\Lambda^{(2)}$.

Then, the simulation experiment steps are as follows:

1. Using the true values for the model parameters, simulate a dataset $\boldsymbol{n}_{1:T}$ of the annual number of events over $T = 20$ years;
2. Obtain correlated MCMC samples from the target posterior distribution after discarding burnin samples, $\{\boldsymbol{\theta}_{\Lambda,l}, \boldsymbol{\lambda}_{1:T,l}, \theta_{\rho,l}\}, l = 1,001, \ldots, 50,000$. Here, we use the slice sampler Algorithm 12.2;
3. Estimate desired posterior quantities such as posterior mean of parameters of interest and posterior standard deviations.

Further analysis can be done by repeating steps 1–3 for independent data realizations and then analyzing average of the results; these can be found in Peters *et al.* (2009).

Results for this simulation experiment as a function of data size are given in Table 12.1. That is, we study the accuracy of the parameter estimates as the number of observations increases. A typical run with 5 years of data and 1 expert in the bivariate case for 50,000 simulations took approximately 50 s and for the case of 10 risk profiles it took approximately 43 min.[1] The standard errors in the estimates (due to finite number of MCMC iterations) were in the range 1–5% and are not presented in the table.

TABLE 12.1 **Posterior estimates for $\Theta_\Lambda^{(1)}, \Theta_\Lambda^{(2)}$ and copula parameter Θ_ρ**

Year	1	2	5	10	15	20
$\mathbb{E}[\Theta_\Lambda^{(1)}]$	2.83	4.49	3.31	4.88	4.36	5.07
$\text{stdev}[\Theta_\Lambda^{(1)}]$	1.74	2.02	1.38	1.29	1.10	1.09
$\mathbb{E}[\Theta_\Lambda^{(2)}]$	10.23	10.85	8.72	8.91	8.58	9.94
$\text{stdev}[\Theta_\Lambda^{(2)}]$	3.92	3.52	2.95	2.12	2.04	1.85
$\mathbb{E}[\Theta_\rho]$	0.21	0.47	0.61	0.66	0.70	0.74
$\text{stdev}[\Theta_\rho]$	0.54	0.39	0.30	0.24	0.19	0.15

In this case, a single data set is generated using Gaussian ($\rho = 0.9$) copula model as specified. Posterior standard deviations are given in brackets next to estimate. Joint estimation was used.

[1] Computing time is quoted for a standard PC, Intel Core 2 with 2.40 GHz CPU and 2.39 GB of RAM.

These results demonstrate that our model and estimation methodology is successfully able to estimate jointly the risk profiles and the correlation parameter. It is also clear that with few observations, for example, $T \leq 5$, and a vague prior for the copula parameter, it will be difficult to accurately estimate the copula parameter. This is largely due to the fact that the posterior distribution in this case is diffuse. However, as the number of observations increases, the accuracy of the estimate improves and the estimates are reasonable in the case of 15 or 20 years of data. Additionally, we could further improve the accuracy of this prediction if we incorporated expert opinions into the prior specification of the copula parameter instead of using a vague prior.

Other results presented in Peters *et al.* (2009) demonstrate that, as expected from credibility theory, the joint estimation is better than the marginal, that is, the posterior standard deviations for $\Theta_\Lambda^{(1)}$ and $\Theta_\Lambda^{(2)}$ are less when joint estimation is used. In addition, the rate of convergence of the posterior mean for Θ_Λ to the true value is faster under the joint estimation and there is a strong correlation between $\Theta_\Lambda^{(1)}$ and $\Theta_\Lambda^{(2)}$. Thus, the standard practice in the industry of performing marginal estimation of risk profiles may lead to incorrect results.

Overall, this example demonstrates how the combination of all the relevant sources of data can be achieved and that a sampling methodology has the ability to estimate jointly all the model parameters, including the copula parameter. One can extend this methodology to more sophisticated and flexible copula-based models with more than one parameter. This should be relatively trivial since the methodology developed applies directly. However, the challenge in the case of a more sophisticated copula model relates to finding a relevant choice of prior distribution on the correlation structure.

12.7.3 PREDICTIVE DISTRIBUTION

Conceptually, quantification of the predictive distribution (accounting both for process and parameter uncertainties) for a bank's annual loss in the case of many risks is similar to the case of single risk considered in Section 13.7. If correlation modeling cannot be done then, as required by Basel II, the 0.999 quantile should be quantified for each risk cell as described in Section 13.7; the total capital is just a sum of these quantiles. In this section, we assume that the dependence model between risks is developed.

Consider the annual loss in a bank over the next year, Z_{T+1}. Denote the density of the annual loss, conditional on parameters $\boldsymbol{\theta}$, as $f(z_{T+1}|\boldsymbol{\theta})$. Typically, practitioners will take point estimates $\hat{\boldsymbol{\theta}}$ of all model parameters; conditional on these point estimates construct the predictive distribution $f(z_{T+1}|\hat{\boldsymbol{\theta}})$. Then, the latter is used to calculate risk measures such as the 0.999 quantile, $Q_{0.999}(\hat{\boldsymbol{\theta}})$. Typically, given observations, the MLEs $\hat{\boldsymbol{\theta}}$ are used as the "best fit" point estimators for $\boldsymbol{\theta}$.

However, the parameters $\boldsymbol{\theta}$ are unknown and it is important to account for this uncertainty when the capital charge is estimated (especially for risks with small datasets) as discussed in Shevchenko (2008). If the Bayesian inference approach is taken, then the parameter $\boldsymbol{\theta}$ is modeled by random variable $\boldsymbol{\Theta}$ and the predictive density (accounting for parameter uncertainty) of Z_{T+1}, given all data $\boldsymbol{Y}$ used in the estimation procedure, is

$$f(z_{T+1}|\boldsymbol{y}) = \int f(z_{T+1}|\boldsymbol{\theta})\pi(\boldsymbol{\theta}|\boldsymbol{y})d\boldsymbol{\theta}. \tag{12.53}$$

Here, $\pi(\boldsymbol{\theta}|\boldsymbol{y})$ is the posterior density for $\boldsymbol{\Theta}$. Also, it is assumed that, given parameters $\boldsymbol{\Theta}$, Z_{T+1} and $\boldsymbol{Y}$ are independent. The 0.999 quantile of the predictive distribution

$$Q_q^P = F_{Z_{T+1}|\boldsymbol{Y}}^{-1}(q) = \inf\{z \in \mathbb{R} : \Pr[Z_{T+1} > z|\boldsymbol{Y}] \le 1 - q\}, \tag{12.54}$$

where $q = 0.999$, can be used as a risk measure for capital calculations; also see formula (13.79).

Another approach under a Bayesian framework to account for parameter uncertainty is to consider a quantile $Q_q(\boldsymbol{\theta})$ of the conditional annual loss density $f(\cdot|\boldsymbol{\theta})$:

$$Q_q(\boldsymbol{\Theta}) = F_{Z_{T+1}|\boldsymbol{\Theta}}^{-1}(q) = \inf\{z \in \mathbb{R} : \Pr[Z_{T+1} > z|\boldsymbol{\Theta}] \le 1 - q\}, \tag{12.55}$$

where we are interested in $q = 0.999$. Then, given that $\boldsymbol{\Theta}$ is distributed from $\pi(\boldsymbol{\theta}|\boldsymbol{y})$, one can find the distribution of $Q_q = Q_q(\boldsymbol{\Theta})$ and form a predictive interval to contain the true value of Q_q with some probability.[2] Under this approach, one can argue that the conservative estimate of the capital charge accounting for parameter uncertainty should be based on the upper bound of the constructed interval. Note that specification of the confidence level is required and it might be difficult to argue that the commonly used confidence level 0.95 is good enough for estimation of the 0.999 quantile.

In OpRisk, it seems that the objective should be to estimate the full predictive distribution (12.53) for the annual loss Z_{T+1} over next year conditional on all available information and then estimate the capital charge as a quantile $Q_{0.999}^P$ of this distribution (12.54).

Consider all risk cells in the bank. Assume that multivariate model is specified. That is, the frequency $p(\cdot|\boldsymbol{\alpha})$ and severity $f(\cdot|\boldsymbol{\beta})$ densities for each cell are chosen and the dependence structure between risks parameterized by some parameter vector $\boldsymbol{\rho}$ is specified. Also, suppose that the posterior $\pi(\boldsymbol{\theta}|\boldsymbol{y}), \boldsymbol{\theta} = (\boldsymbol{\alpha}, \boldsymbol{\beta}, \boldsymbol{\rho})$ is estimated. Then, the predictive distribution (12.53) for the annual loss across all risk cells over next year can be calculated using the Monte Carlo procedure with the following logical steps.

Algorithm 12.3 (Monte Carlo Predictive Distribution for Many Risks)

1. *For* $k = 1, \ldots, K$
   a) *Simulate all model parameters (including the dependence parameters) from their joint posterior* $\pi(\boldsymbol{\theta}|\boldsymbol{y})$. *If the posterior is not known in closed form, then this simulation can be done using MCMC (see Section 7.4). For example, one can run MCMC for K iterations beforehand and simply take the k-th iteration parameter values;*
   b) *Given model parameters* $\boldsymbol{\theta} = (\boldsymbol{\alpha}, \boldsymbol{\beta}, \boldsymbol{\rho})$, *simulate the annual frequencies* $N^{(j)}$ *and severities* $X_s^{(j)}$, $s = 1, \ldots, N^{(j)}$ *for all risks* $j = 1, \ldots, J$ *with a chosen dependence structure. Calculate the bank annual loss* $Z_k = Z^{(1)} + \cdots + Z^{(J)}$, *where* $Z^{(j)} = \sum_{s=1}^{N^{(j)}} X_s^{(j)}$ *is the annual loss due to the j-th risk.*
2. *Next k*

---

[2]This is similar to forming a confidence interval in the frequentist approach using the distribution of $Q_{0.999}(\hat{\boldsymbol{\theta}})$, where $\hat{\boldsymbol{\theta}}$ is treated as random.

Remark 12.5 *Obtained annual losses (total across all risks for next year)* $Z_1, \ldots, Z_K$ *are samples from the predictive density (12.53). A full specification of the dependence model is required. In general, sampling from the joint posterior of all model parameters can be accomplished via MCMC; see Peters et al. (2009) and Dalla Valle (2009). The 0.999 quantile* $Q^P_{0.999}$ *and other distribution characteristics can be estimated using the simulated samples in the usual way; see Section 13.2.*

Note that in the aforementioned Monte Carlo procedure the risk profile $\boldsymbol{\theta}$ is simulated from its posterior for each simulation. Thus, we model both the process uncertainty, which comes from the fact that frequencies and severities are random variables, and the parameter risk (parameter uncertainty), which comes from the fact that we do not know the true values of $\boldsymbol{\theta}$. Using samples from the joint posterior distribution of the model parameters, we can construct the predictive distribution by removing the parameter uncertainty from the model considered, including the uncertainty arising from the dependence parameters.

EXAMPLE 12.2

As an example, consider the Model Assumptions 12.1. Then the predictive density for the annual loss Z_{T+1} is

$$\pi\left(z_{T+1}|\boldsymbol{n}_{1:T}, \boldsymbol{\delta}_{1:K}\right) = \int \pi\left(z_{T+1}|\boldsymbol{\theta}_\Lambda, \theta_\rho\right) \pi\left(\boldsymbol{\theta}_\Lambda, \theta_\rho|\boldsymbol{n}_{1:T}, \boldsymbol{\delta}_{1:T}\right) d\boldsymbol{\theta}_\Lambda d\theta_\rho. \tag{12.56}$$

Here, we used the model assumptions that given $\boldsymbol{\Theta}_\Lambda$ and Θ_ρ we have that Z_{T+1} is independent from the data $(N_{1:T}, \Delta_{1:K})$. To obtain samples from this predictive distribution, add simulation of $(\boldsymbol{\theta}_\Lambda, \boldsymbol{\theta}_\rho)$ from the posterior distribution (e.g., using slice sampler methodology) as an extra step before step 1 in Algorithm 12.1. Specifically, if one wants to simulate K annual losses from the predictive distribution, then this would involve first running the slice sampler for K iterations after *burnin*. Then, for each iteration k, one would use the state of the Markov chain $(\boldsymbol{\theta}_{\Lambda,k}, \theta_{\rho,k})$ in the simulation Algorithm 12.1. ■

12.8 A Note on Negative Diversification and Dependence Modeling

We conclude this chapter with a brief note on dependence modeling and VaR subadditivity properties. As has already been discussed in Chapter 6, VaR is not a coherent risk measure; see Artzner *et al.* (1999). In particular, under some circumstances, VaR measure may fail a subadditivity property

$$\text{VaR}_q[Z] \leq \sum_{j=1}^{J} \text{VaR}_q[Z^{(j)}]; \tag{12.57}$$

see Embrechts *et al.* (2009a,b). That is, dependence modeling could also increase VaR. Note that if there is a perfect positive dependence between risks, that is, $Z^{(j)} = H_j^{-1}(U), j = 1, \ldots, J$, where $U \sim \mathcal{U}(0,1)$ and $H_j(\cdot)$ is a distribution of $Z^{(j)}$, then

$$\text{VaR}_q[Z] = \sum_{j=1}^{J} \text{VaR}_q[Z^{(j)}]. \tag{12.58}$$

That is, the failure of the subadditivity means that the VaR for the sum of risks is larger than the VaR in the case of perfectly dependent risks. This is very counterintuitive given a typical expectation of diversification benefits. In particular, the diversification

$$D_q = 1 - \frac{\text{VaR}_q[\sum_j Z^{(j)}]}{\sum_j \text{VaR}_q[Z^{(j)}]} \tag{12.59}$$

is expected to be positive while the subadditivity failure corresponds to the negative diversification. The latter may occur even for independent risks when the risks are heavy tailed. It was shown and discussed in Nešlehová *et al.* (2006) that if independent risks are Pareto type, $Z^{(j)} \sim F_j(x) = 1 - x^{-\alpha_j} C_j(x)$, with the tail indexes $0 < \alpha_j < 1$, then

$$\text{VaR}_q[Z] > \sum_{j=1}^{J} \text{VaR}_q[Z^{(j)}], \tag{12.60}$$

at least for sufficiently large q. The case of $0 < \alpha_j \leq 1$ corresponds to infinite mean distribution, that is, $\mathbb{E}[Z^{(j)}] = \infty$.

Remark 12.6 *To simplify notation, the index of discrete time (year) is dropped. Implicitly, in the discussion of diversification issues, we refer to the next year.*

EXAMPLE 12.3

Assume that we have two independent risks, $X \sim Pareto(\beta, 1)$ and $Y \sim Pareto(\beta, 1)$, where $Pareto(\beta, a) = 1 - (x/a)^{-\beta}$. Calculating the $\text{VaR}_{0.999}[X + Y]$ using, for example, FFT, we can easily find the diversification D_q as defined in (12.59). Figure 12.3 shows the results for $D_{0.999}$ versus β that demonstrate negative diversification for $\beta < 1$. ■

EXAMPLE 12.4

In the previous example, we found that the diversification is positive for $\beta > 1$. In particular, $D_{0.999} \approx 0.27$ when $\beta = 4$, that is mean, variance, and skewness are finite. It is important to realize that diversification depends on the quantile level. Figure 12.3 shows the results for D_q versus q in the case of $\beta = 4$. One can see that diversification is positive for high level quantiles but may become zero and negative for lower quantiles.

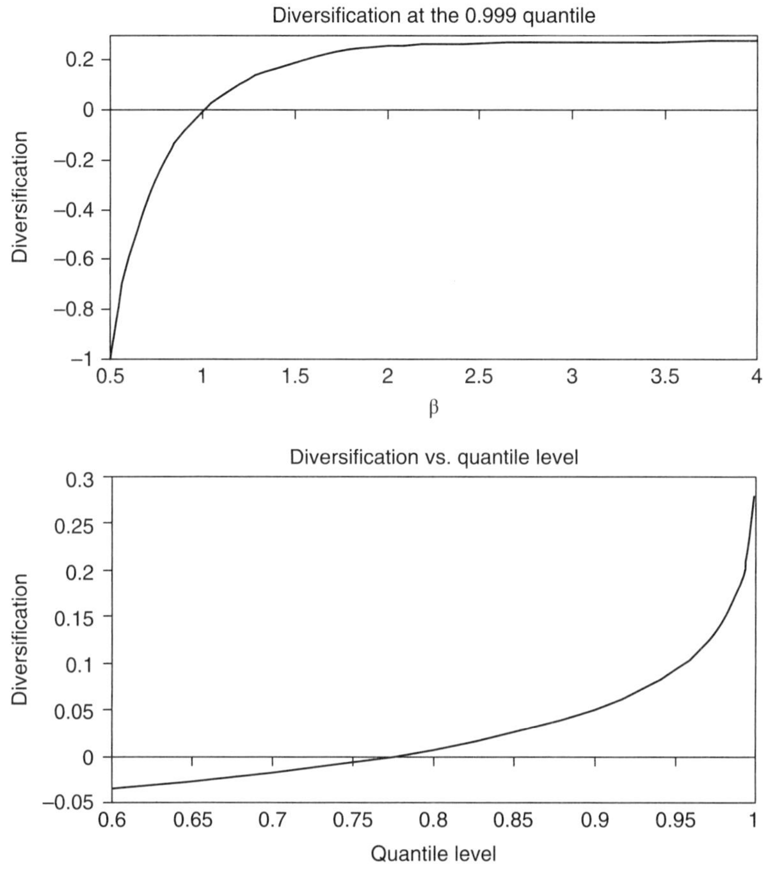

FIGURE 12.3 Upper figure: the diversification for random variables $X \sim \textit{Pareto}(\beta, 1)$ and $Y \sim \textit{Pareto}(\beta, 1)$ versus. β. Bottom figure: the diversification for random variables $X \sim \textit{Pareto}(4, 1)$ and $Y \sim \textit{Pareto}(4, 1)$ versus. quantile level q

■

CHAPTER THIRTEEN

Loss Aggregation

Estimation of the capital under the Loss Distribution Approach (LDA) requires calculation of the distribution for the aggregate (compound) loss

$$Z = X_1 + \cdots + X_N,$$

where the frequency N is a discrete random variable. Closed-form solutions are not available for the distributions typically used in OpRisk and numerical evaluation is required. This is one of the classical problems in risk theory. Before the era of personal computers, it was calculated using approximations such as that based on the asymptotic central limit theory or on ad hoc reasoning using, for example, shifted Gamma approximation. With modern computer processing power, these distributions can be calculated virtually *exactly* using numerical algorithms. The easiest to implement is the Monte Carlo method. However, because it is typically slow, Panjer recursion and Fourier inversion techniques are also widely used. Both have a long history, but their applications to computing very high quantiles of the compound distribution functions with high frequencies and heavy tails are only recent developments and various pitfalls exist. This chapter describes numerical algorithms that can be successfully used for this problem. In particular, Monte Carlo, Panjer recursion, and Fourier transformation methods are presented. Several closed-form approximations are also reviewed.

13.1 Analytic Solution

In general, there are two types of analytic solutions for calculating the compound distribution, denoted hereafter in this chapter by $H(z)$. These are based on convolutions and method of characteristic functions. Typically, the analytic solutions do not have closed form, and numerical methods (such as Monte Carlo, Panjer recursion, Fast Fourier Transform (FFT), or direct integration) are required. These solutions and methods are described in the following sections. To introduce required notation and definitions, consider the following model setup.

Fundamental Aspects of Operational Risk and Insurance Analytics: A Handbook of Operational Risk, First Edition. Marcelo G. Cruz, Gareth W. Peters, and Pavel V. Shevchenko.

Model Assumptions 13.1 *The annual loss in a risk cell is modeled by a compound random variable*

$$Z = \sum_{i=1}^{N} X_i, \tag{13.1}$$

where

- N *is the number of events (frequency) over 1 year modeled as a discrete random variable with probability mass function* $p_k = \mathbb{P}\mathrm{r}[N = k]$, $k = 0, 1, \ldots$. *Note that there is a finite probability of no loss occurring over the considered time period if* $N = 0$ *is allowed, that is* $\mathbb{P}\mathrm{r}[Z = 0] = \mathbb{P}\mathrm{r}[N = 0]$;
- X_i, $i \geq 1$, *are positive severities of the events (loss amounts) modeled as independent and identically distributed random variables from a continuous distribution function* $F(x)$ *with* $x \geq 0$ *and* $F(0) = 0$. *The corresponding density function is denoted as* $f(x)$;
- N *and* X_i *are independent for all i, that is, the frequencies and severities are independent;*
- *The distribution and density functions of the annual loss* Z *are denoted as* $H(z)$ *and* $h(z)$, *respectively;*
- *All model parameters (parameters of the frequency and severity distributions) are assumed to be known. Of course, in reality, the model parameters are unknown and estimated using past data over* T *years.*

The methods described in this chapter can be used to calculate the distribution of compound loss over any time period. For simplicity, only the most relevant case of a 1-year time period is considered here. Extension to the case of other time periods is trivial.

13.1.1 ANALYTIC SOLUTION VIA CONVOLUTIONS

The density and distribution functions of the sum of independent random variables can be calculated via convolution as described in Section 5.5. Thus, for given $N = k$, the distribution of the sum $X_1 + \cdots + X_k$ is just k-fold convolution $F^{(k)*}(z) = \mathbb{P}\mathrm{r}[X_1 + \cdots + X_n \leq z]$ of the severity distribution $F(\cdot)$, and the distribution of the annual loss (13.1) can be calculated as

$$\begin{aligned} H(z) = \mathbb{P}\mathrm{r}[Z \leq z] &= \sum_{k=0}^{\infty} \mathbb{P}\mathrm{r}[Z \leq z | N = k] \mathbb{P}\mathrm{r}[N = k] \\ &= \sum_{k=0}^{\infty} p_k F^{(k)*}(z). \end{aligned} \tag{13.2}$$

The k-fold convolution $F^{(k)*}(z)$ is calculated recursively as

$$F^{(k)*}(z) = \int_0^z F^{(k-1)*}(z - x) f(x) dx$$

with

$$F^{(0)*}(z) = \begin{cases} 1, & z \geq 0, \\ 0, & z < 0. \end{cases}$$

In this method, the integration limits are 0 and z. This is because we consider non-negative severities. The obtained formula is analytic. However, closed-form solutions are rare, an in depth discussion on such models is provided in Peters and Shevchenko (2015), in addition a special discussion on discrete distributions in this class of models is provided briefly at the end of this chapter. Panjer recursion and FFT, discussed in Sections 13.3 and 13.5, are very efficient numerical methods to calculate these convolutions.

13.1.2 ANALYTIC SOLUTION VIA CHARACTERISTIC FUNCTIONS

The characteristic function of the compound distribution can be easily calculated via the characteristic function of the severity and the probability generating function of the frequency. For definitions and results on characteristic functions, see Section 5.5. Then, the inverse transform of the characteristic function can be used to calculate the actual compound distribution. Again, typically, the inverse transform cannot be performed in closed form, and FFT or direct integration methods can be used to calculate the required integrals numerically. To introduce the notation, define as follows:

- The characteristic function of the severity density $f(x)$ is

$$\varphi(t) = \int_{-\infty}^{\infty} f(x)e^{\mathrm{i}tx}dx, \tag{13.3}$$

  where $\mathrm{i} = \sqrt{-1}$ is a unit imaginary number;
- The *probability-generating function* of a frequency distribution with probability mass function $p_k = \mathbb{P}\mathrm{r}[N = k]$ is

$$\psi(s) = \sum_{k=0}^{\infty} s^k p_k. \tag{13.4}$$

The characteristic function of the sum of independent random variables is just a product of their individual characteristic functions. Thus, for given $N = k$, the characteristic function of $X_1 + \cdots + X_k$ is just $(\varphi(t))^k$. Then, the characteristic function of the compound loss Z in model (13.1), denoted by $\chi(t)$, can be expressed through the probability-generating function of the frequency distribution and characteristic function of the severity distribution as

$$\chi(t) = \sum_{k=0}^{\infty} (\varphi(t))^k p_k = \psi(\varphi(t)). \tag{13.5}$$

Given the characteristic function, the density of the annual loss Z can be calculated via the inverse Fourier transform as

$$h(z) = \frac{1}{2\pi} \int_{-\infty}^{\infty} \chi(t) \exp(-\mathrm{i}tz)dt, \quad z \geq 0. \tag{13.6}$$

In the case of non-negative severities, the density and distribution functions of the compound loss can be calculated using the following lemma.

Lemma 13.1 *For a non-negative random variable Z with a characteristic function $\chi(t)$, the density $h(z)$ and distribution $H(z)$ functions, $z \geq 0$, are*

$$h(z) = \frac{2}{\pi} \int_0^\infty \mathcal{R}\mathrm{e}[\chi(t)] \cos(tz) dt, \quad z \geq 0; \tag{13.7}$$

$$H(z) = \frac{2}{\pi} \int_0^\infty \mathcal{R}\mathrm{e}[\chi(t)] \frac{\sin(tz)}{t} dt, \quad z \geq 0. \tag{13.8}$$

Proof: The proof is simple and can be shown by defining a function $\widetilde{h}(z)$ such that $\widetilde{h}(z) = h(z)$ if $z \geq 0$ and $\widetilde{h}(z) = h(-z)$ if $z < 0$; for details, see Shevchenko (2011, lemma 3.1). ■

Changing variable $x = t \times z$, the formula (13.8) can be rewritten as

$$H(z) = \frac{2}{\pi} \int_0^\infty \mathcal{R}\mathrm{e}[\chi(x/z)] \frac{\sin(x)}{x} dx,$$

which is often a useful representation to study limiting properties. In particular, in the limit $z \to 0$, it gives

$$H(z \to 0) = \frac{2}{\pi} \mathcal{R}\mathrm{e}[\chi(\infty)] \int_0^\infty \frac{\sin(x)}{x} dx = \mathcal{R}\mathrm{e}[\chi(\infty)]. \tag{13.9}$$

This leads to a correct limit $H(0) = \mathbb{P}\mathrm{r}[N = 0]$, because the severity characteristic function $\varphi(\infty) \to 0$ (in the case of continuous severity distribution function).

EXAMPLE 13.1 Poisson frequency

Assume that frequency N is distributed from *Poisson*(λ); then characteristic function of aggregate loss $Z = X_1 + \cdots + X_N$ is

$$\chi(t) = \sum_{k=0}^\infty (\varphi(t))^k \frac{e^{-\lambda}\lambda^k}{k!} = \exp(\lambda\varphi(t) - \lambda). \tag{13.10}$$

Substituting $\varphi(\infty) \to 0$, note that in the case of Poisson frequency, formula (13.9) gives an obvious result $H(0) = \exp(-\lambda)$. ■

EXAMPLE 13.2 Negative Binomial frequency

Assume that frequency N is from Negative Binomial distribution $NegBinomial(m, q)$; then characteristic function of aggregate loss $Z = X_1 + \cdots + X_N$ is

$$\begin{aligned}\chi(t) &= \sum_{k=0}^{\infty} (\varphi(t))^k \binom{k+m-1}{k} (1-q)^k q^m \\ &= \left(\frac{q}{1-(1-q)\varphi(t)}\right)^m. \end{aligned} \tag{13.11}$$

Substituting $\varphi(\infty) \to 0$, note that in the case of Negative Binomial frequency, formula (13.9) gives an obvious result $H(0) = q^m$. ■

EXAMPLE 13.3 Binomial frequency

Assume that frequency N is from binomial distribution $Binomial(n, q)$; then characteristic function of aggregate loss $Z = X_1 + \cdots + X_N$ is

$$\begin{aligned}\chi(t) &= \sum_{k=0}^{\infty} (\varphi(t))^k \binom{n}{k} q^k (1-q)^{n-k} \\ &= \frac{1}{(1+q(\varphi(t)-1))^n}. \end{aligned} \tag{13.12}$$

Substituting $\varphi(\infty) \to 0$, note that in the case of Negative Binomial frequency, formula (13.9) gives an obvious result $H(0) = (1-q)^n$. ■

13.1.3 MOMENTS OF COMPOUND DISTRIBUTION

In general, the compound distribution cannot be found in closed form. However, given that characteristic function of the compound distribution can be expressed through the characteristic functions of the severity and frequency via (13.5), the moments (if exist) of the compound distribution can be calculated as

$$M_k = \mathbb{E}[Z^k] = (-\mathrm{i})^k \left.\frac{d^k \chi(t)}{dt^k}\right|_{t=0}, \quad k = 1, 2, \ldots . \tag{13.13}$$

Similarly, the central moments can be found as

$$\begin{aligned}\mu_k &= \mathbb{E}[(Z - \mathbb{E}[Z])^k] \\ &= (-\mathrm{i})^k \left.\frac{d^k \chi(t) \exp(-\mathrm{i}t\mathbb{E}[Z])}{dt^k}\right|_{t=0}, \quad k = 1, 2, \ldots \end{aligned} \tag{13.14}$$

and cumulants (or semi-invariants) can be calculated as

$$\kappa_k = (-\mathrm{i})^k \left. \frac{d^k \ln \chi(t)}{dt^k} \right|_{t=0}. \tag{13.15}$$

The moments can be calculated via the cumulants and vice versa. Here, $\chi(t)$ is characteristic function of compound distribution given by (13.5). Then, one can derive the explicit expressions for all moments of compound distribution via the moments of frequency and severity, noting that $\varphi(0) = 1$ and using the relations

$$\left. \frac{d^k \psi(s)}{ds^k} \right|_{s=1} = \mathbb{E}[N(N-1)\cdots(N-k+1)], \tag{13.16}$$

$$(-\mathrm{i})^k \left. \frac{d^k \varphi(t)}{dt^k} \right|_{t=0} = \mathbb{E}[X_1^k], \tag{13.17}$$

that follow from the definitions of the probability-generating and characteristic functions (13.4) and (13.3), respectively, though the expression is lengthy for high moments.

In application, typically only the first four moments are most often used with the following relations:

$$\mu_2 = \kappa_2 \equiv \mathbb{V}\mathrm{ar}[Z]; \quad \mu_3 = \kappa_3; \quad \mu_4 = \kappa_4 + 3\kappa_2^2. \tag{13.18}$$

Then, closely related distribution characteristics, skewness and kurtosis, are

$$\text{Skewness} = \frac{\mu_3}{(\mu_2)^{3/2}}, \tag{13.19}$$

$$\text{Kurtosis} = \frac{\mu_4}{(\mu_2)^2} - 3. \tag{13.20}$$

These formulas relating characteristic function and moments can be found in many textbooks on probability theory.

Using the expression for characteristic function of the compound distribution (13.5) and formulas (13.16), it is easy to find the explicit expressions for the first four moments of the compound distribution (the calculus is simple but lengthy). Specifically, The first four moments of the compound random variable $Z = X_1 + \cdots + X_N$, where $X_1, \ldots, X_N$ are independent and identically distributed, and independent of N, are given by

$$\mathbb{E}[Z] = \mathbb{E}[N]\mathbb{E}[X_1], \tag{13.21}$$

$$\mathbb{V}\mathrm{ar}[Z] = \mathbb{E}[N]\mathbb{V}\mathrm{ar}[X_1] + \mathbb{V}\mathrm{ar}[N]\mathbb{E}[X_1]^2, \tag{13.22}$$

$$\mathbb{E}[(Z - \mathbb{E}[Z])^3] = \mathbb{E}[N]\mu_3^X + 3\mathbb{V}\mathrm{ar}[N]\mathbb{V}\mathrm{ar}[X_1]\mathbb{E}[X_1] + \mu_3^N \mathbb{E}[X_1]^3,$$

$$\mathbb{E}[(Z - \mathbb{E}[Z])^4] = \mathbb{E}[N]\mu_4^X + 4\mathbb{V}\mathrm{ar}[N]\mu_3^X \mathbb{E}[X_1] \tag{13.23}$$

$$+ 3\mathbb{V}\mathrm{ar}[X_1]^2 \left(\mathbb{V}\mathrm{ar}[N] + \mathbb{E}[N](\mathbb{E}[N]-1)\right)$$

$$+ 6\mathbb{E}[X_1]^2\mathbb{V}\mathrm{ar}[X_1]\left(\mu_3^N + \mathbb{E}[N]\mathbb{V}\mathrm{ar}[N]\right) + \mu_4^N \mathbb{E}[X_1]^4. \tag{13.24}$$

Here, explicitly,

$$\mu_3^X = \mathbb{E}[(X_1 - \mathbb{E}[X_1])^3], \quad \mu_3^N = \mathbb{E}[(N - \mathbb{E}[N])^3],$$
$$\mu_4^X = \mathbb{E}[(X_1 - \mathbb{E}[X_1])^4], \quad \mu_4^N = \mathbb{E}[(N - \mathbb{E}[N])^4].$$

It is also assumed that the required moments of severity and frequency exist.

Many parametric distributions can be used as an approximation for a compound loss distribution using moment matching; some examples are given in Section 13.6. Here, for illustration, we calculate the moments of compound loss in the case of Poisson distributed frequencies in two examples.

EXAMPLE 13.4 Moments of Compound Poisson

If frequencies are Poisson-distributed, $N \sim Poisson(\lambda)$, then

$$\mathbb{E}[N] = \mathbb{V}\mathrm{ar}[N] = \mathbb{E}[(N - \mathbb{E}[N])^3] = \lambda,$$
$$\mathbb{E}[(N - \mathbb{E}[N])^4] = \lambda(1 + 3\lambda),$$

and compound loss moments calculated using formulas (13.21–13.24) are

$$\mathbb{E}[Z] = \lambda\mathbb{E}[X_1], \quad \mathbb{V}\mathrm{ar}[Z] = \lambda\mathbb{E}[X_1^2],$$
$$\mathbb{E}[(Z - \mathbb{E}[Z])^3] = \lambda\mathbb{E}[X_1^3],$$
$$\mathbb{E}[(Z - \mathbb{E}[Z])^4] = \lambda\mathbb{E}[X_1^4] + 3\lambda^2(\mathbb{E}[X_1^2])^2. \tag{13.25}$$

Moreover, if the severities are from LogNormal distribution, $X_1 \sim LogNormal(\mu, \sigma^2)$, then

$$\mathbb{E}[X_1^k] = \exp(k\mu + k^2\sigma^2/2). \tag{13.26}$$

■

EXAMPLE 13.5 Cumulants of Compound Poisson

Consider aggregate loss $Z = X_1 + \cdots + X_N$, where N is from $Poisson(\lambda)$; $X_1, \ldots, X_N$ are independent and identically distributed, and independent of N. The cumulants of Z can be found using the definition of cumulants (13.15) and the characteristic function for compound Poisson (13.10) as follows

$$\kappa_k = (-\mathrm{i})^k \left.\frac{d^k \ln\chi(t)}{dt^k}\right|_{t=0} = \lambda(-\mathrm{i})^k \left.\frac{d^k\varphi(t)}{dt^k}\right|_{t=0} = \lambda\mathbb{E}[X_1^k], \quad k = 1, 2, \ldots.$$

■

13.1.4 VALUE-AT-RISK AND EXPECTED SHORTFALL

Once compound loss distribution is evaluated, the risk measures such as Value-at-Risk (VaR) and expected shortfall (ES) can be calculated. Analytically, VaR of the compound loss is calculated as the inverse of the compound distribution

$$\mathrm{VaR}_{\alpha}[Z] = H^{-1}(\alpha). \tag{13.27}$$

The ES of the compound loss, $\mathrm{ES}_{\alpha}[Z]$, can be calculated using its definition (6.11) in Section 6.2.4, that is,

$$\mathrm{ES}_{\alpha}[Z] = \frac{1}{1-\alpha}\int_{\alpha}^{1} \mathrm{VaR}_p[Z]dp. \tag{13.28}$$

For numerical algorithms such as Panjer recursion or FFT it is often easier to calculate the equivalent expression for ES

$$\mathrm{ES}_{\alpha}[Z] = \frac{1}{1-\alpha}\left(\mathbb{E}[Z] - \int_{0}^{\alpha} \mathrm{VaR}_p[Z]dp\right) \tag{13.29}$$

because, typically, the mean of compound distribution $\mathbb{E}[Z] = \mathbb{E}[N]\mathbb{E}[X_1]$ is known in closed form and the required integration is between 0 and α, instead of tail integration α to 1. Note that $\mathrm{ES}_{\alpha}[Z]$ is defined for a given quantile q_{α}, that is, the quantile $H^{-1}(\alpha)$ has to be computed first (unless we calculate ES as a minimum of cost function using Proposition 6.6).

Assuming that there is no jump in distribution at the quantile $q_{\alpha} = \mathrm{VaR}_{\alpha}[Z]$, the expression for ES can be rewritten as

$$\mathrm{ES}_{\alpha}[Z] = \mathbb{E}[Z|Z \geq q_{\alpha}] = \frac{\mathbb{E}[Z]}{1-\alpha} - \frac{1}{1-\alpha}\int_{0}^{q_{\alpha}} zh(z)dz. \tag{13.30}$$

Using the expression for $h(z)$ via its characteristic function $\chi(t)$ given by formula (13.7), it can be rewritten as

$$\mathrm{ES}_{\alpha}[Z] = \frac{1}{1-H(q_{\alpha})}\left[\mathbb{E}[Z] - H(q_{\alpha})q_{\alpha} + \frac{2q_{\alpha}}{\pi}\int_{0}^{\infty} \mathcal{R}\mathrm{e}\left[\chi(x/q_{\alpha})\right]\frac{1-\cos x}{x^2}dx\right]; \tag{13.31}$$

see also Shevchenko (2011, solution for exercise 3.1). This can be used to calculate ES via direct integration of characteristic function as by Luo and Shevchenko (2009). Strictly speaking, in formulas (13.30) and (13.31), we assumed that there is no jump in distribution at q_{α}. If required, the formula can be easily generalized using Proposition 6.5.

13.2 Monte Carlo Method

The easiest numerical method to calculate the compound loss distribution is the Monte Carlo method with the following steps: (i) simulate the annual number of events N from the frequency distribution; (ii) simulate independent severities $X_1, \ldots, X_N$ from the severity distribution; (iii) calculate $Z = \sum_{i=1}^{N} X_i$. Repeat these steps K times to get $Z_1, \ldots, Z_K$ independent samples of Z from a compound distribution $H(\cdot)$. All random numbers simulated here are independent.

Obtained $Z_1, \ldots, Z_K$ are samples from a compound distribution $H(\cdot)$. Now the distribution $H(\cdot)$ can be estimated by empirical distribution

$$\hat{H}(z) = \frac{1}{n} \sum_{k=1}^{K} \mathbb{I}_{\{Z_k \leq z\}}. \tag{13.32}$$

Distribution characteristics can be estimated using the simulated samples in the usual way. Here, we just mention the quantile and ES which are of primary importance for OpRisk.

13.2.1 QUANTILE ESTIMATE

Consider a sample of i.i.d. random variables $Z_1, \ldots, Z_K$ and corresponding sample sorted into the ascending order (the order statistics) $Z_{(1,K)} \leq \cdots \leq Z_{(K,K)}$. Then a standard estimator of the quantile $q_\alpha = H^{-1}(\alpha)$ is

$$\hat{Q}_\alpha = \hat{H}^{-1}(\alpha) = Z_{(\lceil K\alpha \rceil, K)}, \tag{13.33}$$

where $\lceil . \rceil$ denotes rounding upward, that is, $\lceil K\alpha \rceil$ is the smallest integer larger than or equal to $K\alpha$. Then, for a given realization of the sample $\boldsymbol{Z} = \boldsymbol{z}$, the quantile estimate is $\hat{q}_\alpha = z_{(\lceil K\alpha \rceil, K)}$. It is important to estimate the numerical error (due to the finite number of simulations K) in the quantile estimator. Formally, it can be assessed using the following asymptotic result

$$\frac{h(q_\alpha)\sqrt{K}}{\sqrt{\alpha(1-\alpha)}} \left(\hat{Q}_\alpha - q_\alpha \right) \to \mathit{Normal}(0,1), \quad \text{as} \quad K \to \infty \tag{13.34}$$

(see, e.g., Stuart and Ord 1994, pp. 356–358 and Glasserman 2004, p. 490). This means that the quantile estimator $\hat{Q}_\alpha$ converges to the true value q_α as the sample size K increases and asymptotically $\hat{Q}_\alpha$ is normally distributed with the mean q_α and standard deviation

$$\text{stdev}[\hat{Q}_\alpha] = \frac{\sqrt{\alpha(1-\alpha)}}{h(q_\alpha)\sqrt{K}}. \tag{13.35}$$

Typically, the density $h(q_\alpha)$ is not known and should be estimated itself, and thus the use of the given formula is difficult. It is often easier to estimate the error of the quantile estimator using a nonparametric statistic by forming a conservative confidence interval $[Z_{(r,K)}, Z_{(s,K)}]$ to contain the true quantile value q_α with the probability at least γ:

$$\Pr[Z_{(r,K)} \leq q_\alpha \leq Z_{(s,K)}] \geq \gamma, \quad 1 \leq r < s \leq K. \tag{13.36}$$

Indices r and s can be found by utilizing the fact that the true quantile q_α is located between $Z_{(M,K)}$ and $Z_{(M+1,K)}$ for some M. The number of losses M not exceeding the quantile q_α has a Binomial distribution, $\mathit{Binomial}(K, \alpha)$, because it is the number of successes from K independent and identical attempts with success probability α; denote this distribution as

$\mathbb{P}\mathrm{r}[M \leq m] = Binomial(m; K, \alpha)$. Thus, the probability that the interval $[Z_{(r,K)}, Z_{(s,K)}]$ contains the true quantile is simply

$$\begin{aligned}\mathbb{P}\mathrm{r}[r \leq M \leq s-1] &= \sum_{i=r}^{s-1} \binom{K}{i} \alpha^i (1-\alpha)^{K-i} \\ &= Binomial(s-1; K, \alpha) - Binomial(r-1; K, \alpha). \end{aligned} \tag{13.37}$$

A common practice is to choose r and s that are symmetric around and closest to the index $\lceil K\alpha \rceil$, and such that the probability (13.37) is not less than the desired confidence level γ. The mean and variance of the Binomial distribution are $K\alpha$ and $K\alpha(1-\alpha)$, respectively. For large K, approximating the Binomial by the Normal distribution with this mean and variance leads to a simple approximation for the conservative confidence interval bounds:

$$\begin{aligned} r &= \lfloor l \rfloor, \quad l = K\alpha - F_N^{-1}((1+\gamma)/2)\sqrt{K\alpha(1-\alpha)}, \\ s &= \lceil u \rceil, \quad u = K\alpha + F_N^{-1}((1+\gamma)/2)\sqrt{K\alpha(1-\alpha)}, \end{aligned} \tag{13.38}$$

where $F_N^{-1}(\cdot)$ is the inverse of the standard Normal distribution. This formula works very well for $K\alpha(1-\alpha) \geq 50$.

A priori, the number of simulations required to achieve a specific accuracy is not known. One of the approaches is to continue simulations until a desired numerical accuracy is achieved. Typically, $K \geq 10^5$ should be used to achieve a good numerical accuracy for the 0.999 quantile but it can be much larger in the case of heavy-tailed distributions. If the number of simulations to get acceptable accuracy is very large (e.g., $K > 10^7$), then you might not be able to store the whole array of samples $Z_1, \ldots, Z_K$ when implementing the algorithm, due to computer memory limitations. However, if you need to calculate just the high quantiles, then you need to save only $K - \lceil K\alpha \rceil + 1$ largest samples to estimate the quantiles (13.33). This can be done by using sorting *on the fly* algorithms, where you keep a specified number of largest samples as you generate the new samples; see Press *et al.* (2002, section 8.5). Moments (mean, variance, etc.) can also be easily calculated *on the fly* without saving all samples into the computer memory.

Note that formula (13.38) can be used for estimating the quantile numerical error if Monte Carlo samples $Z_1, \ldots, Z_K$ are independent and identically distributed. If the samples are correlated, for example, generated by Markov Chain Monte Carlo (MCMC), then (13.38) can significantly underestimate the error. In this case, one can use *batch sampling* or *effective sample size* methods described in Section 7.5.2.

EXAMPLE 13.6

Assume that $K = 10^5$ independent samples Z_k were drawn from some distribution. Suppose that we would like to construct a conservative confidence interval to contain the 0.999 quantile with probability at least $\gamma = 0.95$. Then, sort the samples in ascending order and using (13.38) calculate $F_N^{-1}((1+\gamma)/2) \approx 1.96$, $r = 99{,}880$ and $s = 99{,}920$, and $\lceil K\alpha \rceil = 99{,}900$. That is, our best point estimate for the quantile is $Z_{(\lceil K\alpha \rceil, K)}$ and $[Z_{(r,K)}, Z_{(s,K)}]$ contains the true quantile with probability at least 0.95.

Note that for these values of r and s, we get

$$Binomial(s-1;K,\alpha) - Binomial(r-1;K,\alpha) \approx 0.955.$$

For this example, if we would use exact formula (13.37) instead of approximation (13.38), we would get the same result because $Binomial(s-1;K,\alpha) - Binomial(r-1;K,\alpha)$ is less then 0.95 for $(r = 99,881; s = 99,920)$ or $(r = 99,880; s = 99,919)$. ■

EXAMPLE 13.7

Assume that independent losses $X_1, \ldots, X_n$ are sampled (or observed) from the density $f(x)$. Then the quantile q_α at confidence level α is estimated empirically as $\hat{Q}_\alpha = X_{(\lceil n\alpha \rceil, n)}$, where $X_{(1,n)}, \ldots, X_{(n,n)}$ is the data sample $\boldsymbol{X}$ sorted into the ascending order. The standard deviation of this empirical estimate can be estimated using formula (13.35),

$$\text{stdev}[\hat{Q}_\alpha] = \frac{\sqrt{\alpha(1-\alpha)}}{f(q_\alpha)\sqrt{n}}, \tag{13.39}$$

where $f(q_\alpha; \mu, \sigma)$ is the density of $LogNormal(\mu, \sigma^2)$. Thus, to calculate the quantile within relative error $\varepsilon = 2 \times \text{stdev}[\hat{Q}_\alpha]/q_\alpha$, we need

$$n = \frac{4\alpha(1-\alpha)}{\varepsilon^2 (f(q_\alpha) q_\alpha)^2} \tag{13.40}$$

observations. For example, if losses are from LogNormal distribution $LogNormal(\mu = 0, \sigma = 2)$ and we try to estimate quantile at the level $\alpha = 0.999$, then, according to formula (13.40), we need $n = 140,986$ samples to achieve $\varepsilon = 0.1$ accuracy. In the case of $n = 1000$ data points, we get $\varepsilon = 1.18$ accuracy. ■

13.2.2 EXPECTED SHORTFALL ESTIMATE

Consider a sample of i.i.d. random variables $Z_1, \ldots, Z_K$ from some distribution $H(z)$ (with finite mean) and corresponding ordered sample $Z_{(1,K)} \leq \cdots \leq Z_{(K,K)}$. Then, in general (i.e., sample can be from distribution with jumps and there might be repeated values in a sample), ES (6.11) can be estimated as

$$\widehat{\text{ES}}_\alpha = \frac{\sum_{i=k}^{K} Z_{(i,K)}}{K-k+1} \to \text{ES}_\alpha[Z], \quad \text{as} \quad K \to \infty, \tag{13.41}$$

where $k = \lceil K\alpha \rceil$ is the smallest integer larger than or equal to $K\alpha$, that is, $\hat{Q}_\alpha = Z_{(k,K)}$ is the sample estimator of the quantile $q_\alpha = H^{-1}(\alpha)$. Typically, in OpRisk, the distribution is continuous at the quantile q_α, and ES can be calculated as the conditional tail expectation $\mathrm{ES}_\alpha[Z] = \mathbb{E}[Z|Z \geq q_\alpha]$. In this case, it can be estimated as a simple average of losses larger than or equal to q_α.

$$\widehat{\mathrm{ES}}_\alpha = \frac{\sum_{i=1}^{K} Z_i \mathbb{I}_{Z_i \geq \hat{Q}_\alpha}}{\sum_{i=1}^{K} \mathbb{I}_{Z_i \geq \hat{Q}_\alpha}} \to \mathrm{ES}_\alpha[Z], \quad \text{as} \quad K \to \infty. \tag{13.42}$$

The convergence follows from the strong law of large numbers applied to the numerator and denominator and the convergence of the quantile estimator. If we assume that the quantile q_α is known, then in the limit $K \to \infty$, the central limit theorem gives

$$\frac{\sqrt{K}}{\sigma}(\widehat{\mathrm{ES}}_\alpha - \mathrm{ES}_\alpha) \to \mathit{Normal}(0, 1), \tag{13.43}$$

where σ can be estimated as

$$\hat{\sigma}^2 = K \frac{\sum_{k=1}^{K} (Z_k - \widehat{\mathrm{ES}}_\alpha)^2 \mathbf{1}_{Z_k \geq q_\alpha}}{\left(\sum_{k=1}^{K} \mathbf{1}_{Z_k \geq q_\alpha}\right)^2}.$$

Then, the standard deviation of $\widehat{\mathrm{ES}}_\alpha$ is estimated by $\hat{\sigma}/\sqrt{K}$ (see Glasserman 2005). However, it will underestimate the error in ES estimate because the quantile q_α is not known and estimated itself by $\hat{q}_\alpha$. Approximation for asymptotic standard deviation of ES estimate can be found in Yamai and Yoshiba (2002, appendix 1). In general, the standard deviation of the Monte Carlo estimates can always be evaluated by simulating K samples many times; see the batch sampling method described in Section 7.5.2. For heavy-tailed distributions and high quantiles, it is typically observed that the error in quantile estimate is much smaller than the error in ES estimate.

13.3 Panjer Recursion

The calculation of compound distribution via the convolution (13.2) for some class of frequency distributions can be reduced to a simple recursion introduced by Panjer (1981) and referred to as Panjer recursion. Detailed treatment of Panjer recursion and its extensions can be found in Sundt and Vernic (2009); for a good introduction in the context of OpRisk, see Panjer (2006, sections 5 and 6). We present a basic Panjer recursion that is typically good enough for OpRisk calculations.

Panjer recursion has been developed for the case of discrete severities. To use the method for continuous severities, the continuous severity should be approximated by the discrete one. The easiest approach is just to round all loss amounts to the nearest multiple of unit δ, for example, round to the nearest USD 1000. To concentrate severity, whose continuous distribution is $F(x)$, on $\{0, \delta, 2\delta, \ldots\}$, one can choose $\delta > 0$ and use the central difference approximation

$$\begin{aligned} f_0 &= F(\delta/2), \\ f_n &= F(n\delta + \delta/2) - F(n\delta - \delta/2), \quad n = 1, 2, \ldots. \end{aligned} \tag{13.44}$$

Discretization can also be done via the forward and backward differences:

$$\begin{aligned} f_n^U &= F(n\delta + \delta) - F(n\delta); \\ f_n^L &= F(n\delta) - F(n\delta - \delta). \end{aligned} \tag{13.45}$$

These allow for calculation of the upper and lower bounds for the compound distribution. As an example, Table 13.1 gives results of discretization for the *LogNormal*$(\mu = 0, \sigma = 2)$ in the case of step $\delta = \$1$ USD.

In addition to these basic discretization methods, other slightly more sophisticated versions of discretization methods have been developed. For instance, in Gerber (1982) an approach was developed which involves a localization of the standard technique of moment matching which produces a system of non-linear equations that can be solved using the Lagrange formula locally for each set of grid points. Consider an interval $(x\delta, x\delta + 2\delta]$ and associate local masses to be solved for at the left grid points denoted by $\{p_0(x\delta), p_1(x\delta), p_2(x\delta)\}$ in which the following system of equations must be solved for the zeroth, first and second local moments:

$$\begin{aligned}
p_0(x\delta) + p_1(x\delta) + p_2(x\delta) &= \int_{x\delta}^{x\delta+2\delta} dF(x),\\
(x\delta)p_0(x\delta) + (x\delta + \delta)p_1(x\delta) + (x\delta + 2\delta)p_2(x\delta) &= \int_{x\delta}^{x\delta+2\delta} xdF(x),\\
(x\delta)^2 p_0(x\delta) + (x\delta + \delta)^2 p_1(x\delta) + (x\delta + 2\delta)^2 p_2(x\delta) &= \int_{x\delta}^{x\delta+2\delta} x^2 dF(x).
\end{aligned}$$

The solution to this system of equations for the masses $\{p_0(x\delta), p_1(x\delta), p_2(x\delta)\}$ is obtained in closed form according to

$$p_j(x\delta) = \int_{x\delta}^{x\delta+2\delta} \prod_{i \neq j} \frac{\tau - x\delta - i\delta}{(j - i)\delta} dF(\tau), \quad \forall j \in \{0, 1, 2\}. \tag{13.46}$$

The system of equations above can be evaluated in closed form in some cases, in other cases one needs to use for instance a quadrature approximation. For instance, one could approximate the integrals via the trapezoidal (trapezium rule) using the result

$$\begin{aligned}
&\int_{x_0}^{x_0+x\delta} f(x)dx\\
&\approx f(x_0)\delta x + \frac{1}{2} f'(x_0)\delta x^2 + \frac{1}{6} f''(x_0)\delta x^3 + \cdots\\
&= \frac{1}{2}\left(f(x_0) + f(x_0 + \delta x)\delta x + O(\delta x^3)\right).
\end{aligned} \tag{13.47}$$

with the choices $f(x) = f_X(x), f(x) = xf_X(x)$ and $f(x) = x^2 f_X(x)$ utilised.

TABLE 13.1 **Example of discretization of *LogNormal*$(\mu = 0, \sigma = 2)$ distribution using central f_n, backward f_n^L, and forward f_n^U differences with the step $\delta = 1$**

n	f_n	f_n^L	f_n^U
0	0.364455845	0	0.5
1	0.215872117	0.5	0.135544155
2	0.096248034	0.135544155	0.073058159
⋮	⋮	⋮	⋮

Remark 13.1 *Note, we did not explicitly call these probability masses since the solution can actually produce negative results, which are clearly not usable for this discretisation method - making this approach some times untenable, depending on the grid points and model. To obtain the final discretised distribution simply add the masses at each grid point to get the discretised distribution approximation. Once the sequence of n discretized values* $\{p_j\}_{j=0}^{n}$ *are obtained, they can be turned into values* $\left\{\hat{f}(j)\right\}_{j=1}^{n}$ *by making sure they are all positive and normalized. Note, the final values of these discretization methods may need to be enforced to be positive and normalized for some applications.*

Another approach to discretization is known as Loyd's algorithm, which was popularized in the signal processing and engineering literature and named after Stuart Lloyd, see Lloyd (1982). This algorithm is a form of Voronoi iteration or relaxation for finding evenly-spaced sets of points in subsets of Euclidean spaces, and partitions of these subsets into well-shaped and uniformly sized convex cells. In this regard it is closely related to the concept of k-means clustering. It is well known that in one dimension Lloyd's algorithm converges to a centroidal Voronoi diagram. This algorithm is typically applied to settings where data is available and one wants to 'group' or 'cluster' the data by partitioning the convex hull of the data. In the setting we consider here the Lloyd algorithm is applied in a non-standard way, not to data, but instead to the discretization of a known distribution function. In this case, the points are selected as well as the mass at the discrete points to minimise a squared error criterion. Hence, to compute the discretization of the approximate pdf of a severity model X, $\hat{f}_X(z)$. We can achieve this by using the Lloyd algorithm, which minimizes the mean-square error and in the process produces a set of grid points $\{\delta_i\}_{i=1}^{N}$ for which we have probability masses $\left\{\hat{f}_i\right\}_{i=1}^{N}$.

NOTE: under this discretization method the grid points are not required to be uniformly spaced. This will have implications for utilisation of this method in calculating integrals which should be considered. The algorithm is detailed by the following steps for N discrete grid points:

Algorithm 13.1 (Lloyd algorithm)

1. *Initialization.*
   a) *Choose the initial 'quantisation' levels ie. discretization steps* $\delta_i^{(0)}$, $i = 1, \ldots, N$;
   b) *Set the quantisation boundaries* $b_i^{(0)} = \frac{\delta_{i+1}^{(0)} - \delta_i^{(0)}}{2}$, $i = 1, \ldots, N-1$;
   c) *Construct the initial density function:* $\hat{f}_X(x) = \sum_{i=1}^{N} \hat{f}_i \delta_{\delta_i^{(0)}}(x)$;
   d) *Set the initial distortion* $D^{(0)} = \infty$.
2. *Update rule: While* $|D^{(k)} - D^{(k-1)}| > \epsilon$ *perform the following steps:*
   a) *Quantizer levels:* $\delta_i^{(k+1)} = \frac{\int_{b_{i-1}^{(k)}}^{b_i^{(k)}} x\hat{f}_X(x)dx}{\int_{b_{i-1}^{(k)}}^{b_i^{(k)}} \hat{f}_X(x)dx}$;
   b) *Quantizer boundaries:* $b_i^{(k+1)} = \frac{\delta_{i+1}^{(k+1)} - \delta_i^{(k+1)}}{2}$;
   c) *Distortion:* $D^{(k+1)} = \sum_{i=1}^{N} \int_{b_{i-1}^{(k+1)}}^{b_i^{(k+1)}} (x - \delta_i^{(k+1)})^2 \hat{f}_X(x)dx$;
   d) *Increment counter:* $k = k + 1$.

In addition, we observe that the final values of these discretization methods may need to be normalized for some applications.

Given discrete severity $f_k = \mathbb{P}\mathrm{r}[X_i = k\delta]$ and discrete frequency $p_k = \mathbb{P}\mathrm{r}[N = k]$, the compound loss $Z = X_1 + \cdots + X_N$ is also discrete with some $h_k = \mathbb{P}\mathrm{r}[Z = k\delta]$; $k = 0, 1, \ldots$ that will be calculated by Panjer recursion. For simplicity, assume that $f_0 = 0$. Then, the discrete version of convolution (13.2) is

$$\begin{aligned} h_n &= \sum_{k=1}^{n} p_k f_n^{(k)*}, \quad n \geq 1, \\ h_0 &= \mathbb{P}\mathrm{r}[Z = 0] = \mathbb{P}\mathrm{r}[N = 0] = p_0, \end{aligned} \tag{13.48}$$

where $f_n^{(k)*} = \sum_{i=0}^{n} f_{n-i}^{(k-1)*} f_i$ with $f_0^{(0)*} = 1$ and $f_n^{(0)*} = 0$ if $n \geq 1$.

Note that the condition $f_0 = \mathbb{P}\mathrm{r}[X = 0] = 0$ implies that $f_n^{(k)*} = 0$ for $k > n$ and thus the above summation is up to n only. If $f_0 > 0$, then $f_n^{(k)*} > 0$ for all n and k; and the upper limit in summation (13.48) should be replaced by infinity. The number of operations to calculate $h_0, h_1, \ldots, h_n$ using (13.48) explicitly is on the order of n^3. If the maximum value for which the compound distribution should be calculated is large, the number of computations become prohibitive due to $O(n^3)$ operations. Fortunately, if the frequency N belongs to the so-called Panjer classes, (13.48) is reduced to a simple recursion introduced by Panjer (1981) and referred to as Panjer recursion.

Theorem 13.1 (Panjer recursion) *If the frequency probability mass function p_n, $n = 0, 1, \ldots$ satisfies*

$$p_n = \left(a + \frac{b}{n}\right) p_{n-1}, \quad \text{for} \quad n \geq 1 \quad \text{and} \quad a, b \in \mathbb{R}, \tag{13.49}$$

then it is said to be in Panjer class $(a, b, 0)$ and the compound distribution (13.48) satisfies the recursion

$$\begin{aligned} h_n &= \frac{1}{1 - af_0} \sum_{j=1}^{n} \left(a + \frac{bj}{n}\right) f_j h_{n-j}, \quad n \geq 1, \\ h_0 &= \sum_{k=0}^{\infty} (f_0)^k p_k. \end{aligned} \tag{13.50}$$

The initial condition in (13.50) is simply a probability-generating function of N at f_0, that is, $h_0 = \psi(f_0)$; see (13.4). If $f_0 = 0$, then it simplifies to $h_0 = p_0$. The Panjer recursion requires $O(n^2)$ operations to calculate $h_0, \ldots, h_n$ in comparison with asymptotic $O(n^3)$ of explicit convolution. If severity is restricted by a value of the largest possible loss m, then the upper limit in the recursion (13.50) should be replaced by $\min(m, n)$.

It was shown by Sundt and Jewell (1981) that (13.49) is satisfied for the Poisson, Negative Binomial, and Binomial distributions with the parameters (a, b), and starting values h_0 are as follows:

- Poisson frequency, *Poisson*(λ):

$$a = 0, \quad b = \lambda, \quad h_0 = \exp(\lambda(f_0 - 1)). \tag{13.51}$$

- Negative Binomial frequency, *NegBinomial*(r, q):

$$a = 1 - q, \quad b = (1 - q)(r - 1), \quad h_0 = \left(1 + (1 - f_0)\frac{1 - q}{q}\right)^{-r}. \tag{13.52}$$

- Binomial frequency, *Binomial*(m, q):

$$a = -\frac{q}{1 - q}, \quad b = \frac{q(m + 1)}{1 - q}, \quad h_0 = (1 + q(f_0 - 1))^m. \tag{13.53}$$

Strong stability of Panjer recursion was established for the Poisson and Negative Binomial cases (see Panjer and Wang 1993). The accumulated rounding error of the recursion increases linearly in n with a slope not exceeding 1. Serious numerical problems may occur for the case of Binomial distribution. Typically, instabilities in the recursion appear for significantly underdispersed frequencies of severities with a large negative skewness which are not typical in OpRisk. The basic Panjer recursion can be implemented as follows.

Algorithm 13.2 (Panjer recursion)

1. *Initialization: calculate* f_0 *and* h_0, *see (13.51–13.53), and set* $H_0 = h_0$;
2. *For* $n = 1, 2, \ldots$
   a) *Calculate* f_n. *If severity distribution is continuous, then* f_n *can be found using discretization (13.44) or (13.45).*
   b) *Calculate* $h_n = \frac{1}{1 - af_0} \sum_{j=1}^{n} \left(a + \frac{bj}{n}\right) f_j h_{n-j}$;
   c) *Calculate* $H_n = H_{n-1} + h_n$;
   d) *Interrupt the procedure at* $n = \hat{n}$ *if* $H_{\hat{n}}$ *is larger than or equal to the required quantile level* α, *for example,* $\alpha = 0.999$. *Then the estimate of the quantile* q_α *is* $\hat{n} \times \delta$.
3. *Next* n *(i.e. do an increment* $n = n + 1$ *and return to step 2).*

Given calculated $H_0, H_1, \ldots$ as $H_n = \sum_{i=0}^{n} h_i$, the distribution function is just

$$\hat{H}(x) = H_{\lfloor x/\delta \rfloor} \tag{13.54}$$

and the quantile estimator is

$$\hat{q}_\alpha = \hat{H}^{-1}(\alpha) = \hat{n} \times \delta, \quad \hat{n} = \min\{n : H_n \geq \alpha\}. \tag{13.55}$$

A formal calculation of ES by definition (13.28) involves summation over points above the quantile. To avoid unnecessary calculations of distribution H_n for $n\delta$ above the quantile, it is easier to calculate ES using a discretized version of (13.30). Then the ES for compound distribution (13.54) is

$$\widehat{\mathrm{ES}}_\alpha[Z] = \frac{\mathbb{E}[Z]}{1 - \alpha} - \frac{\delta}{1 - \alpha} \sum_{n=0}^{\hat{n}} n h_n + \hat{q}_\alpha \frac{H_{\hat{n}} - \alpha}{1 - \alpha}. \tag{13.56}$$

Here, the summation is over the finite number of points (for distribution with non-negative support, which is the case for OpRisk). We also assume that the mean $\mathbb{E}[Z] = \mathbb{E}[N]\mathbb{E}[X]$ can be found exactly, which is typically the case. The last term in (13.56) is typically small.

As an example, Table 13.2 presents results of Panjer recursion calculations of the $Poisson(100) - LogNormal(\mu = 0, \sigma = 2)$ compound distributions using central difference discretization with the step $\delta = 1$. Of course, the accuracy of the result depends on the step size as shown by the results for the 0.999 quantile and ES versus δ (see Table 13.3). It is, however, important to note that the error of the result is due to discretization only and there is no truncation error (i.e. the severity is not truncated by some large value). Note also that the 0.999 quantile estimate is more accurate than the 0.999 shortfall estimate (for a given time δ).

The use of forward f_n^U and backward f_n^L severity discretizations will produce the upper and lower bounds for the compound distribution. Thus, the lower and upper bounds for a quantile are obtained with f_n^U and f_n^L, respectively (see examples in Shevchenko, 2011, section 3.3.1).

If frequency is large, then underflow may occur in computations of (13.50). Underflow occurs in the case when the numerical calculations produce a number outside the range of representable numbers leading to zero. It is easy to see in the case of $Poisson(\lambda)$ and $f_0 = 0$ when $h_0 = \exp(-\lambda)$. In this case, the underflow will occur for $\lambda \gtrsim 700$ on a 32 bit computer with double-precision calculations. Rescaling h_0 by large factor γ to calculate the recursion (and descaling the result) will not resolve the issue because overflow will occur for $\gamma h(n)$ (i.e., calculations will produce a number outside of the representative range leading to ∞). The following identity helps to overcome this problem in the case of Poisson frequency:

$$H^{(m)*}(z; \lambda/m) = H(z; \lambda). \tag{13.57}$$

That is, calculate the compound distribution $H(z; \lambda/m)$ for some large m to avoid underflow. Then preform m convolutions for the obtained distribution directly or via FFT (see Panjer and Willmot 1986). Similar identity is available for Negative Binomial, $NegBinomial(r, p)$:

$$H^{(m)*}(z; r/m) = H(z; r). \tag{13.58}$$

In the case of Binomial, $Binomial(M, p)$:

$$H^{(m)*}(z; m_1) * H(z; m_2) = H(z; M), \tag{13.59}$$

TABLE 13.2 **Example of Panjer recursion calculating the *Poisson*(100) − *LogNormal*($\mu = 0, \sigma = 2$) compound distributions using central difference discretization with the step $\delta = 1$**

n	f_n	h_n	H_n
0	0.364455845	2.50419×10^{-28}	2.50419×10^{-28}
1	0.215872117	5.40586×10^{-27}	5.65628×10^{-27}
2	0.096248034	6.07589×10^{-26}	6.64152×10^{-26}
⋮	⋮	⋮	⋮
5847	2.81060×10^{-9}	4.44337×10^{-7}	0.998999329
5848	2.80907×10^{-9}	4.44061×10^{-7}	0.998999773
5849	2.80755×10^{-9}	4.43785×10^{-7}	0.999000217

TABLE 13.3 **Convergence of Panjer recursion estimates $\hat{q}_{0.999}$ and $\widehat{\text{ES}}_{0.999}$ of the 0.999 quantile and ES, respectively, for the *Poisson*(100) − *LogNormal*($\mu = 0, \sigma = 2$) compound distributions using central difference discretization versus the step size δ**

δ	N	$\hat{q}_{0.999}$	$\widehat{\text{ES}}_{0.999}$
2	2,921	5842	20,131
1	5,849	5849	13,519
0.5	11,703	5851.5	10,831
0.25	23,411	5852.75	9,873
0.125	46,824	5853	9,575
0.0625	93,649	5853.0625	9,494

Here, $N = \hat{q}_{0.999}/\delta$ is the number of steps required.

where $m_1 = \lfloor M/m \rfloor$ and $m_2 = M - m_1 m$. For numerical efficiency, one can choose $m = 2^k$ so that instead of m convolutions of $H(\cdot)$ only k convolutions are required $H^{(2)*}, H^{(4)*}, \ldots, H^{(2^k)*}$, where each term is the convolution of the previous one with itself.

13.4 Panjer Extensions

The Panjer recursion formula (13.50) can be extended to a class of frequency distributions $(a, b, 1)$.

Definition 13.1 (Panjer class $(a, b, 1)$) *The distribution is said to be in $(a, b, 1)$ Panjer class if it satisfies*

$$p_n = \left(a + \frac{b}{n}\right) p_{n-1}, \quad \textit{for} \quad n \geq 2 \quad \textit{and} \quad a, b \in \mathbb{R}. \tag{13.60}$$

■

Theorem 13.2 (Extended Panjer recursion) *For the frequency distributions in a class $(a, b, 1)$:*

$$h_n = \frac{(p_1 - (a+b)p_0)f_n + \sum_{j=1}^{n} (a + bj/n) f_j h_{n-j}}{1 - af_0}, \quad n \geq 1,$$

$$h_0 = \sum_{k=0}^{\infty} (f_0)^k p_k. \tag{13.61}$$

The distributions of $(a, b, 0)$ class are special cases of $(a, b, 1)$ class. There are two types of frequency distributions in $(a, b, 1)$ class:

- Zero-truncated distributions, where $p_0 = 0$: that is, zero-truncated Poisson, zero-truncated Binomial, and zero-truncated Negative Binomial;
- Zero-modified distributions, where $p_0 > 0$: the distributions of $(a, b, 0)$ with modified probability of zero. It can be viewed as a mixture of $(a, b, 0)$ distribution and degenerate distribution concentrated at zero.

Finally, we would like to mention a generalization of Panjer recursion for the (a, b, l) class

$$p_n = \left(a + \frac{b}{n}\right) p_{n-1}, \quad \text{for} \quad n \geq l + 1. \tag{13.62}$$

For initial values $p_0 = \cdots = p_{l-1} = 0$, and in the case of $f_0 = 0$, it leads to the recursion

$$h_n = p_l f_n^{(l)*} + \sum_{j=1}^{n} (a + bj/n) f_j h_{n-j}, \quad n \geq l.$$

The distribution in this class is, for example, $l - 1$ truncated Poisson. For an overview of high-order Panjer recursions, see Hess *et al.* (2002). Other types of recursions

$$p_n = \sum_{j=1}^{k} (a_j + b_j/n) p_{n-1}, \quad n \geq 1, \tag{13.63}$$

are discussed by Sundt (1992). Application of the standard Panjer recursion in the case of the generalized frequency distributions, such as the extended Negative Binomial, can lead to numerical instabilities. Generalization of the Panjer recursion that leads to numerically stable algorithms for these cases is presented by Gerhold *et al.* (2010). Discussion on a multivariate version of Panjer recursion can be found by Sundt (1999) and bivariate cases are discussed by Vernic (1999) and Hesselager (1996).

In the case of severities from a phase-type distribution (distribution with a rational probability-generating function), the recursion (13.50) is reduced to $O(n)$ operations (see Hipp 2006). Typically, the severity distributions are not phase-type distributions and approximation is required. This is useful for modeling small losses but not suitable for heavy-tailed distributions because the phase-type distributions are light-tailed (see Bladt 2005) for a review.

The analog of Panjer recursion for the case of continuous severities is given by the following integral equation.

Theorem 13.3 (Panjer recursion for continuous severities) *For frequency distributions in* $(a, b, 1)$ *class and continuous severity distributions on positive real line,*

$$h(z) = p_1 f(z) + \int_0^x (a + by/z) f(y) h(z - y) dy. \tag{13.64}$$

The proof is presented by Panjer and Willmot (1992, theorems 6.14.1 and 6.16.1). Note that the integral equation (13.62) holds for $(a, b, 0)$ class because it is a special case of $(a, b, 1)$. The integral equation (13.64) is a Volterra integral equation of the second type. There are different methods to solve it described by Panjer and Willmot (1992). A method of solving this equation using hybrid MCMC (minimum variance importance sampling via reversible jump MCMC) is presented by Peters *et al.* (2007), this is discussed briefly in the advanced section at the end of this chapter.

13.5 Fast Fourier Transform

Another efficient method to calculate compound distributions via the inversion of the characteristic function is FFT. The method originates from the signal-processing field. The existence of the algorithm became generally known in the mid-1960s, but it was independently discovered by many researchers much earlier. A detailed explanation of the method in application to aggregate loss distribution can be found by Robertson (1992). Often, OpRisk practitioners in banking regard the method as difficult. However, in fact, it is a very simple algorithm to implement, although to make it really efficient, especially for heavy-tailed distribution, some improvements are required. We describe the essential steps and theory required for successful implementation of the FFT for OpRisk. The basic FFT algorithm is very simple and its code is short; see, for example, C code provided by Press *et al.* (2002, chapter 12).

FFT works with discrete severity (in the same way as Panjer recursion case), that is, continuous severity distributions should be discretized using central difference (13.44) or backward/forward differences (13.45). The method is based on the discrete Fourier transformation defined as follows.

Definition 13.2 (Discrete Fourier transformation) *For a sequence $f_0, f_1, \ldots, f_{M-1}$, the following transformation*

$$\phi_k = \sum_{m=0}^{M-1} f_m \exp\left(\frac{2\pi \mathrm{i}}{M} mk\right), \quad k = 0, 1, \ldots, M-1 \tag{13.65}$$

is called the discrete Fourier transformation (DFT). Then the inverse transformation to recover the original sequence f_k from ϕ_k is

$$f_k = \frac{1}{M}\sum_{m=0}^{M-1} \phi_m \exp\left(-\frac{2\pi \mathrm{i}}{M} mk\right), \quad k = 0, 1, \ldots, M-1. \tag{13.66}$$

■

The number of operations to calculate M points of ϕ_m are on the order of M^2, that is, $O(M^2)$. If M is a power of 2, then DFT can be efficiently calculated via FFT algorithms with the number of computations $O(M \log_2 M)$. This is because the DFT of length M can be represented as the sum of DFT over even points ϕ_k^e and DFT over odd points ϕ_k^e:

$$\phi_k = \phi_k^e + \exp\left(\frac{2\pi \mathrm{i}}{M} k\right)\phi_k^o,$$

$$\phi_k^e = \sum_{m=0}^{M/2-1} f_{2m} \exp\left(\frac{2\pi \mathrm{i}}{M} mk\right),$$

$$\phi_k^o = \sum_{m=0}^{M/2-1} f_{2m+1} \exp\left(\frac{2\pi \mathrm{i}}{M} mk\right).$$

Subsequently, each of these two DFTs can be calculated as a sum of two DFTs of length $M/4$. For example, ϕ_k^e is calculated as a sum of ϕ_k^{ee} and ϕ_k^{eo}. This procedure is continued until the transforms of the length 1. The latter is simply an identity operation. Thus, every obtained pattern of odd and even DFTs will be f_m for some m, $\phi_k^{eo\cdots ooe} = f_m$. The bit-reversal procedure can be used to find m that corresponds to a specific pattern. That is, set $e = 0$ and $o = 1$, then the reverse pattern of *es* and *os* is the value of m in binary. The code for FFT, where M is integer power of 2, is short and simple. For example, see C code provided by Press *et al.* (2002, chapter 12) with the following logical steps.

Algorithm 13.3 (Simple FFT)

1. *Sort the data in a bit-reversed order. The obtained points are simply one-point transforms;*
2. *Combine the neighbor points into nonoverlapping pairs to get two-point transforms. Then combine two-point transforms into four-point transforms and continue subsequently until the final M point transform is obtained. Thus, there are* $\log_2 M$ *iterations and each iteration involve is on the order of M operations.*

The inverse FFT transformation is calculated in the same way as FFT. The only differences are sign change and division by M (see (13.65) and (13.66)). Once the FFT algorithm is available, then the compound distribution can be calculated via FFT as follows.

Algorithm 13.4 (Compound Distribution via FFT)

1. *Discretize severity to obtain $f_0, f_1, \ldots, f_{M-1}$, where $M = 2^r$ with integer r, and M is the truncation point in the aggregate distribution;*
2. *Using FFT, calculate the characteristic function of the severity $\varphi_0, \ldots, \varphi_{M-1}$;*
3. *Calculate the characteristic function of the compound distribution using (13.5), that is, $\chi_m = \psi(\varphi_m), \quad m = 0, 1, \ldots, M-1$;*
4. *Perform inverse FFT (which is the same as FFT except the change of sign under the exponent and factor $1/M$) applied to $\chi_0, \ldots, \chi_{M-1}$ to obtain the compound distribution $h_0, h_1, \ldots, h_{M-1}$.*

Once the compound distribution $h_0, h_1, \ldots, h_{M-1}$ is calculated, its quantile and ES can be estimated using (13.55) and (13.56), respectively; that is, the same as for Panjer recursion.

If there is no truncation error in the severity discretization, that is, $\sum_{m=0}^{M-1} f_m = 1$, then FFT procedure calculates the compound distribution on $m = 0, 1, \ldots, M$. That is, the mass of compound distribution beyond M is "wrapped" and appears in the range $m = 0, \ldots, M-1$ (the so-called *aliasing error*). This error is larger for heavy-tailed severities. To decrease the error for compound distribution on $0, 1, \ldots, n$, one has to take M much larger than n. If the severity distribution is bounded and M is larger than the bound, then one can put zero values for points above the bound (the so-called padding by zeros). An example is given in Table 13.4, where $(Q_{0.999}^{(1)}, \mathrm{ES}_{0.999}^{(1)})$ are the quantile and ES estimators, respectively, calculated via FFT using the central difference discretization with the tail probability compressed into the last point $f_{M-1} = 1 - F(\delta(M-1) - \delta/2)$; and $(Q_{0.999}^{(2)}, \mathrm{ES}_{0.999}^{(2)})$ are the estimators calculated via FFT using the central difference discretization with the tail probability ignored, that is, $f_{M-1} = F(\delta(M-1) + \delta/2) - F(\delta(M-1) - \delta/2)$. These are compared with the Panjer

TABLE 13.4 **Example of FFT calculating the 0.999 quantile and ES of the *Poisson*(100) – *LogNormal*($\mu = 0, \sigma = 2$) compound distribution using central difference discretization with the step $\delta = 0.5$**

r	$L = \delta \times 2^r$	$Q^{(1)}_{0.999}$	$\mathrm{ES}^{(1)}_{0.999}$	$Q^{(2)}_{0.999}$	$\mathrm{ES}^{(2)}_{0.999}$
14	8,192	5117	12,831	5665.5	11,291
15	16,384	5703.5	11,180	5834	10,872
16	32,768	5828	10,886	5850	10,834
17	65,536	5848.5	10,839	5851.5	10,831
18	131,072	5851.5	10,832	5851.5	10,831
19	262,144	5851.5	10,831	5851.5	10,831

The exact Panjer recursion for this discretization step gives $Q_{0.999} = 5851.5$ and $\mathrm{ES}_{0.999} = 10,831$.

recursion exact results for this discretization. As one can see, the truncation should be large enough to get accurate results for both estimators, although $(Q^{(2)}_{0.999}, \mathrm{ES}^{(2)}_{0.999})$ are a bit more accurate than $(Q^{(1)}_{0.999}, \mathrm{ES}^{(1)}_{0.999})$.

Another way to reduce the error is to apply some transformation to increase the tail decay (the so-called *tilting*). The exponential tilting technique for reducing aliasing error under the context of calculating compound distribution was first investigated by Grubel and Hermesmeier (1999). Many authors suggest the following tilting transformation:

$$\widetilde{f}_j = \exp(-j\theta)f_j, \quad j = 0, 1, \ldots, M-1, \tag{13.67}$$

where $\theta > 0$. This transformation commutes with convolution in a sense that convolution of two functions $f(x)$ and $g(x)$ equals the convolution of the transformed functions $\widetilde{f}(x) = f(x)\exp(-\theta x)$ and $\widetilde{g}(x) = g(x)\exp(-\theta x)$ multiplied by $\exp(\theta x)$, that is,

$$(f * g)(x) = e^{\theta x}(\widetilde{f} * \widetilde{g})(x). \tag{13.68}$$

This can easily be shown using the definition of convolution. Then calculation of the compound distribution is performed using the transformed severity distribution as follows.

Algorithm 13.5 (Compound Distribution via FFT with Tilting)

1. *Apply FFT to a set* $\widetilde{f}_0, \ldots, \widetilde{f}_{M-1}$ *to obtain* $\widetilde{\phi}_0, \ldots, \widetilde{\phi}_{M-1}$;
2. *Apply the inverse FFT to the set* $\widetilde{\chi}_0, \ldots, \widetilde{\chi}_{M-1}$ *to obtain* $\widetilde{h}_0, \widetilde{h}_1, \ldots, \widetilde{h}_{M-1}$;
3. *Untilt by calculating final compound distribution as* $h_j = \widetilde{h}_j \exp(\theta j)$.

This tilting procedure is very effective in reducing the aliasing error. The parameter θ should be as large as possible but not producing under- or overflow that will occur for very large θ. Embrechts and Frei (2009) reported that the choice $M\theta \approx 20$ works well for standard double-precision (8 bytes) calculations. Evaluation of the probability-generating function $\psi(\cdot)$ of the frequency distribution may lead to the problem of underflow in the case of large frequencies that can be resolved using formulas (13.57–13.59).

To demonstrate the effectiveness of the tilting, an example is provided in Table 13.5, where $(Q^{(2)}_{0.999}, \mathrm{ES}^{(2)}_{0.999})$ are the quantile and ES estimators via FFT, respectively,

TABLE 13.5 **Example of FFT calculating the 0.999 quantile of the *Poisson*(100) – *LogNormal*(μ = 0, σ = 2) compound distribution using central difference discretization with the step δ = 0.5**

r	$L = \delta \times 2^r$	$Q^{(2)}_{0.999}$	$ES^{(2)}_{0.999}$	$Q^{(tilt)}_{0.999}$	$ES^{(tilt)}_{0.999}$
14	8,192	5665.5	11,291	5851.5	10,831
15	16,384	5834	10,872	5851.5	10,831
16	32,768	5850	10,834	5851.5	10,831
17	65,536	5851.5	10,831	5851.5	10,831
18	131,072	5851.5	10,831	5851.5	10,831
19	262,144	5851.5	10,831	5851.5	10,831

The exact Panjer recursion for this discretisation step gives $Q_{0.999} = 5851.5$ and $\text{ES}_{0.999} = 10,831$.

using the central difference discretization with the tail probability ignored, that is, $f_{M-1} = F(\delta(M-1)+\delta/2) - F(\delta(M-1)-\delta/2)$; and $(Q^{(tilt)}_{0.999}, \text{ES}^{(tilt)}_{0.999})$ are the estimators via FFT using the central difference discretization with tilting. The tilting, parameter θ is chosen to be $\theta = 20/M$. The results presented in Table 13.5 demonstrate the efficiency of the tilting. If FFT is performed without tilting, then the truncation level for the severity should exceed the quantile significantly. In this particular case, it should exceed by approximately a factor of 10 to get the exact result for this discretization step. The latter is obtained by Panjer recursion, which does not require discretization beyond the calculated quantile. The FFT and Panjer recursion are approximately the same in terms of computing time required for quantile estimate in this case. However, once the tilting is utilized, the cutoff level does not need to exceed the quantile significantly to obtain the exact result—making FFT superior to Panjer recursion. Moreover, in this case, the treatment of the severity tail by ignoring it or absorbing into the last point f_{M-1} does not make any difference when tilting is applied.

For comparison of FFT and Panjer, see Embrechts and Frei (2009) and Bühlmann (1984a). Comprehensive numerical examples comparing MC, Panjer recursion, and FFT are provided by Shevchenko (2011, section 3.6).

13.6 Closed-Form Approximation

There are several well-known approximations for the compound loss distribution. These can be used with different success depending on the quantity to be calculated and distribution types. Even if the accuracy is not good, these approximations are certainly useful from the methodological point of view in helping to understand the model properties. The quantile estimate derived from these approximations can also be used successfully to set a cutoff level for FFT algorithms that will subsequently determine the quantile more precisely.

Many parametric distributions can be used as an approximation for a compound loss distribution by moment matching. This is because the moments of the compound loss can be calculated in closed form. In particular, the first four moments are given by formulas (13.21–13.24). Of course these can only be used if the required moments exist, which is not the case for some heavy-tailed risks with infinite moments. We describe Normal and translated Gamma approximations. In addition, for heavy-tailed severities, we describe efficient closed-form approximation for the tail of compound distribution.

Normal Approximation. If the severities $X_1, X_2, \ldots$ are independent and identically distributed with finite mean and variance, then at very high frequencies the central limit theory is expected to provide a good approximation to the distribution of the annual loss Z. That is, the compound distribution is approximated by the Normal distribution

$$H(z) \approx Normal(\mathbb{E}[Z], \mathbb{V}\text{ar}[Z]), \tag{13.69}$$

where the mean and variance are given by formulas (13.21) and (13.22), respectively. This result is asymptotic and a priori we do not know how well it will perform for specific distribution types and distribution parameter values. Moreover, it cannot be used for the cases where variance or mean are infinite.

EXAMPLE 13.8

If N is distributed from $Poisson(\lambda)$ and $X_1, \ldots, X_N$ are independent random variables from $LogNormal(\mu, \sigma^2)$, then

$$\mathbb{E}[Z] = \lambda \exp(\mu + 0.5\sigma^2), \quad \mathbb{V}\text{ar}[Z] = \lambda \exp(2\mu + 2\sigma^2). \tag{13.70}$$

■

Translated Gamma Approximation. Typically, in OpRisk the compound distribution is positively skewed. For example, in the case of Poisson distributed frequencies, the skewness of the compound distribution (see (13.25)) is

$$\frac{\mathbb{E}[(Z - \mathbb{E}[Z])^3]}{(\mathbb{V}\text{ar}[Z])^{3/2}} = \frac{\lambda \mathbb{E}[X^3]}{(\lambda \mathbb{E}[X^2])^{3/2}} > 0, \tag{13.71}$$

which approaches zero as λ increases but finite positive for finite $\lambda > 0$. To improve the Normal approximation (13.69), the compound loss can be approximated by the shifted Gamma distribution, which has a positive skewness (assuming that the first three moments of compound distribution exist). In this case, Z is approximated as $Y + a$ where a is a shift and Y is a random variable from $Gamma(\alpha, \beta)$. Then the parameters are estimated by matching the mean, variance, and skewness of the approximate distribution and the correct one:

$$a + \alpha\beta = \mathbb{E}[Z]; \quad \alpha\beta^2 = \mathbb{V}\text{ar}[Z]; \quad \frac{2}{\sqrt{\alpha}} = \mathbb{E}[(Z - \mathbb{E}[Z])^3] / (\mathbb{V}\text{ar}[Z])^{3/2}. \tag{13.72}$$

EXAMPLE 13.9

If frequencies are Poisson-distributed, $N \sim Poisson(\lambda)$, then

$$a + \alpha\beta = \lambda \mathbb{E}[X]; \quad \alpha\beta^2 = \lambda \mathbb{E}[X^2]; \quad \frac{2}{\sqrt{\alpha}} = \lambda \mathbb{E}[X^3] / \left(\lambda \mathbb{E}[X^2]\right)^{3/2}. \tag{13.73}$$

■

VaR Closed-Form Approximation. If severities $X_1, \ldots, X_N$ are independent and identically distributed from the subexponential (heavy tail) distribution $F(x)$, and frequency distribution satisfies

$$\sum_{n=0}^{\infty}(1+\epsilon)^n \mathbb{P}\mathrm{r}[N=n] < \infty$$

for some $\epsilon > 0$, then the tail of the compound distribution $H(z)$, of the compound loss $Z = X_1 + \cdots + X_N$, is related to the severity tail as

$$1 - H(z) \sim \mathbb{E}[N](1 - F(z)), \quad \text{as} \quad z \to \infty \tag{13.74}$$

(see Embrechts *et al.* 1997, theorem 1.3.9). This is also discussed in detail by Peters and Shevchenko (2015). The validity of this asymptotic result was demonstrated for the cases when N is distributed from Poisson, Binomial, or Negative Binomial. This approximation can be used to calculate the quantiles of the annual loss as

$$\mathrm{VaR}_\alpha[Z] \approx F^{-1}\left(1 - \frac{1-\alpha}{\mathbb{E}[N]}\right), \quad \text{as} \quad \alpha \to 1. \tag{13.75}$$

For application in the OpRisk context, see Böcker and Klüppelberg (2005). Under the assumption that the severity has a finite mean, Böcker and Sprittulla (2006) derived a correction reducing the approximation error of (13.75). This was further refined by Degen (2010) for heavy-tailed finite mean severity as

$$\mathrm{VaR}_\alpha[Z] = F^{-1}\left(1 - \frac{1-\alpha}{\mathbb{E}[N]}\right) + \mathbb{E}[X]\left(\mathbb{E}[N] + \frac{\mathbb{V}\mathrm{ar}[N]}{\mathbb{E}[N]} - 1\right) + o(1), \quad \alpha \to 1. \tag{13.76}$$

For heavy-tailed infinite mean severity and large α, Degen (2010) derives the following approximation

$$\begin{aligned}\mathrm{VaR}_\alpha[Z] \approx\ & F^{-1}\left(1 - \frac{1-\alpha}{\mathbb{E}[N]}\right)\\ & - (1-\alpha)F^{-1}\left(1 - \frac{1-\alpha}{\mathbb{E}[N]}\right)\frac{c_\xi/\mathbb{E}[N]}{1-1/\xi}\left(\mathbb{E}[N] + \frac{\mathbb{V}\mathrm{ar}[N]}{\mathbb{E}[N]} - 1\right),\end{aligned} \tag{13.77}$$

where $\xi > 1$ is a tail index of severity distribution (i.e., $\lim_{t\to\infty}(1-F(tx))/(1-F(x)) = x^{-1/\xi}$) and $c_\xi = \frac{1}{2}(1-\xi)\Gamma^2(1-1/\xi)/\Gamma(1-2/\xi)$ with $\Gamma(\cdot)$ denoting the standard Gamma function.

EXAMPLE 13.10

Consider a heavy-tailed *Poisson*(λ)–*GPD*(ξ,β) compound distribution. In this case, (13.75) gives

$$\mathrm{VaR}_\alpha[Z] \approx \frac{\beta}{\xi}\left(\frac{\lambda}{1-\alpha}\right)^{\xi}, \quad \text{as} \quad \alpha \to 1. \tag{13.78}$$

This implies a simple scaling, $\mathrm{VaR}_\alpha[Z] \propto \lambda^\xi$, with respect to the event intensity λ for large α. ■

EXAMPLE 13.11

To demonstrate the accuracy of the previous approximations, consider the compound distribution *Poisson*($\lambda = 100$)–*LogNormal*($\mu = 0, \sigma = 2$) with relatively heavy-tailed severity. Calculating moments of the LogNormal distribution $\mathbb{E}[X^m]$ using (13.26) and substituting into (13.25) gives

$$\mathbb{E}[Z] \approx 738.9056, \quad \mathbb{V}\text{ar}[Z] \approx 298095.7987,$$
$$\mathbb{E}[(Z - \mathbb{E}[Z])^3]/(\mathbb{V}\text{ar}[Z])^{3/2} \approx 40.3428.$$

Approximating the compound distribution by the Normal distribution with this mean and variance gives Normal approximation. Approximating the compound distribution by the translated Gamma distribution (13.72) with these mean, variance, and skewness gives

$$\alpha \approx 0.002457, \quad \beta \approx 11013.2329, \quad a \approx 711.8385.$$

Figure 13.1 shows the Normal and translated Gamma approximations for the tail of the compound distribution in comparison with the corrected heavy-tailed approximation (13.76). It is easy to see that the corrected heavy-tailed asymptotic approximation (13.76) converges to the *exact* result for large quantile level $\alpha \to 1$, while the Normal and Gamma approximations perform badly. Comparison of the corrected heavy-tailed approximation (13.76) with the one without correction (13.74) is shown in Figure 13.2, demonstrating that correction significantly improves the results. All the results are compared with the "*exact*" values obtained by FFT.

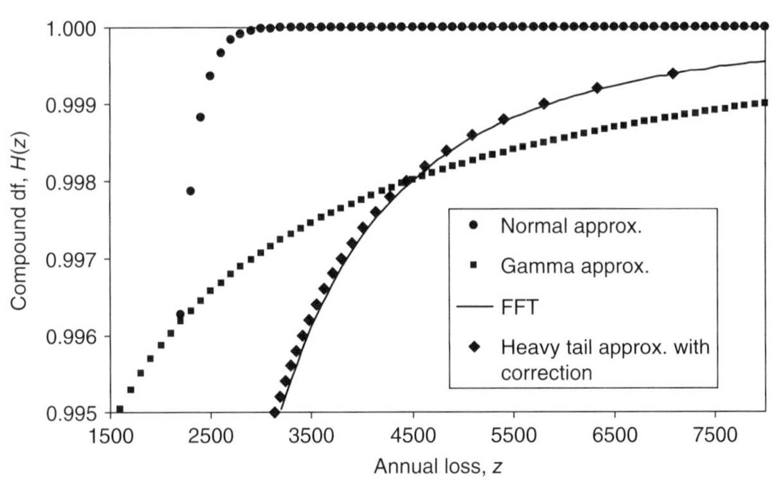

FIGURE 13.1 **Different closed-form approximations for the tail of the *Poisson*(100) – *LogNormal*($\mu = 0, \sigma = 2$) distribution. See Example 13.11 for details**

The results for the case of not-so-heavy tail, when the severity distribution is *LogNormal*(0, 1), are shown in Figure 13.3 and 13.4. Here, the Gamma approximation outperforms Normal approximation and heavy-tailed approximations are very bad. However, note that the corrected heavy-tailed approximation performs much better than the one without correction.

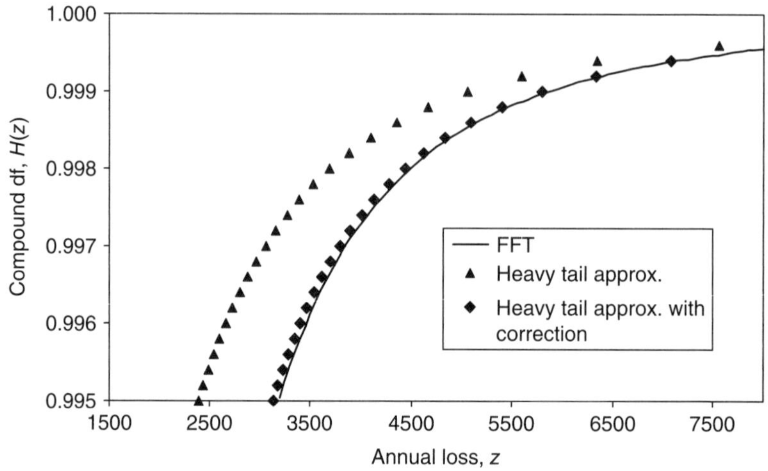

FIGURE 13.2 Heavy-tailed approximation and the corrected heavy-tailed approximation given by formulas (13.75) and (13.76), respectively, for the tail of the $Poisson(100) - LogNormal(\mu = 0, \sigma = 2)$ distribution. See Example 13.11 for details

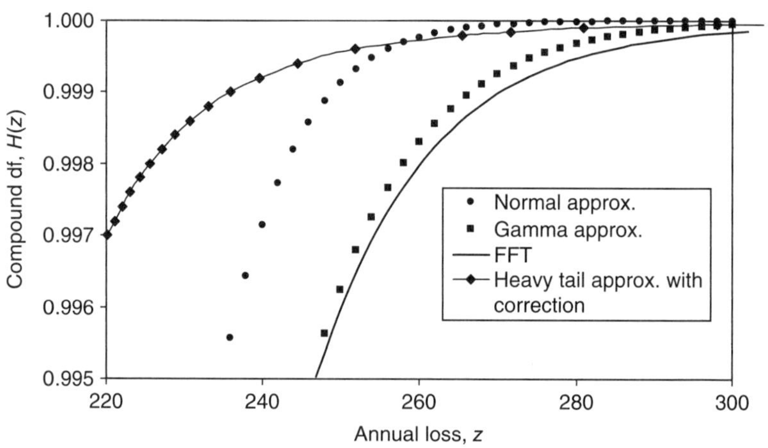

FIGURE 13.3 Different approximations for the tail of the $Poisson(100) - LogNormal(0, 1)$ distribution. See Example 13.11 for details

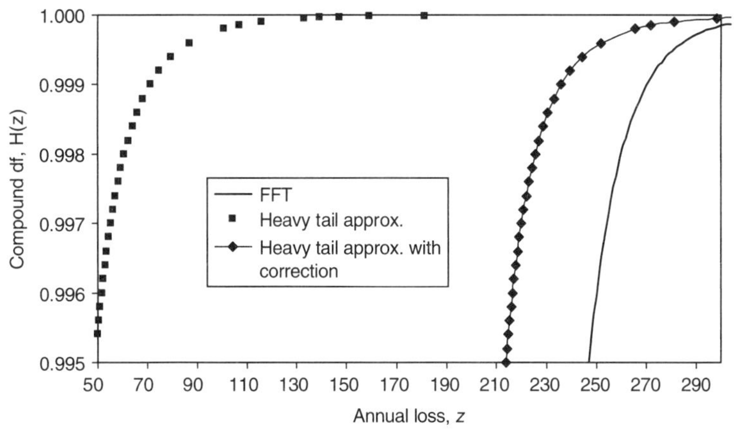

FIGURE 13.4 **Heavy-tailed approximation and the corrected heavy-tailed approximation given by formulas (13.75) and (13.76), respectively, for the tail of the** *Poisson*(100) − *LogNormal*(0, 1) **distribution. See Example 13.11 for details**

The accuracy of the heavy-tailed approximation (13.74) improves for more heavy-tailed distributions, such as GPD with infinite variance or even infinite mean. ■

13.7 Capital Charge Under Parameter Uncertainty

According to the Basel II requirements (BCBS 2006), the final bank capital should be calculated as a sum of the risk measures in the risk cells if the bank's model cannot account for correlations between risks accurately. If this is the case, then one needs to calculate VaR for each risk cell separately and sum VaRs over risk cells to estimate the total bank capital. It is equivalent to the assumption of perfect dependence between risks. Modeling of dependence between risks and aggregation issues were discussed in Chapter 10. In this section, we consider one risk cell, but note that adding quantiles over the risk cells to find the quantile of the total loss distribution is not necessarily conservative. In fact, it can underestimate the capital in the case of heavy-tailed distribution as discussed in Chapter 10.

Under the LDA model, the annual loss in a risk cell over the next year $T+1$ is modeled as a random variable Z_{T+1} with some density $f(z_{T+1}|\boldsymbol{\theta})$, where $\boldsymbol{\theta}$ are model parameters. Given data $\boldsymbol{Y}$ over past T years (frequencies and severities) generated from some distributions parameterized by $\boldsymbol{\theta}$, the main task is to estimate the distribution of Z_{T+1}. The maximum likelihood estimation (MLE) $\hat{\boldsymbol{\theta}}^{\text{MLE}}$ is often used as the "best fit" point estimate for $\boldsymbol{\theta}$. Then, a typical industry practice is to estimate the annual loss distribution for the next year as $f(z_{T+1}|\hat{\boldsymbol{\theta}}^{\text{MLE}})$ and its 0.999 quantile, $Q_{0.999}(\hat{\boldsymbol{\theta}}^{\text{MLE}})$, is used for the capital charge calculation.

However, the parameters $\boldsymbol{\theta}$ are unknown and it is important to account for this uncertainty when the capital charge is estimated, especially for risks with small datasets. The Bayesian inference approach is an elegant and convenient way to accomplish this task.

13.7.1 PREDICTIVE DISTRIBUTIONS

Under the Bayesian approach, the unknown parameters are modeled by random variables $\mathbf{\Theta}$ and their posterior density $\pi(\boldsymbol{\theta}|\boldsymbol{y})$ is calculated. Then, the predictive density of Z_{T+1}, given data $\boldsymbol{Y} = \boldsymbol{y}$, is defined as follows.

Definition 13.3 (Predictive density for annual loss) *Suppose that*

(a) Given $\mathbf{\Theta} = \boldsymbol{\theta}$*, the conditional density of the annual loss* Z_{T+1} *is* $f(z_{T+1}|\boldsymbol{\theta})$*;*

(b) Given data $\boldsymbol{Y} = \boldsymbol{y}$*, the posterior density of* $\mathbf{\Theta}$ *is* $\pi(\boldsymbol{\theta}|\boldsymbol{y})$*;*

(c) Given $\mathbf{\Theta}$*,* Z_{T+1} *and* $\boldsymbol{Y}$ *are independent.*

Then the predictive density of Z_{T+1} *is*

$$f(z_{T+1}|\boldsymbol{y}) = \int f(z_{T+1}|\boldsymbol{\theta})\pi(\boldsymbol{\theta}|\boldsymbol{y})d\boldsymbol{\theta}. \tag{13.79}$$

■

Remark 13.2

- *The predictive distribution accounts for both process and parameter uncertainties;*
- *It is assumed that, given* $\mathbf{\Theta}$*,* Z_{T+1} *and* $\boldsymbol{Y}$ *are independent. If they are not independent, then* $f(z_{T+1}|\boldsymbol{\theta})$ *should be replaced by* $f(z_{T+1}|\boldsymbol{\theta}, \boldsymbol{y})$*;*
- *If a frequentist approach is taken to estimate the parameters, then* $\boldsymbol{\theta}$ *should be replaced by the point estimators* $\hat{\boldsymbol{\theta}}$ *and the integration should be done with respect to the density of* $\hat{\boldsymbol{\theta}}$*.*

The ultimate goal in capital charge calculation is to estimate the 0.999 quantile of the annual loss distribution. It is important to realize that there are two ways to define the required quantile to account for parameter uncertainty.

Definition 13.4 (Quantile of the predictive density $f(z_{T+1}|\boldsymbol{y})$) *The quantile of a random variable with the predictive density (13.79) is*

$$Q_q^P = F^{-1}_{Z_{T+1}|\boldsymbol{Y}}(q) = \inf\{z \in \mathbb{R} : \Pr[Z_{T+1} > z|\boldsymbol{Y}] \leq 1 - q\}, \tag{13.80}$$

where $q \in (0, 1)$ *is a quantile level and* $F^{-1}_{Z_{T+1}|\boldsymbol{Y}}(q)$ *is the inverse of the distribution corresponding to the density (13.79).* ■

Then, $Q^P_{0.999}$ can be used as a risk measure for capital calculations. Here, "P" in the upper script is used to emphasize that this is a quantile of the full predictive distribution.

Another approach under a Bayesian framework to account for parameter uncertainty is to consider a quantile of the annual loss density $f(z_{T+1}|\boldsymbol{\theta})$ conditional on parameter $\mathbf{\Theta} = \boldsymbol{\theta}$, defined in a standard way as follows.

Definition 13.5 (Quantile of the conditional density $f(z|\boldsymbol{\theta})$) *The quantile of a random variable with the density $f(z|\boldsymbol{\theta})$ is*

$$Q_q(\boldsymbol{\theta}) = F^{-1}_{Z_{T+1}|\boldsymbol{\Theta}}(q) = \inf\{z \in \mathbb{R} : \Pr[Z_{T+1} > z|\boldsymbol{\Theta} = \boldsymbol{\theta}] \leq 1 - q\}, \tag{13.81}$$

where $q \in (0, 1)$ is a quantile level and $F^{-1}_{Z_{T+1}|\boldsymbol{\Theta}}(q)$ is the inverse of the distribution corresponding to the density $f(z_{T+1}|\boldsymbol{\theta})$. ∎

That is, the quantile $Q_q(\boldsymbol{\theta})$ is a function of $\boldsymbol{\theta}$ and thus $Q_q(\boldsymbol{\Theta})$ is a random variable with some distribution. Given that $\boldsymbol{\Theta}$ is distributed with the density $\pi(\boldsymbol{\theta}|\mathbf{y})$, one can find the *predictive distribution* of $Q_q(\boldsymbol{\Theta})$ and its characteristics. In particular, the mean of this distribution can be used as a point estimator:

$$\widehat{Q_q(\boldsymbol{\Theta})}^{\text{MMSE}} = \int Q_q(\boldsymbol{\theta})\pi(\boldsymbol{\theta}|\mathbf{y})d\boldsymbol{\theta}. \tag{13.82}$$

Other standard point estimators are the mode and median. A predictive interval $[L, U]$ can be formed to contain the true value with a probability α:

$$\Pr\left[L \leq Q_q(\boldsymbol{\Theta}) \leq U\right] = \alpha \tag{13.83}$$

or one-sided predictive interval

$$\Pr\left[Q_q(\boldsymbol{\Theta}) \leq U\right] = \alpha. \tag{13.84}$$

As before, for capital charge calculations, we are interested in $q = 0.999$. Then one can argue that the conservative estimate of the capital charge accounting for parameter uncertainty should be based on the upper bound of the constructed predictive interval.

Remark 13.3

- *Specification of the confidence level α is required to form a conservative interval for $Q_q(\boldsymbol{\Theta})$. It might be difficult to justify a particular choice of α. For example, it might be difficult to argue that the commonly used confidence level $\alpha = 0.95$ is good enough for estimation of the 0.999 quantile;*
- *This is similar to forming a confidence interval in the frequentist approach using the distribution of $Q_{0.999}(\widehat{\boldsymbol{\theta}}^{MLE})$, where $\widehat{\boldsymbol{\theta}}^{MLE}$ is treated as random.*

In OpRisk, it seems that the objective should be to estimate the full predictive distribution (13.79) for the annual loss Z_{T+1} over next year conditional on all available information. The capital charge should then be estimated as a quantile of this distribution, that is, $Q^P_{0.999}$ given by (13.80).

13.7.2 CALCULATION OF PREDICTIVE DISTRIBUTIONS

Consider a risk cell in the bank. Assume that the frequency $p(\cdot|\boldsymbol{\alpha})$ and severity $f(\cdot|\boldsymbol{\beta})$ densities for the cell are chosen. Suppose also that the posterior density $\pi(\boldsymbol{\theta}|\mathbf{y}), \boldsymbol{\theta} = (\boldsymbol{\alpha}, \boldsymbol{\beta})$ is estimated. Then, the predictive annual loss distribution (13.79) in the cell can be calculated using the Monte Carlo procedure with the following logical steps.

Algorithm 13.6 (Full predictive loss distribution via Monte Carlo)

1. *For* $k = 1, \ldots, K$
   a) *For a given risk simulate the risk parameters* $\boldsymbol{\theta} = (\boldsymbol{\alpha}, \boldsymbol{\beta})$ *from, the posterior* $\pi(\boldsymbol{\theta}|\mathbf{y})$. *If the posterior is not known in closed form then, this simulation can be done using MCMC (see Section 7.4). For example, one can run MCMC for K iterations (after burn-in) beforehand and simply take the k-th iteration parameter values;*
   b) *Given* $\boldsymbol{\theta} = (\boldsymbol{\alpha}, \boldsymbol{\beta})$, *simulate the annual number of events N from* $p(\cdot|\boldsymbol{\alpha})$ *and severities* $X^{(1)}, \ldots, X^{(N)}$ *from* $f(\cdot|\boldsymbol{\beta})$, *then calculate the annual loss* $Z^{(k)} = \sum_{n=1}^{N} X^{(n)}$.
2. *Next k*

Obtained annual losses $Z^{(1)}, \ldots, Z^{(K)}$ are samples from the predictive density (13.79). Extending this procedure to the case of many risks is easy but requires specification of the dependence model; see Chapter 10. In this case, in general, all model parameters (including the dependence parameters) should be simulated from their joint posterior in Step (a). Then, given these parameters, Step (b) should simulate all risks with a chosen dependence structure. In general, sampling from the joint posterior of all model parameters can be accomplished via MCMC (see Peters *et al.* 2009 and Dalla Valle 2009). The 0.999 quantile $Q^P_{0.999}$ and other distribution characteristics can be estimated using the simulated samples in the usual way; see Section 13.2.

This procedure is easily adapted to calculate the predictive distribution of $Q_{0.999}(\boldsymbol{\Theta})$. In particular, in Step (b), one can calculate the quantile $Q_{0.999}(\boldsymbol{\theta})$ of the conditional density $f(z|\boldsymbol{\theta})$, using, for example, FFT. Then the obtained K samples of the quantile can be used to estimate the distribution of $Q_{0.999}(\boldsymbol{\Theta})$ implied by the posterior $\pi(\boldsymbol{\theta}|\mathbf{y})$. To summarize, the logical steps of Monte Carlo procedure are as follows.

Algorithm 13.7 (Posterior distribution of quantile via MC)

1. *For* $k = 1, \ldots, K$
   a) *For a given risk simulate the risk parameters* $\boldsymbol{\theta} = (\boldsymbol{\alpha}, \boldsymbol{\beta})$ *from the posterior* $\pi(\boldsymbol{\theta}|\mathbf{y})$. *If the posterior is not known in closed form, then this simulation can be done using MCMC (see Section 7.4). For example, one can run MCMC for K iterations beforehand and simply take the k-th iteration parameter values;*
   b) *Given* $\boldsymbol{\theta} = (\boldsymbol{\alpha}, \boldsymbol{\beta})$, *calculate the quantile* $Q_q^{(k)}(\boldsymbol{\theta})$ *of* $f(z|\boldsymbol{\theta})$ *using FFT or other methods.*
2. *Next k*

Note that in these Monte Carlo procedures, the risk profile $\boldsymbol{\Theta}$ is sampled from its posterior for each simulation $k = 1, \ldots, K$. Thus, we model both the process uncertainty, which comes from the fact that frequencies and severities are random variables, and the parameter risk (parameter uncertainty), which comes from the fact that we do not know the true values of $\boldsymbol{\theta}$.

EXAMPLE 13.12

The parameter uncertainty is ignored by the estimator $Q_{0.999}(\hat{\boldsymbol{\theta}}^{\text{MLE}})$ but is taken into account by $Q^P_{0.999}$. The following illustrative example is taken from Shevchenko (2008, section 8). Figure 13.5 presents results for the relative bias (averaged over

100 realizations) $\mathbb{E}[Q^P_{0.999} - Q_{0.999}(\hat{\boldsymbol{\theta}}^{\text{MLE}})]/Q^{(0)}_{0.999}$, where $\hat{\boldsymbol{\theta}}^{\text{MLE}}$ is MLE, $Q^{(0)}_{0.999}$ is the quantile of $f(\cdot|\boldsymbol{\theta}_0)$ and $\boldsymbol{\theta}_0$ is the true value of the parameter. The frequencies and severities are simulated from $Poisson(\lambda_0 = 10)$ and $LogNormal(\mu_0 = 1, \sigma_0 = 2)$, respectively. Constant priors are also used for the parameters so that there are closed-form expressions for the posterior; see Sections 15.2.3 and 15.2.4. In this example, the bias induced by parameter uncertainty is large: it is approximately 10% after 40 years (i.e., approximately 400 data points) and converges to zero as the number of losses increases. A similar analysis for a multivariate case was performed by Dalla Valle (2009) with real data. For high-frequency/low-severity risks, where a large amount of data is available, the impact is certainly expected to be small. However, for low-frequency/high-severity risks, where the data are very limited, the impact can be significant.

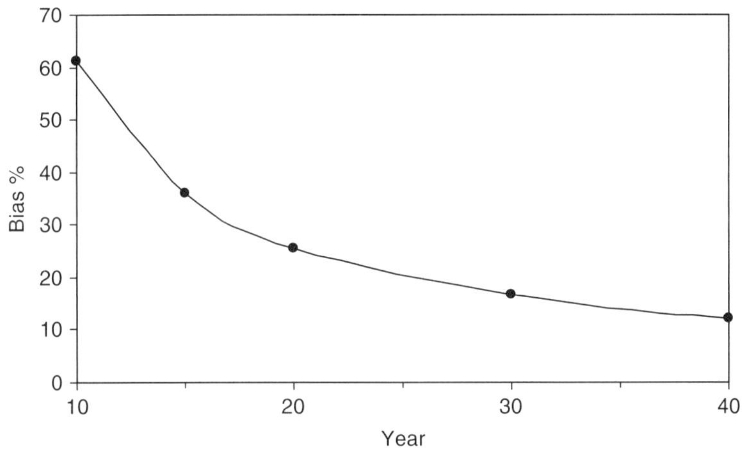

FIGURE 13.5 **Comparison of the estimators of the 0.999 annual loss quantile versus number of observation years. Losses were simulated from *Poisson*(10) and *LogNormal*($\mu = 1, \sigma = 2$). Parameter uncertainty is ignored by $Q_{0.999}(\hat{\boldsymbol{\theta}}^{MLE})$ (MLE) but taken into account by $Q^P_{0.999}$ (Bayesian). Relative bias $\mathbb{E}[Q^P_{0.999} - Q_{0.999}(\hat{\boldsymbol{\theta}}^{MLE})]/Q^{(0)}_{0.999}$ is estimated as an average over 100 realizations**

■

13.8 Special Advanced Topics on Loss Aggregation

The advanced topics discussed in the second half of this chapter include:

- Discretisation Errors and Extrapolation Methods;
- Classes of Discrete Distributions: Discrete Infinite Divisibility and Discrete Heavy Tails. These classes of distributions naturally admit either analytic compound distributions or efficient methods for evaluation of the loss aggregated distribution when used as the severity distribution;

- Recursions for Convolutions (Partial Sums) with Discretised Severity Distributions (Fixed number of losses n);
- Alternatives to Panjer Recursions: Recursions for Compound Distributions with Discretised Severity Distributions;
- Higher Order Recursions for Discretised Severity Distributions in Compound LDA Models;
- Recursions for Discretised Severity Distributions in Compound Mixed Poisson LDA Models; and
- Continuous Versions of the Panjer Recursion.

A foreword on notation contained in this advanced section of the chapter, all distributions and densities will be assumed to be continuous unless otherwise stated, at certain stages in the presentation it will become important to discretise the severity distributions as was done initially when working with the Panjer recursion above, at which point it will be made explicit when such representations are utilised to avoid confusion. We note that when a continuous distribution is discretised onto an equispaced grid of points in its support, given for some interval $\triangle \in \mathbb{R}^+$ by $0, \triangle, 2\triangle, \ldots$, this is equivalent to working on the space of integers, $0, 1, 2, \ldots$, the resulting discretized distribution will be denoted by $\{\hat{f}_n\}_{n\in\mathbb{N}}$, where $\hat{f}_n = f(n\triangle)$ and $\hat{f}_n \geq 0$. Note, in some instances it may also be required to impose a normalization condition on the discretized distribution values where $\sum_{n=1}^{\infty} \hat{f}_n = 1$.

In addition, we will adopt the convention from the actuarial literature, see a summary in Sundt and Vernic (2009) in which $\mathcal{P}_1$ will denote the class of all univariate distributions on the integers and $\mathcal{P}_{1l}$ the class of all distributions $f \in \mathcal{P}_1$ satisfying the condition that $f(x) = 0$ for all integers $x < l$ and finally, the notation $\mathcal{P}_{1\underline{l}}$ which denotes the sub-class of distributions in $\mathcal{P}_{1l}$ with a positive mass at l. Analogously, these definitions will also carry through for functions, that is densities are denoted by the class $\mathcal{P}$ and functions by the label $\mathcal{F}$, for example densities only known up to normalization. To clarify this point, we consider $\mathcal{F}_l$ for all integers l to be the class of all functions on the integers and $\mathcal{F}_{1l}$ as the set of functions $f \in \mathcal{F}_1$ which satisfy $f(x) = 0$ for all integers $x < l$.

13.8.1 DISCRETISATION ERRORS AND EXTRAPOLATION METHODS

This section is of relevance to the study of the approximation error incurred when one makes an approximating discretization to the severity distribution in order to calculate an aggregate distribution of some form, as was discussed in the section on Panjer recursions a the start of this chapter.

In particular, Richardson extrapolation, also known as extrapolation to the limit or acceleration of convergence, can often be performed to reduce discretisation errors. This can be particularly relevant when evaluating risk measures based on these discretised annual loss distributions. The approaches discussed in the two articles by Grubel and Hermesmeier (1999) and Grübel and Hermesmeier (2000) consider how to increase the accuracy for a given discretisation budget through methods that reduce aliasing error and incorporate extrapolation procedures. In particular the method of Richardson extrapolation as developed in Richardson (1911) is discussed. Richardson extrapolation is considered as a sequence acceleration procedure which can improve the rate of convergence of sequences, such as those that may arise in recursive evaluation of convolution distributions on a discretisation grid. Examples of the wide spread use of Richardson extrapolation include the method of Romberg integration Romberg (1955) which combines Richardson extrapolation to speed up a trapezoidal integration rule for definite integrals.

We first briefly recall how Richardson extrapolation works and then consider how it may be utilised to improve the accuracy of recursions for compound process recursions. The generic specification of Richardson Extrapolation if provided in Definition 13.6.

Definition 13.6 (Richardson Extrapolation) *Consider approximating a generic density f, for instance from a convolution, at a particular point x which cannot be evaluated directly. However, one has an approximation that is a function of the discretisation effort denoted generically by h which produces an approximation $\widehat{f}_h(x)$ for any $h > 0$ and relates to the true density $f(x)$ according to*

$$\widehat{f}_h(x) = f(x) + Ch^{\alpha} + o\left(h^{\beta}\right). \tag{13.85}$$

Richardson extrapolation then takes the approximations for different h values and combines them in such a manner as to improve the rate of convergence and therefore the accuracy of the combined approximation. Assuming that rate α is known but the constant C is intractable, then one can define a new approximation with improved convergence rate of $\widehat{f}_h \to f(x)$ by the following combined approximation known as the Richardson extrapolation of $\widehat{f}_h(x)$ to give

$$\widetilde{f}(x; h, k) := \frac{k^{\alpha}\widehat{f}_h(x) - \widehat{f}_{kh}(x)}{k^{\alpha} - 1} \tag{13.86}$$

for some h and k. Where clever choices of h and k can significantly accelerate the convergence of the new approximation. ∎

The first proposal to utilise extrapolation methods to accelerate insurance based recursions and evaluations was in the definite integration of a function with respect to a compound process, such as would be required under a spectral risk measure, see discussion in Embrechts *et al.* (1993). Then it was developed further in the context of compound processes and convolution recursions in the work of Grubel and Hermesmeier (1999) which we discuss briefly below.

The application of Richardson extrapolation as applicable to the recursions in this chapter will in general follow a four stage procedure, which is based on a given selected k, h and selected α for the given approximation method according to Equation 13.90, as follows :

Algorithm 13.8 (Richardson extrapolation)

1. *Discretise the severity distribution input to obtain both $\widehat{f}_{X,h}(x)$ and $\widehat{f}_{X,kh}(x)$ where the h will correspond to the discretisation unit $\triangle$ as discussed in the methods previously mentioned;*
2. *Given the discretised severity distributions, these can then be considered as inputs to a generic non-linear mapping $\Phi(\cdot)$. The mapping corresponds in this chapter to one of the recursions to be presented throughout this chapter that allows one to evaluate for instance, the density, distribution, or tail of the n-fold convolutions of severity distributions or the evaluation of these quantities for the compound process of an annual loss LDA model given by*

$$\hat{f}_{Z,h}(x) = \Phi\left(\hat{f}_{X,h}(x)\right) \tag{13.87}$$

for some mapping $\Phi(\cdot)$ that represents the class of numerical recursion utilised for the evaluation of the intractable density or distribution;

3. *Evaluate the Richardson Extrapolation based on approximations $\hat{f}_{Z,h}(x)$ and $\hat{f}_{Z,kh}(x)$ given by Equation 13.86 according to*

$$\widetilde{f}_Z(z) := \frac{k^\alpha \hat{f}_{Z,h}(z) - \hat{f}_{Z,kh}(z)}{k^\alpha - 1}. \tag{13.88}$$

4. *Evaluate required functionals based on the Richardson extrapolated result for density $\widetilde{f}_Z(z)$ and distribution $\widetilde{F}_Z(z)$ based on the mapped discretised input severity distributions $\hat{f}_{Z,h}(z)$ and $\hat{f}_{Z,kh}(z)$, to obtain functional approximations such as for example:*

$$\begin{aligned} \overline{F}_Z(x) &\approx \int_0^x d\widetilde{F}_Z(z) = \int_0^x \widetilde{f}_Z(z)dz \\ \overline{F}_X^{(n)*}(x) &\approx \int_0^x d\widetilde{F}_X^{(n)*}(z) = \int_0^x \widetilde{f}_X^{(n)*}(z)dz \\ SRM_{Z,h}(x) &\approx \int \psi(z)d\widetilde{F}_Z(z). \end{aligned} \tag{13.89}$$

For example, consider the case of the SRM that will be evaluated as the integral of the risk aversion function $\psi(x)$ with respect to the annual loss distribution that was discussed in Peters and Shevchenko (2015, chapter 6). Then considering the generic representation

$$\hat{f}_h(x) = f(x) + Ch^\alpha + o\left(h^\beta\right). \tag{13.90}$$

in this context the extrapolation to the limit relation of this type for the approximation of the SRM would be given by considering an approximation integral expansion of the form

$$\begin{aligned} SRM_{Z,h}(\alpha) &:= \int \psi(x)dF_{Z,h}(x) \\ &= \int \psi(x)d\Phi\left(F_{X,h}(x)\right) \\ &= \int \phi(x)d\Phi\left(F_X(x)\right) + \zeta\left(\Phi; F_X, \psi\right)h^\alpha + o\left(h^\beta\right) \end{aligned} \tag{13.91}$$

as $h \downarrow 0$ and $\beta \geq \alpha > 0$ and function $\zeta(\Phi; F_X, \psi) \neq 0$. Examples are provided for particular instances of α and β rates in Embrechts *et al.* (1993) in the case of integration results.

In the case that one is primarily interested in just the discretisation density say from an n-fold convolution of the severity $\hat{f}_{X,h}^{(n)*}$ then the key results are developed for Richardson extrapolation methods in Grubel and Hermesmeier (1999). In particular if one considers that a density exists for $\hat{f}_X^{(n)*}$ for the measure $\nu_X^{(n)*}$ then one way of considering the convergence of $\hat{f}_{X,h}^{(n)*}$ to $f_X^{(n)*}$ as discretisation error is diminished ($h \downarrow 0$) is to consider the approximation

$$\hat{f}_{X,h}(kh) = \frac{1}{h}\nu_{X,h}(\{kh\}) + g(kh)h^{\alpha} + O\left(h^{\beta}\right), \tag{13.92}$$

which holds uniformly in integer discretisation steps $k \in \mathbb{J}^+$ as $h \downarrow 0$ for some function $g(\cdot)$ which depends on the mapping $\phi(\cdot)$ and measure input. This could be combined into a Richardson extrapolation procedure as long as $\beta > \alpha$. Details of particular examples of function g and rates α and β are provided in detail in Grubel and Hermesmeier (1999, section 3).

13.8.2 CLASSES OF DISCRETE DISTRIBUTIONS: DISCRETE INFINITE DIVISIBILITY AND DISCRETE HEAVY TAILS

If one considers a discretized continuous or a discrete distribution $\left\{\hat{f}_n\right\}_{n\in\mathbb{N}}$ which satisfies the constraint that $\hat{f}_n \geq 0$ for all $n \in \mathbb{N}$, which is normalized with $\sum_n \hat{f}_n = 1$, then it is first useful to recall the following definition of a non-degenerate discrete distribution.

Definition 13.7 (Non Degenerate Discrete Distribution) *A discrete distribution that is normalized is called non degenerate if $\hat{f}_n < 1$ holds for all $n \in \mathbb{N}$.* ■

In addition, it will be useful to define the class of discrete distributions known as the Panjer class.

Definition 13.8 (General Panjer Class of Discrete Distributions) *A discrete distribution $\left\{\hat{f}_n\right\}_{n\in\mathbb{N}}$ with parameters $a, b \in \mathbb{R}$ and order $k \in \mathbb{N}$ is in the Panjer class if $\hat{f}_n = 0$ for all $n \leq k-1$ and the probabilities satisfy the recursion*

$$\hat{f}_{n+1} = \left(a + \frac{b}{n+1}\right)\hat{f}_n \tag{13.93}$$

for all $n \geq k$. Such a class of discrete distribution that satisfies this condition is known as a Panjer class family with parameters a, b and order k. ■

There will be further discussion on these classes of discrete distributions in the following sections. In this section we will also state a representation of such distributions according to a differential equation in terms of its probability generating function, see discussion in Hess *et al.* (2002). It will be useful to first define the probability generating function given in Definition 13.9.

Definition 13.9 (Probability Generating Function) *A random variable X with discrete probability mass distribution $\left\{\hat{f}_n\right\}_{n\in\mathbb{N}}$ is characterized by the probability generating function given by*

$$m_X(z) := \mathbb{E}\left[z^X\right] = \sum_{n=0}^{\infty}\hat{f}_n z^n, \tag{13.94}$$

where given the p.g.f., denoted $m_X(z)$, one obtains the probabilities by differentiation and evaluation at the origin according to

$$\hat{f}_n = \frac{m_X^{(n)}(0)}{n!}, \ \forall n \in \mathbb{N}. \tag{13.95}$$

■

One can then show that the following representational equivalence for degenerate discrete severity models given in Theorem 13.4 that represents the probability generating function (p.g.f.) in terms of the solution to an ordinary differential equation (o.d.e.).

Theorem 13.4 (Panjer Class *k* and o.d.e. for p.g.f.) *If a random variable X has discrete distribution $\left\{\hat{f}_n\right\}_{n\in\mathbb{N}}$ that is non degenerate, then the following statements are equivalent:*

1. $\left\{\hat{f}_n\right\}_{n\in\mathbb{N}}$ *is in the Panjer a, b class of order k;*
2. *The p.g.f. satisfies the differential equation given by*

$$(1 - az)m_X^{(n+1)}(z) = ((n + 1)a + b)\, m_X^{(n)}(z) \tag{13.96}$$

for $z \in [0, 1)$ and initial condition $m_X^{(j)}(0) = 0$ for all $j \leq k - 1$.

In addition to the Panjer class of distributions, it will sometimes be of relevance to consider lower and upper truncated Poisson distributions and in particular convolutions of such random variables when they are considered independent. The class of upper and lower truncated Poisson n-fold convolution distributions was studied in Huang and Fung (1993) and is denoted as the family of D-distributions. Several properties of this family of distributions are known such as the expressions for integer moments, see Huang and Fung (1993).

Definition 13.10 (D-Distributions: *n*-Fold Convolutions of Upper and Lower Truncated Poisson) *Consider n independent counting random variables each i.i.d. from an upper and lower truncated Poisson distribution such that*

$$N_i \sim Poisson(\lambda_i)\mathbb{I}\left[N_i \in \{n_{min,i}, n_{max,i}\}\right] \tag{13.97}$$

for some positive integers n_{min} and n_{max} satisfying $0 < n_{min,i} < n_{max,i} < \infty$. Then the distribution of the convolution given by

$$N_n = \sum_{k=1}^{n} N_k \sim D - Distribution\,(n, \mathcal{L}, \mathbf{\Lambda}) \tag{13.98}$$

where the D-Distribution is characterized by the discretely supported distribution given by

$$\mathbb{P}\text{r}\,(X = k) = \prod_{i=1}^{n} e\,(N_i, M_i; \lambda_i)^{-1}\, D(x, n; \mathcal{L}\Lambda)\frac{1}{x!} \tag{13.99}$$

with

$$x \in \left\{ \sum_{i=1}^{n} n_{min,i}, \sum_{i=1}^{n} n_{min,i} + 1, \ldots, \sum_{i=1}^{n} n_{max,i} \right\}, \tag{13.100}$$

and the D numbers given by

$$D(x, n; \mathcal{L}, \Lambda) = \sum_{y \in Y} \binom{x}{y_1, y_2, \ldots, y_n} \tag{13.101}$$

with

$$\begin{aligned} \boldsymbol{Y} &= \left\{ \boldsymbol{y} : \boldsymbol{y} = (y_1, y_2, \ldots, y_n), x = \sum_{i=1}^{n} y_i \right\} \\ \mathcal{L} &= \left\{ (n_{min,i}, n_{max,i}) : n_{min,i} < n_{max,i}, n_{min,i} \in \mathbb{N}^+, n_{max,i} \in \mathbb{N}^+ \right\} \\ \Lambda &= \{ \lambda_i : \lambda_i > 0 \}. \end{aligned} \tag{13.102}$$

and $e(\cdot)$ the incomplete exponential function with parameter λ given by

$$e(N, M; \lambda) = \begin{cases} \sum_{i=N}^{M} \frac{\lambda^i}{i!}, & \lambda > 0, 0 \leq N < M, \ N, M \in \mathbb{N} \\ \sum_{i=0}^{M} \frac{\lambda^i}{i!}, & \lambda > 0, N = -1, -2, -3, \ldots, \\ 0, & \text{otherwise}. \end{cases} \tag{13.103}$$

■

It will also be useful in sections of this chapter to define the special class of discrete distributions $\left\{ \hat{f}_n \right\}_{n \in \mathbb{N}}$ which will have the property that they are infinitely divisible, as characterised in Theorem 13.5, see discussions in Steutel and Van Harn (2003).

Theorem 13.5 (Characterizing the Infinitely Divisible Distributions) *The following properties exist for members of the class of infinitely divisible distributions:*

1. *A distribution concentrated on a dirac mass is infinitely divisible;*
2. *The class of infinitely divisible distributions is closed under the operation of convolution;*
3. *The class of infinitely divisible distributions is closed under linear translations;*
4. *The class of infinitely divisible distributions is closed under constant scalings;*
5. *A mixed Poisson distribution is infinitely divisible if the mixing distribution is also infinitely divisible;*
6. *A mixed Poisson distribution with infinitely divisible mixing distribution can be expressed as a compound Poisson distribution with severity distribution in the class $\mathcal{P}_{11}$;*
7. *A compound distribution is infinitely divisible if its counting distribution (frequency distribution) is infinitely divisible and in the class of distributions $\mathcal{P}_{10}$;*
8. *An infinitely divisible distribution in the class $\mathcal{P}_{10}$ has a positive probability in zero;*

9. *A non-degenerate distribution f in the class $\mathcal{P}_{1\underline{0}}$ is infinitely divisible if and only if it can be expressed as a compound Poisson distribution with a severity distribution g which is in the class $\mathcal{P}_{11}$.*

Random variables which are distributed according to a discrete distribution which is infinitely divisible have the advantage that when evaluating sums of such random variables under an independence assumption will result in simple expressions for the form of the resulting compound distribution. This is explored in the following sections through De Pril transforms for such classes of random variable.

To determine if a discrete distribution is infinitely divisible one can utilise the necessary and sufficient condition for a discrete distribution to be infinitely divisible given in Theorem 13.6, see Katti (1967).

Theorem 13.6 (Discrete Infinitely Divisible Distributions: Necessary and Sufficient Conditions) *Consider a discrete distribution $\left\{\hat{f}_i\right\}_{i=1}^{N}$ for $i = 0, 1, 2, \ldots$ with $\hat{f}_0 \neq 0$ and $\hat{f}_1 \neq 0$. Then the necessary and sufficient condition for $\left\{\hat{f}_i\right\}_{i=1}^{N}$ to be infinitely divisible is that it must satisfy the recursion given below, which must be strictly non-negative for all $i \in \mathbb{N}$.*

$$\pi_i = i\frac{\hat{f}_i}{\hat{f}_0} - \sum_{j=1}^{i-1} \pi_{i-j}\frac{\hat{f}_j}{\hat{f}_0} \geq 0. \tag{13.104}$$

A corollary of this result discussed in Bondesson *et al.* (1996, corollary 2.6) is given below.

Corollary 13.1 *If a discrete distribution $\hat{f}_n$ on positive integers n is of the form $\hat{f}_n = (n+1)c_n$ for some sequence c_n which is completely monotone, then the distribution is infinitely divisible.*

To validate that a discrete distribution is in the class of infinitely divisible distributions it was later shown in Warde and Katti (1971) that it is sufficient to ensure the following condition provided in Theorem 13.7 holds.

Theorem 13.7 (Discrete Distribution Infinite Divisibility Sufficient Condition 1) *A discrete distribution $\left\{\hat{f}_n\right\}_{n\in\mathbb{N}}$, with $\hat{f}_0 \neq 0$, $\hat{f}_1 \neq 0$ is infinitely divisible if the ratios $\frac{\hat{f}_i}{\hat{f}_{i-1}}$ for $i = 1, 2, \ldots$ form a monotone increasing sequence.*

A second alternative sufficient condition that can be considered for discrete distributions with support on the non-negative integers, as given in Theorem 13.8, see Steutel (1973).

Theorem 13.8 (Discrete Distribution Infinite Divisibility Sufficient Condition 2) *A discrete distribution $\left\{\hat{f}_n\right\}_{n\in\mathbb{N}}$ on the non-negative integers, with $\hat{f}_0 \neq 0$, is infinitely divisible if it satisfies the recursion*

$$\hat{f}_{n+1} = \sum_{k=0}^{n} q_k \hat{f}_{n-k} \tag{13.105}$$

for $q_k \geq 0$ for all $k = 1, 2, 3, \ldots$.

Remark 13.4 *Distributions that satisfy the above are also known to be of compound geometric distribution form.*

Yet a third sufficient condition can be stated for discrete infinitely distributed random variables on the positive integers, given in Theorem 13.9, see discussion in Steutel (1973).

Theorem 13.9 (Discrete Distribution Infinite Divisibility Sufficient Condition 3) *A discrete distribution $\{\hat{f}_n\}_{n\in\mathbb{N}}$ on the non-negative integers with a distribution which is log-convex, is infinitely divisible if it satisfies that*

$$\hat{f}_{n+1}\hat{f}_{n-1} \geq \hat{f}_n^2, \tag{13.106}$$

for all $n = 1, 2, \ldots$.

It is also worth noting that it was shown in Convolutions (Generalized Gamma) that the class of discrete distributions satisfying Corollary 13.1 are comprised of the class of mixtures of negative binomial distributions of order 2.

Having defined the notion of discrete infinite divisibility it is natural to consider the question, are there any cases in which one can take a distribution which is infinitely divisible with support $(0, \infty)$ and discretize it to a distribution which takes integer support and preserves the infinite divisibility?

In Bondesson *et al.* (1996) they studied this type of question under a particular form of discretization based on simple rounding by taking the integer component. That is consider a random variable X decomposed as its integer component $[X]$ and remainder fractional part $\{X\}$ given by $X = [X] + \{X\}$. In this setting they were able to state the following result in Theorem 13.10. Before stating this result we will briefly recall the definition of a log-concave density function for an OpRisk severity model, see Definition 13.11.

Definition 13.11 (Log-Concave Density Functions) *A continuous loss random variable X has a density $f_X(x)$ which is said to be log-concave if it can be expressed as follows*

$$f_X(x) = \exp\left(\phi(x)\right), \tag{13.107}$$

where $\phi(x)$ is a concave function. ∎

Densities that are log-concave satisfy the conditions:

1. $\ln f_X\left(\lambda x + (1-\lambda)y\right) \geq \lambda \ln f_X(x) + (1-\lambda) f_X(y)$;
2. Analogously, $f_X\left(\lambda x + (1-\lambda)y\right) \geq f_X(x)^{\lambda} f_X(y)^{1-\lambda}$;
3. $f_X\left(\frac{1}{2}(x+y)\right) \geq \sqrt{f_X(x) f_X(y)}$;
4. In Ibragimov (1956) it was shown that a density on $\mathbb{R}$ will be log-concave if and only if when it is convolved with a unimodal density, the resulting convolved density is again unimodal.

Theorem 13.10 (Continuous Infinite Divisibility to Discrete Infinite Divisibility) *If a loss random variable X with support on $(0, \infty)$ has a log-concave density then both X and the integer supported random variable given by $[X]$ are characterized by infinitely divisible distributions.*

In addition it was shown in Bondesson *et al.* (1996, theorem 3.3) that one can characterize the classes of infinitely divisible severity distributions for which the integer rounded discrete distributions will also be infinitely divisible as those satisfying the result in Theorem 13.11.

Theorem 13.11 (Characterising Infinitely Divisible Distributions: Continuous and Discrete) *Consider a continuous loss random variable with a severity density that satisfies the representation given by*

$$f_X(x) = (x + a)h(x), \tag{13.108}$$

where $a \geq 0$ and $h(x)$ is a completely monotone function. In this case the severity distribution is infinitely divisible. In addition, $[X]$ the discretized loss random variable on the integers is infinitely divisible for such a model when $a \geq 1$.

In mentioning the class of discrete infinitely divisible distributions, one can then as what types of models may be suitable to consider in this class for OpRisk modelling of severity. Before proceeding to discuss such models of relevance to OpRisk settings, we first recall the definition of the discrete Sibuya distribution given in Definition 13.12, see discussion in Devroye (1993).

Definition 13.12 (Discrete Sibuya Distribution) *A random variable S is said to be distributed according to a Sibuya distribution with parameter γ if it has a p.g.f. given by*

$$m_S(z) = 1 - (1 - z)^{\gamma} \tag{13.109}$$

for $\gamma \in (0, 1]$. In addition, it can therefore be shown that the discrete probability mass function is given by

$$\Pr(S = n) = \begin{cases} \frac{\gamma(1-\gamma)\cdots(n-1-\gamma)}{n!}, & n > 1, \\ \gamma, & n = 1. \end{cases} \tag{13.110}$$

■

In turns out that there is a class of discrete severity distributions that form the analog of the α-Stable severity model, for the discrete support case as given in Definition 13.13, see discussion in Steutel and Van Harn (2003) and Christoph and Schreiber (1998a).

It is useful to recall that α-stable random variables, that are discussed in Peters and Shevchenko (2015), satisfy the condition that for strictly stable random variables $X_i \in \mathbb{R}$ with $i = 0, 1, 2, \ldots, n$, one has the characterization given by

$$X_1 + \ldots + X_n \stackrel{d}{=} c_n X_0 + d_n, \tag{13.111}$$

for some sequence $c_n = n^{\frac{1}{\alpha}}$ and (strict stability involves $d_n = 0$). The tail index coefficient $\alpha \in [0, 2]$ is the index of stability, dictating how heavy the tails of the distribution will be, with light tails if α is close to 2 and heavy tailed as α decreases towards 0. In addition it is known that α must satisfy the condition

$$X \stackrel{d}{=} kX_1 + (1 - k^{\alpha})^{\frac{1}{\alpha}} X_2, \tag{13.112}$$

for $k \in (0,1)$. It was shown in Christoph and Schreiber (1998a) that one can utilise this condition to re-express the discrete version of the α-Stable family of severity models on the integer support which is non-negative.

Definition 13.13 (Discrete Integer Values Stable Family of Severity Models: Class 1) *A discrete non-negative lattice loss random variable X is stable distributed with stability index $\gamma \in (0,1]$, the discrete analog of tail index α, if the p.g.f. is represented by*

$$m_X(z) = \exp\left(-\lambda(1-z)^{\gamma}\right), \quad |z| \leq 1, \tag{13.113}$$

for some parameter $\lambda > 0$. The probability mass function of a discrete stable loss random variable is given by

$$\mathbb{P}\mathrm{r}\left(X = k\right) = (-1)^k e^{-\lambda} \sum_{m=0}^{k} \sum_{j=0}^{m} \frac{m!(\gamma j)!}{(m-j)!j!(\gamma j - k)!k!} (-1)^j \frac{\lambda^m}{m!}, \quad k = 0, 1, 2, \ldots. \tag{13.114}$$

■

It will also be useful to observe the following recursive evaluation available to probabilities for such a distribution given in Lemma 13.2, see Christoph and Schreiber (1998a, theorem 2).

Lemma 13.2 (Recursions for Discrete Stable Probabilities) *If X is a discrete stable severity loss random variable with discrete stable model parameters $\gamma \in (0,1]$ and λ then the following recursion holds for the probability evaluations*

$$(k+1)\mathbb{P}\mathrm{r}\left(X = k+1\right) = \lambda \sum_{m=0}^{k} \mathbb{P}\mathrm{r}\left(X = k - m\right)(m+1)(-1)^m \frac{\gamma!}{(\gamma - (m+1))!(m+1)!} \tag{13.115}$$

for all $k \in \mathbb{N}^+$ and with $\mathbb{P}\mathrm{r}\left(X = 0\right) = \exp(-\lambda)$.

One can also make the following remarks regarding the discrete stable distribution.

Remark 13.5 *The following properties of the discrete stable severity distribution are known:*

- *One can consider the standard characterization of the α-stable distribution given by Equation 13.112 being modified to obtain the discrete stable analog by replacing kX by the term $k \circ X$ defined by the sum of i.i.d. random variables Z_j satisfying the equality in distribution given by*

$$k \circ X \stackrel{d}{=} \sum_{j=0}^{X} Z_j \tag{13.116}$$

for $k \in (0,1)$ and $\mathbb{P}\mathrm{r}\left(Z_j = 1\right) = 1 - \mathbb{P}\mathrm{r}\left(Z_j = 0\right) = q$, i.e. it admits a mixed Poisson form, where the mixing distribution of the Poisson is known in the literature as a Sibuya distribution;

- *If the discrete Stable random variable has a stability index parameter $\gamma = 1$, then it will correspond to the Poisson distribution with intensity parameter λ;*

- *If the discrete Stable random variable has a stability index parameter $\gamma < 1$ then the resulting random variable has a distribution which is infinitely divisible, discrete self-decomposable, unimodal and normally attracted to a stable law;*
- *Since the discrete Stable distribution is in the domain of attraction of a strictly stable random variable, then it also has the following fractional lower order moment properties*

$$\mathbb{E}[X^r] < \infty \tag{13.117}$$

for $0 \leq r < \gamma < 1$;

- *The simulation of discrete stable random variables are studied in Devroye (1993). In this paper it is shown that since the discrete stable distribution can be written as a compound Poisson distribution, then random variable realizations can be drawn from the model*

$$X = \sum_{i=1}^{Y} Z_i \tag{13.118}$$

with $Y \sim Poisson(\lambda)$ and i.i.d. discrete random variables Z_i given by the Sibuya distribution with p.g.f. given by

$$m_Z(z) = 1 - (1 - z)^{\gamma}. \tag{13.119}$$

- *Hence, one has the discrete stable distributed random variable $X \sim S_{\gamma}(\lambda)$ if it can be represented in law according to a mixed Poisson distribution*

$$X \sim Poisson\left(\lambda^{\frac{1}{\gamma}} S\right) \tag{13.120}$$

with random variable

$$S \sim Sibuya(\gamma, 1). \tag{13.121}$$

A related family of discrete stable distributions are known as the discrete Linnik distributions first studied in Christoph and Schreiber (1998b), see Definition 13.14.

Definition 13.14 (Discrete Stable Laws: Class 2 - Linnik Family) *A discrete severity random variable X has a Linnik Law if it has a p.g.f. given by*

$$m_X(z) = \frac{1}{\left(1 + (1 - z)^{\gamma}\right)^{\beta}}, \tag{13.122}$$

with $\beta > 0$ and $\gamma \in (0, 1]$. ■

Remark 13.6 *It can be shown that the discrete Linnik distributed severity random variable will also admit a mixed Poisson distribution representation given by consideration a Gamma*

$$X \sim Poisson\left(G^{\frac{1}{\gamma}} S\right) \tag{13.123}$$

with random variable

$$S \sim Sibuya(\gamma, 1), \tag{13.124}$$

and independent Gamma random variable

$$G \sim Gamma(\beta, 1). \tag{13.125}$$

One can also demonstrate the following useful asymptotic tail behaviours of the discrete stable distribution (class 1), given in Theorem 13.12, see Christoph and Schreiber (1998a, theorem 3).

Theorem 13.12 (Tail Asymptotic of Discrete Stable Distributions) *Consider a discrete Stable severity loss random variable X with parameters $\gamma \in (0,1)$ and λ. Then the following asymptotics hold as $n \to \infty$*

$$\mathbb{P}\mathrm{r}\,(X = n) = \frac{1}{\pi}\sum_{j=1}^{m}\frac{(-1)^{j+1}}{j!}\lambda^j \sin(\gamma j\pi)B(\gamma j + 1, n - \gamma j) + O\left(n^{-\gamma(m+1)-1}\right), \quad m < n, \tag{13.126}$$

with Beta function $B(x,y) = \frac{\Gamma(x)\Gamma(y)}{\Gamma(x+y)}$. In addition one can show that the right tail probability is given by

$$\mathbb{P}\mathrm{r}\,(X \geq n) = \frac{1}{\pi}\sum_{j=1}^{m}\frac{(-1)^{j+1}}{j!}\lambda^j \sin(\gamma j\pi)B(\gamma j, n - \gamma j) + O\left(n^{-\gamma(m+1)}\right), \quad m < n. \tag{13.127}$$

13.8.3 RECURSIONS FOR CONVOLUTIONS (PARTIAL SUMS) WITH DISCRETISED SEVERITY DISTRIBUTIONS (FIXED *n*)

In the following sections we will discuss and explore recursions for partial sums and compound processes. The three main classes of recursion considered are known as the Panjer, De Pril and the method of Kornya's approximation, see discussions and comparisons in Kuon *et al.* (1987).

Here we consider the recursive evaluation of an n-fold convolution $F_{Z_n}(x)$ in the case in which the severity distribution has been discretised onto an equispaced grid (according to one of the approaches presented previously), w.l.o.g. over integer values $x = 1, 2, 3 \ldots$. In this case it is proven in Sundt and Vernic (2009, theorem 2.8) that a recursion for the evaluation of the discretised density arising from the n-fold convolution distribution $\widehat{F}_{Z_n}(x)$ is achieved according to the result in Theorem 13.13.

Consider a discrete density $\widehat{f} \in \mathcal{P}_{1\underline{0}}$, then the evaluation of the n-fold convolution of such a distribution $\widehat{f}_{Z_n}(x) = \widehat{f}^{(n)*}(x)$ is obtained according to the recursive relationship developed in Theorem 13.13.

Theorem 13.13 (Discretised Severity Distribution *n*-Fold Convolution Recursions) *The n-fold convolution for the density $\widehat{f}_{Z_n}(x) = \widehat{f}^{(n)*}(x)$, with severity distribution satisfying the condition that $\widehat{f}(x) = 0,\ \forall x < 0$, can be evaluated according to the following recursion*

$$\hat{f}_{Z_n}(x) = \frac{1}{\hat{f}(0)} \sum_{y=1}^{x} \left((n+1)\frac{y}{x} - 1 \right) \hat{f}(y) \hat{f}_{Z_n}(x-y), \ \forall x \geq 1, \tag{13.128}$$

with initialization given by

$$\hat{f}_{Z_n}(0) = \hat{f}(0)^n. \tag{13.129}$$

To understand the initialization stage in this recursion, consider the two fold convolution between a distribution $\hat{f} \in \mathcal{P}_{10}$ and itself which is then given by

$$\left(\hat{f} * \hat{f}\right)(z) = \sum_{x=-\infty}^{\infty} \hat{f}(x)\hat{f}(z-x), \ \forall z = 0, 1, 2, \ldots. \tag{13.130}$$

Since one has that $\hat{f}(x) = 0$ for $x < 0$ and $\hat{f}(z-x) = 0$ for $x > z$ due to the membership of this density in the class $\mathcal{P}_{10}$ then one gets the finite sum

$$\left(\hat{f} * \hat{f}\right)(z) = \sum_{x=0}^{\infty} \hat{f}(x)\hat{f}(z-x), \ \forall z = 0, 1, 2, \ldots, \tag{13.131}$$

and in particular at the origin one obtains

$$\left(\hat{f} * \hat{f}\right)(0) = \hat{f}(0)^2. \tag{13.132}$$

When this is extrapolated to the n-fold convolution one obtains the condition, $\hat{f}_{Z_n}(0) = \hat{f}(0)^n$. Next, to understand where the recursive relationship is derived from, we adopt the approach in Sundt and Vernic (2009), where one considers the addition of an auxiliary random variable Y with the density $\hat{f}$ which is independent of the random variable Z_n which has density $\hat{f}^{(n)*}$. Then one can show via an argument of symmetry that the following expression holds for all $z_n \in \{1, 2, \ldots\}$ according to

$$\sum_{y=0}^{z_n} \left((n+1)\frac{y}{z_n} - 1 \right) \hat{f}(y) \hat{f}^{(n)*}(z_n - y) = 0, \tag{13.133}$$

from which one can obtain a recursive expression for the solution for $\hat{f}_{Z_n}(x) = \hat{f}^{(n)*}(x)$ as detailed in Theorem 13.13.

If there is a known upper bound on the severity distribution due to the application of an insurance policy or due to the total liability that one may be exposed to for a given type of OpRisk as discussed in Chapter 17, then in this case one can develop a recursion for the truncated discretised severity distribution given in Theorem 13.14, see details in Sundt and Vernic (2009, theorem 2.9).

Theorem 13.14 (Discretised Truncated Severity *n*-Fold Convolution Recursions) *The n-fold convolution for the density $\hat{f}_{Z_n}(x) = \hat{f}^{(n)*}(x)$, with severity distribution $\hat{f} \in \mathcal{P}_1$ and furthermore there exists a finite k such that $k = \max\left\{x : \hat{f}(x) > 0\right\} < \infty$, then one can evaluate the convolved distribution according to the following recursion*

$$\hat{f}_{Z_n}(x) = \frac{1}{\hat{f}(k)} \sum_{y=1}^{nk-x} \left(\frac{y(n+1)}{nk-x} - 1 \right) \hat{f}(k-y)\hat{f}_{Z_n}(x+y), \;\; \forall x \in \{nk-1, nk-2, \ldots, 0\}, \tag{13.134}$$

with initialization given at the maximum index k by

$$\hat{f}_{Z_n}(k) = \hat{f}(k)^n. \tag{13.135}$$

In cases in which there exist a need to evaluate recursively a sequence of convolutions corresponding to $\left\{\hat{f}^{(j)*}\right\}_{1 \le j \le n}$ for each $j \in \{1, 2 \ldots, n\}$ for a discrete density $\hat{f} \in \mathcal{P}_{1\underline{0}}$. In this case the above convolution identities may be excessively computational compared to a simple recursive evaluation in $j = 1, 2, 3 \ldots, n$ for all $x \in \{0, 1, 2, \ldots\}$ involving,

$$\begin{aligned} \hat{f}^{(j)*}(x) &= \hat{f} * \hat{f}^{(j-1)*}(x) \\ &= \begin{cases} \displaystyle\sum_{y=0}^{x} \hat{f}(y)\hat{f}^{(j-1)*}(x-y), & \forall j \in \{1, 3, 5, 7, \ldots\}, \\ \displaystyle 2 \sum_{y=0}^{(x-1)/2} \hat{f}^{(j/2)*}(y)\hat{f}^{(j/2)*}(x-y) + \left(\hat{f}^{(j/2)*}(x/2)\right)^2 \mathbb{I}_{x \text{ even}}, & \forall j \in \{2, 4, 6, 8, \ldots\}. \end{cases} \end{aligned} \tag{13.136}$$

However, one can improve on the efficiency of evaluation for a sequence of distributions $\left\{\hat{f}^{(j)*}\right\}_{1 \le j \le n}$ if additional information is known about the distributions in the convolution. For instance consider the result in Proposition 13.1, based on Sundt and Vernic (2009, theorem 2.7).

Proposition 13.1 *Consider a density $\hat{f} \in \mathcal{P}_{1\underline{0}}$ that satisfies, for some a and b, the recursive relationship*

$$\hat{f}(x) = \left(a + \frac{b}{n}\right)\hat{f}(x-1), \; \forall x \in \{1, 2, \ldots\}. \tag{13.137}$$

Then the convolution of $\hat{f}$ with itself $\hat{f}^{(2)*}$ also satisfies this recursion in Equation 13.137 with a unchanged and the new $\widetilde{b}$ given by $\widetilde{b} = a + 2b$. As a consequence when extrapolated to the case of evaluation of each distribution in the sequence $\left\{\hat{f}^{(j)*}\right\}_{1 \le j \le n}$ one gets the relationship

$$\hat{f}^{(j)*}(x) = \left(a + \frac{(a+b)j - a}{x}\right)\hat{f}^{(j)*}(x-1), \; \forall j, x \in \{1, 2, \ldots\}. \tag{13.138}$$

Having defined these basic n-fold convolution identities for discretised severity models and the sequential version for sequences of increasing n-fold convolutions, we next present a well known family of transforms that further improve the efficiency of evaluation of convolutions between discretised severity distributions. This leads naturally to the notion of the De Pril Transforms for n-Fold convolutions (partial sums) with discretised severity distributions.

Therefore, in this advanced section we introduce the notion of a De Pril transform for a distribution which can then be utilised to devise an efficient recursive relationship for the evaluation of an n-fold convolution. In De Pril (1986) and Karl-Heinz (1994) developed a range of recursive identities were introduced relating to convolutions for partial sums, after discretization of the severity model. It was then later recognised that these identities were highly efficient methods for evaluation of the distribution of an n-fold convolution comprised of distributions defined over the non-negative integers with a positive probability at the origin, and in Sundt (2005), Sundt (1998) and Dhaene and Vandebroek (1995) one of the most important of the recursions identified in De Pril's early work was named in his honour as the De Pril transform.

In Definition 13.15 we provide the formal representation of the De Pril transform of a discrete probability density on the non-negative integers with a positive probability mass in zero, $\hat{f} \in \mathcal{P}_{1\underline{0}}$.

Definition 13.15 (De Pril Transform of a Density) *The De Pril transform, denoted by φ_f, of a discrete probability density on the non-negative integers with a positive probability mass in zero, $\hat{f} \in \mathcal{P}_{1\underline{0}}$, is given by*

$$\varphi_f(x) = \frac{1}{\hat{f}(0)}\left[x\hat{f}(x) - \sum_{y=1}^{x-1}\varphi_f(y)\hat{f}(x-y)\right], \; \forall x \in \{0, 1, 2, \ldots\}, \tag{13.139}$$

with $\varphi_f(0) = 0$. ■

Remark 13.7 *It can be observed that solving this recursion with respect to $\hat{f}(x)$ produces*

$$\hat{f}(x) = \frac{1}{x}\sum_{y=1}^{x}\varphi_f(y)\hat{f}(x-y), \; \forall x \in \{1, 2, \ldots\}, \tag{13.140}$$

and conversely, one can obtain the expression for the De Pril transform in Equation 13.139 by solving Equation 13.140 with respect to $\varphi_f(x)$. Hence, given φ_f and the initial value $\hat{f}(0)$, then the distribution of $\hat{f}$ can be evaluated recursively. Therefore, one can show using the property of normalization of the probability density (mass) $\hat{f}$,

$$\sum_{x=0}^{\infty}\hat{f}(x) = 1, \tag{13.141}$$

that the De Pril transform φ_f is a unique representation of the distribution $\hat{f}$.

It is also the case that a distribution in the class $\mathcal{P}_{1\underline{0}}$ will be infinitely divisible if and only if the De Pril transform of its density is non-negative.

13.8.3.1 De Pril's First Method. In this section we consider the partial sum given by

$$Z_n = \sum_{i=1}^{n} X_j \tag{13.142}$$

with each X_i independent with distributions $X_i \sim F_i(x)$ and density $f_j(x)$. After discretization one has probability mass functions for each random variable given by $\{\hat{f}_i\}_{i=1}^{n}$. Typically in OpRisk one would consider the case that all X_i loss random variables were i.i.d. If each discretized probability mass function $\hat{f}_j \in \mathcal{P}_{1\underline{0}}$ then one can find the n-fold convolution

$$\hat{f}_{Z_n}(x) = *_{j=0}^{n}\hat{f}_j = \left(\hat{f}_1 * \hat{f}_2 * \cdots * \hat{f}_n\right)(x), \tag{13.143}$$

by using the De Pril transform as displayed below. That is, the real benefit of utilising the De Pril transform in the context of efficiently evaluating the n-fold convolution of a set of n different discretized severity distributions $\hat{f}_j \in \mathcal{P}_{1\underline{0}}$ for all $j \in \{1, 2, \ldots, n\}$, given by $\hat{f}^{(n)*}(x) = *_{j=0}^{n}\hat{f}_j$, is presented by the result in Theorem 13.15, see De Pril (1989). The approach described below for evaluation of the n-fold convolution is known colloquially as De Pril's First Method which involves the steps:

Step 1: For each distribution $\hat{f}_j$ evaluate the De Pril transform $\varphi_{\hat{f}_j}(x)$ in Equation 13.140. **Note:** this step can be simplified by noting that the De Pril transform of the n-fold convolution of a distribution in $\mathcal{P}_{10}$ is n times the De Pril transform of that distribution, see Sundt and Vernic (2009, corollary 6.2).

Step 2: Find the De Pril transform $\varphi_{f^{(n)*}(x)}$ of the convolved distribution $\hat{f}^{(n)*}(x) = *_{j=0}^{n}\hat{f}_j$ by simply summing the n De Pril Transforms

Step 3: Find the evaluation of the n-fold convolved distribution $\hat{f}^{(n)*}(x)$ by using the recursion in Equation 13.140.

Theorem 13.15 (De Pril Transform of an n-Fold Convolution) *The De Pril transform of the convolution of a finite number of discrete densities $\hat{f}_j \in \mathcal{P}_{1\underline{0}}$ for $j \in \{1, 2, \ldots, n\}$ is given by $\hat{f}^{(n)*}(x) = *_{j=1}^{n}\hat{f}_j$ and can be evaluated exactly as the sum of the De Pril transforms of these discrete densities, where*

$$\varphi_{f^{(n)*}}(x) = \sum_{j=1}^{n} \varphi_{\hat{f}_j}(x) = \frac{1}{x}\sum_{j=1}^{n}\sum_{y=1}^{x} \varphi_{\hat{f}_j}(y)\hat{f}_j(x-y), \ \forall x \in \{1, 2, \ldots\}, \tag{13.144}$$

with $\varphi_{\hat{f}_j}(0) = 0$ for all $j \in \{1, 2, \ldots, n\}$.

As a result of this theorem, one observes that the evaluation of the n-fold convolution can be performed exactly through a linear combination of De Pril transforms.

Remark 13.8 *Hence, it is clear that the evaluation of an n-fold convolution of discrete densities in $\mathcal{P}_{1\underline{0}}$ can be performed by first obtaining the De Pril transform of each density and then finding the De Pril transform of the convolution through summation. Then given the De Pril transform of the convolution, one trivially obtains the evaluation of the n-fold density through*

$$\hat{f}^{(n)*}(x) = \frac{1}{x}\sum_{y=1}^{x}\varphi_{f^{(n)*}}(y)\hat{f}^{(n)*}(x-y),\ \forall x \in \{1,2,\ldots\}, \tag{13.145}$$

with initial value given by $\hat{f}^{(n)*}(0) = \prod_{j=1}^{n}\hat{f}_j(0)$

13.8.3.2 De Pril's Second Method.

Consider the case of evaluation of the n-fold convolution given by a set of severity distributions $\hat{f}_j \in \mathcal{P}_{1\underline{0}}$ for all $j \in \{1,2,\ldots,n\}$ according to $\hat{f}^{(n)*}(x) = *_{j=0}^{n}\hat{f}_j$. Under the second method of De Pril the idea is to bypass the utilisation of the recursive evaluation for each distribution given by applying the recursion in Equation 13.139 and instead to utilise a closed form expression for evaluation of the De Pril transform of each density given by $\varphi_{f_j}(x)$.

As described in Sundt and Vernic (2009) this involves the following steps for the De Pril Second Method:

Step 1: For each distribution $\hat{f}_j$ evaluate the De Pril transform $\varphi_{f_j}(x)$ in closed form, avoiding the recursive evaluation.

Find the De Pril transform in closed form by representing each distribution $\hat{f}_j$ in terms of a compound Bernoulli representation. The compound Bernoulli for discretized distribution $\hat{f}_j$ will have a frequency π_j and severity $\hat{f}_j$ component.

- A frequency distribution given by a Bernoulli distribution with probability of success $\pi_j = 1 - \hat{f}_j(0)$;
- A severity distribution $\hat{h}_j \in \mathcal{P}_{11}$ which is given by

$$\hat{h}_j(x) = \frac{\hat{f}_j(x)}{\pi_j}\quad \textit{for}\ \ x = 1,2,\ldots \tag{13.146}$$

Then calculate the De Pril transform via the new representation

$$\varphi_{f_j}(x) = -x\sum_{n=1}^{x}\frac{1}{n}\left(\frac{\pi_j}{\pi_j - 1}\right)^n \hat{h}_j^{(n)*}(x),\ \ x = 1,2,\ldots. \tag{13.147}$$

Step 2: Find the De Pril transform $\varphi_{f^{(n)*}(x)}$ of the convolved distribution $\hat{f}^{(n)*}(x) = *_{j=0}^{n}\hat{f}_j$ by simply summing the n De Pril Transforms.

Step 3: Evaluate the density of the n-fold convolution using the result of Sundt and Vernic (2009) to obtain the density given in closed form by

$$\hat{f}^{(n)*}(x) = -\frac{1}{x}\sum_{n=1}^{x}\hat{f}^{(n)*}(x-n)\sum_{j=1}^{M}\left(\frac{\pi_j}{\pi_j - 1}\right)^n,\ \ x = 1,2,\ldots. \tag{13.148}$$

There are also many approximation based techniques for evaluation of the n-fold convolution and a detailed account of these can be found in Sundt and Vernic (2009, chapter 7).

Next we explore some special sub-families of distributions that have been discussed in previous chapters for their particular useful properties when utilised within an LDA model structure, these are the infinitely divisible distributional families. We discuss what special features the De Pril transform will have when applied to convolutions involving this family of distributions.

13.8.3.3 De Pril Transforms and Convolutions of Infinitely Divisible Distributions. The De Pril transform is especially relevant in the context of infinitely divisible distributions, see details in Sundt and Vernic (2009, chapter 4, theorem 4.1, theorem 4.2, corollary 4.1 and theorem 4.5). There computationally efficient properties that can be developed for evaluation of the De Pril transform of such classes of distributions, as will be explored in the next few results.

Based on the properties of the family of infinitely divisible distributions one can then study the De Pril transform for such a class of distributions, including convolutions of such distributions. The first identity of interest is to recall the properties of compound Poisson distributions discussed in Peters and Shevchenko (2015) in terms of convolutions of such distributions, see Cont and Tankov (2004). This result can clearly be seen as an efficient recursive procedure for the evaluation of aggregation of a number of independent single loss LDA risk processes in financial hierarchical banking structure.

Proposition 13.2 (Convolutions of Compound Poisson Distributions: Multiple LDA Risks) *Consider m independent compound processes, given by annual loss random variables $Z^{(j)} = \sum_{n=0}^{N^{(j)}} X_n^{(j)} \sim F_{Z^{(j)}}$, each representing a single risk process and given by severity distributions $X^{(j)} \sim f_j$ and a Poisson frequency distribution $N^{(j)} \sim Poisson\left(\lambda_j\right)$ for all $j \in \{1, 2, \ldots, m\}$. If one considers the convolution of each of these compound Processes given by*

$$Z_T = \sum_{j=1}^{m} Z^{(j)} \sim F_{Z_T} = *_{j=1}^{m} F_{Z^{(j)}}, \tag{13.149}$$

where $F^{(m)*}$ is a compound Poisson distribution with rate parameter in the frequency distribution λ_T and severity distribution $f_T(x)$ given by

$$\lambda_T = \sum_{j=1}^{m} \lambda_j, \text{ and } f_T(x) = \frac{1}{\lambda_T} \sum_{j=1}^{m} \lambda_j f_j(x). \tag{13.150}$$

Now utilising the properties of infinitely divisible distributions and their unique representation as compound Poisson distributions detailed in point 9 of Theorem 13.5 which states that an infinitely divisible distribution in the class $\mathcal{P}_{1\underline{0}}$ can always be expressed as a compound Poisson distribution with severity distribution in the class $\mathcal{P}_{11}$. Hence, if one considers the m-fold convolution of infinitely divisible discrete distributions $\hat{f}_j \in \mathcal{P}_{1\underline{0}}$ for all $j \in \{1, 2, \ldots, m\}$ then one can utilise this representation for each distribution and re-express the solution as the m-fold convolution of compound Poisson distributions and utilise the results in Proposition 13.2. Alternatively, there is a different approach one may adopt in addressing this problem utilising the De Pril transform, given in Proposition 13.3.

Proposition 13.3 (De Pril Transform and Infinitely Divisible Severity Distributions) *Consider the m-fold convolution of discrete infinitely divisible distributions $\hat{f}_j \in \mathcal{P}_{1\underline{0}}$ for all $j \in \{1, 2, \ldots, m\}$. One could evaluate the compound Poisson distribution representation of each distribution*

$$\hat{f}_j(x) = \sum_{n=0}^{\infty} \frac{\lambda_j^n}{n!} \exp(-\lambda_j)\hat{g}_j(x) \tag{13.151}$$

which would involve determining for the Poisson frequency distribution, the λ_j and the severity distribution $\hat{g}_j(x)$. Then application of Proposition 13.2 and Theorem 13.13 would produce a solution involving evaluation first of the compound Poisson representation of each infinitely divisible distribution and then the m-fold infinitely divisible distribution with total intensity λ_T and severity $\hat{f}_T(x)$ according to

$$\begin{aligned} \lambda_T &= \sum_{j=1}^{m} \lambda_j = \sum_{j=1}^{m} - \ln g_j(0), \\ \hat{f}_T(x) &= \frac{1}{\lambda_T} \sum_{j=1}^{m} \lambda_j \hat{f}_j(x) \\ &= \frac{1}{\lambda_T} \sum_{j=1}^{m} \lambda_j \left[\frac{1}{\hat{g}_j(0)} \left(\frac{\hat{g}_j(x)}{\lambda_j} - \frac{1}{x} \sum_{y=1}^{x-1} y\hat{f}_j(y)\hat{g}_j(x-y) \right) \right], \ \forall x \in \{1, 2, \ldots\}. \end{aligned} \tag{13.152}$$

However, this computation can be performed in an alternative manner via the De Pril transform which avoids the need to evaluate each λ_j and more importantly the recursions for evaluation of $\hat{f}_j(x)$. Instead knowledge of the De Pril transform of each distribution f_j is utilised to evaluate the De Pril transform of the m-fold convolution $\hat{f}^{(m)*}(x) = *_{j=1}^{m}\hat{f}_j(x)$ directly as follows

$$\varphi_{f^{(m)*}} = \sum_{j=1}^{m} \varphi_{f_j} = \sum_{j=1}^{m} \frac{1}{\hat{f}_j(0)} \left(x\hat{f}_j(x) - \sum_{y=1}^{x-1} \varphi_{f_j}(y)\hat{f}_j(x-y) \right), \ \forall x \in \{1, 2, \ldots\}, \tag{13.153}$$

and the resulting distribution is then the solution

$$\hat{f}^{(m)*}(x) = \frac{1}{x} \sum_{y=1}^{x} \varphi_{f^{(m)*}(x)}(y)\hat{f}^{(m)*}(x-y), \ \forall x \in \{1, 2, \ldots\}. \tag{13.154}$$

EXAMPLE 13.13 De Pril Transform of a Poisson Distribution

Consider the Poisson distribution $\hat{f}$ with parameter λ, then the De Pril transform is given trivially by considering the Poisson distribution as a compound Poisson distribution with a severity $\hat{g}$ concentrated on unity, resulting in the expression

$$\varphi_f(y) = \lambda \mathbb{I}[y = 1]. \tag{13.155}$$

■

It is also often convenient in OpRisk LDA models to note the result in Theorem 13.16 which relates a mixed Poisson distribution to a compound distribution, which can be easily evaluated using a De Pril transform.

Theorem 13.16 (Mixed Poisson Distributions) *A mixed Poisson distribution that has a mixing distribution in the class $\mathcal{P}_{10}$ can be re-expressed according to a compound distribution with the mixing distribution making up the frequency distribution and a severity distribution given by a Poisson distribution with unit rate.*

There are several general extensions to such results provided in the actuarial literature, see for example details of the Wilmot class of mixing distributions in Sundt and Vernic (2009, chapter 3).

13.8.4 ALTERNATIVES TO PANJER RECURSIONS: RECURSIONS FOR COMPOUND DISTRIBUTIONS WITH DISCRETISED SEVERITY DISTRIBUTIONS

In some settings it may be advantageous both from a computational efficiency as well as numerical accuracy or stability to consider alternative recursions or higher order recursions for compound processes. There are numerous other recursions available for compound distribution evaluation, some of the more useful variants are given below under assumptions on either the frequency, the severity distribution or both.

An example of this is the recursion for a compound Poisson distribution with severity distribution satisfying that $f(0) = 0$, then one obtains a recursion given in Theorem 13.17.

Theorem 13.17 (Recursions for Compound Poisson Distributions with Discretised Severity) *Consider a compound Poisson distribution for a single risk LDA model in which the severity distribution $f(x)$ is discretised (w.l.o.g.) over the non-negative integers and satisfies $f(0) = 0$ then the compound distribution given by*

$$f_{Z_N}(x) = \sum_{n=1}^{\infty} \frac{\lambda^n}{n!} \exp(-\lambda) f^{(n)*}(x), \tag{13.156}$$

is evaluated recursively according to

$$f_{Z_N}(x) = \frac{\lambda}{x} \sum_{y=1}^{x} y f(y) f_{Z_N}(x-y), \forall x \in \{1, 2, \ldots\}, \tag{13.157}$$

with initialization

$$f_{Z_N}(0) = \exp(-\lambda). \tag{13.158}$$

Additionally, in Sundt and Vernic (2009, section 3.3) an alternative class of recursions for mixed Poisson compound distributions can be utilised if certain conditions are satisfied for the discretised severity distribution. If the severity distribution $f(x)$ after discretisation satisfies that the first derivative of the power series representation of the probability mass function satisfies the relation

$$\frac{d}{ds}\mathbb{E}\left[s^X\right] = \int_0^\infty s^x dF(x) = \frac{\sum_{y=1}^r \eta(y)s^{y-1}}{1-\sum_{y=1}^r \chi(y)s^y} \tag{13.159}$$

for some functions η and χ and r either a positive integer of infinity then the recursion in Theorem 13.18 for the compound Poisson distribution evaluation holds.

Theorem 13.18 *If the frequency distribution is a Poisson distribution with rate λ and the severity distribution f is discretised and takes values with positive probability only on the non-negative integers (w.l.o.g.). Furthermore, assume that there exists function η and χ on $\{1, 2, \ldots, r\}$ and an integer r either a positive integer of infinity that satisfies the that the first derivative of the power series representation of the probability mass function satisfies the relation*

$$\frac{d}{ds}\mathbb{E}\left[s^X\right] = \int_0^\infty s^x dF(x) = \frac{\sum_{y=1}^r \eta(y)s^{y-1}}{1-\sum_{y=1}^r \chi(y)s^y}, \tag{13.160}$$

then the compound distribution is evaluated according to the recursion

$$f_{Z_N}(x) = \sum_{y=1}^{r}\left(\frac{\lambda}{x}\eta(x) + \left(1-\frac{y}{x}\right)\chi(y)\right)f_{Z_N}(x-y), \quad \forall x \in \{1, 2, \ldots\}. \tag{13.161}$$

In addition, in cases in which the frequency distribution satisfies the recursive evaluation given by

$$p_n = \left(a + \frac{b}{n}\right)p_{n-1}, \quad \forall n \in \{l, l+1, \ldots, r\}. \tag{13.162}$$

then in this case the compound distribution can be evaluated recursively according to

$$f_{Z_N}(x) = p_l f^{(l)*}(x) - \left(a + \frac{b}{r+1}\right)p_r f^{(r+1)*}(x) + \sum_{y=1}^{x}\left(a + b\frac{y}{x}\right)f(y)f_{Z_N}(x-y),$$
$$\forall x \in \{l, l+1, \ldots\}. \tag{13.163}$$

To complete this section of recursions for evaluation of the compound process we also mention the framework developed and known as Waldmann's recursion Waldmann (1996, theorem 1) which provides an alternative recursion one can consider. The advantage of the Waldmann recursion is that instead of working with the density of the compound process it considers the distribution function which has the advantage that it is strictly monotone. The recursion proceeds according to Theorem 13.19.

Theorem 13.19 (Waldmann's Recursion for Compound Distributions with Discretised Severity) *Consider the compound process with discretised severity distribution given by $F_{Z_N}(x)$ for all $x \in \mathbb{N}_0$. If the frequency distribution satisfies the condition that*

$$p_n = \left(a + \frac{b}{n}\right)p_{n-1}, \quad \forall n \in \{1, 2, 3 \ldots\}. \tag{13.164}$$

Then the distribution of the compound process can be evaluated recursively according to

$$xF_{Z_N}(x) = r_1(x) + r_2(x), \tag{13.165}$$

where $F_{Z_N}(0) = p_0$ *and for all* $x \in \mathbb{N}_0$ *one has*

$$\begin{aligned} r_1(x) &= r_1(x-1) + F_{Z_N}(x-1), \\ r_2(x) &= a\sum_{i=1}^{x-1} f(i) r_2(x-i) + (a+b)\sum_{i=1}^{x} i f(i) F_{Z_N}(x-i), \end{aligned} \tag{13.166}$$

with $r_1(0) = 0$.

There are also approaches to stabilize and safe guard this recursive algorithm against underflows and overflows, see Waldmann (1996, section 3) for details.

13.8.5 HIGHER ORDER RECURSIONS FOR DISCRETISED SEVERITY DISTRIBUTIONS IN COMPOUND LDA MODELS

The Panjer recursion was introduced for the evaluation of the compound process distribution recursively when the frequency distribution had probabilities satisfying the recursive relationship given by

$$p_n = \left(a + \frac{b}{n}\right) p_{n-1}, \tag{13.167}$$

for some distribution $p \in \mathcal{P}_{10}$. In this section recursions for compound distributions in which the counting distribution satisfies a generalized higher order recursion given by the most general representation

$$p_n = \sum_{i=1}^{k} \left(a_i + \sum_{j=0}^{l \vee i-1} \frac{b_{i,j}}{n-j} \right) p_{n-i}, \quad \forall n \in \{l+1, l+2, \ldots\} \tag{13.168}$$

are considered, where the order is denoted by the "lag" k. The results for the recursive evaluation of a compound distribution of an annual loss random variable $Z = \sum_{n=1}^{N} X_n$ with severity density $f \in \mathcal{P}_{10}$ and frequency distribution $p \in \mathcal{P}_{10}$ that also satisfies the higher order recursion in Equation 13.168 are given in Theorem 13.20 as derived in Sundt and Vernic (2009, chapter 5.1). Note, this results is the most general formulation that also incorporates cases in which $j = 0$ for which the recursion for the probabilities in the frequency distribution satisfy the simplified higher order recursion given by

$$p_n = \sum_{i=1}^{k} \left(a_i + \frac{b_i}{n} \right) p_{n-i}. \tag{13.169}$$

Theorem 13.20 *Consider the compound process LDA model with discretised severity distribution* $f \in \mathcal{P}_{10}$ *and frequency distribution* $p \in \mathcal{P}_{10}$ *that satisfies the recursive evaluation of the probability of* $\mathbb{P}\mathrm{r}[N = n] = p_n$ *given as a linear combination of* k *terms corresponding to the probabilities* $\{\mathbb{P}\mathrm{r}[N = n-1], \ldots, \mathbb{P}\mathrm{r}[N = n-k]\}$ *according to,*

$$p_n = \sum_{i=1}^{k} \left(a_i + \sum_{j=0}^{l \vee i-1} \frac{b_{i,j}}{n-j} \right) p_{n-i}, \quad \forall n \in \{l+1, l+2, \ldots\} \tag{13.170}$$

Then the compound process LDA annual loss distribution $f_Z(x)$ is recursively evaluated according to

$$\begin{aligned} f_Z(x) = \frac{1}{1-\tau_a(f(0))} &\left(\sum_{n=1}^{l} \left(p_n - \sum_{i=1}^{k} a_i p_{n-i} \right) f^{(n)*}(x) - \sum_{i=1}^{k} \sum_{j=0}^{l \vee i-1} \sum_{n=j+1}^{l} \frac{b_{i,j}}{n-j} p_{n-i} f^{(n)*}(x) \right. \\ &+ \sum_{i=1}^{k} \sum_{y=1}^{x} a_i f^{(i)*}(y) f_Z(x-y) \\ &\left. + \sum_{i=1}^{k} \sum_{j=0}^{l \vee i-1} b_{i,j} \left(c_{i-j} f^{(j)*}(x) + \frac{1}{i-j} \sum_{y=0}^{x-1} f^{(j)*}(y) \sum_{z=1}^{x-y} \frac{z}{x-y} f^{(i-j)*}(z) f_Z(x-y-z) \right) \right) \end{aligned}$$

for all $x \in \{1, 2, \ldots\}$, $\tau_a(f(0))$ denoting the probability generating function and coefficients

$$c_i = \sum_{n=1}^{\infty} \frac{1}{n} p_{n-i} f(0)^n, \quad \forall i \in \{1, 2, \ldots\} \tag{13.171}$$

Clearly the evaluation of the coefficients c_i proves an intractable quantity computationally due to the infinite summation. However, as noted in Sundt and Vernic (2009), if one further assumes that the severity distribution f is in the class $\mathcal{P}_{11}$ then the simplified result in Corollary 13.2 applies.

Corollary 13.2 *Consider the compound process LDA model with discretised severity distribution $f \in \mathcal{P}_{11}$ and frequency distribution $p \in \mathcal{P}_{10}$ that satisfies the recursive evaluation of the probability of $\mathbb{P}\mathrm{r}[N=n] = p_n$ given as a linear combination of k terms corresponding to the probabilities $\{\mathbb{P}\mathrm{r}[N=n-1], \ldots, \mathbb{P}\mathrm{r}[N=n-k]\}$ according to,*

$$p_n = \sum_{i=1}^{k} \left(a_i + \sum_{j=0}^{l \vee i-1} \frac{b_{i,j}}{n-j} \right) p_{n-i}, \quad \forall n \in \{l+1, l+2, \ldots\} \tag{13.172}$$

Then the compound process LDA annual loss distribution $f_Z(x)$ is recursively evaluated according to

$$\begin{aligned} f_Z(x) = &\sum_{n=1}^{l} \left(p_n - \sum_{i=1}^{k} a_i p_{n-i} \right) f^{(n)*}(x) - \sum_{i=1}^{k} \sum_{j=0}^{l \vee i-1} \sum_{n=j+1}^{l} \frac{b_{i,j}}{n-j} p_{n-i} f^{(n)*}(x) \\ &+ \sum_{i=1}^{k} \sum_{y=1}^{x} a_i f^{(i)*}(y) f_Z(x-y) \\ &+ \sum_{i=1}^{k} \sum_{j=0}^{l \vee i-1} \frac{b_{i,j}}{i-j} \sum_{y=j}^{x-1} f^{(j)*}(y) \sum_{z=i-j}^{x-y} \frac{z}{x-y} f^{(i-j)*}(z) f_Z(x-y-z) \end{aligned} \tag{13.173}$$

for all $x \in \{1, 2, \ldots\}$.

13.8.6 RECURSIONS FOR DISCRETISED SEVERITY DISTRIBUTIONS IN COMPOUND MIXED POISSON LDA MODELS

In this section we consider generalizing the class of allowable frequency distributions to be mixed types. In particular we briefly detail recursions for the evaluation of compound mixed Poisson distributions, the Wilmot class of mixing distributions and a simplified recursion. Then a recent generalization of the Panjer recursion due to Gerhold *et al.* (2010) will be considered which can be shown to improve numerical stability of the Panjer recursion and extend the application of the recursion to the class mixed type compound distributions.

In this section we consider the class of compound process distributions in which the frequency distribution is of a mixed type, generically represented according to Definition 13.16. Note, these distributions are also known as doubly stochastic processes and Cox processes and have been utilised in numerous applications in the risk and insurance literature, see examples in OpRisk in Peters *et al.* (2011) and in the Bayesian context in Peters *et al.* (2009).

Definition 13.16 (Mixed Poisson Type Frequency Distributions) *Consider Λ as a positive random variable with distribution U. Then define the frequency distribution for the number of losses N annually in the single risk LDA model to be defined conditionally as follows*

$$\begin{aligned}\mathbb{P}\mathrm{r}\left[N=n|\Lambda=\lambda\right]&=\frac{\lambda^n}{n!}\exp\left(-\lambda\right),\quad\text{and,}\\ \mathbb{P}\mathrm{r}\left[N=n\right]&=\int_0^\infty\frac{\lambda^n}{n!}\exp\left(-\lambda\right)dU(\lambda)\ \ \forall n\in\{0,1,2,\ldots\}.\end{aligned}\tag{13.174}$$

■

We note that several examples will be provided for such frequency distributions in the chapters on closed form LDA models and insurance models in Chapter 17. It is also worth noting that the convolution between a finite number of mixed Poisson distributions is equivalent to considering a mixed Poisson distribution in which the mixing distribution is obtained by the convolution between the mixing distributions of these distributions.

Next we detail how to perform recursive evaluation of the mixed Poisson compound distribution with severity distribution f and mixing distribution generically denoted by U given by

$$\begin{aligned}f_{Z_N}(x)&=\sum_{n=0}^{\infty}\mathbb{P}\mathrm{r}\left[N=n\right]f^{(n)*}(x)\\ &=\sum_{n=0}^{\infty}\int_0^\infty\frac{\lambda^n}{n!}\exp\left(-\lambda\right)dU(\lambda)f^{(n)*}(x).\end{aligned}\tag{13.175}$$

If one considers discretising the severity distribution $f(x)$ to take support (w.l.o.g.) on the non-negative integers, then a recursive evaluation of the mixed Poisson compound process, due to Sundt and Jewell (1981, section 3.3) proceeds to evaluate $f_{Z_N}(x)$ for all $x\in\{0,1,2\ldots\}$ as detailed in Proposition 13.4.

To understand this recursion it is beneficial to first consider the basic recursion that one obtains when considering specifically the evaluation of a Poisson compound distribution in a single risk LDA model. As detailed in Sundt and Jewell (1981, theorem 2.2) one can efficiently evaluate the compound Poisson distribution according to Theorem 13.17 presented previously.

Given this recursion in the context of the mixed Poisson compound distributions this recursion can be turned into an alternative recursion with respect to moments of the Laplace transform of the mixing distribution. Consider the recursion

$$f_{Z_N}(x) = \frac{\lambda}{x}\sum_{y=1}^{x} yf(y)f_{Z_N}(x-y), \forall x \in \{1,2,\ldots\}, \tag{13.176}$$

and then multiply this by a power of the mixing random variable and its distribution, $\lambda^i dU(\lambda)$ to obtain in the notation of Sundt and Jewell (1981) the recursion for the i-th integer moment

$$\begin{aligned}\nu_i(x) &= \int_0^\infty \lambda^i f_{Z_N}(x;\lambda)dU(\lambda)\\ &= \frac{1}{x}\sum_{y=1}^{x} yf(y)\nu_{i+1}(x-y), \forall x \in \{1,2,\ldots\} \text{ and } \forall i \in \{0,1,2,\ldots\}.\end{aligned} \tag{13.177}$$

This second recursion for the values of $\nu_i(x)$ can then be utilised to evaluate mixed Poisson Compound distribution as detailed in Proposition 13.4.

Proposition 13.4 (Recursive Evaluation of Mixed Poisson Compound Distributions) *Consider a mixed Poisson compound distribution for a single risk LDA model, with discretised severity density given by $f(x)$ for $x \in \{0,1,2,\ldots\}$ and mixed frequency distribution in Definition 13.16. The evaluation of the compound process annual loss distribution at a point x according to*

$$f_{Z_N}(x) = \sum_{n=0}^{\infty}\int_0^\infty \frac{\lambda^n}{n!}\exp(-\lambda)\, dU(\lambda)f^{(n)*}(x), \tag{13.178}$$

proceeds by evaluation of the initialization $f_{Z_N}(0)$ according to the Laplace transform of the mixing distribution U evaluated at $1-f(0)$ via

$$\nu_0(0) = f_{Z_N}(0) = \mathcal{L}[U(1-f(0))] = \int_0^\infty \exp(-\lambda(1-f(0)))\, dU(\lambda). \tag{13.179}$$

Then for all values of $y \in \{1,2,\ldots,x\}$ evaluate y-th derivative of the Laplace transform of the mixing distribution at $1-f(0)$ according to

$$\begin{aligned}\nu_y(0) &= (-1)^y = \mathcal{L}[\lambda^y U(1-f(0))] = \int_0^\infty \lambda^y \exp(-\lambda(1-f(0)))\, dU(\lambda)\\ &= (-1)^y \frac{d^y}{ds^y}\mathcal{L}[U(s)]|_{s=(1-f(0))}\end{aligned} \tag{13.180}$$

Then for all $y \in \{1,2,\ldots,x\}$ and all $z \in \{1,2,\ldots,y\}$ evaluate $\nu_{y-z}(z)$ via the recursion

$$\begin{aligned}\nu_i(x) &= \int_0^\infty \lambda^i f_{Z_N}(x;\lambda)dU(\lambda)\\ &= \frac{1}{x}\sum_{y=1}^{x} yf(y)\nu_{i+1}(x-y), \forall x \in \{1,2,\ldots\} \text{ and } \forall i \in \{0,1,2,\ldots\}.\end{aligned} \tag{13.181}$$

Finally, the value of the mixed Poisson compound distribution of $f_{Z_N}(y) = \nu_0(y)$.

Remark 13.9 *The computational efficiency and storage requirements for the recursive evaluation presented in Proposition 13.4 will become untenable as the value of* $x \to \infty$. *Unfortunately, this is precisely the situation in which one wishes to consider the utilization of this recursion in many applications related to estimation of the tail of the compound distribution for a single risk LDA model.*

Corollary 13.3 *If the severity distribution f is discretised and takes values with positive probability only on the non-negative integers (w.l.o.g.). Furthermore, assume that there exists function* η *and* χ *on* $\{1, 2, \ldots, r\}$ *and an integer r either a positive integer of infinity that satisfies the that the first derivative of the power series representation of the probability mass function satisfies the relation*

$$\frac{d}{ds}\mathbb{E}\left[s^X\right] = \int_0^\infty s^x dF(x) = \frac{\sum_{y=1}^r \eta(y)s^{y-1}}{1 - \sum_{y=1}^r \chi(y)s^y}, \tag{13.182}$$

then in the recursive evaluation of the mixed compound process, the stage that consider for all $y \in \{1, 2, \ldots, x\}$ *and all* $z \in \{1, 2, \ldots, y\}$ *evaluate* $\nu_{y-z}(z)$ *the recursive evaluation of*

$$\begin{aligned}\nu_i(x) &= \int_0^\infty \lambda^i f_{Z_N}(x; \lambda) dU(\lambda) \\ &= \frac{1}{x}\sum_{y=1}^x y f(y)\nu_{i+1}(x-y), \forall x \in \{1, 2, \ldots\} \text{ and } \forall i \in \{0, 1, 2, \ldots\}\end{aligned} \tag{13.183}$$

can be replaced with an alternative recursion involving

$$\begin{aligned}\nu_i(x) = \sum_{y=1}^r \left(\frac{\eta(y)}{x}\nu_{i+1}(x-y) + \left(1 - \frac{y}{x}\right)\chi(y)\nu_i(x-y)\right), \\ \forall x \in \{1, 2, \ldots\} \text{ and } \forall i \in \{0, 1, 2, \ldots\}.\end{aligned} \tag{13.184}$$

It is well known that one can improve the efficiency of the recursion presented in Proposition 13.4 if additional restrictions on the mixing distribution are satisfied. The class of Wilmot mixing distributions correspond to an important set of such mixing distributions that allow for improved computational efficiency of the evaluation of mixed compound Poisson distributions, as defined in Definition 13.17, see numerous properties and details of this class in Sundt and Vernic (2009, section 3.7).

Definition 13.17 (Wilmot Class of Mixing Distributions) *Consider a continuous mixing distribution U defined according to Definition 13.16. Furthermore, assume it takes a finite support on the interval* $[\lambda_{\min}, \lambda_{\max}]$ *with* $\lambda_{\min} \geq 0$ *and* $\lambda_{\max} \leq \infty$. *A mixing distribution* $U(\lambda)$ *on this support that admits a density* $u(\lambda)$ *that satisfies the condition that on the log scale its derivative can be represented according to*

$$\frac{d}{d\lambda}\ln u(\lambda) = \frac{\sum_{i=0}^k \eta(i)\lambda^i}{\sum_{i=0}^k \chi(i)\lambda^i}, \forall \lambda \in [\lambda_{\min}, \lambda_{\max}], \tag{13.185}$$

belongs to the Wilmot Class of mixing distributions for some functions η *and* χ. ■

If one has a mixed type frequency distribution model satisfying membership of the Wilmot class, then efficient recursions for the evaluation of the compound process with discretised severity proceed as detailed in Theorem 13.21, see derivations in Sundt and Vernic (2009, theorem 3.2).

Theorem 13.21 (Mixed Frequency Compound Recursion: Wilmot Class) *Consider a Poisson mixed frequency distribution in which the random intensity Λ is a positive random variable with distribution U. Then define the frequency distribution for the number of losses N annually in the single risk LDA model to be defined conditionally as follows*

$$\begin{aligned} \mathbb{P}\mathrm{r}\left[N = n|\Lambda = \lambda\right] &= \frac{\lambda^n}{n!}\exp\left(-\lambda\right), \quad \text{and,} \\ \mathbb{P}\mathrm{r}\left[N = n\right] &= \int_0^{\infty} \frac{\lambda^n}{n!}\exp\left(-\lambda\right)dU(\lambda) \;\; \forall n \in \{0,1,2,\ldots\}. \end{aligned} \tag{13.186}$$

Furthermore assume the mixing distribution U is in the Wilmot class and satisfies that it takes support on the interval $[\lambda_{min}, \lambda_{max}]$ with $0 \leq \lambda_{min} < \lambda_{max} \leq \infty$ and admits a differentiable density u satisfying the condition for Wilmot class membership

$$\frac{d}{d\lambda}\ln u(\lambda) = \frac{\sum_{i=0}^{k}\eta(i)\lambda^i}{\sum_{i=0}^{k}\chi(i)\lambda^i}, \forall \lambda \in [\lambda_{\min}, \lambda_{\max}]. \tag{13.187}$$

Then one has the following recursion for the annual loss distribution f_Z given the discretised severity distribution f_X over the non-negative integers $x \in \{1, 2, \ldots\}$ according to

$$\rho(k)\nu_k(x) = \sum_{y=1}^{x} f_X(y)\sum_{i=0}^{k}\chi(i)\nu_i(x-y) - \sum_{i=0}^{k-1}\rho(i)\nu_i(x) + w_{\lambda_{min}}(x) - w_{\lambda_{max}}(x) \tag{13.188}$$

with $\chi(-1) = \nu(-1) = \chi(k+1) = 0$ and

$$\begin{aligned} \rho(i) &= (1 - f_X(0))\,\chi(i) - \nu(i) - (i+1)\chi(i+1), \quad i = -1, 0, 1, \ldots, k \\ w_\lambda(x) &= f_Z(x;\lambda)u(\lambda)\sum_{i=0}^{k}\chi(i)\lambda^i, \quad x = 0,1,2,\ldots; \lambda \in (\lambda_{min}, \lambda_{max}) \\ \nu_i(x) &:= \int \lambda^i f_Z(x;\lambda)dU(\lambda) = \frac{1}{x}\sum_{y=1}^{x} y f_X(y)\nu_{i+1}(x-y). \end{aligned} \tag{13.189}$$

Next we consider the cases of recursions for compound distributions in which the severity distribution is not assumed to be discretised.

13.8.7 CONTINUOUS VERSIONS OF THE PANJER RECURSION

Although, in some settings, the discretisation of a continuous severity distribution might be justifiable, this is not the preferred approach in most OpRisk models. However, the Panjer recursion approach may also be applied in a continuous setting, leading to the recursion for the density given by:

$$f_{Z_N}(x) = p_1 f_X(x) + \int_0^x \left(a + \frac{b\tau}{x}\right) f_X(\tau) f_{Z_N}(x-\tau)d\tau \tag{13.190}$$

where $f_X(x)$ is the severity density and a, b and $p_1 = \mathbb{P}\text{r}[N = 1]$ parameterize the frequency distribution of the compound process. There are many approaches to evaluate this expression. We focus primarily on Importance Sampling based approaches in this chapter. For alternative numerical techniques to a Panjer recursion see works such as inversion transforms (fast Fourier transforms) and series expansions (Bergstrom, 1953); see Shevchenko (2011) and Cruz (2002) and references therein.

13.8.7.1 The Panjer Recursion via Volterra Integral Equations of the Second Kind. One can now observe that the Panjer recursion specified in Equation 13.190 can be recognized as a Volterra equation of the second kind, see discussions in Peters *et al.* (2007) and references therein. In general the Volterra integral equation of the second kind takes the form:

$$f(x) = g(x) + \int_0^x K\left(x, x_1, f\left(x_1\right)\right) dx_1. \tag{13.191}$$

Therefore one can observe that for the Panjer recursion if one selects a linear Volterra equation in which,

$$K\left(x, x_1, f\left(x_1\right)\right) = k\left(x, x_1\right) f\left(x_1\right), \tag{13.192}$$

this will produce

$$f(x) = g(x) + \int_0^x k\left(x, x_1\right) f\left(x_1\right) dx_1, \tag{13.193}$$

allowing one to make an association directly between the linear Volterra equation of the second kind and the Panjer recursion. This involves making the following identifications:

$$\begin{aligned} x_1 &= x - \tau, \\ g(x) &= p_1 f_X(x), \\ k\left(x, x_1\right) &= \left(a + b\frac{x - x_1}{x}\right) f_X\left(x - x_1\right), \\ f\left(x_1\right) &= f_{Z_N}\left(x_1\right). \end{aligned} \tag{13.194}$$

Therefore, one can obtain the following recursive representation for this integral equation, first working with the representation from the Volterra integral equation,

$$\begin{aligned} f(x) &= g(x) + \int_0^x k\left(x, x_1\right) f\left(x_1\right) dx_1 \\ &= g(x) + \int_0^x k\left(x, x_1\right) \left[g(x) + \int_0^{x_1} k\left(x_1, x_2\right) f\left(x_2\right) dx_2\right] dx_1, \end{aligned} \tag{13.195}$$

where $g : [0, x] \mapsto \mathbb{R}$ and $k : [0, x] \times [0, x] \mapsto \mathbb{R}$ are known functions and $f : [0, x] \mapsto \mathbb{R}$ is unknown. Now if one recognises that this equation is also expressed as according to the expression

$$f(x) = g(x) + \int_0^x r\left(x, x_1\right) g\left(x_1\right) dx_1, \tag{13.196}$$

where r is the resolvent kernel for the Volterra equation of the second kind which can therefore be expressed according to a von Neumann series expansion, see Baker (2000). This produces

$$r(x, x_1) = \sum_{n=1}^{\infty} k^n(x, x_1) \tag{13.197}$$

where one defines $k^0(x, x_1) = 1$ $k^1(x, x_1) = k(x, x_1)$ and

$$k^n(x, x_1) = \int_0^x k(x, u) k^{n-1}(u, x_1)\, du, \quad n = 2, 3, 4, \ldots, \tag{13.198}$$

such that the von Neumann series expansion exists under the condition that

$$\sum_{n=0}^{\infty} \int_0^x |k^n(x_0, x_n|\, dx_n < \infty. \tag{13.199}$$

Therefore, if one applies this series expansion to Equation 13.193 one obtains for any point $x_0 \in \mathbb{R}^+$,

$$f(x_0) = g(x_0) + \sum_{n=1}^{\infty} \int_0^{x_0} \cdots \int_0^{x_{n-1}} g(x_n) \prod_{l=1}^{n} k(x_{l-1}, x_l)\, dx_{1:n} \tag{13.200}$$

with notation $x_{1:n}$ denoting an n-tuple $(x_1, x_2, \ldots, x_n)$. Now as in Peters *et al.* (2007) we define the domains of integration according to the following notations where $D_k(x_{k-1}) = [0, x_{k-1}]$ is the conditional one-dimensional domains of integration and the domain of integration of the n-th term in the summation as

$$D_{1:n}(x_0) = \{(x_1, \ldots, x_n) : x_0 > x_1 > \cdots > x_n\},$$

with the convention that $D_{1:0}(x_0) = \{\emptyset\}$. Under this representation one may rewrite the series expansion according to

$$f(x_0) = g(x_0) + \sum_{n=1}^{\infty} \int_{D_{1:n}(x_0)} g(x_n) \prod_{l=1}^{n} k(x_{l-1}, x_l)\, dx_{1:n}. \tag{13.201}$$

Additionally, we define the domain $\hat{D}_{0:n}(D_0) = \{(x_0, x_1, \ldots, x_n) : D_0 \ni x_0 > x_1 > \cdots > x_n\}$ and D_0 corresponds to the region of values over which one wishes to characterize the annual loss distribution, for example an interval $[x_a, x_b]$.

Clearly, one can now start to characterise the tail distribution of the compound process under this framework according to the integral over the domain $[x, \infty)$ using the series representation given by

$$\begin{aligned} \overline{F}_{Z_N}(x) &= \Pr[Z_N > x] = \sum_{n=1}^{\infty} \Pr[N = n] \overline{F}_{Z_n}(x) \\ &= \int_x^{\infty} g(\tau)\, d\tau + \sum_{n=1}^{\infty} \int_x^{\infty} \int_{D_{1:n}(\tau)} g(x_n) \prod_{l=1}^{n} k(x_{l-1}, x_l)\, dx_{1:n} d\tau. \end{aligned} \tag{13.202}$$

In Peters *et al.* (2007) a set of importance sampling estimation methodologies were developed to approximation numerically the recursive integral representations developed. This involved primarily importance sampling based methods on a path space. The choice of the importance sampling distribution was selected in two settings, one for efficiency and one utilising a trans-dimensional Markov chain Monte Carlo proposal which was provably optimal in minimising the variance of the importance sampling weights.

13.8.7.2 Importance Sampling Solutions to the Continuous Panjer Recursion.

To introduce the importance sampling estimation framework to estimate the compound process distribution pointwise or over an interval it will be convenient to introduce the following additional notation:

$$\begin{aligned} f_0\left(x_0\right) &= g\left(x_0\right), \\ f_n\left(x_{0:n}\right) &= g\left(x_n\right)\prod_{l=1}^{n} k\left(x_{l-1}, x_l\right), \end{aligned} \tag{13.203}$$

which then allows for a representation of the evaluation of the density either at a point x_0 by

$$f\left(x_0\right) = f_0\left(x_0\right) + \sum_{n=1}^{\infty} \int_{D_{1:n}(x_0)} f_n\left(x_{0:n}\right) dx_{1:n}, \tag{13.204}$$

or on an interval D_0 using the $\hat{D}_{0:n}\left(D_0\right)$.

Therefore one can frame this quantity of interest according to an expectation with respect to some importance sampling distribution π according to:

$$\begin{aligned} f(x) &= \frac{f_0(x)}{\pi(0)} + \sum_{n=1}^{\infty} \int_{D_{1:n}(x)} \frac{f_n\left(x, x_{1:n}\right)}{\pi\left(n, x_{1:n}\right)} \pi\left(n, x_{1:n}\right) dx_{1:n} \\ &= \mathbb{E}\left[\frac{f_n\left(x, x_{1:n}\right)}{\pi\left(n, x_{1:n}\right)}\right]. \end{aligned} \tag{13.205}$$

As discussed there are then two estimation problems of interest, the estimation of $f(x)$ pointwise and the characterization of $f(x)$ over some interval by obtaining samples from its restriction to that interval.

The space upon which the importance sampling is performed is now a path-space since it either corresponds to:

1. estimation of $f(x)$ pointwise via an importance sampling space $\bigcup_{n=0}^{\infty}\{n\} \times D_{1:n}(x)$, or
2. estimation of $f(x)$ over an interval D_0 via an importance sampling space $\bigcup_{n=0}^{\infty}\{n\} \times \hat{D}_{1:n}\left(\left[x_a, x_b\right]\right)$.

As noted in Peters *et al.* (2007), when one is interested in estimating the function pointwise, as $f_0(x)$ is known, it would be more efficient in the sense that variance would be reduced on both a per sample basis and a per unit of computation basis to instead estimate $f(x) - f_0(x)$

by importance sampling on the smaller space $\bigcup_{n=1}^{\infty} \{n\} \times D_{1:n}(x)$ and this approach introduces no further complications.

It is critical to any importance sampling based procedure to ensure a suitable importance sampling distribution is selected. In this section we detail a simple Markov chain proposal formulation and refer the interested reader to a trans-dimensional Markov chain proposal developed to minimize the variance of the importance sampling weights in Peters *et al.* (2007, section 3.2).

Consider a proposal distribution that is intuitive to understand and simple to simulate from as detailed in Peters *et al.* (2007, section 3.1). The solution in this setting would involve starting with a Markov chain from x (or with some initial distribution μ which covers the region of interest if we wish to characterize f over some interval rather than at a point) and a transition kernel for the Markov chain denoted by $M(x, y)$ which is the probability density for going from state x to state y. The initial distribution μ, when it is used, and transition kernel M are selected such that $\mu(x) > 0$ over the region of interest and $M(x, y) > 0$ if $k(x, y) \neq 0$, which is important to ensure the importance sampling scheme is well defined over the domain of interest, avoiding bias in estimates. In addition, the space explored by M is designed to have an absorbing cemetery state that we denote by d, where $d \notin [0, \infty)$ and $M(x, d) = P_d$ for any x. Therefore, the proposal we consider for the importance sampler over the path space in the case of considering a evaluation at a point x_0 takes the following form $\pi\left(n, x_{1:n}\right) = \pi(n)\pi_n\left(x_{1:n}\right)$ with

$$\begin{aligned} \pi(n) &= \mathbb{P}\mathrm{r}\left[X_{1:n} \in D_{1:n}\left(x_0\right), X_{n+1} = \{d\}\right] = \left(1 - P_d\right)^n P_d, \\ \pi_n\left(x_{1:n}\right) &= \frac{\prod_{k=1}^{n} M\left(x_{k-1}, x_k\right)}{\left(1 - P_d\right)^n}. \end{aligned} \tag{13.206}$$

Note, the dependency of $\pi_n\left(x_{1:n}\right)$ on the point x_0 is not made explicit here but is understood from the construction.

Now using this proposal one can develop an importance sampling based estimation. To present this estimation we introduce the particle notation in which we represent the annual loss distribution $f_{Z_N}(x)$ according to an empirical measure with P independent path samples, either at a point $x = x_0$ according to Equation 13.207

$$\hat{f}_{Z_N}\left(x_0\right) = \frac{1}{P} \sum_{i=1}^{P} W\left(x_0, X_{1:n^{(i)}}^{(i)}\right), \tag{13.207}$$

or over an interval $D_0 = \left[x_a, x_b\right]$ according to Equation 13.208

$$\hat{f}_{Z_N}\left(x_0\right) = \frac{1}{P} \sum_{i=1}^{P} W\left(X_{0:n^{(i)}}^{(i)}\right) \delta\left(x_0 - X_0^{(i)}\right), \tag{13.208}$$

where for the i-th particle (independent sample) we denote the path with $n^{(i)}$ stages, run until absorption, by $X_{0:n^{(i)}}^{(i)}$, the importance weight for the path by $W\left(X_{0:n^{(i)}}^{(i)}\right)$ and the dirac mass $\delta\left(x_0 - X_0^{(i)}\right)$ located at $X_0^{(i)}$.

The importance sampling approximation of the annual loss density $f_{Z_N}(x)$ is given by the following steps:

- Generate P independent Markov chain paths $\left\{X^{(i)}_{0:n^{(i)}+1}\right\}_{i=1}^{P}$ until absorption $X^{(i)}_{n^{(i)}+1} = d$;
- Evaluate the importance weights for each particle on the path space. If evaluation of the annual loss density at a point is desired for a value x_0 then the weight is given by:

$$W\left(X^{(i)}_{0:n^{(i)}}\right) = \begin{cases} \left(\prod_{s=1}^{n^{(i)}} \frac{k\left(X^{(i)}_{s-1}, X^{(i)}_{s}\right)}{M\left(X^{(i)}_{s-1}, X^{(i)}_{s}\right)}\right) \frac{g\left(X^{(i)}_{n^{(i)}}\right)}{P_d}, & \text{if } n^{(i)} \geq 1, \\ \frac{g\left(X^{(i)}_{0}\right)}{\mu\left(X^{(i)}_{0}\right) P_d}, & \text{if } n^{(i)} = 0. \end{cases} \tag{13.209}$$

If X_0 is being sampled from some distribution μ in order to characterize f over some interval, then the importance weight function becomes

$$W\left(X^{(i)}_{0:n^{(i)}}\right) = \begin{cases} \frac{1}{\mu\left(X^{(i)}_{0}\right)} \left(\prod_{n=1}^{n^{(i)}} \frac{k\left(X^{(i)}_{n-1}, X^{(i)}_{n}\right)}{M\left(X^{(i)}_{n-1}, X^{(i)}_{n}\right)}\right) \frac{g\left(X^{(i)}_{n^{(i)}}\right)}{P_d}, & \text{if } n^{(i)} \geq 1, \\ \frac{g\left(X^{(i)}_{0}\right)}{\mu\left(X^{(i)}_{0}\right) P_d}, & \text{if } n^{(i)} = 0. \end{cases} \tag{13.210}$$

Then the empirical measure,

$$\hat{f}_Z(x_0) = \frac{1}{N} \sum_{i=1}^{N} W_1\left(X^{(i)}_{0:n^{(i)}}\right) \delta\left(x_0 - X^{(i)}_{0}\right)$$

forms an unbiased Monte Carlo approximation of the expectation of $f_Z(z)$ for any set D_0 given by $\mathbb{E}\left[\int_{D_0} \hat{f}(x_0) dx_0\right] = \int_{D_0} f(x_0) dx_0$. Furthermore, detailed discussions on the optimal choice with respect to minimizing the variance of the importance weights is developed in Peters *et al.* (2007) and Doucet *et al.* (2010).

CHAPTER FOURTEEN

Scenario Analysis

14.1 Introduction

The Basel II/Basel III Advanced Management Approach (AMA) includes the requirement for an OpRisk system to have four key elements: internal data, external data, scenario analysis, and factors reflecting the business environment and internal control systems (see BCBS, 2006, p. 152). One of these key elements that is particularly subjective is *scenario analysis*. In the final proposals, the Basel committee specified more detailed criteria for each of the four fundamental elements, in particular for scenario analysis (BCBS, 2006, paragraph 675, p. 154):

> "A bank must use scenario analysis of expert opinion in conjunction with external data to evaluate its exposure to high-severity events. This approach draws on the knowledge of experienced business managers and risk management experts to derive reasoned assessments of plausible severe losses. For instance, these expert assessments could be expressed as parameters of an assumed statistical loss distribution. In addition, scenario analysis should be used to assess the impact of deviations from the correlation assumptions embedded in the bank's OpRisk measurement framework, in particular, to evaluate potential losses arising from multiple simultaneous OpRisk loss events. Over time, such assessments need to be validated and reassessed through comparison to actual loss experience to ensure their reasonableness".

Estimation of low-frequency/high-severity risks cannot be done using historically observed losses from one bank only. Typically, in practice, there are insufficient data to estimate high quantiles of the risk distribution. It is also clear that estimation based on historical losses is backward-looking and has limited ability to predict future losses of constantly changing banking environment. Hence, it is important to have scenario analysis/expert judgments incorporated into the model. These judgments may provide valuable information for forecasting and decision making, especially for risk cells lacking loss data. The use of a consistent and comprehensive approach to scenario analysis would allow regulators to compare the use of scenario analysis by banks across the industry and would also allow firms to allocate economic capital according to rigorous scenario analysis of risks in its business lines. The underlying assumption is that scenarios contain important information about severe but plausible future losses that have not yet been seen in historical observations. The incorporation of scenario analysis should bring that information to the quantification process and lead to a more accurate capital assessment process. The overall result would be a more robust banking environment.

Fundamental Aspects of Operational Risk and Insurance Analytics: A Handbook of Operational Risk, First Edition. Marcelo G. Cruz, Gareth W. Peters, and Pavel V. Shevchenko.

In 2003, a working group of internationally active banks identified the main steps in a so-called scenario-based AMA (sbAMA) process for calculating OpRisk capital (see sbAMA Working Group, 2003). The working group defined scenarios as "*potential future events, whose evaluation involves answering two fundamental questions: firstly, what is the potential frequency of a particular scenario occurring and secondly, what is its potential loss severity?*" There are several stages of an sbAMA life cycle:

1. Scenario generation—identify plausible OpRisk scenarios;
2. Scenario assessment—analyze and prioritize potential scenarios;
3. Data quality—review assessment factors, loss data (internal/external);
4. Determination of parameter values—select and combine values in potential loss matrices;
5. Model output—estimate economic or regulatory capital for the quantile we are interested in (e.g., 99.9% of Basel II) from the aggregated loss distribution values.

National banking regulators have been expanding the Basel II rules for local use in their countries. For example, in late 2005, the Australian Prudential Regulatory Authority (APRA) published detailed guidance (APRA, 2005) for Australian banks wishing to be accredited in the use of AMA. This guidance covers all aspects of an AMA and, in particular, specifies slightly more detailed requirement for scenario analysis:

> "Scenario analysis must be incorporated into a bank's OpRisk measurement system to evaluate the bank's exposure to high-severity loss events. The bank must collect scenarios that draw upon the knowledge of experienced business managers and risk management experts to derive reasoned assessments of plausible severe losses. The set of developed scenarios should be comprehensive and capture all material sources of OpRisk across all of a bank's business activities and geographic locations. A bank's process for building a database of scenario-based events must be robust and methodical and is required to be applied consistently across the bank. A bank's OpRisk management framework must include policies and procedures that identify how scenario analysis will be incorporated into the OpRisk measurement system. Scenarios and their use in OpRisk modelling must be independently reviewed and validated. Over time, scenarios must be reassessed through comparison to actual loss experience to assess their reasonableness".

The adaptation of Basel II by the US national supervisory authorities through domestic rule-making procedures, The Final Rule (2007), also requires that large, internationally active financial institutions incorporate scenario analysis into their OpRisk assessment and quantification systems.

Unfortunately, these regulatory documents do not provide any specifics on how the financial institutions should incorporate scenario analysis into the risk framework and what types of clients, products, processes, and models should be covered by scenario analysis in practice. So far, limited success exists in meeting this requirement. A common feature of the current practice is the absence of widely accepted and theoretically well-grounded approaches to incorporating scenarios. As a result, practitioners often use many different ad hoc approaches.

A methodical approach to estimating risk in any scenario analysis exercise is extremely important. Research shows that people (and business managers in particular) are not good at producing accurate estimates of risk, especially of low-probability/high-impact events. This matter is a subject of a relatively new discipline of Behavioral Finance, for which Kahneman and Tversky won the Nobel Prize in Economics in 2002.

The Loss Distribution Approach (LDA) separately models the severity and the frequency distributions of losses and finds the annual total loss distribution through the convolution of the two distributions. Thus expert opinions are used to estimate frequency and severity distributions. However, published studies on the use of expert elicitation for OpRisk LDA are scarce. Among the few that are available are Berkowitz (2000); Frachot *et al.* (2004b), Alderweireld *et al.* (2006), Steinhoff and Baule (2006), Peters and Hübner (2009), and Dutta and Babbel (2013).

Scenarios are hypothetical realizations of inherent risks in an institution. They can certainly be useful in making forward-looking adjustments to the frequency and severity of the risk to account for the latest or expected future changes in the institution's risk profile and business environment that have not been reflected in historical losses yet. Scenario experts build each scenario as a hypothetical realization of a specific risk under specific circumstances. Typically, each scenario is assigned with a duration (the number of years during which this scenario happens only once on average) and severity (some institutions assign loss ranges with lower and upper bounds, while other institutions assign just lower bounds or point estimates of the anticipated loss amounts).

The main problem with scenarios is that no knowledge exists on true loss frequencies. Experts can generate too many scenarios in certain loss regions and too few scenarios in some other regions. As a result, scenario-implied frequencies of losses in different regions might not necessarily follow true loss frequencies. One way of dealing with this challenge is to adjust scenario frequencies by some multiplier to align them with historical loss frequencies. However, such an adjustment is questionable because it implies that true frequencies are those of backward-looking historical losses, which might not necessarily be the case. Dutta and Babbel (2013) propose the replacement of historical frequencies with scenario frequencies whenever the former are lower than the latter. Although this assumption seems conservative, if too many scenarios have been generated from the body relative to the tail of the severity distribution, then this situation might lead to a reduction in capital. It is possible that if managers are responsible for generating scenarios for their business lines, they might avoid generating large scenario losses if it will negatively impact on their performance. Therefore, in this case, whether the reduction in capital happens due to expert opinions or simply because of a disproportionate amount of scenarios is not clear (see discussion in Ergashev, 2012).

If modelers simply pool scenario losses together with historical loss observations to calculate the capital, then, as Dutta and Babbel (2013) rightly point out, it does not take into account scenario frequencies. Alternatively, if modelers treat scenario losses as add-ons to capital, it may lead to the rejection of capital numbers due to their unrealistically high values. In general, both pooling and using scenarios as add-ons should be avoided.

Modelers may also simulate scenarios from their empirical distribution and add them to the pool of historical losses to find the combined severity distribution and calculate scenario-adjusted capital. Berkowitz (2000) proposed such an approach to incorporate scenarios for measuring market risk. This approach is not directly applicable to OpRisk modeling mainly because of the necessity to separately model the severity and frequency of losses. Dutta and Babbel (2013) propose a similar simulation-based approach that takes into account the specifics of modeling OpRisk.

Using scenario analysis, experts express opinions on potential losses and corresponding probabilities in the business line/event type risk cells. Often these opinions/judgments take the following forms:

- Opinions on a distribution family;
- Opinions on distribution parameters;
- Opinions on the number of losses with the amount to be within some ranges;
- Opinions on the quantiles of loss distribution and overall frequency;

- Separate opinions on the frequency of the losses and quantiles of the severity;
- Opinions on how often the losses exceed some threshold level;
- Point estimates of anticipated loss amounts.

In this chapter, we present several approaches discussed in the literature. Some of the concepts and principles of scenario analysis have already been discussed in Section 2.6 briefly. The question of combining scenario analysis with historical data will also be discussed in detail in Chapter 15.

14.2 Examples of Expert Judgments

As a result of scenario analysis workshops, the capital modeling team is provided with expert judgments about frequency and severity of OpRisk for business lines and event types. In this section, we present a few examples of such expert opinions/judgments.

EXAMPLE 14.1 Expert Opinions on Quantiles

Suppose that a bank has not experienced any losses due to fines for improper trade practices, which is part of "clients, products, and business practices", but that it recognizes the potential to incur such losses in the future. An expert (or group of experts) estimates that these losses may occur every 5 years in the "asset management" business line. This gives an estimate for the annual frequency (i.e., if $Poisson(\lambda)$ is selected to model annual frequency, then $\hat{\lambda} = 1/5$. An expert also estimates that if the loss occurs, then the probabilities (likelihood) of the loss exceeding USD 10 million, USD 30 million, USD 50 million, and USD 120 million are 0.9, 0.5, 0.25, and 0.1, respectively, and the maximum possible loss is USD 600 million. That is, a scenario-based severity distribution at points (0, 10, 30, 50, 120, 600) is (0, 0.1, 0.5, 0.75, 0.9, 1). It is presented in Figure 14.1 with linear interpolation between specified distribution points.

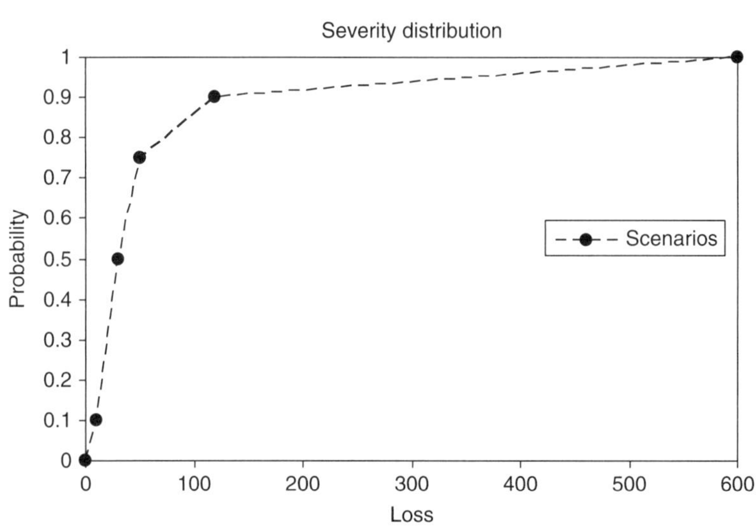

FIGURE 14.1 Scenario analysis example

EXAMPLE 14.2 Expert Estimates in Severity Brackets

Assume that the outcomes of the scenario analysis workshop produce results on the loss frequencies in predefined severity brackets in Table 14.1. Then we can estimate the annual frequency of this risk to be 102 (the sum of all frequencies) and severity distribution is easily calculated from relative frequencies (see Figure 14.2).

TABLE 14.1 **Using scenario analysis outcomes on the loss frequencies in predefined severity brackets**

Loss bracket (in USD thousands)	Annual loss frequency	Relative frequency(%)
>5000	7	6.9
1000–5000	10	9.8
500–1000	15	14.7
100–500	30	29.4
50–100	40	39.2
Total	102	

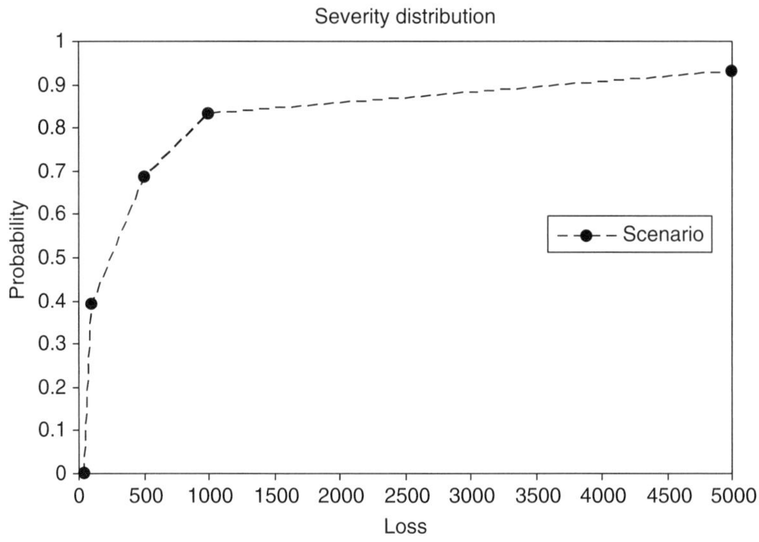

FIGURE 14.2 Severity distribution estimate from expert opinions on the loss frequencies in pre-defined severity brackets

EXAMPLE 14.3 Opinions on How Often the Loss Exceeding Some Level May Occur

Many studies (e.g., Frachot *et al.*, 2004b, Alderweireld *et al.*, 2006, Steinhoff and Baule, 2006, Peters and Hübner, 2009) suggest that questions on "*how often the loss exceeding some level may occur*" are well understood by OpRisk experts. Here, experts express the opinion that a loss of amount L or higher is expected to occur every d years. If there are M experts, then we have M opinions $(L_1, d_1), \ldots, (L_M, d_M)$. These opinions can be used to fit assumed frequency and severity distributions. For example, assume that the frequency is modeled by *Poisson*(λ) and severity is modeled by distribution $F(x|\boldsymbol{\theta})$. Then, the number of losses exceeding level L_i is distributed from *Poisson*$(\lambda(1 - F(L_i|\boldsymbol{\theta})))$. That is, the expected number of losses exceeding L_i per year is

$$\widetilde{\lambda} = \lambda(1 - F(L_i|\boldsymbol{\theta})). \tag{14.1}$$

This is typically interpreted as the loss exceeding L_i occurs (on average) every $1/\widetilde{\lambda}$ years or the expected duration between losses exceeding L_i is $1/\widetilde{\lambda}$. Then the parameters $(\lambda, \boldsymbol{\theta})$ can be estimated as

$$(\hat{\lambda}, \hat{\boldsymbol{\theta}}) = \arg\min_{\lambda, \boldsymbol{\theta}} \sum_{j=1}^{M} w_j \left(d_j - \frac{1}{\lambda(1 - F(L_j|\boldsymbol{\theta}))} \right)^2, \tag{14.2}$$

where w_j is the weight associated with the j-th opinion. The previous literature suggests to use a weight w_j equal to the inverse of the variance estimate of the duration between events exceeding L_j, that is, $w_j = 1/d_j$. If the severity is assumed to be from a two-parameter distribution, then one can fit all three model parameters (frequency and severity) using three or more opinions. However, the previous method does not allow for estimation of parameter uncertainty (prior distribution) if a Bayesian approach is undertaken. For the latter, it is important that experts specify not just the expected duration d_j, but also the uncertainty of their estimates. This will be discussed more in Section 14.3 and Chapter 15. ■

14.3 Pure Bayesian Approach (Estimating Prior)

If a Bayesian approach is taken to estimate frequency and severity distribution parameters, which are denoted by random vector $\boldsymbol{\theta}$, then expert opinions can be used to estimate the prior distributions for the parameters (so-called *pure Bayesian approach*), that is, distribution of the parameters before observing data. These are combined with the likelihood of observed data to find the posterior distributions of the parameters (distribution of the parameters after the data are observed). This approach will be discussed in detail in Chapter 15. In the Bayesian approach, both observations and parameters are considered to be random. Consider random data $\boldsymbol{X} = (X_1, X_2, \ldots, X_n)$ whose joint density, for given parameters $\boldsymbol{\Theta} = (\Theta_1, \Theta_2, \ldots, \Theta_K)$, is $h(\boldsymbol{x}|\boldsymbol{\theta})$. Then the joint density is

$$h(\boldsymbol{x}, \boldsymbol{\theta}) = h(\boldsymbol{x}|\boldsymbol{\theta})\pi(\boldsymbol{\theta}) = \pi(\boldsymbol{\theta}|\boldsymbol{x})h(\boldsymbol{x}) \Rightarrow \pi(\boldsymbol{\theta}|\boldsymbol{x}) = h(\boldsymbol{x}|\boldsymbol{\theta})\pi(\boldsymbol{\theta})/h(\boldsymbol{x}), \tag{14.3}$$

where

- $\pi(\boldsymbol{\theta})$ is the probability density of the parameters, random vector $\boldsymbol{\Theta}$ a so-called prior density function. Typically, $\pi(\boldsymbol{\theta})$ depends on a set of further parameters that are called hyper-parameters, omitted here for simplicity of notation;
- $\pi(\boldsymbol{\theta}|\boldsymbol{x})$ is the density of parameters given data $\boldsymbol{X}$, a so-called posterior density;
- $h(\boldsymbol{x}, \boldsymbol{\theta})$ is the joint density of observed data and parameters;
- $h(\boldsymbol{x}|\boldsymbol{\theta})$ is the density of observations for given parameters. This is the same as a likelihood function if considered as a function of $\boldsymbol{\theta}$, that is, $L_{\boldsymbol{x}}(\boldsymbol{\theta}) = h(\boldsymbol{x}|\boldsymbol{\theta})$;
- $h(\boldsymbol{x})$ is a marginal density of $\boldsymbol{X}$ that can be written as $h(\boldsymbol{x}) = \int h(\boldsymbol{x}|\boldsymbol{\theta})\pi(\boldsymbol{\theta})d\boldsymbol{\theta}$, otherwise known as the model evidence.

In the context of OpRisk, one can follow the proceeding three logical steps:

- The prior distribution $\pi(\boldsymbol{\theta})$ should be estimated by scenario analysis (expert opinions with reference to external data);
- Then the prior distribution should be weighted with the observed data using formula (14.3) to get the posterior distribution $\pi(\boldsymbol{\theta}|\boldsymbol{x})$;
- The predictive distribution of X_{n+1} given the data $\boldsymbol{X}$ can be calculated using formula (15.6) discussed in Chapter 15.

Berger (1985) lists several methods for estimating the prior:

- *Histogram approach.* Split the space of the parameter $\boldsymbol{\theta}$ into intervals and specify the subjective probability for each interval. From this, the smooth density of the prior distribution can be determined;
- *Relative likelihood approach.* Compare the intuitive likelihoods of the different values of $\boldsymbol{\theta}$. Again, the smooth density of prior distribution can be determined. It is difficult to apply this method in the case of unbounded parameters;
- *Distribution function determinations.* Subjectively construct the distribution function for the prior and sketch a smooth curve;
- *Matching a given functional form.* Find the prior distribution parameters assuming some functional form for the prior distribution to match prior beliefs (on the moments, quantiles, etc.) as close as possible.

Matching a given functional is a popular method. The use of a particular method is determined by a specific problem and expert experience. Usually, if the expected values for the quantiles (or mean) and their uncertainties are estimated by the expert, then it is possible to fit the priors.

Often, expert opinions are specified for some quantities such as quantiles or other risk characteristics rather than for the parameters directly. In this case, it might be better to assume some priors for these quantities that will imply a prior for the parameters. In general, given model parameters $\boldsymbol{\theta} = (\theta_1, \ldots, \theta_n)$, assume that there are risk characteristics $d_i = g_i(\boldsymbol{\theta})$, $i = 1, 2, \ldots, n$ that are well understood by experts. These could be some quantiles, expected values, expected durations between losses exceeding high thresholds, etc. Now, if experts specify the joint prior $\pi(d_1, \ldots, d_n)$, then using a standard transformation method, the prior for $\theta_1, \ldots, \theta_n$ is

$$\pi(\boldsymbol{\theta}) = \pi(g_1(\boldsymbol{\theta}), \ldots, g_n(\boldsymbol{\theta})) \left| \frac{\partial (g_1(\boldsymbol{\theta}), \ldots, g_n(\boldsymbol{\theta}))}{\partial (\theta_1, \ldots, \theta_n)} \right|, \tag{14.4}$$

where $|\partial (g_1(\boldsymbol{\theta}), \ldots, g_n(\boldsymbol{\theta})) / \partial (\theta_1, \ldots, \theta_n)|$ is the Jacobian determinant of the transformation. Essentially, the main difficulty in specifying a joint prior is due to a possible dependence between the parameters. It is convenient to choose the characteristics (for specification of the prior) such that independence can be assumed. For example, if the prior for the quantiles $q_1, \ldots, q_n$ (corresponding to probability levels $p_1 < p_2 < \cdots < p_n$) is to be specified, then to account for the ordering it might be better to consider the differences

$$d_1 = q_1, d_2 = q_2 - q_1, \ldots, d_n = q_n - q_{n-1}.$$

Then, it is reasonable to assume independence between these differences and impose constraints $d_i > 0$, $i = 2, \ldots, n$. If experts specify the marginal priors $\pi(d_1), \pi(d_2), \ldots, \pi(d_n)$ (e.g., Gamma priors), then the full joint prior is

$$\pi(d_1, \ldots, d_n) = \pi(d_1) \times \pi(d_2) \times \cdots \times \pi(d_n)$$

and the prior for parameters $\boldsymbol{\theta}$ is calculated by transformation using (14.4). To specify the i-th prior $\pi(d_i)$, an expert may use the approaches listed earlier. For example, if $\pi(d_i)$ is $Gamma(\alpha_i, \beta_i)$, then the expert may provide the mean and variational coefficient for $\pi(d_i)$ (or median and 0.95 quantile), which should be enough to determine α_i and β_i.

Examples of expert estimates for the prior are presented in Section 15.2.3 (also see Example 15.2) for estimation of the Poisson frequency and in Sections 15.2.4 and 15.2.5 for estimating LogNormal and Pareto severities.

14.4 Expert Distribution and Scenario Elicitation: Learning from Bayesian Methods

Other methods to perform scenario analysis and expert elicitation of loss distributional information from experts include approaches adopted for prior elicitation in the Bayesian statistics literature, such as Garthwaite *et al.* (2005), Jacobs (1995), Das *et al.* (2013), and the textbook review by O'Hagan *et al.* (2006).

It is first of all important to realize that the study of expert elicitation is a widely covered science, both in psychology and statistics. In early studies, it was shown by Seidenfeld *et al.* (1989) that the perfect combination of expert opinion is basically impossible; however, one can make good progress with approximations as we will discuss.

If one considers elicitation and scenario analysis as the process of formulating knowledge and beliefs about one or more uncertain loss process quantities into a probability distribution for those quantities, then the natural question that arises is how best to perform such a task and how to judge its success. In this regard, one must distinguish between the quality of an expert's knowledge and the accuracy with which that knowledge is translated into probabilistic form.

It is then clear that one can judge an elicitation as successful if the resulting distribution for the loss process is derived accurately and reflects the true physical process. Before proceeding, it is important to realize that there are many types of expert opinion, but the main focus of this section will be on those opinions on scenarios that can be expressed as quantiles or probability distributions.

The following key steps are generally agreed upon in the literature as key steps in any expert ellicitation and scenario analysis exercise (see discussions in Kadane, 1980, Wolpert, 1989, Winkler, 1986 and Garthwaite *et al.*, 2005).

- Select the experts and prepare training appropriate for the experts' background. Identify the appropriate aspects of the problem that will be required to be elicited and ensure the training covers these aspects;
- Elicit a summary of each expert's distribution on the decided relevant quantities and it is generally agreed that one should not attempt to ask experts to estimate moments of a distribution directly, with the exception of perhaps the mean. It is more widely accepted that a more robust approach will be to elicit probabilities of events or quantiles of the predictive distribution;
- Fit a probability distribution to each of the summaries from each expert. The type of model fitted could be parametric or nonparametric (see different examples in Oakley and O'Hagan, 2007 and Das *et al.*, 2013);
- Perform feedback and make the elicitation an iterative approach where the adequacy of each judgment of the expert is considered relative to the implications of the fitted model.

Ideas from works on expert elicitation processes were implemented in a freely available toolkit known as the Sheffield Elicitation Framework (SHELF),[1] which is covered under copyright when it comes to commercial usage (see details on the associated url). In agreement with the standard industry practice of structured workshops, the SHELF framework is developed to be performed with a group elicitation in mind and comprises a framework for eliciting beliefs of one or more experts as a group. As noted by Das *et al.* (2013), there are three typical approaches to group elicitation exercises:

1. Conduct the elicitation from each expert separately and merge the results through either a linear combination that is, weighted average or a logarithmic opinion pool obtained by a normalized weighted geometric mean;
2. Experts perform their elicitation separately and then discuss their opinions until a reconciliation of some form is met;
3. Consider each expert as providing an opinion from a common population distribution; see, for example, in the OpRisk setting, Peters *et al.* (2009). The expert opinions can then be utilized to infer the location of the distribution via Bayes rule.

As noted by O'Hagan *et al.* (2006), the psychological literature suggests that people are prone to certain heuristics and biases in how they respond to situations involving uncertainty. As a result, some of the ways of asking questions about uncertain quantities are preferable to others, and appear to be more reliable. It is therefore critical that significant consideration

[1] SHELF is available at http://www.tonyohagan.co.uk/shelf/.

be given regarding appropriate experts, and the information that should be elicited to aid in OpRisk quantification.

As a guide, the developers of the SHELF framework suggest that a group should ideally not comprise more than about five experts to avoid lengthy eliciatation workshops that may not achieve a concensus. As noted by these authors in the introduction notes to the SHELF package:

> "Elicitation is the process of capturing expert knowledge about one or more uncertain quantities in the form of a probability distribution. It can be done informally, but when the expert judgements are sufficiently important it is necessary to employ a formal procedure in the interests of quality and defensibility".

In this toolkit, one can find a set of documents that can be modified to help facilitate a distributional eliciation which include the following:

- An overivew of the elicitation process;
- Pre-elicitation briefing notes;
- Elicitation record sheets; and
- Distribution fitting instructions and R-code or performing the fitting once eliciation data are obtained from experts.

The approaches provided in this toolkit include methods for elicitation of scenario and distributional data based on the following.

- **Quantile methods**. This elicitation exercise aims to obtain information about the following quantities, which are then used to fit a set of plausible distributions that are discussed and a feedback process is undertaken;
  1. Plausible range. The range of values for a random unknown quantity such as an annual loss for a particular risk type. It involves extracting expert opinion on the lower and uper bounds, that is, range of the plausible losses $[L, U]$ where these are logical bounds such that it is extremely unlikely that they lie outside this range;
  2. Median loss. The value M that each expert expects has an equal probability of losses below and above this mid value;
  3. Lower and upper quantiles. These correspond to finding for each expert's median, the mid range between the lower and upper bounds such that equal probability of losses lies on either side of the lower/upper quantile in the intervals $[L, M]$ or $[M, U]$, respectively.
- **Quantile and probability methods**. This elicitation exercise aims to obtain information about the following quantities, which are then used to fit a set of plausible distributions that are discussed and a feedback process is undertaken;
  1. Plausible range, median, upper and lower quantiles, which are used to fit plausible loss distributions;
  2. The probability component enters where the facilitator selects loss values X_0, X_1, and X_2 and then for each of the fitted model possibilities, and after discussion on reasoning about the differences in each expert's fitted distribution, an assessment requiring group consensus values for probabilities is undertaken to obtain L and U lower

and upper bounds based on probabilities $\mathbb{P}\text{r}(L < X < X_1)$, $\mathbb{P}\text{r}(X_2 < X < U)$, and $\mathbb{P}\text{r}(L < X < X_0)$.

- **Roulette methods**. See Oakley *et al.* (2010);
- **Roulette and probability-based methods**;
- **Tertile methods**. Plausible range, median, upper and lower tertiles (thirds), which are used to fit plausible loss distributions;
- **Tertile and probability methods**. Same idea as quantile and probability method except tertiles are considered.

As noted by the authors of SHELF, any successful elicitation process will typically require a facilitator who has expertise in the process of elicitation. It is then the role of the facilitator to help guide the experts, in the case of OpRisk, the business managers from different business units and quantitative experts, traders, database managers, etc., through the process. The process generally follows these basic steps:

1. Identify the appropriate experts for a given risk type eliciation;
2. Obtain a suitable fixed date for the workshop and send around preliminary briefing material. It is suggested in SHELF that "[b]riefing material may be short or may include substantial training documents (for instance, concerning probability and its use to represent expert knowledge)".
   From experience, this will depend on the type of experts present and their background;
3. Experts should be provided with feedback lines to the facilitator post elicitation, should things need updating or be partially resolved.

14.5 Building Models for Elicited Opinions: Hierarchical Dirichlet Models

Ideas from works on expert elicitation processes were implemented in a freely available toolkit known as SHELF,[2] which is covered under copyright when it comes to commercial usage. In agreement with the standard industry practice of structured workshops, SHELF is developed to be performed with a group elicitation in mind and comprises a framework for eliciting beliefs of one or more experts as a group. In these SHELF packages, simple distributions are fitted to the data from exponential family models and other common two- and three- parameter shape and scale density models. In general, one may wish to allow for a greater amount of flexibility. In addition, one may wish to factor in particular attributes for each expert in a regression structure; detailed discussion on this is provided in Chapter 16.

In the following, we detail a few basic examples of prior models for multiple experts opinions that are elicited as a well as a collection of information on independent attributes of each expert that may also provide additional strength to weight their opinions. To achieve the modeling of sets of distributions from experts, one could do this purely on a parameter space of a simplex, modeling the sets of elicited probabilities of each expert as i.i.d. draws from, for example, a Dirichlet distribution. Alternatively, one could instead model, at a higher

[2]SHELF is available at http://www.tonyohagan.co.uk/shelf/.

level, a distribution over distributions, that is, a distribution over possible expert opinion–elicited distributions via what is known as a Dirichlet process (see examples in Neal, 2000 and Shahbaba and Neal, 2009).

As a simple example, consider the stylized case in which it is assumed that each expert is asked to produce through an elicitation process a set of k probabilities around scenarios and events of relevance to the modeling goals, denoted here by $\boldsymbol{Y}_j \in \mathbb{S}_{k-1}$ for the j-th expert for probability simplex $\mathbb{S}_{k-1}$ embedded in $\mathbb{R}^k$. It will be assumed that each set of experts produces a set of i.i.d. probability vectors from a population distribution given by a Dirichlet distribution as detailed in Definition 14.1.

Definition 14.1 (Dirichlet distribution) *A k-dimensional random vector $\boldsymbol{Y}$ of probabilities on the $k-1$ simplex follows a Dirichlet distribution if its density is given by*

$$f_Y(\boldsymbol{y};\boldsymbol{\alpha}) = c(\boldsymbol{\alpha}) \prod_{i=1}^{k} y_i^{\alpha_i - 1}, \tag{14.5}$$

with $k \geq 2$, $\alpha_i > 0, y_i > 0, y_1 + \cdots + y_{k-1} < 1, y_k = 1 - y_1 - \cdots - y_{k-1}$ and

$$c(\boldsymbol{\alpha}) = \frac{\Gamma\left(\sum_{j=1}^{k} \alpha_j\right)}{\prod_{j=1}^{k} \Gamma(\alpha_j)}. \tag{14.6}$$

■

Using this model, one can assume n sets of elicited probability vectors from n experts that forms a joint expert prior given in exponential family form

$$f_{Y_1,\ldots,Y_n}(\boldsymbol{y}_1,\ldots,\boldsymbol{y}_n;\boldsymbol{\alpha}) = \exp\left(n \ln c(\boldsymbol{\alpha}) + \sum_{j=1}^{k} \alpha_j \sum_{i=1}^{n} \ln y_{ij} - \sum_{i=1}^{n}\sum_{j=1}^{k} \ln y_{ij} \right) \prod_{i=1}^{n} \mathbb{I}_{\mathbb{S}_k}(\boldsymbol{y}_i). \tag{14.7}$$

As noted by Das *et al.* (2013), one may then consider a hierarchical prior structure for the population distribution of elicited expert opinions, where the prior on the hyperparameters $\boldsymbol{\alpha}$ can be made "conjugate" by selecting the form

$$f(\boldsymbol{\alpha};\boldsymbol{\theta}) \propto \exp\left(\sum_{j=1}^{k} \alpha_j \theta_j + \theta_{k+1} \ln c(\boldsymbol{\alpha}) \right), \tag{14.8}$$

where $\boldsymbol{\theta} \in \mathbb{R}^{k+1}$.

The second modeling example one can consider, which incorporates into the prior distribution additional regression features related to the attributes of each expert, is given by a Hierarchical Dirichlet regression model (see, for instance, Bishop and Nasrabadi 2006 or Das *et al.* 2013, section 3.1).

The model assumes, as before, n experts, each producing an elicited random vector of $\boldsymbol{Y}_i \in \mathbb{S}_{k-1}$ of k probabilities characterizing the likelihood of events or scenarios that the expert judges. Furthermore, assume that in the elicitation process, independent attributes (variables, factors, features) of each expert are recorded. Assume there are m attributes for each elicited probability for each expert given by independent vector $\boldsymbol{x}_{ij} = \left(x_{ij1}, \ldots, x_{ijm}\right)$ for the i-th expert on the j-th elicited probability (event, scenario, etc.). Then the resulting extension to the previously specified Dirichlet model is given by the hierarchical Dirichlet regression prior model with density

$$
\begin{aligned}
&f_{Y_1,\ldots,Y_n}\left(\boldsymbol{y}_1, \ldots, \boldsymbol{y}_n; \boldsymbol{\beta}\right) \\
&= \prod_{i=1}^{n} \exp\left(\ln\left(\frac{\Gamma\left(\sum_{j=1}^{k} g\left(\boldsymbol{x}_{ij}^T \beta_j\right)\right)}{\prod_{j=1}^{k} \Gamma\left(g\left(\boldsymbol{x}_{ij}^T \beta_j\right)\right)}\right) + \sum_{j=1}^{k} g\left(\boldsymbol{x}_{ij}^T \beta_j\right) \ln y_{ij} - \sum_{j=1}^{k} \ln y_{ij}\right) \mathbb{I}_{\mathbb{S}_k}[\boldsymbol{y}_i],
\end{aligned}
\tag{14.9}
$$

where $g(\cdot)$ is the so-called link function. Extensions to this model can be found in discussions of Dey *et al.* (2000) and Das *et al.* (2013). For instance, we can consider the Dirichlet mixed effect regression prior model given by the hierarchical structure

$$
\begin{aligned}
\boldsymbol{Y}_i &\sim \textit{Dirichlet}\left(\boldsymbol{X}_i^T \boldsymbol{\beta} + \boldsymbol{W}_i + \epsilon_i\right), \\
\boldsymbol{W}_i &\sim \textit{Normal}\left(0, \Sigma\right), \\
\epsilon_i &\sim \textit{Normal}\left(\boldsymbol{0}, \tau^{-1} \boldsymbol{I}_{k-1}\right),
\end{aligned}
\tag{14.10}
$$

with hyperparameter priors given by

$$
\begin{aligned}
\boldsymbol{\beta} &\sim \prod_{i=1}^{m} \prod_{j=1}^{k-1} \textit{Normal}\left(\beta_j; 0, \lambda^{-1}\right), \\
\lambda &\sim \textit{Gamma}\left(a_0, b_0\right), \\
\Sigma &\sim \textit{InverseWishart}\left(k-1, \Sigma_0\right), \\
\tau &\sim \textit{Gamma}\left(c_0, d_0\right).
\end{aligned}
\tag{14.11}
$$

14.6 Worst-Case Scenario Framework

Benefits and drawbacks exist of assigning probability distributions to scenarios. If scenarios are provided with their probability distributions, it is easier to incorporate them into the quantification framework. The difficulty is that scenario experts generally have very limited knowledge about probabilistic concepts. The task of assigning probability distributions around scenarios requires a sophisticated process of eliciting expert opinion that brings additional subjectivity to the process of incorporating scenarios. It can be argued that it is better to take a minimalistic approach by eliciting a minimum amount of expert information in very simple terms that does not require additional processing.

In this regard, a framework is proposed by Ergashev (2012) that links scenarios to historical losses. This framework derives from the concept that the focus should be on worst-case scenarios only, because only these scenarios contain valuable information about the tail behavior of operational losses. Following Ergashev (2012), we call a scenario a "once-in-an-M-year" scenario if it has a duration of M years. Although each expert-generated scenario is important from the risk management perspective, one may argue that from the risk quantification perspective the focus should be on worst-case ("worst-in-a-certain-year") scenarios only, to emphasize their duration. The worst-in-an-M-year event (where M is a natural number) is a rare event that results in the largest loss the institution experiences once in every M years on average. The worst-in-an-M_1-year loss must be less than or equal to a worst-in-an-M_2-year loss as long as $M_1 < M_2$. The worst-in-a-certain-year events introduce a natural order to scenario losses and make possible the comparison of these losses with the corresponding quantiles of the severity distribution in the base model, where the base model is the one used to fit the historical OpRisk losses.

From the quantification perspective, the most efficient way of presenting scenarios is to assume that a duration and a lower bound of the loss amount accompany each scenario. For example, if a scenario comes with a range, one can choose the lower bound of that range as the scenario's lower bound for quantification purposes. If only the point estimates of scenario losses are available, one can still conservatively use those point estimates as the lower bounds. Importantly, this proposed framework can still be used in situations where the definition of scenarios involves probability distributions, with the provision that each scenario distribution possesses a unique and strictly positive lower bound.

Ergashev (2012) considers two seemingly different but equivalent definitions of the worst-in-an-certain-year event.

Definition 14.2 *For any natural number M the definition of the loss amount of the worst-in-an-M-year event, V, is*

$$\mathbb{P}\mathrm{r}[\max(X_1, \dots, X_N) > V] = \frac{1}{M}, \tag{14.12}$$

where $X_1, \dots, X_N$ are losses and N is the random variable representing the annual frequency of losses. ■

Definition 14.3 *For any natural number M, the definition of the loss amount of the worst-in-an-M-year event, U, is*

$$\mathbb{P}\mathrm{r}\left[\sum_{i=1}^{N} \mathbb{I}_{\{X_i > U\}} \geq 1\right] = \frac{1}{M}, \tag{14.13}$$

where $X_1, \dots, X_N$ are observed losses and N is the random variable representing annual frequency of losses. ■

Ergashev (2012) shows that these two definitions are equivalent as follows. Suppose U is such that Equation (14.13) is true. Then

$$\begin{aligned}\mathbb{P}\mathrm{r}\left[\sum_{i=1}^{N} \mathbb{I}_{\{X_i > U\}} \geq 1\right] &= 1 - \mathbb{P}\mathrm{r}\left[\sum_{i=1}^{N} \mathbb{I}_{\{X_i > U\}} = 0\right] \\ &= 1 - \mathbb{P}\mathrm{r}[X_1 < U, \dots, X_N < U] \\ &= 1 - \mathbb{P}\mathrm{r}[\max(X_1, \dots, X_N) \leq U] \\ &= \mathbb{P}\mathrm{r}[\max(X_1, \dots, X_N) > U]\end{aligned}$$

and thus $U = V$.

The worst-in-a-certain-year loss severity depends on the annual loss frequency of the risk. For two risks with the same severity distribution but different annual frequencies, if the annual frequency of the first institution is greater than that of the second institution, then the worst-in-a-certain-year loss of the first institution must be greater than that of the second institution.

Formally, if we assume the losses $X_1, X_2, \ldots$ are independent and identically distributed from $F(\cdot)$ and the annual frequency of the losses N is from $Poisson(\lambda)$, then the distribution of the maximum loss is

$$G(x) = \mathbb{P}\mathrm{r}[\max(X_1, \ldots, X_N) \leq x] = \exp(-\lambda(1 - F(x)));$$

for a proof see, for example, Section 5.6.2. It follows that if $M_1 < M_2$, then the loss amount of the worst-in-an-M_1-year event is always less than or equal to that of the worst-in-an-M_2-year event.

The scenario in this framework comprises both duration M and lower bound L of the unknown loss amount and the set of scenarios is denoted by $\boldsymbol{S} = (S_1, \ldots, S_k)$, where $S_i = (M_i, L_i)$. To integrate scenarios into the base model, we begin by first identifying the set of worst-in-a certain-year scenarios, denoted by $\boldsymbol{S}^w = (S_1^w, \ldots, S_r^w)$, from the set of all scenarios $\mathbf{S}$. This can be achieved by using a simple procedure described by Ergashev (2012, section 4.2) with following the steps:

1. Find in $\mathbf{S}$ the scenario with the lower bound and denote it by $\hat{S}$; remove $\hat{S}$ from $\mathbf{S}$ and add it to $\mathbf{S}^w$;
2. Throw away all scenarios in $\mathbf{S}$ with durations that are equal to or greater than the duration of $\hat{S}$ (but which have smaller lower bounds);
3. Repeat steps 1 and 2 until $\mathbf{S}$ is empty.

We would then have a set of worst-in-a-certain-year scenarios $\mathbf{S}^w$ consisting of pairs of lower bounds of maximum loss L_i and associated frequencies M_i, so that $S_i^w = (M_i^w, L_i^w)$. We do not know much about the distribution of S^w, denoted by $G_S(\cdot)$, except that for the unknown loss V_i of scenario S_i^w, $V_i > L_i^w$ and $G_S(V_i) = 1 - 1/M_i^w$. The relation between severity distribution and scenario $S^w = (M^w, L^w)$ is just

$$F^{-1}\left(1 + \frac{1}{\lambda}\ln\left(1 - \frac{1}{M^w}\right)\right) > L^w.$$

Then we try to incorporate this information into the base model (the one that uses internal data) and say that a scenario $S_i^w = (M_i^w, L_i^w)$ is concordant with the base model if

$$G_S^{-1}\left(1 - \frac{1}{M_i^w}\right) > L_i^w.$$

Putting this in words, if the base model's historical maximum loss distribution, at the $(1 - 1/M_i^w)$-th quantile, is greater than L_i^w, then it agrees with the worst-in-a-certain-year scenario projections of potential losses at that point. Concordant scenarios are uninformative in the sense that no apparent reason exists for making any adjustments to the base model's severity distribution at the corresponding quantiles (i.e., the internal losses–driven model is larger than the scenarios losses). The goal of this framework is to identify where the base model is discordant with worst-in-a-certain-year scenario information and adjust the severity distribution accordingly, as this represents a possible shift in the institution's risk profile as well as its true loss distribution (i.e., where scenarios are larger than historical losses). Ergashev (2012) presents five approaches to adjust the base distribution using scenarios with inequality constraints

$$F^{-1}(q_i) \geq L_i^w, \quad q_i = 1 + \frac{1}{\lambda} \ln\left(1 - \frac{1}{M_i^w}\right), \quad i = 1, \ldots, r. \tag{14.14}$$

The first three approaches allow for the direct incorporation of scenario analysis into the quantification framework through the derivation of a scenario-adjusted distribution function, which is then used to calculate scenario-adjusted capital. The first one, the stochastic dominance approach, shifts the distribution of the base model severity distribution of historical losses to the right so that all the inequalities in (14.14) are satisfied, that is, at any quantile, the corresponding loss amount is greater than or equal to the loss amount implied by the base model severity distribution. Therefore, this approach should always result in higher scenario-adjusted capital than the base model capital. With such an approach, the modelers are able to discard scenarios that are not as heavy-tailed as the base model and also have an elegant way to adjust the base model based on the risk profile defined by experts.

The other two approaches, the constraint estimation approach and the constrained Markov chain Monte Carlo approach, incorporate the constraints (14.14) inside the estimation process. The last two approaches incorporate the scenario information into the quantification framework indirectly, because under these approaches there is no need to find a scenario-adjusted distribution. Instead, the curve-fitting approach leads to a scenario-driven distribution as a curve that is the closest to the set of points (q_i, L_i^w), $i = 1, \ldots, r$ under the scenario constraints (14.14). The fitted distribution curve is then used to calculate the scenario-driven capital, which can be considered as a benchmark for the base model's capital number. Ideally, this curve should make the inequalities (14.14) binding by turning them into equalities, although this change might not always be possible. Under the minimum distance approach, it is assumed that several competing severity distributions fit the base model reasonably well. Using the minimum distance approach allows to choose the severity distribution with the smallest deviation from the scenario constraints (14.14). Detailed description of each approach can be found in Ergashev (2012). All these approaches share a property of the built-in conservatism that makes it difficult for the scenario-adjusted capital to fall below the capital implied by the base model even when scenario experts substantially understate the severity of scenario losses.

14.7 Stress Test Scenario Analysis

Stress-testing has been part of the risk manager's toolkit for decades. However, it has always been in poor relation with analytical techniques to control risk. Recent renewed interested in stress-testing has been motivated by the financial crisis of 2007/2009 that has shown the limitations of the purely statistical techniques such as Value-at-Risk (VaR) that were supposed to provide the cornerstones of the financial system; events of once-in-many-thousand-years rarity keep on occurring with disconcerting regularity since the beginning of the crisis. The quote from the article by Aragonés *et al.* (2001) points exactly to the problem: "*[T]raditional stress testing is done on a stand-alone basis, and the results of stress tests are evaluated side-by-side with the results of traditional market risk (or VaR) models. This creates problems for risk managers, who then have to choose which set of risk exposures to 'believe'. Risk managers often don't know whether to believe their stress test results, because the stress tests exercises give them no idea of how likely or unlikely stress-test scenarios might be*".

Stress-testing is the subject of many consultation papers by the Bank for International Settlements (BIS) and other international bodies; one of the recent papers for example, BCBS

(May 2009c). Nowadays, "*risk*" and "*uncertainty*" are often used interchangeably in the risk management literature. However, these concepts are very different: the word "risk" refers to situations where we know for sure the probabilities attached to future events and we know exactly the possible future events while "uncertainty" refers to situations where there is no such probabilistic knowledge but we still know what may hit us tomorrow.

Stress-testing is supposed to examine whether a financial institution would be able to withstand exceptional risk losses. Stress-testing should be used as a complementary approach to the VaR-based LDA approach in order to ensure that a bank would be able to cover its losses even if it faces more severe risk events. As stated by Jorion (2007): "*Whenever the stress tests reveal some weakness, management must take steps to manage the identified risks. One solution could be to set aside enough capital to absorb potential large losses. Too often, however, this amount will be cripplingly large, reducing the return on capital*". Some of the aspects of stress-testing with respect to macroeconomic and financial institution factors have already been discussed in Chapter 4. Here we present an example of scenario analysis for stress-testing.

The results of stress tests are often difficult to interpret because they give us no idea of the probabilities of the events concerned, and in the absence of such information, we often do not know what to do with them. Suppose, for instance, that stress-testing reveals that our firm will become bankrupt under a particular scenario. Should we act on this information? If the scenario is very likely, it would be very unwise not to act on it. However, if the scenario is extremely unlikely, then it becomes almost irrelevant, because we would not usually expect management to take expensive precautions against events that may be too improbable to worry about. As pointed out by Berkowitz (2000), this absence of probabilities puts stress-testing in a statistical purgatory. We have some loss numbers but often it is not clear whether we should be concerned about them or not.

It is important to note that stress-testing methods are not comparable with each other. Moreover, the applications of the same stress tests to different financial institutions are not comparable with each other, because the results are always bound to the specific risk profile of a financial institution. Adopting bad assumptions or using irrelevant scenarios would result in irrelevant losses being considered. Since the stress tests often define events with a very low probability of occurrence, the results become difficult to interpret and it is not clear what actions should be taken by the management in order to mitigate the risks. Quite often, the results of stress tests appear unacceptably large and they are just ignored and dismissed as irrelevant. However, it is valuable to evaluate stress test results at different points in time to assess whether the exposures to severe OpRisk losses have changed. The scenarios can be divided into two groups based on the type of event they define. The first group uses historical events like the 9/11 terrorist attacks or the unauthorized trading in Société Générale in 2007. The second group, more widely used in practice, uses hypothetical scenarios. The scenarios are based on plausible risk events that have not happened yet, but have a nonzero probability to occur. A scenario can also be based on an analysis of a new product a bank is going to implement.

A typical scenario consists of the description of a complex state of the world that would impose an extreme risk event on the financial institution and would include probabilities and frequencies of occurrence of the particular state of the world, business activities impacted by the event, maximum internal and external loss amounts generated by the occurrence of such an event, and also possible mitigation techniques that may be available. Even though such a scenario claims to be realistic, it is not possible to include all possible risk factors and features. However, risk managers must try to define the scenarios, so that they correspond to reality as much as possible; for a discussion of this topic, see Jorion (2007). The aim of using scenarios is to get an overview of low-frequency events that might have severe impact on the financial institution.

TABLE 14.2 Historical scenarios (loss amounts in EUR thousands)

Scenario name	Estimated loss
Unauthorized trading	112,000
Process management failure—software loss	7,300
External fraud—theft	21,180

Source: Taken from Rippel and Teplý (2008)

TABLE 14.3 Hypothetical scenario details employee strike duration: probability, distribution)

Probability (%)	Duration	Estimated Loss (€)
70	1 hour	138,515
25	1 day	3,750,446
4	2–4 days	9,056,450
1	5 days	20,890,382

Source: Taken from Rippel and Teplý (2008)

We present an example of a stress test scenario analysis by Rippel and Teplý (2008), where the three historical scenarios are unauthorized trading, external fraud, and process management failure (see Table 14.2). These scenarios are based on concrete historical events. The estimated losses were quite high and were treated as the worst-case losses.

Rippel and Teplý (2008) also considered the hypothetical scenarios such as employee strike that would hit all the regions. This type of scenario was selected due to evidence of similar events. The frequency of the scenario assessment was estimated to be one per 40 years based on historical data from other regions where the duration of the strike ranged from 1 day to 1 week. It was assumed that the frequency of strikes would be quite low in the region of Central Europe. Usually, the duration of such strikes is limited only to several hours. The other important feature of a strike is its extent—a strike can range from one branch to a national strike. It was also assumed that the employees from all regions would go on strike at the same time. Such a scenario has a very low probability, but if it occurred, it would have significant negative impact on the bank. The severity of this scenario depends on two factors: the extent and the duration of the strike. The extent was set to the whole country. The duration was assumed to range from 1 hour to 5 business days, and the probability for each class was estimated according to the assumptions stated earlier; these are presented in Table 14.3. A strike was assumed to cause four types of losses: (i) the direct loss of revenue from branches was estimated based on the list of bank branches and their revenues per day; (ii) the costs connected with expenses on substitute employees who would be hired in order to maintain the bank's critical operations; (iii) the loss of clients (the most severe loss)—while a 1-hour strike is not considered to have impact on customer satisfaction, in case of a whole, week strike, up to 5% of customers might decide to move to competitors; (iv) the costs connected with commercial disputes (the losses were estimated based on interest costs from nonrealized transactions and estimated amount of dispute penalties). After taking into account all the assumed loss sources, the total loss was computed. It is easy to find from Table 14.3 that the worst-case loss is € 20,890,382 and the average loss of this scenario is € 1,605,733.

The worst-case scenario is a strike that lasts 5 days. In this case, the loss amount reaches € 20 million, which is approximately 2% of the Basel income ratio. Such a strike is considered to cause significant harm to a bank, especially by the loss of 5% customers. Such a scenario would also have a very negative impact on the brand image and the credibility of the bank would be damaged, resulting in counterparty risk.

In total, six tests were run. The aim was to analyze whether a bank would be able to handle particular combinations of events defined in the combination of scenarios. The impact of such joint scenarios was evaluated. It was found that even if all the scenarios were considered, the estimated regulatory capital would not exceed 12% of the Basel income ratio, suggesting that a bank would be able to handle the losses of such high magnitude.

14.8 Bow-Tie Diagram

There are well-developed techniques outside of the finance industry in safety-conscious industries, such as airline maintenance, mining, and defence, that may be applied to scenario analysis in OpRisk. Specifically, concepts have been developed in safety management disciplines such as the *Bow-Tie Diagram*—a graphical model with logical relationship between the causes and consequences of an undesired event. The Bow-Tie technique has been used since the 1970s, and has been incorporated into risk management plans of many companies. The paper by McConnell and Davies (2006) proposes a structured approach based on the Bow-Tie concept for scenario analysis for AMA in Basel II. It describes how such a concept may be used by banks and regulators to satisfy the requirements of Basel II and to improve OpRisk management across the industry. The Bow-Tie diagram is a diagrammatic representation of hazardous events. The name Bow-Tie is due to its bow-tie appearance (for illustration, see Figure 14.3): in the center of the diagram is an *incident* (undesirable event), to its left are possible *causes*, and to its right are possible *consequences* (outcomes) of the incident. Between the causes and the incident are possible *preventative (proactive) controls* and between the incident and consequences are possible *mitigative (reactive) controls*. The Bow-Tie diagram is an effective way to identify and communicate risks and required responses. The diagram displays the links between potential causes, preventative controls, and mitigation controls. It is often used for what-if and root-cause analyses where quantification is not possible or desirable.

In 2004, the US Federal Aviation Authority (FAA) mandated that its regulated entities employ the Bow-Tie diagram as the main mechanism for safety analyses (see FAST, 2004).

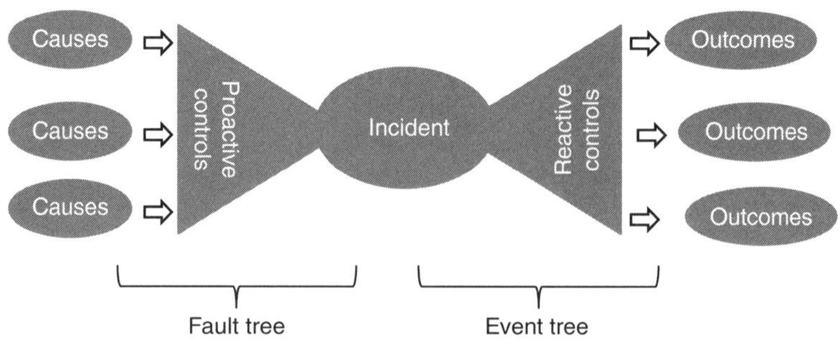

FIGURE 14.3 Bow-Tie diagram

This technique is also recommended by other bodies responsible for safety in air traffic control (EUROCONTROL, 2004) and safety management in hazardous industries (Work Cover, 2001).

Figure 14.3 illustrates the key components of a Bow-Tie diagram:

- **Causes:** Potential causes of an undesirable incident;
- **Proactive controls:** Actions taken to reduce the likelihood of an undesirable incident occurring;
- **Incident:** An event that can cause undesirable outcomes;
- **Reactive controls:** Actions taken to reduce the impact of an undesirable incident;
- **Outcomes:** Potential results of an undesirable incident.

The left-hand side of the diagram is often called a *Fault Tree*, which is a detailed analysis of the combination of causes (faults) that can possibly give rise to an undesirable incident, while the right-hand side is often called an *Event Tree*, which is a detailed analysis of the outcomes or consequences of an undesirable incident. In essence, the diagram attempts to answer the two main questions: what is the potential frequency of a particular scenario occurring (i.e., left side/Fault Tree) and what is its potential loss severity (i.e., right side/Event Tree)? In industrial applications, Bow-Tie analyses are most often employed to identify and assess the potentially disastrous impact of the failure of mechanical components, such as chemical containment vessels or airframe components.

Of course, the Bow-Tie technique is not a panacea, it is merely a way of making risk management assumptions, analyses, and conclusions explicit. It has known weaknesses, including the following:

- The quality of the final analysis will entirely depend on the quality of the analysis process and the analysts and experts taking part;
- The technique does not help in uncovering underlying causes, but merely in making their consequences explicit; therefore, an earlier analysis step (i.e., risk identification) is required;
- It is a semiquantitative methodology and hence requires an additional step of estimating the impact of each outcome numerically as required by Basel II;
- It can be "gamed" by staff members who may have a different agenda, so it requires additional supporting information to be captured such as external data or other documented factors that can suffice as evidence.

The Bow-Tie technique does not offer a new or different way of analyzing risk. The reason for its increased use is that the diagrams that it creates greatly assist in the communication of the hazard analysis process—particularly to nonspecialists. As an illustration, McConnell and Davies (2006) consider an example of a System Security category of External Fraud event type, presented in Figure 14.4.

Once the Bow-Tie diagram is constructed, quantitative analyses can be performed but it requires knowledge about the system and availability of precise data such as probability and interdependence of input events. Quantitative analysis of a Bow-Tie is still a major challenge. The probabilities for the input events are often missing or hard to come by, which introduces data uncertainty. Elicitation of experts' knowledge for the missing data may provide an

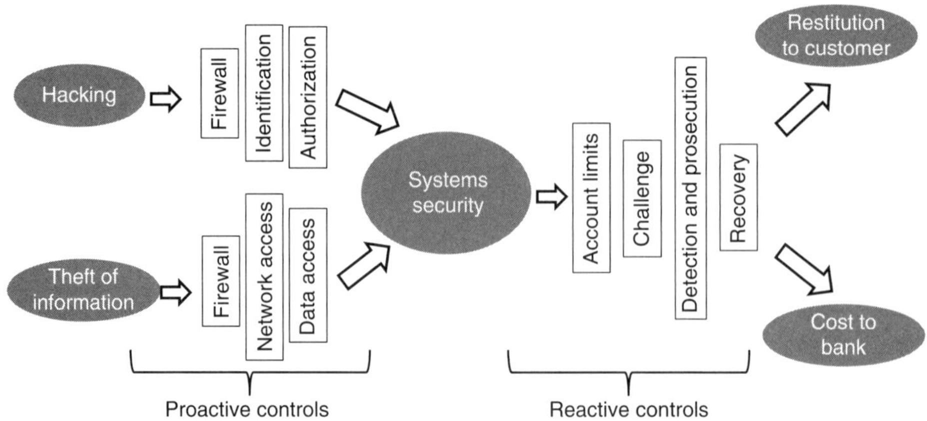

FIGURE 14.4 **Illustration of an example of a Bow-Tie diagram for System Security category of External Fraud event type**

alternative; however, such knowledge incorporates uncertainties and may undermine the credibility of risk analysis. Another major problem with Bow-Tie diagrams is that they remain limited by their restriction to the graphical representation of different scenarios without any consideration to the dynamic aspect of real systems. For recent papers tackling these problems, see Ferdousa *et al.* (2013), Badreddine and Ben Amor (2010), and references therein.

The Bow-Tie approach assumes that causes and consequences are separated by hazards, which is certainly convenient for modeling. However, Fault Tree components might also interact with events in the Event Tree; this can be handled by Bayesian network methods discussed next, where we also discuss risk quantification aspects.

14.9 Bayesian Networks

A purely probabilistic approach to risk management places all the emphasis on the association among variables, rather than on the causal links among them. However, assigning links among variables on a causal rather than associative basis is cognitively much easier and more "natural". A popular way to deal with complex processes via causal models is to build a Bayesian net, which is the subject of this section. Bayesian nets enable reasoning under uncertainty and combine the advantages of an intuitive visual representation with a sound mathematical basis in Bayesian probability. With Bayesian nets, it is possible to articulate dependencies between different variables and to propagate consistently the impact of evidence (observations) on the probabilities of uncertain outcomes. The underlying theory of Bayesian nets combines Bayesian probability theory and uses conditional independence to represent dependencies between variables (see Pearl, 1986, Spegelhalter and Cowell, 1992). Bayesian nets have proven useful in many areas of application such as medical expert systems, diagnosis of failures, pattern matching, speech recognition, and, more relevantly, for the OpRisk community, risk assessment of complex systems in high-stakes environments (see Neil *et al.*, 2001, Ale *et al.*, 2009, Fenton *et al.*, 2004, including financial institutions Alexander, 2003, Neil *et al.*, 2005, Adusei-Poku, 2005, Cowell *et al.*, 2006, Rebonato, 2010).

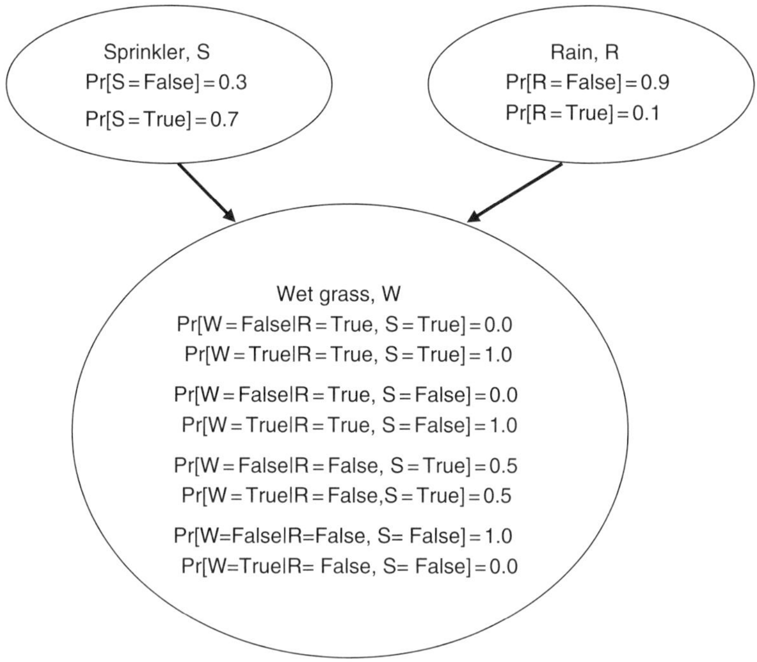

FIGURE 14.5 **Wet grass example**

14.9.1 DEFINITION AND EXAMPLES

Consider a popular simple example where garden grass observed in the morning is either wet or dry, and if the grass is wet then it is due to either the sprinkler or rain. The probability that the sprinkler is on is 0.3 and the probability of rain is 0.1. Denote the corresponding binary variables as W, S, and R that can have only two possible values, True and False. The graph of this example, with probabilities of the grass being wet conditional on the state of the sprinkler and rain, is presented in Figure 14.5. In general, a *graph* is defined as follows.

Definition 14.4 (Graph) *A graph is a set of points (called nodes or vertices) that represent random variables connected by lines (called arcs or edges).* ■

In our example, the variables W, S, and R are *nodes* and connection lines are *arrows* (*directed edge*). The arrow between two nodes indicates a causal relationship between the nodes, for example, the change in S causes a change in W, the change in R causes a change in W. The node where the edge originates is called *parent* and the node to which the edge points is called *child*. This example in Figure 14.5 is a *directed acyclic graph*.

Definition 14.5 (Directed acyclic graph) *A directed acyclic graph is a graph formed by collection of nodes (vertices) and directed edges where each edge connects one node to another one in such a way that moving along edges in the edge direction it is impossible to return to a previous node. This means that, starting from any node, we cannot go back to the same node following the arrows that connect the other nodes.* ■

If there is no edge between some nodes, then these nodes are independent (e.g., R and S are independent in our example). The joint density of W, S, and R can be written as

$$p(W, S, R) = p(W|S, R)p(S|R)p(R),$$

which can be simplified using independence of R and S as

$$p(W, S, R) = p(W|S, R)p(S)p(R),$$

which can be easily calculated using probabilities for R and S and conditional probabilities for W presented in Figure 14.5. It is also a trivial exercise to find marginal (unconditional) distributions for all variables, for example, for $p(W)$ we get $\mathbb{P}\mathrm{r}[W = \mathit{True}] = 0.235$ and $\mathbb{P}\mathrm{r}[W = \mathit{False}] = 0.765$. We can also answer questions about probabilities given some information. For example, if the grass is wet, then was it caused by rain or sprinkler? That is, we can find $\mathbb{P}\mathrm{r}[R = \mathit{True}|W = \mathit{True}]$ and $\mathbb{P}\mathrm{r}[S = \mathit{True}|W = \mathit{True}]$ using Bayes theorem

$$\mathbb{P}\mathrm{r}[R = \mathit{True}|W = \mathit{True}] = \frac{\mathbb{P}\mathrm{r}[R = \mathit{True}, W = \mathit{True}]}{\mathbb{P}\mathrm{r}[W = \mathit{True}]}.$$

We might need to add another variable C representing season as in Figure 14.6 and specify conditional densities $p(R|C)$ and $p(S|C)$. Here, R and S are independent given its parent C. Then, joint density of all variables can be written (using conditional independencies in the structure) as

$$p(C, W, S, R) = p(W|S, R)p(S|C)p(R|C)p(C).$$

Similar and other examples can be found by Pearl (2009). Of course, we could take a purely associative approach, and build all the probability tables. If we are using a purely frequentist approach, we just collect data and calculate all the conditional probabilities from our dataset. However if we have to provide subjective probabilities, our cognitive aptitude in dealing with some questions rather than others does make a big difference. Some of the assignments we are requested to make invoke a causal link among variables, and others are linked by a subtler associative ("diagnostic") kind of relationship. It so happens that our mind works much more effectively in a causal rather than in an associative mode. The associative link between

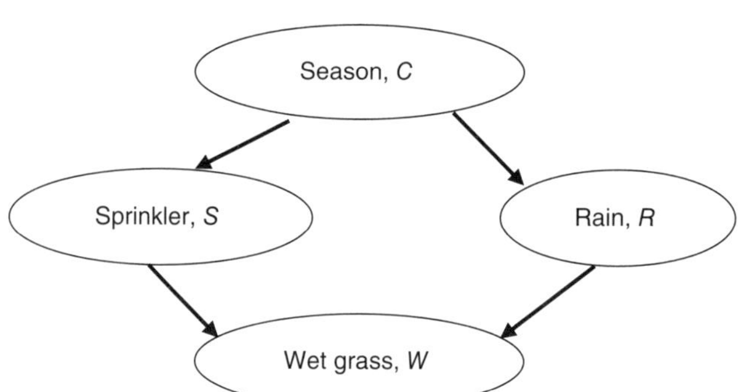

FIGURE 14.6 Wet grass example with variable C representing season

variables contained in a purely probabilistic description can make our interpretation difficult, and this can be true even when we have plenty of relevant data. If understanding, rather than just analyzing, the output of complicated analysis is our final goal, we can follow a causal model such as in our simple wet grass example, where

- The season of the year affects the probability of rain and of the sprinkler being on or off (but is not affected by either);
- The status of the sprinkler (on or off) and whether it rained or not affects the probability of the grass being wet (but sprinkler status and occurrence of rain are not caused by the wetness of the grass).

If we have n events $E_1, \ldots, E_n$, where each is a discrete random variable that can take m values, then there will be m^n joint events. In principle, the knowledge of joint distribution allows to derive everything someone might want from a statistical perspective. However, it is just unrealistic for risk managers to assign all required joint probabilities. *Bayesian nets* (or *Bayesian networks*) are the tools that help experts to specify the required probabilities; causal links among the variables in the network simplify the task.

Definition 14.6 (Bayesian networks) *Bayesian nets (or networks) are directed, acyclical graphs.* ■

Each node (or vertex) in a Bayesian net is associated with a discrete random variable and with a list of numbers called a *conditional probability table*. The examples we considered in Figures 14.5 and 14.6 are simple Bayesian nets. Another general illustration is presented in Figure 14.7; if an arrow points from node A to node B, node A is said to be the *parent* of node B and B is the *child* of node A. The parents, parents of the parents, etc. are called *ancestors*; the children, the children of children, etc. are called the *descendants*. A causal link between A and B means that knowledge of A helps in the assessment of the probability of B happening. When we find a causal link, it is easier for experts to assign conditional probabilities.

Formally defining, for Bayesian net with n random variables $\boldsymbol{E} = \{E_1, \ldots, E_n\}$, each variable E_i with parents $pa(E_i)$ has an associated conditional probability table $p(E_i|pa(E_i))$. The overall joint density of $\boldsymbol{E}$ in the Bayesian net is the product of all conditional probabilities

$$p(\boldsymbol{E}) = \prod_{i=1}^{n} p(E_i|pa(E_i)). \tag{14.15}$$

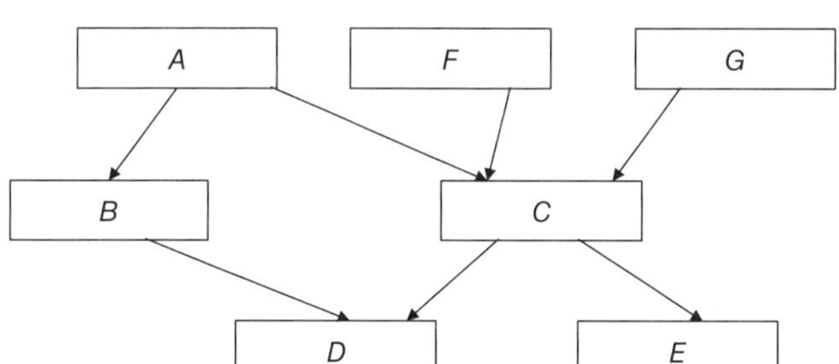

FIGURE 14.7 Bayesian net example: A causes B and C, F causes C, G causes C, B and C cause D and C cause D and E

The power of Bayesian nets comes from the fact that we can interpret the existence of an arrow between two nodes as representing a causal link between the associated random variables. Once all conditional probabilities are given, Bayesian nets provide a full graphical (and, via the tables, numerical) representation of the joint distribution.

14.9.2 CONSTRUCTING AND SIMULATING A BAYESIAN NET

To build Bayesian nets, we must first identify the variables relevant to our application. Next, we associate each variable to a node. In the following step, we draw arcs between the nodes. We do so keeping in mind that whenever we draw an arc (a line) between two nodes, we mean that there is a causal link between the two variables. Now, wherever we find a line, we put an arrow at one end or the other. The causal link goes in the direction of the arrow: from the parent to the child. Note that, in general, one does not consider causalities in which there may be a feedback effect and thus the arrows typically go in one direction.

When we build Bayesian nets, we make the fundamental assumption that, conditional on its parents, any variable is independent of all other variables apart from its descendants. This also means that there is no path dependence in our net, see the example in Figure 14.7, the probability of D happening does not depend on whether C was caused by A or F. This condition of path independence is very closely linked to the Markov condition; a time-ordered process, X_i, is a Markov process if

$$p(X_k|X_{k-1}, X_{k-2}, \dots, X_{k-n}) = p(X_k|X_{k-1}).$$

The conditional independence simplifies finding the joint probability. For example, in the case of the net in Figure 14.7, the joint probability

$$p(A, B, C, D, E, F, G) = p(A)p(F)p(G)p(B|A)p(C|F, A, G)p(D|B, C)p(E|C).$$

In short, for a Bayesian net with at most m parents, we need at-most-m-conditioned probabilities to build all the joint probabilities.

To perform inference/calculations using a Bayesian net, we need to specify prior distributions for each node, that is, unconditional distributions for the nodes without a parent (e.g., $p(A), p(F)$, and $p(G)$ for nodes A, F, and G in Figure 14.7) and conditional distributions for child nodes (e.g., $p(B|A), p(C|A, F, G), p(D|B, C)$, and $p(E|C)$ in Figure 14.7). These distributions can be determined by experts or from data using, for example, maximum likelihood by taking the ratio of the event to the frequency of the parent. Then a simple sampling from a Bayesian net can be done as follows.

Algorithm 14.1 (Logic sampling from a Bayesian Net)

1. *Sample the unconditional nodes (e.g., A, F, and G are sampled from $p(A)$, $p(F)$, and $p(G)$ in Figure 14.7);*
2. *Sample the child nodes using conditional probabilities at these nodes given sample of the parents until all nodes are sampled (e.g., in Figure 14.7, B and C are sampled from $p(B|A)$ and $p(C|A, F, G)$; then D and E are sampled from $p(D|B, C)$ and $p(E|C)$); this process will produce one sample of the Bayesian net;*
3. *Repeat the above steps many times; obtained samples for each node can be used to estimate marginal probabilities for the nodes.*

If we are given a specific information (evidence) about the state of some nodes, then we can use this algorithm but will need to discard Bayesian net samples that have node values inconsistent with the evidence. Of course, this approach is not efficient and there are more efficient algorithms (where no Bayesian net samples are rejected) that can be found in the literature. A very good conceptual treatment of Bayesian nets in general can be found by Williamson (2005). Pearl (2009) provides an excellent discussion of causality. For readers who want to delve much more deeply and rigorously into Bayesian networks, Heckerman *et al.* (1995) provides a very good and solid reference.

14.9.3 COMBINING EXPERT OPINION AND DATA IN A BAYESIAN NET

It is possible to update the prior distributions specified by experts for the nodes in a Bayesian net using information from the data utilizing a Bayesian approach, which has already been discussed in Sections 7.2 and 14.3. Under this approach, if a model is parameterized by $\boldsymbol{\theta}$, then $\boldsymbol{\theta}$ is treated as random with prior density $\pi(\boldsymbol{\theta})$ and the estimate for $\boldsymbol{\theta}$, updated by the vector of new data $\boldsymbol{Y}$, is found from the posterior density $\pi(\boldsymbol{\theta}|\boldsymbol{Y})$, which is proportional to the prior density times the likelihood of the data. For example, as presented by Yoon (2003), assume that the node random variable Y can have values $y_1, \ldots, y_m$ with corresponding probabilities (model parameters) $q_1, \ldots, q_m$. The density of n independent samples with n_1 samples having value y_1, n_2 samples having value y_2, etc. is described by the multinomial density

$$p(n_1, \ldots, n_m|q_1, \ldots, q_m) = \frac{n!}{\prod_{i=1}^{m} n_i!} \prod_{i=1}^{m} q_i^{n_i}. \tag{14.16}$$

A convenient candidate for the prior density of $q_1, \ldots, q_m$ is the Dirichlet distribution $D(\alpha_1, \ldots, \alpha_m)$ with the density

$$\pi(q_1, \ldots, q_m) = \frac{\Gamma(\alpha)}{\prod_{i=1}^{m} \Gamma(\alpha_i)} \prod_{i=1}^{m} q_i^{\alpha_i - 1}, \quad \alpha = \sum_i \alpha_i. \tag{14.17}$$

In the case of binary nodes (two-parameter case), it is just Beta distribution. Using expert opinions about expected value and range of q_i, we can estimate $\alpha_1, \ldots, \alpha_m$ using statistics

$$\mathbb{E}[q_i] = \frac{\alpha_i}{\alpha} \quad \text{and} \quad \mathbb{V}\text{ar}[q_i] = \frac{\alpha_i(\alpha - \alpha_i)}{\alpha^2(\alpha + 1)}. \tag{14.18}$$

The convenience of Dirichlet distribution as the prior comes from the fact that the posterior density $\pi(q_1, \ldots, q_m|n_1, \ldots, n_m)$ can be found in closed form to be Dirichlet distribution $D(\alpha_1 + n_1, \ldots, \alpha_m + n_m)$, that is, it is conjugate prior distribution discussed in Section 7.2.1, with the density

$$\pi(q_1, \ldots, q_m|n_1, \ldots, n_m) \propto \prod_{i=1}^{m} q_i^{\alpha_i + n_i - 1}. \tag{14.19}$$

14.9.4 BAYESIAN NET AND OPERATIONAL RISK

It is important in OpRisk management to be able to create quantitative risk models that provide a basis for fruitful discussions about specific risk processes. Detailed causal modeling at a business process level can absorb organization-specific input, reveal the importance of risk drivers (causal factors), identify the potential lack of controls/barriers, and quantify the overall risk expected and unexpected losses. Such modeling tools have been applied in the safety critical industries for a long time; for example, Fault Trees (Haasl, 1965) and Event Trees (Nielsen, 1971). There are various shortcomings in their use related to time dynamics, subjective probability, and a large number of dependent parameters and multi-state variables rather than binary-state variables.

To handle these challenges, many practitioners in the safety critical industry have adopted Bayesian nets over the last decades. Numerous researchers have highlighted advantages of causal modeling via Bayesian nets when compared to the traditional actuarial techniques. It is argued that a major benefit of Bayesian nets is the structured and scientifically sound way of combining statistical analysis with other information sources such as knowledge of business processes, near misses, and expert opinions using Bayesian methods. Bayesian nets can enable risk practitioners to link the operational conditions of the bank, including control environment, directly to the probability and severity of losses, that is, assess the quantitative effect of risk management decisions.

Many OpRisk practitioners have a view that Bayesian net models are suitable as a tool for risk management, but not as a tool for establishing economic and regulatory capitals. The main challenges include the following:

- Illustration of wide applicability for the full complexity of OpRisk in banks (examples so far have been applied only to simple parts of banking processes);
- Consideration of time dynamics for OpRisk events. Operational losses evolve through series of sequential events in time (e.g., sequential checks performed by a set of controls that may or may not be functional); this is possible to achieve under the Bayesian net framework; see details of dynamic Bayesian nets by Neil *et al.* (2009);
- Implementation of continuous variables for loss severities (Bayesian nets use static discretization of continuous variables that limits the model accuracy for loss severities, especially in the cases of heavy-tailed distributions).

The paper by Neil *et al.* (2009) addresses these problems presenting a hybrid dynamic Bayesian network model. The model considered by Neil *et al.* (2009) is presented in Figure 14.8. The first layer models the potential loss events, E_t, and their evolution in time as they are influenced by controls, C_t, embedded within the business process. This dynamic time-based evolution of an event given the controls is modeled by $p(E_t|E_{t-1}, C_t)$ with time periods $t = 1, \ldots, T$. The performance of each control, C_t, is modeled as a function of a set of operational failure modes, O_j. These failure modes are in turn influenced by a set of causal factors, F_i. The occurrence of a failure mode for a specific control O_{C_t} is dependent, through causal factors, on the occurrence of a failure mode for a previous control in time $O_{C_{t-s}}$, which is modeled using a special kind of causal factor called a dependency factor, D_k. A dependency factor, D_k, is a causal factor conditioned on the occurrence of some operational failure mode specific for a control, $O_{C_{t-s}}$, which in turn influences a failure mode of a secondary control, O_{C_t}, where $s = 1, \ldots, t$.

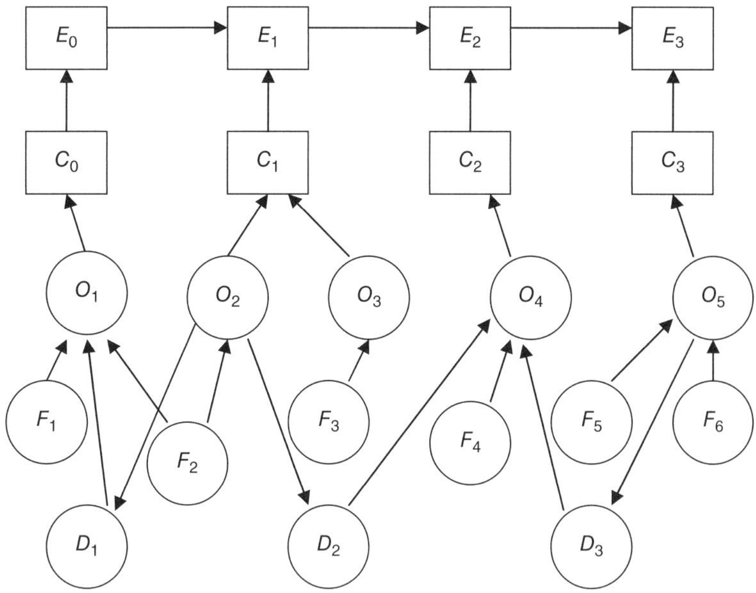

FIGURE 14.8 Illustration of Bayesian Network

TABLE 14.4 **Logical conditional transition probabilities $p(E_t|E_{t-1}, C_t)$**

E_{t-1} = fail				E_{t-1} = ok	
	C_t =fail	C_t =ok		C_t =fail	C_t =ok
E_t =fail	1	0	E_t =fail	0	0
E_t =ok	0	1	E_t =ok	1	1

For this model setup, the operational failure mode is modeled as a function of causal factors, F_i, and dependency factors, D_k; the failures that have no dependency relationship with other failure modes are modeled as a function of F_i only. The business process is then represented by a sequence of discrete time-dependent events such that we have a dynamic Bayesian network as shown in Figure 14.8, which is a graph representation of the full joint distribution:

$$p(\boldsymbol{E}, \boldsymbol{C}, \boldsymbol{O}, \boldsymbol{F}, \boldsymbol{D}) = \prod_{t=1}^{T} \prod_{s=1}^{t} \prod_{j} \prod_{i} \prod_{k} p(E_t|E_{t-1}, C_t)p(C_t|\boldsymbol{O}_{C_t}) \tag{14.20}$$

$$\times\, p(O_j|\boldsymbol{F}_{O_j}, D_{O_j})p(D_k|\boldsymbol{O}_{C_{t-s}})p(F_i)p(C_0), \tag{14.21}$$

where $\boldsymbol{E}, \boldsymbol{C}, \boldsymbol{O}, \boldsymbol{F}, \boldsymbol{D}$ are sets of loss events, controls, operational failure modes, causal factor variables, and dependency factors.

Each of the state transitions $p(E_t|E_{t-1}, C_t)$ is specified by a discrete node probability table for the transition probability of the loss event from an undetected to a detected state, dependent on the control state. If the control operates correctly at time t, the loss event E_{t-1} would transit to a correct operating state at E_t using the logical conditional transition probabilities given in Table 14.4. For the given model specification, the marginal probability of occurrence for each loss event can be written as

$$p(\boldsymbol{E}) = \sum_{\boldsymbol{C},\boldsymbol{O},\boldsymbol{F},\boldsymbol{D}} \prod_{t=1}^{T} \prod_{s=1}^{t} \prod_{j} \prod_{i} \prod_{k} p(E_t|E_{t-1}, C_t)p(C_t|\boldsymbol{O}_{C_t}) \tag{14.22}$$

$$\times\, p(O_j|\boldsymbol{F}_{O_j}, \boldsymbol{D}_{O_j})p(D_k|\boldsymbol{O}_{C_{t-s}})p(F_i)p(C_0). \tag{14.23}$$

The paper, Neil *et al.* (2009) presents the junction tree algorithm for performing the inference and describes extension to modeling severity and calculation of the total annual loss distribution.

14.10 Discussion

Expert elicitation is certainly one of the challenges in operational risk because many managers and employees may not have a sound knowledge of statistics and probability theory. This may lead to misleading results and misunderstanding of the true loss process model. It is important that questions answered by experts are simple and well understood by respondents. There are psychological aspects involved. There is a vast literature on expert elicitation published by statisticians, especially in areas such as security and ecology. For a good review, see O'Hagan (2006). Scenarios come with their own biases, such as anchoring, confirmation, availability, and overconfidence (see discussions in Section 2.6). For example, anchoring is a common human tendency to rely too heavily, or "anchor", on one trait or piece of information when making decisions. Tversky and Kahneman (1974) introduced the concept that usually once the anchor is set, a bias exists toward that value. For descriptions of the other biases, see Kahneman *et al.* (1982).

Merging scenario analysis with historical data will be discussed more in Chapter 15. One of the problems with the required merge is that scenarios are formulated for loss processes while historical data are often collected in Basel II business line/event type risk cells that include many processes; moreover, some processes may contribute to different risk cells.

CHAPTER FIFTEEN

Combining Different Data Sources

This chapter provides a detailed Overview of OpRisk methods to combine data sources, particularly aimed to aid practitioners in forming rigorous and statistically justified methods for combining different data sources. We cover in this chapter:

- Linear Weighted Combining based on the minimum variance principle;
- Bayesian methods for the combining of two data sources with posterior and predictive distributional models developed for frequency and severity models. Special topics include different methods for prior development and hyper-parameter estimation;
- Bayesian methods for combining expert opinion with internal and external data;
- Combining data sources using Linear Bayes or Credibility Theory;
- Non-parametric Bayesian methods for combining of data sources;
- Combining data sources based on other generalized uncertainty methods such as Dempster-Shafer theory and p-boxes.

It is hard to perform a robust estimation of low frequency/ high severity risks using data from a single financial institution. As the number of these large events in a financial institution would be minimal, any statistical analysis would present significant challenges. There is simply not enough data to estimate high quantiles of the risk distribution. Other sources of information that can be used to improve risk estimates and are required by the Basel II for OpRisk Advanced Measurement Approaches (AMA) are internal data, relevant external data, scenario analysis, and factors reflecting the business environment and internal control systems. Specifically, Basel II AMA includes the following requirement[1] (BCBS 2006, p. 152):

> Any OpRisk measurement system must have certain key features to meet the supervisory soundness standard set out in this section. These elements must include the use of internal data, relevant external data, scenario analysis and factors reflecting the business environment and internal control systems.

We discussed scenario analysis in Chapter 14 and fitting historically observed losses in Chapter 7, also risk indicators/factors will be discussed in Chapter 16. Combining these different data sources for model estimation is certainly one of the main challenges in OpRisk.

Fundamental Aspects of Operational Risk and Insurance Analytics: A Handbook of Operational Risk, First Edition. Marcelo G. Cruz, Gareth W. Peters, and Pavel V. Shevchenko.

[1]The original text is available free of charge on the BIS website www.BIS.org/bcbs/publ.htm.

Conceptually, the following ways have been proposed to process different data sources of information; (see e.g., Berger 1985, sections 4.11 and 4.12):

- Numerous ad hoc procedures;
- Parametric and nonparametric Bayesian methods;
- General nonprobabilistic 'uncertainty' based methods such as Dempster–Shafer theory.

Some of the ad hoc procedures will be presented shortly and Bayesian methods are the main focus of this chapter. Finally, we present the basic concepts and methods of Dempster–Shafer theory and closely related ideas of "probability boxes" (referred to as *"p-boxes"*). Dempster–Shafer theory is based on the so-called belief functions and *Dempster's rule* for combining evidence (see Dempster 1968 and Shafer 1976). It is often referred to as a generalization of the Bayesian method. For a good summary on the methods for obtaining Dempster–Shafer structures and "p-boxes", and aggregation methods handling a conflict between the objects from different sources, see Ferson *et al.* (2003).

Often in practice, accounting for factors reflecting the business environment and internal control systems is achieved via scaling of data. Then ad hoc procedures are used to combine internal data, external data, and expert opinions. For example:

- Fit the severity distribution to the combined samples of internal and external data and fit the frequency distribution using internal data only;
- Estimate the Poisson annual intensity for the frequency distribution as $w\lambda_{int} + (1 - w)\lambda_{ext}$, where the intensities λ_{ext} and λ_{int} are implied by the external and internal data, respectively, using expert specified weight w;
- Estimate the severity distribution as a mixture

$$w_1 F_{SA}(x) + w_2 F_I(x) + (1 - w_1 - w_2) F_E(x),$$

where $F_{SA}(x)$, $F_I(x)$ and $F_E(x)$ are the distributions identified by scenario analysis, internal data, and external data, respectively, using expert-specified weights w_1 and w_2.

Statistically sound and consistent methodologies to combine different data sources are the main topics of this chapter discussed next.

15.1 Minimum Variance Principle

Probably the easiest to use and most flexible procedure to combine estimators obtained from different data sources is the minimum variance principle. Under the *minimum variance principle*, the combined estimator is a linear combination of the individual estimators obtained from internal data, external data, and expert opinion separately with the weights chosen to minimize the variance of the combined estimator.

The rationale behind the principle is as follows. Consider two unbiased independent estimators $\widehat{\Theta}^{(1)}$ and $\widehat{\Theta}^{(2)}$ for parameter θ, that is, $\mathbb{E}\left[\widehat{\Theta}^{(k)}\right] = \theta$ and $\mathbb{V}\mathrm{ar}[\widehat{\Theta}^{(k)}] = \sigma_k^2$, $k = 1, 2$.

Then the combined unbiased linear estimator and its variance are

$$\hat{\Theta}_{tot} = w_1\hat{\Theta}^{(1)} + w_2\hat{\Theta}^{(2)}, \quad w_1 + w_2 = 1, \tag{15.1}$$

$$\mathbb{V}\mathrm{ar}[\hat{\Theta}_{tot}] = w_1^2\sigma_1^2 + (1 - w_1)^2\sigma_2^2. \tag{15.2}$$

It is easy to find the weights minimizing $\mathbb{V}\mathrm{ar}[\hat{\Theta}_{tot}]$:

$$w_1 = \frac{\sigma_2^2}{\sigma_1^2 + \sigma_2^2} \quad \text{and} \quad w_2 = \frac{\sigma_1^2}{\sigma_1^2 + \sigma_2^2}.$$

The weights behave as expected in practice. In particular, $w_1 \to 1$ if $\sigma_1^2/\sigma_2^2 \to 0$ (σ_1^2/σ_2^2 is the uncertainty of the estimator $\hat{\Theta}^{(1)}$ over the uncertainty of $\hat{\Theta}^{(2)}$) and $w_1 \to 0$ if $\sigma_2^2/\sigma_1^2 \to 0$. This method can easily be extended to combine three or more estimators using the following theorem.

Theorem 15.1 (Minimum variance estimator) *Assume that we have $\hat{\Theta}^{(i)}$, $i = 1, 2, \ldots, K$ unbiased and independent estimators of θ with variances $\sigma_i^2 = \mathbb{V}\mathrm{ar}[\Theta^{(i)}]$. Then the linear estimator*

$$\hat{\Theta}_{tot} = w_1\hat{\Theta}^{(1)} + \cdots + w_K\hat{\Theta}^{(K)}$$

is unbiased and has a minimum variance if

$$w_i = \frac{1/\sigma_i^2}{\sum_{k=1}^{K}(1/\sigma_k^2)}.$$

In this case, $w_1 + \cdots + w_K = 1$ and

$$\mathbb{V}\mathrm{ar}[\hat{\Theta}_{tot}] = \left(\sum_{k=1}^{K}\frac{1}{\sigma_k^2}\right)^{-1}.$$

Proof: The linear estimator $\hat{\theta}_{tot} = w_1\hat{\theta}_1 + \cdots + w_K\hat{\theta}_K$ is unbiased, that is, $\mathbb{E}[\hat{\theta}_{tot}] = \theta$, if $w_1 + \cdots + w_K = 1$ because $\mathbb{E}[\hat{\theta}_k] = \theta$. Minimization of the variance

$$\mathbb{V}\mathrm{ar}[\hat{\theta}_{tot}] = w_1^2\sigma_1^2 + \cdots + w_K^2\sigma_K^2$$

under the constraint $w_1 + \cdots + w_K$ is equivalent to unconstrained minimization of the

$$\Psi = \mathbb{V}\mathrm{ar}[\hat{\theta}_{tot}] - \lambda(w_1 + \cdots + w_K - 1),$$

which is a well-known method of Lagrange multipliers. Optimization of this method requires solution of the following equations:

$$\frac{\partial\Psi}{\partial w_i} = 2w_i\sigma_i^2 - \lambda = 0, \quad i = 1, \ldots, K;$$

$$\frac{\partial\Psi}{\partial\lambda} = -(w_1 + \cdots + w_K - 1) = 0.$$

This gives

$$\frac{1}{2}\lambda = \left(\sum_{k=1}^{K}(1/\sigma_k^2)\right)^{-1}, \quad w_i = \frac{1/\sigma_i^2}{\sum_{k=1}^{K}(1/\sigma_k^2)}. \tag{15.3}$$

■

It is a simple exercise to extend this principle to the case of unbiased estimators with known linear correlations. Heuristically, minimum variance principle can be applied to almost any quantity, including a distribution parameter or distribution characteristic such as mean, variance, or quantile. The assumption that the estimators are unbiased estimators for θ is probably reasonable when combining estimators from different experts (or from expert and internal data). However, it is certainly questionable if applied to combine estimators from the external and internal data. The following sections focus on the Bayesian inference method that can be used to combine these data sources in a consistent statistical framework.

EXAMPLE 15.1

Assume that there are two independent unbiased estimates of the expected annual number of losses $\hat{\theta}^{(1)} = 10$ and $\hat{\theta}^{(2)} = 15$ with corresponding variances $\sigma_1^2 = 9$ and $\sigma_2^2 = 4$. For example, these could be expert opinions. Then, using Theorem 15.1, the combined unbiased estimate is

$$\hat{\theta}_{tot} = w_1\hat{\theta}^{(1)} + (1 - w_1)\hat{\theta}^{(2)},$$

where $w_1 = \sigma_2^2/(\sigma_1^2 + \sigma_2^2) = 4/13$, that is, $\hat{\theta}_{tot} \approx 13.5$ with the variance estimate $w_1^2\sigma_1^2 + (1 - w_1)^2\sigma_2^2 \approx 2.8$. ■

15.2 Bayesian Method to Combine Two Data Sources

The Bayesian inference method can be used to combine different data sources in a consistent statistical framework. The main concept of the Bayesian approach has already been introduced in Section 7.2. Now we consider the approach in detail in the context of combining.

Consider random data $\boldsymbol{X} = (X_1, X_2, \ldots, X_n)$ whose joint density, for given parameters $\boldsymbol{\Theta} = (\Theta_1, \Theta_2, \ldots, \Theta_K)$, is $h(\boldsymbol{x}|\boldsymbol{\theta})$. In the Bayesian approach, both observations and parameters are considered to be random. Then the joint density is

$$h(\boldsymbol{x}, \boldsymbol{\theta}) = h(\boldsymbol{x}|\boldsymbol{\theta})\pi(\boldsymbol{\theta}) = \pi(\boldsymbol{\theta}|\boldsymbol{x})h(\boldsymbol{x}), \tag{15.4}$$

where

- $\pi(\boldsymbol{\theta})$ is the probability density of the parameters, a so-called prior density function. Typically, $\pi(\boldsymbol{\theta})$ depends on a set of further parameters that are called hyper-parameters, omitted here for simplicity of notation;

- $\pi(\boldsymbol{\theta}|\boldsymbol{x})$ is the density of parameters given data $\boldsymbol{X}$, a so-called posterior density;
- $h(\boldsymbol{x}, \boldsymbol{\theta})$ is the joint density of observed data and parameters;
- $h(\boldsymbol{x}|\boldsymbol{\theta})$ is the density of observations for given parameters. This is the same as a likelihood function if considered as a function of $\boldsymbol{\theta}$, that is, $L_{\boldsymbol{x}}(\boldsymbol{\theta}) = h(\boldsymbol{x}|\boldsymbol{\theta})$;
- $h(\boldsymbol{x})$ is a marginal density of $\boldsymbol{X}$ that can be written as

$$h(\boldsymbol{x}) = \int h(\boldsymbol{x}|\boldsymbol{\theta})\pi(\boldsymbol{\theta})d\boldsymbol{\theta}. \tag{15.5}$$

For simplicity of notation, we consider continuous $\pi(\boldsymbol{\theta})$ only. If $\pi(\boldsymbol{\theta})$ is a discrete probability function, then the integration in the given expression should be replaced by a corresponding summation (see Definition 5.7).

Predictive distribution. The objective (in the context of OpRisk) is to estimate the predictive distribution (frequency and severity) of a future observation X_{n+1} conditional on all available information $\boldsymbol{X} = (X_1, X_2, \ldots, X_n)$. Assume that, conditionally, given $\boldsymbol{\Theta}$, X_{n+1} and $\boldsymbol{X}$ are independent, and X_{n+1} has a density $f(x_{n+1}|\boldsymbol{\theta})$. It is even common to assume that $X_1, X_2, \ldots, X_n, X_{n+1}$ are all conditionally independent (given $\boldsymbol{\Theta}$) and identically distributed. Then the conditional density of X_{n+1}, given data $\boldsymbol{X} = \boldsymbol{x}$, is

$$f(x_{n+1}|\boldsymbol{x}) = \int f(x_{n+1}|\boldsymbol{\theta})\pi(\boldsymbol{\theta}|\boldsymbol{x})d\boldsymbol{\theta}. \tag{15.6}$$

If X_{n+1} and $\boldsymbol{X}$ are not independent, then the predictive distribution should be written as

$$f(x_{n+1}|\boldsymbol{x}) = \int f(x_{n+1}|\boldsymbol{\theta}, \boldsymbol{x})\pi(\boldsymbol{\theta}|\boldsymbol{x})d\boldsymbol{\theta}. \tag{15.7}$$

Posterior distribution. Bayes's theorem (see Theorem 7.3) says that the posterior density can be calculated from (15.4) as

$$\pi(\boldsymbol{\theta}|\boldsymbol{x}) = h(\boldsymbol{x}|\boldsymbol{\theta})\pi(\boldsymbol{\theta})/h(\boldsymbol{x}). \tag{15.8}$$

Here, $h(\boldsymbol{x})$ plays the role of a normalization constant. Thus, the posterior distribution can be viewed as proportional to the product of a prior knowledge with a likelihood function for observed data.

In the context of OpRisk, one considers the following three logical steps:

- The prior distribution $\pi(\boldsymbol{\theta})$ should be estimated by scenario analysis (expert opinions with reference to external data);
- Then the prior distribution should be weighted with the observed data using formula (15.8) to get the posterior distribution $\pi(\boldsymbol{\theta}|\boldsymbol{x})$;
- Formula (15.6) is then used to calculate the predictive distribution of X_{n+1} given the data $\boldsymbol{X} = \boldsymbol{x}$.

Remark 15.1 *Of course, the posterior density can be used to find parameter point estimators. Typically, these are the mean, mode, or median of the posterior (see Section 7.2.3). The use of the posterior mean as the point parameter estimator is optimal in a sense that the mean squared error of prediction is minimized. For more on this topic, see Section 7.3 or Bühlmann and Gisler (2005,* *section 2.3). However, in the case of OpRisk, it is more appealing to use the whole posterior to calculate the predictive distribution (15.6).*

If the data $X_1, X_2, \ldots, X_n$ are conditionally (given $\boldsymbol{\Theta} = \boldsymbol{\theta}$) independent and X_k is distributed with a density $f_k(\cdot|\boldsymbol{\theta})$, then the joint density of the data for given $\boldsymbol{\theta}$ can be written as $h(\boldsymbol{x}|\boldsymbol{\theta}) = \prod_{i=1}^{n} f_i(x_i|\boldsymbol{\theta})$. Denote the posterior density calculated after k observations as $\pi_k(\boldsymbol{\theta}|x_1, \ldots, x_k)$, then using (15.8), observe that

$$\begin{aligned}\pi_k(\boldsymbol{\theta}|x_1, \ldots, x_k) &\propto \pi(\boldsymbol{\theta}) \prod_{i=1}^{k} f_i(x_i|\boldsymbol{\theta}) \\ &\propto \pi_{k-1}(\boldsymbol{\theta}|x_1, \ldots, x_{k-1}) f_k(x_k|\boldsymbol{\theta}).\end{aligned} \tag{15.9}$$

It is easy to see from Equation (15.9) that the updating procedure that calculates the posteriors from priors can be done iteratively. Only the posterior distribution calculated after $k-1$ observations and the k-th observation are needed to calculate the posterior distribution after k observations. Thus, the loss history over many years is not required, making the model easier to understand and manage, and allowing experts to adjust the priors at every step. Formally, the posterior distribution calculated after $k - 1$ observations can be treated as a prior distribution for the k-th observation. In practice, initially, we start with the prior distribution $\pi(\boldsymbol{\theta})$ identified by expert opinions and external data only. Then, the posterior distribution $\pi(\boldsymbol{\theta}|\boldsymbol{x})$ is calculated, using Equation (15.8), when actual data are observed. If there is a reason (e.g., the new control policy introduced in a bank), then this posterior distribution can be adjusted by an expert and treated as the prior distribution for subsequent observations. Examples will be presented in the following sections.

Conjugate prior distributions. So-called conjugate distributions (see Definition 7.10) are very useful in practice when Bayesian inference is applied. We present conjugate pairs (Poisson–Gamma, LogNormal–Normal, Pareto–Gamma) that are good illustrative examples for modeling frequencies and severities in OpRisk. Several other pairs (Binomial–Beta, Gamma–Gamma, Exponential–Gamma) can be found, for example, in Bühlmann and Gisler (2005). In all these cases, the prior and posterior distributions have the same type and the posterior distribution parameters are easily calculated using the prior distribution parameters and observations (or recursively using Equation (15.9)).

15.2.1 ESTIMATING PRIOR: PURE BAYESIAN APPROACH

In general, the structural parameters of the prior distributions can be estimated subjectively using expert opinions (*pure Bayesian approach*) or using data (*empirical Bayesian approach*). Pure Bayesian approach has already been considered in Section 14.3. For convenience of readers, we repeat main points. In a pure Bayesian approach, the prior distribution is specified subjectively (i.e., in the context of OpRisk, using expert opinions). Berger (1985) lists several methods:

- *Histogram approach.* Split the space of the parameter $\boldsymbol{\theta}$ into intervals and specify the subjective probability for each interval. From this, the smooth density of the prior distribution can be determined;

- *Relative likelihood approach.* Compare the intuitive likelihoods of the different values of $\boldsymbol{\theta}$. Again, the smooth density of prior distribution can be determined. It is difficult to apply this method in the case of unbounded parameters;
- *Distribution function determinations.* Subjectively construct the distribution function for the prior and sketch a smooth curve;
- *Matching a given functional form.* Find the prior distribution parameters assuming some functional form for the prior distribution to match prior beliefs (on the moments, quantiles, etc.) as closely as possible.

In this chapter, the method of matching a given functional form will be used often. The use of a particular method is determined by a specific problem and expert experience. Usually, if the expected values for the quantiles (or mean) and their uncertainties are estimated by the expert, then it is possible to fit the priors.

Often, expert opinions are specified for some quantities such as quantiles or other risk characteristics rather than for the parameters directly. In this case, it might be better to assume some priors for these quantities that will imply a prior for the parameters. In general, given model parameters $\boldsymbol{\theta} = (\theta_1, \ldots, \theta_n)$, assume that there are risk characteristics $d_i = g_i(\boldsymbol{\theta})$, $i = 1, 2, \ldots, n$ that are well understood by experts. These could be some quantiles, expected values, expected durations between losses exceeding high thresholds, etc. Now, if experts specify the joint prior $\pi(d_1, \ldots, d_n)$, then using a transformation method, the prior for $\theta_1, \ldots, \theta_n$ is

$$\pi(\boldsymbol{\theta}) = \pi(g_1(\boldsymbol{\theta}), \ldots, g_n(\boldsymbol{\theta})) \left| \frac{\partial\left(g_1(\boldsymbol{\theta}), \ldots, g_n(\boldsymbol{\theta})\right)}{\partial\left(\theta_1, \ldots, \theta_n\right)} \right|, \tag{15.10}$$

where $\left|\partial\left(g_1(\boldsymbol{\theta}), \ldots, g_n(\boldsymbol{\theta})\right) / \partial\left(\theta_1, \ldots, \theta_n\right)\right|$ is the Jacobian determinant of the transformation. Essentially, the main difficulty in specifying a joint prior is due to a possible dependence between the parameters. It is convenient to choose the characteristics (for specification of the prior) such that independence can be assumed. For example, if the prior for the quantiles $q_1, \ldots, q_n$ (corresponding to probability levels $p_1 < p_2 < \cdots < p_n$) is to be specified, then to account for the ordering it might be better to consider the differences

$$d_1 = q_1, \quad d_2 = q_2 - q_1, \ldots, d_n = q_n - q_{n-1}.$$

Then, it is reasonable to assume independence between these differences and impose constraints $d_i > 0$, $i = 2, \ldots, n$. If experts specify the marginal priors $\pi(d_1), \pi(d_2), \ldots, \pi(d_n)$ (e.g., Gamma priors), then the full joint prior is

$$\pi(d_1, \ldots, d_n) = \pi(d_1) \times \pi(d_2) \times \cdots \times \pi(d_n),$$

and the prior for parameters $\boldsymbol{\theta}$ is calculated by transformation using Equation (15.10). To specify the i-th prior $\pi(d_i)$, an expert may use the approaches listed earlier. For example, if $\pi(d_i)$ is $Gamma(\alpha_i, \beta_i)$, then the expert may provide the mean and variational coefficient for $\pi(d_i)$ (or median and 0.95 quantile), which should be enough to determine α_i and β_i.

A very appealing statistical experiment demonstrating the importance of subjective prior information was given by Savage (1961):

1. A lady, who adds milk to her tea, claims to be able to tell whether the tea or milk was poured into the cup first. In all 10 trials, her answer is correct;

2. A music expert claims to be able to distinguish a page of Haydn's score from a page of Mozart's score. In all 10 trials, he makes a correct determination;
3. A drunken friend says that he can predict the outcome of a coin flip. In all 10 trials, his prediction is correct.

In all three cases, the unknown parameter to identify is the probability of the correct answer. Classical statistical approach (based on hypothesis testing), ignoring our prior opinion, would give a very strong evidence that all these claims are correct. We would not doubt this result for situation 2. However, for situation 3, our prior opinion is that this prediction is impossible and we would tend to ignore the empirical evidence. Different people may give different prior opinions for situation 1. Anyway, in all these cases, prior information is certainly valuable.

15.2.2 ESTIMATING PRIOR: EMPIRICAL BAYESIAN APPROACH

Under the empirical Bayesian approach, the parameter $\boldsymbol{\theta}$ is treated as a random sample from the prior distribution. Then using collective data of *similar* risks, the parameters of the prior are estimated using a marginal distribution of observations. Depending on the model setup, the data can be collective industry data, collective data in the bank, etc.

To explain, consider K similar risks where each risk has its own risk profile $\boldsymbol{\Theta}^{(i)}$, $i = 1, \ldots, K$ (see Figure 15.1). Given $\boldsymbol{\Theta}^{(i)} = \boldsymbol{\theta}^{(i)}$, the risk data $X_1^{(i)}, X_2^{(i)}, \ldots$ are generated from the distribution $F(x|\boldsymbol{\theta}^{(i)})$. The risks are different, having different risk profiles $\boldsymbol{\theta}^{(i)}$, but what they have in common is that $\boldsymbol{\Theta}^{(1)}, \ldots, \boldsymbol{\Theta}^{(K)}$ are distributed from the same density $\pi(\boldsymbol{\theta})$. Then, one can find the unconditional distribution of the data $\boldsymbol{X}$ and fit the prior distribution using all data (across all similar risks). This could be done, for example, by the maximum

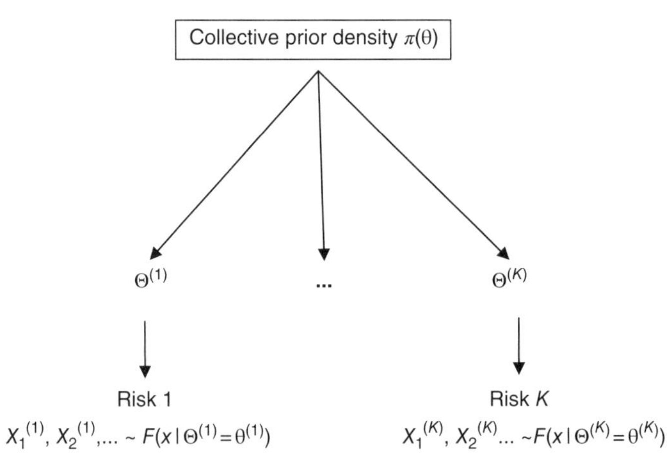

FIGURE 15.1 Empirical Bayes approach – interpretation of the prior density $\pi(\boldsymbol{\theta})$. Here, $\boldsymbol{\Theta}^{(i)}$ is the risk profile of the i-th risk. Given $\boldsymbol{\Theta}^{(i)} = \boldsymbol{\theta}^{(i)}$, the risk data $X_1^{(i)}, X_2^{(i)}, \ldots$ are generated from the distribution $F(x|\boldsymbol{\theta}^{(i)})$. The risks are different having different risk profiles $\boldsymbol{\theta}^{(i)}$, but $\boldsymbol{\Theta}^{(1)}, \ldots, \boldsymbol{\Theta}^{(K)}$ are distributed from the same common density $\pi(\boldsymbol{\theta})$

likelihood method or the method of moments or even empirically. This approach will be discussed in detail in Section 15.3.

15.2.3 POISSON FREQUENCY

Consider the annual number of events for a risk in one bank in year t modeled as a random variable from the Poisson distribution *Poisson* (λ). The intensity parameter λ is not known and the Bayesian approach models it as a random variable Λ. Then the following model for years $t = 1, 2, \ldots, T, T+1$ (where $T+1$ corresponds to the next year) can be considered.

Model Assumptions 15.1

- *Suppose that, given $\Lambda = \lambda$, the data $N_1, \ldots, N_{T+1}$ are independent random variables from the Poisson distribution, Poisson(λ):*

$$\Pr[N_t = n|\Lambda = \lambda] = e^{-\lambda}\frac{\lambda^n}{n!}, \quad \lambda \geq 0. \tag{15.11}$$

- *The prior distribution for Λ is a Gamma distribution, Gamma(α, β), with a density*

$$\pi(\lambda) = \frac{(\lambda/\beta)^{\alpha-1}}{\Gamma(\alpha)\beta}\exp(-\lambda/\beta), \quad \lambda > 0,\ \alpha > 0, \beta > 0. \tag{15.12}$$

That is, λ plays the role of $\boldsymbol{\theta}$ and $\boldsymbol{N} = (N_1, \ldots, N_T)$ the role of $\boldsymbol{X}$ in (15.8).
Posterior. Given $\Lambda = \lambda$, under the Model Assumptions 15.1, $N_1, \ldots, N_T$ are independent and their joint density, at $\boldsymbol{N} = \boldsymbol{n}$, is given by

$$h(\boldsymbol{n}|\lambda) = \prod_{i=1}^{T} e^{-\lambda}\frac{\lambda^{n_i}}{n_i!}. \tag{15.13}$$

Thus, using formula (15.8), the posterior density is

$$\pi(\lambda|\boldsymbol{n}) \propto \frac{(\lambda/\beta)^{\alpha-1}}{\Gamma(\alpha)\beta}\exp(-\lambda/\beta)\prod_{i=1}^{T} e^{-\lambda}\frac{\lambda^{n_i}}{n_i!} \propto \lambda^{\alpha_T-1}\exp(-\lambda/\beta_T), \tag{15.14}$$

which is *Gamma*(α_T, β_T), that is, the same as the prior distribution with updated parameters α_T and β_T given by

$$\alpha \to \alpha_T = \alpha + \sum_{i=1}^{T} n_i, \quad \beta \to \beta_T = \frac{\beta}{1 + \beta \times T}. \tag{15.15}$$

Improper constant prior. It is easy to see that, if the prior is constant (improper prior), that is, $\pi(\lambda|\boldsymbol{n}) \propto h(\boldsymbol{n}|\lambda)$, then the posterior is *Gamma*(α_T, β_T) with

$$\alpha_T = 1 + \sum_{i=1}^{T} n_i, \quad \beta_T = \frac{1}{T}. \tag{15.16}$$

In this case, the mode of the posterior $\pi(\lambda|\boldsymbol{n})$ is

$$\hat{\lambda}_T^{\text{MAP}} = (\alpha_T - 1)\beta_T = \frac{1}{T}\sum_{i=1}^{T} n_i, \tag{15.17}$$

which is the same as the maximum likelihood estimate (MLE) $\hat{\lambda}_T^{\text{MLE}}$ of λ.

Predictive distribution. Given data, the conditional predictive distribution for N_{T+1} is Negative Binomial, *NegBinomial*$(\alpha_T, 1/(1+\beta_T))$:

$$\begin{aligned}
\mathbb{P}\text{r}[N_{T+1} = m|\boldsymbol{N} = \boldsymbol{n}] &= \int f(m|\lambda)\pi(\lambda|\boldsymbol{n})d\lambda \\
&= \int e^{-\lambda}\frac{\lambda^m}{m!}\frac{\lambda^{\alpha_T-1}}{(\beta_T)^{\alpha_T}\Gamma(\alpha_T)}e^{-\lambda/\beta_T}d\lambda \\
&= \frac{(\beta_T)^{-\alpha_T}}{\Gamma(\alpha_T)m!}\int e^{-(1+1/\beta_T)\lambda}\lambda^{\alpha_T+m-1}d\lambda \\
&= \frac{\Gamma(\alpha_T+m)}{\Gamma(\alpha_T)m!}\left(\frac{1}{1+\beta_T}\right)^{\alpha_T}\left(\frac{\beta_T}{1+\beta_T}\right)^m.
\end{aligned} \tag{15.18}$$

It is assumed that given $\Lambda = \lambda$, N_{T+1} and $\boldsymbol{N}$ are independent. The expected number of events over the next year, given past observations, $\mathbb{E}[N_{T+1}|\boldsymbol{N}]$, that is, mean of *NegBinomial*$(\alpha_T, 1/(1+\beta_T))$ (which is also a mean of the posterior distribution in this case), allows for a good interpretation as follows:

$$\begin{aligned}
\mathbb{E}[N_{T+1}|\boldsymbol{N} = \boldsymbol{n}] &= \mathbb{E}[\lambda|\boldsymbol{N} = \boldsymbol{n}] \\
&= \alpha_T\beta_T \\
&= \beta\frac{\alpha + \sum_{i=1}^{T} n_i}{1+\beta T} \\
&= w_T\hat{\lambda}_T^{\text{MLE}} + (1-w_T)\lambda_0.
\end{aligned} \tag{15.19}$$

Here:

- $\hat{\lambda}_T^{\text{MLE}} = \frac{1}{T}\sum_{i=1}^{T} n_i$ is the estimate of λ using the observed counts only;
- $\lambda_0 = \alpha\beta$ is the estimate of λ using a prior distribution only (e.g., specified by expert);
- $w_T = \frac{T\beta}{T\beta+1}$ is the credibility weight in [0,1) used to combine λ_0 and $\hat{\lambda}_T^{\text{MLE}}$.

Remark 15.2

- *As the number of observed years T increases, the credibility weight w_T increases and vice versa. That is, the more observations we have, the greater credibility weight we assign to the estimator based on the observed counts, while the lesser credibility weight is attached to the expert opinion estimate. In addition, the larger the volatility of the expert opinion (larger β), the greater credibility weight is assigned to observations;*
- *Recursive calculation of the posterior distribution is very simple. That is, consider observed annual counts $n_1, n_2, \ldots, n_k, \ldots,$ where n_k is the number of events in the k-th year. Assume*

that the prior Gamma(α, β) *is specified initially, then the posterior* $\pi(\lambda|n_1, \ldots, n_k)$ *after the k-th year is a Gamma distribution, Gamma*(α_k, β_k)*, with* $\alpha_k = \alpha + \sum_{i=1}^{k} n_i$ *and* $\beta_k = \beta/(1 + \beta \times k)$. *Observe that*

$$\alpha_k = \alpha_{k-1} + n_k, \quad \beta_k = \frac{\beta_{k-1}}{1 + \beta_{k-1}}. \tag{15.20}$$

This leads to a very efficient recursive scheme, where the calculation of posterior distribution parameters is based on the most recent observation and parameters of posterior distribution calculated just before this observation.

Estimating prior. Suppose that the annual frequency of the OpRisk losses N is modeled by the Poisson distribution, *Poisson*$(\Lambda = \lambda)$, and the prior density $\pi(\lambda)$ for Λ is *Gamma*(α, β). Then, $\mathbb{E}[N|\Lambda] = \Lambda$ and $\mathbb{E}[\Lambda] = \alpha \times \beta$. The expert may estimate the expected number of events but cannot be certain of the estimate. One could say that the expert's "best" estimate for the expected number of events corresponds to $\mathbb{E}[\mathbb{E}[N|\Lambda]] = \mathbb{E}[\Lambda]$. If the expert specifies $\mathbb{E}[\Lambda]$ and an uncertainty that the "true" λ for next year is within the interval $[a,b]$ with a probability $\Pr[a \le \Lambda \le b] = p$ (it may be convenient to set $p = 2/3$), then the equations

$$\begin{aligned} &\mathbb{E}[\Lambda] = \alpha \times \beta, \\ &\Pr[a \le \Lambda \le b] = p = \int_a^b \pi(\lambda|\alpha, \beta) d\lambda = F_{\alpha,\beta}^{(G)}(b) - F_{\alpha,\beta}^{(G)}(a) \end{aligned} \tag{15.21}$$

can be solved numerically to estimate the structural parameters α and β. Here, $F_{\alpha,\beta}^{(G)}(\cdot)$ is the Gamma distribution, *Gamma*(α, β), that is,

$$F_{\alpha,\beta}^{(G)}(y) = \int_0^y \frac{x^{\alpha-1}}{\Gamma(\alpha)\beta^{\alpha}} \exp\left(-\frac{x}{\beta}\right) dx.$$

In the insurance industry, the uncertainty for the "true" λ is often measured in terms of the coefficient of variation, $\text{Vco}[\Lambda] = \sqrt{\text{Var}[\Lambda]}/\mathbb{E}[\Lambda]$. Given the expert estimates for $\mathbb{E}[\Lambda] = \alpha\beta$ and $\text{Vco}[\Lambda] = 1/\sqrt{\alpha}$, the structural parameters α and β are easily estimated.

EXAMPLE 15.2

If the expert specifies $\mathbb{E}[\Lambda] = 0.5$ and $\Pr[0.25 \le \Lambda \le 0.75] = 2/3$, then we can fit a prior distribution *Gamma*$(\alpha \approx 3.407,\ \beta \approx 0.147)$ by solving (15.21). Assume now that the bank experienced no losses over the first year (after the prior distribution was estimated). Then, using formulas (15.20), the posterior distribution parameters are $\hat{\alpha}_1 \approx 3.407 + 0 = 3.407$, $\hat{\beta}_1 \approx 0.147/(1 + 0.147) \approx 0.128$ and the estimated arrival rate using the posterior distribution is $\hat{\lambda}_1 = \hat{\alpha}_1 \times \hat{\beta}_1 \approx 0.436$. If, during the next year, no losses are observed again, then the posterior distribution parameters are $\hat{\alpha}_2 = \hat{\alpha}_1 + 0 \approx 3.407$, $\hat{\beta}_2 = \hat{\beta}_1/(1 + \hat{\beta}_1) \approx 0.113$, and $\hat{\lambda}_2 = \hat{\alpha}_2 \times \hat{\beta}_2 \approx 0.385$. Subsequent observations will update the arrival rate estimator correspondingly using formulas (15.20). Thus, starting from the expert specified prior, observations regularly update (refine) the posterior distribution. The expert might reassess the posterior distribution at any point in time (the posterior distribution can be treated as a prior distribution for the next period) if new practices/policies are introduced in the bank that affect the frequency of the loss. That is, if we have a new policy at time k, the expert may reassess the parameters and replace $\hat{\alpha}_k$ and $\hat{\beta}_k$ by $\hat{\alpha}_k^*$ and $\hat{\beta}_k^*$, respectively.

In Figure 15.2, we show the posterior best estimate for the arrival rate $\hat{\lambda}_k = \hat{\alpha}_k \times \hat{\beta}_k$, $k = 1, \ldots, 15$ (with the prior distribution as in the previous example), when the annual number of events N_k, $k = 1, \ldots, 25$ are simulated from $Poisson(\lambda = 0.6)$ and the realized samples for 25 years are

$$n_{1:25} = (0, 0, 0, 0, 1, 0, 1, 1, 1, 0, 2, 1, 1, 2, 0, 2, 0, 1, 0, 0, 1, 0, 1, 1, 0).$$

In the same figure, we show the standard MLE of the arrival rate $\hat{\lambda}_k^{\text{MLE}} = \frac{1}{k}\sum_{i=1}^{k} n_i$. After approximately 8 years, the estimators are very close to each other. However, for a small number of observed years, the Bayesian estimate is more accurate as it takes the prior information into account. Only after 12 years do both estimators converge to the true value of 0.6 (this is because the bank was very lucky to have no events during the first 4 years). Note that for this example we assumed the prior distribution with a mean equal to 0.5, which is different from the true arrival rate. Thus, this example shows that an initially incorrect prior estimator is corrected by the observations as they become available. It is interesting to observe

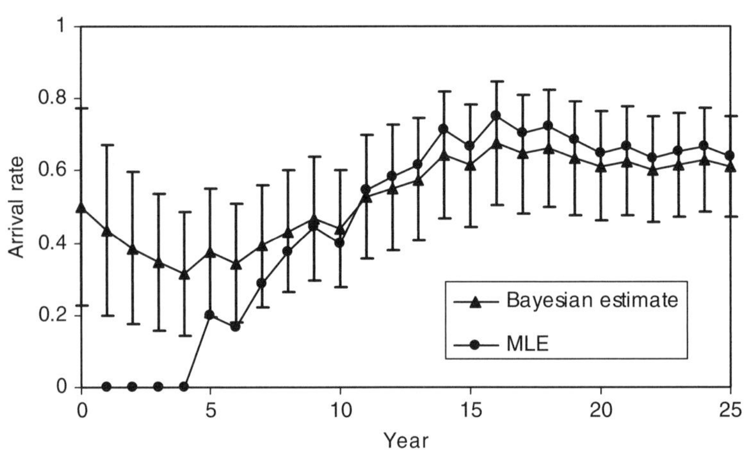

FIGURE 15.2 The Bayesian and the standard maximum likelihood estimates of the arrival rate versus the observation year. The Bayesian estimate is a mean of the posterior distribution when the prior distribution is $Gamma(\alpha, \beta)$ with $\mathbb{E}[\Lambda] = 0.5$; $\alpha \approx 3.41$ and $\beta \approx 0.15$. The MLE is a simple average over the number of observed events. The annual counts were sampled from the Poisson(0.6). Error bars for Bayesian estimates correspond to the posterior standard deviation. See Example 15.2 for details

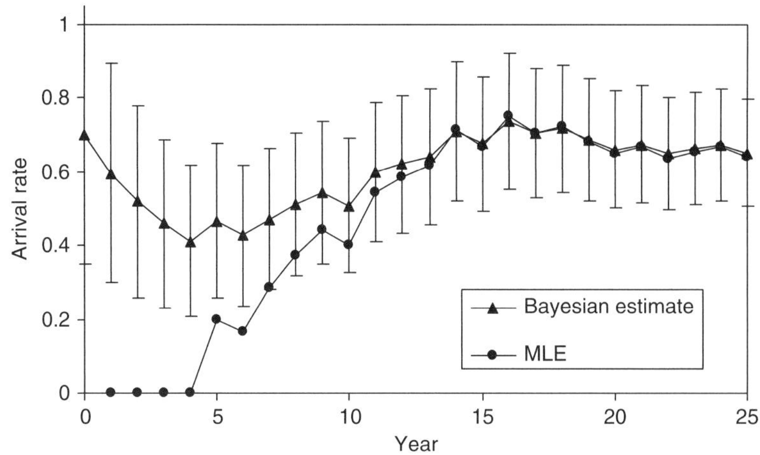

FIGURE 15.3 **The Bayesian and the standard maximum likelihood estimates of the arrival rate versus the observation year. The Bayesian estimate is a mean of the posterior distribution when the prior distribution is *Gamma*(α, β) with $\mathbb{E}[\Lambda] = 0.7$ and Vco$[\Lambda] = 0.5$; $\alpha = 4$ and $\beta = 0.175$. The MLE is a simple average over the number of observed events. The annual counts were sampled from the Poisson(0.6). Error bars for Bayesian estimates correspond to the posterior standard deviation. See Example 15.2 for details**

that, in year 14, the estimators become slightly different again. This is because the bank was unlucky to experience event counts (1, 1, 2) in the years (12, 13, 14). As a result, the MLE becomes higher than the true value, while the Bayesian estimate is more stable (smooth) with respect to the unlucky years. If this example is repeated with different sequences of random numbers, then one would observe quite different MLEs (for small k) and more stable Bayesian estimates.

In Figure 15.3, we show the maximum likelihood and Bayesian posterior estimates for the arrival rates if the prior is *Gamma*$(\alpha = 4, \beta = 0.175)$. The parameters of the prior correspond to the situation when the expert specifies $\mathbb{E}[\Lambda] = \alpha\beta = 0.7$ and Vco$[\Lambda] = 1/\sqrt{\alpha} = 0.5$.

Finally, we note that in both cases, the standard deviation of the posterior distribution *Gamma*(α_k, β_k) is large for small k. It is indicated by the error bars in Figures 15.2 and 15.3 and calculated as $\beta_k\sqrt{\alpha_k}$. ■

15.2.4 THE LOGNORMAL SEVERITY

Assume that the loss severity for a risk in one bank is modeled as a random variable from a LogNormal distribution, *LogNormal*(μ, σ^2), whose density is

$$f(x|\mu, \sigma) = \frac{1}{x\sqrt{2\pi\sigma^2}} \exp\left(-\frac{(\ln x - \mu)^2}{2\sigma^2}\right). \tag{15.22}$$

This distribution often gives a good fit for OpRisk loss data. It also belongs to a class of heavy-tailed distributions (i.e., the distribution tail $1 - F(x)$ decays to 0 slower than any exponential $\exp(-\epsilon x)$, $\epsilon > 0$). The parameters μ and σ are not known and the Bayesian approach models these as random variables Θ_μ and Θ_σ, respectively. We assume that the losses over the years $t = 1, 2, \ldots, T$ are observed and should be modeled for the next year $T + 1$. To simplify notation, we denote the losses over the past T years as $X_1, \ldots, X_n$ and the future losses as $X_{n+1}, \ldots$. Then the model can be structured as follows. For simplicity, assume that σ is known and μ is unknown. The case where both σ and μ are unknown will be treated later.

Model Assumptions 15.2

- *Suppose that, given σ and $\Theta_\mu = \mu$, the data $X_1, \ldots, X_n, \ldots$ are independent random variables from LogNormal(μ, σ^2). That is, $Y_i = \ln X_i$, $i = 1, 2, \ldots$ are distributed from Normal(μ, σ^2);*
- *Assume that parameter σ is known and the prior distribution for Θ_μ is Normal(μ_0, σ_0^2). That is, the prior density is*

$$\pi(\mu) = \frac{1}{\sigma_0\sqrt{2\pi}} \exp\left(-\frac{(\mu - \mu_0)^2}{2\sigma_0^2}\right). \tag{15.23}$$

Denote the losses over the past years as $\boldsymbol{X} = (X_1, \ldots, X_n)$ and the corresponding log losses as $\boldsymbol{Y} = (Y_1, \ldots, Y_n)$. Note that μ plays the role of $\boldsymbol{\theta}$ in Equation (15.8).

Posterior. Under these assumptions, the joint density of the data over past years (conditional on σ and $\Theta_\mu = \mu$) at position $\boldsymbol{Y} = \boldsymbol{y}$ is given by

$$h(\boldsymbol{y}|\mu, \sigma) = \prod_{i=1}^{n} \frac{1}{\sigma\sqrt{2\pi}} \exp\left(-\frac{(y_i - \mu)^2}{2\sigma^2}\right). \tag{15.24}$$

Then, using Equation (15.8), the posterior density can be written as

$$\begin{aligned}\pi(\mu|\boldsymbol{y}) &\propto \frac{\exp\left(-\frac{(\mu-\mu_0)^2}{2\sigma_0^2}\right)}{\sigma_0\sqrt{2\pi}} \prod_{i=1}^{n} \frac{\exp\left(-\frac{(y_i-\mu)^2}{2\sigma^2}\right)}{\sigma\sqrt{2\pi}} \\ &\propto \exp\left(-\frac{(\mu - \mu_{0,n})^2}{2\sigma_{0,n}^2}\right),\end{aligned} \tag{15.25}$$

that corresponds to Normal$(\mu_{0,n}, \sigma_{0,n}^2)$, i.e. the same as the prior distribution with updated parameters

$$\mu_0 \to \mu_{0,n} = \frac{\mu_0 + \omega \sum_{i=1}^{n} y_i}{1 + n\omega}, \tag{15.26}$$

$$\sigma_0^2 \to \sigma_{0,n}^2 = \frac{\sigma_0^2}{1 + n\omega}, \quad \text{where } \omega = \sigma_0^2/\sigma^2. \tag{15.27}$$

The expected value of Y_{n+1} (given past observations), $\mathbb{E}[Y_{n+1}|\boldsymbol{Y}=\boldsymbol{y}]$, allows for a good interpretation, as follows:

$$\mathbb{E}[Y_{n+1}|\boldsymbol{Y}=\boldsymbol{y}] = \mu_{0,n} = \frac{\mu_0 + \omega\sum_{i=1}^{n} y_i}{1+n\omega} = w_n\bar{y}_n + (1-w_n)\mu_0, \tag{15.28}$$

where

- $\bar{y}_n = \frac{1}{n}\sum_{i=1}^{n} y_i$ is the estimate of μ using the observed losses only;
- μ_0 is the estimate of μ using a prior distribution only (e.g., specified by expert);
- $w_n = \frac{n}{n+\sigma^2/\sigma_0^2}$ is the credibility weight in [0,1) used to combine μ_0 and $\bar{y}_n$.

Remark 15.3

- *As the number of observations increases, the credibility weight w increases and vice versa. That is, the more observations we have, the greater weight we assign to the estimator based on the observed counts and the lesser weight is attached to the expert opinion estimate. Moreover, larger uncertainty in the expert opinion σ_0^2 leads to a higher credibility weight for observations and larger volatility of observations σ^2 leads to a higher credibility weight for expert opinions;*
- *The posterior distribution can be calculated recursively as follows. Consider the data $Y_1, Y_2, \ldots, Y_k, \ldots$. Assume that the prior distribution, $Normal(\mu_0, \sigma_0^2)$, is specified initially, then the posterior density $\pi(\mu|y_1, \ldots, y_k)$ after the k-th event is $Normal(\mu_{0,k}, \sigma_{0,k}^2)$, with*

$$\mu_{0,k} = \frac{\mu_0 + \omega\sum_{i=1}^{k} y_i}{1+k\omega}, \quad \sigma_{0,k}^2 = \frac{\sigma_0^2}{1+k\omega},$$

where $\omega = \sigma_0^2/\sigma^2$. It is easy to show that

$$\mu_{0,k} = \frac{\mu_{0,k-1} + \omega_{k-1}y_k}{1+\omega_{k-1}}, \quad \sigma_{0,k}^2 = \frac{\sigma^2\omega_{k-1}}{1+\omega_{k-1}}, \tag{15.29}$$

with $\omega_{k-1} = \sigma_{0,k-1}^2/\sigma^2$. That is, calculation of the posterior distribution parameters can be based on the most recent observation and the parameters of the posterior distribution calculated just before this observation.

Estimating prior. Suppose that X, the severity of operational losses, is modeled by the LogNormal distribution, $LogNormal(\mu, \sigma^2)$, and Model Assumptions 15.2 are satisfied. Then, for given Θ_μ (and σ is known), the expected loss is

$$\Omega = \mathbb{E}[X|\Theta_\mu] = \exp(\Theta_\mu + \tfrac{1}{2}\sigma^2), \tag{15.30}$$

and the quantile at level q is

$$Q_X(q) = \exp(\Theta_\mu + \sigma z_q), \tag{15.31}$$

where $z_q = \Phi^{-1}(q)$ is the inverse of the standard Normal distribution. That is, Ω and $Q_X(q)$ are functions of Θ_μ.

Consider the case when the prior distribution for Θ_μ is $Normal(\mu_0, \sigma_0^2)$. In this case, unconditionally, Ω is distributed from $LogNormal(\mu_0 + \frac{1}{2}\sigma^2, \sigma_0^2)$ and the quantile $Q_X(q)$ is distributed from $LogNormal(\mu_0 + \sigma z_q, \sigma_0^2)$. Then, the expert may specify "the best" estimate of the expected loss $\mathbb{E}[\Omega]$ and uncertainty, that is, the interval $[a,b]$ such that the true expected loss is within the interval with a probability $p = \Pr[a \le \Omega \le b]$. Then, the equations

$$\begin{aligned} p &= \Pr[a \le \Omega \le b] = \Phi\left(\frac{\ln b - \frac{1}{2}\sigma^2 - \mu_0}{\sigma_0}\right) - \Phi\left(\frac{\ln a - \frac{1}{2}\sigma^2 - \mu_0}{\sigma_0}\right), \\ \mathbb{E}[\Omega] &= \exp(\mu_0 + \tfrac{1}{2}\sigma^2 + \tfrac{1}{2}\sigma_0^2), \end{aligned} \tag{15.32}$$

can be solved to find μ_0, σ_0. Here, $\Phi(\cdot)$ is the standard Normal distribution.

EXAMPLE 15.3

For example, assume that $\sigma = 2$ and the expert estimates are $\mathbb{E}[\Omega] = 10$ and $p = \Pr[8 \le \Omega \le 12] = 2/3$. Then, solving (15.32) gives $\mu_0 \approx 0.28$ and $\sigma_0 \approx 0.21$. Finally, using Equation (15.26) we can calculate the posterior parameters $\mu_{0,k}, \sigma_{0,k}$ as observations X_k, $k = 1, 2, \ldots$ become available. ■

One can also try to fit parameters μ_0 and σ_0 using estimates for some quantile and uncertainty by solving

$$\begin{aligned} p &= \Pr[a \le Q_X(q) \le b] = \Phi\left(\frac{\ln b - \sigma z_q - \mu_0}{\sigma_0}\right) - \Phi\left(\frac{\ln a - \sigma z_q - \mu_0}{\sigma_0}\right), \\ \mathbb{E}[Q_q] &= \exp(\mu_0 + \sigma z_q + \tfrac{1}{2}\sigma_0^2). \end{aligned} \tag{15.33}$$

Remark 15.4 *If the uncertainty for Ω or $Q_X(q)$ in (15.32)–(15.33) is measured using the coefficient of variation* $\mathrm{Vco}[X] = \sqrt{\mathbb{V}\mathrm{ar}[X]}/\mathbb{E}[X]$, *then μ_0, σ_0 are easily expressed in the closed form. In the insurance industry,* Vco *is often provided by regulators.*

The $LogNormal(\mu, \sigma^2)$ **Severity with Unknown** μ **and** σ. Now, consider the case of both μ and σ unknown and modeled by random variables Θ_μ and Θ_σ, respectively. Suppose that, given $\Theta_\mu = \mu$ and $\Theta_\sigma = \sigma$, the data $X_1, \ldots, X_n, \ldots$ are independent random variables from $LogNormal(\mu, \sigma^2)$, that is, $Y_i = \ln X_i \sim Normal(\mu, \sigma^2)$. Assume also that the prior distribution of Θ_σ^2 is the inverse Chi-squared distribution, $InvChiSq(\nu, \beta)$, and the prior distribution of Θ_μ (given $\Theta_\sigma = \sigma$) is $Normal(\theta, \sigma^2/\phi)$. Under these assumptions, the joint posterior density can be found in closed form. For details and proofs, see Shevchenko (2011, section 4.3.5). The posterior has the same form as the joint prior distribution with parameters updated as

$$\begin{aligned} \nu_n &= \nu + n, \\ \beta_n &= \beta + \phi\theta^2 + n\bar{y}^2 - (\phi\theta + n\bar{y})^2/(\phi + n), \\ \theta_n &= \phi\theta + n\bar{y}/(\phi + n), \\ \phi_n &= \phi + n, \end{aligned} \tag{15.34}$$

where $\bar{y} = \frac{1}{n}\sum_{i=1}^{n} y_i$ and $\bar{y^2} = \frac{1}{n}\sum_{i=1}^{n} y_i^2$. It is easy to see that if the prior is constant (improper prior), that is, $\pi(\mu, \sigma|\boldsymbol{y}) \propto h(\boldsymbol{y}|\mu, \sigma)$, then the posterior densities $\pi(\sigma^2|\boldsymbol{y})$ and $\pi(\mu|\sigma^2, \boldsymbol{y})$ correspond to the $InvChiSq(\nu_n, \beta_n)$ and $Normal(\theta_n, \sigma_n^2/\phi_n)$, respectively, with

$$\nu_n = n - 3, \quad \beta_n = n\bar{y^2} - n(\bar{y})^2, \quad \theta_n = \bar{y}, \quad \phi_n = n. \tag{15.35}$$

In this case, the mode of the posterior density $\pi(\mu, \sigma|\boldsymbol{y})$ is

$$\hat{\mu}^{\text{MAP}} = \bar{y}, \quad (\hat{\sigma}^2)^{\text{MAP}} = \bar{y^2} - (\bar{y})^2, \tag{15.36}$$

which are the same as MLEs of μ and σ^2.

Estimating prior for both μ and σ. For given Θ_μ and Θ_σ, the loss quantile at the level q is

$$Q_X(q) = \exp(\Theta_\mu + \Theta_\sigma z_q), \tag{15.37}$$

see (15.31). Thus, one can find Θ_σ via two quantiles Q_{q_2} and Q_{q_1} as

$$\Theta_\sigma = \frac{\ln(Q_X(q_2)/Q_X(q_1))}{z_{q_2} - z_{q_1}}. \tag{15.38}$$

Then, one can try to fit the prior distribution for Θ_σ using the expert opinions on

$$\mathbb{E}[\ln(Q_X(q_2)/Q_X(q_1))] \quad \text{and} \quad \Pr[a \leq Q_X(q_2)/Q_X(q_1) \leq b],$$

or the opinions involving several pairs of quantiles. Given σ, the prior distribution for μ can be estimated using Equation (15.32) or (15.33).

15.2.5 PARETO SEVERITY

Another important example of the severity distribution, which is very useful to fit heavy-tailed losses, for a given threshold $L > 0$, is the Pareto distribution, $Pareto(\xi, L)$, with a density

$$f(x|\xi) = \frac{\xi}{L}\left(\frac{x}{L}\right)^{-\xi-1}. \tag{15.39}$$

It is defined for $x \geq L$ and $\xi > 0$. If $\xi > 1$, then the mean is $L\xi/(\xi - 1)$; otherwise, the mean does not exist. The tail parameter ξ is unknown and modeled by a random variable Θ_ξ.

Model Assumptions 15.3

- *Suppose that conditionally, given $\Theta_\xi = \xi$, the data $X_1, \ldots, X_n, \ldots$ are independent random variables from $Pareto(\xi, L)$;*
- *The prior distribution for the tail parameter Θ_ξ is $Gamma(\alpha, \beta)$, that is, the prior density is*

$$\pi(\xi) \propto \xi^{\alpha-1}\exp(-\xi/\beta). \tag{15.40}$$

Denote the losses over past years as $\boldsymbol{X} = (X_1, \ldots, X_n)$.

Posterior. Given $\boldsymbol{X} = \boldsymbol{x}$, under the previous assumptions, the posterior density (using (15.8)) is given by

$$\pi(\xi|\boldsymbol{x}) = \xi^n \exp\left(-(\xi+1)\sum_{i=1}^{n} \ln(x_i/L)\right) \xi^{\alpha-1} \exp(-\xi/\beta)$$
$$\propto \xi^{\alpha_n - 1} \exp(-\xi/\beta_n), \tag{15.41}$$

which is $Gamma(\alpha_n, \beta_n)$, that is, the same as the prior distribution with updated parameters (i.e., a conjugate model)

$$\alpha \to \alpha_n = \alpha + n, \quad \beta^{-1} \to \beta_n^{-1} = \beta^{-1} + \sum_{i=1}^{n} \ln(x_i/L). \tag{15.42}$$

The mean of the posterior distribution for Θ_ξ allows for a good interpretation, as follows:

$$\hat{\xi} = \mathbb{E}[\Theta_\xi|\boldsymbol{X} = \boldsymbol{x}] = \alpha_n\beta_n$$
$$= \frac{\alpha + n}{\beta^{-1} + \sum_{i=1}^{n} \ln(x_i/L)}$$
$$= w_n\hat{\xi}_n^{\text{MLE}} + (1 - w_n)\xi_0, \tag{15.43}$$

where

- $\hat{\xi}_n^{\text{MLE}} = \frac{1}{n}\sum_{i=1}^{n} \ln(x_i/L)$ is the MLE of ξ using the observed losses;
- $\xi_0 = \alpha\beta$ is the estimate of ξ using a prior distribution only (e.g., specified by expert);
- $w_n = \left[\sum_{i=1}^{n} \ln(x_i/L)\right] \times \left[\sum_{i=1}^{n} \ln(x_i/L) + 1/\beta\right]^{-1}$ is the weight in [0,1) combining ξ_0 and $\hat{\xi}_n^{\text{MLE}}$.

The posterior distribution can be easily calculated recursively. Consider the observed losses $x_1, x_2, \ldots, x_k, \ldots$. Assume that the prior, $Gamma(\alpha, \beta)$, is specified initially, then the posterior $\pi(\xi|x_1, \ldots, x_k)$ after the k-th event is $Gamma(\alpha_k, \beta_k)$ with

$$\alpha_k = \alpha + k, \quad \beta_k^{-1} = \beta^{-1} + \sum_{i=1}^{k} \ln(x_i/L).$$

Thus, the parameters of the posterior can be calculated recursively as

$$\alpha_k = \alpha_{k-1} + 1, \quad \beta_k^{-1} = \beta_{k-1}^{-1} + \ln(x_k/L). \tag{15.44}$$

Again, this leads to a very efficient recursive scheme, where the calculation of the posterior distribution parameters is based on the most recent observation and parameters of the posterior distribution calculated just before this observation.

Remark 15.5 *It is important to note that the prior and posterior distributions of* Θ_ξ *are Gamma distributions formally defined for* $\xi > 0$. *Thus, there is a finite probability that* $\Pr[\Theta_\xi \leq 1] > 0$, *which leads to infinite means of predicted distributions, that is,* $\mathbb{E}[X_i] = \infty$ *and* $\mathbb{E}[X_{n+1}|\boldsymbol{X}] = \infty$. *If we do not want to allow for infinite mean behavior, then* ξ *should be restricted to* $\xi > 1$. *See Section 7.2.4 on how to deal with this.*

Improper constant prior. It is easy to see that if the prior is constant (improper prior), that is, $\pi(\xi|\boldsymbol{x}) \propto h(\boldsymbol{x}|\xi)$, then the posterior is *Gamma*(α_n, β_n) with

$$\alpha_n = n + 1, \quad \beta_n^{-1} = \sum_{i=1}^{n} \ln(x_i/L). \tag{15.45}$$

In this case, the mode of the posterior density $\pi(\xi|\boldsymbol{x})$ is

$$\hat{\xi}^{\text{MAP}} = \frac{n}{\sum_{i=1}^{n} \ln(x_i/L)}, \tag{15.46}$$

which is the same as MLE of ξ.

Estimating prior. Suppose that X, the severity of operational losses exceeding threshold L, is modeled by the Pareto distribution, *Pareto*(ξ, L). Then, conditionally on $\Theta_\xi = \xi$, the expected loss

$$\mathbb{E}[X|\Theta_\xi = \xi] = \mu(\xi) = \frac{L\xi}{\xi - 1}, \quad \text{if} \quad \xi > 1, \tag{15.47}$$

and the quantile of the loss distribution at level q is

$$Q_{X|\Theta_\xi=\xi}(q) = f_q(\xi) = L\exp\left(-\frac{\ln(1-q)}{\xi}\right), \quad \xi > 0. \tag{15.48}$$

The mean and quantile of the loss are functions of ξ and thus, unconditionally, are random variables

$$\mu(\Theta_\xi) \quad \text{and} \quad f_q(\Theta_\xi),$$

respectively. If there is a reason to believe that, unconditionally, expected loss is finite, then the tail parameter ξ should satisfy $\xi \geq B > 1$. Now, assume that we choose the prior distribution for Θ_ξ to be *Gamma*(α, β) distribution truncated below B, that is, to have a density

$$\pi(\xi) = \frac{\xi^{\alpha-1}\exp(-\xi/\beta)}{(1 - F^{(G)}_{\alpha,\beta}(B))\Gamma(\alpha)\beta^\alpha}\mathbb{I}_{\{\xi \geq B\}}, \quad \xi \geq B, \ \alpha > 0, \ \beta > 0, \tag{15.49}$$

where $F^{(G)}_{\alpha,\beta}(\cdot)$ is a Gamma distribution, *Gamma*(α, β). If the expert estimates $\mathbb{E}[\Theta_\xi]$ and the uncertainty $\Pr[a \leq \Theta_\xi \leq b] = p$, then the following two equations

$$\begin{aligned} \mathbb{E}[\Theta_\xi] &= \alpha\beta\frac{1 - F^{(G)}_{\alpha+1,\beta}(B)}{1 - F^{(G)}_{\alpha,\beta}(B)}, \\ \Pr[a \leq \Theta_\xi \leq b] &= \frac{F^{(G)}_{\alpha,\beta}(b) - F^{(G)}_{\alpha,\beta}(a)}{1 - F^{(G)}_{\alpha,\beta}(B)} \end{aligned} \tag{15.50}$$

can be solved to estimate the structural parameters: shape α and scale β.

EXAMPLE 15.4

Assume that the lower bound for the tail parameter is $B = 2$ and the expert estimates are $\mathbb{E}[\Theta_\xi] = 5$, $\mathbb{P}\mathrm{r}[4 \leq \Theta_\xi \leq 6] = 2/3$. Then we can fit $\alpha \approx 23.086$, $\beta \approx 0.217$ and can calculate the posterior distribution parameters α_k, β_k when observations $x_1, x_2, \ldots$ become available, using (15.20). In Figure 15.4, we show the subsequent posterior best estimates for the tail parameter

$$\xi_k = \alpha_k \beta_k \frac{1 - F^{(G)}_{\alpha+1,\beta}(B)}{1 - F^{(G)}_{\alpha,\beta}(B)}, \quad k = 1, 2, \ldots, \tag{15.51}$$

when the losses X_k are simulated from *Pareto*$(4, 1)$. The actual simulated loss values are

$$x_{1:15} = (1.089, 1.181, 1.145, 1.105, 1.007, 1.451, 1.187, 1.116, 1.753, 1.383, 2.167, 1.180, 1.334, 1.272, 1.123).$$

In the same figure, we show the standard MLE of the tail parameter

$$\hat{\xi}_k^{\mathrm{MLE}} = \left(\frac{1}{k} \sum_{i=1}^{k} \ln(x_i/L) \right)^{-1}.$$

It is easy to see that the Bayesian estimates are more stable while the MLEs are quite volatile when the number of observations is small. As the number of observations increases, the two estimators become almost the same. As another example,

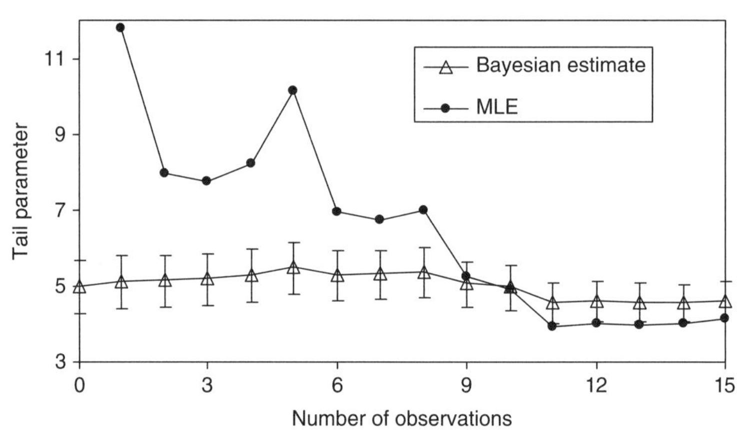

FIGURE 15.4 **The Bayesian and the standard maximum likelihood estimates of the Pareto tail parameter versus the number of observations. The losses were sampled from *Pareto*(4, 1). The prior distribution is *Gamma*(23.1, 0.22), truncated below $B = 2$. See Example 15.4 for details. Error bars correspond to 0.25 and 0.75 quantiles of the posterior distribution**

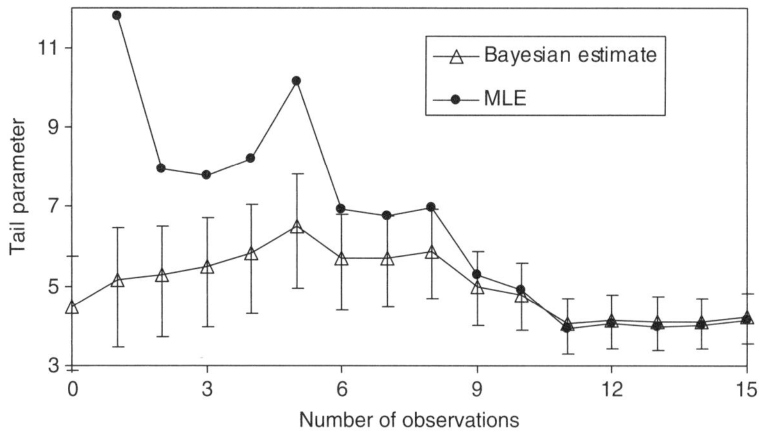

FIGURE 15.5 **The Bayesian and the standard maximum likelihood estimates of the Pareto tail parameter versus the number of observations. The losses were sampled from *Pareto*(4, 1). The prior distribution is *Gamma*(4, 1.125). See Example 15.4 for details. Error bars correspond to 0.25 and 0.75 quantiles of the posterior distribution**

Figure 15.5 compares the Bayesian estimate and MLE when the Gamma prior is specified by an expert who says that $\mathbb{E}[\Theta_\xi] = 4.5$ and $\text{Vco}[\Theta_\xi] = 0.5$. This gives the parameters of the prior $\alpha = 4$ and $\beta = 1.125$. ■

If it is difficult to express opinions on ξ directly, the expert may try to estimate the expected loss, quantile, or their uncertainties. It might be difficult numerically to fit α and β if the expert specifies unconditional expected loss or expected quantile

$$\mathbb{E}[\mu(\Theta_\xi)] = \int_B^\infty \mu(\xi)\pi(\xi)d\xi,$$

$$\mathbb{E}[f_q(\Theta_\xi)] = L\int_B^\infty f_q(\xi)\pi(\xi)d\xi, \tag{15.52}$$

respectively, as these are not easily expressed. Nevertheless, there is no problem in principle. Fitting opinions on uncertainties might be easier. For example, if the expert estimates the interval $[a, b]$ such that the true expected loss is within the interval with the probability $\Pr[a \le \mu(\Theta_\xi) \le b] = p$, then it leads to the equation

$$\Pr[a \le \mu(\Theta_\xi) \le b] = p = \int_{\widetilde{b}}^{\widetilde{a}} \pi(\xi)d\xi = \frac{F^{(G)}_{\alpha,\beta}(\widetilde{a}) - F^{(G)}_{\alpha,\beta}(\widetilde{b})}{1 - F^{(G)}_{\alpha,\beta}(B)}, \tag{15.53}$$

where $\widetilde{a} = \frac{a}{a-L}$, $\widetilde{b} = \frac{b}{b-L}$. Here, the interval bounds should satisfy $L < a < b \le B \times L/(B-1)$. The estimation of the interval $[a, b]$, $L < a < b$, such that the true quantile is within the interval with the probability $\mathbb{P}\mathrm{r}[a \le f_q(\Theta_\xi) \le b] = p$, leads to the equation

$$\mathbb{P}\mathrm{r}[a \le f_q(\Theta_\xi) \le b] = L \int_{C_1}^{C_2} \pi(\xi) d\xi = \frac{F_{\alpha,\beta}^{(G)}(C_2) - F_{\alpha,\beta}^{(G)}(C_1)}{1 - F_{\alpha,\beta}^{(G)}(B)}, \tag{15.54}$$

$$C_1 = -\frac{\ln(1-q)}{\ln(b/L)}, \quad C_2 = -\frac{\ln(1-q)}{\ln(a/L)},$$

where the interval bounds should satisfy

$$L < a < b \le L \exp\left(-\frac{\ln(1-q)}{B}\right).$$

Equations (15.53) and (15.54) or similar ones can be used to fit α and β. If the expert specifies more than two quantities, then one can use, for example, a nonlinear least square procedure to fit the structural parameters.

15.3 Estimation of the Prior Using Data

The prior distribution can be estimated using a marginal distribution of observations. The data can be collective industry data, collective data in the bank, etc. This approach is referred to as *empirical Bayes* (see Section 15.2.2 and Figure 15.1).

15.3.1 THE MAXIMUM LIKELIHOOD ESTIMATOR

Consider, for example, J *similar* risk cells with the data $\{X_k^{(j)}, k = 1, 2, \ldots, j = 1, \ldots, J\}$. This can be, for example, a specific business line/event-type risk cell in J banks. Denote the data over past years as $\boldsymbol{X}^{(j)} = (X_1^{(j)}, \ldots, X_{K_j}^{(j)})$, that is, K_j is the number of observations in bank j over past years. Assume that $X_1^{(j)}, \ldots, X_{K_j}^{(j)}$ are conditionally independent and identically distributed from the density $f(\cdot|\boldsymbol{\theta}^j)$, for given $\boldsymbol{\Theta}^{(j)} = \boldsymbol{\theta}^{(j)}$. That is, the risk cells have different risk profiles $\boldsymbol{\Theta}^{(j)}$. Assume now that the risks are similar, in a sense that $\boldsymbol{\Theta}^{(1)}, \ldots, \boldsymbol{\Theta}^{(J)}$ are independent and identically distributed from the same density $\pi(\boldsymbol{\theta})$. That is, it is assumed that the risk cells are the same a priori (before we have any observations) (see Figure 15.1). Then the joint density of all observations can be written as

$$f(\boldsymbol{x}^{(1)}, \ldots, \boldsymbol{x}^{(J)}) = \prod_{j=1}^{J} \int \left[\prod_{k=1}^{K_j} f(x_k^{(j)}|\boldsymbol{\theta}^{(j)}) \right] \pi(\boldsymbol{\theta}^{(j)}) d\boldsymbol{\theta}^{(j)}. \tag{15.55}$$

The parameters of $\pi(\boldsymbol{\theta})$ can be estimated using the maximum likelihood method by maximizing Equation (15.55). The distribution $\pi(\boldsymbol{\theta})$ is a prior distribution for the j-th cell. Using internal data of the j-th risk cell, its posterior density is calculated from Equation (15.8) as

$$\pi(\boldsymbol{\theta}^{(j)}|\boldsymbol{x}^{(j)}) = \prod_{k=1}^{K_j} f(x_k^{(j)}|\boldsymbol{\theta}^{(j)}) \pi(\boldsymbol{\theta}^{(j)}), \tag{15.56}$$

where $\pi(\boldsymbol{\theta})$ was fitted with MLE using Equation (15.55). The basic idea here is that the estimates based on observations from all banks are better than those obtained using a smaller number of observations available in the risk cell of a particular bank.

15.3.2 POISSON FREQUENCIES

It is not difficult to include a priori known differences (exposure indicators, expert opinions on the differences, etc.) between the risk cells from the different banks. As an example, we consider the case when the annual frequency of the events is modeled by the Poisson distribution with the Gamma prior and estimate structural parameters using the industry data with differences between the banks taken into account.

Model Assumptions 15.4 *Consider J risk cell with the loss frequencies $\{N_{j,k},\ k = 1, 2, \ldots, j = 1, \ldots, J\}$, where $N_{j,k}$ is the annual number of events in the j-th risk cell in the k-th year. Denote the data over past years in risk cell j as $\boldsymbol{N}_j = (N_{j,1}, \ldots, N_{j,K_j})$ and the data over past years in all risk cells as $\boldsymbol{N}_{1:J} = (\boldsymbol{N}_1, \ldots, \boldsymbol{N}_J)$. Assume the following:*

- *Given $\Lambda_j = \lambda_j$, $N_{j,k}$ are independent random variables from Poisson$(\lambda_j V_{j,k})$, with probability mass function denoted as $f(\cdot|\lambda_j)$. Here, $V_{j,k}$ is the known constant (i.e., the gross income or the volume or combination of several exposure indicators) and λ_j is a risk profile of the cell in the j-th bank;*
- *$\Lambda_1, \ldots, \Lambda_J$ are independent and identically distributed from Gamma(α, β) with the density denoted as $\pi(\cdot)$;*
- *Denote $N_j = \sum_{k=1}^{K_j} N_{j,k}$ and $V_j = \sum_{k=1}^{K_j} V_{j,k}$.*

Given Model Assumptions 15.4, the joint density of all data (over all J risk cells) can be written as

$$
\begin{aligned}
f(\boldsymbol{n}_{1:J}) &= \prod_{j=1}^{J} \int \left[\prod_{k=1}^{K_j} f(n_{j,k}|\lambda_j) \right] \pi(\lambda_j) d\lambda_j \\
&= \prod_{j=1}^{J} \int \left[\prod_{k=1}^{K_j} e^{-\lambda_j V_{j,k}} \frac{(V_{j,k}\lambda_j)^{n_{j,k}}}{(n_{j,k})!} \right] \frac{\lambda_j^{\alpha-1} e^{-\lambda_j/\beta}}{\Gamma(\alpha)\beta^{\alpha}} d\lambda_j \\
&= \left[\prod_{j=1}^{J} \prod_{k=1}^{K_j} \frac{(V_{j,k})^{n_{j,k}}}{(n_{j,k})!} \right] \prod_{j=1}^{J} \frac{\Gamma(\alpha + n_j)}{\Gamma(\alpha)\beta^{\alpha}(V_j + 1/\beta)^{\alpha+n_j}}.
\end{aligned} \tag{15.57}
$$

The parameters α and β can now be estimated using the maximum likelihood method by maximizing

$$
\ln f(\boldsymbol{n}_{1:J}) \propto \sum_{j=1}^{J} \left\{ \ln \Gamma(\alpha + n_j) - \ln \Gamma(\alpha) - \alpha \ln \beta - (\alpha + n_j) \ln \left(\frac{1}{\beta} + V_j \right) \right\} \tag{15.58}
$$

over α and β. To avoid the use of numerical optimization required for maximizing Equation (15.58), one could also use a method of moments (see Equations (15.62) and (15.63)). Once the prior distribution parameters α and β are estimated, then, using (15.8), the posterior distribution of λ_j for the j-th risk cell has a density

$$\pi(\lambda_j|\boldsymbol{n}_j) \propto \frac{(\lambda_j/\beta)^{\alpha-1}}{\Gamma(\alpha)\beta} e^{-\lambda_j/\beta} \prod_{k=1}^{K_j} e^{-\lambda_j V_{j,k}} \frac{(V_{j,k}\lambda_j)^{n_{j,k}}}{n_{j,k}!}$$
$$\propto \lambda^{n_j+\alpha-1} \exp\left(-\lambda_j V_j - \frac{\lambda_j}{\beta}\right), \tag{15.59}$$

which is $Gamma(\hat{\alpha}, \hat{\beta})$ with

$$\hat{\alpha} = \alpha + \sum_{k=1}^{K_j} n_{j,k}, \; \hat{\beta} = \beta\left(1 + \beta\sum_{k=1}^{K_j} V_{j,k}\right)^{-1}. \tag{15.60}$$

Assume that the exposure indicator of the cell in the j-th bank for the next year is $V_{j,K_j+1} = V$. Then, the predictive distribution for the annual number of events in the cell (conditional on the past internal data) is Negative Binomial, $NegBinomial(\hat{\alpha}, \hat{p} = 1/(1+V\hat{\beta}))$:

$$\Pr[N_{K_j+1} = n|\boldsymbol{N}_j = \boldsymbol{n}_j] = \int e^{-\lambda V} \frac{(V\lambda)^n}{n!} \frac{\lambda^{\hat{\alpha}-1}}{\Gamma(\hat{\alpha})\hat{\beta}^{\hat{\alpha}}} e^{-\lambda/\hat{\beta}} d\lambda$$
$$= \frac{\Gamma(n+\hat{\alpha})}{\Gamma(\hat{\alpha})n!}(1-\hat{p})^n \hat{p}^{\hat{\alpha}}. \tag{15.61}$$

Remark 15.6 *Observe that we have scaled the parameters for considering a priori differences. This leads to a linear volume relation for the variance function. To obtain different functional relations, it might be better to scale the actual observations. For example, given observations $X_{j,k}, j = 1, \ldots, J$, $k = 1, \ldots, K_j$ (these could be frequencies or severities), consider variables $Y_{j,k} = X_{j,k}/V_{j,k}$. Assume that, for given $\boldsymbol{\Theta}_j = \boldsymbol{\theta}_j$, $\{Y_{j,k}, k = 1, \ldots, K_j\}$ are independent and identically distributed from $f(\cdot|\boldsymbol{\theta}_j)$. Assume also that $\boldsymbol{\Theta}_1, \ldots, \boldsymbol{\Theta}_J$ are independent and identically distributed from $\pi(\cdot)$. Then one can construct the likelihood of $Y_{j,k}$ using (15.55) to fit parameters of $\pi(\cdot)$ or try to use the method of moments.*

Estimating prior using method of moments. To avoid the use of numerical optimization required for maximizing (15.58), one could also use a method of moments. For example, given Model Assumptions 15.4, denote $\lambda_0 = \mathbb{E}[\Lambda_j] = \alpha\beta$, $\sigma_0^2 = \mathbb{V}\mathrm{ar}[\Lambda_j] = \alpha\beta^2$. Then the estimates $\hat{\lambda}_0$ and $\hat{\sigma}_0^2$ for λ_0 and σ_0^2, respectively, are

$$\hat{\lambda}_0 = \frac{1}{J}\sum_{j=1}^{J}\hat{\lambda}_j, \; \hat{\lambda}_j = \frac{1}{K_j}\sum_{k=1}^{K_j}\frac{n_{j,k}}{V_{j,k}}, \quad j = 1, \ldots, J, \tag{15.62}$$

$$\hat{\sigma}_0^2 = \max\left[\frac{1}{J-1}\sum_{j=1}^{J}(\hat{\lambda}_j - \hat{\lambda}_0)^2 - \frac{\hat{\lambda}_0}{J}\sum_{j=1}^{J}\frac{1}{K_j^2}\sum_{k=1}^{K_j}\frac{1}{V_{j,k}}, \; 0\right]. \tag{15.63}$$

These can easily be used to estimate α and β as $\hat{\alpha} = \hat{\lambda}_0/\hat{\beta}$ and $\hat{\beta} = \hat{\sigma}_0^2/\hat{\lambda}_0$ correspondingly. For a proof, see, for example, Shevchenko (2011, proposition 4.1). Alternative unbiased moment estimators can be found in Bühlmann and Gisler (2005, section 4.10).

15.4 Combining Expert Opinions with External and Internal Data

In the previous sections, we showed how to combine two data sources: expert opinions and internal data; or external data and internal data. In order to estimate the risk capital of a bank and to fulfill the Basel II requirements, risk managers have to take into account internal data, relevant external data (industry data), and expert opinions. The aim of this section is to provide an example of methodology to be used to combine these three sources of information. Here, we follow the approach suggested by Lambrigger *et al.* (2007). As in the previous section, we consider one risk cell only. In terms of methodology, we go through the following steps:

- In any risk cell, we model the loss frequency and the loss severity by parametric distributions (e.g., Poisson for the frequency or Pareto, LogNormal, etc. for the severity). For the considered bank, the unknown parameter vector $\boldsymbol{\theta}$ (e.g., the Poisson parameter or the Pareto tail index) of these distributions has to be quantified;
- A priori, before we have any company-specific information, only industry data are available. Hence, the best-prediction of our bank-specific parameter $\boldsymbol{\theta}$ is given by the belief in the available external knowledge such as the provided industry data. This unknown parameter of interest is modeled by a prior distribution (structural distribution) corresponding to a random vector $\boldsymbol{\Theta}$. The parameters of the prior distribution (hyperparameters) are estimated using data from the whole industry by, for example, MLE, as described in Section 15.3. If no industry data are available, the prior distribution could come from a "super expert" that has an overview over all banks;
- The true bank-specific parameter $\boldsymbol{\theta}_0$ is treated as a realization of $\boldsymbol{\Theta}$. The prior distribution of a random vector $\boldsymbol{\Theta}$ corresponds to the whole banking industry sector, whereas $\boldsymbol{\theta}$ stands for the unknown underlying parameter set of the bank being considered. Due to the variability among banks, it is natural to model $\boldsymbol{\theta}$ by a probability distribution. Note that $\boldsymbol{\Theta}$ is random with known distribution, whereas $\boldsymbol{\theta}_0$ is deterministic but unknown;
- As time passes, internal data

  $$\boldsymbol{X} = (X_1, \ldots, X_K)$$

  as well as expert opinions

  $$\boldsymbol{\Delta} = (\Delta_1, \ldots, \Delta_M)$$

  about the underlying parameter $\boldsymbol{\theta}$ become available. This affects our belief in the distribution of $\boldsymbol{\Theta}$ coming from external data only and adjusts the prediction of $\boldsymbol{\theta}_0$. The more information on $\boldsymbol{X}$ and $\boldsymbol{\Delta}$ we have, the better we are able to predict $\boldsymbol{\theta}_0$. That is, we replace the prior density $\pi(\boldsymbol{\theta})$ by a conditional density of $\boldsymbol{\Theta}$ given $\boldsymbol{X}$ and $\boldsymbol{\Delta}$.

In order to determine the posterior density $\pi(\boldsymbol{\theta}|\boldsymbol{x}, \boldsymbol{\delta})$, consider the joint conditional density of observations and expert opinions (given the parameter vector $\boldsymbol{\theta}$):

$$h(\boldsymbol{x}, \boldsymbol{\delta}|\boldsymbol{\theta}) = h_1(\boldsymbol{x}|\boldsymbol{\theta})h_2(\boldsymbol{\delta}|\boldsymbol{\theta}), \tag{15.64}$$

where h_1 and h_2 are the conditional densities (given $\boldsymbol{\Theta} = \boldsymbol{\theta}$) of $\boldsymbol{X}$ and $\boldsymbol{\Delta}$, respectively. Thus, $\boldsymbol{X}$ and $\boldsymbol{\Delta}$ are assumed to be conditionally independent given $\boldsymbol{\Theta}$.

Remark 15.7

- *Notice that, in this way, we naturally combine external data information* $\pi(\boldsymbol{\theta})$ *with internal data* $\boldsymbol{X}$ *and expert opinion* $\boldsymbol{\Delta}$*;*
- *In classical Bayesian inference (as it is used, for example, in actuarial science), one usually combines only two sources of information as described in the previous sections. Here, we combine three sources simultaneously using an appropriate structure, that is, Equation (15.64);*
- *Equation (15.64) is quite a reasonable assumption. Assume that the true bank-specific parameter is* $\boldsymbol{\theta}_0$*. Then, (15.64) says that the experts in this bank estimate* $\boldsymbol{\theta}_0$ *(by their opinion* $\boldsymbol{\Delta}$*) independently of the internal observations. This makes sense if the experts specify their opinions regardless of the data observed. Otherwise, we should work with the joint distribution* $h(\boldsymbol{x}, \boldsymbol{\delta}|\boldsymbol{\theta})$*.*

We further assume that observations as well as expert opinions are conditionally independent and identically distributed, given $\boldsymbol{\Theta} = \boldsymbol{\theta}$, so that

$$h_1(\boldsymbol{x}|\boldsymbol{\theta}) = \prod_{k=1}^{K} f_1(x_k|\boldsymbol{\theta}), \tag{15.65}$$

$$h_2(\boldsymbol{\delta}|\boldsymbol{\theta}) = \prod_{m=1}^{M} f_2(\delta_m|\boldsymbol{\theta}), \tag{15.66}$$

where f_1 and f_2 are the marginal densities of a single observation and a single expert opinion, respectively. We have assumed that all expert opinions are identically distributed, but this can be generalized easily to expert opinions having different distributions.

Here, the unconditional parameter density $\pi(\boldsymbol{\theta})$ is the *prior* density, whereas the conditional parameter density $\pi(\boldsymbol{\theta}|\boldsymbol{x}, \boldsymbol{\delta})$ is the *posterior* density. Let $h(\boldsymbol{x}, \boldsymbol{\delta})$ denote the unconditional joint density of the data $\boldsymbol{X}$ and expert opinions $\boldsymbol{\Delta}$. Then, it follows from the Bayes theorem that

$$h(\boldsymbol{x}, \boldsymbol{\delta}|\boldsymbol{\theta})\pi(\boldsymbol{\theta}) = \pi(\boldsymbol{\theta}|\boldsymbol{x}, \boldsymbol{\delta})h(\boldsymbol{x}, \boldsymbol{\delta}). \tag{15.67}$$

Note that the unconditional density $h(\boldsymbol{x}, \boldsymbol{\delta})$ does not depend on $\boldsymbol{\theta}$ and thus the posterior density is given by

$$\pi(\boldsymbol{\theta}|\boldsymbol{x}, \boldsymbol{\delta}) \propto \pi(\boldsymbol{\theta}) \prod_{k=1}^{K} f_1(x_k|\boldsymbol{\theta}) \prod_{m=1}^{M} f_2(\delta_m|\boldsymbol{\theta}). \tag{15.68}$$

For the purposes of OpRisk, it should be used to estimate the predictive distribution of future losses.

Hereafter, in this section, we assume that the parameters of the prior distribution are known and we look at a single risk cell in one bank. Therefore, the index representing bank or risk cell is not introduced.

15.4.1 CONJUGATE PRIOR EXTENSION

Equation (15.68) can be used in a general setup, but it is convenient to find some *conjugate* prior distributions such that the prior and the posterior distribution have a similar type, or where at least the posterior distribution can be calculated analytically. This type of distribution has been treated in Section 15.2 when two data sources have to be combined. For the case of (15.68), the standard definition of the conjugate prior distribution, Definition 7.10, can be extended as follows.

Definition 15.1 (Conjugate Prior Distribution) *Let F denote the class of density functions $h(\boldsymbol{x}, \boldsymbol{\delta}|\boldsymbol{\theta})$, indexed by $\boldsymbol{\theta}$. A class U of prior densities $\pi(\boldsymbol{\theta})$ is said to be a* conjugate family *for F if the posterior density $\pi(\boldsymbol{\theta}|\boldsymbol{x}, \boldsymbol{\delta}) \propto \pi(\boldsymbol{\theta})h(\boldsymbol{x}, \boldsymbol{\delta}|\boldsymbol{\theta})$ also belongs to the class U for all $h \in F$ and $\pi \in U$.* ■

Again, in general, the posterior distribution cannot be calculated analytically but can be estimated numerically – for instance, by the Markov chain Monte Carlo (MCMC) methods described in Section 7.4.

15.4.2 MODELING FREQUENCY: POISSON MODEL

To model the loss frequency for OpRisk in a risk cell, consider the following model.

Model Assumptions 15.5 (Poisson–Gamma–Gamma) *Assume that a risk cell in a bank has a scaling factor V for the frequency in a specified risk cell (it can be the product of several economic factors such as the gross income, the number of transactions, or the number of staff).*

- **(a)** *Let $\Lambda \sim Gamma(\alpha_0, \beta_0)$ be a Gamma distributed random variable with shape parameter $\alpha_0 > 0$ and scale parameter $\beta_0 > 0$, which are estimated from (external) market data. That is, the density of $Gamma(\alpha_0, \beta_0)$ plays the role of $\pi(\boldsymbol{\theta})$ in (15.68);*
- **(b)** *Given $\Lambda = \lambda$, the annual frequencies, $N_1, \ldots, N_T, N_{T+1}$, where $T + 1$ refers to next year, are assumed to be independent and identically distributed with $N_t \sim Poisson(V\lambda)$. That is, $f_1(\cdot|\lambda)$ in (15.68) corresponds to the probability mass function of a $Poisson(V\lambda)$ distribution;*
- **(c)** *A financial company has M expert opinions Δ_m, $1 \le m \le M$, about the intensity parameter Λ. Given $\Lambda = \lambda$, Δ_m and N_t are independent for all t and m, and $\Delta_1, \ldots, \Delta_M$ are independent and identically distributed with $\Delta_m \sim Gamma(\xi, \lambda/\xi)$, where ξ is a known parameter. That is, $f_2(\cdot|\lambda)$ corresponds to the density of a $Gamma(\xi, \lambda/\xi)$ distribution.*

Remark 15.8

- *The parameters α_0 and β_0 in Model Assumptions 15.5 are hyperparameters (parameters for distributions of the parameters) and can be estimated using the maximum likelihood method or the method of moments (see, e.g., Section 15.3);*
- *In Model Assumptions 15.5, we assume*

$$\mathbb{E}[\Delta_m|\Lambda] = \Lambda, \quad 1 \le m \le M, \tag{15.69}$$

that is, expert opinions are unbiased. A possible bias might only be recognized by the regulator, as he alone has the overview of the whole market.

Note that the *coefficient of variation* of the conditional expert opinion $\Delta_m|\Lambda$ is

$$\mathrm{Vco}[\Delta_m|\Lambda] = (\mathbb{V}\mathrm{ar}[\Delta_m|\Lambda)])^{1/2}/\mathbb{E}[\Delta_m|\Lambda] = 1/\sqrt{\xi},$$

and thus is independent of Λ. This means that ξ, which characterizes the uncertainty in the expert opinions, is independent of the true bank-specific Λ. For simplicity, we have assumed that all experts have the same conditional Vco and thus have the same credibility. Moreover, this allows for the estimation of ξ as

$$\hat{\xi} = (\hat{\mu}/\hat{\sigma})^2, \tag{15.70}$$

where

$$\hat{\mu} = \frac{1}{M}\sum_{m=1}^{M}\delta_m \quad \text{and} \quad \hat{\sigma}^2 = \frac{1}{M-1}\sum_{m=1}^{M}(\delta_m - \hat{\mu})^2, \quad M \geq 2.$$

In a more general framework, the parameter ξ can be estimated, for example, by maximum likelihood.

In the insurance practice, ξ is often specified by the regulator denoting a lower bound for expert opinion uncertainty; for example, Swiss Solvency Test (see Swiss Financial Market Supervisory Authority 2006, appendix 8.4). If the credibility differs among the experts, then $\mathrm{Vco}[\Delta_m|\Lambda]$ should be estimated for all m, $1 \leq m \leq M$. Admittedly, this may often be a challenging issue in practice.

Remark 15.9 *This model can be extended to a model where one allows for more flexibility in the expert opinions. For convenience, it is preferred that experts are conditionally independent and identically distributed, given Λ. This has the advantage that there is only one parameter, ξ, that needs to be estimated.*

Using the notation from Section 15.4, the posterior density of Λ, given the losses up to year K and the expert opinions of M experts, can be calculated. Denote the data over past years as follows:

$$\boldsymbol{N} = (N_1, \ldots, N_T),$$
$$\boldsymbol{\Delta} = (\Delta_1, \ldots, \Delta_M).$$

Denote also the arithmetic means by

$$\overline{N} = \frac{1}{T}\sum_{t=1}^{T} N_t, \quad \overline{\Delta} = \frac{1}{M}\sum_{m=1}^{M}\Delta_m, \quad \text{etc.} \tag{15.71}$$

Then, the posterior density is given by the following theorem.

Theorem 15.2 *Under Model Assumptions 15.5, given loss information $\boldsymbol{N} = \boldsymbol{n}$ and expert opinion $\boldsymbol{\Delta} = \boldsymbol{\delta}$, the posterior density of Λ is*

$$\pi(\lambda|\boldsymbol{n}, \boldsymbol{\delta}) = \frac{(\omega/\phi)^{(\nu+1)/2}}{2K_{\nu+1}(2\sqrt{\omega\phi})}\lambda^{\nu} e^{-\lambda\omega - \lambda^{-1}\phi}, \tag{15.72}$$

with

$$\begin{aligned}\nu &= \alpha_0 - 1 - M\xi + T\overline{n},\\ \omega &= VT + \frac{1}{\beta_0},\\ \phi &= \xi M\overline{\delta},\end{aligned} \tag{15.73}$$

and

$$K_{\nu+1}(z) = \frac{1}{2}\int_0^{\infty} u^{\nu} e^{-z(u+1/u)/2} \mathrm{d}u. \tag{15.74}$$

Here, $K_\nu(z)$ is a modified Bessel function of the third kind (see, e.g., Abramowitz and Stegun 1965, p. 375).

Proof: Model Assumptions 15.5 applied to (15.68) yield

$$\begin{aligned}\pi(\lambda|\boldsymbol{n},\boldsymbol{\delta}) &\propto \lambda^{\alpha_0-1}e^{-\lambda/\beta_0}\prod_{t=1}^{T} e^{-V\lambda}\frac{(V\lambda)^{n_t}}{n_t!}\prod_{m=1}^{M}\frac{(\delta_m)^{\xi-1}}{(\lambda/\xi)^{\xi}}e^{-\delta_m\xi/\lambda}\\ &\propto \lambda^{\alpha_0-1}e^{-\lambda/\beta_0}\prod_{t=1}^{T} e^{-V\lambda}\lambda^{n_t}\prod_{m=1}^{M}(\xi/\lambda)^{\xi}e^{-\delta_m\xi/\lambda}\\ &\propto \lambda^{\alpha_0-1-M\xi+T\overline{n}}\exp\left(-\lambda\left(VT+\frac{1}{\beta_0}\right)-\frac{1}{\lambda}\xi M\overline{\delta}\right).\end{aligned}$$

■

Remark 15.10

- *A distribution with density (15.72) is known as the generalized inverse Gaussian (GIG) distribution GIG(ω, ϕ, ν). This is a well-known distribution with many applications in finance and risk management (see McNeil et al. 2005, pp. 75, 497). The GIG has been analyzed by many authors; see a discussion by Jørgensen (1982). The GIG belongs to the popular class of subexponential (heavy-tailed) distributions; see Embrechts (1983) for a proof and Cruz et al. (2014) for a detailed treatment of subexponential distributions. The GIG with $\nu \leq 1$ is a distribution of the first hitting time for certain time-homogeneous processes (see, e.g., Jørgensen 1982, chapter 6). In particular, the (standard) inverse Gaussian (i.e., the GIG with $\nu = -3/2$) is known by financial practitioners as the distribution function determined by the first passage time of a Brownian motion. The algorithm for generating realizations from a GIG can be found, for example, in Lambrigger et al. (2007);*
- *In comparison with the classical Poisson–Gamma case of combining two sources of information (considered in Section 15.2.3), where the posterior is a Gamma distribution, the posterior $\pi(\lambda|\cdot)$ in (15.75) is more complicated. In the exponent, it involves both λ and $1/\lambda$. Note that expert opinions enter via the term $1/\lambda$ only;*
- *Observe that the classical exponential dispersion family with associated conjugates (Bühlmann and Gisler, 2005, chapter 2.5) allows for a natural extension to GIG-like distributions. In this*

sense, the GIG distributions enlarge the classical Bayesian inference theory on the exponential dispersion family.

For our purposes, it is interesting to observe how the posterior density transforms when new data from a newly observed year arrive. Let ν_k, ω_k, and ϕ_k denote the parameters for the data $(N_1, \ldots, N_k)$ after k accounting years. Implementation of the update processes is then given by the following equalities (assuming that expert opinions do not change).

Recursive calculation for parameters (as a function of sample size):

$$\begin{aligned} \nu_{k+1} &= \nu_k + n_{k+1}, \\ \omega_{k+1} &= \omega_k + V, \\ \phi_{k+1} &= \phi_k. \end{aligned} \tag{15.75}$$

Obviously, the information update process has a very simple form and only the parameter ν is affected by the new observation n_{k+1}. The posterior density (15.75) does not change its type every time new data arrive and, hence, is easily calculated.

The moments of a GIG are not available in a closed form through elementary functions but can be expressed in terms of Bessel functions (see Appendix A.4.13). In particular, the posterior expected number of losses is

$$\mathbb{E}[\Lambda|\boldsymbol{N} = \boldsymbol{n}, \boldsymbol{\Delta} = \boldsymbol{\delta}] = \sqrt{\frac{\phi}{\omega}} \frac{K_{\nu+2}(2\sqrt{\omega\phi})}{K_{\nu+1}(2\sqrt{\omega\phi})}. \tag{15.76}$$

The mode of a GIG has a simple expression (see Appendix A.4.13) that gives the posterior mode

$$\text{mode}(\Lambda|\boldsymbol{N} = \boldsymbol{n}, \boldsymbol{\Delta} = \boldsymbol{\delta}) = \frac{1}{2\omega}(\nu + \sqrt{\nu^2 + 4\omega\phi}). \tag{15.77}$$

It can be used as an alternative point estimator instead of the mean. In addition, the mode of a GIG differs only slightly from the expected value for large $|\nu|$.

We are clearly interested in robust prediction of the bank-specific Poisson parameter and thus the Bayesian estimator (15.76) is a promising candidate within this OpRisk framework. The examples below show that, in practice, (15.76) outperforms other classical estimators. To interpret (15.76) in more detail, we make use of asymptotic properties. Using properties of Bessel functions, it is easy to show that

$$R_{\nu^2}(2\nu) \to \nu \quad \text{as} \quad \nu \to \infty, \tag{15.78}$$

where

$$R_\nu(z) = \frac{K_{\nu+1}(z)}{K_\nu(z)}$$

(see Lambrigger *et al.* 2007, lemma B.1 in appendix B). Using this result, a full asymptotic interpretation of the Bayesian estimator (15.76) can be found as follows.

Theorem 15.3 *Under Model Assumptions 15.5, the following asymptotic relations hold:*

(a) *If* $T \to \infty$*, then* $\mathbb{E}[\Lambda|\boldsymbol{N}, \boldsymbol{\Delta}] \to \mathbb{E}[N_t|\Lambda = \lambda]/V = \lambda$;

(b) *If* $\text{Vco}[\Delta_m|\Lambda] \to 0$*, then* $\mathbb{E}[\Lambda|\boldsymbol{N}, \boldsymbol{\Delta}] \to \Delta_m$*,* $m = 1, \ldots, M$;

(c) *If* $M \to \infty$, *then* $\mathbb{E}[\Lambda|\boldsymbol{N}, \boldsymbol{\Delta}] \to \mathbb{E}[\Delta_m|\Lambda = \lambda] = \lambda$;

(d) *If* $\text{Vco}[\Delta_m|\Lambda] \to \infty$, $m = 1, \ldots, M$, *then*

$$\mathbb{E}[\Lambda|\boldsymbol{N}, \boldsymbol{\Delta}] \to \frac{1}{VT\beta_0 + 1}\mathbb{E}[\Lambda] + \frac{1}{V}\left(1 - \frac{1}{VT\beta_0 + 1}\right)\overline{N}.$$

(e) *If* $\mathbb{E}[\Lambda] = $ *constant and* $\text{Vco}[\Lambda] \to 0$, *then* $\mathbb{E}[\Lambda|\boldsymbol{N}, \boldsymbol{\Delta}] \to \mathbb{E}[\Lambda]$.

Proof: The proof is given Lambrigger *et al.* (2007, appendix C). These asymptotic relations should be understood in a probability sense, that is, true with probability 1 (the so-called P-*almost surely*). ∎

Remark 15.11 *The GIG mode and mean are asymptotically the same for* $\nu \to \infty$*; also* $4\omega\phi/\nu^2 \to 0$ *for* $T \to \infty$, $M \to \infty$, $M \to 0$ *or* $\xi \to 0$. *Then, one can approximate the posterior mode as*

$$\text{mode}(\Lambda|\boldsymbol{N} = \boldsymbol{n}, \boldsymbol{\Delta} = \boldsymbol{\delta}) \approx \frac{\nu}{2\omega}1_{\{\nu \geq 0\}} + \frac{\phi}{|\nu|} \tag{15.79}$$

and obtain the results of Theorem 15.3 in an elementary manner avoiding Bessel functions.

Theorem 15.3 yields a natural interpretation of the posterior density (15.72) and its expected value (15.76):

- As the number of observations increases, we give more weight to them and in the limit $T \to \infty$ (case a), we completely believe in the observations N_k and neglect a priori information and expert opinion;
- On the other hand, the more the coefficient of variation of the expert opinions decreases, the more weight is given to them (case b);
- In Model Assumptions 15.5, we assume experts to be conditionally independent. In practice, however, even for $\text{Vco}[\Delta_m|\Lambda] \to 0$, the variance of $\overline{\Delta}|\Lambda$ cannot be made arbitrarily small when increasing the number of experts, as there is always a positive covariance term due to positive dependence between experts. Since we predict random variables, we never have "perfect diversification", that is, in practical applications we would probably question property c;
- Conversely, if experts become less credible in terms of having an increasing coefficient of variation, our model behaves as if the experts do not exist (case d). The Bayes estimator is then a weighted sum of prior and posterior information with appropriate credibility weights. This is the classical credibility result obtained from Bayesian inference on the exponential dispersion family with two sources of information (see 15.19);
- Of course, if the Vco of the prior distribution (i.e., of the whole banking industry) vanishes, the external data are not affected by internal data and expert opinion (case e).

This interpretation shows that the model behaves as we would expect and require in practice. Thus, there are good reasons to believe that it provides an adequate model to combine internal observations with relevant external data and expert opinions, as required by many risk managers. One can even go further and generalize the results from this section in a natural way to a Poisson–Gamma–GIG model, that is, where the prior distribution is a GIG. Then, the posterior distribution is again a GIG (see also Model Assumptions 15.5).

EXAMPLE 15.5

A simple example, taken from Lambrigger *et al.* (2007, example 3.7) illustrates the described methodology combining three data sources. It also extends Example 15.2, where two data sources are combined using the classical Bayesian inference approach. Assume the following:

- External data (e.g., provided by external databases or regulator) estimate the intensity of the loss frequency (i.e., the Poisson parameter Λ), which has a prior Gamma distribution, $\Lambda \sim Gamma(\alpha_0, \beta_0)$, as $\mathbb{E}[\Lambda] = \alpha_0\beta_0 = 0.5$ and $\mathbb{P}\mathrm{r}[0.25 \leq \Lambda \leq 0.75] = 2/3$. Then, the parameters of the prior are $\alpha_0 \approx 3.407$ and $\beta_0 \approx 0.147$ (see Example 15.2);
- One expert gives an estimate of the intensity as $\delta = 0.7$. For simplicity, we consider in this example one expert only and, hence, the Vco is not estimated using (15.70), but given a priori (e.g., by the regulator): $\mathrm{Vco}[\Delta|\Lambda] = \sqrt{\mathbb{V}\mathrm{ar}[\Delta|\Lambda]}/\mathbb{E}[\Delta|\Lambda] = 0.5$, that is, $\xi = 4$;
- The observations of the annual number of losses $n_1, n_2, \ldots$ are sampled from $Poisson(0.6)$ and are the same as in the Example 15.2.

This means that a priori we have a frequency parameter distributed as $Gamma(\alpha_0, \beta_0)$ with mean $\alpha_0\beta_0 = 0.5$. The true value of the parameter λ for this risk in a bank is 0.6, that is, it does worse than the average institution. However, our expert has an even worse opinion of his institution, namely $\delta = 0.7$. Now, we compare the following:

- The pure MLE

$$\hat{\lambda}_k^{\mathrm{MLE}} = \frac{1}{k}\sum_{i=1}^{k} n_i,$$

- The Bayesian estimate Equation (15.19)

$$\hat{\lambda}_k^{(2)} = \mathbb{E}[\Lambda|N_1 = n_1, \ldots, N_k = n_k], \tag{15.80}$$

  without expert opinion; and
- The Bayesian estimate derived in formula (15.76)

$$\hat{\lambda}_k^{(3)} = \mathbb{E}[\Lambda|N_1 = n_1, \ldots, N_k = n_k, \Delta = \delta], \tag{15.81}$$

  which combines internal data and expert opinions with the prior.

The results are plotted in Figure 15.6. The estimator $\hat{\lambda}_k^{(3)}$ shows a much more stable behavior around the true value $\lambda = 0.6$, due to the use of the prior information (market data) and the expert opinions. Given adequate expert opinions, $\hat{\lambda}_k^{(3)}$ clearly outperforms the other estimators, particularly if only a few data points are available.

One could think that this is only the case when the experts' estimates are appropriate. However, even if experts fairly under- (or over-)estimate the true parameter λ, the method presented here performs better for our dataset than the other mentioned methods, when a few data points are available. Figure 15.7 displays the same estimators, but where the expert's opinion is $\delta = 0.4$, which clearly underestimates the true expected value 0.6.

In Figure 15.6, $\hat{\lambda}_k^{(3)}$ gives better estimates when compared to $\lambda_k^{(2)}$. Observe also that in Figure 15.7, $\hat{\lambda}_k^{(3)}$ gives more appropriate estimates than $\lambda_k^{(2)}$. Though the expert is too optimistic, $\hat{\lambda}_k^{(3)}$ manages to correct $\hat{\lambda}_k^{\text{MLE}}$ ($k \leq 10$), which is clearly too low.

This example yields a typical picture observed in numerical experiments that demonstrates that the Bayes estimator (15.76) is often more suitable and stable than MLEs based on internal data only. Note that in this example the prior distribution as well as the expert opinion do not change over time. However, as soon as new information is available or when new risk management tools are in place, the corresponding parameters may be easily adjusted.

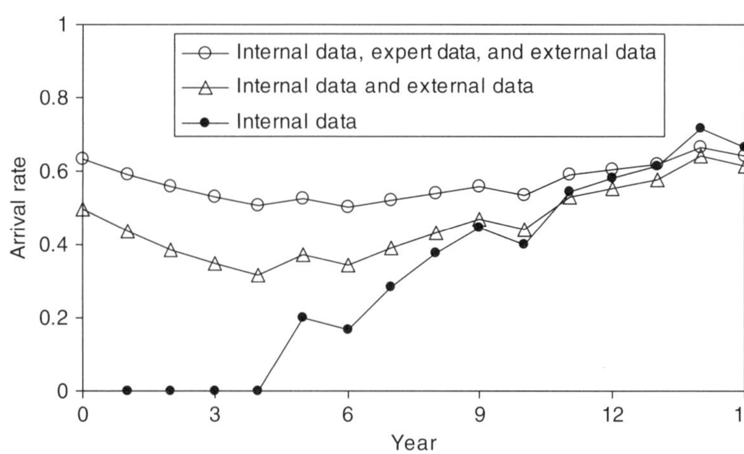

FIGURE 15.6 (o) The Bayes estimate $\hat{\lambda}_k^{(3)}$, $k = 1, \ldots, 15$, combines the internal data simulated from *Poisson*(0.6), external data giving $\mathbb{E}[\Lambda] = 0.5$, and expert opinion $\delta = 0.7$. It is compared with the Bayes estimate $\hat{\lambda}_k^{(2)}$ ($\triangle$), combines external data and internal data, and the classical MLE, $\hat{\lambda}_k^{\text{MLE}}$ ($\bullet$). See Example 15.5 for details

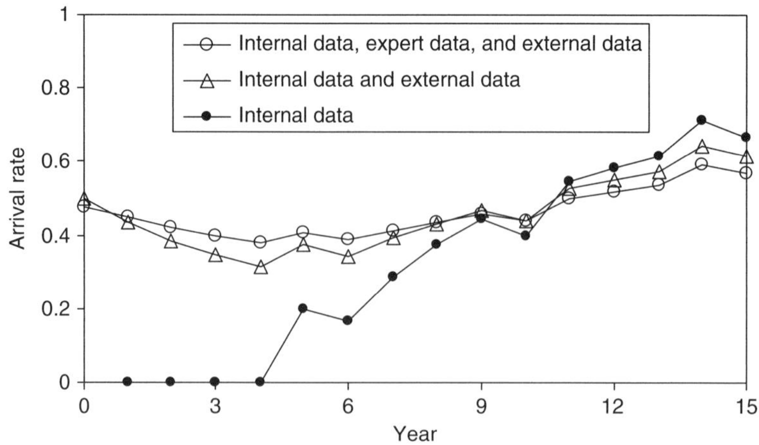

FIGURE 15.7 (○) The Bayes estimate $\hat{\lambda}_k^{(3)}$, $k = 1, \ldots, 15$, combines the internal data simulated from $Poisson(0.6)$, external data giving $\mathbb{E}[\Lambda] = 0.5$, and expert opinion $\delta = 0.4$. It is compared with the Bayes estimate $\hat{\lambda}_k^{(2)}$ (△), which combines external data and internal data, and the classical MLE, $\hat{\lambda}_k^{\mathrm{MLE}}$ (•). See Example 15.5 for details

Remark 15.12 (Modeling Frequency: Poisson with Stochastic Intensity) *In this section, we considered the situation where Λ is the same for all years $t = 1, 2, \ldots$. However, in general, the evolution of Λ_t can be modeled as having deterministic (trend, seasonality) and stochastic components, that is, one may consider a sequence $\Lambda_1, \Lambda_2, \ldots, \Lambda_T, \Lambda_{T+1}$, where $T + 1$ corresponds to the next year. In actuarial mathematics, this is called a mixed Poisson model. That is, Λ_t is not only different for different banks and different risks but also may change from year to year for a risk in the same bank. A simple case, extending Model Assumptions 15.5 to the case when Λ_t is purely stochastic and distributed according to a Gamma distribution, is considered by Peters et al. (2009) also see Shevchenko (2011, section 4.5.3). The model setup with random intensities Λ_t can be utilized by the modeler to introduce a dependence between different risk cells, by introducing dependence between $\Lambda_t^{(1)}, \ldots, \Lambda_t^{(J)}$, where superscript refers to the risk cell (see, e.g., Peters et al. 2009).*

15.4.3 LOGNORMAL MODEL FOR SEVERITIES

In general, one can use the methodology summarized by Equation (15.68) to develop a model combining external data, internal data, and expert opinion for estimation of the severity. For illustration purposes, this section considers the LogNormal severity model; the Pareto severity model is developed in the next section.

Consider modeling severities $X_1, \ldots, X_K, \ldots$ using $LogNormal(\mu, \sigma^2)$, where $\boldsymbol{X} = (X_1, \ldots, X_K)$ are the losses over past T years. Here, we take an approach considered in Section 15.2.4, where μ is unknown and σ is known. The unknown μ is treated under the Bayesian approach as a random variable Θ_μ. Then combining external data, internal data, and expert opinions can be accomplished using the following model.

Model Assumptions 15.6 (LogNormal–Normal–Normal) *Let us assume the following severity model for a risk cell in one bank:*

(a) *Let* $\Theta_\mu \sim Normal(\mu_0, \sigma_0^2)$ *be a normally distributed random variable with parameters* μ_0, σ_0*, which are estimated from (external) market data, that is,* $\pi(\boldsymbol{\theta})$ *in (15.68) is the density of* $Normal(\mu_0, \sigma_0^2)$*;*

(b) *Given* $\Theta_\mu = \mu$*, the losses* $X_1, X_2, \ldots$ *are conditionally independent with a common LogNormal distribution:*

$$X_k \sim LogNormal(\mu, \sigma^2),$$

where σ *is assumed to be known. That is,* $f_1(\cdot|\mu)$ *in (15.68) corresponds to the density of a* $LogNormal(\mu, \sigma^2)$ *distribution;*

(c) *The financial company has* M *experts with opinions* Δ_m*,* $1 \leq m \leq M$*, about* Θ_μ*. Given* $\Theta_\mu = \mu$*,* Δ_m *and* X_k *are independent for all* m *and* k*, and* $\Delta_1, \ldots, \Delta_M$ *are independent with a common Normal distribution:*

$$\Delta_m \sim Normal(\mu, \xi^2),$$

where ξ *is a parameter estimated using expert opinion data. That is,* $f_2(\cdot|\mu)$ *corresponds to the density of a* $Normal(\mu, \xi^2)$ *distribution.*

Remark 15.13

- *For* $M \geq 2$*, the parameter* ξ *can be estimated by the standard deviation of* δ_m*:*

$$\hat{\xi} = \left(\frac{1}{M-1} \sum_{m=1}^{M} (\delta_m - \overline{\delta})^2 \right)^{1/2}. \tag{15.82}$$

- *The hyperparameters* μ_0 *and* σ_0 *are estimated from market data, for example, by MLE or by the method of moments;*
- *In practice, one often uses an ad hoc estimate for* σ*, which is usually based on expert opinion only. However, one could think of a Bayesian approach for* σ*, but then an analytical formula for the posterior distribution in general does not exist and the posterior then needs to be calculated numerically, for example, by MCMC methods.*

Under Model Assumptions 15.6, the posterior density is given by

$$\begin{aligned}
\pi(\mu|\boldsymbol{x}, \boldsymbol{\delta}) &\propto \frac{1}{\sigma_0\sqrt{2\pi}} \exp\left(-\frac{(\mu-\mu_0)^2}{2\sigma_0^2}\right) \\
&\quad \times \prod_{k=1}^{K} \frac{1}{\sigma\sqrt{2\pi}} \exp\left(-\frac{(\ln x_k - \mu)^2}{2\sigma^2}\right) \prod_{m=1}^{M} \frac{1}{\xi\sqrt{2\pi}} \exp\left(-\frac{(\delta_m-\mu)^2}{2\xi^2}\right) \\
&\propto \exp\left[-\left(\frac{(\mu-\mu_0)^2}{2\sigma_0^2} + \sum_{k=1}^{K} \frac{(\ln x_k - \mu)^2}{2\sigma^2} + \sum_{m=1}^{M} \frac{(\delta_m-\mu)^2}{2\xi^2}\right)\right] \\
&\propto \exp\left[-\frac{(\mu-\hat{\mu})^2}{2\hat{\sigma}^2}\right],
\end{aligned} \tag{15.83}$$

with

$$\hat{\sigma}^2 = \left(\frac{1}{\sigma_0^2} + \frac{K}{\sigma^2} + \frac{M}{\xi^2}\right)^{-1},$$

and

$$\hat{\mu} = \hat{\sigma}^2 \times \left(\frac{\mu_0}{\sigma_0^2} + \frac{1}{\sigma^2}\sum_{k=1}^{K} \ln x_k + \frac{1}{\xi^2}\sum_{m=1}^{M} \delta_m\right).$$

In summary, we derived the following theorem (also see Lambrigger *et al.* 2007).

Theorem 15.4 *Under Model Assumptions 15.6, the posterior distribution of* Θ_μ*, given loss information* $\boldsymbol{X} = \boldsymbol{x}$ *and expert opinion* $\boldsymbol{\Delta} = \boldsymbol{\delta}$*, is a Normal distribution, Normal*$(\hat{\mu}, \hat{\sigma}^2)$*, with*

$$\hat{\sigma}^2 = \left(\frac{1}{\sigma_0^2} + \frac{K}{\sigma^2} + \frac{M}{\xi^2}\right)^{-1}$$

and

$$\hat{\mu} = \mathbb{E}[\Theta_\mu | \boldsymbol{X} = \boldsymbol{x}, \boldsymbol{\Delta} = \boldsymbol{\delta}] = \omega_1 \mu_0 + \omega_2 \overline{\ln x} + \omega_3 \overline{\delta}, \tag{15.84}$$

where $\overline{\ln x} = \frac{1}{K}\sum_{k=1}^{K} \ln x_k$ *and the credibility weights are*

$$\omega_1 = \hat{\sigma}^2/\sigma_0^2, \quad \omega_2 = \hat{\sigma}^2 K/\sigma^2, \quad \omega_3 = \hat{\sigma}^2 M/\xi^2.$$

This theorem yields a natural interpretation of the considered model. The estimator $\hat{\mu}$ in (15.84) weights the internal and external data as well as the expert opinion in an appropriate manner. Observe that under Model Assumptions 15.6, the mean of the posterior distribution can be calculated explicitly. This is different from the frequency model in Section 15.4.2, where asymptotic calculations (Theorem 15.3) were required for the interpretation of the terms. However, interpretation of the terms is exactly the same as in Theorem 15.3. The more credible the information, the higher is the credibility weight in Equation (15.84) – as expected from an appropriate model for combining internal observations, relevant external data, and expert opinions.

15.4.4 PARETO MODEL

Consider modeling severities $X_1, \ldots, X_K, \ldots$ using *Pareto*(γ, L) with a density

$$f(x) = \frac{\gamma}{L}\left(\frac{x}{L}\right)^{-\gamma-1}, \quad x \geq L, \quad \xi > 0, \tag{15.85}$$

where $\boldsymbol{X} = (X_1, \ldots, X_K)$ are the losses over past T years. Note that if $\xi > 1$, then the mean is $L\xi/(\xi - 1)$; otherwise, the mean does not exist. Here, we take an approach considered in Section 15.2.5, where γ is unknown and the threshold L is known. The unknown γ is treated under the Bayesian approach as a random variable Θ_γ. Then, combining external data, internal data, and expert opinions can be accomplished using the following model.

Model Assumptions 15.7 (Pareto–Gamma–Gamma) *Let us assume the following severity model for a risk cell in one bank:*

(a) *Let $\Theta_\gamma \sim Gamma(\alpha_0, \beta_0)$ be a Gamma-distributed random variable with parameters α_0 and β_0, which are estimated from (external) market data, that is, $\pi(\boldsymbol{\theta})$ in (15.68) is the density of a $Gamma(\alpha_0, \beta_0)$ distribution;*

(b) *Given, $\Theta_\gamma = \gamma$, the losses $X_1, X_2, \ldots$ in the risk cell are assumed to be conditionally independent and Pareto-distributed:*

$$X_k \sim Pareto(\gamma, L),$$

where the threshold $L \geq 0$ is assumed to be known and fixed. That is, $f_1(\cdot|\gamma)$ in Equation (15.68) corresponds to the density of a $Pareto(\gamma, L)$ distribution;

(c) *A financial company has M experts with opinions Δ_m, $1 \leq m \leq M$, about the parameter Θ_γ. Given $\Theta_\gamma = \gamma$, Δ_m and X_k are independent for all m and k, and $\Delta_1, \ldots, \Delta_M$ are independent and identically distributed with*

$$\Delta_m \sim Gamma(\xi, \gamma/\xi),$$

where ξ is a parameter estimated using expert opinion data (see 15.70). That is, $f_2(\cdot|\gamma)$ corresponds to the density of a $Gamma(\xi, \gamma/\xi)$ distribution.

Theorem 15.5 *Under Model Assumptions 15.7, given loss information $\boldsymbol{X} = \boldsymbol{x}$ and expert opinion $\boldsymbol{\Delta} = \boldsymbol{\delta}$, the posterior distribution of Θ_γ is $GIG(\omega, \phi, \nu)$ with the density*

$$\pi(\gamma|\boldsymbol{x}, \boldsymbol{\delta}) = \frac{(\omega/\phi)^{(\nu+1)/2}}{2K_{\nu+1}(2\sqrt{\omega\phi})}\gamma^{\nu} e^{-\gamma\omega - \gamma^{-1}\phi}, \tag{15.86}$$

where

$$\begin{aligned} \nu &= \alpha_0 - 1 - M\xi_i + K, \\ \omega &= \frac{1}{\beta_0} + \sum_{k=1}^{K} \ln\frac{x_k}{L}, \\ \phi &= \xi M\bar{\delta}. \end{aligned} \tag{15.87}$$

Proof: This is straightforward from the calculation of the posterior density

$$\begin{aligned} \pi(\gamma|\boldsymbol{x}, \boldsymbol{\delta}) &\propto \gamma^{\alpha_0 - 1} e^{-\gamma/\beta_0} \prod_{k=1}^{K} \frac{\gamma}{L}\left(\frac{x_k}{L}\right)^{-\gamma-1} \prod_{m=1}^{M} \frac{(\delta_m)^{\alpha-1}}{\beta^{\alpha}} e^{-\delta_m/\beta} \\ &\propto \gamma^{\alpha_0 - 1 - M\xi + K} \exp\left[-\gamma\left(\frac{1}{\beta_0} + \sum_{k=1}^{K} \ln\frac{x_k}{L}\right) - \frac{\xi M\bar{\delta}}{\gamma}\right]. \end{aligned} \tag{15.88}$$

■

Hence, as in Theorem 15.2 for modeling Poisson frequencies, the posterior distribution is a GIG with the convenient property that the term γ in the exponent in Equation (15.88) is only affected by the internal observations, whereas the term $1/\gamma$ is driven by the expert opinions.

Remark 15.14 *It seems natural to generalize this result to the case of the GIG prior distribution. In particular, changing the assumption (a) in Model Assumptions 15.7 to $\Theta_\gamma \sim GIG(\omega_0, \phi_0, \nu_0)$, with the parameters ν_0, ω_0, ϕ_0, the posterior density $\pi(\gamma|\boldsymbol{x}, \boldsymbol{\delta})$ is $GIG(\omega, \phi, \nu)$ with*

$$\begin{aligned} \nu &= \nu_0 - M\xi + K, \\ \omega &= \omega_0 + \sum_{k=1}^{K} \ln(x_k/L), \\ \phi &= \phi_0 + \xi M\bar{\delta}. \end{aligned} \tag{15.89}$$

Note that for $\phi_0 = 0$, the prior GIG is a Gamma distribution and hence we are in the Pareto–Gamma–Gamma situation of Model Assumptions 15.7.

The posterior mean (that can be used as a Bayesian point estimator for γ) can be calculated as

$$\mathbb{E}[\Theta_\gamma|\boldsymbol{X} = \boldsymbol{x}, \boldsymbol{\Delta} = \boldsymbol{\delta}] = \sqrt{\frac{\phi}{\omega}} \frac{K_{\nu+2}(2\sqrt{\omega\phi})}{K_{\nu+1}(2\sqrt{\omega\phi})} \tag{15.90}$$

(see Appendix A.4.13). The MLE of the Pareto tail index γ is also easily calculated as

$$\hat{\gamma}^{\text{MLE}} = \frac{K}{\sum_{k=1}^{K} \ln(x_k/L)}. \tag{15.91}$$

Then, completely analogous to Theorem 15.3, the following theorem gives a natural interpretation of the Bayesian (posterior mean) estimator.

Theorem 15.6 *Under Model Assumptions 15.7, the following asymptotic relations hold* P*-almost surely:*

(a) *If $K \to \infty$, then $\mathbb{E}[\Theta_\gamma|\boldsymbol{X}, \boldsymbol{\Delta}] \to \mathbb{E}[X_k|\Theta_\gamma = \gamma]/V = \gamma$;*

(b) *If $\text{Vco}[\Delta_m|\Delta_\gamma] \to 0$, then $\mathbb{E}[\Theta_\gamma|\boldsymbol{X}, \boldsymbol{\Delta}] \to \Delta_m$, $m = 1, \ldots, M$;*

(c) *If $M \to \infty$, then $\mathbb{E}[\Theta_\gamma|\boldsymbol{X}, \boldsymbol{\Delta}] \to \mathbb{E}[\Delta_m|\Theta_\gamma = \gamma] = \gamma$;*

(d) *If $\text{Vco}[\Delta_m|\Theta_\gamma] \to \infty$, $m = 1, \ldots, M$, then*

$$\mathbb{E}[\Theta_\gamma|\boldsymbol{X}, \boldsymbol{\Delta}] \to (1 - w)\,\mathbb{E}[\Theta_\gamma] + w\hat{\gamma}^{\text{MLE}},$$

where $w = K\beta_0/(\hat{\gamma}^{\text{MLE}} + K\beta_0)$;

(e) *If $\mathbb{E}[\Theta_\gamma] = $ constant and $\text{Vco}[\Theta_\gamma] \to 0$, then $\mathbb{E}[\Theta_\gamma|\boldsymbol{X}, \boldsymbol{\Delta}] \to \mathbb{E}[\Theta_\gamma]$.*

Remark 15.15

- *Theorem 15.6 basically says that the higher the precision of a particular source of risk information, the higher is its corresponding credibility weight. This means that we obtain the same interpretations as for Theorem 15.3 and Equation (15.84);*
- *Observe that in Sections 15.4.2 and 15.4.3, we have applied Bayesian inference to the expected values of the Poisson and the Normal distribution, respectively. However, Bayesian inference is*

much more general and, basically, can be applied to any reasonable parameter. In this section, it is applied to the Pareto tail index;

- *Observe that Model Assumptions 15.7 lead to an infinite mean model because the Pareto parameter Θ_γ can be less than 1 with positive probability. For finite mean models, the range of possible γ has to be restricted to $\gamma > 1$. This does not impose difficulties (see Section 7.2.4).*

The update process of (15.87) and (15.89) has again a simple linear form when new information arrives. The posterior density (15.86) does not change its type every time a new observation arrives. In particular, only the parameter ω is affected by a new observation.

Recursive calculation for parameters (as a function of sample size):

$$\begin{aligned} \nu_{k+1} &= \nu_k + 1, \\ \omega_{k+1} &= \omega_k + \ln(x_{k+1}/L), \\ \phi_{k+1} &= \phi_k. \end{aligned} \tag{15.92}$$

The following example illustrates the simplicity and robustness of the posterior mean estimator.

EXAMPLE 15.6

Assume that a bank would like to model its risk severity by a Pareto distribution with tail index Θ_γ. The regulator provides external prior data, saying that $\Theta_\gamma \sim Gamma(\alpha_0, \beta_0)$ with $\alpha_0 = 4$ and $\beta_0 = 9/8$, that is, $\mathbb{E}[\Theta_\gamma] = 4.5$ and $\mathrm{Vco}[\Theta_\gamma] = 0.5$. The bank has one expert opinion δ with $\mathrm{Vco}[\Delta|\Theta_\gamma = \gamma] = 0.5$, that is, $\xi = 4$. We then observe the losses sampled from a *Pareto*$(4, 1)$ distribution, the same as in Example 15.4. In Figure 15.8 and Figure 15.9, the following estimators are compared when expert opinion $\delta = 3$ and $\delta = 5$:

- The classical MLE

$$\hat{\gamma}_k^{\mathrm{MLE}} = \frac{k}{\sum_{i=1}^{k} \ln(x_i/L)}. \tag{15.93}$$

- The Bayesian posterior mean estimate Equation (15.43)

$$\gamma_k^{(2)} = \mathbb{E}[\Theta_\gamma | X_1 = x_1, \ldots, X_k = x_k], \tag{15.94}$$

  which does not account for expert opinions;

- The Bayesian Posterior mean estimate, which includes expert opinion

$$\hat{\gamma}_k^{(3)} = \mathbb{E}[\Theta_\gamma | X_1 = x_1, \ldots, X_k = x_k, \Delta = \delta], \tag{15.95}$$

  given by Equation (15.90).

Figures 15.8 and 15.9 show the high volatility of the MLE for small numbers k. It is very sensitive to newly arriving losses. The estimator $\hat{\gamma}_k^{(3)}$ shows a much more stable behavior around the true value $\alpha = 4$, most notably when a few data points

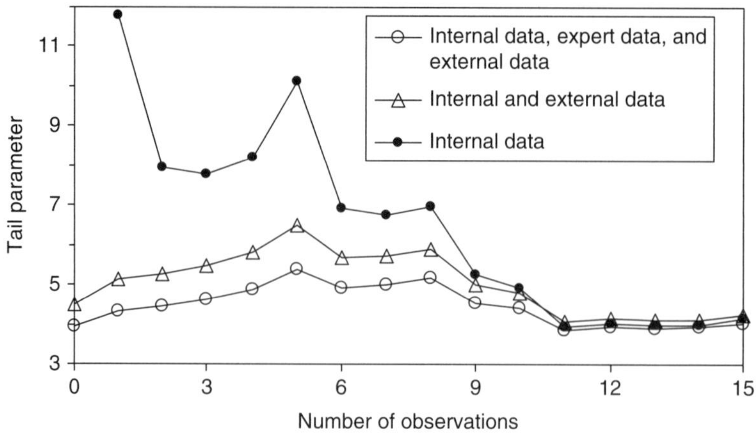

FIGURE 15.8 (○) The Bayes estimate $\hat{\gamma}_k^{(3)}$, $k = 1, \ldots, 15$, combines the internal data simulated from *Pareto*(4, 1), external data giving $\mathbb{E}[\Theta_\gamma] = 4.5$, and expert opinion $\delta = 3$. It is compared with the Bayes estimate $\hat{\gamma}_k^{(2)}$ (△), which combines external data and internal data, and the classical MLE, $\hat{\gamma}_k^{\text{MLE}}$ (•). See Example 15.6 for details

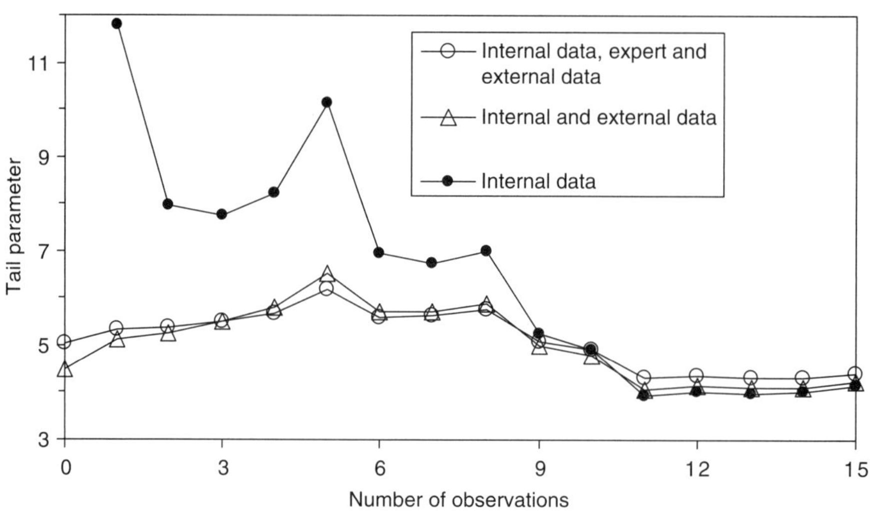

FIGURE 15.9 (○) The Bayes estimate $\hat{\gamma}_k^{(3)}$, $k = 1, \ldots, 15$, combines the internal data simulated from *Pareto*(4, 1), external data giving $\mathbb{E}[\Theta_\gamma] = 4.5$, and expert opinion $\delta = 5$. It is compared with the Bayes estimate $\hat{\gamma}_k^{(2)}$ (△), which combines external data and internal data, and the classical MLE, $\hat{\gamma}_k^{\text{MLE}}$ (•). See Example 15.6 for details

are available. This example also shows that consideration of the relevant external data and well-specified expert opinions stabilizes and smoothes the estimator in an appropriate way. ■

15.5 Combining Data Sources Using Credibility Theory

Quantification of the frequency and severity distributions of the low-frequency/high-severity losses (that typically account for most of the OpRisk capital) is a challenging task. The data are so limited that often full quantification of frequency, severity, and related prior distributions is problematic. In this situation, methods of credibility theory are very useful as they require less information. Credibility theory approach has been successfully used in the insurance industry and actuarial sciences for many decades. It can be used to estimate frequency and severity distributions of the low-frequency large losses in each risk cell by taking into account bank internal data, expert opinions, and industry data. An excellent textbook on credibility theory is the one by Bühlmann and Gisler (2005); also see Kaas *et al.* (2001, section 7.2).

Consider a model parameterized by θ that generates data $X_1, \ldots, X_n, \ldots$. In general, we are interested in estimation of some function of θ (e.g., $\mu(\theta)$) given past data $\boldsymbol{X} = (X_1, \ldots, X_n)$. Under the Bayesian approach, θ is modeled by random variable Θ. Let $\hat{\mu(\Theta)}$ be some estimator of $\mu(\Theta)$. Then the unconditional mean squared (MSEP) error of prediction of an estimator $\hat{\mu(\Theta)}$ is

$$\begin{aligned}
\text{MSEP} &= \mathbb{E}[(\mu(\Theta) - \hat{\mu(\Theta)})^2] \\
&= \mathbb{E}\left[\mathbb{E}\left[\left(\mu(\Theta) - \mathbb{E}[\mu(\Theta)|\boldsymbol{X}] + \mathbb{E}[\mu(\Theta)|\boldsymbol{X}] - \hat{\mu(\Theta)}\right)^2 |\boldsymbol{X}\right]\right] \\
&= \mathbb{E}\left[(\mu(\Theta) - \mathbb{E}[\mu(\Theta)|\boldsymbol{X}])^2\right] + \mathbb{E}\left[\left(\mathbb{E}[\mu(\Theta)|\boldsymbol{X}] - \hat{\mu(\Theta)}\right)^2\right].
\end{aligned}$$

It is easy to see that the posterior mean

$$\hat{\mu(\Theta)} = \mathbb{E}[\mu(\Theta)|\boldsymbol{X}],$$

minimizes MSEP and thus is the best estimator with respect to the quadratic loss function; also see Section 7.3.

In general, the posterior mean cannot be found in closed form. The prior and conditional distributions should also be specified, which is certainly a problem in the case of small datasets. The credibility theory initiated by Bühlmann (1970) considers estimators that are linear in observations $X_1, X_2, \ldots$ and minimize a quadratic loss function. This allows for simple calculation of the estimators, referred to as *credibility estimators* or linear Bayes estimators.

The credibility estimators have already appeared in the previous sections. For example, the estimator for the expected intensity of events (15.19), when frequencies are modeled by *Poisson*$(\Lambda = \lambda)$ and the prior for Λ is *Gamma*(α, β), is

$$\hat{\Lambda} = \mathbb{E}[\Lambda|N_1, \ldots, N_T] = w\overline{N} + (1 - w)\lambda_0,$$

where

- $\overline{N} = \frac{1}{T}\sum_{t=1}^{T} N_t$ is the estimator of λ using the observed counts only;
- $\lambda_0 = \alpha\beta$ is the estimator of λ using a prior distribution only (e.g., specified by expert or from external data);
- $w = \frac{T}{T+1/\beta}$ is the credibility weight in [0,1) used to combine λ_0 and $\overline{N}$.

The estimator $\hat{\Lambda}$ is linear in data $N_1, \ldots, N_T$ and minimizes the MSEP

$$\mathbb{E}[(\hat{\Lambda} - \Lambda)^2].$$

Of course, the estimator $\hat{\Lambda}$ was derived assuming a specific prior distribution. The credibility theory avoids this assumption and requires the knowledge of the first and second moments only. To demonstrate the idea, consider a simplistic credibility model.

Model Assumptions 15.8 (Simple Credibility Model) *Consider the following credibility model assumptions that will admit a linear Bayes estimator of the conditional mean $\mu(\theta)$:*

- *Given $\Theta = \theta$, random variables $X_1, X_2, \ldots$ are independent and identically distributed with*

$$\mu(\theta) = \mathbb{E}[X_j|\Theta = \theta], \quad \sigma^2(\theta) = \mathbb{V}\mathrm{ar}[X_j|\Theta = \theta].$$

- *Θ is a random variable with*

$$\mu_0 = \mathbb{E}[\mu(\Theta)], \quad \tau^2 = \mathbb{V}\mathrm{ar}[\mu(\Theta)].$$

The aim of credibility estimators is to find an estimator of $\mu(\Theta)$ that is linear in $X_1, \ldots, X_n$, that is,

$$\widehat{\mu(\Theta)} = \hat{a}_0 + \hat{a}_1 X_1 + \cdots + \hat{a}_n X_n$$

and minimize quadratic loss function, that is,

$$(\hat{a}_0, \ldots, \hat{a}_n) = \min_{a_0, \ldots, a_n} \mathbb{E}\left[(\mu(\Theta) - a_0 - a_1 X_1 - \cdots - a_n X_n)^2\right].$$

The invariance of the distribution of $X_1, \ldots, X_n$ under permutations of X_j gives $\hat{a}_1 = \hat{a}_2 = \cdots = \hat{a}_n := \hat{b}$. Then, by solving the minimization problem for two parameters a_0 and b by setting corresponding partial derivatives with respect to a_0 and b to zero, one obtains

$$\widehat{\mu(\Theta)} = w\overline{X} + (1 - w)\mu_0,$$

where

$$w = \frac{n}{n + \sigma^2/\tau^2}, \quad \overline{X} = \frac{1}{n}\sum_{i=1}^{n} X_i.$$

For details of the proof and discussion, see Bühlmann and Gisler (2005, section 3.1).

15.5.1 BÜHLMANN–STRAUB MODEL

In OpRisk, we are interested in the Loss Distribution Approach (LDA) model for the annual loss. That is, for a risk cell, the annual number of events $N_1, N_2, \ldots$ are modeled as random variables from some discrete distribution $P(\cdot|\boldsymbol{\lambda})$ and the severities of the events $X_1, X_2, \ldots$ are modeled as random variables from a continuous distribution $F(\cdot|\boldsymbol{\theta})$. Under the Bayesian approach, $\boldsymbol{\lambda}$ and $\boldsymbol{\theta}$ are distribution parameters that are not known and are modeled by random variables $\boldsymbol{\Lambda}$ and $\boldsymbol{\Theta}$, respectively. Often, the credibility approach takes the empirical Bayes setup (see Section 15.2.2). That is, it considers a group of risks, where $\boldsymbol{\Lambda}$ are different for different risks but are drawn from the same distribution (the prior distribution) common across the risks (and similar for $\boldsymbol{\Theta}$). In this framework, we do not consider risks individually but regard each risk as embedded in a group of *similar* risks (*collective*). If a pure Bayesian setup is taken, then the prior distribution is specified by the expert.

Usually, the credibility estimators are used to estimate the expected number of events or the expected loss. However, in general, they can be applied to estimate any square integrable valued random variable Z based on some known random vector $\boldsymbol{Y}$. For example, the elements of $\boldsymbol{Y}$ can be the MLEs, transformed data, quantiles, etc. In particular, the credibility estimators for the severity and frequency distribution parameters can be calculated using the model developed by Bühlmann and Straub (1970); see also Bühlmann and Gisler (2005, model assumptions 4.1 and theorems 4.2, 4.4).

Model Assumptions 15.9 (Bühlmann–Straub Model) *Consider a portfolio of J risks modeled by random variables $Y_{j,k}: \ k = 1, 2, \ldots, j = 1, \ldots, J$. Assume that, for known weights $w_{j,k}$, the j-th risk is characterised by an individual risk profile θ_j, which is itself the realization of a random variable Θ_j, and*

- *Given Θ_j, the data $Y_{j,1}, Y_{j,2}, \ldots$ are independent with*

$$\mathbb{E}[Y_{j,k}|\Theta_j] = \mu(\Theta_j), \quad \mathbb{V}\mathrm{ar}[Y_{j,k}|\Theta_j] = \sigma^2(\Theta_j)/w_{j,k} \tag{15.96}$$

  for all $j = 1, \ldots, J$;
- *The pairs $(\Theta_1, Y_{1,k}; \ k \geq 1), \ldots, (\Theta_J, Y_{J,k}; \ k \geq 1)$ are independent;*
- *$\Theta_1, \ldots, \Theta_J$ are independent and identically distributed with*

$$\mu_0 = \mathbb{E}[\mu(\Theta_j)], \quad \sigma^2 = \mathbb{E}[\sigma^2(\Theta_j)], \quad \tau^2 = \mathbb{V}\mathrm{ar}[\mu(\Theta_j)]$$

  for all j.

Theorem 15.7 (Bühlmann–Straub Credibility Estimators) *Under Model Assumptions 15.9, given the available data $\boldsymbol{Y}_j = (Y_{j,1}, \ldots, Y_{j,K_j}), j = 1, \ldots, J$, the inhomogeneous and homogeneous credibility estimators of $\mu(\Theta_j)$ are given as follows:*

- *The inhomogeneous credibility estimator is*

$$\widehat{\widehat{\mu(\Theta_j)}} = \alpha_j \overline{Y}_j + (1 - \alpha_j)\mu_0. \tag{15.97}$$

- *The homogeneous credibility estimator is*

$$\widehat{\widehat{\mu(\Theta_j)}}^{\mathrm{hom}} = \alpha_j \overline{Y}_j + (1 - \alpha_j)\hat{\mu}_0. \tag{15.98}$$

Here:

$$\hat{\mu}_0 = \sum_{j=1}^{J} \frac{\alpha_j}{\alpha_0} \overline{Y}_j, \quad \overline{Y}_j = \sum_{k=1}^{K_j} \frac{w_{j,k}}{\widetilde{w}_j} Y_{j,k}, \quad \alpha_j = \frac{\widetilde{w}_j}{\widetilde{w}_j + \sigma^2/\tau^2},$$

$$\alpha_0 = \sum_{j=1}^{J} \alpha_j, \quad \widetilde{w}_j = \sum_{k=1}^{K_j} w_{j,k}.$$

Remark 15.16

- *Note that K_j may vary between the risks;*
- *Structural parameters μ_0, σ^2, and τ^2 can be determined using expert opinions (pure Bayes) or using data of all risks (empirical Bayes);*
- *The difference between inhomogeneous and homogeneous credibility estimators is that the latter estimates μ_0 by $\hat{\mu}_0$ using the data for all risks.*

Using these credibility estimators, Bühlmann *et al.* (2007) suggested a "toy" model for OpRisk, where the Pareto and Poisson distributions were used for modeling severity and frequency, respectively. Although the model might be simple, it is a very good illustration of a consistent credibility approach for estimating low-frequency/high-severity OpRisks. We illustrate the use of the model in a simple case of J risks without considering a full hierarchical model.

15.5.2 MODELING FREQUENCY

Consider a collection of J similar risk cells (see Figure 15.10). Let $N_{j,k}$ be the annual number of events, with the event losses exceeding some high threshold L, in the j-th risk cell ($j = 1, \ldots, J$) in the k-th year. That is, the same threshold L is used across all risk cells in a collection (e.g., one can choose the threshold equal to the threshold in the database of external data).

Model Assumptions 15.10 (Poisson Frequency) *Assume the following:*

(a) *Given $\Lambda_j = \lambda_j$, $N_{j,k}$ are independent and distributed from Poisson$(\nu_j\lambda_j)$, that is,*

$$\Pr\left[N_{j,k} = n | \Lambda_j = \lambda_j\right] = \frac{\left(\nu_j\lambda_j\right)^n}{n!} \exp\left(-\nu_j\lambda_j\right) \tag{15.99}$$

and moments

$$\mathbb{E}[N_{j,k}|\Lambda_j] = \nu_j\Lambda_j, \quad \mathbb{V}\mathrm{ar}[N_{j,k}|\Lambda_j] = \nu_j\Lambda_j. \tag{15.100}$$

The arrival rate parameter is defined as $\nu_j\Lambda_j$, where ν_j are the known a priori constants and Λ_j are the risk profiles of the bank cells. The constants ν_j are scaling factors, reflecting differences in frequencies across the risks, discussed later;

(b) *Assume that $\Lambda_1, \ldots, \Lambda_J$ are independent and identically distributed with*

$$\mathbb{E}[\Lambda_j] = \lambda_0 \quad \text{and} \quad \mathbb{V}\mathrm{ar}[\Lambda_j] = (\omega_0)^2,$$

and $(\Lambda_1, N_{1,k};\ k \geq 1), \ldots, (\Lambda_J, N_{J,k}; k \geq 1)$ are independent;

(c) *The available data are* $\{N_{j,1}, \ldots, N_{j,K_j} : j = 1, \ldots, J\}$.

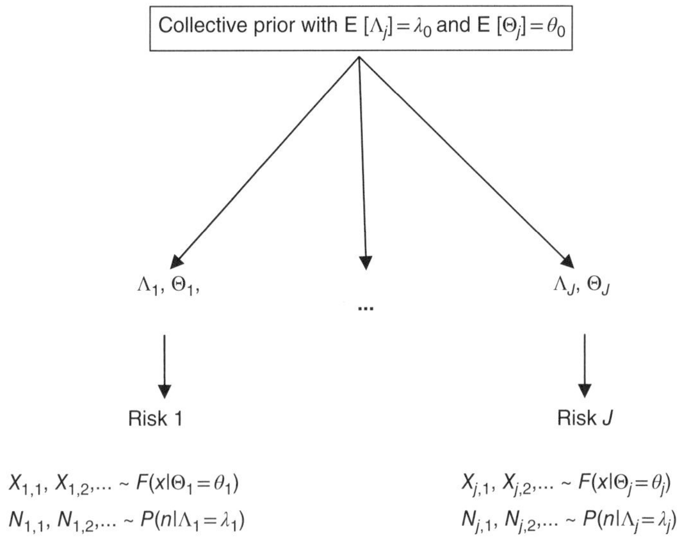

FIGURE 15.10 Example of the credibility model for OpRisk. Given $\Theta_j = \theta_j$ and $\Lambda_j = \lambda_j$, $X_{j,k} \sim Pareto(a_j\theta_j, L_j)$ and $N_{j,k} \sim Poisson(\nu_j\lambda_j)$ are the losses (above threshold L_j and their annual frequencies in risk cells $j = 1, \ldots, J$, respectively. The risk profiles Λ_j are drawn from common distribution with $\mathbb{E}[\Lambda_j] = \lambda_0, \mathbb{V}\text{ar}[\Lambda_j] = (\omega_0)^2$; risk profiles Θ_j are from common distribution with $\mathbb{E}[\Theta_j] = \theta_0$, $\mathbb{V}\text{ar}[\Theta_j] = (\tau_0)^2$. Scaling factors a_j and ν_j for the relative differences between the risks can be specified using expert opinions or known factors

The Arrival Rate MLE Using Data in a Risk Cell

Under the first assumption in Model Assumptions 15.10, $N_{j,k}$, $k = 1, \ldots, K_j$ in the j-th risk cell are conditionally independent. Thus, given $\Lambda_j = \lambda_j$, the standard MLE of λ_j is

$$\hat{\Lambda}_j = \frac{1}{\widetilde{\nu}_j} \sum_{k=1}^{K_j} N_{j,k}, \quad \widetilde{\nu}_j = \nu_j K_j \tag{15.101}$$

with

$$\mathbb{E}[\hat{\Lambda}_j | \Lambda_j = \lambda_j] = \lambda_j,$$
$$\mathbb{V}\text{ar}[\hat{\Lambda}_j | \Lambda_j = \lambda_j] = \lambda_j / \widetilde{\nu}_j.$$

Again, a common situation in OpRisk is that only a few large losses are observed for some risk cells, so the standard MLEs of parameters λ_j will not be reliable. The idea is to use data from a collection of risks to improve the estimates of the arrival rate parameter.

The Arrival Rate Estimator Improved by Bank Data

Under the second assumption in Model Assumptions 15.10, $\Lambda_1, \Lambda_2, \ldots$ are independent and identically distributed with $\mathbb{E}[\Lambda_j] = \lambda_0$ and $\mathbb{V}\text{ar}[\Lambda_j] = (\omega_0)^2$. Observe that the standardized frequencies $F_{j,k} = N_{j,k}/\nu_j$ satisfy

$$\mathbb{E}[F_{j,k}|\Lambda_j] = \Lambda_j \quad \text{and} \quad \mathbb{V}\mathrm{ar}[F_{j,k}|\Lambda_j] = \Lambda_j/\nu_j. \tag{15.102}$$

Thus, $F_{j,k}$ satisfy the Bühlmann–Straub model (15.96)-(15.98) and the credibility estimator for Λ_j is given by

$$\hat{\hat{\Lambda}}_j = \gamma_j \hat{\Lambda}_j + (1 - \gamma_j)\lambda_0, \tag{15.103}$$

where

$$\gamma_j = \frac{\widetilde{\nu}_j}{\widetilde{\nu}_j + \lambda_0/(\omega_0)^2}. \tag{15.104}$$

The structural parameters λ_0 and ω_0 can be estimated using all data from a collection of J risks by solving two nonlinear equations (using, for example, an iterative procedure; see Bühlmann and Gisler (2005, pp. 102–103):

$$(\hat{\omega}_0)^2 = \max\left[c \times \left\{A - \frac{J\hat{\lambda}_0}{\nu_0}\right\}, 0\right], \quad \hat{\lambda}_0 = \frac{1}{\widetilde{\widetilde{\gamma}}} \sum_j \gamma_j \hat{\Lambda}_j, \tag{15.105}$$

where

$$\nu_0 = \sum_{j=1}^{J} \widetilde{\nu}_j, \quad A = \frac{J}{J-1} \sum_{j=1}^{J} \frac{\widetilde{\nu}_j}{\nu_0}(\hat{\Lambda}_j - \overline{F})^2, \quad \widetilde{\gamma} = \sum_j \gamma_j,$$

$$\overline{F} = \frac{1}{J} \sum_{j=1}^{J} \hat{\Lambda}_j, \quad c = \frac{J}{J-1} \left\{ \sum_{j=1}^{J} \frac{\widetilde{\nu}_j}{\nu_0} \left(1 - \frac{\widetilde{\nu}_j}{\nu_0}\right) \right\}^{-1}.$$

Here, the coefficients γ_j are given in Equation (15.104) with λ_0 and ω_0 replaced by $\hat{\lambda}_0$ and, $\hat{\omega}_0$ respectively.

Remark 15.17

- *Based on the cell data and all data in a collection of J risks, the best credibility estimator of the arrival rate parameter in the j-th cell is* $\nu_j \hat{\hat{\Lambda}}_j$;
- *We assumed that the constants* ν_j *are known a priori. Note that these constants are defined up to a constant factor, that is, the coefficients* γ_j *(and the final estimates of the arrival rate parameters) will not change if all* ν_j *are changed by the same factor. Hence, only relative differences between risks play a role. These constants have the interpretation of a priori differences and can be fixed by the expert opinions on expected annual number of losses exceeding threshold L for each risk cell. For example, the expert may estimate the expected annual number of events (exceeding threshold* L_j*) denoted* n_j *in the j-th cell as* $\hat{n}_j$ *and estimate* ν_j *as* $\hat{n}_j/\lambda_0$. *Only relative differences play a role here; thus (without loss of generality),* λ_0 *can be set equal to 1. For an example of using expert opinions for quantification of frequency and severity distributions, see Alderweireld et al. (2006) and Shevchenko and Wüthrich (2006).*

15.5.3 MODELING SEVERITY

Again, consider a collection of J similar risk cells (see Figure 15.10).

Model Assumptions 15.11 (Pareto Severity) *Assume the following:*

- *Given,* $\Theta_j = \theta_j$, *the losses* $X_{j,k}$, $k \geq 1$ *above threshold* L_j *in the j-th risk cell* ($j = 1, \ldots, J$) *are independent and Pareto-distributed, Pareto*$(a_j\theta_j, L)$, *with the density*

$$f(x) = \frac{a_j\theta_j}{L}\left(\frac{x}{L}\right)^{-a_j\theta_j - 1} \tag{15.106}$$

respectively, for $x \geq L$ *and* $a_j\theta_j > 0$. *It is assumed that the threshold L is known and is the same across risk cells. Here* a_j *are known a priori constants (differences) and* θ_j *are the risk profiles of the cells in the bank. The constants* a_j *are scaling factors, reflecting differences in severities across the risks, which can be fixed by experts as discussed;*

- *Assume that* $\Theta_1, \ldots, \Theta_J$ *are independent and identically distributed with*

$$\mathbb{E}[\Theta_j] = \theta_0 \quad \textit{and} \quad \mathbb{V}\mathrm{ar}[\Theta_j] = (\tau_0)^2,$$

and $(\Theta_1, X_{1,k};\ k \geq 1), \ldots, (\Theta_J, X_J, k;\ k \geq 1)$ *are independent. Here,* θ_0 *is a risk profile of the collection.*

- *The available data are denoted as* $\{X_{j,1}, \ldots, X_{j,\widetilde{K}_j} : j = 1, \ldots, J\}$.

Remark 15.18

- *Note that the number of available losses in the j-th risk cell, denoted as* $\widetilde{K}_j$, *is the number of events over* K_j *years. The latter is the number of observed years for modeling annual frequencies in the previous section;*
- *The results in this section are valid if thresholds are different for different risk cells, although in the previous section for modeling frequencies we assumed the same threshold across risk cells;*
- *The Pareto distribution is often used in the insurance industry to model large claims and is a good candidate for modeling large OpRisk losses. It is interesting to note that the conditional distribution of the losses exceeding any higher level* $\widetilde{L}$ *is also a Pareto distribution with parameters* $a_j\theta_j$ *and* $\widetilde{L}$.

The Tail Parameter MLE Using Data in a Risk Cell

Under the first assumption in Model Assumptions 15.11, the losses $X_{j,k}$, $k \geq 1$ in the j-th risk cell are conditionally (given Θ_j) independent and Pareto-distributed. Thus, MLE of θ_j is

$$\hat{\Psi}_j = \left[\frac{a_j}{\widetilde{K}_j}\sum_{k=1}^{\widetilde{K}_j} \ln\left(\frac{X_{j,k}}{L}\right)\right]^{-1}. \tag{15.107}$$

It is easy to show (see Rytgaard 1990) that an unbiased estimator of θ_j is

$$\hat{\Theta}_j = \frac{\widetilde{K}_j - 1}{\widetilde{K}_j}\hat{\Psi}_j, \tag{15.108}$$

with

$$\mathbb{E}[\hat{\Theta}_j|\Theta_j = \theta_j] = \theta_j, \quad \mathbb{V}\text{ar}[\hat{\Theta}_j|\Theta_j = \theta_j] = \frac{(\theta_j)^2}{\widetilde{K}_j - 2}. \tag{15.109}$$

A common situation in OpRisk is that only a few losses are observed for certain risk cells. Thus, the standard MLE $a_j\hat{\Theta}_j$ (based on the data in the j-th risk cell only) for the Pareto tail parameters will not be reliable (this is easy to see from the variance in (15.109)). The idea is to use the collective losses (from bank, industry, etc.) to improve the estimates of the Pareto parameters in the risk cells.

The Tail Parameter Estimator Improved by Collective Data

The tail parameter estimator $a_j\hat{\Theta}_j$ can be improved using all data in the collection of J risks as follows. Under the second assumption in Model Assumptions 15.11, $\Theta_1, \ldots, \Theta_J$ are independent and identically distributed random variables with $\mathbb{E}[\Theta_j] = \theta_0$ and $\mathbb{V}\text{ar}[\Theta_j] = (\tau_0)^2$, where θ_0 is a risk profile for the whole collective. Observe that the unbiased estimators $\hat{\Theta}_j$ (see 15.109), satisfy the assumptions of the Bühlmann–Straub model (15.96)–(15.98) and thus the credibility estimator is given by

$$\hat{\hat{\Theta}}_j = \alpha_j\hat{\Theta}_j + (1 - \alpha_j)\theta_0, \tag{15.110}$$

where

$$\alpha_j = \frac{\widetilde{K}_j - 2}{\widetilde{K}_j - 1 + (\theta_0/\tau_0)^2}.$$

The structural parameters θ_0 and $(\tau_0)^2$ can be estimated using data across all risk cells in the bank by solving two nonlinear equations (using, for example, an iterative procedure; see Bühlmann and Gisler, 2005, pp. 116–117):

$$(\hat{\tau}_0)^2 = \frac{1}{J-1}\sum_{j=1}^{J} \alpha_j(\hat{\Theta}_j - \hat{\theta}_0)^2, \quad \hat{\theta}_0 = \frac{1}{W}\sum_{j=1}^{J} \alpha_j\hat{\Theta}_j, \tag{15.111}$$

where $W = \sum_{j=1}^{J} \alpha_j$.

Here, the coefficients α_j are given in (15.110), with θ_0 and $(\tau_0)^2$ replaced by $\hat{\theta}_0$ and $(\hat{\tau}_0)^2$, respectively. If the solution for $(\hat{\tau}_0)^2$ is negative, then we set $\alpha_j = 0$ and

$$\hat{\theta}_0 = \frac{1}{W}\sum_{j=1}^{J} w_j\hat{\Theta}_j, \quad w_j = K_j - 2, \quad W = \sum_{j=1}^{J} w_j.$$

The best credibility estimate for the tail parameter in the j-th cell (based on the cell data and all data in the collection) is $a_j\hat{\hat{\Theta}}_j$. We assume that constants a_j are known a priori. Note that these constants are defined up to a constant factor, that is, coefficients α_j (and final estimates of tail parameters) will not change if all $a_j, j = 1, \ldots, J$, are changed/scaled by the same factor. Hence, only relative differences between risks play a role. These constants have the interpretation of a priori differences and can be fixed by an expert using opinions on, for example, quantiles of losses exceeding L_j. For example, the expert may estimate the probability q_j, that the loss in the j-th cell will exceed level H_j, as $\hat{q}_j$ and use relations

$$a_j\theta_j = -\ln q_j/\ln(H_j/L_j) \quad \text{and} \quad \mathbb{E}[\Theta_j] = \theta_0$$

to estimate a_j as $-\ln\hat{q}_j/[\theta_0 \ln(H_j/L_j)]$. Only relative differences play a role, so here (without loss of generality) θ_0 can be set equal to 1. Experts may specify several quantiles, then a_j can be estimated using, for example, a least squared method. Ideally, the expert specifying constants a_j has a complete overview of all risk cells in the bank, as only relative differences between risks are important. However, in practice, opinions from experts with special knowledge of business specifics within a risk cell are required. Combining opinions from different experts is one of the problems to be resolved by a practitioner.

15.5.4 NUMERICAL EXAMPLE

To illustrate the previous procedures, consider an example where losses (exceeding USD 1 million) are observed across 10 risk cells as given in Table 15.1 and all risk cells are the same a priori, $a_1 = \cdots = a_{10} = 1$. Using these losses, the MLEs for the tail parameters $\hat{\Theta}_j$, presented in Table 15.1, are calculated by (15.108). Then, using (15.110) and (15.111), we estimate the structural parameters $(\hat{\tau}_0)^2 \approx 1.116$ and $\hat{\theta}_0 \approx 3.157$, and credibility coefficients $\alpha_j \approx 0.446$ (the coefficients are the same because equal number of losses are observed in the cells).

TABLE 15.1 **Losses (in million USD) exceeding USD 1 million observed across 10 risk cells, and corresponding maximum likelihood and credibility estimators for the Pareto tail parameter in the risk cells**

Cell 1	Cell 2	Cell 3	Cell 4	Cell 5	Cell 6	Cell 7	Cell 8	Cell 9	Cell 10
Losses (in million USD) exceeding USD 1 million observed in risk cells									
1.557	9.039	1.166	1.548	1.578	1.201	1.006	1.741	1.364	1.074
1.079	2.138	1.037	1.040	1.282	2.815	1.169	1.165	2.036	1.103
1.047	1.008	1.136	1.045	1.092	3.037	1.215	1.010	1.014	1.664
1.199	1.761	2.104	1.774	1.658	1.001	1.116	1.096	1.217	1.049
1.395	1.654	1.774	1.045	2.025	1.114	1.010	1.060	1.202	1.104
1.060	1.073	1.161	1.856	1.129	1.422	1.560	1.352	1.095	2.924
3.343	2.435	1.080	1.636	1.946	2.397	1.059	1.044	1.348	1.265
2.297	4.357	1.154	1.403	1.831	1.241	1.059	1.678	1.191	1.333
1.297	1.576	1.257	2.522	1.478	1.522	1.050	1.882	1.161	1.424
1.180	1.113	1.231	1.113	1.208	1.243	1.231	1.401	1.017	1.435
Maximum likelihood estimators (MLEs) $\hat{\Theta}_j$, $j = 1, \ldots, 10$									
2.499	1.280	3.688	2.487	2.264	1.992	6.963	3.335	4.194	2.870
Credibility estimators $\hat{\hat{\Theta}}_j$, $j = 1, \ldots, 10$ disregarding industry data									
2.863	2.319	3.394	2.858	2.759	2.637	4.855	3.236	3.620	3.029

The credibility estimators $\hat{\hat{\Theta}}_j$, shown in Table 15.1, are calculated using (15.110). In this example, the MLEs are quite volatile as the number of observations is small. For example, cell 7 has no large losses and thus its MLE is high; cell 10 has one large loss and thus its MLE is smaller. One could easily calculate cell MLEs versus the number of observations in a cell and observe that MLEs are highly volatile for a small number of observations. One important observation may lead to a substantial change in the MLE. The credibility estimators (based on data in the bank) are smoother in comparison with MLEs. This is because a credibility estimator is a weighted average, according to credibility theory, between a risk cell MLE and the estimator of the structural parameter $\hat{\theta}_0$ based on all data in the collection. The credibility weights α_j are approximately 0.45, which means that a risk cell MLE (based on observations in a cell) $\hat{\Theta}_j$ and the a priori estimate $\hat{\theta}_0 \approx 3.157$ are weighted with 0.45, and 0.55, respectively.

15.5.5 REMARKS AND INTERPRETATION

The credibility formulas (15.103) and (15.110) for the frequency and severity parameter estimators, based on a cell and collective data, have a simple interpretation.

- As the number of observations in the j-th cell increases, the larger are the credibility weights γ_j and α_j that are assigned to the estimators $\hat{\Lambda}_j$ and $\hat{\Theta}_j$ (based on the cell observations) and the lesser are the weights that are assigned to the estimators $\hat{\theta}_0$ and $\hat{\lambda}_0$ (based on all observations in a collection of risks), respectively;
- Also, the larger the τ_0 and ω_0 (variance across risk cells in a collection), the larger are the weights that are assigned to $\hat{\Theta}_j$ and $\hat{\Lambda}_j$ correspondingly. For a detailed discussion on the credibility parameters, refer to Bühlmann and Gisler (2005, section 4.4).

It is not difficult to consider a hierarchical model, where the collection of risks is part of another larger collection. For example, one can consider the collection of similar risks in the bank and then consider a collection of banks (i.e., the banking industry). This will further improve the estimates of arrival rate $\nu_j\lambda_j$ and the tail parameter $a_j\theta_j$. This can be done using a hierarchical credibility model (see Bühlmann and Gisler, 2005, chapter 6). In particular, one can consider M banks with bank-specific parameters $\lambda_0^{(m)}$ and $\theta_0^{(m)}$ modeled by random variables $\Lambda_0^{(m)}$ and $\Theta_0^{(m)}$, $m = 1, \ldots, M$, respectively. Then assume the following:

(a) $\Lambda_0^{(m)}$ are independent and identically distributed random variables with

$$\mathbb{E}[\Lambda_0^{(m)}] = \lambda_{coll} \quad \text{and} \quad \mathbb{V}\text{ar}[\Lambda_0^{(m)}] = \omega_{coll}^2.$$

(b) $\Theta_0^{(m)}$ are independent and identically distributed random variables with

$$\mathbb{E}[\vartheta_0^{(m)}] = \vartheta_{coll} \quad \text{and} \quad \mathbb{V}\text{ar}[\vartheta_0^{(m)}] = \tau_{coll}^2.$$

The credibility weights and estimators in such a hierarchical model can be calculated as described by Bühlmann *et al.* (2007).

The Capital Calculations. For the purposes of the regulatory capital calculations of OpRisk, the annual loss distribution, in particular its 0.999 quantile (VaR) as a risk measure, should be quantified for each Basel II risk cell in the matrix of eight business lines times seven risk types and for the whole bank. The credibility model presented here is for modeling low-frequency/high-severity losses exceeding some large threshold L. Given the credibility estimates for the model parameters, the annual loss distribution can be calculated as usual using methods listed in Chapter 13; also see Bühlmann *et al.* (2007, section 5). Of course, modeling of the high-frequency/low-severity losses (below threshold L) should be added to the model before the final OpRisk capital charge is estimated. For a related actuarial literature on this topic, see Sandström (2006) and Wüthrich (2006). That is, one can model the losses above threshold L using credibility theory as described earlier, while the losses below the threshold are modeled separately. Note that typically the low-frequency/high-severity losses give the largest contribution to the final capital charge. The number of high-frequency/low-impact losses recorded in the bank internally is usually large enough to obtain reliable estimates by a standard fitting of the frequency and severity distributions without the use of the external data.

The important assumption in calculating the credibility estimates is that the risk cells are independent. While it is an important (and quite realistic) assumption of the proposed model that the low-frequency/high-severity losses from different risk cells are independent, dependence can be considered between the high-frequency/low-impact losses from different risk cells. Accurate quantification of the dependencies between the risks is a difficult task; this is an open field for future research. The dependence can be introduced using different methods (copula methods, common shocks, etc.), which are discussed in Chapters 10, 11 and 12.

15.6 Nonparametric Bayesian Approach via Dirichlet Process

Typically, under the Bayesian approach, we assume that there is an unknown distribution underlying observations $x_1, \ldots, x_n$ and this distribution is parametrized by $\boldsymbol{\theta}$. Then we place a prior distribution on the parameter $\boldsymbol{\theta}$ and try to infer the posterior of $\boldsymbol{\theta}$ given observations $x_1, \ldots, x_n$. Under the nonparametric approach, we do not make an assumption that the underlying loss process–generating distribution is parametric; we put a prior on the distribution directly and find the posterior of the distribution given data that is a combination of the prior with the empirical data distribution.

One of the most popular Bayesian nonparametric models is based on the Dirichlet process introduced by Ferguson (1973). The Dirichlet process represents a probability distribution of the probability distributions. It can be specified in terms of a base distribution $H(x)$ and a scalar concentration parameter $\alpha > 0$ and denoted as $DP(\alpha, H)$. For example, assume that we model severity distribution $F(x)$, which is unknown and modeled as random at each point x using $DP(\alpha, H)$. Then, the mean value of $F(x)$ is the base distribution $H(x)$ and variance of $F(x)$ is $H(x)(1 - H(x))/(\alpha + 1)$. That is, as the concentration parameter α increases, the true distribution comes closer to the base distribution $H(x)$. Each draw from the Dirichlet process is a distribution function and for $x_1 < x_2 < \cdots < x_k$, the distribution of

$$F(x_1), F(x_2) - F(x_1), \ldots, 1 - F(x_k),$$

is a $k + 1$ multivariate Dirichlet distribution

$$\mathit{Dirichlet}(\alpha H(x_1), \alpha(H(x_2) - H(x_1)), \ldots, \alpha(1 - H(x_k))),$$

formally defined as follows.

Definition 15.2 (Dirichlet Distribution) *A d-variate Dirichlet distribution is denoted as* $Dirichlet(\alpha_1, \alpha_2, \ldots, \alpha_d)$*, where* $\alpha_i > 0$. *The random vector* $(Q_1, Q_2, \ldots, Q_d)$ *has a Dirichlet distribution if its density function is*

$$f(q_1, q_2, \ldots, q_{d-1}) = \frac{\Gamma(\alpha_1 + \cdots + \alpha_d)}{\prod_{i=1}^{d} \Gamma(\alpha_i)} \prod_{i=1}^{d} q_i^{\alpha_i - 1}, \tag{15.112}$$

where $q_i > 0$ *and* $q_1 + \cdots + q_d = 1$. ■

There are several formal definitions of Dirichlet processes; for a detailed description see Ghosh and Ramamoorthi (2003). For the purposes of this book, we just present a few important results here that can be easily adopted for OpRisk. In particular, the i-th marginal distribution of $Dirichlet(\alpha_1, \ldots, \alpha_d)$ is $Beta(\alpha_i, \alpha_0 - \alpha_i)$, where $\alpha_0 = \alpha_1 + \cdots + \alpha_d$. Thus, the marginal distribution of the Dirichlet process $DP(\alpha, H)$ is Beta distribution $F(x) \sim Beta(\alpha H(x), \alpha(1 - H(x)))$, that is, explicitly it has the Beta density

$$\Pr[F(x) \in dy] = \frac{\Gamma(\alpha)}{\Gamma(\alpha H(x))\Gamma(\alpha(1 - H(x)))} y^{\alpha H(x)} (1 - y)^{\alpha(1 - H(x)) - 1} dy, \tag{15.113}$$

where $\Gamma(\cdot)$ is a Gamma function.

If the prior distribution for $F(x)$ is $DP(\alpha, H)$, then after observing $x_1, \ldots, x_n$, the posterior for $F(x)$ is

$$DP\left(\alpha + n, \frac{\alpha}{\alpha + n} H(x) + \frac{n}{\alpha + n} \frac{1}{n} \sum_{i=1}^{n} \mathbb{I}_{x_i \le x}\right). \tag{15.114}$$

In other words, the Dirichlet process is a conjugate prior with respect to empirical sample distribution; in posterior, our unknown distribution $F(x)$ will have updated concentration parameter $\alpha + n$ and updated base distribution

$$\widetilde{H}(x) = \frac{\alpha}{\alpha + n} H(x) + \frac{n}{\alpha + n} \frac{1}{n} \sum_{i=1}^{n} \mathbb{I}_{x_i \le x}, \tag{15.115}$$

which is a weighted sum of the prior base distribution and empirical distribution with the weights $\alpha/(\alpha + n)$ and $n/(\alpha + n)$, respectively.

The modeler can choose $H(x)$ as an expert opinion on distribution $F(x)$, then the posterior estimate of the distribution $F(x)$ after observing data $x_1, \ldots, x_n$ will be given by $\widetilde{H}(x)$ in (15.115).

Remark 15.19

- *As new data are collected, the posterior distribution, converges to the empirical distribution, which itself converges to the true distribution of* $F(x)$*;*
- *The larger the value of* α*, the less impact new data will have on the posterior estimate of* $F(x)$*; if* $\alpha = 0$*, the posterior distribution will simply be the empirical distribution of the data;*

- *The concentration parameter α can be interpreted as an "effective sample size" associated with the prior estimate. In assigning the value of c, the modeler should attempt to quantify the level of information contained in the scenario estimates, as measured by the equivalent amount of data that would provide a similar level of confidence. The modeler can also estimate α from a likely interval range of severities or frequencies (i.e., from the variance of the possible distribution). Cope (2012) suggests that given the rarity of the scenarios considered, the assigned value of α will likely be low, often less than 10 and possibly as low as 1.*

EXAMPLE 15.7 Combining Scenario with Data Using a Dirichlet Process

Assume that an expert provides estimates in USD million for a risk severity as follows. If loss occurs, then the probability to exceed 10, 30, 50, and 120 are 0.9, 0.5, 0.25, and 0.1, respectively, and the maximum possible loss is USD 600 million. That is, probability distribution $H(x)$ at points (0, 10, 30, 50, 120, 600) is (0, 0.1, 0.5, 0.75, 0.9, 1). This is presented in Figure 15.11 with linear interpolation between specified distribution points.

If we choose the prior for the unknown severity distribution $F(x)$ as $DP(\alpha, H(x))$ with concentration parameter $\alpha = 10$, then expected value for $F(x)$ from the prior is $H(x)$ and bounds for $F(x)$ for each x can be calculated from the marginal beta distribution (15.113). For example, the lower and upper bounds in Figure 15.11 correspond to 0.1 and 0.9 quantiles of the Beta distribution $Beta(\alpha H(x), \alpha(1 - H(x)))$, that is, will contain the true value of $F(x)$ with probability 0.8 for each x.

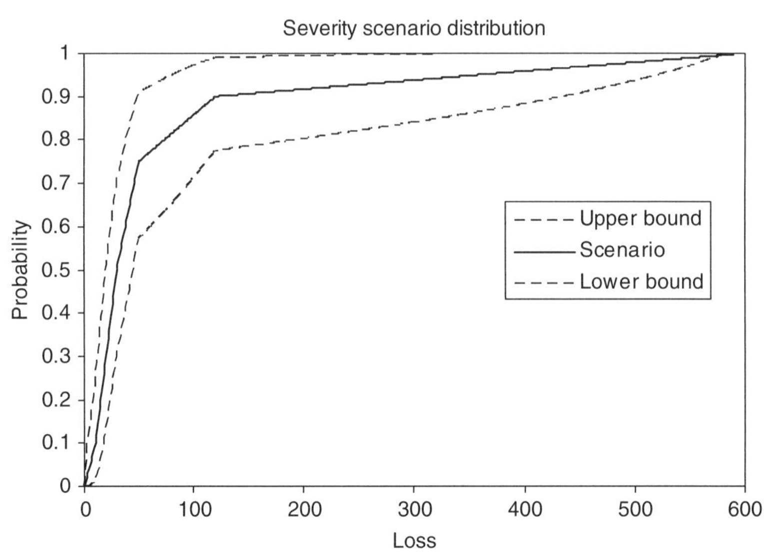

FIGURE 15.11 **Dirichlet marginal bounds for scenario severity distribution; for details, see Example 15.7**

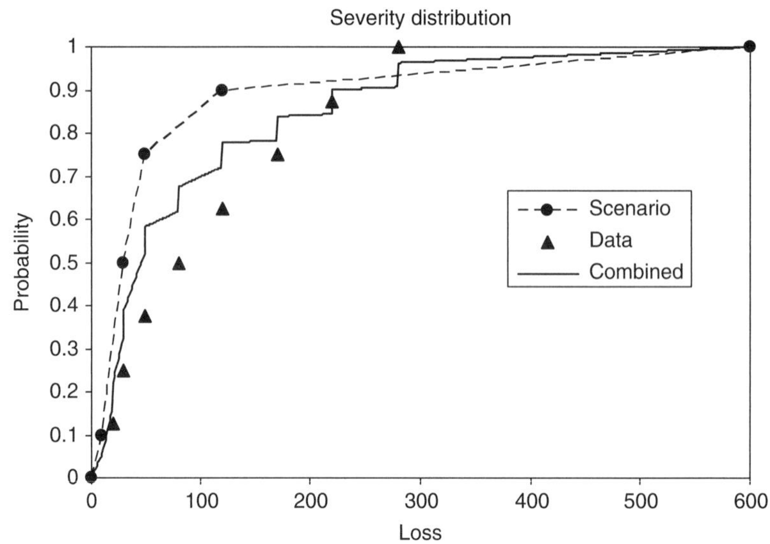

FIGURE 15.12 **Combining scenario severity distribution with empirical distribution of the observed data; for details, see Example 15.7**

Now, assume that we observe the actual losses 20, 30, 50, 80, 120, 170, 220, and 280 all in USD million. The posterior mean of $F(x)$ combining scenario and data is easily calculated using (15.115) and presented in Figure 15.12 along with the empirical data and scenario distribution. ■

15.7 Combining Using Dempster–Shafer Structures and p-Boxes

Often risk assessment includes situations where there is little information on which to evaluate a probability or the information is nonspecific, ambiguous, or conflicting. In this case, one can work with bounds on probability. For example, this idea has been developed by Walley and Fine (1982) and Berleant (1993) and there are suggestions that the idea has its roots from Boole (1854). Williamson and Downs (1990) introduced interval-type bounds on cumulative distribution functions called "probability boxes" or "p-boxes". They also described algorithms to compute arithmetic operations (addition, subtraction, multiplication, and division) on pairs of p-boxes.

The method of reasoning with uncertain information known as Dempster–Shafer theory of evidence was suggested by Dempster (1967, 1968) and Shafer (1976). A special rule to combine the evidence from different sources was formulated by Dempster (1968); it is somewhat controversial and there are many modifications to the rule such as shown by Yager (1986, 1987).

For a good summary on the methods for obtaining Dempster–Shafer structures and "p-boxes", and aggregation methods handling a conflict between the objects from different sources, see Ferson *et al.* (2003). The use of p-boxes and Dempster–Shafer structures in risk analyses offers many significant advantages over a traditional probabilistic approach. Ferson *et al.* (2003) list the following practical problems faced by analysts that can be resolved using these methods:

- Imprecisely specified distributions;
- Poorly known or even unknown dependencies;
- Non-negligible measurement uncertainty;
- Nondetects or other censoring in measurements;
- Small sample size;
- Inconsistency in the quality of input data;
- Model uncertainty; and
- Nonstationarity (nonconstant distributions).

Walley (1991) emphasized that the use of imprecise probabilities does not require one to assume the actual existence of any underlying distribution function. This approach could be useful in risk analyses even when the underlying stochastic processes are nonstationary or could never, even in principle, be identified to precise distribution functions. Oberkampf *et al.* (2001) and Oberkampf (2005) demonstrated how the theory could be used to model uncertainty in engineering applications of risk analysis stressing that the use of p-boxes and Dempster–Shafer structures in risk analyses offers many significant advantages over a traditional probabilistic approach.

These features are certainly attractive for OpRisk, especially for combining expert opinions, and were applied for OpRisk by Sakalo and Delasey (2011). At the same time, some writers consider these methods as unnecessary elaboration that can be handled within the Bayesian paradigm through Bayesian robustness (Berger, 1985, section 4.7). It might also be difficult to justify the application of Dempster's rule (or its other versions) to combine statistical bounds for empirical data distribution with exact bounds for expert opinions.

15.7.1 DEMPSTER–SHAFER STRUCTURES AND P-BOXES

A Dempster–Shafer structure on the real line is similar to a discrete distribution except that the locations where the probability mass resides are sets of real values (*focal elements*) rather than points. The correspondence of probability masses associated with the focal elements is called the basic probability assignment. This is analogous to the probability mass function for an ordinary discrete probability distribution. Unlike a discrete probability distribution on the real line, where the mass is concentrated at distinct points, the focal elements of a Dempster–Shafer structure may overlap one another, and this is the fundamental difference that distinguishes Dempster–Shafer theory from traditional probability theories. Dempster–Shafer theory has been widely studied in computer science and artificial intelligence, but has never achieved complete acceptance among probabilists and traditional statisticians, even though it can be rigorously interpreted as classical probability theory in a topologically coarser space.

Definition 15.3 (Dempster–Shafer Structure) *A finite Dempster–Shafer structure on the real line* $\mathbb{R}$ *is a probability assignment, which is a mapping*

$$m : 2^{\mathbb{R}} \to [0; 1],$$

where $m(\emptyset) = 0$; $m(a_i) = p_i$ *for focal elements* $a_i \subseteq \mathbb{R}$, $i = 1, 2, \ldots, n$; *and* $m(D) = 0$ *whenever* $D \neq a_i$ *for all i, such that* $0 < p_i$ *and* $p_1 + \cdots + p_n = 1$. ∎

For convenience, we will assume that the focal elements a_i are closed intervals $[x_i, y_i]$. Then implementation of a Dempster–Shafer structure will require $3n$ numbers: one for each p_i, and x_i and y_i for each corresponding focal element.

Remark 15.20 *Note that* $2^{\mathbb{R}}$ *denotes a power set. The power set of a set* $\mathbb{S}$ *is the set of all subsets of* $\mathbb{S}$ *including the empty set* $\emptyset$ *and* $\mathbb{S}$ *itself. If* $\mathbb{S}$ *is a finite set with K elements, then the number of elements in its power set is* 2^K. *For example, if* $\mathbb{S}$ *is the set* $\{x, y\}$, *then the power set is* $\{\emptyset, x, y, \{x, y\}\}$.

The upper and lower probability bounds can be defined for a Dempster–Shafer structure. These are called *plausibility* and *belief* functions defined as follows.

Definition 15.4 (Plausibility Function) *The plausibility function corresponding to a Dempster–Shafer structure* $m(A)$ *is the sum of all masses associated with sets that overlap with or merely touch the set* $b \subseteq \mathbb{R}$

$$Pls(b) = \sum_{a_i \cap b \neq \emptyset} m(a_i),$$

which is the sum over i such that $a_i \cap b \neq \emptyset$. ∎

Definition 15.5 (Belief Function) *The belief function corresponding to a Dempster–Shafer structure* $m(A)$ *is the sum of all masses associated with sets that are subsets of* $b \subseteq \mathbb{R}$

$$Bel(b) = \sum_{a_i \subseteq b} m(a_i),$$

which is the sum over i such that $a_i \subseteq b$. ∎

Obviously, $Bel(b) \leq Pls(b)$. Moreover, if one of the structures (either Dempster–Shafer structure, or *Bel* or *Pls*) is known, then the other two can be calculated. Considering sets of all real numbers less than or equal to z, it is easy to get upper and lower bounds for a probability distribution of a random real-valued quantity characterized by a finite Dempster–Shafer structure.

Consider a Dempster–Shafer structure with focal elements that are closed intervals $[x_i, y_i]$. We can specify it by listing the focal elements and their associated probability masses p_i as $\{([x_1, y_1], p_1), ([x_2, y_2], p_2), \ldots, ([x_n, y_n], p_n)\}$. Then the left bound (cumulative plausibility function) is

$$F^U(z) = \sum_{x_i \leq z} p_i \tag{15.116}$$

and the right bound (cumulative belief function) is

$$F^L(z) = \sum_{y_i \le z} p_i. \tag{15.117}$$

These functions are nondecreasing and right continuous functions from real numbers onto the interval $[0, 1]$ and $F^L(z) \le F^U(z)$, that is, proper distribution functions. They define the so-called p-box $[F^L(z), F^U(z)]$ that can be defined without any reference to Dempster–Shafer structure.

Definition 15.6 (Probability Box or p-Box) *p-box is a set of all probability distributions $F(x)$ such that $F^L \le F(x) \le F^U(x)$, where $F^L(x)$ and $F^U(x)$ are nondecreasing functions from the real line into $[0, 1]$. It is denoted as $[F^L, F^U]$.* ■

That is, we say that $[F^L, F^U]$ is a p-box of a random variable X whose distribution $F(x)$ is unknown except that $F^L \le F(x) \le F^U(x)$.

EXAMPLE 15.8

Consider the following Dempster–Shafer structure with three focal elements that have the same probability $1/3$:

$$\text{Structure A} = \begin{cases} [x_1 = 5, y_1 = 20]; & p_1 = 1/3 \\ [x_2 = 10, y_2 = 25]; & p_2 = 1/3 \\ [x_3 = 15, y_3 = 30]; & p_3 = 1/3 \end{cases}$$

Plausibility and belief functions are easily calculated using (15.116) and (15.117), respectively and presented by structure A in Figure 15.13. ■

15.7.2 DEMPSTER'S RULE

The central method in the Dempster–Shafer theory is Dempster's rule for combining evidence (see Shafer 1976 and Dempster 1967). In some situations, this rule produces counterintuitive results, and various alternative versions of the rule have been suggested, for example, by Yager (1987). In this section, we briefly describe only the original Dempster's rule, which is used to combine evidence obtained from two or more independent sources for the same quantity in question (e.g., expert opinions about a specific risk). A considerably more extensive review of this literature is available by Sentz and Ferson (2002).

Definition 15.7 (Dempster's Rule) *The combination of two independent Dempster–Shafer structures $m_1(A)$ and $m_2(B)$ with focal elements a_i and b_j, respectively, is another Dempster-Shafer structure with probability assignment*

$$m(\emptyset) = 0; \quad m(c) = \frac{1}{1 - \mathbb{K}} \sum_{a_i \cap b_j = c} m_1(a_i) m_2(b_j) \quad \textit{for } c \ne \emptyset, \tag{15.118}$$

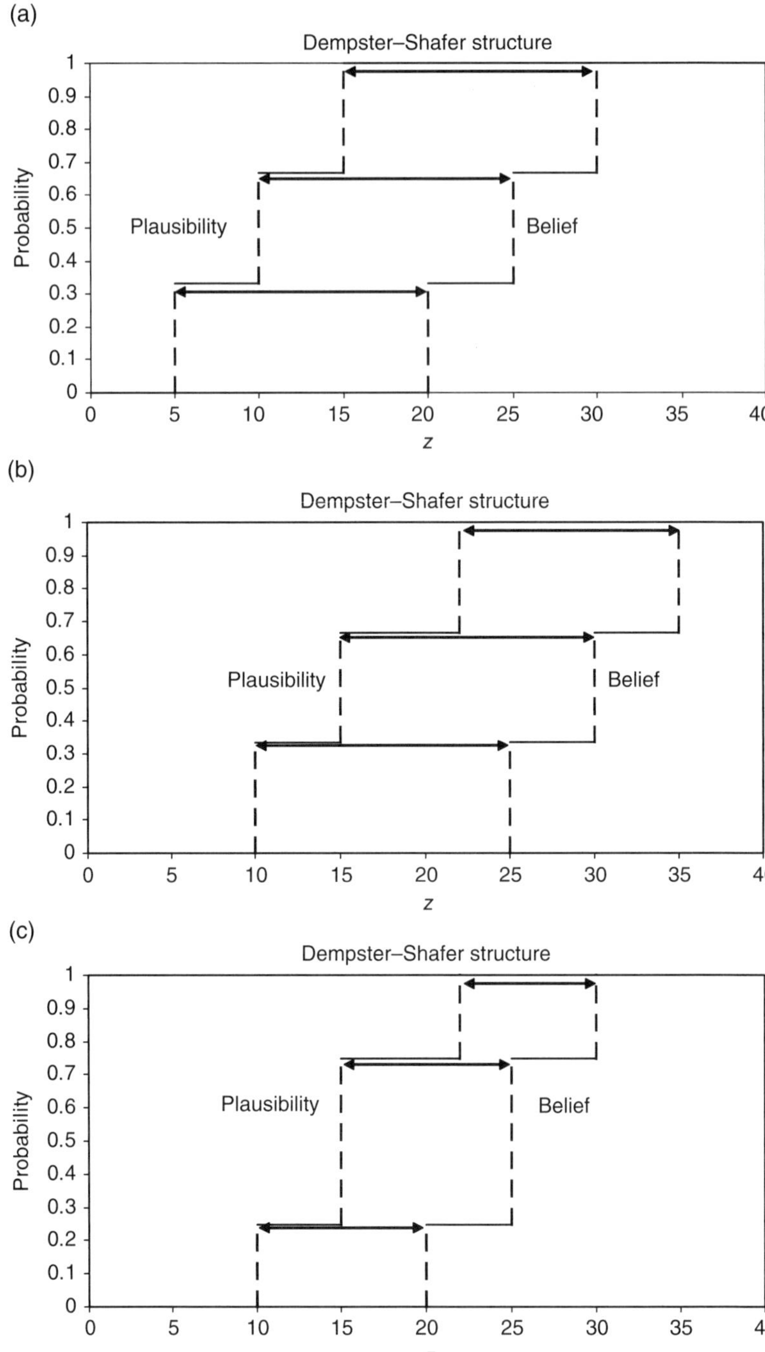

FIGURE 15.13 Plausibility and belief functions for Dempster–Shafer structures in Examples 15.8 and 15.9. Focal elements of the structure are indicated by arrows. Structure C is a result of combining structures A and B via Dempster's rule

that is, the sum over all i and j such that intersection of a_i and b_j is equal to c, where

$$\mathbb{K} = \sum_{a_i \cap b_j = \emptyset} m_1(a_i) m_2(b_j) \tag{15.119}$$

is the mass associated with the conflict present in the combined evidence. ■

EXAMPLE 15.9

Consider two independent Dempster–Shafer structures A and B with focal elements a_i and b_j, respectively:

$$\text{Structure A} = \begin{cases} [5, 20], \frac{1}{3} \\ [10, 25], \frac{1}{3} \\ [15, 30], \frac{1}{3} \end{cases} \qquad \text{and Structure B} = \begin{cases} [10, 25], \frac{1}{3} \\ [15, 30], \frac{1}{3} \\ [22, 35], \frac{1}{3} \end{cases}$$

The only combination of focal elements between these two structures that has no intersection is $a_1 = [5, 20]$ with $b_3 = [22, 35]$. Thus, the conflict of information in (15.119) is $\mathbb{K} = \frac{1}{3}\frac{1}{3} = \frac{1}{9}$. Intersections of other combinations are listed below with probabilities calculated using Dempster rule (15.118):

$$\text{Structure C} = \begin{cases} [10, 20], \frac{1}{8} \\ [15, 20], \frac{1}{8} \\ [10, 25], \frac{1}{8} \\ [15, 25], \frac{1}{4} \\ [22, 25], \frac{1}{8} \\ [15, 30], \frac{1}{8} \\ [22, 30], \frac{1}{8} \end{cases}$$

Note that intersection $c_4 = [15, 25]$ is produced by two combinations: a_2 with b_2; and a_3 with b_1. Thus, c_4 has probability $(\frac{1}{3}\frac{1}{3} + \frac{1}{3}\frac{1}{3})/(1 - \mathbb{K}) = 1/4$ while all other elements of structure C are produced by one combination and have probability $\frac{1}{3}\frac{1}{3}/(1 - \mathbb{K}) = \frac{1}{8}$ each. Plausibility and belief functions of all structures are easily calculated using (15.116) and (15.117), respectively and are presented in Figure 15.13 for all structures. ■

15.7.3 INTERSECTION METHOD

If the estimates to be aggregated represent claims that the quantity has to be within some limits, then the *intersection method* is perhaps the most natural kind of aggregation. The idea is simply to use the smallest region that all estimates agree. For example, if we know for sure that a true value of the quantity a is within the interval $x = [1, 3]$, and we also know from another source of evidence that a is also within the interval $y = [2, 4]$, then we may conclude that a is certainly within the interval $x \cap y = [2, 3]$.

The most general definition of intersection can be specified in terms of p-boxes. If there are K p-boxes $F_1 = [F_1^L, F_1^U], \ldots, F_K = [F_K^L, F_K^U]$, then their intersection is a p-box $[F^L, F^U]$, where

$$F^U = \min(F_1^U, \ldots, F_K^U), \quad F^L = \max(F_1^L, \ldots, F_K^L) \tag{15.120}$$

if $F^L(x) \leq F^U(x)$ for all x. This operation is used when the analyst is highly confident that each of multiple p-boxes encloses the distribution of the quantity in question. This formulation extends to Dempster–Shafer structures easily. The cumulative plausibility and belief functions of such structures form p-boxes.

Despite its several desirable properties, the intersection has only limited application for aggregation in OpRisk because it requires a very strong assumption that the individual estimates are each absolutely correct. It is certainly not recommended for the cases where any of the experts might be wrong. In practice, wrong opinions can be more typical than correct ones. For more detailed discussion and examples, see Ferson *et al.* (2003).

15.7.4 ENVELOPE METHOD

In the previous section on aggregation via intersection, it is assumed that all the estimates to be aggregated are completely reliable. If the analyst is only confident that at least one of the estimates encloses the quantity, but does not know which estimate, the method of *enveloping* can be used to aggregate the estimates into one reliable characterization. In general, when the estimates to be aggregated represent claims about the true value of a quantity and these estimates have uncertain reliability, enveloping is often an appropriate aggregation method. The idea is to identify the region where any estimate might be possible as the aggregation result. In particular, if one expert says that the value is 1 and another expert says that it is 2, we might decide to use the interval $[1, 2]$ as the aggregated estimate. If there are K p-boxes $F_1 = [F_1^L, F_1^U], \ldots, F_K = [F_K^L, F_K^U]$, then their envelope is defined to be a p-box $[F^L, F^U]$ where

$$F^U = \max(F_1^U, \ldots, F_K^U), \quad F^L = \min(F_1^L, \ldots, F_K^L). \tag{15.121}$$

This operation is always defined. It is used when the analyst knows that at least one of multiple p-boxes describes the distribution of the quantity in question. This formulation extends to Dempster–Shafer structures easily. The cumulative plausibility and belief functions of such structures form p-boxes. The result of aggregating these p-boxes can then be translated back into a Dempster–Shafer structure by canonical discretization. However, enveloping is sensitive to claims of general ignorance. This means that if only one expert provides an inconclusive opinion, it will determine the result of the aggregation. The overall result of enveloping will be as broad as the broadest input. The naive approach to omit any inconclusive estimates before calculating the envelope will not be sufficient in practice because any estimate that is not meaningless but just very wide can swamp all other estimates. Again, for more detailed discussion, the reader is referred to Ferson *et al.* (2003).

15.7.5 BOUNDS FOR THE EMPIRICAL DATA DISTRIBUTION

P-boxes and Dempster–Shafer structures can be constructed for empirical data using distribution free bounds around an empirical distribution function (Kolmogorov 1933, 1941, Smith 1939). Similar to the confidence intervals around a single number, these are bounds on a statistical distribution as a whole. As the number of samples increases, these confidence limits would converge to the empirical distribution function. Given independent samples $X_1, \ldots X_n$ from unknown continuous distribution $F(x)$, the empirical distribution of the data is

$$F_n(x) = \frac{1}{n}\sum_{1}^{n} \mathbb{I}_{X_i \le x}.$$

The lower and upper bounds (referred to as Kolmogorov–Smirnov (KS) bounds) for the distribution $F(x)$ can be calculated as

$$\begin{aligned} F_n^L &= \max(0, F_n(x) - D(\alpha, n)), \\ F_n^U &= \min(1, F_n(x) + D(\alpha, n)), \end{aligned} \tag{15.122}$$

where $D(\alpha, n)$ is a critical value for the one-sample KS statistic D_n at the confidence level $100(1 - \alpha)\%$ and sample size n, that is,

$$\mathbb{P}\mathrm{r}[D_n \le D(\alpha, n)] = 1 - \alpha,$$

where

$$D_n = \sup_x |F_n(x) - F(x)|\,.$$

The tabulated values for $D(\alpha, n)$ as well as a numerical approximations can be found in Miller (1956). For example, for sample size $n = 10$ and $\alpha = 0.05$ (i.e., 95% confidence level), $D(\alpha, n) = 0.40925$ and approximation for $0.01 < \alpha < 0.2$

$$D(\alpha, n) \approx \sqrt{\frac{\ln(2/\alpha)}{2n} - \frac{0.16693}{n} - \frac{A(\alpha)}{n^{3/2}}}, \tag{15.123}$$

where

$$A(\alpha) = 0.09037\left(\log_{10}\frac{2}{\alpha}\right)^{3/2} + 0.01515\left(\log_{10}\frac{\alpha}{2}\right)^2 - 0.08467\frac{\alpha}{2} - 0.11143.$$

Note that, typically, Kolmogorov–Smirnov (KS) statistics D_n is used for goodness-of-fit testing to compare a sample with a reference probability distribution. The null hypothesis that the sample is from $F(x)$ is rejected at level α if D_n exceeds a critical value $D(\alpha, n)$.

Theoretically, the left tail of the KS upper limit extends to negative infinity. But, of course, the smallest possible value might be limited by other considerations. For instance, there might be a theoretical lower limit at zero. If so, we could use this fact to truncate the upper (left) bound at zero. The right tail of the lower limit likewise extends to positive infinity. Sometimes it may be reasonable to select some value at which to truncate the largest value of a quantity too.

EXAMPLE 15.10

Assume that we have the following i.i.d. samples

$$(3.5, 4, 6, 8.1, 9.2, 12.3, 14.8, 16.9, 18, 20).$$

Also assume that the lower bound for the samples is zero and the upper bound is 30. Then KS bounds at 80% confidence are calculated using (15.122) and presented in Figure 15.14.

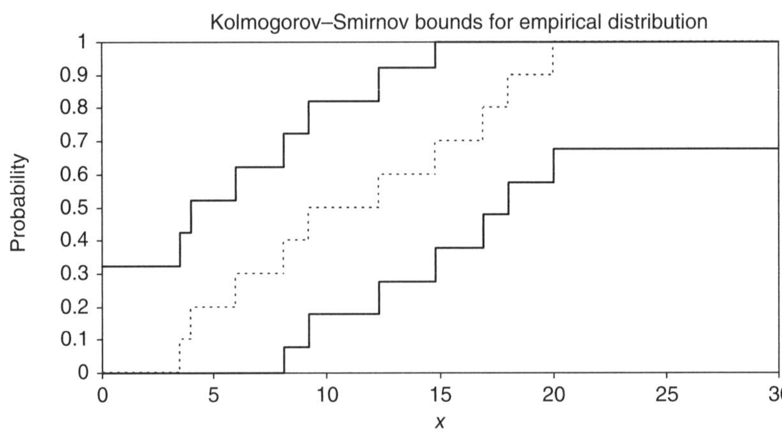

FIGURE 15.14 **Kolmogorov–Smirnov bounds for empirical distribution; for details, see Example 15.10**

The KS bounds make no distributional assumptions, but they do require that the samples are independent and identically distributed. In practice, an independence assumption is sometimes hard to justify. KS bounds are widely used in probability theory and risk analyses, for instance, as a way to express the reliability of the results of a simulation.

Formally, the KS test is valid for continuous distribution functions. In the discrete case too, KS bounds are conservative, that is, these bounds can be used in the case of discrete distributions but may not represent best possible bounds.

The confidence value α should be chosen such that the analyst believes the p-box contains the true distribution. The same hypothesis must also be assumed for the construction of the p-box from expert estimates. However, note that a p-box defined by KS confidence limits is fundamentally different from the sure bounds. The KS bounds are not certain bounds but statistical ones. The associated statistical statement is that 95% (or whatever is specified by α) of the time the true distribution will be within the bounds. It is not completely clear how to combine the KS p-box with the expert specified p-box; the choices of the upper limit and confidence level α for KS bounds can be problematic.

15.8 General Remarks

This chapter described how the parameters of the frequency and severity distributions are estimated using internal data, external data, and expert opinion. Then calculation of VaR (accounting for parameter uncertainty) for each risk cell can easily be done using a simulation approach as described in Section 13.7. The approaches and issues related to modeling dependence and aggregation over many risks are discussed in Chapters 10,11 and 12.

The main motivation for the use of the Bayesian approach is that, typically, the bank's internal data of the large losses in risk cells are so limited that the standard MLEs are not reliable. Overall, the use of the Bayesian inference method for the quantification of the frequency and severity distributions of OpRisks is very promising. The method is based on specifying the prior distributions for the parameters of the frequency and severity distributions using expert opinions or industry data. Then, the prior distributions are weighted with the actual observations in the bank to estimate the posterior distributions of the model parameters. These are used to estimate the annual loss distribution for the next accounting year. The estimation of low-frequency risks using this method has several appealing features such as stable estimators, simple calculations (in the case of conjugate priors), and the ability to take into account expert opinions and industry data. The approach allows for combining all three data sources: internal data, external data, and expert opinions required by Basel II.

If the data are very limited, it might be difficult to specify the prior distributions. Then one can use a closely related credibility theory approach to estimate parameters of the frequency and severity distributions for the low-frequency/high-severity risks, as described in Section 15.5.

The models presented in this chapter give illustrative examples that can be extended to a full-scale application. The approach has a simple structure, which is beneficial for practical use and can engage the bank risk managers, statisticians, and regulators in productive model development and risk assessment.

Several general remarks on the described Bayesian method for OpRisk are worth making:

- Validation of the models in the case of small datasets is problematic. Formally, justification of the model assumptions (such as conditional independence between the losses or common distribution for the risk profiles across the risks) can be based on the analysis of the unconditional properties (e.g., unconditional means and covariances) of the losses and should be addressed during model implementation;
- Presented examples have a simplistic dependence on time but can be extended to the case of a more realistic time component;
- Adding extra levels to the considered hierarchical structure may be required to model the actual risk cell structure in a bank;
- One of the features of the described method is that the variance of the posterior distribution $\pi(\boldsymbol{\theta}|\cdot)$ will converge to zero for a large number of observations. This means that the true value of the risk profile will be known exactly. However, there are many factors (political, economical, legal, etc.) changing in time that might not allow for precise knowledge of the risk profiles. One can model this by limiting the variance of the posterior distribution by some lower levels (e.g., 5%). This has been done in many solvency approaches for the insurance industry, for example, in the Swiss Solvency Test (see Swiss Financial Market Supervisory Authority 2006, formulas (25) and (26));

- For convenience, we have assumed that expert opinions are independent and identically distributed. However, all formulas can easily be generalized to the case of expert opinions modeled by different distributions;
- It would be ideal if the industry risk profiles (prior distributions for frequency and severity parameters in risk cells) are calculated and provided by the regulators to ensure consistency across the banks. Unfortunately, this may not be realistic at the moment. Banks might thus estimate the industry risk profiles using industry data available through external databases from vendors and consortia of banks. The data quality, reporting, and survival biases in external databases are the issues that should be considered in practice.

Finally, in this book, we consider modeling OpRisk but the use of similar Bayesian models is also useful in other areas (such as credit risk, insurance, environmental risk, and ecology) where, mainly due to lack of internal observations, a combination of internal data, external data, and expert opinions is required.

CHAPTER SIXTEEN

Multifactor Modeling and Regression for Loss Processes

In this chapter, we introduce several statistical approaches that may be developed to incorporate Key Risk Indicators (KRIs) into a Loss Distribution Approach (LDA) model structure to add information that will inform parameter estimation and capital measurement. In particular, we demonstrate how one may introduce covariates that will allow one to model capital dynamically and the changes that may occur in capital due to risk factors that may be internal to an organization or external such as macroeconomic and microeconomic factors. We start this chapter with a basic introduction to Generalized Linear Models (GLMs), then introduce regularization concepts and quantile regression. Following this background review, we explain how to use such statistical models in practical OpRisk settings.

16.1 Generalized Linear Model Regressions and the Exponential Family

In this section, we introduce to OpRisk modelling the widely utilized class of regression models known as the GLM structure (see Nelder and Wedderburn 1972) and its hierarchical versions Lee and Nelder (1996). Effectively, the GLM is a flexible generalization of ordinary linear regression that allows for response variables that are distributed from a more general distribution than the standard linear model, which assumes normally distributed responses. Throughout this chapter we will develop a general framework for the introduction of such regression modelling to OpRisk contexts. The GLM generalizes linear regression by allowing the linear model to be related to the response variable via a link function and by allowing the magnitude of the variance of each measurement to be a function of its predicted mean value. Hence, such a regression model framework provides a very flexible class of models that allow us to specify a relationship between some variable of interest, the observations Y_i, and a collection of potential explanatory variables given by the observed p covariates in the p-dimensional vectors $\boldsymbol{x}_i$.

Fundamental Aspects of Operational Risk and Insurance Analytics: A Handbook of Operational Risk, First Edition. Marcelo G. Cruz, Gareth W. Peters, and Pavel V. Shevchenko.

Hence, the n data points consist of pairs $(Y_1, \boldsymbol{x}_1), \ldots, (Y_n, \boldsymbol{x}_n)$, where Y_i is the response for the i-th case in the dataset and $\boldsymbol{x}_i = \left(x_{i,1}, \ldots, x_{i,p}\right)^T$ is the corresponding vector of explanatory variables, note the response may also be vector valued, though for convenience here we treat the univariate case. When specifying GLM regression, one must consider two aspects: the distribution for the response and the mean and variance relationships in terms of the covariates.

16.1.1 BASIC COMPONENTS OF A GENERALIZED LINEAR MODEL REGRESSION IN THE EXPONENTIAL FAMILY

There is a large literature on GLM modeling and such regression structures represent a widely developed class of regression models (see discussions by McCullagh and Nelder 1989 and Denison *et al.* 2002). In brief, the key components of a GLM specification involve the following:

- A GLM structure considers that given a covariate vector $\boldsymbol{x}_i$, the response Y_i has some probability distribution with mean μ_i, such that $g(\mu_i) = \boldsymbol{x}_i^T \boldsymbol{\beta} = \sum_{j=1}^{p} \beta_j x_{i,j}$ $(= \eta_i$, for example) for some coefficient vector $\boldsymbol{\beta} = (\beta_1, \ldots, \beta_p)^T$ and monotonic, differentiable function $g(\cdot)$ (the link function). η_i is the linear predictor for Y_i.;
- If the distributions of the responses are considered in the exponential family, then Y_i has density of the form

$$f(y_i; \theta_i, \phi) = \exp\left[\frac{y_i\theta_i - b(\theta_i)}{\phi} + c(y_i, \phi)\right], \tag{16.1}$$

  for some parameters θ_i and ϕ, and functions $b(\cdot)$ and $c(\cdot, \cdot)$. This family contains many standard distributions. The mean is $\mu_i = \partial b / \partial \theta|_{\theta=\theta_i}$ and the variance is $V_i = \phi \partial^2 b / \partial \theta^2|_{\theta=\theta_i} = \phi \partial \mu / \partial \theta|_{\theta=\theta_i}$. ϕ is the dispersion parameter, and is the same for all i.;
- If $g(\cdot)$ is the identity, and we specify a Normal distribution for Y_i, we end up with a standard linear regression model. If we write the *Normal*(μ, σ^2) in "exponential family" form, we find $\theta = \mu$ and $\phi = \sigma^2$, so in this case, the dispersion parameter is just the variance and the assumption of a common ϕ is just the usual assumption of constant variance, homoskedascity in the standard linear regression model.

Remark 16.1 *We note that the exponential family of GLM structures produces a family of models that contains many standard distributions, allowing for continuous response distributions as well as discrete response distributions such as the Normal, Exponential, Gamma, Chi-squared, Beta, Dirichlet, Bernoulli, categorical, Poisson, Wishart, Inverse Wishart, and many others.*

As some examples of the GLM regression model that will be of relevance to OpRisk modelling we present first the Binomial, Logistic, Poisson and multinomial regression models as simple examples. We will simply present the model structure and then show the generic framework for parameter estimation under likelihood and Bayesian models in future sections of this chapter.

EXAMPLE 16.1 Binomial Regression and the GLM for OpRisk Settings

Assume that the observational data follows a Binomial distribution, that is the random variables $\{Y_i\}_{i=1:n}$ take values in the set $\{0, 1\}$ and are distributed i.i.d. conditional on the p-dimensional vectors of covariates $\{\boldsymbol{x}_i\}_{i=1:n}$ according to the distribution

$$Y_i \sim Bernoulli\left(\pi_i\right), \quad \text{for } \pi \in [0, 1], \tag{16.2}$$

such that

$$\mathbb{P}\text{r}\left(Y_i = y_i | \boldsymbol{x}_i\right) = \pi_i^{y_i}\left(1 - \pi_i\right)^{1-y_i}. \tag{16.3}$$

It will also be useful to recall the relationship between the mean and variance of this random variables response,

$$\begin{aligned} \mathbb{E}\left[Y_i | \boldsymbol{x}_i\right] &= \mu_i = \pi_i, \\ \mathbb{V}\text{ar}\left[Y_i | \boldsymbol{x}_i\right] &= \sigma_i^2 = \pi_i\left(1 - \pi_i\right). \end{aligned} \tag{16.4}$$

From these relationships we would like to develop the regression relationship, specified for instance through the relationship between the success probability π_i and the regressor $\boldsymbol{x}_i$. A standard approach to achieve this is to consider the logit or log-odds link function which allows one to map the probabilities π_i from the range $(0, 1)$ to the entire real line $\mathbb{R}$ according to the relationship

$$\nu_i = \text{logit}\left(\pi_i\right) = \ln \frac{\pi_i}{1 - \pi_i} \tag{16.5}$$

so that now ν_i can be written in terms of say a linear model such as

$$\nu_i = \sum_{j=1}^{p} \beta_j \boldsymbol{x}_{i,j}. \tag{16.6}$$

■

EXAMPLE 16.2 Logistic Regression and the GLM for OpRisk Settings

Assume that one has k independent observations $y_1, \ldots, y_k$ and that the i-th observation is the number of successes in n_i Bernoulli trials such that the random variables $\{Y_i\}_{i=1:n}$ take values in the set $\{0, 1, 2, \ldots, n_i\}$ and are distributed i.i.d. conditional on the p-dimensional vectors of covariates $\{\boldsymbol{x}_i\}_{i=1:n}$ according to the distribution

$$Y_i \sim Binomial\left(n_i, \pi_i\right), \quad \text{for } \pi_i \in [0, 1], \tag{16.7}$$

such that

$$\mathbb{P}\mathrm{r}\left(Y_i = y_i \middle| \boldsymbol{x}_i\right) = C_{y_i}^{n_i} \pi_i^{y_i} \left(1 - \pi_i\right)^{n_i - y_i}. \tag{16.8}$$

It will also be useful to recall the relationship between the mean and variance of this random variables response,

$$\begin{aligned} \mathbb{E}\left[Y_i \middle| \boldsymbol{x}_i\right] &= \mu_i = n_i \pi_i, \\ \mathbb{V}\mathrm{ar}\left[Y_i \middle| \boldsymbol{x}_i\right] &= \sigma_i^2 = n_i \pi_i \left(1 - \pi_i\right). \end{aligned} \tag{16.9}$$

From these relationships we would like to develop the regression relationship, specified for instance through the relationship between the success probability π_i and the regressor $\boldsymbol{x}_i$. A standard approach to achieve this is to consider the logit or log-odds link function which allows one to map the probabilities π_i from the range $(0, 1)$ to the entire real line $\mathbb{R}$ according to the relationship

$$\nu_i = \operatorname{logit}\left(\pi_i\right) = \ln \frac{\pi_i}{1 - \pi_i} \tag{16.10}$$

so that now ν_i can be written in terms of say a linear model such as

$$\nu_i = \sum_{j=1}^{p} \beta_j \boldsymbol{x}_{i,j}. \tag{16.11}$$

■

Next, if the observation data is counts then the appropriate regression structure may come from a Poisson regression model as follows in Example 16.3.

EXAMPLE 16.3 Poisson Regression and the GLM for OpRisk Settings

Assume that one has independent observations $y_1, \ldots, y_n$ and that the i-th observation is a count of an event in a pre-specified time period, such that the random variables $\{Y_i\}_{i=1:n}$ take values in the set $\{0, 1, 2, \ldots\}$ and are distributed i.i.d. conditional on the p-dimensional vectors of covariates $\{\boldsymbol{x}_i\}_{i=1:n}$ according to the distribution

$$Y_i \sim Poisson\left(\lambda_i\right), \quad \text{for } \lambda_i \geq 0, \tag{16.12}$$

such that

$$\mathbb{P}\mathrm{r}\left(Y_i = y_i \middle| \boldsymbol{x}_i\right) = \frac{\lambda_i^{y_i}}{y_i!} \exp\left(-\lambda_i\right). \tag{16.13}$$

It will also be useful to recall the relationship between the mean and variance of this random variables response,

$$\begin{aligned} \mathbb{E}\left[Y_i \middle| \boldsymbol{x}_i \right] &= \mu_i = \lambda_i, \\ \mathbb{V}\mathrm{ar}\left[Y_i \middle| \boldsymbol{x}_i \right] &= \sigma_i^2 = \lambda_i. \end{aligned} \tag{16.14}$$

From these expressions we would like to develop the regression relationship, specified for instance through the relationship between the intensity λ_i and the regressor $\boldsymbol{x}_i$. A standard approach to achieve this is to consider the class of log-linear models where the link function allows one to map the intensity λ_i from the range $\mathbb{R}^+$ to the entire real line $\mathbb{R}$ according to the relationship

$$\nu_i = \ln\left(\lambda_i\right) \tag{16.15}$$

so that now ν_i can be written in terms of say a linear model such as

$$\nu_i = \sum_{j=1}^{p} \beta_j \boldsymbol{x}_{i,j}. \tag{16.16}$$

■

In some cases the response of a survey of business managers regarding their risk profiles may be quantified initially according to a discrete score from say $1, 2, \ldots, J$, then one may wish to understand relationships between such responses and other covariates related to the risk process that formed the survey questions. In this case one requires the use of a multinomial regression structure as specified in the following example.

EXAMPLE 16.4 Multinomial Regression and the GLM for OpRisk Settings

Assume that one has independent observations $y_1, \ldots, y_n$ and that the i-th observation is a discrete valued response from a set of possible responses according to a discrete score such that the random variables $\{Y_i\}_{i=1:n}$ take values in the set $\{0, 1, 2, \ldots, J\}$ and are distributed i.i.d. conditional on the p-dimensional vectors of covariates $\{\boldsymbol{x}_i\}_{i=1:n}$ according to the distribution

$$Y_i \sim \mathit{Multinomial}\left(\pi_{i,1}, \pi_{i,2}, \ldots, \pi_{i,J}\right), \quad \text{for } \pi_{i,j} \in [0, 1] \tag{16.17}$$

with $\sum_{i=1}^{J} \pi_{i,j} = 1$ such that

$$\mathbb{P}\mathrm{r}\left(Y_i = j \middle| \boldsymbol{x}_i \right) = \pi_{i,j}. \tag{16.18}$$

If one now records the total number of responses from all participants as n, then treating $Y_i = j$ as a random variable for the indicator of the i-th response in the

j-th possible item response, then the $\sum_{i=1}^{n} Y_i = n$ and one has the multinomial distribution with mass function for number of respondents n_i given by

$$\mathbb{Pr}\left(Y_{i,1} = y_{i,1}, Y_{i,2} = y_{i,2}, \ldots, Y_{i,J} = y_{i,J}; \boldsymbol{x}_i\right) = C^{n_i}_{y_{i,1},\ldots,y_{i,J}} \prod_{j=1}^{J} \pi_{i,j}^{y_{i,j}}. \tag{16.19}$$

From these expressions we would like to develop the regression relationship, specified for instance through the relationship between the probabilities $\pi_{i,j}$ and the regressor $\boldsymbol{x}_i$. A standard approach to achieve this is to consider the class of multinomial logit model we assume that the log-odds of each response follow a linear model with

$$\nu_{i,j} = \ln\left(\frac{\pi_{i,j}}{\pi_{i,J}}\right) = \alpha_j + \sum_{s=1}^{p} \boldsymbol{x}_{i,s}\beta_{j,s}. \tag{16.20}$$

for $j = 1, \ldots, J-1$. This model is analogous to a logistic regression model, except that the probability distribution of the response is multinomial instead of Binomial and we have $J-1$ equations instead of one. ■

16.1.2 BASIS FUNCTION REGRESSION

In some settings, it is also sensible to consider functions of the input independent variables in the regression structure (the covariates) as this will give additional flexibility to the regression's explanatory power. These classes of models are typically called basis function regression models. In such basis regression models, one has two components to consider when performing model selection: selection of the most explanatory covariates in the regression and the most appropriate choice of basis function model choice. Rather than discussing the many approaches to model selection in regression modeling, we refer the reader to detailed discussions on such items in texts such as Hastie *et al.* (2009) and Yuan and Lin (2006). Instead, we will discuss briefly approaches that will produce parsimonious model structures through approaches based on regularization, with a particular focus on the relevance to OpRisk based on Bayesian GLM regularization models.

Consider a general basis function regression structure in which we need to perform model selection to assess the most suitable class of basis functions and we will jointly perform regularization of the regression coefficients on the basis functions to remove bases (transformed covariates) that are not explanatory of the variation in the response in a given model structure.

Consider for the k-th basis function model a function of the GLM regression mean given by

$$g(\mu_i) = \boldsymbol{\Phi}_k\left(\boldsymbol{x}_i\right)^T \boldsymbol{\beta} = \sum_{j=1}^{p} \beta_j \Phi_k\left(x_{i,j}\right) \tag{16.21}$$

($= \eta_i$, for example) for some coefficient vector $\boldsymbol{\beta} = (\beta_1 \ \ldots \ \beta_p)^T$, a basis function regression design matrix $\boldsymbol{\Phi}_k$ where each element contains the basis function applied to the covariate corresponding to the i-th column $\Phi_k(x_{i,j})$ in which $\Phi_k : \mathbb{R} \mapsto \mathbb{R}$, and the first column containing the normalized basis function $\Phi(1) = 1$ corresponding to the intercept. In specifying the function of the mean $g(\cdot)$ (known as the link function), we must ensure it is selected to be strictly monotonic and differentiable, we saw a few examples above including for instance the logit transform. Then one can say that after application of the link function to the basis function linear model structure, the result, which we term η_i, forms the *linear predictor* for Y_i.

Having presented basic details of the GLM structure, one needs to consider how to perform basic parameter estimation. This can follow two standard structures: the likelihood-based estimation or a Bayesian model structure; we will first discuss Maximum Likelihood Estimation (MLE) in the GLM structure.

16.2 Maximum Likelihood Estimation for Generalized Linear Models

When fitting a GLM to data, the aim is to estimate the coefficient vector $\boldsymbol{\beta}$ and, if necessary, the dispersion parameter ϕ. In the context of exponential families, $\boldsymbol{\beta}$ can be regarded as a function of θ (and vice versa), since both are related to the mean of the distribution. The log likelihood for $\boldsymbol{\beta}$ and ϕ in exponential family models is therefore given by

$$\ell\left(\boldsymbol{\beta}, \phi|\mathbf{y}\right) = \sum_{i=1}^{n} \left[\frac{y_i\theta_i - b(\theta_i)}{\phi} + c(y_i, \phi)\right]. \tag{16.22}$$

Differentiating with respect to β_j $(j = 1, \ldots, p)$ and setting to zero the derivative yields the p score equations

$$U_j = \frac{\partial \ell}{\partial \beta_j} = \frac{1}{\phi}\sum_{i=1}^{n} \frac{\partial}{\partial \beta_j}\left[y_i\theta_i - b(\theta_i)\right] = 0, \qquad (j = 1, \ldots, p). \tag{16.23}$$

It can be shown that

$$U_j = \sum_{i=1}^{n}\left[\frac{y_i - \mu_i}{V_i}\left(\frac{\partial \mu_i}{\partial \eta_i}\right)x_{i,j}\right] = \sum_{i=1}^{n}\left[\frac{y_i - \mu_i}{V_i g'\left(\mu_i\right)}x_{i,j}\right]. \tag{16.24}$$

where μ_i and V_i are defined as specified earlier.

The solution of the score equations (which, from Equation (16.23), clearly does not depend on ϕ) yields the MLE of $\boldsymbol{\beta}$. Assembling all p equations into vector form, we seek the solution of

$$\mathbf{U}\left(\boldsymbol{\beta}\right) = \mathbf{0}, \tag{16.25}$$

where $\mathbf{U}\left(\boldsymbol{\beta}\right) = \left(U_1, \ldots, U_p\right)^T$ is the score vector of the log likelihood derivatives.

16.2.1 ITERATED WEIGHTED LEAST SQUARES MAXIMUM LIKELIHOOD FOR GENERALISED LINEAR MODELS

However, the solution must be obtained numerically in all but the simplest cases. One possible numerical approach used widely in practice is the Newton–Raphson algorithm, which may be used to solve these equations. One must start with an initial guess at the solution, $\boldsymbol{\beta}^{(0)}$, and then successively calculate updates given by

$$\boldsymbol{\beta}^{(t+1)} = \boldsymbol{\beta}^{(t)} - \left[\left. \frac{\partial \mathbf{U}}{\partial \boldsymbol{\beta}} \right|_{\boldsymbol{\beta}^{(t)}} \right]^{-1} \mathbf{U}\left(\boldsymbol{\beta}^{(t)}\right) \tag{16.26}$$

until convergence is achieved. Note that the notation $\partial \mathbf{U} / \partial \boldsymbol{\beta}$ represents a $p \times p$ matrix of second derivatives of the log likelihood with respect to $\boldsymbol{\beta}$. This is actually Newton–Raphson in p dimensions.

Conventionally, when fitting GLMs, one would typically replace the matrix of second derivatives in Equation (16.26) by its expected value $-\mathbf{I}(\boldsymbol{\beta})$, where $\mathbf{I}(\boldsymbol{\beta})$ is the Fisher information matrix. The resulting algorithm is called the method of scoring. The method of scoring is a reliable alternative to Equation (16.26), so long as the initial value $\boldsymbol{\beta}^{(0)}$ is chosen sensibly. Moreover, if the link function $g(\cdot)$ is such that $g(\mu_i) = \theta_i$ in the exponential family, it can be shown that the scoring method is exactly the same as Equation (16.26). Such a link function is called a canonical link.

The scoring algorithm is particularly convenient because the information matrix has an alternative representation as the covariance matrix of the score vector. Together with Equation (16.24), this can be used to show that the information matrix is given by the form

$$\mathbf{I}(\boldsymbol{\beta}) = \mathbf{X}^T \mathbf{W} \mathbf{X}, \tag{16.27}$$

where $\boldsymbol{X}$ is an $n \times p$ matrix whose (i, j)-th entry is $x_{i,j}$, and $\mathbf{W}$ is a diagonal $n \times n$ matrix with elements $w_{ii} = \left(\partial \mu_i / \partial \eta_i\right)^2 / V_i = \left[g'(\mu_i)\right]^{-2} / V_i$ (which depend on $\boldsymbol{\beta}$). The iterative scheme for the scoring method is therefore transformed into the following updating steps:

$$\boldsymbol{\beta}^{(t+1)} = \boldsymbol{\beta}^{(t)} + \left[\mathbf{X}^T \mathbf{W}^{(t)} \mathbf{X}\right]^{-1} \mathbf{U}\left(\boldsymbol{\beta}^{(t)}\right). \tag{16.28}$$

Multiplying both sides by $\mathbf{X}^T \mathbf{W}^{(t)} \mathbf{X}$ gives

$$\mathbf{X}^T \mathbf{W}^{(t)} \mathbf{X} \boldsymbol{\beta}^{(t+1)} = \mathbf{X}^T \mathbf{W}^{(t)} \mathbf{X} \boldsymbol{\beta}^{(t)} + \mathbf{U}\left(\boldsymbol{\beta}^{(t)}\right). \tag{16.29}$$

Noting that $\mathbf{X}\boldsymbol{\beta}^{(t)} = \boldsymbol{\eta}^{(t)}$, the vector of linear predictors at iteration t, and that $\mathbf{U}\left(\boldsymbol{\beta}^{(t)}\right)$ can itself be written as a vector product involving the matrix $\mathbf{X}^T \mathbf{W}^{(t)}$ (from 16.24), the scoring algorithm can finally be written in matrix form as

$$\left[\mathbf{X}^T \mathbf{W}^{(t)} \mathbf{X}\right] \boldsymbol{\beta}^{(t+1)} = \mathbf{X}^T \mathbf{W}^{(t)} \mathbf{z}^{(t)} \Rightarrow \boldsymbol{\beta}^{(t+1)} = \left[\mathbf{X}^T \mathbf{W}^{(t)} \mathbf{X}\right]^{-1} \mathbf{X}^T \mathbf{W}^{(t)} \mathbf{z}^{(t)}, \tag{16.30}$$

where $\mathbf{z}^{(t)}$ is an $n \times 1$ vector whose i-th element is given by

$$z_i^{(t)} = \eta_i^{(t)} + \left(y_i - \mu_i^{(t)}\right) \left(\left. \frac{\partial \eta}{\partial \mu} \right|_{\mu_i^{(t)}} \right) = \eta_i^{(t)} + \left(y_i - \mu_i^{(t)}\right) g'\left(\mu_i^{(t)}\right).$$

Equation (16.30) gives the solution of the weighted least-squares regression of $\mathbf{z}^{(t)}$ upon $\mathbf{X}$, with weights contained in the diagonal elements of $\mathbf{W}^{(t)}$. For this reason, $\mathbf{z}^{(t)}$ is sometimes called the adjusted dependent variate.

The need for iteration arises because $\mathbf{z}$ and $\mathbf{W}$ both depend, in general, upon $\boldsymbol{\beta}$. Expressed in this form, the algorithm for fitting GLMs is referred to as iterative weighted least squares (IWLS). A side effect of the algorithm is that for large samples, the covariance matrix of the parameter estimates can be calculated easily from the information matrix (16.27); this enables us to derive standard errors for the estimates. Hence, to summarize the steps involved in the GLM coefficient estimation via an MLE procedure, perform the stages given in Algorithm 16.1.

Algorithm 16.1 (Maximum Likelihood for GLM Regression in Exponential Family)

1. *Assemble your explanatory variables into an $n \times p$ matrix $\boldsymbol{X}$;*
2. *Choose a sensible starting value $\boldsymbol{\beta}^{(0)}$;*
3. *Repeat the following stages:*
   a) *For the current estimate of $\boldsymbol{\beta}$, calculate the vector $\boldsymbol{\eta}$ of linear predictors, the vector $\boldsymbol{\mu}$ of means, and the vector $\boldsymbol{V}$ of variances. If the model involves an unknown dispersion parameter ϕ, set it to unity. Note the MLE of $\boldsymbol{\beta}$ is independent of ϕ;*
   b) *Calculate the diagonal elements of $\boldsymbol{W}$ where the i-th element is $[g'(\mu_i)]^{-2}/V_i$;*
   c) *Calculate the vector $\mathbf{z}$, with the i-th element $\eta_i + (y_i - \mu_i)\, g'(\mu_i)$;*
   d) *Calculate the $p \times n$ matrix $\boldsymbol{X}^T\boldsymbol{W}$ by multiplying each row of $\boldsymbol{X}$ by the corresponding diagonal element of $\boldsymbol{W}$ and then taking the transpose;*
   e) *Calculate the $p \times p$ matrix $\boldsymbol{X}^T\boldsymbol{W}\boldsymbol{X}$, and invert it to obtain $\left[\mathbf{X}^T\mathbf{W}\mathbf{X}\right]^{-1}$. Also calculate the $p \times 1$ vectors $\boldsymbol{X}^T\boldsymbol{W}\boldsymbol{z}$ and $\boldsymbol{U} = \boldsymbol{X}^T\boldsymbol{W}(\boldsymbol{z} - \boldsymbol{\eta})$;*
   f) *Note that $\boldsymbol{U}$ is the score vector, so one can set a tollerance for convergence and stop the repeat when the score vector $\boldsymbol{U}$ is within this tollerance or zero. Otherwise, recalculate $\boldsymbol{\beta}$ as $\left[\mathbf{X}^T\mathbf{W}\mathbf{X}\right]^{-1}\mathbf{X}^T\mathbf{W}\mathbf{z}$ and go back to the start of this loop.*
4. *Estimate the dispersion parameter via a method of moments procedure, to obtain $\hat{\phi}$;*
5. *Extract the diagonal elements of $\left[\mathbf{X}^T\mathbf{W}\mathbf{X}\right]^{-1}$, multiply by $\hat{\phi}$, and take the square roots, to obtain standard errors for the parameter estimates.*

16.2.2 MODEL SELECTION VIA THE DEVIANCE IN A GLM REGRESSION

Note that there is an extensive section on model selection methodology in Chapter 8; here there is a brief introduction to the use of the deviance information criterion particularly in the GLM regression context. In GLM regression model selection, it is common to compare and perform model selection based on either the likelihood ratio or via the deviance hypothesis tests. The deviance for a model is the equivalent of the residual sum of squares in a linear model, and is defined according to

$$D = 2\phi\left[\ell(\boldsymbol{y};\boldsymbol{y}) - \ell(\boldsymbol{\mu};\boldsymbol{y})\right], \tag{16.31}$$

where $\boldsymbol{y}$ is the vector of observed responses; $\boldsymbol{\mu}$ is the corresponding vector of modeled means and $\ell(\boldsymbol{m};\boldsymbol{y})$ is the log likelihood for $\boldsymbol{y}$ when the mean vector is $\boldsymbol{m}$.

D is also sometimes called the residual deviance. If you have two models with deviances D_1 and D_2, respectively, and model 2 is an extension of model 1 containing p extra parameters, then a formal test of model 1 against model 2 can be carried out by comparing the quantity $\frac{(D_1-D_2)/p}{D_2/\nu}$ to an $F_{p,\nu}$ distribution, where ν is the residual degrees of freedom for model 2 and this distribution is characterized by the F-distribution according to

$$
\begin{aligned}
f(x; d_1, d_2) &= \frac{\sqrt{\frac{(d_1 x)^{d_1} d_2^{d_2}}{(d_1 x + d_2)^{d_1+d_2}}}}{x \mathrm{B}\left(\frac{d_1}{2}, \frac{d_2}{2}\right)} \\
&= \frac{1}{\mathrm{B}\left(\frac{d_1}{2}, \frac{d_2}{2}\right)} \left(\frac{d_1}{d_2}\right)^{\frac{d_1}{2}} x^{\frac{d_1}{2}-1} \left(1 + \frac{d_1}{d_2} x\right)^{-\frac{d_1+d_2}{2}}
\end{aligned}
\tag{16.32}
$$

for real $x \geq 0$. Here B is the beta function. In many applications, the parameters d_1 and d_2 are positive integers, but the distribution is well-defined for positive real values of these parameters. One can also write the cumulative distribution function as a special function also,

$$
F(x; d_1, d_2) = I_{\frac{d_1 x}{d_1 x + d_2}} \left(\frac{d_1}{2}, \frac{d_2}{2}\right), \tag{16.33}
$$

where $I(\cdot, \cdot)$ is the regularized incomplete beta function.

This is the exact equivalent of the F-test for comparing nested linear models and hence one may construct an "analysis of deviance" table in the same way as we would construct an analysis of variance (ANOVA) table for linear models. The need for an F-test arises from the unknown dispersion parameter. If the dispersion parameter is known (as in the Poisson case, where we know that $\phi = 1$), an alternative is to base tests on the χ^2 distribution for the likelihood ratio statistic $\Lambda = 2(\ell_2 - \ell_1) = \phi^{-1}(D_1 - D_2)$, where ℓ_1 and ℓ_2 are the log likelihoods for the two models and the χ^2 density and distribution is given by

$$
f(x;\ k) = \begin{cases} \frac{x^{(k/2-1)} e^{-x/2}}{2^{k/2}\Gamma\left(\frac{k}{2}\right)}, & x \geq 0; \\ 0, & \text{otherwise}, \end{cases} \tag{16.34}
$$

where $\Gamma(k/2)$ denotes the Gamma function, which has closed form values for integer k. Its cumulative distribution function is

$$
F(x;\ k) = \frac{\gamma\left(\frac{k}{2}, \frac{x}{2}\right)}{\Gamma\left(\frac{k}{2}\right)} = P\left(\frac{k}{2}, \frac{x}{2}\right), \tag{16.35}
$$

where $\gamma(s, t)$ is the lower incomplete Gamma function and $P(s, t)$ is the regularized Gamma function.

If D is the deviance for a model, $\phi^{-1}D$ is sometimes called the scaled deviance.

The deviance for the Poisson regression model is given by considering the Poisson log likelihood for data $\boldsymbol{y}$ with fitted values $\boldsymbol{m}$ given by

$$
\ell(\boldsymbol{m}; \boldsymbol{y}) = \sum_{i=1}^{n} [y_i \ln m_i - m_i - \ln y_i!]\,. \tag{16.36}
$$

Substitute $\ell(\boldsymbol{y}; \boldsymbol{y})$ and $\ell(\boldsymbol{\mu}; \boldsymbol{y})$ into Equation (16.31), to get an expression for D.

16.3 Bayesian Generalized Linear Model Regressions and Regularization Priors

For a detailed discussion on Bayesian modeling and methodology, see the discussion in the section on estimation in Chapter 7. In a Bayesian GLM regression model, we still utilize an exponential family likelihood model as described previously; however, we supplement this with prior distributions for the unknown parameters in the model, typically the regression coefficients and the dispersion parameter. These represent a widely developed class of regression models (see discussions by McCullagh and Nelder 1989 and Denison *et al.* 2002). We will discuss a particular class of Bayesian regression models, those subject to regularization prior structures, which allows for the joint posterior parameter estimation as well as parsimonious model selection through shrinking the nonexplanatory covariates to be insignificant.

Sparse regression analysis initially studied in the context of penalized least squares or likelihood has gained increasing popularity since the seminal paper on the least absolute shrinkage and selection operator (LASSO) Tibshirani (1996). Since this work, many approaches under both frequentist and Bayesian have been proposed to extend these sparsity-inducing regression frameworks. Most approaches consider modification to an L2 least-squares criterion via addition of an L_q penalty on the coefficients that form the argument of the optimization. The downside of this is a nonconvex optimization problem and a bias in the estimate for the regression coefficients $\boldsymbol{\beta} \in \mathbb{R}^p$ and the upside is an induced sparsity through shrinkage of less consequential regressor coefficients to 0. Therefore, the idea is to incur a bias in estimated coefficients so that a sparse solution will deliver other important attributes such as parsimony. The coefficients of the regression are therefore estimated to minimize the following general penalized criterion

$$\arg\min_{\boldsymbol{\beta}} \sum_{i=1}^{n} \left(y_i - \sum_{j=1}^{p} \beta_j x_{i,j} \right)^2 + \gamma \nu(\boldsymbol{\beta}), \tag{16.37}$$

for some form of sparsity-inducing penalty $\nu(\boldsymbol{\beta})$ and a regularization strength γ.

In a frequentist setting, the most common choice is the L_1 regularization known as LASSO, that is, a penalty term $\gamma \sum_{i=1}^{p} |\beta_i|$, where the coefficients of the regression are estimated to minimize the following penalized criterion

$$\arg\min_{\boldsymbol{\beta}} \sum_{i=1}^{n} \left(y_i - \sum_{j=1}^{p} \beta_j x_{i,j} \right)^2 + \gamma \sum_{i=1}^{p} |\beta_i|, \tag{16.38}$$

where the penalty and strength of the penalty γ penalize large values of the parameters β_i. The constant γ is typically chosen via a cross-validation procedure and it controls the trade-off between least-squares $\gamma = 0$ and more shrunken estimates of $\boldsymbol{\beta}$, that is, it therefore trades off bias and variance.

As the popularity of the LASSO penalization method grew, it was natural to explore Bayesian regression alternatives and which classes of priors would induce such sparsity-inducing constraints on the regression model structures. To proceed in this direction, we will first make the assumption that the likelihood is a simple multivariate Gaussian structure given for responses and independent covariate pairs $\{(y_n, \boldsymbol{x}_n)\,; n = 1, \ldots, N\}$ by the density

$$\pi\left(y_n|\boldsymbol{x}_n, \boldsymbol{\beta}\right) = \left(\frac{1}{2\pi\sigma^2}\right)^{\frac{N}{2}} \exp\left(-\frac{1}{2\sigma^2}\sum_{i=1}^{N}\left(f\left(\boldsymbol{x}_n, \boldsymbol{\beta}\right) - y_n\right)^2\right), \tag{16.39}$$

where $f\left(\boldsymbol{x}_n, \boldsymbol{\beta}\right)$ is the regression mean function.

One can then begin constructing for this linear Gaussian regression model structure different prior structures. The most widely used in practice is the Gaussian–Inverse Gamma prior structure for the coefficients and error variance given in Definition 16.1.

Definition 16.1 (Gaussian–Inverse-Gamma Bayesian Regression Prior) *Consider the prior on the coefficients in a GLM regression, $\boldsymbol{\beta}$, conditional on the observation noise variance σ^2, given by the multivariate Gaussian distribution*

$$\pi\left(\boldsymbol{\beta}|\sigma^2\right) = \frac{1}{(2\pi)^{\frac{p}{2}}|\sigma^2\Sigma|^{\frac{1}{2}}} \exp\left(-\frac{2}{\sigma^2}\boldsymbol{\beta}^T\Sigma^{-1}\boldsymbol{\beta}\right) \tag{16.40}$$

with zero mean (for regularization) and hyper parameters in the $p \times p$ covariance matrix Σ. The joint prior for coefficients and noise variance is then given, assuming i.i.d. observations and hetroskedascity by the prior Generalized Inverse Gaussian (GIG) structure

$$\pi\left(\boldsymbol{\beta}, \sigma^2|\Sigma, a, b\right) = \frac{b^a}{\Gamma(a)(2\pi)^{\frac{p}{2}}|\sigma^2\Sigma|^{\frac{1}{2}}}\sigma^{-a-1} \exp\left(-\frac{2}{\sigma^2}\boldsymbol{\beta}^T\Sigma^{-1}\boldsymbol{\beta} - \frac{b}{\sigma}\right). \tag{16.41}$$

■

Remark 16.2 *One can then adjust the amount of shrinkage induced on the Maximum a Posteriori (MAP) estimator for the coefficients $\hat{\boldsymbol{\beta}}^{MAP}$ by varying the prior covariance matrix Σ.*

Note that one may adopt other choices of prior for σ^2 such as uninformative priors. Priors such as the GIG will produce a conjugate model, whereas other priors may not and will result in requirements for inference via Markov chain Monte Carlo (MCMC) methods etc. (see Chapter 7).

Under this joint prior structure for $\boldsymbol{\beta}$ and σ^2, one has to develop a Bayesian conjugate model in which the joint posterior is given by combining the likelihood in Equation (16.39) with a heirarichical prior such as the GIG, which produces the joint posterior given by

$$\pi\left(\boldsymbol{\beta}, \sigma^2|\boldsymbol{y}, \boldsymbol{X}\right) \propto \left(\sigma^2\right)^{-\frac{N+p}{2-a-1}} \exp\left(-\frac{1}{2\sigma^2}\left[(\boldsymbol{y} - \boldsymbol{X}\boldsymbol{\beta})^T(\boldsymbol{y} - \boldsymbol{X}\boldsymbol{\beta}) + \boldsymbol{\beta}^T\Sigma^{-1}\boldsymbol{\beta}^T + 2b\right]\right). \tag{16.42}$$

One may now observe that such a posterior structure admits certain conjugate structures, such as the marginal posterior distribution of $\boldsymbol{\beta}$ given by the Student-t distribution with $N + 2a$ degrees of freedom and parameters for location and covariance

$$\widetilde{\mu} = \left(\Sigma^{-1} + \boldsymbol{X}^T\boldsymbol{X}\right)^{-1} \left(\boldsymbol{X}^T\boldsymbol{X}\right)\hat{\boldsymbol{\beta}}$$
$$\widetilde{\Sigma} = \frac{2b + s^2 + \hat{\boldsymbol{\beta}}^T \left(\Sigma + (\boldsymbol{X}^T\boldsymbol{X})^{-1}\right)^{-1} \hat{\boldsymbol{\beta}}}{N + 2a} \left(\Sigma^{-1} + (\boldsymbol{X}^T\boldsymbol{X})\right)^{-1} \tag{16.43}$$

with $\hat{\boldsymbol{\beta}} = \left(\boldsymbol{X}^T\boldsymbol{X}\right)^{-1} \boldsymbol{X}^T\boldsymbol{y}$ and $s^2 = (\boldsymbol{y} - \boldsymbol{X}\boldsymbol{\beta})^T (\boldsymbol{y} - \boldsymbol{X}\boldsymbol{\beta})$.

Remark 16.3 *It is clear from this marginal posterior distribution that the hyperparameters Σ can have a significant influence on posterior parameter estimates for $\boldsymbol{\beta}$. Hence, it is common to include a prior for these hyperparameters from one of the following choices:*

- ***Ridge Regression**. Here the prior for Σ is based on the assumption of a scaled identity matrix $\Sigma = c\boldsymbol{I}$;*
- ***Zellner's g-Prior**. Here the prior is set to the entropy prior given by $\Sigma = g\left(\boldsymbol{X}^T\boldsymbol{X}\right)^{-1}$.*

Under a Bayesian modeling paradigm, in which the regression coefficients are treated as a random vector, one may recover the LASSO estimates from the MAP point estimator of the coefficients via a choice of prior on the coefficients given by the multivariate Laplace distribution given in Definition 16.2.

Definition 16.2 (Bayesian LASSO and Multivariate Laplace Prior) *The multivariate Laplace prior on the coefficients is given by*

$$p(\boldsymbol{\beta}; \lambda, \sigma) = \left(\frac{\lambda}{2\sqrt{\sigma^2}}\right)^p \exp\left(-\frac{\lambda\sum_{i=1}^p |\beta_i|}{\sqrt{\sigma^2}}\right), \tag{16.44}$$

where this prior is parameterized with $\frac{\lambda}{\sqrt{\sigma^2}}$ as suggested by Park and Casella (2008). ∎

It will often be of direct use to also recognize the mixture representation of the univariate Laplace distribution by an infinite mixture of Gaussians, as it is a special case of the Scaled Mixture of Normals in the α-stable family (see discussion in Peters and Shevchenko 2015). This infinite Gaussian mixture has the form given in Lemma 16.1.

Lemma 16.1 (Gaussian Infinite Scaled Mixture Representation for Laplace Distributions) *Consider the Laplace-distributed random varaible X; then its density has the following scale mixture of Gaussian representation,*

$$\begin{aligned} f_X(x; \lambda, \sigma) &:= \frac{\lambda}{2\sqrt{\sigma^2}} \exp\left(-\frac{\lambda}{\sqrt{\sigma^2}}|x|\right) \\ &= \int_0^\infty \frac{1}{\sqrt{2\pi z}} \exp\left(-\frac{x^2}{2z}\right) \frac{\lambda^2}{2\sigma^2} \exp\left(-\frac{\lambda^2}{2\sigma^2}\right) dz. \end{aligned} \tag{16.45}$$

Unlike the garrotte or the ridge penalties (see discussions by Yuan and Lin 2007, Breiman 1995 and Hoerl and Kennard 1970), the Laplace prior will produce truly sparse solutions as $\gamma = \frac{\lambda}{\sqrt{\sigma^2}}$ increases.

When the LASSO multivariate Laplace prior is combined with the inverse Gamma prior for the observation variance, one obtains the joint posterior given by

$$\pi\left(\boldsymbol{\beta},\sigma^2|\boldsymbol{y},\boldsymbol{X}\right)\propto\left(\sigma^2\right)^{-\frac{N+p}{2-a-1}}\lambda^p\exp\left(-\frac{1}{2\sigma^2}\left[(\boldsymbol{y}-\boldsymbol{X\beta})^T(\boldsymbol{y}-\boldsymbol{X\beta})+2b\right]-\frac{\lambda\sum_{j=1}^p|\beta_j|}{\sqrt{\sigma^2}}\right). \tag{16.46}$$

One can then introduce the scale mixture of Gaussian representation and a corresponding set of p auxiliary variables $\{\tau_i^2\}_{i=1}^p$ to the posterior distribution to obtain the following conjugate posterior distribution structure for the Bayesian LASSO framework:

$$\begin{aligned}
&\boldsymbol{\beta}|\sigma^2,\tau_1^2,\ldots,\tau_p^2,\boldsymbol{y},\boldsymbol{X}\sim Normal\left(\left(\boldsymbol{D}_\tau^{-1}+\boldsymbol{X}^T\boldsymbol{X}\right)^{-1}(\boldsymbol{X}^T\boldsymbol{X})\hat{\boldsymbol{\beta}},\sigma^2\left(\boldsymbol{D}_\tau^{-1}+\boldsymbol{X}^T\boldsymbol{X}\right)^{-1}\right)\\
&\sigma^2|\boldsymbol{\beta},\tau_1^2,\ldots,\tau_p^2,\boldsymbol{y},\boldsymbol{X}\sim InverseGamma\left(\frac{N}{2}+\frac{p}{2}+a,\,b+\frac{1}{2}\sum_{n=1}^N(f(\boldsymbol{x}_n,\boldsymbol{\beta})-y_n)^2+\frac{\lambda}{2}\boldsymbol{\beta}^T\boldsymbol{D}_\tau^{-1}\boldsymbol{\beta}\right)\\
&\tau_j^{-2}|\boldsymbol{\beta},\sigma^2,\boldsymbol{y},\boldsymbol{X}\sim InverseGaussian\left(\sqrt{\frac{\lambda^2\sigma^2}{\beta_j^2}},\lambda^2\right)
\end{aligned} \tag{16.47}$$

with $\boldsymbol{D}_\tau=\text{diag}\left(\tau_1^2,\ldots,\tau_p^2\right)$.

A limitation in this approach is the use of identical penalization on each regression coefficient. This can lead to unacceptable bias in the resulting estimates (Fan and Li 2001). For example, the classical L_1 regularization can lead to an overshrinkage of large regression coefficients even in the presence of many zeros. This has resulted in sparsity-inducing non-convex penalties that use different penalty coefficients on each regression coefficient, that is, $\sum_{i=1}^p\gamma_i|\beta_i|$ have been proposed as have grouping regularization constraints; see adaptive and sequential estimation approaches by Zou (2006), Zou and Li (2008), Candès *et al.* (2008), and Chartrand and Yin (2008). Alternative, nonconvex approaches include the bridge regression framework, that is, $\gamma\sum_{i=1}^p|\beta_i|^q$ with $q\in(0,1)$, which leads to the L_q regularization problem (Polson *et al.* 2013). Compared to the previous nonconvex prior, the latter possesses the advantage of not introducing additional variables that need to be tuned.

Therefore, in addition to the L1 penalty LASSO prior obtained from the Laplace distribution, another regularization prior that has been popular in the Bayesian context involves the L_q prior, especially when it is used in what is known as an elastic net prior structure (see, e.g., Bornn *et al.* 2010 and Nguyen *et al.* 2013), and Definition 16.3.

Definition 16.3 (L_q Prior) *The multivariate Lq prior on the coefficients is given by*

$$\pi(\boldsymbol{\beta};\lambda,\sigma)=\left(\frac{\lambda}{\sigma^2}\right)^{\frac{p}{2}}\exp\left(-\frac{\lambda\sum_{i=1}^p|\beta_i|^q}{\sigma^2}\right),\quad q\in(1,2), \tag{16.48}$$

where this prior is parameterized with $\frac{\lambda}{\sigma^2}$. ∎

When the L_q prior is combined with the multivariate Gaussian prior, one obtains the Elastic Net prior given by

$$\pi\left(\boldsymbol{\beta};\sigma^2,\lambda_1,\lambda_2\right)\propto\exp\left(-\frac{\lambda_1}{\sqrt{\sigma^2}}\sum_{j=1}^{p}|\beta_j|-\frac{\lambda_2}{2\sigma^2}\boldsymbol{\beta}^T\Sigma^{-1}\boldsymbol{\beta}\right), \tag{16.49}$$

which acts like a combination of L_2 and L_q penalizations providing a compromise between the Laplace and the Gaussian priors.

The resulting posterior full conditionals for the elastic net prior can be obtained as follows:

$$\begin{aligned}
&\boldsymbol{\beta}|\sigma^2,\tau_1^2,\ldots,\tau_p^2,\boldsymbol{y},\boldsymbol{X}\\
&\quad\sim \textit{Normal}\left(\left(\boldsymbol{D}_\tau^{-1}+\boldsymbol{X}^T\boldsymbol{X}+\lambda_2\mathbb{I}\right)^{-1}(\boldsymbol{X}^T\boldsymbol{X})\hat{\boldsymbol{\beta}},\sigma^2\left(\boldsymbol{D}_\tau^{-1}+\boldsymbol{X}^T\boldsymbol{X}+\lambda_2\mathbb{I}\right)^{-1}\right),\\
&\sigma^2|\boldsymbol{\beta},\tau_1^2,\ldots,\tau_p^2,\boldsymbol{y},\boldsymbol{X}\\
&\quad\sim \textit{InverseGamma}\left(\frac{N}{2}+\frac{p}{2}+a,b+\frac{1}{2}\sum_{n=1}^{N}\left(f(\boldsymbol{x}_n,\boldsymbol{\beta})-y_n\right)^2+\frac{\lambda_1}{2}\boldsymbol{\beta}^T\boldsymbol{D}_\tau^{-1}\boldsymbol{\beta}+\frac{\lambda_2}{2}\boldsymbol{\beta}^T\boldsymbol{\beta}\right),\\
&\tau_j^{-2}|\boldsymbol{\beta},\sigma^2,\boldsymbol{y},\boldsymbol{X}\sim \textit{InverseGaussian}\left(\sqrt{\frac{\lambda_1^2\sigma^2}{\beta_j^2}},\lambda_1^2\right)
\end{aligned} \tag{16.50}$$

To complete the discussion on regularization priors we also note the following classes of priors deverloped by Nguyen *et al.* (2013). In their study, they consider the use of two other forms of regularization priors: the exponential power distribution used in the Bayesian bridge regression (see Polson *et al.*, 2013) and the symetric α-stable distribution as an alternative family of regularizing priors.

- **Prior Class 1: Exponential Power Distribution: Bridge**
  The exponential power (*EP*) distribution with zero mean is defined as:

$$f(x;\gamma,q)=\frac{q}{2\gamma\Gamma(1/q)}\exp\left(-|x/\gamma|^q\right). \tag{16.51}$$

- **Prior Class 2: α-Stable Distribution**
  The symetric α-stable distribution was presented as a new class of prior distributions for the regression coefficients. The α-stable distribution with characteristic exponent $0<\alpha<2$, dispersion parameter $\zeta>0$, location parameter μ, and skewness parameter $\beta\in[-1;1]$ is only defined through its characteristic function:

$$\ln\phi(t)=\begin{cases}\mathrm{i}\mu t-\zeta^\alpha|t|^\alpha\left[1-\mathrm{i}\beta\,\mathrm{sgn}(t)\tan\left(\frac{\alpha\pi}{2}\right)\right], & \alpha\neq 1,\\ \mathrm{i}\mu t-\zeta|t|\left[1+\mathrm{i}\beta\,\mathrm{sgn}(t)\frac{2}{\pi}\ln|t|\right], & \alpha=1.\end{cases} \tag{16.52}$$

  In this section, we are particularly interested in a regularization prior with the symetric α-stable (α) distribution ($\mu=0$, $\beta=0$).

To understand the behavior of these two different prior choices of Bayesian GLM regularization we present a few comparisons of the influence of the prior with respect to the type of penalty, that is, the shrinkage effect that each choice may impose. To achieve this we present plots of the negative log densities (i.e., penalty function) for the α-stable distribution and the exponential power distribution; (see Figure 16.1) for different values of q.

- For $q = 2$, these two distributions are equivalent to the Normal distribution, producing a convex penalty (Ridge regression).;
- For $q < 1$, the penalty function from the exponential power distribution is nonconvex whereas the one from the symmetric α-stable distribution is nonconvex when the characteristic exponent of the distribution is $0 < q < 2$. In particular, for $q = 1$, we can see the greater kurtosis and heavier tails provided by the stable distribution. As mentioned previously, the relatively light tails of the exponential power distribution prior are unattractive as they tend to shrink large values of the coefficients even when there is clear evidence from the likelihood that they correspond to large values. This is an important motivation for the class of α-stable priors we introduce in this chapter.

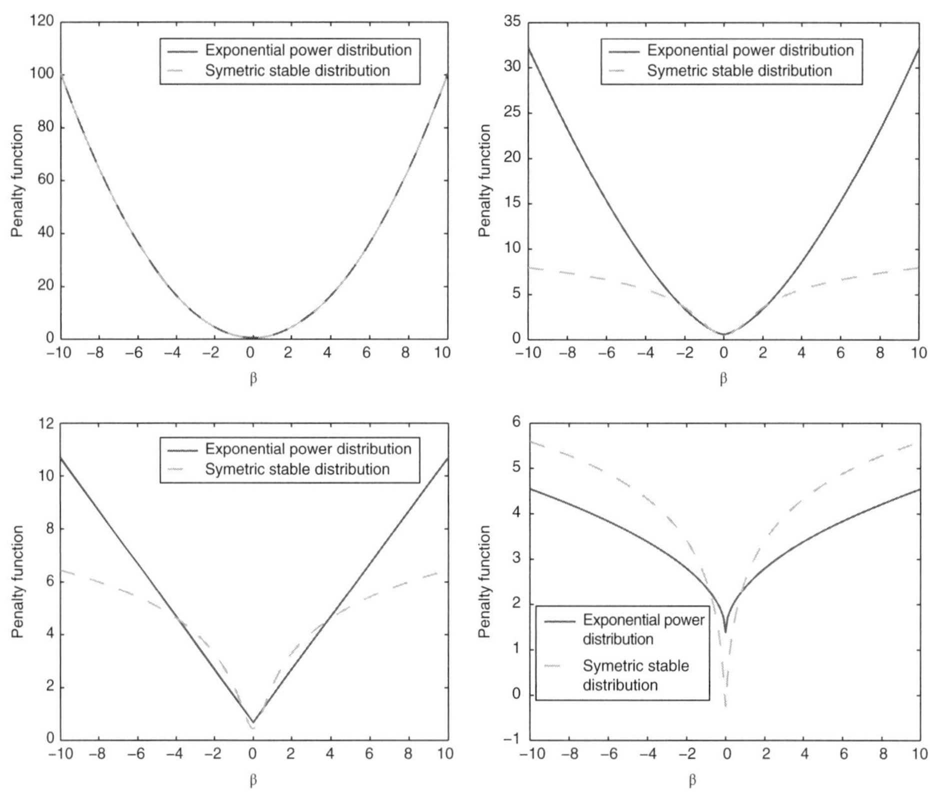

FIGURE 16.1 **Comparison of the penalty term induced by the log prior of the regression coefficient to be either the exponential power distribution or the α-stable distribution ($\gamma_{EP} = 2\gamma_\alpha = 1$). Top left $q = 2$, Top right $q = 1.5$, Bottom Left $q = 1$, Bottom Right $q = 0.5$. (For color detail, please see color plate section.)**

Finally, we note that from a Bayesian perspective, the use of MAP estimates is not exploiting the full posterior information; see Tibshirani (2011) and Park and Casella (2008), who explore the of full posterior analysis of Laplace prior via MCMC or SMC samplers; see discussion in Chapter 7.

Therefore, in the remainder of this section, we focus on the design of efficient algorithms to fully explore the regression coefficient posterior distribution when nonconvex penalty functions with the same penalty coefficient for each regression term are used. Polson *et al.* (2013) propose an MCMC algorithm for the Bayesian Bridge regression problem. In previous Bayesian approaches, they commonly assume that the possibly nonlinear basis function(s) required to link the input variables (exploratory independent variables) to the observed response $\boldsymbol{y}$ is perfectly known. In general, this may not be the case and so we consider an efficient Bayesian algorithm for the joint model selection of these basis functions as well as the regressor coefficients under a nonconvex penalized regression model. Finally, we also introduce a new class of priors based on the α-stable family. We contrast their performance w.r.t. regularization against L_q priors. In doing so, we will particularly consider two simple choices that OpRisk practitioners may encounter in practice. The two choices considered from this GLM basis regression structure previously discussed under the exponential model family setting in a Bayesian regularized regression framework will involve the Normal family and the Poisson family.

- **Normal Regression Model**
  We consider the standard generalized linear basis regression model involving a link function $g(\cdot)$ given by the identity, as well as specifying a Normal distribution for the responses Y_i. To achieve this in the "exponential family" form, we find $\theta = \mu$ and $\phi = \sigma^2$. Here, the dispersion parameter is just the variance and the assumption of a common ϕ is just the usual assumption of constant variance (homoskedascity).;
- **Poisson Regression Model**
  We consider a discrete response model for observations that correspond to counts which, in any given time or space increment, are independent and distributed according to a Poisson distribution. Under the GLM structure our aim is to explain the observed counts with regard to the intensity function constructed via a link function in terms of a linear basis regression. To construct our Poisson regression model we consider the canonical link function in which $g(\cdot)$ is the logarithmic transformation of the linear basis function regression. In the exponential family formulation specified, the Poisson distribution is obtained by considering $\theta = \ln\lambda$, $b(\theta) = \lambda$, $a(\phi) = 1$, and $c(y, \phi) = -\ln y!$ where we denote the Poisson distribution intensity (mean) by λ.

16.3.1 BAYESIAN MODEL SELECTION FOR REGULARLIZED GLM REGRESSION

In OpRisk settings, it will often be relevant to consider several families of non-nested regression models, each specified by the choice of basis function transforming the covariates. In this context, one will utilize regularization to remove nonexplanatory regressors and model selection for the most suitable choice of basis. To achieve this one may perform Bayesian model selection in which the aim is to approximate $\pi(\mathcal{M}_k|\boldsymbol{y})$ for each of the models $k \in \{1, 2, \ldots, K\}$ which corresponds to the posterior model probability by using Bayes' theorem,

$$\pi(\mathcal{M}_k|\boldsymbol{y}) \propto \pi(\boldsymbol{y}|\mathcal{M}_k)\pi(\mathcal{M}_k), \tag{16.53}$$

where $\pi(\boldsymbol{y}|\mathcal{M}_k)$ denotes the marginal likelihood under model $\mathcal{M}_k$, also known as Bayes evidence, and $\pi(\mathcal{M}_k)$ corresponds to the model prior. Moreover, one may also be interested in estimating the parameters that define each model through the parameter posterior $\pi(\boldsymbol{\theta}|\boldsymbol{y}, \mathcal{M}_k)$. In the two examples considered in this study, the parameter is defined as

$$\boldsymbol{\theta} = \begin{cases} \left\{\beta, \sigma_y^2, \gamma\right\}, & \text{Normal model,} \\ \{\beta, \gamma\}, & \text{Poisson model.} \end{cases} \tag{16.54}$$

In these models, the two distributions of interest, that is, the conditional parameter posterior $\pi(\boldsymbol{\theta}|\boldsymbol{y}, \mathcal{M}_k)$ and the associated marginal likelihood $\pi(\boldsymbol{y}|\mathcal{M}_k)$, are intractable. Therefore, we resort to an Importance Sampling (IS)-based Monte Carlo solution to jointly approximate these two quantities. This is a challenge due to the high dimension of the parameter $\boldsymbol{\theta}$, so classical IS methods will be inefficient and produce high variance estimators. Consequently, we utilize a special class of algorithms known as "SMC samplers".

16.4 Bayesian Estimation and Model Selection via SMC Samplers

In this section, we briefly recall the SMC Samplers algorithm discussed in the estimation Chapter 7 and we describe briefly this special class of SMC algorithms specifically designed to work in settings in which the sequence of target distributions to be sampled from are all defined on the same fixed support (see discussions by Del Moral *et al.* 2006, Peters 2005, Peters *et al.* 2009a). This is different from standard SMC algorithms for state space models (particle filtering) in which the sequence of distributions evolves on a product space, and as a result requires modification to the incremental importance sampling (IS) weight expressions.

In short, the SMC Sampler generates weighted samples (termed *particles*) from a sequence of arbitrary distributions π_t, for $t = 1, \ldots, T$, where π_T may be of particular interest and is referred as the target distribution. Procedurally, this involves mutation (or move), correction (or importance weighting), and selection (or resampling). The final weighted particles at distribution π_T are considered weighted samples from the target distribution π.

In more detail, suppose that at time $t-1$, the distribution π_{t-1} can be approximated empirically using N-weighted particles. These particles are first propagated to the next distribution π_t using a mutation kernel $K_t(\boldsymbol{\theta}_{t-1}; \boldsymbol{\theta}_t)$, and then assigned new weights $w_t = w_{t-1} W_t(\boldsymbol{\theta}_1, \ldots \boldsymbol{\theta}_t)$, where w_{t-1} is the weight of a particle at time $t-1$ and W_t is the incremental weight given by

$$W_t(\boldsymbol{\theta}_1, \ldots, \boldsymbol{\theta}_t) = \frac{\pi_t(\boldsymbol{\theta}_t) L_{t-1}(\boldsymbol{\theta}_t; \boldsymbol{\theta}_{t-1})}{\pi_{t-1}(\boldsymbol{\theta}_{t-1}) K_t(\boldsymbol{\theta}_{t-1}; \boldsymbol{\theta}_t)}. \tag{16.55}$$

There is a range of possible things to consider when designing an SMC Sampler algorithm: the appropriate sequence of distributions, the choice of mutation kernel, and then the optimal choice of backward mutation kernel $L_{t-1}(\cdot;\cdot)$ (for a given mutation kernel); see discussion on the optimal choices for these components by Del Moral *et al.* (2006) and Peters (2005). In the context of the modeling undertaken in this chapter, we will utilize the SMC Sampler algorithm to also perform model selection for the basis function choices as detailed later.

16.4.1 PROPOSED SMC SAMPLER SOLUTION

Consider the use of the SMC Sampler on the *artificial* sequence of distributions $\{\pi_t(\boldsymbol{\theta})\}_{t=1}^{T}$ as follows:

$$\pi_t(\boldsymbol{\theta}) \propto \pi(y|\boldsymbol{\theta}, \mathcal{M}_k)^{\phi_t}\pi(\boldsymbol{\theta}|\mathcal{M}_k), \tag{16.56}$$

where conditioned on a specific model, $\mathcal{M}_k$, $\pi(\boldsymbol{\theta}|\mathcal{M}_k)$ is the prior of the model parameters and ϕ_t is a nondecreasing temperature schedule with $\phi_1 = 0$ and $\phi_T = 1$. We thus sample initially from $\pi_1(\boldsymbol{\theta}) = \pi(\boldsymbol{\theta}|\mathcal{M}_k)$ directly and introduce the effect of the likelihood gradually in order to obtain at this end ($t = T$) an approximation of the conditional parameter posterior $\pi(\boldsymbol{\theta}|\boldsymbol{y}, \mathcal{M}_k)$. As shown by Del Moral *et al.* (2006), the marginal likelihood of interest to make a decision regarding which basis function to use can be approximated with SMC Samplers as

$$Z_T = Z_1 \prod_{t=2}^{T} \frac{Z_t}{Z_{t-1}} \approx \prod_{t=1}^{T}\left(\sum_{i=1}^{N_p} w_t^i\right), \tag{16.57}$$

where $Z_t = \int \pi(y|\boldsymbol{\theta}, \mathcal{M}_k)^{\phi_t}\pi(\boldsymbol{\theta}|\mathcal{M}_k)d\boldsymbol{\theta}$ corresponds to the normalizing constant of the target distribution at iteration t. As a consequence, the following procedure is performed:

1. For each model $\mathcal{M}_k$, $k \in 1, \ldots, K_{\max}$, approximate the conditional parameter posterior distribution $\pi(\boldsymbol{\theta}|\boldsymbol{y}, \mathcal{M}_k)$ as well as the marginal likelihood $\pi(\boldsymbol{y}|\mathcal{M}_k)$.;
2. Approximate the model posterior $\pi(\mathcal{M}_k|\boldsymbol{y})$, which is the model posterior, via the approximation of $\pi(\boldsymbol{y}|\mathcal{M}_k)$; and model prior $\pi(\mathcal{M}_k)$.

Successively Random Walk Metropolis Hastings within Gibbs proposal kernels is used for the mutation step of the algorithm $K_t(.;.)$ by randomly partitioning the parameter vector $\boldsymbol{\theta}$ into B blocks. The SMC Sampler algorithm for model $\mathcal{M}_k$ proceeds according to the following steps:

- Initialize particle system from the prior

$$\{\boldsymbol{\theta}_1^i\}_{i=1}^{N_p} \sim \pi(\boldsymbol{\theta}|\mathcal{M}_k) \tag{16.58}$$

  and set $\{\tilde{w}_1^i\}_{i=1}^{N_p} = 1/N_p$.;
- For the sequence $t = 1, \ldots, T$ perform the following steps:
  - Perform evaluation of the particle weights for each particle $i = 1, \ldots, N_p$ to evaluate the un-normalized weights

$$w_t^i = \tilde{w}_{t-1}^i \frac{\pi_t(\boldsymbol{\theta}_{t-1}^i)}{\pi_{t-1}(\boldsymbol{\theta}_{t-1}^i)} = \tilde{w}_{t-1}^i \frac{\pi(y|\boldsymbol{\theta}_{t-1}, \mathcal{M}_k)^{\phi_t}}{\pi(y|\boldsymbol{\theta}_{t-1}, \mathcal{M}_k)^{\phi_{t-1}}}. \tag{16.59}$$

  - Renormalize the particle weights for each particle $i = 1, \ldots, N_p$ via

$$\tilde{w}_t^i = w_t^i \left[\sum_{j=1}^{N_p} w_t^j\right]^{-1}. \tag{16.60}$$

- Perform the selection stage if effective sample size (ESS) $< N_p/2$, which involves a resample stage.;
- Peform mutation for each of the particles $i = 1, \ldots, N_p$, which involves sampling.;
- $\theta_t^i \sim K_t(\boldsymbol{\theta}_{t-1}^i; \cdot)$ where $K_t(\cdot; \cdot)$ is a $\pi_t(\cdot)$ invariant Markov kernel.

  The resulting weighted particle system $\left\{\boldsymbol{\theta}_t^i, \tilde{w}_t^i|\mathcal{M}_k\right\}_{i=1}^{N_p}$ approximates. $\pi_t(\boldsymbol{\theta}) \propto \pi(y|\boldsymbol{\theta}, \mathcal{M}_k)^{\phi_t}\pi(\boldsymbol{\theta}|\mathcal{M}_k)$.

16.5 Illustrations of SMC Samplers Model Estimation and Selection for Bayesian GLM Regressions

Here we present a few examples to illustrate the performance of the SMC sampler discussed in the previous section to perform joint model selection and parameter estimation in the two different GLMs of Gaussian and Poisson discussed previously. All results have been obtained using the approach of Section 16.4 with the following settings: $N_p = 500$ particles and $T = 50$ iterations have been used to approximate the sequence of distributions. A piece-wise linear tempering schedule $\{\phi_t\}$ has been selected. The sequence increased uniformly from 0 to 7/50 for the first 10 time points, then from 7/50 to 20/50 for the next 20, and finally from 20/50 to 1 for the last 20 time points.

The different basis functions considered in this chaper are described in Table 16.1 with equally spaced centers c_i on some chosen bounded support of interest for the univariate input variable $x \in [-4; 12]$. The following priors have been used: $p(\mathcal{M}_k) = 1/K_{\max}$ and an Inverse-Gamma prior for both γ and σ_y^2. Finally, in order to validate the proposed algorithm, the results will be compared with the frequentist LASSO implemented using the coordinate descent algorithm (Friedman *et al.* 2010) and for which the tuning parameter is obtained by a 10-fold cross-validation procedure.

16.5.1 NORMAL REGRESSION MODEL

To illustrate the example for the Normal regression model, consider $n = 40$ observations under model $\mathcal{M}_2$ ($\sigma_y^2 = 2$) with regression coefficients set to zero except $\beta_0 = 1, \beta_3 = \beta_{15} = \beta_{24} = 5$,

TABLE 16.1 **Description of the different basis functions used in the numerical simulation section — $r_i = ||x - c_i||$ defines the L_1-norm between the input univariate variable and the i-th center of the current basis — ρ_i is a scale factor**

	Model	Expression Φ_k^i
$\mathcal{M}_1$	Linear	x
$\mathcal{M}_2$	Gaussian	$\exp(-(\rho_i r_i)^2)$
$\mathcal{M}_3$	Inverse quadratic	$[1 + (\rho_i r_i)^2]^{-1}$
$\mathcal{M}_4$	Sigmoidal	$[1 + \exp(-r_i/\rho_i)]^{-1}$
$\mathcal{M}_5$	B-spline	See Lee (1982) Order=2
$\mathcal{M}_6$	Mollifier	$\begin{cases} \exp\left(-\frac{1}{1-(\rho_i r_i)^2}\right) & \text{if } \lvert\rho_i r_i\rvert < 1 \\ 0 & \text{otherwise} \end{cases}$

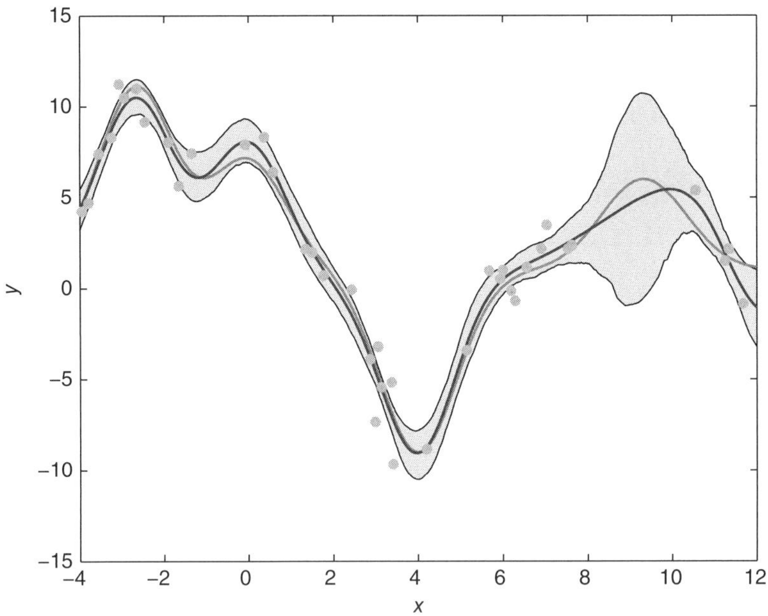

FIGURE 16.2 **Normal regression with *EP* prior ($q = 1$): true function in blue, observed responses in green-filled circles, posterior mean from SMC under model $\mathcal{M}_3$ in red, and confidence region in gray (5–95% percentiles). (For color detail, please see color plate section.)**

$\beta_8 = \beta_{20} = -5$, and $\beta_5 = \beta_{17} = 3$. For the basis functions ($\mathcal{M}_2$ to $\mathcal{M}_6$), we have 12 equally spaced centers c_i with 2 different scale parameters. As a consequence, the dimension of the parameter vector $\boldsymbol{\theta}$ to estimate is 28 for $\mathcal{M}_2$ to $\mathcal{M}_6$ and 4 for $\mathcal{M}_1$. As shown in Figure 16.2, the SMC Sampler is able to efficiently predict the unknown function even if only a few observations are available. As opposed to frequentist LASSO, the proposed approach can give a confidence interval on the predicted curve, which is of great interest in many applications. Table 16.2 clearly show the ability of the proposed method to give an accurate estimate of the regression coefficients. The mean squared error (MSE) on $\boldsymbol{\beta}$ is indeed divided by a factor of approximately 4.5 compared to to the one obtained with the frequentist LASSO. We can also see that a slightly lower MSE is given by the use of the proposed prior for regularization, that is, the symmetric stable distribution. From Figure 16.3, we can see the shrinkage effect on the approximate marginal posterior distribution on some coefficient which is 0 in reality. As expected, as q decreases, this marginal posterior distribution is shrunk around 0, a bit more rapidly with the use of α. Finally, from Figure 16.4, we study the model choice given by the proposed SMC Sampler. From these results, we can see that the algorithm is able to give a good model with high probability. There is some uncertainty with model $\mathcal{M}_5$ owing to the similarity of the two models. Again, the posterior model probability for the correct model is higher when the α is used as a prior for the regression coefficient.

16.5.2 POISSON REGRESSION MODEL

In the second example, a Poisson regression model is considered. A total of $n = 100$ observations have been generated with $\mathcal{M}_3$. For the basis functions, 25 equally spaced centers c_i

TABLE 16.2 Median of the mean squared error (MSE) between true regression coefficients and the estimated ones under the true model $\mathcal{M}3$ based on 50 replications

	EP	α	LASSO
$q = 1$	29.13	28.98	133.69
$q = 0.8$	29.37	29.06	

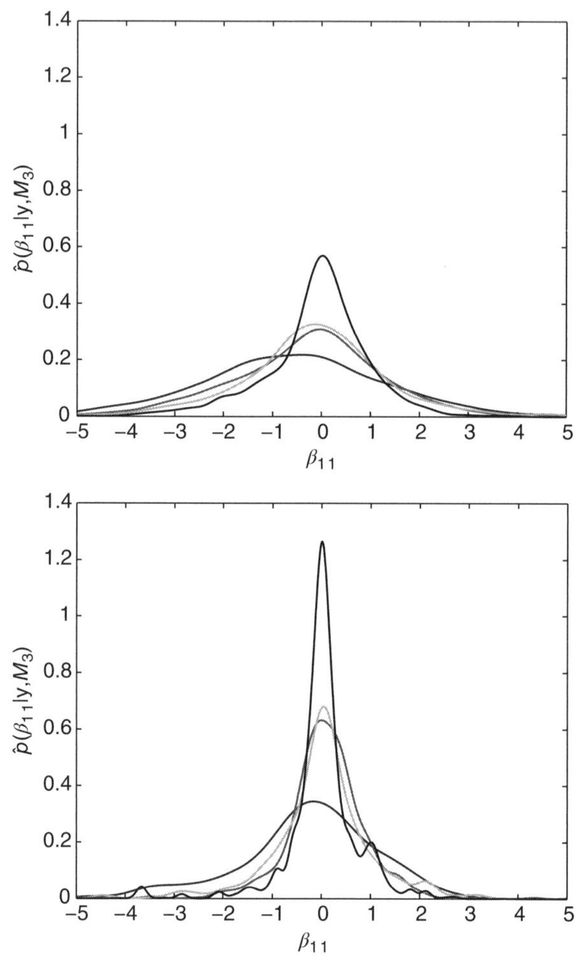

FIGURE 16.3 Comparison of the shrinkage results obtained with the two different priors as q decreases (blue: $q = 1.5$, red: $q = 1$, green: $q = 0.8$, black: $q = 0.5$). Top plot is EP prior and bottom plot is a symmetric α-Stable prior. (For color detail, please see color plate section.)

have been used with the same scale parameter. Figure 16.5 shows the resulting mean predicted curve (and associated confidence interval) obtained by using the proposed SMC Sampler under the true model. As in the previous case, the true curve is always within

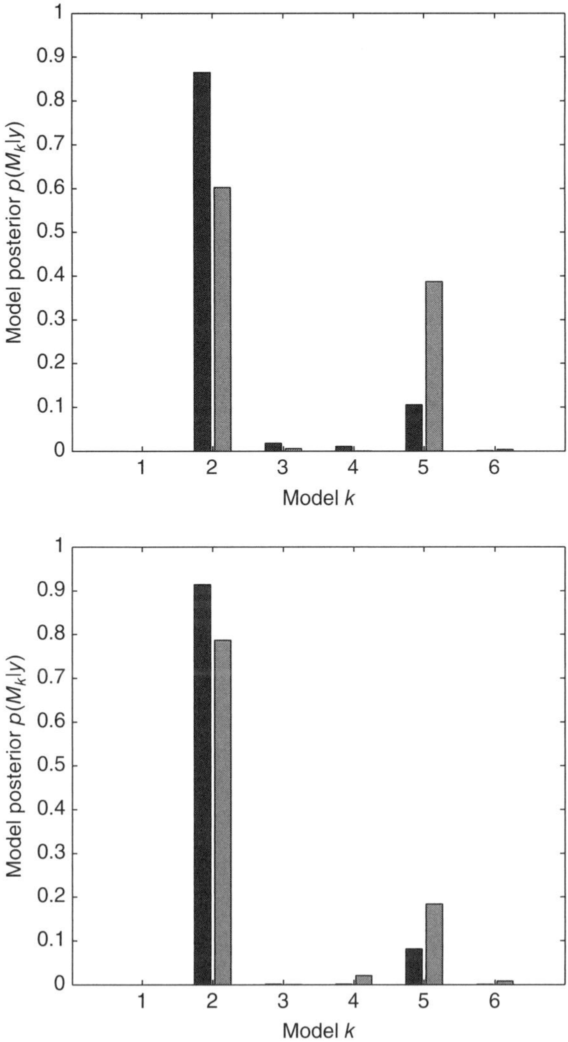

FIGURE 16.4 **Comparison of the approximation of the model posterior (blue: α, red: *EP*) (For color detail, please see color plate section.)**

the confidence region. Table 16.3 presents the mean squared prediction errors obtained by the proposed approach by using the two different priors as well as the ones obtained by the frequentist LASSO. The SMC Sampler with the α and $q = 1$ slightly outperforms the others.

Hence, these examples illustrate that the proposed algorithm is efficient for model selection and parameter estimation in penalized regression models. Moreover, the proposed class of priors based on α-stable family distribution represents an alternative to EP distribution commonly used in L_q regularization. There is a detailed exploration of these ideas by Nguyen *et al.* (2013).

TABLE 16.3 Median of the mean squared prediction error for the proposed approach using the different priors as well as the LASSO estimate, based on 50 replications. Best fitting results are in bold font

	EP		α		LASSO
	$q = 1$	$q = 0.5$	$q = 1$	$q = 0.5$	
$\mathcal{M}_1$	17.4087	17.4370	**17.2961**	17.4117	17.3391
$\mathcal{M}_2$	2.7994	2.4939	**2.3589**	2.5132	3.4125
$\mathcal{M}_3$	2.6987	2.5305	**2.4089**	2.6344	2.5451
$\mathcal{M}_4$	3.0428	3.0087	**2.9484**	3.3708	3.3917
$\mathcal{M}_5$	3.3182	3.6683	**3.2428**	6.4821	4.3435
$\mathcal{M}_6$	3.0162	3.0104	**2.9131**	3.4356	3.1100

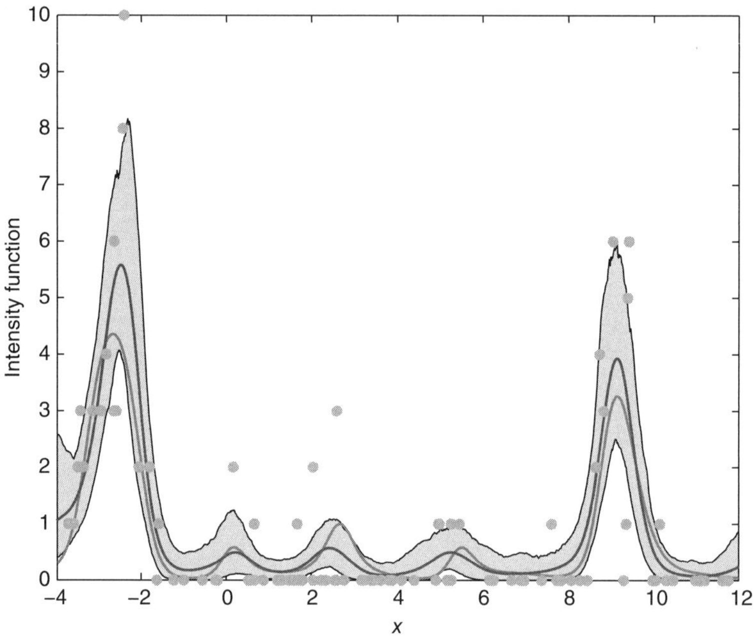

FIGURE 16.5 Poisson regression with α prior ($q = 1$): true function in blue, observed count responses in green-filled circles, posterior mean from SMC under model $\mathcal{M}_3$ in red, and confidence region in gray (5–95% percentiles). (For color detail, please see color plate section.)

16.6 Introduction to Quantile Regression Methods for OpRisk

In this section, we briefly introduce the notion of percentile-based regression methods, which will involve the estimation of quantiles via a quantile regression structure. Quantile regression is a statistical technique intended to estimate, and conduct inference about, conditional quantile functions that can be suitably used to serve the purpose of establishing the sensitivity of capital and risk measure estimations to different input covariates, which may be internal risk indicators, external factors, and micro- and macroeconomic factors related to the current operating environment of the bank.

Just as classical linear regression methods discussed earlier typically minimize sums of squared residuals (perhaps with regularization constraints on the coefficients) enabling one to estimate models for conditional mean functions, quantile regression methods offer a mechanism for estimating models for the conditional median function and the full range of other conditional quantile functions.

Such regression structures and models allow one to study the effect of explanatory variables on the entire conditional distribution of the response variable and not only on its center. By supplementing the estimation of conditional mean functions with techniques for estimating an entire family of conditional quantile functions, quantile regression is capable of providing a more complete statistical analysis of the stochastic relationships among random variables.

Quantile regression has been applied to a wide range of applications in economics and finance. In quantitative investment, least square regression–based analysis is extensively used for analyzing factor performance, assessing the relative attractiveness of different firms, and monitoring the risks in their portfolios. Engle and Manganelli (2004) consider the quantile regression for the Value-at-Risk (VaR) model. They construct a conditional autoregressive Value-at-Risk model (CAVaR), and employ quantile regression for the estimation. Recall that this is achievable since VaR is defined as a quantile of the loss distribution of a portfolio within a given time period and a confidence level. Accurate VaR estimation can help financial institutions maintain appropriate capital levels to cover the risk from the corresponding portfolio. In this section, we consider both parametric and nonparametric quantile regressions, which, under a Bayesian paradigm, may be estimation via MCMC strategies; see details of such sampling methods in Chapter 7.

Initially, we present basic concepts of quantile regression models for loss modeling and capital estimation. We begin with the classical introduction to quantile-based regression and the associated loss function utilized in the parameter estimation in quantile regression (Figure 16.6). There are many excellent introductions to quantile regression that the reader may consult for more detailed discussions such as Koenker and Hallock (2001), Koenker (2001, 2005), Fitzenberger *et al.* (2002), Buchinsky (1998), Yu and Moyeed (2001), and Gilchrist (2002).

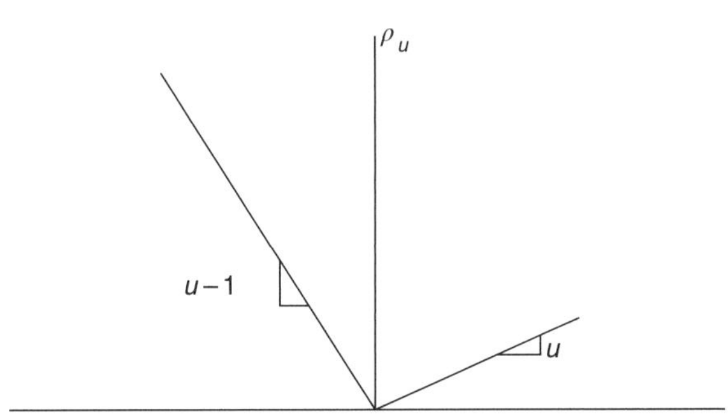

FIGURE 16.6 **Quantile regression loss function**

16.6.1 NONPARAMETRIC QUANTILE REGRESSION MODELS

Let $Y_i > 0$ be a set of observed losses and $\boldsymbol{x}_i = (1, x_{i,1}, \ldots, x_{i,m})$ be a vector of covariates that describe Y_i. The quantile function for the log-transformed data $Y_i^* = \ln Y_i \in \Re$ is

$$Q_{Y^*}(u|\boldsymbol{x}_i) = \beta_{0,u} + \sum_{k=1}^{m} \beta_{k,u}, x_{i,k}, \tag{16.61}$$

where $u \in (0, 1)$ is the quantile level, $\boldsymbol{\beta}_u = (\beta_{0,u}, \ldots, \beta_{k,u})$ are estimated by solving

$$\min_{\beta_{0,u},\ldots,\beta_{m,u}} \sum_{i=1}^{n} \rho_u(\epsilon_i) = \sum_{i=1}^{n} \epsilon_i[u - I(\epsilon_i < 0)], \tag{16.62}$$

and $\epsilon_i = y_i^* - \beta_{0,u} - \sum_{k=1}^{m} \beta_{k,u} x_{i,k}$. Koenker and Hallock (2001) provide an illustration of the loss function ρ_u for quantile regression, which we consider later Koenker and Machado (1999) and Yu and Moyeed (2001) show that parameters $\beta_{k,u}$ can be estimated via an asymmetric Laplace distribution (ALD) with density given by

$$f(y_i^*|\mu_i, \sigma_i^2, p) = \frac{p(1-p)}{\sigma_i} \exp\left(-\frac{(y_i^* - \mu_i)}{\sigma_i}[p - I(y_i^* \leq \mu_i)]\right), \tag{16.63}$$

where the skew parameter $0 < p < 1$ gives the quantile level u, $\sigma_i > 0$ is the scale parameter, and $-\infty < \mu_i < \infty$ is the location parameter. Since the pdf (16.63) contains the loss function (16.62), it is clear that parameter estimates that maximize (16.63) will minimize (16.62).

Essentially, we assume Y_i^* follows an ALD model family denoted by $Y_i^* \sim AL(\mu_i, \sigma_i^2, p)$. Then

$$Y_i^* = \beta_0 + \sum_{k=1}^{m} \beta_k x_{i,k} + \epsilon_i \sigma_i, \tag{16.64}$$

where $\epsilon_i \sim AL(0, 1, p)$ and one considers m-dimensional covariate vectors of covariates to explain the location or mean given by $\boldsymbol{x}_i$ and ν dimensional vectors of covariates $\boldsymbol{s}_i$ to explain the scale or variance according to,

$$\begin{aligned} \mu_i &= \beta_0 + \sum_{k=1}^{m} \beta_k x_{i,k}, \\ \sigma_i^2 &= \exp(\alpha_0 + \sum_{k=1}^{\nu} \alpha_k s_{i,k}). \end{aligned} \tag{16.65}$$

As a tool to estimate quantile regression, the ALD is a three-parameter distribution that has established a direct link to the estimation of quantiles and quantile regression (see discussions by Yu and Zhang 2005). The properties of the distribution are given in more detail in Chapter 9 and we also refer the interested reader to the work of Yu and Moyeed (2001), who apply the ALD model for quantile regression.

We note that one may consider a generalization of the ALD to incorporate a dynamic mean, variance, and shape parameter, which provides a direct link to investigating the effect of the covariates on quantiles. Using the ALD family provides a mechanism for Bayesian inference of quantile regression models. The benefit of using a Bayesian procedure lies in the adoption of available prior information and the provision of a complete predictive distribution for the required capital and risk measures.

16.6.2 PARAMETRIC QUANTILE REGRESSION MODELS

Alternatively, under a parametric approach, we may assume $Y_i^* \sim F(y^*|\boldsymbol{\theta})$, where $F(y^*|\boldsymbol{\theta})$ is the conditional cumulative distribution function and $\boldsymbol{\theta} \in \boldsymbol{\Theta}$ is a vector of model parameters. The quantile function for the conditional distribution of Y_i^* given $\boldsymbol{x}_i$ is

$$Q_{Y^*}(u|\boldsymbol{x}_i) \equiv \inf_{y^*} F(y^*|\boldsymbol{\theta}) \geq u = \arg\min_{\boldsymbol{\theta}\in\boldsymbol{\Theta}} E[\rho_u(\epsilon_i)], \tag{16.66}$$

where the loss function is

$$\rho_u(\epsilon) = \epsilon\left(u - \mathbb{I}[\epsilon < 0]\right), \tag{16.67}$$

and $\epsilon_i = y_i^* - \beta_{0,u} - \sum_{k=1}^m \beta_{k,u} x_{i,k}$ and u is a quantile level between $(0, 1)$. Then the quantile function in (16.66) can be written according to a linear regression model, where for notational convenience we drop the additional index of u for the quantile level, giving

$$Q_{Y^*}(u|\boldsymbol{x}_i) = \beta_0 + \sum_{k=1}^{m} \beta_k x_{i,k} + Q_\epsilon(u)\sigma_i, \tag{16.68}$$

where $Q_\epsilon(u) = F_z^{-1}(u)$ is the inverse distribution for the standardized variable $Z_i = \frac{Y_i - \mu_i}{\sigma_i}$ and the quantile function for Y_i is $Q_Y(u|\boldsymbol{x}_i) = \exp(Q_{Y^*}(u|\boldsymbol{x}_i))$.

Next we will provide three illustrative families of models that one may adopt in practice for parametric quantile regression modeling: ALD, polynomial power Pareto model of Cai (2010), and the Generalized Beta family of models; see discussions on aspects of these models in Chapter 9.

16.6.2.1 Asymmetric Laplace Distribution.

If one models the residuals ϵ_i by an ALD family, the quantile function for the observed data Y_i^* is given by (16.68), where $F_z^{-1}(u)$ is the inverse distribution function

$$F_{AL}^{-1}(u|\mu, \sigma^2, p) = \begin{cases} \mu + \frac{\sigma}{1-p}\ln\left(\frac{u}{p}\right), & \text{if } 0 \leq u \leq p, \\ \mu - \frac{\sigma}{p}\ln\left(\frac{1-u}{1-p}\right), & \text{if } p < u \leq 1, \end{cases} \tag{16.69}$$

with $\mu = 0$ and $\sigma^2 = 1$. The mean, variance, skewness γ, and kurtosis κ of the ALD family are respectively given by

$$\begin{aligned}\mathbb{E}(Y) &= \mu + \frac{\sigma(1-2p)}{p(1-p)},\\ \mathbb{V}\mathrm{ar}(Y) &= \frac{\sigma^2(1-2p+2p^2)}{(1-p)^2p^2},\\ \gamma &= \frac{2[(1-p)^3-p^3]}{((1-p)^2+p^2)^{3/2}},\\ \kappa &= \frac{9p^4+6p^2(1-p)^2+9(1-p)^4}{(1-2p+2p^2)^2}.\end{aligned}$$

Note that the true shape parameter p of the ALD family and the estimated p gives an indication of the magnitude and direction of skewness.

ALD(μ_i, σ_i, p) is skewed to left when $p > 0.5$ and skewed to right when $p < 0.5$, which corresponds to the fact that most log-transformed loss data are skewed to the left and the risk measure or captial estimation that quantile levels are typically significantly larger than, say, 50% (the median). Figure 16.7 show a variety of ALD densities, its skewness, and kurtosis, respectively.

16.6.2.2 Polynomial Power Pareto Model.

Cai (2010) presents a polynomial power Pareto (PP) quantile function model and a Bayesian method for parameter estimation. This model combines a power law model into a Pareto distribution for additional flexibility in the tail behavior of the observed quantity, which enables one to model both the main body and the tails of a distribution. In considering the PP model, the conditional quantile function of the response (reserve in each cell) is comprised of two components:

- Component 1. A power distribution $F_1(y) = y^{\frac{1}{\gamma_1}}$ where $y \in [0, 1]$ and $\gamma_1 > 0$ with a corresponding quantile function given by $Q_1(u; \gamma_1) = u^{\gamma_1}$ for $u \in [0, 1]$;
- Component 2. A Pareto distribution function $F_2(y) = 1 - y^{-\frac{1}{\gamma_2}}$ where $y \geq 1$ and $\gamma_2 > 0$ with a corresponding quantile function given by $Q_2(u; \gamma_2) = (1-u)^{-\gamma_2}$.

One may use the fact that the product of the two quantile functions will remain a strictly valid quantile function producing the new quantile function family known as the polynomial power Pareto model. The resulting structural form given by the inverse of the Pareto distribution with an additional polynomial power term is

$$F_{PP}^{-1}(u|\gamma_1, \gamma_2) = u^{\gamma_1}(1-u)^{-\gamma_2}. \tag{16.70}$$

Hence, the quantile function is again given by (16.68), where $Q_\epsilon(u) = F_{PP}^{-1}(u)$ and $Q_Y(u) = \exp(Q_{Y^*}(u))$.

The pdf of PP distribution for $Y_i^* = \ln Y_i$ is

$$f_{PP}(y_i^*|\gamma_1, \gamma_2) = \frac{u_i^{1-\gamma_1}(1-u_i)^{\gamma_2+1}}{\sigma_i[\gamma_2 u_i + \gamma_1(1-u_i)]},$$

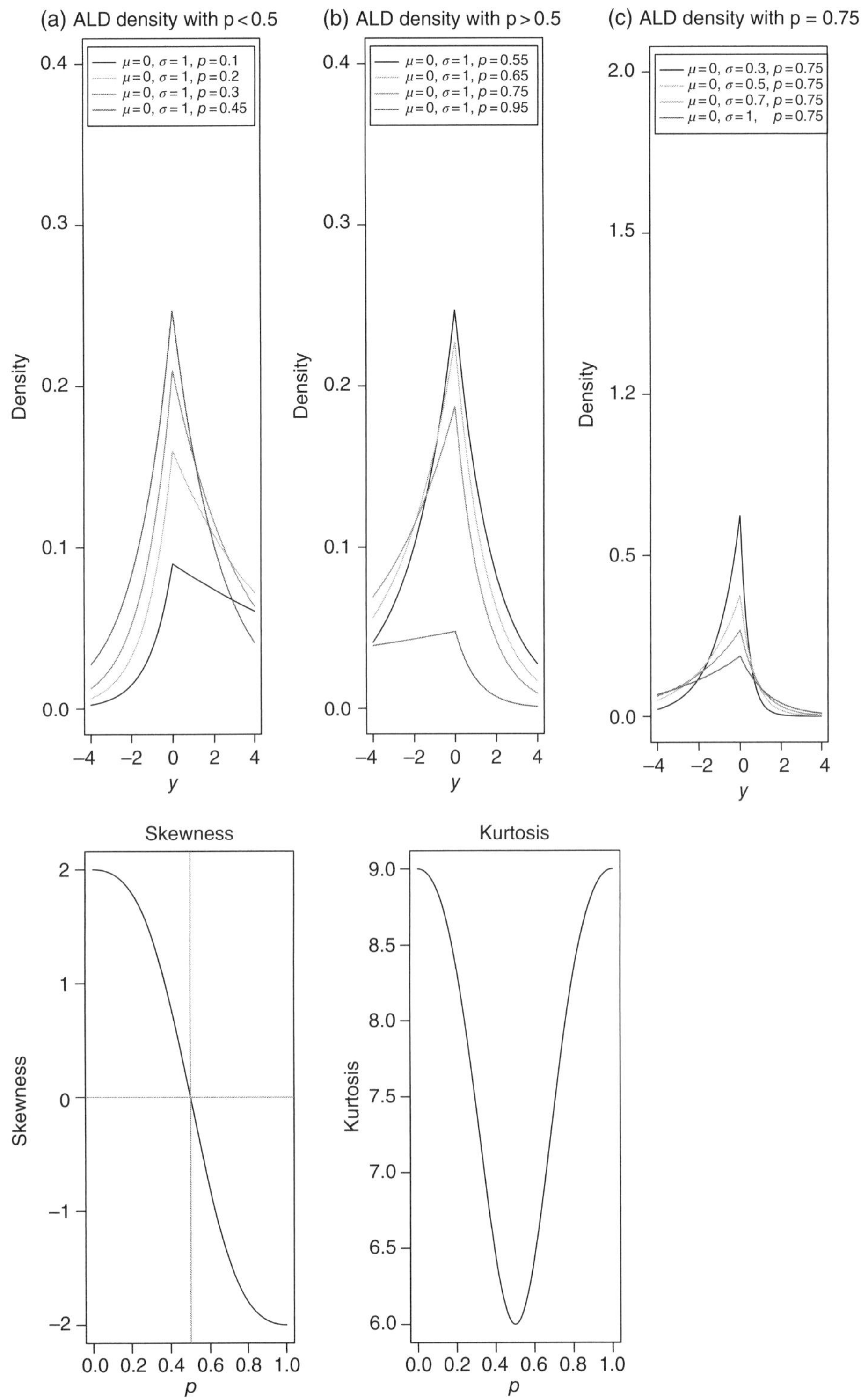

FIGURE 16.7 Top: Assymetric Laplace densities for a range of parameter values. Bottom: ALD skewness and kurtosis

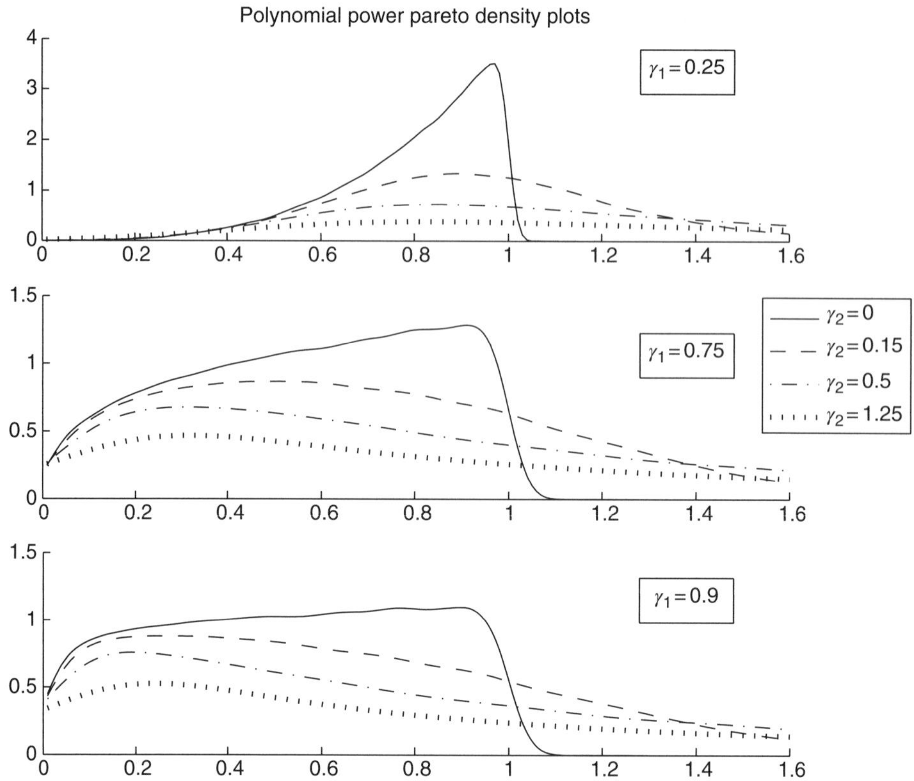

FIGURE 16.8 **The pdf of power Pareto distribution**

where $y_i^* = \beta_0 + \sum_{k=1}^{m} \beta_k x_{ik} + u_i^{\gamma_1}(1-u_i)^{-\gamma_2}\sigma_i$ and the inverse distribution is

$$F_{PP}^{-1}(u|\gamma_1, \gamma_2) = u^{\gamma_1}(1-u)^{-\gamma_2}.$$

Hence, the quantile function is again given by (16.68), where $Q_\epsilon(u) = F_{PP}^{-1}(u)$ and $Q_Y(u) = \exp(Q_{Y^*}(u))$.

To complete the specification of the polynomial power Pareto plots we demonstrate the shape of the density that can be obtained for a range of different power parameters for the power and Pareto components with a unit scale factor $\sigma = 1$. The plots in Figure 16.8 demonstrate the flexible skew, kurtosis, and tail features that can be obtained from such a model by varying the parameters γ_1 and γ_2.

When constructing the Bayesian model for this quantile regression model, one must take care to ensure the posterior support is well defined; this is carefully explained as well as an efficient MCMC sampling scheme in the paper by Cai (2010).

To complete the posterior distribution specification, that is, to define $\pi(\boldsymbol{\beta}, \boldsymbol{\sigma}, \gamma_1, \gamma_2|y_{1,1}, \ldots, y_{n,n})$, it suffices to consider the representation of two components: the likelihood of the data for the regression structure (i.e., the density, not the quantile function), and the prior specifications for the model parameters. We adopt the Bayesian model proposed by Cai (2010) with a support on the parameters of the quantile regression for which the posterior is defined uniquely characterized by the three sets of parameter space constraints Ω_1,Ω_2, and Ω_3 given by

$$\Omega_1 = \left\{ (\beta_0, \ldots, \beta_m) \; : \; \beta_0 + \sum_{k=1}^{m} \alpha_k x_{i,k} < y_i, \;\; \forall i \in \{1, 2, \ldots, N\} \right\}$$
$$\Omega_2 = \left\{ (\sigma_0, \ldots, \sigma_\nu) \; : \; \sigma_0 + \sum_{k=1}^{\nu} \sigma_k s_{i,k} > \epsilon > 0, \;\; \forall i \in \{1, 2, \ldots, N\} \right\}$$
$$\Omega_3 = (0, M] \times (0, \infty), \;\; M \in \mathbb{R}_+.$$

Under these parameter space restrictions, the resulting posterior for the polynomial power Pareto model can be shown to be well defined as a proper density (see Cai 2010, theorem 1).

16.6.2.3 Generalized Beta Distribution of the Second-Type Family. Loss data often exhibit heavy-tailed behavior as has been discussed in numerous places throughout the book. To incorporate this feature into a quantile regression one may adopt the family of generalized beta distributions of the second kind (GB2 models), which have several attractive features for modeling heavy-tailed loss data. The GB2 model nests a number of important distributions as special cases. The GB2 distribution has four parameters, which allow it to be expressed in various flexible densities. The density function is specified for a random variable $Y_i \sim GB2(a, b_i, p, q)$ on positive support, that is, $Y_i \in \mathbb{R}^+$ with shape parameters a, p, q and a density given by

$$f_{GB2}(y_i|a, b_i, p, q) = \frac{\frac{a}{b_i}(y_i/b_i)^{ap-1}}{B(p, q)[1 + (y_i/b_i)^a]^{p+q}}, \quad \text{for } y_i > 0, \tag{16.71}$$

where

$$b_i = \frac{\mu_i B(p, q)}{B(p + 1/a, q - 1/a)} \tag{16.72}$$

and

$$\mathbb{E}(Y_i) = \mu_i = \exp\left(\beta_0 + \sum_{k=1}^{m} \beta_k x_{i,k} \right). \tag{16.73}$$

GB2 distribution can also be written as a beta distribution

$$f_B(z_i|p, q) = \frac{1}{B(p, q)} z_i^{p-1} (1 - z_i)^{p+q} \tag{16.74}$$

via the transformation $z_i = \dfrac{(y_i/b_i)^a}{1 + (y_i/b_i)^a}$. Hence, the distribution function of GB2 is given by

$$F_{GB2}(y_i|a, b_i, p, q) = \int_0^{z_i} \frac{t^{p-1}(1 - t)^{(q-1)}}{B(p, q)} dt = \frac{B(z_i|p, q)}{B(p, q)} = F_B(z_i|p, q), \tag{16.75}$$

where $B(z_i|p, q)$ is the incomplete beta function.

Then the quantile function for the GB2 model is given by

$$Q_Y(u) = \frac{\exp\left(\beta_0 + \sum_{k=1}^{m} \beta_k x_{i,k}\right) B(p,q)}{B(p+1/a, q-1/a)} \left(\frac{F_B^{-1}(u|p,q)}{1 - F_B^{-1}(u|p,q)}\right)^{\frac{1}{a}}. \tag{16.76}$$

When $q = \infty$, GB2 distribution becomes a Generalized Gamma (GG) distribution.

16.6.2.4 Two Special Cases of GB2. In this section, we consider a special subfamily of the GB2 family discussed, the GG, introduced by Stacy (1962); see discussions in Chapter 9. This distribution is a three-parameter distribution with density function given by

$$\begin{aligned} f_{GG}(y_i|a, b_i, p) &= \lim_{q\to\infty} \frac{\frac{a}{b_i}(y_i/b_i)^{ap-1}}{B(p,q)[1 + (y_i/b_i)^a]^{p+q}} \\ &= \frac{a(y_i/b_i)^{ap} \exp[-(y_i/b_i)^a]}{y_i\Gamma(p)}, \quad \text{for } y_i > 0, \end{aligned} \tag{16.77}$$

where

$$b_i = \frac{\mu_i \Gamma(p)}{\Gamma(p+1/a)} \tag{16.78}$$

and

$$\mathbb{E}(Y_i) = \exp\left(\beta_0 + \sum_{k=1}^{m} \beta_k x_{ik}\right) = \frac{b_i \Gamma(p+1/a)}{\Gamma(p)}. \tag{16.79}$$

It is a special case of the GB2 distribution when $b = q^{1/q}\beta$ and $q \to \infty$ (Cummins *et al.* 1990 and McDonald 1984).

The distribution function is

$$F_{GG}(y_i|a, b_i, p) = \int_0^{z_i} \frac{t^{p-1}e^{-t}}{\Gamma(p)} dt = \frac{\gamma_1(z_i|p)}{\Gamma(p)} = F_G(z_i|1, p), \tag{16.80}$$

where $\gamma_1(z_i|p)$ is the lower incomplete Gamma function and $z_i = (y_i/b_i)^a$. Hence, the quantile function is given by

$$Q_Y(u) = \frac{\exp\left(\beta_0 + \sum_{k=1}^{m} \beta_k x_{i,k}\right) \Gamma(p)}{\Gamma(p+1/a)} \left(F_G^{-1}(u|1,p)\right)^{1/a}. \tag{16.81}$$

Gamma is a special case of GG and GB2 when $a = 1$. Its density is well known and can be expressed using Equation (16.71) by replacing a with 1.

The estimation of quantile regression models is straightforward using Bayesian method with MCMC and Gibbs sampling algorithms; see discussions in Chapter 7.

16.7 Factor Modeling for Industry Data

Most low-frequency/high-impact OpRisks have very limited datasets for calibration. As a result, the uncertainty in distribution parameters can be very large. For example, De Fontnouvelle *et al.* (2007) present results for several large international banks with a large variation in the shape parameter of the Generalized Pareto Distribution (GPD) for each bank such that possible capital may vary from about USD 200 million to USD 4300 million. To handle this issue Basel II requires the use of external data to complement internal datasets. There are several possible biases from the use of external data that should be handled, such as reporting bias (reporting threshold can be different for different banks); control bias (data are collected from banks with different control systems); and scale bias (data are collected from banks of different sizes). Here we consider the scaling bias issue.

There were several attempts in the past to find a relationship between OpRisk severity and bank size starting from Shih *et al.* (2000), who tested the relationship between bank size indicators of revenue, assets, and number of employees with OpRisk severities. Their study used the OpVar database and reported a strong nonlinear relationship with all variables (with largest correlation to the revenue). They tested a simple model

$$\ln X_i^{(j)} = \beta_0 + \beta_1 \ln y^{(j)} + \epsilon_i^j,$$

where $X_i^{(j)}$ is OpRisk losses in bank j, $y^{(j)}$, is revenue of bank j, and ϵ_i^j are zero mean i.i.d. random variables. The conclusion was that size accounts for only 5% of the variability in the severity. However, their study ignored reporting bias in the OpVar database, which is too large to be ignored as was shown by De Fontnouvelle *et al.* (2006). Another study by Na and *et al.* (2006) considered scaling of aggregate losses with respect to factors such as macroeconomic, geopolitical, cultural, business environment, and bank size. They found that mean and standard deviation of aggregate losses scale similarly but also reported that the results are not convincing. More recent papers by Dahen and Dionne (2010) considered regression

$$\ln X_i = \beta_0 + \beta_1 \ln y_i + \sum_j \alpha_j BL_{ij} + \sum_k \gamma_k ET_{ik} + \epsilon_i,$$

where y_i is total assets in a bank where the loss X_i occured, BL_{ij} is a business line indicator, ET_{ik} is an event-type indicator, and ϵ_i are i.i.d. Normal random variables (error term). The adjusted R^2 value of 29% was reported for overall fit to the data. The low value of R^2 might indicate that Normal distribution assumption for log losses is not appropriate; various previous studies have found evidence that operational losses are (extremely) heavy-tailed.

Most of the studies focused on scaling of the mean of the log losses with respect to the size of a bank. However, the mean of the log losses does not relate to any meaningful statistic of the (raw) loss distribution. Hence, results are difficult to interpret in most circumstances. Second, the OpRisk capital charge for a bank is determined by the 99.9th percentile of the loss distribution. The information provided by a scaling model fitted to the mean of the log losses provides very little information on how the 99th percentile may scale (see discussion by Ganegoda and Evans 2013). In general, the mean of a loss distribution provides very little information when losses are heavy-tailed. The second issue that Rootzén and Klüppelberg (1999) raised is that the mean may be quite unstable for losses that follow a heavy-tailed distribution. For example, for a widely used Pareto model for heavy tailed-losses $\mathbb{P}\mathrm{r}[X > x] = 1 - (x/x_0)^{-\xi}$, for $x > x_0$, the mean is infinite for $0 < \xi \leq 1$ and finite for $\xi > 1$. When the mean is finite but ξ is close to 1, a similar model can have very different means. To illustrate this, Ganegoda and Evans (2013) consider two banks where OpRisk follows a Pareto distribution with $x_0 = 1$ and $\xi = 1.01$ for bank I and $\xi = 1.001$ for bank II. Both distributions are very similar with the loss level at the 99.9th percentile being 933 for Bank I and 993 for Bank II (i.e., the capital amounts for these banks should be very similar). However, the mean operational loss at Bank I is 101, whereas the mean loss at Bank II is 1001, which is about 10 times larger. This motivates one to seek scaling properties not only for the scale but also for the shape parameter of the severity. Specifically, Ganegoda and Evans (2013) considered shape and scale parameters modeled via explanatory variables using the recently introduced "generalized additive models for location scale and shape" (GAMLSS) framework by Rigby and Stasinopoulos (2005). They considered several distributional assumptions for severity and found that a log-gamma distribution provides the best fit. Their results suggest that the tail index of the operational loss distribution and the size of a bank have a negative relationship. In other words, the tail of the operational loss distribution in large banks is heavier than in small banks.

GAMLSS is a very general class of regression model that incorporates popular GLM, Generalized Additive models (GAMs), Generalized Linear Mixed Models (GLMMs), and Generalized Additive Mixed Models (GAMMs) together. However, GAMLSS is more general than these models, since it relaxes the assumption that the response variable belongs to the natural exponential family. It is a convenient framework to test various distributional assumptions such as Gumbel, Weibull, and Student-t, in addition to the standard natural exponential family distributions. The second advantage of the GAMLSS framework is that it does not limit the modeling to the location of the distribution as in GLM and the other similar frameworks. The standard GLM setup (similar to ordinary least squares (OLS)) cannot model distributional parameters other than the location parameter explicitly.

In the GAMLSS framework, all distributional parameters can be explicitly modeled using both fixed and random effects (see Rigby and Stasinopoulos, 2005). Furthermore, each distributional parameter can be modeled as linear and/or nonlinear, parametric and/or smooth nonparametric functions of explanatory variables and/or random effects.

Consider a random sample of independent observations (e.g., losses or log losses) $\boldsymbol{y} = (y_1, y_2, \ldots, y_n)$. Let $f(y_i; \boldsymbol{\theta})$ be the density function conditional on the parameter vector $\boldsymbol{\theta}$. The parameter vector can have any number of distributional parameters $\boldsymbol{\theta} = (\theta_1, \ldots, \theta_K)$ the each of which can be modeled by explanatory variables. Then the modeler should define link functions $g_k(\cdot)$, which specifies the relationship between the linear predictor and the distributional parameters as

$$g_k(\theta_k) = \beta_1^{(k)} X_{i,1}^{(k)} + \beta_2^{(k)} X_{i,2}^{(k)} + \cdots + \beta_J^{(k)} X_{i,J}^{(k)}, \tag{16.82}$$

where $X_{i,j}^{(k)}$ is the value of the j-th explanatory variable relating to observation y_i in the k-th distributional parameter. In matrix notation, it can be written as $g(\theta_k) = \boldsymbol{X}^{(k)} \boldsymbol{\beta}^{(k)}$ with $\boldsymbol{X}^{(k)}$ the design matrix of the k-th parameter. Parameters $\boldsymbol{\beta}^{(k)}$ can be estimated using likelihood

$$\prod_{i=1}^{n} \ln f(y_i; \boldsymbol{\beta}^{(1)}, \ldots, \boldsymbol{\beta}^{(K)}) \tag{16.83}$$

using the maximum likelihood method or Bayesian MCMC approach. For example, Ganegoda and Evans (2013) carried out maximum likelihood parameter estimation for OpRisk losses in AlgoOpData using the software R package GAMLSS (see Stasinopoulos and Rigby 2007). They found a good fit for log-Gamma distribution when log losses are modeled by Gamma distribution $Gamma(\alpha, \gamma)$ with log link functions for shape α and scale γ parameters, that is, $g_k(\cdot) = \ln(\cdot)$ in (16.82).

16.8 Multifactor Modeling under EVT Approach

Under the EVT approach, the frequency N_t of loss exceedances X_i over large enough threshold u follows a Poisson process with intensity λ and loss exceedances X_i follow $GPD(\xi, \theta)$, whose distribution function is

$$G_{\xi,\theta}(x) = \begin{cases} 1 - (1 + \xi x/\theta)^{-1/\xi}, & \xi \neq 0, \\ 1 - \exp(-x/\theta), & \xi = 0, \end{cases} \tag{16.84}$$

where $x \geq 0$ when $\xi \geq 0$ and $0 \leq x \leq -\theta/\xi$ when $\xi < 0$.

In practice, the assumption of i.i.d. about the number of exceedances and severity is typically not appropriate. OpRisk losses might depend on covariates such as economic factors, business lines, event types, and time. Consider the observed vectors $z_i = (t_i, \boldsymbol{y}_i, x_i)$, $i = 1, \ldots, n$, where $0 \leq t_1 \leq \cdots \leq t_n \leq T$ are exceedance times, $\boldsymbol{y}_i$ is observed vector of covariates (explanatory variables) at time t_i, and x_i is observed loss exceedance over the threshold u. In the study of OpRisk Willis dataset by Chavez-Demoulin *et al.* (2013), the model parameters λ,ξ, and θ are considered to be functions of time and covariates as follows.

The Poisson intensity λ is modeled as

$$\lambda(\boldsymbol{y}, t) = \exp(\boldsymbol{y}^T \boldsymbol{\beta}_\lambda + h_\lambda(t)), \tag{16.85}$$

where $\boldsymbol{\beta}_\lambda$ is a vector of parameters, $h_\lambda(t)$ is a general measurable function of time that does not depend on specific parameters.

It is assumed that $\xi > 0$ and GPD parameters (ξ, θ) are replaced with (ξ, ν), where $\nu = \ln((1 + \xi)\theta)$ to make parameters orthogonal with respect to the Fisher information metric (which is important for convergence of the fitting procedure) that are modelled as

$$\xi(\boldsymbol{y}, t) = \boldsymbol{y}^T \boldsymbol{\beta}_\xi + h_\xi(t), \tag{16.86}$$

$$\nu(\boldsymbol{y}, t) = \boldsymbol{y}^T \boldsymbol{\beta}_\nu + h_\nu(t), \tag{16.87}$$

where $h_\xi(t)$ and $h_\nu(t)$ are general measurable functions of time that do not depend on specific parameters. After estimating $\xi(y, t)$ and $\nu(y, t)$, one can estimate θ using

$$\theta(\boldsymbol{y}, t) = \frac{\exp(\nu(\boldsymbol{y}, t))}{1 + \xi(\boldsymbol{y}, t)}.$$

The likelihood function of loss exceedances can be written as

$$L_{\boldsymbol{X}}(\xi, \theta) = \prod_{i=1}^{n} \frac{1}{\beta} \left(1 + \xi \frac{x_i}{\theta}\right)^{-(1+1/\xi)} \tag{16.88}$$

and reparameterized likelihood is

$$\widetilde{L}_X(\xi,\nu) = L_X\left(\xi, \frac{\exp(\nu)}{1+\xi}\right).$$

The model for Poisson process (16.85) is a standard GAM that can be estimated using available functions, for example, in *R*-software package. However, to estimate all severity parameters (β_ξ, β_ν) and functions $(h_\xi(t), h_\nu(t))$, Chavez-Demoulin *et al.* (2013) had to introduce a special backfitting algorithm because the severity factor model (16.84) with (16.86,16.87) does not lie directly within any standard GAM. The described methodology was applied by Chavez-Demoulin *et al.* (2013) to a dataset of OpRisk losses provided by Willis; the considered covariates were business lines and event types. To fit smooth functions $h_\xi(t), h_\nu(t)$, the penalized MLE method was utilized. Details of the fitting procedure including links to detailed R implementation are provided by Chavez-Demoulin *et al.* (2013).

CHAPTER SEVENTEEN

Insurance and Risk Transfer: Products and Modeling

In this chapter, we address the following components of OpRisk insurance modeling:

1. What is the incentive and motivation for undertaking risk transfer?
2. What types of insurance products can one consider in the OpRisk setting and how might they apply to the LDA model structure?
3. What types of models can one develop for understanding analytically the impact of insurance in OpRisk settings?
4. What impact does each of the different insurance policies have on capital estimation (risk measures)?

In discussing these questions, we provide a detailed introduction to the basic properties of insurance products, structuring, and modeling. We detail what types of losses are insurable in a classical sense and then set the tone for the more advanced aspects in Chapter 18, which deals with catastrophe bonds and bespoke insurance products. We discuss the important aspects of insurance product structuring such as moral hazard and how this may influence OpRisk modeling and applications. After these basic definitions, the majority of this chapter deals with aspects of single peril insurance product structures with deterministic and stochastic features, followed by a detailed analysis of how one may construct quantitative LDA models incorporating such insurance properties.

17.1 Motivation for Insurance and Risk Transfer in OpRisk

The notion of risk transfer and insurance has been studied for a long period, and in particular there is a well-established economic model for the transfer of such risk; see an overview in Gollier (2005). Under the standard economic framework for risk transfer, it is argued that competition in insurance markets will result in a Pareto-efficient allocation of risk in the economy,

Fundamental Aspects of Operational Risk and Insurance Analytics: A Handbook of Operational Risk, First Edition. Marcelo G. Cruz, Gareth W. Peters, and Pavel V. Shevchenko.

where the notion of Pareto optimality is provided in Definition 17.1. In addition, this economic reasoning also concludes that all diversifiable risks in the economy are removed through the mechanism of mutual risk-sharing arrangements that involves pooling and diversification of risk in a deep insurance/reinsurance market. It also involves the conclusion that all risks are insurable since any residual systematic risk will be attributed to agents with a competitive advantage with regard to risk management, such as insurers. In practice, this is not the case since there are still significant diversifiable risks that are borne by individuals that Gollier (2005) attributes to an inefficient risk-sharing ex ante.

Definition 17.1 (Pareto Optimality) *Pareto efficiency or Pareto optimality refers to an allocation of assets or resources in which it is impossible to make any one better off without making at least one individual worse off.* ■

As detailed in Doherty (1997b), the 1980s and 1990s witnessed significant changes in the insurance landscape, especially in the property-liability insurance market that they listed as

1. Withdrawal of commercial business into alternative risk management vehicles and strategies;
2. Crises and coverage changes in liability insurance;
3. Integration of insurer asset and liability management;
4. Emergence of innovative reinsurance instruments such as financial reinsurance;
5. Experiments with radical regulation;
6. Corporate reorganization and reassembly, such as merger activity among brokers leading to increased concentration;
7. The securitization of catastrophe risk through instruments such as exchange traded catastrophe options and future as well as OTC catastrophe bonds.

With regard to the establishment of the catastrophe insurance linked securities (ILS), it was argued by Babbel and Santomero (1996) and Santomero and Babbel (1997) that this was by no means coincidental. In this section, we are interested in how such growth in the ILS markets can be beneficially utilized in the OpRisk context. Modeling the impact of insurance mitigation for different risk cells and business units is an important challenge in the setting of OpRisk (OpRisk) management yet to be fully understood and therefore adopted in practice. The slow uptake of insurance policies in OpRisk for capital mitigation can be partially attributed to the limited understanding of their impact in complex multi-risk, multiperiod scenarios, under heavy-tailed losses and the fair premium to charge for such policies, as well as a relatively conservative Basel II regulatory cap of 20% for capital reduction due to insurance. Therefore, although OpRisk models are maturing, OpRisk insurance mitigation is still in its infancy (Bazzarello *et al.* 2006, Brandts 2004, Peters *et al.* 2011a).

The Basel II OpRisk regulatory requirements for the advanced measurement approach (BCBS, 2006, p. 148) state that "Under the AMA, a bank will be allowed to recognise the risk mitigating impact of insurance in the measures of OpRisk used for regulatory minimum capital requirements. The recognition of insurance mitigation will be limited to 20% of the total OpRisk capital charge calculated under the AMA". Therefore, from the perspective of a financial institution, such as a bank, there is a strong incentive to understand the effect of insurance mitigation on the OpRisk capital.

From the insurer's perspective, a quantitative understanding of the impact of insurance in OpRisk extreme loss scenarios will allow for accurate pricing of insurance premiums and an understanding of what aspects of OpRisk loss processes are actually insurable. In addition by studying the risk transfer from bank to insurer, this will aid in modeling of the required capital for an insurer under Solvency 2. As discussed in the initiatives developed by the International Association of Insurance Supervisors (Kawai 2005, Linder and Ronkainen 2004), the Solvency 2 framework was developed as a similar system to the Basel II three-pillar system. It specifies the financial resources that a company must hold to be considered solvent. In Sandström (2006), the IAIS guidance under Principle 8 discusses minimum capital in the following non prescriptive manner: "A minimum level of capital has to be specified", it is therefore a quantitative challenge to decide how to model such capital. In the second phase of the EU project Solvency 2, the commission introduced two distinct levels of solvency: these are measured according to an upper level, the solvency capital requirement (SCR), and lower level, the minimum capital requirement (MCR); see Sandström (2006). For more discussion on the relationship between the OpRisk banking sector claims process as viewed by an insurer and the insurance mitigation as viewed by a bank or financial institution, see the study in Peters *et al.* (2011a).

Thinking purely of the OpRisk and Basel II/Basel III context, we note that the Basel III policy for application of insurance mitigation requires that a bank must have a detailed framework for recognizing insurance. In addition, this framework must be made available to the regulator to assess whenever insurance mitigation is applied as stated in Pillar III of the (BCBS, 2006, p. 155): "... risk mitigation calculations must reflect the bank's insurance coverage in a manner that is transparent in its relationship to, and consistent with, the actual likelihood and impact of loss used in the bank's overall determination of its OpRisk capital".

Finally, as noted in Chernobai *et al.* (2007, p. 52), for a bank to consider applying insurance mitigation to their risk processes and ultimately to their capital measure, they must already be of a suitable credit worthiness as measured by the following requirements:

- The insurance company (whether it be self-insurance from an insurance branch of the bank or external insurance) must hold a credit rating that is at least A-grade;
- The coverage from the insurance product must be in alignment with the LDA models and resulting actual likelihood of loss events that are utilized by the bank in calculation of the OpRisk capital.

To proceed with the understanding of the role of insurance and risk transfer in OpRisk, we note that throughout this chapter we will work with application of insurance under the context of an LDA model structure, involving an annual loss in a risk cell (business line/event type) modeled as a compound random variable,

$$Z_t^{(j)} = \sum_{s=1}^{N_t^{(j)}} X_s^{(j)}(t) . \tag{17.1}$$

Here, $t = 1, 2, \ldots, T, T+1$ in our framework is discrete time (in annual units) with $T+1$ corresponding to the next year. The superscript j is used to identify the risk cell. The annual number of events $N_t^{(j)}$ is a random variable distributed according to a frequency counting distribution $P^{(j)}(\cdot)$, typically Poisson. The severities in year t are represented by random variables $X_s^{(j)}(t)$,

$s \geq 1$, distributed according to a severity distribution $F^{(j)}(\cdot)$. Severities represent actual loss amounts per event. The total bank's loss in year t is calculated as

$$Z_t = \sum_{j=1}^{J} Z_t^{(j)}, \tag{17.2}$$

where formally for OpRisk under the Basel II requirements $J = 56$ (seven event types times eight business lines). However, this may differ depending on the financial institution and type of problem. In general throughout this section, we will drop the upper risk cell (business line/event type) index unless explicitly required such as in multiple loss insurance products.

17.2 Fundamentals of Insurance Product Structures for OpRisk

It is important to realize that the need for insurance in OpRisk settings becomes clearer when one considers the different types of risk processes present in a large banking institution. In a general classification, one can consider the risk exposures to be classified into two sources of risk: those that are **controllable** and can be managed through improvement to weaknesses or lack of compliance in internal controls, and those that are external and **noncontrollable**. It is precisely in the context of noncontrollable risks that risk transfer can be a critical tool to help manage exposures.

We start with a very basic definition of what insurance is and some basic components of insurance.

Definition 17.2 (Insurance Policy) *At a fundamental level, one can consider insurance to be the fair transfer of risk associated with a loss process between two financial entities. The transfer of risk is formalized in a legal insurance contract that is facilitated by the financial entity taking out the insurance mitigation making a payment to the insurer offering the reduction in risk exposure. The contract or insurance policy legally sets out the terms of the coverage with regard to the conditions and circumstances under which the insured will be financially compensated in the event of a loss. As a consequence, the insurance contract policy holder assumes a guaranteed and often known proportionally small loss in the form of a premium payment corresponding to the cost of the contract in return for the legal requirement for the insurer to indemnify the policy holder in the event of a loss.* ■

Under this definition, one can then interpret the notion of insurance as a risk management process in which a financial institution may hedge against potential losses from a given risk process or group of risk processes. Mehr *et al.* (1980) and Berliner (1982) discuss at a high level the fundamental characteristics of what it means to be an insurable loss or risk processes, which we review in Definition 17.3. Of course, this is intended to be a simple guide and is not to be considered definitive; in addition, it provides the standard actuarial view on insurability. This is not directly equivalent to the economists view as we note subsequently.

Definition 17.3 (Insurable Losses) *Mehr et al. (1980) and Chernobai et al. (2007, chapter 3) define insurable risks as those that should have the following common characteristics:*

1. *The risks must satisfy the "Law of Large Numbers". In other words, there should be a large number of similar exposures;*
2. *The loss must take place at a known recorded time, place, and from a reportable cause, known as a definite loss process;*
3. *The loss process must be considered subject to randomness. That is, the events that result in the generation of a claim should be random or at a minimum outside the control of the policy holder;*
4. *The loss amounts generated by a particular risk process must be commensurate with the charge premium, and associated insurer business costs such as claim analysis, contract issuance and processing. Therefore, risk processes that are to be insured must be sufficiently large;*
5. *The estimated premium associated with a loss process must be affordable. In other words, the likelihood of an insured event causing a catastrophic loss so large that to insure it would result in a premium charge so large that no one would purchase such a contract must be considered. This is particularly important in high-consequence rare-event settings; see discussions in Peters et al. (2011a), who consider this question in a general setting;*
6. *The probability of a loss should be able to be estimated for a given risk process. In addition, one should be able to estimate some statistic characterizing the typical, average, median, etc., loss amount;*
7. *The final requirement typically is that either the risk process has a very limited chance of a catastrophic loss that would bankrupt the insurer and in addition the events that occur to create a loss occur in a nonclustered fashion or the insurer will cap the total exposure.* ■

Gollier (2005) argues that there is also a need to consider the economic ramifications for insurable risks. In particular, the adds to this definition of insurable risks the need to consider the economic market for such risk transfers. In particular, Gollier (2005) discusses uninsurable and partially insurable losses, where an uninsurable loss occurs when "'..., *given the economic environment, no mutually advantageous risk transfer can be exploited by the consumer and the supplier of insurance*". Gollier (2005) defines a partially uninsurable loss as one that arises when the two parties to the risk transfer exchange can only partially benefit or exploit the mutually advantageous components of the risk transfer; this has been studied in numerous studies, such as Aase (1993), Arrow (1964), Arrow (1965), Borch (1962), and Raviv (1979).

As noted in Gollier (2005), from the economist's perspective, the basic model for risk transfer involves a competitive insurance market in which the Law of Large Numbers is utilized as part of the evaluation of the social surplus of the transfer of risk. However, unlike the actuarial view presented earlier, the maximum potential loss and the probabilities associated with this loss are not directly influential when it comes to assessing the size of risk transfers at market equilibrium. In addition, the economic model adds factors related to the degree of risk aversion of market participants "agents" and their degree of optimism when assessing the insurability of risks in the economy. Classically these features are all captured by the economic model known as the Arrow–Borch–Raviv model of perfect competition in insurance markets; see a good review in Gollier (2005, section 2) and Ghossoub (2012).

Under the standard Arrow–Borch–Raviv model, the demand for insurance contracts involves an optimal contract for a buyer taking the form of a deductible contract. This result is obtained under the assumption (among others) that the insurer is a risk-neutral expected utility maximizer, and the buyer is a risk-averse expected utility maximizer. It is also assumed that both parties to the risk transfer share the same probabilistic beliefs about the realizations of the underlying insurable loss. In particular, the distribution function for losses is common knowledge and can depend on several factors such as prevention efforts by the agents; in addition, the efforts associated with such prevention are assumed costless. Clearly in practice, there will be an asymmetry in the information available to the sponsors of the insurance compared to the investors depending on the type of peril covered and the sophistication of the investor. It is interesting to note that in the context of catastrophe bonds, some aspects of the information asymmetry traditionally present are starting to dissipate with the emergence of improved statistical modeling in some perils that are made available by specialized consulting firms for a fee.

This feature of information asymmetry was noted in Ghossoub (2012), where he argues that there is a heterogeneity of beliefs in the classical insurance model, and consequently considers instead a setting where the investor and the insurer have preferences yielding different subjective beliefs. The investor then demands an insurance contract that will maximize the (subjective) expected utility of terminal wealth with respect to an associated subjective probability measure. Alternatively, the insurer sets premiums on the basis of their subjective probability measure. In this setting, they were able to show that under the condition they term vigilance (a consistency requirement on the insurer's subjective probability), then there exists an event for which the investor assigns their subjective probability measure completely and the consequent optimal insurance contract is a "generalized deductible contract".

In addition, it is noted in Lakdawalla and Zanjani (2012) the "puzzling" nature of the incompleteness of the catastrophe risk transfer. They claim that "*the price of risk transfer seems high, risk is not spread evenly among insurers ... and reinsurance consumers do not purchase coverage for high layers of risk*" which is directly contradictory to the findings from the framework of the Arrow–Borch–Raviv model. This is also documented in Froot (2001), and several plausible explanations are offered relating to inefficiencies in the insurance market as causes. For instance, in Froot (2001), it is argued that catastrophe bonds (a type of ILS) serve a well-defined economic role in risk transfer markets, which is based on their full collateralization. This allows them to avoid exposure to counterparty defaults, and Froot (2001) further notes that as frictional costs in catastrophe bonds such as issuance cost, secondary market illiquidity decrease relative to more traditional reinsurance, then there will be an increase in the utilization of catastrophe bonds for risk transfer. This has already started to occur and can be beneficial to OpRisk.

Returning to the notion of uninsurable risks, whether defined from an actuarial or an economic perspective, we note that in OpRisk several risks are not directly insurable and this can arise for many reasons. In cases in which a loss process is not directly insurable due to no existing product structure for the type of peril considered, or it may be partially insurable and then an insurer can still develop products for OpRisk through use of deductibles, exclusions, conditions, attachment points, and total cover limits as defined below.

Definition 17.4 (Insurance Policy Deductibles) *In an insurance policy, the deductible is the amount of expenses that must be paid out of pocket before an insurer will pay any expenses.* ■

Definition 17.5 (Insurance Policy Exclusions) *An exclusion in an insurance policy is a contractually legal exception in which an insurer will not cover a loss. In other words, exclusions remove a portion of coverage from the insurance contract and they are typically stated through descriptions of property, perils, hazards, or losses that may arise in the loss process being insured under specific causes* *that are deemed to be the aspects of the loss process that result in the loss process being noninsurable and are therefore not covered by the policy.* ■

Definition 17.6 (Insurance Policy Conditions) *In an insurance policy, the conditions list a set of rules of conduct and provisions as well as the duties and obligations required by the policy holder in order for them to make a claim and therefore obtain coverage under the policy.* ■

Definition 17.7 (Attachment Point) *In an insurance policy, an attachment point corresponds to the value at which excess insurance or reinsurance limits apply. If, for example, a captive's retention is USD x, then this is the "attachment point" at which excess reinsurance limits would apply. In other words, it corresponds to the amount of money an insurer pays until the point at which supplemental insurance (perhaps from another provider) begins to provide coverage.* ■

Definition 17.8 (Insurance Policy Total Cover Limits) *In an insurance policy, the insurer may specify a total liability in the form of a total coverage limit that can be claimed by the policy holder in the event of a catastrophic loss.* ■

From a general perspective, one can consider the role of insurance within the three pillars of Basel III OpRisk modeling as stylized by the representation in Figure 17.1.

Now we would like to make some comments on the role that insurance mitigation can play in OpRisk and its context, while highlighting the need for risk managers to be vigilant of an apparent side effect of loss coverage known as moral hazard. Here, we treat this issue at a high level before discussing in some more detail in the section on catastrophe bonds and reinsurance.

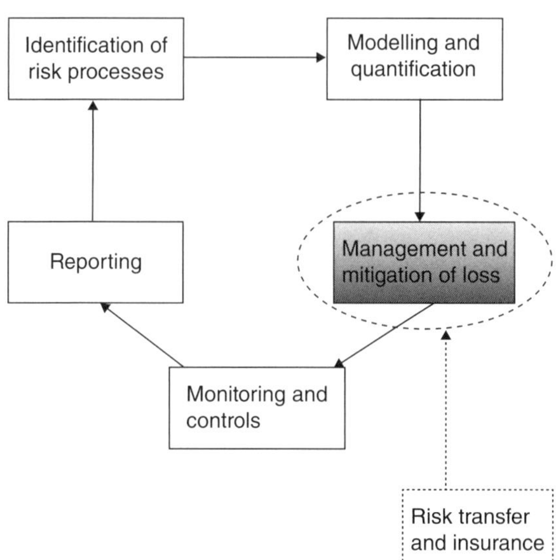

FIGURE 17.1 A picture depictsing a highly stylized view of the risk management process from Pilar I to Pilar III under the Basel II/Basel III accord. It is intended to be a continuous review cycle in which components of each stage are reinvestigated and revised over time under changing banking environments, and importantly for this chapter it demonstrates where we perceive the insurance products playing a role

When consideration is made as to whether the concept of insurance for OpRisk loss processes will further stabilize the banking environment with respect to OpRisk loss exposures, it is important to understand the component of moral hazard present as defined generically in the OpRisk context in Definition 17.9.

Definition 17.9 (Moral Hazard in OpRisk Arising from Insurance) *Traditionally, a moral hazard describes a situation in which a bank becomes less risk averse under certain conditions that transfer the cost borne by its risk taking to an outside party. Put simply it is a tendency to be more willing to take a higher level of risk than usual under the knowledge that the potential costs of taking such risk will be borne, in whole or in part, by others.* ■

This could conceivably arise in the setting of losses covered by insurance in OpRisk settings. However, this would of course have the opposite of the intended effect of the insurance policy of stabilizing the business exposure and banking sector. In addition, this transfer of risk would be counterproductive from the perspective of the Pilar III on risk reporting and refinement of risk management practices. There can be two possible manifestations of moral hazard in the OpRisk context, the *exante* moral hazard arising from increase risk taking due to transfer of risk and the *exposte* moral hazard resulting from biased reporting that could arise due to the complications involved with modeling and reporting the loss mitigations associated to insurance deductions in OpRisk loss processes. It is therefore important for risk managers and regulators to instill controls in the business process to reduce the possibilities of occurrence of such moral hazards.

17.3 Single Peril Policy Products for OpRisk

In this section, we discuss different options that banks and financial institutions covered under Basel II/Basel III have at their disposal to consider when evaluating the cost–benefit analysis of different risk transfer decisions.

Traditionally there are many insurance products that could be utilized for coverage of aspects of OpRisk loss processes. Therefore, in general in this chapter, we will not be specific with particular types of insurance products that can be applied for particular risk processes in OpRisk. We simply note that in general the following forms of insurance product can be considered as the components that would go into the construction of the generic insurance products we describe later, which include the following nonexhaustive list (some of these groups will overlap). We begin with a discussion on peril-specific products that are available for specific categories of risk.

1. **Casualty insurance**. This class of products mainly involve liability coverage of an organization for negligent acts or omissions. In BCBS (2003), it is stated that to apply casualty insurance the bank must be insurable and it refers to items 1, 2, 3 and 4 in Definition 17.3 of an insurable loss process;
2. **Property insurance**. This class of products mainly involves coverage for the wide class of property related items, which are usually classified by either "real property" or "personal property", for items such as buildings; contents of buildings; money and securities; motor vehicles and trailers; property in transit; ships and their cargo; and boilers and machinery. The class of insurance products available for each of these aspects of property can include

boiler insurance, builder risk insurance, earthquake insurance, fidelity bond insurance, flood insurance, land lord insurance, terrorism insurance, and windstorm insurance;

3. **Liability insurance**. This class of products can include coverage of aspects such as public liability, directors and officers liability, environmental liability, errors and omissions, and professional liability insurance;
4. **Umbrella liability insurance**. Even if a bank holds a policy for General Liability Insurance, it may face a claim, settlement, or judgment that exceeds the cover limit. In such cases, an Umbrella Liability Insurance contract will cover the uncovered expenses from the original liability insurance product;
5. Other specific categories such as **business interruption insurance, collateral protection insurance, legal expense insurance**, and **pollution insurance**.

In general, there is no direct mapping between different risk processes under a Basel II/Basel III risk and business unit framework (such as the 56 risk cells in the Basel accord). However, there will be combinations of different policies that will be combined to mitigate a risk process. It is precisely this combination of coverages that we will consider on aggregate to develop theory for the modeling of OpRisk insurance products.

When thinking of specific types of insurance products that are directly applicable to OpRisk loss processes, it is argued in Chernobai *et al.* (2007), Scott and Jackson (2002), and Lewis and Lantsman (2005) in their studies on transfer of risk for rogue trading in OpRisk that the OpRisk loss process classes are covered by the insurance products as shown in Table 17.1.

Next, we present some generic insurance product structures and the role they play on the LDA loss process structure.

TABLE 17.1 **OpRisk processes and corresponding insurance products**

OpRisk risk process	Insurance product
Fraudulent and dishonest acts committed by employees—e.g., rogue trading	Fidelity Bonds, which is a form of insurance (not a bond), is also known as: in Australia—employee dishonesty insurance coverage; and in UK–fidelity guarantee insurance coverage
Natural disasters, fire, and theft	Property insurance
Failed IT infrastructure protections to prevent malicious and accidental IT crime	Electronic and computer crimes insurance
Losses from fraudulent activity by directors and executives that may result, for example, from alleged errors in judgment, breaches of duty, or wrongful acts.	Director's and officers' liability coverage
Coverage for financial losses that include bankers' professional indemnity, electronic & computer crime, unauthorized trading, and credit card exposures	Financial Institutions Crime Coverage (Swiss Re)
Losses arising from liabilities to third parties for claims arising from employee negligence when providing professional services such as investment advice to clients	Personal indemnity insurance
Unauthorized financial transaction and trading activities	Unauthorized trading insurance

17.4 Generic Insurance Product Structures for OpRisk

In a financial institution, typically there are numerous insurance policies against different risk types, some affecting single risks and some affecting several risk cells at once. These insurance policies can be considered per risk type and per business unit; however, in OpRisk, all policies must satisfy the Basel II regulatory requirements. In the following examples, we present several possible generic insurance product structures and we detail how these may effect the OpRisk LDA process. Later, we will then explain how one may construct a portfolio of available insurance products to replicate these simple mitigation policies.

Therefore, in this section, we present different versions of insurance model specified by top cover limits (TCL) under multiple risk modeling scenarios. These are selected to be fundamental insurance models that provide information about the building blocks for more advanced policy structures. Building on these we also consider several advanced insurance policy structures, the first involving a Basel II haircut with a linearly increasing TCL over the duration of the year (Bazzarello *et al.*, 2006). The second policy involves a stochastic banding loss structure for the TCL, which can be considered an extension of the model proposed in Bazzarello *et al.* (2006), to a stochastic insurance structure. The third is a proportional deterministic or stochastic annual policy. In particular, we will be able to show that under such policy types, several interesting closed-form LDA models can be adopted that also incorporate the insurance policy and, as a consequence, one can also solve for optimal purchase strategies for such products in finite and infinite time horizons. A subset of the insurance policies we discuss in this chapter may be found in detail in Peters *et al.* (2011a).

The following list presents a brief summary of the policies (and acronyms for these policy structures) presented in detail later:

- Individual Loss Policy Uncapped (ILPU);
- Individual Loss Policy Capped (number of events) (ILPCn);
- Individual Loss Policy Capped (maximum total accumulated compensation) (ILPCa);
- Accumulated Loss Policy (ALP);
- Combined Loss Policy (two variants) (CLP1 and CLP2);
- Accumulated Loss Policy (with d risk exposures) (ALPd);
- Proportional Individual Loss Policy (PILP);
- Haircut Individual Loss Policy with Top Cover Limit (HILP-TCL);
- Haircut Individual Loss Policy (HILP);
- Stochastic Banding Policy (BILP).

17.4.1 GENERIC DETERMINISTIC POLICY STRUCTURES

Here, we consider the basic deterministic policy structures that can form the building blocks of more complex policy structures and admit interpretable results and intuition for the impact of such policies on the mitigation of the LDA loss process and the effect on capital risk measures.

We begin with individual loss processes, followed by a policy applicable to combined risk processes, which may be used in practice to exploit knowledge of dependence properties of particular risk processes in Basel II. For studies of the impact of dependence models in OpRisk, see Peters *et al.* (2009). In addition, we assume without loss of generality a simple setting in which deductible excess is zero. We start by defining the per event policy that we term the ILP given in Definition 17.10.

Definition 17.10 (Individual Loss Policy Uncapped (ILPU)) *The ILPU provides a specified maximum compensation, a TCL on a per event basis for a given year at a cost of purchase of C, though there is no upper limit on the number of events the TCL is applied too. Therefore, one can write the ILPU risk mitigated loss LDA model from the perspective of the banking institution according to the loss process*

$$Z_t^{(ILPU)} = C + \sum_{s=1}^{N_t} \max \{X_s(t) - TCL, 0\} . \tag{17.3}$$

■

To understand this insurance product, we consider the very simple illustration given in Example 17.1.

EXAMPLE 17.1 Application of the ILPU to a Single Risk LDA Model

Consider the ILPU policy for a single risk process in which we consider a year consisting of five OpRisk losses of {6, 10, 8, 2, 5} in USD million. For each loss, the insurer will provide compensation on the loss up to the value TCL, as illustrated in Figure 17.2, where the solid lines represent the ILP policies TCL for this risk process.

As can be seen in losses 2 and 3, the value of the loss exceeds the TCL of USD 7 million, hence the insurer provides compensation of USD 7 million (highlighted in gray) and the bank still incurs the loss above this value (highlighted in black).

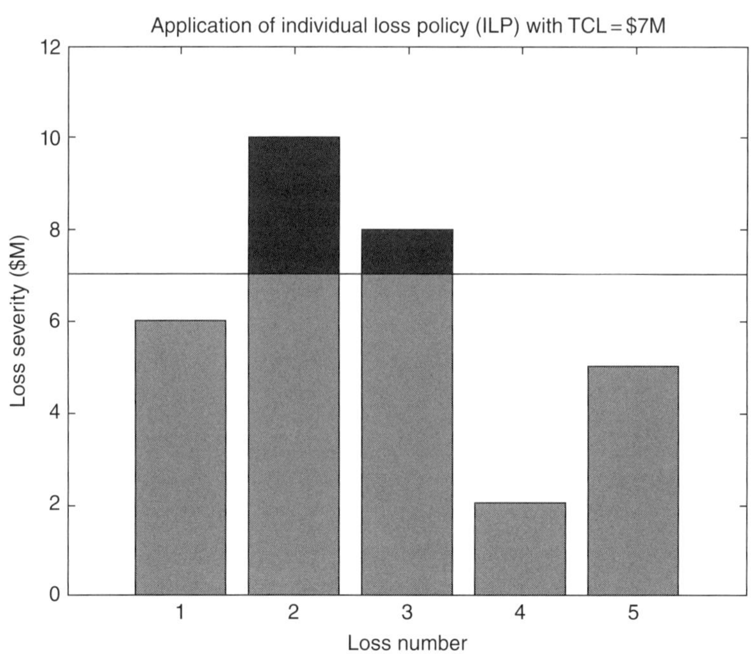

FIGURE 17.2 Individual loss policy and the application of the top cover limit

■

A variant of this policy involves the ILP policy in which a cap is imposed on the total claims that can be covered by the policy, depending on the context of the claim process this may be on a number of different factors such as total number of events covered or total amount of claims covered. These two variants are detailed later and denoted by ILPCn, and ILPCa respectively.

Definition 17.11 (Individual Loss Policy Capped (ILPCn)) *The ILPCn provides a specified maximum compensation, a TCL on a per event basis for a given year at a cost of purchase of C, and there is an upper limit on the number of events that the TCL is applied to that is denoted by n_T. Therefore, one can write the resulting ILPCn risk mitigated loss LDA model from the perspective of the banking institution according to the loss process*

$$Z_t^{(ILPCn)} = C + \sum_{s=1}^{N_t} \max\{X_s(t) - TCL, 0\} \mathbb{I}[s \leq n_T] + \sum_{s=1}^{N_t} X_s(t) \mathbb{I}[s > n_T], \tag{17.4}$$

where $\mathbb{I}[\cdot]$ is the indicator function. ■

Definition 17.12 (Individual Loss Policy Capped (ILPCa)) *The ILPCa provides a specified maximum compensation, a TCL on a per event basis for a given year at a cost of purchase of C, and there is an upper limit on the total accumulated loss (AL) coverage applied denoted by AL, such that the loss event that takes the accumulated claim above threshold coverage AL is also covered up to TCL. Therefore, one can write the ILPCa risk mitigated loss LDA model from the perspective of the banking institution according to the loss process*

$$\begin{aligned} Z_t^{(ILPCa)} &= C + \sum_{s=1}^{N_t} \max\{X_s(t) - TCL, 0\} \mathbb{I}\left[\sum_{k=1}^{s} \max\{X_k(t) - TCL, 0\} \leq AL\right] \\ &\quad + \sum_{s=1}^{N_t} X_s(t) \mathbb{I}\left[\sum_{k=1}^{s} \max\{X_k(t) - TCL, 0\} > AL\right], \end{aligned} \tag{17.5}$$

where $\mathbb{I}[\cdot]$ is the indicator function. ■

A second type of generic policy one may adopt is the ALP given in Definition 17.13.

Definition 17.13 (Accumulated Loss Policy (ALP)) *The ALP provides a specified maximum compensation on losses experienced over a one-year insurance period at a cost of purchase of C. The resulting risk mitigated loss can be expressed according to Equation (17.6). This formulation can be simplified, though this representation is particularly useful when application of the policy on an event by event basis that may be required in simulation studies and also if ACL is time dependent.*

$$\begin{aligned} Z_t^{(ALP)} &= C + \sum_{s=1}^{N_t} X_s(t) \times \mathbb{I}\left(\sum_{k=1}^{s-1} X_k(t) \geq ACL\right) \\ &\quad + \left(\left(\sum_{k=1}^{s} X_k(t)\right) - ACL\right) \times \mathbb{I}\left(0 < ACL - \sum_{k=1}^{s-1} X_k(t) < X_s(t)\right), \end{aligned} \tag{17.6}$$

with $\mathbb{I}(\cdot)$ denoting the indicator function. ■

To understand this insurance product, we consider the very simple illustration given in Example 17.2.

EXAMPLE 17.2 Application of the ALP to a Single Risk LDA Model

As for the ILP model, for comparison purposes we also illustrate the application of the ALP policy with the annual loss example presented in the ILP Definition 17.10 and ACL = USD 25 million. In this case, the insurer will provide complete compensation of all losses over the year until the value of compensation reaches the limit ACL, as depicted in Figure 17.3. In this setting, the insurer compensates the bank for losses 1 to 3. However, the fourth loss brings the total claim value to USD 26 million, which will exceed the ACL cap of USD 25 million, hence the insurer only compensates the bank for the first USD 1 million of the fourth loss and the bank is exposed to the entirety of the fifth loss.

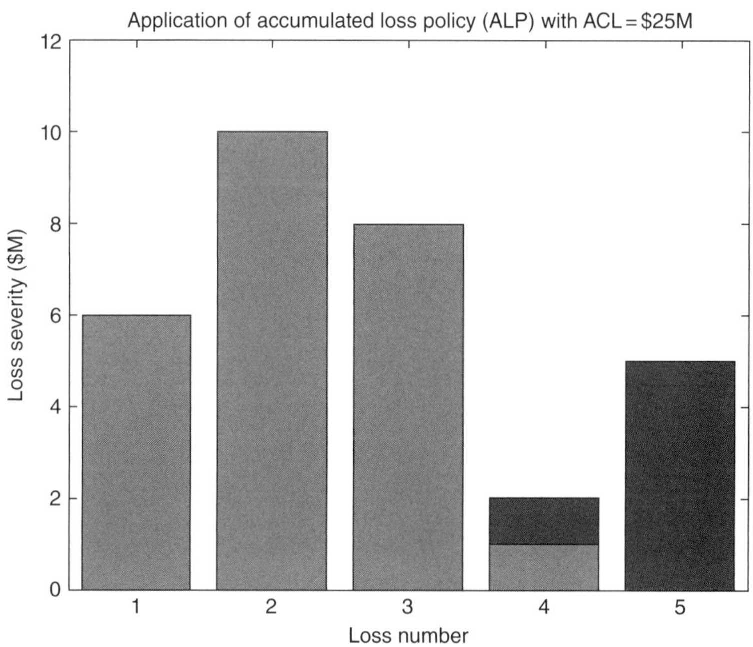

FIGURE 17.3 Accumulated loss policy and the application of the accumulated cover limit

■

As a third generic policy type, one may adopt the CLP1 or CLP2 policies that are simple variants of the ILPCa approach where the coverage of the loss event that results in an exceedance above an accumulation threshold is only partially covered. This case is presented in Definitions 17.14 and 17.15.

Definition 17.14 (Combined Loss Policy (CLP1)) *A CLP1 insurance contract provides a specified maximum compensation, TCL, on a per event basis up to a maximum per year loss, ACL at a cost of purchase of C. The resulting risk mitigated loss process can be expressed as*

$$Z_t^{(CLP1)} = C + \sum_{s=1}^{N_t} \left[X_s(t) \times \mathbb{I}\left(\sum_{k=1}^{s-1} \min(X_k(t), TCL) \geq ACL \right) \right.$$
$$+ \max(X_s(t) - TCL, 0) \times \mathbb{I}\left(\sum_{k=1}^{s} \min(X_k(t), TCL) \leq ACL \right)$$
$$+ \left(X_s(t) - \left(ACL - \sum_{k=1}^{s-1} \min(X_k(t), TCL) \right) \right)$$
$$\left. \times \mathbb{I}\left(ACL - \sum_{k=1}^{s-1} \min(X_k(t), TCL) < \min(X_s(t), TCL) \right) \right]. \quad (17.7)$$

■

To understand this insurance product, we consider a very simple illustration given in Example 17.3.

EXAMPLE 17.3 Application of the CLP1 to a Single Risk LDA Model

Considering again the example in Section 17.10, we illustrate the application of such a policy in Figure 17.4. Under the CLP1, the insurer will provide compensation on the loss up to the value TCL. However, once the total value of claims exceeds the aggregate limit ACL, the insurer will not provide any further compensation.

As can be seen, the insurer only compensates the bank for the first USD 7 million of losses 2 and 3. In addition, the total value of claims is exceeded by the fifth loss. Hence, the insurer will only compensate the first USD 3 million of the fifth loss and the remaining exposure is incurred fully by the bank.

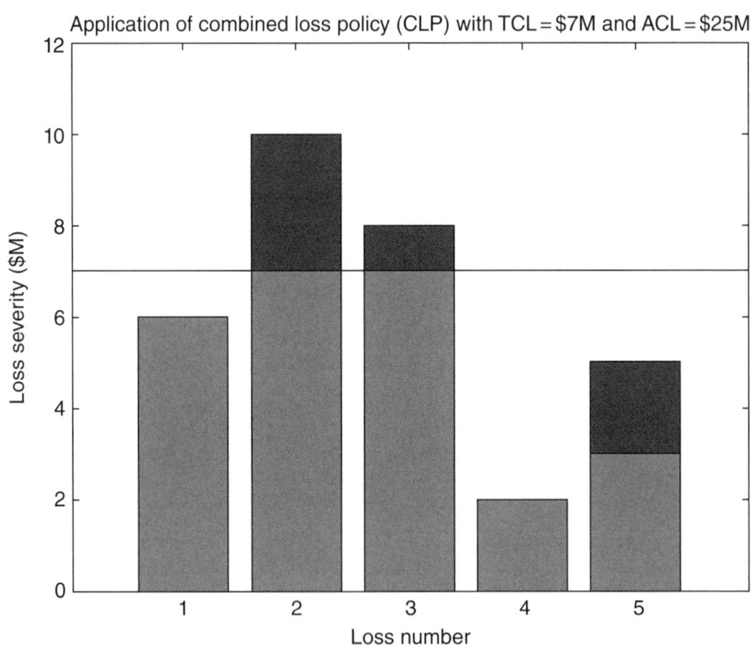

FIGURE 17.4 Combined loss policy and the application of the ILP and ACL

Under the generic policy type specified by CLP2 in Definition 17.15, the consideration of coverage is reversed compared to the CLP1 policy. The structure of policy CLP2, though it is presented generically later, it has strong similarity with the structure of certain types of catastrophe insurance products, as will be discussed later.

Definition 17.15 (Combined Loss Policy (CLP2)) *A CLP2 insurance contract provides a specified maximum compensation, TCL, on a per event basis that only takes effect on the next loss once a maximum per year loss ACL is reached, at a cost of purchase of C. The resulting risk mitigated loss process can be expressed as*

$$Z_t^{(CLP2)} = C + \sum_{s=1}^{N_t} \left[X_s(t) \times \mathbb{I}\left(\sum_{k=1}^{s} X_k(t) \leq ACL \right) + \max\{X_s(t) - TCL, 0\} \times \mathbb{I}\left(\sum_{k=1}^{s} X_k(t) > ACL \right) \right]. \tag{17.8}$$

■

To understand this insurance product, we consider a very simple illustration given in Example 17.4.

EXAMPLE 17.4 Application of the CLP2 to a Single Risk LDA Model

Considering again the example in Section 17.10, we illustrate the application of such a policy in Figure 17.5. Under the CLP2, the insurer will provide compensation on the loss up to the value TCL. This will only occur once the total value of claims exceeds the aggregate limit ACL, before this aggregation (trigger level) the insurer will not provide any compensation. If the ACL is set at USD 15 million, then there will be no coverage for loss 1 and loss 2 under the policy CLP2. In addition, with a TCL in this example given by USD 7 million, we see the remaining losses after this trigger point of ACL will in this example be partially covered for loss 3 and completely covered for losses 4 and 5.

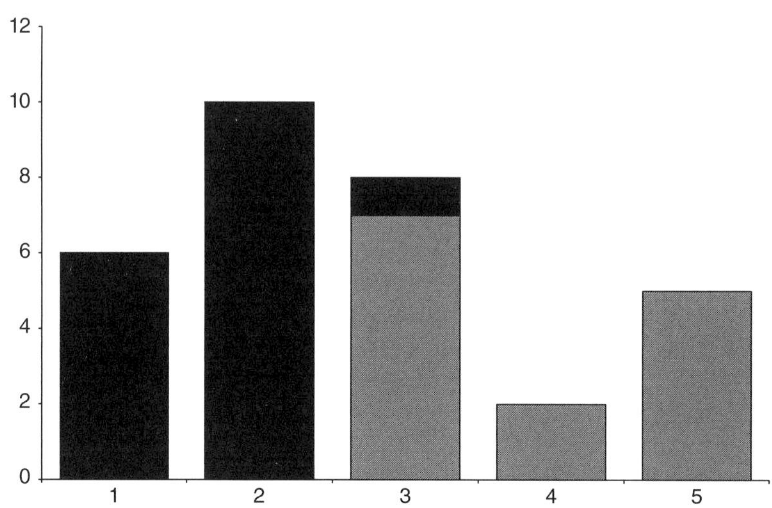

FIGURE 17.5 Combined Loss Policy and the application of the ILP and ACL in CLP2

■

One may also consider policies that are coupled or contingent on the behavior of d different risk processes in a given year. To illustrate this, we consider a fourth example that one may adopt, termed the ALPd given in Definition 17.16.

Definition 17.16 (Accumulated Loss Policy—*d* Risk Exposures (ALPd)) *Consider the multiple risk setting for a d-variate risk process, where a cap on compensation for accumulated losses across d loss processes over the year is imposed, denoted ALPd at a cost of purchase of C. Under this approach, the insurer will provide compensation to the bank with the ACL limit placed over claims on the combined d risk exposures. As applied to the basic LDA model, a simplified risk mitigated loss can be expressed for the j-th risk process according to*

$$Z_t^{(ALPd)} = C + \max\left[\left(\sum_{j=1}^{d}\sum_{s=1}^{N_t^{(j)}} X_s^{(j)}(t)\right) - ACL, 0\right]. \tag{17.9}$$

■

17.4.2 GENERIC STOCHASTIC POLICY STRUCTURES: ACCOUNTING FOR COVERAGE UNCERTAINTY

It is required in Basel II/Basel III specifications to take reasonable consideration of payment uncertainties. These can arise from legal disputes regarding individual claims against particularly large loss events, uncertainty in default of insurers providing coverage in the fact of a catastrophic loss event, or lengthy delays and claim run-offs for payment of total coverage which factor in the fact that large claims will typically be payed in proportions over time, meaning that coverage for any given year is uncertain. Hence, having defined these basic insurance product structures, we note that Basel II/Basel III requires other insurance modeling conditions as outlined in BCBS (2006, p. 155). Two of these being residual term of a policy and payment uncertainty to be considered, which we model through stochastic policy structures.

All the policies discussed so far have a deterministic deduction applied either per loss or aggregated over the annual loss for the year. To address the residual term aspect and the payment uncertainty component, we consider three advanced insurance models. The first is based on a random coverage per loss modeled as a proportion of the loss amount. This may arise in a simple model for such payment uncertainties, which is factoring in litigation costs and legal challenges for risk process known to have infrequent but high consequence losses. Claims arising from such loss process would likely lead to challenge from an insurer and may result in proportional coverage actually applying in a given year. The second follows guidelines proposed in Basel II/Basel III relating to the insurance premium haircut and the third is based on a stochastic banding structure. In particular, the third stochastic model extends the model of Bazzarello *et al.* (2006), allowing one to capture the notion of payment uncertainty. This is a critical aspect of both Basel II and Solvency 2 modeling; see BCBS (2006, p. 155).

In the following first example, we still consider deterministic policy specifications; however, the amount deducted for a given loss is now a random variable as detailed under a PILP given in Definition 17.17.

Definition 17.17 (Proportional Individual Loss Policy (PILP)) *The PILP provides a specified maximum compensation, given by a proportional top cover limit (PTCL) on a per event basis* *for a given year at a cost of purchase of C. Therefore, one can write the PILP risk mitigated loss LDA model from the perspective of the banking institution according to the loss process*

$$Z_t^{(PILP)} = C + \sum_{s=1}^{N_t} \max\left\{ X_s(t) - \underbrace{\theta X_s(t)}_{PTCL}, 0 \right\}. \tag{17.10}$$

for some policy specified $\theta \in [0, 1)$, *which defines the PTCL of %x coverage.* ■

To understand this insurance product, we consider a very simple illustration given in Example 17.5.

EXAMPLE 17.5 Application of the PILP to a Single Risk LDA Model

Consider the PILP policy for a single risk process in which we consider a year consisting of five OpRisk losses of $\{6, 10, 8, 2, 5\}$ in USD million. For each loss, the insurer will provide compensation on the loss given by the value PTCL, which in this example is set at 40%, as illustrated in Figure 17.6, where the solid lines represents the PILP policies PTCL for each loss event in the risk process. As can be seen in each loss, the value of the compensation is at 40% of each individual loss (highlighted in gray) and the bank still incurs the loss above this value (highlighted in black).

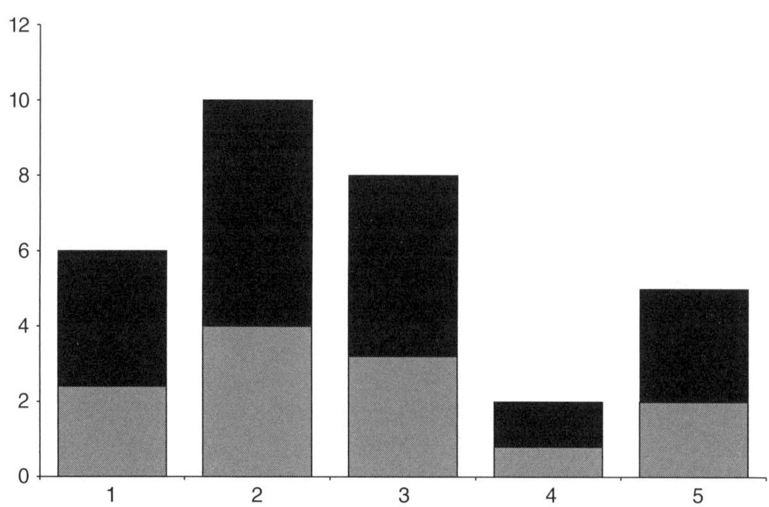

FIGURE 17.6 **Individual Loss Policy and the application of the Top Cover Limit** ■

If one wants to model explicitly the residual term of a policy, this can be considered through a HILP product. Under the Basel II framework, it is clearly specified that "*for policies with a residual term of less than one year, the bank must make appropriate haircuts*" (BCBS, 2006, section 678, p. 155). Modeling the haircut complicates the LDA model since now one requires explicit knowledge of the arrival time process. In this section, we model the interarrival time of losses in a year as exponentially distributed, such as would be relevant for a simple LDA model with a Poisson frequency distribution assumption. Under this model, we consider the simplest scenario in which a basic haircut is applied to the single risk LDA model as discussed in Definition 17.18.

Definition 17.18 (Haircut Individual Loss Policy (HILP-TCL)) *Under the HILP, insurance is applied to the loss process; however, each compensation amount is specified by a proportion of the TCL. In particular, the insurance mitigation follows a discounted time-sensitive factor, or haircut factor. For simplicity, we consider a linear function increasing from 0% insurance mitigation at the beginning of the year up to 100% of the TCL at the end of the year. We can therefore write the risk mitigated loss process according to*

$$Z_t^{(HILP)} = \sum_{s=1}^{N_t} \max\left(X_s\left(t\right) - \alpha\left(t\right) TCL, 0\right). \tag{17.11}$$

■

A second variant of the haircut coverage involves a time increasing deterministic function for the proportion of coverage of a loss as detailed Definition 17.19.

Definition 17.19 (Haircut Individual Loss Policy (HILP)) *Under the HILP, insurance is applied to the loss process. However, each compensation amount is specified by a proportion of the loss. In particular, the insurance mitigation follows a discounted time-sensitive factor, or haircut factor. For simplicity, we consider a linear function increasing from 0% insurance mitigation at the beginning of the year up to 100% of the losses at the end of the year. We can therefore write the risk mitigated loss process according to*

$$Z_t^{(HILP)} = \sum_{s=1}^{N_t} \max\left(X_s\left(t\right) - \alpha\left(t\right) X_s\left(t\right), 0\right). \tag{17.12}$$

■

To model insurer payment uncertainty for OpRisk, we consider a banding model. Payment uncertainty, as discussed in Brandts (2004), generally arises as a result of disagreements between a bank or financial institution and its insurer as to the true value of loss that will be realized. As such, when modeling the resulting processes for both annual loss from the banking perspective and claims from the insurer's perspective, it is important to account for such uncertainty to ensure appropriate capitalization and solvency.

As proposed in Bazzarello *et al.* (2006), a banded structure for payment uncertainty allows for accounting for the fact that severe losses will typically attract more disagreement from insurers and are more likely to be affected by payment delays on such claims on these losses. Therefore, severe losses may be more realistically modeled as being discounted by larger values due to their heightened likelihood of payment uncertainty arising from counter party disputes on larger claims. Previously such models were deterministic, we extend these models to treat payment uncertainty as a stochastic process. To achieve this, we consider a stochastic banding structure across different levels of severity in which we can reflect higher probabilities of reductions in total coverage of losses as severity of such losses increases.

Definition 17.20 (Stochastic Banding Policy (BILP)) *Under the BILP policy, at a cost of C, the risk mitigated loss process can be expressed according to Equation (17.13):*

$$Z_t^{(BILP)} = C + \sum_{s=1}^{N_t} \max\left(X_s(t) - CL\left(X_s(t)\right), 0\right). \tag{17.13}$$

We will segment the top level of cover for the policy as quantified by the TCL into D bands of equal length L (possibly unequal length depending on the application). Under this segmentation, we can define a function that identifies in which band X is located, denoted by the indicator function $d(X)$ for a given band and defined according to

$$d(X) = \min\left(D, \lfloor X/L \rfloor + 1\right). \tag{17.14}$$

According to the definition of $d(X)$, we can now define $CL(X)$ as

$$CL(X) = (d(X) - 1)L + \delta_{d(X)} \min\left(L, X - (d(X) - 1)L\right), \tag{17.15}$$

where $\delta_X \sim Beta(\alpha(X), \beta(X))$ is a random variable from Beta distribution with

$$\begin{aligned}
\alpha(X) &= \mathbb{I}\left(d(X) \geq \lceil (D+1)/2 \rceil\right) \\
&\quad + \left(\lceil (D+1)/2 \rceil - d(X)\right) \left[\frac{2}{\left(D - \lceil (D+1)/2 \rceil\right)}\right] \times \mathbb{I}\left(d(X) < \lceil (D+1)/2 \rceil\right), \\
\beta(X) &= \mathbb{I}\left(d(X) \leq \lfloor (D+1)/2 \rfloor\right) \\
&\quad + \left(d(X) - \lfloor (D+1)/2 \rfloor\right) \left[\frac{2}{\left(D - \lfloor (D+1)/2 \rfloor\right)}\right] \times \mathbb{I}\left(d(X) > \lfloor (D+1)/2 \rfloor\right).
\end{aligned}$$

However, in actual application of payment uncertainty to an observed claims processes, it is unlikely that the bands of cover limit would have equal length. Therefore, to account for this, we will convert the i-th band on the $[0, 1]$ basic scale with length $l = \frac{L}{TCL}$ to the i-th band on the $[0, 1]$ log-scale with length B_i via the transformation

$$B_i = \frac{\exp(il) - \exp((i-1)l)}{\exp(1) - 1}, \quad \text{for } i = 1, \ldots, D. \tag{17.16}$$

Hence, the band identifying function in Equation (17.14), $d(X)$, can be redefined in the log-scale case to $b(X)$, which identifies the log-band in which X is located

$$b(X) = \mathbb{I}\left(X \leq B_1 TCL\right) + \sum_{i=2}^{D-1} \left[i\, \mathbb{I}\left(\sum_{j=1}^{i-1} B_i < \frac{X}{TCL} \leq \sum_{j=1}^{i} B_i \right) \right] + D\, \mathbb{I}\left(X > B_{D-1} TCL\right). \tag{17.17}$$

From here, the calculation of $CL(X)$ is the same, where $d(X)$ is replaced by $b(X)$. ■

In the following, we illustrate the BILP policy coverage under two different banding structure assumptions.

EXAMPLE 17.6 Understanding BILP Coverage

To illustrate the calculation of $CL(X)$ under the two different banding structures, refer to Figure 17.7. As can be seen, this figure illustrates the application of the 3-banded insurance structure to the same loss of value USD 8 million. Under the basic banded policy, $L = 5$ and hence the loss is categorized into the second band. This means the insurer will provide complete compensation of the first band USD 5 million, plus a proportion of the remaining loss USD 3 million (= USD 8 million − USD 5 million) as determined by $\delta_2 \sim Beta(1, 1)$. However, under the log banded policy (*Note:* for simplicity the bandwidths have been selected as integer values), the loss is categorized into the third band. As such, the insurer will provide complete compensation for the first two bands USD 6 million (= USD 2 million − USD 4 million), plus a proportion of the remaining loss USD 2 million (= USD 8 million − USD 6 million) as determined by $\delta_3 \sim Beta(1, 3)$.

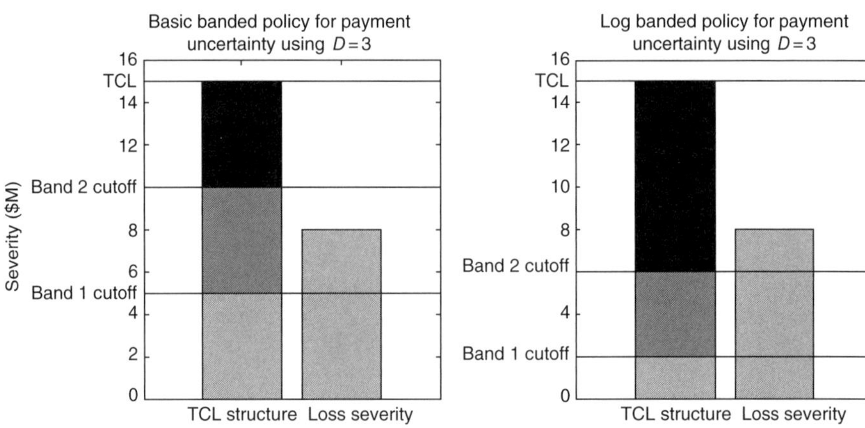

FIGURE 17.7 Linear and logarithmic stochastic banding policies

It should also be mentioned that in the reinsurance industry the aforementioned banded structure may look a little like the notion of layering as detailed in Definition 17.21.

Definition 17.21 (Reinsurance Layering) *In the reinsurance industry, one often divides a large risk into several layers. A layer in reinsurance is comparable to a tranche in catastrophe bond series. Given a random loss variable $X \sim F(x)$, then a layer with limit h and attachment point a, denoted by $X_{(a,a+h]}$, is defined according to*

$$X_{(a,a+h]} = \begin{cases} 0, & \text{if } X < a, \\ X - a, & \text{if } X \in (a, a+h], \\ h, & \text{if } X \geq a+h. \end{cases} \tag{17.18}$$

Under a layer scheme, the expected loss in a given layer with attachment a and limit h is given by

$$\mathbb{E}\left[X_{(a,a+h]}\right] = \int_{a}^{a+h} [1 - F(x)]\, dx. \tag{17.19}$$

■

In the next section, we will illustrate how certain LDA model families that are of particular relevance to the context of insurance in OpRisk can be developed to obtain closed-form expressions for the annual loss distribution under the insurance mitigation developed in the ILPU, ILPCn, ILPCa, ALP, PILP, HLP and BILP policy types. This will involve development of two families of models that have heavy-tailed severity distributions that are both closed under convolution and also invariant to translation by the insurance mitigation on the loss process.

17.5 Closed-Form LDA Models with Insurance Mitigations

In this section, we illustrate that under flexible families of LDA loss process models, one may characterize the resulting insurance mitigated loss process distribution and density in closed form. To develop closed-form LDA models under the setting of insurance mitigation, we will utilize properties of particular families of severity models that are infinitely divisible. In particular, we will consider the following classes of severity family:

- Inverse Gaussian;
- α-stable—with strict positive support (perfect right skew $\beta = 1$);
- Large claim number approximations via geometric stable representations;
- Insured loss process Gamma and Beta basis series expansion representations.

In terms of the frequency distributions we consider, we will look at both Poisson, doubly stochastic Poisson-Gamma and Poisson-generalized-hyper-geometric (Sichel) process models in the definition of the following insurance mitigated processes.

In developing this section, we simply state without extensive detail the closed-form LDA-Insured models. For the purpose of this section, it will be relevant to recall the following basic definitions.

17.5.1 INSURANCE MITIGATION UNDER THE POISSON-INVERSE-GAUSSIAN CLOSED-FORM LDA MODELS

If the severity model is taken to be the inverse-Gaussian family of models, then one has a density and distribution function defined over a support $(0, \infty)$ given by Definition 17.22. It is interesting to note that such models have been proposed for LDA loss models in the actuarial literature in Ter Berg (1994) and Hadwiger (1942).

Definition 17.22 (Inverse-Gaussian (Wald) Severity Model) *If the severity model is given by $X \sim InverseGaussian(\mu, \psi)$ with $x > 0$ and mean $\mu > 0$ and shape parameter $\psi > 0$, then one has the following density and distribution functions:*

$$
\begin{aligned}
f_X(x;\mu,\psi) &= \left[\frac{\psi}{2\pi x^3}\right]^{\frac{1}{2}} \exp\left(\frac{-\psi(x-\mu)^2}{2\mu^2 x}\right),\\
F_X(x;\mu,\psi) &= \Phi\left(\sqrt{\frac{\psi}{x}}\left(\frac{x}{\mu}-1\right)\right) + \exp\left(\frac{2\psi}{\mu}\right)\Phi\left(-\sqrt{\frac{\psi}{x}}\left(\frac{x}{\mu}+1\right)\right),
\end{aligned}
\tag{17.20}
$$

with $\Phi(x) = \frac{1}{\sqrt{2\pi}}\int_{-\infty}^{x}\exp\left(-\frac{1}{2}t^2\right)dt$ the standard Normal distribution. ■

Under this class of severity model, we can obtain the following results for closed-form expressions for the insured LDA risk process.

Theorem 17.1 (Poisson-Inverse-Gaussian LDA Under PILP Coverage) *Consider the PILP coverage characterized by the insurance mitigated loss process*

$$
Z_t^{(PILP)} = C + \sum_{s=1}^{N_t} \max\left\{X_s(t) - \underbrace{\theta X_s(t)}_{PTCL}, 0\right\},
\tag{17.21}
$$

for some policy specified $\theta \in [0,1)$, which defines the (proportional TCL) PTCL or percentage of each individual losses relative coverage. Then the distribution of the annual loss process Z represented by a compound process model with LDA structure in which the frequency is $N_t \sim Poisson(\lambda)$ and the severity model $X_i(t) \sim InverseGaussian(\mu, \psi)$ can be re-expressed as the annual loss process of the insured process $Z_t^{(PILP)}$ via mixture density comprising inverse-Gaussian components with Poisson mixing weights for $N_t > 0$,

$$
f_{Z^{PILP}}(z) = \sum_{n=1}^{\infty} \exp(-\lambda)\frac{\lambda^n}{n!}\left[\frac{\psi_n}{2\pi z^3}\right]^{\frac{1}{2}} \exp\left(\frac{-\psi_n(z-\mu_n)^2}{2\mu_n^2 z}\right),
\tag{17.22}
$$

with

$$
\begin{aligned}
\mu_n &= (1-\theta)n\mu,\\
\psi_n &= (1-\theta)^2 n^2 \psi,
\end{aligned}
$$

and $f_Z(0) = \mathbb{P}r[N_t = 0] = \exp(-\lambda)$. The exact form of the annual loss cumulative distribution function is also expressible in closed form,

$$
\begin{aligned}
\Pr(Z < z) &= F_{Z^{PILP}}(z)\\
&= \sum_{n=1}^{\infty}\exp(-\lambda)\frac{\lambda^n}{n!}\left[\Phi\left(\sqrt{\frac{\psi_n}{z}}\left(\frac{z}{\mu_n}-1\right)\right) + \exp\left(\frac{2\psi_n}{\mu_n}\right)\Phi\left(-\sqrt{\frac{\psi_n}{z}}\left(\frac{z}{\mu_n}+1\right)\right)\right]\\
&\quad + \exp(-\lambda)\times\mathbb{I}[z=0].
\end{aligned}
\tag{17.23}
$$

In addition, one can develop models in which the intensity of the number of losses per year is treated as a random variable, in which case we are considering the insured process being modeled by a doubly stochastic process as specified in Theorem 17.2.

Theorem 17.2 (Doubly Stochastic Poisson-Inverse-Gaussian LDA Under PILP Coverage) *Consider the PILP coverage characterized by the insurance mitigated loss process*

$$Z_t^{(PILP)} = C + \sum_{s=1}^{N_t} \max\left\{ X_s(t) - \underbrace{\theta X_s(t)}_{PTCL}, 0 \right\} \tag{17.24}$$

for some policy specified $\theta \in [0, 1)$, *which defines the PTCL or percentage of each individual losses relative coverage. Then the distribution of the annual loss process Z represented by a doubly stochastic compound process model with LDA structure in which the frequency is* $N_t \sim$ *Poisson*(Λ) *and the intensity of the number of loss events each year is a random variable with one of the following two possible models.*

- ***Gamma Intensity***. *Here, we assume that the intensity parameter for the mean number of losses in a year is a random variable given by*

$$\Lambda \sim \Gamma(\lambda; \alpha, \beta) = \frac{\beta^\alpha}{\Gamma(\alpha)} \exp(-\beta\lambda)\lambda^{\alpha-1}, \tag{17.25}$$

where $\alpha > 0$ *and* $\beta > 0$, *which results in the mixed Poisson model involving the probability of the event* $\{N = n\}$ *being given by*

$$\begin{aligned} \mathbb{P}\mathrm{r}[N = n] &= \int_0^\infty \exp(-\lambda)\frac{\lambda^n}{n!}\frac{\beta^\alpha}{\Gamma(\alpha)} \exp(-\beta\lambda)\lambda^{\alpha-1} d\lambda \\ &= \frac{(\alpha+n-1)!}{(\alpha-1)!n!}\left(\frac{\beta}{1+\beta}\right)^\alpha \left(\frac{1}{1+\beta}\right)^n, \end{aligned} \tag{17.26}$$

see details in Peters et al. (2011, theorem 6). Here, $\Gamma(\alpha)$ *is the gamma function;*

- ***Generalised Inverse-Gaussian Intensity***. *In this case, we consider a generalized version of the Sitchel model where we assume that the intensity parameter for the mean number of losses in a year is a random variable given by*

$$\begin{aligned} \Lambda \sim GIG(\lambda; \alpha, \beta, \gamma) = &\left(\frac{2\sqrt{(1-\beta)}}{\alpha\beta}\right)^\gamma \frac{\lambda^{\gamma-1}}{\left[2K_\gamma\left(\alpha\sqrt{(1-\beta)}\right)\right]} \\ &\times \exp\left(-\left[\frac{1}{\beta}-1\right]\lambda - \frac{\alpha^2\beta}{4\lambda}\right), \end{aligned} \tag{17.27}$$

where $\gamma \in \mathbb{R}$, $\alpha > 0$ *and* $\beta \in [0, 1]$, *which results in the mixed Poisson model involving the probability of the event* $\{N = n\}$ *given by*

$$\begin{aligned}
&\mathbb{P}\mathrm{r}\,(N = n) \\
&= \int_0^\infty \exp(-\lambda)\frac{\lambda^n}{n!}\left(\frac{2\sqrt{(1-\beta)}}{\alpha\beta}\right)^{\gamma}\frac{\lambda^{\gamma-1}}{\left[2K_\gamma\left(\alpha\sqrt{(1-\beta)}\right)\right]} \\
&\quad\times \exp\left(-\left[\frac{1}{\beta}-1\right]\lambda - \frac{\alpha^2\beta}{4\lambda}\right)d\lambda \\
&= \frac{(1-\beta)^{\frac{\gamma}{2}}}{K_\gamma\left(\alpha\sqrt{(1-\beta)}\right)}\frac{(\alpha\beta)^n}{2^n n!}K_{n+\gamma}(\alpha)\,.
\end{aligned} \tag{17.28}$$

See details in Sichel (1982, equation 2.4).

For examples of such models in the doubly stochastic binomial, negative binomial processes, see the details in Peters et al. (2011). If under the considered LDA model one utilizes a severity model $X_i(t) \sim InverseGaussian(\mu, \psi)$, *the PILP insured annual loss process* $Z_t^{(PILP)}$ *can be expressed via a mixture density comprising inverse-Gaussian components with weights specified by* $\mathbb{P}\mathrm{r}[N = n]$ *under one of the models provided earlier, giving*

$$f_{Z^{PILP}}(z) = \sum_{n=1}^{\infty}\mathbb{P}\mathrm{r}[N = n]\left[\frac{\psi_n}{2\pi z^3}\right]^{\frac{1}{2}}\exp\left(\frac{-\psi_n(z-\mu_n)^2}{2\mu_n^2 z}\right), \tag{17.29}$$

with

$$\begin{aligned}
\mu_n &= (1-\theta)n\mu, \\
\psi_n &= (1-\theta)^2 n^2\psi,
\end{aligned}$$

and $f_Z(0) = \mathbb{P}\mathrm{r}\,[N_t = 0] = \exp(-\lambda)$ *for* $N = 0$. *The exact form of the annual loss cumulative distribution function is also expressible in closed form,*

$$\begin{aligned}
\mathbb{P}\mathrm{r}\,[Z < z] &= F_{Z^{PILP}}(z) \\
&= \sum_{n=1}^{\infty}\mathbb{P}\mathrm{r}[N = n]\left[\Phi\left(\sqrt{\frac{\psi_n}{z}}\left(\frac{z}{\mu_n}-1\right)\right) + \exp\left(\frac{2\psi_n}{\mu_n}\right)\Phi\left(-\sqrt{\frac{\psi_n}{z}}\left(\frac{z}{\mu_n}+1\right)\right)\right] \\
&\quad + \exp(-\lambda)\times\mathbb{I}\,[z = 0]\,.
\end{aligned} \tag{17.30}$$

This presents a very generic and highly flexible class of risk processes with a simple and easily interpretable insurance mitigation that admits closed-form representations, allowing for efficient study and understanding of such models in practice. Furthermore, we note the following features of the GIG-based model.

Remark 17.1 *We note that in the case of the generalized Inverse Gaussian (GIG) mixing distribution for the number of losses in a year, we can characterize special subfamilies of the GIG model as follows: if* $\gamma > 0$ *and* $\alpha \to 0$, *then the family reduces to the Pearson Type III (gamma) model; if* $\gamma < 0$ *and* $\beta \to 1$, *then the family reduces to the Pearson Type V family.*

In the case that one considers a haircut policy as specified under the HILP coverage, then the following closed-form LDA models can be obtained as detailed in Theorem 17.3.

Theorem 17.3 (Poisson-Inverse-Gaussian LDA Under HILP Coverage) *Consider the HILP coverage characterized by the insurance mitigated loss process*

$$Z_t^{(HILP)} = \sum_{s=1}^{N_t} \max\left(X_s(t) - \alpha(t) X_s(t), 0\right) \tag{17.31}$$

for some functional form of the haircut specified according to $\alpha(t)$, *which is an increasing percentage as a function of time* t *within the year of coverage. Then the distribution of the annual loss process* Z *represented by a compound process model with LDA structure in which the frequency is* $N_t \sim \text{Poisson}(\lambda)$ *and the severity model* $X_i(t) \sim \text{InverseGaussian}(\mu, \psi)$ *can be re-expressed as the annual loss process of the insured process* $Z_t^{(PILP)}$ *via mixture density comprising Inverse-Gaussian components with Poisson mixing weights for* $N_t > 0$,

$$f_{Z^{PILP}}(z) = \sum_{n=1}^{\infty} \exp(-\lambda)\frac{\lambda^n}{n!}\left[\frac{\psi_n}{2\pi z^3}\right]^{\frac{1}{2}} \exp\left(\frac{-\psi_n(z-\mu_n)^2}{2\mu_n^2 z}\right), \tag{17.32}$$

with

$$\mu_n = \mu \sum_{i=1}^{n} \left[1 - \alpha(t_i)\right],$$

$$\psi_n = \psi \left(\sum_{i=1}^{n} \left[1 - \alpha(t_i)\right]\right)^2,$$

where $\{t_i\}_{i=1}^n$ *are the times at which the* n *losses occurred during the year and* $f_Z(0) = \mathbb{P}\text{r}(N_t = 0) = \exp(-\lambda)$ *for* $N = 0$. *The exact form of the annual loss cumulative distribution function is also expressible in closed form,*

$$\begin{aligned}
\mathbb{P}\text{r}(Z < z) &= F_{Z^{PILP}}(z) \\
&= \sum_{n=1}^{\infty} \exp(-\lambda)\frac{\lambda^n}{n!}\left[\Phi\left(\sqrt{\frac{\psi_n}{z}}\left(\frac{z}{\mu_n} - 1\right)\right)\right. \\
&\quad \left. + \exp\left(\frac{2\psi_n}{\mu_n}\right)\Phi\left(-\sqrt{\frac{\psi_n}{z}}\left(\frac{z}{\mu_n} + 1\right)\right)\right] \\
&\quad + \exp(-\lambda) \times \mathbb{I}[z = 0].
\end{aligned} \tag{17.33}$$

17.5.1.1 Insurance Mitigation and the Coefficient of Variation. Before proceeding, we recall the basic notion of the coefficient of variation for an LDA annual loss random variable, provided in Definition 17.23.

Definition 17.23 (Coefficient of Variation for an LDA Model) *In the simplest form, the coefficient of variation of a loss process with a compound process LDA model* $Z = \sum_{i=1}^{N} X_i$, *denoted by* $c_v(Z)$ *is defined according to the ratio given by*

$$c_v(Z) = \frac{\sqrt{\mathbb{V}\text{ar}[Z]}}{\mathbb{E}[Z]}. \tag{17.34}$$

■

This measure of variation is often considered as it may contain the desirable property that it is independent of the unit in which the measurement has been taken, so it is a dimensionless number. It is standard practice to consider this metric as a comparative measure of variability when assessing data sets with different units or widely different means; in such cases, one should use the coefficient of variation instead of the standard deviation.

EXAMPLE 17.7 **Coefficient of Variation for Poisson-Inverse-Gaussian LDA Model**

If one considers the LDA model comprising Poisson frequency model such that $N_t \sim Poisson(\lambda)$ and the severity model utilized is an inverse-Gaussian model such that $X_i(t) \sim InverseGaussian(\mu, \psi)$ with a resulting annual loss given by

$$Z_t = \sum_{s=1}^{N_t} X_s(t). \tag{17.35}$$

Then one can find the coefficient of variation by first finding the first and second moments of the LDA annual loss model according to

$$\begin{aligned}
\mathbb{E}[Z_t] &= \mathbb{E}[\mathbb{E}[Z_t|N_t = n]] \\
&= \mathbb{E}[N_t X_1(t)] \\
&= \lambda\mathbb{E}[X_1(t)] \\
&= \lambda\mu, \\
\mathbb{V}\text{ar}[Z_t] &= \mathbb{E}[\mathbb{V}\text{ar}[Z_t|N_t = n]] + \mathbb{V}\text{ar}[\mathbb{E}[Z_t|N_t = n]] \\
&= \lambda\mathbb{E}\left[(X_1(t))^2\right] \\
&= \lambda\left(\frac{\mu^3}{\psi} + \mu^2\right),
\end{aligned} \tag{17.36}$$

which gives a coefficient of variation for the Poisson-inverse-Gaussian LDA model according to

$$c_v(Z_t) = \sqrt{\frac{\mu + \psi}{\lambda\psi}}. \tag{17.37}$$

■

It is interesting to also consider the coefficient of variation in the case in which the LDA model has insurance mitigation applied. In such cases, it was noted in Ter Berg (1994) and detailed in Sterk (1979, chapter 5) that when one applies insurance to the loss process, the insurance will naturally have a loss eliminating effect. This is somewhat obvious and is indeed the sole reason for purchase of insurance policies in the first instance; however, what is perhaps not so obvious is the fact that if one considers the coefficient of variation of the LDA model prior to application of insurance mitigation versus that of the insurance mitigated insurance process, the coefficient of variation will increase as a result of insurance mitigation. One way to understand why this is the case is that the insurance policy will reduce the total mean loss. Then it is true that if one considers the coefficient of variation, when the mean value is close to zero, the coefficient of variation will approach infinity and is therefore sensitive to small changes in the mean. As the mean annual loss decreases, then one will naturally observe an increased coefficient of variation. We can verify this is the case by developing an expression for the Poisson-Inverse-Gaussian LDA model after considering insurance mitigation under the PILP structure as detailed in Example 17.8.

EXAMPLE 17.8 Coefficient of Variation for Poisson-Inverse-Gaussian LDA Model with PILP Insurance Mitigation

Consider the PILP coverage characterized by the insurance mitigated loss process

$$Z_t^{(PILP)} = C + \sum_{s=1}^{N_t} \max\left\{ X_s(t) - \underbrace{\theta X_s(t)}_{PTCL}, 0 \right\} \tag{17.38}$$

for some policy specified $\theta \in [0, 1)$, which defines the (proportional TCL) PTCL or percentage of each individual losses relative coverage. Then the distribution of the annual loss process Z represented by a compound process model with LDA structure in which the frequency is $N_t \sim Poisson(\lambda)$ and the severity model $X_i(t) \sim InverseGaussian(\mu, \psi)$. The resulting coefficient of variation of the PILP insurance mitigated process is then given by first finding the mean and variance of the resulting insurance mitigated process given by

$$\begin{aligned}
\mathbb{E}\left[Z_t^{(PILP)}\right] &= \mathbb{E}\left[\mathbb{E}\left[Z_t | N_t = n\right]\right] \\
&= \sum_{n=0}^{\infty} \exp(-\lambda)\frac{\lambda^n}{n!}\mu_n \\
&= (1-\theta)\mu \sum_{n=1}^{\infty} \exp(-\lambda)\frac{\lambda^n}{(n-1)!} \\
\mathbb{V}\text{ar}\left[Z_t^{(PILP)}\right] &= \mathbb{E}\left[\mathbb{V}\text{ar}\left[Z_t | N_t = n\right]\right] + \mathbb{V}\text{ar}\left[\mathbb{E}\left[Z_t | N_t = n\right]\right] \qquad (17.39) \\
&= \mathbb{E}\left[\frac{\mu_{N_t}^3}{\psi_{N_t}}\right] + \mathbb{V}\text{ar}\left[\mu_{N_t}\right] \\
&= (1-\theta)\frac{(\mu)^3}{\psi}\mathbb{E}\left[N_t\right] + (1-\theta)^2 (\mu)^2\, \mathbb{V}\text{ar}\left[N_t\right] \\
&= \lambda(1-\theta)(\mu)^2\left[\frac{\mu}{\psi} + (1-\theta)\right],
\end{aligned}$$

where μ_n and ψ_n is notation used to denote the parameters of the Inverse Gaussian of the partial sum with n losses, which produces an insurance mitigated coefficient of variation given by

$$c_v\left(Z_t^{(PILP)}\right) = \frac{\sqrt{\lambda(1-\theta)\left[\frac{\mu}{\psi} + (1-\theta)\right]}}{(1-\theta)\sum_{n=1}^{\infty} \exp(-\lambda)\frac{(\lambda)^n}{(n-1)!}}. \tag{17.40}$$

From this analysis we see that when we take the first-order approximation of the coefficient of variation given by

$$c_v\left(Z_t^{(PILP)}\right) \approx \frac{\sqrt{\lambda(1-\theta)\left[\frac{\mu}{\psi} + (1-\theta)\right]}}{(1-\theta)\exp(-\lambda)(\lambda)}, \tag{17.41}$$

noting that $\theta \in [0, 1]$, if $\theta << \frac{\mu+\psi}{\psi}$, then one has the approximate relationship between the coefficient of variation for the insured process and the uninsured process given by

$$c_v\left(Z_t^{(PILP)}\right) = \frac{c_v\left(Z_t\right)}{\sqrt{1-\theta}\exp(-\lambda)}. \tag{17.42}$$

This relationship shows that as $\theta \rightarrow 1$, that is, it approaches full insurance coverage per loss under the PILP coverage, the coefficient of variation of the insured loss process indeed is increasing. ■

17.5.2 INSURANCE MITIGATION AND POISSON-α-STABLE CLOSED-FORM LDA MODELS

There are a number of different parameterizations that have been developed for the four-parameter α-stable characteristic function; a detailed discussion is provided in Peters and Shevchenko (2015). We first present some results in regard to two different series expansions for the density and distribution function of the α-stable severity model under the B-type parameterisations of Zolotarev.

Definition 17.24 (Zolotarev's B-Type Stable Parameterizations) *A random variable X with α-stable distribution, denoted by $X \sim S_\alpha(x; \beta_B, \gamma_B, \delta_B; B)$, denotes the univariate four-parameter stable distribution family under parameterizations B type of Zolotarev (1986, theorem C.3) with characteristic function in the following form:*

$$\ln \Phi_X(\theta) = \ln \mathbb{E}\left[\exp(\mathrm{i}\theta X)\right] = \begin{cases} \gamma_B\left(\mathrm{i}\theta\delta_B - |\theta|^\alpha \exp\left(-\mathrm{i}(\pi/2)\beta_B K(\alpha)\mathrm{sgn}\theta\right)\right) & \alpha \neq 1 \\ \gamma_B\left(\mathrm{i}\theta\delta_B - |\theta|^\alpha\left(\pi/2 + \mathrm{i}\beta_B \ln|\theta|\mathrm{sgn}\theta\right)\right) & \alpha = 1 \end{cases} \tag{17.43}$$

with $K(\alpha) = \alpha - 1 + \operatorname{sgn}(1 - \alpha)$ *and stability index* $\alpha \in (0\ ,\ 2]$*, skewness* $\beta_B \in [-1, 1]$*, "rate"* $\gamma_B > 0$*, and location* $\delta_B \in \mathbb{R}$*. Here,* $\mathrm{i} = \sqrt{-1}$ *is a unit complex number.* ■

Lemma 17.1 (α-Stable Severity Density and Distribution Representations) *Consider the standardized B-type α-stable random variable, standardized such that $\gamma = 1$ and $\delta = 0$. Then the density function can be evaluated pointwise according to the following series expansions (Zolotarev, 1986, equation 2.4.6, p. 89):*

$$
f_X(x; \alpha, \beta, 1, 0; B)
= \begin{cases}
\frac{1}{\pi} \sum_{n=1}^{\infty} (-1)^{n-1} \frac{\Gamma(\frac{n}{\alpha}+1)}{\Gamma(n+1)} \sin\left(n\pi \frac{1+\beta K(\alpha)}{2\alpha}\right) x^{n-1}, & \text{if } \alpha > 1, \beta \in [-1, 1], x \in \mathbb{R}, \\
\frac{1}{\pi} \sum_{n=1}^{\infty} (-1)^{n-1} n b_n x^{n-1}, & \text{if } \alpha = 1, \beta \in (0, 1], x \in \mathbb{R}, \\
\frac{1}{\pi} \sum_{n=1}^{\infty} (-1)^{n-1} \frac{\Gamma(n\alpha+1)}{\Gamma(n+1)} \sin\left(n\pi\alpha \frac{1+\beta K(\alpha)}{2\alpha}\right) x^{-n\alpha-1}, & \text{if } \alpha < 1, \beta \in [-1, 1], x \in \mathbb{R}^+,
\end{cases}
\tag{17.44}
$$

where the coefficients b_n are given by

$$
b_n = \frac{1}{\Gamma(n+1)} \int_0^{\infty} \exp(-\beta u \ln u)\, u^{n-1} \sin\left[(1+\beta)u\frac{\pi}{2}\right] du. \tag{17.45}
$$

In addition, the distribution function of an α-stable severity model can be evaluated pointwise according to the convergent series expansion given by

$$
F_X(x; \alpha, \beta, 1, 0; B)
= \begin{cases}
1 - \frac{1}{\pi\alpha} \sum_{n=1}^{\infty} (-1)^{n-1} \frac{\Gamma(n\alpha+1)}{n\Gamma(n+1)} \sin\left(n\pi\alpha \frac{1+\beta K(\alpha)}{2\alpha}\right) (x)^{-\frac{n}{\alpha}}, \\
\qquad \text{if } \alpha < 1, \beta \in [-1, 1], x \in \mathbb{R}, \\
1 + \frac{1}{2}\left(1 + \frac{\beta K(\alpha)}{\alpha}\right) + \frac{1}{\pi} \sum_{n=1}^{\infty} (-1)^{n-1} \frac{\Gamma(\frac{n}{\alpha}+1)}{n\Gamma(n+1)} \sin\left(n\pi \frac{1+\beta K(\alpha)}{2\alpha}\right) x^n, \\
\qquad \text{if } \alpha > 1, \beta \in [-1, 1], x \in \mathbb{R}, \\
1 - \frac{1}{\pi} b_0 + \frac{1}{\pi} \sum_{n=1}^{\infty} (-1)^{n-1} b_n x^n, \qquad \text{if } \alpha = 1, \beta \in (0, 1], x \in \mathbb{R}^+,
\end{cases}
$$

In all other cases, it suffices to utilize the duality principle of infinitely divisible stable distributions, which has the consequence that

$$
F_X(-x; \alpha, \beta, 1, 0; B) + F_X(x; \alpha, -\beta, 1, 0; B) = 1. \tag{17.46}
$$

In addition, it may be of interest in some cases to consider instead a special basic function expansion for the distribution and density of the α-stable severity density specifically for the case of positive support of the losses, given in terms of Laguerre polynomials, see Definition 17.25.

Definition 17.25 (Laguerre Polynomials) *The generalized Laguerre polynomials are defined as the solutions to the second-order linear differential equation, for integers n according to*

$$xy'' + (\alpha + 1 - x)y' + ny = 0.$$

The solutions form a sequence of polynomials $L_0, L_1, \ldots$ which are also an orthonormal sequence given by several different representations. First, it will be useful to consider the case of the standard Laguerre polynomials in which $\alpha = 0$ and then the following recursive relationship for their definition is considered for $x \in \mathbb{R}$ and for all $k \in \mathbb{N}^+$

$$\begin{aligned} L_0(x) &= 1, \\ L_1(x) &= 1 - x, \\ L_{k+1}(x) &= \frac{1}{k+1}\left((2k + 1 - x)L_k(x) - kL_{k-1}(x)\right). \end{aligned} \tag{17.47}$$

Now the generalized Laguerre polynomials, for cases in which $\alpha \neq 0$, are given by

$$\begin{aligned} L_n^{(\alpha)}(x) &= \frac{x^{-\alpha}\exp(x)}{n!}\frac{d^n}{dx^n}\left(\exp(-x)x^{n+\alpha}\right) \\ &= \sum_{i=0}^{n}(-1)^i \begin{pmatrix} n+\alpha \\ n-i \end{pmatrix} \frac{x^i}{i!}, \end{aligned}$$

or for $x \in \mathbb{R}$ according to the special functions known as the confluent hypergeometric functions or Kummer's functions by

$$\begin{aligned} L_n^{(\alpha)}(x) &= \binom{n+\alpha}{n} M(-n, \alpha + 1, x) \\ &= \frac{(-1)^n}{n!} U(-n, \alpha + 1, x), \end{aligned} \tag{17.48}$$

where one defines

$$\begin{aligned} M(a, b, z) &= \sum_{n=0}^{\infty} \frac{(a)_{(n)} z^n}{(b)_{(n)} n!}, \\ U(a, b, z) &= \frac{\Gamma(1-b)}{\Gamma(a-b+1)} M(a, b, z) + \frac{\Gamma(b-1)}{\Gamma(a)} z^{1-b} M(a - b + 1, 2 - b, z), \end{aligned} \tag{17.49}$$

with Pochhammer symbols $(a)_{(n)} = a(a+1)(a+2)\ldots(a+n-1)$. ■

These generalized Laguerre polynomials can then be utilized to obtain a series expansion for the α-stable severity density as follows.

Lemma 17.2 (α-Stable Strictly Positive Support Severity Density Representations) *The standardized perfectly skewed α-stable severity density $S_\alpha(x; 1, 1, 0; B)$ can be represented in the heavy-tailed cases by the following Laguerre polynomial series expansion for $x > 0$; see Zolotarev (1986, theorem 2.4.4)*

$$f_X(x;\alpha,1,1,0;B) = \begin{cases} x^{\frac{1-\alpha^2}{\alpha}} \exp(-x^{-\alpha}) \sum\limits_{n=0}^{\infty} k_n^{(s)}(\alpha)\, L_n^{(s)}(x^{-\alpha}), & \forall x > 0, 0 < \alpha < 1 \\ \dfrac{1}{x} \exp(-x) \sum\limits_{n=0}^{\infty} k_n^{(s)} \left(\dfrac{1}{\alpha}\right) L_n^{(s)}(x), & \forall x > 0, 1 < \alpha < 2 \end{cases} \tag{17.50}$$

with the coefficient functions defined by

$$k_n^{(s)}(\alpha) = \alpha \left(\frac{\Gamma(n+1)}{\Gamma(n+1+s)} \right)^{\frac{1}{2}} \sum_{m=0}^{n} \frac{(-1)^m \Gamma(1+s+n)}{\Gamma(m+1)\Gamma(n-m+1)\Gamma(1+\alpha(s+m))} \tag{17.51}$$

and s is any fixed number greater than -1.

Furthermore, it is worth observing that one can "standardize" these loss random variables according to the two following affine transformations, for the case of $\alpha = 1$, by

$$\frac{2\,(X - \gamma_B \delta_B)}{\pi \gamma_B} \sim S\,(1, \beta_B, 1, 0; 0)$$

and in the second case when $\alpha \neq 1$ by

$$\frac{2\,(X - \gamma_B \delta_B)}{\cos\left(\frac{\pi}{2} \beta_B K(\alpha)\right) \gamma_B} \sim S\,(\alpha, \beta_B, 1, 0; 0)\,,$$

where $K(\alpha) = \alpha - 1 + \text{sgn}(1-\alpha)$.

The support of a random variable $X \sim S\,(\alpha, \beta_B, \gamma_B, \delta_B)$ as a function of the distribution parameters is given by the results in Lemma 17.3.

Lemma 17.3 *Denote the sample space S_X or support of the distribution of a univariate random variable $X \sim S\,(\alpha, \beta_B, \gamma_B, \delta_B)$ as follows:*

$$S_X = Supp\,(S\,(\alpha, \beta_B, \gamma_B, \delta_B)) = \begin{cases} [\gamma_B \delta_B, \infty)\,, & \alpha < 1 \text{ and } \beta_B = 1, \\ (-\infty, \gamma_B \delta_B]\,, & \alpha < 1 \text{ and } \beta = -1, \\ (-\infty, \infty)\,, & \text{otherwise.} \end{cases} \tag{17.52}$$

Hence, we will consider setting $\beta = 1$ and restricting $\delta_B \geq 0$ to ensure the support of the stable model is strictly positive.

17.5.2.1 Closed-Form Poisson-α-Stable LDA Models with Insurance Mitigation.

Given this class of models, we can obtain the following analytic results for the insured LDA model under the ILP structure detailed in Theorem 17.4.

Theorem 17.4 (Poisson-α-Stable LDA Under PILP Coverage) *Consider the ILP coverage characterized by the insurance mitigated loss process*

$$Z_t^{(ILPU)} = C + \sum_{s=1}^{N_t} \max\left\{X_s(t) - \theta X_s(t), 0\right\}, \tag{17.53}$$

for some policy specified $\theta \in [0,1)$, which defines the PTCL of %x coverage. If each loss $X_i(t) \sim F_X(x;\alpha,\beta,\gamma,\delta;B)$, then one has the transformed loss following the distribution

$$X_s(t)(1-\theta) \sim F_X\left(x;\alpha,\tilde{\beta}_s,\tilde{\gamma}_s,\tilde{\delta}_s;B\right) \tag{17.54}$$

with transformed parameters

$$\tilde{\beta}_s = \begin{cases} \operatorname{sgn}\left[(1-\theta)\right]\beta_B, & \alpha = 1, \\ \operatorname{sgn}\left[(1-\theta)\right]\cot\left(\frac{\pi}{2}\alpha\right)\tan\left(\frac{\pi}{2}\beta_B K(\alpha)\right), & \alpha \neq 1, \end{cases}$$

$$\tilde{\gamma}_s = \begin{cases} |(1-\theta)|\,\frac{\pi}{2}\gamma_B, & \alpha = 1, \\ |(1-\theta)|\cos\left(\frac{\pi}{2}\beta_B K(\alpha)\right)\gamma_B, & \alpha \neq 1, \end{cases}$$

$$\tilde{\delta}_s = \begin{cases} (1-\theta)\gamma_B\delta_B, & \alpha = 1, \\ (1-\theta)\gamma_B\delta_B - \frac{2}{\pi}\cot\left(\frac{\pi}{2}\alpha\right)\tan\left(\frac{\pi}{2}\beta_B K(\alpha)\right)(1-\theta)\ln\left[(1-\theta)\right], & \alpha \neq 1. \end{cases}$$

Then the distribution of the annual loss process Z represented by a compound process model with LDA structure in which the frequency is $N_t \sim Poisson(\lambda)$ and the severity model $X_i(t) \sim F_X(x;\alpha,\beta_B,\gamma_B,\delta_B;B)$ can be re-expressed as the annual loss process of the insured process $Z_t^{(ILPU)}$ via mixture density comprising α-stable components with Poisson mixing weights given generically by

$$f_{Z^{ILPU}}(z) = \sum_{n=0}^{\infty}\exp(-\lambda)\frac{\lambda^n}{n!}S_n(z;\alpha,\widetilde{\beta}_n,\widetilde{\gamma}_n,\widetilde{\delta}_n;B), \tag{17.55}$$

where the parameters for the partial sum mixture component parameters α, $\widetilde{\beta}(n)$, $\widetilde{\gamma}(n)$, $\widetilde{\delta}(n)$ are expressed with respect to the parameters α, $\widetilde{\beta}_s$, $\widetilde{\gamma}_s$, $\widetilde{\delta}_s$ as follows:

$$\widetilde{\gamma}(n)^{\alpha} = \begin{cases} \sum_{i=1}^{n}\left|\frac{\pi}{2}\widetilde{\gamma}_i\right|, & \alpha = 1, \\ \sum_{i=1}^{n}\left|\cos\left(\frac{\pi}{2}\widetilde{\beta}_i K(\alpha)\right)\widetilde{\gamma}_i\right|^{\alpha}, & \alpha \neq 1, \end{cases}$$

$$\widetilde{\beta}(n) = \begin{cases} \dfrac{\sum_{i=1}^{n}\widetilde{\beta}_i\left|\frac{\pi}{2}\widetilde{\gamma}_i\right|}{\widetilde{\gamma}(n)^{\alpha}}, & \alpha = 1, \\ \dfrac{\sum_{i=1}^{n}\cot\left(\frac{\pi}{2}\alpha\right)\tan\left(\frac{\pi}{2}\widetilde{\beta}_i K(\alpha)\right)\left|\cos\left(\frac{\pi}{2}\widetilde{\beta}_i K(\alpha)\right)\widetilde{\gamma}_i\right|^{\alpha}}{\widetilde{\gamma}(n)^{\alpha}}, & \alpha \neq 1, \end{cases}$$

$$\widetilde{\delta}(n) = \begin{cases} \sum_{i=1}^{n}\widetilde{\delta}_i\widetilde{\gamma}_i - \frac{2}{\pi}\sum_{i=1}^{n}\cot\left(\frac{\pi}{2}\alpha\right)\sin\left(\frac{\pi}{2}\widetilde{\beta}_i K(\alpha)\right)\widetilde{\gamma}_i, & \alpha = 1, \\ \sum_{i=1}^{n}\widetilde{\delta}_i\widetilde{\gamma}_i, & \alpha \neq 1. \end{cases}$$

Note that the distribution and density functions for the PILP insured LDA model comprising the Poisson-α-stable model are then easily derived from these results for a desired Stable series representation. An example of the Laguerre series expansion is provided next.

EXAMPLE 17.9 Poisson-α-Stable PILP Laguerre Series Representation

In this example, we develop a mixture representation of the LDA Poisson-α-stable (B parameterization) model after PILP insurance is applied via a strictly positively support Laguerre series representation presented in Lemma 17.2. This is achieved via the general result presented in Theorem 17.4 after considering the case in which the severity model has a skewness parameter constraint involving considering $\beta_B = 1$, for any other $\alpha_B \in [0, 2]$, $\gamma_B > 0$ and $\delta_B \in \mathbb{R}$. To obtain a closed-form expression for the insurance mitigated loss process given by

$$f_{Z^{PILP}}(z) = \sum_{n=0}^{\infty} \exp(-\lambda) \frac{\lambda^n}{n!} S_n(z; \alpha, \widetilde{\beta}_n, \widetilde{\gamma}_n, \widetilde{\delta}_n; B), \tag{17.56}$$

where the parameters for the partial sum mixture component parameters α, $\widetilde{\beta}(n)$, $\widetilde{\gamma}(n)$, $\widetilde{\delta}(n)$ are given in Theorem 17.4 one first considers each partial sum given $N = n$ and performs standardization. The resulting standardized insurance mitigated annual loss $(\widetilde{Z}_n)$ for $N = n$ losses is given by considering the transformations

$$\widetilde{Z}_n = \begin{cases} \dfrac{2\left(Z_n - \tilde{\gamma}_B(n)\tilde{\delta}_B(n)\right)}{\pi \tilde{\gamma}_B(n)}, & \alpha = 1, \\ \dfrac{2\left(Z_n - \tilde{\gamma}_B(n)\tilde{\delta}_B(n)\right)}{\cos\left(\frac{\pi}{2}\tilde{\beta}_B(n)K(\alpha)\right)\tilde{\gamma}_B(n)}, & \alpha \neq 1. \end{cases} \tag{17.57}$$

Now, using the Laguerre polynomial basis series representation of the stable mixture components, one obtains the density representation given for any $s > -1$ by

$$f_{Z^{PILP}}(z) = \begin{cases} \sum_{n=1}^{\infty} \exp(-\lambda)\frac{\lambda^n}{n!} z^{\frac{1-\alpha^2}{\alpha}} \exp(-z^{-\alpha}) \sum_{r=0}^{\infty} k_r^{(s)}(\alpha)\, L_r^{(s)}(x^{-\alpha}), & \forall x > 0, 0 < \alpha < 1, \\ \sum_{n=1}^{\infty} \exp(-\lambda)\frac{\lambda^n}{n!}\frac{1}{x} \exp(-x) \sum_{r=0}^{\infty} k_r^{(s)}\left(\frac{1}{\alpha}\right) L_r^{(s)}(x), & \forall x > 0, 1 < \alpha < 2. \end{cases} \tag{17.58}$$

■

In the following, we also present a second simplified closed-form representation of the LDA family of risk process for the PILP insurance mitigated Poisson-α-stable model. We cover the representation from Nolan's $S1$ parameterization to the Zolotarev B-Type parameterization. If one considers setting $\alpha_1 = \alpha_B = 0.5$ and

$$\beta_1 = \begin{cases} \beta_B, & \alpha = 1, \\ \cot\left(\frac{\pi}{2}\alpha\right)\tan\left(\frac{\pi}{2}\beta_B K(\alpha)\right), & \alpha \neq 1, \end{cases} \tag{17.59}$$

for the severity model, then one has the two parameter subfamily of models given by the Poisson-Lévy family of LDA models as detailed in Peters *et al.* (2011). In this case according to Samorodnitsky and Taqqu (1994, section 1.2, property 1.2.1) and Nolan (2015, proposition 1.17), one has the severity model closed-form representation given in Lemma 17.4.

Lemma 17.4 (Lévy Severity Model (B-Parameterizations)) *Given a severity model that is $\frac{1}{2}$-Stable (Lévy), such that $X \sim S(0.5, \frac{1}{2K(0.5)}, \frac{\sqrt{2}}{2}\gamma_B, \gamma_B\delta_B; B)$, this model specifies the subfamily of α-stable models with positive real support $x \in [\gamma_B\delta_B, \infty)$. The density and distribution functions are analytic and given, respectively, for $\gamma_B\delta_B < x < \infty$ by*

$$f_X(x) = \sqrt{\frac{\frac{\sqrt{2}}{2}\gamma_B}{2\pi}} \frac{1}{(x-\gamma_B\delta_B)^{3/2}} \exp\left(-\frac{\frac{\sqrt{2}}{2}\gamma_B}{2(x-\gamma_B\delta_B)}\right),$$

$$F_X(x) = erfc\left(\sqrt{\frac{\frac{\sqrt{2}}{2}\gamma_B}{2(x-\gamma_B\delta_B)}}\right).$$

Under this result, we may state the following representation for the B-type parameterized PILP insurance mitigated LDA model for the Poisson–Lévy family given in Example 17.10 (Figure 17.8).

EXAMPLE 17.10 PILP Insurance Mitigated Poisson–Lévy LDA Family

The distribution of the annual loss process Z^{PILP} represented by a compound process model with LDA structure in which the frequency is $N_t \sim Poisson(\lambda)$ and the severity model $X_i(t) \sim S(0.5, \frac{1}{2K(0.5)}, \frac{\sqrt{2}}{2}\gamma_B, \gamma_B\delta_B; B)$, then the exact density of the annual loss process can be expressed analytically as a mixture density comprising α-stable components with Poisson mixing weights for $N_t > 0$,

$$f_Z(z) = \sum_{n=1}^{\infty} \exp(-\lambda)\frac{\lambda^n}{n!}\left[\sqrt{\frac{\tilde{\gamma}_n}{2\pi}} \frac{1}{\left(z-\tilde{\delta}_n\right)^{3/2}} \exp\left(-\frac{\tilde{\gamma}_n}{2\left(z-\tilde{\delta}_n\right)}\right)\right] \times \mathbb{I}\left[\tilde{\gamma}_n\tilde{\delta}_n < z < \infty\right] \tag{17.60}$$

with

$$\tilde{\gamma}_n^{0.5} = n\left|\frac{\sqrt{2}}{2}\gamma_B\right|^{0.5}, \quad \tilde{\beta}_n = \frac{n\left|\frac{\sqrt{2}}{2}\gamma_B\right|^{0.5}}{2K(0.5)\tilde{\gamma}_n^{0.5}}, \quad \text{and} \quad \tilde{\delta}_n = n\gamma_B\delta_B,$$

and $f_Z(0) = \mathbb{P}\text{r}[N_t = 0] = \exp(-\lambda)$ for $N = 0$. The exact form of the annual loss cumulative distribution function is also expressible in closed form:

$$\mathbb{P}\text{r}[Z < z] = F_Z(z) = \sum_{n=1}^{\infty} \exp(-\lambda)\frac{\lambda^n}{n!}\text{erfc}\left(\sqrt{\frac{\tilde{\gamma}_n}{2\left(z-\tilde{\delta}_n\right)}}\right) \times \mathbb{I}\left[\tilde{\delta}_n < z < \infty\right] + \exp(-\lambda) \times \mathbb{I}[z = 0]. \tag{17.61}$$

■

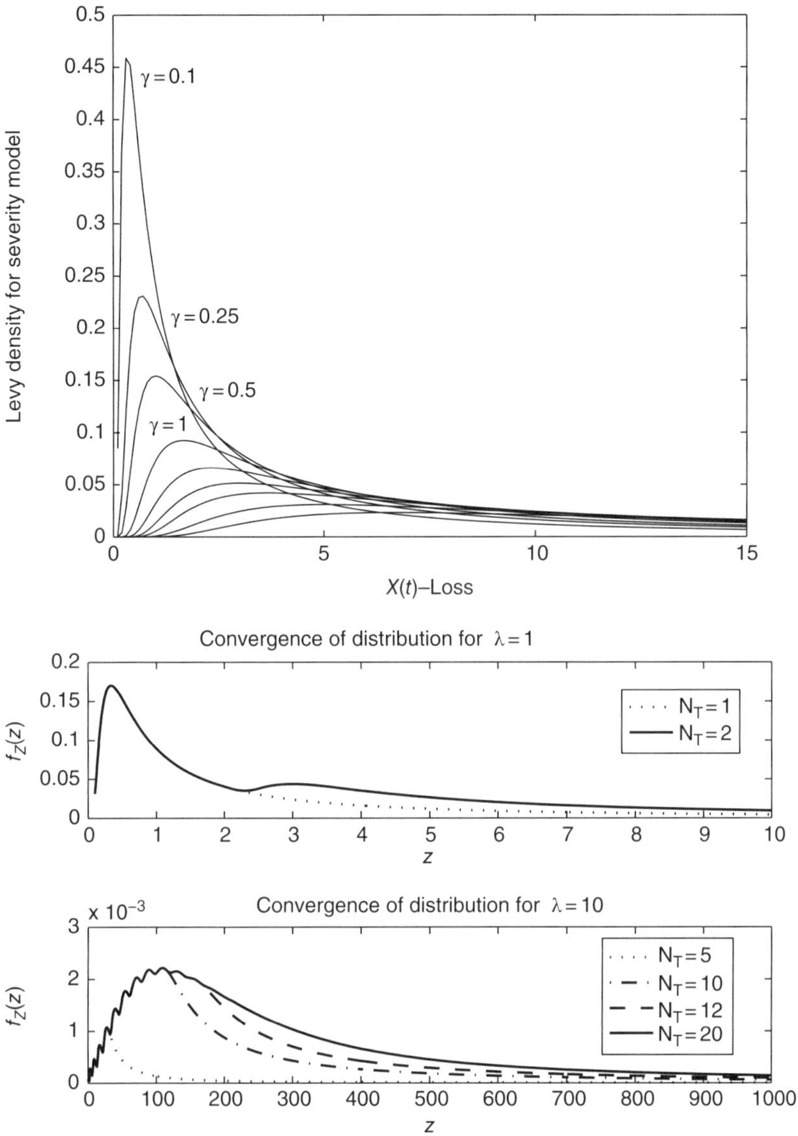

FIGURE 17.8 (**Top**) Lévy severity model as a function of scale parameter, with location $\delta = 0$. (**Bottom**) Truncated sum annual loss distribution approximations

17.5.3 LARGE CLAIM NUMBER LOSS PROCESSES: GENERIC CLOSED-FORM LDA MODELS WITH INSURANCE MITIGATION

The interest in geometric stable models lies in the fact that they act as exact asymptotic representations to geometric compound sums of i.i.d. random variables. Under this class of models, one may obtain closed-form results for any LDA model and any insurance policy asymptotically in the limit of large mean number of annual losses. Therefore, the application of this closed-form model is particularly of interest in risk process model settings in which one has high numbers of claims arriving each year. In OpRisk, this would typically arise in settings such as credit card fraud. An alternative place where this may be of interest is in approximation of an overall business unit or even financial institutions LDA model, which really comprises many loss processes and many different insurance policies; this can also be a setting in which we may adopt a geometric stable approximation for the overall grouped data loss process.

In these cases, we may obtain a closed-form expression (asymptotically) for the annual loss distribution of the insured process. To proceed, we first present a basic characterization of the geometric stable limiting model and then present the representations that are available for closed-form expressions of the resulting insured annual loss density and distribution, as well as tail asymptotics.

17.5.3.1 Characterizing Geometric Stable Approximations for Insured Loss Processes. To describe the main results utilized in this section to perform approximations of large sample loss processes (with insurance), it is beneficial to first recall the results developed in Klebanov *et al.* (1985), which characterize what has become known as geometric infinite Divisibility. We will then move to the notion of geometric stability and finish by showing the relationships between these two properties and important models that will satisfy both properties that can be considered in practical applications in OpRisk LDA modeling. Though geometric infinite divisibility is not strictly required for the approximations in this section to be applied, it is useful to consider as several subclasses of well-studied models considered in the geometric stable family are also geometrically infinitely divisible, which provides additional insight into their representational properties when combining, that is, forming linear combinations of two or more loss process approximations under this class.

The concept of geometric infinite divisibility originally arose from a problem proposed by Zolotarev that was addressed by Klebanov and basically involved the characterization of all random variables Y that satisfy the property that for any $p \in (0,1)$ there is a random variable X_p such that the following holds in distribution

$$Y \stackrel{d}{=} X_p + \epsilon_p Y, \tag{17.62}$$

with $Y \perp X_p \perp \epsilon_p$ and $\mathbb{P}\mathrm{r}\left[\epsilon_p = 0\right] = p$ and $\mathbb{P}\mathrm{r}\left[\epsilon_p = 1\right] = 1 - p$.

While at first sight this may seem a little disjoint from the insurance modeling context being studied in this chapter, in the following sections after defining formally the notion of geometric infinite divisibility in Definition 17.26, it will be shown that one can utilize this concept to aid in the specification of important families of geometrically stable families of distributions that can be used to approximate the annual loss of LDA models with generic insurance policies.

Definition 17.26 (Geometric Infinite Divisibility in OpRisk) *A real-valued random variable corresponding to the annual loss in an LDA model with insurance, denoted $\widetilde{Z}$, is geometrically infinitely divisible (g.i.d.) iff for any $p \in (0,1)$ there exists a sequence of i.i.d., real-valued random variables $\left\{X_{(i,p)}\right\}$ such that the following holds*

$$\widetilde{Z} \stackrel{d}{=} \sum_{i=1}^{N_p} X_{(i,p)}, \tag{17.63}$$

where $N_p \sim Geometric(p)$ with mass function

$$\mathbb{P}\mathrm{r}\left[N_p = n\right] = p(1-p)^{n-1}, \quad n = 1, 2, \ldots. \tag{17.64}$$

The class of g.i.d. distributions is denoted by $F \in \mathcal{F}_{gid}$. It is also equivalent to stating that $Z \sim F \in \mathcal{F}_{gid}$ iff its characteristic function $\Phi_{\widetilde{Z}}(\theta)$ produces a transformed characteristic function, denoted generically by $\Phi(\theta)$, given by the transformation

$$\Phi(\theta) = \exp\left(1 - \frac{1}{\Phi_{\widetilde{Z}}(\theta)}\right), \tag{17.65}$$

which under this transformation becomes infinitely divisible (in the standard sense) and therefore satisfies under log transformation the canonical Kolmogorov representation

$$\ln\left(\Phi(\theta)\right) = \mathrm{i}\delta\theta + \int \left(\exp(\mathrm{i}\theta u) - 1 - \mathrm{i}\theta u\right) \frac{1}{u^2} dK(u), \tag{17.66}$$

for some nondecreasing measure $K(u)$ such that $K(-\infty) = 0$. ■

Remark 17.2 *It is also worth noting that if $\Phi(\theta)$ is an infinitely divisible characteristic function, then so is $|\Phi(\theta)|$; see discussions in Kawata and Maejima (1977).*

Next we discuss a related concept of geometric stability (which we need for the majority of models we consider in this section). To explain the notion of geometric stability, it will be informative to first recall the following basic properties of α-stable models that will help to inform the properties of the geometric stable models to follow.

For a given number of annual losses n, in the case of the α-stable severity model, one has the property that if $X_1, X_2, \ldots, X_n$ are i.i.d. α-stable distributed random variables, then for suitably selected constant functions a_n and b_n one has that

$$S_n = a_n\left(X_1 + X_2 + \cdots + X_n\right) + b_n \tag{17.67}$$

is also α-stable distributed.

Under the geometric stable limiting model, we can shift from the fixed number of losses n to the setting of interest in OpRisk modeling, namely, the compound process setting where the number of losses is also a random variable N. That is, we will obtain a single distribution approximation for the LDA compound process, rather than the mixture representations obtained under the α-stable models. To consider specifically the domain of attraction of the geometric stable distribution, we consider that we may approximate the frequency distribution by a geometric frequency distribution with parameter p and we denote the resulting approximation for the annual loss random variable by Z_{N_p} and the insured annual loss under this framework by $\widetilde{Z}_{N_p}$ throughout the following section.

Under the geometric stability results, one may adopt a broader set of base LDA models and insurance products while still obtaining classes of closed-form distributional results for the compound process annual loss under an LDA modeling framework. In particular, once we have that the average number of losses is large we may approximate any insured LDA model with the following limiting result in Lemma 17.5 and Definition 17.27.

Lemma 17.5 (Asymptotic Approximation of Compound Loss Distributions) *An annual loss distribution under an LDA model framework has a geometric stable limiting distribution for the compound sum iff the number of elements in the sum N_p is a random variable, which is geometrically distributed with parameter p (frequency distribution approximation) and i.i.d. losses with any severity distribution $X_i \sim F_X$ such that one has that the compound annual loss*

$$Z_{N_p} = a_{N_p}\left(X_1 + X_2 + \cdots + X_{N_p}\right) + b_{N_p}, \tag{17.68}$$

which will have a limiting geometric stable distribution as $p \to 0$ iff the appropriate normalizing sequence $a_{N_p} > 0$ and $b_{N_p} \in \mathbb{R}$ exists.

Remark 17.3 *Informally, this statement shows that a geometric stable random variable has the property that for any $p \in (0, 1)$ it can be represented as a sum of a random number (geometric frequency distribution) of i.i.d. loss random variables.*

The formal definition of the geometric stable model is given in Definition 17.27 according to the representation of Kozubowski (1999). It should be noted that it is common to distinguish between geometric stability and strict geometric stability (as is also the case with α-Stable models); see Klebanov *et al.* (1985).

Definition 17.27 (Geometric Stable LDA Model Distribution Limits) *Consider an LDA model with a sequence of i.i.d. loss random variables $X_i \sim F_X(x)$ and the number of annual losses being modeled by a geometric distribution, $N_p \sim Geometric(p)$ with mean $\mathbb{E}\left[N_p\right] = \frac{1}{p}$. Then one has a geometric stable distribution asymptotically as $p \to 0$ and $\mathbb{E}\left[N_p\right] \to \infty$ iff the compound process for the resulting LDA model converges in distribution to the geometric stable distribution,*

$$Z_{N_p} = a_{N_p}\sum_{i=1}^{N_p} X_i + b_{N_p} \sim F_{Z_{N_p}}(z) \xrightarrow{d} GS\left(z; \alpha_{GS}, \beta_{GS}, \gamma_{GS}, \delta_{GS}\right) \tag{17.69}$$

with tail index $\alpha_{GS} \in [0, 2]$, skewness parameter $\beta_{GS} \in [-1, 1]$, scale parameter $\gamma_{GS} > 0$, and location parameter $\delta_{GS} \in \mathbb{R}$ where the log-characteristic function is given by

$$\begin{aligned}
\Phi_{Z_{N_p}}(\theta) &= \mathbb{E}\left[\exp(i\theta Z_{N_p})\right] \\
&= \left[1 + \gamma_{GS}|\theta| w\left(\theta, \alpha_{GS}, \beta_{GS}\right) - i\delta_{GS}\theta\right]^{-1} \\
&= \begin{cases} \left[1 + \gamma_{GS}|\theta|\left(1 + i\beta_{GS}(2\pi) sgn(\theta)\ln|\theta|\right) - i\delta_{GS}\theta\right]^{-1}, & \alpha = 1 \\ \left[1 + \gamma_{GS}^{\alpha_{GS}}|\theta|^{\alpha_{GS}}\left(1 - i\beta_{GS} sgn(\theta)\tan\left(\frac{\pi\alpha}{2}\right)\right) - i\delta_{GS}\theta\right]^{-1}, & \alpha \neq 1. \end{cases}
\end{aligned} \tag{17.70}$$

The class of geometrically stable distributions is denoted by $F \in \mathcal{F}_{gs}$. The subclass of strictly geometrically Stable distributions will be denoted by $F \in \mathcal{F}_{gs}^{*}$, which corresponds to the models with parameters in which $\delta_{GS} = 0$ and $\alpha_{GS} \neq 1$ or $\beta_{GS} = 0$ and $\alpha_{GS} = 1$. ■

Just like the case of the α-stable models, there are also several parameterizations of the characteristic function of the geometric stable model that are available; here we also note the parameterizations of Kozubowski (2000b). Hence, in addition to the general geometric stable model, one can also define the class of strictly geometrically stable models typically reparameterized as shown in Definition 17.28.

Definition 17.28 (Geometric Stable Parameterisations of Kozubowski (Strictly Stable Laws)) *A strictly geometric stable random variable can also have a characteristic function given by the three-parameter representation*

$$\Phi_Z(\theta) = \left[1 + \lambda_{GS_K}|\theta|^{\alpha_{GS_K}} \exp\left(-\mathrm{i}\frac{\pi}{2}\alpha_{GS_K}\tau_{GS_K}\mathrm{sgn}(\theta)\right)\right]^{-1}, \tag{17.71}$$

where the subscript on the parameters GS_K denotes that we are considering the geometric stable model under the parameterization of Kozubowski. In this case, the parameters have support for the tail index $\alpha_{GS_K} \in (0, 2]$, the scale $\lambda_{GS_K} > 0$, and the skewness $|\tau_{GS_K}| \leq \min\left(1, \frac{2}{\alpha_{GS_K}} - 1\right)$. ■

In addition, we have the geometric stable characterization given in Theorem 17.5; see Klebanov *et al.* (1985, theorem 4).

Theorem 17.5 *A random variable for the annual loss Z is geometrically strictly stable iff for some $\alpha_{GS} \in (0, 2]$ and some $p_1 \in (0, 1)$ and $p_2 \in (0, 1)$ such that the quotient $\frac{\ln p_1}{\ln p_2}$ is irrational, and the following equality in distribution holds:*

$$Z \stackrel{d}{=} p_1^{\frac{1}{\alpha_{GS}}} \sum_{n=1}^{N_{p_1}} X_n \stackrel{d}{=} p_2^{\frac{1}{\alpha_{GS}}} \sum_{n=1}^{N_{p_2}} X_n. \tag{17.72}$$

In the context of this chapter and the closed-form expressions for the insured OpRisk loss processes being derived in this section, the aforementioned asymptotic result is more interesting if we consider the distribution of the *insured* severity model. Consider a generic severity model in the LDA structure and apply a generic insurance policy, denoting the resulting insurance mitigated loss severity model by $\widetilde{X}_i \sim \widetilde{F}(x)$. Then the aforementioned result also allows one to characterize in closed form an approximation of the distribution for the annual loss of the insurance mitigated loss process.

Remark 17.4 *Having provided a formal definition of the notion of geometric infinite divisibility and geometric stability, we note that in an OpRisk setting, any sets of LDA models (with insurance mitigation or without) that satisfy these two conditions will not only result in the individual loss process LDA models being uniquely characterized by an annual loss model from the geometric stable family, but in addition the joint aggregation of these annual loss processes for the institution-wide insured loss process will also be uniquely characterized by the family of geometric stable models.*

There are well-known families of distributions and LDA compound processes that satisfy that they are both geometrically infinitely Divisible as well as being geometrically stable. These include the compound geometric-exponential distributions, the Mittag-Leffler distributions (see Pillai and Jayakumar 1995), and the wider class of distributions constructed from Berstein functions that satisfy these two conditions in Fujita (1993). In fact in Aly and Bouzar (2000), it is stated that any positively supported random variable $X \in \mathbb{R}^+$ is geometrically strictly stable iff it is also necessarily geometrically infinitely divisible as stated in Lemma 17.6.

Lemma 17.6 (Geometric Stability and Geometric Infinite Divisibility) *Consider the real-valued positively supported random variable $Z \in \mathbb{R}^+$ with Laplace Stiegjtles transform denoted $\psi_Z(u)$. Then random variable $Z \sim F$ is geometrically strictly Stable ($F \in \mathcal{F}^*_{gs}$) iff $\psi_Z(u)$ satisfies*

$$\psi_Z(u) = \frac{p\psi_Z(u)}{1 - q\psi_Z(u)} \tag{17.73}$$

for any $p \in (0,1)$. In addition, all annual loss random variables Z satisfying this condition of geometric strict stability with only positive support are necessarily geometrically infinitely divisible ($F \in \mathcal{F}_{gid}$) as well as being infinitely divisible ($F \in \mathcal{F}_{id}$).

One can further state the following properties that are equivalent in Proposition 17.1; see discussion in Aly and Bouzar (2000, proposition 4.4).

Proposition 17.1 (Positive Geometrically Stable Random Variable Equivalencies) *Consider the real-valued positively supported annual loss random variable $Z \in \mathbb{R}^+$, then the following statements are equivalent:*

1. *$Z \sim F$ is g.i.d. with $F \in \mathcal{F}_{gid}$;*
2. *Z is g.i.d. and has a compound Poisson mixture representation denoted by $N_\lambda(Z)$, which is also g.i.d. for all Poisson intensities $\lambda > 0$;*
3. *Z is g.s. and has a compound Poisson mixture representation denoted by $N_\lambda(Z)$, which is also g.s. for all Poisson intensities $\lambda > 0$;*
4. *Z is compound exponential;*
5. *If the distribution of Z has an atom at the origin, then these are also equivalent to saying that r.v. Z satisfies the stability equation $Z \stackrel{d}{=} B(Z + S)$ for some Bernoulli random variable B and some independent positive real-valued random variable $S \in \mathbb{R}^+$.*

To understand how to work with and study the features of such an asymptotic result as geometric stability, we first present some details of the characterization of this family of models. We note the following relationship between a geometric stable model and the α-Stable model in Lemma 17.7; see Kozubowski (1994).

Lemma 17.7 (Geometric Stable and α-Stable Characteristic Functions) *The geometric stable characteristic function is expressed in terms of the log of the α-stable characteristic function (Φ_Y) under Zolotarev's B-Type parameterizations according to the relationship*

$$\begin{aligned}\Phi_{Z_{N_p}}(\theta; \alpha_{GS}, \beta_{GS}, \gamma_{GS}, \delta_{GS}) &= [1 - \ln \Phi_Y(\theta; \alpha_B, \beta_B, \gamma_B, \delta_B)]^{-1} \\ &= \left[1 + \gamma_{GS}^{\alpha_{GS}} |\theta|^{\alpha_{GS}} \omega_{\alpha_{GS},\beta_{GS}}(\theta) - i\delta_{GS}\theta\right]^{-1},\end{aligned} \tag{17.74}$$

where

$$\omega_{\alpha_{GS},\beta_{GS}}(\theta) = \begin{cases} 1 - i\beta_{GS}\operatorname{sgn}(\theta)\tan\left(\frac{\pi\alpha_{GS}}{2}\right), & \alpha_{GS} \neq 1, \\ 1 + i\beta_{GS}\frac{2}{\pi}\operatorname{sgn}(\theta)\ln|\theta|, & \alpha_{GS} = 1, \end{cases} \tag{17.75}$$

and with the following subsequent relationships between the parameters

$$\begin{aligned} \alpha_{GS} = \alpha_B, \quad \beta_{GS} &= \begin{cases} \cot\left(\frac{\pi}{2}\alpha_B\right)\tan\left(\frac{\pi}{2}\beta_B K(\alpha_B)\right), & \alpha \neq 1, \\ \beta_B, & \alpha = 1, \end{cases} \\ \delta_{GS} = \delta_B \gamma_B, \quad \gamma_{GS} &= \begin{cases} \cos\left(\frac{\pi}{2}\beta_B K(\alpha_B)\right)\gamma_B, & \alpha \neq 1, \\ \frac{\pi}{2}\gamma_B, & \alpha = 1, \end{cases} \end{aligned} \tag{17.76}$$

that produces the equivalence in distribution for $Z_{N_p} \sim GS_{\alpha_{GS}}(\beta_{GS}, \gamma_{GS}, \delta_{GS})$ *and a B-Type parameterized stable random variable according to* $X \sim S_{\alpha_B}(\beta_B, \gamma_B, \delta_B)$ *according to*

$$Z_{N_p} \stackrel{d}{=} \begin{cases} \delta_{GS} W + W^{\frac{1}{\alpha_{GS}}} \gamma_{GS} X, & \alpha_{GS} \neq 1, \\ \delta_{GS} W + W \gamma_{GS} X + \frac{2}{\pi} W \ln\left(W \gamma_{GS}\right), & \alpha_{GS} = 1, \end{cases} \tag{17.77}$$

with $W \sim$ *Exponential*(1) *and* $X \perp W$.

Remark 17.5 *Having specified explicitly the relationship between the characterization of the geometric stable law and the* α*-stable law, one may utilize representations developed for stable to characterize distributional properties.*

17.5.3.2 Developing a Geometric Stable Approximation of an Insured LDA Model. The results on asymptotic convergence of suitably scaled compound loss processes to Geometric stable models mean that in practice one can fit the geometric stable model as an approximation to any compound process with a large number claim numbers. This can be achieved under many methods depending on what data are available. If the actual annual loss amounts are obtained as a sample over several years, one may utilize parameter estimation for the geometric stable model followed by a compound hypothesis test for the goodness of fit of such an approximation model. Of course, in the context of this chapter, this would be done for the insured annual loss amounts. Alternatively, one may have access to the LDA model frequency and severity (fitted) models as well as a particular type of insurance policy, in which case the approximation of the process by a geometric stable model could be undertaken in a number of ways: moment matching, quantile matching either explicitly or alternatively via simulation and estimation.

Remark 17.6 (Geometric Stable Approximations for Generic Insured LDA Models) *The most generic approach to this problem would involve taking the desired LDA model, simulating many realizations of the annual losses from the LDA model, in the process also applying the required insurance policy to the loss process. This will allow one to obtain a sample for the insured annual loss amounts (of any desired size for a given computational budget), which could then be fitted to a geometric stable model.*

Assuming access to a set of i.i.d. realizations of the insured annual losses for T years (either observed or simulated from an LDA model with insurance applied), denoted by $\left\{\widetilde{Z}_t\right\}_{t=1}^{T}$, one can perform fitting of the resulting geometric stable approximation. In Kozubowski (1999), several approaches are presented to perform estimation of a geometric stable distributions parameters from a data sample. These approaches are based on the estimation literature in the α-stable setting, modified for the context of the geometric stable case. We will consider one such case that has been found to perform well and is efficient to implement in the α-stable setting, based on a method of moments type estimation from the sample characteristic function. This method is an adaption of the approach proposed in α-stable settings by Press (1972). We summarize the required implementation steps (assuming that only the LDA model is known for the loss process and the insurance policy specifications):

Performing a Geometric Stable Approximation of an Insured OpRisk Loss Process

1. Generate under the desired LDA model T years worth of loss counts $\{N_t\}_{t=1}^{T}$;
2. For each year t, generate the losses $\{X_i(t)\}_{i=1}^{N_t}$ for year t;
3. Apply the given insurance policy structure to the annual loss process to obtain transformed annual losses $\left\{\widetilde{Z}_t\right\}_{t=1}^{T}$;
4. Estimate the approximation for the annual loss via a geometric stable distribution function $\widetilde{Z}_t \sim F_{\widetilde{Z}_t}(z; \alpha_{GS}, \beta_{GS}, \gamma_{GS}, \delta_{GS})$ by estimation of the parameters as follows
   a) Using the fact that the characteristic function of $F \in \mathcal{F}_{gs}$ satisfies

$$\frac{1}{\Phi(\theta)} = 1 + \gamma_{GS}|\theta| w(\theta, \alpha_{GS}, \beta_{GS}) - \mathrm{i}\delta_{GS}\theta, \tag{17.78}$$

   therefore if one considers the real and imaginary parts separately a system of equations can be obtained to solve for the parameters by substitution. In each case, it will of course involve the estimation of the empirical characteristic function given for a value of θ and the i.i.d. samples $\widetilde{Z}_1, \widetilde{Z}_1, \ldots, \widetilde{Z}_T$ by

$$\hat{\Phi}(\theta) = \frac{1}{T}\sum_{t=1}^{T} \exp\left(\mathrm{i}\theta\widetilde{Z}_t\right) \tag{17.79}$$

   such that $\hat{\Phi}(\theta) \to \Phi(\theta)$ almost surely as $T \to \infty$.

   b) **Real Components (Solving for α_{GS} and γ_{GS}).** First, considering the real component given for any value of θ by

$$\nu(\theta) := \left|\mathcal{R}\mathrm{e}\left[\frac{1}{\Phi(\theta)}\right] - 1\right| = \gamma_{GS}\,|\theta|^{\alpha_{GS}} \tag{17.80}$$

   then by considering two distinct nonzero values of θ denoted by θ_1 and θ_2, one can solve the system of equations to obtain

$$\begin{aligned} \alpha_{GS} &= \ln\left[\frac{\nu(\theta_1)}{\nu(\theta_2)}\right]\left\{\ln\left[\frac{\theta_1}{\theta_2}\right]\right\}^{-1} \\ \gamma_{GS} &= \exp\left(\frac{\ln|\theta_1|\ln[\nu(\theta_2)] - \ln|\theta_2|\ln[\nu(\theta_1)]}{\ln\left[\frac{\theta_1}{\theta_2}\right]}\right), \quad \text{if } \alpha_{GS} \neq 1. \end{aligned} \tag{17.81}$$

c) **Imaginary Components (Solving for β_{GS} and δ_{GS})**. Considering the imaginary component given for any value of θ and parameters α_{GS} and γ_{GS} by

$$\eta(\theta) := -\mathcal{I}\text{m}\left[\frac{1}{\Phi(\theta)}\right] = \delta_{GS}\theta + \gamma_{GS}\left|\theta\right|^{\alpha_{GS}}\beta_{GS}\tan\left(\frac{\pi\alpha_{GS}}{2}\right)\text{sgn}\left(\theta\right), \quad \text{if } \alpha_{GS} \neq 1. \tag{17.82}$$

Then by considering two distinct nonzero values of θ denoted by θ_3 and θ_4 one can solve the system of equations to obtain

$$\begin{aligned}\beta_{GS} &= \frac{\frac{\eta(\theta_3)}{\theta_3} - \frac{\eta(\theta_4)}{\theta_4}}{\frac{2\gamma_{GS}}{\pi}\ln\left|\frac{\theta_4}{\theta_3}\right|}\\ \delta_{GS} &= \frac{\eta\left(\theta_3\right)\frac{\ln|\theta_4|}{\theta_3} - \eta\left(\theta_4\right)\frac{\ln|\theta_3|}{\theta_4}}{\ln\left|\frac{\theta_4}{\theta_3}\right|}.\end{aligned} \tag{17.83}$$

Note. One may obtain confidence intervals for the parameter estimates using the results discussed in (Kozubowski, 1999, appendix 1);

5. Perform a goodness-of-fit compound hypothesis test to assess the approximation accuracy of the geometric stable model assumption for a finite sample; see omnibus tests for such an analysis in Chapter 8.

For the purpose of analysis of the geometric stable model approximation, both for parametric bootstrap estimations of the p-value in the GOF test and estimation of capital measures under this model, it will be relevant to know how to simulate from such an approximation model given the estimated parameters. To achieve this, we can utilize several approaches as described in Kozubowski (2000a). For instance, given the ability to simulate α-stable random variables, one could perform transformations to obtain draws from the geometric stable model. However, as noted in Kozubowski (2000a) there is an alternative equivalent exact simulation method for strictly geometric stable models that avoids the assumption one can draw from the α-stable model equivalent.

In the case of the strictly geometric stable random variable $Z \sim F_Z \in \mathcal{F}^*_{gs}$ with parameters α_{GS_K}, λ_{GS_K}, and τ_{GS_K}, one can perform simulation as follows:

Algorithm 17.1 (Simulating from a Strictly Geometric Stable Distribution)

1. *Set* $p = \frac{1+\tau_{GS_K}}{2}$;
2. *Generate an exponential random variable* $E \sim Exp(1)$;
3. *Generate a uniform random variable* $U \sim Uniform(0, 1)$;
4. *If* $U \leq p$
   Then set $\rho = \alpha_{GS_K} p$ *and set* $I = 1$;
5. *Else*
   Then set $\rho = \alpha_{GS_K}(1 - p)$ *and set* $I = -1$;
6. *If* $\rho = 1$
   Then set $W = 1$;

7. *Else*
   Generate a uniform random variable $V \sim Uniform(0, 1)$ *and set*
   $W = \sin(\pi\rho)\cot(\pi\rho V) - \cos(\pi\rho)$;
8. *Set* $Z = IZ\left(\lambda_{GS_K} W\right)^{\frac{1}{\alpha_{GS_K}}}$.

Alternatively, in the case of $F \sim \mathcal{F}_{gs}$ with parameters α_{GS}, βGS, γ_{GS}, and δ_{GS} one can perform simulation as follows:

Algorithm 17.2 Simulating from a Geometric Stable Distribution (via an α-Stable Model)

1. *Generate a standard* α*-stable random variable* $S \sim \mathcal{S}_{\alpha_{GS}}\left(\beta_{GS}, 1, 0\right)$;
2. *Generate a standard exponential random variable* $E \sim Exp(1)$;
3. *If* $\alpha_{GS} = 1$
   Then set $Z = \delta_{GS}E + \gamma_{GS}ES + \frac{2}{\pi}\gamma_{GS}\beta_{GS}E\ln\left(E\gamma_{GS}\right)$;
4. *Else*
   Then set $Z = \delta_{GS}E + \gamma_{GS}E^{\frac{1}{\alpha_{GS}}}S$.

17.5.3.3 Large Claim Number Insured LDA Model Approximations: Densities, Distributions, and Tails. A geometric stable distribution or geo-stable distribution is a type of leptokurtic probability distribution. The geometric stable distribution may be symmetric or asymmetric. In general, one can develop a generic Geometric Stable distribution series representation such as those proposed in Kozubowski (1999, section 3), where both symmetric and skewed geometric stable series representations are presented. Representations of this form are most of interest with regard to understanding the characterization of such a loss process, though they add little in practical value with regard to evaluation of the annual loss distribution. We briefly present these for general geometric stable models before presenting more specific density and distribution representations.

The series of Kozubowski (1999, section 3) is based on the characterization of univariate infinitely divisible random variables by a LePage series; see LePage *et al.* (1997). In Theorem 17.6, we state the relationship between the series representation of a symmetric α-stable random variable and the equivalent form for the geometric stable random variable as well as the nonsymmetric case for the geometric stable; see details in Kozubowski (1999, theorem 3.1).

Theorem 17.6 (Series Representations for Geo-Stable Laws for Insured LDA Models) *Consider the uninsured LDA model with frequency distribution* $N_p \sim Geometric(p)$ *and severity distribution* $X_i \sim F_{X_i}(x)$ *and the resulting generic insured distribution for each i.i.d. loss will be denoted generically by* $\widetilde{X}_i \sim \widetilde{F}(x)$ *(the distribution of the transformed severity random variable under a given insurance mitigation product). Now consider the compound process for the insured annual loss given by*

$$\widetilde{Z}_{N_p} = \sum_{i=1}^{N_p} \widetilde{X}_i. \tag{17.84}$$

In the asymptotic limit as $\lim_{p\to 0}$, *one can consider the suitably scaled annual compound process random variable* $\widetilde{Z}_{N_0}$, *which can have a symmetric or asymmetric asymptotic distribution limit:*

1. ***Symmetric Geometric Stable Representation:***
In the symmetric case $\widetilde{Z}_{N_0} \sim GS_{\alpha_{GS}}(0, \gamma_{GS}, \delta_{GS})$, *one has the representation given according to a transformation of the* α*-stable LePage series representation,*

$$\lim_{p\to 0} \widetilde{Z}_{N_p} \stackrel{d}{=} W^{\frac{1}{\alpha_{GS}}} \underbrace{\sum_{i=1}^{\infty} D_i \Gamma_i^{-\frac{1}{\alpha_{GS}}}}_{\text{Symmetric } \alpha\text{-Stable}} \tag{17.85}$$

with random variable $W \sim Exp(1)$ *such that for all i one has* $W \perp \Gamma_i \perp D_i$, *a sequence of i.i.d. random variables* $\{\Gamma_i\}$, *where* Γ_i *are arrival times of a poisson point process with unity intensity and an independent Rademacher sequence of random variables* $\{D_i\}$ *(an i.i.d. sequence of random variables taking the values* $+1$ *and* -1 *with probabilities* p_0 *and* q_0 *respectively each);*

2. ***Nonsymmetric Geometric Stable Representation:***
To define the nonsymmetric case, we first define the following distribution function notations for the severity of the insured process $\widetilde{X}_i$:

$$\begin{aligned} G_+(x) &= 1 - \widetilde{F}_{X_i}(x|X_i \geq 0) = \mathbb{P}\mathrm{r}[X_i \geq x | X_i \geq 0] \\ G_-(x) &= 1 - \widetilde{F}_{-X_i}(x| - X_i \geq 0) = \mathbb{P}\mathrm{r}[-X_i \geq x | - X_i \geq 0] \\ G(x) &= \mathbb{P}\mathrm{r}(|X_i| \geq x), \quad p_0 = \mathbb{P}\mathrm{r}[X_i \geq 0], \quad q_0 = \mathbb{P}\mathrm{r}[-X_i > 0]. \end{aligned} \tag{17.86}$$

In the nonsymmetric case with Z_{N_p} *is asymptotically geometric stable as* $Z_{N_0} \sim GS_{\alpha_{GS}}(\beta_{GS}, \gamma_{GS}, \delta_{GS})$, *one has the series representation*

$$\lim_{p\to 0} \widetilde{Z}_{N_p} \stackrel{d}{=} \sum_{i=1}^{N_p^*} D_i G_{D_i}^{-1}\left(\frac{\Gamma_i}{\Gamma_{N_p^*+1}}\right) \tag{17.87}$$

with $N_p^* = G_p^{-1}(U)$ *such that* $G_p^{-1}(x) = \inf\{y : \mathbb{P}\mathrm{r}(N_p \leq y) \geq x\}$ *and* $U \sim Uniform(0,1)$ *with the condition that for all i one has independence* $U \perp D_i \perp \Gamma_i$. *In addition, one has the limiting behavior for the geometric stable domain of attraction in the non symmetric case, therefore producing the almost sure convergence to the following related series representation for the nonsymmetric case:*

$$\begin{aligned} &\lim_{p\to 0} \frac{1}{a_p}\left[\sum_{i=1}^{N_p^*} D_i G_{D_i}^{-1}\left(\frac{\Gamma_i}{\Gamma_{N_p^*+1}}\right) - b_p\right] \\ &\stackrel{a.s.}{=} W^{\frac{1}{\alpha_{GS}}} \sum_{i=1}^{\infty}\left(Z_i \Gamma_i^{-\frac{1}{\alpha_{GS}}} - C_i\right) + f(W, \alpha_{GS}, A, B, p_0, q_0) \end{aligned} \tag{17.88}$$

with $Z_i = A^{\frac{D_i+1}{2}} + B^{\frac{D_i-1}{2}}$ *in which A and B are given by*

$$A = \lim_{n\to\infty} \frac{G_+^{-1}\left(\frac{1}{n}\right)}{G^{-1}\left(\frac{1}{n}\right)}, \qquad B = \lim_{n\to\infty} \frac{G_-^{-1}\left(\frac{1}{n}\right)}{G^{-1}\left(\frac{1}{n}\right)} \tag{17.89}$$

and the function f is given by

$$f\left(W, \alpha_{GS}, A, B, p_0, q_0\right) = \begin{cases} \dfrac{\alpha_{GS}}{\alpha_{GS}-1}\left(p_0A^2 - q_0B^2\right)\left(W - W^{\frac{1}{\alpha_{GS}}}\right), & \alpha_{GS} \neq 1, \\ \left(p_0A^2 - q_0B^2\right) W(\ln W - 1), & \alpha_{GS} = 1 \end{cases} \tag{17.90}$$

and the series normalizing constants involve

$$\begin{aligned} a_p &= G^{-1}(p), \\ b_p &= p_0 A \int_p^1 G_+^{-1}(X)dx - q_0 B \int_p^1 G_-^{-1}(x)dx. \end{aligned} \tag{17.91}$$

Remark 17.7 *The aforementioned representations are instructive of the properties of the geometric stable model approximation as well as being very general in that they relate the geometric stable representation directly to the properties of the severity model density. In addition, they provide a guide to simulation of random variables from the geometric stable process via simple transformations of independent random variables.*

However, in practice, it will be important to be able to evaluate the density pointwise. In the following, we also provide representations that admit density and distribution function representations of the geometric stable approximation for an insured LDA model, under appropriate scaling.

Before presenting representations of the distribution and density of the geometric stable families, we can first characterize the family of geometrically infinitely divisible distributions. This is important as a special subset of these distributions will also correspond to a subclass of the geometric stable family that happen to admit a closed-form representation.

A general representation of the distributions in the class $\mathcal{F}_{gid}$ is provided by Fujita (1993), which characterizes all random variables $Z \in \mathbb{R}^+$ such that $Z \sim F \in \mathcal{F}_{gid}$ via the class of Bernstein functions, given in Definition 17.29; see details in Berg and Forst (1975, definition 9.1).

Definition 17.29 (Bernstein Functions) *A* C^∞*-function* $f : (0, \infty) \mapsto \mathbb{R}$ *is a Bernstein function if* $f(z) \geq 0$ *for all* $z > 0$ *and it satisfies the derivative conditions for all* $n \in \mathbb{J}^+$

$$(-1)^n \frac{d^n f(z)}{dz^n} \leq 0, \tag{17.92}$$

which results in $f(z)$ *having a first derivative that is a completely monotone function. This allows the class of such functions* $f(z)$ *to be characterized exactly by the representation*

$$f(z) = a + bz + \int_0^\infty (1 - \exp(-sz))\,\mu(ds), \quad z > 0 \tag{17.93}$$

for constants $a > 0$, $b > 0$ and positive measure $\mu(ds)$ on support $(0, \infty)$ that satisfies

$$\int_0^\infty \frac{s}{1+s}\mu(ds) < \infty. \tag{17.94}$$

∎

One can then utilize two constraints on the function $f(z)$ proposed by Fujita (1993) given by the left and right limiting values of the function to produce a unique representation of all positively supported geometrically stable models, as detailed in Theorem 17.7.

Theorem 17.7 (Characterization the Geometrically Infinitely Divisible Loss Processes) *Any random variable $Z \in \mathbb{R}^+$ with a distribution function satisfying $Z \sim F \in \mathcal{F}_{gid}$ is uniquely represented by the series*

$$F(z) = -\sum_{n=1}^{\infty}(-1)^n W^{(n)*}([0, z]), \quad z > 0, \tag{17.95}$$

where $W(dz)$ is a positive measure that satisfies for some Bernstein function f with the constraints $\lim_{z\downarrow 0} f(z) = 0$ and $\lim_{z\to\infty} f(z) = \infty$ the relationship given by

$$\frac{1}{f(z)} = \int_0^\infty \exp(-sz)W(dz), \quad z > 0. \tag{17.96}$$

Remark 17.8 *This is a general characterization of all positively supported geometrically infinitely divisible distributions, though it only provides a representation and in general does not provide a constructive manner to obtain a representation for a general distribution, via an approach to directly solve for the measure $W(dz)$ by finding the appropriate Bernstein function.*

Now returning to the focus of geometrically stable distributions and densities, we can define two important subfamilies: those that have distributions satisfying $F \in \mathcal{F}_{gs}$ and those that have both $F \in \mathcal{F}_{gs}$ and $F \in \mathcal{F}_{gid}$.

An important subfamily with $F \in \mathcal{F}_{gs}$ is generally referred to by an alternative name: Linnik and generalized Linnik laws. A symmetric geometric stable distribution is also referred to as a Linnik distribution (Kotz *et al.* 1995b, Klebanov *et al.* 1996, Kotz and Ostrovskii 1996 and Erdogan and Ostrovskii 1998). In the case of the Kozubowski parameterizations setting, $\tau_{GS_K} = 0$ produces a symmetric distribution corresponding to the Linnik distribution. There are also generalized versions of the geometric stable model Linnik class that are nonsymmetric as presented in Erdogan (1999).

An important subfamily with $F \in \mathcal{F}_{gs}$ as well as $F \in \mathcal{F}_{gid}$ is generally referred to by the alternative name of Mittag-Leffler laws; see discussion in Kozubowski (1999). This subclass of models of the geometric stable family is known in the physics literature as the Mittag-Leffler function distributions, developed by Pillai and Sandhya (1990). This subfamily admits a positive support and is directly based on the Mittag-Leffler function, given in Definition 17.30. Under the parameterizations of Kozubowski, this subfamily is characterized by setting $\alpha_{GS_K} \leq 1$ making it heavy tailed when $\alpha_{GS_K} << 1$ and exponential when $\alpha_{GS_K} = 1$ and perfectly skewed through $\tau_{GS_K} = 1$.

Definition 17.30 (Geometric Stable Subfamily: Mittage-Leffler Distributions) *The annual loss random variable* $Z \in \mathbb{R}^+$ *has a Mittag-Leffler (geo-stable distribution) denoted by* $Z \sim ML(\alpha_{GS})$ *if it has a distribution* F_Z *given by*

$$F_Z(z) = \sum_{k=1}^{\infty} \frac{(-1)^{k-1} z^{k\alpha_{GS_K}}}{\Gamma(1 + k\alpha_{GS_K})}. \tag{17.97}$$

This distribution has the property $F \in \mathcal{F}^*_{gid}$ *and in addition* $F \in \mathcal{F}_{id}$. ■

More generally, there have also been several series expansions and integral representations developed for the family of distributions $F \in \mathcal{F}_{gs}$ that we discuss later.

A second characterization of a generalized family of the distributions $F \in \mathcal{F}^*_{gs}$ was expressed by a popular integral representation discussed in Erdogan and Ostrovskii (1998), Klebanov *et al.* (1996), and the two part series of Kotz *et al.* (1995a,b). This is presented in Theorem 17.8; see Erdogan (1999, theorem 1).

Theorem 17.8 (Generalized Nonsymmetric Linnik (Geo-Stable) Densities) *Consider the Kozubowski parameterizations of strictly stable models in which, without loss of generality, the scale is set such that* $\lambda_{GS_K} = 1$. *Then the density of an insured annual loss process with insurance mitigation can be approximately represented according to the density denoted by* $p(z; \alpha_{GS_K}, \tau_{GS_K})$. *This is a two-parameter form that will admit several different representations given by the following three cases:*

1. *if* $\alpha_{GS_K} \in (0,1)$, $0 \le \tau_{GS_K} \le \frac{\pi}{2}$ *or* $\alpha_{GS_K} \in [1,2)$, $0 \le \tau_{GS_K} \le \frac{\pi}{\alpha_{GS_K}} - \frac{\pi}{2}$ *then for* $z \in \mathbb{R}$ *one has*

$$p(z; \alpha_{GS_K}, \tau_{GS_K}) = \frac{\sin\left(\frac{\pi\alpha_{GS_K}}{2} + \alpha_{GS_K}\tau_{GS_K} sgn(z)\right)}{\pi} \int_0^{\infty} \frac{\exp(-sgn(z)y)\, y^{\alpha_{GS_K}}\, dy}{\left|1 + \exp(i\alpha_{GS_K}\tau_{GS_K} sgn(z))\, y^{\alpha_{GS_K}} \exp\left(\frac{i\pi\alpha_{GS_K}}{2}\right)\right|^2}, \tag{17.98}$$

2. *if* $\alpha_{GS_K} \in [1,2)$ *and* $\tau_{GS_K} = \frac{\pi}{\alpha_{GS_K}} - \frac{\pi}{2}$ *then*

$$p(z; \alpha_{GS_K}, \tau_{GS_K}) = \begin{cases} -\dfrac{\sin(\pi\alpha_{GS_K})}{\pi} \displaystyle\int_0^{\infty} \frac{\exp(yz)\, y^{\alpha_{GS_K}}\, dy}{|1 - \exp(i\pi\alpha_{GS_K})\, y^{\alpha_{GS_K}}|^2}, & z < 0, \\ \dfrac{1}{\alpha_{GS_K}} \exp(-z), & z > 0. \end{cases} \tag{17.99}$$

3. *if* $\tau_{GS_K} \in \left(-\frac{\pi}{\alpha_{GS_K}}, 0\right)$, *then one has*

$$p(z; \alpha_{GS_K}, \tau_{GS_K}) = p(-z; \alpha_{GS_K}, -\tau_{GS_K}). \tag{17.100}$$

Now, in special cases, one can even get closed-form series expansions with bounded approximation error due to truncation of the series; see detailed results for a range of values of α_{GS_K} in Erdogan (1999, theorems 5, 6 and 7). In Theorem 17.9, we present one of these results of direct relevance to practitioners who may wish to work with such geometric stable approximations.

Theorem 17.9 (Finite Series Characterization of Nonsymmetric Linnik Densities) *Consider the Kozubowski parameterizations of strictly stable models in which, without loss of generality, the scale is set such that $\lambda_{GS_K} = 1$. Then the density of an insured annual loss process with insurance mitigation can be approximately represented according to the density denoted by $p\left(z;\alpha_{GS_K},\tau_{GS_K}\right)$. If one considers the parameter ranges $\alpha_{GS_K} \in (0,2)$ and $|\tau_{GS_K}| < \min\left(\frac{\pi}{2}, \frac{\pi}{\alpha_{GS_K}} - \frac{\pi}{2}\right)$, then the following finite sum density approximation for the annual loss applies:*

$$
\begin{aligned}
&p\left(z;\alpha_{GS_K},\tau_{GS_K}\right)\\
&= \frac{1}{\pi}\sum_{n=1}^{N}(-1)^{n+1}\Gamma\left(1+\alpha_{GS_K}n\right)\sin\left(\frac{\pi\alpha_{GS_K}n}{2} + n\alpha_{GS_K}\tau_{GS_K}sgn(z)\right)|z|^{-1-\alpha_{GS_K}n} + R_{N,\alpha_{GS_K}}(z),
\end{aligned}
\tag{17.101}
$$

with remainder term bounded in absolute value as follows:

$$
\left|R_{N,\alpha_{GS_K}}(z)\right| \le \frac{\alpha_{GS_K}\Gamma\left(1+\alpha_{GS_K}(N+1)\right)}{\pi\left|\sin\left(\frac{\pi\alpha_{GS_K}}{2} + \alpha_{GS_K}\tau_{GS_K}sgn(z)\right)\right|}|z|^{-1-\alpha_{GS_K}(N+1)}. \tag{17.102}
$$

We note that this model will be suitable for transformed loss processes under insurance mitigation for any of the aforementioned policy types. Where it is recognized that as $p \to 0$, the number of independent (not necessarily identically distributed) losses will dominate the number of modified insured losses in cases in which there are two distributions such as the ILPCn and ILPCa insurance policies. In addition, this closed-form result will be suitable for any choice of severity model in the LDA setup.

To conclude the discussion on the approximation of insured loss processes with large claim numbers, it is interesting to note the following asymptotic result for the representation of the tail of the geometric stable density as the annual loss amount gets large $\widetilde{Z} \to \infty$ given in Theorem 17.10; see details in Klebanov *et al.* (1996, section 3). Other results of this form for the Linnik subfamily are provided in the two combined works of Kotz *et al.* (1995a,b).

Theorem 17.10 (Tail Asymptotics for Geometric Stable Densities) *Consider the density of a geometric stable random variable denoted by $p\left(z;\alpha_{GS},\beta_{GS},\gamma_{GS},\delta_{GS}\right)$, then the following tail asymptotic representations apply:*

$$
\begin{aligned}
&p\left(z;\alpha_{GS},\beta_{GS},\gamma_{GS},\delta_{GS}\right) \sim\\
&\begin{cases}
\frac{1}{\pi}\sum_{s=0}^{\infty}\sum_{k=0}^{s}\sin\left(\frac{\pi\alpha_{GS}(1+\beta_{GS})(s-k)}{2}\right)\frac{(-1)^k s!}{k!(s-k)!}\delta_{GS}^k\gamma_{GS}^{s-k}\Gamma\left((s-k)\alpha_{GS}+k+1\right)\\
\quad\times z^{-(s-k)\alpha_{GS}-k-1}, \quad z\to\infty,\\
\\
\frac{1}{\pi}\sum_{s=0}^{\infty}\sum_{k=0}^{s}\sin\left(\frac{\pi\alpha_{GS}(1-\beta_{GS})(s-k)}{2}\right)\frac{ks!}{k!(s-k)!}\delta_{GS}^k\gamma_{GS}^{s-k}\Gamma\left((s-k)\alpha_{GS}+k+1\right)\\
\quad\times z^{-(s-k)\alpha_{GS}-k-1}, \quad z\to-\infty.
\end{cases}
\end{aligned}
$$

17.5.4 GENERIC CLOSED-FORM APPROXIMATIONS FOR INSURED LDA MODELS

In this section, we consider classes of models for which we do not have elegant properties of closure under convolution either in the original LDA formulation or as a result of the application of a particular insurance policy, which alters or removes the infinite divisibility for the resulting insured process. In particular, we will demonstrate that for insured processes that have finite d-th order moments, we can obtain a d-th order series representation for insured processes in one of two cases, those with infinite support on the real line, and secondly, those with support that is on a truncated domain of the real line due to the application of an insurance policy (such as under an ILPU policy). In particular, the series expansions we will derive will have the following important features:

- The density representation obtained will have a strictly positive support;
- The value of the density at all points in this support will be positive;
- The resulting finite truncated density representation will be normalized to 1 on this support with known approximation accuracy.

This makes the representation developed a valid density representations up to a known order of approximation. To achieve this, we will utilize a very important property of orthogonality of the Laguerre polynomials with a Gamma kernel (density). We begin this section by discussing the case in which the insured loss process denoted by annual loss random variable $\widetilde{Z}$ takes support on the whole positive real line, then detailing the results for bounded support cases. In each case, the series expansions considered utilize properties of orthogonality between particular kernel functions (probability density functions) that will act together with a polynomial basis to form a series representation. In this regard, it will be relevant to recall some basic properties of orthogonality between particular density functions and classes of polynomials; for many details on these matters, the interested reader is referred to Jackson (1941) and Osilenker (1999). In particular, it is of relevance to consider what are known as the Askey–Sheme of polynomials; see Askey and Wilson (1985). This class presents a wide array of subfamilies of polynomial bases that are orthogonal to a range of different density functions, making them directly useful for series representations of stochastic processes such as loss process (such expansions are also related to Wiener–Askey Chaos expansions).

In Theorem 17.11, we make explicit the notion of orthogonality as used throughout this section see details in Xiu and Karniadakis (2002) and references therein.

Theorem 17.11 *A polynomial class denoted generically by $\{P_n(x), n \in \mathbb{N}\}$ where $P_n(x)$ is a polynomial of exact degree $n \in \mathbb{N} = \{0, 1, 2, \dots\}$ is an orthogonal system with respect to a real positive measure μ, on support Ω, if the following integral identity holds for some nonzero constants a_n and a Dirac delta mass when $m = n$ denoted by $\delta_n(m)$:*

$$\int_{\Omega} P_n(x)P_m(x)d\mu(x) = a_n^2\delta_n(m), \quad n, m \in \mathbb{N}. \tag{17.103}$$

If the measure $\mu(dx)$ admits a density denoted generically by $w(x)$ (called the weighting function) then the orthogonality condition holds if

$$\int_{\Omega} P_n(x)P_m(x)w(x)dx = a_n^2\delta_n(m), \quad n, m \in \mathbb{N}. \tag{17.104}$$

In addition, it is often useful when performing evaluation of such polynomials in the Askey scheme, to note that all orthogonal polynomials $\{P_n(x)\}$ satisfy the recurrence relationship given by

$$-xP_n(x) = a_n P_{n+1}(x) - (a_n + c_n) P_n(x) + c_n P_{n-1}(x), \quad n \geq 1 \tag{17.105}$$

with $a_n, c_n \neq 0$, $\frac{c_n}{a_{n-1}} > 0$ and the initial values $P_{-1}(x) = 0$ and $P_0(x) = 1$.

Remark 17.9 *The Askey scheme of polynomials involves the hypergeometric orthogonal polynomials that satisfy some type of differential or difference equation and provides the limit relations between them. The orthogonal polynomials in the Askey scheme each have different weighting functions for the orthogonality relationship developed, which leads to the following orthogonality relationships between subfamilies of Askey scheme polynomials and density functions:*

1. *Hermite polynomials are associated with the Gaussian distribution;*
2. *Laguerre polynomials with the Gamma distribution;*
3. *Jacobi polynomials with the Beta distribution;*
4. *Charlier polynomials with the Poisson distribution;*
5. *Meixner polynomials with the Negative Binomial distribution;*
6. *Krawtchouk polynomials with the Binomial distribution;*
7. *Hahn polynomials with the Hypergeometric distribution.*

The two examples of direct interest to this chapter involve the Laguerre polynomials and Gamma distribution and the Jacobi polynomials with the Beta distribution. See further discussions in Xiu and Karniadakis (2002).

The remainder of these closed-form results to be derived will consider the generic representations of the insured loss process annual loss density and distribution can be approximated via the following weighted polynomial approximations of order d given generically by

$$\begin{aligned} \hat{f}_{\widetilde{Z},d}(z) &= \sum_{k=0}^{d} a_k P_k(x), \\ \hat{F}_{\widetilde{Z},d}(z) &= \sum_{k=0}^{d} A_k P_k(x), \end{aligned} \tag{17.106}$$

where one can show that the optimal choice of coefficients is given for the k-th term by the Fourier coefficients of the density $f_{\widetilde{Z}}(z)$ and distribution $F_{\widetilde{Z}}(z)$ with respect to the desired polynomial basis given by

$$\begin{aligned} a_k &= \int f_{\widetilde{Z}}(z) P_k(z) w(z) dz, \\ A_k &= \int F_{\widetilde{Z}}(z) P_k(z) w(z) dz. \end{aligned} \tag{17.107}$$

The weight function can be defined over a finite support or the whole real line. In the context of this chapter, two possible supports are of interest; the whole positive real line and an interval $[0, E]$ for some $E > 0$.

Remark 17.10 *It is well known that these series representations correspond to Fourier series-type expansions in different bases (compared to the classical cosine basis). It is therefore no surprise to find that the selection of the coefficients follows a similar procedure. However, in this application, one cannot evaluate easily the insured loss distribution; however, samples are available from the insured LDA model either through observations or via simulation methods of the LDA model giving samples* $\left\{\widetilde{Z}_t\right\}_{t=1}^{T}$*, which can be used to estimate the coefficients* $\hat{a}_k$ *and* $\hat{A}_k$*. Typically one may wish to preserve the property of unbiasedness in which* $\mathbb{E}\left[\hat{a}_k\right] = a_k$ *and* $\mathbb{E}\left[\hat{a}_k\right] = A_k$ *and each coefficient* $\hat{a}_k \to a_k$ $\hat{A}_k \to A_k$ *converges as* $T \to \infty$ *with a known rate; see discussions in Kronmal and Tarter (1968).*

Note that there are associated implications with the unbiasedness of each particular coefficient estimate, namely, the most trivial being that the resulting density and distribution function estimators will be by definition biased since one will have

$$\mathbb{E}\left[f_{\widetilde{Z}}(z) - \hat{f}_{\widetilde{Z},d}(z)\right] = \sum_{k=d+1}^{\infty} a_k P_k(x), \tag{17.108}$$

which can only be zero when all a_k are zero for $k \in \{d+1, d+2, \ldots\}$. The bias will be small if the a_k and A_k decrease quickly as k increases, that is, the leading coefficients dominate.

It is also worth noting that much of the literature of Fourier series approximations to distributions and densities has revolved around choice of the weighting function and focussing functions, with the most popular criterion for selection of these choices of functions being the mean integrated square error (M.I.S.E.) criterion; see discussions in Rosenblatt (1956), Parzen (1962), and Whittle (1958). This is not the case in the applications considered here since the support considered will be of a particular form that would lend itself to the choice of particular polynomial and kernel weight choices.

Algorithm 17.3 Generic Approach to Series Approximation of Insured Annual Loss Density

1. *Given an LDA frequency model (fitted to data), simulate* T *total years of loss counts* $\{N_t\}_{t=1}^{T}$ *from the frequency distribution;*
2. *Given an LDA severity model (fitted to data), simulate for each of the* T *total years of loss amounts* $\{X_i(t)\}_{i=1}^{N_t}$ *from the severity distribution;*
3. *Given a particular insurance policy, obtain the resulting insured loss amounts* $\left\{\hat{Z}_t\right\}_{t=1}^{T}$ *by applying the insurance policy to the simulated loss amounts and aggregating the uninsured amounts;*
4. ***IF*** *using a support* $\widetilde{Z} \in \mathbb{R}^+$
   - **a)** *Transform the simulated annual loss data (via an invertible mapping) to obtain equal mean and variance such that* $T(\widetilde{Z}) = \beta\widetilde{Z}$ *with* β *selected to make* $\mathbb{E}\left[T(\widetilde{Z})\right] = \mathbb{V}\mathrm{ar}\left(T(\widetilde{Z})\right)$*;*
   - **b)** *Utilize the Gamma weight function (density) and the resulting orthogonal polynomial basis given by Laguerre polynomials and fix the order of approximation* d *and the coefficient* α *(Note: Selecting* $\alpha > 1$ *weights the errors in the tails strongly and for* $\alpha \leq 1$ *the weight function has little effect near the origin and strongly weights error for large* z*);*

c) *Estimate using the simulated data the coefficients* $\{a_k\}_{k=1}^{d}$ *via*

$$\hat{a}_k = \frac{1}{T}\frac{\Gamma(\alpha)}{k!\Gamma(\alpha+k)}\sum_{t=1}^{T} L_k^{\alpha}\left(T\left(\widetilde{Z}_t\right)\right) \tag{17.109}$$

5. ***ELSEIF*** *using a support* $\widetilde{Z} \in [0, E]$

a) *Transform the simulated annual loss data (via an invertible mapping) such that* $T(\widetilde{Z}) \in [-1, 1]$;

b) *Utilize the standard Beta weight function (density) and the resulting orthogonal polynomial basis given by Jacobi polynomials and fix the order of approximation* d *and the coefficients* $\widetilde{\alpha} = \alpha - 1$ *and* $\widetilde{\beta} = \beta - 1$;

c) *Estimate using the simulated data the coefficients* $\{a_k\}_{k=1}^{d}$ *via*

$$\hat{a}_k = \frac{1}{T}\frac{1}{B(\widetilde{\alpha}+1, \widetilde{\beta}+1)2^{\widetilde{\alpha}+\widetilde{\beta}+1}}\sum_{t=1}^{T} P_k^{\widetilde{\alpha},\widetilde{\beta}}\left(T\left(\widetilde{Z}_t\right)\right) \tag{17.110}$$

Remark 17.11 *The generic approaches given earlier can be applied in any setting for approximation of the density and distribution functions. Note that in some cases it will be convenient to modify the representation in order to obtain a series say in terms of linear combinations of a certain density such as the example presented next with the gamma density.*

17.5.4.1 Series Expansions for General LDA Insured Processes (Support $\mathbb{R}^{+}$).

In this case, we will demonstrate how to obtain closed-form expressions for the insured annual loss process that we will denote by process $\widetilde{Z}$ as a function of properties of the LDA model for the uninsured LDA model annual loss denoted by Z. To proceed, we will first introduce an important orthogonality property of the Laguerre polynomial with a gamma density kernel in Theorem 17.12; see derivation in Osilenker (1999, p. 184).

Theorem 17.12 (Orthogonality of Laguerre Polynomials and Gamma Density Kernels) *Consider the Gamma density kernel given by*

$$g(z; \alpha, \beta = 1) = \frac{z^{\alpha-1}}{\Gamma(\alpha)}\exp(-z) \tag{17.111}$$

with scale parameter $\beta = 1$ *and shape parameter* α *such that the mean and variance are given by* $\mathbb{E}[Z] = \mathbb{V}\mathrm{ar}[Z] = \alpha$, *then the following orthogonality condition holds for generalized Laguerre polynomials* $L_n^{(\alpha)}(x)$ *such that*

$$\int_0^{\infty} \frac{z^{\alpha-1}}{\Gamma(\alpha)}\exp(-z)L_n^{(\alpha)}(z)L_m^{(\alpha)}(z)dz = \begin{cases} 0, & m \neq n, \\ \dfrac{n!\Gamma(\alpha+n)}{\Gamma(\alpha)}, & m = n. \end{cases} \tag{17.112}$$

Remark 17.12 *Note that the orthogonality result utilizes the representation of the generalized Lagurre polynomials according to the relationship*

$$L_n^{(\alpha)}(x) = (-1)^n x^{1-\alpha}\exp(-x)\frac{d^n}{dx^n}\left(x^{n+\alpha-1}\exp(-x)\right). \tag{17.113}$$

The more standard representation of these polynomials is via the Rodrigues' formula to obtain the representation

$$\widetilde{L}_n^{(\alpha)}(x) = \frac{1}{n!} x^{-\alpha} \exp(-x) \frac{d^n}{dx^n} \left(x^{n+\alpha} \exp(-x) \right). \tag{17.114}$$

The reason for this slight change in notation is associated with the form of the Gamma density parameterizations used and the resulting orthogonality result. To see the relationship between the two, we note the following:

$$L_n^{(\alpha)}(x) = n!(-1)^n \widetilde{L}_n^{(\alpha-1)}(x). \tag{17.115}$$

To illustrate the forms of the first few generalized Laguerre polynomials under the representation adopted in this section, they are given by the following coefficient representations with respect to shape parameter α:

$$\begin{aligned}
L_0^{(\alpha)}(x) &= 1, \\
L_1^{(\alpha)}(x) &= x - \alpha, \\
L_2^{(\alpha)}(x) &= x^2 - 2(\alpha+1)x + (\alpha+1)\alpha, \\
L_3^{(\alpha)}(x) &= x^3 - 3(\alpha+2)x^2 + 3(\alpha+2)(\alpha+1)x - (\alpha+2)(\alpha+1)\alpha, \\
L_4^{(\alpha)}(x) &= x^4 - 4(\alpha+3)x^3 + 6(\alpha+3)(\alpha+2)x^2 - 4(\alpha+3)(\alpha+2)(\alpha+1)x \\
&\quad + (\alpha+3)(\alpha+2)(\alpha+1)\alpha, \\
L_5^{(\alpha)}(x) &= x^5 - 5(\alpha+4)x^4 + 10(\alpha+4)(\alpha+3)x^3 - 10(\alpha+4)(\alpha+3)(\alpha+2)x^2 \\
&\quad + 5(\alpha+4)(\alpha+3)(\alpha+2)(\alpha+1)x - (\alpha+4)(\alpha+3)(\alpha+2)(\alpha+1)\alpha.
\end{aligned}$$

We can now assume that the manner in which we obtain the series expansion for the annual loss density and distribution of the insured loss process $\widetilde{Z}$ is via a series expansion that will exploit the orthogonality condition and utilize the Gamma basis functions to ensure that support is strictly positive, as shown in Theorem 17.13. This approach is based on the proposed series expansions of Bowers and Newton (1966) and Smith (1992) which utilises directly the orthogonality between the Gamma density kernel and the Laguerre polynomials to obtain an efficient series representation.

Theorem 17.13 (Generic Series Representation of Insured Loss Process) *Under an arbitrary specification of the severity and frequency distributions in the model for the LDA of the loss process being modeled and for an arbitrary choice of insurance product, the resulting insured annual loss $\widetilde{Z}$ can be represented under a simple transformation $T\left(\widetilde{Z}\right) = \zeta \widetilde{Z}$ such that ζ is selected such that $\mathbb{E}\left[T\left(\widetilde{Z}\right)\right] = \mathbb{V}\text{ar}\left(T\left(\widetilde{Z}\right)\right)$, up to d-th order by the Gamma density series*

$$\hat{f}_{T(\widetilde{Z}),d}(z) = \underbrace{\frac{z^{\alpha-1}}{\Gamma(\alpha)} \exp(-z)}_{\text{Gamma density shape } \alpha} \sum_{i=0}^{d} a_i L_i^{(\alpha)}(z) \tag{17.116}$$

assuming the first d moments of the insured process $\widetilde{Z}$ are finite, where

$$a_i = \frac{\Gamma(\alpha)}{i!\Gamma(\alpha+i)} \int_0^\infty f_{T(\widetilde{Z})}(z) L_i^{(\alpha)}(z) dz. \tag{17.117}$$

Remark 17.13 *Note that in the aforementioned series representation, the series is expanded in Laguerre polynomials; however, each polynomial term is multiplied by the Gamma density kernel. Since this is a slightly different representation to the most familiar approach presented in Equation (17.106), it will result in a slightly different representation of the optimal choice of coefficients.*

It is always good practice to consider the accuracy of the series approximation, and perhaps to consider the appropriate length of the series approximation (truncation point) that will ensure a certain accuracy in the series approximation to the actual loss process density or distribution being approximated. One way to perform such an analysis for the aforementioned series approximation for a finite order of approximation is to adopt a similar analysis as is performed when assessing the accuracy of a finite truncation in a Fourier series expansion. In particular, the expansion of a function according to a finite truncated Fourier series can be represented at any order by a trigonometric polynomial, which maximizes a certain integral; see Jackson (1941). This can also be seen to be the case with the aforementioned series expansion and was exploited in Bowers and Newton (1966) to study the accuracy and optimality of the finite order series approximation given earlier; see details in Theorem 17.14. This theorem demonstrates the optimality of the choice of form for the coefficients a_i in the earlier representation; though it does not provide the rate of convergence of the series representation; for these types of results, the interested reader is referred to the study of Kronmal and Tarter (1968).

Theorem 17.14 (Optimality of the Generic Gamma-Laguerre Series Representation) *The accuracy of the series representation can be studied for a given shape parameter α in the Gamma basis by consideration of the integral of the squared error between the exact annual loss density and a series approximation given by*

$$\begin{aligned} R &= \Gamma(\alpha) \int_0^\infty \frac{1}{z^{\alpha-1}\exp(-z)} \left[f_{T(\widetilde{Z})}(z) - f_{T(\widetilde{Z}),d}(z) \right]^2 dz \\ &= \Gamma(\alpha) \int_0^\infty \frac{1}{z^{\alpha-1}\exp(-z)} \left[f_{T(\widetilde{Z})}(z) - \frac{z^{\alpha-1}}{\Gamma(\alpha)} \exp(-z) \sum_{i=0}^{d} a_i L_i^{(\alpha)}(z) \right]^2 dz, \end{aligned} \tag{17.118}$$

which if minimized, for a given α, with respect to a_i coefficients via $\frac{\partial R}{\partial a_i} = 0$ yields the relationship (via orthogonality between the Laguerre and Gamma kernel),

$$a_i = \frac{\Gamma(\alpha)}{i!\Gamma(\alpha+i)} \int_0^\infty f_{T(\widetilde{Z})}(z) L_i^{\alpha}(z) dz. \tag{17.119}$$

It should be noted that this representation, while it normalizes to 1 when integrated over the support $[0, \infty)$, may not be guaranteed to be strictly a density as it can have terms that oscillate in the expansion and therefore can produce negative values of the density for some values of $z \in [0, \infty)$. This is also well known to happen in other areas of series expansions for distributions and densities such as the Gramm-Charlier type A series expansion classes and their variants. Therefore, we need to propose some additional constraints to ensure the resulting series expansion is also a valid density with strictly positive density function values. First, we note the following remark regarding the assumption of the existence of the moments of the loss process.

Remark 17.14 *The original uninsured LDA loss process model for Z need not have* $\mathbb{E}\left[(Z)^d\right] < \infty$, *it is required under this particular series representation to have that the resulting insured process does satisfy the condition* $\mathbb{E}\left[\left(\widetilde{Z}\right)^d\right] < \infty$. *In many insurance settings described earlier, this is not an overly restrictive criterion to require and allows in many cases for subexponential uninsured annual loss LDA models (even infinite mean LDA models) to still satisfy this condition for the insured process.*

In Bowers and Newton (1966), this type of series is utilized for an insurance setting and they re-express conveniently the series representation for the fourth-order case in the following manner for the density and distribution functions, respectively:

$$\begin{aligned}\hat{f}_{T(\widetilde{Z}),5}(z) &= \gamma(z;\alpha,1)\,(1-A+B-C) + \gamma(z;\alpha+1,1)(3A-4B+5C)\\ &\quad + \gamma(z;\alpha+2,1)(-3A+6B-10C) + \gamma(z;\alpha+3,1)(A-4B+10C)\\ &\quad + \gamma(z;\alpha+4,1)(B-5C) + \gamma(z;\alpha+5,1)C,\end{aligned} \tag{17.120}$$

$$\begin{aligned}\hat{F}_{T(\widetilde{Z}),5}(z) &= \Gamma(z;\alpha,1)\,(1-A+B-C) + \Gamma(z;\alpha+1,1)(3A-4B+5C)\\ &\quad + \Gamma(z;\alpha+2,1)(-3A+6B-10C) + \Gamma(z;\alpha+3,1)(A-4B+10C)\\ &\quad + \Gamma(z;\alpha+4,1)(B-5C) + \Gamma(z;\alpha+5,1)C,\end{aligned}$$

with $\gamma(z;\alpha,1)$ the density of a Gamma random variable with shape α and unit scale, $\Gamma(z;\alpha,1)$ the distribution function and the coefficients given by

$$\begin{aligned}A &= \frac{\mu_3 - 2\alpha}{3!}, \quad B = \frac{\mu_4 - 12\mu_3 - 2\alpha^2 + 18\alpha}{4!},\\ C &= \frac{\mu_5 - 20\mu_4 - (10\alpha - 120)\,\mu_3 + 60\alpha^2 - 144\alpha}{5!},\end{aligned} \tag{17.121}$$

with μ_n the n-th moment about the mean of the annual loss $T\left(\widetilde{Z}\right)$. Note that for convenience (Bartlett, 1965, p. 451), the LDA moments about the mean, that is, central moments about

$$\mu_1 = \mathbb{E}\left[T\left(\widetilde{Z}\right)\right] = \zeta\mathbb{E}\left[\widetilde{Z}\right] = \zeta\mathbb{E}\left[\widetilde{X}\right]\mathbb{E}\left[N\right]$$

for μ_2 to μ_4 of the compound process are provided according to the expressions for the mean of the insured severity model (r.v. $\widetilde{X}$) and the mean of the frequency distribution as follows:

$$\mu_2 = \mathbb{E}\left[\left(T\left(\widetilde{Z}\right) - \mu_1\right)^2\right] = \zeta^2\mathbb{E}\left[\left(\widetilde{Z} - \mathbb{E}\left[\widetilde{Z}\right]\right)^2\right] = \zeta^2\mathbb{E}\left[\widetilde{X}^2\right]\mathbb{E}\left[N\right],$$

$$\mu_3 = \mathbb{E}\left[\left(T\left(\widetilde{Z}\right) - \mu_1\right)^3\right] = \zeta^3\mathbb{E}\left[\widetilde{X}^3\right]\mathbb{E}\left[N\right],$$

$$
\begin{aligned}
\mu_4 &= \mathbb{E}\left[\left(T\left(\widetilde{Z}\right) - \mu_1\right)^4\right] = \zeta^4\left(\mathbb{E}\left[\widetilde{X}^4\right]\mathbb{E}\left[N\right] + 3\mathbb{E}\left[\widetilde{X}^2\right]^2\mathbb{E}\left[N\right]^2\right), \\
\mu_5 &= \mathbb{E}\left[\left(T\left(\widetilde{Z}\right) - \mu_1\right)^5\right] = \zeta^5\left(\mathbb{E}\left[\widetilde{X}^5\right]\mathbb{E}\left[N\right] + 10\mathbb{E}\left[\widetilde{X}^2\right]\mathbb{E}\left[\widetilde{X}^3\right]\mathbb{E}\left[N\right]^2\right).
\end{aligned}
\tag{17.122}
$$

Although in the majority of examples considered in OpRisk the aforementioned series expansion will be well behaved in the sense that the density will remain strictly positive in evaluation and not oscillate over the range of the support typically utilized in practice, it is still important to be certain this is the case for any give LDA model and insurance policy combination. To achieve this one may adopt the approach considered in Jondeau and Rockinger (1999) for the Gram-Charlier series. Next, a similar method as proposed in this work for the Gamma series expansion will be developed in Theorem 17.15 for the case of a series expansion of the insured loss process with respect to moments 1–4 (i.e., up to capturing the skew and kurtosis of the insured loss process). The basic idea behind this approach is to obtain an analytic functional representation of the domain of the skew and kurtosis of the resulting insured LDA process annual loss distribution that will admit a guaranteed strictly positive density value. Characterizing this via the compound process skewness and kurtosis is a natural and intuitive method that can be interpretable in terms of the constraints on the family of LDA models that will be imposed to enforce this condition of a strict density representation, which can then also be directly mapped into suitable parameter ranges (domains) of the LDA model that will result in such skew–kurtosis regions.

To summarize, we note that the fourth order series expansion approximation $\hat{f}_{T(\widetilde{Z}),4}(z)$ has coefficients that are functions directly of the skew and kurtosis of the LDA model being approximated; to see this we make an explicit representation of the coefficients a_0, $a_1, \ldots, a_4$ next:

$$
\begin{aligned}
a_0 &= \int_0^\infty f_{T(\widetilde{Z})}(z)dz = 1, \\
a_1 &= \frac{1}{\alpha}\int_0^\infty f_{T(\widetilde{Z})}(z)(z-\alpha)dz = 0, \\
a_2 &= \frac{\Gamma(\alpha)}{2!\Gamma(\alpha+2)}\int_0^\infty f_{T(\widetilde{Z})}(z)\left[z^2 - 2(\alpha+1)z + (\alpha+1)\alpha\right]dz = 0, \\
a_3 &= \frac{\Gamma(\alpha)}{3!\Gamma(\alpha+3)}\left(\mu_3 - 2\alpha\right), \\
a_4 &= \frac{\Gamma(\alpha)}{4!\Gamma(\alpha+4)}\left(\mu_4 - 12\mu_3 - 3\alpha^2 + 18\alpha\right).
\end{aligned}
\tag{17.123}
$$

This shows that the resulting fourth order approximation of the density (and distribution) $\hat{f}_{T(\widetilde{Z}),4}(z)$ will be parameterized with respect to the LDA model by the excess skewness (denoted by s) and excess kurtosis (denoted by k), which are functions of μ_3 and μ_4. Namely, one may now write

$$\begin{aligned}\hat{f}_{T(\widetilde{Z}),4}(z) &= \frac{z^{\alpha-1}}{\Gamma(\alpha)}\exp(-z)\sum_{i=0}^{4} a_i L_i^{(\alpha)}(z) \\ &= \gamma(z;\alpha,1)\left[1+\frac{\Gamma(\alpha)}{3!\Gamma(\alpha+3)}\left(\mu_3-2\alpha\right)L_3^{(\alpha)}(z)\right. \\ &\quad\left.+\frac{\Gamma(\alpha)}{4!\Gamma(\alpha+4)}\left(\mu_4-12\mu_3-3\alpha^2+18\alpha\right)L_4^{(\alpha)}(z)\right] \\ &= \gamma(z;\alpha,1)\left[1+\frac{\Gamma(\alpha)}{3!\Gamma(\alpha+3)}\left(s\mu_2^{\frac{3}{2}}-2\alpha\right)L_3^{(\alpha)}(z)\right. \\ &\quad\left.+\frac{\Gamma(\alpha)}{4!\Gamma(\alpha+4)}\left((k+3)\mu_2^2-12s\mu_2^{\frac{3}{2}}-3\alpha^2+18\alpha\right)L_4^{(\alpha)}(z)\right].\end{aligned} \tag{17.124}$$

Hence, we now wish to consider the regions (sets) of values of (s,k) that will produce strictly valid densities since for some (s,k) pairs, the pdf $\hat{f}_{T(\widetilde{Z}),4}(z)$, can be negative for some values of z while for other pairs of (s,k) the pdf approximation may be strictly positive but multimodal (due to alternating signs in the series expansion).

In Theorem 17.15, the region $\mathcal{D}$ in the (s,k)-plane is identified that ensures the density approximation $\hat{f}_{T(\widetilde{Z}),4}(z)$ is positive definite. This will be characterized by considering the following notions from analytic geometry, which are modifications developed for the aforementioned class of series expansion, which are analogous to those proposed in Jondeau and Rockinger (1999) for Gram-Charlier series expansions.

It will be sufficient to consider two conditions that will allow one to characterize the set $\mathcal{D}$ of values for (s,k) that produce the "envelope" for the density approximation in which it will remain positive. The aim will be to find an analytic expression for the curve characterizing the boundary of $\mathcal{D}$ for the skewness and kurtosis as a function of z solved for the condition that for all values of z the density approximation remains positive definite. This involves parametric specification of the boundary where for a given value of z one has the density approximation reaching zero. One can then find the subregion defined by $\hat{f}_{T(\widetilde{Z}),4}(z)=0$ for all values of z, characterized with regard to the space of (s,k) values in $\mathcal{D}$. In deriving the following expressions, it will be convenient to recall the following result in Lemma 17.8 regarding derivatives of generalized Laguerre polynomials.

Lemma 17.8 (Derivatives of Generalized Laguerre Polynomials) *Consider the generalized Laguerre polynomial representation of Rodrigues $\widetilde{L}_n^{\alpha}(x)$ given in a different representation by*

$$\widetilde{L}_n^{(\alpha)}(z)=\sum_{i=0}^{n}(-1)^i\frac{\Gamma(n+\alpha+1)}{\Gamma(\alpha+i+1)\Gamma(n-i+1)}\frac{z^i}{\Gamma(i+1)}. \tag{17.125}$$

Then this will produce for the p-th derivative the expression

$$\frac{d^p}{dz^p}\widetilde{L}_n^{(\alpha)}(z)=(-1)^p\widetilde{L}_{n-p}^{(\alpha+p)}(z). \tag{17.126}$$

Therefore, to obtain the derivative of the polynomial representation adopted in this section, we recall the relationship

$$L_n^{(\alpha)}(x) = n!(-1)^n \widetilde{L}_n^{(\alpha-1)}(x), \tag{17.127}$$

which gives the derivative

$$\begin{aligned} \frac{1}{n!(-1)^n}\frac{d^p}{dz^p}\left[L_n^{(\alpha+1)}(z)\right] &= \frac{1}{(n-p)!(-1)^n}L_{n-p}^{(\alpha+p+1)}(z) \\ \Rightarrow \frac{d^p}{dz^p}\left[L_n^{(\alpha)}(z)\right] &= \frac{n!}{(n-p)!}L_{n-p}^{(\alpha+p)}(z). \end{aligned} \tag{17.128}$$

We first state the two conditions that will characterize this envelope for a given value of z according to the following:

The first condition states that the density must be non-negative.

Condition 1

$$\begin{aligned} &\hat{f}_{T(\widetilde{Z}),4}(z) \geq 0 \\ &\quad\Rightarrow \frac{\Gamma(\alpha)}{3!\Gamma(\alpha+3)}\left(s\mu_2^{3/2} - 2\alpha\right)L_3^{(\alpha)}(z) \\ &\quad+ \frac{\Gamma(\alpha)}{4!\Gamma(\alpha+4)}\left((k+3)\mu_2^2 - 12s\mu_2^{3/2} - 3\alpha^2 + 18\alpha\right)L_4^{(\alpha)}(z) + 1 \geq 0 \\ &\quad\Rightarrow s\mu_2^{3/2}c_1(z) + k\mu_2^2 c_2(z) + c_3(z) \geq 0. \end{aligned} \tag{17.129}$$

The second condition specifies the points at which the maximum and minimum occur in the oscillating tails, for a given pair (s, k).

Condition 2

$$\begin{aligned} &\frac{d}{dz}\hat{f}_{T(\widetilde{Z}),4}(z) = 0 \\ &\quad\Rightarrow \gamma(z;\alpha-1,1) - \gamma(z;\alpha,1) + \frac{\Gamma(\alpha)}{3!\Gamma(\alpha+3)}\left(s\mu_2^{3/2} - 2\alpha\right) \\ &\quad\times\left[L_3^{(\alpha)}(z)\left(\gamma(z;\alpha-2,1) - \gamma(z;\alpha,1)\right) + 3L_2^{(\alpha+1)}(z)\gamma(z;\alpha,1)\right] \\ &\quad+\left\{\frac{\Gamma(\alpha)}{4!\Gamma(\alpha+4)}\left((k+3)\mu_2^2 - 12s\mu_2^{3/2} - 3\alpha^2 + 18\alpha\right)\right. \\ &\quad\left.\times\left[L_4^{(\alpha)}(z)\left(\gamma(z;\alpha-2,1) - \gamma(z;\alpha,1)\right) + 4L_3^{(\alpha+1)}(z)\gamma(z;\alpha,1)\right]\right\} = 0 \\ &\quad\Rightarrow s\mu_2^{3/2}c_4(z) + k\mu_2^2 c_5(z) + c_6(z) = 0 \end{aligned} \tag{17.130}$$

with

$$
\begin{aligned}
c_1(z) &:= \frac{\Gamma(\alpha)}{3!\Gamma(\alpha+3)} L_3^{(\alpha)}(z) - 12\frac{\Gamma(\alpha)}{4!\Gamma(\alpha+4)} L_4^{(\alpha)}(z),\\
c_2(z) &:= \frac{\Gamma(\alpha)}{4!\Gamma(\alpha+4)} L_4^{(\alpha)}(z),\\
c_3(z) &:= \frac{\Gamma(\alpha)}{4!\Gamma(\alpha+4)} \left(3\mu_2^2 - 3\alpha^2 + 18\alpha\right) L_4^{(\alpha)}(z) - \frac{\Gamma(\alpha)}{3!\Gamma(\alpha+3)} 2\alpha L_3^{(\alpha)}(z) + 1,\\
c_4(z) &:= \left[\frac{\Gamma(\alpha)}{3!\Gamma(\alpha+3)} \left(L_3^{(\alpha)}(z)\left(\gamma(z;\alpha-2,1) - \gamma(z;\alpha,1)\right) + 3L_2^{(\alpha+1)}(z)\gamma(z;\alpha,1)\right)\right]\\
&\quad - 12\frac{\Gamma(\alpha)}{4!\Gamma(\alpha+4)} \left[L_4^{(\alpha)}(z)\left(\gamma(z;\alpha-2,1) - \gamma(z;\alpha,1)\right) + 4L_3^{(\alpha+1)}(z)\gamma(z;\alpha,1)\right],\\
c_5(z) &:= \frac{\Gamma(\alpha)}{4!\Gamma(\alpha+4)} \left[L_4^{(\alpha)}(z)\left(\gamma(z;\alpha-2,1) - \gamma(z;\alpha,1)\right) + 4L_3^{(\alpha+1)}(z)\gamma(z;\alpha,1)\right],\\
c_6(z) &:= \gamma(z;\alpha-2,1) - \gamma(z;\alpha,1)\\
&\quad - 2\alpha\frac{\Gamma(\alpha)}{3!\Gamma(\alpha+3)} \times \left[L_3^{(\alpha)}(z)\left(\gamma(z;\alpha-2,1) - \gamma(z;\alpha,1)\right) + 3L_2^{(\alpha+1)}(z)\gamma(z;\alpha,1)\right]\\
&\quad + \frac{\Gamma(\alpha)}{4!\Gamma(\alpha+4)} \left(3\mu_2^2 - 3\alpha^2 + 18\alpha\right)\\
&\quad \times \left[L_4^{(\alpha)}(z)\left(\gamma(z;\alpha-2,1) - \gamma(z;\alpha,1)\right) + 4L_3^{(\alpha+1)}(z)\gamma(z;\alpha,1)\right].
\end{aligned}
$$

One now notes that these two conditions specify two linear equations with respect to (s, k), which can be solved by substitution to find the curve in the (s, k) plane as a function of z that guarantees strictly positive definite density representations for the approximation, as detailed in Theorem 17.15.

Theorem 17.15 (Generic Strict Density Representation Fourth-Order Insured Loss Processes) *Under an arbitrary specification of the severity and frequency distributions in the model for the LDA of the loss process being modeled and for an arbitrary choice of insurance product, the resulting insured annual loss* $\widetilde{Z}$ *can be represented under a simple transformation* $T\left(\widetilde{Z}\right) = \zeta\widetilde{Z}$ *such that* ζ *is selected such that* $\mathbb{E}\left[T\left(\widetilde{Z}\right)\right] = \mathbb{V}\text{ar}\left(T\left(\widetilde{Z}\right)\right)$, *up to d-th order by the Gamma density series*

$$
\hat{f}_{T(\widetilde{Z}),d}(z) = \underbrace{\frac{z^{\alpha-1}}{\Gamma(\alpha)}\exp(-z)}_{\textit{Gamma density shape } \alpha} \sum_{i=0}^{4} a_i L_i^{(\alpha)}(x) \tag{17.131}
$$

assuming the first four moments of the insured process $\widetilde{Z}$ *are finite, where*

$$
a_i = \frac{\Gamma(\alpha)}{i!\Gamma(\alpha+i)} \int_0^\infty f_{T(\widetilde{Z})}(z) L_i^{(\alpha)}(z)dz. \tag{17.132}
$$

This will be a strictly positive definite density approximation with positive support iff the following conditions for the skewness and kurtosis of the transformed annual loss $T\left(\widetilde{Z}\right)$ *satisfy that the*

LDA models excess skewness and kurtosis are in the domain $(s, k) \in \mathcal{D}$ *defined analytically by the boundary curves given by*

$$s(z) = \frac{1}{\mu_2^2 c_1(z)} \left[\left(c_6(z) - \frac{c_4(z)}{c_1(z)} c_3(z) \right) \left(c_5(z) - \frac{c_4(z)}{c_1(z)} c_2(z) \right)^{-1} c_2(z) - c_3(z) \right],$$

$$k(z) = \left(\frac{c_4(z)}{c_1(z)} c_3(z) - c_6(z) \right) \left[\mu_2^2 \left(c_5(z) - \frac{c_4(z)}{c_1(z)} c_2(z) \right) \right]^{-1}. \tag{17.133}$$

Remark 17.15 *It may also be relevant in practice to note that one can also make such series expansions robust through the utilization of truncated L-central moments in place of standard central moments used earlier. These will be more robust in the sense that the resulting excess skewness and kurtosis will not be as strongly affected by changes in the observed losses or parameter values estimated for the underlying LDA model, each time the model is actually estimated and the series expansion applied by estimation of the sample moments.*

17.5.4.2 Series Expansions for General LDA Insured Processes (Truncated Support on $[0, E]$).

In the case of insured loss processes, it may also be possible that the insurance mitigation has the effect of truncating the support of the annual loss random variable via an upper bound on the total loss incurred on each loss event, as in the case of the ILPU policy structure. In such cases, it will be relevant to perform a series expansion that should not only satisfy that it takes a positive support, but now a support on an interval say $[0, E]$, where E bay be, for example, a *TCL* for the policy.

To achieve this requirement, we discuss two possible approaches, the first is a simple extension to the aforementioned generic series representations based on a basis that is constructed from a Beta kernel rather than a gamma kernel. We first recall the representation of the two-parameter Beta distribution used as a kernel in this series expansion on finite support as shown in Definition 17.31.

Definition 17.31 (Beta Density) *A random variable X has a general Beta density with shape and scale parameters* $\alpha > 0$ *and* $\beta > 0$ *and with location and range parameters a and* $b \geq a$ *iff the density has the form*

$$f_X(x) = \frac{(x-a)^{\alpha-1}(b-x)^{\beta-1}}{B(\alpha,\beta)(b-a)^{\alpha+\beta-1}}, \quad a \leq x \leq b, \tag{17.134}$$

with $B(\alpha, \beta)$ *representing the Beta function given by*

$$B(\alpha,\beta) = \int_0^1 t^{\alpha-1}(1-t)^{\beta-1} dt. \tag{17.135}$$

■

Typically in OpRisk settings one would be interested in setting $a = 0$ and for example $b = TCL$. In the first approach discussed it will be shown how one may maintain the representation aforementioned with a further approximation to the orthogonality condition, which is specified according to Lemma 17.10, where the choice $a = 0$ and $b = 1$ are used, which would typically mean that one would transform the insured loss process $\widetilde{Z}$ with support $[0, E]$ to the interval $[0, 1]$ via an invertible transform, denoted by $T(\widetilde{Z})$. Then the series expansion would be performed on the transformed insured process and inference on capital from the resulting representation can be mapped back to the original space via the invertible mapping. The best linear invertible mapping to be considered is given by the result in Lemma 17.9.

Lemma 17.9 *Given a random variable X with support $[a, b]$ distributed according to a general Beta distribution $X \sim GB(x; \alpha, \beta, a, b)$, the transformed random variable*

$$T(X) = \frac{X - a}{b - a} \tag{17.136}$$

will result in $T(X) \sim Beta(x; \alpha, \beta) = GB(x; \alpha, \beta, 0, 1)$.

One can then state the following result for the orthogonality conditions that will be of interest in this section.

Lemma 17.10 (Approximate Asymptotic Orthogonality of Beta Kernel to Laguerre Polynomials) *The Laguerre polynomial basis is approximately asymptotically orthogonal to a Beta kernel (weight function) since the following holds*

$$\int_0^\infty \frac{x^{\alpha-1}(1-x)^{\beta-1}}{B(\alpha,\beta)} L_n^{(\alpha)}(x) L_m^{(\alpha)}(x) dx \approx \begin{cases} 0, & m \neq n \\ \beta \dfrac{n!\Gamma(\alpha+n)}{\Gamma(\alpha)}, & m = n, \end{cases} \tag{17.137}$$

which uses the fact that $\lim_{\beta\to\infty} \beta B(\alpha, \beta) \to \Gamma(\alpha, 1)$.

If one really takes $\beta \to \infty$, it would be clear that this result would diverge to infinity, hence the notion of the approximate asymptotic orthogonality. In other words in practice for a finite value of β one can utilize the approximate orthogonality condition to obtain a series representation as presented earlier. Perhaps a more suitable approach would be to undertake a series expansion with a different subfamily of the Askey scheme of polynomials, namely, the truly orthogonal basis given by the Jacobi polynomials as detailed in Definition 17.32.

Definition 17.32 *Jacobi polynomials (hypergeometric polynomials) form a class of orthogonal polynomials with respect to the weight*

$$(1-x)^\alpha(1+x)^\beta, \tag{17.138}$$

on the interval $[-1, 1]$. *They are defined by hypergeometric function as follows:*

$$P_n^{(\alpha,\beta)}(x) = \frac{(\alpha+1)_n}{n!} {}_2F_1\left(-n, 1+\alpha+\beta+n; \alpha+1; \frac{1-x}{2}\right)$$
$$= \frac{\Gamma(\alpha+n+1)}{\Gamma(\alpha+\beta+n+1)} \sum_{m=0}^{n} \frac{\Gamma(\alpha+\beta+n+m+1)}{\Gamma(n-m+1)\Gamma(\alpha+m+1)m!} \left(\frac{x-1}{2}\right)^m, \tag{17.139}$$

with $(\alpha+1)_n$ *representing the Pochhammer's symbol (rising factorial).* ■

This family of orthogonal polynomials also contains other well-known families of polynomials, namely, the Gegenbauer, Legendre, Zernike, and Chebyshev polynomials. In Theorem 17.16, we present the orthogonality result for the general beta density and the Jacobi polynomials. This result makes use of the symmetry property of the Beta density in which $f(x; \alpha, \beta) = f(1-x; \beta, \alpha)$.

Theorem 17.16 (Orthogonality of Jacobi Polynomials and Beta Density Kernels) *Consider the Beta density kernel on support* $[-1, 1]$ *given by*

$$g(x; \alpha, \beta) = \frac{(x+1)^{\alpha-1}(1-x)^{\beta-1}}{B(\alpha,\beta)2^{\alpha+\beta-1}}. \tag{17.140}$$

Then relabelling the variables $\widetilde{\alpha} = \alpha - 1$ *and* $\widetilde{\beta} = \beta - 1$, *one has the following orthogonality condition for Jacobi polynomials* $P_n^{(\widetilde{\alpha},\widetilde{\beta})}(x)$ *such that*

$$\frac{1}{B(\widetilde{\alpha}+1,\widetilde{\beta}+1)2^{\widetilde{\alpha}+\widetilde{\beta}+1}} \int_{-1}^{1} (1-x)^{\widetilde{\alpha}} (1+x)^{\widetilde{\beta}} P_n^{(\widetilde{\alpha},\widetilde{\beta})}(x) P_m^{(\widetilde{\alpha},\widetilde{\beta})}(x) dx$$
$$= \begin{cases} 0, & m \neq n, \\ \dfrac{1}{B(\widetilde{\alpha}+1,\widetilde{\beta}+1)2^{\widetilde{\alpha}+\widetilde{\beta}+1}} \dfrac{2^{\widetilde{\alpha}+\widetilde{\beta}+1}}{2n+\widetilde{\alpha}+\widetilde{\beta}+1} \dfrac{\Gamma(n+\widetilde{\alpha}+1)\Gamma(n+\widetilde{\beta}+1)}{\Gamma(n+\widetilde{\alpha}+\widetilde{\beta}+1)n!}, & m = n. \end{cases} \tag{17.141}$$

Hence, again one may utilize the orthogonality results for the Beta polynomials to obtain the representation of the d-th order series expansion given with respect to Beta density weight function and Jacobi polynomials on a restricted support after transformation from $[0, E]$ to $[-1, 1]$ by the result in Theorem 17.17. In this case, we may present two different representations (which are directly related): the first is the standard Fourier series-type representation with respect to Jacobi polynomials and the second representation is with respect to a liner combination of Beta densities.

Theorem 17.17 (Generic Fourier Series Representation of Insured Loss Process (Bounded Support)) *Under an arbitrary specification of the severity and frequency distributions in the model for the LDA of the loss process being modeled and for an arbitrary choice of insurance product, the resulting insured annual loss* $\widetilde{Z} \in [0, E]$ *can be represented, under a simple* *transformation to the interval* $T\left(\widetilde{Z}\right) \in [-1, 1]$, *up to d-th order by the Jacobi polynomial series with Beta weights given by*

$$\hat{f}_{T(\widetilde{Z}),d}(z) = \sum_{i=0}^{d} a_i P_i^{(\alpha,\beta)}(z), \tag{17.142}$$

where

$$a_i = \int_0^\infty f_{T(\widetilde{Z})}(z) P_i^{(\alpha,\beta)}(z)(z+1)^{\alpha-1}(1-z)^{\beta-1} dz. \tag{17.143}$$

In the case that we choose to expand the insured annual loss density in terms of Beta densities, one can adopt the following representation. Note that this choice is optimal in the sense that the coefficients are selected to minimize the integrated weighted squared error criterion given in Theorem 17.18, which is the analog to the Gamma results derived previously.

Theorem 17.18 (Optimality of the Generic Beta–Jacobi Series Representation) *The accuracy of the series representation can be studied for a given shape parameter* α *in the Gamma basis by consideration of the integral of the squared error between the exact annual loss density and a series approximation given by*

$$\begin{aligned} R &= B(\alpha,\beta)2^{\alpha+\beta-1} \int_{-1}^{1} \frac{1}{(z+1)^{\alpha-1}(1-z)^{\beta-1}} \left[f_{T(\widetilde{Z})}(z) - f_{T(\widetilde{Z}),d}(z)\right]^2 dz \\ &= B(\alpha,\beta)2^{\alpha+\beta-1} \int_{-1}^{1} \frac{1}{(z+1)^{\alpha-1}(1-z)^{\beta-1}} \\ &\quad \times \left[f_{T(\widetilde{Z})}(z) - \frac{(z+1)^{\alpha-1}(1-z)^{\beta-1}}{B(\alpha,\beta)2^{\alpha+\beta-1}} \sum_{i=0}^{d} a_i P_i^{(\alpha,\beta)}(z)\right]^2 dz \end{aligned} \tag{17.144}$$

which if minimized, for a given α *and* β *combination, with respect to* a_i *coefficients via* $\frac{\partial R}{\partial a_i} = 0$ *yields the relationship (via orthogonality between the Jacobi and Beta kernel),*

$$a_i = \frac{B(\alpha,\beta)\left[2i+\alpha+\beta-1\right]\Gamma(i+\alpha+\beta-1)i!}{\Gamma(i+\alpha)\Gamma(i+\beta)} \int_{-1}^{1} f_{T(\widetilde{Z})}(z) P_i^{\alpha,\beta}(z) dz. \tag{17.145}$$

Theorem 17.19 (Generic Beta Density Series Representation of Insured Loss Process (Bounded Support)) *Under an arbitrary specification of the severity and frequency distributions in the model for the LDA of the loss process being modeled and for an arbitrary choice of* *insurance product, the resulting insured annual loss* $\widetilde{Z} \in [0, E]$ *can be represented, under a simple transformation to the interval* $T\left(\widetilde{Z}\right) \in [-1, 1]$, *up to d-th order by the beta density series*

$$\hat{f}_{T(\widetilde{Z}),d}(z) = \underbrace{\frac{(z+1)^{\alpha-1}(1-z)^{\beta-1}}{B(\alpha,\beta)2^{\alpha+\beta-1}}}_{\text{Beta density}} \sum_{i=0}^{d} a_i P_i^{(\alpha,\beta)}(z), \tag{17.146}$$

where

$$a_i = \frac{B(\alpha,\beta)\left[2i+\alpha+\beta-1\right]\Gamma(i+\alpha+\beta-1)i!}{\Gamma(i+\alpha)\Gamma(i+\beta)} \int_{-1}^{1} f_{T(\widetilde{Z})}(z) P_i^{\alpha,\beta}(z) dz. \tag{17.147}$$

CHAPTER EIGHTEEN

Insurance and Risk Transfer: Pricing Insurance-Linked Derivatives, Reinsurance, and CAT Bonds for OpRisk

This chapter deals with more advanced aspects of OpRisk insurance mitigation and covers three main topics: catastrophe bonds (CAT bonds) and insurance-linked derivatives for extreme loss (low frequency, high consequence) risk transfer, insurance portfolio selection, and purchase strategies for OpRisk insurance products (including optimal decision rules for when to purchase insurance under a multiple stopping time framework).

In this chapter, we address the following components of OpRisk insurance modeling.

1. How does one manage the transfer of risk with regard to issues such as adverse selection, moral hazard, counterparty risk and withdrawal, payment uncertainty (in amount and time), and systemic risk?
   We particularly address this question in reference to more advanced risk transfer mechanisms such as insurance-linked derivatives like CAT bonds.
2. What are CAT bonds, how are they structured and priced, and how are they relevant to OpRisk?
3. How can one optimize a coverage of Basel III allowable insurance mitigations using linear portfolio of CAT bonds and insurance products via optimal portfolio theory based on tail functional risk measures for generic OpRisk loss processes that do not have a direct insurance product available?
4. What are the optimal multiple times to purchase or construct such insurance mitigation portfolios given an LDA model structure under finite or infinite time horizons?

We begin this chapter with a detailed coverage of aspects of incorporating insurance-linked derivatives into OpRisk risk transfer strategies, detailing the features and mechanisms available.

Fundamental Aspects of Operational Risk and Insurance Analytics: A Handbook of Operational Risk, First Edition. Marcelo G. Cruz, Gareth W. Peters, and Pavel V. Shevchenko.

18.1 Insurance-Linked Securities and CAT Bonds for OpRisk

The development of risk transfer products for OpRisk settings by insurers is a relatively new and growing field in both academic research and industry, where new products are developed with greater understanding of catastrophe and high-consequence low-frequency loss processes. Early and influential work on catastrophe modeling was undertaken by Rene Thom and Sir Christopher Zeemen; see, for instance, the widely acclaimed book Thom (1977); Thom and Zeeman (1974), and the reviews in Zeeman (1977, 1979). As a result of this early work in mathematics, the theory of catastrophes and chaos is well established; however, we would argue that the statistical theory of risk processes and modeling of such stochastic processes is still a burgeoning field that is highly relevant for study in structuring and issuing insurance products for such disasters. We observe that in many cases the approaches adopted to study these catastrophic events in the applied mathematics literature are very different to the modeling and estimation-based approaches adopted in statistics and actuarial research. In addition, the question of how best to structure these products and price these products is still an active area of research.

Remark 18.1 *We note that the majority of products considered in this section are primarily of interest in OpRisk for uncontrollable risk processes that are typically of low frequency and very high consequence.*

The focus of this section will be primarily on CAT bonds such as those mentioned in Grossi and Kunreuther (2005) and Woo (1999). There are however multiple mechanisms and products available to perform risk transfer for nature risks (generically natural disasters). The types of products include insurance, insurance derivatives, CAT bonds, industry loss warranties (contracts triggered by a given industry loss in a defined jurisdiction), sidecars, and captives; see detailed discussions in Barrieu and Albertini (2010, chapter 1). The most prominent form of insurance-linked security (ILS) is the CAT bond that was designed to facilitate the direct transfer of catastrophe insurance risk from insurers, reinsurers, and financial institutions (known as sponsors) to the investors in the capital markets. These capital markets have sufficient depth to absorb such disasters should a CAT bond default.

Remark 18.2 *Insurance-Linked Securities are unlike traditional corporate bonds or other fixed income instruments in that they have a primary risk derived from one or more adverse insurance-related events, often linked to a particular peril or group of perils directly driven by nature risk. The primary risks in the majority of ILS arise from perils such as earthquakes, wind storms and hurricanes, extreme mortality, terrorism, and other mixed policies (comprising several of these perils). They are often specific to particular jurisdictions prone to such natural disasters.*

Before proceeding, it is useful to recall what a captive is in Definition 18.1.

Definition 18.1 (Captive Insurers) *The term captive was first coined by Frederic M. Reiss to refer to the situation in which a policyholder owns the insurance company that wrote the insurance policy (hence the name captive). There are several types of captive insurers.*

- *Pure captives. These involve the captive insuring its own parent and affiliates (subsidiary companies or financial institutions);*
- *Homogeneous captives. These involve the captive insuring only a single type of industry or for a single type of peril;*
- *Heterogeneous captives. These involve captives that insure a group of diverse companies or multiple perils.* ■

Given this wide range of products available, it is interesting to think about how the CAT bond market can be beneficial to the transfer of risk in an OpRisk setting. In particular, there seems to be three basic mechanisms available under this class of assets.

- **Approach 1.** The first is of relevance to larger financial institutions and involves the issuance of CAT bonds to cover exposure to nature risk that would cover losses they would be exposed to from perils arising from natural disasters: wind storms, hurricanes, tsunami, wild fires, flooding, earthquake, mass loss of life, and terrorism.
- **Approach 2.** The second approach involves financial institutions who own their own insurance arms often known as captives that provide their own insurance cover within the financial corporation; these captives can also be involved with insurance linked securities, reinsurance, and CAT bonds.
- **Approach 3.** The third approach is to construct multiple peril insurance portfolio in which a financial institution purchases a combination of insurance products for certain perils affecting particular lines of OpRisk exposures that are covered by reinsurers who are fully funded through a CAT bond issuance, as well as perhaps shorting CAT bonds. In particular, short selling of a CAT bond (naked or covered) is a type of strategy in which one obtains capital to cover potential OpRisk loss exposures to natural disasters such that if the natural disaster occurs to trigger the bond default, then the actual asset need never be delivered.

The third strategy mentioned earlier is possible if one believes there is sufficient depth in the secondary CAT bond markets to provide volume for such strategies. One would argue that this has not always been the case, but is significantly improved these days. As discussed in Araya (2005), the development of the market for insurance-linked derivatives and especially CAT bonds has not been smooth. There are different opinions of what was the first CAT bond issued, some say it was the “Act of God” bond completed by Nationwide Insurance Co. of Columbus in Ohio in 1994, whereas others refer to the first incarnation of ILS on the Chicago Board of Trade (CBOT) in 1992, which provided a market for catastrophe insurance, options and futures (technically not bonds and not OTC). Since these early days of catastrophe insurance, there has been a steady increase in market growth and a widening of the pool of market participants in these ILS products from traditional reinsurers to speculators from the hedge fund industry. As some CAT bonds mature and new ones are issued, the distribution of covered perils changes year on year; however, it was stated in Araya (2005) that a significant share of the perils covered (outstanding CAT bond exposures) in any given year are attributed to US hurricane and earthquake perils. Historically, the market for CAT bonds has grown such that it was reported in Barrieu and Albertini (2010, chapter 2) that in 1998–2001 there were at least USD 1–2 billion issuances per year, then following the attack on the World Trade Center issuances went up to over USD 2 billion per year from 200 to 2005, then from 2006 issuances jumped up to USD 4 billion per year following Hurricane Katrina in New Orleans, with a record issuance of USD 7 billion in 2007. Following the GFC, there has been a softening in the market, though the last year or two has seen a steady increase again.

In Figure 18.1, an aggregate total issuance in millions of USD is provided for particular types of CAT bonds grouped as earthquakes, wind storms and hurricane, extreme mortality, medical benefits, mixed policies. It was stated in Barrieu and Albertini (2010, chapter 2) that CAT bond risk principal in the capital markets in 2009 comprised around 8% globally of the entire estimated property limits, which would rise to an even larger percent when other ILS products were accounted for such as those related to sidecars.

The website http://www.artemis.bm/ provides a detailed list of CAT bond issuances globally, which includes details in the "Deal Directory" such as the issuer, cedent, risks/perils covered by the CAT bond, the size of the bond and the data of issuance since 1996. In Figure 18.2 a plot of time series for issuance sizes for the bonds in USD (converted 25/06/2013) is provided from 1996 to 2013 for a range of risks/perils that include earthquakes; windstorms and hurricanes; extreme mortality; medical benefits; and mixed coverage of several natural disasters/perils or specialty bonds (e.g., including bonds such as the Hoplon Insurance Ltd. and MyLotto24 Lottery winnings bond, the Kortis Capital Ltd. and Swiss

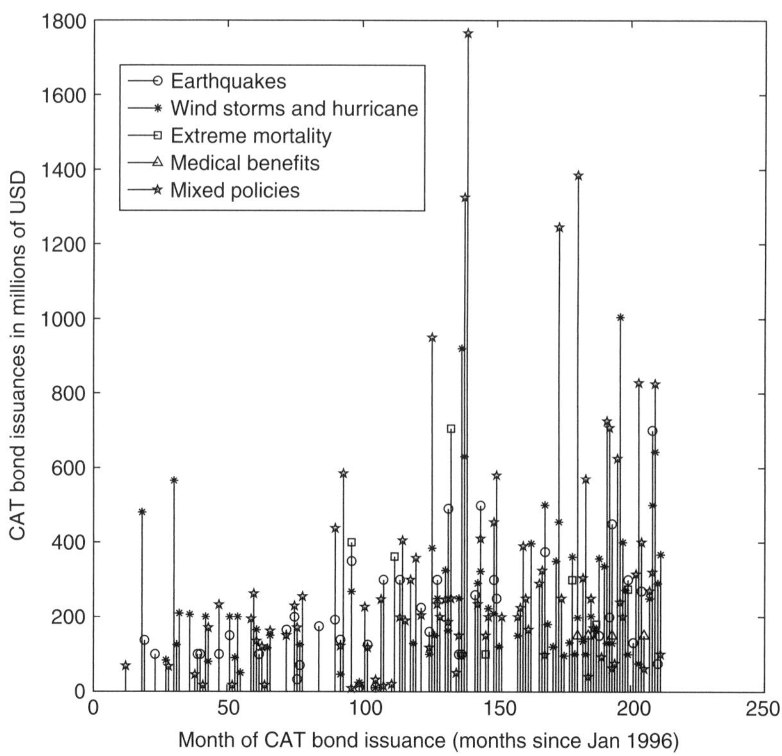

FIGURE 18.1 **CAT bond issuances (worldwide) in USD (data source: http://www.artemis.bm/. Note for some currencies such as the German Deutsche Mark (DEM) it became obsolete with the € and so was converted via 1 DEM = 0.511292 EUR)**

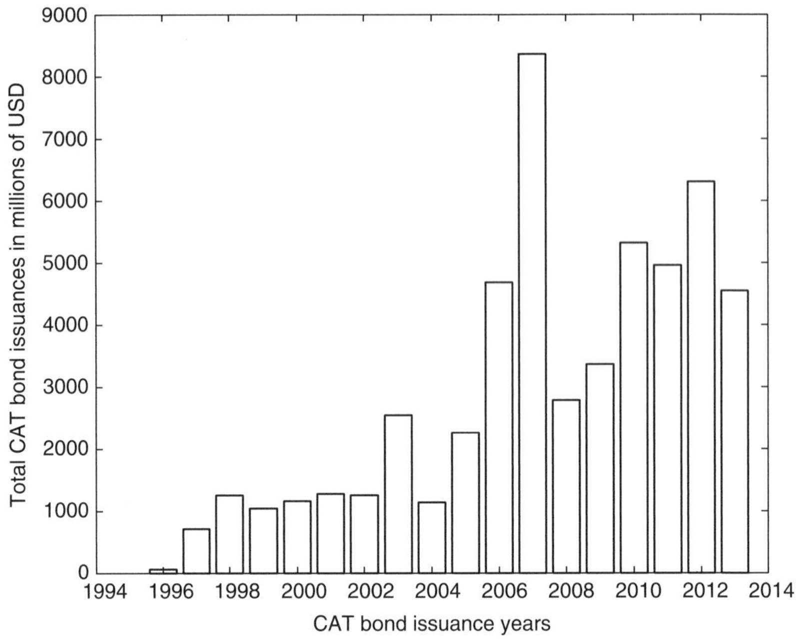

FIGURE 18.2 **Total CAT bond issuances (worldwide) in USD (data source: http://www.artemis.bm/. Note for some currencies such as the German Deutsche Mark (DEM) it became obsolete with the € and so was converted via 1 DEM = 0.511292 EUR)**

Re longevity risk bond, the Kelvin Ltd. and Koch Energy Trading, Inc. temperature risk bond, the Javelin Re Ltd. and Arrow Capital Re worldwide all risks bond and the multiperil bonds of "Merna Reinsurance Ltd. and State Farm" and "Gamut Re Ltd. and Nephila Capital").

As discussed in Loubergé *et al.* (1999), the first incarnation of CAT bonds were not actual bonds—instead, they were futures and options on futures that were issued on the CBOT in 1992 as CAT insurance options and futures and were developed around four indices related to natural disasters. These were developed by the Insurance Services Office (ISO) and constructed based on a pool of 30 insurers in the US. Each index covered a different geographical region of the US and were updated only quarterly. The first line of products were simple futures products and vanilla European options on the futures were available. These products were then modified in 1995 to be based on indices produced by Property Claims Services (PCS), which included up to 80% of the US CAT bond market; see discussions in Schradin (1996). The resulting PCS options were exchange traded and cash-based financial derivatives on an underlying that were the loss indices (standardized in their objective, regional and temporal dimensions). The PCS indices were updated daily instead of quarterly as was the case with the ISO indices. The PCS indices reflect the dollar cumulative amounts of CAT claims greater than USD 5 million of insured property damages in a specified US region and time; see discussion in Schradin (1996). Then following these early developments, the futures contracts available were discontinued and the vanilla options were replaced with European call spreads in which a call is purchased along with the simultaneous writing of a call at a higher strike price, creating the ability for market participants (typically insurers and reinsurers) to buy coverage with layers of aggregate excess of loss reinsurance.

When discussing CAT options traded on the CBOT, it has been noted by several authors that the market in the late 1990s and early 2000s was not deep enough or liquid enough to attract widespread uptake of these products by investors. In general, several authors have posed different arguments for this either based on the exposure to basis risk for the investors, the different regulation standards, and also the lack of supply of such products. In general, it is clear that CAT derivatives cannot involve arbitrage trading since the underlying index cannot be replicated. In addition, speculative trading required to enhance liquidity and efficiency of the market cannot easily take place since such traders have a limited information set about the possibility of a trigger event. Hence, CAT spreads conceived as a useful hedging tool have struggled to grow as a market. For background-specific details on the PCS CAT options on the CBOT as they were originally developed, see Schradin (1996). Then in Aase (1999) an equilibrium-based valuation model for CAT futures contracts and derivatives from such contracts was developed, where the underlying delivery value was an insurance index. In Embrechts and Meister (1997), a study of the pricing approaches for CAT futures in the context of incomplete markets and the impact this has on the pricing where the underlying index is based on the ISO.

The index-linked catastrophic loss futures as well as the more liquid call option spreads that superseded them on the CBOT were eventually delisted as a result of low trading volume. In their place, new classes of assets based on over-the-counter (OTC) products are enjoying a rapid and successful development. In the following sections, we focus instead on the pricing and valuation of CAT bonds as they are currently developed. These alternative OTC issued CAT bonds currently developed are significantly more successful products in terms of the liquidity and uptake as was demonstrated in the previous issuance plots.

18.1.1 BACKGROUND ON INSURANCE-LINKED DERIVATIVES AND CAT BONDS FOR EXTREME RISK TRANSFER

Cummins *et al.* (2002) pose the provocative question: "Can insurers pay for the big one?". Then in Peters *et al.* (2011a) a study is performed on a related question in the context of OpRisk where they consider "Impact of insurance for OpRisk: is it worthwhile to insure or be insured for severe losses?". As a result of such studies of the capacity of reinsurance markets to bear the cost of catastrophic disasters such as earthquakes, hurricanes, floods, fires, etc. and the resulting claims over multiple years arising from losses from such events could easily reach into the USD 50–USD 100 billion for single events; see discussion in Lee and Yu (2007) and Froot (2007). Such severe losses would stress the capacity of the insurance industry and threaten the credit risk of many reinsurers. This would in turn impact on the capital mitigation offered under OpRisk settings through heightened payment uncertainties. As a result, there have been a number of private and public policy proposals; see discussions in Loubergé *et al.* (1999), where two alternatives to traditional reinsurance for such catastrophe coverage are noted:

1. Mandatory public provision of coverage; versus
2. Nonmandatory use of government intervention.

In the first case of mandatory public provision, it would rely upon the ability of the government to dissipate losses across the population of the country or state over time. This type of approach is adopted, for instance, by the National Flood Insurance Program in the US and in France under a surplus on all property-liability insurance contracts that goes to a public fund for natural catastrophes; see discussion in Magnan (1995). In the case of nonmandatory use of government intervention it is noted in Loubergé *et al.* (1999) and Lewis and Murdock (1996) that one could utilize a combination of CAT bonds on the Chicago Board of Trade (CBOT) with additional coverage in higher levels using contracts supplied by a federal authority, though such an approach has not been implemented widely in practice.

In the commercial sector, the CAT bond market is a relatively new market in terms of insurance products (see issuances worldwide in Figure 18.1) and has been discussed in detail in works such as Coval *et al.* (2009) and Lee and Yu (2007). However, as noted in Cox and Pedersen (2000), there were specific examples of such products discussed in the early 1970s. Goshay and Sandor (1973) and a product specifically targeting Japanese earthquakes was on the market as early as the mid-1980s; see discussion in Ollard (1985). However, the mainstream establishment of such products occurred in the mid-1990s when the CBOT introduced first exchange-traded futures, then later options based on industry-wide loss indices. More recently, the approach that is popular, and we will discuss, has involved privately placed CAT bonds (CAT bonds), also sometimes known as "Act of God or insurance-linked bonds", which have been developed in a number of ways. In general, they all have the basic principle involved that they were developed to ease the transfer of catastrophe-based insurance risk from insurers, reinsurers, and sponsors to a much larger pool of market participants corresponding to the capital market investors. A running history of such bonds is kept by ARTEMIS,[1] and it was noted in Zhu (2011) that the market for such CAT bonds (insurance-linked derivatives) is in excess of USD 10 billion issuances.

Definition 18.2 (CAT Bond) *A CAT bond is an ILS, which acts as a mechanism by which catastrophe risk is transferred from one party known as the sponsor directly to investors in the capital markets. Hence, one can consider CAT bonds as risk-linked securities that transfer a specified set of "nature" risks (natural catastrophes or man-made events such as terrorism) from a sponsor to investors in the capital markets in the form of "credit risk". It is also useful to think of a CAT bond as a reinsurance contract between the sponsor (parties seeking protection) and a special purpose vehicle. The bond issuance can be understood according to the following four basic participant groups:*

1. *Sponsor: government agency, insurer, reinsurer;*
2. *Issuer: structuring agent, investment bank, special purpose vehicle SPV;*
3. *Collateral: swap counterparty, bank with high credit rating;*
4. *Investors: insurers, institutional investors, hedge funds, reinsurers.*

Typically, CAT bonds are issued by an insurance company (sponsor) via a structuring agent or special purpose vehicle (SPV) (issuer) such as an investment bank, which then sells the bonds to capital

[1] http://www.artemis.bm/

market to investors (investors) and proceeds of the sale are placed in a special account (collateral account). The premium and the earnings from the account are collected and paid to the issuer. Three typical cases arise:

- *Case 1. Contract stipulates that the principle provided at issuance is secure, coupon payments are subject to forfeit upon trigger;*
- *Case 2. Contract stipulates that the principle provided at issuance is fractionally secure (perhaps depending on time to maturity) with portion of principle subject to forfeit upon a trigger, coupon payments are subject to forfeit upon trigger;*
- *Case 3. Contract stipulates that the principle provided at issuance as well as coupon payments are subject to forfeit upon trigger.*

The issuer typically makes a floating rate coupon/dividend payment quarterly to investors to compensate their risk. In the advent that there is no catastrophe during the life of the bond, it is most common that the issuer will return the principle to the investors. If a catastrophe occurs before the maturity of the bond, the issuer pays the sponsor according to the reinsurance contract policy specifications and provides the remaining principle to investors.
Catastrophe Losses Covered. The losses covered under CAT bonds are typically characterized by a loss exceedance curve of $\overline{F}(x) = 1 - F(x)$, *where* $F(x)$ *is the distribution of the loss amount from the catastrophe that is obtained in one of two common ways:*

- *Utilize commercially available catastrophe modeling software (statistical models) with a company exposure data;*
- *Use parametric indicators based on peril-specific features (Richter scale thresholds in a given geographic location, aggregate industry loss indexes, etc.) to design payout functions.*

The loss exceedance curve contains two important components: the expected frequency of default, and the recovery rate given default. ■

The following remarks further clarify the typical view that market participants take on the CAT bond products.

Remark 18.3 *Note that since the bond is issued directly by the structuring agent or SPV, it means that the bond is not affected by the sponsor's credit rating(s) and is also not considered a debt of the sponsors. Typically these bonds are low rated and risky but provide multiple year coverage, with the most widely used issuance periods offering 3-year maturity. However, in the event of a catastrophe before the maturity of the bond, the remaining bond principal would be forgiven and the insurance company would use this money to pay their claim-holders.*

Hence, CAT bonds are typically structured as a floating rate bond in which the principal will be forfeit when a prespecified trigger condition occurs. If triggered, the principal is paid to the sponsor. The triggers are linked to major natural catastrophes and perils associated with the intended coverage of the bond.

Therefore, in the market, the CAT bond appeals to sponsors and investors and providing a general economic benefit. Sponsors benefit since an alternative flexible source of risk financing is available to produce greater coverage capacity with a more stable price discovery mechanism

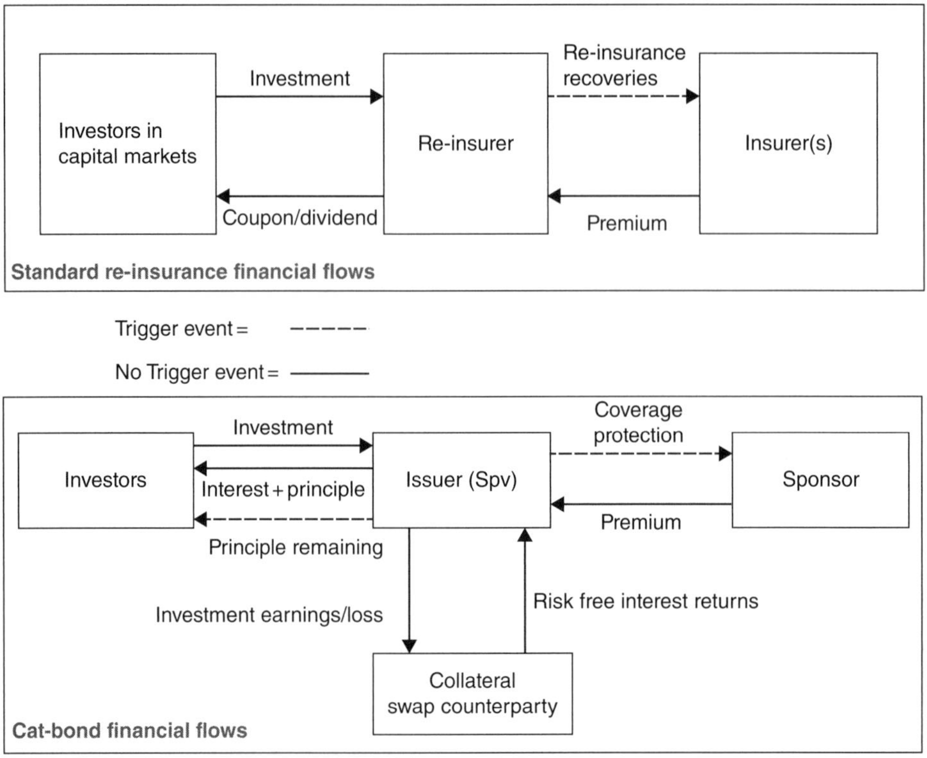

FIGURE 18.3 **Picture depicting a highly stylized version of the flows between participants under a standard reinsurance contract versus those typically undertaken under a CAT bond**

driven by the larger depth and diversification benefits of the capital markets that such bonds are sold into. Investors benefit due to additional diversity in their portfolios, with assets that are uncorrelated largely with financial securities and their associated risk exposures, while also obtaining attractive spreads and yields often offered by CAT bonds. An excellent overview of the economic rational for the insurance and reinsurance markets such as CAT bonds and in general ILS is provided in Gollier (2005).

To understand the difference between a traditional reinsurance contract and a CAT-bond, we illustrate the relationships between each part involved in Figure 18.3; see discussions in Härdle and Cabrera (2010).

Then in Figure 18.4 we plot the data for several key aspects of CAT bond issuances between 1997 and 2000 as detailed in the detailed analysis of CAT bond hedges in Cummins *et al.* (2004). The data when presented in early 2000 were at a county level and originated from the Florida Insurance Commissioner. It comprised insured residential property values for 255 of the 264 insurers writing property coverage in Florida in 1998. The insurers in the sample were representative of 93% of the total insured residential property values.

One immediately observes the following features from this plot, the spread premium of CAT bonds (corresponding to the annual coupon rate above the one-year LIBOR) tends to be significantly higher than the expected loss of the bond principle. This led to the observation made by Zhu (2011) as summarized in Remark 18.4.

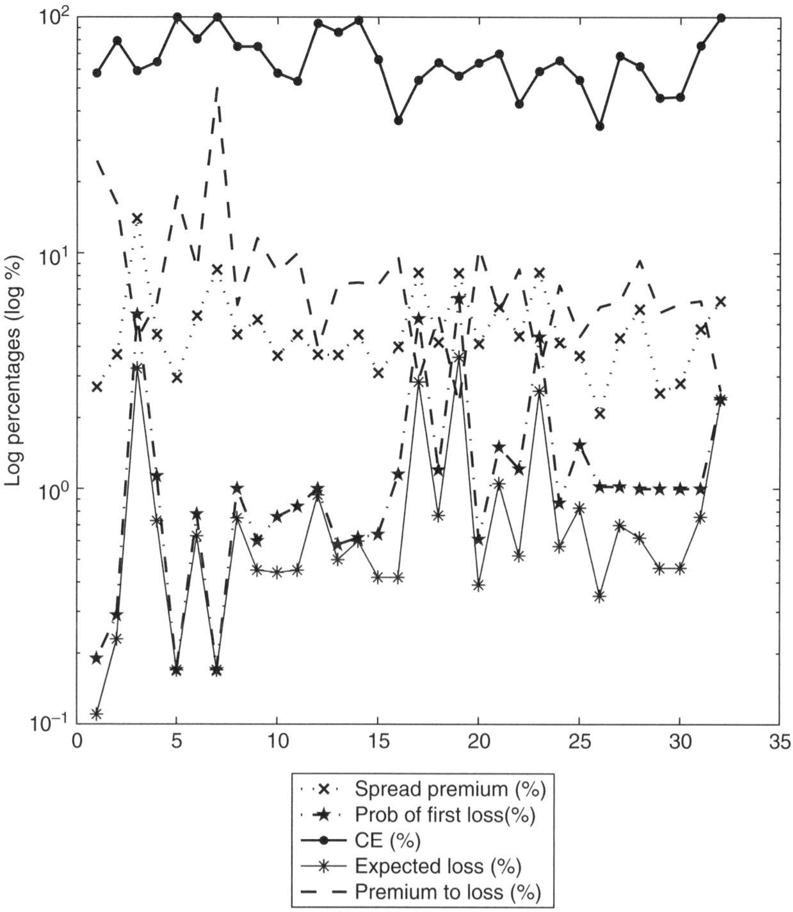

FIGURE 18.4 The plots contained are issuances from 1997 to 2000 based on Table 1.B in Cummins *et al.* (2004). It depicts the spread premium corresponding to the annual coupon rate above the one-year LIBOR; the probability of a trigger under the bond; (CE) the expected principle payment proportion to the issuing insurer, conditional on a loss that triggers payment under the bond, expressed as a percentage of the principle of the bond; the expected loss given by the probability of a loss times the expected principle payment proportion to the issuing insurer; the premium to expected loss, which is the ratio of the spread premium to the expected loss of principle of the bond

Remark 18.4 (CAT Bond Premium Puzzle) *One typically considers catastrophe risk assets as only a small amount of the total wealth in the economy and typically uncorrelated with the capital market; see Zhu (2011). If this is the case, then one ought to expect that CAT bond premiums would be priced at close to the risk-free interest rate, where actuarial pricing would suggest that the fair premium today would correspond to actuarial fair discounted losses that are covered by the bond. However, in practice the data suggest that CAT bond spread is significantly above the expected loss of the bond's principle, which is in contradiction with arbitrage pricing (equilibrium) theory for financial markets. In fact, it was reported in Cummins et al. (2004) that during the period 1997 to 2000, 32 of the CAT bonds issued in this period had an average spread yield of 9.09 times the expected loss, which is a substantial spread premium relative to the expected principle loss. In* *addition, it can be observed that premium spreads are more pronounced for CAT bonds with lower probabilities of trigger, which is also surprising and at odds with standard capital market theories.*

Understanding the cause and behavior of this feature of CAT bonds related to the premium puzzle is an ongoing theoretical concern in economics and financial mathematics.

In addition to these observed features and the properties specified relating to the process undertaken if a forfeiture of the bond occurs, it is also important to consider what mechanism is stipulated to quantify or result in such an event as forfeiture. Therefore, it is important to understand how the trigger event for what will result in forfeiture of the principle will arise (also known as *debt forgiveness*).

18.1.2 TRIGGERS FOR CAT BONDS AND THEIR IMPACT ON RISK TRANSFER

Together, the sponsor and the structuring agent select the mechanism by which the bond will pay out, or trigger. Following this, a quantitative modeling agent will typically assess the risk associated with the product via a loss distribution for possible exposures in order to estimate the amount of protection offered by the bond. Then jointly the modeling agent and the structuring agent write the offering documentation specifying the terms of the investment product for the market and investors, which is provided to the rating agency for a rating assessment.

The first CAT bond was rated by Moody's Investors Service in 1997 and corresponded to the Residential Re Limited issuance, since this first rating up to the period of 2005 it was reported in Araya (2005, chapter 2.1) that USD 6.2 billion of rated securities are available in the market, this number would have substantially grown since 2005 with the steady growth of the market.

It should also be noted that a rating for a CAT bond differs to the typical rating based on the default probability of the issuer going into bankruptcy, in a CAT bond the rating is instead based on estimation of the probability of default due to a trigger and subsequent forfeit of the principle. The probability of such triggers is then estimated using different catastrophe modeling approaches both mathematical and statistical in nature. Primarily there are a few companies that provide assessments of CAT bond structures and triggers settings and models to the rating agencies, the dominant firms specializing in this field are Applied Insurance Research (AIR), EQE International, and Risk Management Solutions. Typically these consulting firms produce model analysis that provides a summary of their proprietary disaster models in the form of an annual probability of loss exceedance that corresponds to the particular peril covered by the CAT bond. As summarized in Araya (2005), the models developed to assess this probability involve probabilistic descriptions of the natural disaster (hazard analysis), the performance of the assets in the book of business (portfolio modelling and the vulnerability analysis), and the resulting loss analysis involving the convolution of the two.

CAT bonds are typically sponsored by insurance and reinsurance companies, governments, and catastrophe-exposed corporations that have an influence on the rating attained. CAT bonds are issued to protect against almost any peril for which an established catastrophe model has been developed. It may also be common to write reinsurance policies where in this case they act as a cedant, which helps offer some level of protection to investors of the CAT bond; see the Definition 18.3. Such relationships provide much needed leverage since insurance companies are regulated so that they may not write policies in excess of a certain percentage of their collateral. However, insurance companies do not have to hold collateral against policies that are reinsured. Most insurance companies employ some sort of reinsurance program in order to more efficiently manage their operations. This can also be performed in OpRisk settings by captives within a financial institution and sidecars.

Definition 18.3 (Cedent) *A cedent is a party to an insurance contract who passes financial obligation for certain potential losses to an insurer. As a result of bearing a particular risk of loss, the cedent pays an insurance premium.* ■

CAT bonds are classified according to the following features:

- The number of perils covered, where options typically include single-peril (one natural disaster such as earthquake in one region of the world) and multiperil classifications that involve multiple tranches (multiple natural disasters such as earthquake, windstorms, hurricanes in multiple regions of the world). In the case of multiple tranches, each may vary in payment terms, coupon rates and event credit ratings; see Barrieu and Albertini (2010, chapters 2 and 4);
- The type of losses they cover (per event or aggregate losses), which is detailed in the offering contract and documentation;
- The trigger mechanism (clause that results in a default of the CAT bond and subsequent forfeiture of captial/coupons). These can be based on many features as detailed later and can involve *first-event* or *second/third* events.

Though a vast array of natural and man-made disasters exist and can be covered by CAT bonds, in general one may classify CAT bonds into one of four categories based on their trigger types; see details in Araya (2005).

CAT Bond Trigger Classifications

1. **Indemnity**. This trigger involves an issuer's actual losses in which the sponsor is indemnified in the same manner as if they had simply purchased a reinsurance. Where the contracts will state a layer in the CAT bond of x million excess over y million in losses. Then if the total claims accumulated from a catastrophic event exceed y million, the bond is triggered. Typically indemnity-based trigger CAT bonds also include provisions to extend the maturity of the security in order to allow for development periods of the claim process in the event of a natural disaster for the given peril(s) covered that result in a trigger;
2. **Modelled Loss**. This is an alternative to considering actual accumulated claims. In this case, the CAT bonds have an associated exposure portfolio constructed that is linked to a computer model (statistical model) for the given catastrophe. Then when an actual catastrophic event occurs, the observed/measured/estimated details of the event are utilized in the simulation model to compare to the exposure database for the CAT model. Then the CAT bond is triggered if the modeled losses under the event parameters exceed a specified threshold;
3. **Industry Loss Indexation**. This is an alternative to considering actual accumulated claims; instead, the trigger is based on an insurance industry-wide index for a given peril, where the exceedance of a given peril results in the triggering of the CAT bond;
4. **Peril Based (or Parametric)**. This is an alternative to considering actual accumulated claims; instead, the trigger is "parameterized" by specification of physical properties/parameters

of a given peril such as Richter scale recordings for an earthquake or wind speeds, etc., measured at multiple locations. Under a parametric trigger, losses to the CAT bond holders are triggered when parameters that define the peril covered exceed certain prespecified thresholds. Often these parametric trigger thresholds are modeled to correlate well in magnitude to actual losses (insurance claims) that may have been experienced historically from such perils.

As discussed in Loubergé *et al.* (1999), when the trigger condition is a fixed proportion of the losses, then a development period for the claims process must be incorporated into the risk period in order to determine the exact total loss amount. The actual trigger amount can be based on the insurer's losses or on industry-wide losses as discussed earlier in the classifications. To understand the effects of different trigger choices, we first define the notion of basis risk.

18.1.2.1 Triggers for CAT Bonds: Basis Risk versus Moral Hazard.

It is discussed in Doherty (1997a) among others that different choices for the construction of the triggers in CAT bonds may result in different trade-offs between credit risk, basis risk, and moral hazard.

Definition 18.4 (Basis Risk) *Basis risk involves the risk that offsetting investments in a hedging strategy will not experience price changes in entirely opposite directions from each other. In other words, when the two investments are not perfectly correlated in the hedging strategy, there is the potential for additional risk of losses (or gains) from the hedged position; this is known as basis risk. In the context of CAT bonds the basis risk is defined as the potential differences between actual losses in the sponsor's portfolio of assets in the event of a covered natural hazard and the losses predicted by the catastrophe modeling analysis.* ■

For instance, if the trigger is based purely on the individual insurers losses (an indemnity trigger), then in this construction of a CAT bond there will be no basis risk present for the sponsor; however, there will be the possibility of a moral hazard for the investors due to inflation of reported losses by the insurer. Conversely, if the trigger is based on an index of multiple insurers losses or "industry losses", then there is clearly the possibility of a basis risk. Furthermore, the amount of exposure to such a basis risk would be a function of the strength of the dependence between the industry losses and the firms losses, with weaker dependence relationships increasing the basis risk. There is also another interesting strategy that is underutilized and could be considered in the CAT bond asset class in future, as noted in Doherty (1997a) and summarized in Remark 18.5.

Remark 18.5 *If the trigger index for the CAT bond is allowed to be flexibly selected by the primary, then they would be afforded the possibility to trade-off basis risk and moral hazard. To understand this, consider the case in which the primary has the ability to select the indices upon which the trigger is defined. Then if the primary can find an industry portfolio with a similar exposure profile to their own, they may select this index. This would minimize the basis risk and the moral hazard risk will be controlled as they will not have the ability to overquote claims.*

Traditionally the class of trigger most favored by sponsors was the indemnity trigger as these types of CAT bond were perceived to be similar in character to the ultimate net loss type coverage offered in insurance industries and therefore they carry minimal basis risk. However, the disadvantages of the indemnity trigger structures relative to other choices of trigger include a less competitive risk spread premiums, disclosure requirements, perceived legal exposure to payment uncertainty, and claims processing uncertainty, as well as the time and cost involved in verification of claims by external auditors (often requested by the investors in the event of a trigger and subsequent default). Hence, even though originally the indemnity triggers were most popular, as the market for CAT bonds has grown, this class of trigger has reduced in popularity, being replaced by the parametric trigger type. The primary reason for this is due to three main developments: the acceptance by the sponsor of basis risk involved with such instruments; the demand by investors for greater transparency of the claims reporting, processing, and payment processes and trigger conditions offered by the less ambiguous parametric triggers provide this transparency; and the reduction in payment uncertainty periods offered by parametric triggers, where settlements tend to be more rapid, rather than extending the lifetime of the bond to cover the claims development process.

18.1.3 RECENT TRENDS IN CAT BONDS

The following recent developments have occurred in CAT bond structuring.

1. **Hybrid Triggers**. Recently, there has been a move to constructing triggers that are based on multiple events/losses. Examples include products such as Atlas Re's first, second event CAT bond and Atlas II's first, third event bond to name a few. This is one example of what are known in the industry of hybrid triggers that generally involve multiple trigger types and offer features such as granular disaggregation of industry triggers in order to minimize basis risk to sponsors;
2. **Takedowns**. There are also recent developments known as shelf-programmes that involve a sponsor incorporating into the CAT bond contract the option to issue additional bonds known as "takedowns". This additional flexibility has the advantage that the issuance cost to the sponsor is significantly reduced as well as allowing for long-term and short-term planning scenarios to be accommodated;
3. **Resets**. CAT bonds have been issued that offer multiyear coverage. In such settings, under a nonparametric bond class, the risk is estimated via analysis of exposures either based on the sponsor or the industry, depending on the type of trigger specified. It has been recognized that over longer time frames such as multiple year coverage the exposures faced at issuance (and therefore used in the calculation of liability coverage, etc.) may have increased significantly. This would expose the sponsor to a heightened basis risk over the duration of the maturity of the CAT bond. Therefore, an approach to tackling this is known as a **reset**, in which the CAT bond is allowed to be annually remodeled based on current exposures. This is typically achieved by modifying the trigger conditions to ensure a constant probability of loss corresponding to the level specified at issuance.

To complete the basic introduction to CAT bonds, it should be noted that these products continue to develop and to round off this section we also note there is a different payout timing and a different loss verification process in CAT bonds compared with standard insurance products; see details in Barrieu and Albertini (2010, chapter 4).

18.1.4 MANAGEMENT STRATEGIES FOR UTILIZATION OF INSURANCE-LINKED DERIVATIVES AND CAT BONDS IN OPRISK

We start this section by highlighting the different approaches that have been proposed as generic strategies for an insurer or a financial institution to provide coverage of some or all of the losses they may be exposed to with regard to natural disasters. Under Basel II/Basel III, financial institutions are required to provide capital to cover losses from process that may contain elements of catastrophic risks such as natural disasters (wind, storm, rain, floods, earthquake, fire, riots, disease, etc.). Such loss processes can be rare and of a potentially high consequence; therefore, it is particularly, important to consider the possibility of insurance mitigations for such loss processes.

Doherty (1997a) summarizes the different generic strategies that financial institutions may consider in order to attempt to manage their exposure to catastrophe risk according to four categories: asset hedges; liability hedges; post-loss equity re-capitalization; and leverage management. In general, they note that catastrophe hedging instruments will inherently face design choices, such as the trigger choices discussed earlier for CAT bonds that will ultimately involve trade-off between aspects such as credit risk, basis risk, and moral hazard. In what follows, we first briefly discuss some of the suggestions made in Doherty (1997a) and the implications.

- **Asset Hedge**. This is defined to be an asset that hedges against risk in holding an alternative asset. In the insurance setting, a reinsurance policy is a basic form of asset hedge. In OpRisk, this may be relevant, for example, in the situation of a financial institution with both a banking function and an insurance function often known as captives. When this insurance function provides coverage for certain losses in insurance products utilized for OpRisk coverage by the same financial institution, such a reinsurance contract could be obtained as an asset hedge;
- **Liability Hedge**. Here, instead of taking an asset that will hedge against risk, it instead involves a portfolio with a liability (opposite side of the balance sheet). The idea is that when the asset being hedged against has a loss in value, the corresponding liability (which if carefully selected to be correlated to the price behavior of the asset) should also see a subsequent reduction in the liability offsetting the loss;
- **PostLoss Equity ReCapitalization**. The idea of this strategy is to recapitalize the bank or portions of the bank after a significant loss has occurred that could threaten the viability of the banking business and solvency. Recapitalization is a form of corporate reorganization involving change in a company's capital structure. There are several ways that recapitalization may occur such as leveraged recapitalization (where the bank issues bonds to raise money, and then buys back its own shares); leveraged buyout; or partial or complete nationalization. The idea of such strategies is to generate a revenue stream at times in which liquidity is most needed without further increase in leverage. This is typically achieved by unlocking in some fashion the illiquid assets present in a bank. Examples of two such approaches discussed in Doherty (1997a) include postloss equity financing in which the price at which new equity is issued is reduced by the loss and the second approach suggested involves the bank or insurer purchasing a put option on its own stock that can be exercised should a catastrophic loss be realized of a given magnitude;
- **Leverage Management**. This strategy is utilized to manage catastrophic losses ex-post.

As an example of a liability hedge, we note that one may develop a reverse convertible approach; see Definition 18.5. In the case of CAT bonds, one can conceive of a reverse convertible being used as a liability hedge in which it acts as an alternative mechanism to debt forfeiture, where the debt is instead converted into another asset, in this equity. Hence, one may construct a conversion option into the debt, where the option is then exercised by the issuer and not the bond holder; this is termed reverse convertible debt (RCD) in Doherty (1997a).

Definition 18.5 (Reverse Convertibles) *A reverse convertible security involves a short-term note that is linked to an underlying stock. Typically the security provides a constant income stream through payment of a high coupon rate. Then at maturity the owner receives either 100% of the par value if the stock value has stayed above a predetermined level; alternatively, they may be delivered a predetermined number of shares of the underlying stock if the value of the stock price drops below a prespecified trigger level.* ■

An example of RCD could involve converting the bond into a fixed number of shares for each bond. Hence, a fall in the share price will result in the option being in the money. A second approach would be to base an RCD for the primary issuer on a trigger of a similar form to the CAT bond.

All of the aforementioned strategies rely on the ability to answer the important question of pricing such insurance-linked derivatives. The next section provides some insight and pointers as to why this is not such a straightforward endeavor.

18.2 Basics of Valuation of ILS and CAT Bonds for OpRisk

There is a rich literature on different mathematical approaches to developing the notion of a fair value or price for financial instruments; an interesting technical review of such mechanism of particular relevance to this section is provided by Platen and Heath (2006, chapters 9 and 14). In this chapter, the authors provide a means of studying in a common framework several pricing mechanisms under a unified framework known as the benchmark approach, details of which will be discussed in this section.

In the following sections, we provide a nontechnical review of basic and fundamental aspects of relevance to valuation and pricing and we do so by considering differences and relationships between pricing (valuation) in the context of the following aspects: traded versus nontraded assets or liabilities; complete versus incomplete markets, arbitrage versus nonarbitrage pricing; actuarial versus financial pricing; and real-world versus risk-neutral pricing. Each of these concepts will be discussed in the next few sections in basic detail and relevant key references will be provided.

To begin the discussion, we consider several ideal market assumptions in order to obtain concise statements regarding fair value pricing that sometimes even produce closed-form expressions for these prices. To begin we consider the notion of the Efficient Market Hypothesis; see Malkiel and Fama (1970) and Definition 18.6.

Definition 18.6 (Efficient Market Hypothesis) *Assumptions or hypotheses about market efficiency are assertions regarding the degree to which stock prices reflect all available relevant information. The efficient market hypothesis is a theory of investment that states as its core principle that it is not possible for an investor to outperform the market because all available information is already built into all stock prices. Therefore, stock market efficiency causes existing share prices to always reflect all relevant information and, as a consequence of the EMH, stocks always trade at their fair value.* ■

Under the EMH, it is possible to replicate the cash flows of an asset and the value of the replicating assets provide a unique value for the asset. Due to the assumption of market efficiency, one may assume that any other value would have been exploited by arbitrageurs trading. Under these market conditions, one may start to think about arbitrage-free pricing and valuation. In this context, one would typically perform valuation of a financial asset under the actuarial framework via a "deflator" or "pricing kernel" $f(t)$, which is given by

$$f(t) = \mathbb{E}_t\left[\frac{\xi_T}{\xi_t}f(T)\right], \tag{18.1}$$

where the deflator ξ_t is concerned with achieving market-consistent valuations of assets and liabilities. In addition to using deflators, the arbitrage-free market-consistent value can also be obtained in the mathematical finance audience by considering the notion of risk-neutral pricing according to a change of measure. These methods involve using probabilistic expectations of discounted present values of future cash flows, but in a world where all investors are risk neutral. Being risk neutral, these investors will be indifferent toward risk and hence they do not require riskier assets to have a greater expected return than the risk-free return when making investment decisions. Under these market conditions, all assets would provide an expected return that is equal to the return from a risk-free benchmark, irrespective of the risks associated with the cash flows. Hence, a risk-neutral valuation is achieved by calibrating a cash flow projection and an asset valuation model. The underlying asset classes will have the same expected return equal to the risk-free rate. Therefore, the present fair value would be given by a discounted expected value

$$f(t) = \mathbb{E}_{\mathbb{Q}}\left[\exp(-r(T-t))f(T)\right], \tag{18.2}$$

where r is the risk-free interest rate and $\mathbb{Q}$ denotes the discounted expected present fair value is taken under the risk neutral pricing measure generically denoted in this chapter by $\mathbb{Q}$.

We will discuss how in the complete market setting the deflator and risk-neutral settings can yield the same unique result in general; however if one moves to incomplete markets, this starts to differ. Typically, in the context of risk of an insurance modeling, the incomplete market framework arises naturally from the types of assets and liabilities under consideration. At the risk of stereotyping, it is typically the perspective of actuarial valuation that risk-neutral valuations are somewhat unrealistic since one could argue that in practice most investors require some form of compensation for the risk they take on in a particular financial contract or investment and they would otherwise not invest in more risky assets unless it had a higher expected return to compensate for this additional risk.

Under actuarial valuation via deflator methods, one may project future cash flows under realistic scenarios that incorporate volatile risky assets with expected returns exceeding the risk-free rate. However, the final valuation result involves an enforced arbitrage free value that corresponds to the risk-neutral valuation. For this to occur, the interest rate used to discount future cash flows under the deflator method is constrained. In this fashion, the deflator method can be considered as a class of stochastic discount factors that are coupled with a realistic scenario generator to achieve the same valuation result as the market-consistent risk-neutral valuation result.

Given this basic contextual background, we first begin with a high-level discussion of the classical complete market setting, starting with a key celebrated result of the mathematical finance investigations into pricing, the Black–Scholes option pricing formulation. Then we proceed to actuarial deflator methods and real-world pricing. In this regard, we consider a key component of a complete market, that it is the notion of frictionless markets under the Black–Scholes formulation. In the seminal paper on the pricing of options and corporate liabilities, see Black and Scholes (1973), an arbitrage-free pricing formulation that depends on observable quantities and did not require knowledge regarding investor's preferences or beliefs about their expected returns on the underlying stock. The conditions required for such a pricing framework involved what Merton (1976) referred to as "ideal conditions" in the market for the underlying stock and the option. Before presenting a brief overview of these conditions, it is useful to recall Definition 18.7 for a frictionless market.

Definition 18.7 (Frictionless Market) *A frictionless market is one in which all costs and restraints associated with transactions are nonexistent, that is, there are no trading costs, transaction costs, or differential taxes. This is a theoretical trading environment.* ■

It should be noted that friction is a type of market incompleteness and that every complete market is frictionless, but the converse does not hold. Now, one is in a position to consider the ideal conditions referred to by Merton for the pricing framework of Black and Scholes to hold included originally:

1. Frictionless markets;
2. Trading on the markets takes place continuously in time;
3. Borrowing and short selling are admitted without restriction and with full proceeds available;
4. Borrowing rates equal to lending rates;
5. The short-term interest rate is known and constant through time;
6. The stock pays no dividends or other distributions during the life of the option;
7. The option is only exercised at expiry; and
8. *The stock follows a "geometric" Brownian motion in time, giving a LogNormal distribution for stock prices between time points.*

This framework was relaxed and extended in works by Merton (1973) to allow for stochastic interest rates and other features. However, still of direct importance to this pricing framework were the assumptions that trading takes place continuously in time and the *price dynamics of the stock have a continuous sample path with probability one*. It is precisely this continuity condition in which over small time intervals only small price changes are admissible, which implies a form of local Markov property that is contentious in the context of insurance-linked derivatives. For example, if the asset price process follow a continuous time stochastic jump process for example, a Lévy process, then with some nontrivial probability there exists the possibility that the asset price may change significantly even over short time periods. Therefore, then the price process follows such a jump-type process one must rethink how to perform valuation.

In this section, will first provide a high-level discussion on the approaches to pricing such products. The valuation and study of the associated properties of such insurance-linked derivatives such as CAT bonds is not such a straightforward problem. The reason for this is that it involves some particular features of insurance-linked derivatives that distinguishes such products from other more well-studied traded financial derivative assets. Insurance-linked derivatives are primarily established to transfer catastrophic "nature" risks. In general, such risks have been proposed to be modeled by jump diffusions, Poisson processes, inhomogeneous Poisson processes, and doubly stochastic Poisson (Cox) processes in a compound process framework; see discussions in Embrechts and Meister (1997) and an interesting calibration example for earthquakes in Härdle and Cabrera (2010).

As carefully detailed in Embrechts and Meister (1997), Lane (2000), Pelsser (2011), and Wüthrich (2010), the pricing of insurance-linked derivatives such as CAT bonds requires the framework of incomplete markets. There are multiple approaches that one may adopt when developing a pricing framework for insurance-linked derivatives; several of these have been considered in the literature for CAT bonds and this section aims to summarize the features of each to provide a clear overview of each and the associated valuation models proposed in the literature. Fortunately, this area is still a growing field and so reasonable coverage can be expected in this endeavor.

We will first present a general discussion on the intricacies of such a valuation setting. Then we will present two general cases the settings in which people have made idealized frameworks to perform valuation based on assuming that a complete market framework can be approximately considered and the insights gained from such approximation. This will be followed by the more realistic framework of incomplete markets the implications of the fair valuation and the introduction of risk aversion and utility to perform unique pricing in an arbitrage-free setting. In both cases, we will present a range of complexity in the models that have been proposed based on simplifications of the behavior of interest rates ranging from constant interest rates, simple diffusion models for the interest rates, through to term structure–based approaches.

When dealing with natural catastrophes, one may consider a model for their occurrence and their severity under a standard actuarial framework involving a compound Poisson process or a mixed jump-diffusion framework for the underlying risk index. As noted in Vaugirard (2003), under such a model framework the pricing of insurance-linked derivatives involves an incomplete market and so the well-known methodology of replicating portfolios does not apply. Instead, it is noted that the five fundamental approaches one could consider involve as follows.

1. Assume risk that is associated with jumps is inconsequential due to diversification—Merton framework; see Merton (1976);
2. Construct a variance minimizing hedge strategy in such a way that it would allow one to select a unique martingale measure from the many possible equivalent martingale measures arising from the pricing based on a discontinuous process to obtain the valuation of the CAT bond; see examples in Föllmer and Sondermann (1986), Föllmer and Schweizer (1991), Sondermann (1991), and Colwell and Elliott (1993);
3. Specify additional information such as the utility function of the investors in the CAT bond and determine the valuation based on the historical empirical price process measure; see examples in Pliska (1997), Kallsen (2003), and Hugonnier *et al.* (2005);
4. Assume the pricing framework is approximately one in which the market is complete by restricting the problem to one that does not allow for random jump amplitudes; see examples in Cox and Ross (1976);
5. Assume the pricing framework is approximately one in which the market is complete by embedding the incomplete market into a complete market through the introduction of an artificial asset; see examples in Shirakawa (1991).

While talking about pricing approaches, one can note that it is common practice to determine the price of an asset through reference to the underlying economic value of the asset. This leads one to the classes of "general equilibrium models" that aim to explain through economic reasoning why the value of financial assets change as a result of changing economic forces or factors. Examples of such approaches can be attributed to Merton (1976) through examples of pricing under what is known as the Intertemporal Capital sset Pricing Model (ICAPM) or the common insurance pricing framework of Bühlmann (1970) known as actuarial pricing. An alternative approach to pricing that does not endeavor to explain the economic reasons for price changes or to attribute such changes to particular changes in macro- or microeconomic factors is known as the benchmark approach; see, for example, Long (1990) and the book-length review of Platen and Heath (2006). Under a benchmark approach one is focused on a pricing framework based on marking to market.

At this stage, it will be useful to recall the following informal Definition 18.8 for arbitrage pricing theory of Ross (1976), Roll and Ross (1980), and Ross (1973).

Definition 18.8 (Arbitrage Pricing Theory) *Under arbitrage pricing theory (APT), one adopts a model in which an asset's returns can be predicted using the relationship between that same asset and many common risk factors. Put another way, the APT predicts a relationship between the returns of a portfolio and the returns of a single asset via a linear combination of many independent macroeconomic variables. Under an arbitrage opportunity, one has a trading strategy that is costless at time t, has no negative value in times $T > t$, and a positive probability of strictly positive values.* ■

Hence, the assumption of no arbitrage means that it is impossible to develop a strategy or portfolio that will guarantee a riskless sure profit in the market considered. Put another way, one can consider a no-arbitrage condition as stating that there is no possible strategy that involves making money with no initial investment and without any possibility of loss.

Remark 18.6 *The APT is often considered an alternative to the capital asset pricing model since it has more flexible requirements on assumptions. Under the CAPM formula, one requires the market's expected return; however, under the APT, the risky asset's expected return and the risk premium of a number of macroeconomic factors are required.*

Returning to the notion of pricing risky assets, under the APT framework, one of the most widely used methods in valuation involves the notion of risk-neutral pricing as defined next.

Definition 18.9 (Risk Neutral) *Risk neutral refers explicitly to an indifference to risk. In terms of investors, the risk-neutral investor is primarily concerned with their expected return on investment, not so much about the level of risk they are taking on.* ■

The modern day pricing approach replaces the classical framework that involved expectations of discounted quantities, known as the "the present value principle", by the concept of deflators, numeraires (which are basically inverse deflators), or the application of the present value principle after a change of measure. One of the most popular pricing techniques involves the risk-neutral pricing framework under APT as it allows one to develop a convenient fair value pricing that involves a change of measure, accompanied by the change of reference numeraire. Here, the term numeraire, taken from the French for money, coinage, or face value, is used in economics to refer to a unit of account. For example, in general, a numeraire could be applied to a single good, which becomes the base good or reference good. Using this numeraire good, all similar goods are then valued and priced against the base good. This reference or numeraire good allows one to make comparisons between similar goods in order to identify which goods are worth more than others. It is known that when the value process of a numeraire portfolio is used as a discount process, the relative value processes of all other portfolios with respect to it will be martingales or at least supermartingales; see discussions in Long (1990) and Korn and Schäl (2009) and the references therein. For any equivalent martingale measure obtained, one may work with respect to this measure under a consistent pricing system, see Harrison and Kreps (1979). However, ensuring the existence of a unique equivalent martingale measure or selection of such a measure from an infinite number of possibilities (as will be discussed further after) for pricing ILS assets can be problematic, overly restrictive on the desired price process models and arises primarily due to the price process models typically adopted, which involve jumps.

Without changing the price process models (which typically admit jumps or discontinuities), one can overcome this challenge using a benchmark approach. It is shown in Platen and Heath (2006, chapter 9) exactly how the benchmark approach to valuation adopts as numeraire the growth optimal portfolio (GOP), defined to be the portfolio that maximizes the expected logarithmic utility from terminal wealth. The GOP was first proposed by Kelly (1956) and Latane (1959) and later developed for applications by Breiman (1961), Christensen and Larsen (2007), and Platen (2005). In Long (1990), it was shown how the GOP is the natural numeraire portfolio when pricing contingent claims with the real-world probability measure used as the pricing measure. An interesting discussion and development in the discrete time application of the GOP to pricing is provided in Korn and Schäl (2009). However, unlike the standard pricing frameworks that require the existence of an equivalent martingale measure, this is not required when pricing under the benchmark approach. Hence, the main advantage that the benchmark approach has over other more classical pricing methods is that as soon as there is existence of a GOP, then one can perform pricing under a real-world measure. This therefore allows one to broaden the class of prices process models. In the pricing discussions given later, a unique martingale measure $\mathbb{Q}$ is defined by the concept of the numeraire portfolio. The particular choice of measure $\mathbb{Q}$ is then justified through a change of numeraire, which is in place of the typical change of measure. Then one applies the important fact that uniqueness in the pricing measure is obtained by the fact that the equivalent martingale measure after the change of numeraire may be the original real-world probability measure so long as the appropriate numeraire is selected. Hence, instead of finding a change of measure to move from the real-world process to a risk-neutral pricing measure, one instead seeks a numeraire change that allows one to perform pricing under a martingale measure given by the real-world process. In many general cases, one may obtain a required numeraire portfolio for such an endeavor via the GOP. Note: The change of numeraire that results in the real-world process being a martingale measure ensures that derivatives priced under this real-world measure represent the best forecast of the future values.

Hence, in the following sections, we will consider the concept of *real-world pricing* under the benchmark approach. As demonstrated by Platen and Heath (2006), it contains the standard actuarial pricing framework, the CAPM and ICAPM frameworks as special cases. In general, the real-world pricing approach allows one to obtain prices for payoffs to be obtained via conditional expectations under the real-world probability measure (i.e., the probability measure that models the market as it evolves and as one observes it through empirical observations). In the following section, we present the basic technical details for the real-world pricing framework discussed; for a comprehensive presentation, see the book-length discussion in Platen and Heath (2006).

18.2.1 PROBABILISTIC PRICING FRAMEWORKS: COMPLETE AND INCOMPLETE MARKETS, REAL-WORLD PRICING, BENCHMARK APPROACH, AND ACTUARIAL VALUATION

There exist a wide range of different approaches, that areas of financial mathematics, economics, and insurance have sought to explore, in order to justify the valuation of financial instruments. This chapter does not aim to present any detailed level of presentation of these approaches; instead, it will illustrate in the context of insurance-linked derivatives what has been achieved under some of these different approaches when valuing CAT bonds. In addition, a selection of influential works (which is by no means to be considered comprehensive) will be provided for each of the topics touched upon.

As discussed earlier, in financial mathematics, one of the most traditional approaches of pricing involves the assumption of a complete, efficient market hypothesis with the existence of a real-world price process that admits (under a change of measure) a unique risk-neutral pricing measure that is an equivalent martingale measure. This risk-neutral pricing measure can be used to perform discounted expected value calculations. As discussed in the previous section, this highly utilized framework will not typically apply in the context of insurance-linked derivative pricing. The reason for the need to consider an alternative pricing framework will be discussed in detail later. Fortunately, there are many other approaches one may adopt to overcome this issue when pricing insurance-linked derivatives. Next, we primarily concentrate on the benchmark approach and the actuarial pricing approach; see the book-length coverages in Bühlmann (1970), Gerber (1990), and recently Wüthrich (2010), which is common in insurance and accounting, providing another important example in this direction. To proceed, it will be beneficial to define a few basic quantities that will be used later in the illustrations and examples of pricing CAT bonds.

We remind the reader that when we refer to bonds we consider them as a security that establishes a creditor relationship between the purchaser (creditor) and the issuer (debtor). Under this contractual relationship, the issuer is entitled to receive a certain amount of money in return for the bond, and in return the issuer is then obliged to repay the principal at the end of the lifetime of the bond (maturity). In general, bonds will make coupon or interest payments that are determined as part of the product structuring (hence the name, fixed income securities).

Definition 18.10 (Money Market Account) *A money market account is a fictitious bank account whose balance grows at the random spot rate, denoted by r_t, over time. Therefore, one dollar deposited into a money market account at time 0 results in a bank balance at time t given by*

$$C_t = \exp\left(\int_0^t r_u du\right). \tag{18.3}$$

■

Definition 18.11 (Zero-Coupon and Defaultable Bonds) *A zero-coupon bond is a fictitious financial security that pays one dollar (the face value or par value) at a fixed time T (the maturity time). For $t \in [0, T]$ and $r_t \geq 0$, the price B_t of this pure discount bond fluctuates randomly in $[0, 1]$, but $B_T = 1$ with certainty. Hence, a zero-coupon bond is simply a discounted bond bought at a price lower than its face value, with the face value repaid at the time of maturity.*

A defaultable bond is a pure discount bond that promises to pay one dollar at its fixed expiry (maturity) date; however, this payment is subject to a risk that may result in a default at some random time $\tau \in [0, T)$. *If a default occurs, it pays a random recovery rate* $R \in [0, 1)$ *at the default time* τ. ■

We remind the reader that typically bonds have two main sources of risk that affect the bonds investment value: credit risk (default) and interest rate risk (rate fluctuations). However, for CAT bonds, the main source of risk for a default is not credit risk but instead what could loosely be termed nature risk.

Definition 18.12 (Bond Yield) *In general, a yield refers to the income return on an investment and can be generically used to refer to the interest, dividends received from a security or other sources. Typically it is expressed annually as a percentage based on the following components: investment cost; current market value or face value. In the context of bonds, there are three yields:*

1. *Coupons (the bond interest rate fixed at issuance);*
2. *Current yield (the bond interest rate as a percentage of the current price of the bond); and*
3. *Yield to maturity (an estimate of what an investor will receive if the bond is held to its maturity date)*

In some special classes of bonds such as nontaxable municipal bonds, there will be also a fourth yield corresponding to a tax-equivalent yield determined by the investor's tax bracket. ■

If one considers how the price of a bond with fixed cash flows changes in price, the two factors that drive the price in standard settings correspond to two sources:

1. A predictable riskless source corresponding to the passage of time (convergence towards par); and
2. A change in the yield either through changes in the benchmark yield and/or changes in the yield spread.

The yield–price relationship is inverse, and we would like to have a measure of how sensitive the bond price is to yield changes. A first attempt, which does not really directly capture this desired feature, is the Macaulay duration, a later definition known as the modified duration provides the required price sensitivity measure with respect to changes in yield. In particular, it is popular to consider either a linear approximation, known as the modified duration that is a measure of the price sensitivity to yields, or for larger yield changes one may consider a higher-order approximation such as the quadratic approximation known as the convexity.

Definition 18.13 (Bond Duration) *The duration of a financial asset consisting of fixed cash flows, for example, a bond, is the weighted average of the times until those fixed cash flows are received. When an asset is considered as a function of yield, duration also measures the price sensitivity to yield, the rate of change of price with respect to yield or the percentage change in price for a parallel shift in yields. Typically calculation of the duration involves the present value, the yield, the coupon cash flow to investors, and the final maturity and call features. In all definitions of* *duration, the larger the duration, the greater the interest rate risk or reward for bond prices. There are typically two commonly considered duration measures:*

1. *Frederick Macaulay (1938) proposed that duration be determined by the weighted average specified according to*

$$D = \frac{\sum_{j=1}^{T} (1+r)^{-t_j} t_j c_j}{\sum_{j=1}^{T} (1+r)^{-t_j} c_j}, \tag{18.4}$$

   where r is the periodic yield (for a single period), t_j is the time until the j-th cashflow and c_j is the j-th periods cash flow. Hence, Macaulay duration is the name given to the weighted average time until cash flows are received and is measured in years;

2. *Modified duration: The name given to the price sensitivity and is the percentage change in price for a unit change in yield given by*

$$DM = \frac{D}{(1+r)} = -\frac{1}{B(r)} \frac{\partial B(r)}{\partial r}, \tag{18.5}$$

   where D is the Macaulay duration, r is the periodic yield, and $B(r)$ represents the price of the bond at yield r. ■

Remark 18.7 *When yields are continuously compounded, then the Macaulay duration and the modified duration will be equal in value. If, however, yields are periodically compounded (i.e., interest rates are compounded), then the Macaulay and the modified durations will differ. Though Macaulay duration cannot be directly utilized to determine true price sensitivity to yield changes, it can be used to show the following:*

1. *The duration of a zero coupon bond is equal to its time to maturity;*
2. *The duration of a coupon bearing bond is less than its time to maturity;*
3. *Given two bonds with the same coupon rate and yield, then the bond with the greater maturity has a higher duration; and*
4. *Given two bonds that have the same yield and maturity, then the one with the lower coupon rate has a higher duration.*

Having defined these basic concepts relating to bonds and their properties, we now proceed to illustrate some examples of pricing CAT bonds by first establishing an appropriate pricing framework for CAT bonds. Explaining why standard approaches to valuation need to be reconsidered when valuing insurance-linked derivatives and special products such as CAT bonds.

In particular, we will discuss the "no-arbitrage" pricing framework in the context of insurance-linked derivatives. This has been studied in a number of papers such as the works of Harrison and Kreps (1979), Harrison and Pliska (1981), Föllmer (1991), Delbaen and Schachermayer (1994), and Embrechts and Meister (1997). We will recapitulate the discussions provided from Embrechts and Meister (1997), which itself is a selection of ideas based on the work of Föllmer (1991). We first consider the price process $(X_t)_{0 \le t \le T}$, which is associated with a probability triple $(\Omega, \mathcal{F}, \mathbb{P})$, where Ω is the sample space of possible outcomes,

$\mathcal{F}$ represents the sigma algebra, and $\mathbb{P}$ the relevant probability measure for the process. Furthermore, assume a contingent claim cashflow H is an $\mathcal{F}_T$ measurable random variable on the probability space $(\Omega, \mathcal{F}, \mathbb{P})$. As a reminder, we note the definition of a measurable function.

Definition 18.14 (Measurable Function) *A measurable function is a mapping that is "structure-preserving" between two measurable spaces, in our case probability spaces. That is, the function is measurable if the preimage of each measurable set is measurable. Given two measurable spaces $(\Omega_1, \mathcal{F}_1)$ and $(\Omega_2, \mathcal{F}_2)$, in which Ω_1 and Ω_2 are sample spaces with their corresponding sigma algebras $\mathcal{F}_1$ and $\mathcal{F}_2$. Then a function $f : \Omega_1 \mapsto \Omega_2$ is measurable if $f^{-1}(E) \in \mathcal{F}_1$ for all $E \in \mathcal{F}_2$.* ■

Measurable functions have the following useful mathematical properties.

Remark 18.8 (Properties of Measurable Functions) *The following properties preserve measurability.*

1. *The sum of two measurable functions is a measurable function;*
2. *The product (or quotient if no division by 0) for two measurable functions is a measurable function;*
3. *The composition of two measurable functions is a measurable function; and*
4. *The pointwise supremum, infimum, lim sup, and lim inf of a sequence of measurable functions are each measurable.*

In addition, it will be valuable to recall the definition of a predictable process, see Definition 18.15, as well as the definition of a martingale, see Definition 18.16.

Definition 18.15 (Predictable Process) *A predictable process can be considered as a stochastic process whose value is knowable at a prior time. The predictable processes form the smallest class that is closed under taking limits of sequences and contains all adapted left-continuous processes. Formally, a continuous time stochastic process $(X_t)_{t\geq 0}$ on a filtered probability space $\left(\Omega, \mathcal{F}, (\mathcal{F}_t)_{t\geq 0}, \mathbb{P}\right)$ is predictable if X_t is measurable with respect to the sigma algebra $\mathcal{F}_{t^-}$ for any time t.* ■

Definition 18.16 (Martingale) *A martingale is a stochastic process (i.e., a sequence of random variables) such that at a particular time in the realized sequence, the expectation of the next value in the sequence is equal to the present observed value. This will be true even if knowledge of all prior observed values is available at the current time. Formally, a stochastic process $X : T \times \Omega \mapsto S$ is a martingale with respect to a filtration $\mathcal{F}_t$ and probability measure $\mathcal{P}$ if the following hold:*

- *X_t is adapted;*
- *$\mathbb{E}\,|X_t| < +\infty$ for all time t; and*
- *$\mathbb{E}\left[X_t \middle| \mathcal{F}_{t-1}\right] = X_{t-1}$ for all $t \geq 1$.* ■

Using these probabilistic definitions allows one to also define the notion of a complete market as detailed next.

Definition 18.17 (Complete Market) *A complete market is one in which the complete set of possible gambles on future states of the world can be constructed with existing assets without friction. This is a situation in which every agent has the ability to exchange every good either directly or indirectly with every other agent without transaction costs. Put another way one could state that completeness means that the underlying price process* $(X_t)_{0 \leq t \leq T}$ *is such that every contingent claim may be replicated by a self-financing strategy. Formally, the market denoted by the probabilistic construction* $\left\{ \left(\Omega, \mathcal{F}, (\mathcal{F}_T)_{0 \leq t \leq T}, \mathbb{Q} \right), (X_t)_{0 \leq t \leq T} \right\}$ *is complete if for every contingent claim that is square integrable* $H \in L^2$*, then the probability space* $(\Omega, \mathcal{F}_T, \mathbb{Q})$ *admits an Ito representation with respect to the initial investment value at time* $t = 0$ *denoted* H_0 *and a predictable process* (ζ_t)*, which is given by,*

$$H = H_0 + \int_0^T \zeta_s dX_s, \tag{18.6}$$

in terms of the process $(X_t)_{0 \leq t \leq T}$. ■

Now given these basic fundamental definitions, we reiterate the discussion that there is a rich and well-developed literature on asset pricing in complete and arbitrage-free markets that relies on these definitions. First, we will introduce a few key aspects of the standard pricing theory before following the real-world pricing framework discussed earlier. We follow this path as it is informative to consider the differences between the standard pricing approach of APT combined with change of measure to a risk-neutral framework versus the real-world pricing approach that can also be developed under an APT framework except now a change in numeraire is applied in order to make the real-world process into the appropriate equivalent martingale measure for pricing.

18.2.1.1 Basics of Arbitrage-Free Risk-Neutral Pricing. The basic probabilistic definitions allow one to consider the context of working with pricing under a "risk-neutral measure", also known as an equivalent martingale measure that forms for traditional exchange-traded assets the key process for pricing. To understand at a basic level how this risk-neutral measure plays a role in pricing, we first discuss risk preferences and then proceed to the definition of two key theorems known as the Fundamental Theorems of Asset Pricing I and II. Together these two theorems link the role of the risk-neutral measure to the assumptions of arbitrage-free and completeness to guarantee that the fair value of a financial derivative is given by the discounted expected value of the future payoff under the unique risk-neutral measure.

So why one would like to consider a risk-neutral measure? To answer this question, one can first postulate that the prices of assets depend on their risk and furthermore that investors will expect a greater profit if they are exposed to more uncertainty or risk. Hence, if the price today of a claim on a risky amount realized tomorrow differs from its expected value and furthermore if one believes that investors in this asset are risk averse, then in order to make such risky a investment, said investors, will need to be rewarded with a risk premium in order for them to be willing to bear the risk associated with a potential loss.

Given these assumptions, then the pricing of an asset that involves the calculation of the fair value today, via an expectation, should be adjusted according to the investor's risk preference. That is, one would first take the expected value today and then adjust for the investor's risk preference. If this were attempted, it would be very difficult to obtain a unique price since the discounted rates would vary as a function of a given individuals risk preference. The key results that arise from the fundamental theorems of asset pricing allow one to avoid such complications since they show that in a complete market with no arbitrage opportunities there is an alternative way to perform this pricing.

Basically, instead of worrying about how to adjust for any given investor's risk preferences, one can instead perform the pricing by a single once of change of probability measure. This measure change adjusts the probabilities of future outcomes in such a way that they incorporate all investors' risk premia. Then the present value fair price is obtained by taking the expectation under this new probability distribution, the risk-neutral measure. This approach, when applicable, has the benefit that once the risk-neutral measure is obtained (if it exists) every asset can be priced by simply taking its expected payoff. Whereas without this risk-neutral measure in this context, if we had worked with the real-world measure, then pricing each asset would require a different adjustment for their different levels of risk.

At this stage, it would be instructive to recall two key fundamental theorems of arbitrage (critical to this standard pricing framework), which provide necessary and sufficient conditions for a market to be considered arbitrage free and complete. In the simplest case of a discrete finite state market, one can state the following two fundamental theorems of asset pricing. The first fundamental theorem relates to the existence of an arbitrage-free market and crucially depends on the existence of at least one risk-neutral probability measure, while the second fundamental theorem relates to the completeness of a market that in turn can be shown to directly depend on the uniqueness of a single risk-neutral pricing measure; see detailed discussions in Musiela and Rutkowski (2005, section 2.6.6) and the references therein.

Theorem 18.1 (First Fundamental Theorem of Asset Pricing) *A discrete market (finite state) with a probability space $(\Omega, \mathcal{F}, \mathbb{P})$ is arbitrage free if and only if there exists at least one risk-neutral probability measure $\mathbb{Q}$ that is equivalent to the original probability measure $\mathbb{P}$, denoted $\mathbb{P} \sim \mathbb{Q}$.*

Theorem 18.2 (Second Fundamental Theorem of Asset Pricing) *Given a market with base components: a collection of S stocks and B risk-free bonds, denoted by (S, B), then such a market is said to be complete if and only if there exists a unique risk-neutral measure that is equivalent to $\mathbb{P}$ and has numeraire B.*

The assertion of these two fundamental theorems of asset pricing then leads one to the definition of the risk-neutral pricing measure, given in Definition 18.18.

Definition 18.18 (Unique Martingale Measure) *A market will be arbitrage free and complete (frictionless) iff there exists a unique probability measure under which the prices of all traded assets, when divided by an appropriate numeraire, will be martingales.* ■

Under the existence and uniqueness of a risk-neutral measure for pricing of an asset, it is then standard to proceed as follows. Consider a maturity (future time) T that the derivative on the asset price process $(X_t)_{0\leq t\leq T}$ pays an amount H_T, which is a $\mathcal{F}_T$ measurable random variable. Assume that the discount factor from the current time ($t = 0$) to the maturity future time T is denoted by $P(0, T)$. Then the fair value of the derivative today ($t = 0$) is given by

$$H_0 = P(0, T)\mathbb{E}_{\mathbb{P}}\left[\frac{d\mathbb{Q}}{d\mathbb{P}}H_T\right] = P(0, T)\mathbb{E}_{\mathbb{Q}}\left[H_T\right] \tag{18.7}$$

with $\frac{d\mathbb{Q}}{d\mathbb{P}}$ the standard Radon–Nikodym derivative.

Remark 18.9 *We now see the direct link between the existence and uniqueness of the risk-neutral pricing measure and the assumptions of an arbitrage-free and complete market (at least in the discrete market case). These features are directly linked to the pricing of the contingent claim at $t = 0$ and the representation of the change of measure from the real world to the risk-neutral pricing measure to perform the valuation uniquely.*

So far we have simply discussed the possibility of a fair pricing framework and the associated requirements from an economic perspective on the market and prices of assets that will ensure the existence and uniqueness of such a measure. We have not commented on how one may construct or obtain such a pricing measure; for an in-depth technical analysis of the results discussed later, it is best to refer to a comprehensive text such as Karatzas and Shreve (1991). In the following discussion, we aim to provide a very basic introduction to the steps involved with obtaining such a risk-neutral pricing measure. To proceed, we consider a general stochastic process for the price of a traded asset given by stochastic diffusion process or stochastic differential equation (s.d.e.)

$$dX_t = \mu(t, X_t)\, dt + \sigma(t, X_t)\, dW_t, \tag{18.8}$$

for some drift function $\mu(t, X_t)$, and volatility function $\sigma(t, X_t)$, which is driven by Brownian motion W_t with the properties that $\mathbb{E}[dW_t] = 0$ and $\mathbb{E}\left[(dW_t)^2\right] = dt$. Note that the Brownian motion temporal path is a continuous function in time that is nowhere differentiable.

It will be often important when undertaking pricing and valuation of assets or liabilities to consider transforms of the price process X_t such as those generically represented previously by payoffs of a derivative of the price process at some future time T, denoted by H_T. More, explicitly we may consider a generic transformed process $Y_t = f(t, X_t)$ and as what is the resulting process for the new diffusion Y_t? To answer this question, under suitable technical conditions, one may apply Ito's Lemma in Theorem 18.3.

Theorem 18.3 (Ito's Lemma) *Consider the generic s.d.e. for the process X_t according to*

$$dX_t = \mu(t, X_t)\, dt + \sigma(t, X_t)\, dW_t, \tag{18.9}$$

for some drift function $\mu(t, X_t)$, and volatility function $\sigma(t, X_t)$, which is driven by Brownian motion W_t. Then the new transformed process $Y_t = f(t, X_t)$ for a twice continuously differential function f has s.d.e. given by

$$\begin{aligned} dY_t = {} & \left[\frac{\partial}{\partial t} f(t, X_t) + \mu(t, X_t)\frac{\partial}{\partial x} f(t, X_t) + \frac{1}{2}\sigma^2(t, X_t)\frac{\partial^2}{\partial x^2} f(t, X_t)\right] dt \\ & + \sigma(t, X_t)\frac{\partial}{\partial x} f(t, X_t)\, dW_t. \end{aligned} \tag{18.10}$$

Returning to the previously defined notion of a martingale, one may note that a martingale will be a stochastic process (s.d.e.) that stays on average at the same level and therefore is a differential equation with the dt component or term that is zero. The significance of this observation will be made more apparent in the following discussion on risk-neutral pricing and Girsanov's theorem.

Hence, given this price process for the financial asset, the expected value of a payoff for a financial derivative, which is some functional of the price process (derived from the underlying asset) that is, made explicit by the function $f(X_T)$ at some time $T > t$, is given with respect to the value of the underlying asset at time t by the function $v(t,x)$, which is the conditional expectation

$$v(t,x) = \mathbb{E}\left[Y_T \mid Y_t = f(x)\right] = \mathbb{E}\left[f(X_T) \mid X_t = x\right]. \tag{18.11}$$

The form of the expected payoff $v(t,x)$ is a solution to the Kolmogorov backward equation (KBE) given by the partial differential equation

$$\frac{\partial}{\partial t}v(t,x) + \mu(t,x)\frac{\partial}{\partial x}v(t,x) + \frac{1}{2}\sigma^2(t,x)\frac{\partial^2}{\partial x^2}v(t,x) = 0, \tag{18.12}$$

for $t < T$ and a boundary condition given by $u(T,x) = f(x)$.

In addition to understanding how to derive the s.d.e., for a transform of another s.d.e., it will also be important to understand how to obtain a new s.d.e. under a change of measure. This is a natural point to discuss the fundamental result provided by Girsanov's theorem, given in Theorem 18.4. This result describes how the dynamics of stochastic processes change when the original measure is changed to an equivalent probability measure. It is therefore at the core of the change of measure required for pricing when one moves from a real-world price process (physical process measure) to the risk-neutral pricing measure for the underlying asset price or interest rate. That is, Girsanov's theorem will be important as it enables the key result that if $\mathbb{Q}$ is a measure absolutely continuous with respect to $\mathbb{P}$ then every $\mathbb{P}$-semimartingale is a $\mathbb{Q}$-semimartingale.

Theorem 18.4 (Girsanov's Theorem) *For any stochastic process G_t (Girsanov kernel) that satisfies the condition $\int_0^t G_s^2 ds < \infty$ with probability one and has the Radon–Nikodym derivative representation given by,*

$$R_t = \exp\left(\int_0^t G_s dW_s - \frac{1}{2}\int_0^t G_s^2 ds\right), \tag{18.13}$$

where W_t is Brownian motion under a probability measure $\mathbb{P}$. Then if one defines the probability measure $\mathbb{Q}$ according to $d\mathbb{Q} = R_t d\mathbb{P}$ then under the probability measure $\mathbb{Q}$ with associated process $\widetilde{W}_t$ given by,

$$\widetilde{W}_t = W_t - \int_0^t R_s ds, \tag{18.14}$$

is also a Brownian motion.

Now, utilizing Ito's Lemma, one can show that the process R_t given earlier is a martingale (with dt term set to zero) under probability measure $\mathbb{Q}$, characterized by the s.d.e. $dR_t = G_t R_t dW_t$. Hence, given some price process s.d.e., one can utilize the following basic steps to find the risk-neutral s.d.e. representation by combining Ito's Lemma with Girsanov's theorem:

1. Consider a s.d.e. process for the price of an asset denoted by $(X_t)_{t\geq 0}$ given by the generic s.d.e.

$$dX_t = \mu(t, X_t)\,dt + \sigma(t, X_t)\,dW_t, \tag{18.15}$$

driven by a Brownian motion $(W_t)_{t \geq 0}$ which is defined with respect to a measure $\mathbb{P}$ (real-world process measure). If one applies a change in probability measure to the process $(X_t)_{t \geq 0}$ to take it from being defined with respect to measure $\mathbb{P}$ (real-world measure) to measure $\mathbb{Q}$ (risk-neutral pricing measure), then this implies a Radon–Nikodym derivative R_t with respect to the process X_t (some times denoted $\frac{d\mathbb{Q}}{d\mathbb{P}}$). Note, this change of measure is associated to appropriate selection of a base unit (numeraire) asset that has a strictly positive price process that when used to scale the price process of the asset of interest will produce a corresponding price process that is a martingale under measure $\mathbb{Q}$;

2. Apply Ito's Lemma to the Radon–Nikodym derivative R_t to obtain its s.d.e. representation and hence infer its Girsanov kernel G_t. Note, the process R_t will be a martingale with respect to the new probability measure $\mathbb{Q}$. In addition, one notes that the only choice for the Girsanov kernel that will nullify the differential in time component dt, thereby turning the process X_t into a martingale under measure $\mathbb{Q}$ is given by the functional form

$$G_t = -\frac{\mu(t, X_t)}{\sigma(t, X_t)}, \tag{18.16}$$

which corresponds to the market price of risk;

3. Obtain the s.d.e. for the measure changed price process by substituting the new s.d.e. process for the measure changed Brownian motion (with $\widetilde{W}_t$ defined w.r.t. measure $\mathbb{Q}$) given by,

$$dW_t = d\widetilde{W}_t + G_t dt, \tag{18.17}$$

to obtain

$$dX_t = [\mu(t, X_t) + \sigma(t, X_t) G_t] dt + \sigma(t, X_t) d\widetilde{W}_t. \tag{18.18}$$

To further understand this process, there is a good discussion on this valuation in Embrechts and Meister (1997), where the following stages are detailed along with a discussion on the valuation under the risk-neutral measure based on the existence of a complete arbitrage-free market. Returning to the generic notation for a derivative payoff or contingent claim, denoted by $H$and given the existence of an Ito decomposition of the contingent claim $H = H_0 + \int_0^T \zeta_s dX_s$ one may construct a riskless portfolio replication of the claim H using the premium H_0. This elementary procedure follows the following stages.

Riskless Portfolio Construction for Replication of a Claim H using Premium H_0.

- At a time t set up a portfolio holding an amount ζ_t in the risky asset (stock) X_t and hold an amount $\nu_t = \left(H_0 + \int_0^t \zeta_s dX_s\right) - \zeta_t X_t$ in the riskless asset (risk-free bond). At time t, the value of the portfolio is given by $v(t) = \zeta_t X_t + \nu_t = H$;
- To calculate H_0, the value of the contingent claim at time $t = 0$, we use the following facts (by construction):
  - The process (X_t) is a $\mathbb{Q}$-martingale. Note it need not be a $\mathbb{P}$-martingale;
  - The process (ζ_t) is predictable such that the process denoted by $(I_t)_{0 \leq t \leq T}$ and given by $\int_0^t \zeta_s dX_s$ is a $\mathbb{Q}$-martingale;

- The expectation of the process $(I_t)_{0\leq t\leq T}$, w.r.t. the measure $\mathbb{Q}$ has the property that $\mathbb{E}_{\mathbb{Q}}[I_t] = \mathbb{E}_{\mathbb{Q}}[I_0] = 0$.

- Then the expected value of the contingent claim w.r.t. the measure $\mathbb{Q}$ given by $\mathbb{E}_{\mathbb{Q}}[H] = H_0$ and given knowledge of H and measure $\mathbb{Q}$ one knows the fair present value of the contingent claim H_0.

It turns out that once one moves away from the previously described pricing in an ideal market to more realistic settings, then it is not so straightforward to obtain results that will guarantee the existence and uniqueness of the risk-neutral pricing measure.

Remark 18.10 *In particular, the existence of a unique risk-neutral pricing measure will crucially depend on the properties of the price process of the underlying asset* $(X_t)_{0\leq t\leq T}$. *For example it is well known that under certain established pricing frameworks such as the Cox–Ross–Rubenstein binomial tree model, the Bachelier Brownian motion model or the Markovitz–Black–Scholes geometric Brownian motion models one may prove the existence and uniqueness of a risk-neutral pricing measure. Other processes for the price are not so straightforward or may not even admit a unique risk-neutral pricing measure.*

Hence, we conclude by noting that traditionally the "financial pricing" approach (or option pricing) approach works under a change of measure in the equivalent martingale world of the $\mathbb{Q}$ measure (via risk-adjusted probabilities), while we will discuss in the next section the actuarial and real-world pricing frameworks that operate traditionally in the real-world $\mathbb{P}$ measure via objective probabilities and observed data consisting of projected losses and a likelihood of such losses that are converted to an actuarial fair value. In practice, this means that traditionally financial pricing has involved pricing say an option based on the minimal cost of setting up a hedging portfolio, whereas actuarial pricing of an insurance contract involves actuarial present value of costs and additional risk premiums for uncertainty associated with correlations, parameter estimation, and capital costs.

It should also be noted that under the actuarial deflator approach utilized when pricing assets or liabilities that are not traded on the market, such is the case for many contingent claims and long-dated cash flows considered in actuarial risk and insurance applications will also no longer be unique since the market price of risk is "unknown". Examples of such settings will be discussed in more detail later with respect to OpRisk, but in general they may include longevity, long-dated cashflows and wage inflation as well as certain types of reinsurance. This leads one to the frameworks discussed earlier that will be based on real-world pricing.

18.2.1.2 Real-world Pricing: Benchmark and Actuarial Frameworks (Dispersion Measures and Deflators). ILS that are constructed to transfer catastrophic risk from nature to the capital markets require a particular framework for valuation. In general, natural catastrophes can be incorporated into models for the assets and liabilities of an insurer, reinsurer, or a financial institution by considering a class of models involving jump-diffusion processes for an underlying risk index. In this context, we noted previously that financial markets are incomplete and as a consequence the methodology of replicating portfolios is not applicable. In the paper by Gerber and Shiu (1994), they consider the family of dispersion measures known as the Esscher transform, that they utilize for option pricing (as the deflator methodology equivalent of risk-neutral pricing in financial mathematics). Under the Esscher transform, they are able to show that one may obtain an efficient technique for valuing derivative securities if the logarithms of the prices of the underlying security come from a particular class that follows a stochastic processes with independent and stationary increments. Furthermore, they show that popular processes in this class include the popular families of models given by the Wiener process, the Poisson process, the Gamma process, and the inverse Gaussian process. Therefore, under this family of dispersion measures, one may select the parameter of this transform such that when it is applied to a security price process of the previously mention forms this would produce an equivalent probability measure. This resulting equivalent probability measure for the specially selected Esscher transform parameter (corresponding to the market price of risk) will produce a martingale for the discounted price of any underlying security with respect to the new transformed measure. Hence, in terms of valuation, one may calculate the value of any derivative or contingent claim future cashflows as the expectation, with respect to the equivalent martingale measure, of the discounted payoffs.

For the incomplete market case, particularly in the context of insurance, there is an excellent article detailing the properties of pricing in incomplete settings such as under compound Poisson processes, mixed and doubly stochastic compound Poisson processes, typically encountered in insurance applications; see Bühlmann *et al.* (1996). These processes posses a "jump" structure that distinguishes them from the more standard diffusion processes utilized in pricing in finance. The consequence of this manifests itself typically in two related outcomes: the first is that the risk cannot be completely hedged, and secondly in the existence not of a unique martingale measure but a potentially infinite number of such measures, naturally leading to the question of which to select in the pricing?

Before discussing this point, we first note that the following conditions under which there will be a martingale measure(s) for compound Poisson processes, mixed compound Poisson processes and doubly stochastic compound Poisson processes given in Theorem 18.5; see Meister (1995, proposition 2.11) and Embrechts and Meister (1997, theorem 1 and theorem 2).

Theorem 18.5 *Consider a mixed compound Poisson process on* $(\Omega, \mathcal{F})$ *under measures* $\mathbb{P}$ *and* $\mathbb{Q}$ *and denoted by* $(Z_t)_{t\geq 0}$. *Furthermore, the compound Poisson process has severity loss processes* $\{X_n\}_{n=1}^{N_t}$ *with each* X_n *being i.i.d. and associated to measures* $\mathbb{P}_{X_1}$ *and* $\mathbb{Q}_{X_1}$ *and the count process* $(N_t)_{t\geq 0}$ *is associated with measures* $\mathbb{P}_\Lambda$ *and* $\mathbb{Q}_\Lambda$. *Then the following are equivalent:*

1. *The measures* $\mathbb{P}_{X_1}$ *and* $\mathbb{Q}_{X_1}$ *are equivalent* $\mathbb{P}_{X_1} \sim \mathbb{Q}_{X_1}$ *for all* $s \geq 0$*;*
2. *Conditional on the filtration* $\mathcal{F}_s$ *the measures* $\mathbb{P}$ *and* $\mathbb{Q}$ *are conditionally equivalent* $\mathbb{Q}|_{\mathcal{F}_s} \sim \mathbb{P}|_{\mathcal{F}_s}$ *for all* $s \geq 0$*;*
3. *There exists a function* $\gamma : \mathbb{R} \mapsto \mathbb{R}$, *which is measurable such that*

$$\mathbb{E}_{\mathbb{P}_{X_1}}\left[\exp\left(\gamma\left(X_1\right)\right)\right] = 1,$$

the Novikov condition is satisfied

$$\mathbb{E}_{\mathbb{P}_{X_1}}\left[X_1^2 \exp\left(\gamma\left(X_1\right)\right)\right] < \infty,$$

and the resulting Radon–Nikodym derivative of measure $\mathbb{Q}$ *with respect to measure* $\mathbb{P}$ *conditional on filtration* $\mathcal{F}_s$ *can be expressed as a function of* γ *as follows:*

$$\left.\frac{d\mathbb{Q}}{d\mathbb{P}}\right|_{\mathcal{F}_s} = \exp\left(\sum_{n=1}^{N_s} \gamma\left(X_n\right)\right) \frac{\mathbb{E}_{\mathbb{Q}_\Lambda}\left[\Lambda^{N_s} \exp\left(-\Lambda_s\right)\right]}{\mathbb{E}_{\mathbb{P}_\Lambda}\left[\Lambda^{N_s} \exp\left(-\Lambda_s\right)\right]}. \tag{18.19}$$

Furthermore, if one considers the process $(Z_t - pt)_{t\geq 0}$, *it can be shown to be an* $\mathcal{F}_t$*-Martingale under the measure* $\mathbb{Q}$ *for all* $s \geq 0$ *if and only if the following hold:*

1. *There exists* $\lambda > 0$ *and function* $\beta : \mathbb{R}^+ \mapsto \mathbb{R}$, *which is measurable, such that* $\mathbb{E}_{\mathbb{P}}\left[\exp\left(\beta\left(X_1\right)\right)\right] = \lambda$ *and* $\mathbb{E}_{\mathbb{P}}\left[X_1^2 \exp\left(\beta\left(X_1\right)\right)\right] < \infty$ *and*

$$\left.\frac{d\mathbb{Q}}{d\mathbb{P}}\right|_{\mathcal{F}_s} = \exp\left(\sum_{n=1}^{N_s} \beta\left(X_n\right) - \lambda s\right)\left(\mathbb{E}_{\mathbb{P}}\left[\Lambda^{N_s}\exp\left(-\Lambda s\right)\right]\right)^{-1}, \tag{18.20}$$

2. $p = \mathbb{E}_{\mathbb{P}}\left[X_1 \exp\left(\beta\left(X_1\right)\right)\right]$.

A very flexible example model for such an insurance process that encompasses as special cases several other models is the generalized inverse Gaussian mixed compound Poisson model (also known in statistics as the compound Sichel process after the work in Sichel (1974)), treated here as a continuous time process, see Example 18.1. This model family also then naturally contains the mixed Poisson gamma models as a subfamily of processes; see discussion in the insurance context in the book Panjer and Willmot (1992). There is a very detailed pricing framework developed for processes of this type in the context of stop-loss catastrophe reinsurance contracts in Dassios and Jang (2003), where they consider Cox processes (doubly stochastic Poisson processes) with shot noise driving the stochastic intensity of the Poisson process for the model of the claim arrival process. Following this work, Jaimungal and Wang (2006) considered the pricing of the now delisted CBOT catastrophe options with stochastic interest rates and compound Poisson losses.

EXAMPLE 18.1 Compound Sichel Process

Consider a mixed compound Poisson process on $(\Omega, \mathcal{F})$ under measures $\mathbb{P}$ and $\mathbb{Q}$ and denoted by $(Z_t)_{t\geq 0}$. Furthermore, the compound Poisson process has severity loss processes $\{X_n\}_{n=1}^{N_t}$ with each X_n being i.i.d. and associated to measures $\mathbb{P}_{X_1}$ and $\mathbb{Q}_{X_1}$ and the count process $(N_t)_{t\geq 0}$ is associated with measures $\mathbb{P}_\Lambda$ and $\mathbb{Q}_\Lambda$. Assume that one models the count process according to the following specifications involving setting $\mathbb{P}_\Lambda \sim GIG(\mu_1, \beta_1, \lambda_1)$ and $\mathbb{Q}_\Lambda \sim GIG(\mu_2, \beta_2, \lambda_2)$, where the generalized inverse Gamma distribution is given by the density

$$f(x; \mu, \beta, \lambda) = \frac{\left(\frac{\mu}{\beta}\right)^{\frac{\lambda}{2}}}{2K_\lambda\left(\sqrt{\mu\beta}\right)} x^{\lambda-1} \exp\left(-\frac{1}{2}\left(\mu x + \frac{\beta}{x}\right)\right), \tag{18.21}$$

where K_λ denotes the modified Bessel function of the third kind and the density has a strictly positive support ($x > 0$) and parameter restrictions $\mu > 0, \beta > 0, \lambda \in \mathbb{R}$. Note the GIG distribution family was first discovered by Etienne Halphen; see discussion in Perreault *et al.* (1999a) and later popularized by Ole Barndorff-Nielsen, where it became known as the generalized inverse Gaussian distribution, see Barndorff-Nielsen *et al.* (1992). Furthermore, the following properties of the model are known such as the existence of the moments given by the moment generating function having derivatives that are well defined at the origin, with the MGF given by

$$M_X(t) = \mathbb{E}\left[\exp(tX)\right] = \left(\frac{\mu}{\mu - 2t}\right)^{\frac{\lambda}{2}} \frac{K_{\lambda+2}\left(\sqrt{\beta(\mu - 2t)}\right)}{K_\lambda\left(\sqrt{\mu\beta}\right)}. \tag{18.22}$$

The resulting Radon–Nikodym derivative of measure $\mathbb{Q}$ with respect to measure $\mathbb{P}$ when conditioned on the filtration $\mathcal{F}_t$ is then given for any $t \geq 0$ by

$$\begin{aligned}\left.\frac{d\mathbb{Q}}{d\mathbb{P}}\right|_{\mathcal{F}_t} &= \exp\left(\sum_{n=1}^{N_t} \gamma\left(X_n\right)\right)\left(\frac{\mu_2}{\mu_1}\right)^{N_t}\left(\frac{1+\beta_1 t}{1+\beta_2 t}\right)^{\frac{N_t}{2}} \\ &\times \frac{K_{\lambda_2+N_t}\left(\mu_2\beta_2^{-1}\sqrt{1+2\beta_2 t}\right)}{K_{\lambda_1+N_t}\left(\mu_1\beta_1^{-1}\sqrt{1+2\beta_1 t}\right)} \frac{\left(1+2\beta_1 t\right)^{\frac{\lambda_1}{2}}}{\left(1+2\beta_2 t\right)^{\frac{\lambda_2}{2}}} \frac{K_{\lambda_1}\left(\frac{\mu_1}{\beta_1}\right)}{K_{\lambda_2}\left(\frac{\mu_2}{\beta_2}\right)}.\end{aligned} \tag{18.23}$$

In addition, this process can still be characterized under the change of measure as a mixed compound Poisson process. ■

Remark 18.11 *It can be shown from Theorem 18.5 that in any case in which the compound Poisson process $(Z_t)_{t\geq 0}$ has a jump size that is not constant, then the equivalent Martingale measure for $(Z_t - pt)_{t\geq 0}$ cannot be unique.*

It turns out that all is not lost, one can extend the framework of assumptions made about the pricing model in order to restrict the possible martingale measures (premium principles) to allow one to obtain a fair price. To proceed in this direction, it will be useful to recall the notion of Pareto optimality, see Definition 18.19, which has incidently been widely studied in insurance modelling, see examples in Borch (1960, 1962), Bühlmann (1980), Bühlmann (1984a), Bühlmann and Jewell (1978), Gerber (1978), and the book-length review of Gollier (1992). The reason such a notion from game theory and economics is being considered here is that it provides a way of understanding the notion of a replicating portfolio in a complete market setting. In addition, when in the setting of an incomplete market, one may adopt a framework of utility theory to select a unique pricing measure.

Definition 18.19 (Pareto Optimality) *Pareto efficiency or Pareto optimality refers to an allocation of assets or resources in which it is impossible to make any one better off without making at least one individual worse off.* ■

Remark 18.12 *One can show that in a continuous market setting by using continuous trading in a selected set of commodities one can span a continuum of states of the world, thereby allowing for the ability to achieve Pareto optimal outcomes in portfolio replication and hedging.*

In the case of incomplete markets, due to the price process admitting jumps, one can turn to a well-known actuarial approach to fair pricing (selection of a risk-neutral pricing measure) with which to obtain the discounted present value, via the work of the Swedish actuary Esscher (1932) and utilized in a more modern treatment in Gerber and Shiu (1996). It will be useful at this stage to recall the definition of a discrete and continuous Esscher transform given in Definition 18.20, also commonly known in statistics as an "exponential tilting" and used widely in developing asymptotic series expansions for distributions and densities such as the Edgeworth or saddle point types (discussed earlier); see Small (2010) and references therein. The family of Esscher transformations includes the result presented in Theorem 18.5 and Example 18.1.

Definition 18.20 (Esscher Transform or Exponential Tilting) *Consider a continuous random variable X defined with respect to a probability measure $\mathbb{P}$ and a nonzero constant real number h such that $\mathbb{E}_{\mathbb{P}}[\exp(hX)]$ exists. One can then define the Esscher transform, denoted $\mathcal{E}_h[\cdot]$, of the original probability measure $\mathbb{P}$ for X in terms of an equivalent new probability measure $\mathbb{Q} \sim \mathbb{P}$ (same null sets) with the following properties:*

1. $\mathcal{E}_{h_1}\mathcal{E}_{h_2}[\mathbb{P}] = \mathcal{E}_{h_1+h_2}[\mathbb{P}]$;
2. $\mathcal{E}_h^{-1}[\mathbb{P}] = \mathcal{E}_{-h}[\mathbb{P}]$.

If the measure $\mathbb{P}$ that characterizes random variable X admits a Radon–Nikodym derivative with respect to a suitable measure ν given by a density $f_X(x)$, then we see that the Esscher transform of the density, denoted by $\mathcal{E}_h[f_X(x)]$, will be a new density given by

$$f(x;h) := \mathcal{E}_h[f_X(x)] = \frac{\exp(hx)f_X(x)}{\int_{-\infty}^{\infty}\exp(hx)f_X(x)dx}. \tag{18.24}$$

■

Next we discuss how this Esscher transform can be used to obtain a risk-neutral Esscher pricing measure. Using this notion of an Esscher transformed (distorted) measure, we may consider the continuous process setting. For $t \geq 0$, consider a price process (e.g., an insurance futures process, contingent claim, or alternatively a CAT bond price) denoted by process $(F_t)_{t\geq 0}$ on a futures market characterized probabilistically by $\left(\Omega, \mathcal{F}, (\mathcal{F}_t)_{0\geq t\geq T}, \mathbb{P}\right)$. Assume that this price process is derived from an underlying Lévy stochastic process $(X_t)_{t\geq 0}$ with independent and stationary increments and a moment generating function $M_X(t)$ that exists such that the asset price at time t, given with regard to the process X_t and the initial price F_0, is obtained by the transformation

$$F_t = F_0 \exp(X_t), \tag{18.25}$$

where one assumes that $F_t - rt$ can be positive or negative with r the risk free force of interest.

In this setting, the process $\left(\exp(hX_t)\left(\mathbb{E}[\exp(hX_t)]\right)^{-1}\right)_{t\geq 0}$ is a positive martingale that can be utilized to develop a change of measure. To see this, consider the original price process measure $\mathbb{P}$, then under the change of measure w.r.t. this martingale, the resulting measure $\mathbb{Q}$ is the Esscher transform with parameter h, known in actuarial science as the "risk neutral Esscher measure". Then one can select a unique parameter h^* such that the process $(\exp(-rt)F_t)_{t\geq 0}$ is a martingale.

Remark 18.13 (Risk-Neutral Esscher Pricing Principle) *Hence, the Esscher principle of pricing allows one obtain a unique Martingale measure for the pricing of insurance futures that, for the right continuous price process denoted $(F_t)_{t\geq 0}$, can be used to perform the pricing under discounting with regard to the new measure $\mathbb{Q}$ according to the expectation $F_0 = \mathbb{E}_{\mathbb{Q}}[F_t|\mathcal{F}_t]$ for some unique $h = h^*$ such that the risk-neutral Esscher measure is given by*

$$\frac{d\mathbb{Q}}{d\mathbb{P}} = \frac{\exp(hF_t)}{\mathbb{E}_{\mathbb{P}}[\exp(hF_t)]}, \tag{18.26}$$

and the process (F_t) is a $\mathbb{Q}$-Martingale.

Remark 18.14 (Motivating the Need to Consider a Generalized Esscher Transform) *It is noted in Zhu (2011) that when pricing CAT bonds, perhaps the specification of the Esscher transform with only a single parameter (degree of freedom), in defining the change of measure, is too restrictive. They note that from an economic perspective the parameter h in the stochastic discount factor, for the pricing under the standard Esscher risk-neutral measure is related to the risk-aversion coefficient under the assumed subjective expected utility (EU) framework. In its earliest conception this EU framework comprised a core set of five axioms of individual risk preferences; see Von Neumann and Morgenstern (2007). However, it has been suggested in Zhu (2011) that the existence of a CAT Bond Premium Puzzle may result in the need to change or generalize the measure change when performing the risk-neutral pricing since subjective EU fails to explain such features. The justification they offer for this is based on arguments made in Bantwal and Kunreuther (2000) and Froot (2001) who each attempt an economic reasoning for the presence of a premium puzzle explained with regard to individual risk preferences. They utilize analysis of behavioral economic factors that could be invoked to explain this premium puzzle not accounted for under the standard expected utility framework specification of risk preferences. These include ambiguity aversion, myopic loss aversion, selection bias, and threshold behaviors to name a few. Therefore, the result of applying the standard Esscher pricing measure change that is appropriate under an assumption of an ideal expected utility framework may need to be relaxed for the pricing of CAT bonds.*

As a consequence of the ideas presented in Remarks 18.4 and 18.14, it was proposed in Zhu (2011) to consider pricing CAT bonds under a generalized Esscher transform to help explain a more appropriate premium that will capture the features that are empirically observed in CAT bond spreads: high spreads and when moving from large probability of trigger (CAT) bonds to those with lower probabilities of trigger, the corresponding premium spreads increase. One way to attempt to reconcile this puzzle and perform appropriate pricing under a modified risk-neutral pricing measure, accounting for the agent behaviors such as ambiguity aversion, is to utilize the generalized Esscher transform in Proposition 18.1. The difference between the standard Esscher transform and that proposed for use in pricing CAT bonds in Zhu (2011) is that their pricing kernel is a form of augmented Esscher transform that applies also to the Poisson process (frequency distribution of losses in an LDA model structure) to represent the ambiguity aversion of the agents.

Proposition 18.1 (Pricing CAT Bonds Under the Generalized Esscher Transform)
Consider the standard compound Poisson risk process, under real-world measure $\mathbb{P}$, given by

$$Z_t = \sum_{i=1}^{N_t} X_i(t) \tag{18.27}$$

with i.i.d. random losses $X_i(t) \sim F_X(x)$ *and* $N_t \sim \text{Poisson}(\lambda)$. *Then for the risk process* $(Z_t)_{t\geq 0}$ *the generalized (augmented) Esscher transform of Zhu (2011) is given for the compound process* Z_t *at time t by the Radon–Nikodym derivative of measure* $\mathbb{Q}$ *with respect to measure* $\mathbb{P}$ *by*

$$\frac{d\mathbb{Q}}{d\mathbb{P}} = \frac{\exp\left(hZ_t + gN_t\right)}{\mathbb{E}\left[\exp\left(hZ_t + gN_t\right)\right]}. \tag{18.28}$$

Therefore, the generalized Esscher transformed distribution F_{Z_t} *at time t is obtained by the distortion transformation*

$$F_Z(z,t;g,h) := \mathcal{E}_{g,h}\left[F_{Z_t}\right] = \mathbb{E}\left[\frac{\exp\left(hZ_t + gN_t\right)}{\mathbb{E}\left[\exp\left(hZ_t + gN_t\right)\right]}\mathbb{I}\left(Z_t \leq z\right)\right], \tag{18.29}$$

where $F_Z(z,t;g,h)$ *is the generalised Esscher transformed distribution now with respect to measure* $\mathbb{Q}$. *Furthermore, the resulting moment generating function is found to be*

$$M_Z(z,t;g,h) = \exp\left(\lambda M_X(h)\exp(g)\left[\frac{M_X(h+z)}{M_X(h)} - 1\right]t\right) \tag{18.30}$$

with M_X *the moment generating function of the loss random variable X.*

Remark 18.15 *Under this generalized Esscher transform, one preserves the loss process according to a compound Poisson process with modified Poisson rate parameters given by* $\lambda M_X(h)\exp(g)$ *and new loss amount random variables given by transformation* $\frac{M_X(h+z)}{M_X(h)}$.

Remark 18.16 *Under this generalized Esscher transform applied to a simple compound Poisson process, one achieves a risk-neutral pricing that is able to capture ambiguity aversion in the agents preferences. It can be shown that such a transform produces a stochastic discount factor that has the form that it will provide an equilibrium economy in which the agents are averse to both risk and uncertainty with respect to loss occurrence (due to the augmented transform applying to the frequency of occurrence of losses).*

Other alternative developments that generalize the Esscher transform dispersion measure have also been proposed, for example, see the second-order Esscher transform developed in Monfort and Pegoraro (2012) for discrete time pricing models. This two-parameter Esscher transform is a exponential-quadratic version of a stochastic discounting factor (i.e., deflator), which is based on the idea of a developing a second-order Laplace transform of the security process being studied. The resulting discrete time Esscher transform introduced is argued to be the analog of the continuous time Girsanov-type change of measure in which a diffusion component in the real-world process, different from the risk-neutral one, implies mutually singular real-world and risk-neutral probabilities. Next we briefly present the two-parameter generalization in Definition 18.21. We present the univariate case as this is of relevance to the models considered in this chapter; however, it is straightforward to consider this framework also in multivariate settings.

Definition 18.21 (Second-Order Esscher Transform) *Consider a random variable X characterized by the probability measure* $\mathbb{P}$ *on support* $\mathbb{R}$ *that admits a density f with respect to a measure* ν. *The second-order Esscher transform of the density f is a new density given by*

$$f(x;\theta_1,\theta_2) = \mathcal{E}_{\theta_1,\theta_2}[f(x)] = \frac{f(x)\exp\left(\theta_1 x + \theta_2 x^2\right)}{\mathcal{L}_2[f(x);\theta_1,\theta_2]}, \tag{18.31}$$

where $\mathcal{L}_2[f(x);\theta_1,\theta_2]$ *corresponds to the second-order Laplace transform defined by*

$$\mathcal{L}_2[f(x);\theta_1,\theta_2] = \int f(x)\exp\left(\theta_1 + \theta_2 x^2\right) d\nu(x). \tag{18.32}$$

■

The main difference between this extended Esscher transform family and the original formulation is that this transform allows one to modify the first- and second-order moments of the distribution being transformed (such as mean and covariance in the Gaussian case).

The Esscher transform pricing measure has also been extended to a randomized Esscher transform given in Definition 18.22 as derived by Siu *et al.* (2001). This is particularly relevant in the context of Bayesian risk analysis. In particular, the Randomized Esscher transform has been utilized under a Bayesian framework for measuring coherently risk associated with derivatives under the Gerber-Shiu pricing framework for complete and incomplete markets. Importantly the random Esscher transform can be shown to preserve the property of coherency of a given coherent risk measure for price processes that include Wiener, multiplicative binomial, Poisson, gamma, and inverse gaussian.

In the case of ILS pricing in incomplete markets for any of these aforementioned price processes this randomized Esscher transform also allows one to develop "Bayesian Esscher scenarios". These Bayesian Esscher scenario analysis is a generalization of the notion of financial scenario analysis and stress testing with the important new feature that it allows for consistent incorporation of historical data as well as subjective expert opinions on possible outcomes of, for example, default on a CAT bonds; see details of this framework in Siu *et al.* (2001, section 2).

In order to define the random Esscher transform, consider the objective measure $\mathbb{P}$ for the price process X_t at time t on a space $(\Omega, \mathcal{F})$. Furthermore, assume that this measure admits a distribution $F(x,t)$, which can be practically selected based on a given subset of historical data and furthermore all agents in the economy agree on this reference distribution (they have the same historical information on which they condition their knowledge of this distribution). Then, define the set of Esscher parameters at some time $t_0 > 0$ according to the λ distortion parameter values that satisfy the conditions of membership of the following set

$$\chi = \left\{ \lambda \in \mathbb{R} \ : \ \int_{-\infty}^{\infty} \exp(\lambda x) dF(x,t) < \infty \right\}. \tag{18.33}$$

Using this particular set of Esscher parameters, one has a definition of the random Esscher transform given later.

Definition 18.22 (Random Esscher Transform) *Consider the random variable* $\Lambda \in \chi$ *for the Esscher parameter corresponding to the market price of risk, which has an a priori distribution based on beliefs or risk preferences from agents' subjective views (Bayesian Esscher scenarios) denoted for the i-th such risk preference by the prior* $\pi_i(\lambda)$. *Then the random Esscher measure* $\mathbb{Q}_\Lambda$ *that is equivalent to measure* $\mathbb{P}$ *on* $(\Omega, \mathcal{F})$, *which is associated to the random Esscher market price of risk* Λ, *can be defined by the family of Esscher measures* $\{\mathbb{Q}_\lambda \sim \mathbb{P} : \Lambda = \lambda \in \chi\}$. *Then under any member of this family, the random distribution of* X_t *under the new measure* $\mathbb{Q}_\Lambda$ *is given by the* *random Esscher transform*

$$
\begin{aligned}
F(x,t;\Lambda) := \mathcal{E}_{\Lambda}\left[F(x,t)\right] &= \mathbb{E}_{\mathbb{P}}\left[\frac{\exp(\Lambda X_t)}{\mathbb{E}_{\mathbb{P}}\left[\exp(\Lambda X_t)\right]}\mathbb{I}\left[X_t \leq x\right]\right] \\
&= \frac{\int_{-\infty}^{x}\exp(\Lambda y)dF(y,t)}{\mathbb{E}_{\mathbb{P}}\left[\exp(\Lambda X_t)\right]}.
\end{aligned}
\tag{18.34}
$$

■

The randomized Esscher transform can then be used for pricing under the following two steps:

1. First performing a prior elicitation that involves collecting different sets of prior probabilities assigned to a given set of generalized scenarios according to different subjective views of agents risk preferences;
2. Utilize Bayes theorem to combine in a consistent mathematical manner the prior beliefs and the observations of stock prices and market data.

Hence, the posterior will determine the probability weighting given to a particular combination of market observations and subjective risk preferences.

Remark 18.17 (Random Esscher Transform Applications: Bayesian Pricing) *A range of applications can be developed in pricing under this randomized distortion measure. The main application discussed in Siu et al. (2001) is the development of a form of financial scenario analysis under a Bayesian paradigm that will lead to a posterior pricing framework. This involves specification of a Bayesian Esscher scenario, which is characterized with respect to the product probability space* $(\Omega \times \chi, \mathcal{F} \otimes \mathcal{N}, \pi_i)$ *where the sample space is the product space of* Ω *and the set of Esscher parameters* χ*; the sigma algebra (event space) is the tensor product of the sigma algebra that the price process is defined with respect to* $\mathcal{F}$ *and a new sigma algebra, denoted* $\mathcal{N}$*, which corresponds to the collection of all measurable subsets of Esscher parameters* χ*; the product measure* π_i*, for the i-th set of subjective views (risk preferences) in the Esscher scenarios, is given for each* $D \in \mathcal{F}$ *and* $N \in \mathcal{N}$ *by the posterior measure:*

$$
\pi_i(D \times N) = \int_N \mathbb{Q}_{\lambda}(D)d\pi_i(\lambda). \tag{18.35}
$$

Under this Bayesian Esscher scenario defined by $(\Omega \times \chi, \mathcal{F} \otimes \mathcal{N}, \pi_i)$*, Siu et al. (2001) show that the following useful properties hold when considering pricing under such a framework:*

1. *The process* $\{X_t\}_{t\geq 0}\Big| \{\Lambda = \lambda\}$ *has stationary and independent increments if* $\{X_t\}_{t\geq 0}$ *also has these properties;*
2. *For each time* $t \geq 0$*, the conditional distribution function of* X_t *conditional on* $\{\Lambda = \lambda\}$ *is given by*

$$
F(x,t;\lambda) = \mathbb{Q}_{\lambda}\left(\{\omega \in \Omega; X_t(\omega) \leq x\}\right), \tag{18.36}
$$

which is the data generating distribution.

In the remainder of this section, we also comment on some other well-known actuarial pricing methods that involve some form of probability transformation or distortion. Those most widely used include the variance loading, technique, the standard deviation loading, and the previously presented Esscher transform. Before completing this introductory section on real-world pricing and actuarial pricing, we also mention the approach proposed in the actuarial community by Wang (2002), titled "A universal framework for pricing financial and insurance risks". The approach developed by Wang was derived from the work of Bühlmann (1984a) and focuses on the transform given in Definition 18.24. In this procedure, Wang introduced a class of transforms (in response partly to the Basel Accords) to find fair value pricing approaches to all assets and liabilities irrespective of whether they were traded or not (as required in the Basel Accords), where this transform was based on earlier works by Venter (1991) and Butsic (1999).

This class of probability transform (or distortion measure) has been shown to recover an equivalent pricing solution as the CAPM model (under an assumption of returns having a Normal distribution) and the Black–Scholes economy (under the assumption of a LogNormal distribution). The motivation for the development of this distortion measure comes from the idea of Venter (1991), where it was argued that no-arbitrage pricing assumptions always imply a distributional transformation. In the context of reinsurance with layering, one can explain this notion well. Then one can understand this argument based on the following simplified example; see details in Wang (2004).

Consider the loss $X \sim F(x)$ that is layerd and represented by $X_{(a,a+h]}$ with an attachment point a and a limit h that is very close to the attachment point. In this case, the expected loss to the layer is given by the approximation

$$\begin{aligned} \mathbb{E}\left[X_{(a,a+h]}\right] &= \int_a^{a+h} [1 - F(x)]\, dx \\ &= \int_a^{a+h} S(x) dx \\ &\approx hS(a), \quad \text{as } h \downarrow a. \end{aligned} \tag{18.37}$$

Then if one observes the price for a thin layer given by $E^*\left[X_{(a,a+h]}\right]$, then one can use this observed price to make inference on the approximate price implied loss exceedance probability given by

$$S^*(a) \approx \frac{1}{h}\mathbb{E}^*\left[X_{(a,a+h]}\right]. \tag{18.38}$$

In general, one would always expect that $S^*(a) \geq S(a)$ since the layer price will typically contain a risk loading that is additional to $\mathbb{E}\left[X_{(a,a+h]}\right]$, which in turn implies that looking at the observed market price one can see empirically a direct transform of the loss exceedance curve from $S(x)$ to $S^*(x)$.

A more mathematical approach to verifying the relationship between the Wang transform and alternative pricing mechanism now proceeds as follows. First, we note that in Platen and Heath (2006) it was shown that the benchmark approach also can be shown to contain as a special subclass the large family of pricing methods known as capital asset pricing model (CAPM) and its variants. To understand the applicability of the approach of Wang and its relationship to benchmark approach and real-world pricing, we briefly mention some details regarding CAPM.

It will be sufficient here to just consider the basic linear regression model version of the classical CAPM model in its original form that involves obtaining predictions regarding the equilibrium expected returns on assets, under the assumption that all investors have a one-period horizon and returns follow a multivariate Gaussian distribution that has sufficient statistics given by the mean vector of returns and the covariance matrix (correlations and variances). Under this simple CAPM model, one assumes the following linear model relationship:

$$\mathbb{E}[R_i] = r_f + \beta_i \left(\mathbb{E}[R_M] - r_f\right) \tag{18.39}$$

with R_i and R_M the rates of return on asset i and a market portfolio (index, etc.), respectively, r_f is the risk-free rate of return (e.g., Libor), and the β-coefficient given for asset i represents the sensitivity of the expected excess asset returns to the expected excess market returns. Under the assumption that R_i and R_M are distributed according to a Gaussian distribution, one can then define the market price of risk for asset i according to the expression:

$$\lambda_i = \frac{\mathbb{E}[R_i] - r_f}{\sigma_i} = \rho_{i,M}\lambda_M \tag{18.40}$$

which is also known as the Sharpe ratio (as detailed in Definition 18.27) and where $\rho_{i,M}$ is the linear correlation coefficient between returns on the asset and the index and λ_M is the market price of risk of the index. The conclusions from this idealized model are that only systematic risk will acquire an additional risk premium in an efficient market. However, when it comes to non-Gaussian and incomplete markets, this model has major deficiencies that are well documented.

To proceed, it will be useful to recall the definition of comonotonicity of two measurable functions given in Definition 18.23.

Definition 18.23 (Comonotonicity of Measurable Functions) *Assume a sample space Ω and a corresponding sigma algebra $\mathcal{F}$, then consider a $\mathcal{F}$-measurable family of functions $(X_i)_{i\in I}$, which will be comonotonic functions iff*

$$[X_i(\omega) - X_i(\omega')]\left[X_j(\omega) - X_j(\omega')\right] \geq 0, \quad \forall i,j \in I, \quad \forall \omega,\omega' \in \Omega. \tag{18.41}$$

■

In the following, we present briefly the basic principle behind the approach proposed by Wang for actuarial valuation. This approach extends the CAPM to allow one to price all types of assets and liabilities, with any type of distribution (process), traded or underwritten, in finance and insurance markets. The key to this transform is the introduction of a parameter known as the "market price of risk", which is utilized to obtain a "risk-adjusted" fair valuation price. It is assumed that the market price of risk is a continuously increasing function of duration. This pricing method can be applied to any contingent claim or payoff so long as it is co-monotone with its underlying assets or liabilities; see an excellent review of properties of comonotonicity of functions in economic and risk applications in Chateauneuf *et al.* (1997).

Definition 18.24 (Wang Transform) *Consider a financial asset or liability, with value denoted by X_t, over time horizon $[0, T]$. Assume X_t has distribution $F(x)$, then the Wang transformation of this distribution is given by the "risk-adjusted distribution"*

$$F^*(x) := \mathcal{W}_\lambda [F(x)] = \Phi\left(\Phi^{-1}(F(x)) + \lambda\right) \tag{18.42}$$

with $\Phi(\cdot)$ representing the standard Gaussian distribution and λ representing the market price of risk that indicates the associated amount of systematic risk. Under this transform, the expected value of $\mathbb{E}^*[X]$ under $F^*(x)$ will correspond to the risk-adjusted actuarial fair value of the asset or liability at time T, which can then be discounted to any time via the risk-free rate. ∎

Then one may link the Wang transform back to the financial pricing framework of the CAPM model if one assumes a competitive market in which the risk-adjusted return for all assets in the portfolio are equal to the risk-free rate, then the market price of risk in the classical CAPM model will be equivalent to the market price of risk defined in the Wang transform.

Remark 18.18 *If the Wang transform is applied to a Normal or LogNormal distributed random variable, then under the Wang transformation the distribution form is invariant, meaning that the Normal and LogNormal distributions are retained for the transformed distribution function. Furthermore, the Wang transform is the same as the Esscher transform in the case of a Gaussian distribution.*

To obtain the correlation in the CAPM model when the original price process (asset or liability value distribution) is non-Gaussian means one must also modify the correlation coefficient and this is trivially achieved by transforming marginally each random variable to a Gaussian distribution via $U_i = \Phi^{-1}(F(x))$ and taking the correlation between the transformed random variables via the Pearson linear correlation coefficient.

Remark 18.19 *It can be shown that the Wang transform and the Esscher transform are the only two distortion measures that are able to recover the CAPM model and the Black–Scholes option pricing formula.*

Given an underlying risk X and a function h that maps this risk to a payoff $Y = h(X)$, that is, a derivative or contingent payoff, then Wang provides two methods that, though they involve different mathematical steps at each stage, will produce equivalent fair value prices; see Wang (2002, p. 218).

1. First take the Wang transform $F_Y^*(y;\lambda) = \mathcal{W}_\lambda[F_X(x)]$ of the underlying risk random variable (processes) X with respect to its distribution. Then derive the distribution of the risk-adjusted derivative or contingent payoff $F_Y^*(y;\lambda)$ as a function of $F_X^*(x)$ using change of variable or probability transforms with the relationship $Y^* = h(X^*)$ to obtain

$$F_Y^*(y;\lambda) = \mathbb{P}\mathrm{r}[Y^* \le y] = \mathbb{P}\mathrm{r}[h(X^*) \le y] = \mathbb{P}\mathrm{r}\left[X^* \le h^{-1}(y)\right] = F_X^*(h^{-1}(y)). \tag{18.43}$$

2. Alternatively, one can first derive the distribution of F_Y using the relationship $Y = h(x)$,

$$F_Y(y) = \mathbb{P}\mathrm{r}[Y \le y] = \mathbb{P}\mathrm{r}[h(X) \le y] = \mathbb{P}\mathrm{r}\left[X \le h^{-1}(y)\right] = F_X(h^{-1}(y)). \tag{18.44}$$

   Then apply the Wang transformation $F_Y^*(y;\lambda) = \mathcal{W}_\lambda[F_Y(y)]$.

This equivalence in relationship will hold if the transformation h is monotone.

Remark 18.20 (Criticism of Wang Transform) *The Wang transform has been criticised in the literature since it fails to match heavy-tailed or fat-tailed features often observed in financial returns series. In this regard, Wang (2002) also proposed a two-parameter version based on the Student's t distribution; however, this transform loses the nice theoretical properties present in the original Wang transform as it is no longer consistent with Bühlmann's economic premium principle and also fails to reproduce the Black–Scholes and CAPM formulations.*

As noted in Wang (2004) and Kijima and Muromachi (2008), the justification of the Wang transform via the classical CAPM, while important from the perspective of understanding the properties of pricing under the Wang transform distortion measure, it does not always produce a distorted measure that fits well historical data in practical performance. This is not such a surprise; the main reason for this is that the standard definition of the Wang transform is equivalent to the two-moment CAPM. Under the setting of the CAPM family, there have been significant enhancements to the classical formulation where in addition to risk premium associated with volatility (second moments) it is common in practice to also consider risk premiums associated to higher moments that also play a role in the required distortion measure; see discussion in Kozik and Larson (2001). In this study, and they note that the rate of return distribution should take into account higher-order moments, associated to skewness and kurtosis, especially in the context of catastrophe insurance products where some of the most extremely skewed distributions occur. In addition to the need to account for risk associated with higher moments, there is also risk associated to statistical estimation of models. From the statistical perspective, this manifests itself typically in the form of model and parameter uncertainty that also needs to be accounted for when performing pricing under such model frameworks. Empirically, it was shown by Kozik and Larson (2001) that taking into account a third moment in a modified CAPM framework can significantly improve the fits for empirical financial data. It turns out that it is relatively straightforward to modify the Wang transform to also account for such features as higher moments (heavy tails, skewness, and asymmetry), as well as parameter uncertainty.

In Wang (2004), a modification to the standard Wang transform is proposed based on accounting for parameter uncertainty in which the modified Wang transform takes the form of a two-parameter transform based on a Student's t transform given in Definition 18.25.

Definition 18.25 (Two-Parameter Wang Transform: Accounting for Parameter Uncertainty) *Consider a financial asset or liability, with value denoted by X_t, over time horizon $[0, T]$. Assume X_t has distribution $F(x)$, which has possible features such as nontrivial parameter uncertainty or significant skewness or kurtosis. Accounting for such features can be achieved partially by the modified two-parameter Wang transformation of this distribution given by the "risk-adjusted distribution"*

$$F^{*,(2)}(x) := \mathcal{W}^{(2)}_{\lambda,k}\left[F(x)\right] = \mathcal{Q}\left(\Phi^{-1}\left(F(x)\right) + \lambda\right) \tag{18.45}$$

with $\mathcal{Q}$ representing the Student's t distribution with location parameter $\mu = 0$ and degrees of freedom parameter k, $\Phi(\cdot)$ represents the standard Gaussian distribution and λ representing the market price of risk which indicates the associated amount of systematic risk. Here, we utilize the upper script (2) to denote the modification to account for the two-parameter Wang transform. ■

While this modification to the Wang transform is no longer consistent with the risk-neutral arbitrage-free classical CAPM model, it has been shown to produce improved empirical fits for the distorted risk measure when compared with real-world prices for ILS such as CAT bonds. The reason for this, as noted by Wang (2004), is that the Student's t adjustment "captures two opposing forces that often distort investors' rational behavior, namely greed and fear. Although investors may fear unexpected large losses, they desire unexpected large gains. As a result the tail probabilities are often inflated at both tails; and the magnitude of distortion normally increases at the extreme tails".

Other extensions that generalize the standard Wang transform have been developed in Kijima and Muromachi (2008), which maintain the theoretical property of the Wang transform with regard to its consistency with Bühlmann's principle of premium calculation, making it an interesting extension to consider in practice. This modified Wang transform, which we denote as the generalised Wang transform, is given in Definition 18.26.

Definition 18.26 (Generalized Wang Transforms) *Consider a financial asset or liability, with value denoted by X_t, over time horizon $[0, T]$. Assume X_t has distribution $F(x)$, then the generalized Wang transformation of this distribution is given by the "risk-adjusted distribution"*

$$F^*(x) := \mathcal{GW}_{\lambda,G,Y}\left[F(x)\right] = \mathbb{E}_Y\left[\Phi\left(G^{-1}\left(F(x)\right)\right) Y + \lambda\right] \tag{18.46}$$

with $\Phi(\cdot)$ representing the standard Gaussian distribution, λ representing the market price of risk, which indicates the associated amount of systematic risk, Y is any positive valued random variable, and $G(x)$ is the distribution of the random variable corresponding to the ratio of U/Y, where $U \perp Y$ and U has a standard Gaussian distribution. ■

Note that this generalized Wang transform recovers the standard Wang transform when the random variable Y is a Dirac mass on the event $\{Y = 1\}$ almost surely. In addition, the following distributional bound was obtained by Kijima and Muromachi (2008), which relates the standard Wang transform distorted distribution function and the Generalized Wang transform distorted distribution; see Theorem 18.6.

Theorem 18.6 *When $\lambda \geq 0$, then for all x such that $F(x) < \frac{1}{2}$, the following inequality holds*

$$\mathcal{W}_\lambda[F(x)] \geq \mathcal{GW}_{\lambda,G,Y}\left[F(x)\right] \tag{18.47}$$

for all $Y > 0$ and $G(x)$ is the distribution of the random variable corresponding to the ratio of U/Y, where $U \perp Y$ and U has a standard Gaussian distribution.

In the continuous time setting where one considers the change of measure for a diffusion process, there have also been studies of the associated relationships and consistency of the Wang transform and risk-neutral arbitrage-free pricing. Pelsser (2008, section 4) discuss the relationship between the Wang transform and risk-neutral arbitrage-free pricing for the case in which one considers a traded asset X_t with transition distribution $F(t, x; T, y)$. Under the distortion measure corresponding to the Wang transform, one obtains the new measure

$$F^*(t, x; T, y) = \Phi\left(\Phi^{-1}\left(F(t, x; T, y)\right) - \lambda(t, T)\right) \tag{18.48}$$

for some market price of risk deterministic function $\lambda(t, T)$, which they point out is the unique solution to the integro differential equation that ensures the choice of $\lambda(t, T)$ is selected to enforce the martingale condition $\mathbb{E}_{X^*}\left[X_T | X_t = x\right] = x$ for all T, given $\forall T > t$ by

$$
\begin{aligned}
\mathbb{E}_X\left[X_T \mid X_t=x\right] &= \int_{-\infty}^{\infty} X_t dF\left(t, x; T, X_T\right) \\
&= \int_{-\infty}^{\infty} X_t d\Phi\left(\Phi^{-1}\left(F\left(t, x; T, X_T\right)\right)-\lambda(t, T)\right) \\
&= \mathbb{E}_{X^*}\left[X_T \mid X_t=x\right] \\
&= x
\end{aligned}
\tag{18.49}
$$

with initial condition $\lambda(t,t) = 0$. This insight then allowed Pelsser (2008) to identify the conditions under which the Wang distortion measure would reproduce the required change of measure for arbitrage-free risk-neutral pricing. After some stochastic calculus using the standard theorems of Girsanov stated previously, they obtain the following result presented in Proposition 18.2.

Proposition 18.2 (Equivalence Between R-N Arbitrage-Free and Wang Distortion Pricing) *The Wang transform is equivalent to the risk-neutral arbitrage-free pricing (R-N AFP) framework in a continuous price process setting iff the following conditions on the drift and volatility of the process are satisfied, with $\phi(\cdot)$ and $\Phi(\cdot)$ the standard Gaussian density and distributions respectively, giving conditions:*

$$
\frac{\partial}{\partial x}\left[\mu(t, x)\left(\phi\left(\Phi^{-1}(F(t, x; T, y))\right)\right)^{-1} \frac{\partial}{\partial x} F(t, x; T, y)\right]=0 \tag{18.50}
$$

and

$$
\frac{\partial}{\partial x}\left[\sigma(t, x)\left(\phi\left(\Phi^{-1}(F(t, x; T, y))\right)\right)^{-1} \frac{\partial}{\partial x} F(t, x; T, y)\right]=0. \tag{18.51}
$$

We also note that there have been studies performed in Goovaerts and Laeven (2008) that also consider the stochastic process generalization of the Esscher transform for pricing, termed the Esscher–Girsanov distortion measures. This is the distortion measure analog of the change of measure achieved by the Girsanov theorem.

To summarize, in the context of pricing CAT bonds (incomplete markets where risk-neutral arbitrage-free methods are not directly applicable), the way to think about this form of dispersion measure pricing is to consider the default risk for CAT bonds as directly transferrable to catastrophe insurance contracts, then the excess yield spreads over say the risk-free rate for the CAT bond may be translated into a risk premium in dollars for an equivalent insurance contract providing equivalent protection. This can then be utilized to obtain estimates of the market price of risk to proceed with the pricing under the Wang transform.

18.2.2 RISK ASSESSMENT FOR REINSURANCE: ILS AND CAT BONDS

Since CAT bonds are exposed to nature risk, just like corporate bonds that are exposed to credit defaults, they naturally are required to offer investors a higher yield than standard risk-free rates (LIBOR). For the market to be attractive to investors, the excess yields spread over the risk-free rate should be such that it provides sufficient compensation for the expected default rate as well as a risk loading associated with uncertainty in default risk.

The measure of risk in investments in a traded or a nontraded asset is typically considered through the specification and evaluation of a risk measure; see Chapter 6 for details. In Artzner *et al.* (1997, 1999), four desirable and now well-established properties of risk measures were developed that would characterize the so-called "coherent" classes of risk measures; the properties to be satisfied were invariance, positive homogeneity, monotonicity, and subadditivity. In the context of insurance (non-traded assets) a similar characterization was developed in Wang *et al.* (1997). Wang *et al.* (1997) assumed that individual insurers were operating in a competitive market and were therefore price takers; they then developed four axioms that could be used to describe the behavior of insurance prices. It is not the intention of this section to detail these comprehensively for either of the frameworks of Artzner or Wangs; instead, the reader is referred to Chapter 6, which has a detailed discussion on risk measures of relevance to OpRisk. Instead, this section aims to succinctly provide some related discussion on risk measures for distortion measures such as the Esscher and Wang transforms, not covered in other chapters, while also making connections to popular mathematical finance measures of risk such as the Sharpe ratio discussed in Sharpe (1998) and Sharpe and Sharpe (1970).

The ability to measure the riskiness of an ILS such as a CAT bond was considered in the works of Wang (2004, 2002). In these works, a probability transform approach is adopted to modify the standard Sharpe ratio. The standard definition of the Sharpe ratio, which measures the risk adjusted performance in mutual funds, is given by Definition 18.27.

Definition 18.27 (Sharpe Ratio) *The Sharpe ratio is a measure of how well the return of an asset compensates an investor for risk taken. The Sharpe ratio is a deviation risk measure that measures the excess return (or risk premium) per unit of deviation in an investment asset or a trading strategy, typically referred to as risk, see Sharpe (1998). The Sharpe ratio is defined by the expression*

$$S = \frac{\mathbb{E}\left[R - R_I\right]}{\sqrt{\mathrm{Var}\left[R - R_I\right]}}, \tag{18.52}$$

where R is the asset return and R_I is the return on a benchmark asset or index, such as the risk-free rate or the S&P 500. The ex ante version uses the expected returns on these assets, while the ex-post version uses the actual realized returns. When comparing two assets versus a common benchmark, the one with a higher Sharpe ratio provides better return for the same risk (or, equivalently, the same return for lower risk). ■

It can be shown that one can relate the Sharpe ratio measure of risk to the family of distortion measures characterized by the Wang transform. In doing so, the market price of risk obtained through the Wang transform can be directly interpreted as the ILS equivalent of the Sharpe ratio for returns on traded market securities; this is summarized in the Remark 18.2.

Remark 18.21 *Consider a financial asset or liability, with value denoted by X_t, over time horizon $[0, T]$ and X_t has distribution $F(x)$, which is Gaussian. Then the Wang transform $F^*(x) := \mathcal{W}_\lambda\left[F(x)\right]$ will produce a distorted distribution $F^*(x)$, which is also Gaussian and the parameter λ will correspond to exactly the Sharpe ratio. This is a direct consequence of the Wang transforms equivalence with the CAPM model discussed previously, since in the classic CAPM where* *asset returns are assumed to follow multivariate Normal distributions, the market price of risk is the Sharpe ratio, which represents the excess return per unit of volatility.*

In the work of Wang, the standard definition of the Sharpe ratio, which works well for elliptical distributions with finite first and second moment such as the Gaussian distribution or under log-returns the LogNormal distribution, is modified to account for cases in which skew and heavy tails are present. This is particularly important when assessing the riskiness of assets such as CAT bond issues, where one cannot straightforwardly apply the standard Sharpe ratio concept due to the fact that the asset return is skewed and with jumps. In such situations, one would find that most of the probability mass will be centered at the atom corresponding to a zero loss and there is a small probability of potentially large negative returns; these features are taken into account in different ways under each of the different classes of Wang transform measures of risk, obtained from the Wang transform distortion parameters in each case denoted generically by λ. For instance, it turns out that the class of distortion measures corresponding to the standard Wang transform can be utilized to extend the Sharpe ratio concept to credit risks with skewed return distributions. This will then allow one to evaluate the risk-adjusted performance of a CAT bond asset. This is also achieved by considering the market price of risk associated with the two-parameter and generalized Wang transforms.

18.3 Applications of Pricing ILS and CAT Bonds

To begin this application section, we first present a basic probabilistic framework for the market for say CAT bonds. Then we consider two classes of example, simple idealistic examples, which illustrate features of pricing CAT bonds, followed by some more sophisticated examples that reflect the real-world actuarial pricing principles discussed earlier.

18.3.1 PROBABILISTIC FRAMEWORK FOR CAT BOND MARKET

Next we establish the probabilistic framework for the CAT bond market, assets, and underlying products; this is based on the framework developed in Cox and Pedersen (2000, section 5). We separate the structure into both "financial market variables" and "catastrophic risk variables" as detailed next:

- **Financial Market Variables**. Contained on a filtered probability space denoted $\left(\Omega^{(1)}, \mathcal{F}^{(1)}, \mathbb{P}^{(1)}\right)$ in which the sample space $\Omega^{(1)}$ can be finite or infinite depending on the application and encompasses all the sample paths (stochastic trajectories) that financial variables can have in times $t = 1, 2, \ldots, T$; $\mathcal{F}^{(1)}$ represents the filtration that captures how information in the financial market evolves that comprises an increasing sequence of sets of events such that $\left\{\mathcal{F}_0^{(1)} \subseteq \mathcal{F}_1^{(1)} \subseteq \cdots \subseteq \mathcal{F}_T^{(1)}\right\}$ with $\mathcal{F}_t^{(1)}$ denoting securities prices or other investment information in the market at time t; and the probability measure $\mathbb{P}^{(1)}$ is defined over the sigma-algebra $\mathcal{F}_T^{(1)}$;
- **Catastrophic Risk Variables**. Contained on a filtered probability space denoted $\left(\Omega^{(2)}, \mathcal{F}^{(2)}, \mathbb{P}^{(2)}\right)$ in which the sample space $\Omega^{(2)}$ can be finite or infinite depending on the application and encompasses all the sample paths (stochastic trajectories) that the

actual "nature stochastic process" can take in terms of the catastrophe being covered by the CAT bond (e.g., a spatial-temporal stochastic process for wind speeds); $\mathcal{F}^{(2)}$ represents the filtration that captures how information in the nature stochastic process evolves; and the probability measure $\mathbb{P}^{(2)}$ is defined over the sigma-algebra $\mathcal{F}_T^{(2)}$ and determines the chance of particular catastrophic events (which will have particular models depending on the application);

- **Joint Model**. The full model used for the pricing will then comprise tuples $\left(\omega^{(1)}, \omega^{(2)}\right)$ that include the state of the financial market and the nature catastrophic risk variables, which therefore jointly form elements of a product space formulation in which the sample space is given by $\Omega = \Omega^{(1)} \times \Omega^{(2)}$;
- It will be assumed that the resulting measure on the product sample space Ω is given by an assumption of independence for events in nature versus events that depend only on economic risk variables and is therefore given by $\mathbb{P}\left(\omega^{(1)}, \omega^{(2)}\right) = \mathbb{P}^{(1)}\left(\omega^{(1)}\right)\mathbb{P}^{(2)}\left(\omega^{(2)}\right)$;
- It will be assumed that the resulting filtration adopted is the product space filtration given at each time t by the product space $\mathcal{F}_t = \mathcal{F}_t^{(1)} \times \mathcal{F}_t^{(2)}$;
- As in Cox and Pedersen (2000), we consider the two new filtrations defined for all times $t = 1, 2, \ldots, T$ by

$$\begin{aligned} \mathcal{A}_t^{(1)} &= \mathcal{F}_t^{(1)} \times \left\{\emptyset, \Omega^{(2)}\right\}, \\ \mathcal{A}_t^{(2)} &= \mathcal{F}_t^{(2)} \times \left\{\emptyset, \Omega^{(1)}\right\}, \end{aligned} \tag{18.53}$$

and note that any random variable or function on the joint probability space $(\Omega, \mathcal{F}, \mathbb{P})$ that is measurable with respect to $\mathcal{A}_t^{(1)}$ (resp. $\mathcal{A}_t^{(2)}$) is dependent only on the financial risk (resp. nature catastrophe risk) variables. Furthermore, it is simple to show these two filtrations are by construction independent of each other, as was shown in Cox and Pedersen (2000, lemma 5.1). As noted by these authors, one must be careful not to refer specifically to measures $\mathbb{P}^{(1)}$ or $\mathbb{P}^{(2)}$ as being independent under $\mathbb{P}$ because strictly speaking neither of these measures is defined on the space given by $(\Omega, \mathcal{F}, \mathbb{P})$. Hence, the need for these additional filtrations to formalize this intuitive notion.

Having presented a general probabilistic framework in which to consider undertaking the pricing of the CAT bonds, we proceed by presenting a range of approaches that researchers have adopted to solve the pricing under different assumptions. To achieve this, we develop two example frameworks proposed in the literature for pricing CAT bond products.

- **Framework 1.** This involves making a simplifying assumption of a complete market with either deterministic or stochastic discounting interest rate;
- **Framework 2.** This involves working with an incomplete market hypothesis; again we can consider deterministic or stochastic interest rate models.

Before proceeding, we also note the following useful property of CAT bonds that are not present in corporate bonds when it comes to valuation.

Remark 18.22 (Value Additivity for CAT Bonds) *In modeling a CAT bond, it should also be noted that when a trigger occurs with subsequent loss of the coupon payments, these coupon payments* *will have no influence on the contingency of default. This is unlike a corporate bond in which the coupon payments affect the solvency status of the indebted firm. As a consequence of this feature of CAT bonds, it is noted in Vaugirard (2003) that an insurance coupon bond can be modeled by a portfolio of zero-coupon bonds with weights in the portfolio designed to match the coupon payments. This makes the valuation framework for such assets simpler to construct.*

18.3.2 FRAMEWORK 1: ASSUMING COMPLETE MARKET AND ARBITRAGE-FREE PRICING

We will start this section by using an illustration of a simple illustration of a single period analysis in which catastrophe reinsurance is developed under a framework of a high-yield bond. In Example 18.2, a model is developed to demonstrate a simple framework for securitization under a CAT bond, though it is too simplistic in practice to utilize for a real-world pricing scenario; several variations of it were used effectively by Cox and Pedersen (2000), Tilley (1995), and Froot *et al.* (1995) to illustrate the interesting properties of catastrophe reinsurance in ideal settings. The variation we present is from Cox and Pedersen (2000), which is detailed in toy model 1 and toy model 2.

EXAMPLE 18.2 Toy Model 1 (Single Period): Securitized Catastrophe Reinsurance as a High Yield Bond

In this example, we consider the development of a single period catastrophe reinsurance product in terms of a high-yield bond. We make the following definitions of the model parameters and assumptions:

1. In this model, a reinsurer will pay a fixed amount S at the end of the period if a catastrophic event occurs as defined in the contract. Alternatively, if no event occurs of the prescribed severity or intensity of the CAT bond contract, then no payment is made. It is assumed that S is specified in the contract;
2. Denote by p_r the probability that a catastrophic event occurs during the period as estimated by the reinsurance markets assessment of the probability of a catastrophe;
3. Denote by p_b the default probability of the junk bonds issued by the reinsurer to raise capital to provide full funding in the event of a catastrophic event (in order to make payment of claims against the CAT bond). The probability p_b is the assessment of the bond holders/bond market.
   Note 1. In this toy model, one may assume for simplicity that the risk of default on the junk bonds is synonymous with the risk of default of the CAT bond due to a catastrophic event. Of course in practice this may not be the case, as the issuer of the junk bonds faces other financial pressures from the markets they participate in that are directly uncorrelated with nature risk. The bond owners are exposed directly to credit risk of the reinsurer (though indirectly one may assume that the primary source of this risk of default stems from the risk associated with the catastrophic event).

Note 2. This toy model assumes unrealistically that the reinsurer losses for a single exposure (catastrophe covered by the CAT bond) are the primary source of losses for the investors; however, in practice, this could be diversified across a portfolio or exposures.
Note 3. Even if we assume the risk of defaults are the same for the junk bonds and the CAT-bond, it is still expected to be the case that p_r and p_b are different due to information asymmetries between each participating agent;

4. Denote by B_r the fair value price of purchase of the reinsurance (CAT bond value) from the perspective of the reinsurer offering the CAT bond; and denote by B_b the implied price of the bond from the perspective of the bond market who purchase the junk bonds issued by the reinsurer in order to provide full funding for the payouts in the event of a catastrophe.
Note: These two CAT-bond purchase prices B_r and B_b will generally differ due to the different assessments of the probability of default;
5. Denote by r the single period effective default-free interest rate.

Reinsurance Market. The fair value of the reinsurance as determined by the reinsurance market is then specified according to the following relationship between the bond price and the probability of a catastrophe given by

$$B_r = \frac{1}{1+r} p_r S. \tag{18.54}$$

It is noted in Cox and Pedersen (2000) that the capital that is set aside to cover losses through the payout of S is derived in this context via full funding and not the typical mechanism of reinsurance that involves diversification of a portfolio. Therefore, to obtain the required full funding for the catastrophic payout, the reinsurer borrows the required capital through a mechanism involving issuance of defaultable junk bonds that offer high returns with a commensurate risk of default. This issuance raises an amount of capital in cash C given by the following expression that ensures that suitable capital is available to cover any default on the CAT bond due to insurance claims arising from this coverage,

$$(B_r + C)(1 + r) = S. \tag{18.55}$$

Bond Market. If during the single period there is no catastrophic event, the investors get back their principle invested and a coupon payment for their risk exposure, which is given by the difference between the amount paid for the bond and the capital $G = S - C$. The price per unit of face value is determined by the bond market as a discounted expected cashflow given by

$$\frac{1}{1+r}\left(1 + \frac{G}{S}\right)(1 - p_b), \tag{18.56}$$

where $\frac{G}{S}$ represents the coupon rate.

Determining the Probability of Default of Capital Raising Junk Bonds and the Implied Fair Value. If the bond is designed to sell at face value, then this means that effectively investors pay a unit 1 to receive $1 + \frac{G}{S}$ at the end of the period if no catastrophe occurs that implies the following relationship if no catastrophe occurs:

$$1 = \frac{1}{1+r}\left(1 + \frac{G}{S}\right)(1 - p_b). \tag{18.57}$$

This allows one to determine the probability of default of the junk bonds according to the perspective of the bond market as given by

$$p_b = \frac{\frac{G}{S} - r}{1 + \frac{G}{S}}. \tag{18.58}$$

Then this implies a price for reinsurance from the perspective of the bond market (capital raising junk bonds valuation) denoted by B_b given by

$$B_b = \frac{1}{1+r}. \tag{18.59}$$

Profitability of CAT Bond. This provides a lower bound for the price premium that the reinsurer should charge B_r in order to not make any total losses, that is, B_r is greater than some function of B_b. This will be satisfied in cases in which

$$p_r \geq \frac{\frac{G}{S} - r}{1 + \frac{G}{S}}. \tag{18.60}$$

■

Having illustrated through a single period toy model how the basic pricing and securitization of a CAT bond can be achieved, it is worth noting some additional research that has been performed on assessing the relationship between the risk associated with the default of the junk bonds issued by the reinsurer and the default risk of the CAT bond.

We have seen in this toy model that one can think of the CAT bond when a trigger occurs (catastrophic event) as similar in behavior to a defaultable corporate bond (some high-yield bond). Although they may behave in a similar fashion, there is a fundamental difference in the features of CAT bonds in that the default risk of the CAT bond is not correlated with underlying financial market variables such as interest rates. This was empirically studied to test the validity of this zero correlation claim in Canter *et al.* (1997) where it was shown that the correlation coefficient between annual returns on the S&P500 index and the PCS index during the period of 1994–1995 was statistically insignificant from zero. If one ignores the intricacies of CAT bonds, then in Kielholz and Durrer (1997) it is argued that introducing such products into a diversified portfolio will improve the performance by translating the efficient frontier to the left. However, as discussed in Briys (1997) and Loubergé *et al.* (1999), this is not quiet the case since CAT bond-specific details can effect this result; for instance, they highlight the following three features to be considered cautiously.

1. Forecasts of risk return on CAT bonds in diversified portfolios are based purely on in sample analysis and typically do not account for nonstationarity and time-varying parameter and model uncertainty;
2. Optional features present in CAT bond contract specifications result in the analysis of such products under a standard mean-variance portfolio theory framework cannot be directly applied;
3. CAT bonds have been shown in Briys (1997) to have unconventional durations, where the duration of a CAT bond consisting of fixed cash flows (coupons) corresponds to the weighted average of the times until those fixed cash flows are received.

Next, by way of a second toy model, one may extend the framework presented in the first toy model representation of a catastrophe reinsurance product pricing framework in terms of a high-yield bond, by making a multiperiod analysis as presented in Example 18.3. This second example is derived from the results of Cox and Pedersen (2000, section 3) as it provides a means to analyze the following additional features.

1. Multiple periods in which coupons are paid at the end of each period in addition to the principle payment at the maturity of the bond;
2. In the event of a default due to catastrophic events, one can incorporate fractional coupon payments and fractional principle payments if the events occurred within the coupon period. Some form of prorata payment can be applied as a function of how the interest rate is modeled.

The following example again makes the simplifying assumption that one may price such products in an arbitrage-free complete market under a unique risk-neutral measure denoted by $\mathbb{Q}$. In addition, we will consider the discrete time setting to help simplify the mathematics and illustrate the main features.

EXAMPLE 18.3 **Toy Model 2 (Multiple Period): Securitized Catastrophe Reinsurance as a High-Yield Bond**

In this example, we consider the development of a multiple period catastrophe reinsurance product in terms of a high-yield bond. We make the following definitions of the model parameters and assumptions:

1. Assume a face value for the multiperiod CAT bond of 1 and a schedule of coupon payments $\{c_k\}_{k=1}^{T}$ at the end of each period until maturity, which is at the T-th period from issuance, at which time a payment of $1 + c_T$ is made if no catastrophe has occurred. It will be assumed for simplicity that the coupon schedule and amounts are specified in the contract;
2. If a catastrophe occurs during the k-th coupon period, then the bond makes a prorata payment for the coupon and the principle corresponding to fractional

payment of $f(1+c)$ at the termination time due to catastrophe trigger and subsequent default;

3. Assume that the market is complete and arbitrage free so that a unique risk-neutral pricing measure $\mathbb{Q}$ can be obtained so that the fair value for the discounted price of each coupon payment at the issuance date can be obtained;
4. Denote by $\{r_k\}_{k=1}^{T-1}$ the stochastic process corresponding to the one period interest rates during the lifetime of the bond;
5. Denote by P_k the price of a riskless zero-coupon bond with face value of 1 dollar and maturity at time k;
6. Assume that the risk-neutral pricing measure for the complete arbitrage-free market is independent of the probability and timing of when a catastrophe may occur in the interval $[0, T]$.

Discounted Expected Price at Issuance (t = 0). Under the assumption of a unique risk-neutral pricing measure, one may calculate the fair value of the discounted coupon payment schedule at the issuance time of the bond using the following expectation, which we also rewrite in terms of a sequence of nondefaultable zero coupon bonds $\{P_k\}$

$$
\begin{aligned}
B_0 &= \mathbb{E}_{\mathbb{Q}}\left[\sum_{k=1}^{T}\left[\prod_{s=0}^{k-1}(1+r_0)\right]^{-1} c_k\right] \\
&= \sum_{k=1}^{T} c_k P_k,
\end{aligned}
\tag{18.61}
$$

where it is assumed here that there is no risk of default making the sequence $\{c_k\}_{k=1}^{T}$ deterministic.

Default due to Catastrophe Event ($t = \tau$). Now we introduce uncertainty into the coupon cash flow making the sequence $\{c_k\}_{k=1}^{T}$ stochastic due to the possibility of a catastrophe that may trigger default. If a catastrophic event occurs that results in a trigger of the default clause of the CAT bond, then one can define the following cash flow stream to investors by the additional modification to the coupon payment sequence given by one of the following two options.

Coupon and Principle at Risk (forgiven in default event):

$$
c_k = \begin{cases} c\mathbb{I}[\tau > k] + f(c+1)\mathbb{I}[\tau = k], & k = 1, 2, \ldots, T-1, \\ (c+1)\mathbb{I}[\tau > T] + f(c+1)\mathbb{I}[\tau = T], & k = T, \end{cases}
\tag{18.62}
$$

where it is assumed each period has a fixed common coupon payment of c.

Coupon at Risk (forgiven in default event):

$$c_k = \begin{cases} c\mathbb{I}\left[\tau > k\right] + fc\mathbb{I}\left[\tau = k\right], & k = 1, 2, \ldots, T-1, \\ 1 + c\mathbb{I}\left[\tau > T\right] + fc\mathbb{I}\left[\tau = T\right], & k = T, \end{cases} \tag{18.63}$$

where it is assumed each period has a fixed common coupon payment of c.

In this case, one may write the modified fair value under the risk-neutral pricing framework according to the discounted expectation

$$B_0 = c\sum_{k=1}^{T-1} P_k\mathbb{Q}\left(\tau > k\right) + (c+1)P_T\mathbb{Q}\left(\tau > T\right) + f(c+1)\sum_{k=1}^{T} P_k\mathbb{Q}\left(\tau = k\right), \tag{18.64}$$

where $\mathbb{Q}\left(\tau > k\right)$ represents the probability (under the risk-neutral pricing measure) that the catastrophe does not occur in the first k periods. The distribution of the random variable τ will depend on how the CAT bond is constructed.

Coupon Rate to Recover Principle of CAT Bond. The coupon rate can now be calculated to ensure the principle of the CAT bond is recovered as a function of the default exposure according to

$$c = \frac{1 - P_T\mathbb{Q}(\tau > T) - f\sum_{k=1}^{T} P_k\mathbb{Q}(\tau = k)}{\sum_{k=1}^{T} P_k\mathbb{Q}(\tau > k) + f\sum_{k=1}^{T} P_k\mathbb{Q}(\tau = k)}. \tag{18.65}$$

■

In the second example presented, it is clear that one would need to make some additional assumptions or alternatively undertake some modeling or empirical analysis to assess the probability of a catastrophic event in the interval $[0, T]$. With this distribution, one would need to relate it back to the risk-neutral pricing measure, as specified in the introduction to this pricing section where the probabilistic framework is developed for the financial risk and the nature risk. Next, we will present another basic pricing example for the case of a parametric trigger bond.

EXAMPLE 18.4 **Parametric Trigger CAT Bond: Valuation**

Consider a generic peril, based CAT bond issued for a period $[0, T]$, which has the attributes that it will pay for the catastrophic event, I dollars at the next coupon payment period, if the monitored parameters θ of the peril (e.g., seismic activity, wind speeds, ocean heights, etc., at a set of k locations) exceed a specified threshold L within the period of coverage, triggering the forfeit of capital in the bond. If the catastrophic event does not occur and the trigger clause is not exercised, then a payment of a fixed coupon c_t dollars is paid out systematically at specified times $t \in \{t_1, t_2, \ldots, t_J\}$ with $0 < t_1 < t_2 < \cdots < t_J < T$ such that $t_i = iT/J$. Denote by τ the random time at which a catastrophe may occur in period $[0, T]$ with a probability of such an event in interval $[t_{i-1}, t_i]$ denoted by $\tilde{p}_{t_i} = \mathbb{P}\mathrm{r}\left(\tau \in [t_{i-1}, t_i]\right)$. Furthermore, assume for instance that the probability of the trigger, denoted at time τ by $p_t = \mathbb{P}\mathrm{r}\left(\theta_1(t) > L, \theta_2(t) > L, \ldots, \theta_k(t) > L \middle| \tau = t\right)$ is defined by the probability that the given parameters, at time t, at the k cites $\{\theta_i(t)\}_{i=1}^{k}$ that characterize the peril, exceeds the threshold L at each specified monitoring cite and is constant over time increment $[t_{i-1}, t_i]$. Assuming a constant interest rate r compounded at a rate given on the interval times T/J, then in this simple case one can obtain the minimum fair value of the total coupon payments $\{c_{t_i}\}_{i=1}^{J}$ (in dollars) made to investors and the minimum required total principle capital V required to be raised from initial investment in the CAT bond issue that corresponds to these coupons. These are obtained as functions of the probability of a trigger in a given period and a given prespecified insurance coverage. To see this, consider the trivial solutions to the system of equations created by evaluating each possible event outcome.

- **Catastrophe occurs in period $[0, t_1]$ and trigger is activated creating payout of L from the principle, with no coupons paid,**

$$
\begin{aligned}
&V(1+r)^J = I(1+r)^J \mathbb{P}\mathrm{r}\left(\theta_1(\tau) > L, \ldots, \theta_k(\tau) > L \middle| \tau \in [0, t_1]\right) \mathbb{P}\mathrm{r}\left(\tau \in [0, t_1]\right) \\
&\quad\Rightarrow \quad V = I p_{t_1} \tilde{p}_{t_1}.
\end{aligned}
$$

- **Catastrophe occurs in period $[t_1, t_2]$ and trigger is activated creating payout of L from the principle, with one coupon paid,**

$$
\begin{aligned}
&V(1+r)^J - c_{t_1}(1+r)^{J-1}\left[1 - \tilde{p}_{t_1}\mathbb{P}\mathrm{r}\left(\theta_1(\tau) > L, \ldots, \theta_k(\tau) > L \middle| \tau \in [0, t_1]\right)\right] \\
&\quad = I(1+r)^{J-2}\mathbb{P}\mathrm{r}\left(\theta_1(\tau) > L, \ldots, \theta_k(\tau) > L \middle| \tau \in [t_1, t_2]\right) \tilde{p}_{t_2} \\
&\quad \Rightarrow \quad c_{t_1} = \frac{\left(V(1+r)^2 - I p_{t_2}\tilde{p}_{t_2}\right)}{\left(1 - p_{t_2}\tilde{p}_{t_2}\right)(1+r)}.
\end{aligned}
$$

$\vdots$

- **Catastrophe occurs in period $[t_{k-1}, t_k]$ and trigger is activated creating payout of L from the principle, with $k-1$ coupons paid,**

$$
\begin{aligned}
&V(1+r)^J - \sum_{i=1}^{k-1} c_{t_i}(1+r)^{J-i} \\
&\qquad \times \left[1 - \tilde{p}_{t_i}\mathbb{P}\mathrm{r}\left(\theta_1(\tau) > L, \ldots, \theta_k(\tau) > L \middle| \tau \in [t_{i-1}, t_i]\right)\right] \\
&\quad = I(1+r)^{J-k}\mathbb{P}\mathrm{r}\left(\theta_1(\tau) > L, \ldots, \theta_k(\tau) > L \middle| \tau \in [t_{k-1}, t_k]\right) \mathbb{P}\mathrm{r}\left(\tau \in [t_{k-1}, t_k]\right) \\
&\quad \Rightarrow c_{t_{k-1}} = \frac{V(1+r)^J - \sum_{i=1}^{k-2} c_{t_i}(1+r)^{J-i}\left[1 - p_{t_i}\tilde{p}_{t_i}\right] - I(1+r)^{J-k} p_{t_k}\tilde{p}_{t_k}}{(1+r)^{J-k-1}\left[1 - p_{t_{k-1}}\tilde{p}_{t_{k-1}}\right]}.
\end{aligned}
$$

$\vdots$

- **Catastrophe never occurs in any period in $[0, T]$ and so trigger is never activated and J coupons are therefore paid,**

$$V(1+r)^J - \sum_{i=1}^{J} c_{t_i}(1+r)^{J-i} \times \left[1 - \tilde{p}_{t_i}\mathbb{P}\mathrm{r}\left(\theta_1(\tau) > L, \ldots, \theta_k(\tau) > L \middle| \tau \in [t_{i-1}, t_i]\right)\right] = 0$$

$$\Rightarrow c_{t_J} = \frac{V(1+r)^J - \sum_{i=1}^{J-1} c_{t_i}(1+r)^{J-i}\left[1 - p_{t_i}\tilde{p}_{t_i}\right]}{\left[1 - p_{t_J}\tilde{p}_{t_J}\right]}.$$

with the convention that $t_0 = 0$. Note, though these assumptions are simplistic in nature, they provide insight into the CAT bond parametric coupon values.

■

Of course, this illustrative example could be modified trivially to include cases where the coupon payments are specified or the value of the initial principle is specified along with the payout amount, etc. In addition, with the advent of new products such as multiple event (multiple trigger CAT bonds), one could also trivially solve these equations in an analogous fashion for such contracts.

Increasingly, the CAT bond coverage is also being packaged into products that industries can access for coverage such as the Example 18.5 we provide from ACE Insured TM the product "ACE Catastrophe Management 2.5SM". This is just a representative set of details of the typical specifications of a contract offered for catastrophe risk to supplement other policies they offer.

EXAMPLE 18.5

Under the ACE Insured TM[2] product "ACE Catastrophe Management 2.5SM" coverage is offered for managing threats to an insured reputation and other expenses directly related to catastrophic events. The specified benefits they list that are of relevance to OpRisk coverage include

- Coverage for costs associated with the disaster scene;
- Coverage for the impacted third-party funeral, psychological counseling, and temporary living costs;
- Coverage for the costs of the employment of engineers, scientists, or other professionals for the purposes of rescue or attempted rescue;
- Coverage for travel-related costs for directors, officers, and others to manage the repercussions of the catastrophic event.

The coverage is stated as "triggered when there is a catastrophic event resulting in traumatic body injury or property damage that is likely to result in damages covered by the lead excess policy".

The amounts of coverage provided and eligibility for the purchase of such an insurance product include

1. Specifically designed for clients who have a general liability attachment point of USD 5 million or higher;
2. Provides expanded catastrophe management services coverage for a maximum of 10% of a USD 25 million policy limit up to USD 2.5 million;
3. Requires a minimum of 10% coinsurance (insureds can select up to 50%);
4. Requires a minimum premium of USD 50,000.

■

There have also been pricing frameworks developed in the continuous time setting such as the contingent claim analysis adopted by Loubergé *et al.* (1999), which is present in brief in Example 18.6. In this framework, it is again assumed, under a simplification, that the market is complete and that one may therefore obtain a risk-neutral pricing measure. The example in Loubergé *et al.* (1999) is of interest here as it provides a simplified framework to perform valuation and duration calculations of CAT bonds as detailed later. Though the assumption of a complete market and the existence of a unique risk-neutral pricing measure in this setting is unrealistic, it does provide a simple insight into an understanding of an idealized relationship between the bond value and sensitivity as a function of the trigger in the contract (in this case, an index-based trigger). In particular, when considering the sensitivity, it will be measured for a bond by what is known as the duration and under (strong) simplifying assumptions; it will be shown that the duration of a CAT bond is such that it will remain greater than the maturity for the entire life of the bond. This is important to consider since typically the literature on CAT bonds pays little attention to the evaluation of their duration, though an industry paper out of the now defunct Lehmann Brothers firm by Briys (1997) pointed out that CAT bonds have unconventional durations and this was further illustrated under strict simplifying assumptions in Loubergé *et al.* (1999).

EXAMPLE 18.6 **Index Triggered CAT Bonds: Valuation and Duration**

Loubergé *et al.* (1999) consider the development of a simple contingent claim model for CAT bond pricing in which the following assumptions are made.

1. Assume the model evolves according to a continuous time pricing framework;
2. Assume the interest rate is denoted by r and is constant over time;

3. Assume that the bond is a zero-coupon bond that at issuance is characterized by a face value V and maturity T. There are no coupon payments assumed in this example to simplify the analysis;
4. Assume that the bond payoff is contingent on a trigger that is only evaluated at maturity based on an accumulated index $I(t)$ of claims resulting from natural catastrophes occurring during the life of the bond.

Stochastic Process for Trigger Index
For simplicity, one may assume a stochastic process for the index, $I(t)$ which is a simple geometric Brownian motion

$$dI(t) = \mu I(t)dt + \sigma I(t)dW. \tag{18.66}$$

Fair Value at Issuance
Under this simple example, the final payoff is contingent upon the value of the index at maturity $I(T)$ and a trigger threshold L producing one of two outcomes.

1. If $I(T) \leq L$, then the payoff is the face value V;
2. Otherwise if the accumulated index triggers the bond $I(T) > L$, then the payoff is treated as $\min(V_0, V - (I(T) - L))$, where V_0 is a specified minimum payoff based on the maturing face value V.

It can then be shown that under this simple model one can decompose the value at maturity $v(T)$ according to a linear combination of three components:

1. A long position in a riskless zero-coupon bond;
2. A short position in a catastrophe call with strike L;
3. A long position in a catastrophe call with strike $V + L - V_0$.

Where the decomposed value at maturity of the CAT bond is now expressed by the following linear combination

$$v(T) = V - \max(0, I(T) - L) + \max(0, I(T) - (L + V - V_0)). \tag{18.67}$$

Note that decomposing the maturing value in terms of these simple option structures is beneficial as it allows one to apply, under the specified market assumptions, a standard pricing framework based on Black–Scholes. Therefore, under the constant interest rate assumption, the present value of the bond (fair value under a risk-neutral pricing measure Q simply based on the index process $I(t)$) is given by the expected discounted prices to the issuance time. Under the simple GBM model for the index process, the closed-form solution is based on the well-known Black–Scholes formula:

$$\begin{aligned} v(0) &= \mathbb{E}_Q\left[V\exp(-rT)\right] - \mathbb{E}_Q\left[\max(0, I(T)-L)\exp(-rT)\right] \\ &\quad + \mathbb{E}_Q\left[\max\left(0, I(T)-(L+V-V_0)\right)\exp(-rT)\right] \\ &= V\exp(-rT) - \underbrace{C_E(I(0), L, T)}_{\text{European call option}} + \underbrace{C_E\left(I(0), V+L-V_0, T\right)}_{\text{European call option}} \\ &= V\exp(-rT) + I(0)\Phi\left(d_1(I(0), T)\right) - L\exp(-rT)\Phi\left(d_2(I(0), T)\right) \\ &\quad + I(0)\Phi\left(d_3(I(0), T)\right) - (V+L-V_0)\exp(-rT)\Phi\left(d_4(I(0), T)\right) \end{aligned} \tag{18.68}$$

with standard values

$$\begin{aligned} d_1(x, T) &= \frac{1}{\sigma\sqrt{T}}\ln\left(\frac{x}{L\exp(-rT)}\right) + \frac{\sigma\sqrt{T}}{2}, \\ d_2(x, T) &= d_1(x, T) - \sigma\sqrt{T}, \\ d_3(x, T) &= \frac{1}{\sigma\sqrt{T}}\ln\left(\frac{x}{(V+L-V_0)\exp(-rT)}\right) + \frac{\sigma\sqrt{T}}{2}, \\ d_4(x, T) &= d_3(x, T) - \sigma\sqrt{T}. \end{aligned} \tag{18.69}$$

In addition, one can write the value of the CAT bond under this formulation using the aforementioned expression where the final time T is replaced with the time to maturity.

Modified Duration/Price Sensitivity of CAT Bond

Since we consider in this case the CAT bond is directly a function of the yield (i.e., the return you get on the bond), the sensitivity of the CAT bond price can also be calculated according to the modified duration. The modified duration is a price sensitivity measure that measures the percentage derivative of price with respect to yield. Using the fact that we have a value for the bond at time t given (as given earlier) by

$$v(t) = V\exp(-r(T-t)) - C_E(I(t), L, T-t) + C_E\left(I(t), V+L-V_0, T-t\right), \tag{18.70}$$

then we get a duration given by

$$\begin{aligned} D(t) &= \frac{d\ln(v(t))}{dr} = -\frac{1}{v(t)}\frac{dv(t)}{dr} \\ &= -\frac{1}{v(t)}\frac{d}{dr} \\ &\quad \times \left[V\exp(-r(T-t)) - C_E(I(t), L, T-t) + C_E\left(I(t), V+L-V_0, T-t\right)\right] \end{aligned}$$

$$
\begin{aligned}
&= -\frac{1}{v(t)}\frac{d}{dr}\left[\, V\exp(-r(T-t)) + I(t)\Phi\left(d_1(I(t),T-t)\right)\right.\\
&\quad - L\exp(-r(T-t))\Phi\left(d_2(I(t),T-t)\right) + I(t)\Phi\left(d_3(I(t),T-t)\right)\\
&\quad \left.- (V+L-V_0)\exp(-r(T-t))\Phi\left(d_4(I(t),T-t)\right)\right]\\
&= -\frac{1}{v(t)}\left[-(T-t)V\exp(-r(T-t))\right.\\
&\quad + (T-t)L\exp(-r(T-t))\Phi\left(d_2(I(t),T-t)\right)\\
&\quad \left.+ (T-t)\left(V+L-V_0\right)\exp(-r(T-t))\Phi\left(d_4(I(t),T-t)\right)\right]\\
&= (T-t)\left[1+\frac{I(t)}{v(t)}\left(\Phi\left(d_2(I(t),T-t)\right) - \Phi\left(d_4(I(t),T-t)\right)\right)\right]
\end{aligned}
\tag{18.71}
$$

Outcome. This shows that the modified duration is strictly greater than 1, meaning that the duration is greater than the time to maturity. Loubergé *et al.* (1999) find under the simplifying assumptions made that the duration of the CAT bond at any time t exceeds the time to maturity. *This shows that such a bond has a greater exposure to interest rate risk than a typical zero coupon bond.* ■

Note, that Loubergé *et al.* (1999, p. 134) also present a similar analysis in the less restrictive model assumptions in which they consider a stochastic interest rate model.

18.3.3 FRAMEWORK 2: ASSUMING INCOMPLETE ARBITRAGE-FREE PRICING

In addition to these features of CAT bonds it is also well acknowledged in the literature that CAT bond payments cannot be hedged, that is, there is no exact replicating portfolio comprising primitive assets or bonds that can perfectly reproduce the CAT bond payments; see discussion in Cox and Pedersen (2000). Therefore, the simplified models presented previously act purely as informative examples of how such CAT bond products behave under ideal stylized conditions in which strong assumptions regarding the existence of a unique martingale measure for fair value pricing and a lack of arbitrage opportunities are assumed given. As discussed previously, it is argued by Embrechts and Meister (1997) and Cox and Pedersen (2000) that the appropriate framework required to price CAT bonds involves an incomplete markets approach. We will discuss the pricing and valuation of CAT bonds in a few different settings under an incomplete market hypothesis. We will highlight the key components of different approaches that several authors have advocated in this regard.

We begin this section with a discrete time multiperiod example of valuation of CAT bonds under the benchmark financial economics framework developed for CAT bond valuation in an incomplete market setting by Cox and Pedersen (2000); see details in Platen (2006) and the book-length presentation in Platen and Heath (2006). We work again with the examples of Cox and Pedersen (2000) as they correspond to the framework developed and provide clear and comprehensive approach to the pricing steps required for practitioners.

EXAMPLE 18.7 Valuation of CAT Bonds in Incomplete Markets (Discrete Time, Multiperiod Model)

The model and valuation for CAT bonds proposed in Cox and Pedersen (2000, section 5) are summarized as follows.

1. Assume the market for the CAT bond is incomplete and the pricing of uncertain cash flows is achieved via a technique known as the representative agent. In this approach, one assumes a representative utility function and an aggregate consumption process, then the agent utilizes this utility function to make decisions regarding possible investments in cash flows from the assets in the market that will be termed consumption streams (paths or trajectories in time);
2. The resulting generic "consumption streams" are assumed to be adapted processes that are therefore only dependent on observable information and are denoted by the process $\{c_k\}_{k=0}^{T}$, where T denotes the CAT bond maturity and k denotes the index for the discrete time units;
3. The total consumption available in the economy at any given time and state of the world is defined by the "aggregate consumption process" that they denoted by $\left\{C_k^*\right\}_{k=0}^{T}$, and at time $t = 0$ this aggregate consumption of the economy is known exactly, at all other times it is stochastic and unknown. In addition it is assumed that the aggregate consumption is only dependent on financial risk variables such that $C_k^*(\omega) = C_k^*\left(\omega^{(1)}, \omega^{(2)}\right) = C_k^*\left(\omega^{(1)}\right)$ for all $k \in \{1, 2, \ldots, T\}$ so that the process $\left\{C_k^*\right\}_{k=0}^{T}$ is adapted to the filtration $\mathcal{A}^{(1)}$;
4. The representative agent's utility is assumed additively seperable and therefore the price $V_k(d_k)$ at $k = 0$ of a generic future cash flow $\{d_k\}_{k=1}^{T}$ is given by the expectation

$$V_k(d_k) = \mathbb{E}_{\mathbb{P}}\left[\sum_{k=1}^{T} \frac{u_k\left(C_k^*\right)}{u_0\left(C_0^*\right)} d_k\right], \tag{18.72}$$

   where $u_k(\cdot)$ are representative utility functions for the representative agent and the generic cashflow is assumed to depend on both financial risk variables and catastrophe risk variables such that $d_k(\omega) = d_k\left(\omega^{(1)}, \omega^{(2)}\right)$ for all $k \in \{1, 2, \ldots, T\}$;
5. Assume the one period interest rates, denoted by $\{r_k\}_{k=1}^{T}$, are defined by the conditional expectations (see justification for this definition in Cox and Pedersen (2000, p. 68))

$$\frac{1}{1+r_k} := \frac{1}{u_k\left(C_k^*\right)} \mathbb{E}_{\mathbb{P}}\left[u_{k+1}\left(C_{k+1}^*\right) \middle| \mathcal{F}_k\right]. \tag{18.73}$$

   Under this assumption, they are able to remove the form of the utility function and the aggregate endowment process from the pricing framework by associating the price relation to the valuation measure approach of arbitrage-free pricing;

6. Define the change of measure from $\mathbb{P}$ to $\mathbb{Q}$ such that all resulting prices are discounted expectations with respect to the previously defined one-period interest rate process,

$$\left.\frac{d\mathbb{Q}}{d\mathbb{P}}\right|_{\mathcal{F}_T}(\omega) := (1+r_0)\prod_{k=1}^{T-1}(1+r_k(\omega))\,\frac{u_T\left(C_T^*(\omega)\right)}{u_0\left(C_0^*\right)} \tag{18.74}$$

which results in the valuation of the generic future cash flow under the new "risk-neutral pricing" measure $\mathbb{Q}$ according to

$$V_k(d_k) = \mathbb{E}_{\mathbb{Q}}\left[\sum_{k=1}^{T} d_k \prod_{s=0}^{k-1}(1+r_s)^{-1}\right]. \tag{18.75}$$

One can also obtain the marginalized cash flow over the nature risk (i.e., the catastrophe risk variables) as given by

$$\overline{d}_k(\omega^{(1)}) = \mathbb{E}_{\mathbb{Q}}\left[d_k\left(\omega^{(1)},\omega^{(2)}\right)\middle|\,\mathcal{A}_k^{(1)}\right], \tag{18.76}$$

which allows one to obtain the valuation of the future cash-flows purely with respect to financial risk variables according to

$$V_k(d_k) = \mathbb{E}_{\mathbb{Q}}\left[\sum_{k=1}^{T} \overline{d}_k \prod_{s=0}^{k-1}[1+r_s]^{-1}\right]. \tag{18.77}$$

From this probabilistic pricing framework, one can then approximately evaluate a CAT bonds values in settings in which the coupons (cash flows) depend exclusively on the catastrophe risk variables (through the trigger) or they depend on both financial and catastrophe risk variables with a known functional dependence. Then in either case one can basically in practice select an arbitrage-free interest rate model, then for each interest rate state of the world, one calculates the bonds expected cash-flows conditionally on the interest rate path. Finally, one may use Equation (18.77) to evaluate the expected price of cashflows. ■

Other approaches have been adopted to performing pricing in the incomplete market setting such as the model of Lee and Yu (2007), where the valuation of catastrophe reinsurance with CAT bonds under a contingent claim model is undertaken. However, for simplicity as was undertaken in the previously present model, these authors also assume the existence of a risk-neutral process that they parameterize for use in pricing.

EXAMPLE 18.8 Valuation of CAT Bonds in Incomplete Markets (Continuous Time Model)

The model and valuation for CAT bonds proposed in Lee and Yu (2007) is summarized as follows for the main features of interest. The model proposed by these authors includes the specification of the asset dynamics value of a reinsurance company according to a LogNormal diffusion that also incorporates explicitly the effect of a stochastic interest rate to reflect the exposure that most reinsurers will have to fixed income markets.

Asset Dynamics. The asset dynamics are then given by

$$dV_t = \mu_V V_t dt + \phi_V V_t dr_t + \sigma_V V_t dW_{V,t} \tag{18.78}$$

with $(r_t)_{t\geq 0}$ the diffusion process for the instantaneous interest rate r_t's dynamics, which are assumed to follow a square root diffusion of Cox *et al.* (1985) and $dW_{V,t}$ increments of a driving Brownian motion representing the credit risk associated with the assets. Furthermore, they assume a parametric model for the risk-neutral process for the interest rates.

Liability Dynamics (Noncatastrophe Insurance Lines). It is also assumed that the liability dynamics follows the process

$$dL_t = (r_t + \mu_L) L_t dt + \phi_L L_t dr_t + \sigma_L L_t dW_{L,t} \tag{18.79}$$

with L_t representing the present value of liabilities for the reinsurer not associated with a catastrophe (i.e., claims arising from reinsurance coverage provided for other insurance lines), ϕ_L is the instantaneous interest rate elasticity of the reinsurers liabilities, μ_L is the risk premium of small short time shocks, and $W_{L,t}$ is a driving Brownian motion representing small idiosyncratic shocks to the capital market (day-to-day).

Catastrophe Liability Dynamics (Catastrophe Insurance Lines). The catastrophe loss dynamics are modeled according to a loss dynamic model given by the compound Poisson process

$$C_t = \sum_{n=1}^{N_t} c_n(t), \tag{18.80}$$

where C_t represents the covered catastrophe loss at time t. In addition, the basis risk impact on the reinsurance valuation is taken into account through a second compound process representing an index of catastrophe losses, given by dynamics

$$C_{\text{index},t} = \sum_{n=1}^{N_t} c_{\text{index},n}(t) \tag{18.81}$$

with the process $\{N_t\}_{t\geq 0}$ representing the counting process for the number of losses and $c_n(t)$ is the loss amount for the n-th catastrophe loss event at time t. Lee and Yu (2007) then proceed to present two scenarios, the case where a primary insurer has no CAT bonds and then the case where they have purchased CAT bonds and the calculation of the rate on line (ROL) is performed numerically. The ROL represents the premium rate per dollar covered by the catastrophe reinsurance; see Lee and Yu (2007, equation 13) for details. ■

In addition to the examples presented earlier, there have been several other interesting models proposed, such as the approach adopted by Vaugirard (2003), which involves development of an approach to the valuation of insurance-linked derivatives that can account for interest rate dynamics and catastrophic events while taking into account a framework of nontraded underlyings. In addition, this valuation framework is of interest as it involves the ability to incorporate the existence of arbitrage prices for CAT bonds in an incomplete market with nontraded underlying state variables for CAT bonds that are OTC traded.

EXAMPLE 18.9 **Valuation of CAT Bonds in Incomplete Markets with Arbitrage (Continuous Time Model Jump-Diffusion Process)**

In this example, a brief highlight of the model and pricing approach proposed in Vaugirard (2003) is provided. In particular, when developing this pricing approach Vaugirard (2003) adopt Merton's approach in Merton (1976), which assumes any risk associated with jumps is diversifiable and can therefore be ignored. That is, the β in the CAPM model for portfolios that only include the nonsystematic risk is insignificant from zero with the sum given by the risk-free rate. In addition, Vaugirard (2003) argues that Merton's approach is well suited since it is directly applicable to cases in which underlying state variables are noninvestment assets. The key features of this model are as follows.

1. Consider a CAT bond modeled like a corporate bond with insurance-linked risk instead of credit default risk. Assume the bond holder (investor or cedent) stands to lose coupons and possibly a fraction of the principle if a trigger based on a natural risk index at time t, denoted by I_t, exceeds a predefined threshold K. If the index does not exceed K in the time interval $[0, T]$, then payment of the face value F is made to the investor, otherwise the investor receives the face value minus write down. Assume at time $t = 0$ that $I_0 < K$;
2. Define by T the risk exposure period and the bond maturity is $T' > T$ to allow for lags in the risk index assessment at maturity;
3. Define the stochastic process sources of randomness for the catastrophes as follows: Brownian motion $\{W_t\}_{t\in[0,T]}$—noncatastrophic nature risk; Poisson process with intensity λ_t denoted $\{N_t\}_{t\in[0,T]}$—occurrence of catastrophes; severity random variables (i.i.d.) $\{U_j\}_{j\geq 0}$; and Brownian motion $\{W_{2t}\}_{t\in[0,T]}$—interest rate fluctuations;

4. Assume that the risk attitude for investors is separated into two components: nature risk and market risk.
 Nature Risk. Assume that investors are neutral with regard to jump risk arising from catastrophes (nature risk).
 This assumption is argued to be reasonable based on the ability to diversify catastrophe risk from an investment due to the fact that nature risk is uncorrelated with market risks. Which is in alignment with Merton's stance that jump risk is not systematic.
 Market Risk. Assume any additional variation in the risk index not attributed to catastrophes can be replicated through existing exchange-traded securities. In addition, it is assumed that variation in interest rates can also be replicated through exchange traded assets.
 Each component of this assumption is argued to be reasonable since domestic interest rates can be replicated using risk-free bonds and noncatastrophe-related changes in the risk index can be replicated by instruments such as energy and power derivatives, weather derivatives and contingent claims on multiple commodities.

The model proposed involves the following features.

Interest Rate Model. Interest rates follow a simple mean reversion as in the Vasicek model, see Vasicek (1977), given by

$$dr_t = a\left(b - r_t\right)dt + \sigma_r dW_{2t}. \tag{18.82}$$

Risk Index Model. The model for the risk index process I_t is given by a Poisson jump diffusion process with three components: the expected instantaneous index change conditional on no catastrophe occurring; unanticipated instantaneous fluctuations in the index not due to catastrophes; and instantaneous index changes attributed to a catastrophic event. The resulting jump diffusion with these three features is then given by

$$dI_t = \mu_t I_{t^-} dt + \sigma_t I_{t^-} dW_t + J_t dN_t, \tag{18.83}$$

with I_{t^-} the index value just before t, μ_t the drift (that can be stochastic), σ the volatility that is deterministic, J_t is the stochastic size of the jumps given by

$$J_t = \sum_{n=1,+\infty} U_n \mathbb{I}_{[\tau_{n-1},\tau_n]}(t), \tag{18.84}$$

where at time τ_j the jump in I_t of size $\triangle I_{\tau_j} = I_{t^-} U_j$, with $(1 + U_j) \sim LogNormal(\mu, \sigma^2)$. Vaugirard (2003, proposition 1) utilize Girsanov's theorem to find a risk-neutral pricing measure to evaluate the expected value of the discounted contingent claim at time t under the jump diffusion and interest rate diffusion models presented. Then Vaugirard (2003, proposition 2) develop a generic expression for valuation of the pure discount CAT bond is provided in terms of the first passage time of the loss index through the barrier (trigger level). ■

To conclude we note that there have been also other noteworthy studies such as the model proposed by Lee and Yu (2002) that prices default risky CAT bonds incorporating into the model aspects of moral hazard and basis risk. There is also an interesting hybrid CAT bond structure proposed in Barrieu and Loubergé (2009), which aims to improve market efficiency for CAT bonds. In particular, it considers an alteration to the between the sponsor, the SPV, and the investor such that depending on two events, a catastrophe and independently a market crash, different payouts to the investor and to the sponsor are made. The aim being to develop catastrophe risk transfer with protection against stock market declines, which it is argued will result in increasing volume in the CAT bond markets.

18.4 Sidecars, Multiple Peril Baskets, and Umbrellas for OpRisk

In this section, we discuss alternative insurance products for multiple perils that could also be considered in the context of OpRisk, starting with a brief mention of the concept of a sidecar given in Definition 18.28.

Definition 18.28 (Reinsurance Sidecars) *A reinsurance sidecar is a financial structure created to allow investors to take on the risk and return of a group of reinsurance or insurance policies (a "book of business") written by an insurer or reinsurer that in turn earn the risk and return that arises from that business. The insurer or reinsurer will only cede the premiums associated with the book of business for such a sidecar vehicle if the investors make sufficient investment in the vehicle so as to ensure that claims can be met should they arise. In general, the investor liability is then limited to this invested capital.* ∎

Hence, the notion of an insurance sidecar can be utilized in OpRisk as a product structure for a given financial institution that may have exposure to multiple perils that need coverage and they may wish, through the portfolio of a captive to that financial institution, to raise additional capital for coverage of losses, should they arise from such exposures, through the notion of a sidecar. Such a strategy could bring investor capital into the institution to help mitigate against potential insurance losses in exchange for profits from a portion of the captives insurance premiums and investments such as CAT bonds.

In the following section, we will consider a generalized framework, which will focus on two settings in which the class of multiple peril basket insurance products will be of interest. The first is in the case of a single risk process that has exposure to multiple perils and the second is in the case of an insurance policy for multiple risk processes each exposed to one or more perils. In addition, there is the situation that the perils with which a risk process or processes are exposed may not all have direct relationships to existing insurance products currently available on the market in the given jurisdiction of operation of the bank seeking insurance.

To address such situations, there has been the development of multiple risk or multiple peril type insurance products that can be generically defined according to a following portfolio products in Definition 18.31. These could be constructed in two basic approaches: the first is internally within the bank seeking coverage on their risk process(s) with exposure to multiple perils, via a portfolio of sufficiently highly rated insurers offering products for each of the perils considered for the risk process. Alternatively, the second approach is to purchase one of the specialized products available for OpRisk that are offered by large insurers. We note that in this second case there can be strict eligibility requirements for purchase of the insurance product such as minimum thresholds on the total OpRisk capital of the bank. In addition, the size of the required premium for such products can also make these products either prohibitive or uneconomical for a large number of banking institutions. Hence, a larger portion of banks seeking insurance for OpRisk may be in the first category.

In either of these cases, one can present a common framework and definition of insurance coverage that is constructed from a multiple peril basket. To illustrate this, we consider the case of an OpRisk loss process that is affected by multiple sources of peril, and where the single risk processes under question has losses that do not have a direct insurance product available to provide coverage for all exposures. To achieve this, we need to first define the notions of peril and CAT bonds.

Definition 18.29 (Peril and Hazard in Insurance) *Peril and hazard are related to the cause of losses, where a peril can be defined as giving rise to losses, while a hazard is defined as influencing the operation of the peril. Typically one would classify hazards according to physical or moral: a physical hazard will relate to the physical characteristics of the risk, while a moral hazard will relate to the attitude and conduct of people.* ■

Particular examples of Perils and Hazards will be discussed further in the context of OpRisk CAT bonds. Before proceeding with a formal definition of multiple peril baskets, we will discuss first the notion of umbrella insurance.

18.4.1 UMBRELLA INSURANCE

In Definition 18.30, we define the notion of an umbrella insurance product.

Definition 18.30 (Umbrella Insurance) *An umbrella insurance policy in its most basic form is constructed to act in conjunction with an existing base insurance policy for a given loss process. Typically, umbrella insurance is purchased as a liability insurance policy that acts as an additional protection of assets and future income of insured party which is in addition to the primary policies coverage. It is distinct from what is known as excess insurance, which only covers claims once all underlying policies are exhausted, in that umbrella insurance is capable of what is known as the "drop down" feature. This allows umbrella insurance policies to provide coverage for underlying policy gaps, meaning that in some cases the umbrella insurance contract may eventuate as the primary insurance contract on a particular risk. The term umbrella refers to the more general coverage options of this policy compared to peril-specific policies. In addition, such umbrella policies may in particular cases provide coverage for claims that would otherwise have been excluded from primary policies.* ■

There are several examples of commercial umbrella coverages that are of relevance to OpRisk settings such as the product known as the "ACE Umbrella PlusSM", which is offered as a commercial umbrella liability insurance for US national accounts. The same complay also offers a different product known as the "ACE USA Excess Casualty" which specializes in Fortune 2000 US corporations and privately owned company equivalents. These policies include aspects such as minimum attachment points, which may be individually evaluated on a per risk basis, and aggregate coverage limits. Example details can include aspects such as general liability of USD 1 million per occurrence and USD 2 million general aggregate and USD 2 million products aggregate; employer's liability: USD 1 million; general limits of USD 50 million per occurrence or aggregate are available.

18.4.2 OPRISK LOSS PROCESSES COMPRISED OF MULTIPLE PERILS

In this section we consider the setting in which we have an OpRisk loss process structure that has the individual LDA risk process models comprising losses that may arise from multiple sources of peril, some of which may be directly insurable and others may not. The number of such perils and the sources of exposures for a given risk process will of course depend on the chosen mapping of the banks business unit structure to the Basel II/Basel III event types. In such settings, under a standard LDA framework, one defines a risk process $\{Z_t\}$ for the annual loss corresponding to the particular business unit in the chosen hierarchy mapping and an official Basel II defined event type such as, for instance, internal fraud; external fraud; employment practices and workplace safety; damage to physical assets; business disruption and systems failures to name a few.

Remark 18.23 *Hence, the focus of this section is on OpRisk loss processes that are subject to loss events that can come from several sources of peril and are associated with multiple sources of hazards.*

For such risk process $\{Z_t\}$, there may be multiple sources of peril that result in the losses in the year $\{X_i(t)\}_{i=1}^{N_t}$. For instance, consider d sources of peril that impact the given risk process Z_t in year t, then one may consider the actual process as comprising a random number of $N_t = \sum_{i=1}^{d} N_t(i)$ total losses in the year t comprising of $N_t(i)$ losses from the i-th peril. Then each individual loss $X_i(t)$ would be comprising a contribution attributed to $X_{i,j}(t)$ arising from the loss from the j-th peril in the given event such that $X_i(t) = \sum_{j=1}^{d} X_{i,j}(t)$. Note that for any given loss event, the contribution from any one of the d perils may range from 0 to 100%.

Remark 18.24 *Typically, when modeling such risk process in OpRisk under an LDA framework we do not need to distinguish explicitly the source of the individual component losses from each of the perils that contributed to the total loss in a given event. In other words, we would not decompose each loss amount $X_i(t)$ into components from each peril; instead, we would focus on modeling the distribution for $X_i(t)$. However, when insurance is considered where some of the d contributing perils, there will be coverage from particular policies available to these portions of the loss events, say for the j-th peril in the i-th event there would be some coverage for the loss amount $X_{i,j}(t)$ depending on the policy specifications.*

Hence, when we wish to consider such features, we have a loss process in year t denoted by annual loss Z_t, which is modeled under an LDA framework in which each of the individual loss events could be considered as arising wholly or in part from one of d different sources of insurable perils that fall under the given risk process categorization in the OpRisk structure of the institution. In this case we may wish to consider the decomposition of the standard LDA loss process comprising $N_t \sim F_N(n)$ and $\{X_i(t)\}_{i=1}^{N_t}$ with i.i.d. $X_i(t) \sim F_X(x)$ into the following model:

$$Z_t = \sum_{i=1}^{N_t} X_i(t) = \sum_{i=1}^{N_t} \sum_{j=1}^{d} X_{i,j}(t), \tag{18.85}$$

with the allowance that any given peril j the amount $X_{i,j}(t)$ could be zero for the i-th loss event such that there would be $N_t(i)$ total random number of nonzero losses (could be comprised of portions of losses from larger losses) in the year for peril i. In this class of OpRisk process case, one could aim to cover a portion of the aggregate loss in a given year either on a composite level or on a per loss event level. In either case, a special insurance basket would need to be constructed with components of relevance to coverage for each of d sources of peril represented by the OpRisk loss process, some of which may not have a direct insurance product to cover this form of loss.

To illustrate how this may arise, consider the classes of Basel II/Basel III event types such as damage to physical assets that could be attributed from multiple perils in any particular geographic region such as natural disasters — wind damage, storm damage, flooding, earthquake, fire, hurricane; and terrorism; vandalism. A second example to consider where a single risk process in a given business unit and event type hierarchy may be exposed to multiple perils is when one considers the event type given by business disruption & systems failures that can have contributing perils coming from infrastructure failures; utility disruptions, software failures, hardware failures, etc., some of which will be covered under particular policies and others will not.

Therefore, for risk processes that are exposed to multiple perils, we define for OpRisk the single risk process multiple peril basket insurance that is generically given as detailed in Definition 18.31. In specification of this insurance structure, it will be assumed that there will be an insurance policy, CAT bond, or umbrella coverage that can be purchased to provide some amount of coverage for all perils that the loss process is exposed to and furthermore that the amount of coverage one may purchase from the market is not bounded (of course, these are simplifying assumptions in practice). In general, this may not be the case, but we assume this for simplicity of notation and development later.

We will first define the insurance portfolio followed by specification of the analog of a modern portfolio theory solution to portfolio selection that we adapt to the insurance setting. In this regard, we will consider the adaption of the solution based on Markowitz's approach (see Markowitz 1959 and Rubinstein 2002).

Definition 18.31 (OpRisk Single Risk Process Multiple Peril Baskets) *A multiple peril basket in its simplest form is defined as a linearly weighted portfolio of insurance contracts, umbrella contracts, and CAT bonds combined to mitigate a predefined portion of each of the sources of peril that contribute to a given risk process, thereby transferring the associated risk of the loss process up to a prespecified coverage limit. Consider the following parameters that would be considered in the definition of such a basket for covering the loss processes, with m combined insurance products and k CAT bonds in order to cover $m + k$ perils that the given loss process is exposed too:*

1. ***Total Top Cover Limit*** *(TCL). This would be comprising a combination of coverage from several insurance products and CAT bonds constructed to cover several perils that may result in losses under the particular OpRisk loss process. The total coverage that is purchased from each product for each peril comprised m different insurance products defined by coverage limits $\{ICL_j\}_{j=1}^{m}$ and effective coverage from k different bonds (e.g., CAT bonds) defined by coverage $\{BCL_j\}_{j=1}^{k}$ such that the resulting TCL for the OpRisk loss process covered by the multiple peril basket is given by*

$$TCL = \sum_{j=1}^{m} ICL_j + \sum_{j=m+1}^{m+k} BCL_j. \tag{18.86}$$

2. ***Individual Peril Minimum Coverage Limits*** *(MCL_i). One may assume that a minimum amount of coverage for the i-th peril is required, producing the set of constraints for the insurance products given by $ICL_i \geq MCL_i$ and the bond products (CAT-bonds) given by constraints $BCL_j \geq MCL_j$;*
3. ***Individual Peril Insurance Product Premium*** *P_i. For the i-th peril, the required premium for the insurance product to cover the losses attributed to this peril, for inclusion in the portfolio is given by P_i. It is assumed that P_i is strictly increasing as the required amount of coverage ICL_i increases;*
4. ***Individual Peril Bond Prices*** *B_i. For the i-th peril the required bond price for inclusion in the portfolio is given by B_i;*
5. ***Individual peril Bond specifications*** *$\boldsymbol{\theta}_i$. For the i-th peril the bond is specified by a vector of parameters $\boldsymbol{\theta}_i$ comprising elements: maturity T_i, coupon dates subject to no-trigger (default) $\{\tau_1, \tau_2, \ldots\}$ with coupon amounts $\{c_1, c_2, \ldots\}$, type of trigger (to indicate default) and required specifications of trigger event such as threshold, modeled probability of trigger p_i, and exposure coverage in the event of trigger given by BCL_i.* ■

Given this very general specification of the total coverage provided by a multiple peril basket for a given loss process, it is also important to note the following practical aspects of the coverage.

Remark 18.25

- *Formally in an ideal setting the ICL for each policy and the BCL for each bond asset would be deterministically fixed in the contract and then one would simply have the option of determining the total number of policies of a particular type, subject to attachment points, the total umbrella coverage required, and the types of bond coverage that could be considered to provide an overall total coverage for a particular OpRisk loss process. Then given a requirement for a particular mitigation level, for example, 20% of the calculated capital for the risk process as specified in OpRisk settings, one could deterministically optimize the number of policies and bonds required such that the contribution from each came as close as possible to the required overall TCL for the particular loss process while minimizing the total cost of the coverage. This could ideally be solved by a deterministic search through the space of portfolio cost and coverage to find the optimal combination that reproduces the TCL required for the given loss process;*
- *In addition, we will assume for simplicity that ICL and BCL are continuous and unbounded positive quantities, though in practice we may expect upper bounds on these quantities. In practice, the ICL on each policy and BCL are typically subject to potentially substantial payment uncertainties due to several factors: litigation and challenge of individual claims, default of the insurer, and payment time uncertainties. This makes the effective coverages inherently stochastic in practice and therefore it is perhaps prudent to consider these quantities as random variables (even though they are formally specified in a contract). It is therefore possible that the ICL (BCL) may effectively be less than the premium (purchase price of the bond) in any given year. Hence, given coverage of ICL (BCL) in the contract, we define the effective coverage for insurance policy (bond) i, at the end of the contract (yearly), to be $\widetilde{ICL}_i = ICL_i - UPC_i$, where $UPC_i \in [0, ICL_i]$ corresponds to a random variable for the dollar amount at the end of the contract of all forms of unpaid or outstanding*

claims that were disputed. The analogous amount for bond products that default or fail to make claim payments immediately at the time of trigger are denoted by effective coverage $\widetilde{BCL}_i$.

To proceed with the optimal portfolio selection, which aims to address the question of how much coverage should one purchase from each insurance product type in order to maximize the expected total coverage "return" while minimizing the variance or some other measure of "risk" associated with the coverage return. In Definition 18.32, we define what we mean by return via the notion of a coverage on investment (COI).

Definition 18.32 (Coverage on Investment) *We define the random variable corresponding to the amount of effective annual insurance coverage (in dollars) for insurance policies by* $\widetilde{ICL}_i = ICL_i - UPC_i$ *and bonds by* $\widetilde{BCL}_i = BCL_i - UPC_i$. *Then the coverage on investment (COI) is given by the ratio of the effective coverage per year to the dollar amount invested for the coverage of* ICL_i (BCL_i) *from the insurance product (insurance policy or CAT bond) according to the random variable* R_i *given by*

$$R_i = \begin{cases} \dfrac{\widetilde{ICL}_i}{P_i}, & \textit{Insurance Policy } i, \\ \\ \dfrac{\widetilde{BCL}_i}{P_i}, & \textit{Bond (CAT bond } i), \end{cases} \tag{18.87}$$

where $\widetilde{ICL}_i$ *or* $\widetilde{BCL}_i$ *correspond to the coverage offered for the premium/bond price* P_i *paid on the contract coverage of* ICL_i (BCL_i). *To distinguish the case of an insurance policy versus a bond, one may adopt* R_i^I *or* R_i^B, *respectively.* ■

We will now assume that for each insurance policy or CAT bond product we may model the random variable R_i by a LogNormal distribution with $R_i \sim LogNormal\left(\mu_i, \sigma_i^2\right)$. Of course, in practice, this model assumption can be tested and other families of distribution can be considered; however, the choice of an elliptic family (in this case on the log scale) makes the following portfolio optimization conveniently a convex optimization programme.

One can now aim to perform portfolio selection to address the question:

> Given a particular loss process exposed to $m + k$ perils, what is the optimal amount of coverage for each insurance policy and CAT bond to ensure the expected total effective coverage is maximized, whilst the risk (variance) in the total effective coverage is minimized.

To address this question, we establish the following modern portfolio theory (MPT) portfolio selection framework, the first approach will be based on the trade-off between the expected total effective COI for the basket of insurance products versus the variance in the total effective COI. We will adopt a standard Markowitz portfolio selection framework here that will be a convex optimization framework if we assume that the log total effective COI is a random variable with a distribution function with is in the elliptic family and that each individual log effective COIs are also distributed according to an elliptic family, perhaps with correlation induced between the prices or the amounts of coverage that go into each policy in the basket.

Markowitz MPT for Insurance Basket Portfolio Selection

- Define the log total effective coverage on investment random variable according to the linear combination of individual policy and bond returns (log COIs) as follows:

$$\ln\left(R_T\right) = \sum_{i=1}^{m} w_i^I \ln\left(R_i^I\right) + \sum_{i=m+1}^{m+k} w_i^B \ln\left(R_i^B\right) \tag{18.88}$$

  with w_i^I corresponding to the weighting of component asset insurance policy i that corresponds to the proportion of insurance policy asset i in the portfolio insurance basket, and w_i^B corresponds to the equivalent proportion of bond asset i in the basket;
- Assume the insurance portfolio total effective coverage on investment "return" is the proportion-weighted combination of the constituent insurance products effective coverage on investments "returns";
- Assume the portfolio total effective coverage on investment has a "risk/uncertainty/volatility", which is a function of the correlations σ_{ij} between the effective COI's of the component assets, for all asset pairs (i,j);
- Then one may define the expected total effective coverage:

$$\mathbb{E}\left[\ln\left(R_T\right)\right] = \sum_{i=1}^{m} w_i^I \mathbb{E}\left[\ln\left(R_i^I\right)\right] + \sum_{i=m+1}^{m+k} w_i^B \mathbb{E}\left[\ln\left(R_i^B\right)\right]. \tag{18.89}$$

- Then one may define the variance of the total effective coverage:

$$\mathbb{V}\text{ar}\left[\ln\left(R_T\right)\right] = \sum_{j=1}^{m}\sum_{i=1}^{m} w_i^I w_j^I \sigma_i^I \sigma_j^I \rho_{i,j}^I + \sum_{j=1}^{k}\sum_{i=1}^{k} w_i^I w_j^I \sigma_i^B \sigma_j^B \rho_{i,j}^B + \sum_{j}^{m}\sum_{i}^{k} w_i^I w_j^B \sigma_i^I \sigma_j^B \rho_{i,j}^{I,B} \tag{18.90}$$

  with $\rho_{i,j}^{\cdot} = 1$ if $i = j$;
- The portfolio selection can now be performed by considering a desired risk tolerance (associate with how certain one would like to be about the total effective coverage) denoted by $q \in [0, \infty)$, and the resulting efficient frontier is obtained by minimizing the objective function

$$\boldsymbol{w}^T \Sigma \boldsymbol{w} - q \boldsymbol{R}^T \boldsymbol{w}, \tag{18.91}$$

  where

  1. $\boldsymbol{w}$ corresponds to the vector of insurance portfolio weights satisfying the constraint $\sum_i w_i = 1$. In the context of an insurance basket, it will be only possible that $w_i^I \geq 0$ though one may short the CAT bond assets so one may have $w_i^B \in \mathbb{R}$;
  2. Σ represents the covariance matrix of effective coverages on investments for insurance products and bonds in the basket portfolio;
  3. $q \geq 0$ is the risk tolerance that ranges between $q = 0$ for a portfolio with minimal risk (i.e., minimum uncertainty regarding the total effective coverage on investment) through to $q \to \infty$ for a progressively higher expected effective total coverage

on investment with progressively higher unbounded uncertainty associated with the expected coverage;

4. $\boldsymbol{R}$ is a vector of expected effective coverages on investment for each insurance product of bond in the portfolio;
5. $\boldsymbol{w}^T \Sigma \boldsymbol{w}$ represents the variance associated with the portfolios effective total coverage on investment;
6. $\boldsymbol{R}^T \boldsymbol{w}$ represents the expected total effective coverage from the insurance portfolio.

Note that one could also incorporate constraints related to minimum ICL_i or BCL_i amounts for a given product covering the i-th peril exposure as well as budget constraints.

Remark 18.26 *Since this application of MPT is nonstandard, there are a few aspects of the aforementioned problem to be considered:*

1. *One could factor into the MPT framework the practical reality that several insurance policies and CAT bonds in the basket (portfolio) are not necessarily continuously divisible, i.e., there may be fixed predefined ICL_i amounts for a given policy in the possible products one may purchase;*
2. *The discrete nature of some insurance contract coverages and bond coverages could be incorporated in the optimization routine adopted;*
3. *The assets of the insurance portfolio may not be highly liquid, especially for some specialised insurance products. Therefore opportunities for purchasing new insurance contracts may be limited and may occur in limited windows of time. In addition, insurance contracts of some CAT-bonds that have already been purchased may not be abandoned without a loss of a sunk cost. Incorporating this feature would require additional constraint parameterizations in the optimizations objective function and optimization solver.*

We conclude this section with some indications of some of the more specialized products available for consideration when constructing a basket of insurance products for a particular loss process exposed to multiple perils. In terms of specialty products, a large reinsurance company that offers a number of products in the space of OpRisk loss processes to a global market is Swiss Re. They have teams such as in the US the Excess and Surplus market Casualty group that specializes in "U.S-domiciled surplus lines wholesale brokers with primary, umbrella and follow-form excess capacity for difficult-to-place risks in the Excess and Surplus market". This group aims to seek coverage solutions for challenging risks not in the standard/admitted market. The types of coverage limits offered are quoted as being of the range: USD 10 million limits in umbrella and follow form excess; USD 5 million CGL limits for each occurrence; USD 5 million general aggregate limit; USD 5 million products/completed operations; and USD 5 million personal and advertising injury. There is also groups like the professional and management liability team in Swiss Re for example that provide bespoke products for "protection for organisations and their executives, as well as other professionals, against allegations of wrongdoing, mismanagement, negligence, and other related exposures". In addition, as discussed in Van den Brink (2002), there are some specialty products that are available for OpRisk insurance coverage offered by Swiss Re and known as the Financial Institutions OpRisk Insurance (FIORI) that covers OpRisk causes such as liability, fidelity, and unauthorized activity, technology risk, asset protection, and external fraud. It is noted in Chernobai *et al.* (2007) that the existence of such specialised products is limited in scope and market since the resulting premium one may be required to pay for such an insurance product can typically run into very significant costs, removing the actual gain from obtaining the insurance contract in terms of capital mitigation in the first place.

18.5 Optimal Insurance Purchase Strategies for OpRisk Insurance via Multiple Optimal Stopping Times

We begin this section with a brief overview of multiple optimal stopping time theory required for the development of optimal purchase strategies for OpRisk settings.

Assume an agent sequentially observes a process given by $\{W(t)\}_{t=1}^{T}$, for a fixed $T < +\infty$ and wants to choose $k < T$ of these observations in order to maximize (or minimize) the expected sum of these chosen observations. For $k = 1$, this problem is known in the literature as the house selling problem (see Sofronov 2013 for an updated literature review) since one of its interpretations is as follows. If the agent is willing to sell a house and assume that at most T bids will be observed, he wants to choose the optimal time τ such that the house will be sold for the highest possible value. The extension of this problem for $k > 1$ is known as the multiple house selling problem, where the agent wants to sell k identical houses.

Formally, the mathematical framework for such a problem consists of a filtered probability space $(\Omega, \mathcal{F}, \{\mathcal{F}_t\}_{t\geq 0}, \mathbb{P}\mathrm{r})$, where $\mathcal{F}_t = \sigma(W(t))$ is the sigma-algebra generated by $W(t)$. Within this framework, where we assume the flow of information is given only by the observed values of W, it is clear that any decision at time t should take into account only values of the process W up to time t. It is also required that two actions cannot take place at the same time, that is, we do not allow two stopping times to occur at the same discrete time instant. These assumptions are precisely stated in the following definition, but for further details on the theory of multiple optimal stopping rules we refer the reader to Nikolaev and Sofronov (2007) and Sofronov (2013).

Definition 18.33 (Multiple Stopping Rules) *A collection of integer-valued random variables $(\tau_1, \ldots, \tau_i)$ is called an i-multiple stopping rule if the following conditions hold:*

1. $\{\omega \in \Omega : \tau_1(\omega) = m_1, \ldots, \tau_j(\omega) = m_j\} \in \mathcal{F}_{m_j}$, $\forall m_j > m_{j-1} > \cdots > m_1 \geq 1$, $j = 1, \ldots, i$;
2. $1 \leq \tau_1 < \tau_2 < \ldots < \tau_i < +\infty$, *a.s.* ∎

Given the mathematical definition of a stopping rule, the notion of optimality of these rules can be made precise in Definitions 18.34, 18.35 and 18.36.

Definition 18.34 (Gain Function for a Multiple Stopping Rule) *For a given multiple stopping rule $\boldsymbol{\tau} = (\tau_1, \ldots, \tau_k)$, the gain function utilized in this chapter takes the following additive form:*

$$g(\boldsymbol{\tau}) = W(\tau_1) + \cdots + W(\tau_k).$$ ∎

Definition 18.35 (Value Function for Multiple Stopping Rule) *Let $\mathcal{S}_m$ be the class of multiple stopping rules $\boldsymbol{\tau} = (\tau_1, \ldots, \tau_k)$ such that $\tau_1 \geq m$ a.s. The function*

$$v_m = \sup_{\boldsymbol{\tau} \in \mathcal{S}_m} \mathbb{E}[g(\boldsymbol{\tau})],$$

is defined as the m-value of the game and, in particular, if $m = 1$ *then* v_1 *is the value of the game.* ■

Definition 18.36 (Optimal Multiple Stopping Rule) *A multiple stopping rule* $\boldsymbol{\tau}^* \in \mathcal{S}_m$ *is called an optimal multiple stopping rule in* $\mathcal{S}_m$ *if* $\mathbb{E}[W(\boldsymbol{\tau}^*)]$ *exists and* $\mathbb{E}[W(\boldsymbol{\tau}^*)] = v_m$. ■

The following result of Nikolaev and Sofronov (2007, theorem 3) provides the optimal multiple stopping rule that maximizes the expectation of the sum of the observations.

Theorem 18.7 *Let* $W(1), W(2), \ldots, W(T)$ *be a sequence of independent random variables with known distribution functions* $F_1, F_2, \ldots, F_T$*, and the gain function* $g(\boldsymbol{\tau}) = \sum_{j=1}^{k} W(\tau_j)$*. Let* $v^{L,l}$ *be the value of a game where the agent is allowed to stop* l *times* ($l \leqslant k$) *and there are* L ($L \leqslant T$) *steps remaining. If there exist* $\mathbb{E}[W(1)], \mathbb{E}[W(2)], \ldots, \mathbb{E}[W(T)]$*, then the value of the game is given by*

$$
\begin{aligned}
v^{1,1} &= \mathbb{E}[W(T)],\\
v^{L,1} &= \mathbb{E}\big[\max\{W(T-L+1), v^{L-1,1}\}\big], \quad 1 < L \le T,\\
v^{L,l+1} &= \mathbb{E}\big[\max\{v^{L-1,l} + W(T-L+1), v^{L-1,l+1}\}\big], \quad l+1 < L \le T,\\
v^{l,l} &= \mathbb{E}\left[v^{l-1,l-1} + W(T-l+1)\right].
\end{aligned}
$$

If we put

$$
\begin{aligned}
&\tau_1^* = \min\{m_1 : 1 \leqslant m_1 \leqslant T-k+1, W(m_1) \geqslant v^{T-m_1,k} - v^{T-m_1,k-1}\};\\
&\tau_i^* = \min\{m_i : \tau_{i-1}^* < m_i \leqslant T-k+i, W(m_i) \geqslant v^{T-m_i,k-i+1} - v^{T-m_i,k-i}\}, \quad i = 2, \ldots, k-1;\\
&\tau_k^* = \min\{m_k : \tau_{k-1}^* < m_k \leqslant T, W(m_k) \geqslant v^{T-m_k,1}\};
\end{aligned}
\tag{18.92}
$$

then $\boldsymbol{\tau}^* = (\tau_1^*, \ldots, \tau_k^*)$ *is the optimal multiple stopping rule.*

In the examples explored in this chapter, it will always be optimal to stop the process exactly k times, but this may not be true, for example, if some reward is given to the product holder for less than k years of claims of insurance. In the absence of such considerations, one may proceed with assuming always k years of claims will be made. In Theorem 18.7, we can see that the value function for $L > l$ is artificial and $v^{0,1}$, for example, has no interpretation (Figure 18.5). On the other hand, $v^{1,1}$ cannot be calculated using the general formula (it would depend on $v^{0,1}$). With one stop remaining and one step left, from the reasons given earlier, we are obliged to stop, and, therefore, there is no maximization step when calculating $v^{1,1}$, that is, $v^{1,1} = \mathbb{E}[W(T-1+1)]$. The same argument is valid for $l > 1$ and, in this case,

$$
v^{L,l} = \mathbb{E}\left[\max\{v^{L-1,l-1} + W(T-L+1),\ v^{L-1,l}\}\right], \ 1 \le l \le T
$$

and, if we have $l \le (T-1)$ steps left and also l stops, we must stop in all the steps remaining. So,

$$
v^{l,l} = \mathbb{E}\left[v^{l-1,l-1} + W(T-l+1)\right].
$$

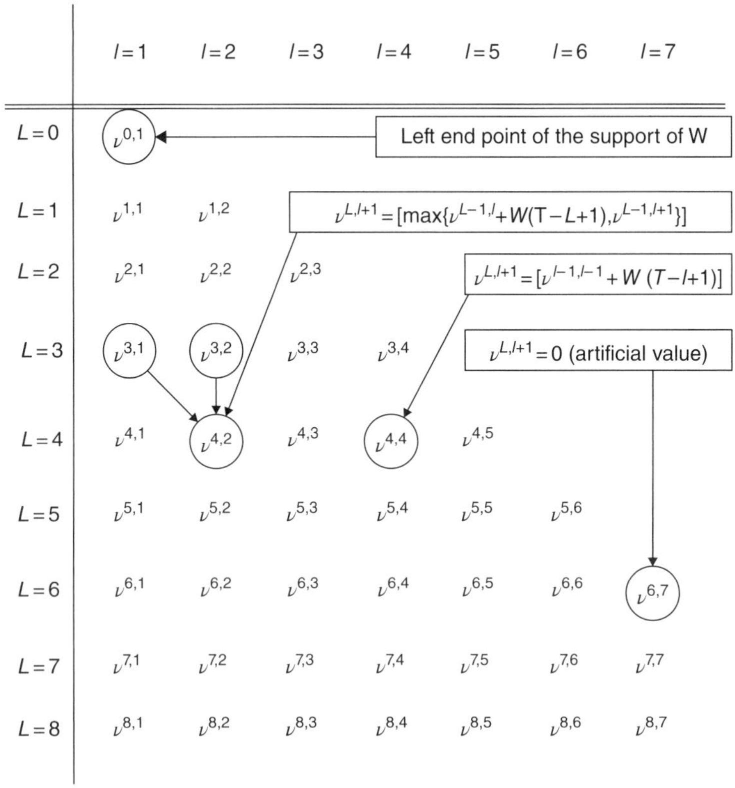

FIGURE 18.5 **Schematic representation of the value function iteration**

From Theorem 18.7 and the assumption of independence of the annual losses, one can see that to be able to calculate the optimal rule we only need to calculate (unconditional) expectations like $\mathbb{E}[W]$ and $\mathbb{E}[\max\{c_1 + W,\ c_2\}]$, for different values of c_1 and c_2. In addition, since $0 \leq v^{L-1,l} \leq v^{L-1,l+1}$, one actually only needs to calculate $\mathbb{E}[\max\{c_1 + W,\ c_2\}]$ for $0 \leq c_1 \leq c_2$.

Having presented the basic results required for developing an multiple optimal stopping time formulation, the insurance application may be developed. In this regard, the remaining section is based on the work developed in Targino *et al.* (2013) for OpRisk insurance purchase strategies. In that work an interesting general question is posed relating to how one may construct insurance products satisfying the axioms and definitions earlier while allowing a sufficiently general class of policies that may actually be suitable for a wider range of financial institutions and banks than those specialized products currently on offer.

More specifically, this section explores aspects of an insurance product that provides its owner several opportunities to decide which annual OpRisk loss(es) to insure. This product can be thought of as a way to decrease the cost paid by its owner to the insurance company in a similar way to what occurs with swing options in energy markets (see, e.g., Jaillet *et al.* (2004) and Carmona and Touzi (2008)): *instead of buying T yearly insurance policies over a period of T years, the buyer can negotiate with the insurance company a contract that covers only k of the T years (to be chosen by the owner).* This type of structured product will result in a reduction in the cost of insurance or partial insurance for OpRisk losses and this aspect is highlighted in Allen *et al.* (2009, p. 188), where they note that "even without considering the cost of major catastrophes, insurance coverage is very expensive". In addition, it may be interesting to explore such structures if the flexibility they provide results in an increased uptake of such products for OpRisk coverage, further reducing insurance premiums and resulting perhaps in greater competition in the market for these products.

A focus on three basic generic "building block" policies (see Definitions 18.37 to 18.39) that can be combined to create more complex types of protection. For these three basic policies, a "moderate-tailed" model for annual risks is developed that leads to closed-form usage strategies of the insurance product, answering the question: When is it optimal to ask the insurance company to cover the annual losses?

Suppose the company holds an insurance product that lasts for T years and grants the company the right to mitigate k of its T annual losses through utilization of its insurance claims. To clarify consider a given year $t \leq T$ where the company will incur N_t losses adding up to $Z_t = \sum_{n=1}^{N_t} X_n(t)$, assuming it has not yet utilized all its k insurance mitigations it then has the choice to make an insurance claim or not. If it utilizes the insurance claim in this year, the resulting annual loss will be denoted by $\widetilde{Z}_t$. Such a loss process model structure is standard in OpRisk and insurance and is typically referred to as the loss distributional approach (LDA).

In this context, the company's aim is to choose k distinct years out of the T in order to minimize its expected operational loss over the time interval $[0, T]$, where it is worth noting that if $Z > \widetilde{Z}$ that is, if the insurance is actually mitigating the company's losses, all its k rights should be exercised. The question that then must be addressed is what is the optimal decision rule, that is define the multiple optimal stopping times for making the k sets of insurance claims.

Since these closed-form results rely upon the stochastic loss model considered, we also provide a general framework applicable for any loss process. Therefore, in Section 18.5.4, we discuss a method based on series expansions of unknown densities to calculate the optimal rules when the combination of insurance policy and severity density does not lead to analytical results.

18.5.1 EXAMPLES OF BASIC INSURANCE POLICIES

In the remainder of this chapter, if a process $\{Z_t\}_{t=1}^{T}$ is a sequence of i.i.d. random variables, the time index will be dropped and one can denote a generic r.v. from this process by Z. For the rest of the chapter, $\mathbb{I}_A$ will denote the indicator function on the event A, that is, $\mathbb{I}_A = 1$ if A is valid and zero otherwise.

Definition 18.37 (Individual Loss Policy (ILP)) *This policy applies a constant haircut to the loss process in year t in which individual losses experience a TCL as specified by*

$$\widetilde{Z} = \sum_{n=1}^{N} \max\left(X_n - TCL,\ 0\right). \qquad \blacksquare$$

Definition 18.38 (Accumulated Loss Policy (ALP)) *The ALP provides a specified maximum compensation on losses experienced over a year. If this maximum compensation is denoted by ALP, then the annual insured process is defined as*

$$\widetilde{Z} = \left(\sum_{n=1}^{N} X_n - ALP \right) \mathbb{I}_{\left\{ \sum_{n=1}^{N} X_n > ALP \right\}}.$$ ■

Definition 18.39 (Postattachment Point Coverage (PAP)) *The attachment point is the insured's retention point after which the insurer starts compensating the company for accumulated losses at point PAP*

$$\widetilde{Z} = \sum_{n=1}^{N} X_n \times \mathbb{I}_{\left\{ \sum_{k=1}^{n} X_k \leq PAP \right\}}.$$ ■

In the remainder of this section, an example LDA model that leads to closed-form solutions for the annual loss distribution under each insurance policy is considered. In particular, a Poisson frequency distribution and an inverese Gaussian severity model are considered. Following the nomenclature in Franzetti (2011, table 3.3), the inverse Gaussian distribution possess a "moderate tail", which makes it a reasonable model for OpRisk losses for many risk process types and is often used in practice. This family of distributions also has the advantage of being closed under convolution, and this characteristic is essential if closed-form solutions for the multiple optimal stopping problem are to be obtained.

For the model considered in this section, it will be useful to recall the following properties of the family of inverse Gaussian severity models. The inverse Gaussian distribution and its relationship with the generalized inverse Gaussian distribution will be of direct us in the remainder of this chapter, especially when evaluating the value function in closed form for the optimal stopping rules; see additional details in Folks and Chhikara (1978), Jørgensen (1982), and Tweedie (1957, section 2).

Consider a sequence of i.i.d. inverse Gaussian random variables $X_1, \ldots, X_n$ with distribution parameters $\mu, \lambda > 0$, i.e.,

$$f_X(x;\ \mu, \lambda) = \left(\frac{\lambda}{2\pi} \right)^{1/2} x^{-3/2} \exp\left\{ \frac{-\lambda(x-\mu)^2}{2\mu^2 x} \right\}, \quad x > 0.$$

In addition, denote G to be a generalized inverse Gaussian (GIG) r.v. with parameters $\alpha, \beta > 0$, $p \in \mathbb{R}$, that is,

$$f_G(x;\ \alpha, \beta, p) = \frac{(\alpha/\beta)^{p/2}}{2K_p(\sqrt{\alpha\beta})} x^{p-1} \exp\left\{ -\frac{1}{2}(\alpha x + \beta/x) \right\}, \quad x > 0,$$

where K_p is a modified Bessel function of the third kind (sometimes called modified Bessel function of the second kind), defined as

$$K_p(z) = \frac{1}{2} \int_0^{+\infty} u^{p-1} e^{-z(u+1/u)/2} du.$$

Lemma 18.1 (Closure Under Convolution for Inverse Gaussian Losses) *The inverse Gaussian family of random variables is closed under convolution and the distribution of its sum is given by*

$$S_n := \sum_{l=1}^{n} X_l \sim IG(n\mu,\ n^2\lambda). \tag{18.93}$$

Lemma 18.2 (Embedding an Inverse Gaussian in a Generalized Inverse Gaussian) *Any inverse Gaussian random variable can be represented as a generalized inverse Gaussian, and for the particular case of Lemma 18.1 the relationship is*

$$f_{S_n}(x;\ n\mu, n^2\lambda) \equiv f_G(x;\ \lambda/\mu^2, n^2\lambda, -1/2). \tag{18.94}$$

Lemma 18.3 *Modified Bessel functions of the third kind are symmetric around zero in the parameter p. In particular, when $p = 1/2$,*

$$\frac{K_{1/2}\left(\frac{n\lambda}{\mu}\right)}{K_{-1/2}\left(\frac{n\lambda}{\mu}\right)} = 1. \tag{18.95}$$

Lemma 18.4 *The density of an inverse Gaussian r.v. has the following property (which clearly holds for any power of x, with the proper adjustment in the last parameter of the GIG in the RHS):*

$$xf_G(x;\ \lambda/\mu^2, n^2\lambda, -1/2) \equiv n\mu\, f_G(x;\ \lambda/\mu^2, n^2\lambda, 1/2). \tag{18.96}$$

The symmetry in Lemma 18.3 can be seen through the following characterization of modified Bessel functions of the third kind

$$K_p(x) := \int_0^{+\infty} \exp\{-x\cosh(t)\}\cosh(pt)dt,$$

(see Watson (1922), p. 181) and the fact that $\cosh(-p) = \cosh(p)$. The last result, Lemma 18.4, follows from Lemma 18.3 and a simple comparison of the densities.

Having defined a few basic examples of insurance policies as well as a loss process model based on a Poisson-inverse-Gaussian LDA model structure, one now has to develop appropriate objective functions for the definition of the multiple optimal stopping rule strategies to maximize (minimize).

18.5.2 OBJECTIVE FUNCTIONS FOR RATIONAL AND BOUNDEDLY RATIONAL INSUREES

One may consider two possible general populations for the potential insuree. The first group are those that are perfectly rational, meaning that they will always act in an optimal fashion when given the chance and, more importantly, are capable (i.e., have the resources) of figuring out what is the optimal behavior. In this case, we will consider a global objective function to be optimized.

The second group represent boundedly rational insurees who act suboptimally. This group represents firms who are incapable or lack the resources/knowledge to understand how to act optimally when determining their optimal behaviors/actions and will be captured by local behaviors. Hence, these two populations will be encoded in two objective functions: one that is optimal (globally) and one that represents a suboptimal (local strategy) the boundedly rational population would likely adopt. These behaviors can be made precise through the following exercising strategies, for the first and second groups, respectively.

1. **Global Risk Transfer Strategy**. Minimizes the (expected) total loss over the period $[0, T]$;
2. **Local Risk Transfer Strategy**. Minimizes the (expected) sum of the losses at the insurance times (i.e., stopping times).

These two different groups can be understood as, for example, large corporations, with employees dedicated to fully understand the mathematical nuances of this kind of contract and small companies, with limited access to information. The group with "bounded rationality" may decide (heuristically, without the usage of any mathematical tool) to follow the so-called local risk transfer strategy, which will produce smaller gain in the period $[0, T]$.

As studied in Targino *et al.* (2013), these two different objective functions can lead to completely different exercising strategies, and therefore an insurance company who sells such a contract should be aware of these different behaviours.

For the first loss function, the formal objective is to minimize

$$\sum_{\substack{t=1\\ t\notin\{\tau_1,\ldots,\tau_k\}}}^{T} Z(t) + \sum_{j=1}^{k} \widetilde{Z}(\tau_j) = \sum_{t=1}^{T} Z(t) - \sum_{\substack{t=1\\ t\in\{\tau_1,\ldots,\tau_k\}}}^{T} \left\{Z(t) - \widetilde{Z}(t)\right\}.$$

Since $\sum_{t=1}^{T} Z(t)$ does not depend on the choice of $\tau_1, \ldots, \tau_k$, this is, in fact, equivalent to maximize

$$\sum_{j=1}^{k} W(\tau_j) = \sum_{j=1}^{k} \left\{Z(\tau_j) - \widetilde{Z}(\tau_j)\right\},$$

where the process W is defined as $W(t) = Z(t) - \widetilde{Z}(t)$.

For the second objective function, the company aims to minimize the total loss not over period $[0, T]$ but instead only at times at which the decisions are taken to apply insurance and therefore claim against losses in the given year,

$$\sum_{j=1}^{k} \widetilde{Z}(\tau_j)$$

and, in this case, the process W should be viewed as $W(t) = -\widetilde{Z}(t)$.

Remark 18.27 *As noted by Targino et al. (2013), if the agent is trying to maximize the first loss function (using $W = Z - \widetilde{Z}$), then W is non-negative stochastic process, and only one kind of expectation is required to be calculated, since if $c_1 = c_2 = 0$, then $\mathbb{E}[\max\{c_1 + W, c_2\}] = \mathbb{E}[W]$.*

Remark 18.28 *If the agent is trying to minimize the expected gain of the sum of* $\widetilde{Z}(t)$ *random variables (instead of maximizing it), one can rewrite the problem as follows. Define a process* $\widetilde{W}(t) = -\widetilde{Z}(t)$ *and note that* $\min \mathbb{E}[\sum_{j=1}^{k} \widetilde{Z}(\tau_j)] = \max \mathbb{E}[\sum_{j=1}^{k} W(\tau_j)]$. *Therefore, the optimal stopping times that maximize the expected sum of the process* W *are the same that minimize the expected sum of the process* $\widetilde{Z}$.

Having defined the two different objective functions, one may now proceed to develop the multiple optimal stopping rules in closed form for the three insurance policies and the Poisson-inverse-Gaussian LDA model. That is, under the Poisson-inverse-Gaussian LDA model, where $X_n \sim InverseGaussian(\lambda, \mu)$ and $N \sim Poisson(\lambda_N)$, the optimal times (years) to exercise or make claims on the insurance policy for the accumulated loss policy (ALP) and the post attachment point coverage policy (PAP) can be calculated analytically regardless of where the global or local gain (objective) functions are considered. For the solution to the individual loss policy (ILP), when using the gain function as the local objective function given by the sum of the losses at the stopping times (insurance claim years) it was proposed by Targino *et al.* (2013) to model the losses after the insurance policy is applied. In the following sections, we consider the ALP and PAP policies to illustrate the concepts developed.

18.5.3 CLOSED-FORM MULTIPLE OPTIMAL STOPPING RULES FOR MULTIPLE INSURANCE PURCHASE DECISIONS

Since we assume the annual losses $Z_1, \ldots, Z_T$ are identically distributed, we will denote by Z a r.v. such that $Z \stackrel{d}{=} Z_1$, where $\widetilde{Z}$ is the insured process; $S_n = \sum_{k=1}^{n} X_k$ is the partial sum up to the n-th loss; $p_m = \Pr[N = m]$ is the probability of observing m losses in 1 year. The gain W will be defined as either $-\widetilde{Z}$, when the objective is to minimize the loss at the times the company uses the insurance policy (local optimality), or $Z - \widetilde{Z}$, in case the function to be minimized is the total loss over the time horizon $[0, T]$, that is, (global optimality).

18.5.3.1 Accumulated Loss Policy (ALP).

In the case of the ALP insurance model given in Definition 18.38, one can model the severity of the losses before applying the insurance policy. Conditional upon the fact that $\sum_{n=1}^{m} X_n > ALP$, the annual loss after the application of the insurance policy will be $\sum_{n=1}^{m} X_n - ALP$. With this in mind, one can then calculate the distribution of the insured process, $\widetilde{Z}$, and also of the random variable $Z - \widetilde{Z}$.

Targino *et al.* (2013) prove that under the local risk transfer case one may obtain closed-form results for the insured loss process, as detailed in Proposition 18.3.

Proposition 18.3 (Local Risk Transfer Case) *The distribution and density of the insured process are given, respectively, by*

$$F_{\widetilde{Z}}(z) = \sum_{m=1}^{+\infty} F_{IG}(z + ALP;\ m\mu, m^2\lambda)C_m + C_0, \tag{18.97}$$

$$f_{\widetilde{Z}}(z) = \sum_{m=1}^{+\infty} \left\{ f_{IG}(z + ALP;\ m\mu, m^2\lambda)C_m \right\} \mathbb{I}_{\{z>0\}} + C_0 \mathbb{I}_{\{z=0\}}, \tag{18.98}$$

where the constants $C_0, C_1, C_2, \ldots$ *are defined as*

$$C_0 := \sum_{m=1}^{+\infty} F_{IG}(ALP;\; m\mu, m^2\lambda)p_m + p_0$$

$$C_m := \overline{F}_{IG}(ALP;\; m\mu, m^2\lambda)p_m, \quad m = 1, 2, \ldots$$

Targino *et al.* (2013) then demonstated that after calculating the distribution of $\widetilde{Z}$ one can also calculate expectations of the form $\mathbb{E}\left[\max\left\{c_1 + W, c_2\right\}\right]$ with respect to the loss process Z, and therefore one can consequently obtain the multiple optimal stopping rules under the accumulated loss policy via direct application of Theorem 18.7.

Theorem 18.8 (Local Risk Transfer Case) *Using the notation of Theorem 18.7 and defining* $W(t) = -\widetilde{Z}(t)$, *for* $t = 1, \ldots, T$, *the multiple optimal stopping rule is given by the set of equations in (18.92), where*

$$v^{1,1} = -\sum_{m=1}^{+\infty} C_m\left(m\mu\overline{F}_{GIG}(ALP;\; \lambda/\mu^2, m^2\lambda, 1/2) - ALP \times \overline{F}_{GIG}(ALP;\; \lambda/\mu^2, m^2\lambda, -1/2)\right),$$

$$\begin{aligned}
v^{L,1} = -\sum_{m=1}^{+\infty} C_m\Bigg[\Bigg(&m\mu\Big(F_{GIG}(v^{L-1,1} + ALP;\; \lambda/\mu^2, m^2\lambda, 1/2)\\
&- F_{GIG}(ALP;\; \lambda/\mu^2, m^2\lambda, 1/2)\Big)\\
&+ ALP\Big(F_{GIG}(v^{L-1,1} + ALP;\; \lambda/\mu^2, m^2\lambda, -1/2) - F_{GIG}(ALP;\; \lambda/\mu^2, m^2\lambda, -1/2)\Big)\Bigg)\\
&+ v^{L-1,1}\overline{F}_{GIG}(v^{L-1,1} + ALP;\; \lambda/\mu^2, m^2\lambda, -1/2)\Bigg],
\end{aligned}$$

$$\begin{aligned}
v^{L,l+1} = -\sum_{m=1}^{+\infty} C_m\Bigg[\Bigg(&m\mu\Big(F_{GIG}(v^{L-1,l+1} - v^{L-1,l} + ALP;\; \lambda/\mu^2, m^2\lambda, 1/2)\\
&- F_{GIG}(ALP;\; \lambda/\mu^2, m^2\lambda, 1/2)\Big)\\
&- (v^{L-1,l} - ALP)\Big(F_{GIG}(v^{L-1,l+1} - v^{L-1,l} + ALP;\; \lambda/\mu^2, m^2\lambda, -1/2)\\
&- F_{GIG}(ALP;\; \lambda/\mu^2, m^2\lambda, -1/2)\Big)\Bigg)\\
&+ v^{L-1,l+1}\overline{F}_{GIG}(v^{L-1,l+1} - v^{L-1,l} + ALP;\; \lambda/\mu^2, m^2\lambda, -1/2)\Bigg] - v^{L-1,l}C_0,
\end{aligned}$$

$$\begin{aligned}
v^{l,l} = v^{l-1,l-1} - \sum_{m=1}^{+\infty} C_m\Big(&m\mu\overline{F}_{GIG}(ALP;\; \lambda/\mu^2, m^2\lambda, 1/2)\\
&- ALP\overline{F}_{GIG}(ALP;\; \lambda/\mu^2, m^2\lambda, -1/2)\Big).
\end{aligned}$$

Now, considering the global objective, it was then observed by Targino *et al.* (2013) that if we assume the company wants to minimize its total loss over the period $[0, T]$ the gain achieved through the accumulated loss policy (ALP) is given by

$$
\begin{aligned}
W &= Z - \widetilde{Z} \\
&= \sum_{n=1}^{N} X_n - \left(\sum_{n=1}^{N} X_n - ALP \right) \mathbb{I}_{\{\sum_{n=1}^{N} X_n > ALP\}} \\
&= ALP\mathbb{I}_{\{\sum_{n=1}^{N} X_n > ALP\}} + \left(\sum_{n=1}^{N} X_n \right) \mathbb{I}_{\{\sum_{n=1}^{N} X_n > ALP\}} \\
&= \min \left\{ ALP, \sum_{n=1}^{N} X_n \right\}.
\end{aligned}
$$

For notational convenience, one may denote by $W_m = \min \left\{ ALP, \sum_{n=1}^{m} X_n \right\}$ the annual gain conditional on the fact that m losses were observed.

Targino *et al.* (2013) were able to again prove that under the global risk transfer case one may obtain closed-form results for the insured loss process, as detailed in Proposition 18.4.

Proposition 18.4 (Global Risk Transfer Case: ALP) *The distribution and density of the gain process are given, respectively, by*

$$
F_W(w) = \mathbb{I}_{\{w \geq ALP\}} + F_{S_m}(w)\mathbb{I}_{\{w < ALP\}}, \tag{18.99}
$$

$$
f_W(w) = \sum_{m=1}^{N} \left\{ \left(\overline{F}_{S_m}(ALP)\mathbb{I}_{\{w=ALP\}} + f_{S_m}(w)\mathbb{I}_{\{0<w<ALP\}} \right) p_m \right\} + p_0\mathbb{I}_{\{w=0\}}. \tag{18.100}
$$

As in the local case, again for the global case Targino *et al.* (2013) demonstrated that after calculating the distribution of the gain, W, we can calculate expectations w.r.t. it and, therefore, the multiple optimal stopping rule under the ALP is then obtained via direct application of Theorem 18.7.

Theorem 18.9 (Global Risk Transfer Case: ALP) *Defining $W(t) = Z(t) - \widetilde{Z}(t)$, for $t = 1, \ldots, T$, the multiple optimal stopping rule is given by (18.92), where*

$$
\begin{aligned}
v^{1,1} &= \sum_{m=1}^{+\infty} p_m \left\{ \overline{F}_{S_m}(ALP)ALP + m\mu F_{GIG}(ALP;\ \lambda/\mu^2, m^2\lambda, 1/2) \right\}, \\
v^{L,1} &= \sum_{m=1}^{+\infty} p_m \Big\{ \overline{F}_{S_m}(ALP) \max\{ALP,\ v^{L-1,1}\} + m\mu\big(F_{GIG}(ALP;\ \lambda/\mu^2, m^2\lambda, 1/2) \\
&\quad - F_{GIG}(v^{L-1,1};\ \lambda/\mu^2, m^2\lambda, 1/2)\big) + v^{L-1,1} F_{S_m}(\min\{v^{L-1,1},\ ALP\}) \Big\} + p_0 v^{L-1,1}, \\
v^{L,l+1} &= \sum_{m=1}^{+\infty} p_m \Big\{ \overline{F}_{S_m}(ALP) \max\{v^{L-1,l} + ALP,\ v^{L-1,l+1}\} \\
&\quad + v^{L-1,l}(F_{S_m}(ALP) - F_{S_m}(v^{L-1,l+1} - v^{L-1,l})) + m\mu\big(F_{GIG}(ALP;\ \lambda/\mu^2, m^2\lambda, 1/2)
\end{aligned}
$$

$$
\begin{aligned}
&- F_{GIG}(v^{L-1,l+1} - v^{L-1,l};\ \lambda/\mu^2, m^2\lambda, 1/2)) \\
&+ v^{L-1,l+1} F_{S_m}(\min\{v^{L-1,l+1} - v^{L-1,l},\ ALP\})\Big\} + p_0 v^{L-1,l+1},
\end{aligned}
$$

$$
v^{l,l} = \sum_{m=1}^{+\infty} p_m \Big\{ \overline{F}_{S_m}(ALP)ALP + m\mu F_{GIG}(ALP;\ \lambda/\mu^2, m^2\lambda, 1/2) \Big\}.
$$

18.5.3.2 Postattachment Point Coverage (PAP).

As in the ALP case, for the post attachment point coverage policy one can also model the intraannual losses before applying the insurance policy, but to calculate the distribution and density of both the insured process $\widetilde{Z}$ and the gain process $Z - \widetilde{Z}$ it will be necessary to consider an additional conditioning step. Since the insured loss is given by Definition 18.39, it will be convenient to define the first time the aggregated loss process exceeds the threshold *PAP*, which can be formally defined as the following stopping time

$$
M_m^* = \inf \left\{ n \le m : \sum_{k=1}^{n} X_k > PAP \right\} \tag{18.101}
$$

and, $M_m^* = +\infty$, if $\sum_{k=1}^{n} X_k \le PAP$.

Targino *et al.* (2013) prove that under the local risk transfer case one may obtain closed-form results for the insured loss process, as detailed in Proposition 18.5.

Proposition 18.5 (Local Risk Transfer Case: PAP) *The distribution and the density of the insured process are given, respectively, by*

$$
F_{\widetilde{Z}}(z) = \sum_{m=1}^{+\infty} \left\{ \sum_{m^*=1}^{m} \left(F_{IG}(z;\ m^*\mu, (m^*)^2\lambda) D_{m^*,m} \right) + F_{IG}(z;\ m\mu,\ m^2\lambda) D_m \right\} + p_0
$$

$$
f_{\widetilde{Z}}(z) = \sum_{m=1}^{+\infty} \left\{ \sum_{m^*=1}^{m} \left(f_{IG}(z;\ m^*\mu, (m^*)^2\lambda) D_{m^*,m} \right) + f_{IG}(z;\ m\mu,\ m^2\lambda) D_m \right\} \mathbb{I}_{z>0} + p_0 \mathbb{I}_{z=0},
$$

where

$$
\begin{aligned}
D_{m^*,m} &= \Pr\left[M_m^* = m^*\right] p_m, \quad m = 1, 2, \ldots,\ m^* = 1, \ldots, m, \\
D_m &= F_{IG}(PAP,\ m\mu,\ m^2\lambda) p_m, \quad m = 1, 2, \ldots
\end{aligned}
$$

Targino *et al.* (2013) then demonstated that after calculating the distribution of the insured annual loss one can also calculate expectations required to obtain the multiple optimal stopping rules under the PAP Policy via direct application of Theorem 18.7.

Theorem 18.10 (Local Risk Transfer Case: PAP) *Defining* $W(t) = -\widetilde{Z}(t)$, *for* $t = 1, \ldots, T$ *the multiple stopping rule is given by (18.92), where*

$$
v^{1,1} = -\sum_{m=1}^{+\infty} \sum_{m^*=1}^{m} m^* \mu D_{m^*,m} + \sum_{m=1}^{+\infty} m\mu D_m,
$$

$$
\begin{aligned}
v^{L,1} = \sum_{m=1}^{+\infty} \sum_{m^*=1}^{m} &\Big\{ m\mu F_{GIG}(v^{L-1,1};\ \lambda/\mu^2, (m^*)^2\lambda, 1/2) \\
&+ v^{L-1,1} \overline{F}_{IG}(v^{L-1,1};\ m^*\mu, (m^*)^2\lambda) \Big\} D_{m^*,m}
\end{aligned}
$$

$$
\begin{aligned}
&+\sum_{m=1}^{+\infty}\Big(m\mu F_{GIG}(v^{L-1,1};\ \lambda/\mu^2, m^2\lambda, 1/2) + v^{L-1,1}\overline{F}_{IG}(v^{L-1,1};\ m\mu, m^2\lambda)\Big)D_m,\\
v^{l,l+1} &= \sum_{m=1}^{+\infty}\sum_{m^*=1}^{m}\Bigg\{v^{L-1,l}F_{IG}(v^{L-1,l+1} - v^{L-1,l};\ m^*\mu, (m^*)^2\lambda)\\
&+ m\mu F_{GIG}(v^{L-1,l+1} - v^{L-1,l};\ \lambda/\mu^2, (m^*)^2\lambda, 1/2)\\
&+ v^{L-1,l+1}\overline{F}_{IG}(v^{L-1,l+1} - v^{L-1,l};\ m^*\mu, (m^*)^2\lambda)\Bigg\}D_{m^*,m}\\
&+\sum_{m=1}^{+\infty}\Big(v^{L-1,l}F_{IG}(v^{L-1,l+1} - v^{L-1,l};\ m\mu, m^2\lambda)\\
&+ m\mu F_{GIG}(v^{L-1,l+1} - v^{L-1,l};\ \lambda/\mu^2, m^2\lambda, 1/2)\\
&+ v^{L-1,l+1}\overline{F}_{IG}(v^{L-1,l+1} - v^{L-1,l};\ m\mu, m^2\lambda)\Big)D_m + v^{L-1,l}p_0,\\
v^{l,l} &= v^{l-1,l-1} - \sum_{m=1}^{+\infty}\sum_{m^*=1}^{m} m^*\mu D_{m^*,m} + \sum_{m=1}^{+\infty} m\mu D_m.
\end{aligned}
$$

In the case of the global objective, Targino *et al.* (2013) showed that the gain process in the PAP–total loss case takes the following form:

$$
\begin{aligned}
W &= Z - \widetilde{Z}\\
&= \sum_{n=1}^{N} X_n - \sum_{n=1}^{N} X_n \times \mathbb{I}_{\{\sum_{k=1}^{n} X_k \le PAP\}}\\
&= \sum_{n=1}^{N} X_n \mathbb{I}_{\{\sum_{k=1}^{n} X_k > PAP\}}.
\end{aligned}
$$

Targino *et al.* (2013) prove that the resulting annual loss distribution and density are then given by Proposition 18.6.

Proposition 18.6 (Global Risk Transfer Case: PAP) *The distribution and the density functions of the insured process are given, respectively, by*

$$
\begin{aligned}
F_W(w) &= \sum_{m=1}^{+\infty}\left\{\sum_{m^*=1}^{m}\left(\mathbb{P}\mathrm{r}\left[\sum_{n=m^*}^{m} X_n \le w\right]\mathbb{P}\mathrm{r}\left[M_m^* = m^*\right]p_m\right)\right\}\\
&+\sum_{m=1}^{+\infty}\left\{\mathbb{P}\mathrm{r}\left[\sum_{k=1}^{m} X_k < PAP\right]p_m\right\} + p_0,\\
f_W(w) &= \left(\sum_{m=1}^{+\infty}\left\{\sum_{m^*=1}^{m}\left(f_{IG}(w;\ (m-m^*+1)\mu, (m-m^*+1)^2\lambda)\mathbb{P}\mathrm{r}\left[M_m^* = m^*\right]p_m\right)\right\}\right)\mathbb{I}_{\{w>0\}}\\
&+ \mathbb{P}\mathrm{r}[W=0]\mathbb{I}_{\{w=0\}},
\end{aligned}
$$

where $\mathbb{P}\mathrm{r}[W=0] = \sum_{m=1}^{+\infty}\left\{\mathbb{P}\mathrm{r}\left[\sum_{k=1}^{m} X_k < PAP\right]p_m\right\} + p_0.$

Targino *et al.* (2013) then demonstrated that after calculating the distribution of the insured annual loss one can also calculate expectations required to obtain the multiple optimal stopping rules under the global objective for the PAP policy via direct application of Theorem 18.7.

Theorem 18.11 (Global Risk Transfer Case: PAP) *Defining* $W(t) = Z(t) - \widetilde{Z}(t)$, *for* $t = 1, \ldots, T$ *the multiple stopping rule is given by (18.92), where*

$$
\begin{aligned}
v^{1,1} &= \sum_{m=1}^{+\infty} \sum_{m^*=1}^{m} \mathbb{P}\mathrm{r}\left[M_m^* = m^*\right] p_m (m - m^* + 1)\mu, \\
v^{L,1} &= \sum_{m=1}^{+\infty} \sum_{m^*=1}^{m} \mathbb{P}\mathrm{r}\left[M_m^* = m^*\right] p_m \Big\{ \overline{F}_{GIG}(v^{L-1,1};\ \lambda/\mu^2, (m-m^*+1)^2\lambda, 1/2)(m-m^*+1)\mu, \\
&\quad + v^{L-1,1} F_{GIG}(v^{L-1,1};\ \lambda/\mu^2, (m-m^*+1)^2\lambda, -1/2) \Big\} + v^{L-1,1}\mathbb{P}\mathrm{r}[W=0] \\
v^{l,l+1} &= \sum_{m=1}^{+\infty} \sum_{m^*=1}^{m} \mathbb{P}\mathrm{r}\left[M_m^* = m^*\right] p_m \Big\{ v^{L-1,l} \overline{F}_{GIG}\left(v^{L-1,l+1} - v^{L-1,l};\ \frac{\lambda}{\mu^2}, (m-m^*+1)^2\lambda, -\frac{1}{2}\right) \\
&\quad + \overline{F}_{GIG}(v^{L-1,l+1} - v^{L-1,l};\ \lambda/\mu^2, (m-m^*+1)^2\lambda, 1/2)(m-m^*+1)\mu \\
&\quad + v^{L-1,l+1} F_{GIG}(v^{L-1,l+1} - v^{L-1,l};\ \lambda/\mu^2, (m-m^*+1)^2\lambda, -1/2) \Big\} \\
&\quad + v^{L-1,l+1}\mathbb{P}\mathrm{r}[W=0], \\
v^{l,l} &= v^{l-1,l-1} + \sum_{m=1}^{+\infty} \sum_{m^*=1}^{m} \mathbb{P}\mathrm{r}\left[M_m^* = m^*\right] p_m (m - m^* + 1)\mu.
\end{aligned}
$$

In cases where the LDA model does not admit the required distribution and density for the compound process, or the finite closed-form representations for the value function one can still obtain approximations that are closed form as detailed in the following section. This builds on the results developed in Chapter 17.

18.5.4 ASKI-POLYNOMIAL ORTHOGONAL SERIES APPROXIMATIONS

When one is working with models in which analytical solutions cannot be found, one possible alternative is to create a series expansion of the density of the insured process $\widetilde{Z}$ such that all the expectations necessary in Theorem 18.7 can be analytically calculated. If one can assume that the first n moments of the distribution of the insured process $\widetilde{Z}$ are known, then one can proceed with the following approximations based on a Gamma density basis approximation.

If the n-th first moments of the insured process $\widetilde{Z}$ can be calculated (either algebraically or numerically) and the support of the insured random variable is $[0, +\infty)$, one can use a series expansion of the density of $\widetilde{Z}$ in a gamma basis. For notational convenience, define a new random variable $U = b\widetilde{Z}$, where $b = \mathbb{E}[\widetilde{Z}]/\mathrm{Var}[\widetilde{Z}]$ and set $a = \mathbb{E}[\widetilde{Z}]^2/\mathrm{Var}[\widetilde{Z}]$. Denoting by f_U the density of U the idea is to write f_U as

$$f_U(u) = g(u;a)\left[A_0 L_0^{(a)}(u) + A_1 L_1^{(a)}(u) + A_2 L_2^{(a)}(u) + \dots\right]. \tag{18.102}$$

Since $supp(U) = supp(\widetilde{Z}) = [0, +\infty)$, we assume the kernel $g(\cdot\,;a)$ also has positive support. If $g(u;a) = u^{a-1}e^{-u}/\Gamma(a)$ that is, a Gamma kernel with *shape* $= a$ and *scale* $= 1$, then the orthonormal polynomial basis (with respect to this kernel) is given by the Laguerre polynomials defined as

$$L_n^{(a)}(u) = (-1)^n u^{1-a} e^{-u} \frac{d^n}{du^n}\left(u^{n+a-1}e^{-u}\right). \tag{18.103}$$

Remark 18.29 *Note that the definition of the Laguerre polynomials on Equation (18.103) is slightly different from the usual one, that is, the one based on Rodrigues' formula*

$$\widetilde{L}_n^{(a)} = \frac{u^{-a}e^x}{n!}\frac{d^n}{du^n}\left(e^{-x}x^{n+a}\right),$$

but it is easy to check that

$$L_n^{(a)}(u) = n!(-1)^n \widetilde{L}_n^{(a-1)}.$$

From the orthogonality condition (see, e.g., Jackson 1941, p. 184),

$$\int_0^{+\infty} \frac{x^{a-1}e^{-x}}{\Gamma(a)} L_n^{(a)}(x) L_m^{(a)}(x)dx = \begin{cases} \dfrac{n!\Gamma(a+n)}{\Gamma(a)}, & n = m, \\ 0, \quad n \neq m \end{cases}$$

and using the fact that f_U can be written in the form of Equation (18.102) we find that

$$A_n = \frac{\Gamma(a)}{n!\Gamma(a+n)} \int_0^{+\infty} f_U(x) L_n^{(a)}(x)dx. \tag{18.104}$$

Then, using the characterization of A_n in (18.104) and the fact that $\mathbb{E}[U] = \mathrm{Var}[U] = a$ we can see that

$$A_0 = \int_0^{+\infty} f_U(x) L_0^{(a)}(x)dx = \int_0^{+\infty} f_U(x)dx = 1,$$

$$A_1 = \int_1^{+\infty} f_U(x) L_1^{(a)}(x)dx = \int_0^{+\infty} f_U(x)(z-a)dx = 0,$$

$$A_2 = \int_1^{+\infty} f_U(x) L_2^{(a)}(x)dx = \int_0^{+\infty} f_U(x)(z^2 - 2(a+1)z + (a+1)a)dx = 0.$$

Similar but lengthier calculations show that for $\mu_n = \mathbb{E}\left[(U - \mathbb{E}[U])^n\right]$, $n = 3, 4$,

$$A_3 = \frac{\Gamma(a)}{3!\Gamma(a+3)}(\mu_3 - 2a), \tag{18.105}$$

$$A_4 = \frac{\Gamma(a)}{4!\Gamma(a+4)}(\mu_4 - 12\mu_3 - 3a^2 + 18a). \tag{18.106}$$

Therefore, matching the first four moments, the density of the original random variable $\widetilde{Z}$ can be approximated as

$$f_{\widetilde{Z}}(z) = bf_U(u) \approx b\frac{u^{a-1}e^{-u}}{\Gamma(a)}\left[1 + A_3L_3^{(a)}(u) + A_4L_4^{(a)}(u)\right],$$

where $u = bz$, A_3 and A_4 are given, respectively, by (18.105) and (18.106) and the Laguerre polynomials can be found in Table 18.1. For additional details on the Gamma expansion, we refer the reader to Bowers and Newton (1966).

Remark 18.30 (Ensuring Positivity) *Since this expansion does not ensure positivity of the density at all points (it can be negative for particular choices of skewness and kurtosis), we will adopt the approach discussed in Jondeau and Rockinger (2001) for the Gauss-Hermite Gramm-Charlier case modified to the Gamma-Laguerre setting. To find the region on the (μ_3, μ_4)-plane where $f_U(u)$ is positive for all u, we will first find the region where $f_U(u) = 0$, that is,*

$$\frac{u^{a-1}e^{-u}}{\Gamma(a)}\left(1 + A_3L_3^{(a)}(u) + A_4L_4^{(a)}(u)\right) = 0. \tag{18.107}$$

For a fixed value u, we now want to find the set (μ_3, μ_4) as a function of u such that (18.107) remains zero for small variations on u. This set is given by (μ_3, μ_4) such that

$$\frac{d}{du}\left[\frac{u^{a-1}e^{-u}}{\Gamma(a)}\left(1 + A_3L_3^{(a)}(u) + A_4L_4^{(a)}(u)\right)\right] = 0. \tag{18.108}$$

We can then rewrite Equations (18.107) and (18.107) as the following system of algebraic equations

$$\begin{cases} \mu_3B_1(u) + \mu_4B_2(u) + B_3(u) = 0, \\ \mu_3B_1'(u) + \mu_4B_2'(u) + B_3'(u) = 0, \end{cases}$$

TABLE 18.1 **The first five Laguerre polynomials**

$L_0^{(a)}(u) = 1$
$L_1^{(a)}(u) = u - a$
$L_2^{(a)}(u) = u^2 - 2(a+1)u + (a+1)a$
$L_3^{(a)}(u) = u^3 - 3(a+2)u^2 + 3(a+2)(a+1)u - (a+2)(a+1)a$
$L_4^{(a)}(u) = u^4 - 4(a+3)u^3 + 6(a+3)(a+2)u^2 - 4(a+3)(a+2)(a+1)u + (a+3)(a+2)(a+1)a.$

where

$$
\begin{aligned}
B_1(u) &= \frac{u^{a-1}e^{-u}}{\Gamma(a)}\left(\frac{\Gamma(a)}{3!\Gamma(a+3)}L_3^{(a)}(u) - 12\frac{\Gamma(a)}{4!\Gamma(a+4)}L_4^{(a)}(u)\right);\\
B_2(u) &= \frac{u^{a-1}e^{-u}}{\Gamma(a)}\frac{\Gamma(a)}{4!\Gamma(a+4)}L_4^{(a)}(u);\\
B_3(u) &= \frac{u^{a-1}e^{-u}}{\Gamma(a)}\left(1-2a\frac{\Gamma(a)}{3!\Gamma(a+3)}L_3^{(a)}(u) + \left(-3a^2+18a\right)\frac{\Gamma(a)}{4!\Gamma(a+4)}L_4^{(a)}(u)\right);\\
B_1'(u) &= \left((a-1)u^{-1}-1\right)B_1(u)\\
&\quad + \frac{u^{a-1}e^{-u}}{\Gamma(a)}\left(\frac{\Gamma(a)}{3!\Gamma(a+3)}\frac{dL_3^{(a)}}{du}(u) - 12\frac{\Gamma(a)}{4!\Gamma(a+4)}\frac{dL_4^{(a)}}{du}(u)\right);\\
B_2'(u) &= \left((a-1)u^{-1}-1\right)B_2(u) + \frac{u^{a-1}e^{-u}}{\Gamma(a)}\left(\frac{\Gamma(a)}{4!\Gamma(a+4)}\frac{dL_4^{(a)}}{du}(u)\right);\\
B_3'(u) &= \left((a-1)u^{-1}-1\right)B_3(u)\\
&\quad + \frac{u^{a-1}e^{-u}}{\Gamma(a)}\left(-2a\frac{\Gamma(a)}{3!\Gamma(a+3)}\frac{dL_3^{(a)}}{du}(u) + \left(-3a^2+18a\right)\frac{\Gamma(a)}{4!\Gamma(a+4)}\frac{dL_4^{(a)}}{du}(u)\right);\\
\frac{dL_3^{(a)}}{du}(u) &= 3u^2 - 6(a+2)u + 3(a+2)(a+1);\\
\frac{dL_4^{(a)}}{du}(u) &= 4u^3 - 12(a+3)u^2 + 12(a+3)(a+2)u - 4(a+3)(a+2)(a+1).
\end{aligned}
$$

Therefore, one can solve this system to show that the curve where the approximation will stay positive for all u is given by

$$
\begin{cases}
\mu_4(u) = \left(\frac{B_1'B_3}{B_1} - B_3'\right)\left(B_2' - \frac{B_1'B_2}{B_1}\right)^{-1} \\
\mu_3(u) = -\frac{1}{B_1}\left(\mu_4(u)B_2 + B_3\right)
\end{cases}
\quad \textit{for } u \in [0, +\infty). \tag{18.109}
$$

As an illustration, Figure 18.6 presents (on the left) the histogram of the loss process $Z = \sum_{n=1}^{N} X_n$ for $X \sim LogNormal(\mu = 1, \sigma = 0.8)$ and $N \sim Poisson(\lambda_N = 2)$ and with a solid grey line the Gamma approximation using the first four moments of Z. On the right it is presented the graph of the region where the density is positive for all values of u, given by Equation (18.109). The grey area was calculated numerically, for all combinations in a fine grid on the plane (μ_3, μ_4) it was tested if the density became negative in some point z. Grey points indicate the density is strictly positive. The solid point indicates the third and fourth moments in the LogNormal example and since it lies inside the positivity are we can ensure this approximation is strictly positive for all values of z.

If the the third and fourth moments of the chosen model lied outside the permitted area, one could chose $\hat{\mu}_3$ and $\hat{\mu}_4$ as the estimates that minimize some constrained optimization problem, for instance, the maximum likelihood estimator using

$$f_U(u;\mu_3,\mu_4) = \frac{u^{a-1}e^{-u}}{\Gamma(a)}\left[1 + A_3L_3^{(a)}(u) + A_4L_4^{(a)}(u)\right]$$

as the likelihood. The constrained region is clearly given by a segment of the curve in Equation (18.109) and the endpoints can be found using a root-search method checking for which values of u (the red curve in Figure 18.6) touch the grey area.

18.5.4.1 Series Approximations and the Value Function.

Given the approximation of f_U, and consequently of $f_{\widetilde{Z}}$, one can easily calculate the optimal multiple stopping rule, since $\mathbb{E}[\widetilde{Z}]$ is assumed to be known and $\mathbb{E}[\min\{c_1 + \widetilde{Z},\ c_2\}]$ can be calculated as follows.

Lemma 18.5 *If $G \sim Gamma(a, 1)$, i.e, $f_G(x) = \frac{x^{a-1}e^{-x}}{\Gamma(a)}$, then, similarly to Lemma 18.4, the following property holds*

$$xf_G(x;\ a, 1) \equiv af_G(x;\ a + 1, 1). \tag{18.110}$$

Using this notation, we can rewrite the approximation of $\widetilde{Z}$ as

$$\begin{aligned} f_{\widetilde{Z}}(z) \approx f_G(bz; a, 1)A_1^* &+ f_G(bz; a+1, 1)A_2^* + f_G(bz; a+2, 1)A_3^* \\ &+ f_G(bz; a+3, 1)A_4^* + f_G(bz; a+4, 1)A_5^*, \end{aligned}$$

where

$$\begin{aligned}
A_1^* &= \left(1 - \frac{\Gamma(a+3)}{\Gamma(a)}A_3 + \frac{\Gamma(a+4)}{\Gamma(a)}A_4\right) b, \\
A_2^* &= \left(3\frac{\Gamma(a+3)}{\Gamma(a)}A_3 - 4\frac{\Gamma(a+4)}{\Gamma(a)}A_4\right) b, \\
A_3^* &= \left(-3\frac{\Gamma(a+3)}{\Gamma(a)}A_3 + 6\frac{\Gamma(a+4)}{\Gamma(a)}A_4\right) b, \\
A_4^* &= \left(\frac{\Gamma(a+3)}{\Gamma(a)}A_3 - 4\frac{\Gamma(a+4)}{\Gamma(a)}A_4\right) b, \\
A_5^* &= \left(\frac{\Gamma(a+4)}{\Gamma(a)}A_4\right) b.
\end{aligned}$$

Then, one can calculate the other main ingredient of Theorem 18.7 given by

$$\begin{aligned}
\mathbb{E}[\min\{c_1 + \widetilde{Z},\ c_2\}] &= \int_0^{+\infty} \min\{c_1 + z,\ c_2\} f_{\widetilde{Z}}(z)dz \\
&= \int_0^{+\infty} \left((c_1 + z)\mathbb{I}_{\{c_1+z<c_2\}} + c_2\mathbb{I}_{\{c_1+z\geq c_2\}}\right) f_{\widetilde{Z}}(z)dz \\
&= \int_0^{c_2-c_1} zf_{\widetilde{Z}}(z)dz + c_1 \int_0^{c_2-c_1} f_{\widetilde{Z}}(z)dz + c_2 \int_{c_2-c_1}^{+\infty} f_{\widetilde{Z}}(z)dz
\end{aligned}$$

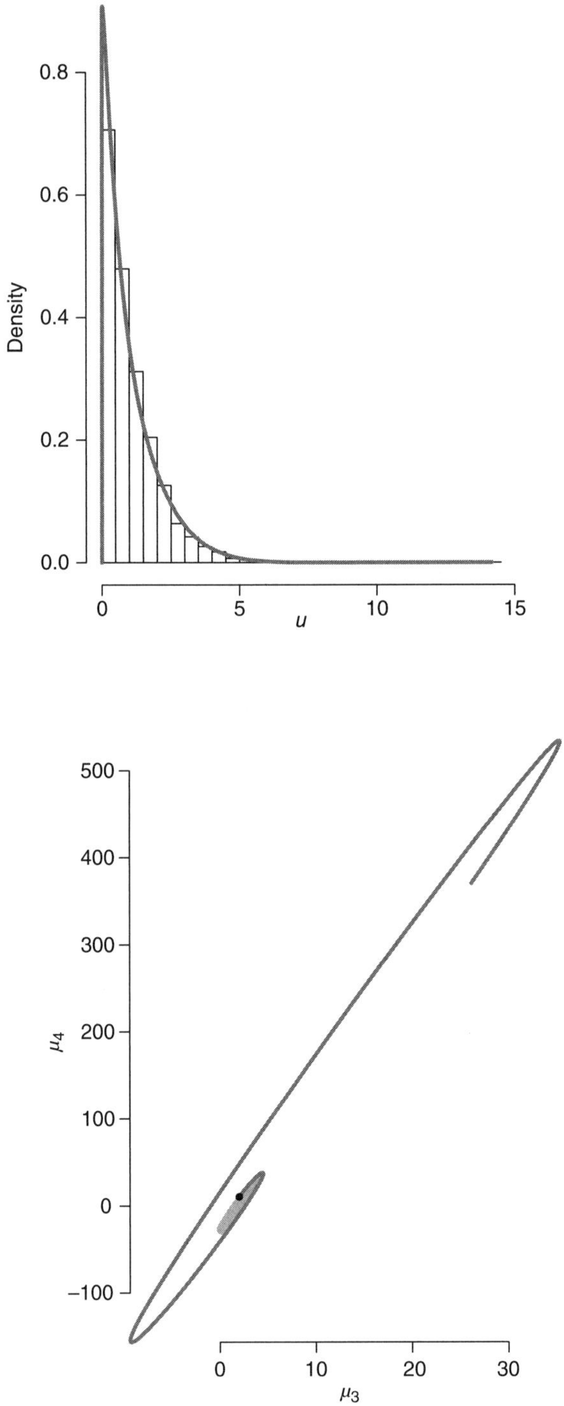

FIGURE 18.6 Approximation using the first four moments for a LogNormal example

$$
\begin{aligned}
&= a\sum_{k=1}^{5} F_G(b(c_2 - c_1); a + k, 1)A_k^* + c_1 \sum_{k=1}^{5} F_G(b(c_2 - c_1); a - 1 + k, 1)A_k^* \\
&\quad + c_2 \sum_{k=1}^{5} \overline{F}_G(b(c_2 - c_1); a - 1 + k, 1)A_k^*.
\end{aligned}
$$

Targino *et al.* (2013) developed a wide range of examples to illustrate the optimal stopping rules in different scenarios; the interested reader is referred to this paper for illustrations and code.

APPENDIX A

Miscellaneous Definitions and List of Distributions

A.1 Indicator Function

The often used indicator symbol $\mathbb{I}_{\{.\}}$ is defined as

$$\mathbb{I}_{\{\cdot\}} = \begin{cases} 1, & \text{if condition in } \{\cdot\} \text{ is true,} \\ 0, & \text{otherwise.} \end{cases} \tag{A.1}$$

In addition on occasion we will also utilise $\mathbb{I}[\cdot]$.

A.2 Gamma Function

The standard gamma function $\Gamma(\alpha)$ is defined as

$$\Gamma(\alpha) = \int_0^\infty t^{\alpha-1} e^{-t} dt, \quad \alpha > 0. \tag{A.2}$$

A.3 Discrete Distributions

A.3.1 POISSON DISTRIBUTION

A Poisson distribution function is denoted as $Poisson(\lambda)$. The random variable N has a Poisson distribution, denoted $N \sim Poisson(\lambda)$, if its probability mass function is

$$p(k) = \mathbb{P}\mathrm{r}[N = k] = \frac{\lambda^k}{k!} e^{-\lambda}, \quad \lambda > 0 \tag{A.3}$$

Fundamental Aspects of Operational Risk and Insurance Analytics: A Handbook of Operational Risk,
First Edition. Marcelo G. Cruz, Gareth W. Peters, and Pavel V. Shevchenko.

for all $k \in \{0, 1, 2, \ldots\}$. Expectation, variance, and variational coefficient of a random variable $N \sim Poisson(\lambda)$ are

$$\mathbb{E}[N] = \lambda, \; \mathbb{V}\mathrm{ar}[N] = \lambda, \quad \mathrm{Vco}[N] = \frac{1}{\sqrt{\lambda}}. \tag{A.4}$$

A.3.2 BINOMIAL DISTRIBUTION

The Binomial distribution function is denoted as $Binomial(n, p)$. The random variable N has a Binomial distribution, denoted $N \sim Binomial(n, p)$, if its probability mass function is

$$p(k) = \mathbb{P}\mathrm{r}[N = k] = \binom{n}{k} p^k (1-p)^{n-k}, \quad p \in (0, 1), \; n \in 1, 2, \ldots \tag{A.5}$$

for all $k \in \{0, 1, 2, \ldots, n\}$. Expectation, variance, and variational coefficient of a random variable $N \sim Binomial(n, p)$ are

$$\mathbb{E}[N] = np, \; \mathbb{V}\mathrm{ar}[N] = np(1-p), \quad \mathrm{Vco}[N] = \sqrt{\frac{1-p}{np}}. \tag{A.6}$$

Remark A.1 *N is the number of successes in n independent trials, where p is the probability of a success in each trial.*

A.3.3 NEGATIVE BINOMIAL DISTRIBUTION

A Negative Binomial distribution function is denoted as $NegBinomial(r, p)$. The random variable N has a Negative Binomial distribution, denoted $N \sim NegBinomial(r, p)$, if its probability mass function is

$$p(k) = \mathbb{P}\mathrm{r}[N = k] = \binom{r+k-1}{k} p^r (1-p)^k, \quad p \in (0, 1), \; r \in (0, \infty) \tag{A.7}$$

for all $k \in \{0, 1, 2, \ldots\}$. Here, the generalized Binomial coefficient is

$$\binom{r+k-1}{k} = \frac{\Gamma(k+r)}{k!\Gamma(r)}, \tag{A.8}$$

where $\Gamma(r)$ is the Gamma function.

Expectation, variance, and variational coefficient of a random variable $N \sim NegBinomial(r, p)$ are

$$\mathbb{E}[N] = \frac{r(1-p)}{p}, \quad \mathbb{V}\mathrm{ar}[N] = \frac{r(1-p)}{p^2}, \quad \mathrm{Vco}[N] = \frac{1}{\sqrt{r(1-p)}}. \tag{A.9}$$

Remark A.2 *If r is a positive integer, N is the number of failures in a sequence of independent trials until r successes, where p is the probability of a success in each trial.*

A.3.4 DOUBLY STOCHASTIC POISSON PROCESS (COX PROCESS)

Let $(\Omega, F, \mathbb{P})$ be a probability space with information structure (filtration) given by $F = \{\mathcal{F}_t, t \in [0, T]\}$. Let N_t be a point process adapted to F. Let λ_t be a non-negative process adapted to F such that

$$\int_0^t \lambda_s ds < \infty, \quad a.s. \tag{A.10}$$

If for all $0 \leq t_1 \leq t_2$ and $u \in \mathbb{R}$ one has

$$\mathbb{E}\left[e^{\mathrm{i}u\left(N_{t_2} - N_{t_1}\right)} \middle| \mathcal{F}_{t_2} \right] = \exp\left\{ \left(e^{\mathrm{i}u} - 1\right) \int_{t_1}^{t_2} \lambda_s ds \right\}, \tag{A.11}$$

then N_t is called a $\mathcal{F}_t$-doubly stochastic Poisson process with intensity λ_t where $\mathcal{F}_t = \sigma\{\lambda_s; s \leq t\}$. One has the following probabilities

$$\Pr\left[N_{t_2} - N_{t_1} = k \middle| \lambda_s; t_1 \leq s \leq t_2\right] = \frac{\exp\left(-\int_{t_1}^{t_2} \lambda_s ds\right) \left[\int_{t_1}^{t_2} \lambda_s ds\right]^k}{k!} \tag{A.12}$$

and in addition one has for τ_k the length of time interval between the $(k-1)$-th and the k-th point the following distribution

$$\Pr\left[\tau_k > t \middle| \lambda_s; t_k \leq s \leq t_k + t\right] = \exp\left(-\int_{t_k}^{t_k+t} \lambda_s ds\right). \tag{A.13}$$

A.4 Continuous Distributions

A.4.1 UNIFORM DISTRIBUTION

A uniform distribution function is denoted as *Uniform*(a, b). The random variable X has a uniform distribution, denoted $X \sim$ *Uniform*(a, b), if its probability density function is

$$f(x) = \frac{1}{b-a}, \quad a < b \tag{A.14}$$

for $x \in [a, b]$. Expectation, variance, and variational coefficient of a random variable $X \sim$ *Uniform*(a, b) are

$$\mathbb{E}[X] = \frac{a+b}{2}, \quad \mathbb{V}\mathrm{ar}[X] = \frac{(b-a)^2}{12}, \quad \mathrm{Vco}[X] = \frac{b-a}{\sqrt{3}(a+b)}. \tag{A.15}$$

A.4.2 NORMAL (GAUSSIAN) DISTRIBUTION

A Normal (Gaussian) distribution function is denoted as *Normal*(μ, σ^2). The random variable X has a Normal distribution, denoted $X \sim$ *Normal*(μ, σ^2), if its probability density function is

$$f(x) = \frac{1}{\sqrt{2\pi\sigma^2}} \exp\left(-\frac{(x-\mu)^2}{2\sigma^2}\right), \quad \sigma^2 > 0, \ \mu \in \mathbb{R} \tag{A.16}$$

for all $x \in \mathbb{R}$. The standard Normal distribution corresponds to $Normal(0, 1)$ and is denoted as $\Phi(\cdot)$. Expectation, variance, and variational coefficient of a random variable $X \sim Normal(\mu, \sigma^2)$ are

$$\mathbb{E}[X] = \mu, \quad \mathbb{V}\text{ar}[X] = \sigma^2, \quad \text{Vco}[X] = \sigma/\mu. \tag{A.17}$$

A.4.3 INVERSE GAUSSIAN DISTRIBUTION

An Inverse Gaussian distribution function is denoted as $InverseGaussian(\mu, \gamma)$. The random variable X has an Inverse Gaussian distribution, denoted $X \sim InverseGaussian(\mu, \gamma)$, if its probability density function is

$$f(x) = \left(\frac{\gamma}{2\pi x^3}\right)^{\frac{1}{2}} \exp\left(-\frac{\gamma(x-\mu)^2}{2\mu^2 x}\right), \quad x > 0, \tag{A.18}$$

where parameters $\mu > 0$ and $\gamma > 0$. The corresponding distribution function is

$$F(x) = \Phi\left(\sqrt{\frac{\gamma}{x}}\left(\frac{x}{\mu} - 1\right)\right) + \exp\left(\frac{2\gamma}{\mu}\right)\Phi\left(-\sqrt{\frac{\gamma}{x}}\left(\frac{x}{\mu} + 1\right)\right), \tag{A.19}$$

where $\Phi(\cdot)$ is the standard Normal distribution. Expectation and variance of $X \sim InverseGaussian(\mu, \lambda)$ are

$$\mathbb{E}[X] = \mu, \quad \mathbb{V}\text{ar}[X] = \frac{\mu^3}{\gamma}.$$

If $X_1, \ldots, X_n$ are independent and $X_i \sim InverseGaussian(\mu w_i, \gamma w_i^2)$, then

$$S_n = \sum_{i=1}^{n} X_i \sim InverseGaussian\left(\mu \sum_{i=1}^{n} w_i, \gamma\left(\sum_{i=1}^{n} w_i\right)^2\right). \tag{A.20}$$

A.4.4 LOGNORMAL DISTRIBUTION

A LogNormal distribution function is denoted as $LogNormal(\mu, \sigma^2)$. The random variable X has a LogNormal distribution, denoted $X \sim LogNormal(\mu, \sigma^2)$, if its probability density function is

$$f(x) = \frac{1}{x\sqrt{2\pi\sigma^2}} \exp\left(-\frac{(\ln x - \mu)^2}{2\sigma^2}\right), \quad \sigma^2 > 0, \ \mu \in \mathbb{R} \tag{A.21}$$

for $x > 0$. Expectation, variance, and variational coefficient of a random variable $X \sim LogNormal(\mu, \sigma^2)$ are

$$\mathbb{E}[X] = e^{\mu + \frac{1}{2}\sigma^2}, \quad \mathbb{V}\text{ar}[X] = e^{2\mu + \sigma^2}(e^{\sigma^2} - 1), \quad \text{Vco}[X] = \sqrt{e^{\sigma^2} - 1}. \tag{A.22}$$

A.4.5 STUDENT'S *t* DISTRIBUTION

A *t* distribution function is denoted as $\mathcal{T}(\nu, \mu, \sigma^2)$. The random variable X has a *t* distribution, denoted $X \sim \mathcal{T}(\nu, \mu, \sigma^2)$, if its probability density function is

$$f(x) = \frac{\Gamma((\nu+1)/2)}{\Gamma(\nu/2)} \frac{1}{\sqrt{\nu\pi}} \left(1 + \frac{(x-\mu)^2}{\nu\sigma^2}\right)^{-(\nu+1)/2} \tag{A.23}$$

for $\sigma^2 > 0$, $\mu \in \mathbb{R}$, $\nu = 1, 2, \ldots$, and all $x \in \mathbb{R}$. Expectation, variance, and variational coefficient of a random variable $X \sim \mathcal{T}(\nu, \mu, \sigma^2)$ are

$$\begin{aligned} \mathbb{E}[X] &= \mu \quad \text{if } \nu > 1, \\ \mathbb{V}\text{ar}[X] &= \sigma^2 \frac{\nu}{\nu-2} \quad \text{if } \nu > 2, \\ \text{Vco}[X] &= \frac{\sigma}{\mu}\sqrt{\frac{\nu}{\nu-2}} \quad \text{if } \nu > 2. \end{aligned} \tag{A.24}$$

A.4.6 GAMMA DISTRIBUTION

A Gamma distribution function is denoted as *Gamma*(α, β). The random variable X has a gamma distribution, denoted as $X \sim$ *Gamma*(α, β), if its probability density function is

$$f(x) = \frac{x^{\alpha-1}}{\Gamma(\alpha)\beta^\alpha} \exp(-x/\beta), \quad \alpha > 0,\ \beta > 0 \tag{A.25}$$

for $x > 0$. Expectation, variance, and variational coefficient of a random variable $X \sim$ *Gamma*(α, β) are

$$\mathbb{E}[X] = \alpha\beta, \quad \mathbb{V}\text{ar}[X] = \alpha\beta^2, \quad \text{Vco}[X] = 1/\sqrt{\alpha}. \tag{A.26}$$

A.4.7 WEIBULL DISTRIBUTION

A Weibull distribution function is denoted as *Weibull*(α, β). The random variable X has a Weibull distribution, denoted as $X \sim$ *Weibull*(α, β), if its probability density function is

$$f(x) = \frac{\alpha}{\beta^\alpha} x^{\alpha-1} \exp(-(x/\beta)^\alpha), \quad \alpha > 0,\ \beta > 0 \tag{A.27}$$

for $x > 0$. The corresponding distribution function is

$$F(x) = 1 - \exp\left(-(x/\beta)^\alpha\right), \quad \alpha > 0,\ \beta > 0. \tag{A.28}$$

Expectation and variance of a random variable $X \sim$ *Weibull*(α, β) are

$$\mathbb{E}[X] = \beta\Gamma(1 + 1/\alpha), \quad \mathbb{V}\text{ar}[X] = \beta^2\left(\Gamma(1 + 2/\alpha) - (\Gamma(1 + 1/\alpha))^2\right).$$

A.4.8 INVERSE CHI-SQUARED DISTRIBUTION

An Inverse Chi-squared distribution is denoted as $InvChiSq(\nu, \beta)$. The random variable X has an Inverse Chi-squared distribution, denoted as $X \sim InvChiSq(\nu, \beta)$, if its probability density function is

$$f(x) = \frac{(x/\beta)^{-1-\nu/2}}{\beta\Gamma(\nu/2)2^{\nu/2}} \exp\left(-\frac{\beta}{2x}\right) \tag{A.29}$$

for $x > 0$ and parameters $\nu > 0$ and $\beta > 0$. Expectation and variance of $X \sim InvChiSq(\nu, \beta)$ are

$$\begin{aligned} \mathbb{E}[X] &= \frac{\beta}{\nu - 2}, \quad \text{for} \quad \nu > 2. \\ \mathbb{V}\text{ar}[X] &= \frac{2\beta^2}{(\nu-2)^2(\nu-4)} \quad \text{for} \quad \nu > 4 \end{aligned}$$

A.4.9 PARETO DISTRIBUTION (ONE-PARAMETER)

A one-parameter Pareto distribution function is denoted as $Pareto(\xi, x_0)$. The random variable X has a Pareto distribution, denoted as $X \sim Pareto\ (\xi, x_0)$, if its distribution function is

$$F(x) = 1 - \left(\frac{x}{x_0}\right)^{-\xi}, \quad x \geq x_0, \tag{A.30}$$

where $x_0 > 0$ and $\xi > 0$. The support starts at x_0, which is typically known and not considered as a parameter. Therefore, the distribution is referred to as a single-parameter Pareto. The corresponding probability density function is

$$f(x) = \frac{\xi}{x_0}\left(\frac{x}{x_0}\right)^{-\xi-1}. \tag{A.31}$$

Expectation, variance, and variational coefficient of $X \sim Pareto(\xi, x_0)$ are

$$\begin{aligned} \mathbb{E}[X] &= x_0\frac{\xi}{\xi-1} \quad \text{if } \xi > 1, \\ \mathbb{V}\text{ar}[X^2] &= x_0^2\frac{\xi}{(\xi-1)^2(\xi-2)} \quad \text{if } \xi > 2, \\ \text{Vco}[X] &= \frac{1}{\sqrt{\xi(\xi-2)}} \quad \text{if } \xi > 2. \end{aligned}$$

A.4.10 PARETO DISTRIBUTION (TWO-PARAMETER)

A two-parameter Pareto distribution function is denoted as $Pareto_2(\alpha, \beta)$. The random variable X has a Pareto distribution, denoted as $X \sim Pareto_2(\alpha, \beta)$, if its distribution function is

$$F(x) = 1 - \left(1 + \frac{x}{\beta}\right)^{-\alpha}, \quad x \geq 0, \tag{A.32}$$

where $\alpha > 0$ and $\beta > 0$. The corresponding probability density function is

$$f(x) = \frac{\alpha\beta^\alpha}{(x+\beta)^{\alpha+1}}. \tag{A.33}$$

The moments of a random variable $X \sim Pareto_2(\alpha, \beta)$ are

$$\mathbb{E}[X^k] = \frac{\beta^k k!}{\prod_{i=1}^k (\alpha - i)}, \qquad \alpha > k.$$

A.4.11 GENERALIZED PARETO DISTRIBUTION

A GPD distribution function is denoted as $GPD(\xi, \beta)$. The random variable X has a GPD distribution, denoted as $X \sim GPD(\xi, \beta)$, if its distribution function is

$$H_{\xi,\beta}(x) = \begin{cases} 1 - (1 + \xi x/\beta)^{-1/\xi}, & \xi \neq 0, \\ 1 - \exp(-x/\beta), & \xi = 0, \end{cases} \tag{A.34}$$

where $x \geq 0$ when $\xi \geq 0$ and $0 \leq x \leq -\beta/\xi$ when $\xi < 0$. The corresponding probability density function is

$$h(x) = \begin{cases} \frac{1}{\beta}(1 + \xi x/\beta)^{-\frac{1}{\xi}-1}, & \xi \neq 0, \\ \frac{1}{\beta}\exp(-x/\beta), & \xi = 0. \end{cases} \tag{A.35}$$

Expectation, variance, and variational coefficient of $X \sim GPD(\xi, \beta)$, $\xi \geq 0$, are

$$\begin{aligned} \mathbb{E}[X^n] &= \frac{\beta^n n!}{\prod_{k=1}^n (1 - k\xi)}, \quad \xi < \frac{1}{n}; \ \mathbb{E}[X] = \frac{\beta}{1-\xi}, \quad \xi < 1; \\ \mathbb{V}\text{ar}[X^2] &= \frac{\beta^2}{(1-\xi)^2(1-2\xi)}, \quad \text{Vco}[X] = \frac{1}{\sqrt{1-2\xi}}, \quad \xi < \frac{1}{2}. \end{aligned} \tag{A.36}$$

A.4.12 BETA DISTRIBUTION

A Beta distribution function is denoted as $Beta(\alpha, \beta)$. The random variable X has a Beta distribution, denoted as $X \sim Beta(\alpha, \beta)$, if its probability density function is

$$f(x) = \frac{\Gamma(\alpha+\beta)}{\Gamma(\alpha)\Gamma(\beta)} x^{\alpha-1}(1-x)^{\beta-1}, \quad 0 \leq x \leq 1, \tag{A.37}$$

for $\alpha > 0$ and $\beta > 0$. Expectation, variance, and variational coefficient of a random variable $X \sim Beta(\alpha, \beta)$ are

$$\mathbb{E}[X] = \frac{\alpha}{\alpha+\beta}, \quad \mathbb{V}\text{ar}[X] = \frac{\alpha\beta}{(\alpha+\beta)^2(1+\alpha+\beta)}, \quad \text{Vco}[X] = \sqrt{\frac{\beta}{\alpha(1+\alpha+\beta)}}.$$

A.4.13 GENERALIZED INVERSE GAUSSIAN DISTRIBUTION

A GIG distribution function is denoted as $GIG(\omega, \phi, \nu)$. The random variable X has a GIG distribution, denoted as $X \sim GIG(\omega, \phi, \nu)$, if its probability density function is

$$f(x) = \frac{(\omega/\phi)^{(\nu+1)/2}}{2K_{\nu+1}(2\sqrt{\omega\phi})} x^{\nu} e^{-x\omega - x^{-1}\phi}, \quad x > 0, \tag{A.38}$$

where $\phi > 0, \omega \geq 0$ if $\nu < -1$; $\phi > 0, \omega > 0$ if $\nu = -1$; $\phi \geq 0, \omega > 0$ if $\nu > -1$; and

$$K_{\nu+1}(z) = \frac{1}{2} \int_0^{\infty} u^{\nu} e^{-z(u+1/u)/2} du.$$

$K_{\nu}(z)$ is called a modified Bessel function of the third kind (see, e.g., Abramowitz and Stegun 1965, p. 375).

The moments of a random variable $X \sim GIG(\omega, \phi, \nu)$ are not available in a closed form through elementary functions but can be expressed in terms of Bessel functions:

$$\mathbb{E}[X^{\alpha}] = \left(\frac{\phi}{\omega}\right)^{\alpha/2} \frac{K_{\nu+1+\alpha}(2\sqrt{\omega\phi})}{K_{\nu+1}(2\sqrt{\omega\phi})}, \quad \alpha \geq 1, \ \phi > 0, \ \omega > 0.$$

Often, using notation $R_{\nu}(z) = K_{\nu+1}(z)/K_{\nu}(z)$, it is written as

$$\mathbb{E}[X^{\alpha}] = \left(\frac{\phi}{\omega}\right)^{\alpha/2} \prod_{k=1}^{\alpha} R_{\nu+k}(2\sqrt{\omega\phi}), \quad \alpha = 1, 2, \ldots$$

The mode is easily calculated from $\frac{\partial}{\partial x} x^{\nu} e^{-(\omega x + \phi/x)} = 0$ as

$$\mathrm{mode}(X) = \frac{1}{2\omega}(\nu + \sqrt{\nu^2 + 4\omega\phi}),$$

which differs only slightly from the expected value for large ν, that is,

$$\mathrm{mode}(X) \to \mathbb{E}[X] \quad \text{for} \quad \nu \to \infty.$$

A.4.14 *d*-VARIATE NORMAL DISTRIBUTION

A d-variate Normal distribution function is denoted as $Normal(\boldsymbol{\mu}, \boldsymbol{\Sigma})$, where $\boldsymbol{\mu} = (\mu_1, \ldots \mu_d)^T \in \mathbb{R}^d$ and $\boldsymbol{\Sigma}$ is a positive definite matrix ($d \times d$). The corresponding probability density function is

$$f(\boldsymbol{x}) = \frac{1}{(2\pi)^{d/2}\sqrt{\det \boldsymbol{\Sigma}}} \exp\left(-\frac{1}{2}(\boldsymbol{x} - \boldsymbol{\mu})^T \boldsymbol{\Sigma}^{-1} (\boldsymbol{x} - \boldsymbol{\mu})\right), \quad \boldsymbol{x} \in \mathbb{R}^d, \tag{A.39}$$

where $\boldsymbol{\Sigma}^{-1}$ is the inverse of the matrix $\boldsymbol{\Sigma}$. Expectations and covariances of a random vector $\boldsymbol{X} = (X_1, \ldots, X_d)^T \sim Normal(\boldsymbol{\mu}, \boldsymbol{\Sigma})$ are

$$\mathbb{E}[X_i] = \mu_i, \quad \mathrm{Cov}[X_i, X_j] = \Sigma_{i,j}, \quad i, j = 1, \ldots, d. \tag{A.40}$$

A.4.15 *d*-VARIATE *t*-DISTRIBUTION

A d-variate t-distribution function with ν degrees of freedom is denoted as $\mathcal{T}_d(\nu, \boldsymbol{\mu}, \boldsymbol{\Sigma})$, where $\nu > 0$, $\boldsymbol{\mu} = (\mu_1, \ldots, \mu_d)^T \in \mathbb{R}^d$ is a location vector and $\boldsymbol{\Sigma}$ is a positive definite matrix $(d \times d)$. The corresponding probability density function is

$$f(\boldsymbol{x}) = \frac{\Gamma\left(\frac{\nu+d}{2}\right)}{(\nu\pi)^{d/2}\Gamma\left(\frac{\nu}{2}\right)\sqrt{\det \boldsymbol{\Sigma}}}\left(1 + \frac{(\boldsymbol{x}-\boldsymbol{\mu})^T\boldsymbol{\Sigma}^{-1}(\boldsymbol{x}-\boldsymbol{\mu})}{\nu}\right)^{-\frac{\nu+d}{2}}, \tag{A.41}$$

where $\boldsymbol{x} \in \mathbb{R}^d$ and $\boldsymbol{\Sigma}^{-1}$ is the inverse of the matrix $\boldsymbol{\Sigma}$. Expectations and covariances of a random vector $\boldsymbol{X} = (X_1, \ldots, X_d)^T \sim \mathit{Normal}(\boldsymbol{\mu}, \boldsymbol{\Sigma})$ are

$$\begin{aligned} \mathbb{E}[X_i] &= \mu_i, \quad \text{if } \nu > 1,\ i = 1, \ldots, d; \\ \mathbb{C}\mathrm{ov}[X_i, X_j] &= \nu\Sigma_{i,j}/(\nu - 2), \quad \text{if } \nu > 2,\ i,j = 1, \ldots, d. \end{aligned} \tag{A.42}$$

Bibliography

Aase, K. 1993. Equilibrium in a reinsurance syndicate: Existence, uniqueness and characterization. *ASTIN Bulletin*, **23** (2), 185–211.

Aase, K. 1999. An equilibrium model of catastrophe insurance futures and spreads. *The Geneva Papers on Risk and Insurance Theory*, **24** (1), 69–96.

Abramowitz, M., & Stegun, I. A. 1965. *Handbook of Mathematical Functions*. New York: Dover Publications.

Acerbi, C. 2002. Spectral measures of risk: A coherent representation of subjective risk aversion. *Journal of Banking & Finance*, **26** (7), 1505–1518.

Acerbi, C., & Tasche, D. 2002. On the coherence of expected shortfall. *Journal of Banking & Finance*, **26** (2), 1487–1503.

Acharya, V., Engle, R., & Pierret, D. 2014. Testing macroprudential stress tests: The risk of regulatory risk weights. *Journal of Monetary Economics*, **65**, 36–53.

Adusei-Poku, K. 2005. *Operational Risk Management—Implementing a Bayesian Network for Foreign Exchange and Money Market Settlement*. Ph.D. thesis, University of Gottingen, Gottingen, Germany.

Akaike, H. 1981. Likelihood of a model and information criteria. *Journal of Econometrics*, **16** (1), 3–14.

Akaike, H. 1983. Information measure and model selection. *Bulletin of International Statistics Institute*, **50**, 277–290.

Alderweireld, T., Garcia, J., & Léonard, L. 2006. A practical operational risk scenario analysis quantification. *Risk Magazine*, **19** (2), 93–95.

Ale, B. J. M., Bellamy, L. J., van der Boom, R., Cooper, J., Cooke, R. M., Goossens, L. H. J., Hale, A. R., Kurowicka, D., Morales, O., Roelen, A. L. C., & Spouge, J. 2009. Further development of a Causal model for Air Transport Safety (CATS); building the mathematical heart. *Reliability Engineering and System Safety*, **94** (9), 1433–1441.

Alexander, C. 2003. Managing operational risks with Bayesian networks. In Alexander, C. (ed.), *Operational Risk, Regulation, Analysis and Management*. Englewood Cliffs, NJ: Prentice Hall, pp. 285–294.

Allen, L., & Bali, T. 2004. *Cyclicality in Catastrophic and Operational Risk Measurements*. Technical report. Baruch College, New York.

Allen, L., Boudoukh, J., & Saunders, A. 2005. *Understanding Market, Credit and Operational Risk: The Value-at-Risk Approach*. Oxford, UK: Blackwell Publishing.

Allen, L., Boudoukh, J., & Saunders, A. 2009. *Understanding Market, Credit, and Operational Risk: The Value at Risk Approach*. Hoboken, NJ: Wiley.

Fundamental Aspects of Operational Risk and Insurance Analytics: A Handbook of Operational Risk, First Edition. Marcelo G. Cruz, Gareth W. Peters, and Pavel V. Shevchenko.

Allen, D., & Satchell, S. 2013. The Four Horsemen: Heavy-tails, Negative Skew, Volatility Clustering, Asymmetric Dependence, Discussion Paper: 2014–004, The University of Sydney, Australia,

Aly, E.-E. A. A., & Bouzar, N. 2000. On geometric infinite divisibility and stability. *Annals of the Institute of Statistical Mathematics*, **52** (4), 790–799.

Amemiya, T. 1984. Tobit models: A survey. *Journal of Econometrics*, **24** (1), 3–63.

Ames, M., Bagnarosa, G., & Peters, G. W. 2013. Reinvestigating the uncovered interest rate parity puzzle via analysis of multivariate tail dependence in currency carry trades. Preprint arXiv:1303.4314, available at http://arxiv.org. Accessed July 1, 2014.

Anders, U., & Sandstedt, M. 2003. An operational risk scorecard approach. *Risk Magazine*, **16** (1), 47–50.

Anderson, T. W., & Darling, D. A. 1952. Asymptotic theory of certain "goodness of fit" criteria based on stochastic processes. *The Annals of Mathematical Statistics*, **23** (2), 193–212.

Andrieu, C., & Thoms, J. 2008. A tutorial on adaptive MCMC. *Statistics and Computing*, **18** (4), 343–373.

Andrieu, C., De Freitas, N., Doucet, A., & Jordan, M. I. 2003. An introduction to MCMC for machine learning. *Machine Learning*, **50** (1), 5–43.

APRA. 2005. *Guidance Note AGN 115.2 (Draft)—Advanced Measurement Approaches to Operational Risk: Quantitative Standards*. Australian Prudential Regulation Authority, Sydney, Australia.

APRA. 2008. *Capital Adequacy: Advanced Measurement Approaches to Operational Risk*. Prudential Standard APS 115, Australian Prudential Regulation Authority.

Aragonés, J. R., Blanco, C., & Dowd, K. 2001. Incorporating stress tests into market risk modeling. *Derivatives Quarterly*, **7** (3), 44–50.

Araujo, A., and Evarist G. 1980. *The central limit theorem for real and Banach valued random variables*. Vol. 431. New York: Wiley.

Araya, R. 2005. Catastrophic risk securitization: Moody's perspective. *Proceedings Policy Issues in Insurance: Catastrophic Risks and Insurance*, **8** (11), 171–182.

Arrow, K. J. 1964. The role of securities in the optimal allocation of risk-bearing. *The Review of Economic Studies*. **31** (2), 91–96.

Arrow, K. J. 1965. *Aspects of the Theory of Risk-Bearing*. Helsinki, Finland: Yrjö Jahnssonin Säätiö.

Artzner, P., Delbaen, F., Eber, J. M., & Heath, D. 1997. Thinking coherently: Generalised scenarios rather than VAR should be used when calculating regulatory capital. *Risk Magazine*, **10** (11), 68–71.

Artzner, P., Delbaen, F., Eber, J. M., & Heath, D. 1999. Coherent measures of risk. *Mathematical Finance*, **9** (3), 203–228.

Askey, R., & Wilson, J. A. 1985. Some basic hypergeometric orthogonal polynomials that generalize Jacobi polynomials. *Memoirs of the American Mathematical Society*, **54** (319), 55pp.

Asmussen, S., & Rojas-Nandayapa, L. 2008. Asymptotics of sums of lognormal random variables with Gaussian copula. *Statistics & Probability Letters*, **78** (16), 2709–2714.

Atchadé, Y. F., & Rosenthal, J. S. 2005. On adaptive Markov chain Monte Carlo algorithms. *Bernoulli*, **11** (5), 815–828.

Atchadé, Y., Fort, G., Moulines, E., & Priouret, P. 2011. Adaptive Markov chain Monte Carlo: Theory and methods. In Barber, D., Cemgil, A. T., & Chiappa, S. (eds.), *Bayesian Time Series Models*. Cambridge, UK: Cambridge University Press, Chapter 2, pp. 32–51.

Atkinson, A. C. 1982. The simulation of generalized inverse Gaussian and hyperbolic random variables. *SIAM Journal on Scientific and Statistical Computing*, **3** (4), 502–515.

Aue, F., & Klakbrener, M. 2006. LDA at work: Deutsche Bank's approach to quantify operational risk. *The Journal of Operational Risk*, **1** (4), 49–95.

Azzalini, A. 1985. A class of distributions which includes the normal ones. *Scandinavian Journal of Statistics*, **12** (2), 171–178.

Babbel, D. F., & Santomero, A. M. 1996. *Risk Management by Insurers: An Analysis of the Process*. Philadelphia, PA: Wharton Financial Institutions Center, Wharton School of the University of Pennsylvania.

Badreddine, A., & Ben Amor, N. 2010. A new approach to construct optimal bow tie diagrams for risk analysis. In Garcia-Pedrajas, N., Herrera, F., Fyfe, C., Benitez, J. M., & Ali, M. (eds.), *Trends in Applied Intelligent Systems*. Lecture Notes in Computer Science, Vol. 6097. Berlin/Heidelberg, Germany: Springer, pp. 595–604.

Bai, Y., Roberts, G. O., & Rosenthal, J. S. 2009. *On the Containment Condition for Adaptive Markov Chain Monte Carlo Algorithms*. Working paper. Centre for Research in Statistical Methodology, University of Warwick, Warwick, UK.

Baker, C. T. H. 2000. A perspective on the numerical treatment of Volterra equations. *Journal of Computational and Applied Mathematics*, **125** (1), 217–249.

Balanda, K. P., & MacGillivray, H. L. 1988. Kurtosis: A critical review. *The American Statistician*, **42** (2), 111–119.

Balanda, K. P., & MacGillivray, H. L. 1990. Kurtosis and spread. *Canadian Journal of Statistics*, **18** (1), 17–30.

Bannister, J. E., & Bawcutt, P. A. 1981. *Practical Risk Management*. London, UK: Whiterby.

Bantwal, V. J., & Kunreuther, H. C. 2000. A cat bond premium puzzle? *The Journal of Psychology and Financial Markets*, **1** (1), 76–91.

Barbe, P., Fougères, A., Genest, C. 2006. On the tail behavior of sums of dependent risks. *ASTIN Bulletin* **36** (2), 361–373.

Barbe, P., Genest, C., Ghoudi, K., & Rémillard, B. 1996. On Kendall's process. *Journal of Multivariate Analysis*, **58** (2), 197–229.

Barndorff-Nielsen, O. E. 1977. Exponentially decreasing distributions for the logarithm of particle size. *Proceedings of the Royal Society of London. A, Mathematical and Physical Sciences*, **353** (1674), 401–419.

Barndorff-Nielsen, O. E. 1978a. Hyperbolic distributions and distributions on hyperbolae. *Scandinavian Journal of Statistics*, **5** (3), 151–157.

Barndorff-Nielsen, O. E. 1978b. *Information and Exponential Families: In Statistical Theory*, Vol. 1978. New York: Wiley.

Barndorff-Nielsen, O. E. 1997. Processes of normal inverse Gaussian type. *Finance and Stochastics*, **2** (1), 41–68.

Barndorff-Nielsen, O. E., & Blaesild, P. 1981. *Hyperbolic Distributions and Ramifications: Contributions to Theory and Application*. Dordrecht, the Netherlands: Springer.

Barndorff-Nielsen, O. E., & Halgreen, C. 1977. Infinite divisibility of the hyperbolic and generalized inverse Gaussian distributions. *Probability Theory and Related Fields*, **38** (4), 309–311.

Barndorff-Nielsen, O. E., & Lindner, A. M. 2004. Some aspects of Lévy copulas. Sonderforschungsbereich 386, Paper 388. Technische Universitat Munchen, München, Germany, available at http://epub.ub.uni-muenchen.de/. Accessed July 1, 2014.

Barndorff-Nielsen, O. E., & Shephard, N. 2001. Modelling by Lévy processes for financial econometrics. In Barndorff-Nielsen, O. E., Mikosch, T., & Resnick, S. (eds.), *Lévy Processes—Theory and Applications*. Boston, MA: Birkhäuser, pp. 283–318.

Barndorff-Nielsen, O. E., & Stelzer, R. 2005. Absolute moments of generalized hyperbolic distributions and approximate scaling of normal inverse Gaussian Lévy processes. *Scandinavian Journal of Statistics*, **32** (4), 617–637.

Barndorff-Nielsen, O. E., Blaesild, P., & Seshadri, V. 1992. Multivariate distributions with generalized inverse Gaussian marginals, and associated Poisson mixtures. *Canadian Journal of Statistics*, **20** (2), 109–120.

Barrieu, P., & Albertini, L. 2010. *The Handbook of Insurance-Linked Securities*, Vol. 525. Hoboken, NJ: Wiley.

Barrieu, P., & Loubergé, H. 2009. Hybrid cat bonds. *Journal of Risk and Insurance*, **76** (3), 547–578.

Bartlett, D. K. 1965. Excess ratio distribution in risk theory. *Transactions of the Society of Actuaries*, **17** (PT 1, 49), 435–453.

Baud, N., Frachot, A., & Roncalli, T. 2002. *Internal Data, External Data and Consortium Data for Operational Risk Measurement: How to Pool Data Properly*? Working paper. Lyonnais, France: Groupe de Recherche Operationnelle, pp. 1–18.

Baud, N., Frachot, A., & Roncalli, T. February 2003. How to avoid over-estimating capital charge for operational risk? *OperationalRisk—Risk's Newsletter*. Available at SSRN http://ssrn.com/abstract=1032591. Accessed July 1, 2014.

Bawa, V. S. 1975. Optimal rules for ordering uncertain prospects. *Journal of Financial Economics*, **2** (1), 95–121.

Bayes, T. 1763. An essay towards solving a problem in the doctrine of chances. *Philosophical Transactions of the Royal Society of London*, **53**, 370–418.

Bazzarello, D., Crielaard, B., Piacenza, F., & Soprano, A. 2006. Modeling insurance mitigation on operational risk capital. *Journal of Operational Risk*, **1** (1), 57–65.

BCBS. 1988. *International Convergence of Capital Measurement and Capital Standards*. Basel, Switzerland: Basel Committee on Banking Supervision, Bank for International Settlements.

BCBS. 1996. *Amendment to the Capital Accord to Incorporate Market Risks*. Basel, Switzerland: Basel Committee on Banking Supervision, Bank for International Settlements.

BCBS. September 2001. *Working Paper on the Regulatory Treatment of Operational Risk*. Basel, Switzerland: Basel Committee on Banking Supervision, Bank for International Settlements.

BCBS. 2002. *Quantitative Impact Study for Operational Risk: Overview of Individual Loss Data and Lessons Learned*. Basel, Switzerland: Basel Committee on Banking Supervision, Bank for International Settlements.

BCBS. 2003. *Operational Risk Transfer across Financial Sectors*. Basel, Switzerland: Basel Committee on Banking Supervision, Bank for International Settlements.

BCBS. June 2004. *International Convergence of Capital Measurement and Capital Standards: A Revised Framework*. Basel, Switzerland: Basel Committee on Banking Supervision, Bank for International Settlements, Basel.

BCBS. June 2006. *International Convergence of Capital Measurement and Capital Standards: A Revised Framework (Comprehensive Version)*. Basel, Switzerland: Basel Committee on Banking Supervision, Bank for International Settlements.

BCBS. July 2009a. *Observed Range of Practice of Key Elements of the Advanced Measurement Approaches*. Basel, Switzerland: Basel Committee on Banking Supervision, Bank for International Settlements.

BCBS. July 2009b. *Results from the 2008 Loss Data Collection Exercise for Operational Risk*. Basel, Switzerland: Basel Committee on Banking Supervision, Bank for International Settlements.

BCBS. May 2009c. *Principles for Sound Stress Testing Practices and Supervision Approaches*. Basel, Switzerland: Basel Committee on Banking Supervision, Bank for International Settlements, Basel.

BCBS. 2011. *Basel III: A Global Regulatory Framework for More Resilient Banks and Banking System*. Basel, Switzerland: Basel Committee on Banking Supervision, Bank for International Settlements.

BCBS. 2012. *Fundamental Review of the Trading Book*. Basel, Switzerland: Basel Committee on Banking Supervision, Bank for International Settlements.

BCBS. 2013. *Basel III: The Liquidity Coverage Ratio and Liquidity Risk Monitoring Tools*. Basel, Switzerland: Basel Committee on Banking Supervision, Bank for International Settlements.

Beaumont, M. A., Zhang, W., & Balding, D. J. 2002. Approximate Bayesian computation in population genetics. *Genetics*, **162** (4), 2025–2035.

Beaumont, M. A., Cornuet, J.-M., Marin, J.-M., & Robert, C. P. 2009. Adaptive approximate Bayesian computation. *Biometrika*, **96** (4), 983–990.

Bedard, M., & Rosenthal, J. S. 2008. Optimal scaling of Metropolis algorithms: Heading towards general target distributions. *The Canadian Journal of Statistics*, **36** (4), 483–503.

Bee, M. 2005a. *Copula-Based Multivariate Models with Applications to Risk Management and Insurance*. Working paper. Dipartimento di Economia, Universita degli Studi di Trento, Trento, Italy.

Bee, M. 2005b. *On Maximum Likelihood Estimation of Operational Loss Distributions*. Discussion paper no. 3. Dipartimento di Economia, Universita degli Studi di Trento, Trento, Italy.

Beirlant, J., Goegebeur, Y., Teugels, J., Segers, J., Waal, D. D., & Ferro, C. 2004. *Statistics of Extremes: Theory and Applications*. Chichester, UK: John Wiley & Sons Ltd.

Bellini, F., & Bignozzi, V. 2013. Elicitable risk measures. Preprint SSRN 2334746. Available on URL: http://ssrn.com.

Beran, R., & Millar, P. W. 1987. Stochastic estimation and testing. *The Annals of Statistics*, **15** (3), 1131–1154.

Berg, D. 2009. Copula goodness-of-fit testing: An overview and power comparison. *The European Journal of Finance*, **15** (7–8), 675–701.

Berg, C., & Forst, G. 1975. *Potential Theory on Locally Compact Abelian Groups*. New York/Heidelberg, Germany: Springer-Verlag.

Berg, D., & Bakken, H. 2005. A goodness-of-fit test for copulae based on the probability integral transform. Preprint series. *Statistical Research Report*. Available at http://urn. nb. no/URN:NBN:no-23420. Accessed July 1, 2014.

Berger, J. O. 1985. *Statistical Decision Theory and Bayesian Analysis*, 2nd edn. New York: Springer.

Bergstrom, H. 1953. On some expansions of stable distributional functions. *Arkiv för Matematik*, **2** (5), 375–378 and 463–474.

Berkowitz, J. 2000. A coherent framework for stress-testing. *Journal of Risk*, **2** (2), 1–11.

Berleant, D. 1993. Automatically verified reasoning with both intervals and probability density functions. *Interval Computations*, **2** (1993), 48–70.

Berliner, B. 1982. *Limits of Insurability of Risks*. Englewood Cliff, NJ: Prentice-Hall.

Bernanke, B. S. 2013. Stress Testing Banks: What Have We Learned? Intervento alla conferenza Maintaining Financial Stability: Holding a Tiger by the Tail, Stone Mountain (Ge). Vol. 8.

Bickel, P. J., & Rosenblatt, M. 1973. On some global measures of the deviations of density function estimates. *The Annals of Statistics*, **1** (6), 1071–1095.

Bingham, N. H., Goldie, C. M., & Teugels, J. L. 1989. *Regular Variation*, vol. 27. Cambridge University Press.

Birkhoff, G. D. 1931. Proof of the ergodic theorem. *Proceedings of the National Academy of Sciences of the United States of America*, **17** (12), 656.

Birnbaum, Z. W., & McCarty, R. C. 1958. A distribution-free upper confidence bound for Pr $[Y < X]$, based on independent samples of X and Y. *The Annals of Mathematical Statistics*, **29** (2), 558–562.

Bishop, C. M., & Nasrabadi, N. M. 2006. *Pattern Recognition and Machine Learning*, Vol. 1. New York: Springer.

Black, F., & Scholes, M. 1973. The pricing of options and corporate liabilities. *The Journal of Political Economy*, **81** (3), 637–654.

Bladt, M. 2005. A review of phase-type distributions and their use in risk theory. *ASTIN Bulletin*, **35** (1), 145–167.

Block, H. W., Savits, T. H., & Shaked, M. 1982. Some concepts of negative dependence. *The Annals of Probability*, **10** (3), 765–772.

Blomqvist, N. 1950. On a measure of dependence between two random variables. *The Annals of Mathematical Statistics*, **21** (4), 593–600.

Bluhm, C., Overbeck, L., & Wagner, C. 2002. *An Introduction to Credit Risk Modeling*. Boca Raton, FL: CRC Press.

Blundell-Wignall, A., & Atkinson, P. 2010. Thinking beyond Basel III: Necessary solutions for capital and liquidity. *OECD Journal: Financial Market Trends*, **2010** (1), 5–6.

Blunden, T. 2003. Scorecard approaches. *Operational Risk: Regulation, Analysis and Management*. Prentice Hall-Financial Times.

Böcker, K., & Klüppelberg, C. 2005. Operational VaR: A closed-form approximation. *Risk Magazine*, **12**, 90–93.

Böcker, K., & Klüppelberg, C. 2008. Modeling and measuring multivariate operational risk with Lévy copulas. *The Journal of Operational Risk*, **3** (2), 3–27.

Böcker, K., & Klüppelberg, C. 2009. First-order approximations to operational risk dependence and consequences. In Gregoriou, G. N. (ed.), *Operational Risk towards Basel III: Best Practices and Issues in Modeling, Management and Regulation*. New York: Wiley, Chapter 11, pp. 219–245.

Böcker, K., & Klüppelberg, C. 2010. Multivariate models for operational risk. *Quantitative Finance*, **10** (8), 855–869.

Böcker, K., & Sprittulla, J. 2006. Operational VAR: Meaningful means. *Risk Magazine*, **12**, 96–98.

Bondesson, L., Kristiansen, G. K., & Steutel, F. W. 1996. Infinite divisibility of random variables and their integer parts. *Statistics & Probability Letters*, **28** (3), 271–278.

Bookstabber, R. M., & McDonald, J. B. 1987. A general distribution for describing security price returns. *The Journal of Business*, **60** (3), 401–424.

Boole, G. 1854. *An Investigation of the Laws of Thought: On Which Are Founded the Mathematical Theories of Logic and Probability*. London, UK: Walton and Maberly.

Borch, K. 1960. Reciprocal reinsurance treaties seen as a two-person co-operative game. *Scandinavian Actuarial Journal*, **1960** (1–2), 29–58.

Borch, K. 1962. Equilibrium in a reinsurance market. *Econometrica: Journal of the Econometric Society*, **30** (3), 424–444.

Bornn, L., Gottardo, R., & Doucet, A. 2010. Grouping priors and the Bayesian elastic net. Preprint arXiv:1001.4083, available at http://arxiv.org. Accessed July 1, 2014.

Borokov, A. A., & Sycheva, N. M. 1968. On asymptotically optimal non-parametric criteria. *Theory of Probability & Its Applications*, **13** (3), 359–393.

Bouyé, E., Durrleman, V., Nikeghbali, A., Riboulet, G., & Roncalli, T. 2000. Copulas for finance-a reading guide and some applications. Preprint SSRN: 1032533, available at http://www.ssrn.com/en/. Accessed July 1, 2014.

Bowers, L. N., Jr., & Newton, L. 1966. Expansion of probability density functions as a sum of Gamma densities with applications in risk theory. *Transactions of Society of Actuaries*, **18** (PT 1, 52), 125–137.

Brandts, S. 2004. *Operational Risk and Insurance: Quantitative and Qualitative Aspects*. Working paper. Goethe University, Frankfurt, Germany.

Brazauskas, V. 2002. Fisher information matrix for the Feller–Pareto distribution. *Statistics & Probability Letters*, **59** (2), 159–167.

Breiman, L. 1961. *Optimal Gambling Systems for Favorable Games*. Berkeley, CA: University of California Press.

Breiman, L. 1995. Better subset regression using the nonnegative garrote. *Technometrics*, **37** (4), 373–384.

Brewer, M. J., Aitken, C. G. G., & Talbot, M. 1996. A comparison of hybrid strategies for Gibbs sampling in mixed graphical models. *Computational Statistics and Data Analysis*, **21** (3), 343–365.

Breymann, W., Dias, A., & Embrechts, P. 2003. Dependence structures for multivariate high-frequency data in finance. *Quantitative Finance*, **3** (1), 1–14.

Brigo, D., & Chourdakis, K. 2012. Consistent single-and multi-step sampling of multivariate arrival times: A characterization of self-chaining copulas. Preprint arXiv:1204.2090, available at http://arxiv.org. Accessed July 1, 2014.

Briys, E. 1997. From Genoa to Kobe: Natural hazards, insurance risks and the pricing of insurance-linked bonds. Working paper. Lehman Brothers International, London, UK.

Brooks, S. 1998. Markov chain Monte Carlo method and its application. *Journal of the Royal Statistical Society: Series D (The Statistician)*, **47** (1), 69–100.

Brooks, S. P., & Gelman, A. 1998. General methods for monitoring convergence of iterative simulations. *Journal of Computational and Graphical Statistics*, **7** (4), 434–455.

Brown, B. M. 1982. Cramér-von Mises distributions and permutation tests. *Biometrika*, **69** (3), 619–624.

Brunel, V. 2013. Solvable models of operational risk and new results on the correlation problem. Preprint arXiv:1308.5064, available at http://arxiv.org. Accessed July 1, 2014.

Buchinsky, M. 1998. Recent advances in quantile regression models: A practical guideline for empirical research. *Journal of Human Resources*, **33** (1), 88–126.

Bühlmann, H. 1970. *Mathematical Methods in Risk Theory*. New York: Springer.

Bühlmann, H. 1980. An economic premium principle. *ASTIN Bulletin*, **11** (11), 52–60.

Bühlmann, H. 1984a. The general economic premium principle. *ASTIN Bulletin*, **14** (1), 13–21.

Bühlmann, H. 1984b. Numerical evaluation of the compound Poisson distribution: Recursion or Fast Fourier transform? *Scandinavian Actuarial Journal*, **1984** (2), 116–126.

Bühlmann, H., & Gisler, A. 2005. *A Course in Credibility Theory and Its Applications*. Berlin, Germany: Springer.

Bühlmann, H., & Jewell, W. S. 1978. *Optimal Risk Exchanges*. Technical report no. ORC-78-10. DTIC Document. California University Berkeley Operations Research Center, Berkeley, CA.

Bühlmann, H., & Straub, E. 1970. Glaubwürdigkeit für Schadensätze. *Bulletin of the Swiss Association of Actuaries*, **70**, 111–133.

Bühlmann, H., Delbaen, F., Embrechts, P., & Shiryaev, A. N. 1996. No-arbitrage, change of measure and conditional Esscher transforms. *CWI Quarterly*, **9** (4), 291–317.

Bühlmann, H., Shevchenko, P. V., & Wüthrich, M. V. 2007. A "toy" model for operational risk quantification using credibility theory. *The Journal of Operational Risk*, **2** (1), 3–19.

Burnham, K. P., & Anderson, D. R. 2002. *Model Selection and Multi-model Inference: A Practical Information-Theoretic Approach*. New York: Springer.

Butsic, R. P. 1999. Capital allocation for property-liability insurers: A catastrophe reinsurance application. *Casualty Actuarial Society Forum*, **Spring**, 1–70.

Cai, Y. 2010. Polynomial power-Pareto quantile function models. *Extremes*, **13** (3), 291–314.

Cam, L., & Morlat, G. 1949. Les lois des débits des rivières françaises. *La Houille Blanche*, **Special B**, 733–740.

Candès, E. J., Wakin, M. B., & Boyd, S. P. 2008. Enhancing sparsity by reweighted l_1 minimization. *Journal of Fourier Analysis and Applications*, **14** (5–6), 877–905.

Cantelli, F. P. 1933. Sulla determinazione empirica delle leggi di probabilita. *Giornale dell'Istituto Italiano degli Attuari*, **4**, 421–424.

Canter, M. S., Cole, J. B., & Sandor, R. L. 1997. Insurance derivatives: A new asset class for the capital markets and a new hedging tool for the insurance industry. *Journal of Applied Corporate Finance*, **10** (3), 69–81.

Capéraà, P., Fougeres, A.-L., & Genest, C. 1997. A nonparametric estimation procedure for bivariate extreme value copulas. *Biometrika*, **84** (3), 567–577.

Cappé, O., Guillin, A., Marin, J. M., & Robert, C. P. 2004. Population Monte Carlo. *Journal of Computational and Graphical Statistics*, **13** (4), 907–929.

Carlin, B. P., & Chib, S. 1995. Bayesian model choice via Markov chain Monte Carlo methods. *Journal of the Royal Statistical Society. Series B (Methodological)*, **57** (3), 473–484.

Carmona, R., & Touzi, N. 2008. Optimal multiple stopping and valuation of swing options. *Mathematical Finance*, **18** (2), 239–268.

Carter, M., & Van Brunt, B. 2000. *The Lebesgue-Stieltjes Integral: A Practical Introduction.* New York: Springer.

Casella, G., & Berger, R. L. 2002. *Statistical Inference*. Pacific Grove, CA: Duxbury.

Casella, G., & George, E. I. 1992. Explaining the Gibbs sampler. *The American Statistician*, **46** (3), 167–174.

Chan, K. S. 1993. Asymptotic behavior of the Gibbs sampler. *Journal of the American Statistical Association*, **88** (Issue 421), 320–326.

Charpentier, A. 2003. Tail distribution and dependence measures. *XXXIV International ASTIN Colloquium*, Berlin, Germany, August, pp. 24–27.

Charpentier, A., & Segers, J. 2007. Lower tail dependence for Archimedean copulas: Characterizations and pitfalls. *Insurance: Mathematics and Economics*, **40** (3), 525–532.

Chartrand, R., & Yin, W. 2008. Iteratively reweighted algorithms for compressive sensing. *Proceedings of ICASSP*, Las Vegas, NV.

Chateauneuf, A., Cohen, M., & Kast, R. 1997. *Comonotone Random Variables in Economics: A Review of Some Results*. Paris, France: Université de Paris I.

Chavez-Demoulin, V., Embrechts, P., & Nešlehová, J. 2006. Quantitative models for operational risk: Extremes, dependence and aggregation. *Journal of Banking & Finance*, **30** (10), 2635–2658.

Chavez-Demoulin, V., Embrechts, P., & Hofert, M. 2013. An extreme value approach for modeling operational risk losses depending on covariates. Working paper. Department of Mathematics, ETH Zurich, Zurich, Switzerland.

Chebana, F., El Adlouni, S., & Bobée, B. 2008. Method of moments of the Halphen distribution parameters. *Stochastic Environmental Research and Risk Assessment*, **22** (6), 749–757.

Chebana, F., El Adlouni, S., & Bobée, B. 2010. Mixed estimation methods for Halphen distributions with applications in extreme hydrologic events. *Stochastic Environmental Research and Risk Assessment*, **24** (3), 359–376.

Chen, Q. 2008. *Dependence Structure for Lévy Processes and Its Application in Finance*. Dissertation, the University of Maryland, College Park, MD.

Chernobai, A., Menn, C., Trück, S., & Rachev, S. T. 2006. A note on the estimation of the frequency and severity distribution of operational losses. *Mathematical Scientist*, **30** (2), 87–97.

Chernobai, A. S., Rachev, S. T., & Fabozzi, F. J. 2007. *Operational Risk: A Guide to Basel II Capital Requirements, Models, and Analysis*. Wiley Finance series, Hoboken, NJ: John Wiley & Sons.

Chernoff, H., & Lehmann, E. L. 1954. The use of maximum likelihood estimates in χ^2 tests for goodness of fit. *The Annals of Mathematical Statistics*, **25** (3), 579–586.

Cherubini, U., Luciano, E., & Vecchiato, W. 2004. *Copula Methods in Finance*. Hoboken, NJ: John Wiley & Sons.

Chib, S. 1995. Marginal likelihood from the Gibbs output. *Journal of the American Statistical Association*, **90** (432), 1313–1321.

Chib, S., & Greenberg, E. 1995. Understanding the Metropolis-Hastings algorithm. *The American Statistician*, **49** (4), 327–335.

Chicheportiche, R., & Bouchaud, J.-P. 2012. Weighted Kolmogorov-Smirnov test: Accounting for the tails. *Physical Review E*, **86** (4), 041115.

Chopin, N. 2002. A sequential particle filter method for static models. *Biometrika*, **89** (3), 539–552.

Chopin, N. 2004. Central limit theorem for sequential Monte Carlo methods and its application to Bayesian inference. *The Annals of Statistics*, **32** (6), 2385–2411.

Christensen, M. M., & Larsen, K. 2007. No arbitrage and the growth optimal portfolio. *Stochastic Analysis and Applications*, **25** (1), 255–280.

Christoph, G., & Schreiber, K. 1998. Discrete stable random variables. *Statistics & Probability Letters*, **37** (3), 243–247.

Christoph, G., & Schreiber, K. 1998. *The Generalized Discrete Linnik Distributions*. Springer.

Cohen, L. W. 1940. On the mean ergodic theorem. *The Annals of Mathematics*, **41** (3), 505–509.

Coles, S., Heffernan, J., & Tawn, J. 1999. Dependence measures for extreme value analyses. *Extremes*, **2** (4), 339–365.

Colwell, D. B., & Elliott, R. J. 1993. Discontinuous asset prices and non-attainable contingent claims. *Mathematical Finance*, **3** (3), 295–308.

Congdon, P. 2006. *Bayesian Statistical Modelling*, 2nd edn. Chichester, UK: John Wiley & Sons, Ltd.

Constantine, G. M., & Savits, T. H. 1996. A multivariate Faa di Bruno formula with applications. *Transactions of the American Mathematical Society*, **348** (2), 503–520.

Cont, R. 2001. Empirical properties of asset returns: Stylized facts and statistical issues. *Quantitative Finance*, **1** (2), 223–236.

Cont, R., & Tankov, P. 2004. *Financial Modelling with Jump Processes*. Boca Raton, FL: Chapman & Hall/CRC Press.

Cont, R., Deguest, R., & Scandolo, G. 2010. Robustness and sensitivity analysis of risk measurement procedures. *Quantitative Finance*, **10** (6), 593–606.

Convolutions, Generalized Gamma. 1992. *Generalized Gamma Convolutions and Related Classes of Distributions and Densities*. Lecture notes in Statistics, Vol. 76. Springer.

Cook, R. D., & Johnson, M. E. 1981. A family of distributions for modelling non-elliptically symmetric multivariate data. *Journal of the Royal Statistical Society. Series B (Methodological)*, **43** (2), 210–218.

Cope, E. W. 2012. Combining scenario analysis with loss data in operational risk quantification. *The Journal of Operational Risk*, **7** (1), 39–56.

Cope, E. W., Antonini, G., Mignola, G., & Ugoccioni, R. 2009. Challenges and pitfalls in measuring operational risk from loss data. *The Journal of Operational Risk*, **4** (4), 3–27.

Coval, J. D., Jurek, J. W., & Stafford, E. 2009. Economic catastrophe bonds. *The American Economic Review*, **99** (3), 628–666.

Cowell, R. J., Verral, R. J., & Yoon, Y. K. 2006. Modelling operational risk with Bayesian networks. *The Journal of Risk and Insurance*, **74** (4), 795–827.

Cowles, M. K., & Carlin, B. P. 1996. Markov chain Monte Carlo convergence diagnostics: A comparative review. *Journal of the American Statistical Association*, **91** (434), 883–904.

Cox, J. C., & Ross, S. A. 1976. The valuation of options for alternative stochastic processes. *Journal of Financial Economics*, **3** (1), 145–166.

Cox, S. H., & Pedersen, H. W. 2000. Catastrophe risk bonds. *North American Actuarial Journal*, **4** (4), 56–82.

Cox, J. C., Ingersoll Jr., J. E., & Ross, S. A. 1985. A theory of the term structure of interest rates. *Econometrica: Journal of the Econometric Society*, **53** (2), 385–407.

Crisan, D., & Doucet, A. 2002. A survey of convergence results on particle filtering methods for practitioners. *IEEE Transactions on Signal Processing*, **50** (3), 736–746.

Crockford, G. N. 1982. The bibliography and history of risk management: Some preliminary observations. *The Geneva Papers on Risk and Insurance*, 7 (23), 169–179.

Cruz, M. G. 2002. *Modeling, Measuring and Hedging Operational Risk.* New York: John Wiley & Sons.

Csillery, K., Blum, M. G. B., Gaggiotti, O. E., & Francois, O. 2010. Approximate Bayesian computation (ABC) in practice. *Trends in Ecology & Evolution*, **25** (7), 410–418.

Csorgo, S., & Faraway, J. J. 1996. The exact and asymptotic distributions of Cramér-von Mises statistics. *Journal of the Royal Statistical Society. Series B (Methodological)*, **58** (1), 221–234.

Cummins, J. D., Dionne, G., McDonald, J. B., & Pritchett, B. M. 1990. Applications of the GB2 family of distributions in modeling insurance loss processes. *Insurance: Mathematics and Economics*, **9** (4), 257–272.

Cummins, J. D., Doherty, N., & Lo, A. 2002. Can insurers pay for the big one? Measuring the capacity of the insurance market to respond to catastrophic losses. *Journal of Banking & Finance*, **26** (2), 557–583.

Cummins, J. D., Lalonde, D., & Phillips, R. D. 2004. The basis risk of catastrophic-loss index securities. *Journal of Financial Economics*, **71** (1), 77–111.

Dahen, H., & Dionne, G. 2010. Scaling models for the severity and frequency of external operational loss data. *Journal of Banking & Finance*, **34** (7), 1484–1496.

Dalkey, O., & Helmer, O. 1963. An experimental application of the Delphi method to the use of the experts. *The Management Science*, **9** (3), 458–472.

Dall'Aglio, G., Kotz, S., & Salinetti, G. 1991. Advances in probability distributions with given marginals: Beyond the copulas, Lectures presented at a *Symposium on Distributions with Given Marginals*, Department of Statistics, University La Sapienza, Rome, Italy, April 1990, Vol. 67.

Dalla Valle, L. 2009. Bayesian copulae distributions, with application to operational risk management. *Methodology and Computing in Applied Probability*, **11** (1), 95–115.

Daniels, H. E. 1954. Saddlepoint approximations in statistics. *The Annals of Mathematical Statistics*, **25** (4), 631–650.

Daniélsson, J., Embrechts, P., Goodhart, C., Keating, C., Muennich, F., Renault, O., & Shin, H. S. 2001. *An Academic Response to Basel II.* Special paper no. 130. LSE Financial Markets Group, London, UK.

Das, S., Yang, H., & Banks, D. 2013. Synthetic priors that merge opinion from multiple experts. *Statistics, Politics, and Policy*, **4** (1), 1–81.

Dassios, A., & Jang, J.-W. 2003. Pricing of catastrophe reinsurance and derivatives using the Cox process with shot noise intensity. *Finance and Stochastics*, **7** (1), 73–95.

Daszykowski, M., Kaczmarek, K., Vander Heyden, Y., & Walczak, B. 2007. Robust statistics in data analysis—A review: Basic concepts. *Chemometrics and Intelligent Laboratory Systems*, **85** (2), 203–219.

Daul, S., De Giorgi, E., Lindskog, F., & McNeil, A. 2003. The grouped t-copula with an application to credit risk. *RISK*, **16**, 73–76.

Decamps, J.-P., Rochet, J.-C., & Roger, B. 2004. The three pillars of Basel II: Optimizing the mix. *Journal of Financial Intermediation*, **13** (2), 132–155.

De Fontnouvelle, P., DeJesus-Rueff, V., Jordan, J. S., & Rosengren, E. S. 2006. Capital and risk: New evidence on implications of large operational losses. *Journal of Money, Credit, and Banking*, **38** (7), 1819–1846.

De Fontnouvelle, P., Rosengren, E. S., & Jordan, J. S. 2007. Implications of alternative operational risk modeling techniques. In Carey, M., & Stulz, R. M. (eds.), *The Risks of Financial Institutions.* Chicago, IL: University of Chicago Press, pp. 475–512.

Degen, M. 2010. The calculation of minimum regulatory capital using single-loss approximations. *Journal of Operational Risk*, **5** (4), 1–15.

Degen, M., Embrechts, P., & Lambrigger, D. D. 2007. The quantitative modeling of operational risk: Between g-and-h and EVT. *ASTIN Bulletin*, **37** (2), 265–291.

Deheuvels, P. 1979. La fonction de dépendance empirique et ses propriétés. Un test non paramétrique d'indépendance. *Bulletin de la Classe des Sciences. 5e Série. Académie Royale de Belgique, Bruxelles*, **65** (6), 274–292.

Delbaen, F., & Schachermayer, W. 1994. A general version of the fundamental theorem of asset pricing. *Mathematische Annalen*, **300** (1), 463–520.

Del Moral, P. 2004. *Feynman-Kac Formulae: Genealogical and Interacting Particle Systems with Applications.* New York: Springer.

Del Moral, P., Doucet, A., & Jasra, A. 2006. Sequential Monte Carlo samplers. *Journal of the Royal Statistical Society: Series B (Statistical Methodology)*, **68** (3), 411–436.

Del Moral, P., Doucet, A., & Jasra, A. 2012. An adaptive sequential Monte Carlo method for approximate Bayesian computation. *Statistics and Computing*, **22** (5), 1009–1020.

Del Moral, P., Peters, G. W., & Vergé, C. 2013. An introduction to particle integration methods: With applications to risk and insurance. In Dick, J., Kuo, F. Y., Peters, G. W., & Sloan, I. H. (eds.), *Monte Carlo and Quasi-Monte Carlo Methods 2012.* Heidelberg, Germany: Springer.

De Luca, G., & Rivieccio, G. 2012. Multivariate tail dependence coefficients for Archimedean copulae. *Advanced Statistical Methods for the Analysis of Large Data-Sets.* New York: Springer, pp. 287–296.

Demarta, S., & McNeil, A. 2005. The t copula and related copulas. *International Statistical Review*, **73** (1), 111–129.

Dempster, A. P. 1967. Upper and lower probabilities induced by a multi-valued mapping. *Annals of Mathematical Statistics*, **38** (2), 325–339.

Dempster, A. P. 1968. A generalization of Bayesian inference. *Journal of the Royal Statistical Society, Series B*, **30** (2), 205–247.

Dempster, A. P., Laird, N. M., & Rubin, D. B. 1977. Maximum likelihood from incomplete data via the EM algorithm. *Journal of the Royal Statistical Society, Series B*, **39** (1), 1–38.

Denault, M. 2001. Coherent allocation of risk capital. *Journal of Risk*, **4**, 1–34.

Denison, D. G. T., Holmes, C. C, Mallick, B. K., & Smith, A. F. M. 2002. *Bayesian Methods for Nonlinear Classification and Regression*, Vol. 386. New York: Wiley.

Dentcheva, D., Penev, S., & Ruszczyński, A. 2010. Kusuoka representation of higher order dual risk measures. *Annals of Operations Research*, **181** (1), 325–335.

Denuit, M., Dhaene, J., Goovaerts, M. J., & Kaas, R. 2005. *Actuarial Theory for Dependent Risks.* Chichester, UK: Wiley.

De Pril, N. 1986. On the exact computation of the aggregate claims distribution in the individual life model. *ASTIN Bulletin*, **16** (2), 109–112.

De Pril, N. 1989. The aggregate claims distribution in the individual model with arbitrary positive claims. *ASTIN Bulletin*, **19** (1), 9–24.

De Vries, C. G., & Zhou, C. 2006. Discussion of "Copulas: Tales and facts", by Thomas Mikosch. *Extremes*, **9** (1), 23–25.

Devroye, L. 1993. A triptych of discrete distributions related to the stable law. *Statistics & Probability Letters*, **18** (5), 349–351.

Dey, D., Ghosh, S. K., & Mallick, B. K. 2000. *Generalized Linear Models: A Bayesian Perspective*, Vol. 5. Boca Raton, FL: CRC Press.

Dhaene, J. & Vandebroek, M. 1995. Recursions for the individual model. *Insurance: Mathematics and Economics*, **16** (1), 31–38.

Dhaene, J., Goovaerts, M. J., & Kaas, R. 2003. Economic capital allocation derived from risk measures. *North American Actuarial Journal*, **7**, 44–59.

Diaconis, P., Holmes, S., & Neal, R. M. 2000. Analysis of a nonreversible Markov chain sampler. *Annals of Applied Probability*, **10** (3), 726–752.

Dobriæ, J., & Schmid, F. 2007. A goodness of fit test for copulas based on Rosenblatt's transformation. *Computational Statistics & Data Analysis*, **51** (9), 4633–4642.

Doherty, N. A. 1997a. Financial innovation in the management of catastrophe risk. *Journal of Applied Corporate Finance*, **10** (3), 84–95.

Doherty, N. A. 1997b. Innovations in managing catastrophe risk. *The Journal of Risk and Insurance*, **64** (4), 713–718.

Dolati, A., & Úbeda-Flores, M. 2006. On measures of multivariate concordance. *Journal of Probability and Statistical Science*, **4** (2), 147–164.

Dorota, K. 2010. *Dependence Modeling: Vine Copula Handbook*. Singapore: World Scientific.

Doucet, A., & Johansen, A. M. 2009. A tutorial on particle filtering and smoothing: Fifteen years later. *Handbook of Nonlinear Filtering*, **12**, 656–704.

Doucet, A., Godsill, S., & Andrieu, C. 2000. On sequential Monte Carlo sampling methods for Bayesian filtering. *Statistics and Computing*, **10** (3), 197–208.

Doucet, A., De Freitas, N., Gordon, N., *et al.* 2001. *Sequential Monte Carlo Methods in Practice*, Vol. 1. New York: Springer.

Doucet, A., Johansen, A. M., & Tadić, V. B. 2010. On solving integral equations using Markov Chain Monte Carlo methods. *Applied Mathematics and Computation*, **216** (10), 2869–2880.

Dowd, K., & Blake, D. 2006. After VaR: The theory, estimation, and insurance applications of quantile-based risk measures. *Journal of Risk and Insurance*, **73** (2), 193–229.

Duane, S., Kennedy, A. D., Pendleton, B. J., & Roweth, D. 1987. Hybrid Monte Carlo. *Physics Letters B*, **195** (2), 216–222.

Durante, F., & Sempi, C. 2010. Copula theory: An introduction. In Jaworski, P., Durante F., Karl Hardle, W., & Rychlik, T. (eds.), *Copula Theory and Its Applications*. Dordrecht, the Netherlands: Springer, pp. 3–31.

Durbin, J. 1973. Weak convergence of the sample distribution function when parameters are estimated. *The Annals of Statistics*, **1** (2), 279–290.

Dutta, K. K., & Babbel, D. F. 2002. On measuring skewness and kurtosis in short rate distributions: The case of the US dollar London inter bank offer rates. Technical report, Working papers 02-25. Wharton School Center for Financial Institutions, Philadelphia, PA.

Dutta, K., & Babbel, D. June 2014. Scenario analysis in the measurement of operational risk capital: A change of measure approach. *Journal of Risk and Insurance*, **18** (2), 303–334.

Dutta, K., & Perry, J. 2006. *A Tale of Tails: An Empirical Analysis of Loss Distribution Models for Estimating Operational Risk Capital.* Working paper no. 06-13. Federal Reserve Bank of Boston, Boston, MA.

Dvoretzky, A., Kiefer, J., & Wolfowitz, J. 1956. Asymptotic minimax character of the sample distribution function and of the classical multinomial estimator. *The Annals of Mathematical Statistics*, **27** (3), 642–669.

Eberlein, E. 2001. Application of generalized hyperbolic Lévy motions to finance. In Barndorff-Nielsen, O. E., Resnick S. I., & Mikosch, M. (eds.), *Lévy Processes: Theory and Applications.* Springer, pp. 319–336.

Eberlein, E., & Keller, U. 1995. Hyperbolic distributions in finance. *Bernoulli*, **1** (3), 281–299.

Efron, B. F., & Hinkley, D. V. 1978. Assessing the accuracy of the maximum likelihood estimator: observed versus expected Fisher information. *Biometrika*, **65** (3), 457–487.

Efron, B. F., & Tibshirani, R. J. 1993. *An Introduction to the Bootstrap*. London, UK: Chapman and Hall.

Eichengreen, B. 2008. *Globalizing Capital: A History of the International Monetary System.* 2nd edn. Princeton, NJ: Princeton University Press.

El Adlouni, S., & Bobée, B. 2007. Sampling techniques for Halphen distributions. *Journal of Hydrologic Engineering*, **12** (6), 592–604.

El Adlouni, S., Chebana, F., & Bobée, B. 2009. Generalized extreme value versus Halphen system: Exploratory study. *Journal of Hydrologic Engineering*, **15** (2), 79–89.

Embrechts, P. 1983. A property of the generalized inverse Gaussian distribution with some applications. *Journal of Applied Probability*, **20**, 537–544.

Embrechts, P. 2006. Discussion of "Copulas: Tales and facts", by Thomas Mikosch. *Extremes*, **9** (1), 45–47.

Embrechts, P. 2009. Copulas: A personal view. *Journal of Risk and Insurance*, **76** (3), 639–650.

Embrechts, P., & Frei, M. 2009. Panjer recursion versus *FFT* for compound distributions. *Mathematical Methods of Operations Research*, **69** (3), 497–508.

Embrechts, P., & Meister, S. 1997. Pricing insurance derivatives, the case of CAT-futures. *Proceedings of the 1995 Bowles Symposium on Securitization of Risk*, Georgia State University, Atlanta, GA. Society of Actuaries, Monograph M-FI97-1, pp. 15–26.

Embrechts, P., & Puccetti, G. 2006. Aggregating risk capital, with an application to operational risk. *The Geneva Risk and Insurance Review*, **31** (2), 71–90.

Embrechts, P., & Puccetti, G. 2008. Aggregation operational risk across matrix structured loss data. *The Journal of Operational Risk*, **3** (2), 29–44.

Embrechts, P., Grübel, R., & Pitts, S. M. 1993. Some applications of the fast Fourier transform algorithm in insurance mathematics This paper is dedicated to Professor WS Jewell on the occasion of his 60th birthday. *Statistica Neerlandica*, **47** (1), 59–75.

Embrechts, P., Klüppelberg, C., & Mikosch, T. 1997. *Modelling Extremal Events for Insurance and Finance*. Berlin, Germany: Springer Verlag.

Embrechts, P., McNeil, A., & Straumann, D. 2002. Correlation and dependence in risk management: Properties and pitfalls. In Dempster, M., & Moffatt, H. (eds.), *Risk Management: Value at Risk and Beyond*. Cambridge, UK: Cambridge University Press, pp. 176–223.

Embrechts, P., Lindskog, F., & McNeil, A. 2003. Modelling dependence with copulas and applications to risk management. *Handbook of Heavy Tailed Distributions in Finance*, **8** (329–384), 1.

Embrechts, P., Nešlehová, J., & Wüthrich, M. V. 2009a. Additivity properties for value-at-risk under Archimedean dependence and heavy-tailedness. *Insurance: Mathematics and Economics*, **44** (2), 164–169.

Embrechts, P., Lambrigger, D. D., & Wüthrich, M. V. 2009b. Multivariate extremes and the aggregation of dependent risks: Examples and counter-examples. *Extremes*, **12** (2), 107–127.

Engle, R. F., & Manganelli, S. 2004. CAViaR: Conditional autoregressive value at risk by regression quantiles. *Journal of Business & Economic Statistics*, **22** (4), 367–381.

Erdogan, M. B. 1999. Analytic and asymptotic properties of non-symmetric Linnik's probability densities. *Journal of Fourier Analysis and Applications*, **5** (6), 523–544.

Erdogan, M. B., & Ostrovskii, I. V. 1998. Analytic and asymptotic properties of generalized Linnik probability densities. *Journal of Mathematical Analysis and Applications*, **217** (2), 555–578.

Ergashev, B. 2012. A theoretical framework for incorporating scenarios into operational risk modeling. *Journal of Financial Services Research*, **41** (3), 145–161.

Ergashev, B., Pavlikov, K., Uryasev, S., & Sekeris, E. 2012. *Estimation of Truncated Data Samples in Operational Risk Modeling*. Working paper. Technical report available at SSRN 2193493. Accessed July 1, 2014.

Eriksson, A., Ghysels, E., & Forsberg, L. 2004. *Approximating the Probability Distribution of Functions of Random Variables: A New Approach*. Montreal, Canada: Cirano.

Eriksson, A., Ghysels, E., & Wang, F. 2009. The normal inverse Gaussian distribution and the pricing of derivatives. *The Journal of Derivatives*, **16** (3), 23–37.

Esary, James D., and Frank Proschan. 1972. "Relationships among some concepts of bivariate dependence". The Annals of Mathematical Statistics 43, no. 2: 651–655.

Esscher, F. 1932. On the probability function in the collective theory of risk. *Skandinavisk Aktuarietidskrift*, **15**, 175–195.

EUROCONTROL. 2004. *Review of Techniques to Support the EATMAP Safety Assessment Methodology*, Vol. 4. Brussels, Belgium: European Organization for the Safety of Air Navigation.

Faa di Bruno, C. F. 1857. Note sur une nouvelle formule de calcul différentiel. *Quarterly Journal of Pure and Applied Mathematics*, **1**, 359–360.

Fan, J., & Li, R. 2001. Variable selection via nonconcave penalized likelihood and its oracle properties. *Journal of the American Statistical Association*, **96** (456), 1348–1360.

Fang, H.-B., Fang, K.-T., & Kotz, S. 2002. The meta-elliptical distributions with given marginals. *Journal of Multivariate Analysis*, **82** (1), 1–16.

FAST. 2004. *Toolsets/System Safety Management Program—Section 4*. Federal Aviation Authority Acquisition System Toolset. U.S. Department of Transportation, Federal Aviation Administration, Washington, DC.

FED. April 2009. *The Supervisory Capital Assessment Program: Design and Implementation*. Federal Reserve Bank, Washington, DC.

Feigin, P. D., & Heathcote, C. R. 1976. The empirical characteristic function and the Cramer-von Mises statistic. *Sankhyâ: The Indian Journal of Statistics, Series A*, **38** (4), 309–325.

Feller, W. 1966. *An Introduction to Probability Theory and Its Applications*, Vol. 2. New York: John Wiley & Sons.

Feller, W. 2008. *An Introduction to Probability Theory and Its Applications*, Vol. 2, New York: John Wiley & Sons.

Feng, J., Li, J., Gao, L., & Hua, Z. 2012. A combination model for operational risk estimation in a Chinese banking industry case. *Journal of Operational Risk*, **7** (2), 17.

Fenton, N. E., Marsh, W., Neil, M., Cates, P., Forey, S., & Tailor, M. 2004. Making resource decisions for software projects. *Proceedings of 26th International Conference on Software Engineering (ICSE 2004)*, Edinburgh, UK. New York: IEEE Computer Society Press, pp. 397–406.

Ferdousa, R., Khana, F., Sadiqb, R., Amyotte, P., & Veitcha, B. 2013. Analyzing system safety and risks under uncertainty using a bow-tie diagram: An innovative approach. *Process Safety and Environmental Protection*, **91** (1–2), 1–18.

Ferguson, T. S. 1973. A Bayesian analysis of some nonparametric problems. *Annals of Statistics*, **1** (2), 209–230.

Ferguson, T. S. 1996. *A Course in Large Sample Theory*. London, UK: Chapman and Hall.

Fermanian, J.-D., Radulovic, D., & Wegkamp, M. 2004. Weak convergence of empirical copula processes. *Bernoulli*, **10** (5), 847–860.

Ferson, S., Kreinovich, V., Ginzburg, L., Myers, D. S., & Sentz, K. January 2003. *Constructing Probability Boxes and Dempster-Shafer Structures*. Sandia National Laboratories, Albuquerque, NM/Livermore, CA. SAND report: SAND2002-4015.

Feuerverger, A., & Mureika, R. A. 1977. The empirical characteristic function and its applications. *The Annals of Statistics*, **5** (1) 88–97.

Field, C., & Genton, M. G. 2006. The multivariate g-and-h distribution. *Technometrics*, **48** (1), 104.

Fischer, T. 2003. Risk capital allocation by coherent risk measures based on one-sided moments. *Insurance: Mathematics and Economics*, **32** (1), 135–146.

Fischer, M. 2010. Generalized Tukey-type distributions with application to financial and teletraffic data. *Statistical Papers*, **51** (1), 41–56.

Fischer, M., & Klein, I. 2004. Kurtosis modelling by means of the J-transformation. *Allgemeines Statistisches Archiv*, **88** (1), 35–50.

Fischer, M., Horn, A., & Klein, I. 2007. Tukey-type distributions in the context of financial data. *Communications in Statistics Theory and Methods*, **36** (1), 23–35.

Fischhoff, B., Slovic, P., & Lichtenstein, S. 1978. Fault trees: Sensitivity of estimated failure probabilities to problem representation. *Journal of Experimental Psychology: Human Perception and Performance*, **4** (2), 330.

Fisher, N. I. 1997. Copulas. In Kotz, S., Read, C. B., & Banks, D. L. (eds.), *Encyclopedia of Statistical Sciences*, Update Vol. 1. New York: Wiley.

Fitzenberger, B., Koenker, R., & Machado, J. A. F. 2002. *Economic Applications of Quantile Regression*. New York: Springer.

Flegal, J. M., Jones, G. L., & Neath, R. C. 2012. Markov chain Monte Carlo estimation of quantiles. Preprint arXiv:1207.6432, available at http://arxiv.org. Accessed July 1, 2014.

Fleming, T. R., O'Fallon, J. R., O'Brien, P. C., & Harrington, D. P. 1980. Modified Kolmogorov-Smirnov test procedures with application to arbitrarily right-censored data. *Biometrics*, **36** (4), 607–625.

Folks, J. L., & Chhikara, R. S. 1978. The inverse Gaussian distribution and its statistical application—A review. *Journal of the Royal Statistical Society. Series B (Methodological)*, **40** (3), 263–289.

Föllmer, H. 1991. Probabilistic aspects of finance. *Bernoulli*, **19** (4), 1306–1326.

Föllmer, H., & Schweizer, M. 1991. Hedging of contingent claims. *Applied Stochastic Analysis*, **5** (1991), 389.

Föllmer, H., & Sondermann, D. 1986. Hedging of contingent claims under incomplete information. In Davis, M. H. A., & Elliott, R. J. (eds.), *Applied Stochastic Analysis, Stochastics Monographs*, Vol. 5. Amsterdam, the Netherlands: North-Holland, pp. 389–414.

Föllmer, H., & Schied, A. 2002. Convex measures of risk and trading constraints. *Finance and Stochastics*, **6** (4), 429–447.

Frachot, A., Roncalli, T., & Salomon, E. 2004a. The correlation problem in operational risk. Working paper. Groupe de Recherche Opérationnelle, Paris, France. Preprint, available at SSRNhttp://ssrn.com/abstract=1032594. Accessed July 1, 2014.

Frachot, A., Moudoulaud, O., & Roncalli, T. 2004b. Loss distribution approach in practice. In Ong, M. (ed.), *The Basel Handbook: A Guide for Financial Practitioners*. London, UK: Risk Books.

Frahm, G., Junker, M., & Schmidt, R. 2005. Estimating the tail-dependence coefficient: Properties and pitfalls. *Insurance: Mathematics and Economics*, **37** (1), 80–100.

Franzetti, C. 2011. *Operational Risk Modelling and Management*. Boca Raton, FL: Taylor & Francis Group.

Fredricks, G. A., & Nelsen, R. B. 2007. On the relationship between Spearman's rho and Kendall's tau for pairs of continuous random variables. *Journal of Statistical Planning and Inference*, **137** (7), 2143–2150.

Frees, E. W., & Valdez, E. A. 1998. Understanding relationships using copulas. *North American Actuarial Journal*, **2** (1), 1–25.

Friedman, J., Hastie, T., & Tibshirani, R. 2010. Regularization paths for generalized linear models via coordinate descent. *Journal of Statistical Software*, **33** (1), 1–22.

Frittelli, M., Maggis, M., & Peri, I. 2014. Risk measures on P(R) and value at risk with probability/loss functions. *Mathematical Finance*, **24** (3), 442–463.

Froot, K. A. 2001. The market for catastrophe risk: A clinical examination. *Journal of Financial Economics*, **60** (2), 529–571.

Froot, K. A. 2007. *The Financing of Catastrophe Risk*. Chicago, IL: University of Chicago Press.

Froot, K. A., Murphy, B., Stern, A., and Usher, S. Summer 1995. The emerging asset class: Insurance risk. *Viewpoint*, **24**, (3), 19–28.

Fujita, Y. 1993. A generalization of the results of Pillai. *Annals of the Institute of Statistical Mathematics*, **45** (2), 361–365.

Gagan, P. 2008. Operational risk—What lies beneath: Operational risk issues underlying the subprime crisis—While outsourcing its credit risk, Countrywide financial created huge operational risks through its business practices and strategy. *RMA Journal*, **91** (1), 96.

Gallant, A. R., & Tauchen, G. 1996. Which moments to match? *Econometric Theory*, **12** (4), 657–681.

Galton, F. 1889. *Natural Inheritance*, Vol. 42. London, UK: Macmillan.

Ganegoda, A., & Evans, J. 2013. A scaling model for severity of operational losses using generalized additive models for location scale and shape (GAMLSS). *Annals of Actuarial Science*, **7** (1), 61–100.

Garcia, R., Renault, E., & Veredas, D. 2011. Estimation of stable distributions by indirect inference. *Journal of Econometrics*, **161** (2), 325–337.

Garson, D. 1991. Interpreting neural-network connection strengths. *AI Expert*, **April**, 47–51.

Garthwaite, P. H., Kadane, J. B., & O'Hagan, A. 2005. Statistical methods for eliciting probability distributions. *Journal of the American Statistical Association*, **100** (470), 680–701.

Gelfand, A. E. 2000. Gibbs sampling. *Journal of the American Statistical Association*, **95** (452), 1300–1304.

Gelfand, A., & Dey, D. 1994. Bayesian model choice: Asymptotic and exact calculations. *Journal of the Royal Statistical Society, Series B (Methodological)*, **56** (3), 501–514.

Gelman, A., & Meng, X. L. 1998. Simulating normalizing constants: From importance sampling to bridge sampling to path sampling. *Statistical Science*, **13** (2), 163–185.

Gelman, A., & Rubin, D. B. 1992. Inference from iterative simulation using multiple sequences. *Statistical Science*, **7** (4), 457–472.

Gelman, A., Gilks, W. R., & Roberts, G. O. 1997. Weak convergence and optimal scaling of random walks Metropolis algorithm. *Annals of Applied Probability*, **7** (1), 110–120.

Geluk, J., & Tang, Q. 2009. Asymptotic tail probabilities of sums of dependent subexponential random variables. *Journal of Theoretical Probability*, **22** (4), 871–882.

Geman, S., & Geman, D. 1984. Stochastic relaxation, Gibbs distributions, and the Bayesian restoration of images. *IEEE Transactions on Pattern Analysis and Machine Intelligence*, **PAMI-6** (6), 721–741.

Genest, C., & Favre, A. C. 2007. Everything you always wanted to know about copula modeling but were afraid to ask. *Journal of Hydrologic Engineering*, **12** (4), 347–368.

Genest, C., & Mackay, J. 1986. The joy of copulas: Bivariate distributions with uniform marginals. *The American Statistician*, **40** (4), 280–283.

Genest, C., & Neslehova, J. 2007. A primer on copulas for count data. *ASTIN Bulletin*, **37** (2), 475.

Genest, C., & Rémillard, B. 2006. Discussion of "Copulas: Tales and facts", by Thomas Mikosch. *Extremes*, **9** (1), 27–36.

Genest, C., & Rémillard, B. 2008. Validity of the parametric bootstrap for goodness-of-fit testing in semiparametric models. *Annales de l'Institut Henri Poincaré: Probabilités et Statistiques*, **44** (6), 1096–1127.

Genest, C., Ghoudi, K., & Rivest, L.-P. 1995. A semiparametric estimation procedure of dependence parameters in multivariate families of distributions. *Biometrika*, **82** (3), 543–552.

Genest, C., Quessy, J. F., & Rémillard, B. 2006. Goodness-of-fit procedures for copula models based on the probability integral transformation. *Scandinavian Journal of Statistics*, **33** (2), 337–366.

Genest, C., Gendron, M., & Bourdeau-Brien, M. 2009a. The advent of copulas in finance. *The European Journal of Finance*, **15** (7–8), 609–618.

Genest, C., Rémillard, B., & Beaudoin, D. 2009b. Goodness-of-fit tests for copulas: A review and a power study. *Insurance: Mathematics and Economics*, **44** (2), 199–213.

Genest, C., Alberto Carabarín-Aguirre, & Fanny H. 2013. Copula parameter estimation using Blomqvist's beta. *Journal de la Société Française de Statistique* **154** (1), 5–24.

Genton, M. G., & Ronchetti, E. 2003. Robust indirect inference. *Journal of the American Statistical Association*, **98** (461), 67–76.

Gerber, H. U. 1978. Pareto-optimal risk exchanges and related decision problems. *ASTIN Bulletin*, **10** (1), 25–33.

Gerber, H. U. 1982. On the numerical evaluation of the distribution of aggregate claims and its stop-loss premiums. *Insurance: Mathematics and Economics*, **1** (1), 13–18.

Gerber, H. U. 1990. *Life Insurance Mathematics*. Berlin, Germany: Springer-Verlag.

Gerber, H. U., & Shiu, E. S. W. 1994. Option pricing by Esscher transforms. *Transactions of the Society of Actuaries*, **46** (99), 140.

Gerber, H. U., & Shiu, E. S. W. 1996. Actuarial bridges to dynamic hedging and option pricing. *Insurance: Mathematics and Economics*, **18** (3), 183–218.

Gerhold, S., Schmock, U., & Warnung, R. 2010. A generalization of Panjers recursion and numerically stable risk aggregation. *Finance and Stochastics*, **14** (1), 81–128.

Geweke, J. 1991. *Evaluating the Accuracy of Sampling-Based Approaches to the Calculation of Posterior Moments*. Research Department, Federal Reserve Bank of Minneapolis, Minneapolis, MN.

Geyer, C. J., & Thompson, E. A. 1995. Annealing Markov chain Monte Carlo with applications to ancestral inference. *Journal of the American Statistical Association*, **90** (431), 909–920.

Ghosh, M. 1981. Multivariate negative dependence. *Communications in Statistics-Theory and Methods*, **10** (4), 307–337.

Ghosh, J., & Ramamoorthi, R. 2003. *Bayesian Nonparametrics*. New York: Springer.

Ghosh, S., & Resnick, S. 2010. A discussion on mean excess plots. *Stochastic Processes and Their Applications*, **120** (8), 1492–1517.

Ghossoub, M. 2012. Belief heterogeneity in the Arrow-Borch-Raviv insurance model. Preprint: SSRN 2028550, available at http://www.ssrn.com/en/. Accessed July 1, 2014.

Giacometti, R., Rachev, S. T., Chernobai, A., & Bertocchi, M. 2008. Aggregation issues in operational risk. *The Journal of Operational Risk*, **3** (3), 3–23.

Gil, A., Segura, J., & Temme, N. M. 2002. Evaluation of the modified Bessel function of the third kind of imaginary orders. *Journal of Computational Physics*, **175** (2), 398–411.

Gilchrist, W. 2002. *Statistical Modelling with Quantile Functions*. Boca Raton, FL: CRC Press.

Giles, D. E. A. 2001. A saddlepoint approximation to the distribution function of the Anderson-Darling test statistic. *Communications in Statistics-Simulation and Computation*, **30** (4), 899–905.

Gilks, W. R., & Berzuini, C. 2002. Following a moving target Monte Carlo inference for dynamic Bayesian models. *Journal of the Royal Statistical Society: Series B (Statistical Methodology)*, **63** (1), 127–146.

Gilks, W. R., & Wild, P. 1992. Adaptive rejection sampling for Gibbs sampling. *Applied Statistics*, **41** (2), 337–348.

Gilks, W. R., Roberts, G. O., & George, E. I. 1994. Adaptive direction sampling. *Journal of the Royal Statistical Society. Series D (The Statistician)*, **43** (1), 179–189.

Gilks, W. R., Best, N. G., & Tan, K. C. 1995. Adaptive rejection Metropolis sampling within Gibbs sampling. *Applied Statistics*, **44** (4), 455–472.

Gilks, W. R., Richardson, S., & Spiegelhalter, D. J. 1996. *Markov Chain Monte Carlo in Practice*. London, UK/Boca Raton, FL: Chapman & Hall/CRC Press (Interdisciplinary Statistics).

Girolami, M., & Calderhead, B. 2011. Riemann manifold Langevin and Hamiltonian Monte Carlo methods. *Journal of the Royal Statistical Society: Series B (Statistical Methodology)*, **73** (2), 123–214.

Givens, G. H., & Raftery, A. E. 1996. Local adaptive importance sampling for multivariate densities with strong nonlinear relationships. *Journal of the American Statistical Association*, **91** (433), 132–141.

Glasserman, P. 2004. *Monte Carlo Methods in Financial Engineering*. New York: Springer.

Glasserman, P. 2005. Measuring marginal risk contributions in credit portfolios. *Journal Computational Finance*, **9** (2), 1–41.

Gneiting, T. 2011. Making and evaluating point forecasts. *Journal of the American Statistical Association*, **106** (494), 746–762.

Godsill, S. 2000. Inference in symmetric alpha-stable noise using MCMC and the slice sampler. *Proceedings of the IEEE International Conference on Acoustics, Speech and Signal Processing*, **6**, 3806–3809.

Gollier, C. 1992. Economic theory of risk exchanges: A review. *Contributions to Insurance Economics*, **13**, 3–23.

Gollier, C. 2005. *Some Aspects of the Economics of Catastrophe Risk Insurance*. Technical report. CESifo working paper series no. 1409, available at SSRN http://ssrn.com/abstract=668384. Accessed July 1, 2014.

Good, I. J. 1953. The population frequencies of species and the estimation of population parameters. *Biometrika*, **40** (3–4), 237–264.

Goovaerts, M. J., & Laeven, R. J. A. 2008. Actuarial risk measures for financial derivative pricing. *Insurance: Mathematics and Economics*, **42** (2), 540–547.

Goshay, R. C., & Sandor, R. 1973. An inquiry into the feasibility of a reinsurance futures market. *Journal of Business Finance*, **5** (2), 56–66.

Götze, F. 1979. Asymptotic expansions for bivariate von Mises functionals. *Probability Theory and Related Fields*, **50** (3), 333–355.

Gourieroux, C., & Monfort, A. 1997. *Simulation-Based Econometric Methods*. New York: Oxford University Press.

Gourieroux, C., Monfort, A., & Renault, E. 2006. Indirect inference. *Journal of Applied Econometrics*, **8** (S1), S85–S118.

Gramacy, R. B., Samworth, R., & King, R. 2010. Importance tempering. *Statistics and Computing*, **20** (1), 1–7.

Green, P. 1995. Reversible jump MCMC computation and Bayesian model determination. *Biometrika*, **82** (4), 711–732.

Greenwood, J. A., Landwehr, J. M., Matalas, N. C., & Wallis, J. R. 1979. Probability weighted moments: Definition and relation to parameters of several distributions expressable in inverse form. *Water Resources Research*, **15** (5), 1049–1054.

Gregoriou, G. N. 2009. *Operational Risk toward Basel III: Best Practices and Issues in Modeling, Management, and Regulation*. New York: Wiley.

Grønneberg, S., & Hjort, N. L. July 2008. *The Copula Information Criterion*. Statistical Research Report No. 7. Department of Mathematics, University of Oslo, Oslo, Norway.

Grossi, P., & Kunreuther, H. 2005. *Catastrophe Modeling: A New Approach to Managing Risk*, Vol. 25. New York: Springer.

Grothe, O., & Nicklas, S. 2013. Vine constructions of Lévy copulas. *Journal of Multivariate Analysis*, **119** (August), 1–15.

Grubel, R., & Hermesmeier, R. 1999. Computation of compound distributions I: Aliasing errors and exponential tilting. *ASTIN Bulletin*, **29** (2), 197–214.

Grübel, R. & Hermesmeier, R. 2000. Computation of compound distributions II: Discretization errors and Richardson extrapolation. *ASTIN Bulletin*, **30** (2), 309–31.

Gudendorf, G., & Segers, J. 2010. Extreme-value copulas. In Jaworski, P., Durante, F., Härdle, W., & Rychlik, T. (eds.), *Copula Theory and Its Applications*. Berlin, Germany: Springer, pp. 127–145.

Guillot, P. 1964. Une extension des lois *A* de Halphen comprenant comme cas limite la loi de Galton-Gibrat. *Revue de Statistique Appliquée*, **12** (1), 63–73.

Gumbel, E. J. 1960. Bivariate exponential distributions. *Journal of the American Statistical Association*, **55** (292), 698–707.

Gupta, A. K., & Nadarajah, S. 2004. *Handbook of Beta Distribution and Its Applications*, Vol. 175. New York: CRC Press.

Haario, H., Saksman, E., & Tamminen, J. 2001. An adaptive Metropolis algorithm. *Bernoulli*, **7** (2), 223–242.

Haario, H., Laine, M., Mira, A., & Saksman, E. 2006. DRAM: Efficient adaptive MCMC. *Statistics and Computing*, **16** (4), 339–354.

Haasl, D. F. 1965. Advanced concepts in fault tree analysis. *System Safety Symposium*, The Boeing Company, Seattle, WA.

Hadwiger, H. 1942. Wahl einer Naherungsfunktion fur Verteilungen auf Grund einer funktionalgeichung. *Blatter fur Versicherungsmathematik*, **5**, 345–352.

Hafner, C. M., & Manner, H. 2010. Dynamic stochastic copula models: Estimation, inference and applications. *Journal of Applied Econometrics*, **27** (2), 269–295.

Halphen, E. 1941. Sur un nouveau type de courbe de frequence. *Comptes Rendus de l'Academie des Sciences*, **213**, 633–635 (due to war constraints, published under the name Dugue).

Halphen, E. 1953. Un exemple d'application des méthodes statistiques: Le problème du plan pour l'équipement électrique français. *Revue de Statistique Appliquée*, **1** (1), 39–46.

Halphen, E. 1955. *Les fonctions factorielles*. Paris, France: Institut de Statistique de l'Université de Paris.

Hanssen, A., & Oigard, T. A. 2001. The normal inverse Gaussian distribution: A versatile model for heavy-tailed stochastic processes. *2001 IEEE International Conference on Acoustics, Speech, and Signal Processing (ICASSP*'01), Salt Lake City, UT, May 7–11, 2001, Vol. 6, pp. 3985–3988.

Härdle, W. K., & Cabrera, B. L. 2010. Calibrating CAT bonds for Mexican earthquakes. *Journal of Risk and Insurance*, **77** (3), 625–650.

Harrison, J. M., & Kreps, D. M. 1979. Martingales and arbitrage in multiperiod securities markets. *Journal of Economic Theory*, **20** (3), 381–408.

Harrison, J. M., & Pliska, S. R. 1981. Martingales and stochastic integrals in the theory of continuous trading. *Stochastic Processes and Their Applications*, **11** (3), 215–260.

Hastie, T., Tibshirani, R., & Friedman, J. 2009. *Linear Methods for Regression*. New York: Springer.

Hastings, W. K 1970. Monte Carlo sampling methods using Markov chains and their applications. *Biometrika*, **57** (1), 97–109.

Headrick, T. C., & Pant, M. D. 2012a. Characterizing Tukey h and hh-distributions through L-moments and L-correlation. *ISRN Applied Mathematics*, **2012**, Article ID 980153, 20pp.

Headrick, T. C., & Pant, M. D. 2012b. A logistic L-moment-based analog for the Tukey g-h, g, h, and h-h system of distributions. *ISRN Probability and Statistics*, **2012**, Article ID 245986, 23pp.

Headrick, T. C., Kowalchuk, R. K., & Sheng, Y. 2008. Parametric probability densities and distribution functions for Tukey g-and-h transformations and their use for fitting data. *Applied Mathematical Sciences*, **2** (9), 449–462.

Heathcote, C. E. 1972. A test of goodness of fit for symmetric random variables. *Australian Journal of Statistics*, **14** (2), 172–181.

Heckerman, D., Geiger, D., & Chickering, D. M. 1995. Learning Bayesian networks: The combination of knowledge and statistical data. *Machine Learning*, **20** (3), 197–243.

Hess, K. T., Liewald, A., & Schmidt, K. D. 2002. An extension of Panjer's recursion. *ASTIN Bulletin*, **32** (2), 283–297.

Hesselager, O. 1996. Recursions for certain bivariate counting distributions and their compound distributions. *ASTIN Bulletin*, **26** (1), 35–52.

Higdon, D. M. 1998. Auxiliary variable methods for Markov chain Monte Carlo with applications. *Journal of the American Statistical Association*, **93** (442), 585–595.

Hipp, C. 2006. Speedy convolution algorithms and Panjer recursions for phase-type distributions. *Insurance: Mathematics and Economics*, **38** (1), 176–188.

Hjort, N. L., Holmes, C., Müller, P., & Walker, S. G. 2010. *Bayesian Nonparametrics*, Vol. 28. Cambridge, UK: Cambridge University Press.

Hoaglin, D. C. 1985. Summarizing shape numerically: The g-and-h distributions. In Hoaglin D. C., Mosteller, F., & Tukey, J. W. (eds.), *Exploring Data Tables, Trends, and Shapes*. New York: Wiley, pp. 461–513.

Hoeffding, W. 1994a. Scale—Invariant correlation theory. In Fisher, N. I., & Sen P. K. (eds.), *The Collected Works of Wassily Hoeffding*. New York: Springer, pp. 57–107.

Hoeffding, W. 1994b. Scale—Invariant correlation theory. In Fisher, N. I. and Sen P. K. (eds.), *The Collected Works of Wassily Hoeffding*. New York: Springer, pp. 109–133.

Hoerl, A. E., & Kennard, R. W. 1970. Ridge regression: Biased estimation for nonorthogonal problems. *Technometrics*, **12** (1), 55–67.

Hofert, M. 2008. Sampling Archimedean copulas. *Computational Statistics & Data Analysis*, **52** (12), 5163–5174.

Hofert, M., Mächler, M., & Mcneil, A. J. 2012. Likelihood inference for Archimedean copulas in high dimensions under known margins. *Journal of Multivariate Analysis*, **110** (September), 133–150.

Hoffman, D. G. 2002. Managing operational risk: 20 firmwide best practice strategies. John Wiley & Sons.

Hogarth, R. M., & Einhorn, H. J. 1992. Order effects in belief updating: The belief-adjustment model. *Cognitive Psychology*, **24** (1), 1–55.

Hosack, G. R., Peters, G. W., & Hayes, K. R. 2012. Estimating density dependence and latent population trajectories with unknown observation error. *Methods in Ecology and Evolution*, **3** (6), 1028–1038.

Hossack, G., Peters, G. W., & Ludsin, S. A. 2014. Interspecific relationships and environmentally driven catchabilities estimated from fisheries data. *Canadian Journal of Fisheries and Aquatic Sciences*, **71** (3), 447–463.

Hua, L., & Joe, H. 2011. Tail order and intermediate tail dependence of multivariate copulas. *Journal of Multivariate Analysis*, **102** (10), 1454–1471.

Huang, M. L., & Fung, K. Y. 1993. D-distribution and its applications. *Statistical Papers*, **34** (1), 143–159.

Huber, P. J. 1972. The 1972 Wald lecture robust statistics: A review. *The Annals of Mathematical Statistics*, **43** (4), 1041–1067.

Hugonnier, J., Kramkov, D., & Schachermayer, W. 2005. On utility-based pricing of contingent claims in incomplete markets. *Mathematical Finance*, **15** (2), 203–212.

Ibragimov, I. A. 1956. On the composition of unimodal distributions. *Theory of Probability & Its Applications*, **1** (2), 255–260.

Jackson, D. 1941. *Fourier Series and Orthogonal Polynomials*. New York: Courier Dover Publications.

Jacobs, R. A. 1995. Methods for combining experts' probability assessments. *Neural Computation*, **7** (5), 867–888.

Jaillet, P., Ronn, E. I., & Tompaidis, S. 2004. Valuation of commodity-based swing options. *Management Science*, **50** (7), 909–921.

Jaimungal, S., & Wang, T. 2006. Catastrophe options with stochastic interest rates and compound Poisson losses. *Insurance: Mathematics and Economics*, **38** (3), 469–483.

Jasra, A., Stephens, D. A., & Holmes, C. C. 2007. On population-based simulation for static inference. *Statistics and Computing*, **17** (3), 263–279.

Jasra, A., Doucet, A., Stephens, D. A., & Holmes, C. C. 2008. Interacting sequential Monte Carlo samplers for trans-dimensional simulation. *Computational Statistics & Data Analysis*, **52** (4), 1765–1791.

Javid, A. A. 2009. Copulas with truncation-invariance property. *Communications in Statistics? Theory and Methods*, **38** (20), 3756–3771.

Jeffreys, H. 1961. *Theory of Probability*, 3rd edn. London, UK: Oxford University Press.

Jiménez, J. A., & Arunachalam, V. 2011. The Use of the Tukey's g-h family of distributions to calculate value at risk and conditional value at risk. *Journal of Risk*, **13** (4), 95–116.

Joag-Dev, K., & Proschan, F. 1983. Negative association of random variables with applications. *The Annals of Statistics*, **11** (1), 286–295.

Joe, H. 1990. Multivariate concordance. *Journal of Multivariate Analysis*, **35** (1), 12–30.

Joe, H. 1997. *Multivariate Models and Dependence Concepts*. London, UK: Chapman & Hall.

Joe, H. 2005. Asymptotic efficiency of the two-stage estimation method for copula-based models. *Journal of Multivariate Analysis*, **94** (2), 401–419.

Joe, H. 2006. Discussion of "Copulas: Tales and facts", by Thomas Mikosch. *Extremes*, **9** (1), 37–41.

Joe, H., Smith, R. L., & Weissman, I. 1992. Bivariate threshold methods for extremes. *Journal of the Royal Statistical Society. Series B (Methodological)*, **54** (1), 171–183.

Joe, H., Li, H., & Nikoloulopoulos, A. K. 2010. Tail dependence functions and vine copulas. *Journal of Multivariate Analysis*, **101** (1), 252–270.

Johansen, A. M. 2009. SMCTC: Sequential Monte Carlo in C++. *Journal of Statistical Software*, **30** (6), 1–41.

Johnson, N. L., Kotz, S., & Balakrishnan, N. 1970. *Distributions in Statistics: Continuous Univariate Distributions*, Vol. 2. New York: Houghton Mifflin.

Johnson, N. L., Kotz, S., & Balakrishnan, N. 1997. *Discrete Multivariate Distributions*. New York: John Wiley & Sons.

Johnson, N. L., Kotz, S., & Balakrishnan, N. 2002. *Continuous Multivariate Distributions*, Vol. 1, *Models and Applications*. New York: John Wiley & Sons.

Jondeau, E., & Rockinger, M. January 1999. *Estimating Gram-Charlier Expansions with Positivity Constraints*. Preprint SSRN 146743, available at http://ssrn.com. Accessed July 1, 2014.

Jondeau, E., & Rockinger, M. 2001. Gram–Charlier densities. *Journal of Economic Dynamics and Control*, **25** (10), 1457–1483.

Jones, G. L. 2004. On the Markov chain central limit theorem. *Probability Surveys*, **1**, 299–320.

Jorge, M., & Boris, I. 1984. Some properties of the Tukey g and h family of distributions. *Communications in Statistics-Theory and Methods*, **13** (3), 353–369.

Jørgensen, B. 1982. *Statistical Properties of the Generalized Inverse Gaussian Distribution*. Lecture Notes in Statistics, Vol. 9. New York: Springer-Verlag.

Jørgensen, B., & Lauritzen, S. L. 2000. Multivariate dispersion models. *Journal of Multivariate Analysis*, **74** (2), 267–281.

Jorion, P. 2007. *Value at Risk: The New Benchmark for Managing Financial Risk*. New York: McGraw-Hill.

JP Morgan. 1996. *Riskmetrics—Technical Document*, 4th edn. New York: Morgan Guaranty Trust Company, Risk Management Advisory.

Kaas, R., Goovaerts, M. J., Dhaene, J., & Denuit, M. 2001. *Modern Actuarial Risk Theory*. Dordrecht, the Netherlands: Kluwer Academic Publishers.

Kadane, J. B. 1980. Predictive and structural methods for eliciting prior distributions. In Zellner, A. (ed.), *Bayesian Analysis in Econometrics and Statistics*. Amsterdam, the Netherlands: North-Holland, pp. 89–93.

Kahneman, D., Slovic, P., & Tversky, A. 1982. *Judgment under Uncertainty: Heuristics and Biases*. Cambridge, UK: Cambridge University Press.

Kalbfleisch, J. D., & Prentice, R. L. 2011. *The Statistical Analysis of Failure Time Data*, Vol. 360. New York: John Wiley & Sons.

Kalkbrener, M. 2005. An axiomatic approach to capital allocation. *Mathematical Finance*, **15** (3), 425–437.

Kalkbrener, M., Lotter, H., & Overbeck, L. 2004. Sensible and efficient capital allocation for credit portfolios. *Risk*, **17** (1), S19–S24.

Kallsen, J. 2002. Utility-based derivative pricing in incomplete markets. *Mathematical Finance–Bachelier Congress*. Berlin, Germany: Springer, pp. 313–338.

Kallsen, J., & Tankov, P. 2006. Characterization of dependence of multidimensional Lévy processes using Lévy copulas. *Journal of Multivariate Analysis*, **97** (7), 1551–1572.

Karamata, J. 1933. Sur un mode de croissance régulière. Théorèmes fondamentaux, *Bull. Soc. Math. France*, **61**, 55–62.

Karatzas, I., & Shreve, S. E. 1991. *Brownian Motion and Stochastic Calculus*, Vol. 113. New York: Springer.

Karl-Heinz, B. Y. 1994. On the exact calculation of the aggregate claims distribution in the individual life model. *ASTIN Bulletin*, **16** (2), 109–112.

Kashyap, A. K., & Stein, J. C. 2004. Cyclical implications of the Basel II capital standards. *Economic Perspectives-Federal Reserve Bank of Chicago*, **28** (1), 18–33.

Kass, R., & Raftery, A. 1995. Bayes factor. *Journal of the American Statistical Association*, **90** (430), 773–792.

Kass, R. E., Carlin, B. P., Gelman, A., & Neal, R. M. 1998. Markov chain Monte Carlo in practice: A roundtable discussion. *The American Statistician*, **52** (2), 93–100.

Katti, S. K. 1967. Infinite divisibility of integer-valued random variables. *The Annals of Mathematical Statistics*, **38** (4), 1306–1308.

Kawai, Y. 2005. IAIS and recent developments in insurance regulation. *The Geneva Papers on Risk and Insurance-Issues and Practice*, **30** (1), 29–33.

Kawata, T., & Maejima, M. 1977. Remarks on an infinitely divisible characteristic function. *Sankhyâ: The Indian Journal of Statistics, Series A*, **39** (2), 130–137.

Kelly Jr., J. 1956. A new interpretation of information rate. *IRE Transactions on Information Theory*, **2** (3), 185–189.

Kendall, M. G. 1938. A new measure of rank correlation. *Biometrika*, **30** (1–2), 81–93.

Kendall, M. G., Stuart, A., & Ord, J. K. 1994. *Distribution Theory*, Vol. 1. London, UK: Arnold.

Kielholz, W., & Durrer, A. 1997. Insurance derivatives and securitization: new hedging perspectives for the US cat insurance market. *The Geneva Papers on Risk and Insurance-Issues and Practice*, **22** (1), 3–16.

Kijima, M., & Muromachi, Y. 2008. An extension of the Wang transform derived from Bühlmann's economic premium principle for insurance risk. *Insurance: Mathematics and Economics*, **42** (3), 887–896.

Kim, G., Silvapulle, M. J., & Silvapulle, P. 2007. Comparison of semiparametric and parametric methods for estimating copulas. *Computational Statistics & Data Analysis*, **51** (6), 2836–2850.

Kimeldorf, George, and Allan R. Sampson. 1989. "A framework for positive dependence." Annals of the Institute of Statistical Mathematics 41, no. 1: 31–45.

Kimberling, C. H. 1974. A probabilistic interpretation of complete monotonicity. *Aequationes Mathematicae*, **10** (2), 152–164.

King, J. L. 2001. *Operational Risk: Measurements and Modelling*. Chichester, UK: John Wiley & Sons.

Klebanov, L. B., Maniya, G. M., & Melamed, I. A. 1985. A problem of Zolotarev and analogs of infinitely divisible and stable distributions in a scheme for summing a random number of random variables. *Theory of Probability & Its Applications*, **29** (4), 791–794.

Klebanov, L. B., Melamed, J. A., Mittnik, S., & Rachev, S. T. 1996. Integral and asymptotic representations of geo-stable densities. *Applied Mathematics Letters*, **9** (6), 37–40.

Klugman, S., & Parsa, A. 1999. Fitting bivariate distributions with copulas. *Insurance: Mathematics and Economics*, **24** (1–2), 139–148.

Klugman, S. A., Panjer, H. H., & Willmot, G. E. 1998. *Loss Models: From Data to Decisions*. New York: John Wiley & Sons.

Klüppelberg, C., Kuhn, G., & Peng, L. 2008. Semi-parametric models for the multivariate tail dependence function—The asymptotically dependent case. *Scandinavian Journal of Statistics*, **35** (4), 701–718.

Knott, M. 1974. The distribution of the Cramér-von Mises statistic for small sample sizes. *Journal of the Royal Statistical Society. Series B (Methodological)*, **36** (3), 430–438.

Ko, B., & Tang, Q. 2008. Sums of dependent nonnegative random variables with subexponential tails. *Journal of Applied Probability*, **45** (1), 85–94.

Koenker, R. 2001. Quantile regression. In El-Shaarawi, A. H. & Piegorsch W. W. (eds.), *Encyclopedia of Environmetrics*. John Wiley & Sons, Ltd.

Koenker, R. 2005. *Quantile Regression*. Cambridge, UK: Cambridge University Press.

Koenker, R., & Machado, J. A. F. 1999. Goodness of fit and related inference processes for quantile regression. *Journal of the American Statistical Association*, **94** (448), 1296–1310.

Koenker, R., & Hallock, K. 2001. Quantile regression: An introduction. *Journal of Economic Perspectives*, **15** (4), 43–56.

Kokoszka, P. S., & Taqqu, M. S. 1994. Infinite variance stable ARMA processes. *Journal of Time Series Analysis*, **15** (2), 203–220.

Kolmogorov, A. N. 1933. Sulla determinazione empirica di una legge di distribuzione. *Giornale dell'Istituto Italiano degli Attuari*, **4** (1), 83–91.

Kolmogorov, A. N. 1941. Confidence limits for an unknown distribution function. *Annals of Mathematical Statistics*, **12** (4), 461–463.

Korn, R., & Schäl, M. 2009. The numeraire portfolio in discrete time: existence, related concepts and applications. *De Gruyter, Advanced Financial Modelling*, Editors: Albrecher H., Runggaldier W.J., Schachermayer W., **8**, 303.

Kornfeld, I. P., Fomin, S. V., & Sinai, Ya. G. 1982. *Ergodic Theory*, Vol. 245. New York: Springer-Verlag.

Korostil, I. A., Peters, G. W., Cornebise, J., & Regan, D. G. 2012. Adaptive Markov chain Monte Carlo forward projection for statistical analysis in epidemic modelling of human papillomavirus. *Statistics in Medicine*, **32** (11), 1917–1953.

Kotz, S., & Ostrovskii, I. V. 1996. A mixture representation of the Linnik distribution. *Statistics & Probability Letters*, **26** (1), 61–64.

Kotz, S., Ostrovskii, I. V., & Hayfavi, A. 1995a. Analytic and asymptotic properties of Linnik's probability densities. II. *Journal of Mathematical Analysis and Applications*, **193** (2), 497–521.

Kotz, S., Ostrovskii, I. V., & Hayfavi, A. 1995b. Analytic and asymptotic properties of Linnik's probability densities, I. *Journal of Mathematical Analysis and Applications*, **193** (1), 353–371.

Koutrouvelis, I. A. 1980. A goodness-of-fit test of simple hypotheses based on the empirical characteristic function. *Biometrika*, **67** (1), 238–240.

Koutrouvelis, I. A., & Kellermeier, J. 1981. A goodness-of-fit test based on the empirical characteristic function when parameters must be estimated. *Journal of the Royal Statistical Society. Series B (Methodological)*, **43** (2), 173–176.

Kozik, T. J., & Larson, A. M. 2001. The n-moment insurance CAPM. *Proceedings of the Casualty Actuarial Society*, **88** (168), 39–63.

Kozubowski, T. J. 1994. Representation and properties of geometric stable laws. In Anastassiou, G., & Rachev, S. T. (eds.), *Approximation, Probability, and Related Fields*. New York: Springer, pp. 321–337.

Kozubowski, T. J. 1999. Geometric stable laws: Estimation and applications. *Mathematical and Computer Modelling*, **29** (10), 241–253.

Kozubowski, T. J. 2000a. Computer simulation of geometric stable distributions. *Journal of Computational and Applied Mathematics*, **116** (2), 221–229.

Kozubowski, T. J. 2000b. Exponential mixture representation of geometric stable distributions. *Annals of the Institute of Statistical Mathematics*, **52** (2), 231–238.

Kratz, M., & Resnick, S. I. 1996. The QQ-estimator and heavy tails. *Stochastic Models*, **12** (4), 699–724.

Krengel, U., & Brunel, A. 1985. *Ergodic Theorems*. Cambridge, UK: Cambridge University Press.

Krokhmal, P. A. 2007. Higher moment coherent risk measures. *Quantitative Finance*, **7** (4), 373–387.

Kronmal, R., & Tarter, M. 1968. The estimation of probability densities and cumulatives by Fourier series methods. *Journal of the American Statistical Association*, **63** (323), 925–952.

Krugman, P., & Wells, R. 2012. *Microeconomics*, 3rd edn. New York: Worth Publishers.

Kruskal, W. H. 1958. Ordinal measures of association. *Journal of the American Statistical Association*, **53** (284), 814–861.

Kumar, P. 2010. Probability distributions and the estimation of Ali-Mikhail-Haq copula. *Applied Mathematical Statistics*, **4** (14), 657–666.

Künsch, H. R. 2005. Recursive Monte Carlo filters: Algorithms and theoretical analysis. *The Annals of Statistics*, **33** (5), 1983–2021.

Kuon, S., Reich, A., & Reimers, L. 1987. Panjer vs Kornya vs De Pril: A Comparison from a Practical Point of View. *ASTIN Bulletin*, **17**, 183.

Kusuoka, S. 2001. On law invariant coherent risk measures. In Kusuoka, S. & Maruyama, T. (eds.), *Advances in Mathematical Economics*. Springer, Springer-Verlag, Tokyo, pp. 83–95.

Lakdawalla, D., & Zanjani, G. 2012. Catastrophe bonds, reinsurance, and the optimal collateralization of risk transfer. *Journal of Risk and Insurance*, **79** (2), 449–476.

Lambert, N. S., Pennock, D. M., & Shoham, Y. 2008. Eliciting properties of probability distributions. *Proceedings of the 9th ACM Conference on Electronic Commerce*. New York: ACM Press, pp. 129–138.

Lambrigger, D. D., Shevchenko, P. V., & Wüthrich, M. V. 2007. The quantification of operational risk using internal data, relevant external data and expert opinions. *The Journal of Operational Risk*, **2** (3), 3–27.

Lane, M. N. 2000. Pricing risk transfer transactions. *ASTIN Bulletin*, **30** (2), 259–293.

Latane, H. A. 1959. Criteria for choice among risky ventures. *The Journal of Political Economy*, **67** (2), 144–155.

Łatuszyński, K., Roberts, G. O., & Rosenthal, J. S. 2013. Adaptive Gibbs samplers and related MCMC methods. *The Annals of Applied Probability*, **23** (1), 66–98.

Lavin, M., & Scherrish, M. 1999. Bayes factors: What they are and what they are not. *The American Statistician*, **53** (2), 119–122.

Lawless, J. F. 1980. Inference in the generalized gamma and log gamma distributions. *Technometrics*, **22** (3), 409–419.

LeCam, L., Mahan, C., & Singh, A. 1983. An extension of a theorem of H. Chernoff and EL Lehmann. In Rizvi, M. H., Rustagi, J. S., & Siegmund, D. (eds.), *Recent Advances in Statistics*. New York: Academic Press, pp. 303–332.

Lee, E. T. Y. 1982. A simplified B-spline computation routine. *Computing*, **29** (4), 365–371.

Lee, Y., & Nelder, J. A. 1996. Hierarchical generalized linear models. *Journal of the Royal Statistical Society. Series B (Methodological)*, **58** (4), 619–678.

Lee, J.-P., & Yu, M.-T. 2002. Pricing default-risky CAT bonds with moral hazard and basis risk. *Journal of Risk and Insurance*, **69** (1), 25–44.

Lee, J.-P., & Yu, M.-T. 2007. Valuation of catastrophe reinsurance with catastrophe bonds. *Insurance: Mathematics and Economics*, **41** (2), 264–278.

Lehmann, E. L. 1966. Some concepts of dependence. *The Annals of Mathematical Statistics*, **37** (5), 1137–1153.

Lehmann, E. L. 1983. *Theory of Point Estimation*. New York: John Wiley & Sons.

Lehmann, E. L., & Casella, G. 1998. *Theory of Point Estimation*, 2nd edn. New York: Springer.

Leipnik, R. B., & Pearce, C. E. M. 2007. The multivariate Faa di Bruno formula and multivariate Taylor expansions with explicit integral remainder term. *Anziam Journal*, **48** (3), 327.

LePage, R., Podgórski, K., & Ryznar, M. 1997. Strong and conditional invariance principles for samples attracted to stable laws. *Probability Theory and Related Fields*, **108** (2), 281–298.

Lewis, P. A. W. 1961. Distribution of the Anderson-Darling statistic. *The Annals of Mathematical Statistics*, **32**, (4), 1118–1124.

Lewis, C. M., & Murdock, K. C. 1996. The role of government contracts in discretionary reinsurance markets for natural disasters. *Journal of Risk and Insurance*, **63** (4), 567–597.

Lewis, C. M., & Lantsman, Y. February 11, 2005. What is a fair price to transfer the risk of unauthorised trading? A case study on operational risk. Technical report, available at SSRN http://ssrn.com/abstract=667103 or http://dx.doi.org/10.2139/ssrn.667103. Accessed July 1, 2014.

Li, H. 2009. Orthant tail dependence of multivariate extreme value distributions. *Journal of Multivariate Analysis*, **100** (1), 243–256.

Lin, S.-W., & Bier, V. M. 2008. A study of expert overconfidence. *Reliability Engineering & System Safety*, **93** (5), 711–721.

Linden, W. J., & Hambleton, R. K. 1997. *Handbook of Modern Item Response Theory*. New York.

Linder, U., & Ronkainen, V. 2004. Solvency II: Towards a new insurance supervisory system in the EU. *Scandinavian Actuarial Journal*, **2004** (6), 462–474.

Lindner, A. 2006. Discussion of "Copulas: Tales and facts", by Thomas Mikosch. *Extremes*, **9** (1), 43–44.

Lindskog, F., & McNeil, A. 2003. Common Poisson shock models: Application to insurance and credit risk modelling. *ASTIN Bulletin*, **33** (2), 209–238.

Ling, C.-H. 1964. *Representation of Associative Functions*. Ph.D. thesis, Illinois Institute of Technology, Chicago, IL.

Linstone, H. A., & Turoff, M. 1975. *The Delphi Method*. New York: Addison-Wesley Publishing Company.

Litterman, R. 1996. Hot spots^{TM} and hedges. *The Journal of Portfolio Management*, **27** (3), 52–75.

Liu, J. S. 2008. *Monte Carlo Strategies in Scientific Computing*. New York: Springer.

Liu, J. S., & Chen, R. 1998. Sequential Monte Carlo methods for dynamic systems. *Journal of the American Statistical Association*, **93** (443), 1032–1044.

Liu, J. S., Wong, W. H., & Kong, A. 1995. Covariance structure and convergence rate of the Gibbs sampler with various scans. *Journal of the Royal Statistical Society. Series B (Methodological)*, **57** (1), 157–169.

Liu, J. S., Chen, R., & Wong, W. H. 1998. Rejection control and sequential importance sampling. *Journal of the American Statistical Association*, **93** (443), 1022–1031.

Lloyd, S. 1982. Least squares quantization in PCM. *Information Theory, IEEE Transactions on* **28** (2), 129–137.

Long, J. B. 1990. The numéraire portfolio. *Journal of Financial Economics*, **26** (1), 29–69.

Loubergé, H., Kellezi, E., & Gilli, M. 1999. Using catastrophe-linked securities to diversify insurance risk: A financial analysis of CAT bonds. *Journal of Insurance Issues*, **22** (2), 125–146.

Lozier, D. W., & Olver, F. W. J. 1994. Numerical evaluation of special functions. *AMS Proceedings of Symposia in Applied Mathematics*, **48**, 79–125.

Lu, J., Guo, L., & Liu, X. 2013. Measuring the operational risk of Chinese commercial banks using the semilinear credibility model. *The Journal of Operational Risk*, **8** (2), 3–34.

Lu, J.-C., & Bhattacharyya, G. K. 1991. Inference procedures for a bivariate exponential model of Gumbel based on life test of component and system. *Journal of Statistical Planning and Inference*, **27** (3), 383–396.

Lugannani, R., & Rice, S. 1980. Saddle point approximation for the distribution of the sum of independent random variables. *Advances in Applied Probability*, **12**, 475–490.

Luo, X., & Shevchenko, P. V. 2009. Computing tails of compound distributions using direct numerical integration. *The Journal of Computational Finance*, **13** (2), 73–111.

Luo, X., & Shevchenko, P. V. 2010. The t copula with multiple parameters of degrees of freedom: bivariate characteristics and application to risk management. *Quantitative Finance*, **10** (9), 1039–1054.

Luo, X., & Shevchenko, P. V. 2012. Bayesian model choice of grouped t-copula. *Methodology and Computing in Applied Probability*, **14** (4), 1097–1119.

Luo, X., Shevchenko, P. V., & Donnelly, J. 2007. Addressing impact of truncation and parameter uncertainty on operational risk estimates. *The Journal of Operational Risk*, **2** (4), 3–26.

MacEachern, S. N., & Berliner, L. M. 1994. Subsampling the Gibbs sampler. *The American Statistician*, **48** (3), 188–190.

MacNeill, I. B. 1974. Tests for change of parameter at unknown times and distributions of some related functionals on Brownian motion. *The Annals of Statistics*, **2** (5), 950–962.

Maddala, G. S. 1983. *Limited Dependent and Qualitative Variables in Econometrics*. Cambridge, UK: Cambridge University Press.

Magnan, S. 1995. Catastrophe insurance system in France. *Geneva Papers on Risk and Insurance-Issues and Practices*, **20** (77), 474–480.

Mahbubul, M., Majumder, A., & Ali, M. M. 2008. A comparison of methods of estimation of parameters of Tukey's gh family of distributions. *Pakistan Journal of Statistics*, **24** (2), 135–144.

Malevergne, Y., & Sornette, D. 2003. Testing the Gaussian copula hypothesis for financial assets dependencies. *Quantitative Finance*, **3** (4), 231–250.

Malkiel, B. G., & Fama, E. F. 1970. Efficient capital markets: A review of theory and empirical work. *The Journal of Finance*, **25** (2), 383–417.

Mari, D. D., & Kotz, S. 2001. *Correlation and Dependence*, Vol. 2. Singapore: World Scientific.

Marin, J.-M., Pudlo, P., Robert, C. P., & Ryder, R. J. 2012. Approximate Bayesian computational methods. *Statistics and Computing*, **22** (6), 1167–1180.

Marinari, E., & Parisi, G. 1992. Simulated tempering: A new Monte Carlo scheme. *Europhysics Letters*, **19** (6), 451–458.

Marjoram, P., Molitor, J., Plagnol, V., & Tavaré, S. 2003. Markov chain Monte Carlo without likelihoods. *Proceedings of the National Academy of Sciences*, **100** (26), 15324–15328.

Markowitz, H. 1959. *Portfolio Selection: Efficient Diversification of Investments*. Cowles Foundation Monograph No. 16. New York: John Wiley & Sons, Inc.

Marshall, C. L. 2001. *Measuring and Managing Operational Risks in Financial Institutions*. Singapore: John Wiley & Sons.

Marshall, A. W., & Olkin, I. 1967. A multivariate exponential distribution. *Journal of the American Statistical Association*, **62** (317), 30–44.

Marshall, A. W., & Olkin, I. 1988. Families of multivariate distributions. *Journal of the American Statistical Association*, **83** (403), 834–841.

Marshall, T., & Roberts, G. 2012. An adaptive approach to Langevin MCMC. *Statistics and Computing*, **22** (5), 1041–1057.

Mason, D. M., & Schuenemeyer, J. H. 1983. A modified Kolmogorov-Smirnov test sensitive to tail alternatives. *The Annals of Statistics*, **11** (3), 933–946.

Massey, Jr., F. J. 1951. The Kolmogorov-Smirnov test for goodness of fit. *Journal of the American Statistical Association*, **46** (253), 68–78.

McConnell, P., & Davies, M. May 2, 2006. Safety first–scenario analysis under Basel II. Technical report of Operational Risk & Regulation, available at http://www.risk.net/data/basel_article_free/april06technical.pdf. Accessed July 1, 2014.

McCullagh, P., & Nelder, J. A. 1989. *Generalized Linear Models*, Vol. 37. London, UK/Boca Raton, FL: Chapman & Hall/CRC Press.

McDonald, J. B. 1984. Some generalized functions for the size distribution of income. *Econometrica: Journal of the Econometric Society*, **52** (3), 647–663.

McDonald, J. B. 1996. Probability distributions for financial models. In G.S. Maddala and C.R. Rao (eds.) *Handbook of Statistics*, **14**, *Statistical Methods in Finance*, Amsterdam: Elsevier Science, 427–461.

McDonald, J. B., & Xu, Y. J. 1995. A generalization of the beta distribution with applications. *Journal of Econometrics*, **66** (1), 133–152.

McLachlan, G. J., & Jones, P. N. 1988. Fitting mixture models to grouped and truncated data via the EM algorithm. *Biometrics*, **44** (2), 571–578.

McLachlan, G. J., & Krishnan, T. 1997. *The EM Algorithm and Extensions*. New York: John Wiley & Sons.

McNeil, A. J. 2008. Sampling nested Archimedean copulas. *Journal of Statistical Computation and Simulation*, **78** (6), 567–581.

McNeil, A. J., & Nešlehová, J. 2009. Multivariate Archimedean copulas, d-monotone functions and L_1-norm symmetric distributions. *The Annals of Statistics*, **37** (5B), 3059–3097.

McNeil, A. J., Frey, R., & Embrechts, P. 2005. *Quantitative Risk Management: Concepts, Techniques and Tools*. Princeton, NJ: Princeton University Press.

Mehr, R. I., & Hedges, B. A. 1963. *Risk Management in the Business Enterprise*. Homewood, IL: Irwin.

Mehr, R. I., Cammack, E., & Rose, T. 1980. *Principles of Insurance*, Vol. 8. Homewood, IL: Irwin, R. D.

Meister, S. 1995. Contributions to the mathematics of catastrophe insurance futures. Unpublished Diplomarbeit, ETH Zürich, Zurich, Switzerland.

Meng, X., & Wong, W. 1996. Simulating ratios of normalizing constants via a simple identity. *Statistical Sinica*, **6**, 831–860.

Mengersen, K. L., Robert, C. P., & Guihenneuc-Jouyaux, C. 1999. MCMC convergence diagnostics: A review. *Bayesian Statistics*, **6**, 415–440.

Merton, R. C. 1973. Theory of rational option pricing. *The Bell Journal of Economics and Management Science*, **4** (1), 141–183.

Merton, R. C. 1976. Option pricing when underlying stock returns are discontinuous. *Journal of Financial Economics*, **3** (1), 125–144.

Metropolis, N., Rosenbluth, A. W., Rosenbluth, M. N., Teller, A. H., & Teller, E. 1953b. Equation of state calculations by fast computing machines. *Journal of Chemical Physics*, **21** (6), 1087–1091.

Meucci, A. May 20, 2011. A short, comprehensive, practical guide to copulas. *GARP Risk Professional*, October 22–27, 2011. Technical report, available at SSRN http://ssrn.com/abstract=1847864 orhttp://dx.doi.org/10.2139/ssrn.1847864. Accessed July 1, 2014.

Meyn, S. P., Tweedie, R. L., & Glynn, P. W. 2009. *Markov Chains and Stochastic Stability*, Vol. 2. Cambridge, UK: Cambridge University Press.

Miazhynskaia, T., & Dorffner, G. 2006. A comparison of Bayesian model selection based on MCMC with an application to GARCH-type models. *Statistical Papers*, **47** (4), 525–549.

Mignola, G., & Ugoccioni, R. October 2005. Tests for extreme value theory. *Operational Risk & Compliance*, **6**, 32–35.

Mignola, G., & Ugoccioni, R. 2006. Effect of a data collection threshold in the loss distribution approach. *The Journal of Operational Risk*, **1** (4), 35–47.

Mihoubi, M. 2008. Bell polynomials and binomial type sequences. *Discrete Mathematics*, **308** (12), 2450–2459.

Mikosch, T. 2006a. Copulas: Tales and facts. *Extremes*, **9** (1), 3–20.

Mikosch, T. 2006b. Copulas: Tales and facts rejoinder. *Extremes*, **9** (1), 55–62.

Miller, L. H. 1956. Table of percentage points of Kolmogorov statistics. *Journal of the American Statistical Association*, **51** (273), 111–121.

Mira, A., & Tierney, L. 2003. Efficiency and convergence properties of slice samplers. *Scandinavian Journal of Statistics*, **29** (1), 1–12.

Mira, A., Møller, J., & Roberts, G. O. 2002. Perfect slice samplers. *Journal of the Royal Statistical Society: Series B (Statistical Methodology)*, **63** (3), 593–606.

Mises, R. V. 1947. On the asymptotic distribution of differentiable statistical functions. *The Annals of Mathematical Statistics*, **18** (3), 309–348.

Monfort, A., & Pegoraro, F. 2012. Asset pricing with second-order Esscher transforms. *Journal of Banking & Finance*, **36** (6), 1678–1687.

Morgenthaler, S., & Tukey, J. W. 2000. Fitting quantiles: Doubling, HR, HQ, and HHH distributions. *Journal of Computational and Graphical Statistics*, **9** (1), 180–195.

Morlat, G. 1951. Note sur l'estimation des débits de crues. *La Houille Blanche*, N° spécial B, 663–681, http://dx.doi.org/10.1051/lhb/1951021

Moscadelli, M. 2004. *The Modelling of Operational Risk: Experiences with the Analysis of the Data Collected by the Basel Committee.* Working paper no. 517. Bank of Italy, Rome, Italy. Technical report, available at SSRN http://ssrn.com/abstract=557214 orhttp://dx.doi.org/10.2139/ssrn.557214. Accessed July 1, 2014.

Murray, I., Adams, R. P., & MacKay, D. J. C. 2010. Elliptical slice sampling. *The Proceedings of the 13th International Conference on Artificial Intelligence and Statistics (AISTATS), JMLR W&CP*, **9**, 541–548.

Musiela, M., & Rutkowski, M. 2005. *Martingale Methods in Financial Modelling*, Vol. 36. Berlin, Germany: Springer.

Na, H., Van Den Berg, J., Miranda, L. C., & Leipoldt, M. 2006. An econometric model to scale operational losses. *The Journal of Operational Risk*, **1** (2), 11–31.

Nair, K. R. M., & Nair, N. U. 1988. On characterizing the bivariate exponential and geometric distributions. *Annals of the Institute of Statistical Mathematics*, **40** (2), 267–271.

Neal, R. M. 1993. *Probabilistic Inference Using Markov Chain Samplers.* Technical Report, Department of Computer Science, University of Toronto, Toronto, Ontario, Canada.

Neal, R. M. 1996. Sampling from multimodal distributions using tempered transitions. *Statistics and Computing*, **6** (4), 353–366.

Neal, R. M. 2000. Markov chain sampling methods for Dirichlet process mixture models. *Journal of Computational and Graphical Statistics*, **9** (2), 249–265.

Neal, R. M. 2001. Annealed importance sampling. *Statistics and Computing*, **11** (2), 125–139.

Neal, R. M. 2003. Slice sampling. *Annals of Statistics*, **31** (3), 705–741.

Neal, R. M. 2010. MCMC using Hamiltonian dynamics. *Handbook of Markov Chain Monte Carlo*, Brooks S., Gelman A., Jones G., and Meng X. (eds.), CRC Press, **54**, 113–162.

Neil, M., Fenton, N. E., Forey, S., & Harris, R. 2001. Using Bayesian belief networks to predict the reliability of military vehicles. *IEE Computing and Control Engineering*, **12** (1), 11–20.

Neil, M., Fenton, N. E., & Tailor, M. 2005. Using Bayesian networks to model expected and unexpected operational losses. *Risk Analysis*, **25** (4), 963–972.

Neil, M., Malcolm, B., & Shaw, R. 2003. Modelling an air traffic control environment using Bayesian belief networks. *21st International System Safety Conference*, Ottawa, Canada, August 4–8, 2003, 1, pp. 11–20.

Neil, M., Hager, D., & Andersen, L. B. 2009. Modeling operational risk in financial institutions using hybrid dynamic Bayesian networks. *Journal of Operational Risk*, **4** (1), 3–33.

Nelder, J. A., & Wedderburn, R. W. M. 1972. Generalized linear models. *Journal of the Royal Statistical Society. Series A (General)*, **135** (3), 370–384.

Nelsen, R. B. 1997. Dependence and order in families of Archimedean copulas. *Journal of Multivariate Analysis*, **60** (1), 111–122.

Nelsen, R. B. 1999. *An Introduction to Copulas*. New York: Springer.

Nelsen, R. B. 2002. Concordance and copulas: A survey. In Cuadras, C. M., Fortiana, J., & Rodriguez-Lallena, J. (eds.), *Distributions with Given Marginals and Statistical Modelling*. Dordrecht, the Netherlands: Springer, pp. 169–177.

Nelsen, R. B., & Úbeda-Flores, M. 2012. Directional dependence in multivariate distributions. *Annals of the Institute of Statistical Mathematics*, **64** (3), 677–685.

Nešlehová, J., Embrechts, P., & Chavez-Demoulin, V. 2006. Infinite mean models and the LDA for operational risk. *Journal of Operational Risk*, **1** (1), 3–25.

Newey, W. K., & Powell, J. L. 1987. Asymmetric least squares estimation and testing. *Econometrica: Journal of the Econometric Society*, **55** (4), 819–847.

Newson, R. 2002. Parameters behind "nonparametric" statistics: Kendall's tau, Somers' D and median differences. *Stata Journal*, **2** (1), 45–64.

Newton, M., & Raftery, A. 1994. Approximate Bayesian inference by the weighted likelihood bootstrap. *Journal of the Royal Statistical Society, Series B (Methodological)*, **56** (1), 1–48.

Nguyen, T. L. T., Septier, F., Peters, G. W., & Delignon, Y. September 2013. Bayesian model selection and parameter estimation in penalized regression model using SMC samplers. *21st European Signal Processing Conference (EUSIPCO)*, Marrakech, Morocco, pp. 1–5.

Nickl, R., & Pötscher, B. M. 2010. Efficient simulation-based minimum distance estimation and indirect inference. *Mathematical Methods of Statistics*, **19** (4), 327–364.

Niederhausen, H. 1981. Sheffer polynomials for computing exact Kolmogorov-Smirnov and Rényi type distributions. *The Annals of Statistics*, **9** (5), 923–944.

Nielsen, D. S. 1971. *Use of Cause-Consequence Charts in Practical Systems Analysis Reliability and Fault Tree Analysis*. Philadelphia, PA: SIAM.

Nikolaev, M. L., & Sofronov, G. 2007. A multiple optimal stopping rule for sums of independent random variables. *Discrete Mathematics and Applications DMA*, **17** (5), 463–473.

Noé, M., & Vandewiele, G. 1968. The calculation of distributions of Kolmogorov-Smirnov type statistics including a table of significance points for a particular case. *The Annals of Mathematical Statistics*, **39** (1), 233–241.

Nolan, J. P. 2015. *Stable Distributions: Models for Heavy Tailed Data*. Boston, MA: Birkhauser. In progress, Chapter 1 online at academic2.american.edu/~jpnolan. Accessed July 1, 2014.

Nowicka-Zagrajek, J., & Wyłomańska, A. 2008. Measures of dependence for stable AR (1) models with time-varying coefficients. *Stochastic Models*, **24** (1), 58–70.

Oakes, D. 2005. On the preservation of copula structure under truncation. *Canadian Journal of Statistics*, **33** (3), 465–468.

Oakley, J. E., & O'Hagan, A. 2007. Uncertainty in prior elicitations: A nonparametric approach. *Biometrika*, **94** (2), 427–441.

Oakley, J. E., Daneshkhah, A., & O'Hagan, A. 2010. Nonparametric prior elicitation using the Roulette method. Technical report, available at http://www.tonyohagan.co.uk/academic/pdf/elic-roulette.pdf. Accessed July 1, 2014.

Oberkampf, W. L. August 2005. Uncertainty quantification using evidence theory. *Advanced Simulation and Computing Workshop Error Estimation, Uncertainty Quantification, and Reliability in Numerical Simulations*, Stanford University, Stanford, CA.

Oberkampf, W. L., Helton, J. C., & Sentz, K. April 2001. *Mathematical Representation of Uncertainty*. Paper No. 2001-1645. American Institute of Aeronautics and Astronautics Non-Deterministic Approaches Forum, Seattle, WA.

Oh, M. S., & Berger, J. O. 1993. Integration of multimodal functions by Monte Carlo importance sampling. *Journal of the American Statistical Association*, **88** (422), 450–456.

O'Hagan, A. 1998. Eliciting expert beliefs in substantial practical applications. *Journal of the Royal Statistical Society: Series D (The Statistician)*, **47** (1), 21–35.

O'Hagan, T. 2005. Elicitation. *Significance*, **2** (2), 84–86.

O'Hagan, A. 2006. *Uncertain Judgements: Eliciting Expert's Probabilities (Statistics in Practice)*. Hoboken, NJ: Wiley.

O'Hagan, A., Buck, C. E., Daneshkhah, A., Eiser, J. R., Garthwaite, P. H., Jenkinson, D. J., Oakley, J. E., & Rakow, T. 2006. *Uncertain Judgements: Eliciting Experts' Probabilities*, Vol. 412. Chichester, UK: Wiley.

Ollard, W. 1985. How SEK Borrows Et 50 Below: SEK Thrills International Banks ; They Know It's the Sharpest Borrower in the World. *Euromoney*, **16**, 13–23.

Olver, F. W. J. 1960. *Royal Society Mathematical Tables*. Cambridge, UK: Cambridge University Press.

Onken, A., Grünewälder, S., Munk, M. H., & Obermayer, K. 2009. Analyzing short-term noise dependencies of spike-counts in macaque prefrontal cortex using copulas and the flashlight transformation. *PLoS Computational Biology*, **5** (11), e1000577.

Oppenheim, A. V., Schafer, R. W., Buck, J. R., *et al.* 1989. *Discrete-Time Signal Processing*, Vol. 2. Englewood Cliffs, NJ: Prentice Hall.

Osband, K., & Reichelstein, S. 1985. Information-eliciting compensation schemes. *Journal of Public Economics*, **27** (1), 107–115.

Osilenker, B. 1999. *Fourier Series in Orthogonal Polynomials*. Singapore: World Scientific Publishing Company Incorporated.

Overbeck, L. 2000. Allocation of economic capital in loan portfolios. In Frank, W. H. and Stahl, G. (eds.), *Measuring Risk in Complex Stochastic Systems*. New York: Springer, pp. 1–17.

Panjer, H. H. 1981. Recursive evaluation of a family of compound distribution. *ASTIN Bulletin*, **12** (1), 22–26.

Panjer, H. H. 2006. *Operational Risks: Modeling Analytics*. New York: Wiley.

Panjer, H. H., & Willmot, G. E. 1986. Computational aspects of recursive evaluation of compound distributions. *Insurance: Mathematics and Economics*, **5** (1), 113–116.

Panjer, H. H., & Willmot, G. 1992. *Insurance Risk Models*. Chicago, IL: Society of Actuaries.

Panjer, H. H., & Wang, S. 1993. On the stability of recursive formulas. *ASTIN Bulletin*, **23** (2), 227–258.

Park, T., & Casella, G. 2008. The Bayesian lasso. *Journal of the American Statistical Association*, **103** (482), 681–686.

Parzen, E. 1962. On estimation of a probability density function and mode. *The Annals of Mathematical Statistics*, **33** (3), 1065–1076.

Paulson, A. S. 1973. A characterization of the exponential distribution and a bivariate exponential distribution. *Sankhyâ: The Indian Journal of Statistics, Series A*, **35** (1), 69–78.

Pearl, J. 1986. Fusion, propagation, and structuring in belief networks. *Artificial Intelligence*, **29** (3), 241–288.

Pearl, J. 2009. *Causality*, 2nd edn. New York: Cambridge University Press.

Pearson, K. 1894. Contributions to the mathematical theory of evolution. *Philosophical Transactions of the Royal Society of London A*, **185**, 71–110.

Pearson, K. 1895. Contributions to the mathematical theory of evolution. II. Skew variation in homogeneous material. *Philosophical Transactions of the Royal Society of London A*, **186**, 343–414.

Pearson, K. 1896. Mathematical contributions to the theory of evolution. On a form of spurious correlation which may arise when indices are used in the measurement of organs. *Proceedings of the Royal Society of London*, **60** (359–367), 489–498.

Pelsser, A. 2008. On the applicability of the Wang transform for pricing financial risks. *ASTIN Bulletin*, **38** (1), 171.

Pelsser, A. May 30, 2011. Pricing in incomplete markets. Technical report, available at SSRN http://ssrn.com/abstract=1855565 or http://dx.doi.org/10.2139/ssrn.1855565. Accessed July 1, 2014.

Peng, L. 2006. Discussion of "Copulas: Tales and facts", by Thomas Mikosch. *Extremes*, **9** (1), 49–50.

Perreault, L., Bobée, B., & Rasmussen, P. F. 1999a. Halphen distribution system. I: Mathematical and statistical properties. *Journal of Hydrologic Engineering*, **4** (3), 189–199.

Perreault, L., Bobée, B., & Rasmussen, P. F. 1999b. Halphen distribution system. II: Parameter and quantile estimation. *Journal of Hydrologic Engineering*, **4** (3), 200–208.

Peters, G. W. 2005. Topics in sequential Monte Carlo samplers. M.Sc., Department of Engineering, University of Cambridge, Cambridge, UK.

Peters, G. W., Johansen, A. M., & Doucet, A. 2007. Simulation of the annual loss distribution in operational risk via Panjer recursions and Volterra integral equations for value-at-risk and expected shortfall estimation. *The Journal of Operational Risk*, **2** (3), 29–58.

Peters, J. P., & Hübner, G. 2009. Modeling operational risk based on multiple experts' opinions. In Gregoriou, G. N. (ed.), *Operational Risk toward Basel III: Best Practices and Issues in Modeling, Management, and Regulation*. Hoboken, NJ: Wiley.

Peters, G. W., Shevchenko, P. V., & Wüthrich, M. V. 2009. Dynamic operational risk: modeling dependence and combining different sources of information. *The Journal of Operational Risk*, **4** (2), 69–104.

Peters, G. W., Briers, M., Shevchenko, P., & Doucet, A. 2013. Calibration and filtering for multi factor commodity models with seasonality: incorporating panel data from futures contracts. *Methodology and Computing in Applied Probability*, **15** (4), 841–874.

Peters, G. W., Shevchenko, P. V., Young, M., & Yip, W. 2011. Analytic loss distributional approach models for operational risk from the α-stable doubly stochastic compound processes and implications for capital allocation. *Insurance Mathematics and Economics*, **49** (3), 565–579.

Peters, G. W., & Shevchenko, P. V. 2015. *Advances in Heavy Tailed Risk Modeling: A Handbook of Operational Risk*. Hoboken, NJ: Wiley.

Peters, G. W., & Sisson, S. A. 2006. Bayesian inference, Monte Carlo sampling and operational risk. *The Journal of Operational Risk*, **1** (3), 27–50.

Peters, G. W., Johansen, A. M., & Doucet, A. 2007. Simulation of the annual loss distribution in operational risk via Panjer recursions and Volterra integral equations for value-at-risk and expected shortfall estimation. *The Journal of Operational Risk*, **2** (3), 29–58.

Peters, G. W., Fan, Y., & Sisson, S. A. 2008. *On Sequential Monte Carlo, Partial Rejection Control and Approximate Bayesian Computation*. Technical report. UNSW, Sydney, Australia.

Peters, G. W., Shevchenko, P. V., & Wüthrich, M. V. 2009a. Model uncertainty in claims reserving within Tweedie's compound Poisson models. *ASTIN Bulletin*, **39** (1), 1–33.

Peters, G. W., Wüthrich, M. V., & Shevchenko, P. V. 2010. Chain ladder method: Bayesian bootstrap versus classical bootstrap. *Insurance: Mathematics and Economics*, **47** (1), 36–51.

Peters, G. W., Byrnes, A. D., & Shevchenko, P. V. 2011a. Impact of insurance for operational risk: Is it worthwhile to insure or be insured for severe losses? *Insurance: Mathematics and Economics*, **48** (2), 287–303.

Peters, G. W., Kannan, B., Lasscock, B., Mellen, C., & Godsill, S. 2011b. Bayesian cointegrated vector autoregression models incorporating alpha-stable noise for inter-day price movements via approximate Bayesian computation. *Bayesian Analysis*, **6** (4), 755–792.

Peters, G. W., Fan, Y., & Sisson, S. A. 2012a. On sequential Monte Carlo, partial rejection control and approximate Bayesian computation. *Statistics and Computing*, **22** (6), 1209–1222.

Peters, G. W., Dong, A. X. D., & Kohn, R. 2012b. A copula based Bayesian approach for paid-incurred claims models for non-life insurance reserving. Preprint arXiv:1210.3849, available at http://arxiv.org. Accessed July 1, 2014.

Peters, G. W., Sisson, S. A., & Fan, Y. 2012c. Likelihood-free Bayesian inference for α-stable models. *Computational Statistics & Data Analysis*, **56** (11), 3743–3756.

Petrella, G., & Resti, A. 2013. Supervisors as information producers: Do stress tests reduce bank opaqueness? *Journal of Banking & Finance*, **37** (12), 5406–5420.

Pflug, Georg Ch.Pflug, G. C. 2000. Some remarks on the value-at-risk and the conditional value-at-risk. In Uryasev, S. P. (ed.), *Probabilistic Constrained Optimization*. New York: Springer, Vol. 49, pp. 272–281.

Pickands, J. 1981. Multivariate extreme value distributions. *Proceedings 43rd Session International Statistical Institute*, Amsterdam, the Netherlands, Vol. 49, pp. 859–878.

Pillai, R. N. & Jayakumar, K. 1995. Discrete Mittag-Leffler distributions. *Statistics & Probability Letters*, **23** (3), 271–274.

Pillai, R. N. & Sandhya, E. 1990. Distributions with complete monotone derivative and geometric infinite divisibility. *Advances in Applied Probability*, **22** (3), 751–754.

Platen, E. 2005. On the role of the growth optimal portfolio in finance. *Australian Economic Papers*, **44** (4), 365–388.

Platen, E. 2006. A benchmark approach to finance. *Mathematical Finance*, **16** (1), 131–151.

Platen, E. & Heath, D. 2006. *A Benchmark Approach to Quantitative Finance*. Berlin, Germany: Springer.

Pliska, S. R. 1997. *Mathematics of Derivative Securities*, Vol. 15. Cambridge, UK: Cambridge University Press.

Pollard, H. 1944. The Bernstein-Widder theorem on completely monotonic functions. *Duke Mathematical Journal*, **11** (3), 427–430.

Polson, N. G., Scott, J. G. and Windle, J. (2014), The Bayesian bridge. *Journal of the Royal Statistical Society: Series B (Statistical Methodology)*, **76** (4), 713–733.

Powojowski, M. R., Reynolds, D., & Tuenter, H. J. H. 2002. Dependent events and operational risk. *ALGO Research Quarterly*, **5** (2), 65–73.

Prause, K. 1999. *The Generalized Hyperbolic Model: Estimation, Financial Derivatives, and Risk Measures*. Ph.D. thesis, University of Freiburg, Freiburg, Germany.

Press, S. J. 1972. Estimation in univariate and multivariate stable distributions. *Journal of the American Statistical Association*, **67** (340), 842–846.

Press, W. H., Teukolsky, S. A., Vetterling, W. T., & Flannery, B. P. 2002. *Numerical Recipes in C*. New York: Cambridge University Press.

Prokhorov, Y. V. 1968. An extension of SN Bernstein's inequalities to multidimensional distributions. *Theory of Probability & Its Applications*, **13** (2), 260–267.

Pugachev, V. S. 1965. *Theory of Random Functions and Its Applications to Control Problems*, 1st edn. London, UK: Pergamon Press.

Raoult, J.-P., & Worms, R. 2003. Rate of convergence for the generalized Pareto approximation of the excesses. *Advances in Applied Probability*, **35** (4), 1007–1027.

Raviv, A. 1979. The design of an optimal insurance policy. *The American Economic Review*, **69** (1), 84–96.

Rayner, G. D., & MacGillivray, H. L. 2002. Numerical maximum likelihood estimation for the g-and-k and generalized g-and-h distributions. *Statistics and Computing*, **12** (1), 57–75.

Rebonato, R. 2010. *Coherent Stress Testing: A Bayesian Approach to the Analysis of Financial Stress*. Chichester, UK: John Wiley & Sons.

Reeves, R. W., & Pettitt, A. N. 2005. A theoretical framework for approximate Bayesian computation. *Statistical Solutions to Modern Problems: Proceedings of the 20th International Workshop on Statistical Modelling*, Sydney, Australia, July 10–15, 2005, pp. 393–396.

Reserve, Federal. 2012. Comprehensive Capital Analysis and Review 2012: Methodology and results for stress scenario projections. Federal Reserve, Washington DC. URL http://www.federalreserve.gov/newsevents/press/bcreg/bcreg20120313a1.pdf.

Resnick, S., & Others. 2004. On the foundations of multivariate heavy-tail analysis. *Journal of Applied Probability*, **41A** (2004), 191–212.

Richardson, L. F. 1911. The approximate arithmetical solution by finite differences of physical problems involving differential equations, with an application to the stresses in a masonry dam. *Philosophical Transactions of the Royal Society of London. Series A, Containing Papers of a Mathematical or Physical Character*, **210**, 307–357.

Rigby, R. A., & Stasinopoulos, D. M. 2005. Generalized additive models for location, scale and shape. *Journal of the Royal Statistical Society: Series C (Applied Statistics)*, **54** (3), 507–554.

Riordan, J. 1946. Derivatives of composite functions. *Bulletin of the American Mathematical Society*, **52** (8), 664–667.

Ripley, B. D. 1987. *Stochastic Simulation*. New York: Wiley.

Rippel, M. & Teplý, P. 2008. Operational risk-scenario analysis. *No. 15/2008. IES Working Paper.*

Ristic, B., Arulampalam, S., & Gordon, N. 2004. *Beyond the Kalman Filter: Particle Filters for Tracking Applications*. Boston, MA: Artech House Publishers.

Robert, C. P. 2001. *The Bayesian Choice*. New York: Springer.

Robert, C. P., & Casella, G. 2004. *Monte Carlo Statistical Methods*, 2nd edn. New York: Springer Texts in Statistics.

Roberts, G. O. 1995. Markov chain concepts related to sampling algorithms. In Gilks, W. R. (ed.), *Markov Chain Monte Carlo in Practice*. Dordrecht, the Netherlands: Springer, pp. 45–57.

Roberts, G. O., & Rosenthal, J. S. 2001. Optimal scaling for various Metropolis-Hastings algorithms. *Statistical Science*, **16** (4), 351–367.

Roberts, G. O., & Rosenthal, J. S. 2002. The polar slice sampler. *Stochastic Models*, **18** (2), 257–280.

Roberts, G. O., & Rosenthal, J. S. 2007. Coupling and ergodicity of adaptive Markov chain Monte Carlo algorithms. *Journal of Applied Probability*, **44** (2), 458–475.

Roberts, G. O., & Rosenthal, J. S. 2009. Examples of adaptive MCMC. *Journal of Computational and Graphical Statistics*, **18** (2), 349–367.

Robertson, J. 1992. The computation of aggregate loss distributions. *Proceedings of the Casualty Actuarial Society*, **79**, 57–133.

Rockafellar, R. T., & Uryasev, S. 2002. Conditional value-at-risk for general loss distributions. *Journal of Banking & Finance*, **26** (7), 1443–1471.

Rockafellar, R. T., Uryasev, S., & Zabarankin, M. 2006. Generalized deviations in risk analysis. *Finance and Stochastics*, **10** (1), 51–74.

Roll, R., & Ross, S. A. 1980. An empirical investigation of the arbitrage pricing theory. *The Journal of Finance*, **35** (5), 1073–1103.

Roman, S. 1980. The formula of Faa di Bruno. *The American Mathematical Monthly*, **87** (10), 805–809.

Romberg, W. 1955. Vereinfachte numerische integration. *Det Kongelige Norske Videnskabers Selskab Forhandlinger*, **28** (7), 30–36.

Rootzén, H., & Klüppelberg, C. 1999. A single number can't hedge against economic catastrophes. *Ambio*, **28** (6), 550–555.

Rosenblatt, M. 1956. Remarks on some nonparametric estimates of a density function. *The Annals of Mathematical Statistics*, **27** (3), 832–837.

Rosenthal, J. S. 2007. AMCMC: An R interface for adaptive MCMC. *Computational Statistics and Data Analysis*, **51** (12), 5467–5470.

Rosenthal, J. S. 2009. Optimal proposal distributions and adaptive MCMC. In Gelman, A., Jones, G., Meng, X. L., & Brooks, S. (eds.), *Handbook of Markov Chain Monte Carlo: Methods and Applications*. Boca Raton, FL: Chapman & Hall/CRC Press.

Rosiński, J. 2007. "Tempering stable processes." Stochastic processes and their applications 117, no. 6: 677–707.

Ross, S. A. 1973. *Return, Risk and Arbitrage*. Technical report. Wharton School Rodney L. White Center for Financial Research, Philadelphia, PA.

Ross, S. A. 1976. The arbitrage theory of capital asset pricing. *Journal of Economic Theory*, **13** (3), 341–360.

Rubinstein, M. 2002. Markowitz's portfolio selection: A fifty-year retrospective. *The Journal of Finance*, **57** (3), 1041–1045.

Rukhin, A. L. 1974. Strongly symmetric families and statistical analysis of their parameters. *Zapiski Nauchnykh Seminarov POMI*, **43**, 59–87.

Rytgaard, M. 1990. Estimation in Pareto distribution. *ASTIN Bulletin*, **20** (2), 201–216.

Saerens, M. 2000. Building cost functions minimizing to some summary statistics. *IEEE Transactions on Neural Networks*, **11** (6), 1263–1271.

Sakalo, T., & Delasey, M. 2011. A framework for uncertainty modeling in operational risk. *The Journal of Operational Risk*, **6** (4), 21–57.

Samorodnitsky, G., & Taqqu, M. S. 1994. *Stable Non-Gaussian Processes*. New York: Chapman & Hall.

Samorodnitsky, G., & Taqqu, M. S. 1997. Stable non-Gaussian random processes. *Econometric Theory*, **13** (1), 133–142.

Sandström, A. 2006. *Solvency: Models, Assessment and Regulation*. Boca Raton, FL: Chapman & Hall/CRC Press.

Sansom, J., & Thompson, P. J. 1998. Detecting components in censored and truncated meteorological data. *Environmetrics*, **9** (6), 673–688.

Santomero, A. M., & Babbel, D. F. 1997. Financial risk management by insurers: An analysis of the process. *Journal of Risk and Insurance*, **64** (2), 231–270.

Savage, L. J. 1961. *The Subjective Basis of Statistical Practice*. Technical report. Department of Statistics, University of Michigan, Ann Arbor, MI.

Savu, C., & Trede, M. May 2006. Hierarchical Archimedean copulas. *International Conference on High Frequency Finance*, Konstanz, Germany.

sbAMA Scenario Based AMA Working Group. 2003. *Scenario Based AMA*. sbAMA Working Group, Technical report, Federal Reserve Bank of New York. Final Version 1.0 available at http://www.newyorkfed.org/newsevents/events/banking/2003/con0529d.pdf. Accessed July 1, 2014.

Scarsini, M. 1984. On measures of concordance. *Stochastica: Revista de Matemática Pura y Aplicada*, **8** (3), 201–218.

Schmidt, R. 2002. Tail dependence for elliptically contoured distributions. *Mathematical Methods of Operations Research*, **55** (2), 301–327.

Schmidt, T. 2006. Coping with copulas. In Rank, J. (ed.), *Copulas from Theory to Applications in Finance*. London, UK: Incisive Media Risk Books, pp. 3–34.

Schoelzel, C., & Friederichs, P. 2008. Multivariate non-normally distributed random variables in climate research—Introduction to the copula approach. *Nonlinear Processes in Geophysics*, **15** (5), 761–772.

Schradin, H. R. 1997. PCS catastrophe insurance options—A new instrument for managing catastrophe risk. *Contribution to the 6th AFIR International Colloquium*, Nurnberg, October 1–3, *British Actuarial Journal*, **3** (I), 241.

Schuermann, T. 2013. Stress testing banks. *International Journal of Forecasting*. Elsevier.

Schwarz, G. 1978. Estimation the dimension of a model. *Annals of Statistics*, **6** (2), 461–464.

Schweizer, B. 1991. Thirty years of copulas. In Dall'Aglio, G., Kotz, S., and Salinetti G. (eds.), *Advances in Probability Distributions with Given Marginals*. Dordrecht, the Netherlands: Kluwer Academic Publishers, pp. 13–50.

Schweizer, B., & Wolff, E. F. 1981. On nonparametric measures of dependence for random variables. *The Annals of Statistics*, **9** (4), 879–885.

Scott, H., & Jackson, H. Spring 2002. Operational risk insurance: Treatment under the New Basel accord. *Aino Bunge International Finance Seminar*. Technical report. Harvard Law School, Cambridge, MA.

Segers, J. 2006. Discussion of "Copulas: Tales and facts", by Thomas Mikosch. *Extremes*, **9** (1), 51–53.

Seidenfeld, T., Kadane, J. B., & Schervish, M. J. 1989. On the shared preferences of two Bayesian decision makers. *The Journal of Philosophy*, **86** (5), 225–244.

Sentz, K, & Ferson, S. 2002. *Combination of Evidence in Dempster-Shafer Theory*. Sandia National Laboratories, Albuquerque, NM/Livermore, CA. SAND report: SAND2002-0835.

Seshadri, V. 2004. Halphen's laws. In *Encyclopedia of Statistical Sciences*. John Wiley & Sons, Inc., pp. 302–306.

Shafer, G. 1976. *A Mathematical Theory of Evidence*. Princeton, NJ: Princeton University Press.

Shahbaba, B., & Neal, R. 2009. Nonlinear models using Dirichlet process mixtures. *The Journal of Machine Learning Research*, **10**, 1829–1850.

Shaked, M. 1977. "A concept of positive dependence for exchangeable random variables." The Annals of Statistics 5, no. 3: 505–515.

Sharpe, W. F. 1998. The Sharpe ratio. In Bernstein, P. L., & Fabozzi, F. J. (eds.), *Streetwise: The Best of the Journal of Portfolio Management*. Princeton, NJ: Princeton University Press, pp. 169–185.

Sharpe, W. F., 1970. *Portfolio Theory and Capital Markets*, Vol. 217. New York: McGraw-Hill.

Shevchenko, P. V. 2008. Estimation of operational risk capital charge under parameter uncertainty. *The Journal of Operational Risk*, **3** (1), 51–63.

Shevchenko, P. V. 2011. *Modelling Operational Risk Using Bayesian Inference*. Berlin, Germany: Springer.

Shevchenko, P. V., & Wüthrich, M. V. 2006. The structural modeling of operational risk via Bayesian inference: combining loss data with expert opinions. *Journal of Operational Risk*, **1** (3), 3–26.

Shevchenko, P. V., & Temnov, G. 2009. Modeling operational risk data reported above a time-varying threshold. *The Journal of Operational Risk*, **4** (2), 19–42.

Shih, J. H., & Louis, T. A. 1995. Inferences on the association parameter in copula models for bivariate survival data. *Biometrics*, **51** (4), 1384–1399.

Shih, J., Khan, A. S., & Medapa, P. 2000. Is the size of an operational loss related to firm size? *Operational Risk Magazine*, **2** (1), 1–2.

Shirakawa, H. 1991. Interest rate option pricing with Poisson-Gaussian forward rate curve processes. *Mathematical Finance*, **1** (4), 77–94.

Sichel, H. S. 1974. On a distribution representing sentence-length in written prose. *Journal of the Royal Statistical Society. Series A (General)*, **137** (1), 25–34.

Sichel, H. S. 1982. Repeat-buying and the generalized inverse Gaussian-Poisson distribution. *Applied Statistics*, **31** (3), 193–204.

Sinclair, C. D., & Spurr, B. D. 1988. Approximations to the distribution function of the Anderson-Darling test statistic. *Journal of the American Statistical Association*, **83** (404), 1190–1191.

Sinclair, C. D., Spurr, B. D., & Ahmad, M. I. 1990. Modified Anderson darling test. *Communications in Statistics-Theory and Methods*, **19** (10), 3677–3686.

Sisson, S. A. & Fan, Y. 2011. Likelihood-free MCMC. In Brooks, S., Gelman, A., Jones, G. L., & Meng, X.-L. (eds.), *Handbook of Markov Chain Monte Carlo*. Boca Raton, FL: Taylor & Francis Group, p. 313.

Sisson, S. A., Fan, Y., & Tanaka, M. M. 2007. Sequential Monte Carlo without likelihoods. *Proceedings of the National Academy of Sciences*, **104** (6), 1760–1765.

Sisson, S. A., Peters, GW, Briers, M, & Fan, Y. 2010. A note on target distribution ambiguity of likelihood-free samplers. Preprint arXiv:1005.5201, available at http://arxiv.org. Accessed July 1, 2014.

Siu, T. K., Tong, H., & Yang, H. 2001. Bayesian risk measures for derivatives via random Esscher transform. *North American Actuarial Journal*, **5** (3), 78–91.

Skilling, J. 2006. Nested sampling for general Bayesian computation. *Bayesian Analysis*, **1** (4), 833–859.

Sklar, A. 1959. Fonctions de répartition à *n* dimensions et leurs marges. *Publications de l'Institut de Statistique l'Université de Paris*, **8**, 229–231.

Sklar, A. 1996. Random variables, distribution functions, and copulas: A personal look backward and forward. *Lecture Notes Monograph Series*, **28**, 1–14.

Small, C. G. 2010. *Expansions and Asymptotics for Statistics*, Vol. 115. London, UK/Boca Raton, FL: Chapman and Hall/CRC Press.

Smirnov, N. V. 1936. Sur la distribution de w2. *Comptes Rendus de l'Académie. des Sciences Paris*, **202**, 449.

Smirnov, N. 1948. Table for estimating the goodness of fit of empirical distributions. *The Annals of Mathematical Statistics*, **19** (2), 279–281.

Smith, R. L. 1939. On the estimation of the discrepancy between empirical curves of distribution for two independent samples. *Bulletin de l'Université de Moscou, Série internationale (Mathématiques)* **2**: 3–16.

Smith, C. E. 1992. A Laguerre series approximation to the probability density of the first passage time of the Ornstein Ullenbeck process. *Noise in Physical Systems and 1/f Fluctuations*. Kyoto, Japan: IOS Press, p. 389.

Smith, A. A., Jr. 2008. Indirect inference. In Durlauf, S. N., & Blume, L. E. (eds.), *The New Palgrave Dictionary of Economics*, 2nd edn. Palgrave Macmillan. Available at http://www.dictionaryofeconomics.com/article?id=pde2008_I000259> doi:10.1057/9780230226203.0778. Accessed July 1, 2014.

Smith, A. F. M., & Roberts, G. O. 1993. Bayesian computation via the Gibbs sampler and related Markov chain Monte Carlo methods. *Journal of the Royal Statistical Society. Series B (Methodological)*, **55** (1), 3–23.

Snedecor, G. W., & Cochran, W. G. 1989. *Statistical Methods*. Ames, IA: Iowa State University Press.

Sofronov, G. 2013. An optimal sequential procedure for a multiple selling problem with independent observations. *European Journal of Operational Research*, **225** (2), 332–336.

Sondermann, D. 1991. Reinsurance in arbitrage-free markets. *Insurance: Mathematics and Economics*, **10** (3), 191–202.

Spearman, C. 1904. "General intelligence", objectively determined and measured. *The American Journal of Psychology*, **15** (2), 201–292.

Spegelhalter, D. J., & Cowell, R. J. 1992. *Learning in Probabilistic Expert Systems (Bayesian Statistics)*, Vol. 4. Oxford, UK: Oxford University Press.

Stacy, E. W. 1962. A generalization of the gamma distribution. *The Annals of Mathematical Statistics*, **33** (3), 1187–1192.

Stasinopoulos, D. M., & Rigby, R. A. 2007. Generalized additive models for location scale and shape (GAMLSS) in R. *Journal of Statistical Software*, **23** (7), 1–46.

Steinbrecher, G., & Shaw, W. T. 2008. Quantile mechanics. *European Journal of Applied Mathematics*, **19** (2), 87–112.

Steinhoff, C., & Baule, R. 2006. How to validate op risk distributions. *OpRisk&Compliance*, August, 36–39.

Sterk, H.-P. 1979. *Selbstbeteiligung unter risikotheoretischen Aspekten*. Karlsruhe, Germany: Verlag Versicherungswirtschaft.

Steutel, F. W. 1973. Some recent results in infinite divisibility. *Stochastic Processes and Their Applications*, **1** (2), 125–143.

Steutel, F. W. & Van Harn, K. 2003. *Infinite Divisibility of Probability Distributions on the Real Line*. CRC Press.

Stramer, O., & Tweedie, R. L. 1999. Langevin-type models II: Self-targeting candidates for MCMC algorithms*. *Methodology and Computing in Applied Probability*, **1** (3), 307–328.

Stuart, A., & Ord, J. K. 1994. *Kendall's Advanced Theory of Statistics*, Vol. 1, *Distribution Theory*, 6th edn. London, UK/Melbourne, Australia/Auckland, New Zealand: Edward Arnold.

Stuart, A., Ord, J. K., & Arnold, S. 1999. *Advanced Theory of Statistics*, Vol. 2A, *Classical Inference and the Linear Models*, 6th edn. London, UK: Oxford University Press.

Sundt, B. 1992. On some extensions of Panjer's class of counting distributions. *ASTIN Bulletin*, **22** (1), 61–80.

Sundt, B. 1999. On multivariate Panjer recursions. *ASTIN Bulletin*, **29** (1), 29–45.

Sundt, B. 1998. A generalisation of the De Pril transform. *Scandinavian Actuarial Journal*, **1998** (1), 41–48.

Sundt, B. 2005. On some properties of De Pril transforms of counting distributions. *ASTIN Bulletin*, **25** (1), 19–31.

Sundt, B., & Jewell, W. S. 1981. Further results on recursive evaluation of compound distributions. *ASTIN Bulletin*, **12** (1), 27–39.

Sundt, B., & Vernic, R. 2009. *Recursions for Convolutions and Compound Distributions with Insurance Applications*. Berlin, Germany: Springer.

Sungur, E. A. 1999. Truncation invariant dependence structures. *Communications in Statistics-Theory and Methods*, **28** (11), 2553–2568.

Swiss Financial Market Supervisory Authority. 2006. *Swiss Solvency Test, Technical Document*. Bern, Switzerland: Swiss Financial Market Supervisory Authority (FINMA).

Kallsen, J. and Tankov, P. 2006. Characterization of dependence of multidimensional Lévy processes using Lévy copulas. *Journal of Multivariate Analysis*, **97** (7), 1551–1572.

Tanner, M. A., & Wong, W. H. 1987. The calculation of posterior distributions by data augmentation. *Journal of the American Statistical Association*, **82** (398), 528–540.

Targino, R. S., Peters, G. W., Sofronov, G., & Shevchenko, P. V. 2013. Optimal insurance purchase strategies via optimal multiple stopping times. Preprint arXiv:1312.0424, available at http://arxiv.org. Accessed July 1, 2014.

Tasche, D. 1999. Risk contributions and performance measurement. Preprint, Department of Mathematics, TU München, München, Germany.

Tasche, D. 2002. Expected shortfall and beyond. *Journal of Banking and Finance*, **26** (7), 1519–1533.

Tasche, D. 2008. *Euler Allocation: Theory and Practice*. Preprint arXiv:0708.2542v2, available at http://arxiv.org. Accessed July 1, 2014.

Tavaré, S., Marjoram, P., Molitor, J., & Plagnol, V. 2003. Markov chain Monte Carlo without likelihoods. *Proceedings of the National Academy of Science, USA*, **100** (26), 15324–15328.

Taylor, M. D. 2007. Multivariate measures of concordance. *Annals of the Institute of Statistical Mathematics*, **59** (4), 789–806.

Ter Berg, P. 1994. Deductibles and the inverse Gaussian distribution. *ASTIN Bulletin*, **24** (2), 319–323.

The Final Rule. 2007. *Risk-Based Capital Standards: Advanced Capital Adequacy Framework—Basel II, Final Rule*. The Office of the Comptroller of Currency, The Federal Reserve System, The Federal Deposit Insurance Corporation, and The Office of Thrift Supervision, Washington, DC.

Thom, R. 1977. Structural stability, catastrophe theory, and applied mathematics. *SIAM Review*, **19** (2), 189–201.

Thom, R., & Zeeman, E. C. 1974. Catastrophe theory: Its present state and future perspectives. *Dynamical Systems-Warwick*, **468**, 366–389.

Thompson, M. B. 2011. *Slice Sampling with Multivariate Steps*. Ph.D. thesis, University of Toronto, Toronto, Canada.

Thompson, M. B., & Neal, R. M. 2010a. Covariance-adaptive slice sampling. Technical report no. 1002, Department of Statistics, University of Toronto, Toronto, Canada, 17pp. Preprint arXiv:1003.3201, available at http://arxiv.org. Accessed July 1, 2014.

Thompson, M. B., & Neal, R. M. 2010. Slice sampling with adaptive multivariate steps: The shrinking-rank method. *JSM 2010, Section on Statistical Computing*, Vancouver, Canada, pp. 3890–3896.

Thomson, W. 1979. Eliciting production possibilities from a well-informed manager. *Journal of Economic Theory*, **20** (3), 360–380.

Tibbits, M. M., Haran, M., & Liechty, J. C. 2011. Parallel multivariate slice sampling. *Statistics and Computing*, **21** (3), 415–430.

Tibshirani, R. 1996. Regression shrinkage and selection via the lasso. *Journal of the Royal Statistical Society. Series B (Methodological)*, **58** (1), 267–288.

Tibshirani, R. 2011. Regression shrinkage and selection via the lasso: A retrospective. *Journal of the Royal Statistical Society. Series B (Statistical Methodology)*, **73** (3), 273–282.

Tilley, J. A. 1995. *The Latest in Financial Engineering: Structuring Catastrophe Reinsurance as a High-Yield Bond*. New York: Morgan Stanley.

Tong, B., & Wu, C. 2012. Asymptotics for operational risk quantified with a spectral risk measure. *Journal of Operational Risk*, **7** (3), 91.

Trivedi, P. K., & Zimmer, D. M. 2007. *Copula Modeling: An Introduction for Practitioners*. Hanover, MA: Now Publishers Inc.

Tsukahara, H. 2005. Semiparametric estimation in copula models. *Canadian Journal of Statistics*, **33** (3), 357–375.

Tukey, J. W. 1977a. *Exploratory Data Analysis*. Reading, MA: Addison Wesley Longman Inc., 231pp.

Tukey, J. W. 1977b. Modern techniques in data analysis. *NSF Sponsored Regional Research Conference*, Southern Massachusetts University, North Dartmouth, MA.

Tversky, A., & Kahneman, D. 1974. Judgment under uncertainty: Heuristics and biases. *Science (New Series)*, **185** (4157), 1124–1131.

Tweedie, M. C. K. 1957. Statistical properties of inverse Gaussian distributions. *The Annals of Mathematical Statistics*, **28** (2), 362–377.

Van den Brink, G. J. 2002. *Operational Risk: The New Challenge for Banks*. London, UK: Palgrave Macmillan.

Van der Vaart, A. W. 2000. *Asymptotic Statistics*, Vol. 3. Cambridge, UK: Cambridge University Press.

Vasicek, O. 1977. An equilibrium characterization of the term structure. *Journal of Financial Economics*, **5** (2), 177–188.

Vaugirard, V. E. 2003. Pricing catastrophe bonds by an arbitrage approach. *The Quarterly Review of Economics and Finance*, **43** (1), 119–132.

Venter, G. G. 1983. Transformed beta and gamma distributions and aggregate losses. *Proceedings of the Casualty Actuarial Society*, **70**, 156–193.

Venter, G. G. 1991. Premium calculation implications of reinsurance without arbitrage. *ASTIN Bulletin*, **21** (2), 223–230.

Vernic, R. 1999. Recursive evaluation of some bivariate compound distributions. *ASTIN Bulletin*, **29** (2), 315–325.

Von Neumann, J., & Morgenstern, O. 2007. *Theory of Games and Economic Behavior* (commemorative edition). Princeton, NJ: Princeton University Press.

Waldmann, K.-H. 1996. Modified recursions for a class of compound distributions. *ASTIN Bulletin*, **26** (2), 213–224.

Walley, P. 1991. *Statistical Reasoning with Imprecise Probabilities*. London, UK: Chapman and Hall.

Walley, P., & Fine, T. L. 1982. Towards a frequentist theory of upper and lower probability. *Annals of Statistics*, **10** (3), 741–761.

Wang, S. 1996. Premium calculation by transforming the layer premium density. *ASTIN Bulletin*, **26** (1), 71–92.

Wang, S. 2002. A universal framework for pricing financial and insurance risks. *ASTIN Bulletin*, **32** (2), 213–234.

Wang, S. S. 2004. Cat bond pricing using probability transforms. Geneva Papers: *Etudes et Dossiers, special issue on Insurance and the State of the Art in Cat Bond Pricing*, **278**, 19–29.

Wang, S. S., Young, V. R., & Panjer, H. H. 1997. Axiomatic characterization of insurance prices. *Insurance: Mathematics and Economics*, **21** (2), 173–183.

Wasserman, L. 1997. *Bayesian Model Selection and Model Averaging*. Technical report. Statistics Department, Carnegie Mellon University, Pittsburgh, PA.

Warde, W. D., & Katti, S. K. 1971. Infinite divisibility of discrete distributions, II. *The Annals of Mathematical Statistics*, **42** (3), 1088–1090.

Watson, G. N. 1922. *A Treatise on the Theory of Bessel Functions*. Cambridge, UK: Cambridge University Press.

West, M. 1987. On scale mixtures of normal distributions. *Biometrika*, **74** (3), 646–648.

Whittle, P. 1958. On the smoothing of probability density functions. *Journal of the Royal Statistical Society. Series B (Methodological)*, **20** (2), 334–343.

Wilk, M. B., & Gnanadesikan, R. 1968. Probability plotting methods for the analysis for the analysis of data. *Biometrika*, **55** (1), 1–17.

Williamson, J. 2005. *Bayesian Nets and Causality: Philosophical and Computational Foundations*. Oxford Scholarship Online. Oxford, UK: OUP, 252pp.

Williamson, R. E. 1956. Multiply monotone functions and their Laplace transforms. *Duke Mathematical Journal*, **23** (2), 189–207.

Williamson, R. C., & Downs, T. 1990. Probabilistic arithmetic I: Numerical methods for calculating convolutions and dependency bounds. *International Journal of Approximate Reasoning*, **4** (2), 89–158.

Winkler, R. L. 1986. *On Good Probability Appraisers*. London, UK: Elsevier.

Wirch, Julia L and Hardy, Mary R. 2001. Distortion risk measures. Coherence and stochastic dominance. *International Congress on Insurance: Mathematics and Economics*, 15–17.

Wittsiepe, R. 2008. IAS 37 provisions, contingent liabilities and contingent assets. *IFRS for Small and Medium-Sized Enterprises: Structuring the Transition Process*, pp. 173–181.

Wolpert, R. L. 1989. Eliciting and combining subjective judgements about uncertainty. *International Journal of Technology Assessment in Health Care*, **5** (4), 537–557.

Wolpert, R. L., & Schmidler, S. C. 2012. α-Stable limit laws for harmonic mean estimators of marginal likelihoods. *Statistica Sinica*, **22** (3), 1233.

Woo, G. 1999. *The Mathematics of Natural Catastrophes*. London, UK: Imperial College Press.

Work Cover. 2001. *Major Hazard Facilities Regulations Guidance Note GN-10 Control Measures*. Victorian Workcover Authority, Technical report, available at http://www.psyfactors.com/ohs_regs/mhf_guidelines.pdf. Accessed July 1, 2014.

Wright, W. F., & Anderson, U. 1989. Effects of situation familiarity and financial incentives on use of the anchoring and adjustment heuristic for probability assessment. *Organizational Behavior and Human Decision Processes*, **44** (1), 68–82.

Wüthrich, M. V. 2006. Premium liability risks: Modelling small claims. *Bulletin of the Swiss Association of Actuaries*, **1**, 27–38.

Wüthrich, M. V. 2010. *Market-Consistent Actuarial Valuation*. Springer, Berlin/Heidelberg, Germany Springer-Verlag EAA Series.

Wüthrich, M. V., & Merz, M. 2008. *Stochastic Claims Reserving Methods in Insurance*. Hoboken, NJ: John Wiley & Sons.

Xiu, D., & Karniadakis, G. E. 2002. The Wiener–Askey polynomial chaos for stochastic differential equations. *SIAM Journal on Scientific Computing*, **24** (2), 619–644.

Xue-Kun Song, P. 2000. Multivariate dispersion models generated from Gaussian copula. *Scandinavian Journal of Statistics*, **27** (2), 305–320.

Yager, R. R. 1986. Arithmetic and other operations on Dempster-Shafer structures. *International Journal of Man-Machine Studies*, **25** (4), 357–366.

Yager, R. R. 1987. On the Dempster-Shafer framework and new combination rules. *Information Sciences*, **41** (2), 93–137.

Yamai, Y., & Yoshiba, T. January 2002. Comparative analyses of expected shortfall and value-at-risk: Their estimation error, decomposition, and optimization. *Monetary and Economic Studies*, **20** (1), 87–121.

Yoon, Y. K. 2003. *Modelling Operational Risk in Financial Institutions Using Bayesian Networks*. Master of Science Dissertation, Cass Business School, London, UK.

Yu, K., & Moyeed, R. A. 2001. Bayesian quantile regression. *Statistics & Probability Letters*, **54** (4), 437–447.

Yu, K., & Zhang, J. 2005. A three-parameter asymmetric Laplace distribution and its extension. *Communications in Statistics: Theory and Methods*, **34** (9–10), 1867–1879.

Yuan, M., & Lin, Y. 2006. Model selection and estimation in regression with grouped variables. *Journal of the Royal Statistical Society: Series B (Statistical Methodology)*, **68** (1), 49–67.

Yuan, M., & Lin, Y. 2007. On the non-negative Garrotte estimator. *Journal of the Royal Statistical Society: Series B (Statistical Methodology)*, **69** (2), 143–161.

Zeeman, E. C. 1977. *Catastrophe Theory: Selected Papers*. Reading, MA: Addison-Wesley.

Zeeman, E. C. 1979. Catastrophe theory. In Giittinger, W., & Eikemeier, H. (eds.), *Structural Stability in Physics*. Springer Series in Synergetics. Berlin, Germany: Springer, Vol. 4, pp, 12–22.

Zhu, W. 2011. Ambiguity aversion and an intertemporal equilibrium model of catastrophe-linked securities pricing. *Insurance: Mathematics and Economics*, **49** (1), 38–46.

Zolotarev, V. M. 1961. Concerning a certain probability problem. *Theory of Probability & Its Applications*, **6** (2), 201–204.

Zolotarev, V. M. 1986. One-dimensional stable distributions. In *Translations of Mathematical Monographs*. Providence, RI: American Mathematical Society, Vol. 65, 284pp.

Zou, H. 2006. The adaptive lasso and its oracle properties. *Journal of American Statistical Association*, **101** (476), 1418–1429.

Zou, H., & Li, R. 2008. One-step sparse estimates in nonconcave penalized likelihood models. *The Annals of Statistics*, **36** (4), 1509–1533.

Index

Fundamental Aspects of Operational Risk and Insurance Analytics: A Handbook of Operational Risk, First Edition. Marcelo G. Cruz, Gareth W. Peters, and Pavel V. Shevchenko.

Wiley Handbooks in

FINANCIAL ENGINEERING AND ECONOMETRICS

Advisory Editor
Ruey S. Tsay
The University of Chicago Booth School of Business, USA

The dynamic and interaction between financial markets around the world have changed dramatically under economic globalization. In addition, advances in communication and data collection have changed the way information is processed and used. In this new era, financial instruments have become increasingly sophisticated and their impacts are far-reaching. The recent financial (credit) crisis is a vivid example of the new challenges we face and continue to face in this information age. Analytical skills and ability to extract useful information from mass data, to comprehend the complexity of financial instruments, and to assess the financial risk involved become a necessity for economists, financial managers, and risk management professionals. To master such skills and ability, knowledge from computer science, economics, finance, mathematics and statistics is essential. As such, financial engineering is cross-disciplinary, and its theory and applications advance rapidly.

The goal of this Handbook Series is to provide a one-stop source for students, researchers, and practitioners to learn the knowledge and analytical skills they need to face today's challenges in financial markets. The Series intends to introduce systematically recent developments in different areas of financial engineering and econometrics. The coverage will be broad and thorough with balance in theory and applications. Each volume will be edited by leading researchers and practitioners in the area and covers state-of-the-art methods and theory of the selected topic.

Published Wiley Handbooks in Financial Engineering and Econometrics

Bauwens, Hafner, and Laurent · *Handbook of Volatility Models and Their Applications*
Brandimarte · *Handbook in Monte Carlo Simulation: Applications in Financial Engineering, Risk Management, and Economics*
Chan and Wong · *Handbook of Financial Risk Management: Simulations and Case Studies*
Cruz, Peters, and Shevchenko · *Fundamental Aspects of Operational Risk and Insurance Analytics: A Handbook of Operational Risk*
James, Marsh, and Sarno · *Handbook of Exchange Rates*
Peters and Shevchenko · *Advances in Heavy Tailed Risk Modeling: A Handbook of Operational Risk*
Viens, Mariani, and Florescu · *Handbook of Modeling High-Frequency Data in Finance*
Szylar · *Handbook of Market Risk*

Forthcoming Wiley Handbooks in Financial Engineering and Econometrics

Bali and Engle · *Handbook of Asset Pricing*
Chacko · *Handbook of Credit and Interest Rate Derivatives*
Florescu, Mariani, Stanley, and Viens · *Handbook of High-Frequency Trading and Modeling in Finance*
Jacquier · *Handbook of Econometric Methods for Finance: Bayesian and Classical Perspecitves*

Longin · *Handbook of Extreme Value Theory and Its Applications to Finance and Insurance*
Starer · *Handbook of Equity Portfolio Management: Theory and Practice*
Szylar · *Handbook of Hedge Fund Risk Management and Performance: In a Challenging Regulatory Environment*
Szylar · *Handbook of Macroeconomic Investing*
Veronesi · *Handbook of Fixed-Income Securities*